D1698727

Modern Alkaloids

Edited by
Ernesto Fattorusso and
Orazio Taglialatela-Scafati

Related Titles

Tietze, Lutz F. / Eicher, Theophil / Diederichsen, Ulf / Speicher, Andreas

Reactions and Syntheses

in the Organic Chemistry Laboratory

2007

ISBN: 978-3-527-31223-8

Hudlicky, Tomas / Reed, Josephine W.

The Way of Synthesis

Evolution of Design and Methods for Natural Products

2007

ISBN: 978-3-527-31444-7

Sarker, Satyajit / Nahar, Lutfun

Chemistry for Pharmacy Students

2007

ISBN: 978-0-470-01780-7

Kayser, Oliver / Quax, Wim J. (eds.)

Medicinal Plant Biotechnology

From Basic Research to Industrial Applications

2006

ISBN: 978-3-527-31443-0

Eicher, Theophil / Hauptmann, Siegfried

The Chemistry of Heterocycles

Structure, Reactions, Syntheses, and Applications

2003

ISBN: 978-3-527-30720-3

van de Waterbeemd, Han / Lennernäs, Hans / Artursson, Per (Eds.)

Drug Bioavailability

Estimation of Solubility, Permeability, Absorption and Bioavailability

2003

ISBN: 978-3-527-30438-7

Hesse, Manfred

Alkaloids

Nature's Curse or Blessing?

2002

ISBN: 978-3-906390-24-6

Modern Alkaloids

Structure, Isolation, Synthesis and Biology

Edited by
Ernesto Fattorusso and Orazio Taglialatela-Scafati

WILEY-VCH Verlag GmbH & Co. KGaA

The Editors

Prof. Ernesto Fattorusso
Univ. Federico II Dipto. di
Chimica delle Sost. Naturali
Via D. Montesano 49
80131 Napoli
Italien

Prof. O. Taglialatela-Scafati
Univ. Federico II, Dipto. di
Chimica delle Sost. Naturali
Via D. Montesano 49
80131 Napoli
Italien

Library of Congress Card No.:
applied for

British Library Cataloguing-in-Publication Data
A catalogue record for this book is available from the British Library.

Bibliographic information published by the Deutsche Nationalbibliothek
Die Deutsche Nationalbibliothek lists this publication in the Deutsche Nationalbibliografie; detailed bibliographic data are available in the Internet at http://dnb.d-nb.de.

Typesetting Thomson Digital, Noida
Printing betz-druck GmbH, Darmstadt
Binding Litges & Dopf GmbH, Heppenheim
Cover Design Grafik-Design Schulz, Fußgönheim

Printed in the Federal Republic of Germany
Printed on acid-free paper

ISBN: 978-3-527-31521-5

Contents

Modern Alkaloids: Structure, Isolation, Synthesis and Biology. Edited by E. Fattorusso and O. Taglialatela-Scafati

ISBN: 978-3-527-31521-5

Preface

Alkaloids constitute one of the widest classes of natural products, being synthesized practically by all phyla of both marine and terrestrial organisms, at any evolutionary level. The extraordinary variety (and often complexity) of alkaloid structures and biological properties have long intrigued natural product chemists (for structure determination and biosynthetic studies), analytical chemists, and synthetic organic chemists. Toxicologists, pharmacologists and pharmaceutical companies have used and will certainly continue to use alkaloids as biological tools and/or as lead compounds for development of new drugs.

When we started our project of a handbook on alkaloid science, we were faced with an impressive number of papers describing the structures and activities of alkaloids, and also with an intense review activity, published in excellent book series or in single books covering specific classes of alkaloids. Consequently, we decided to organize our handbook to present the different aspects of alkaloid science (e.g. the structure and pharmacology of bioactive alkaloids; recent advances in isolation, synthesis, and biosynthesis) in a single volume, aiming to provide representative examples of more recent and promising results as well as of future prospects in alkaloid science. Obviously, the present handbook cannot be regarded as a comprehensive presentation of alkaloid research, but we feel that the diversity of topics treated, ranging from bitterness to the anticancer activity of alkaloids, can provide a good idea of the variety of active research in this field.

In particular, Section I describes the structures and biological activities of selected classes of alkaloids. Almost half of the chapters focus their attention on terrestrial alkaloids (Chapters 1–5). The other half (Chapters 7–11) describe recent results in the field of marine alkaloids, while Chapter 6 is focused on neurotoxic alkaloids produced by cyanobacteria, microorganisms living in both marine and terrestrial environments. The particular emphasis on marine alkaloids undoubtedly reflects our long-standing research activity on marine metabolites, but it is also a result of the impressive amount of work carried out in the last few decades on marine natural product chemistry. Section II (Chapters 12–15) gives an account of modern techniques used for the detection and structural elucidation of alkaloids, while Section III is divided into two parts: different methodologies for the synthesis of alkaloids and accounts of modern biosynthetic studies.

Modern Alkaloids: Structure, Isolation, Synthesis and Biology. Edited by E. Fattorusso and O. Taglialatela-Scafati

ISBN: 978-3-527-31521-5

Finally, we should point out that even today the term alkaloid is ambiguous (a discussion on the definition of alkaloid is presented in Chapter 4). The initial definition of Winterstein and Trier (1910) ("nitrogen-containing basic compounds of plant or animal origin") has obviously been superseded. The most recent definition of alkaloid can be attributed to S. W. Pelletier (1984): "compound containing nitrogen at a negative oxidation level characterized by a limited distribution in Nature". In the preparation of this handbook we have decided to follow this last definition and, thus, to include "borderline" compounds such as capsaicins and non-ribosomal polypeptides.

We cannot conclude without thanking all the authors who have made their expert contributions to the realization of this volume, which we hope will stimulate further interest in one of the most fascinating branches of natural product chemistry.

Naples, July 2007

Ernesto Fattorusso
Orazio Taglialatela-Scafati

List of Contributors

Anna Aiello
Università di Napoli "Federico II"
Dipartimento di Chimica delle
Sostanze Naturali
Via D. Montesano, 49
80131 Napoli
Italy

Raymond J. Andersen
University of British Columbia
Biological Sciences 1450
Vancouver BC, V6T 1Z1
Canada

Giovanni Appendino
Università del Piemonte Orientale
Largo Donegani, 2
28100 Novara
Italy

Prabhat Arya
National Research Council of Canada
Steacie Institute for Molecular Sciences
100 Sussex Drive,
Ottawa, Ontario, K1A 0R6,
Canada

Naoki Asano
Hokuriku University
Faculty of Pharmaceutical Sciences
Ho-3 Kanagawa-machi
Kanazawa, 920-1181
Japan

Christian Bailly
INSERM U-524, Centre Oscar
Lambret
Place de Verdun
59045 Lille
France

Angela Bassoli
Università di Milano
Dipartimento di Scienze Molecolari
Agroalimentari
Via Celoria, 2
20133 Milano
Italy

Roberto G.S. Berlinck
University of Sao Paulo
CP 780, CEP 13560-970
3566590 - Sao Carlos, SP
Brazil

Stefan Bieri
Official Food Control
Authority of Geneva
20, Quai Ernest-Ansermet
1211 Geneva 4
Switzerland

Modern Alkaloids: Structure, Isolation, Synthesis and Biology. Edited by E. Fattorusso and O. Taglialatela-Scafati

ISBN: 978-3-527-31521-5

Gigliola Borgonovo
Università di Milano
Dipartimento di Scienze Molecolari Agroalimentari
Via Celoria, 2
20133 Milano
Italy

Gilberto Busnelli
Università di Milano
Dipartimento di Scienze Molecolari Agroalimentari
Via Celoria, 2
20133 Milano
Italy

Yeun-Mun Choo
University of Mississippi
Department of Pharmacognosy
Mississippi, MS 38677
USA

Philippe Christen
University of Lausanne
School of Pharmaceutical Science EPGL
30, Quai Ernest Ansermet
1211 Genèva 4
Switzerland

Steven M. Colegate
CSIRO Livestock Industries
Private Bag 24
East Geelong, Victoria 3220
Australia

Muriel Cuendet
Gerald P. Murphy Cancer Foundation
3000 Kent Ave, Suite E 2-400
West Lafayette, IN 47906
USA

Sabrina Dallavalle
Università di Milano
Dipartimento di Scienze Molecolari Agroalimentari
Via Celoria, 2
20133, Milano
Italy

Aleksej Dansiov
Department of Biochemistry
McGill University
3655 Promenade Sir William Osler
Montreal, Quebec H3G IV6
Canada

Ana R. Diaz-Marrero
Instituto de Productos Naturales y Agrobiología del CSIC,
Avda Astrofisico F. Sánchez 3
Apdo 195
38206 La Laguna
Tenerife
Spain

Ernesto Fattorusso
Università di Napoli "Federico II"
Dipartimento di Chimica delle Sostanze Naturali
Via D. Montesano, 49
80131 Napoli
Italy

Trina L. Foster
Apoptosis Research Centre
Children's Hospital of Eastern Ontario (CHEO)
401 Smyth Road
Ottawa K1H 8L1
Canada

Dale R. Gardner
Poisonous Plant Research Lab
USDA, Agricultural Research Service
1150 E 1400 N
Logan
Utah, 84341
USA

Kalle Gehring
Department of Biochemistry
McGill University
3655 Promenade Sir William Osler
Montreal
Quebec H3G IV6
Canada

William Gerwick
University of California at San Diego
Scripps Institution of Oceanography
9500 Gilman Drive
La Jolla, CA 92093-0210
USA

Christopher A. Gray
University of British Columbia
Chemistry of Earth and Ocean
Sciences
2146 Health Sciences Mall
Vancouver
British Columbia V6T 1Z1
Canada

Gordon W. Gribble
Dartmouth College
Department of Chemistry
6128 Burke Laboratory
Hanover, NH 03755
USA

Rashel V. Grindberg
University of California, San Diego
Center for Marine Biotechnology and
Biomedicine
Scripps Institution of Oceanography
and The Skaggs School of Pharmacy
and Pharmaceutical Sciences,
La Jolla, California 92093
USA

Mark T. Hamann
University of Mississippi
Department of Pharmacognosy
Mississippi, MS 38677
USA

Jerome Kluza
INSERM U-524, Centre Oscar
Lambret
Place de Verdun
59045 Lille
France

Hans-Joachim Knölker
University of Dresden
Institut für Organische Chemie
Bergstrasse 66
01069 Dresden
Germany

Jun'ichi Kobayashi
Hokkaido University
Graduate School of Pharmaceutical
Sciences
Sapporo 060-0812
Japan

Robert G. Korneluk
National Research Council of Canada
Steacie Institute for Molecular Sciences
100 Sussex Drive,
Ottawa, Ontario, K1A 0R6,
Canada

Miriam H. Kossuga
Instituto de Química de São Carlos
Universidade de São Paulo
CP 780
CEP 13560–970
São Carlos
Brazil

Philippe Marcetti
INSERM U-524, Centre Oscar Lambret
Place de Verdun
59045 Lille
France

Gary E. Martin
Schering - Plough Research Institute
Pharmaceutical Science
556 Morris Avenue
Summit, NJ 07901
USA

Lianne McHardy
University of British Columbia
Biological Sciences 1450
Vancouver BC, V6T 1Z1
Canada

Carmen Mendez
Universidad de Oviedo
Departamento de Biología Funcional
C/. Julián Claveria, s/n
33006 Oviedo
Spain

Marialuisa Menna
Università di Napoli "Federico II"
Dipartimento di Chimica delle Sostanze Naturali
Via D. Montesano, 49
80131 Napoli
Italy

Lucio Merlini
Università di Milano
Dipartimento di Scienze Molecolari Agroalimentari
Via Celoria, 2
20133, Milano
Italy

Hiroshi Morita
Hokkaido University
Graduate School of Pharmaceutical Sciences
Sapporo 060-0812
Japan

Mohammed Naim
Biotechnology Research Institute
National Research Council of Canada
6100 Royalmount Avenue
Montréal, Quebec, H4P 2R2
Canada

John M. Pezzuto
University of Hawaii
Hilo College of Pharmacy
60 Nowelo St., Suite
Hilo, Hawaii 96720
USA

Michael Prakesch
National Research Council of Canada
Steacie Institute for Molecular Sciences
100 Sussex Drive,
Ottawa, Ontario, K1A 0R6,
Canada

Jangnan Peng
University of Mississippi
Department of Pharmacognosy
Mississippi, MS 38677
USA

Karumanchi V. Rao
University of Mississippi
Department of Pharmacognosy
Mississippi, MS 38677
USA

Michel Roberge
University of British Columbia
2146 Health Sciences Mall
Vancouver BC, V6T 1Z3
Canada

Jose A. Salas
Universidad de Oviedo
Departamento de Biología Funcional
C/. Julián Claveria, s/n
33006 Oviedo
Spain

Cesar Sanchez
Universidad de Oviedo
Departamento de Biología Funcional
C/. Julián Claveria, s/n
33006 Oviedo
Spain

Cynthia F. Shumann
University of California, San Diego
Center for Marine Biotechnology and Biomedicine
Scripps Institution of Oceanography and The Skaggs School of Pharmacy and Pharmaceutical Sciences,
La Jolla, California 92093
USA

Marina Solntseva
ACD Limited
Bakuleva 6, Str 1
117513 Moscow
Russia

Carla M. Sorrels
University of California, San Diego
Center for Marine Biotechnology and Biomedicine
Scripps Institution of Oceanography and The Skaggs School of Pharmacy and Pharmaceutical Sciences,
La Jolla, California 92093
USA

Traian Sulea
Biotechnology Research Institute
National Research Council of Canada
6100 Royalmount Avenue
Montréal, Quebec, H4P 2R2
Canada

Orazio Taglialatela-Scafati
Università di Napoli "Federico II"
Dipartimento di Chimica delle Sostanze Naturali
Via D. Montesano, 49
80131 Napoli
Italy

Jean-Luc Veuthey
University of Geneve
Faculty of Sciences
20, Bd d'Yvoy
1211 Genèva 4
Switzerland

Kaoru Warabi
University of British Columbia
Chemistry and Earth and Ocean Sciences
2146 Health Sciences Mall
Vancouver
British Columbia V6T1Z1
Canada

Anthony J. Williams
Chem Zoo
904 Tamaras Circle
Wake Forest, North Carolina 27587
USA

Josh Wingerd
University of California, San Diego
Center for Marine Biotechnology and
Biomedicine
Scripps Institution of Oceanography
and The Skaggs School of Pharmacy
and Pharmaceutical Sciences,
La Jolla, California 92093
USA

Michael Wink
University of Heidelberg,
Institute of Pharmacy and Molecular
Biotechnology
Im Neuenheimer Feld 364
69120 Heidelberg
Germany

I
Bioactive Alkaloids: Structure and Biology

Modern Alkaloids: Structure, Isolation, Synthesis and Biology. Edited by E. Fattorusso and O. Taglialatela-Scafati

ISBN: 978-3-527-31521-5

1
Ecological Roles of Alkaloids

Michael Wink

1.1
Introduction: Defense Strategies in Plants

Plants are autotrophic organisms and serve as both a major and the ultimate source of food for animals and microorganisms. Plants cannot run away or fight back when attacked by a herbivore, nor do they have an immune system to protect them against pathogenic bacteria, fungi, viruses, or parasites. Plants struggle for life, as do other organisms, and have evolved several strategies against herbivorous animals, parasites, microorganisms, and viruses. Plants also compete with neighboring plants for space, light, water, and nutrients [1–8].

Apparently plants have evolved both physical and chemical defense measures, similar to the situation of sessile or slow moving animals. Among physical defense strategies we find [8]

- formation of indigestible cell walls containing cellulose, lignin, or callose;
- presence of a hydrophobic cuticle as a penetration barrier for microbes and against desiccation;
- formation of a thick bark in roots and stems against water loss, microbes, and herbivores;
- development of spines, thorns, hooks, trichomes, and glandular and stinging hairs (often filled with noxious chemicals) against herbivores;
- formation of laticifers and resin ducts (filled with gluey and noxious fluids);
- a high capacity for regeneration so that parts that have been browsed or damaged by infection can be readily replaced (so-called open growth).

Secondly, plants are masters of chemical defense, with a fascinating ability to produce a high diversity of chemical defense compounds, also known as secondary metabolites or allelochemicals [1–17]. Chemical defense involves macromolecular compounds, such as diverse defense proteins (including chitinase [against fungal cell

Modern Alkaloids: Structure, Isolation, Synthesis and Biology. Edited by E. Fattorusso and O. Taglialatela-Scafati

ISBN: 978-3-527-31521-5

walls], β-1,3-glucanases [against bacteria], peroxidase, and phenolase, lectins, protease inhibitors, toxalbumins, and other animal-toxic peptides), polysaccharides, and polyterpenes. More diverse and more prominent are low molecular weight secondary metabolites, of which more than 100 000 have been identified in plants (Figure 1.1).

Among the secondary metabolites that are produced by plants, alkaloids figure as a very prominent class of defense compounds. Over 21 000 alkaloids have been identified, which thus constitute the largest group among the nitrogen-containing secondary metabolites (besides 700 nonprotein amino acids, 100 amines, 60 cyanogenic glycosides, 100 glucosinolates, and 150 alkylamides) [2,3,18,19]. However, the class of secondary metabolites without nitrogen is even larger, with more than 25 000 terpenoids, 7000 phenolics and polyphenols, 1500 polyacetylenes, fatty acids, waxes, and 200 carbohydrates.

1.2
Ecological Roles of Alkaloids

Alkaloids are widely distributed in the plant kingdom, especially among angiosperms (more than 20 % of all species produce alkaloids). Alkaloids are less common but present in gymnosperms, club mosses (*Lycopodium*), horsetails (*Equisetum*), mosses, and algae [1–5,17]. Alkaloids also occur in bacteria (often termed antibiotics), fungi, many marine animals (sponges, slugs, worms, bryozoa), arthropods, amphibians (toads, frogs, salamanders), and also in a few birds, and mammals [1–5,13,17,20].

Alkaloids are apparently important for the well-being of the organism that produces them (Figures 1.1–1.3). One of the main functions is that of chemical defense against herbivores or predators [2,3,8,18]. Some alkaloids are antibacterial, antifungal, and antiviral; and these properties may extend to toxicity towards animals. Alkaloids can also be used by plants as herbicides against competing plants [1,3,8,18]. The importance of alkaloids can be demonstrated in lupins which – as wild plants – produce quinolizidine alkaloids ("bitter lupins"), that are strong neurotoxins (Table 1.1) [21,22]. Since lupin seeds are rich in protein, farmers were interested in using the seeds for animal nutrition. This was only possible after the alkaloids (seed content 2–6 %) had been eliminated. Plant breeders created so-called sweet lupins with alkaloid levels below 0.02 %. If bitter and sweet lupins are grown together in the field it is possible to study the importance of alkaloids for defense. For example, Figure 1.3 shows that rabbits strongly discriminate between sweet and bitter lupins and prefer the former. This is also true for insects, as aphids and mining flies always favor sweet lupins. In the wild, sweet lupins would not survive because of the lack of an appropriate chemical defense [8,21].

Secondary metabolites are not only mono- but usually multifunctional. In many cases, even a single alkaloid can exhibit more than one biological function. During evolution, the constitution of alkaloids (that are costly to produce) has been modulated so that they usually contain more than one active functional group, allowing them to interact with several molecular targets and usually more than one group of enemies [3,18,19,21–24]. Many plants employ secondary metabolites (rarely alka-

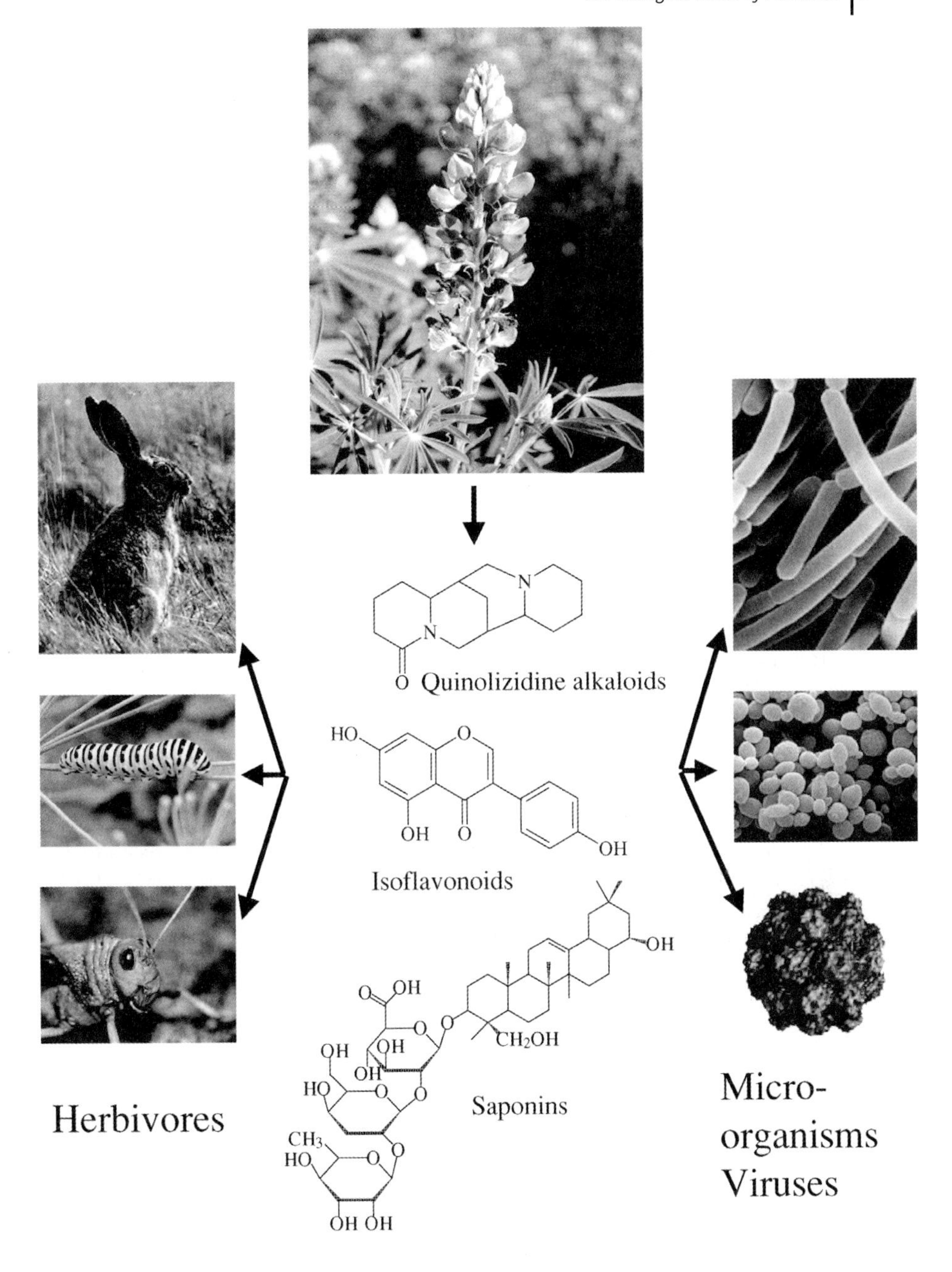

Fig. 1.1 Relationships between plants, their secondary metabolites, and potential enemies (herbivores, microorganisms, and viruses). Example: Lupins produce quinolizidine alkaloids, isoflavonoids, and saponins as main defense compounds.

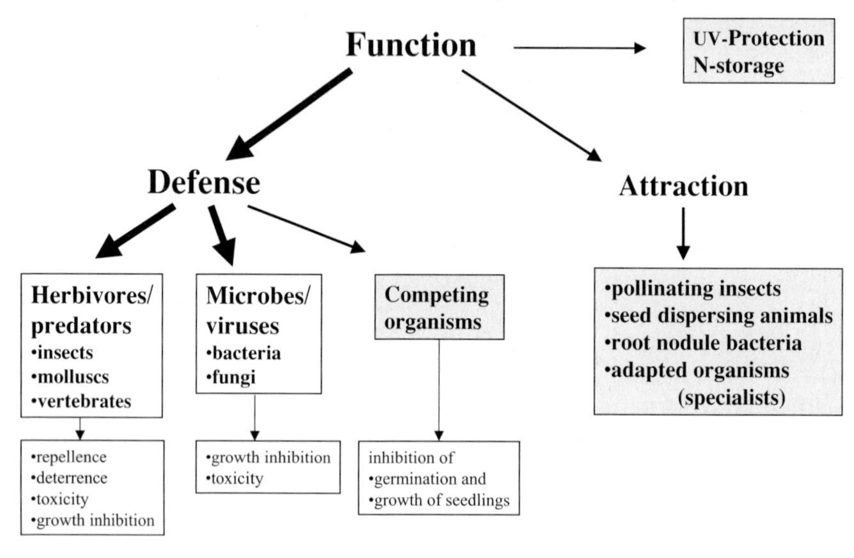

Fig. 1.2 Overview of the ecological functions of secondary metabolites.

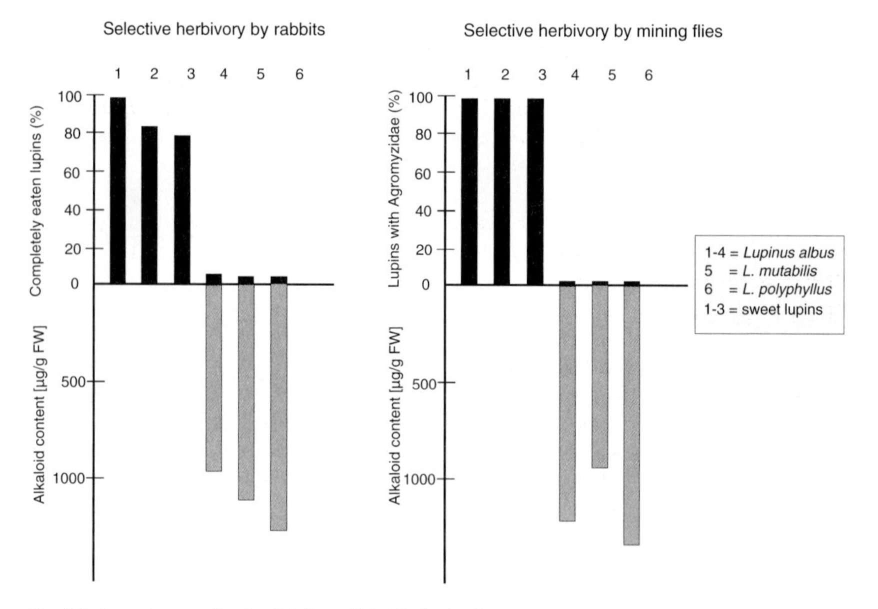

Fig. 1.3 Importance of quinolizidine alkaloids for lupins against herbivores.In this experiment, lupins with or without alkaloids were grown in the field. When rabbits got into the field, they preferentially consumed the sweet, alkaloid-free lupins. Also larvae of mining flies preferred sweet lupins.

Tab. 1.1 Molecular targets of alkaloids in neuronal signal transduction [2,3,19].

Target	Selected alkaloids
Neuroreceptor	
Muscarinic acetylcholine receptor	Hyoscyamine, scopolamine, and other tropane alkaloids (AA); acetylheliosupine and some other pyrrolizidine alkaloids; arecoline (A); berbamine, berberine, and other isoquinoline alkaloids; dicentrine and other aporphine alkaloids; strychnine, brucine; cryptolepine (AA); sparteine and other quinolizidine alkaloids (A); pilocarpine (A); emetine; himbacine and other piperidine alkaloids (A); imperialine (AA); muscarine (A)
Nicotinic acetylcholine receptors	Nicotine and related pyridine alkaloids (A); Ammodendrine (A); anabasine (A); arborine (AA); boldine and other aporphine alkaloids (AA); berberine and related protoberberine alkaloids; C-toxiferine (AA); coniine and related piperidine alkaloids (A); cytisine, lupanine, and other quinolizidine alkaloids (A); tubocurarine (AA); codeine (A); erysodine and related Erythrina alkaloids (AA); histrionicotoxin (AA); lobeline (A); methyllycaconitine (AA); pseudopelletierine (A)
Adrenergic receptors	Acetylheliosupine and related pyrrolizidine alkaloids; ajmalicine, reserpine (AA); arecoline; berbamine, berberine, laudanosine, and other isoquinoline alkaloids (AA); boldine, glaucine, and other aporphine alkaloids (AA); cinchonidine and other quinoline alkaloids; corynanthine, yohimbine, and other indole alkaloids (AA); emetine; ephedrine; ergometrine, ergotamine, and related ergot alkaloids (A/AA); ephedrine and related phenylethylamines (A); higenamine (A); *N*-methyldopamine, octopamine (A)
Dopamine receptor	Agroclavine, ergocornine, and related ergot alkaloids (A); bulbocapnine and related aporphine alkaloids (AA); anisocycline, stylopine, and related protoberberine alkaloids; salsolinol and related isoquinolines (A); tyramine and derivatives (A)
GABA receptor	Bicuculline (AA), cryptopine, hydrastine, corlumine, and related isoquinoline alkaloids (AA); securinine; harmaline and related β-carboline alkaloids (A); muscimol (A); securinine (AA)
Glycine receptor	Corymine, strychnine, and related indole alkaloids (AA)
Glutamate receptor	Histrionicotoxin and related piperidines (AA); ibogaine and related indole alkaloids (AA); nuciferine and related aporphine alkaloids (AA)
Serotonine receptor	Akuaminine and related indole alkaloids (A); annonaine, boldine, liriodenine and related aporphine alkaloids (AA); berberine and related protoberberine alkaloids; ergotamine, ergometrine, and related ergot alkaloids (AA); psilocin, psilocybine (A); bufotenine, *N,N*-dimethyltryptamine, and related indoles (A); harmaline and related β-carboline alkaloids (A); kokusagine and related furoquinoline alkaloids (AA); mescaline (A); ibogaine and other monoterpene indole alkaloids (A); gramine; *N,N*-dimethyltryptamine and derivates (AA)
Adenosine receptor	Caffeine, theobromine, and other purine alkaloids (AA)
Opiate receptor	Morphine and related morphinan alkaloids (A); akuammine, mitragynine (A), ibogaine and related indole alkaloids

(continued)

Tab. 1.1 *(Continued)*

Target	Selected alkaloids
Acetylcholine esterase	Galanthamine (AA); physostigmine and related indole alkaloids (AA); berberine and related protoberberine alkaloids (AA); vasicinol and related quinazolines (AA); huperzine (AA); harmaline and related β-carboline alkaloids (AA); demissine and related steroidal alkaloids (AA)
Monoamine oxidase	Harmaline and related β-carbolines (AA); carnegine, salsolidine, *O*-methylcorypalline, and related isoquinolines (AA); *N,N*-dimethyltryptamine and related indoles (AA);
Neurotransmitter uptake (transporter)	Ephedrine and related phenylalkyl amines (AA); reserpine, ibogaine, and related indole alkaloids (AA); cocaine (AA); annonaine and related aporphine alkaloids (AA); arecaidine (AA); norharman and related β-carboline alkaloids (AA); salsolinol and related isoquinolines (AA)
Na^+, K^+ channels	Aconitine and related diterpene alkaloids (A); veratridine, zygadenine, and related steroidal alkaloids (A); ajmaline, vincamine, ervatamine, and other indole alkaloids (AA); dicentrine and other aporphine alkaloids (AA); gonyautoxin (AA); paspalitrem and related indoles (AA); phalloidin (AA); quinidine and related quinoline alkaloids (AA); sparteine and related quinolizidine alkaloids (AA); saxitoxin (AA); strychnine (AA); tetrodotoxin (AA)
Ca^{2+} channels	Ryanodine (A); tetrandrine, berbamine, antioquine, and related bis-isoquinoline alkaloids (AA); boldine, glaucine, liriodenine, and other aporphine alkaloids (AA); caffeine and related purine alkaloids (A/AA); cocaine (AA); corlumidine, mitragynine, and other indole alkaloids (A/AA); bisnordehydrotoxiferine (AA)
Adenylate cyclase	Ergometrine and related ergot alkaloids (AA); nuciferine and related aporphine alkaloids (AA)
cAMP phosphodiesterase	Caffeine and related purine alkaloids (AA); papaverine (AA); chelerythrine, sanguinarine, and related benzophenanthridine alkaloids (AA); colchicines (AA); infractine and related indole alkaloids (AA)
Protein kinase A (PKA)	Ellipticine and related indole alkaloids (AA)
Protein kinase C (PKC)	Cepheranthine and related bis-isoquinoline alkaloids (AA); michellamine B and related isoquinoline alkaloids (AA); chelerythrine and related benzophenanthridine alkaloids (AA); ellipticine and related indole alkaloids (AA)
Phospholipase (PLA_2)	Aristolochic acid and related aporphine alkaloids (AA); berbamine and related bis-isoquinoline alkaloids (AA)

A = agonist; AA = antagonist.

loids, mostly colored phenolics and fragrant terpenoids) to attract pollinating and seed-dispersing animals; the compounds involved are usually both attractant and feeding deterrents. Attracted animals are rewarded by nectar or fleshy fruit tissues but should leave seeds or flowers undamaged. Hence, a multifunctional or pleiotropic effect is a common theme in alkaloids and other secondary metabolites.

An alkaloid never occurs alone; alkaloids are usually present as a mixture of a few major and several minor alkaloids of a particular biosynthetic unit, which differ in functional groups. Furthermore, an alkaloid-producing plant often concomitantly accumulates mixtures of other secondary metabolites, mostly those without nitrogen, such as terpenoids and polyphenols, allowing them to interfere with even more targets in animals or microorganisms. When considering the total benefits to a plant from secondary metabolites or the pharmacological activities of a drug, the potential additive or even synergistic effect of the different groups of secondary metabolites should be taken into account [10,25].

The multiple functions that alkaloids can exhibit concomitantly include a few physiological tasks: sometimes, alkaloids also serve as toxic nitrogen storage and nitrogen transport molecules [3,8]. Plants that produce few and large seeds, nearly always invest in toxic defense compounds (often alkaloids) that are stored together with proteins, carbohydrates, or lipids [8]. Since nitrogen is a limiting factor for plant growth, nitrogen apparently is a valuable asset for plants. In many species that store nitrogen in proteins and/or secondary metabolites in seeds or tubers, a remobilization has been observed after germination or regrowth in spring [2]. In plants that shed their leaves, alkaloids are usually exported to storage organs prior to leaf fall [2]. Alkaloids are definitely not waste products as had previously been assumed.

Aromatic and phenolic compounds can mediate UV-protecting activities, which might be favorable for plants living in UV-rich environments, such as high altitudes [1]. Alkaloids (such as isoquinoline, quinoline, and indole alkaloids) that derive from aromatic amino acids, such as phenylalanine, tyrosine, and tryptophan, may have UV-absorbing properties, besides antiherbivoral and antimicrobial activities.

Only the defensive properties of alkaloids will be discussed in more detail in this chapter.

1.3 Modes of Action

In order to deter, repel, or inhibit the diverse set of potential enemies, ranging from arthropods and vertebrates to bacteria, fungi, viruses, and competing plants, alkaloids must be able to interfere with important cellular and molecular targets in these organisms. A short overview of these potential targets is given in Figure 1.4a and b. The modulation of a molecular target will negatively influence its communication with other components of the cellular network, especially proteins (cross-talk of proteins) or elements of signal transduction. As a consequence, the metabolism and function of cells, tissues, organs, and eventually the whole organism will be affected and an overall physiological or toxic effect achieved. Although we know the structures of many secondary metabolites, our knowledge of their molecular modes of action is largely fragmentary and incomplete. Such knowledge is, however, important for an understanding of the functions of secondary metabolites in the producing organism, and for the rational utilization of secondary metabolites in medicine or plant protection [10,25].

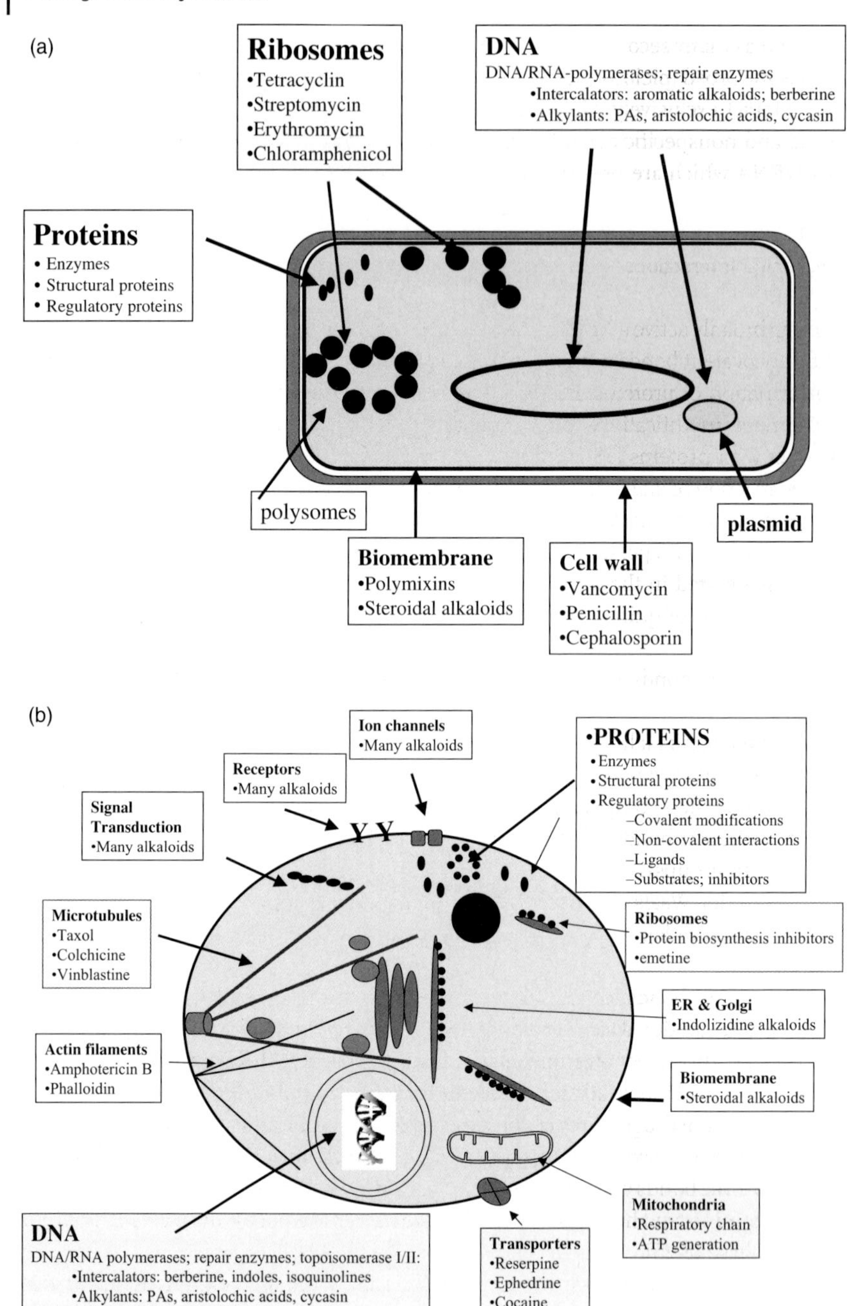

Fig. 1.4 Molecular targets for secondary metabolites, especially alkaloids. (a) Targets in bacterial cells, (b) targets in animal cells.

Whereas many secondary metabolites interact with multiple targets, and thus have unspecific broad (pleiotropic) activities, others, especially alkaloids, are more specific and interact exclusively with a single particular target. Secondary metabolites with broad and nonspecific activities interact mainly with proteins, biomembranes, and DNA/RNA which are present in all organisms.

1.3.1
Unspecific Interactions

Among broadly active alkaloids, a distinction can be made between those that are able to form covalent bonds with proteins and nucleic acids, and those that modulate the conformation of proteins and nucleic acids by noncovalent bonding.

Covalent modifications are the result when the following functional groups interact with proteins [18,25]:

- reaction of aldehyde groups with amino and sulfhydryl groups;
- reaction of exocyclic methylene groups with SH groups;
- reaction of epoxides with proteins and DNA (epoxides can be generated in the liver as a detoxification reaction);
- reaction of quinone structures with metal ions (Fe^{2+}/Fe^{3+}).

Noncovalent bonds are generated when the following groups interact with proteins [18,25]:

- ionic bonds (alkaloids with phenolic hydroxyl groups, that can dissociate as phenolate ions; alkaloid bases that are present as protonated compounds under physiological conditions);
- hydrogen bonds (alkaloids with hydroxyl groups, carbonyl, or keto groups);
- van der Waals and hydrophobic interactions (lipophilic compounds).

Noncovalent bonds, especially hydrogen bonds, ionic bonds, hydrophobic interactions, and van der Waals forces are weak individually, but can be powerful if they work cooperatively. For example, alkaloids with phenolic properties (found in several isoquinoline and indole alkaloids) usually have two or more phenolic hydroxyl groups that can form hydrogen bonds with proteins and nucleic acids. Furthermore, these OH groups may dissociate under physiological conditions to form phenate ions that can form ionic bonds with positively charged amino acid residues, such as those from lysine, arginine, and histidine. These OH groups are crucial for the biological activity of phenolics [18,25].

Molecules of nitrogen-containing compounds, such as alkaloids, amines, and peptides, usually contain (under physiological conditions) positively charged N-atoms that can form ionic bonds with negatively charged amino acid residues of glutamic and aspartic acid in proteins. Both the covalent and the noncovalent interactions will modulate the three-dimensional protein structure, that is, the conformation that is so important for the bioactivities of proteins (enzymes, receptors, transcription factors, transporters, ion channels, hormones, cytoskeleton). A conformational change is usually associated with a loss or reduction in the activity of a protein, leading to inhibition of enzyme or receptor activity or interference with the very important protein–protein interactions [17,18,25].

Lipophilic compounds, such as the various terpenoids, tend to associate with other hydrophobic molecules in a cell; these can be biomembranes or the hydrophobic core of many proteins and of the DNA double helix [10,18,24,25]. In proteins, such hydrophobic and van der Waals interactions can also lead to conformational changes, and thus protein inactivation. A major target for terpenoids, especially saponins, is the biomembrane. Saponins (and, among them, the steroid alkaloids) can change the fluidity of biomembranes, thus reducing their function as a permeation barrier. Saponins can even make cells leaky, and this immediately leads to cell death. This can easily be seen in erythrocytes; when they are attacked by saponins these cells burst and release hemoglobin (hemolysis) [1,6,17]. Among alkaloids, steroidal alkaloids (from Solanaceae) and other terpenoids have these properties.

These pleiotropic multitarget bioactivities are not specific, but are nevertheless effective, and this is critical in an ecological context. Compounds with pleiotropic properties have the advantage that they can attack any enemy that is encountered by a plant, be it a herbivore or a bacterium, fungus, or virus. These classes of compounds are seldom unique constituents; quite often plants produce a mixture of secondary metabolites, often both phenolics and terpenoids, and thus exhibit both covalent and noncovalent interactions. These activities are probably not only additive but synergistic [10,25].

1.3.2 Specific Interactions

Plants not only evolved allelochemicals with broad activities (see Section 1.3.1) but also some that can interfere with a particular target [3,6,17–19,25]. Targets that are present in animals but not in plants are nerve cells, neuronal signal transduction, and the endocrinal hormone system. Compounds that interfere with these targets are usually not toxic for the plants producing them. Plants have had to develop special precautions (compartmentation: resin ducts, trichomes, laticifers) in order to store the allelochemicals with broad activities that could also harm the producer.

Many alkaloids fall into the class of specific modulators and have been modified during evolution in such a way that they mimic endogenous ligands, hormones, or substrates [1,3,18,19]. We have termed this selection process "evolutionary molecular modeling" [12,13,19,23]. Many alkaloids are strong neurotoxins that were selected for defense against animals [2,3,19]. Table 1.1 summarizes the potential neuronal targets that can be affected by alkaloids. Extensive reviews on this topic have been published [2,3,19].

Neurotransmitters derive from amino acids; most of them are amines that become protonated under physiological conditions. Since alkaloids also derive from amino acids (often the same ones as neurotransmitters) it is no surprise that several alkaloids have structural similarities to neurotransmitters. They can be considered as neurotransmitter analogs (Figure 1.5a–c).

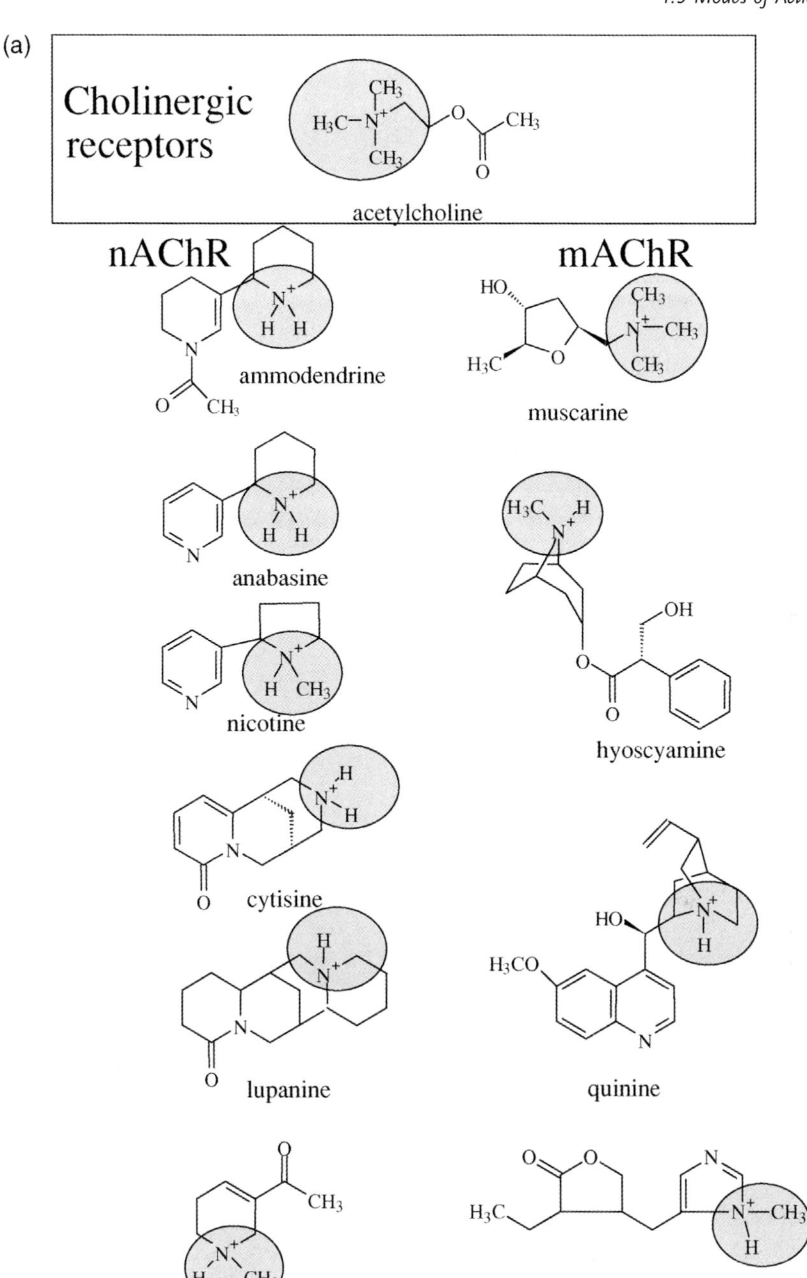

Fig. 1.5 Agonistic or antagonistic modulation of neuroreceptors by alkaloids that mimic neurotransmitters. (a) Interaction at cholinergic neurotransmitters that bind acetylcholine: nicotinic acetylcholine receptor (nAChR) and muscarinic acetylcholine receptors (mAChR), (b) interaction at adrenergic receptors that bind noradrenaline and adrenaline, (c) interaction at serotonergic receptors that bind serotonin.

(b)

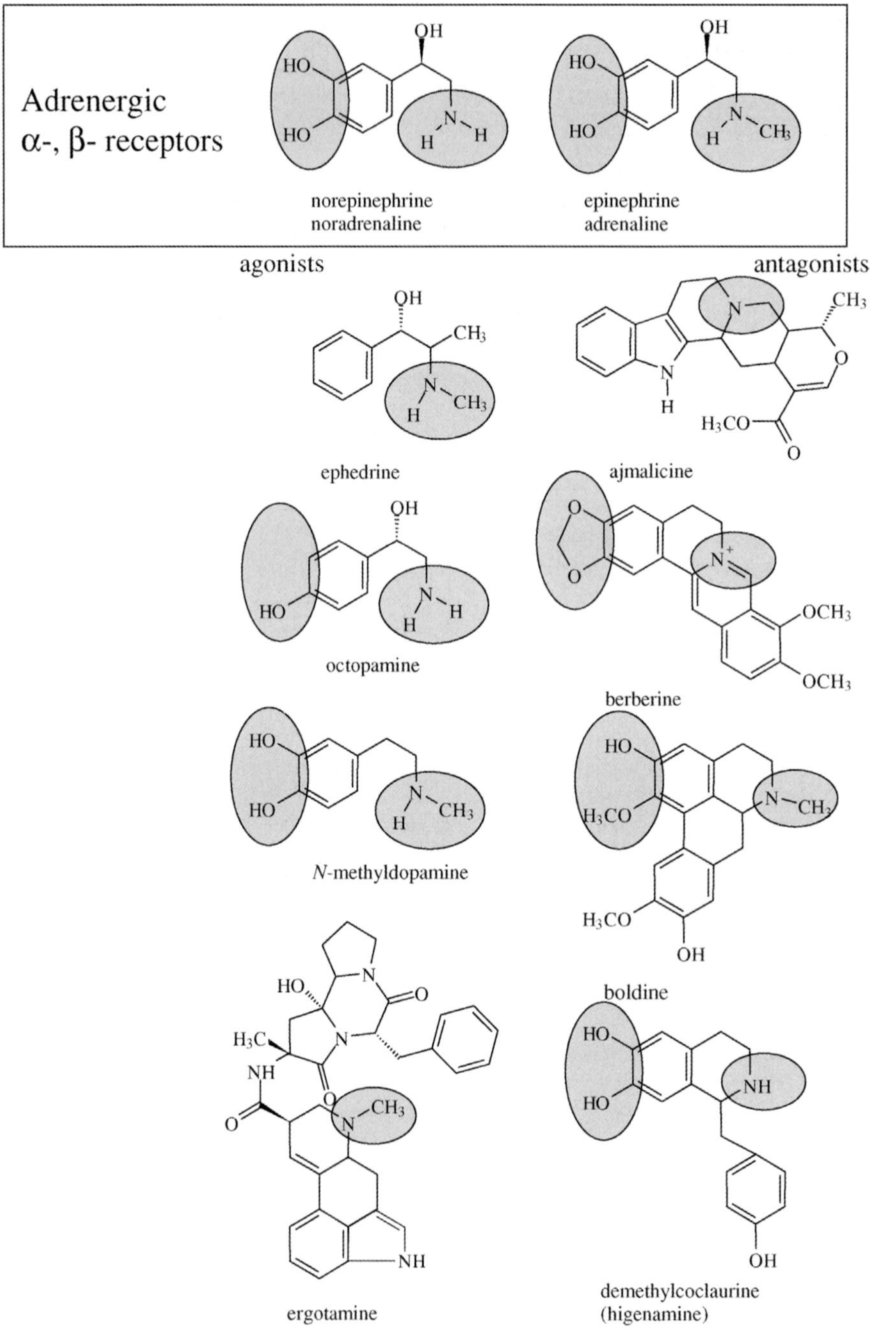

Adrenergic α-, β- receptors

norepinephrine
noradrenaline

epinephrine
adrenaline

agonists

antagonists

ephedrine

ajmalicine

octopamine

berberine

N-methyldopamine

boldine

ergotamine

demethylcoclaurine
(higenamine)

Fig. 1.5 (*Continued*)

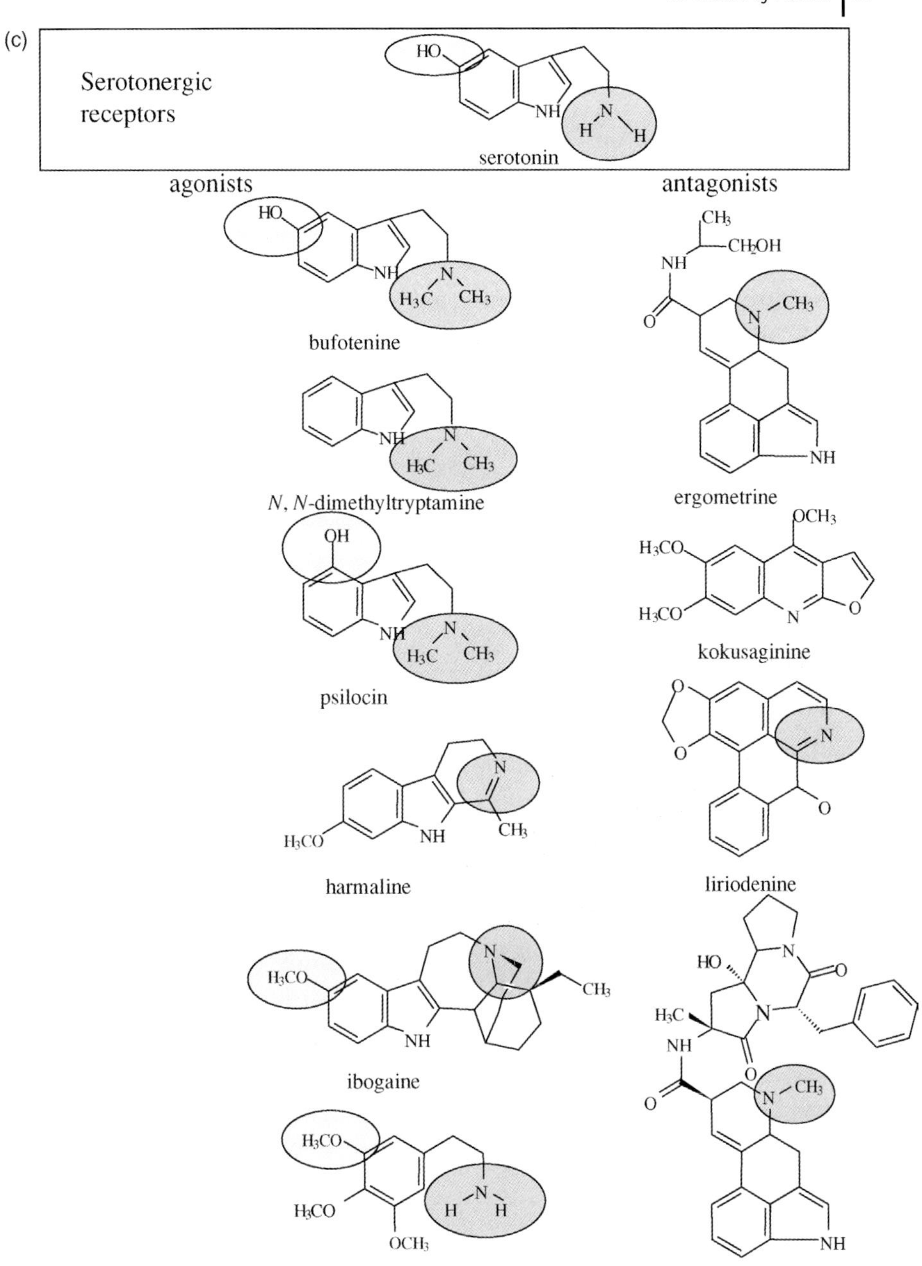

Fig. 1.5 *(Continued)*

Alkaloids that structurally mimic neurotransmitters can bind to neuroreceptors and either activate (agonists) or inactivate (antagonists) them (Table 1.1). Additional important targets are ion channels, such as the Na^+, K^+, and Ca^{2+} channels; several alkaloids are known that inhibit or activate these ion channels (Table 1.1).

Neuronal signal transduction is a very critical target in animals, since all organs are controlled by either the parasympathetic or the sympathetic nervous system. Its disturbance stops organ function (heart and circulation, respiration), mobility, orientation, and ability for flight in most animals. Many alkaloids are indeed strong (even deadly) neurotoxins or have mind-altering and hallucinogenic properties [1–3,17,19].

1.3.3
Cytotoxicity of Alkaloids

Many alkaloids are infamous for their strong toxicity towards animals and humans. Most of the deadly alkaloids fall into the class of neurotoxins (see above). The others have cytotoxic properties (Table 1.2). A cytotoxic effect can be generated when cell membranes are made leaky (as by saponins or steroidal alkaloids), or when elements of the cytoskeleton are inhibited. The spindle poisons vinblastine, vincristine, colchicine, and taxol are particularly famous. Actin filament formation is blocked by fungal poisons such as phalloidin from *Amanita phalloides.*

DNA can also be a target for alkaloids: planar and lipophilic alkaloids, such as berberine and sanguinarine (Figure 1.6) are intercalating compounds that assemble between the stacks of paired nucleotides in the DNA double helix [2,3,18,23]. DNA intercalation can disturb replication, DNA repair, and DNA topoisomerases. Frameshift mutations are one of the adverse consequences of intercalating compounds. Some alkaloids, such as pyrrolizidine alkaloids, aristolochic acids, cycasin, and furoquinoline alkaloids, are known to form covalent adducts with DNA bases. Mutations and tumor formation can be the result of such interactions. DNA alkylation occurs in some alkaloids only after activation by liver enzymes, such as cytochrome p450 oxidases (pyrrolizidine alkaloids, aristolochic acids) [17,18,24].

Ribosomal protein biosynthesis is often inhibited by alkaloids that interact with nucleic acids [23]. There are also more specific inhibitors, such as emetine.

Disturbances of the cytoskeleton, DNA replication, and DNA topoisomerase, or DNA alkylation and intercalation usually lead to cell death by apoptosis [18] (Table 1.2). The cytotoxic properties are usually not specific for animals but also affect bacteria, fungi, other plants, and even viruses. Alkaloids thus defend plants against a wide diversity of enemies. They have the disadvantage that a producing plant could theoretically kill itself by its own poison. Compartmentation, target-site insensitivity, and other mechanisms (which are largely unknown) must have evolved to overcome such problems.

Tab. 1.2 Molecular modes of action of selected cytotoxic alkaloids [3,18].

Alkaloid	Toxicity (animals)	Cyto-toxicity	Apoptosis	Micro-tubules	DNA topoiso-merase	Telo-merase	Membrane lysis	DNA inter-calation	DNA alkylation	Mutagenic	Inhibition of protein biosynthesis
Alkaloids derived from tryptophan											
Camptothecin	X	X	X		X			X			
Cinchonine, cinchonidine	X	X						X			
Cryptolepine	X	X	X		X	X		X			
Dictamnine	X	X						X	X	X	
Ellipticine	X				X			X		X	
Ergotamine	X		X					X			
Evodiamine	X	X		X							
Fagarine	X	X						X		X	
Harmine	X	X	X		X					X	
Quinine	X	X	X					X			
Vincristine	X	X	X	X				X			
Alkaloids derived from phenylalanine, tyrosine											
Aristolochic acids	X	X							X	X	
Berbamine	X							X			
Berberine	X	X	X		X	X		X			
Chelerythrine	X	X	x					X	X		
Chelidonine	X	X	X	X				X			
Colchicine	X	X	X	X							
Coralyne	X	X			X			X			
Dicentrine	X	X			X					X	

(continued)

Tab. 1.2 (*Continued*)

Alkaloid	Toxicity (animals)	Cyto-toxicity	Apoptosis	Micro-tubules	DNA topoiso-merase	Telo-merase	Membrane lysis	DNA inter-calation	DNA alkylation	Mutagenic	Inhibition of protein biosynthesis
Emetine	X	X	X					X			X
Fagaronine	X	X			X			X			
Liriodenine	X	X	X		X					X	
Lycorine	X	X	X								
Noscapine	X	X	X	X							
Piperine		X	X								
Salsolinol	X	X									
Sanguinarine	X	X	X					X	X	x	
Tetrandrine	X	X	X				X				
Alkaloids derived from ornithine, arginine											
Pyrrolizidine alkaloids	X	X	X						X	X	
Miscellaneous alkaloids											
Acronycine	X	X	X					X		X	
Cycasin	X								X	X	
Cyclopamine	X	X	X							X	
Lobeline	X							X			
Maytansine	X	X		X							
Paclitaxel	X	X	X	X							
Solarmargine	X	X	X				X				

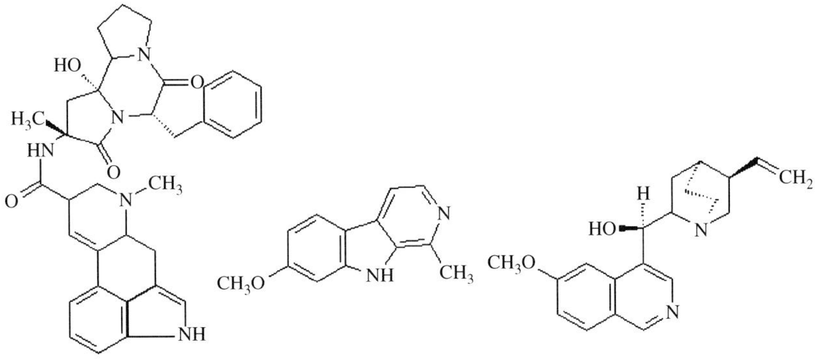

ergotamine (13.7 °C) harmine (16.1 °C) quinidine (8 °C)

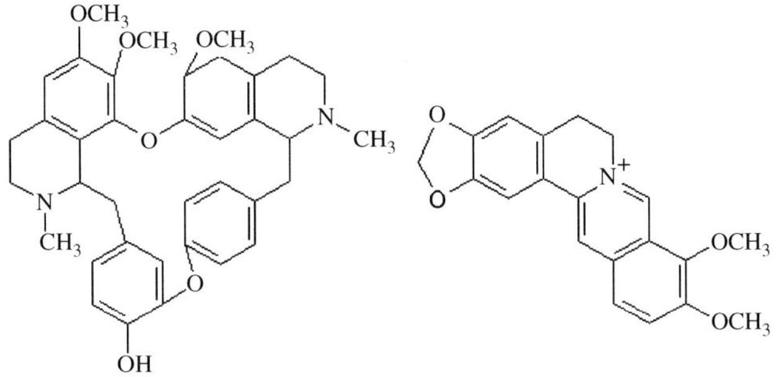

berbamine (13.2 °C) berberine (15.0 °C)

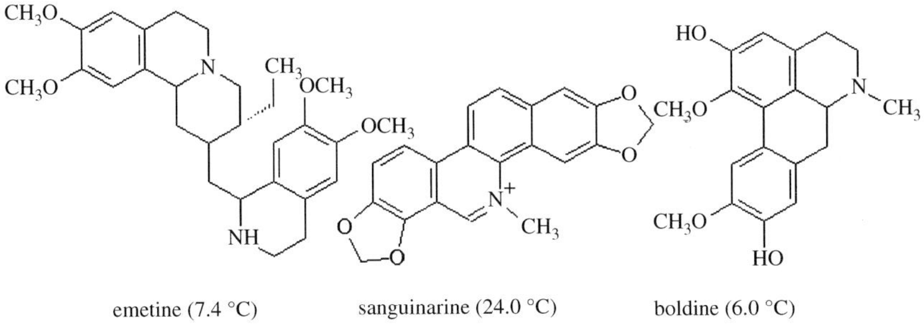

emetine (7.4 °C) sanguinarine (24.0 °C) boldine (6.0 °C)

Fig. 1.6 Examples of alkaloids that intercalate DNA. Intercalation increases the melting temperature of DNA; relevant T_m values are shown in parentheses.

1.4
Evolution of Alkaloidal Defense Systems

Alkaloids are apparently well-adapted molecules that can serve plants as potent defense chemicals which are used on their own or together with other mostly

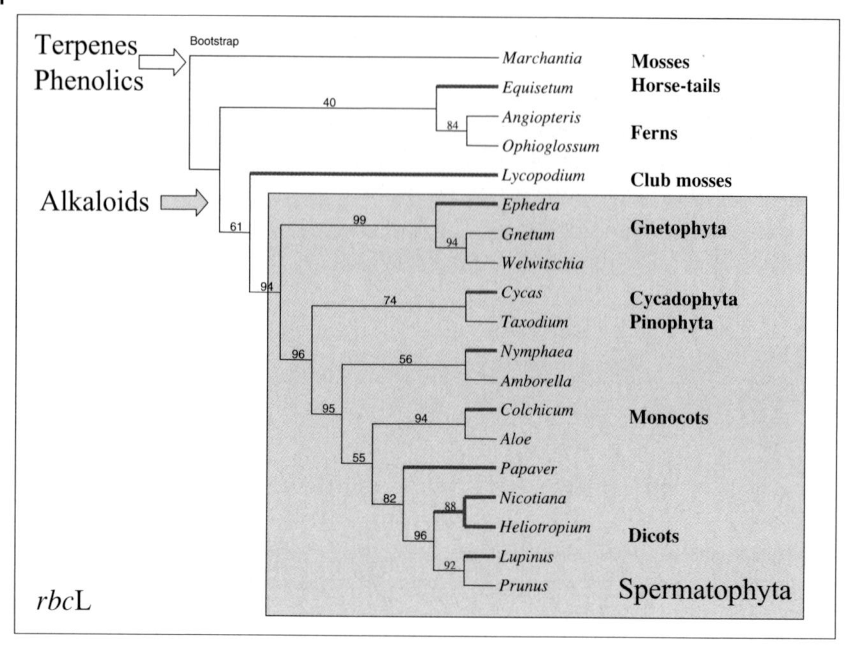

Fig. 1.7 Evolution of alkaloids in the phylogeny of plants. Using nucleotide sequences of the chloroplast gene *rbc*L a phylogenetic tree was computed with Maximum Parsimony. A bootstrap cladogram is shown with bootstrap values shown at the nodes. Branches leading to taxa that accumulate alkaloids are shown in bold.

co-occurring secondary metabolites, such as phenolics or terpenoids. We can speculate on the question of when alkaloid defense systems first arose during the evolution of plants. Figure 1.7 maps the character "alkaloid defense" on a phylogenetic tree that covers the whole plant kingdom. Alkaloids are present already in early branches of land plants, such as lycopods and horse-tails. They are also present in some members of the gymnosperms, especially in the Gnetales (i.e., *Ephedra*) and Cycadales (i.e., families Cycadaceae, Zamiaceae). Only a few conifers produce alkaloids (e.g., *Taxus, Harringtonia*). Within the angiosperms, however, alkaloid formation is a widely distributed trait and especially abundant in the families Solanaceae, Convolvulaceae, Fabaceae, Strychnaceae, Loganiaceae, Apocynaceae, Asclepiadaceae, Ranunculaceae, Papaveraceae, Berberidaceae, Fumariaceae, Buxaceae, Punicaceae, Celastraceae, Erythroxylacae, Zygophyllaceae, Rutaceae, Gelsemiaceae, Colchicaceae, Iridaceae, Boraginaceae, and Asteraceae. It can be speculated that the ancestors of present day plants developed alkaloids as defense chemicals early on because they had to face the attack of herbivorous animals (which were present already in the Cambrian) [2,12,26]. This would mean that genes for the biosynthesis of alkaloids are not only present in plants that actually produce the particular alkaloid but that they may be much more widely distributed among the plant kingdom [2,12,26].

Secondary metabolites with similar structural types and pharmacophoric groups can be seen in several bacteria (where they are often termed antibiotics if they have antimicrobial or cytotoxic properties). Since eukaryotic cells had taken up α-proteobacteria (which became mitochondria) and cyanobacteria (which became chloroplasts), they also inherited a number of genes that encode enzymes for pathways leading to secondary metabolites. Therefore, we may speculate that early plants already had the capacity of building defense compounds and that alkaloids were among the first. Since the numbers and types of herbivores and other enemies have increased within the last 100 million years, angiosperms have had to face more enemies and as a consequence have developed a more complex pattern of defense and signal compounds.

Alkaloids can assume their defense role if they are present at the right place, concentration, and time. Since alkaloids are costly for the plants to produce [2,12], usually only the most important plant tissues and organs (such as young leaves, flowers, seeds, and storage organs, such as roots and tubers) are heavily defended by them. For the same reason, alkaloids are not discarded with falling leaves or senescing tissues but remobilized and stored in seeds, roots, or tubers. Many secondary metabolites accumulate in special cells and tissues.

Several plants produce milk juice sequestered in laticifers; in several plant genera alkaloids are mainly stored in latex vesicles, such as isoquinoline alkaloids in *Papaver* and *Chelidonium*, or piperidine alkaloids in *Lobelia*. If herbivores wound such a plant, the latex will spill out and the herbivore will immediately be confronted with alkaloids. Since most of them are strong poisons, a deterrent effect is usually achieved. Another strategic way to store alkaloids is their sequestration in epidermal vacuoles or in trichomes. These tissues have to ward off not only herbivores (especially small ones) but also microorganisms in the first place. Several classes of alkaloids have been found in epidermal tissues, such as quinolizidine and tropane alkaloids [2,3].

Most plants produce an alkaloid in one organ and transport the alkaloids after synthesis, via either xylem or phloem (Table 1.3), to other plant tissues in which the alkaloids are stored for defense or signaling [2,11,21].

- Xylem transport has been reported for tropane alkaloids and nicotine, which are synthesized in roots but accumulate in aerial parts.

Tab. 1.3 Transport of alkaloids in plants [2].

Alkaloid	Type	Occurrence	Phloem	Xylem
Lupanine	Quinolizidine	*Lupinus, Genista, Cytisus Laburnum, Spartium*	Yes	No
Senecionine	Pyrrolizidine	*Senecio, Petasites, Adenostyles*	Yes	No
Aconitine	Diterpene	*Aconitum*	Yes	?
Swainsonine	Indolizidine	*Astragalus*	Yes	No
Nicotine	Pyrrolidine	*Nicotiana*	No	Yes
Hyoscyamine	Tropane	*Atropa, Datura, Hyoscyamus*	No	Yes
Rutacridone	Quinoline	*Ruta*	No	Yes

- Phloem transport is known for quinolizidine, pyrrolizidine alkaloids, and aconitine.

The transport of toxic alkaloids in the phloem can be an advantage for plants against phloem-feeding insects, such as aphids [8]. For example: alkaloid-rich lupins are avoided by aphids, whereas sweet lupins with very low alkaloid contents are preferred by polyphagous aphids [8,22].

No defense system is 100 % safe. This is also true for alkaloidal defense against herbivores. A few herbivores, mostly insects, have overcome the chemical defense system of their host plants by adapting to it [1,20,27]. A common theme in mono- and oligophagous insects is their ability not only to tolerate alkaloids but also to store them in their body (often the integuments). Alkaloids sequestered by insects include pyrrolizidine, quinolizidine alkaloids, and aconitine [1,3–5,20,27–29]. These specialized insects, which are often aposematically colored, employ the acquired alkaloids for their own defense against predators. The mechanisms by which these specialized insects overcome alkaloid toxicity remain open questions. It could be a target-site modification as observed for cardiac glycosides at the Na^+/K^+-ATPase in Monarch butterflies [30] or simply sequestration in tissues or cells without corresponding targets.

A comparable situation to insect specialists can be found in parasitic and hemiparasitic plants, another example of multitrophic interactions. In several instances it can be shown that the parasites can tap the xylem or phloem of their host plants and sequester the host alkaloids into their own system [2,31]. The parasites would gain chemical defense against herbivores by such a process (Table 1.4). In *Osyris alba* it can be shown that plants exist that can sequester the alkaloids of more than one host plant: that is, pyrrolizidine and quinolizidine alkaloids [32]. The situation of *Lolium* is even more complex [33]. If the grass *Lolium temulentum* is infected by an endophytic

Tab. 1.4 Transfer of alkaloids from host plants to parasitic and hemiparasitic plants [30–32].

Alkaloid	Type	Host plant	Hemiparasite/parasite
Sparteine	Quinolizidine	*Cytisus scoparius*	*Orobanche rapum-genistae*
Retamine	Quinolizidine	*Retama shaerocarpa*	*Viscum cruciatum*
Cytisine	Quinolizidine	*Genista acanthoclada*	*Cuscuta palaestina*
Lupanine	Quinolizidine	*Lupinus* spp.	*Cuscuta reflexa*; *Castilleja integra*
		Lupinus texensis	*Castilleja indivisa*
Thermopsine	Quinolizidine	*Lupinus argenteus*	*Castilleja miniata*
N-Methylcytisine	Quinolizidine	*Spartium junceum*	*Osyris alba*
Isolupanine	Quinolizidine	*Lupinus falcata*	*Pedicularis semibarbata*
Anagyrine	Quinolizidine	*Lupinus* spp.	*Pedicularis semibarbata*
Senecionine	Pyrrolizidine	*Senecio triangularis*	*Pedicularis semibarbata*
Senecionine	Pyrrolizidine	*Senecio* spp.; *Liatris punctata*	*Castilleja integra*
cis-Pinnidol	Piperidine	*Picea engelmannii*	*Arceutholobium microcarpum*
Norditerpene		*Delphinium occidentale*	*Castilleja sulphurea*
Loline	Pyrrolizidine	*Lolium temulentum*	*Rhinanthus minor*

fungus, it acquires the fungal toxins (the pyrrolizidine alkaloid loline). If *Lolium* was parasitized by *Rhinanthus minor*, a second transfer was observed into the hemiparasite, which could increase its fitness through this sequestration [33].

1.5 Conclusions

Alkaloids are not waste products but have evolved mainly as defense compounds against herbivores, but also against microbes, competing other plants, and even viruses. Although the production and storage of alkaloids is costly for plants, they are apparently a good investment against enemies. Alkaloid structures have been shaped during evolution so that they can interfere with critical targets in potential enemies. A disturbance of DNA/RNA and related enzymes, of the cytoskeleton, of ribosomal protein biosynthesis, and of membrane permeability by several alkaloids can be interpreted as a defense against all types of organism, ranging from bacteria to animals. The interference of alkaloids with neuroreceptors, ion channels, and other elements of the neuronal signal transduction chain is more specific and certainly a measure against animal herbivores.

Alkaloids do not only serve as poisons against herbivores and microorganisms; they can also be interesting and important in medicine as pharmaceutical agents. Given at a lower dose (than the plants use for defense) these alkaloids no longer work as poisons but can mediate useful pharmacological activities, such as reducing blood pressure, relieving pain and spasms, stimulating circulation and respiration, or killing tumor cells [10,14,25].

References

1 Harborne, J.B. (1993) *Introduction to Ecological Biochemistry*, 4th edn Academic Press, London.
2 Roberts, M.F. and Wink, M. (1998) *Alkaloids – Biochemistry, Ecological Functions and Medical Applications*, Plenum Press, New York.
3 Wink, M. (1993) Allelochemical properties and the raison d'etre of alkaloids. In Cordell G. (ed.): *The Alkaloids: Chemistry and Biology*, Vol. **43** Academic Press, San Diego, 1–118.
4 Rosenthal, G.A. and Berenbaum, M.R. (1991) Herbivores: their interactions with secondary plant metabolites, *The Chemical Participants*, Vol. **1** Academic Press, San Diego.
5 Rosenthal, G.A. and Berenbaum, M.R. (1992) Herbivores: their interactions with secondary plant metabolites, *Ecological and Evolutionary Processes*, Vol. **2** Academic Press, San Diego.
6 Seigler, D.S. (1998) *Plant Secondary Metabolism*, Kluwer, Boston.
7 Swain, T. (1977) *Annual Review of Plant Physiology and Plant Molecular Biology*, **28**, 479–501.
8 Wink, M. (1988) *Theoretical and Applied Genetics*, **75**, 225–233.
9 Dewick, P.M. (2002) Medicinal natural products, *A Biosynthetic Approach*, Wiley, New York.
10 van Wyk, B.-E. and Wink, M. (2004) *Medicinal Plants of the World*, Briza, Pretoria.
11 Luckner, M. (1990) *Secondary Metabolism in Microorganisms, Plants and Animals*, Springer, Heidelberg.

12 Wink, M. (1999) Biochemistry of plant secondary metabolism, *Annual Plant Reviews*, Vol. **2** Sheffield Academic Press, Sheffield.

13 Wink, M. (1999) Function of plant secondary metabolites and their exploitation in biotechnology, *Annual Plant Reviews*, Vol. **3** Sheffield Academic Press, Sheffield.

14 Mann, J. (1992) *Murder, Magic and Medicine*, Oxford University Press, London.

15 Hegnauer, R. and Hegnauer, M. (1994) *Chemotaxonomie der Pflanzen-Leguminosae*, Birkhäuser Verlag, Basel. Hegnauer, R. and Hegnauer, M. (1996) *Chemotaxonomie der Pflanzen-Leguminosae*, Birkhäuser Verlag, Basel. Hegnauer, R. and Hegnauer, M. (2001) *Chemotaxonomie der Pflanzen-Leguminosae*, Birkhäuser Verlag, Basel.

16 Hegnauer, R. (1962–1996) *Chemotaxonomie der Pflanzen*, Vols. **I–XI** Birkhäuser, Basel.

17 Teuscher, E. and Lindequist, U. (1998) *Biogene Gifte*, Wissenschaftliche Verlagsgesellschaft, Stuttgart.

18 Wink, M. *The Alkaloids* (ed. G., Cordell), Academic Press, in press.

19 Wink, M. (2000) Interference of alkaloids with neuroreceptors and ion channels. In Rahman A. U. (ed.): *Bioactive Natural Products*, Vol. **11** Elsevier, Amsterdam, pp. 3–129.

20 Blum, M.S. (1981) *Chemical Defenses of Arthropods*, Academic Press, New York.

21 Wink, M. (1992) The Role of quinolizidine alkaloids in plant insect interactions. In Bernays E.A. (ed.): *Insect–Plant Interactions*, Vol. **IV** CRC Press, Boca Raton, pp. 133–169.

22 Wink, M. (1993) Quinolizidine alkaloids. In Waterman P. (ed.): *Methods in Plant Biochemistry*, Vol. **8** Academic Press, London, pp. 197–239.

23 Wink, M., Schmeller, T., Latz-Brüning, B. (1998) *Journal of Chemical Ecology*, **24**, 1881–1937.

24 Wink, M. and Schimmer, O. (1999) Modes of action of defensive secondary metabolites. In Wink M. (ed.): *Function of Plant Secondary Metabolites and Their Exploitation in Biotechnology*, Vol. **3** Sheffield Academic Press and CRC Press, Annual Plant Reviews, pp. 17–133.

25 Wink, M. (2005) *Zeitschr Phytother*, teitschrift für Phytotherapie **26**, 271–274.

26 Wink, M. (2003) *Phytochemistry*, **64**, 3–19.

27 Eisner, T. (2004) *For Love of Insects*, Harvard University Press.

28 Eisner, T., Eisner, M., Siegler, M. (2005) *Secret Weapons: Defenses of Insects, Spiders, Scorpions, and Other Many-Legged Creatures*, Harvard University Press.

29 Brown, K.S. and Trigo, J.L. (1995) The ecological activity of alkaloids. Cordell G. (ed.): *The Alkaloids*, Vol. **47** Academic Press, San Diego, pp. 227–354.

30 Holzinger, F. and Wink, M. (1996) *Journal of Chemical Ecology*, **22**, 1921–1937.

31 Stermitz, F.R. (1998) Planta parasites. Roberts M. F. and Wink M. (eds.): *Alkaloids – Biochemistry, Ecological Functions and Medical Applications*, Plenum Press, New York, pp. 327–336.

32 Woldemichael, G. and Wink, M. (2002) *Biochemical Systematics and Ecology*, **30**, 139–149.

33 Lehtonen, P., Helander, M., Wink, M., Sporer, F. and Saikkonen, K. (2006) *Ecology letters*, **8**, 1256–1263.

2
Antitumor Alkaloids in Clinical Use or in Clinical Trials

Muriel Cuendet, John M. Pezzuto

2.1
Introduction

Cancer is a group of diseases characterized by uncontrolled growth and spread of abnormal cells. The estimated worldwide incidence of cancer is about 6 million new cases per year [1]. In the United States, cancer is now the leading cause of death for people younger than 85 years, and it is only exceeded by heart disease in older individuals [2]. The treatment of many diseases is highly dependent on natural products, and this is especially true for the treatment of cancer [3,4]. Unique classes of natural products, such as alkaloids, have shown significant antitumor action. In this chapter, we will focus attention on plant-derived alkaloids used in the clinic, including the *Vinca* alkaloids, analogs of camptothecin, and taxanes, as well as alkaloids currently in clinical trials used to treat, reverse multidrug resistance (MDR), or prevent cancer.

2.2
Antitumor Alkaloids in Clinical Use

2.2.1
Vinca Alkaloids

The vinca alkaloids, isolated from the Madagascar periwinkle, *Catharantus roseus* G. Don., comprise a group of about 130 terpenoid indole alkaloids [5]. Their clinical value was clearly identified as early as 1965 and so this class of compounds has been used as anticancer agents for over 40 years and represents a true lead compound for drug development [6]. Today, two natural compounds, vinblastine (VLB, **1**) and vincristine (VCR, **2**), and two semisynthetic derivatives, vindesine (VDS, **3**) and vinorelbine (VRLB, **4**), have been registered (Figure 2.1). Owing to the pharmaceutical importance and the low content of VLB and the related alkaloid VCR, *Catharanthus roseus* became one of the best-studied medicinal plants. Using super-acidic chemistry, a new family of such compounds was synthesized and vinflunine (VFL, **5**), a difluorinated derivative, was selected for clinical testing [7].

Modern Alkaloids: Structure, Isolation, Synthesis and Biology. Edited by E. Fattorusso and O. Taglialatela-Scafati

ISBN: 978-3-527-31521-5

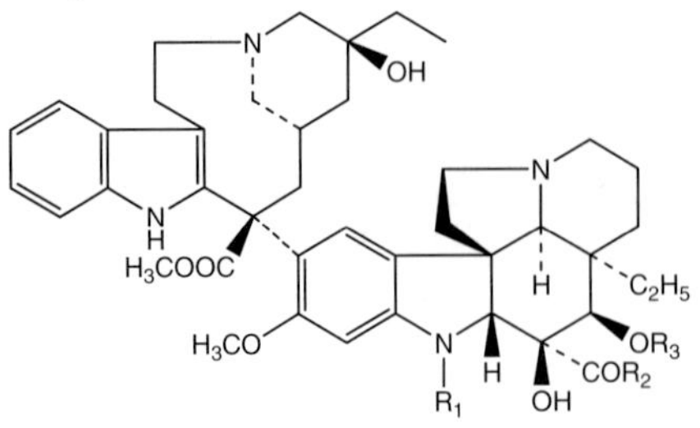

Compound	R_1	R_2	R_3
1	CH_3	OCH_3	$COCH_3$
2	CHO	OCH_3	$COCH_3$
3	CH_3	NH_2	H

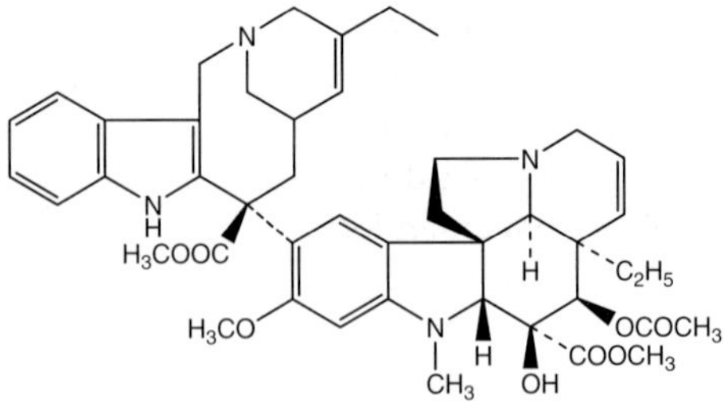

Fig. 2.1 Chemical structures of vinca alkaloids.

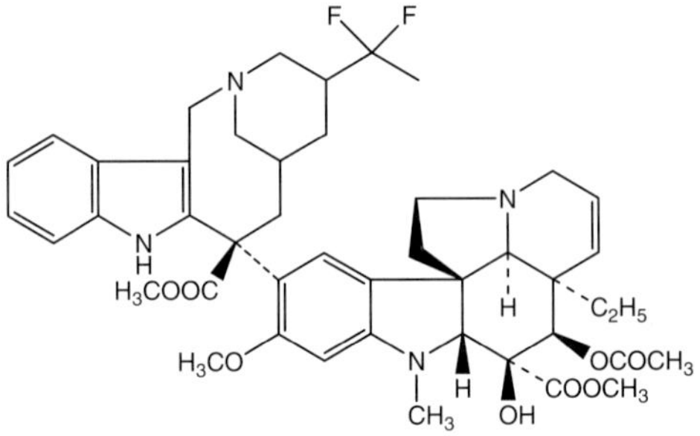

Among the many biochemical effects observed after exposure of cells and tissues to the vinca alkaloids are disruption of microtubules, inhibition of synthesis of proteins and nucleic acids, elevation of oxidized glutathione, alteration of lipid metabolism and the lipid content of membranes, elevation of cyclic adenosine monophosphate (cAMP), and inhibition of calcium-calmodulin-regulated cAMP phosphodiesterase [8–16]. Despite these multiple biochemical actions, the antineoplastic activity of the vinca alkaloids is usually attributed to disruption of microtubules, resulting in dissolution of mitotic spindles and metaphase arrest in dividing cells [9,17–20]. Microtubules are involved in many cellular processes in addition to mitosis, and exposure to vinca alkaloids gives rise to diverse biological effects, many of which could impair essential functions, both in dividing and in nondividing cells [21–23].

The effects of the vinca alkaloids on the organization and function of microtubules have been extensively characterized [8]. It appears that each heterodimer of α-β-tubulin possesses a single "vinca-specific" site of high intrinsic affinity and an unknown number of nonspecific sites of low affinity [24]. The relative strength of drug binding to vinca-specific sites on the α- and β-heterodimers of tubulins is VCR > VDS > VLB > VRLB > VFL [25–27]. From the several effects of VLB on the *in vitro* assembly of microtubules, it is generally assumed that the vinca alkaloids disrupt microtubules by more than one mechanism [28,29]. At low concentrations, VLB inhibits microtubule formation in a "substoichiometric" fashion in that assembly is blocked by binding of only a few molecules to high-affinity sites on tubulin heterodimers located at the ends of microtubules. At higher concentrations, disassembly results from binding of VLB to tubulin heterodimers located along the microtubule surface, through stoichiometric interaction with vinca-specific sites of reduced affinity and/or nonspecific ionic interaction.

Although there is no question that the vinca alkaloids disrupt microtubules, the biological mechanisms underlying the antineoplastic activity of these drugs are less certain. In actively proliferating cells, the mechanism of cytotoxicity of the vinca alkaloids is usually considered to be disruption of the mitotic spindle, resulting in metaphase arrest and, ultimately, cell death [9,17–20]. Among anticancer drugs, the vinca alkaloids are classified as mitotic inhibitors, with their primary site of action being M phase of the cell cycle, but it is not certain that mitotic inhibition is the predominant *in vivo* cytotoxic mechanism. Recent studies suggest that disruption of the cell cycle may lead to cell death through apoptosis [30,31]. The many biological actions of the clinically active vinca alkaloids are seen over a wide range of drug concentrations, and there are selective effects in various normal and neoplastic tissues. The difference in structure of these drugs does not alter the mechanism of action and binding to tubulin in any fundamental way, but structure is of considerable significance with regard to their clinical spectrum of antitumor efficacy and clinical toxicity. *In vivo* differences in biological activity must be due either to heterogeneity of expression of various tubulin isoforms in different tissues, or to differences in processes that influence interaction with tubulin by affecting drug binding or by limiting the availability of the drug [32–35]. A key determinant of the pharmacological activity of the vinca alkaloids in different tissue types appears to be cellular retention of the drug.

Resistance of tumor cells to the cytotoxic actions of the vinca alkaloids has been well-described experimentally and appears to have clinical correlates [36–38]. Although the vinca alkaloids bind to tubulin and disrupt microtubules, some tumor cells express resistance to these agents through a mechanism that does not appear to involve alterations in tubulin binding. When studied with cultured tumor cells, resistance to the vinca alkaloids appears to be due primarily to their decreased accumulation and retention [39]. The altered cellular pharmacology is mediated by the action of P-glycoprotein (Pgp), which is expressed in the plasma membrane of the drug-resistant tumor cells [36]. This protein, encoded by the MDR1 gene, spans the membrane 12 times and forms a pore or channel in the membrane through which drugs are transported [40]. P-Glycoprotein appears to bind the vinca alkaloids and extrudes them from the tumor cell through a process that requires energy [41,42]. This is the most likely mechanism leading to cross-resistance with other compounds such as colchicines, etoposide, or doxorubicin, although alternative mechanisms have been proposed [36]. Novel drug resistance-associated proteins, such as MPR and LRP/MVP, that may have an impact on the cellular and clinical resistance to the vinca alkaloids, have been identified [43,44]. The degree of resistance between vinca alkaloids may differ, and the dose and intensity of therapy may determine the response rate in vinca-resistant disease [45]. It is known that many of the drugs that can circumvent vinca-alkaloid resistance or MDR also bind to Pgp and compete with the anticancer drug for binding to this protein [46,47]. As a consequence of this competition, the efflux of the anticancer drug from the resistant tumor cell is blocked, and concentrations in the cell rise to levels that are apparently cytotoxic [48]. P-Glycoprotein, which is central to vinca-alkaloid resistance and MDR, is also expressed in normal tissues. Drugs that bind to and inhibit tumor cell Pgp will also do the same to the Pgp of these normal tissues, consequently increasing their levels of the anticancer drug and causing unacceptable tissue-toxic effects [49,50]. Several important clinical trials of Pgp modulators have been performed [39,49]. The key finding is that the modulator can increase the plasma levels and half-lives of the anticancer agents, possibly through inhibition of hepatic Pgp, showing that modulators of Pgp have potential pharmacokinetic actions that may necessitate decreasing the dose of the anticancer drug administered [51,52]. Vinca alkaloids have been incorporated into combination chemotherapy protocols, based not only on their lack of cross-resistance with drugs that alkylate DNA but also on their different mechanisms of action.

2.2.1.1 Vinblastine (VLB, 1)

VLB is approved as a component of combination therapy for use in Hodgkin's lymphoma, non-Hodgkin's lymphomas (including lymphocytic, mixed cell, histiocytic, undifferentiated, nodular, and diffuse), advanced mycosis fungoides, advanced testicular carcinoma, Kaposi's sarcoma, and histiocytosis [53–57]. The dose-limiting toxicity of VLB is myelosuppression, with the nadir of leucopenia 5–9 days after administration and recovery occurring within 14–21 days. The usual doses are 4–5 mg/m^2 every week by i.v. injection, although, because of variable leucopenia, an escalating schedule has been suggested.

2.2.1.2 Vincristine (VCR, 2)

VCR has a similar spectrum of activity for Hodgkin's and non-Hodgkin's lymphomas to VLB and has been used in rhabdomyosarcoma of childhood, neuroblastoma, Wilms' tumor (nephroblastoma), Ewing sarcoma, melanoma, and small-cell lung cancer (SCLC) [58–63]. The dose-limiting toxic effect of VCR is neurological, with extensive peripheral neuropathy occurring at higher doses [64]. Early symptoms of neurotoxicity are numbness and painful paresthesias in the fingers and toes, and depression of the Achilles tendon reflex, followed, if treatment continues, by severe muscle weakness and uncoordinated movements. The usual dose in adults is 1.4 mg/m^2 every week by i.v. injection [65].

2.2.1.3 Vindesine (VDS, 3)

The spectrum of antitumor activity of VDS is similar to that of VCR, but with milder experimental neurotoxicity. Its terminal half-life is 24 h and its plasma clearance is intermediate between those of VLB and VCR. The maximal tolerated dose is 4–5 mg/m^2/week, the dose-limiting toxicity being myelosuppression (nadir by days 7–8 and recovery by days 11–13) [66]. Vindesine has already been demonstrated as efficient in childhood acute lymphoid leukemia (ALL), non-Hodgkin's lymphoma, blastic crisis of chronic myeloid leukemia, and esophageal carcinoma. It has also shown activity in Hodgkin's disease, breast and germ cell carcinomas, and melanoma. Intolerance is mainly neurological, with paresthesias, without motor impairment, or hematological, with leukopenia, and sometimes alopecia, asthenia, and muscle pains. There appears to be no cross-resistance with its parent, VCR, as documented in ALL [67].

2.2.1.4 **Vinorelbine (VRLB, 4)**

VRLB is a semisynthetic derivative of VLB (5′-noranhydrovinblastine), structurally distinguished from other members of its class by the modification of the catharanthine nucleus rather than the vindoline ring. This alteration is probably responsible for differences in its antitumor activity and tolerability profile compared with other vinca alkaloids [68]. VRLB is effective as monotherapy or in combination with a platinum derivative in patients with advanced NSCLC and advanced breast cancer [69,70]. Myelosuppression is the major dose-limiting toxicity. VRLB is administered weekly; nadirs are usually reached within 14 days and patients recover within the next two weeks [71]. VRLB is also well absorbed orally. Oral and i.v. forms show similar interindividual variability, the same metabolism pattern, reproducible intrapatient blood exposure, and the same pharmacokinetic–pharmacodynamic relationship. Given at 60 mg/m^2/week for the first three administrations and then increased to 80 mg/m^2/week it achieved the same efficacy as i.v. VRLB (30 mg/m^2) in terms of progression-free survival, overall survival, and objective response [70].

2.2.1.5 **Vinflunine (VFL, 5)**

VFL is a novel vinca alkaloid obtained by semisynthesis using super-acidic chemistry to introduce two fluorine atoms selectively at the 20′ position of VRLB. The preclinical evaluations of the new derivative VFL have already suggested that certain *in vitro* assays, in addition to *in vivo* experiments, could be proposed to select more rationally newer generation *Vincas*. Moreover, recent studies have demonstrated that certain newly identified properties, such as antiangiogenic activities, could enlarge the therapeutic usage of natural and semisynthetic vinca alkaloids. VFL is presently in phase III experimentation for treatment of bladder cancer and nonsmall-cell lung cancer (NSCLC) [72].

2.2.2 **Camptothecin and Analogs**

Camptothecin (CPT, **6**, Figure 2.2) was first isolated from the Chinese ornamental tree *Camptotheca acuminata* Decne, also known as the "tree of joy" and "tree of love." It occurs in different plant parts such as the roots, twigs, and leaves. It is a member of the quinoline alkaloid group and consists of a pentacyclic ring structure that includes a pyrrole quinoline moiety and one asymmetric center within the α-hydroxy lactone

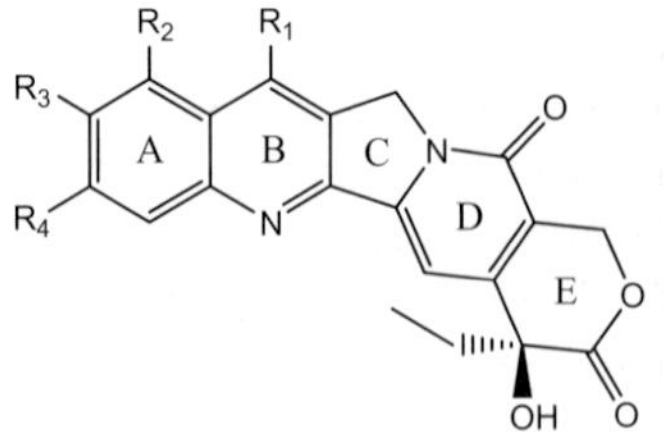

Compound	R_1	R_2	R_3	R_4
6	H	H	H	H
7	$-CH_2CH_3$	H	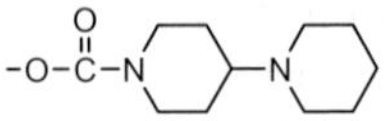 $-O-C(=O)-N$(piperidine ring)$-N$(piperidine ring)	H
8	H	$-CH_2N(CH_3)_2$	OH	H
9	$-CH(NH_3)-CH_2\ CH_2-$		CH_3	F
10	$-CH{=}NOC(CH_3)_3$	H	H	H
11	H	$-CH_2CH_2Si(CH_3)_3$	H	H
12	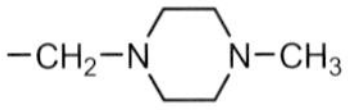 $-CH_2-N$(piperazine ring)$N-CH_3$	H	$-OCH_2CH_2O-$	
13	H	NO_2	H	H

Fig. 2.2 Chemical structures of camptothecin and analogs.

ring with 20(*S*) configuration (ring E). (See Volume 2 for synthesis of camptothecin and analogs.)

In the early 1960s, the discovery of CPT by Wall and Wani as an anticancer drug added an entirely new dimension to the field of chemotherapy [73,74]. Preliminary studies revealed a substantial antitumor activity in standard *in vitro* test systems as well as in mouse leukemia cells. Interest in CPTand analogs remained low until 1985 when it was discovered that CPT, by a unique mechanism, inhibited the enzyme topoisomerase I [75–77]. Topoisomerase I involves the transient single strand cleavage of duplex DNA, followed by unwinding and relegation. These actions facilitate the occurrence of essential cellular processes such as DNA replication, recombination, and transcription [78]. The camptothecins bind to and stabilize the normally transient DNA–topoisomerase I cleavage complex [75]. Collision of the DNA replication fork with the ternary drug–enzyme–DNA complex produces an irreversible double-strand break that ultimately leads to cell death [79]. The camptothecins are, therefore, S phase-specific drugs, because ongoing DNA synthesis is a necessary condition to induce the above sequence of events leading to cytotoxicity. This has important implications for the clinical use of these agents, because optimal therapeutic efficacy of S phase-specific cytotoxic drugs generally requires prolonged exposure of the tumor to concentrations exceeding a minimum threshold.

A variety of mechanisms of resistance to topoisomerase I-targeted agents have been characterized with *in vitro* models, although relatively little is known about their significance in the clinical setting. These mechanisms involve either pretarget events, such as drug accumulation, metabolism, and intracellular drug distribution, or drug–target interactions [80–83]. More recently, post-target events, such as DNA synthesis or repair, cell cycle progression, and regulation of cell death, have also been shown to play an important role in the sensitivity to these drugs [84–88].

2.2.2.1 **Camptothecin (CPT, 6)**

Camptothecin was approved by the US Food and Drug Administration (FDA) in the 1970s against colon carcinoma and, thus, it was evaluated as a possible drug in the treatment of human cancer in phase I and II studies [89,90]. Although CPT showed strong antitumor activity among patients with gastrointestinal cancer, it also caused unpredictable and severe adverse effects including myelosuppression, vomiting, diarrhea, and severe hemorrhagic cystitis. In 1972, these findings resulted in the discontinuation of phase II trials. Since CPT could not be used as a drug of choice owing to its severe toxicity, several groups have tried to synthesize derivatives with lower toxicity. The planar pentacyclic ring structure (rings A–E) was suggested to be one of the most important structural features. Earlier, it was reported that the complete pentacyclic ring system is essential for its activity, but other reports have shown that the E-ring lactone is not essential. However, this ring, in the present lactone form with specific C-20 configuration, is required for better activity [91]. The most successful derivatives of CPT have been obtained by modifications of rings A and B. To date, the only CPT analogs approved for clinical use are irinotecan and topotecan, which were obtained by modification of these rings. Modifications can involve additions to the quinoline ring or the complete replacement of the quinoline ring with an alternative ring system, but the quinoline ring system was found to be the most potent [92].

2.2.2.2 **Irinotecan (CPT-11)**

Irinotecan (7, Figure 2.2) is a water-soluble prodrug designed to facilitate parenteral administration of the biologically active 7-ethyl-10-hydroxy analog of camptothecin (SN-38) [93]. Irinotecan is now widely used, especially for colorectal and lung cancers [94,95]. The main dose-limiting toxicities of irinotecan therapy are myelosuppression and delayed-type diarrhea [96]. Unfortunately, as a result of its complicated and highly variable expressed metabolic pathways, irinotecan and its metabolites are subject to extensive interindividual pharmacokinetic and pharmacodynamic variability [97]. Many attempts have been made to understand the complex pharmacokinetic profile of irinotecan. However, up to now, this has not resulted in a better prediction of occurrence and severity of adverse effects in clinical practice. The approved administration schedule of irinotecan in the United States is 125 mg/m^2 given as a 90-min i.v. infusion once weekly for four of six weeks [96]. In Europe, the most widely used dosing regimen is 350 mg/m^2 given as a 60-min i.v. infusion once every three weeks [98], whereas in Japan, 100 mg/m^2 every week or 150 mg/m^2 every other week are the schedules more commonly used [99].

2.2.2.3 Topotecan

Topotecan (**8**, Figure 2.2) is a semisynthetic derivative of camptothecin with a basic *N,N*-dimethylaminomethyl functional group at C-9 that confers water solubility on the molecule. Topotecan is approved in more than 70 countries for the second-line treatment of metastatic ovarian cancer after failure of initial or subsequent chemotherapy, and in more than 30 countries for the treatment of patients with chemotherapy-sensitive SCLC after failure of first-line chemotherapy. The approved schedule is a 30-min i.v. infusion at a starting dose of 1.5 mg/m^2 per day on days 1–5 of a 21-day course [100]. The dose-limiting toxicity for topotecan is myelosuppression, with the most common severe adverse event being noncumulative grade 4 neutropenia [101]. Thus, toxicity is generally managed by treatment delays and dose reductions.

2.2.2.4 Exatecan

Exatecan (**9**, Figure 2.2) is a synthetic hexacyclic water-soluble derivative of camptothecin that potently inhibits the growth of human tumor cell lines and tumor xenografts, including tumors resistant to irinotecan or topotecan [102]. Phase II clinical trials suggest that exatecan has modest activity against pancreatic, gastric, and breast cancers [103–105]. Neutropenia proved to be dose limiting [106].

2.2.2.5 Gimatecan

The realization that position 7 of camptothecin allows several options in chemical manipulation of the drug has stimulated a systematic investigation of a variety of substituents in this position. These efforts resulted in the identification of a novel series of 7-oxyiminomethyl derivatives. Among compounds of this series, gimatecan (**10**, Figure 2.2), a lipophilic derivative, was selected for further development. Using different schedules and dosing durations, gimatecan exhibited an acceptable toxicity profile, with myelotoxicity being the dose-limiting toxic effect. The clinical development of gimatecan is currently ongoing, with phase II studies in diverse tumor types (colon, lung, breast carcinoma, and pediatric tumors) [107].

2.2.2.6 Karenitecin

Karenitecin (**11**, Figure 2.2) is a very lipophilic compound that exhibits more potent cytotoxic activity than camptothecin with both *in vitro* and *in vivo* scenarios. In addition to superior *in vitro* potency, its increased lactone stability and enhanced oral bioavailability are potential clinical advantages [108]. Karenitecin showed significant activity in metastatic melanoma and the dose-limiting toxicity consisted of neutropenia and thrombocytopenia [109].

2.2.2.7 Lurtotecan

The liposome encapsulated form of lurtotecan (**12**, Figure 2.2), which is a water-soluble analog of camptothecin, has an *in vivo* potency similar to topotecan [110]. Pharmacology/toxicology studies in tumor xenograft models show that encapsulation of lurtotecan results in a significant increase in therapeutic index (percentage tumor growth inhibition) and in plasma concentrations higher than free lurtotecan and topotecan

[110]. Dose-limiting toxicity is myelosuppression: primarily thrombocytopenia, although neutropenia was also noted [111]. So far, liposomal lurtotecan has shown limited activity in phase II studies of ovarian, and head and neck tumors [112–114].

2.2.2.8 Rubitecan (9-nitrocamptothecin)

Rubitecan (**13**, Figure 2.2), a potent but poorly soluble camptothecin analog, is an orally available camptothecin analog that also has potential for delivery transdermally or by inhalation. Rubitecan exists in equilibrium as 9-nitro-camptothecin (9-NC) and 9-amino-camptothecin (9-AC), a metabolite that is thought to be active although it failed in clinical trials [115]. Preclinically, rubitecan has shown activity against a broad spectrum of tumor types with *in vitro* and *in vivo* human tumor xenograft models [116]. Frustratingly, the level of activity of an agent in preclinical models has not always translated into similar activity against human tumors in clinical trials. To date, with the exception of pancreatic and possibly ovarian cancer, rubitecan has shown disappointing activity against a number of other solid tumors in relatively small phase I/II trials; however, it has shown some activity against pancreatic cancer, a malignancy that remains difficult to treat, so clinical trials for this indication will continue to be evaluated [117].

2.2.3 Taxanes

Taxanes are a class of structurally complex yet homogenous diterpene alkaloids that occur in the genus *Taxus*, commonly known as the yew. This family of diterpenoids has long been known for its toxicity as well as for other biological activities. The first chemical study of the metabolites of the yew dates backs to the mid-nineteenth century, when a mixture of taxanes was obtained by the German pharmacist Lucas in 1856 [118]. The structure characterization of these compounds, which were named taxine by Lucas was, however, extremely slow, owing to the complexity of the substance and the lack of modern spectroscopic techniques. In 1963, the constitution of the taxane nucleus was established for the first time by independent work of Lythgoe's group, Nakanishi's group, and Uyeo's group as tricyclic polyalcohols esterified with acids, such as taxinine [119,120], whose stereochemistry was established three years later [121]. More importantly, in 1971, Wani and Wall discovered the highly potent anticancer agent paclitaxel (**14**, Figure 2.3) [122,123]. This remarkable accomplishment not only shifted the attention of the scientific community to paclitaxel itself, but also attracted extensive studies on various species of yew tree that led to the isolation of many new taxane family members. To date, over 100 taxanes have been isolated and structurally elucidated. Paclitaxel, and an analog, docetaxel, are currently regarded as very useful anticancer chemotherapeutic agents.

2.2.3.1 Paclitaxel

Paclitaxel (**14**, Figure 2.3), a complex diterpenoid alkaloid, occurs widely in plants of *Taxus* species. Among various *Taxus* species and various parts of taxus trees, the

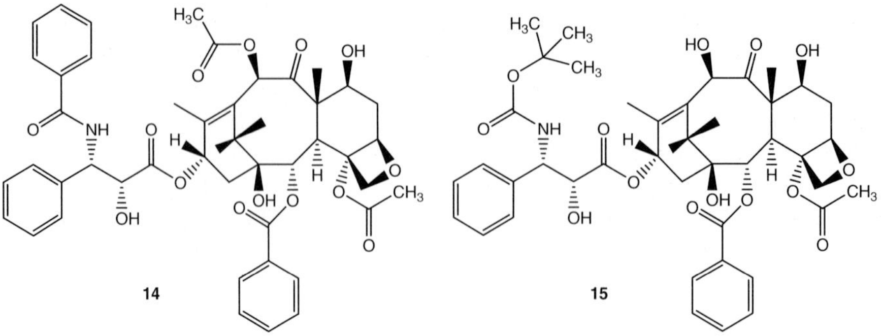

14

15

Fig. 2.3 Chemical structures of taxanes.

bark of *T. brevifolia* Nutt. has the highest content of paclitaxel [124]. However, even this content is relatively low – only about 0.01 % on a dry weight basis. Moreover, yew trees grow very slowly. Paclitaxel isolation is therefore restricted as a result of the relative scarcity of the Pacific yew tree and the low yield of its bark. However, other species and other parts of taxus trees also produce paclitaxel and related taxanes [125]. The major barrier to early clinical development of paclitaxel was its low abundance in the yew trees, since chemical synthesis had not been possible; there simply was not enough paclitaxel to perform the appropriate trials. This situation has changed because of the development of novel semisynthetic methods and the identification of new sources of taxanes. An important biosynthetic precursor of paclitaxel, 10-deacetylbaccatin III, was isolated in 1981 [126] in reasonably good yield from the leaves of *Taxus baccata* L., as a renewable source, as well as from the bark of *Taxus brevifolia* in 1982 [127]. It serves as the starting material for the production of paclitaxel through a coupling reaction with an appropriately protected side chain that can be prepared synthetically. The total synthesis of paclitaxel has also been a challenging target for a number of research groups. Both the Holton group and the Nicolaou group synthesized paclitaxel almost simultaneously in 1994, but both syntheses are too cumbersome to have commercial value [128,129].

Among antineoplastic drugs that interfere with microtubules, paclitaxel exhibits a unique mechanism of action [130–133]. Paclitaxel promotes assembly of microtubules by shifting the equilibrium between soluble tubulin and microtubules toward assembly, reducing the critical concentration of tubulin required for assembly. The result is stabilization of microtubules, even in the presence of conditions that normally promote disassembly of microtubules. The remarkable stability of microtubules induced by paclitaxel is damaging to cells because of the perturbation in the dynamics of various microtubule-dependent cytoplasmic structures that are required for such functions as mitosis, maintenance of cellular morphology, shape changes, neurite formation, locomotion, and secretion [134–137]. The binding site for paclitaxel is distinct from the binding sites for the vinca alkaloids, colchicine or podophyllotoxin [132]. Other natural products, such as epothilone and discodermolide, have also been reported to share the mechanism of paclitaxel in promoting microtubule assembly and have shown potential anticancer activity. Cells treated with pharmacologic levels of paclitaxel are arrested in the G_2 and M phases of the cell cycle and contain disorganized arrays of microtubules, often aligned in parallel bundles [133].

Paclitaxel was found to be highly active in numerous preclinical tumor models, and in 1981 it entered phase I clinical trials, which established toxicity profiles and dose schedules for further trials. A high prevalence of acute hypersensitivity reactions necessitated discontinuation of many trials. However, concomitant administration of steroids histamine H_1 and H_2 receptor antagonists and prolonged infusion (6–24 h) have been used successfully. Partial or minor responses were observed in patients with melanoma, refractory ovarian carcinoma, breast carcinoma, NSCLC, and head and neck carcinoma [138]. Phase II trials began in 1985, and during those trials paclitaxel showed remarkable activity against refractory and advanced ovarian cancer, which was confirmed by other investigators [139,140]. Activities have also been reported in metastatic breast cancer, as well as in NSCLC, urothelial, and head and neck cancers [141–143]. In December of 1992, the FDA approved paclitaxel for the treatment of drug refractory metastatic ovarian cancer [144]. Today, the drug is used for a variety of cancers, including ovarian, breast, small-cell and large-cell lung cancers, and Karposi's sarcoma. The dose recommendations are generally in the range of 200–250 mg/m^2. A predominant focus is now the evaluation of paclitaxel in combination with other agents, such as cisplatin, carboplatin, or vinorelbine [145,146]. Paclitaxel is tolerated better than any other anticancer drug used today, but it also has some side effects, such as nausea, numbness, and a reduction in the white blood cells, which can be prevented by the administration of a granulocyte colony-stimulating factor [147].

Resistance to paclitaxel has been described in cultured cells, and it appears in several forms. In one, resistance is associated with altered expression or mutation in α- and β-tubulins [148,149], whereas in another, paclitaxel resistance is associated with overexpression of Pgp and the MDR phenotype [150].

2.2.3.2 **Docetaxel**

Docetaxel (**15**, Figure 2.3), a semisynthetic side-chain analog of paclitaxel, also has potent antitumor activities. It was first synthesized by Potier's group in 1984, from one of the yew tree taxanes, 10-deacetyl-baccatin III, found in the leaves of *Taxus baccata* L. This has the advantage of being a renewable source [151]. Although the mechanisms of action of the taxanes docetaxel and paclitaxel are identical, docetaxel has almost a twofold higher binding affinity for the target site, β-tubulin. Docetaxel, initially developed for the treatment of breast cancer, has a high degree of activity in lung cancer. In clinical trials, individuals previously treated with paclitaxel benefited from docetaxel. Docetaxel is the standard of care in second-line therapy of advanced NSCLC and is effective, alone and in combination, in first-line treatment of advanced NSCLC [152]. The dose recommendations are generally in the range of 75–100 mg/m^2 every three weeks. The most toxic effect seen was neutropenia occurring in the majority of patients. For most patients neutropenia was of short duration, allowing re-treatment on schedule. Patients receiving more than four cycles of docetaxol had fluid retention, which could be largely controlled through the use of diuretics, and were less frequent in patients premedicated with glucocorticoids [153].

2.3 Antitumor Alkaloids in Clinical Trials

2.3.1 Ecteinascidin-743 (Yondelis, Trabectedin)

Ecteinascidin-743 (ET-743, **16**, Figure 2.4) is a tetrahydroisoquinolone alkaloid isolated from *Ecteinascidia turbinata* Herdman, a tunicate that grows on mangrove roots throughout the Caribbean Sea [154]. Most likely, the compound is produced by the marine tunicate as a defense mechanism to survive in its natural environment [155].

ET-743 binds to the N2 position of guanine in the minor groove of DNA with some degree of sequence specificity, altering the transcription regulation of induced genes [156]. ET-743 was selected for clinical development because of its original mode of action involving DNA repair machinery [157] and its cytotoxic activity against a variety of solid tumor cell lines, including sarcoma cell lines, even those resistant to many other cytotoxic agents [158]. In phase I clinical trials, ET-743 was generally well tolerated with noncumulative hematological and hepatic toxicities being the most commonly reported adverse events. The dose-limiting toxicities were neutropenia and fatigue. A dose-related asymptomatic and reversible rise in transaminase levels was prevalent, but not a treatment-limiting toxicity [159]. Patients with any baseline liver-function test exceeding the upper limit of the normal ranges have a significantly

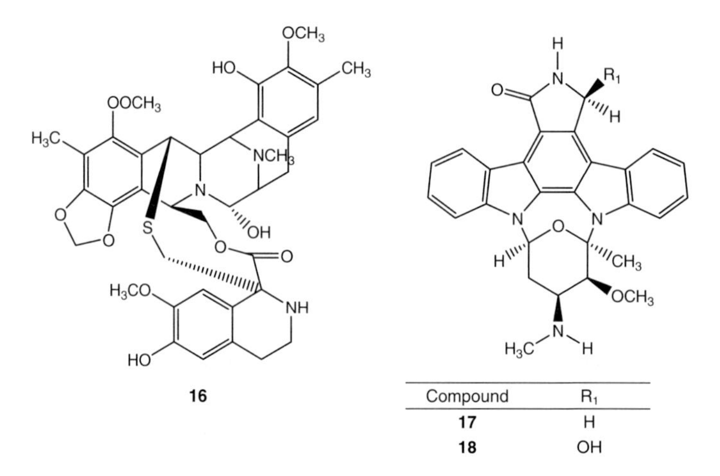

Compound	R_1
17	H
18	OH

Fig. 2.4 Chemical structures of ecteinascidin-743 and staurosporines.

greater incidence of severe hepatic toxicity [160]. This should be used as a reference to identify patients eligible to receive full doses of ET-743. The phase II data for ET-743 administered as a single agent has established a clinical role for the compound in advanced soft tissue sarcoma, and promising potential in pretreated ovarian cancer [154,161]. Owing to its original mechanism of action, ET-743 may act synergistically in combination with other cytotoxic agents. Several preclinical or phase I studies explored this possibility. ET-743 combined with other drugs (i.e. cisplatin, paclitaxel, or doxorubicin) showed more than additive effects in several preclinical systems and initial clinical results (e.g. a combination of ET-743 with cisplatin) appear to confirm the preclinical findings [162,163]. Phase III studies comparing conventional therapy with ET-743 are necessary to fully elucidate the therapeutic potential of this agent.

2.3.2 7-Hydroxystaurosporine (UCN-01)

Staurosporine (**17**, Figure 2.4), an alkaloid isolated from *Streptomyces* spp., is a potent nonspecific protein and tyrosine kinase inhibitor. Thus, efforts to find analogs of staurosporine have identified compounds specific for protein kinases [164]. UCN-01 (**18**, Figure 2.4) is a potent inhibitor of protein kinase C [165], and has antiproliferative activity in several human tumor cell lines. Work from several groups supports the hypothesis that UCN-01 promotes its antitumor activity through induction of apoptosis by either modulation of cyclin-dependent kinases or inhibition of cell cycle checkpoints. UCN-01 may act by abrogating the G_2 block often induced by cellular damage, thus causing inappropriate progression to G_2 and subsequent apoptosis [166]. Also, synergistic effects of UCN-01 have been observed with many chemotherapeutic agents, including mitomycin C, 5-fluorouracil, and camptothecin [167–169]. Phase I clinical studies showed that the drug had an unexpectedly long half-life (about 30 days) [170]. Dose-limiting toxicities were nausea, vomiting, hyperglycemia, and hypoxia when UCN-01 was given as a prolonged 72h infusion, and hypotension when the compound was administered over 1 h. Phase I trials of UCN-01 in combination with topotecan or 5-fluorouracil were relatively well tolerated and showed some preliminary evidence of efficacy [171,172].

2.3.3 Ellipticine and Analogs

Ellipticine (**19**) and its other two naturally occurring analogs, 9-methoxyellipticine (**20**) and olivacine (**21**), showed promising activity as potential anticancer drugs (Figure 2.5). Ellipticine was first isolated in 1959 from the leaves of the evergreen tree *Ochrosia elliptica* Labill., but its biological activities were only recognized in 1967 [173]. Since then, the design, synthesis, and structure–activity relationships of this class of compounds have been studied by a number of laboratories [174,175].

Studies on the mechanism of cytotoxicity and anticancer activity of the ellipticine analogs suggest a complex set of effects, including DNA intercalation, inhibition of topoisomerase, covalent alkylation of macromolecules, and generation of cytotoxic-free radicals [176,177]. Owing to cardiovascular toxicity and hemolysis observed during preclinical toxicity studies, development of the parent compound ellipticine was halted. Interest then shifted to the 9-substituted derivatives, including 9-hydroxyellipticine, 9-methoxyellipticine, and elliptinium. Only limited activity was observed in clinical trials with 9-hydroxyellipticine and 9-methoxyellipticine. Phase II clinical trials of elliptinium yielded moderately promising results. S16020, a cytotoxic agent derived from 9-hydroxyolivacine, was tested in phase I studies, and demonstrated improved toxicities by showing no hemolysis and no detection of anti-S16020 antibodies [178]. None of the ellipticine derivatives have reached clinical practice, but the ellipticine family may still yield clinically useful anticancer drugs.

2.3.4
Acronycine and Analogs

The pyranoacridone alkaloid, acronycine (**22**, Figure 2.5), isolated from *Acronychia baueri* Schott. [179], was shown to exhibit promising activity against a broad spectrum of solid tumors [180]. However, its moderate potency and poor solubility in aqueous solvents severely hampered subsequent clinical trials, which were rapidly discontinued, owing to modest therapeutic effects and dose-limiting gastrointestinal toxicity after oral administration [181]. Consequently, the development of structural analogs with increased potency and/or better water solubility was highly desirable. Efforts to design more potent derivatives were guided by the hypothesis of *in vivo* bioactivation of the 1,2-double bond of acronycine into the corresponding epoxide. The high reactivity of acronycine 1,2-epoxide, which readily reacts with water to give the corresponding *cis* and *trans* diols, suggested that this compound could be the active metabolite of acronycine, able to alkylate some nucleophilic target within the tumor cell. Accordingly, significant improvements in terms of potency were obtained with derivatives modified in the pyran ring, which had a similar reactivity toward nucleophilic agents such as acronycine epoxide but improved chemical

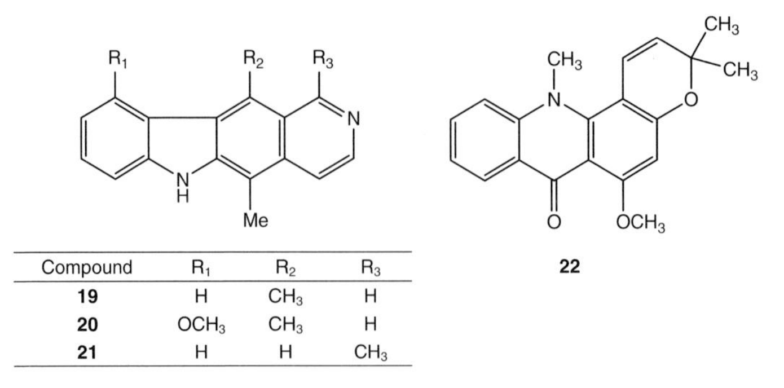

Compound	R1	R2	R3
19	H	CH_3	H
20	OCH_3	CH_3	H
21	H	H	CH_3

Fig. 2.5 Chemical structures of ellipticines and acronycine.

stability [182,183]. The benzo[*b*]acronycine derivative S23906-1 demonstrated a marked antitumor activity in human orthotopic models of lung, ovarian, and colon cancers xenografted in nude mice, and is currently in phase I clinical trials. The mechanism of its action implies alkylation of the 2-amino group of DNA guanine residues by the carbocation resulting from the elimination of the ester leaving group at position 1 of the drug [184].

2.3.5
Colchicine and Analogs

Colchicine (**23**, Figure 2.6) is a three-ring amine alkaloid derived from the corms of *Colchicum autumnale* L. Like the anticancer indole alkaloids, vinblastine and vincristine, it depolymerizes microtubules at high concentrations and stabilizes microtubule dynamics at low concentrations. It was recognized as having damaging effects on tumor vasculature as long ago as the 1930s and 1940s, causing hemorrhage and extensive necrosis in both animal and human tumors [185,186]. Although its toxicity precluded further clinical development, sporadic reports of tumor vascular damage induced by related compounds, such as podophyllotoxin, continued to emerge [187]. Since the late 1990s, the combretastatins and *N*-acetylcolchinol-*O*-phosphate (ZD6126, **24**, Figure 2.6), compounds that resemble colchicine and bind to the colchicine domain on tubulin, have undergone extensive development as antivascular agents. ZD6126 is a phosphate prodrug of the tubulin-binding agent *N*-acetylcolchinol. Profound disruption of the tumor blood vessel network has been noted in a wide variety of preclinical tumor models. The observed effects include vascular shutdown, and reduced tumor blood flow. Histologic assessment has revealed central tumor necrosis extending to within a few cell layers from the tumor margins [188]. Following treatment, a ring of viable tumor cells is invariably found on the tumor periphery [189]. Since the residual tumor tissue can serve as a foundation for tumor regrowth, the vascular disrupting agent could be combined with other treatment options so that the entire tumor cell population can be completely eradicated. Several studies involving the use of ZD6126 in conjunction with conventional chemotherapeutic agents led to enhanced responses in a wide variety of tumor models [188]. Each phase I clinical trial reported similar toxicity profiles, including anemia, constipation, and fatigue [190]. It is noteworthy that this toxicity profile is

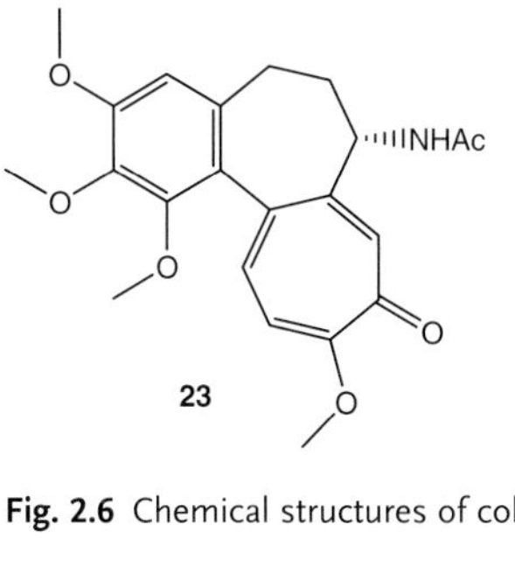

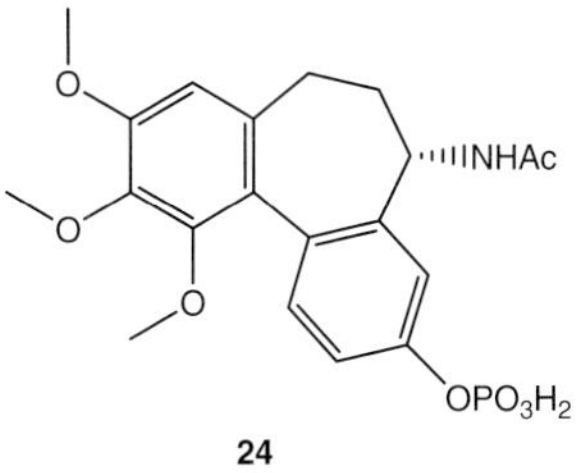

Fig. 2.6 Chemical structures of colchicines and ZD6126.

distinct from the toxicity profile of conventional anticancer agents. These phase I trials revealed some signs of clinical activity, and ZD6126 will be evaluated as a single agent for the treatment of renal carcinoma [191]. However, there is little doubt that the greatest potential utility of vascular damaging agents lies in their combination with other treatment options and this should be evaluated in further clinical trials.

2.3.6 Ukrain

Ukrain (NSC-631570) has been described by Novicky Pharma (Vienna, Austria) as a semisynthetic compound derived from *Chelidonium majus* L., which contains a range of alkaloids, most notably chelidonine. Ukrain consists of one molecule of thiophosphoric acid conjugated to three molecules of chelidonine. A mass-spectrometric analysis of Ukrain components failed to demonstrate the presence of the semisynthetic thiothepa derivative. Instead, analysis of the pharmaceutical preparation Ukrain revealed well-known alkaloids from *Chelidonium majus*, including chelidonine, chelerythrine, sanguinarine, protopine, and allocryptopine [192]. Since 1990, preclinical investigations have pointed to the promising antineoplastic activity of Ukrain. These studies suggested that Ukrain exerts selective cytotoxic effects on tumor cells without adverse side effects on normal cells and tissues [193,194]. However, observations on the selective cytotoxicity of Ukrain are still subject to controversy [195]. In addition to the above mentioned promising preclinical data, some clinical investigations, predominantly from Eastern Europe, suggested beneficial effects of Ukrain in the treatment of patients suffering from various cancers when given as a single drug or in combination with standard chemotherapeutic drugs or ionizing radiation [196]. However, the random clinical trials reported in the literature have serious methodological limitations and the mechanism of action of Ukrain as an anticancer drug remains elusive [196]. Before positive recommendations can be issued, independent replication with definitive trials and larger sample sizes seem mandatory.

2.4 Alkaloids Used for MDR Reversal

2.4.1 *Cinchona* Alkaloids

Cinchona alkaloids isolated from the bark of several species of cinchona trees have been extensively studied for their antimalarial properties [197]. Later, quinine (**25**) and cinchonine (**26**) (Figure 2.7) were shown to have a potential use in reversing multidrug resistance in cancer patients, and are considered as first-generation blockers. In addition to their known role as inhibitors of Pgp, these chemomodulators have been suggested to have intracellular protein targets that may be involved in drug distribution [198–200]. Phase I/II clinical trials demonstrated that quinine

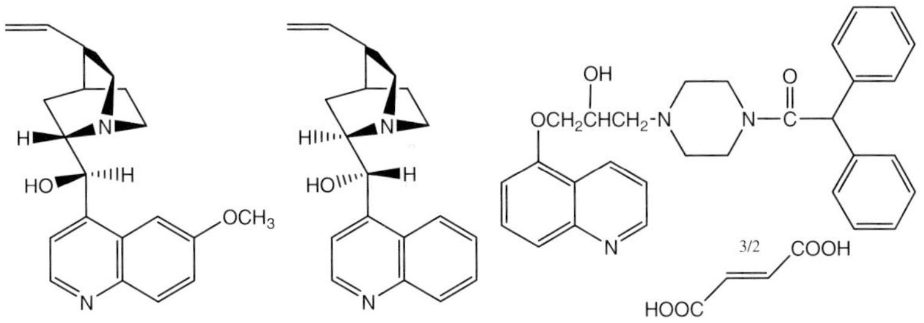

Fig. 2.7 Chemical structures of cinchona alkaloids and MS-209.

could be used safely in combination with various chemotherapeutic agents, including mitoxantrone, vincristine, adriamycin, or paclitaxel for the treatment of clinically resistant acute leukemias, advanced breast cancer, or non-Hodgkin's lymphomas [201–203]. Phase III clinical trials in patients with acute leukemia showed modest results [204,205]. In preclinical studies, cinchonine was reported to be more efficient than quinine as an anti-MDR agent [200,206]. A Phase I trial showed that cinchonine in combination with doxorubicin, vinblastine, cyclophosphamide, and methylprednisolone could be safely administered in patients with malignant lymphoid diseases and an MDR reversing activity was identified in the serum from every patient [207]. The development of third-generation drugs, which are highly specific for Pgp and which seem to have only modest effects on drug clearance, include dofequidar fumarate, tariquidar, zosuquidar, and laniquidar, which are in phase I/II trials and show promise for future treatment.

2.4.2 Dofequidar Fumarate (MS-209)

MS-209 (**27**, Figure 2.7), a quinoline derivative, was developed specifically as a selective Pgp inhibitor [208]. MS-209 alone has no antitumor activity and did not show serious toxicity at doses up to 2000 mg/kg. In preclinical models, MS-209 enhanced the antitumor effects of anticancer agents including adriamycin, vincristine, paclitaxel, and docetaxel in multidrug-resistant tumor cell lines [209]. It also enhanced the efficacy of anticancer agents against cancer cells sensitive to anticancer agents. MS-209 in combination with adriamycin was more effective than adriamycin alone against transplanted murine tumors, multidrug-resistant murine tumors, and human tumors transplanted to nude mice [210]. A phase I trial designed to determine the safety and tolerability of MS-209 when given in combination with docetaxel showed no significant differences in docetaxel-induced nadir neutrophil and platelet counts in the absence or presence of MS-209 [211]. A clinical trial including MS-209 in combination with cyclophosphamide, doxorubicin, and fluorouracil therapy for patients with advanced or recurrent breast cancer showed promising efficacy in patients who had not received prior therapy [212]. By March 2003, phase II trials in breast and NSCLC had been initiated [213].

2.5
Alkaloids Used for Cancer Prevention

Unlike cancer chemotherapy, cancer chemoprevention involves the prevention or delay of carcinogenesis through the ingestion of pharmaceutical or dietary agents [214]. Only few compounds have been approved by the FDA for use as chemopreventive agents. They include the cyclooxygenase-2 specific inhibitor celecoxib (**28**, Figure 2.8) to reduce the number of adenomatous colorectal polyps in familial adenomatous polyposis, as an adjunct to usual care, as well as tamoxifen (**29**) and three aromatase inhibitors, anastrozole (**30**), letrozole (**31**), and exemestane (**32**), for

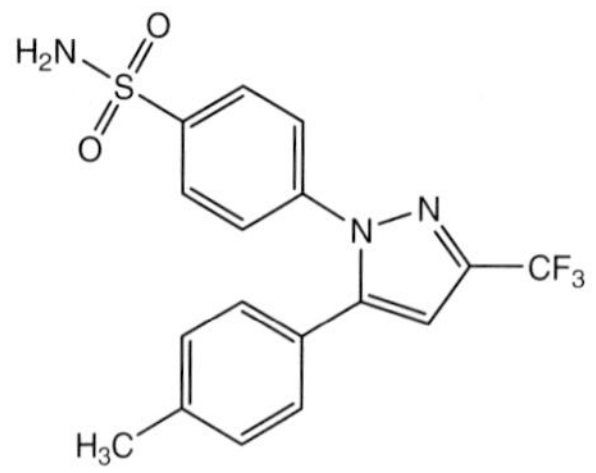

28

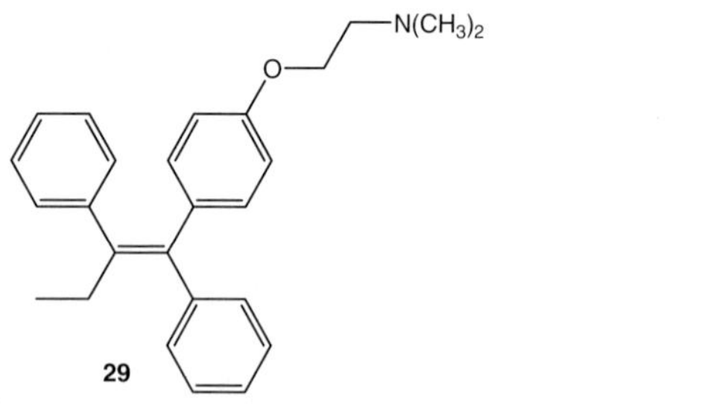

29

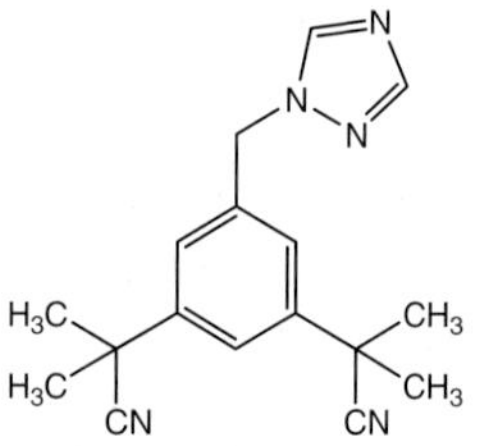

30

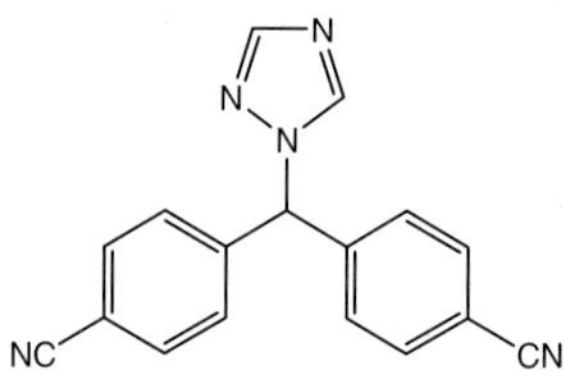

31

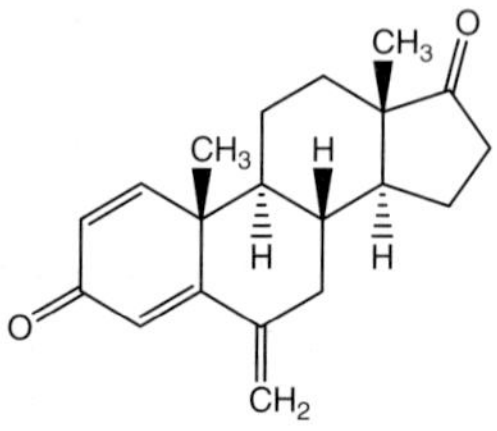

32

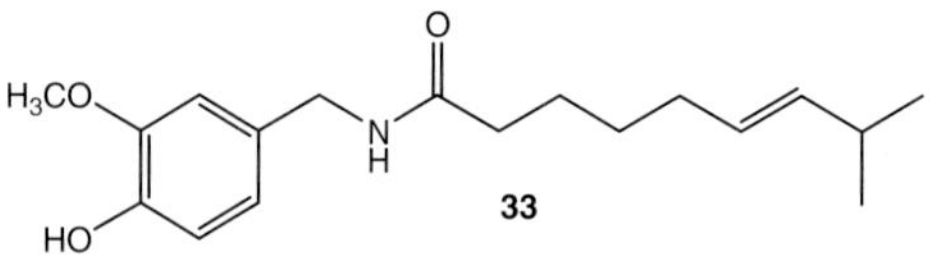

33

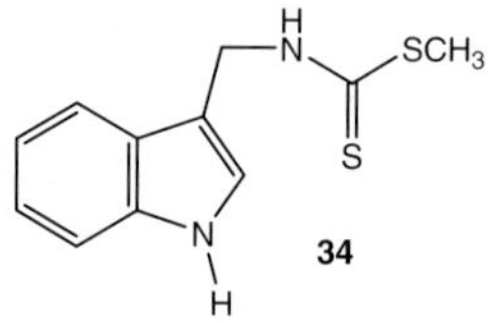

34

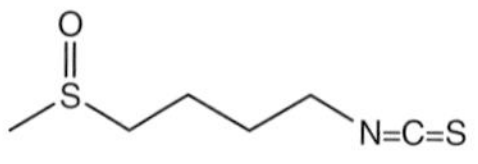

35

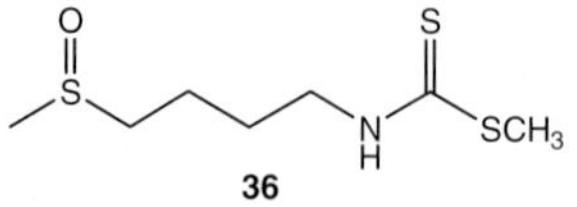

36

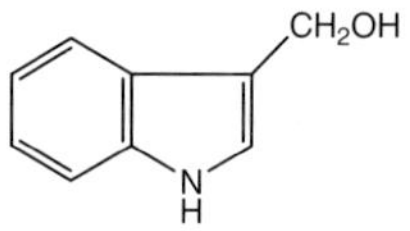

37

Fig. 2.8 Chemical structures of alkaloids with promising cancer chemopreventive activity.

the adjuvant treatment of postmenopausal women with hormone receptor-positive early breast cancer. Obviously, these drugs are not alkaloids.

Capsaicin (**33**), an alkaloid and the main ingredient of hot chili pepper, is found in most *Capsicum* spp. (see Chapter 4). Preclinical studies have demonstrated that capsaicin inhibits mutagenicity and DNA binding of some chemical carcinogens, possibly by suppressing their metabolic activation [215]. With cells in culture, capsaicin inhibited proliferation by decreasing NADH oxidase activity [216]. Capsaicin also alters the expression of tumor forming-related genes by mediating the overexpression of *p53* and/or *c-myc* genes as well as induces apoptosis in cancer cell lines [217,218]. With *in vivo* studies, capsaicin inhibited the growth of xenograft prostate tumors in mice [218]. Promising data suggest the antitumor properties of capsaicin and this compound should be explored additionally for clinical use.

Even though not alkaloids, some nitrogen-containing natural products have shown promising *in vitro* and *in vivo* cancer chemopreventive activities. Cruciferous vegetables provide the indole dithiocarbamate brassinin (**34**) and several natural and synthetic analogs, which were shown to induce quinone reductase and glutathione *S*-transferase activities with *in vitro* and *in vivo* models [219]. Sulforaphane (**35**), an isothiocyanate isolated from broccoli, has potent monofunctional phase 2 enzyme-inducing activity [220]. A novel analog, suforamate (**36**), exhibited similar potency to sulforaphane, with one-third of the toxicity. Sulforamate also increased GSH levels [221]. Sulforaphane and sulforamate have shown similar dose–response patterns for inhibiting 7,12-dimethylbenz(*a*)anthracene (DMBA)-induced lesions in mouse mammary organ culture, and significant inhibition of DMBA-induced mammary carcinogenesis in Sprague–Dawley rats has been demonstrated [222]. Indole-3-carbinol (**37**), a compound found in high concentrations in broccoli, cauliflower, Brussels sprouts, and cabbage, has a number of potential mechanisms of action that are associated with cancer chemoprevention. Indole-3-carbinol has antiestrogenic activity by competing with estrogen for binding sites [223] and reduces the activity of the tumor-promoting enzymes ornithine decarboxylase and tyrosine kinase [224], as well as inducing apoptosis in various cell types through inactivation of the nuclear factor-kappaB [225]. Preclinical and phase I trials in breast, prostate and cervical cancers showed some efficacy and very low toxicity [226,227].

2.6 Conclusions

Natural products, especially alkaloids, have been a source of highly effective conventional drugs for the treatment of many forms of cancer. While the actual compounds isolated from a natural source may not serve as the drugs, they provide leads for the development of potential novel agents to treat or prevent cancer, as well as to modulate MDR. As new technologies are developed, some of the agents which were abandoned or failed earlier clinical studies are now stimulating renewed interest.

2.7 Acknowledgments

The authors are grateful to the National Cancer Institute for support provided under the auspices of program project P01 CA48112 entitled "Natural Inhibitors of Carcinogenesis."

References

1 Stewart, B. W. and Kleihues, P. (2003) *World Cancer Report*, IARC Press, Lyon.
2 American Cancer Society (2006) *Cancer Facts and Figures 2006*, American Cancer Society, Atlanta.
3 Pezzuto, J. M. (1997) *Biochemical Pharmacology*, **53**, 121–133.
4 Cragg, G. M., Newman, D. J., Snader, K. M. (1997) *Journal of Natural Products*, **60**, 52–60.
5 Svoboda, G. H. and Barnes, A. J., Jr. (1964) *Journal of Pharmaceutical Sciences*, **53**, 1227–1231.
6 Johnson, I. S., Armstrong, J. G., Gorman, M., Burnett, J. P., Jr. (1963) *Cancer Research*, **23**, 1390–1427.
7 van Der Heijden, R., Jacobs, D. I., Snoeijer, W., Hallard, D., Verpoorte, R. (2004) *Current Medicinal Chemistry*, **11**, 607–628.
8 Wilson, L., Bamburg, J. R., Mizel, S. B., Grisham, L. M., Creswell, K. M. (1974) *Federation Proceedings*, **33**, 158–166.
9 Malawista, S. E., Sato, H., Bensch, K. G. (1968) *Science*, **160**, 770–772.
10 Watanabe, K., Williams, E. F., Law, J. S., West, W. L. (1981) *Biochemical Pharmacology*, **30**, 335–340.
11 Schroeder, F., Fontaine, R. N., Feller, D. J., Weston, K. G. (1981) *Biochimica et Biophysica Acta*, **643**, 76–88.
12 Kotani, M., Koizumi, Y., Yamada, T., Kawasaki, A., Akabane, T. (1978) *Cancer Research*, **38**, 3094–3099.
13 Kennedy, M. S. and Insel, P. A. (1979) *Molecular Pharmacology*, **16**, 215–223.
14 Creasey, W. A. and Markiw, M. E. (1965) *Biochimica et Biophysica Acta*, **103**, 635–645.
15 Cline, M. J. (1968) *British Journal of Haematology*, **14**, 21–29.
16 Beck, W. T. (1980) *Biochemical Pharmacology*, **29**, 2333–2337.
17 Bruchovsky, N., Owen, A. A., Becker, A. J., Till, J. E. (1965) *Cancer Research*, **25**, 1232–1237.
18 George, P., Journey, L. J., Goldstein, M. N. (1965) *Journal of the National Cancer Institute*, **35**, 355–375.
19 Krishan, A. (1968) *Journal of the National Cancer Institute*, **41**, 581–595.
20 Lengsfeld, A. M., Schultze, B., Maurer, W. (1981) *European Journal of Cancer*, **17**, 307–319.
21 Mareel, M. M., Storme, G. A., De Bruyne, G. K., Van Cauwenberge, R. M. (1982) *European Journal of Cancer Clinical Oncology*, **18**, 199–210.
22 Zakhireh, B. and Malech, H. L. (1980) *Journal of Immunology*, **125**, 2143–2153.
23 Chan, S. Y., Worth, R., Ochs, S. (1980) *Journal of Neurobiology*, **11**, 251–264.
24 Na, G. C. and Timasheff, S. N. (1986) *Biochemistry*, **25**, 6214–6222.
25 Na, G. C. and Timasheff, S. N. (1986) *Biochemistry*, **25**, 6222–6228.
26 Owellen, R. J., Owens, A. H., Jr., Donigian, D. W. (1972) *Biochemical and Biophysical Research Communications*, **47**, 685–691.
27 Fellous, A., Ohayon, R., Vacassin, T., Binet, S., Lataste, H., Krikorian, A., Couzinier, J. P., Meininger, V. (1989) *Seminars in Oncology*, **16**, 9–14.

28 Lobert, S., Ingram, J. W., Hill, B. T., Correia, J. J. (1998) *Molecular Pharmacology*, **53**, 908–915.
29 Jordan, M. A., Margolis, R. L., Himes, R. H., Wilson, L. (1986) *Journal of Molecular Biology*, **187**, 61–73.
30 Tsukidate, K., Yamamoto, K., Snyder, J. W., Farber, J. L. (1993) *American Journal of Pathology*, **143**, 918–925.
31 Kerr, J. F., Winterford, C. M., Harmon, B. V. (1994) *Cancer*, **73**, 2013–2026.
32 Bowman, L. C., Houghton, J. A., Houghton, P. J. (1986) *Biochemical and Biophysical Research Communications*, **135**, 695–700.
33 Ferguson, P. J. and Cass, C. E. (1985) *Cancer Research*, **45**, 5480–5488.
34 Houghton, J. A., Meyer, W. H., Houghton, P. J. (1987) *Cancer Treatment Report*, **71**, 717–721.
35 Owellen, R. J., Hartke, C. A., Dickerson, R. M., Hains, F. O. (1976) *Cancer Research*, **36**, 1499–1502.
36 Beck, W. T. (1987) *Biochemical Pharmacology*, **36**, 2879–2887.
37 Dalton, W. S., Grogan, T. M., Rybski, J. A., Scheper, R. J., Richter, L., Kailey, J., Broxterman, H. J., Pinedo, H. M., Salmon, S. E. (1989) *Blood*, **73**, 747–752.
38 Goldstein, L. J., Fojo, A. T., Ueda, K., Crist, W., Green, A., Brodeur, G., Pastan, I., Gottesman, M. M. (1990) *Journal of Clinical Oncology*, **8**, 128–136.
39 Beck, W. T. (1984) *Advances in Enzyme Regulation*, **22**, 207–227.
40 Gerlach, J. H., Endicott, J. A., Juranka, P. F., Henderson, G., Sarangi, F., Deuchars, K. L., Ling, V. (1986) *Nature*, **324**, 485–489.
41 Beck, W. T., Cirtain, M. C., Lefko, J. L. (1983) *Molecular Pharmacology*, **24**, 485–492.
42 Cornwell, M. M., Pastan, I., Gottesman, M. M. (1987) *The Journal of Biological Chemistry*, **262**, 2166–2170.
43 Zaman, G. J., Flens, M. J., van Leusden, M. R., de Haas, M., Mulder, H. S., Lankelma, J., Pinedo, H. M., Scheper, R. J., Baas, F., Broxterman, H. J. F, Borst, P. (1994) *Proceedings of the National Academy of Sciences of the United States of America*, **91**, 8822–8826.
44 Scheffer, G. L., Wijngaard, P. L., Flens, M. J., Izquierdo, M. A., Slovak, M. L., Pinedo, H. M., Meijer, C. J., Clevers, H. C., Scheper, R. J. (1995) *Nature Medicine*, **1**, 578–582.
45 Lowenbraun, S., DeVita, V. T., Serpick, A. A. (1970) *Cancer*, **25**, 1018–1025.
46 Qian, X. D. and Beck, W. T. (1990) *The Journal of Biological Chemistry*, **265**, 18753–18756.
47 Safa, A. R. (1988) *Proceedings of the National Academy of Sciences of the United States of America*, **85**, 7187–7191.
48 Tsuruo, T., Iida, H., Tsukagoshi, S., Sakurai, Y. (1981) *Cancer Research*, **41**, 1967–1972.
49 Arceci, R. J. (1993) *Blood*, **81**, 2215–2222.
50 Horton, J. K., Thimmaiah, K. N., Houghton, J. A., Horowitz, M. E., Houghton, P. J. (1989) *Biochemical Pharmacology*, **38**, 1727–1736.
51 Sonneveld, P., Suciu, S., Weijermans, P., Beksac, M., Neuwirtova, R., Solbu, G., Lokhorst, H., van der Lelie, J., Dohner, H., Gerhartz, H., Segeren, C. M., Willemze, R., Lowenberg, B. (2001) *British Journal of Haematology*, **115**, 895–902.
52 Sikic, B. I., Fisher, G. A., Lum, B. L., Halsey, J., Beketic-Oreskovic, L., Chen, G. (1997) *Cancer Chemotherapy and Pharmacology*, **40** (Suppl.), S13–S19.
53 Volberding, P. A., Abrams, D. I., Conant, M., Kaslow, K., Vranizan, K., Ziegler, J. (1985) *Annals of Internal Medicine*, **103**, 335–338.
54 Gobbi, P. G., Levis, A., Chisesi, T., Broglia, C., Vitolo, U., Stelitano, C., Pavone, V., Cavanna, L., Santini, G., Merli, F., Liberati, M., Baldini, L., Deliliers, G. L., Angelucci, E., Bordonaro, R., Federico, M. (2005)

Journal of Clinical Oncology, **23**, 9198–9207.
55 Canellos, G. P. (2004) *Seminars in Hematology*, **41**, 26–31.
56 Koynov, K. D., Tzekova, V. I., Velikova, M. T. (1993) *International Urology and Nephrology*, **25**, 389–394.
57 McClain, K. L. (2005) *Expert Opinion on Pharmacotherapy*, **6**, 2435–2441.
58 Seeber, S., Siemers, E., Hoffken, K., Schmidt, C. G., Holfeld, H., Schmitt, G., Scherer, E. (1979) *Deutsche medizinische Wochenschrift*, **104**, 804–807.
59 Breitfeld, P. P. and Meyer, W. H. (2005) *Oncologist*, **10**, 518–527.
60 Schleiermacher, G., Rubie, H., Hartmann, O., Bergeron, C., Chastagner, P., Mechinaud, F., Michon, J. (2003) *British Journal of Cancer*, **89**, 470–476.
61 Gommersall, L. M., Arya, M., Mushtaq, I., Duffy, P. (2005) *Nature Clinical Practice. Oncology*, **2**, 298–304.
62 Vuoristo, M. S., Hahka-Kemppinen, M., Parvinen, L. M., Pyrhonen, S., Seppa, H., Korpela, M., Kellokumpu-Lehtinen, P. (2005) *Melanoma Research*, **15**, 291–296.
63 Thatcher, N., Qian, W., Clark, P. I., Hopwood, P., Sambrook, R. J., Owens, R., Stephens, R. J., Girling, D. J. (2005) *Journal of Clinical Oncology*, **23**, 8371–8379.
64 Rosenthal, S. and Kaufman, S. (1974) *Annals of Internal Medicine*, **80**, 733–737.
65 Desai, Z. R., Van den Berg, H. W., Bridges, J. M., Shanks, R. G. (1982) *Cancer Chemotherapy and Pharmacology*, **8**, 211–214.
66 Toso, C. and Lindley, C. (1995) *American Journal of Health-System Pharmacy*, **52**, 1287–1304.
67 Bayssas, M., Gouveia, J., de Vassal, F., Misset, J. L., Schwarzenberg, L., Ribaud, P., Musset, M., Jasmin, C., Hayat, M., Mathe, G. (1980) *Recent Results Cancer Research*, **74**, 91–97.
68 Potier, P. (1989) *Seminars in Oncology*, **16**, 2–4.
69 Curran, M. P. and Plosker, G. L. (2002) *Drugs and Aging*, **19**, 695–721.
70 Gebbia, V. and Puozzo, C. (2005) *Expert Opinion on Drugs Safety*, **4**, 915–928.
71 Romero, A., Rabinovich, M. G., Vallejo, C. T., Perez, J. E., Rodriguez, R., Cuevas, M. A., Machiavelli, M., Lacava, J. A., Langhi, M., Romero Acuna, L. (1994) *Journal of Clinical Oncology*, **12**, 336–341.
72 Bennouna, J., Campone, M., Delord, J. P., Pinel, M. C. (2005) *Expert Opinion on Investigational Drugs*, **14**, 1259–1267.
73 Wall, M., Wani, M., Cook, C., Palmer, K., McPhail, A., Sim, G. (1966) *Journal of the American Chemical Society*, **88**, 3889.
74 Wall, M. E. (1998) *Medicinal Research Reviews*, **18**, 299–314.
75 Hsiang, Y. H., Hertzberg, R., Hecht, S., Liu, L. F. (1985) *The Journal of Biological Chemistry*, **260**, 14873–14878.
76 Redinbo, M. R., Stewart, L., Kuhn, P., Champoux, J. J., Hol, W. G. (1998) *Science*, **279**, 1504–1513.
77 Staker, B. L., Hjerrild, K., Feese, M. D., Behnke, C. A., Burgin, A. B., Jr., Stewart, L. (2002) *Proceedings of the National Academy of Sciences of the United States of America*, **99**, 15387–15392.
78 Gupta, M., Fujimori, A., Pommier, Y. (1995) *Biochimica et Biophysica Acta*, **1262**, 1–14.
79 Tsao, Y. P., Russo, A., Nyamuswa, G., Silber, R., Liu, L. F. (1993) *Cancer Research*, **53**, 5908–5914.
80 Hoki, Y., Fujimori, A., Pommier, Y. (1997) *Cancer Chemotherapy and Pharmacology*, **40**, 433–438.
81 Chen, Z. S., Furukawa, T., Sumizawa, T., Ono, K., Ueda, K., Seto, K., Akiyama, S. I. (1999) *Molecular Pharmacology*, **55**, 921–928.
82 Maliepaard, M., van Gastelen, M. A., de Jong, L. A., Pluim, D., van Waardenburg, R. C., Ruevekamp-Helmers, M. C., Floot, B. G., Schellens, J. H. (1999) *Cancer Research*, **59**, 4559–4563.
83 Danks, M. K., Garrett, K. E., Marion, R. C., Whipple, D. O. (1996) *Cancer Research*, **56**, 1664–1673.

84 Eng, W. K., McCabe, F. L., Tan, K. B., Mattern, M. R., Hofmann, G. A., Woessner, R. D., Hertzberg, R. P., Johnson, R. K. (1990) *Molecular Pharmacology*, **38**, 471–480.

85 Beidler, D. R. and Cheng, Y. C. (1995) *Molecular Pharmacology*, **47**, 907–914.

86 Goldwasser, F., Shimizu, T., Jackman, J., Hoki, Y., O'Connor, P. M., Kohn, K. W., Pommier, Y. (1996) *Cancer Research*, **56**, 4430–4437.

87 Shao, R. G., Cao, C. X., Shimizu, T., O'Connor, P. M., Kohn, K. W., Pommier, Y. (1997) *Cancer Research*, **57**, 4029–4035.

88 Zwelling, L. A., Slovak, M. L., Doroshow, J. H., Hinds, M., Chan, D., Parker, E., Mayes, J., Sie, K. L., Meltzer, P. S., Trent, J. M. (1990) *Journal of the National Cancer Institute*, **82**, 1553–1561.

89 Moertel, C. G., Schutt, A. J., Reitemeier, R. J., Hahn, R. G. (1972) *Cancer Chemotherapy Report*, **56**, 95–101.

90 Muggia, F. M., Creaven, P. J., Hansen, H. H., Cohen, M. H., Selawry, O. S. (1972) *Cancer Chemotherapy Report*, **56**, 515–521.

91 Jaxel, C., Kohn, K. W., Wani, M. C., Wall, M. E., Pommier, Y. (1989) *Cancer Research*, **49**, 1465–1469.

92 Lackey, K., Besterman, J. M., Fletcher, W., Leitner, P., Morton, B., Sternbach, D. D. (1995) *Journal of Medicinal Chemistry*, **38**, 906–911.

93 Garcia-Carbonero, R. and Supko, J. G. (2002) *Clinical Cancer Research*, **8**, 641–661.

94 Vanhoefer, U., Harstrick, A., Achterrath, W., Cao, S., Seeber, S., Rustum, Y. M. (2001) *Journal of Clinical Oncology*, **19**, 1501–1518.

95 Langer, C. J. (2003) *Oncology*, **17**, 30–40.

96 Rothenberg, M. L., Kuhn, J. G., Burris, H. A., 3rd, Nelson, J., Eckardt, J. R., Tristan-Morales, M., Hilsenbeck, S. G., Weiss, G. R., Smith, L. S., Rodriguez, G. I. (1993) *Journal of Clinical Oncology*, **11**, 2194–2204.

97 Mathijssen, R. H., van Alphen, R. J., Verweij, J., Loos, W. J., Nooter, K., Stoter, G., Sparreboom, A. (2001) *Clinical Cancer Research*, **7**, 2182–2194.

98 Rivory, L. P. and Robert, J. (1995) *Cancer Chemotherapy and Pharmacology*, **36**, 176–179.

99 Negoro, S., Fukuoka, M., Masuda, N., Takada, M., Kusunoki, Y., Matsui, K., Takifuji, N., Kudoh, S., Niitani, H., Taguchi, T. (1991) *Journal of the National Cancer Institute*, **83**, 1164–1168.

100 van Warmerdam, L. J., Verweij, J., Schellens, J. H., Rosing, H., Davies, B. E., de Boer-Dennert, M., Maes, R. A., Beijnen, J. H. (1995) *Cancer Chemotherapy and Pharmacology*, **35**, 237–245.

101 Creemers, G. J., Bolis, G., Gore, M., Scarfone, G., Lacave, A. J., Guastalla, J. P., Despax, R., Favalli, G., Kreinberg, R., Van Belle, S., Hudson, I., Verweij, J. (1996) *Journal of Clinical Oncology*, **14**, 3056–3061.

102 Joto, N., Ishii, M., Minami, M., Kuga, H., Mitsui, I., Tohgo, A. (1997) *International Journal of Cancer*, **72**, 680–686.

103 Kuppens, I. E., Beijnen, J., Schellens, J. H. (2004) *Clinical Colorectal Cancer*, **4**, 163–180.

104 Ajani, J. A., Takimoto, C., Becerra, C. R., Silva, A., Baez, L., Cohn, A., Major, P., Kamida, M., Feit, K., De Jager, R. (2005) *Investigational New Drugs*, **23**, 479–484.

105 Esteva, F. J., Rivera, E., Cristofanilli, M., Valero, V., Royce, M., Duggal, A., Colucci, P., DeJager, R., Hortobagyi, G. N. (2003) *Cancer*, **98**, 900–907.

106 Rowinsky, E. K., Johnson, T. R., Geyer, C. E., Jr., Hammond, L. A., Eckhardt, S. G., Drengler, R., Smetzer, L., Coyle, J., Rizzo, J., Schwartz, G., Tolcher, A., Von Hoff, D. D., De Jager, R. L. (2000) *Journal of Clinical Oncology*, **18**, 3151–3163.

107 Pratesi, G., Beretta, G. L., Zunino, F. (2004) *Anticancer Drugs*, **15**, 545–552.

108 Schilsky, R., Hausheer, F., Bertucci, D., Berghorn, E., Kindler, H., Ratain, M. (2000) *Proceedings of the American Society of Clinical Oncology*, **19**, 195.

109 Daud, A., Valkov, N., Centeno, B., Derderian, J., Sullivan, P., Munster, P., Urbas, P., Deconti, R. C., Berghorn, E., Liu, Z., Hausheer, F., Sullivan, D. (2005) *Clinical Cancer Research*, **11**, 3009–3016.

110 Emerson, D. L., Besterman, J. M., Brown, H. R., Evans, M. G., Leitner, P. P., Luzzio, M. J., Shaffer, J. E., Sternbach, D. D., Uehling, D., Vuong, A. (1995) *Cancer Research*, **55**, 603–609.

111 Gelmon, K., Hirte, H., Fisher, B., Walsh, W., Ptaszynski, M., Hamilton, M., Onetto, N., Eisenhauer, E. (2004) *Investigational New Drugs*, **22**, 263–275.

112 Dark, G. G., Calvert, A. H., Grimshaw, R., Poole, C., Swenerton, K., Kaye, S., Coleman, R., Jayson, G., Le, T., Ellard, S., Trudeau, M., Vasey, P., Hamilton, M., Cameron, T., Barrett, E., Walsh, W., McIntosh, L., Eisenhauer, E. A. (2005) *Journal of Clinical Oncology*, **23**, 1859–1866.

113 Duffaud, F., Borner, M., Chollet, P., Vermorken, J. B., Bloch, J., Degardin, M., Rolland, F., Dittrich, C., Baron, B., Lacombe, D., Fumoleau, P. (2004) *European Journal of Cancer*, **40**, 2748–2752.

114 Seiden, M. V., Muggia, F., Astrow, A., Matulonis, U., Campos, S., Roche, M., Sivret, J., Rusk, J., Barrett, E. (2004) *Gynecologic Oncology*, **93**, 229–232.

115 Pantazis, P., Harris, N., Mendoza, J., Giovanella, B. (1995) *European Journal of Haematology*, **55**, 211–213.

116 Clark, J. W. (2006) *Expert Opinion on Investigational Drugs*, **15**, 71–79.

117 Burris, H. A., 3rd, Rivkin, S., Reynolds, R., Harris, J., Wax, A., Gerstein, H., Mettinger, K. L., Staddon, A. (2005) *Oncologist*, **10**, 183–190.

118 Lucas, H. (1856) *Archiv der Pharmazie*, **85**, 145.

119 Kurono, M., Nakadaira, Y., Onuma, S., Sasaki, K., Nakanishi, K. (1963) *Tetrahedron Letters*, **30**, 2153–2160.

120 Ueda, K., Uyeo, S., Yamamoyo, Y., Maki, Y. (1963) *Tetrahedron Letters*, **30**, 2167–2171.

121 Eyre, D., Harrison, J., Lythgoe, B. (1967) *Journal of Chemical Society*, **6**, 452–462.

122 Wani, M. C., Taylor, H. L., Wall, M. E., Coggon, P., McPhail, A. T. (1971) *Journal of the American Chemical Society*, **93**, 2325–2327.

123 Suffness, M. (1995) *Taxol: Science and Applications*, CRC Press, Boca Raton, FL.

124 Vidensek, N., Lim, P., Campbell, A., Carlson, C. (1990) *Journal of Natural Products*, **53**, 1609–1610.

125 Witherup, K. M., Look, S. A., Stasko, M. W., Ghiorzi, T. J., Muschik, G. M., Cragg, G. M. (1990) *Journal of Natural Products*, **53**, 1249–1255.

126 Chauviere, G., Guenard, D., Picot, F., Senilh, V., Potier, P. (1981) *Comptes Rendus Hebdomadaires des seances de l'Academic des Sciences*, **293**, 501– 503.

127 Kingston, D. G., Hawkins, D. R., Ovington, L. (1982) *Journal of Natural Products*, **45**, 466–470.

128 Holton, R., Somoza, C., Kim, H., Liang, F., Biediger, R., Boatman, P., Shindo, M., Smith, C., Kim, S., Nadizadeh, H., Suzuki, Y., Tao, C., Vu, P., Tang, S., Zhang, P., Murthi, K., Gentile, L., Liu, J. (1994) *Journal of the American Chemical Society*, **116** (4), 1597–1598.

129 Nicolaou, K. C., Yang, Z., Liu, J. J., Ueno, H., Nantermet, P. G., Guy, R. K., Claiborne, C. F., Renaud, J., Couladouros, E. A., Paulvannan, K., Sorensen, E. J. (1994) *Nature*, **367**, 630–634.

130 Manfredi, J. J. and Horwitz, S. B. (1984) *Pharmacology and Therapeutics*, **25**, 83–125.

131 Schiff, P. B., Fant, J., Horwitz, S. B. (1979) *Nature*, **277**, 665–667.

132 Kumar, N. (1981) *The Journal of Biological Chemistry*, **256**, 10435–10441.

133 Thompson, W. C., Wilson, L., Purich, D. L. (1981) *Cell Motility*, **1**, 445–454.

134 Baum, S. G., Wittner, M., Nadler, J. P., Horwitz, S. B., Dennis, J. E., Schiff, P. B., Tanowitz, H. B. (1981)

Proceedings of the National Academy of Sciences of the United States of America, **78**, 4571–4575.
135 Letourneau, P. C. and Ressler, A. H. (1984) *The Journal of Cell Biology*, **98**, 1355–1362.
136 Rowinsky, E. K., Donehower, R. C., Jones, R. J., Tucker, R. W. (1988) *Cancer Research*, **48**, 4093–4100.
137 Roytta, M., Laine, K. M., Harkonen, P. (1987) *Prostate*, **11**, 95–106.
138 Rowinsky, E. K., Cazenave, L. A., Donehower, R. C. (1990) *Journal of the National Cancer Institute*, **82**, 1247–1259.
139 McGuire, W. P., Rowinsky, E. K., Rosenshein, N. B., Grumbine, F. C., Ettinger, D. S., Armstrong, D. K., Donehower, R. C. (1989) *Annals of Internal Medicine*, **111**, 273–279.
140 Einzig, A. I., Wiernik, P. H., Sasloff, J., Runowicz, C. D., Goldberg, G. L. (1992) *Journal of Clinical Oncology*, **10**, 1748–1753.
141 Holmes, F. A., Walters, R. S., Theriault, R. L., Forman, A. D., Newton, L. K., Raber, M. N., Buzdar, A. U., Frye, D. K., Hortobagyi, G. N. (1991) *Journal of the National Cancer Institute*, **83**, 1797–1805.
142 Murphy, W. K., Fossella, F. V., Winn, R. J., Shin, D. M., Hynes, H. E., Gross, H. M., Davilla, E., Leimert, J., Dhingra, H., Raber, M.N. Krakoff, J.H., Hong, W.K. (1993) *Journal of the National Cancer Institute*, **85**, 384–388.
143 Ozols, R. F. (1995) *Seminars in Oncology*, **22**, 1–6.
144 Eisenhauer, E. A. and Vermorken, J. B. (1998) *Drugs*, **55**, 5–30.
145 Neijt, J. P. and du Bois, A. (1999) *Seminars in Oncology*, **26**, 78–83.
146 Grunberg, S. M., Dugan, M. C., Greenblatt, M. S., Ospina, D. J., Valentine, J. W. (2005) *Cancer Investigations*, **23**, 392–398.
147 Srivastava, V., Negi, A. S., Kumar, J. K., Gupta, M. M., Khanuja, S. P. (2005) *Bioorganic and Medicinal Chemistry*, **13**, 5892–5908.
148 Haber, M., Burkhart, C. A., Regl, D. L., Madafiglio, J., Norris, M. D., Horwitz, S. B. (1995) *The Journal of Biological Chemistry*, **270**, 31269–31275.
149 Gonzalez-Garay, M. L., Chang, L., Blade, K., Menick, D. R., Cabral, F. (1999) *The Journal of Biological Chemistry*, **274**, 23875–23882.
150 Roy, S. N. and Horwitz, S. B. (1985) *Cancer Research*, **45**, 3856–3863.
151 Gueritte-Voegelein, F., Guenard, D., Potier, P. (1992) *Comptes rendus des séances de la Société de biologie et de ses filiales*, **186**, 433–440.
152 Belani, C. P. and Eckardt, J. (2004) *Lung Cancer*, **46**, S3–S11.
153 Aapro, M. S. (1995) *Seminars in Oncology*, **22**, 1–2.
154 Verweij, J. (2005) *Journal of Clinical Oncology*, **23**, 5420–5423.
155 Minuzzo, M., Marchini, S., Broggini, M., Faircloth, G., D'Incalci, M., Mantovani, R. (2000) *Proceedings of the National Academy of Sciences of the United States of America*, **97**, 6780–6784.
156 Pommier, Y., Kohlhagen, G., Bailly, C., Waring, M., Mazumder, A., Kohn, K. W. (1996) *Biochemistry*, **35**, 13303–13309.
157 Damia, G., Silvestri, S., Carrassa, L., Filiberti, L., Faircloth, G. T., Liberi, G., Foiani, M., D'Incalci, M. (2001) *International Journal of Cancer*, **92**, 583–588.
158 van Kesteren, C., de Vooght, M. M., Lopez-Lazaro, L., Mathot, R. A., Schellens, J. H., Jimeno, J. M., Beijnen, J. H. (2003) *Anticancer Drugs*, **14**, 487–502.
159 van Kesteren, C., Cvitkovic, E., Taamma, A., Lopez-Lazaro, L., Jimeno, J. M., Guzman, C., Mathot, R. A., Schellens, J. H., Misset, J. L., Brain, E., Hillebrand, M. J., Rosing, H., Beijnen, J. H. (2000) *Clinical Cancer Research*, **6**, 4725–4732.
160 Puchalski, T. A., Ryan, D. P., Garcia-Carbonero, R., Demetri, G. D., Butkiewicz, L., Harmon, D., Seiden, M. V., Maki, R. G., Lopez-Lazaro, L., Jimeno, J., Guzman, C., Supko, J. G. (2002) *Cancer Chemotherapy and Pharmacology*, **50**, 309–319.
161 Sessa, C., De Braud, F., Perotti, A., Bauer, J., Curigliano, G., Noberasco,

C., Zanaboni, F., Gianni, L., Marsoni, S., Jimeno, J., D'Incalci, M., Dall'o, E., Colombo, N. (2005) *Journal of Clinical Oncology*, **23**, 1867–1874.

162 D'Incalci, M., Colombo, T., Ubezio, P., Nicoletti, I., Giavazzi, R., Erba, E., Ferrarese, L., Meco, D., Riccardi, R., Sessa, C., Cavallini, E., Jimeno, J., Faircloth, G. T. (2003) *European Journal of Cancer*, **39**, 1920–1926.

163 Takahashi, N., Li, W. W., Banerjee, D., Scotto, K. W., Bertino, J. R. (2001) *Clinical Cancer Research*, **7**, 3251–3257.

164 Takahashi, I., Asano, K., Kawamoto, I., Tamaoki, T., Nakano, H. (1989) *The Journal of Antibiotics*, **42**, 564–570.

165 Seynaeve, C. M., Kazanietz, M. G., Blumberg, P. M., Sausville, E. A., Worland, P. J. (1994) *Molecular Pharmacology*, **45**, 1207–1214.

166 Wang, Q., Worland, P. J., Clark, J. L., Carlson, B. A., Sausville, E. A. (1995) *Cell Growth Differentiation*, **6**, 927–936.

167 Akinaga, S., Nomura, K., Gomi, K., Okabe, M. (1993) *Cancer Chemotherapy and Pharmacology*, **32**, 183–189.

168 Hsueh, C. T., Kelsen, D., Schwartz, G. K. (1998) *Clinical Cancer Research*, **4**, 2201–2206.

169 Jones, C. B., Clements, M. K., Wasi, S., Daoud, S. S. (2000) *Cancer Chemotherapy and Pharmacology*, **45**, 252–258.

170 Sausville, E. A., Arbuck, S. G., Messmann, R., Headlee, D., Bauer, K. S., Lush, R. M., Murgo, A., Figg, W. D., Lahusen, T., Jaken, S., Jing, X., Roberge, M., Fuse, E., Kuwabara, T., Senderowicz, A. M. (2001) *Journal of Clinical Oncology*, **19**, 2319–2333.

171 Hotte, S. J., Oza, A., Winquist, E. W., Moore, M., Chen, E. X., Brown, S., Pond, G. R., Dancey, J. E., Hirte, H. W. (2006) *Annals of Oncology*, **17**, 334–340.

172 Kortmansky, J., Shah, M. A., Kaubisch, A., Weyerbacher, A., Yi, S., Tong, W., Sowers, R., Gonen, M., O'Reilly, E., Kemeny, N., Ilson, D. I., Saltz, L. B., Maki, R. G., Kelsen, D. P., Schwartz, G. K. (2005) *Journal of Clinical Oncology*, **23**, 1875–1884.

173 Goodwin, S., Smith, A. F., Horning, E. C. (1959) *Journal of the American Chemical Society*, **81**, 1903–1908.

174 Acton, E. M., Narayanan, V. L., Risbood, P. A., Shoemaker, R. H., Vistica, D. T., Boyd, M. R. (1994) *Journal of Medicinal Chemistry*, **37**, 2185–2189.

175 Jurayj, J., Haugwitz, R. D., Varma, R. K., Paull, K. D., Barrett, J. F., Cushman, M. (1994) *Journal of Medicinal Chemistry*, **37**, 2190–2197.

176 Kohn, K. W., Waring, M. J., Glaubiger, D., Friedman, C. A. (1975) *Cancer Research*, **35**, 71–76.

177 Froelich-Ammon, S. J., Patchan, M. W., Osheroff, N., Thompson, R. B. (1995) *The Journal of Biological Chemistry*, **270**, 14998–15004.

178 Awada, A., Giacchetti, S., Gerard, B., Eftekhary, P., Lucas, C., De Valeriola, D., Poullain, M. G., Soudon, J., Dosquet, C., Brillanceau, M. H., Giroux, B., Marty, M., Bleiberg, H., Calvo, F., Piccart, M. (2002) *Annals of Oncology*, **13**, 1925–1934.

179 Macdonald, P. L. and Robertson, A. V. (1966) *Australian Journal of Chemistry*, **19**, 275–281.

180 Svoboda, G. H., Poore, G. A., Simpson, P. J., Boder, G. B. (1966) *Journal of Pharmaceutical Sciences*, **55**, 758–768.

181 Scarffe, J. H., Beaumont, A. R., Crowther, D. (1983) *Cancer Treatment Report*, **67**, 93–94.

182 Elomri, A., Mitaku, S., Michel, S., Skaltsounis, A. L., Tillequin, F., Koch, M., Pierre, A., Guilbaud, N., Leonce, S., Kraus-Berthier, L., Rolland, Y., Atassi, G. (1996) *Journal of Medicinal Chemistry*, **39**, 4762–4766.

183 Costes, N., Le Deit, H., Michel, S., Tillequin, F., Koch, M., Pfeiffer, B., Renard, P., Leonce, S., Guilbaud, N., Kraus-Berthier, L., Pierre, A., Atassi, G. (2000) *Journal of Medicinal Chemistry*, **43**, 2395–2402.

184 David-Cordonnier, M. H., Laine, W., Kouach, M., Briand, G., Vezin, H., Gaslonde, T., Michel, S., Doan Thi

Mai, H., Tillequin, F., Koch, M., Leonce, S., Pierre, A., Bailly, C. (2004) *Bioorganic and Medicinal Chemistry*, **12**, 23–29.

185 Boyland, E. and Boyland, M. E. (1937) *Biochemistry Journal*, **31**, 454–460.

186 Seed, L., Slaughter, D. P., Limarzi, L. R. (1940) *Surgery*, **7**, 696–709.

187 Algire, G. H., Legallais, F. Y., Anderson, B. F. (1954) *Journal of the National Cancer Institute*, **14**, 879–893.

188 Blakey, D. C., Westwood, F. R., Walker, M., Hughes, G. D., Davis, P. D., Ashton, S. E., Ryan, A. J. (2002) *Clinical Cancer Research*, **8**, 1974–1983.

189 Siemann, D. W. and Rojiani, A. M. (2002) *International Journal of Radiation Oncology, Biology, Physics*, **53**, 164–171.

190 Gadgeel, S. M., LoRusso, P. M., Wozniak, A. J., Wheeler, C. (2002) *Proceedings of the American Society of Clinical Oncology*, **21**, 438.

191 Siemann, D. W., Chaplin, D. J., Horsman, M. R. (2004) *Cancer*, **100**, 2491–2499.

192 Habermehl, D., Kammerer, B., Handrick, R., Eldh, T., Gruber, C., Cordes, N., Daniel, P. T., Plasswilm, L., Bamberg, M., Belka, C., Jendrossek, V. (2006) *BMC Cancer*, **6**, 14.

193 Hohenwarter, O., Strutzenberger, K., Katinger, H., Liepins, A., Nowicky, J. W. (1992) *Drugs under Experimental and Clinical Research*, **18** (Suppl.), 1–4.

194 Cordes, N., Plasswilm, L., Bamberg, M., Rodemann, H. P. (2002) *International Journal of Radiation Biology*, **78**, 17–27.

195 Panzer, A., Hamel, E., Joubert, A. M., Bianchi, P. C., Seegers, J. C. (2000) *Cancer Letters*, **160**, 149–157.

196 Ernst, E. and Schmidt, K. (2005) *BMC Cancer*, **5**, 69.

197 Warhurst, D. C. (1987) *Acta Leidensia*, **55**, 53–64.

198 Lehnert, M., Dalton, W. S., Roe, D., Emerson, S., Salmon, S. E. (1991) *Blood*, **77**, 348–354.

199 Bennis, S., Ichas, F., Robert, J. (1995) *International Journal of Cancer*, **62**, 283–290.

200 Genne, P., Duchamp, O., Solary, E., Magnette, J., Belon, J. P., Chauffert, B. (1995) *Anticancer Drug Design*, **10**, 103–118.

201 Solary, E., Caillot, D., Chauffert, B., Casasnovas, R. O., Dumas, M., Maynadie, M., Guy, H. (1992) *Journal of Clinical Oncology*, **10**, 1730–1736.

202 Taylor, C. W., Dalton, W. S., Mosley, K., Dorr, R. T., Salmon, S. E. (1997) *Breast Cancer Research and Treatment*, **42**, 7–14.

203 Miller, T. P., Chase, E. M., Dorr, R., Dalton, W. S., Lam, K. S., Salmon, S. E. (1998) *Anticancer Drugs*, **9**, 135–140.

204 Solary, E., Witz, B., Caillot, D., Moreau, P., Desablens, B., Cahn, J. Y., Sadoun, A., Pignon, B., Berthou, C., Maloisel, F., Guyotat, D., Casassus, P., Ifrah, N., Lamy, Y., Audhuy, B., Colombat, P., Harousseau, J. L. (1996) *Blood*, **88**, 1198–1205.

205 Solary, E., Drenou, B., Campos, L., de Cremoux, P., Mugneret, F., Moreau, P., Lioure, B., Falkenrodt, A., Witz, B., Bernard, M., Hunault-Berger, M., Delain, M., Fernandes, J., Mounier, C., Guilhot, F., Garnache, F., Berthou, C., Kara-Slimane, F., Harousseau, J. L. (2003) *Blood*, **102**, 1202–1210.

206 Genne, P., Duchamp, O., Solary, E., Pinard, D., Belon, J. P., Dimanche-Boitrel, M. T., Chauffert, B. (1994) *Leukemia*, **8**, 160–164.

207 Solary, E., Mannone, L., Moreau, D., Caillot, D., Casasnovas, R. O., Guy, H., Grandjean, M., Wolf, J. E., Andre, F., Fenaux, P., Canal, P., Chauffert, B., Wotawa, A., Bayssas, M., Genne, P. (2000) *Leukemia*, **14**, 2085–2094.

208 Suzuki, T., Fukazawa, N., San-nohe, K., Sato, W., Yano, O., Tsuruo, T. (1997) *Journal of Medicinal Chemistry*, **40**, 2047–2052.

209 Nakanishi, O., Baba, M., Saito, A., Yamashita, T., Sato, W., Abe, H.,

Fukazawa, N., Suzuki, T., Sato, S., Naito, M., Tsuruo, T. (1997) *Oncology Research*, **9**, 61–69.

210 Naito, M. and Tsuruo, T. (1997) *Cancer Chemotherapy and Pharmacology*, **40**, S20–S24.

211 Dieras, V., Bonneterre, J., Laurence, V., Degardin, M., Pierga, J. Y., Bonneterre, M. E., Marreaud, S., Lacombe, D., Fumoleau, P. (2005) *Clinical Cancer Research*, **11**, 6256–6260.

212 Saeki, T., Tsuruo, T., Sato, W., Nishikawsa, K. (2005) *Cancer Chemotherapy and Pharmacology*, **56** (Suppl. 1), 84–89.

213 Robert, J. (2004) *Current Opinion in Investigational Drugs*, **5**, 1340–1347.

214 Pezzuto, J. M., Kosmeder, J. W., II, Park, E. J., Lee, S. K., Cuendet, M., Gills, J., Bhat, K., Grubjesic, S., Park, H. S., Mata-Greenwood, E., Tan, Y. M., Yu, R., Lantvit, D. D., Kinghorn, A. D. (2005) Characterization of natural product chemopreventive agents, In Kelloff, G. J., Hawk, E. T., Sigman C. C. (Eds.) *Cancer Chemoprevention: Strategies for Cancer Chemoprevention*, Vol. **2**, Humana Press, Totowa, NJ.

215 Teel, R. W. (1991) *Nutrition Cancer*, **15**, 27–32.

216 Morre, D. J., Chueh, P. J., Morre, D. M. (1995) *Proceedings of the National Academy of Sciences of the United States of America*, **92**, 1831–1835.

217 Kim, J. D., Kim, J. M., Pyo, J. O., Kim, S. Y., Kim, B. S., Yu, R., Han, I. S. (1997) *Cancer Letters*, **120**, 235–241.

218 Sanchez, A. M., Sanchez, M. G., Malagarie-Cazenave, S., Olea, N., Diaz-Laviada, I. (2006) *Apoptosis*, **11**, 89–99.

219 Mehta, R. G., Liu, J., Constantinou, A., Thomas, C. F., Hawthorne, M., You, M., Gerhauser, C., Pezzuto, J. M., Moon, R. C., Moriarty, R. M. (1995) *Carcinogenesis*, **16**, 399–404.

220 Zhang, Y., Talalay, P., Cho, C. G., Posner, G. H. (1992) *Proceedings of the National Academy of Sciences of the United States of America*, **89**, 2399–2403.

221 Gerhauser, C., You, M., Liu, J., Moriarty, R. M., Hawthorne, M., Mehta, R. G., Moon, R. C., Pezzuto, J. M. (1997) *Cancer Research*, **57**, 272–278.

222 Zhang, Y., Kensler, T. W., Cho, C. G., Posner, G. H., Talalay, P. (1994) *Proceedings of the National Academy of Sciences of the United States of America*, **91**, 3147–3150.

223 Yuan, F., Chen, D. Z., Liu, K., Sepkovic, D. W., Bradlow, H. L., Auborn, K. (1999) *Anticancer Research*, **19**, 1673–1680.

224 Hudson, E. A., Howells, L., Ball, H. W., Pfeifer, A. M., Manson, M. M. (1998) *Biochemical Society Transactions*, **26**, S370.

225 Takada, Y., Andreeff, M., Aggarwal, B. B. (2005) *Blood*, **106**, 641–649.

226 Reed, G. A., Peterson, K. S., Smith, H. J., Gray, J. C., Sullivan, D. K., Mayo, M. S., Crowell, J. A., Hurwitz, A. (2005) *Cancer Epidemiology, Biomarkers and Prevention*, **14**, 1953–1960.

227 Bell, M. C., Crowley-Nowick, P., Bradlow, H. L., Sepkovic, D. W., Schmidt-Grimminger, D., Howell, P., Mayeaux, E. J., Tucker, A., Turbat-Herrera, E. A., Mathis, J. M. (2000) *Gynecologic Oncology*, **78**, 123–129.

3
Alkaloids and the Bitter Taste

Angela Bassoli, Gigliola Borgonovo, Gilberto Busnelli

3.1
Introduction

Bitter taste perception is innate in man and usually induces aversive reactions: neonates react with stereotypic rejection responses [1]; all animals, with the exception of some herbivores, react with aversive or repulsive behavior to compounds that humans consider bitter [2,3]. Substances bitter to humans are also often repulsive to invertebrates [4,5]. Since numerous harmful compounds, including xenobiotics, salts, rancid fats, fermentation products, and secondary plant metabolites including alkaloids taste bitter, this taste is usually considered as a defense mechanism against the ingestion of potentially toxic compounds. Incidentally, human responses to bitterness show remarkable diversity, ranging from absolute rejection to strong acceptance. Moreover, the perception of the bitter taste changes significantly during ageing; it is known that some foodstuffs that are usually rejected by infants and children become acceptable to adults, when a certain degree of bitterness in food and beverages starts to become pleasant and contributes to attraction and palatability. This is for instance the case with coffee or beer, or the bitter taste of nicotine in cigarette smoke. This phenomenon undoubtedly has an evolutionary significance. Many natural bitter compounds also have a certain degree of toxicity, therefore the bitter taste receptors have been designed to recognize and avoid them, particularly when the organism is potentially more susceptible to intoxication, as during childhood. Similarly, bitter taste perception changes significantly during pregnancy [6] when the threshold of perception of bitter compounds such as quinine is usually lowered.

A bitter taste has long been associated with alkaloids and in fact many of them have always been known as bitter. Actually, in most cases the bitter taste was initially associated with plant extracts containing alkaloids such as cinchona, belladonna, or aconitum, and subsequently with the pure compounds after isolation. This association is so strong that the bitter taste of food is still sometimes attributed to the presence of alkaloids, even when this has not been demonstrated by the isolation of any active principle and the sensory evaluation of single compounds. There are hundreds of bitter compounds from many other chemical classes, such as terpenes, aminoacids, salts, flavonoids, and a number of synthetic chemicals, and often the bitterness associated

Modern Alkaloids: Structure, Isolation, Synthesis and Biology. Edited by E. Fattorusso and O. Taglialatela-Scafati

ISBN: 978-3-527-31521-5

with these compounds is even stronger than that of alkaloids. Some prototypes of bitter compounds for sensory studies are for instance PROP (6-*n*-propyl-2-thiouracil), PTC (phenylthiocarbamide), denatonium benzoate, sucrose octaacetate, and magnesium sulfate. Even in food, many bitter principles have a nonalkaloid structure: the flavonoid naringin, which gives bitterness to grapefruit, humulone, found in beer, or lactucine, a sesquiterpene lactone found in lettuce, are only some examples. This diversity in the chemical structure of bitter compounds has a parallel in the large variability of the bitter taste receptors: we now recognize many bitter taste receptors, compared to the, probably single, sweet taste receptor. Another significant difference between the bitter taste and the other basic taste chemoreception mechanisms is the sensitivity: the detection threshold for the bitter taste of quinine is in the micromolar range, whereas sucrose or sodium chloride can only be detected at millimolar concentrations. The evolution and adaptation mechanisms seem therefore to have built a single receptor with low affinity for the positive selection of the sugar nutrients, and a family of multiple receptors with higher sensitivity to recognize and avoid potentially dangerous substances from many different food sources, especially plants.

Despite this diversity, the alkaloids are still considered as reference standards for bitter taste: in sensory evaluation methodologies caffeine and quinine solutions are used to train panelists to recognize the bitter taste and to give comparative evaluations of other bitter principles. This is obviously also due to the fact that these two alkaloids are among the most important for their use in food, not only as contaminants but as "active principles" selected for their taste profile and biological activity.

Besides these, a number of other alkaloids are known to have a bitter taste, which was generally discovered accidentally, as many are toxins, poisons, and other biologically active compounds. Finally, the bitter taste of alkaloids also has some importance in the interaction between plants and animals, especially insects, which have a gustative system quite different from that of vertebrates; these interactions have some important applications in ecology and agriculture.

3.2
The Bitter Taste Chemoreception Mechanism

The primary structures of taste reception are the taste buds, which are found in all vertebrates except hagfish [7]. Taste buds are assemblies of approximately 100 cells, which, in humans and other mammals, are present on the tongue, the soft palate, epiglottis, pharynx, and larynx. In insects, taste cells are present also in legs, ovipositor organs, and inside the stomach. There are four types of taste receptor cells (TRCs), called type I to type IV, sometimes subdivided into subclasses [8]. Recently, several signal transduction molecules have been detected in most type II cells and a small subset of type III cells, suggesting that the TRCs for umami, sweet, and bitter taste are among these cells [9,10]. TRCs are not neurons but specialized epithelial cells, and are accordingly designated as secondary sensory cells. In addition to their organization in taste buds, TRCs may also exist as solitary chemosensory cells in the developing mammalian gustatory epithelium [11,12] and chemosensory clusters in the larynx [13]. On the tongue, taste buds are organized in specialized folds and protrusions, the gustatory papillae. The three types of chemosensory papillae are the fungiform, foliate, and vallate papillae. In humans the latter are called circumvallate papillae, because they are completely surrounded by moats. Branches of the three cranial nerves innervate the taste buds, such that individual axons supply several TRCs, which may be located in different taste buds [14–17]. Most of the gustatory information is transmitted by the chorda tympani of the seventh cranial or facial nerve and the glossopharyngeal nerve. In humans the first-order pseudounipolar ganglionic neurons make contact with second-order neurons in the rostral part of the nucleus tractus solitarius (NTS), in a region referred to as the gustatory nucleus, located in the medulla, whereas some differences exists for animals such as rodents and, of course, insects. The response of cultured rat trigeminal ganglion neurons to bitter tastants has been studied [18]. The authors investigated the responses of rat chorda tympani and glossopharnygeal neurons to a variety of bitter-tasting alkaloids, including nicotine (**1**) yohimbine, quinine (**2**), strychnine (**3**), and caffeine (**4**), as well as capsaicin (**5**), the pungent ingredient in hot pepper.

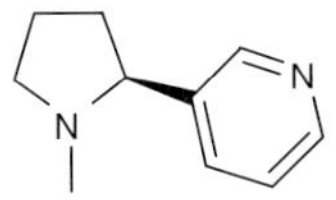

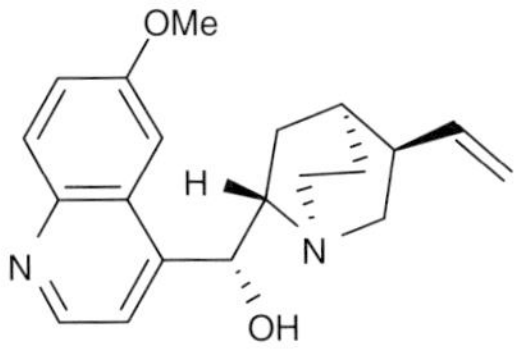

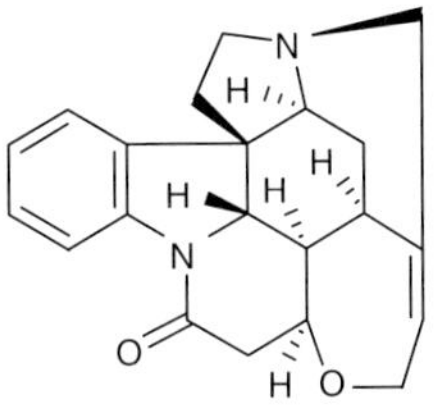

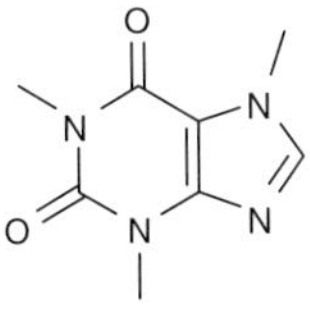

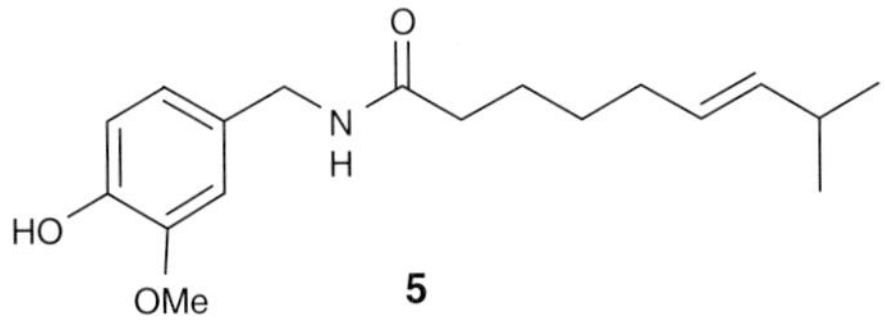

Of the 89 neurons tested, 34 % responded to 1 mM nicotine, 7 % to 1 mM caffeine, 5 % to 1 mM denatonium benzoate, 22 % to 1 mM quinine hydrochloride, 18 % to 1 mM strychnine, and 55 % to 1 μM capsaicin. These data suggest that neurons from the trigeminal ganglion respond to the same bitter-tasting chemical stimuli as do taste receptor cells and are likely to contribute information sent to the higher central nervous system regarding the perception of bitter/irritating chemical stimuli. The neural responses of bitter alkaloid compounds in rats have also been noticed by physiology experiments involving recording the neural activity from rat glossopharyngeal and chorda tympani neurons. Responses to several bitter alkaloids were obtained: 10 mM quinine-HCl, 50 mM caffeine, and 1 mM each nicotine, yohimbine and strychnine, plus a number of nonalkaloid bitter-tasting compounds and some other tastants (NaCl, HCl and capsaicin). It was found that individual neurons in both glossopharyngeal and chorda tympani nerves differed in their relative sensitivities to the various bitter stimuli. To determine relationships among these stimuli, the differences in the evoked responses between each stimulus pair were summarized in a multidimensional scaling space. In these analyses no nerve showed any obvious similarity between the placements of quinine and the other bitter stimuli. Such data suggest that first-order gustatory neurons can discriminate among the above bitter stimuli. For glossopharyngeal neurons, a certain similarity to quinine was found only for nicotine and denatonium, and for chorda tympani neurons, some similarity to quinine was found only for KCl and $MgCl_2$. Of the bitter compounds tested, quinine evoked the greatest response from glossopharyngeal neurons, suggesting that quinine can activate taste receptors cells by more transduction mechanisms than other bitter stimuli [19].

The initial stage of the bitter taste stimulation is the binding of bitter molecules to a protein receptor on the taste buds. T2Rs are a multigene family of the G protein-Coupled Receptors and have a key role in the detection of bitter tastants [20,21]. According to the latest reports the T2Rs gene repertoire in humans comprises 25 full-length genes and 11 pseudogenes [22]. With the exception of *TAS2R1* on chromosomes 5, 9, and 15 TAS2R genes are present in extended clusters on both chromosomes 7 and 12. The mouse genome contains slightly more T2R genes, namely 35 genes and 6 pseudogenes located on chromosomes 2, 6, and 15 [20–23]. All but two genes, which are found on chromosomes 2 and 15 respectively, are located in two clusters on chromosome 6 [24]. As reported above, T2Rs have a key role in the detection of bitter tastants together with α-gustducin, a G protein α-subunit expressed in about 25% of TRCs. Indeed, molecularly cloned bitter-responsive taste receptors have been shown to couple and activate heterotrimetric gustuducin. G protein-coupled receptors mediate the cellular responses to an enormous diversity of signaling molecules, including hormones, neurotransmitters, and local mediators, varied in structure and in function. Despite this, all of the GPCRs have a similar structure and are almost certainly evolutionarily related. All GPCRs have an extracellular N-terminal segment, seven α-helical transmembrane (TM) domains that form the TM core, three exoloops, three cytoloops, and a C-terminal segment. N-terminal segments (7–595 amino acids), loops (5–230 amino acids), and C-terminal segments (12–359 amino acids) vary in size. TAS2Rs are structurally diverse as they display approximately 17–90% sequence identity at the amino acid level, suggesting that different family members may recognize chemicals with widely different structures [22]. In general, the cytoplasmic part of the putative TM segments and the intracellular loops are comparably well conserved, consistent with the idea that intracellular coupling to signal transduction proteins is of limited variability. The extracellular loops and upper parts of the TM segments are less conserved. This is not surprising, since these parts of the TAS2Rs are likely to form heterogeneous binding motifs for the numerous and structurally diverse bitter compounds. On the other hand, the carboxyl-terminal part of TAS2Rs is also highly variable, reflecting possible differences between TAS2Rs in receptor regulation and trafficking. It has been estimated that 6–10 cells per taste bud express

TAS2Rs, corresponding to approximately 15 % of the cells. This estimate applies to rodent taste organs, where almost every taste bud contains TAS2R-positive cells [21]. A similar estimate of approximately 20 % has been reached for TAS2R gene expression in human circumvallate taste buds [25]. If any of the T2Rs is expressed in approximately 15 % of the cells of circumvallate, foliate, and palate taste buds and given that there are over 25 T2Rs in humans, a taste cell should express more than one receptor. Through two-color double-hybridization experiments using a collection of differentially labeled cDNA probes, showed that the majority of the taste receptor cells expressed nearly, if not all, the full complement of T2Rs. The observations are consistent with the fact that mammals are capable of recognizing a wide range of bitter substances but not of distinguishing between them. Bitter compounds activate T2Rs, which activate gustducin heterotrimers [26]. Dissociation of the heterotrimeric gustducin splits the signal into two parts. Activated α-gustducin stimulates a phosphodiesterase (PDE) to hydrolyze cAMP thus reducing the cytosolic concentration of cNMP. The consequences of the α-gustducin-induced decrease in cyclic nucleotide levels are still not well understood. At the same time, the β/γ heterodimer (i.e. β3/γ13) complex, released from activated α-gustducin, activates a phospholipase C (PLCβ2) to generate inositole triphosphate (IP3) which leads to the release of Ca^{2+} from internal stores. Calcium ions in turn stimulate TRPM5 channels, resulting in changes of membrane currents leading to a receptor potential, which translates into action potentials and release of neurotransmitter. Compared to wild-type mice, the responses of α-gustducin knockout mice to bitter stimuli are reduced but not totally abolished, thus suggesting that other G proteins or different pathways are involved. $G\alpha_{i\text{-}3}$, Ga14, Ga15, Gaq, α-transducin are possible candidates as they are also expressed in TRCs. Quench flow studies with murine taste tissue indicate that the alkaloids caffeine and theophylline inhibit PDE to raise TRC cGMP levels [27]. Soluble guanylyl cyclase (GC) is the presumed source of the cGMP; both GC and nitric oxide synthase (NOS) are known to be present in TRCs. Quinine is known to block K^+ channels and to cause TRC depolarization. Bullfrog TRC membrane patches contain a cation channel that is apparently gated directly by quinine and denatonium benzoate and other compounds that humans find bitter. A complete understanding of the physiology of bitter taste requires exact knowledge about the interactions of TAS2Rs with bitter substances. There is a vast excess of numerous, structurally diverse bitter compounds over the about 30 mammalian TAS2R bitter taste receptors [28,29]. The TAS2Rs are therefore highly unlikely to interact with their bitter stimuli in a simple one-to-one ratio. Instead they should be broadly tuned to multiple bitter compounds. Great efforts are being directed toward the deorphanization of all the known receptors and to the elucidation of their structure–activity relationships. The only established correlation between a bitter alkaloid and a TAS2R is that of strychnine, which activates hTAS2R10 [30]. Research into the ligands of the bitter taste receptors is at an initial stage and most of the T2Rs have yet to be analyzed, and it is therefore very likely that in the near future other alkaloids will be identified as ligands for specific receptors.

3.3 Bitter Alkaloids in Food

Despite the supposed universality of bitter taste rejection, many commonly consumed foods and beverages such as fruits, tea, coffee, chocolate, and alcohol have bitterness as a major sensory attribute which, in the overall taste profile of a food, is often appreciated by the consumers. Some alkaloids are certainly responsible for the bitter taste of known food: the taste threshold is available only for atropine (0.1 mM), cocaine (0.5 mM), and morphine (0.5 mM) (Table 3.1).

Caffeine is perhaps the most popular of these compounds. It is moderately bitter in taste, having a taste threshold in water of 0.8–1.2 mM [31]. Major sources of caffeine are the seeds of *Coffea* sp. plants. The caffeine content in raw Arabica coffee is 0.8–2.5 %, while in the Robusta variety it can be as high as 4 %. Caffeine is only partly decomposed during roasting, and can be reduced artificially by chemical extraction during decaffeination processes. The physiological effects of caffeine are well known: stimulation of the central nervous system, increased blood circulation and respiration, and many others. Caffeine is present also in the leaves of tea (*Camellia sinensis*) in various amounts (2.5–5.5 % in dry leaves) according to the origin, age, and

Tab. 3.1 Some bitter alkaloids and their CAS numbers.

Compound	CAS	Compound	CAS
Brucine	357-57-3	Pilocarpine	92-13-7
Atropine	51-55-8	Cathine	492-39-7
Cocaine	50-36-2	Cathinine	5265-18-9
L-Ephedrine	299-42-3	Isoberberine	477-62-3
Morphine	57-27-2	Coumingine	26241-81-6
Heroin	561-27-3	Norcassaidine	26296-41-3
Senecionine	130-01-8	Norcassamidine	36150-73-9
Monocrotaline	315-22-0	Cassaine	468-76-8
Methylergonovine	113-42-8	Retrorsine	480-54-6
Anisotropine methylbromide	80-50-2	Delsoline	509-18-2
Propantheline	298-50-0	Delcosine	545-56-2
Dyphylline	479-18-5	Rodiasine	6391-64-6
Emetine	483-18-1	Infractopicrine	91147-09-0
Vincamine	1617-90-9	Yohimbine	146-48-5
Berberine sulfate	316-41-6	Solasodine	126-17-0
Cincophen	132-60-5	Pistillarin	89647-69-8
Noscapine	128-62-1	Stemonine	27498-90-4
Chloroquine	54-05-7	Glomerine	7471-65-0
Cytisine	485-35-8	Gentianine	439-89-4
Caffeine	58-08-02	Theobromine	83-67-0
Theophylline	58-55-9	Capsaicin	404-86-4
Quinine	130-95-0	Quinidine sulfate	50-54-4
α-Solanine	20562-02-1	β-2-chaconine	469-14-7
α-Chaconine	20562-03-2	Solanidine	80-78-4

processing. Together with theobromine (**6**) and theophylline (**7**), which are present in much lower percentage, it contributes to the taste of tea.

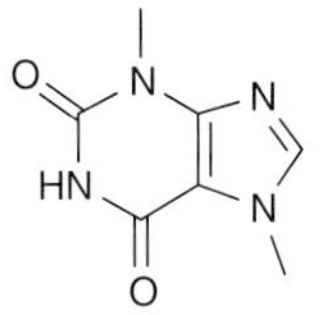

6 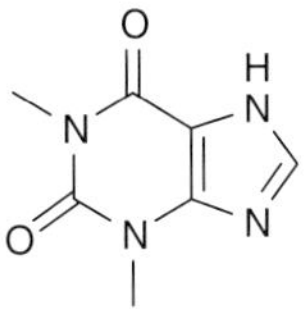 **7**

Caffeine is also present in matè (Paraguayan tea), the infusion from the leaves of *Ilex paraguariensis*. The content of caffeine in matè is 0.5–1.5 % and this beverage is known to stimulate the appetite; this plant has long been the most important alkaloid-containing brewed plant in South America. Yerba matè has been investigated for its caffeine content and the time perception of bitterness of the extract infusions [32]. Caffeine and its stimulating action are also present in other traditional beverages such as those based on cola nuts, the seeds of a tree of the Sterculiaceae family, genus *Cola* sp., growing in West Africa, Madagascar, Sri Lanka, and Central and South America. The caffeine content is about 2.2 % and theobromine is 0.05 %. Theobromine is contained in larger amounts, about 1.2 %, in cocoa beans where caffeine is also present (0.2 %) and therefore these two alkaloids are present in all chocolate derivatives.

Both theobromine and theophylline have a bitter taste similar to that of caffeine, and are responsible for the bitter taste of cacao and black chocolate, but in this case the contribution of other compounds has been demonstrated. Sensory-guided decomposition of roasted cocoa nibs revealed that, besides theobromine and caffeine, a series of bitter-tasting 2,5-diketopiperazines and flavan-3-ols were the key inducers of the bitter taste as well as the astringent mouth-feel imparted upon consumption of roasted cocoa. Actually, the flavanol-3-glycosides not only impart a velvety astringent taste sensation to the oral cavity but also contribute to the bitter taste of tea infusions by amplifying the bitterness of caffeine [33]. The mechanism of action of caffeine and other xanthines on taste receptors is not known in detail. The effects of caffeine in stimulating taste receptor cells have been studied by calcium imaging techniques using intestinal STC-1 cells [34]. These cells are stimulated by caffeine in a dose-dependent manner.

As for many other bitter compounds, the stimulus of caffeine bitterness is transduced through multiple mechanisms and complex pathways. To investigate this idea more thoroughly, caffeine was studied by patch-clamp and radiometric imaging techniques on dissociated rat taste receptor cells. At behaviorally relevant concentrations, caffeine produced strong inhibition of outwardly and inwardly rectifying K currents. Caffeine addition inhibited Ca current, produced a weaker inhibition of Na current, and had no effect on Cl current. Consistent with its effects on voltage-dependent currents, caffeine caused a broadening of the action potential and an increase of the input resistance. Caffeine was an effective stimulus for elevation of intracellular Ca. This elevation was concentration dependent, independent of extracellular Ca or ryanodine, and dependent on intracellular stores as evidenced by thapsigargin treatment. These dual actions on voltage-activated ionic currents and intracellular calcium levels suggest that a single taste stimulus, caffeine, utilizes multiple transduction mechanisms [35].

Quinine and quinidine (**8**) are also bitter. The bitter taste of quinine can still be detected in a dilution of 1 : 184 175 (unit of bitterness) [36]. Quinine is therefore much more bitter than caffeine; 1 mM quinine hydrochloride gives the same bitter stimuli as 30–100 mM caffeine [37]. quinine, as quinine salts or extracts from cinchona bark (*Cinchona officinalis* L., Rubiaceae), is used as a bittering agent in tonic-type drinks, at a concentration of approximately 80 mg/L quinine hydrochloride. Quinine has also been used in some bitter alcoholic beverages and to a small extent in flour confectionery. The major dietary source of quinine are soft drinks of the tonic or bitter lemon type. Some consumption studies to explore some potential toxic effect from quinine consumption have been conducted among users of quinine-containing soft drinks in the United Kingdom, France, and Spain. These studies demonstrated that the current levels used are not of toxicological concern, but recommended that the consumer should be informed by appropriate labeling of the presence of quinine in foods and beverages in which it is used.

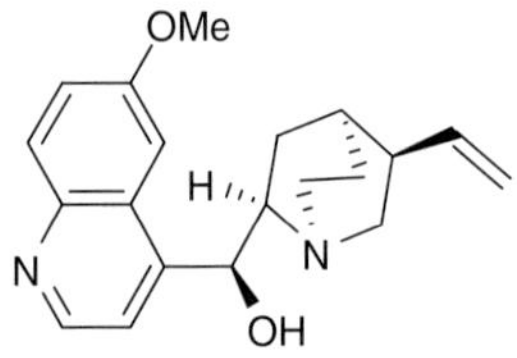

8

Quinine has been also used as a standard compound to study taste discrimination behavior in nonhuman primates, in order to establish a bitter taste (quinine sulfate) as a cue for lever selection and food reward in rhesus monkeys. After training with quinine, all of the animals acquired the discrimination, and the lowest quinine concentration that maintained consistent behavior was 0.3 mg/mL. To assess the specificity of the discrimination, compounds from other human taste categories were tested. A series of compounds that are detected as bitter by humans (caffeine, strychnine, PTC, denatonium benzoate, and urea) produced full generalization to the quinine sulfate discriminative stimulus, while sweet (as sucrose), and salty (as sodium chloride) stimuli did not. There was individual variation among animals in response to "sour" compounds; acetic acid did not generalize to quinine, but hydrochloric acid produced full generalization in one of three animals. These results suggest that a bitter taste cue is controlling the quinine discrimination [38].

Quinine also has pharmacological activity as an antimalarial, and some studies have been made in order to find a method to diminish or suppress its bitter taste. This can be achieved by adding sweet compounds such as sucrose or aspartame or nonspecific bitter taste inhibitors such as NaCl or phosphatidic acid and tannic acid [39]. The effects have been studied by sensory evaluation tests in human volunteers, a binding study, and using an artificial taste sensor. It is difficult to understand to what extent the inhibition is due to adsorption effects or competition for the receptor binding, since the chemoreception mechanism of the bitter taste of alkaloids is not yet clearly understood. In order to suppress the bitter taste of quinine some molecularly imprinted polymers (MIPs) have been devised [40]. L-Arginine, L-ornithine, L-lysine, and L-citrulline were tested as bitterness-suppressant candidates. In an HPLC study using a uniformly sized MIP for cinchonidine, which has a very similar structure to quinine, the retention factor of quinine was significantly shortened by the addition of L-Arg or L-Orn to the mobile phase, whereas a slight or no decrease was observed when L-Ctr and L-Lys were added. This suggests that L-Arg and L-Orn may compete with quinine in the cinchonidine-imprinted space. Finally, the results of human gustatory sensation tests correlated well with the MIP data and are supported also by NMR and molecular-modeling studies. The proposed method using MIPs seems to have a potential for screening bitterness-suppressant agents for quinine and also give some interesting information on the possible interaction of quinine with amino acids in its taste receptor. A specific bitter taste receptor for quinine has not yet been identified; moreover, some experiments recently demonstrated [41] that there is evidence for both receptor-dependent and -independent transduction mechanisms for a number of bitter stimuli, including quinine hydrochloride.

Another bitter tasting alkaloid in food is α-solanine (**9**). This compound and some analogs (α-chaconine (**10**), β-2-chaconine (**11**), and solanidine (**12**)) are responsible for the bitterness of greened potatoes and also for their known toxicity. Organoleptic evaluation of three major potato glycoalkaloids disclosed two distinct taste stimuli: a bitter caffeine-like taste and an astringent pain sensation, characterized as burning and peppery. At higher concentrations the two stimuli merged leaving a persistent burning sensation lasting up to 2 h. The absolute bitter taste thresholds for α-solanine, α-chaconine, and β-2-chaconine are 0.313, 0.078, and 0.078 mg, respectively, with corresponding values for the pain stimulus at 0.625, 0.323, and 0.156 mg. Bitterness thresholds for caffeine and solanidine measured in the same conditions are respectively 1.25 and 0.313 mg, without any pain stimulus up to 1000 ppm [42]. α-Solanine was also identified as the bitter principle of eggplant fruits (*Solanum melongena*) [43].

Capsaicin is the principal chemesthetic compound able to generate the hot sensation, and its activity on the so-called vanilloid receptor has been studied for a long time (for capsaicin see Chapter 4). Interestingly, the taste of capsaicin (and of many varieties of hot pepper) is also described as bitter. Many studies have indicated that capsaicin, traditionally considered to be a pure chemesthetic stimulus, can evoke a bitter taste and might also cross-desensitize the tastes of some bitter and sour tastants. The scope and nature of capsaicin effects on bitter taste have therefore been studied by sensory evaluation techniques. Subjects rated the taste and burning/stinging of quinine sulfate (QSO4) (0.32 and 1.0 mM), saccharin (1.0 and 3.2 mM), urea (3.2 and 10 M), $MgCl_2$, (0.18 and 0.56 M), PROP (0.32 mM), and sucrose (0.32 and 1.0 M) applied to the tongue tip with cotton swabs before and after 10 applications of 300 μM capsaicin. Capsaicin initially evoked a weak bitterness in some subjects, which quickly diminished over repeated exposures. Following capsaicin treatment, the bitterness of QSO4, urea, $MgCl_2$, and PROP was reduced, as was the burning sensation produced by $MgCl_2$ and urea. In another experiment, 29 subjects were examined in the circumvallate region of the tongue using the same general procedure. Capsaicin induced a weak but persistent bitterness in a subset of subjects but failed to desensitize its own bitterness or that of any other tastant. These experiments show that capsaicin can both stimulate and desensitize bitter taste, but in amounts that vary for different bitter stimuli and between the front and back of the tongue [44]. Interestingly, a similar effect has been noticed also for menthol, a cooling chemesthetic compound. Overall, the results suggest that capsaicin and menthol are capable of stimulating a subset of taste neurons that respond to bitter substances, perhaps via receptor-gated ion channels like those recently found in capsaicin- and menthol-sensitive trigeminal ganglion neurons, and that the glossopharyngeal nerve may contain more such neurons than the chorda tympani nerve. The fact that some people fail to perceive bitterness from capsaicin further implies that the incidence of capsaicin-sensitive taste neurons varies between people as well as between gustatory nerves [45].

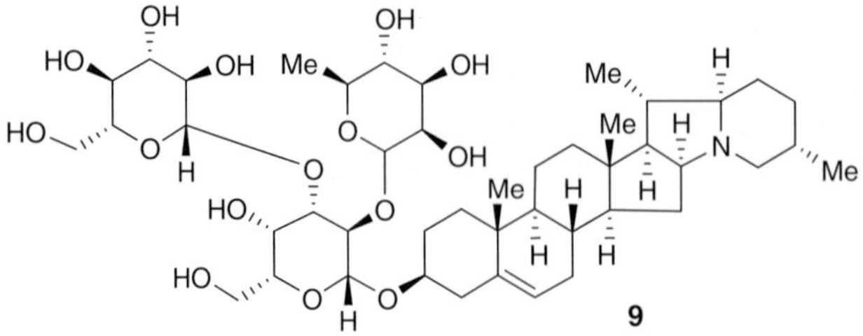

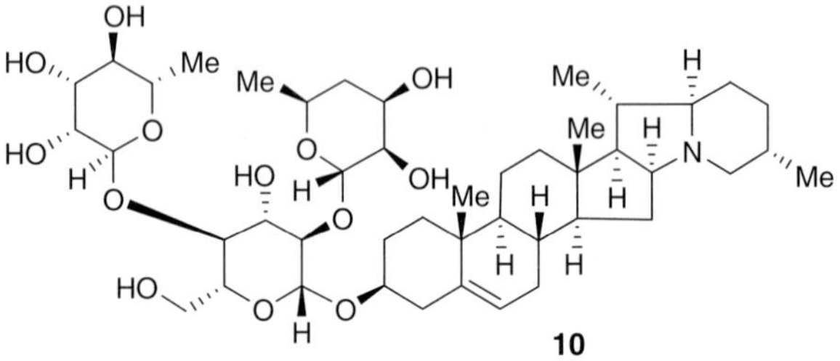

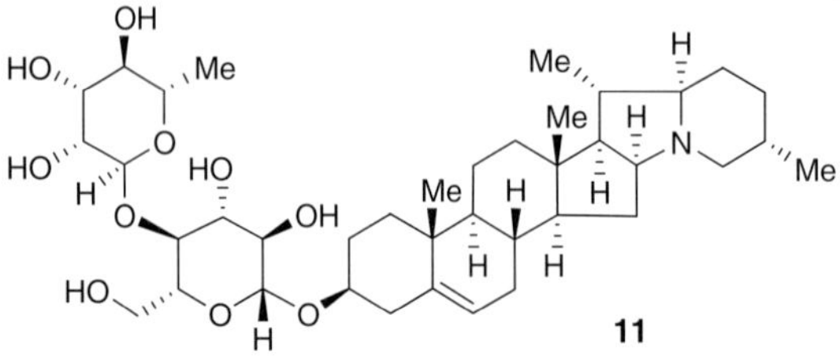

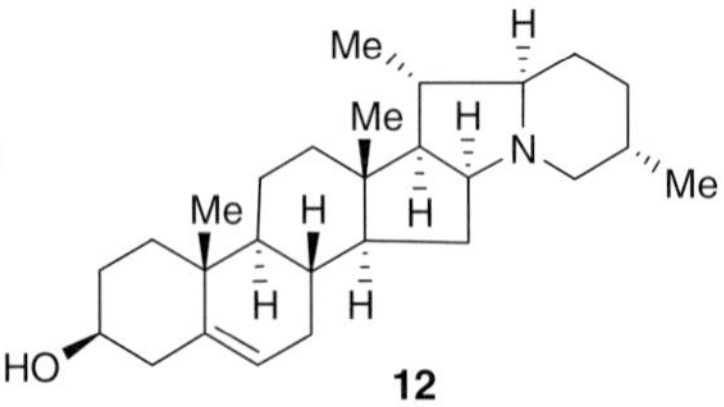

It is known that some cyclic ketopiperazines coming from the Maillard reaction, especially in proline rich foodstuffs, have an intense bitter taste. Some new interesting cyclic nitrogen compounds have been identified by taste dilution analysis from mixtures coming from the Maillard reaction during food processing. Despite these compounds not being alkaloids strictly speaking, they have some interesting structural analogy with alkaloid compounds. One of these compounds, named quinizolate (**13**), exhibits an intense bitter taste at an extraordinarily low detection threshold of 0.00025 mmol/kg of water. This novel taste compound was found to have 2000- and 28-fold lower threshold concentrations than the standard bitter compounds caffeine and quinine hydrochloride, respectively, and, therefore, it is claimed to be one of the most intensely bitter compounds reported so far [46]. It is important to note that in sensory evaluation there is a difference between "detection threshold" and "recognition threshold"; the first determines the concentration of a substance that makes one able to say: this is not pure water; the second means that the panelist is able to state: this is bitter. Therefore these reported data on the taste of new compounds must to be taken carefully especially in comparison with others.

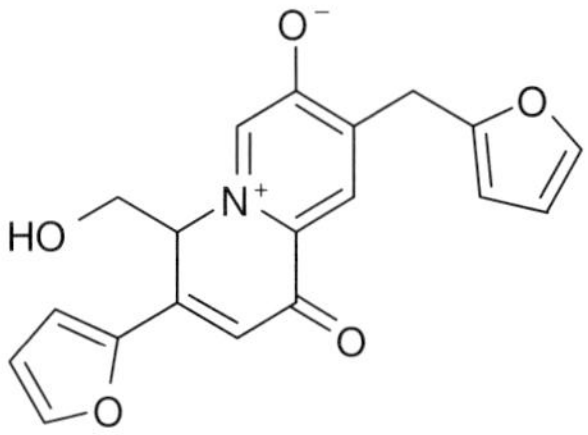

13

3.4 The Bitter Taste of Alkaloids in Other Drugs and Poisons

Most of the alkaloids having psychotropic activity or other general toxicity are described as bitter. Nevertheless, this attribute is very often reported as obvious or "taken for granted" even when there are no exhaustive studies, for the obvious reason that no one is willing to use sensory analysis on toxic compounds. Ancient and modern literature report many anecdotes about the bitter taste of alkaloid poisons. The taste of the hemlock potion offered to Socrates to induce his "suicide" should have been very bitter, but no comments about this are reported by Plato in the last chapter of *Fedone*, describing Socrates death, where he is reported to drink the poison "peacefully," maybe thanks to his interior consciousness. If the bitter taste is indeed a poorly relevant feature of poisons used for suicide, it is a very negative and undesirable characteristic for poisons used in homicides. A nice example of this concept is given in a novel by Agatha Christie [47], where the famous detective Poirot demonstrates that he has some knowledge of taste chemistry:

> "Poirot," I cried, "I congratulate you! This is a great discovery."
> "What is a great discovery?" "Why, that it was the cocoa and
> not the coffee that was poisoned. That explains everything!
> Of course it did not take effect until the early morning, since
> the cocoa was only drunk in the middle of the night." "So you
> think that the cocoa – mark well what I say, Hastings, the
> cocoa – contained strychnine?"

> "Mrs. Inglethorp was in the habit of drinking a cup of cocoa in the middle of the night. Could the strychnine have been administered in that?" "No, I myself took a sample of the cocoa remaining in the saucepan and had it analysed. There was no strychnine present." I heard Poirot chuckle softly beside me. "How did you know?" I whispered. "Listen." "I should say" – the doctor was continuing – "that I would have been considerably surprised at any other result."
>
> "Why?" "Simply because strychnine has an unusually bitter taste. It can be detected in a solution of 1 in 70,000, and can only be disguised by some strongly flavoured substance. Cocoa would be quite powerless to mask it." One of the jury wanted to know if the same objection applied to coffee. "No. Coffee has a bitter taste of its own which would probably cover the taste of the strychnine."

The bitter taste is therefore reported by occasional and accidental tasting or by the observation of rejection behavior by animals. The equation alkaloid = toxic = bitter is so strong that sometimes the presence of some bitter alkaloids was assumed in some bitter plants and never demonstrated by the isolation of any active principle of bitter taste. It is worth noting that many toxic alkaloids are reported as bitter not only in the popular tradition but in scientific literature. Table 3.2 lists some of these compounds, reporting the common name, CAS number and if the compound is listed on Toxnet.

Since these compounds are mostly used as weapons or poisons, the bitter taste is generally not a problem in their use so, with few exceptions, it has seldom been studied systematically. One exception is strychnine, which is used as a potent rodenticide and for which the bitter taste is a limitation since it prevents ingestion by the animals. Some studies on the bitter taste of this alkaloid have been done with the aim of masking the taste of strychnine [48]. The threshold for the detection of strychnine in distilled water is 5.4 μg, in tap water 6.5 μg. Various substances were added to diluted solutions of strychnine to ascertain whether the bitter taste could be masked. It was masked to some extent by certain salts, sucrose, and extracts of yerba santa. It was also noticed that the cation was the significant factor in the masking

Tab. 3.2 Toxic alkaloids referred to as bitter.

Name	CAS	Detection threshold	Toxnet
Strychnine	57-24-9	0.002 mM	X
Nicotine	54-11-5	0.0019 mM	
Sparteine	90-39-1	0.00 085 g/100 mL	
Lupinine	486-70-4	0.0038 g/100 mL	
Hydroxylupanine	15358-48-2	0.0017 g/100 mL	
Lupanine	550-90-3		
Angustifoline	550-43-6		
Epilupinine	486-71-5		

action of salts. The masking efficiency of cations decreased in the order: Mg 78 %, Ca 25 %, Na 8 %, and K 2 %. $NaHCO_3$ increased the bitterness of strychnine. Five percent yerba santa extracts increased the threshold from 5.4 to 36.4 μg. This could be interpreted as a possible competition at molecular level between caffeine contained in mate and strychnine for the same receptors but we still have no evidence about this.

More recently, the interaction of strychnine with cloned taste receptors of the T2Rs family has been studied. It has been demonstrated that TAS2R16 and TAS2R10 are receptors for various bitter glycosides and the alkaloid strychnine, which activates the TAS2R10 receptor at a concentration of 0.2 mM. These data, compared with those for other alkaloids will certainly be very useful in predicting structural requirements for receptor–agonist interactions and in the search for bitter-blocking activities [49].

The bitter taste of nicotine is known and is an important component of the flavor of tobacco and cigarette smoke. Nicotine and its taste have been studied for their effect on taste preferences and therefore on nutrition. There is a well-known relationship between the smoking–nonsmoking habits and the dietary preferences of people; in particular it has been noted that smokers usually consume less sweet food and less vegetables and fruit. This has some consequences on health: the insufficient consumption of food rich in vitamins and antioxidants lead to a major exposure of smokers to cardiovascular diseases; on the other hand, they usually gain body weight when they stop smoking since they start to eat more sweetened food, and this is sometimes a deterrent to stopping smoking. Of course, this is in part due to the anorectic activity of nicotine, which is in part recognized. The taste behavior of smokers may also be related to a different perception of acid and sweet tastes that are representative of different ripening stages of fruit or of polyphenol astringency. The effects of smoking in modifying some sensory perceptions in humans are not well known, although some studies suggest an effect of nicotine on two main tastes, sweet [50–52] and bitter [53]. It is therefore difficult to find data on the influence of smoking on sensory perception and the existing data are quite old and often contradictory. Some authors reported that smoking had no significant effect on the taste receptors [54–57]. Others hypothesized that smoking could affect taste preferences via the taste mechanism [58]. It has been reported for a long time that sugar and salt thresholds are higher among smokers than among nonsmokers but it has also been demonstrated [59] that there are no differences between smokers and nonsmokers in their thresholds for sweet, salty, and sour, but only for bitter (significantly higher for smokers); the authors suggested that the nicotine and other alkaloids in cigarette smoke fatigue the mechanism for perception of bitter. They also concluded that the decrease in sensitivity is progressive with age and thus it appears to be the result of prolonged addiction. Interestingly, smokers appear to drink more coffee than nonsmokers [60,61] and this could also possibly be due to the elevation of the bitter taste threshold, besides other pharmacological effects. It has been demonstrated that nicotine induces taste aversion for sweet substances such as saccharine in rats when preadministered by injection [62]. It is obviously very complicated to distinguish between the effects of nicotine acting as a drug and those generated directly on the taste chemoreception apparatus, and some research is ongoing in order to clarify this point.

In the case of other compounds listed in Table 3.2, the data on bitter taste are reported occasionally and seem not to have been studied in detail.

3.5
Alkaloids and Taste in Insects

In insects the perception of taste is particularly interesting both from the theoretical point of view and for practical applications. Insects have very efficient gustative organs not only in the mouth but also in legs, antennae, in the abdominal region, and even in the internal stomach.

Many excellent books and reviews [63–67] describe this complex system and the role of insect taste in ecology, agriculture, and in the discovery of new molecules to be used as antifeedants, including many alkaloids, and only some relevant points will be summarized here.

Obviously, the gustative system of *Drosophila melanogaster* have been studied in details [68]. The main taste tissue in *Drosophila* spp. is composed of the two labial palps located at the distal end of the proboscis. From taste sensilla located on the prothoracic leg of females (18 sensilla) and males (28 sensilla), six sensilla have been identified that house a neuron activated by bitter compounds. These six sensilla fall into two groups: four were activated when stimulated with quinine but not berberine, and two were activated by berberine but not quinine. All six sensilla showed similar responses to denatonium and strychnine. Most interestingly, the bitter-sensing cell within these six sensilla was found to correspond to the L2 cell, known to be activated by high concentrations of NaCl (a repulsive stimulus). Thus, the L2 cell is a widely tuned neuron that responds to chemically diverse repulsive compounds. Many other prothoracic taste bristles, however, did not appear to house an L2 cell activated by bitter-tasting compounds used in this study; interestingly, however, the firing pattern of the W and S cells in these sensilla – stimulated by water or sugar – was significantly inhibited in the presence of quinine. These findings indicate that the detection of chemical compounds avoided by the fly are mediated through (at least) two different mechanisms, one that leads to the activation of an avoidance neuron (L2 cell) and one that leads to the inhibition of neurons (S cell) involved in the detection of attractive substrates, such as sugars. Hiroi *et al.* [69] recently investigated the firing pattern of labellar neurons stimulated by bitter-tasting compounds, focusing on i-type sensilla, which have only two neurons, facilitating experimental interpretation of spike patterns. Previous investigations indicated that i-type sensilla contain an S cell and an L2 cell, responding to sugars and high salt concentrations, respectively. The L2 cell in these sensilla is also activated by very low concentrations of various bitter compounds, including strychnine, berberine, quinine, and caffeine. In conclusion, the L2 cells of many, but not all, taste bristles located on the labial palps and the legs appear to be broadly tuned and respond to various repulsive stimuli including chemically diverse bitter-tasting compounds such as alkaloids as well as high concentrations of salts.

Since the genome of *Drosophila* spp. is completely known, taste in this insect has also been investigated by using flies with some genes – corresponding to the neurons

Gr5a and Gr66a – ablated by genetic manipulation. Taste responsiveness was assessed by performing proboscis extension reflex assays in control and DTI expressing animals. The proboscis extension reflex is one of the best-studied taste behaviors: when the leg encounters sugar, the proboscis extends. The probability of extension increases as a function of sugar concentration and decreases as increasing concentrations of a bitter substance are added to a fixed concentration of sugar. *Gr5a-Gal4, UAS-DTI* flies show a severe decrease in proboscis extension to trehalose, sucrose, glucose, and low salt, all substances that a fly finds palatable. At high concentrations of sucrose, *Gr5a-Gal4, UAS-DTI* flies extend their proboscis with high probability. Extension probability is reduced when bitter substances were added, with dose sensitivity comparable to wild-type flies. Conversely, *Gr66a-Gal4, UAS-DTI* flies show normal proboscis extension to sugars but show a 10-fold decrease in sensitivity to the noxious substance denatonium, and to the alkaloids berberine, caffeine, and quinine. Both *Gr66a-Gal4, UAS-DTI* and *Gr5a-Gal4, UAS-DTI* flies show normal responses to high salt that are indistinguishable from wild-type. These results seem to suggest that Gr5a cells mediate sugar detection while Gr66a cells participate in the recognition of bitter compounds.

A certain number of alkaloids have been tested toward herbivorous insects and it is observed that many alkaloids act as feeding deterrents at concentrations $>1\%$ w/w. Given the choice, insects tend to select a diet with no or only a small dose of alkaloids. These data indicate that under natural conditions, plants with a high content of alkaloid should be safe from most herbivorous insects. This shows the importance of alkaloids for the well-being of the plant, including the presence of alkaloids at the right concentration at the right time at the right place. Only if the insects are particularly hungry or have no choice do they lower the deterrence threshold and feed on a diet rich in alkaloids they would normally avoid, and in this case toxic effects may appear. Even if most alkaloids are famous poisons, when compared to the deterrence thresholds the lethal concentrations are much higher. This indicates that the function of alkaloids in plants is not primarily to bring about the death of the host herbivore but to avoid a potential danger.

As an example, *Manduca sexta* caterpillars exhibit an aversive behavioral response to many plant-derived compounds that taste bitter to humans, including caffeine and aristolochic acid [70]. This aversive behavioral response is mediated by three pairs of bitter-sensitive taste cells: one responds vigorously to aristolochic acid alone, and the other two respond vigorously to both caffeine and aristolochic acid. There is a peripheral mechanism for behavioral adaptation to specific "bitter" taste stimuli, in fact 24 h of exposure to a caffeinated diet desensitized all of the caffeine-responsive taste cells to caffeine but not to aristolochic acid. In addition, it was found that dietary exposure to caffeine adapted the aversive behavioral response of the caterpillar to caffeine, but not to aristolochic acid. Glendinning et al. [70] propose that the adapted aversive response to caffeine was mediated directly by the desensitized taste cells and that the adapted aversive response did not generalize to aristolochic acid because the signaling pathway for this compound was isolated from that for caffeine. Some compounds that are bitter-tasting to humans, both alkaloid (quinine, quinidine, atropine, and caffeine) and nonalkaloid (denatonium benzoate, sucrose octaacetate, and naringin), deterred feeding and oviposition by *Heliothis virescens* [71]. Preliminary electrophysiology studies of gustatory sensilla on the ovipositor of *H. virescens* provided evidence of three neurons, one of which is responsive to sucrose. Responses of this neuron may be inhibited by quinine and denatonium benzoate.

Some quinolizidine alkaloids such as sparteine (**14**), lupinine (**15**), and their analogs **16–19** are abundant in some varieties of lupin and other Leguminosae and have been recently studied both for their taste and for their possible use as natural pesticides. These compounds are in fact produced by the plant as a defense from predator attacks [72]. While the so called "bitter" lupin species are rich in quinolizidine alkaloids, there are "sweet" species in which the total alkaloids are very low and these are more susceptible to herbivores. As far as we know, the sweet varieties only differ from the bitter forms in their degree of alkaloid concentration. Leaf consumption in sweet (Rumbo) and bitter (El Harrach) *Lupinus albus* varieties by *Anticarsia gemmatalis* increased total alkaloid concentration. Inductive responses in bitter *Lupinus albus* accounted for a 70 % increase in alkaloid level after damage, while concentration in sweet *Lupinus albus* was 130 % greater in treated plants than in controls. The higher induction found in the sweet variety Rumbo was mainly because of an increase in lupanine, sparteine, and tigloiloxilupanine in consumed leaves. Inductive responses found in El Harrach, were also related to an increase in the major alkaloid lupanine and to a lesser extent to increases in tigloiloxilupanine and angustifoline. Lupanine accounted for 73 % and 74 % of the total alkaloid concentration in El Harrach and for 60 % and 68 % in Rumbo, in control and treated plants respectively. Unlike Rumbo, El Harrach plants had albine and angustifoline as part of their alkaloids, although they were present in very low concentrations in both control and treated plants.

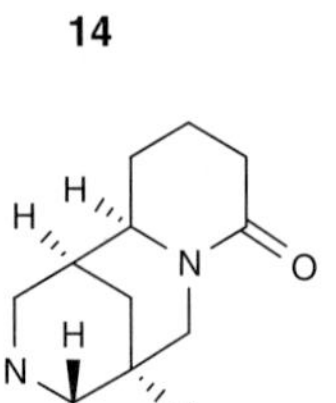

14 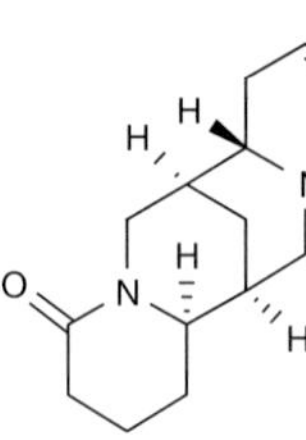 **15** 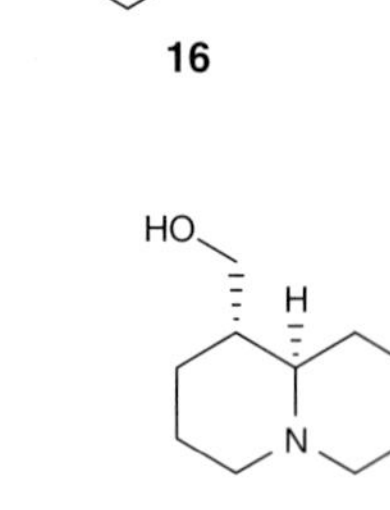 **16**

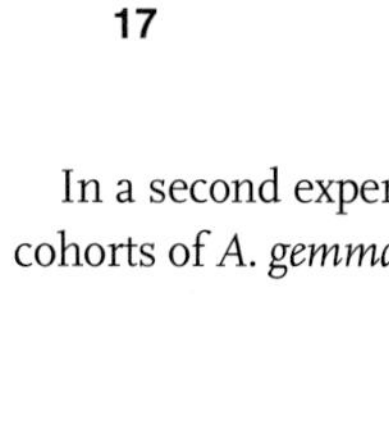

17 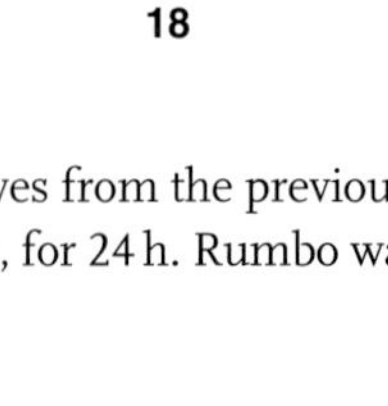 **18** 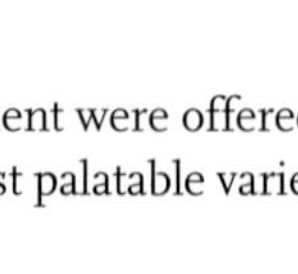 **19**

In a second experiment, leaves from the previous experiment were offered to new cohorts of *A. gemmatalis* larvae, for 24 h. Rumbo was the most palatable variety when leaves were from control plants; however, when Rumbo leaves had experienced previous herbivore attack, subsequent consumption by other anticarsia caterpillars strongly diminished, while preference for El Harrach remained unchanged. *A. gemmatalis* caterpillars consumed an average of 4.0 cm^2 of Rumbo leaves collected from control plants and 1.28 cm^2 (68 % decrease), when leaves were from plants that had experienced previous herbivore attack, while the mean consumption of El Harrach control and previously damaged leaves did not change. Therefore a palatability threshold, that is the level of alkaloid needed for any insect to detect the bitterness of the tissue, might have been overcome after induction in the sweet variety. A similar situation was recorded from insects such as aphids, beetles, and flies: the sweet forms were attacked while the alkaloid-rich ones were largely protected.

The same conclusions were also recorded for vertebrate herbivores. For example rabbits (*Cuniculus europaeus*) and hares (*Lepus europaeus*) clearly prefer the sweet plants and leave the bitter plants almost untouched, at least as long as there is an alternative food source. In conclusion, although taste perception in mammals and insects differs in many aspects, there also some similarities both in anatomy and in the function of the bitter taste perception. A comparison of the effects of alkaloids, as well as of other bitter compounds, will be assisted by further advances in the knowledge of the structure of taste genes and receptors.

3.6
The Bitter Taste of Alkaloids: Should We Avoid, Mask, or Understand?

Understanding of the bitter (and other) taste mechanism at the molecular level is very recent and raises many questions in this aspect of biology. In the case of alkaloids, there are several links between taste and biological function. Indeed, these compounds have a "two-faced" behavior: defensive–offensive compounds that should be avoided, and compounds with important biological activities, features that have always interested humans. This is only an apparent contradiction, since Nature is very skilful in giving us signals that we should heed, if we are to optimize the interactions between our organism and the surrounding world.

There is an interesting debate about this point. Some extreme positions in vegetarianism state that every food with a bitter taste should be avoided since it is toxic. This seems to be quite absurd, since many natural compounds, whose benefits on health have been demonstrated, possess a bitter taste. On the other hand, there is a growing tendency in the scientific community to argue that the bitter taste of many phyto-nutrients should be artificially removed or masked, in order to improve the consumption of these active principles by those consumers who usually avoid them just because of their taste. However, the idea of altering in any way the natural taste perception of individuals should be carefully considered, and history has something to teach us on this point. It is now recognized that the relatively recent practice of elevating sugar, salt, and fat concentrations in foods to improve their palatability has the disadvantage of inducing a change in the taste preferences of consumers which is now recognized as unhealthy, especially for young consumers. In terms of biological activity, this practice can also have a secondary effect. Bitter alkaloids such as caffeine, theobromine or quinine also have well-known pharmacological properties that are the basis of the increase in their consumption during the centuries; coffee, tea, and infusions made from coca leaves have always been used by people to improve endurance during hard work or prolonged walking, or to diminish the sensation of hunger, so they have been selected for their stimulating properties. In a certain sense, the bitter taste of these biologically active substances acts as a "natural deterrent" to their excessive ingestion; we would not ingest large amounts of pure caffeine extract because of its taste. But when the bitterness is masked by the addition of sugar or intensive sweeteners, as in caffeine-based soft drinks, for example, this natural limit is easily overcome. Since caffeine, like many other alkaloids, produces psychotropic effects and dependence, the demand for further ingestion follows: this could be one of the explanations for the great commercial success of this kind of beverage. The most-recent generation of caffeine-based beverages, so-called "energy drinks", have in fact a remarkably elevated caffeine content, in order to satisfy this "caffeine craving" in consumers. In fact there is evidence that the consumption of caffeine is mainly related to caffeine's pharmacological properties, although the influence of flavor has not been eliminated [73]. Today the food industry routinely employs strategies to reduce bitterness in food, such as selective breeding of plants with lower bitter principles content and/or commercial de-bittering processes [74]. Recent advances in the understanding of the receptor mechanisms of bitterness perception will also probably lead to the design of specific compounds able to mask or block the bitter taste of some useful nutrients; but the same compounds could also be employed to induce unnatural consumption of some compounds such as alkaloids, altering the natural regulatory mechanisms that taste operates in nature. Therefore, it is to be hoped that progress in clarifying the mechanisms of the bitter taste of alkaloids will help us to understand, rather than simply avoid or mask, this complex phenomenon and its importance in biology and life.

3.7 Acknowledgments

We thank Maria Teresa Cascella for suggesting the citations on bitter taste taken from Plato and Agatha Christie and for finding the original texts.

References

1 Steiner, J.E. (1994) *Olfaction and Taste XI*, Springer, Tokyo, pp. 284–287.
2 Nolte, D.L., Mason, J.R., Lewis, S.L. (1994) *Journal of Chemical Ecology*, **20**, 303–308.
3 Lindemann, B. (1996) *Physiological Reviews*, **76**, 718–766.
4 Matsunami, H. and Amrein, H. (2003) *Genome Biology*, **4**, 220.
5 Hilliard, M.A., Bergamasco, C., Arbucci, S., Plasterk, R.H.,

Bazzicalupo, P. (2004) *The EMBO Journal*, **23**, 1101–1111.

6 Sipiora, M.L., Murtaugh, M.A., Gregoire, M.B., Duffy, V.B. (2000) *Physiology and Behavior*, **69**, 259–67.

7 Northcutt RG, R.G. (2004) *Brain, Behavior and Evolution*, **64**, 198–206.

8 Clapp, T.R., Yang, R., Stoick, C.L., Kinnamon, S.C., Kinnamon, J.C. (2004) *The Journal of Comparative Neurology*, **468**, 311–321.

9 Yang, H., Wanner, I.B., Roper, S.D., Chaudhari, N. (1999) *The Journal of Histochemistry and Cytochemistry*, **47**, 431–446.

10 Clapp, T.R., Yang, R., Stoick, C.L., Kinnamon, S.C., Kinnamon, J.C. (2004) *The Journal of Comparative Neurology*, **468**, 311–321.

11 Sbarbati, A., Crescimanno, C., Bernardi, P., Osculati, F. (1999) *Chemical Senses*, **24**, 469–472.

12 Sbarbati, A., Merigo, F., Benati, D., Tizzano, M., Bernardi, P., Osculati, F. (2004) *Chemical Senses*, **29**, 683–692.

13 Sbarbati, A. and Osculati, F. (2003) *Cells, Tissues, Organs*, **175**, 51–55.

14 Martin, J.H. (1989) *Neuroanatomy*, Prentice Hall International, London.

15 Smith, D.V. and Frank, M.E. (1993) *Mechanisms of Taste Transduction*, CRC Press, Boca Raton, FL, pp. 295–338.

16 Miller, I.J., Jr. (1995) *Handbook of Olfaction and Gustation*, Marcel Dekker, New York, pp. 521–547.

17 Whitehead, M.C., Ganchrow, J.R., Ganchrow, D., Yao, B. (1999) *Neuroscience*, **93**, 931–941.

18 Liu, L. and Simon SA, S.A. (1998) *Chemical Senses*, **23**, 125–130.

19 Dahl, M., Erickson, R.P., Simon, S.A. (1997) *Brain Research*, **756**, 22–34.

20 Adler, E., Hoon, M., Mark, A., Mueller, K.L., Chandrashekar, J., Ryba, N.J.P., Zuker, C.S. (2000) *Cell*, **100**, 693–702.

21 Chandrashekar, J., Mueller, K.L., Hoon, M.A., Adler, E., Feng, L., Guo, W., Zuker, C.S., Ryba, N.J.P. (2000) *Cell*, **100**, 703–711.

22 Matsunami, J.P., Montmayeur, P., Buck, L.B. (2000) *Nature*, **404**, 601–604.

23 Conte, C., Ebeling, M., Marcuz, A., Nef, P., Andres-Barquin, P.J. (2003) *Physiological Genomics*, **14**, 73–82.

24 Andres-Barquin, P.J. and Conte, C. (2004) *Cell Biochemistry and Biophys*, **41**, 99–112.

25 Meyerhof, W., Kuhn, C., Brockhoff, A., Winnig, M., Bufe, B., Schöley-Pohl, E., Behrens, M. (2005) *Proceedings of the 8th Wartburg Symposium. State-of-the-art in Flavour Chemistry and Biology*, Deutsche Forschungsanstalt für Lebensmittelchemie, Garching.

26 Meyerhof, W. (2005) *Reviews in Physiology, Biochemistry and Pharmacology*, **154**, 37–72.

27 Rosenzweig, S., Yan, W., Dasso, M., Spielman, A. (1999) *Journal of Neurophysiology*, **81**, 1661–1665.

28 Belitz, H.D. and Wieser, H. (1985) *Food Reviews International*, **1**, 271–354.

29 Keast, R.S. and Breslin, P.A. (2002) *Chemical Senses*, **27**, 123–131.

30 Bufe, B., Hofmann, T., Krautwurst, D., Raguse, J.D., Meyerhof, W. (2002) *Nature Genetics*, **32**, 397–401.

31 Belitz, H.D. and Grosh, W. (1999) *Food Chemistry*, Springer, Berlin.

32 Calvino, A.M., Tamasi, O.P., Ciappini, M.C. (2005) *Food Science and Technology International*, **11**, 401–407.

33 Scharbert, S. and Hofmann, T. (2005) *Journal of Agricultural and Food Chemistry*, **53**, 5377–5384.

34 Masuho, I., Tateyama, M., Saitoh, O. (2005) *Chemical Senses*, **30**, 281–290.

35 Zhao, F., Lu, S., Herness, S. (2002) *American Journal of Physiology*, **283**, R115–R129.

36 Hrynakowski, K., Chwojka, S., Zochowski, A. (1938) *Kron Farm*, **36**, 317–21.

37 Frank, M.E., Bouverat, B.P., MacKinnon, B.I., Hettinger, T.P. (2004) *Physiology and Behavior*, **80**, 421–43.

38 Aspen, J., Gatch, M.B., Woods, J.H. (1999) *Psychopharmacology*, **141**, 251–7.

39 Nakamura, T., Tanigake, A., Miyanaga, Y., Ogawa, T., Akiyoshi, T., Matsuyama, K., Uchida, T. (2002) *Chemical and Pharmaceutical Bulletin*, **50**, 1589–1593.

40 Ogawa, T., Hoshina, K., Haginaka, J., Honda, C., Tanimoto, T., Uchida, T. (2005) *Journal of Pharmaceutical Sciences*, **94**, 353–362.

41 Nelson, T.M., Munger, S.D., Boughter, J.D., Jr. (2005) *BMC Genetics.*

42 Zitnak, A. and Filadelfi, M.A. (1985) *Canadian Institute of Food Science and Technology Journal*, **18**, 337–9.

43 Minasyan, S.M. (1947) *Biokhimiya (Moscow)*, **12**, 298–302.

44 Green, B.G. and Hayes, J. (2003) *Physiology and Behavior*, **79**, 811–821.

45 Green, B.G. and Schullery, M.T. (2003) *Chemical Senses*, **28**, 45–55.

46 Frank, O., Ottinger, H., Hofmann, T. (2001) *Journal of Agricultural and Food Chemistry*, **49**, 231–238.

47 Christie, A. *The Mysterious Affair at Styles: a detective story*, Bodley Head, London, 1920.

48 Ward, J.C. and Munch, J.C. (1930) *Journal of the American Pharmaceutical Association*, 1912–1977.

49 Bufe, B, Scholey-Pohl, E., Kratwurst, D., Hofmann, T., Meyerhof, W. (2004) *Challenges in Taste Chemistry and Biology*, ACS Symposium Series 867, USA, p. 45–59.

50 Grunberg, N.E., Bowen, D.J., Maycock, V.A., Nespor, S.M. (1985) *Psychopharmacology*, **87** (2), 198–203.

51 Etscor, F., Moore, G.A., Hagen, L.S., Caton, T.M., Sanders, D.L. (1986) *Pharmacology, Biochemistry, and Behavior*, **24** (3), 567–570.

52 Villanueva, H.F., Arezo, S., James, J.R., Rosecrans, J.A. (1990) *Pharmacology, Biochemistry, and Behavior*, **37** (1), 59–61.

53 Iiyama, S., Toko, K., Yamafuji, K. (1986) *Agricultural and Biological Chemistry*, **50** (11), 2709–2714.

54 Hopkins, J.W. (1946) *Canadian Journal Research*, **24F**, 203–214.

55 Tilgner, D.J. and Barylko-Pikielna, N. (1959) *Acta Physiologica Polonica*, **10**, 741–754.

56 Aubek, J.P. (1959) *Medical Service Journal Canada*, **15**, 731–733.

57 Cooper, R.M., Bilash, I., Zubek, J.P. (1959) *Journal of Gerontology*, **14**, 6–58.

58 Bronte-Steward, B. (1956) *British Medical Journal*, **4968**, 659.

59 Krut, L.H., Perrin, M.J., Bronte-Steward, B. (1961) *British Medical Journal*, **5223**, 384–387.

60 Beser, E., Baytan, S.H., Akkoyunlu, D., Gul, M. (1995) *Ethiopian Medical Journal*, **33** (3), 155–162.

61 Jessen, A., Buemann, B., Toubro, S., Skovgaard, I.M., Astrup A, A. (2005) *Diabetes, Obesity and Metabolism*, **7** (4), 327–333.

62 Iwamoto, E.T. and Williamson, E.C. (1984) *Pharmacology, Biochemistry, and Behavior*, **21**, 527–32.

63 Barratt-Fornell, A. and Drewsnowski, A. (2002) *Nutrition Today*, **37** (4), 144–150.

64 Arnold, G.W. and Hill, J.L. (1972) *Phytochemical Ecology*, New York Academic Press, London.

65 Wink, M. (1993) *Proceedings of the Phytochemical Society of Europe*, 34, 171–213.

66 Cordell, G.A., Quinn-Beattie, M.L., Farnsworth, N.R. (2001) *Phytotherapy Research*, **15**, 183–205.

67 Wink, M. (1998) *Alkaloids: Biochemistry, Ecology and Medicinal Applications*, Plenum, New York.

68 Amrein, H. and Thorne, N. (2005) *Current Biology*, **15**, R673–R684.

69 Hiroi, M., Meunier, N., Marion-Poll, F., Tanimura, T. (2004) *Journal of Neurobiology*, **61**, 333–342.

70 Glendinning, J.I., Brown, H., Capoor, M., Davis, A., Gbedemah, A., Long, E. (2001) *The Journal of Neuroscience*, **21**, 3688–3696.

71 Ramaswamy, S.B., Cohen, N.E., Hanson, F.E. (1992) *Entomologia Experimentalis et Applicata*, **65**, 81–93.

72 del Pilar Vilarino, M., Mareggiani, G., Grass, M.Y., Leicach, S.R., Ravetta, D.A. (2005) *Journal of Applied Entomology*, **129** (5), 233–238.

73 Newland, M.C. and Brown, K. (1992) *Pharmacology, Biochemistry and Behavior*, **42**, 651–9.

74 Drewnowski, A. and Gomez-Carneros, C. (2000) *The American Journal of Clinical Nutrition*, **72**, 1424–1435.

4
Capsaicin and Capsaicinoids

Giovanni Appendino

4.1
Introduction

Capsaicin (CPS, **1a**), the pungent principle of hot pepper, is one of the best-known natural products, boasting well over 16 000 entries in the SciFinder database [1], and even deserving inclusion in the Webster Dictionary of English [2]. Capsaicin is an extraordinarily versatile agent, as testified by almost 1000 patents covering the use of the natural product, its synthetic analogs, and capsicum oleoresin in fields ranging from pharmacology and nutrition to chemical weapons and shark repellence [1]. Reviews covering various aspects of the biomedical relevance of capsaicin appear regularly in the scientific literature [3]. On the other hand, the seminal review by Suzuki and Iwai on the chemistry, distribution, biochemistry, and ecological aspects of capsaicin has remained un-updated for over two decades [4]. This contribution tries to fill this gap in the light of recent spectacular advances in the molecular characterization of the biological targets, biosynthesis, and physiological role of capsaicin. On account of the huge market of capsaicin-flavored food (over 500 million $/year in the US alone) [5] and the observation that 25 % of the human population enjoys chili or chili-flavored food daily [6], an attempt will also be made to review the molecular gastronomy [7] of chili pepper in the light of these recent advances.

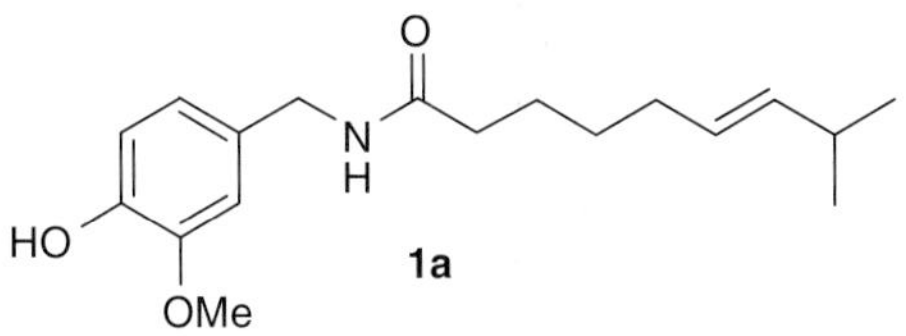

4.2
What Is an Alkaloid? Is Capsaicin an Alkaloid?

Like many other idioms in the language of organic chemistry, the term alkaloid is poorly defined. As an anonymous chemist once remarked "an alkaloid is like my wife.

Modern Alkaloids: Structure, Isolation, Synthesis and Biology. Edited by E. Fattorusso and O. Taglialatela-Scafati

ISBN: 978-3-527-31521-5

I can recognize her when I see her, but I can't define her" [8]. The name "alkaloid" was coined by the German pharmacist Carl Friedrich Wilhelm Meissner in 1819 to refer to plant natural products (the only organic compounds known at that time) showing basic properties similar to those of the inorganic alkalis [9]. The ending "-oid" (from the Greek ειδω, appear) is still used today to suggest similarity of structure or activity, as is evident in names of more modern vintage such as terpenoid, peptoid, or vanilloid. Given the limited structural information on organic compounds available in the early nineteenth century, the definition by Meissner was necessarily vague. As sometimes happens, vague terms have a bright future ahead of them, and even in the absence of a clear definition, the name "alkaloid" became firmly entrenched in the chemical and pharmacological literature of the nineteenth century. "Alkaloid" meant different things to different people. While some chemists such as Königs had a rather narrow view, reserving the name alkaloid for plant bases related to pyridine, others such as Guareschi considered it synonymous with an organic base, applicable to all organic compounds, including synthetic ones, showing basic properties [10]. An important attempt to define alkaloids in structural terms was made by Winterstein and Trier in 1910, almost a century after the name was first coined. In their seminal treatise [11], these authors considered as alkaloids *sensu lato* all nitrogen-containing basic compounds of plant and animal origin, making, however, a clear distinction between *true alkaloids* and *alkaloid-related* bases. To be classified as a true alkaloid, a natural product must meet, besides basicity, four additional requirements:

(1) The nitrogen atom must be included in a heterocyclic system.
(2) The structure must be complex.
(3) The pharmacological activity must be potent.
(4) The distribution must be restricted to the plant kingdom.

Several problems were soon evident with this definition of alkaloids. Many compounds commonly perceived as alkaloids do not show basic properties, bearing amide nitrogen(s) (e.g. colchicine, **2**) or being ammonium salts (e.g. sanguinarine, **3**) or amine *N*-oxides (indicine *N*-oxide, **4**). Furthermore, factors such as structural complexity and bioactivity are relative, while origin as a criterion can sometimes be ambiguous, since certain natural products have a broad distribution in Nature. For instance, the indole derivative bufotenin (**5**) was originally isolated from a toad (*Bufo bufo bufo*), but was later obtained from plants from the genus *Piptadenia*, from the mushroom *Amanita mappa*, and has even been detected in human urine [8].

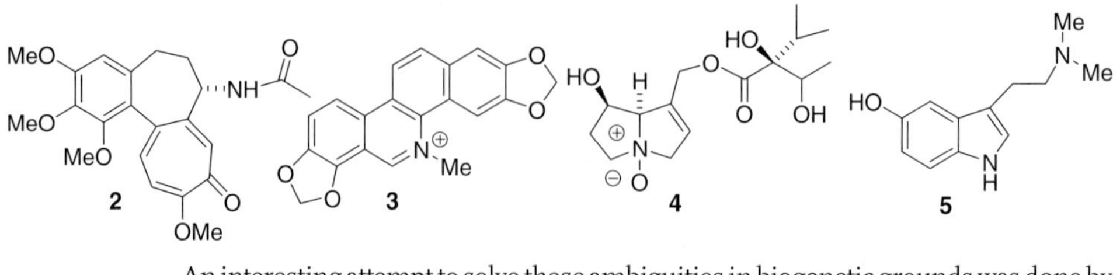

An interesting attempt to solve these ambiguities in biogenetic grounds was done by Hegnauer in the 1960s [12]. While retaining an emphasis on the basic properties, Hegnauer divided alkaloids into three classes: true alkaloids, pseudoalkaloids, and protoalkaloids. True alkaloids are compounds derived from the condensation, generally in a Mannich fashion, between a decarboxylate amino acid and a non-nitrogenous complementary partner. Benzylisoquinoline alkaloids such as papaverine (**6**) and indole alkaloids such as reserpine (**7**) exemplify this biogenetic pattern of formation. Pseudoalkaloids are compounds unrelated biogenetically to amino acids, and whose cyclic nitrogen derives from the formal incorporation of ammonia into a carbon skeleton, generally of terpenoid or polyketide origin. Aconitine (**8**) and solanidine (**9**) exemplify the incorporation of ammonia into a nor-diterpenoid or a steroid skeleton, while coniine (**10**) is the archetypal pseudoalkaloid whose nitrogen is incorporated into a polyketide framework. Finally, protoalkaloids are amino acid-related compounds whose nitrogen atom is not part of a heterocyclic system, but has remained biogenetically "inert." Alkaloid building blocks such as biogenic amines, phenylalkylamines such as ephedrine (**11**) and mescaline (**12**), and esters of unusual amino acids such as taxol (**13**) and meliamine A (**14**) are examples of this type of compound. It is curious to note that, according to the biogenetic classification, neither the steroid alkaloid veratridine (**15**), the very first compound to which the term alkaloid was applied [13], nor the aminated polyketide coniine (**10**), the first alkaloid to be synthesized [14], would actually be alkaloids, instead being pseudoalkaloid.

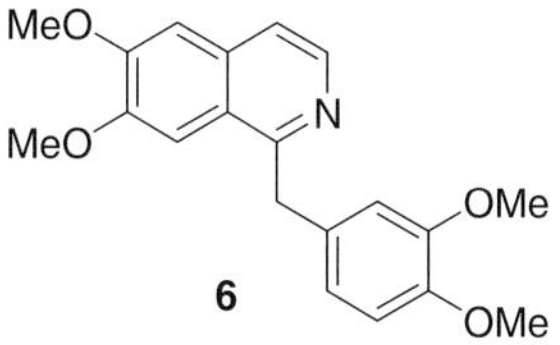

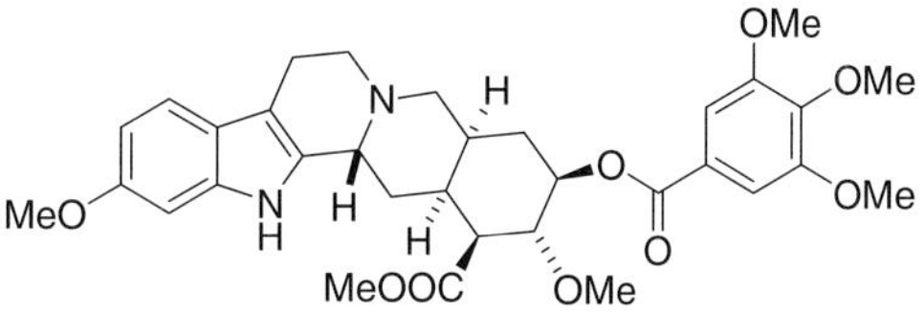

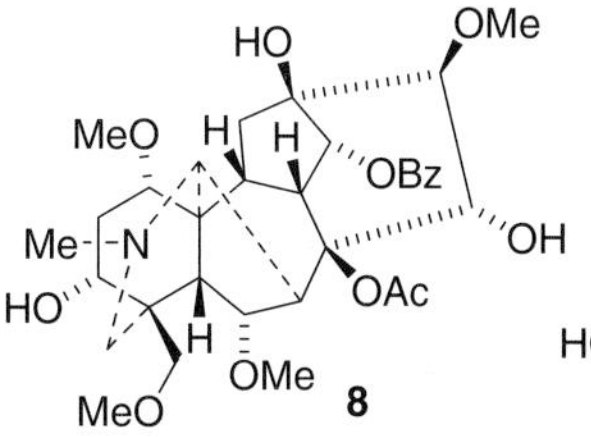

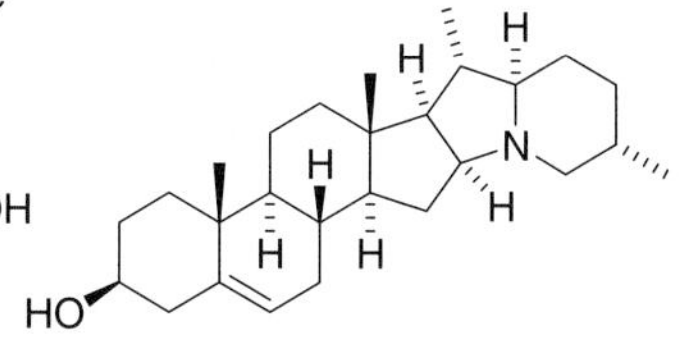

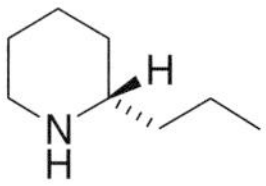

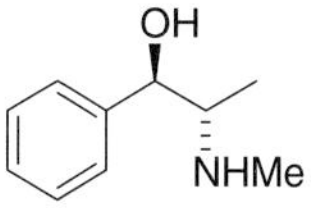

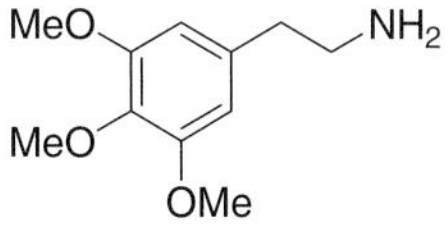

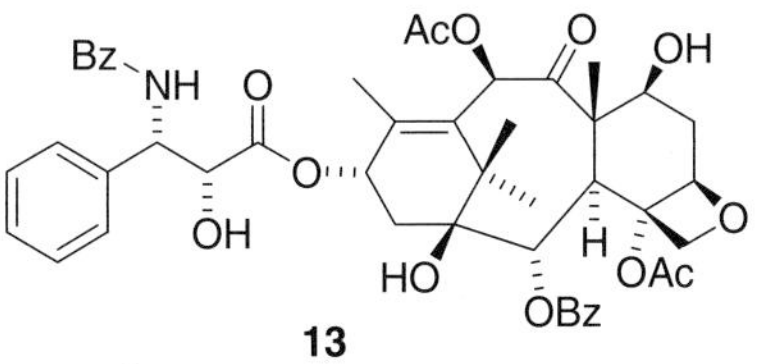

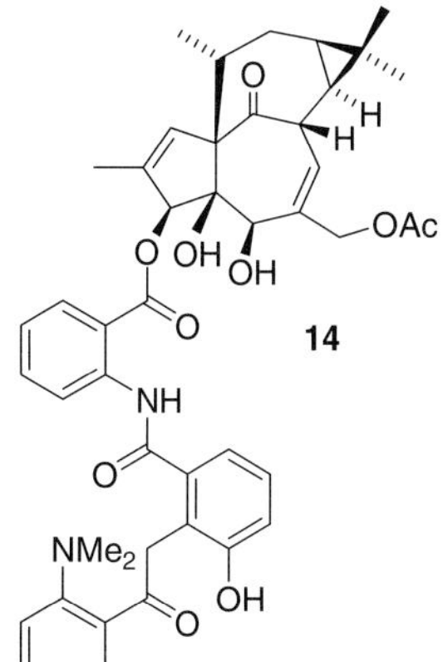

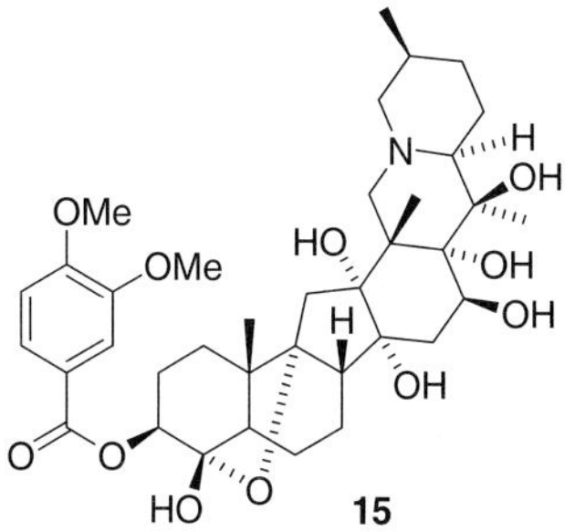

The distinction between true alkaloids, pseudoalkaloids, and protoalkaloids is often difficult to apply. For instance, how should purines be classified? In these compounds, the nitrogen atoms are not implanted in a carbocycle, but rather make up the skeletal framework typical of these compounds. It would be clumsy to consider the purine skeleton as a hydrindane isosterically modified by the inclusion of four nitrogen atoms, and similar difficulties exist for porfirins. In practice, while the biogenetic classification by Hegnauer has merits, especially in the field of chemotaxonomy, its application assumes biogenetic knowledge that, especially for new compounds, can be missing or difficult to guess [15]. For instance, the secondary amide nitrogen of colchicine (**2**) and taxol (**13**) is not included in a cyclic structure, and, without previous biogenetic knowledge, both compounds would be classified as protoalkaloids. However, in biogenetic terms, the two compounds are rather different. The amide nitrogen of colchicine is extruded from an isoquinoline framework [16] while the one of taxol derives from the diotopic enzymatic isomerization of phenylalanine to β-phenylalanine [17]. Thus, colchicine should be an alkaloid and taxol a protoalkaloid.

While the name alkaloid was originally defined in terms of biological and/or biogenetic origin and chemical and pharmacological behavior, its current usage is largely dissociated from these properties, and defined essentially on structural bases. A similar semantic transition was undergone by the adjective "aromatic," exemplifying the tendency of modern organic chemistry to transcend the sensory and chemical properties of compounds. An appropriate modern definition of alkaloid was proposed in 1984 by S.W. Pelletier: an alkaloid is a compound containing nitrogen at a negative oxidation level (−3 for amines, amides, and ammonium salts, −1 for amine oxides) characterized by a limited distribution in Nature [8]. Compounds from various sources can be included, and one speaks therefore of fungal-, anuran-, arthropod-, and mammalian alkaloids when referring to compounds obtained from sources different from plants.

Even with these punctilios, gray areas remains, and the line between alkaloids and other naturally occurring nitrogen compounds is often thin. For instance, in the realm of natural products from sea organisms, where the actual producer is often unknown, the name "marine alkaloids" has become popular, referring to the ecological rather than the biological source of the product. Furthermore, within microbial compounds, antibiotics such as β-lactams are not considered alkaloids, nor are aminoglycosides, while compounds devoid of antibiotic activity such as lysergic acid derivatives are. Since the term antibiotic is no less ambiguous than alkaloid, the matter is still unsettled, as is the inclusion of peptides within alkaloids. Thus, while styrylamine-based cyclopeptides such as lotusine A (**16**) are regarded as alkaloids, cyclic fungal peptides such as amatoxins and phallotoxins are not. Finally, assessing how limited the distribution of a compound should be to deserve inclusion in the alkaloid family is somewhat arbitrary. Few would object to the exclusion from alkaloids of widespread and structurally simple "biogenic amines" such as ethanolamine and dopamine, and of ubiquitous polyamines such as spermidine and putrescine, but these moieties can nevertheless be incorporated into more complex structures of limited distribution in Nature, and originate compounds referred to as alkaloids. An example is homaline (**17**), the bis-cinnamoyl adduct of spermine and cinnamic acid, a compound isolated from the plant *Homalium pronyense* [18]. Both spermine and cinnamic acid are ubiquitous, but the way they are combined in homaline is unique, and it seems therefore logical to consider this compound as an alkaloid.

Returning to the question in the title of this section, capsaicin does not fall into any of the three classic types of nitrogen-bearing plant natural products, being neither a true alkaloid, a protoalkaloid, or a pseudoalkaloid. Capsaicin is of limited distribution in Nature and shows pharmacological activity, but is non-basic, structurally unsophisticated, and not directly derived from an amino acidic precursor. On the other hand, the lack of attributes such as basicity, complexity, and an "amino acidic pedigree" can also be found in compounds commonly perceived as alkaloids. Thus, colchicine is neutral, ephedrine is structurally unsophisticated, and the nitrogen atom of the potato alkaloid solanine is not derived from an amino acid, but rather incorporated into as non-amino acidic framework by a transamination reaction. For the sake of clarity and consistency, it seems therefore convenient to adopt the modern definition of alkaloids, and consider capsaicin, as well as alkylamides such as piperine (**18**) and pellitorine (**19**), as such.

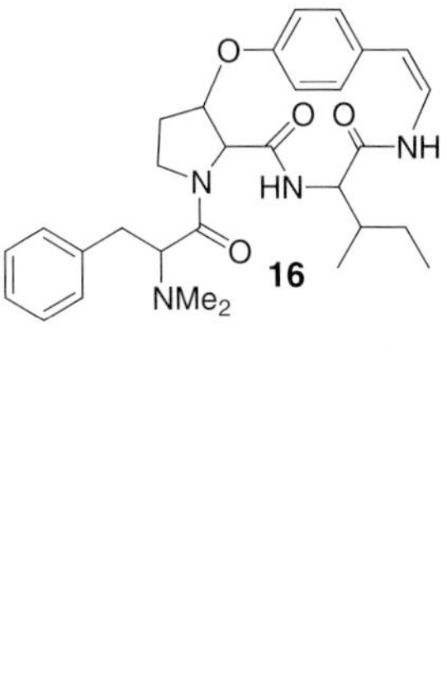

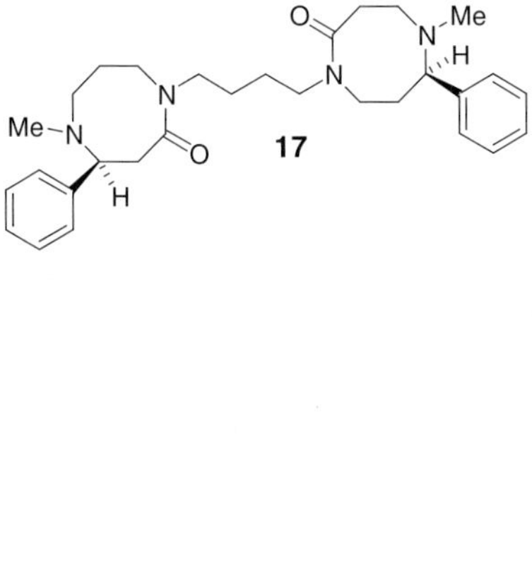

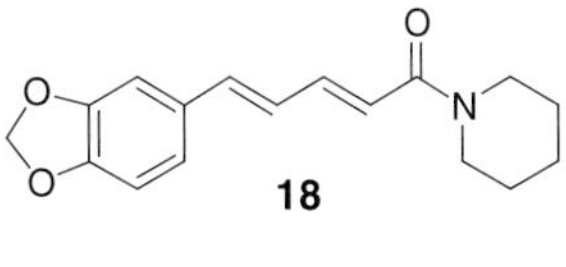

4.3
Diversity, Biosynthesis, and Metabolism of Capsaicinoids

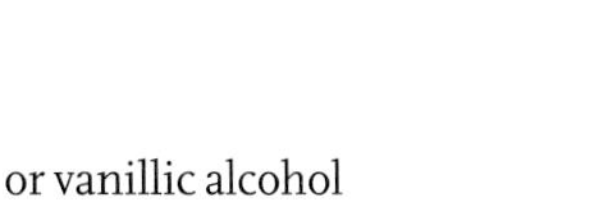

The formation of acyl conjugates of vanillamine (capsaicinoids, **20**) or vanillic alcohol (capsinoids, **21**) with various C_8/C_{13} alkenoic and alkanoic acids is a unique chemical trait of plants from the genus *Capsicum*. Hot peppers are characterized by the presence of vanillyl conjugates of the amide type, absent or replaced by their nonpungent ester isosters (capsinoids) in bell (sweet) peppers [19]. Indeed, the difference between the sensory properties of capsaicin (**1a**) and its naturally occurring ester analog capsiate (**22**) is a remarkable example of the biological relevance of isosterism.

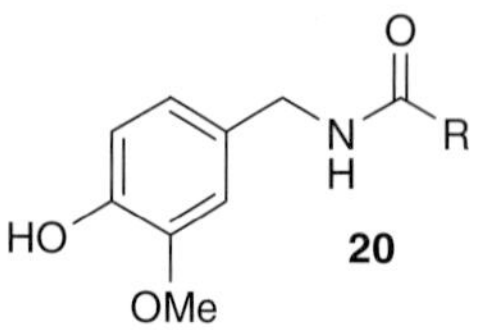

The taxonomy of the genus *Capsicum* is controversial. Twenty-two wild species and five domesticated species are recognized, along with over 2000 cultivars derived from them [20]. The wild species have been relatively poorly investigated, and most data on the distribution of vanillyl conjugates in peppers refer to cultivars. The genus *Capsicum* is apparently endemic to the highlands of Bolivia and Peru [21], but peppers are nowadays cultivated all over the temperate and hot regions of the world. Indeed, it has been estimated that the agricultural land devoted to the cultivation of peppers corresponds to the size of a country such as Switzerland [21].

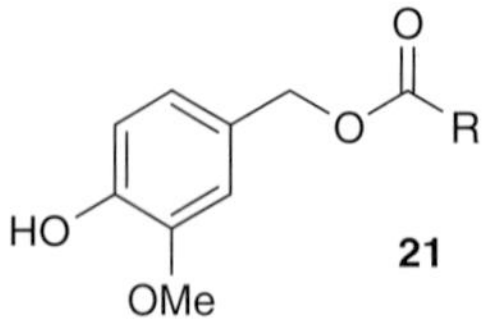

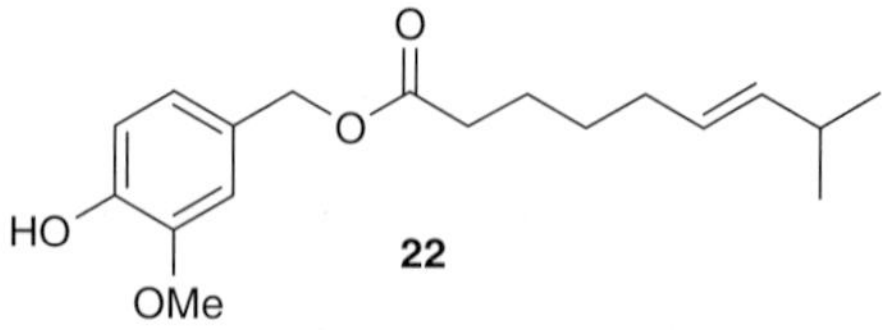

Capsaicinoids occur as complex mixtures of analogs, whose profile is under epigenetic as well as genetic control. Within a single species, it also changes with the organ under investigation, since transport from the site of synthesis is apparently more efficient for some structural types of capsaicinoids than for others. The taxonomic value of the capsaicinoids is, in any case, difficult to assess, since the pattern of distribution of these compounds is inconsistent within a single species, while it is possible to identify clusters characterized by a common capsaicinoid signature in cultivars of different species [22].

Over a dozen capsaicinoids have been detected in capsicum oleoresin [4], but, since most natural species of hot peppers have never been investigated chemically, the number of naturally occurring capsaicinoids might well exceed the current number. Furthermore, some natural capsaicinoids have only been tentatively identified without being actually isolated (Table 4.1). Thus, conclusive evidence based on isolation and structure elucidation is missing for many "natural" capsaicinoids, and therefore confusion exists in the literature. For instance, three homocapsaicins have been reported, but the one elongated on the carbonyl side of the double bond (**1q**) is not a constituent of capsicum oleoresin, and its occurrence in this source is due to confusion with the naturally occurring analogs homologated from the other side of the double bond, namely homocapsaicin I and II [23].

Differences within capsaicinoids depend mainly on their acyl moieties, and three structural elements are involved, namely (a) the length of the acyl chain (C_8–C_{13}), (b) the way it terminates (linear-, iso-, or anteiso-series), and (c) the presence or absence of unsaturation at the ω-3 (capsaicin type) or ω-4 carbon (homocapsaicin I and II type) position. Furthermore, oxidation of the terminal methyls can also occur, as well as phenolic coupling between two vanillyl moieties [24] and glycosidation [25]. The constitutional descriptors *nor* and *homo* are used to describe analogs of capsaicin having extra or fewer carbon compared to the parent compound. Owing to the existence of branching, the use of these descriptors can create confusion. Indeed, *nor*

Tab. 4.1 Naturally occurring capsaicinoids.

Compound	n_1	Δ	n_2	R	Trivial name
1a	1	+	0	CH_3	Capsaicin
1b	1	–	0	CH_3	Dihydrocapsaicin
1c	3	+	0	CH_3	Bis-homocapsaicin[a]
1d	4	+	0	CH_3	Tris-homocapsaicin[a]
1e	0	+	0	CH_3	Norcapsaicin
1f	0	–	0	CH_3	Nordihydrocapsaicin
1g	1	+	1	CH_3	Homocapsaicin I
1h	1	–	1	CH_3	Homodihydrocapsaicin I[a]
1i	1	+	0	CH_3CH_2	Homocapsaicin II
1j	1	–	0	CH_3CH_2	Homodihydrocapsaicin II[a]
1k	0	–	0	CH_3CH_2	Homonordihydrocapsaicin II[a]
1l	1	–	0	H	Nonivamide
1m	2	–	0	H	Decivamide[a]
1n	3	–	0	H	Undecivamide[a]
1o	4	–	0	H	Dodecivamide[a]
1p	1	+	0	CH_2OH	ω-Hydroxycapsaicin
1q	2	+	0	Me	Homocapsaicin[b]

[a]Detected in *Capsicum* oleoresin but not yet isolated.
[b]Wrongly reported as a *Capsicum* constituent owing to confusion with homocapsaicins I and II.

is generally employed to indicate the removal of an alkyl branching, and therefore an uninitiated reader would assume that norcapsaicin (**1e**) is actually nonivamide (**1l**).

The combination of these elements generates the diversity of capsaicinoids reported to date. Generally, the major constituent of the "capsaicinoid soup" are capsaicin (**1a**) and its dihydroderivative (**1b**). Commercial capsaicin powder is an approximately 5 : 1 mixture of capsaicin and dihydrocapsaicin, while analytical (>95 %) capsaicin contains mainly nonivamide as impurity. Despite its trivial name of "synthetic capsaicin," nonivamide is a natural trace constituent of capsicum oleoresin, and concentration >3 % are indicative of adulteration [26]. The addition of nonivamide to capsicum oleoresin has been detected in products from both the food and the pharmaceutical markets. Some of them have been found to contain exclusively nonivamide, even though capsaicin is the only individual constituent of capsicum oleoresin to be approved by the FDA for human use [26].

From a biogenetic standpoint, capsaicin is an acylated degraded phenylpropanoid. Both its aromatic and its acyl moiety are the result of unique metabolic processes that, though simple in principle, are still poorly characterized in terms of enzymology and regulation. Since vanillamine (**23**) is abundant in placental tissues of *Capsicum*, the only site of biosynthesis of capsaicinoids, the limiting factor for the synthesis of capsaicin is the availability of (*E*)-8-methyl-6-nonenoic acid (**24**, MNA) [27]. It is assumed that the availability of the corresponding acids (**25**, **26**) is the limiting factor for the biosynthesis of homocapsaicins. The acylase responsible for the amidation of vanillamine with MNA has not yet been characterized enzymatically, but a gene (Pun 1) encoding for it has been identified [28]. Its product, a putative capsaicin synthase, shows great similarity with acyltransferase of the BAHD super-family [28]. Lack of pungency is associated to a specific deletion, which spans the promoter and the first exon of the predicted coding region, and the resulting allele (pun1) is recessive. This is accordance with the early observations that the production of capsaicin is controlled by a single dominant factor [29].

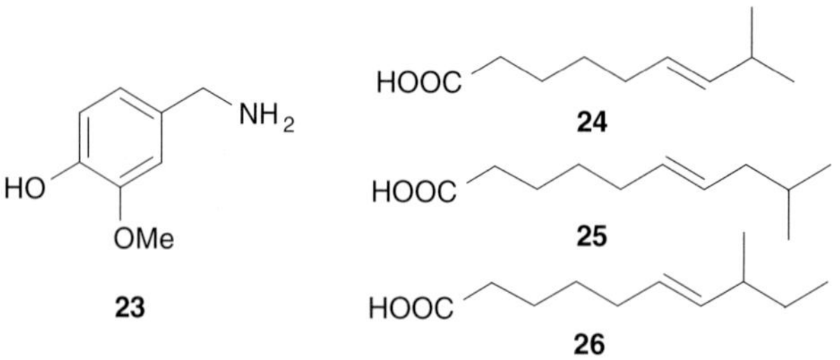

HO OMe NH_2 **23**

HOOC **24**

HOOC **25**

HOOC **26**

Linear capsiacinoids with a C_9/C_{12} chain are only trace constituents of capsicum oleoresin, which mainly contains branched capsaicinoids. The acyl moiety of these compounds is produced by the branched chain fatty acids pathway (Scheme 4.1) [30]. Depending on the nature of the amino acid that acts as the acyl starter precursor, different capsaicinoids are formed. Thus, capsaicinoids of the *iso* series such as CPS and homocapsaicin I are derived from valine and leucine via isobutyrylCoA and isovalerylCoA, respectively, while those from the *anteiso* series such as homocapsaicin II originate from isoleucine via 2-methylbutyrylCoA (Scheme 4.1) [31]. The polymethylene moiety of norcapsaicin has one less carbon than capsaicin. The

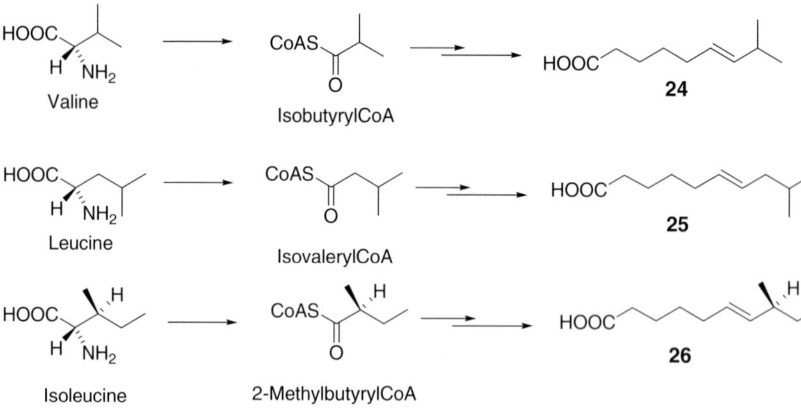

HOOC H NH_2 Valine → CoAS O IsobutyrylCoA → → HOOC **24**

HOOC H NH_2 Leucine → CoAS O IsovalerylCoA → → HOOC **25**

HOOC H H NH_2 Isoleucine → CoAS H O 2-MethylbutyrylCoA → → HOOC H **26**

Scheme 4.1 Biogenetic derivation of capsaicinoid acids **24–26**.

biogenetic origin of this acyl group has not been investigated, nor has that of capsaicinoids with odd linear chains, which are presumably derived from a propionyl starter. A keto acyl synthase (KAS) expressed during the ripening of the fruits is apparently responsible for the formation of the acyl chain of capsaicinoids from the pool of their corresponding starting AcylCoAs [27]. After various cycles of elongation with malonylCoA (two for bisnorcapsaicin, three for CPS, and four for bishomo-capsaicin), capsaicinoid acids are eventually obtained. In accordance with this biogenetic derivation, treatment of placental tissues of *Capsicum* with cerulenine, a known inhibitor of fatty acids synthase, shuts down the production of CPS and leads to the accumulation of vanillamine [27]. Variation in capsaicinoid acids also derives from the way the carbonyl group of the acyl starter is processed after the first step of elongation. Following ketone-to-alcohol reduction and dehydration, the double bond can either be retained (compounds of the unsaturated series) or reduced (compounds of the saturated series).

Vanillamine (**23**), a compound unique to *Capsicum*, is produced from vanillin (**30**) by a pyrophosphate-dependent aminotransferase reaction, sensitive to inhibition by amino oxyacetate [27]. In accordance with the hypothesis that MNA and not vanillamine is the limiting factor for capsaicin synthase, the inhibition degree of the aminotransferase reaction is poorly correlated with the inhibition of capsaicin production. For instance, an 80 % inhibition of the aminotransferase reaction caused only a 20 % reduction of the production of capsaicin, further confirming that MNA and not vanillamine is the limiting factor for the final acylation [27]. The biosynthesis of vanillin in *Capsicum* is presumably similar to that occurring in vanilla pods, starting from phenylalanine (**27**) and proceeding, through the intermediary of cinnamic acid (**28**), coumaric acid (**29**), and caffeic acid (**30**), to ferulic acid (**31**), next oxidized to vanillin (**32**, Scheme 4.2). Regulation seems to exist also at the level of vanillamine, and precisely at the level of caffeate O-methylation, since the concentration of cinnamate and coumarate is independent of that of capsaicin, while a clear correlation was found between pungency and the accumulation of ferulic acid [32]. An *ortho*-diphenol-*O*-methyltransferase responsible for the transformation of coumarate to ferulate has been cloned from pepper, and found to bear similarity with a similar enzyme from tobacco [32].

The amide-to-ester isosteric shift observed in some cultivars of sweet pepper producing capsinoids has not yet been characterized biochemically, but at least two

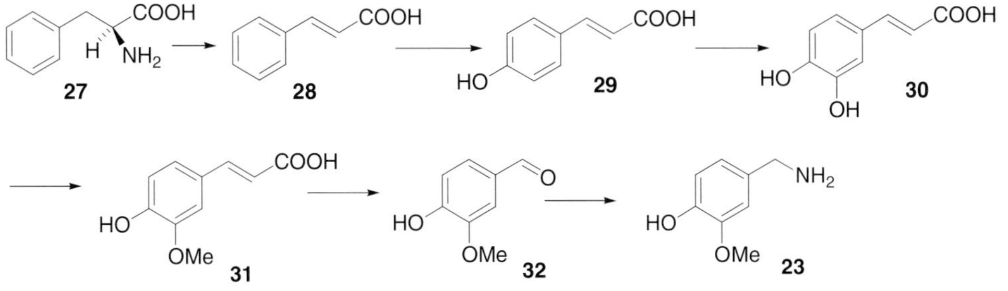

Scheme 4.2 Biogenetic derivation of vanillamine (**23**).

modifications seem necessary, namely the inactivation of the putative vanillin aminotransferase, and a modification of capsaicin-synthase to accept an alcohol rather than an amine substrate.

Capsaicinoids are metabolized in plant tissues mainly via oxidative phenol coupling. In hot pepper fruits, capsaicin biosynthesis competes with the accumulation of lignin-like polymeric structures, and pepper placenta cells contain an exclusive peroxidase, which colocalizes with capsaicin in vacuoles [33]. The lignin-like compounds represent the major metabolites of capsaicin in peppers [34]. Dimeric compounds resulting from 5,5′-phenolic coupling have also been isolated, both from peppers and from the reaction of dihydrocapsaicin with pepper peroxidase. Thus, the homodimer **33**, resulting from 5,5′coupling between capsaicin and its dihydroderivative has been obtained from the Banshou variety of *C. annuum* [24], while treatment of dihydrocapsaicin (**1b**) with crude pepper peroxidase and H_2O_2 afforded, along with polymeric material, 5,5′-bis dihydrocapsaicin (**34a**) and 4-*O*,5′-bis-dihydrocapsaicin (**35**), both in unreported yield [34]. Biomimetic oxidative dimerization of nonivamide with $K_3[Fe(CN)_6]$ afforded in good yield 5,5′-dinonivamide (**34b**) [35a], while the reaction of dihydrocapsaicin with 2,2-diphenyl-1-picrylhydrazyl (DPPH) afforded mainly degraded products, some of them of the 4-*O*,5′-dimeric type (**36**, **37**) [35b].

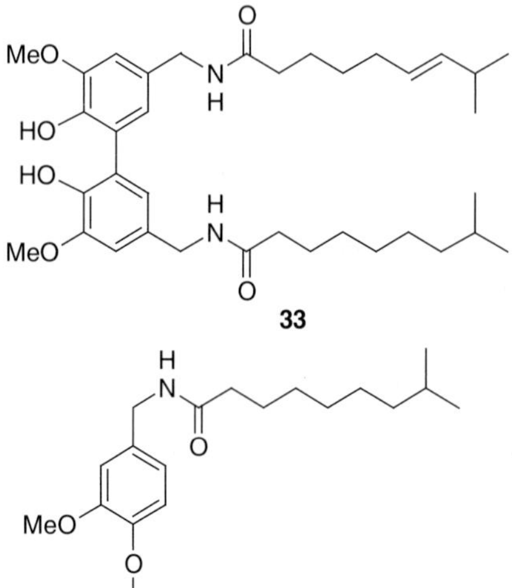

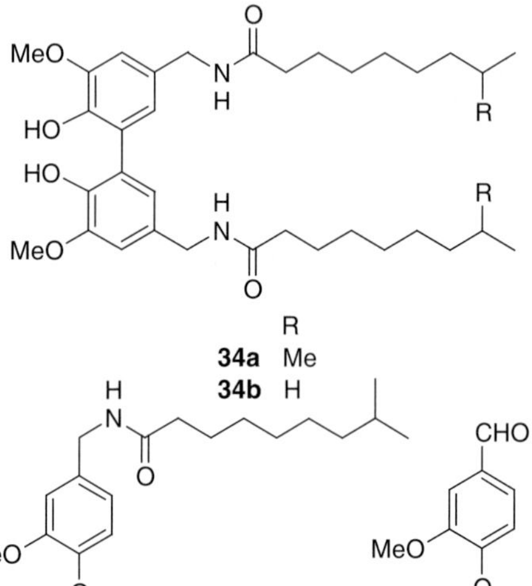

	R
34a	Me
34b	H

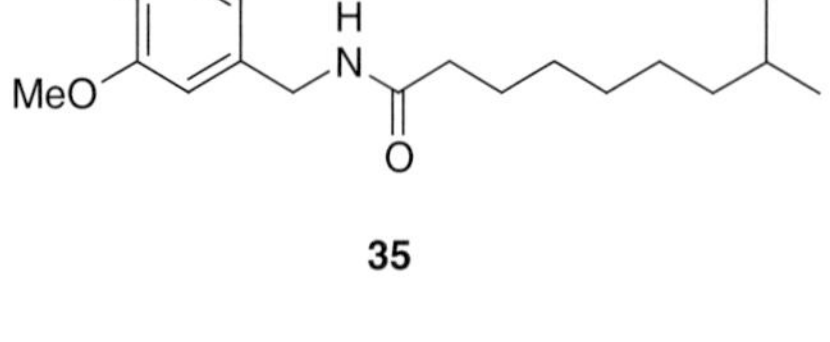

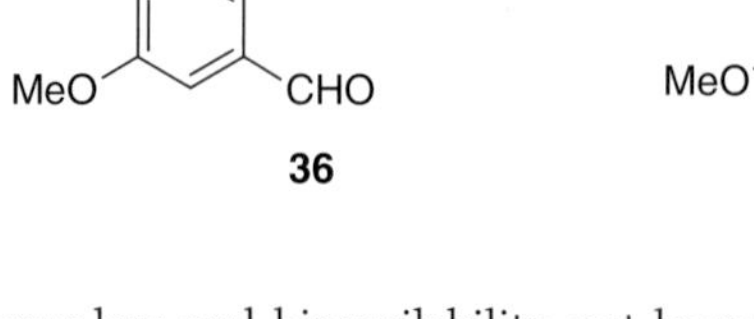

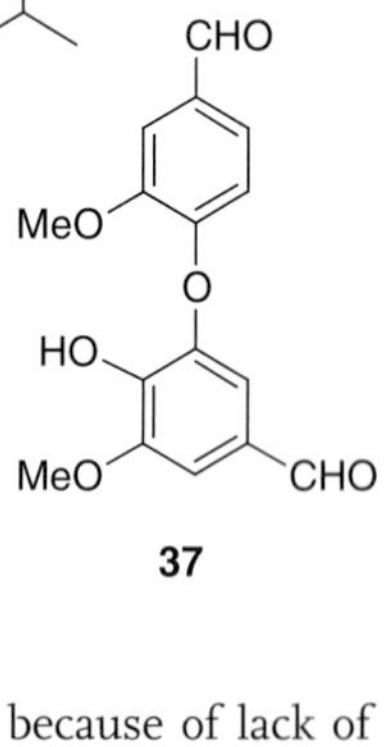

Capsaicin in humans has a very low oral bioavailability, not because of lack of absorption, but because it is almost completely metabolized in the liver before reaching the general circulation, where it exists almost exclusively as metabolites. The very poor oral bioactivity is also responsible for the large difference in LD_{50} between oral and dermal administration of capsaicin (LD_{50} about 190 and >510 mg/kg in mice) compared to intravenous or intraperitoneal administration (LD_{50} about 0.56 and 7.7 mg/kg, respectively, in mice) [36,37].

The metabolism of capsaicin in humans has remained unknown until recently, but is now well characterized, thanks to experiments in cell cultures [38]. Capsaicinoids are metabolized by P450 enzymes, found throughout the body, but especially in liver, lungs, and kidney. The rate of metabolization is significantly lower in lung microsomes than in liver microsomes, suggesting that respiratory tissues are ill-equipped to metabolize capsaicin, and are therefore especially sensitive to its toxic effects. The metabolism of capsaicin is, as expected, essentially oxidative, given its lipophilic character and the presence of the metabolically vulnerable vanillyl moiety. Three major sites of oxidation were recognized, namely the phenolic ring, subjected to O-demethylation and aromatic hydroxylation ortho to the phenolic hydroxyl, the benzylic position, and the terminal isopropyl group. In addition to the compounds of ω-hydroxylation (**1p**) and ω-1 hydroxylation (**42**), dehydrogenation to 8,9-dehydrocapsaicin (**43**) and oxidative *N*, ω-1 macrocyclic cyclization (**44**) were observed. Scheme 4.3 summarizes the results of the human metabolism investigations. 9-Hydroxycapsaicin (**1p**) was previously shown to be a metabolite of CPS in rabbit and in microorganisms [24], but has more recently also been isolated from the Banshou variety of *C. annuum* [24], suggesting that the metabolism of capsaicin shows a certain similarity between different living organisms. 9-Hydroxycapsaicin, and the products of dimeric oxidative coupling of dihydrocapsaicin are nonpungent [24], while the product of phenolic coupling of nonivamide showed only marginal vanilloid activity [40].

4.4 Quantization of Capsaicinoids and Their Distribution in Chili Pepper

Since hot pepper is important for the food and the pharmaceutical industries, a range of different methods have been developed for the analysis of capsaicinoids in plant material and finished products. The separation of CPS (**1a**) and nonivamide (**1l**) is especially challenging, since these compounds have similar behavior in many chromatographic conditions. Since synthetic nonivamide is the most common adulterant of capsicum oleoresin, various strategies have been suggested to overcome this problem. Capillary GC does not require previous derivatization of capsaicinoids, but its separatory power seems lower than that of HPLC, currently the most popular technique for the quantization of capsaicinoids. GC is, however, the method of choice for the analysis of the acyl moieties of capsaicinoids as methyl esters. These can be directly produced from capsaicinoids by oxidative N-dealkylation with DDQ (2,3-dichloro-5,6-dicyanobenzoquinone), followed by alcoholysis of the resulting amides with methanol in the presence of an acidic resin (Scheme 4.4) [41].

The separation of capsaicinoid mixtures is important to assess the Scoville value of food spiced with chili pepper [42]. This "Richter scale" of pungency was devised by the Arizona pharmacist Scoville in 1902 to measure the pungency of peppers (see also Section 4.8). In its original version, the Scoville scale was sensory, being

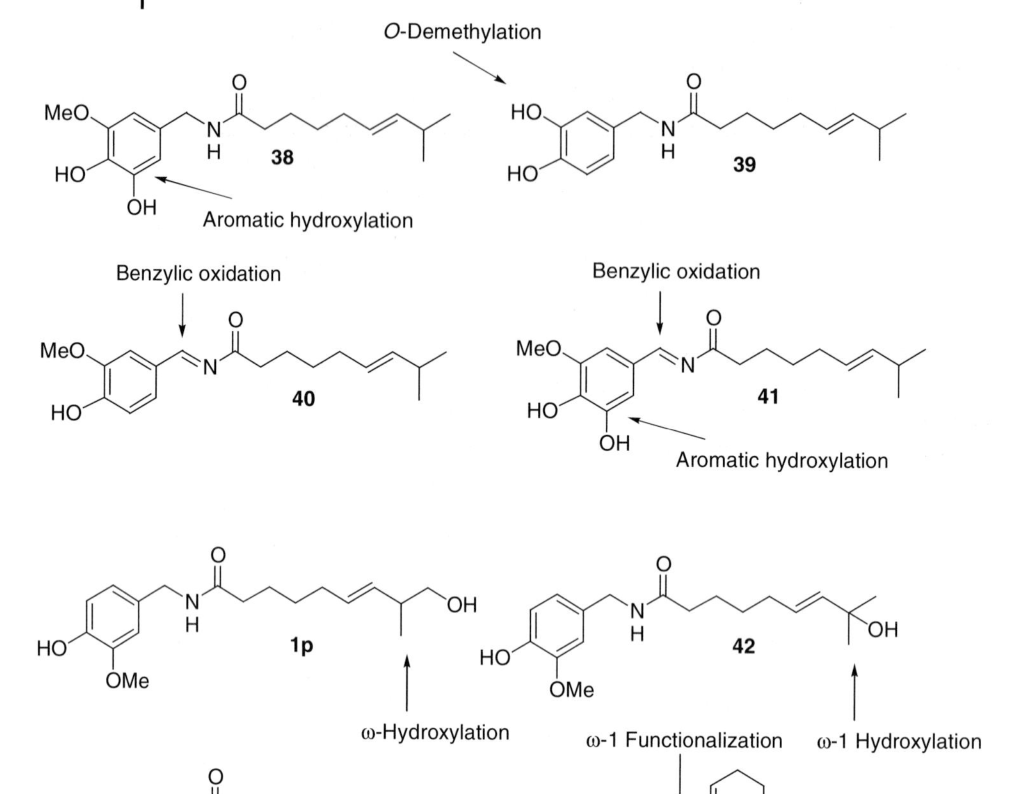

Scheme 4.3 Microsomal metabolites of capsaicin.

based on the stepwise dilution of an extract to the point at which pungency was no longer detectable. Nowadays, the Scoville value is assigned on the basis of the HPLC capsaicinoid profile of a pepper or product containing it, assigning a fixed pungency value to each capsaicinoid, and then averaging these values to their concentration [43]. While HPLC under standard conditions can separate mixtures of up to 10 capsaicinoids, the characterization of the minor constituents generally requires

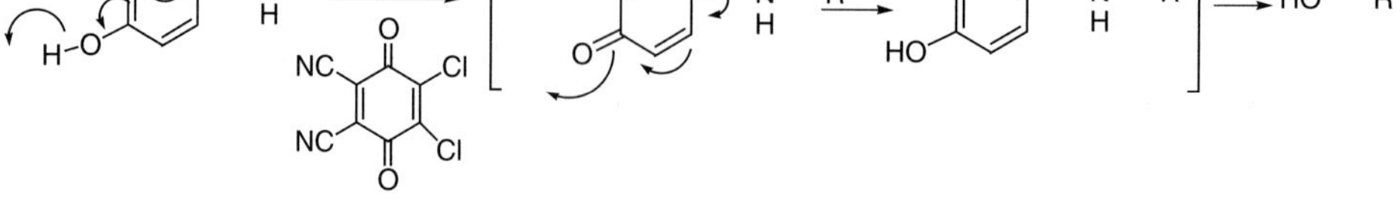

Scheme 4.4 Oxidative *N*-dealkylation of capsaicinoids.

more than one technique. Capillary electrophoreses was found to nicely complement the separation achieved by HPLC [44]. The US Pharmacopoeia reports an official HPLC method for the separation of capsaicinoids in capsicum oleoresin. This uses a C18 RP silica gel column and isocratic conditions (elution with methanol and 2 % aqueous acetic acid in a 56 : 44 ratio) [45]. A systematic investigation on the HPLC separation of capsaicinoids concluded that a mixed phenyl-cation-exchange stationary phase charged with silver ions was the best method to resolve mixtures of minor capsaicinoids [46], but the field of capsaicinoid analysis continues to be vigorously explored, since the separation of these compounds is challenging. Owing to the great variation in the capsaicinoids profile of peppers, our incomplete knowledge of its minor constituents, and the differences in potency between natural capsaicinoids, the pharmacological properties of capsicum oleoresin show a high degree of variability.

The availability of sophisticated analytical methods to separate and quantify capsaicinoids, has made it possible to study their distribution and kinetics of formation. Capsaicinoids are unevenly distributed in pepper plants and fruits. The placenta is their only site of biosynthesis, and they are then translocated to other plant parts (pericarp, seeds, stem, leaves) [47]. This is in striking contrast with what is observed in other solanaceous plants, where alkaloids are produced in the roots and then translocated to the aerial parts. Simple leakage might be responsible for the detection of capsaicin in seeds, owing to the strict anatomical connection with the placenta. Indeed, capsaicinoids are synthesized in the epidermal cells of the placenta, and are then accumulated, just under the cuticle of the placenta surface, in droplets that can be easily broken by pressure, releasing their contents onto the seeds and the inner part of the fruits. Typically, over 85 % of the capsaicinoids content of pungent peppers is located in the placenta, while about 6 % is found in the fleshy pericarp, and about 8 % in the seeds [47]. Several studies have shown a well-defined gradient of concentration in the fruits, with the highest concentration occurring in their basal and apical parts [48]. Differences have also been noticed within the composition of the capsaicinoid mixture from various plant parts, suggesting a subtle, but obscure, physiological meaning in the capsaicinoid signature of an organ. Thus, it was observed that, while capsaicin was the major capsaicinoid in the fruits of some varieties of *C. annuum*, dihydrocapsain was more abundant in their vegetative organs such as leaves and stem [49]. Within a single fruit, the contents and profile of capsaicinoids depends mainly on the maturation stage. Capsaicinoids can already be detected one week after flowering, but differences in the kinetics of production exist, with the highest concentration being reached after two to four weeks, depending on the cultivar. Fifty days after flowering, capsaicinoid concentrations start to decrease through oxidative metabolization (see Section 4.3) [50]. In practice, the maximum concentration of capsaicinoids is attained when the green fruits just begin to change color. Interestingly, within a single plant, fruits collected at the same time from flowering show the same capsaicinoid pattern, but different contents of capsaicinoids, whose concentration follows a spatial gradient along the stem, with the highest concentration of capsaicinoids in apical fruits compared to middle and basal fruits [47,48].

The accumulation of capsaicinoids is also subjected to environmental variables, such as temperature, light, soil moisture, and fertilization level [51]. In general, drought and high night temperatures promote the synthesis of capsaicinoids, explaining the major hotness of peppers grown in dry tropical areas compared to those cultivated in more temperate or humid climates.

4.5
Isolation and Synthesis of Capsaicin

Several methods for the isolation of capsaicin from capsicum oleoresin have been described in the literature or covered by patents. Capsicum oleoresin can be prepared from hot peppers using a variety of organic solvents, but ethanol is the only one suitable for obtaining pharmaceutical-grade material [45]. The purification of CPS from the oleoresin is difficult because of the presence of analogs with similar polarity, chromatographic behavior, and solubility, and because of the unpleasant properties of capsaicinoids. A further disadvantage is the high affinity of capsaicinoids for charcoal, which prohibits the use of this material to remove the abundant pigments (carotenoids, chlorophylls) present in the oleoresin. The classic method to isolate capsaicin from capsicum oleoresin is the one developed by Nelson almost a century ago [52]. In this lengthy procedure, barium chloride is used to remove fatty acids from an acidified extract of the oleoresin, and silver nitrate to precipitate unsaturated capsaicinoids from their saturated analogs. Further purification is achieved by a series of partitions between alkaline water solutions and ether, followed by a final crystallization from ether and washing with boiling petroleum ether (60–110 °C). The final yield depends on the pungency of the starting pepper, and its capsaicinoid profile. Starting from lyophilized ripe Jalapeno, a mild variety of pepper, a yield of over 1 % was reported using a simplified version of the Nelson procedure [53].

Even in its simplified form, the Nelson protocol will deter the most enthusiast natural-product chemist. Unsurprisingly, alternative protocols have been described in the proprietary literature, and their sheer number testify to the ingenuity and commitment of chemists to solve what is still today a daunting task. Some modern modifications to the Nelson method from the proprietary literature include the use of supercritical carbon dioxide to reduce the extraction of pigments in the preparation of the oleoresin [54], and the recourse to macroporous adsorption resins [55] or to repeated extraction with aqueous silver nitrate to trap capsaicinoids and reduce the number of partition steps [56].

The synthesis of capsaicin shows nicely how progress in basic methodologies has simplified the production of natural products, to the point of making synthesis more convenient than isolation. The synthesis of capsaicinoids has practical relevance because the purification of minor capsaicinoids from *Capsicum* oleoresin is very difficult. The first syntheses by Späth [57] in 1930 and by Crombie [58] in 1955 have only historical relevance. Surprisingly, modern syntheses did not appear until the 1980s, probably because capsaicin was wrongly considered a trivial target. The major challenge in the synthesis of CPS is the generation of the (*E*)-double bond of the acyl

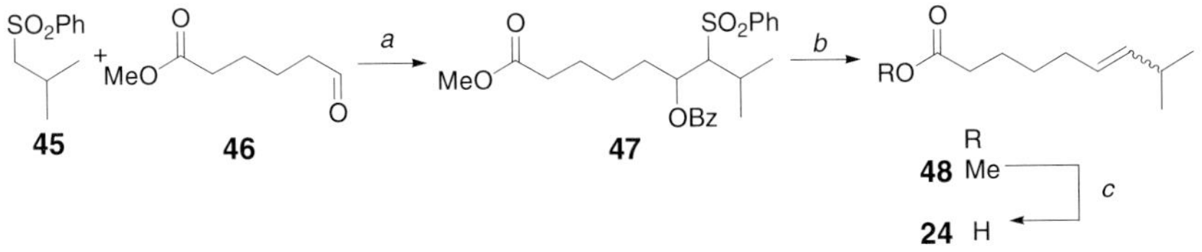

Scheme 4.5 Synthesis of (*E*)-MNA (**24**) by Gannet. a: (i): *n*-BuLi, THF, −78 °C; (ii): BzCl, −78 °C; b: Na(Hg), MeOH, −20 °C, 70–80 % from **45**; c: KOH, EtOH, Δ.

moiety, since Wittig reaction of unstabilized ylides, such as those requested for the synthesis of capsaicin, mainly generates (*Z*)-double bonds. (*Z*)-Capsaicin (zucapsaicin) has never been detected in capsicum oleoresin, and its detection is considered an indicator of sophistication [26]. Since the biological profile of zucapsaicin is poorly defined, the diastereomeric purity of capsaicin of synthetic origin is important for biological investigations. Various strategies have been developed to obtain pure (*E*)-capsaicin by synthesis. These synthetic efforts are of more than academic relevance, as they are important for obtaining the minor capsaicinoids, whose purification from capsicum oleoresin is exceedingly difficult. In 1988, Gannett developed a synthesis based on the Kocienski–Lythgo–Julia olefination (Scheme 4.5) [59]. Condensation of the lithium anion of isobutylphenylsulfone (**45**) with the methyl ester of ω-oxohexanoic acid (**46**) afforded, after trapping with benzoyl chloride, an α-benzoyloxysulfone (**47**), next subjected to reductive elimination with sodium amalgam to afford a 9 : 1 mixture of (*E*/*Z*) methyl esters of MNA. By variation in the nature of the ω-oxoester, various capsaicinoids of the iso-series were obtained. ω-Oxoesters are readily available from their corresponding lactones, and the preparation of isobutylphenylsulfone is straightforward.

In 1989, a different approach was published by Orito [60], in which elaidinization ((*Z*) → (*E*) double bond isomerization) is used to obtain (*E*)-MNA from a (*Z*,*E*)-mixture of diastereomers (Scheme 4.6). Gannet had observed that the iodine-induced photoisomerization of the methyl ester of MNA (**48**) gave only a 7 : 3 (*E*/*Z*) mixture [59], but Orito obtained a better diastereomeric ratio (8 : 1) using nitrous acid. Remarkably, no double-bond migration to form the more stable trisubstituted olefin was observed. This discovery paved the way to a very simple and general synthesis of the acidic component of capsaicinoids. Thus, a Wittig reaction of the phosphonium salt of a 6-bromohexanoic acid (**49**) with isobutyraldehyde (**50**) afforded a 1 : 11

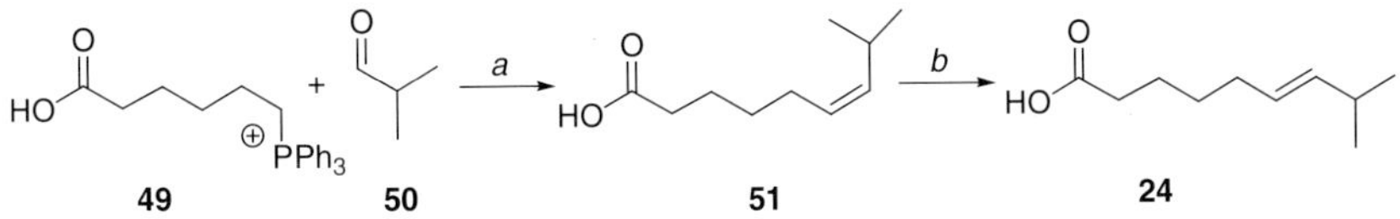

Scheme 4.6 First-generation synthesis of (*E*)-MNA (**24**) by Orito. a: KO^{t}Bu, DMF, 0 °C, 74 % (11 : 1 (*Z*/*E*)); b: $NaNO_2$, HNO_3, 70 °C, 77 %, (1 : 8 (*Z*/*E*)).

mixture of (*E,Z*) diastereomers of MNA, which was next elaidinized with HNO_2, converted to the corresponding chloride, and coupled with vanillamine to afford an 8 : 1 mixture of (*E*/*Z*) capsaicin. Further purification by fractional crystallization from hexane–ether afforded (*E*)-CPS in a rewarding 25 % yield from 6-bromohexanoic acid. This synthesis is easily amenable to the preparation of capsaicinoids of the iso-series by variation of the aldehyde component (isovaleraldehyde for compounds of the homocapsaicin I type) or the ω-bromoacid (compound of the norcapsaicin type), while the obtaining of capsaicinoids from the homocapsaicin II series would require the use of 2-methylbutyraldehyde.

Orito also developed a variation of this synthesis, based on the Wittig reaction of the phosphonium salt from isobutylbromide with a series of lactols. After alcohol-to-acid oxidation, a mixture of MNA isomers is obtained, then isomerized with nitrous acid [61,62].

The two syntheses developed in the late 1980s (Gannett and Orito) are based on a $C_4 + C_6$ assembly strategy. A conceptually different and more stereoselective synthesis was developed by Orito in 1996 [63], based on a $C_6 + C_2 + C_2$ strategy, as first explored by Vig (Scheme 4.7) [64]. The key step is the Claisen orthoester rearrangement of 4-methyl-1-penten-3-ol, in turn obtained by reaction of vinyl magnesium bromide with isobutyraldehyde. Heating this alcohol with triethyl orthoacetate in the presence of catalytic amounts of propionic acid afforded in 73 % yield the ethyl ester of 6-methyl-4-heptenoic acid (that is, bis-norMNA ethyl ester) exclusively as an (*E*)-diastereomer. This compound is a general precursor for all capsaicinoids of the ω-3 iso-type. Two-carbon homologation to MNA was achieved, after ester-to-alcohol reduction with $LiAlH_4$ and mesylation, with a malonic synthesis. Treatment of the

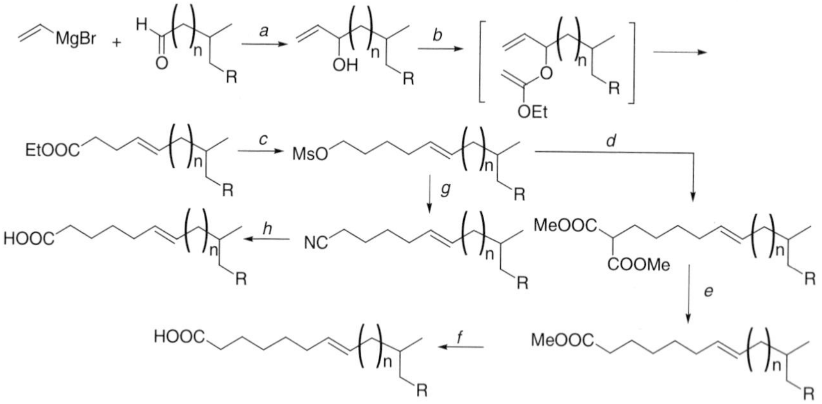

Scheme 4.7 Second-generation synthesis of *E*-MNA (**24**, $n = 0$; R = H) and its higher homologs ($n = 1$, R = H; $n = 0$, R = Me) by Orito (yield for the synthesis of MNA). a: THF, 0 °C, 73 %; b: $CH_3C(OEt)_3$, cat. propionic acid, 138 °C, 3 h, 73 %; c: (i): Li AlH_4, ether, 86 %; (ii): MsCl, TEA, DCM, quantitative; d: sodium diethylmalonate, THF, KI, 80 °C, 3.5 h, 83 %; e: NaCl, DMSO, water, 170 °C, 3 h, 91 %; f: NaOH, MeOH, Δ, 3 h, 89 %; g: NaCN, DMSO, 140 °C; 3 h, quantitative; h: NaOH, MeOH, Δ, 15 h, 91 %.

mesylate with the sodium salt of dimethylmalonate in the presence of potassium iodide, and demethoxycarbonylation according to Krapcho's method (heating in DMSO with NaCl), uneventfully afforded the methyl ester of MNA. Alternatively, one-carbon homologation to nor-MNA was achieved from the key mesylate intermediate by treatment with sodium cyanide in DMSO and hydrolysis (Scheme 4.7). These homologations steps could be carried out in an iterative way, giving access to homo-MNA from the one carbon homologation of the mesylated form of MNA, and to bishomo-MNA from its two-carbon elongation via malonic synthesis. This versatility, coupled to the possibility to carry out the starting sequence starting from the aldehydes corresponding to the terminal moieties of homocapsaicin I and homocapsaicin II, led to the synthesis of all branched natural capsaicinoids, whose physical, chromatographic, and spectroscopic data were reported, sometimes for the first time [63]. The synthesis of capsaicinoids from the dihydro series is much easier, and the key catalytic hydrogenation can be accomplished at any of the various post-olefination steps of the three modern syntheses.

A new synthesis of CPS appeared in the proprietary literature in 2004 (Scheme 4.8) [65]. An Algox Pharmaceutical group reported a preparation of MNA capitalizing on a $C_5 + C_5$ strategy, and based on the alkylation of bromovaleric acid (**52**) with the lithium anion of 3-methyl-1-butyne (**53**), followed by stereoselective reduction of 8-methyl-6-noninoic acid (**54**) to (*E*)-MNA by dissolving-metal reduction.

In all these protocols, chemoselective amidation of vanillamine with NMA was achieved in satisfactory yield (about 80 %) by activation of the acid to its corresponding chloride, followed by condensation with free vanillamine. Free vanillamine is much less stable than its corresponding chloride, and, owing to its polarity, it is difficult to extract quantitatively from water solution, with loss of about 20 % of the product routinely observed in the desalification step [59]. Furthermore, the preparation of chlorides from polyunsaturated fatty acids is not trivial, and therefore alternative coupling protocols were investigated in a series of systematic studies dedicated to the structure–activity relationships of capsaicinoids.

The amidation of phenolic amines with condensing agent does not requires *ex situ* activation of acids, but is troublesome, since carbodiimides are basic enough to deprotonate phenolic hydroxyls and make their acylation competitive with that of an amino group [66]. To solve this problem, protected (benzyl, ethoxyethyl) vanillamine was employed (Scheme 4.9) [67,68]. For amidation with cheap acids, a different protocol was used, directly condensing vanillamine with an excess of acid, and next chemoselectively O-deacylating the *N,O*-diacyl derivative [68]. A more recent protocol is based on the use of the condensing agent PPAA (propylphosphonic acid anhydride = T3P) to chemoselectively acylate a suspension of vanillamine hydrochloride in CH_2Cl_2. After evaporation and filtration over silica gel, pure vanillamides are

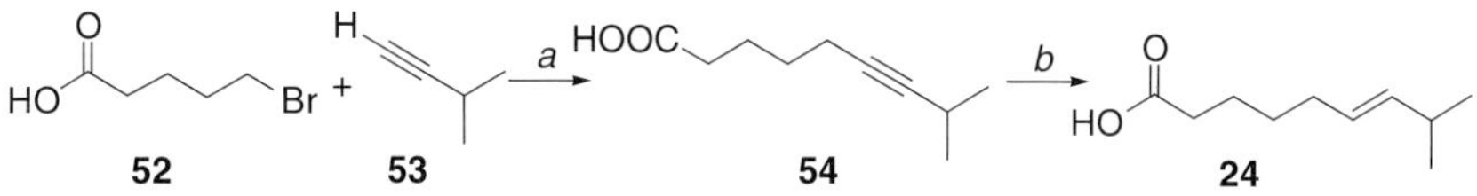

Scheme 4.8 Algox synthesis of (*E*)-MNA (**24**). a: BuLi; b: Na, NH_3.

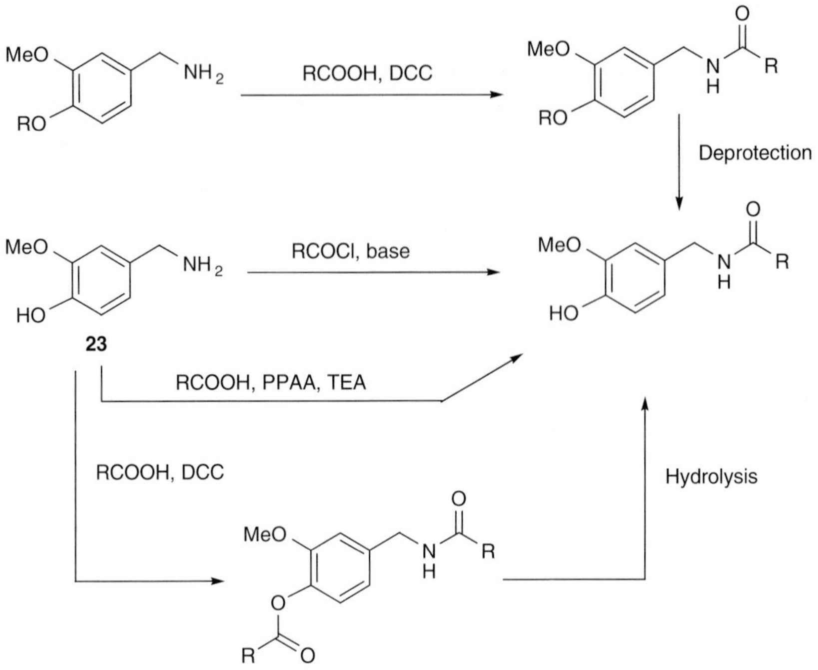

Scheme 4.9 General syntheses of vanillamides. (R = ethoxyethyl or benzyl).

obtained in yields comparable to the other methods. This protocol has the added benefit of being directly applicable to vanillamine hydrochloride, reducing the whole acylation step to a single laboratory operation [66].

Enzymatic syntheses of CPS have also been reported, using various lipase-catalyzed transacylations [69]. Interestingly, a biotechnological process to obtain "natural" vanillin from CPS has been developed, capitalizing on the enzymatic hydrolysis of CPS and the oxidation of vanillamine with a flavoprotein vanillyl alcohol oxidase [70].

4.6
TRV1 as the Biological Target of Capsaicin and the Ecological Raison d'être of Capsaicinoids: A Molecular View

Despite the severe subjective sensation of intolerable burning and inflammation caused by capsaicin, this compound is quite different from obnoxious irritants such as croton oil, mineral acids, and mustards, since it does not produce skin reddening, blistering, edema, or tissue damage. Furthermore, repeated treatment with capsaicin inhibits the perception of pain, leading to desensitization, a unique form of analgesia characterized by a selective impairment of pain sensation. In general, any external stimulus that triggers a response also triggers a process designed to inhibit response to further exposure to the same stimulus. Desensitization is a general feature of biological sensors, and can occur in seconds (flash of light), minutes (odors), or days (tobacco smoke, caffeine). Nevertheless, few sensory stimuli can induce desensitization so effectively as capsaicin. The physiological bases of the peculiar activity of capsaicin have long remained elusive, but their clarification since the mid-1980s has paved the way to important new avenues of investigation for drug discovery, shedding light on the complex mechanism that turn normal pain into the chronic misery of naturopathic pain.

Capsaicin is a lipophilic compounds, and it was long assumed that it could only evoke nonspecific responses by incorporation in cell membranes and perturbation of their organization. In retrospect, this theory sounds totally untenable, but, curiously, similar views were held also on Δ^9-THC, the psychotropic principle of cannabis. The erroneous report that the two enantiomers of Δ^9-THC showed similar bioactivity could have induced many medicinal chemists to dispel the existence of a cannabinoid receptor [71], but for capsaicin it was clear from the outset that strict structure–activity relationships existed, and a receptor model for capsaicin was proposed in 1975 by the Hungarian researcher Szolcsányi [72]. So, apart from its unpleasant properties, there seems to be no explanation as to why capsaicin was ignored for so long by the biomedical community. Support for the existence of a specific receptor was eventually given in 1989 with the discovery that the daphnane diterpenoid resiniferatoxin (RTX, **55**) behaves as an ultrapotent analog of capsaicin [73]. In the following years, the availability of labeled RTX produced overwhelming evidence for the existence of a specific receptor for capsaicin and RTX, setting in motion the race for its cloning, eventually achieved by Julius in 1997 [74]. The capsaicin receptor was named the vanilloid receptor because CPS and RTX share a vanillyl moiety that was considered essential for their activity. More than a decade of intense research has now firmly established the capsaicin receptor (TRPV1) as a druggable target for a host of conditions whose treatment options are currently limited [75]. Thus, malfunctioning of TRPV1 has been suggested for conditions such as pain of various origin (chronic, neuropathic, oncological), urinary incontinence, cough, inflammatory bowel disease, and migraine. TRPV1, the founding member of the vanilloid receptor-like family of transition receptor potential channels, is a nonselective, heat-sensitive cation channel that acts as a polymodal nociceptor to integrate multiple pain stimuli of thermal and chemical (protons, endogenous activators) origin [75]. Under normal conditions, the channel is in the closed state, but the threshold of activation, normally at about 41 °C, is lowered to physiological temperature by acids and by ligands, normally known as vanilloids [75]. The expression of TRPV1 is typical, but not exclusive, of Aδ and C sensory fibers. These dipolar neurons have their somata in sensory ganglia that innervate skin, mucosal membranes, and internal organs, and are involved in the perception of pain, in various reflex responses (cough, micturition), and in neuropeptide (SP, CGRP)-mediated local inflammation. TRPV1 is also expressed in keratinocytes [75], in a variety of internal organs (prostate, pancreas, bladder), as well as in the central nervous system, where its physiological role is unknown [75]. While potentially opening new avenues of biomedical exploitation, this broad distribution cautions against the manipulation of TRPV1 as a selective pharmaceutical strategy.

The regulation of TRPV1 is very complex, but there is a certain agreement that, under resting conditions, a multitude of mechanisms, possibly acting in synergy, contribute to maintaining the channel in an inactive (closed) form [76]. The molecular bases for this baseline inactivation are controversial, but phosphorylation seems critical, being necessary for the insertion of TRPV1 into the cell membrane. Thus, under normal conditions, TRPV1 appears to be mostly sequestered in intracellular compartments as a tetrameric homomer, or associated to cytoplasmatic proteins that, like β-tubulin, have receptor sites for various TRP-channels. TRPV1 was long assumed to be essentially under the inhibitory control of PIP_2 (phosphatidylinositol (4,5)-bisphosphate) and subjected to reversible PKC- and PKA-mediated phosphorylative activation, but more recent studies suggest a more complex scenario of regulation, involving, apart from phospholipases acting on PIP_2 such as phospholipase C, also a diversified kinasic component. In molecular terms, phosphorylation of a single tyrosine residue (Y200) underlies insertion of TRPV1 into the plasma membrane, and this process is controlled by a specific Src tyrosine-kinase that acts as a downstream element of various signaling pathways where PI_3 kinase plays a crucial early role [77]. Binding of NGF to TrkA receptors activates one of these pathways, and other endogenous sensitizers from the "inflammatory soup" (bradichinin, ATP, pro-inflammatory chemokines such as CCL3), might well act through similar, but yet to be characterized, signaling pathways. Complex interactions also exist with the endocannabinoid system, since various endocannabinoids (anandamide, NADA, see below) can also activate TRPV1 [78].

One of the most remarkable features of TRPV1 is desensitization, which is the apparent loss of function following repeated stimulation. The molecular details of TRPV1 desensitization are still largely unknown. Dephosphorylation by calcineurin or binding to calmodulin seem critical for the rapid tachyphylaxis of the receptor, but nothing is known about the long-lasting functional impairment caused by agonists [76].

The complex picture of regulation of TRPV1 offers a host of possibilities for manipulation, but the discovery of receptor agonists and antagonists has traditionally been perceived as the most selective way to exploit the pharmacological potential of TRPV1 [75]. Early studies were centered on the development of agonists capable of desensitizing the receptor while causing minimum side effects such as pain and irritation. More recent investigations have focused instead on the discovery of antagonists capable of making TRPV1 irresponsive to activation by agonists such as capsaicin (vanilloids), protons, and heat. Over the past few years, there has been a remarkable interest in vanilloid antagonists as analgesic and anti-inflammatory agents. This research was spurred by the paucity of clinical options for the treatment of chronic pain, and was made urgent by the demise of COX-2 inhibitors.

Apart from rat [74] and human [79a] sources, TRPV1 has also been cloned from guinea pig [79b], rabbit [79c], chicken [79d], mouse [79e], and dog [79f]. Cloning of the avian version of TRPV1 and the discovery that it is insensitive to CPS [79d] has given a molecular basis to the long-standing observation that birds are not deterred by the pungency of peppers, leading to a satisfactory clarification of the evolutionary significance of capsaicinoids [80]. Plants from the genus *Capsicum* produce fleshy and colored fruits that attract vertebrates and avian consumers. Using nonpungent peppers that are consumed by both mammals and birds, it was found that fruit ingestion by vertebrates inhibits seed germination, while consumption by birds did not damage seed viability, rather promoting it [80]. Birds swallow the fruits and promote the dispersion of seeds, while mammals chew the fruits with their teeth, physically damaging the seeds. Hence, mammals behave as seed predators, while birds are seed dispersers, acting as living "vessels" to carry chilies to new turf. Owing to the selective sensitivity of the mammalian version of TRPV1 to CPS, capsaicinoids function as selective inhibitors of seed predation. Chilies influence the feeding preferences of their potential consumers, deterring consumption of the ripe fruits by mammals. The hotness of peppers might well be an evolutionary adaptation to prevent mammals from ingesting their fruits, but it is instructive to ponder how humans have "fallen in love with this anti-mammalian weapon and spread the chilies much further than any bird ever did" [81].

The immunity of birds to the pungency of capsaicinoids underlies the development of pepper-laced bird seeds that are avoided by squirrels and other competing foraging animals. Since CPS has an interesting activity against *Salmonella* sp. [82], the insensitivity of birds to the pungency of peppers has also served as the basis to develop capsicum-based products as an alternative to antibiotics to prevent salmonella infection in poultry. CPS is apparently not absorbed, since the flesh of the animals remain nonpungent.

4.7 Naturally Occurring Analogs and Antagonists of Capsaicin and Endogenous Vanilloids

The TRPV1 receptor is characterized by a great structural diversity of ligands [83]. Capsaicin and RTX represent the most important first-generation activators, but many more have been discovered, mainly from spices.

The piperidyl amide piperine (**18**) is the major pungent constituent of black pepper (from *Piper* sp.) Compared to capsaicin, piperine is less potent in terms of TRPV1 activation, but shows better desensitizing properties, making it an important lead for the development of anti-inflammatory drugs [84]. Eugenol (**56**) from clove and [6]-gingerol (**57**) from ginger can also activate TRPV1 [83b]. Hot pepper and clove have been traditionally used to treat toothache, and eugenol is used as an antiseptic and analgesic in odontoiatry. Interestingly, TRPV1 was found to be expressed in odontoblasts, the cells responsible for dentin formation [85]. Pungent sesquiterpene dialdehydes such as polygodial (**58**) are contained in plants and sea animals used to flavor food [86]. Polygodial is the major dialdehyde from water pepper (*Polygonum hydropiper* L.), an ancient cheap replacement of black pepper widely cultivated in Italy during the Roman times. Another natural activator of TRPV1 of dietary origin is scutigeral (**59**), a triterpenylated phenol isolated from an edible *Albatrellus* sp.

mushroom [86], while the meroterpenoid cannabidiol (**60**) and the indole alkaloid evodiamine (**61**) are examples of biological analogs of capsaicin isolated from medicinal plants [83].

Garlic's characteristic burning and prickling taste has been traced to the presence of the thiosulfinate allicin (**62**). Unlike capsaicin, allicin shows promiscuity of TRP-activation, activating not only TRPV1, but also TRPA1, the target of isothiocyanates from cruciferous plants [87]. Allicin is not a genuine constituent of garlic, but is produced by the action of allinase on the amino acid alliin (*S*-allylcysteinyl sulfoxide, **63**) [88]. Since alliin and alliinase are contained in distinct anatomical compartments or cells, the reaction is triggered by crushing or mincing a garlic clove, and does not occur to a significant extent when alliinase is inactivated by heat. This is the reason why baked garlic, containing mainly alliin, is nonpungent.

Very few natural products can antagonize the activity of capsaicin, an observation that might have evolutionary relevance, since pungent compounds are presumably produced to deter predation. The SERCA-inhibitor sesquiterpene lactone thapsigargin (**64**) from the Mediterranean plant *Thapsia garganica* L. [89] and the indolylpolyamines **65a** and **65b** from a spider venom [90] are the only TRPV1 antagonists of natural origin identified so far.

TRPV1 is also expressed in the central nervous system, suggesting the existence of endogenous activators. A series of fatty acid amides show endovanilloid activity, namely the endocannabinoid anandamide (**66**), and the dopamine conjugates NADA (*N*-arachidonoyldopamine, **67a**) and OLDA (*N*-oleoyldopamine, **67b**) [91]. Also a variety of polyamines (spermine, spermidine, putrescine) can activate TRPV1, although the high concentration (500 μM) required for binding might have significance only in inflamed tissues, where TRPV1 is overexpressed [92]. Quite remarkably, TRPV1 was also identified as the target of physiologically produced hydrogen sulfide [93].

4.8 Structure–Activity Relationships of Capsaicinoids

Various studies have evidenced strict structure–activity relationships within capsaicinoids, and the existence of these relationships provided a hint for the existence of a specific receptor [72]. On the other hand, different end-points have been used in the literature, and it is difficult to draw a consistent picture. *In vivo* assays such as pungency and ocular irritancy, functional assays such as the measurement of calcium currents in dorsal root ganglion neurons, and receptor assays such as the displacement of labeled RTX have been used. The relationship between the results of these assays is, at best, approximate [75].

Natural capsaicinoids show differences in their pungency, with capsaicin (**1a**) and dihydrocapsaicin (**1b**) being approximately twice as pungent as nordihydrocapsaicin (**1f**) and homocapsaicins (**1g** and **1i**) [43]. The pungency of capsaicinoids and pepper-containing preparations is expressed in Scoville Heat Units (SHU). The SHU is the maximum dilution at which the pungency of a sugar water solution of the substance

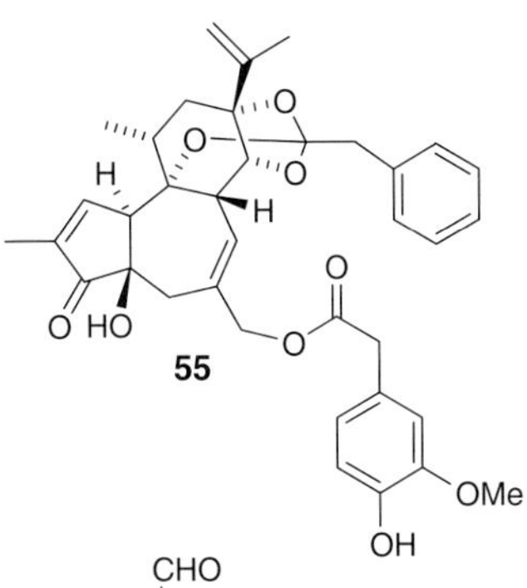

MeO HO **56**

CHO CHO H **58**

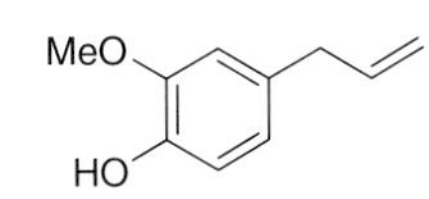

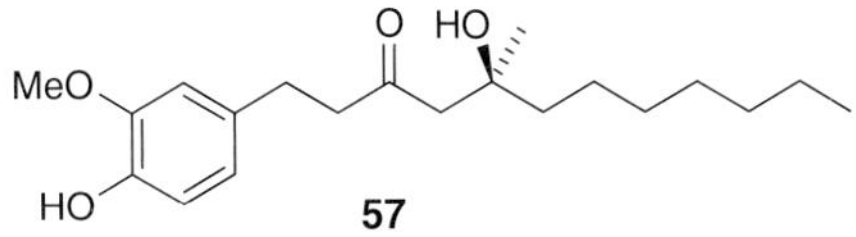

O HO O O O H H O HO **55** O O OMe OH

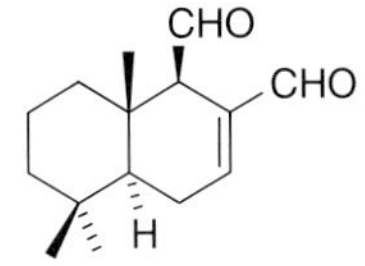

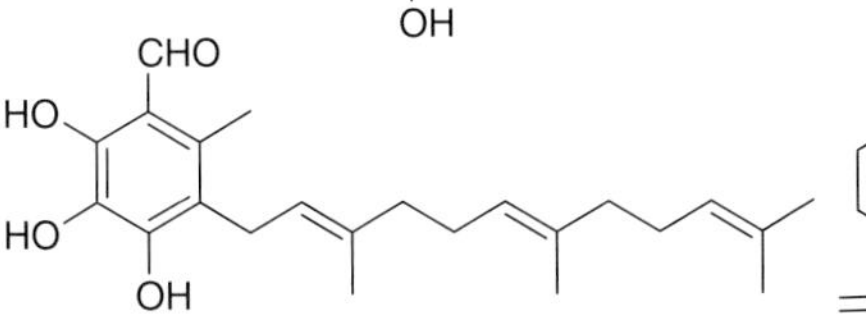

MeO O HO HO **57**

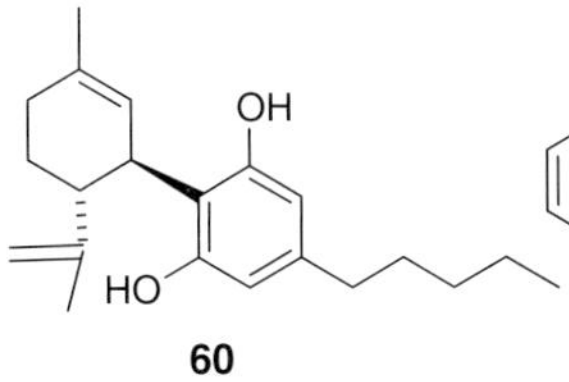

CHO HO HO OH **59**

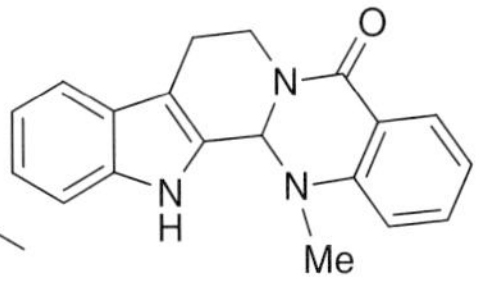

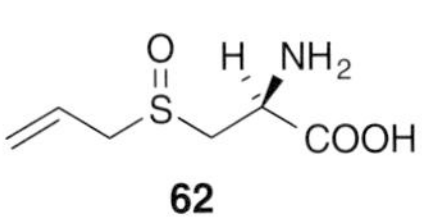

OH HO **60**

O N N H N Me **61**

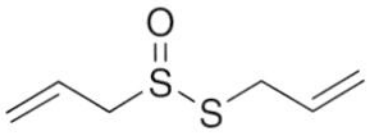

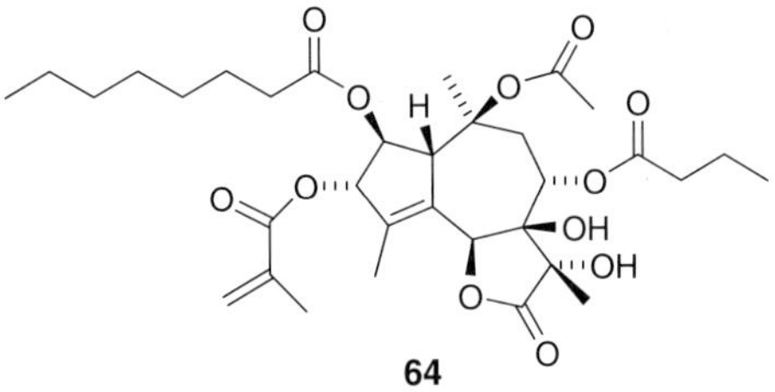

O H NH_2 S COOH **62**

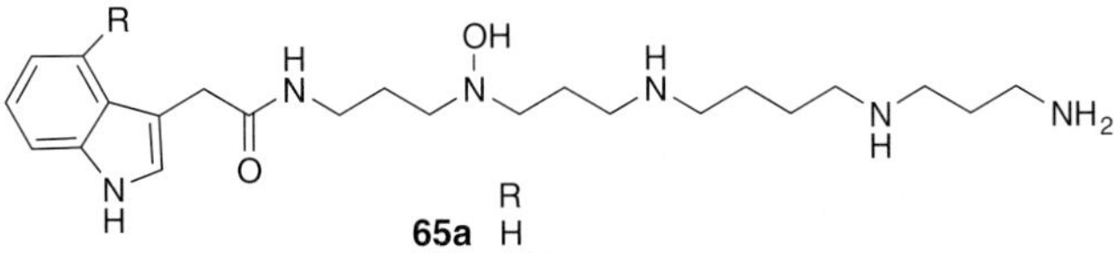

O S S **63**

O O O H O O O OH OH O O O **64**

R H OH N H N N H N H NH_2 O N H

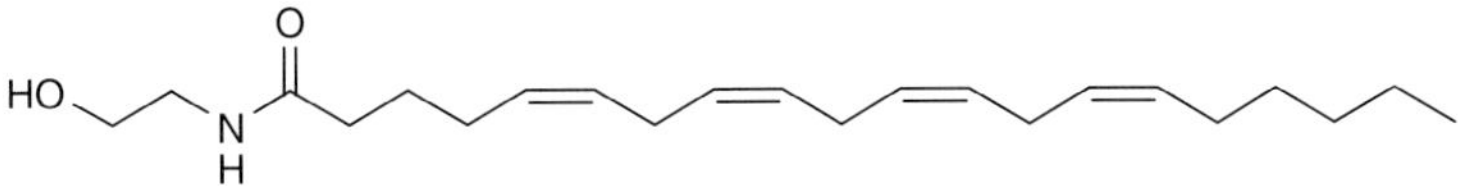

R
65a H
65b OH

O HO N H **66**

H HO N R O HO

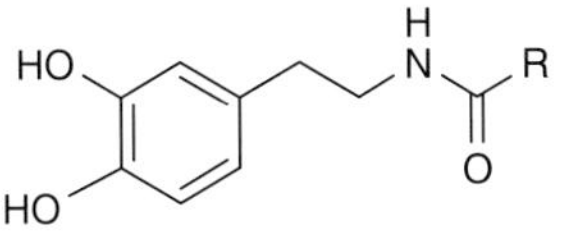

R
67a
67b

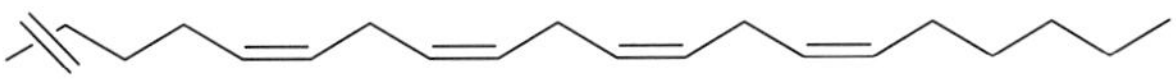

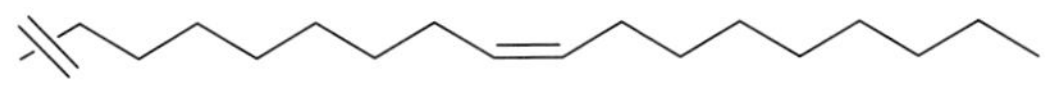

being examined can still be perceived. CPS has a SHU of $16.35 \pm 2.28 \times 10^6$, meaning that CPS can still be perceived as pungent at the dilution of 1 to 16 000 000 (1 mg in 16 L of water, corresponding to a concentration about 200 nM) [42]. The Scoville scale, a sort of Richter scale of pungency, is extensively employed to assess the pungency of pepper-based products. A series of averaged SHU values has been assigned to each major capsaicinoid, and from the HPLC profile of the capsaicinoid fraction, an averaged SHU is calculated. Pure capsaicinoids have SHU $> 8 \times 10^6$. The range of pepper sprays range from 5×10^6 (police-grade) to 2×10^6 (pepper sprays for personal defense). Edible peppers have SHU < 600 000, with the habanero cultivar topping the chart, jalapeno and tabasco in the middle section at around 5000 and 500 SHU, respectively, and sweet pepper having SHU zero [42]. SHU can be translated into approximate concentrations assuming that 15 SHU correspond to one part per million (1 mg/kg) capsaicin. Thus, a habanero pepper with SHU of around 600 000 contains about 4 % capsaicin in dried weight.

The pungency of saturated, linear vanillamides depends on the length of the acyl chain, being maximal at C_9 (nonivamide, **1l**), and decreasing rapidly with shortening or lengthening of the acyl chain. So, lauryl vanillamide (dodecivamide, **1o**) is not pungent, and nor are higher homologs of nonivamide [93]. A still unsettled issue is whether mixtures of capsaicinoids can exhibit synergistic effects, as demonstrated for other natural products occurring in mixtures of closely related compounds. Few systematic studies have been done regarding the activity of natural capsaicinoids in TRPV1 functional or binding assays, but receptor-based structure–activity relationships have been extensively investigated in compounds with a C_{18} side chain. Stearoyl vanillamide (**68**) is neither pungent nor active in assays of TRPV1 activity, but the introduction of a double bond at C-9 affords a compound (*N*-oleoylvanillamine, olvanil, **69a**) only marginally pungent, but more potent than capsaicin on TRPV1 [66]. Since capsaicin and dihydrocapsaicin have very similar properties, the effect of a double bond on the bioactivity of C_{18} fatty acid amides is remarkable. Also remarkable is the minimal oral pungency of olvanil. A possible explanation for this is that highly lipophilic compounds such as olvanil might be bound tightly to salivary proteins and activate TRPV1 slowly, failing to induce an oral sensory response before being swallowed. Indeed, much of the TRPV1 is localized into the interior of cell membranes, and also the capsaicin binding site is localized in the cytoplasmic side of the receptor [94]. To interact with TRPV1, vanilloids must therefore cross the cell membrane and penetrate cells, and all factors interfering with their mobility in cellular membranes are expected to affect their kinetics of TRPV1 activation. The view that a slow uptake might be responsible for the minimal pungency of olvanil is in accordance with the apparent increased affinity for TRPV1 observed after longer incubation time of olvanil with cells expressing TRPV1, with ED_{50} decreasing sevenfold from 29.5 nM to 4.3 nM, in 1 h [94].

The acyl moiety of capsaicin contains only one functional group, a double bond that is redundant for activity. On the contrary, the presence of the double bond is critical for the activity of olvanil, and interesting structure–activity relationships were discovered for this compound. Thus, an analog hydroxylated on the homoallylic carbon is easily available from the amidation of commercial and cheap ricinoleic acid with vanillamine [66]. This compound, named rinvanil (**69b**), retained most of the vanilloid potency of olvanil, but generated an ultra-potent capsaicinoid (phenylacetylrinvanil, PhAR, **69c**) with a two-digit picomolar affinity for TRPV1 by esterification with phenylacetic acid, a key element of the RTX pharmacophore [95]. In accordance with similar observations done on RTX and nonivamide, iodination of the vanillyl moiety of PhAR generated an ultrapotent vanilloid antagonist, while replacement of the vanillyl moiety with aliphatic headgroups, as in **70a** and **70b**) afforded dual agents, capable of activating TRPV1 and behaving as a reverse agonist for CB2, the peripheral cannabinoid receptors [96]. Compounds endowed with this mixed vanilloid–cannabinoid profile are of remarkable interest as anti-inflammatory agents.

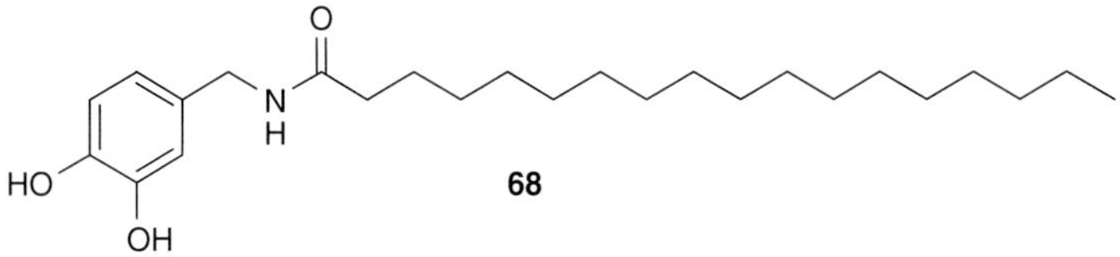

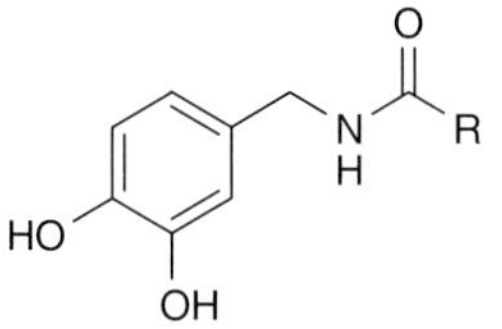

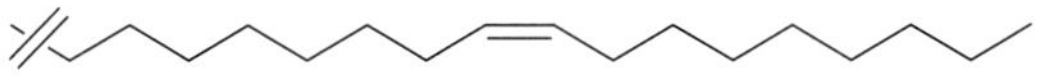

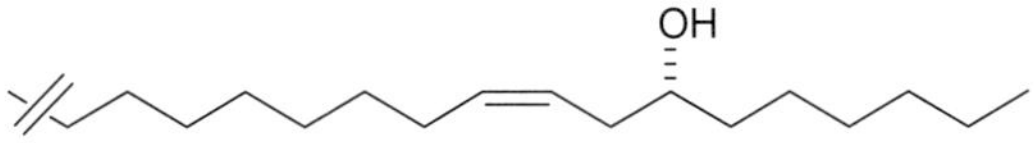

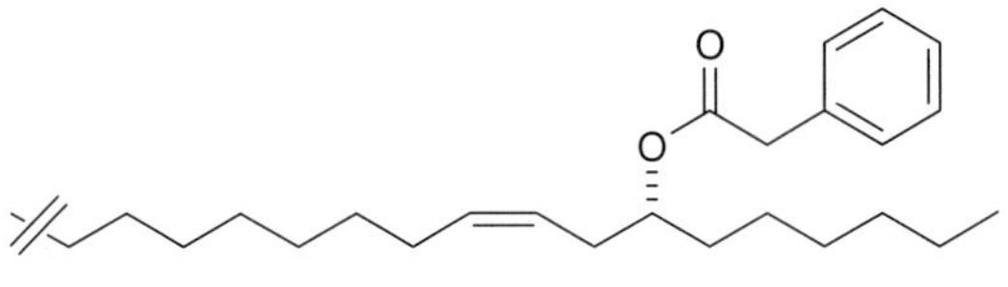

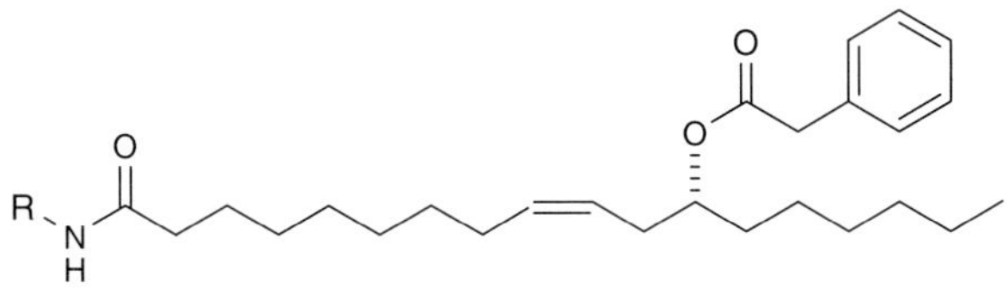

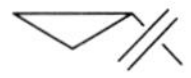

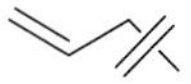

4.9
Molecular Gastronomy of Hot Food

Edible peppers rarely contain more than 1 % capsaicinoids. Since in Western cuisine chili is consumed mainly as a spice, only limited amounts of capsaicinoids are assumed with the diet, with an estimated per capita consumption of 1.5 mg/day [97]. While these are clearly insufficient to produce any systemic effects, the trigeminal stimulation they cause can nevertheless trigger sensory reflexes potentially capable of being translated into pharmacological responses. In some countries (Mexico, India, Thailand) chilies are consumed as a vegetable, and the estimated daily consumption can be higher than 200 mg, corresponding to pharmacological doses whose effects might go beyond gustatory responses [97].

4.9.1
Biomedical Relevance of Capsaicin-Induced Trigeminal Responses

Trigeminal stimulation has long been known to exert anti-inflammatory activity. The molecular mechanism of this reflex remained elusive until the discovery that the trigeminal neurotransmitter acetylcholine binds to the $\alpha7$ nicotinic receptors of macrophages near the vagal endings, inhibiting immune-cell-triggered inflammation [98]. The seminal discovery of a cross-talk between the nervous system and the immune system is currently a very hot topic of biomedical research, and has interesting dietary implications. Thus, binding of capsaicin to TRPV1 activates the vagal endings of the oral cavity, and can potentially trigger a "gustatory" anti-inflammatory trigeminal response, well in accordance with one of the major beneficial effects traditionally ascribed to hot cuisine.

4.9.2
Effect of Capsaicin on Taste

Activation of TRPV1 in the oral cavity enhances gustatory responses to salty compounds, modifying the salt transduction process. The major human transducer of salt taste is an amiloride-insensitive and nonspecific salt receptor, surprisingly identified in a variant of TRPV1 [99]. Unlike TRPV1, this channel is constitutively active at physiological temperature, and is sensitive to capsaicin activation and to inhibition by vanilloid antagonists. Thus, capsaicin increases salt taste sensitivity by lowering the activation threshold of the TRPV1 variant that acts as the main mammalian salt receptor, rationalizing the long-standing culinary observation that spices makes food "more salty" [99]. Salt laced with spices is commercially available to reduce dietary sodium intake. Interestingly, another thermosensitive cation channel from the TRP super-family, TRPM5, is highly expressed in taste buds, where it plays a key role in the perception of sweet, umami, and bitter tastes. The temperature sensitivity of TRPM5 presumably underlies the observation that heat enhances taste perception [100].

4.9.3
Gustatory Sweating

Chili pepper can cause a profuse perspiration, known as gustatory sweating and particularly evident in warm climates. This sweating is clearly distinct from thermo-regulatory and emotional sweating and, despite being first reported over 50 years ago, still lacks a clear physiological basis. Also acidic food and other spices can induce gustatory sweating, while inhibition by cold temperatures is commonly observed. Since the evaporation of sweat has a cooling effect, this might rationalize why hot cuisine is so popular in warm climates. Gustatory sweating might be related to the effects of capsaicin on thermo-regulation. Spicy food makes us feel hotter than we actually are, inducing sweating as a cooling response. TRPV1 is not involved in basal thermal homeostasis, since TRPV1-null animals maintain a normal resting temperature [75], but the systemic administration of capsaicin leads to a decrease of body temperature, a property known for well over 150 years [102]. The very poor oral bioavailability of capsaicin makes it unlikely that chili pepper can induce hypothermia.

4.9.4
Gustatory Rhinitis

Hot pepper induces massive expulsion of mucus from the nose, in a process cogently dubbed "gustatory rhinitis." Indeed, Aztecs suffering from stuffy nose used to eat chilies to expel the troublesome mucus, using them as a nasal decongestants [101]. The physiological bases of gustatory rhinitis from capsaicin are poorly understood, but the process seems to be connected to the sensory properties of capsaicin, leading to the activation of nasal reflexes that facilitate the expulsion of mucus. Also mustard can potently induce gustatory rhinitis, but allyl isothiocyanate (**70**), its active ingredient, is relatively volatile, and therefore can directly stimulate the nasal trigeminal endings by binding to TRPA1, another TRP receptor stimulated by offensive compounds [103].

4.9.5
Hot Food Mitridatism

Trained people can consume large amounts of chili peppers without any adverse effects. The molecular basis for the tolerance to hot food is the facility by which TRPV1 can be desensitized by repeated stimulation. Thus, our tongue can easily become desensitized to capsaicin by repeated (10 times) application of a 1 % solution. Sensitivity to other pungent compounds from spices (piperine (**18**) from pepper, gingerol (**57**) from ginger, allyl isothiocyanate (**71**) from mustard oil) is also lost, and the effect lasts approximately one day [104].

4.9.6 Effect of Capsaicin on Digestion

TRPV1 is expressed in gastric mucosal epithelial cells, where it plays an important role in gastric defense. Its activation by capsaicin stimulates the secretion of bicarbonate and gastric mucus, and directly contributes to the maintenance of mucosa integrity, counteracting the activity of alcohols and other damaging agents [105]. Furthermore, capsaicin shows potent antibacterial activity toward *Helicobacter pylori*, even against metrodinazol-resistant strains, and hot pepper is therefore of potential use for the treatment of gastric and duodenal ulcers, since the active concentrations of CPS (25 μg/mL) can be achieved in the stomach by consumption of hot food [106]. Contrary to common belief, capsaicin and hot food does not harm the stomach, but, on the contrary, protects it against ulcerogenic challenges. Furthermore, TRPV1 receptors play an important role in the protection of the colon against chronic inflammation, and TRPV1-null mice have an increased susceptibility to colon inflammation [107]. Taken together, these studies clearly support the view that capsaicin has a protective role in the gastrointestinal tract.

4.9.7 Capsaicin and Stomach Cancer

Some epidemiological studies support the view that capsaicin has chemopreventive and chemoprotective effects [108], but capsaicin has also been suspected as a potential carcinogen or co-carcinogen [108], and a worrying correlation between the consumption of hot pepper and stomach cancer, at least in some ethnic groups that consume it as a vegetable, has been reported [109]. This correlation is surprising, since peppers contain large amounts of ascorbic acid, a known protective agent for stomach cancer, while capsaicin can block the growth of various human cancer cells with little if any toxicity for primary cells [110], an effect mediated by the inhibition of mitochondrial respiration and the induction of apoptosis [110]. These contrasting data indicate a need for further studies, and a thorough review from the European Scientific Committee on Food concluded that "the available data do not allow to establish a safe exposure level of capsaicin in food" [97].

4.9.8 The Effect of Age and Sex on the Sensitivity to Capsaicin

The activation of TRPV1 from capsaicin can be modulated by steroid hormones. Thus, the female hormone estradiol (**72**) augments the activation of TRPV1 by capsaicin, rationalizing the long-standing and curious observation that woman are more sensitive to hot food than men [111,112]. On the other hand, dehydroepiandrosterone (DHEA, **73**), the most abundant blood steroid, behaves as a vanilloid antagonists [112]. Since the production of DHEA peaks between 20 and 30 years of age and then gradually fades, young people tolerate hot food better that elderly people.

4.9.9
Capsaicin as a Slimming Agent

The anorectic properties of hot pepper have been investigated in the framework of studies on the apparent protection against obesity provided by certain food cultures [113]. Hot food leads to the consumption of body calories, and health beverages flavored with capsicum oleoresin have been developed as slimming agents [114]. In human experiments, hot pepper has been shown to decrease appetite and increase energy expenditure, mainly by lipid oxidation [115]. The satiety-inducing properties of capsaicin were investigated after oral and gastrointestinal (pills) intake with contrasting results. In one study, the effect was similar, suggesting that sensory reflexes are not involved [115a], while an opposite conclusion was reached in a related study [115b]. The addition of 10 g of hot pepper (unreported SHU value) to a high-fat or high-carbohydrate breakfast increased the oxidative breakdown of fats but not that of carbohydrates, a surprising observation since, unlike carbohydrates, fats are unable to increase oxidation rate in humans [116], and a pepper-laced breakfast significantly reduced the desire to eat, and hunger before lunch [116]. Capsaicin seems to be able to adjust fat oxidation to fat intake, triggering a "burn calories and stop eating" message quite appealing in a society where obesity is such a severe problem. On the other hand, the slimming properties of capsaicin have never been investigated in large clinical trials, its mechanistic bases are poorly understood, and its widespread use as a dietary slimming agent would not be appealing to consumers because of its marked sensory properties. The slimming properties of capsaicin are apparently retained by capsiate (**22**) [117], the nonpungent ester isoster of capsaicin found in sweet pepper, and capsaicin *O*-glucoside [118]. These nonpungent compounds have been investigated as an anti-obesity drug in Japan, and biotechnological processes for their production have been developed, no doubt to advertise these compounds, difficult to obtain by isolation, as "natural" [119,120].

4.9.10
Quenching Capsaicin

The burning sensation from capsaicin can be quenched by cold water, ice cubes, or a cold yogurt [81]. As discussed in Section 4.6, capsaicin does not actually activate TRPV1, but only lowers its firing threshold, making it sensitive to lower physiological temperatures. Therefore, its effects can simply be reversed by cooling the oral cavity. Since TRPV1 is also sensitive to alcohol [121] and acids [75], which both potentiate its sensitivity to capsaicin, carbonated drinks and spirits just make the irritation worst, even though capsaicin is more soluble in alcohol than in water. It has been reported that solid and rough materials such as a cube of sugar or a cracker can alleviate capsaicin burning [81]. The rationale for this effect is unknown, but, since many TRP receptors act also as mechanosensors, their activation might interfere with that of thermo-TRPs such as TRPV1.

4.9.11
Chilies and Olive Oil

Olive oil aromatized with chili pepper is a common condiment in the Mediterranean area. Under these conditions, transamidation of capsaicin to *N*-oleoylvanillamine (olvanil, **69a**) has been demonstrated [122]. Olvanil is nonpungent, but more potent as an anti-inflammatory agent than capsaicin, and one therefore wonders if olvanil might also contribute to the beneficial health effects traditionally attributed to olive oil flavored with hot pepper.

4.9.12
Who Should Avoid Chilies?

Capsaicin causes transient bronchoconstriction and induces coughing, especially in individual with severe asthma, potentially triggering fatal crises [37]. These adverse respiratory effects are probably due to the limited capacity of respiratory tissues to metabolize capsaicin (see Section 4.3) [38], and are a major problem with the use of pepper sprays as antiriot agents [37]. Smokers are less sensitive to the respiratory effects of capsaicin, but asthmatic patients should avoid chilies and hot cuisine, as should people using drugs such as ACE-inhibitors, which have an intrinsic capacity to induce cough.

4.9.13
How can the Pungency of Chilies be Moderated?

Pepper lovers know well what chemists have now demonstrated, namely that capsaicinoids are accumulated in the inner placental part of peppers and especially in their basal and apical sections. Therefore, pungency can be moderate by removing the inner part of the pod, which contains the placenta, and cutting away its tip and base.

4.9.14
Psychology of Pepper Consumption

Humans are the only mammals that deliberately consume hot peppers and seem to enjoy their aversive effects, eluding a system designed to keep us from eating food containing potentially harmful compounds. Several theories have been put forward to explain this behavior. The most interesting one is the one that consider hot cuisine as the culinary equivalent of benign masochistic activities such as making a parachute drop, taking a hot bath or an icy shower, or going to a horror movie [123]. We apparently experience a pleasant thrill when confronted with a "constrained risk", that is, a risk typical of a life-threatening event that we nevertheless manage to keep under control by evoking it in a safe setting. Thus, capsaicin mimics potentially damaging heat, but, while consuming chilies, we are well aware that they cannot set us on fire! The trigeminal sensation evoked by capsaicin might disrupts the normal span of attention in a brief, nonadditive, and non-hallucinogenic way, just like a "rollercoaster" drive [123].

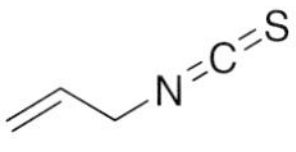

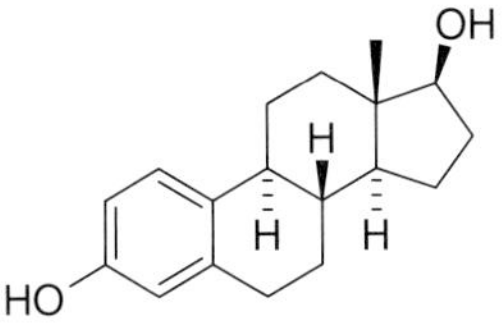

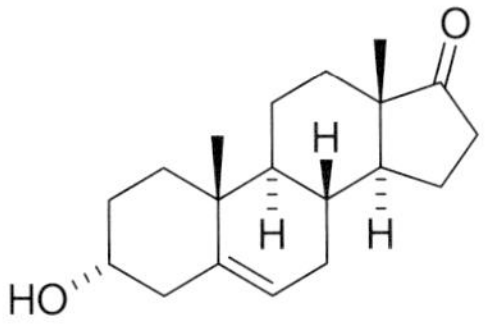

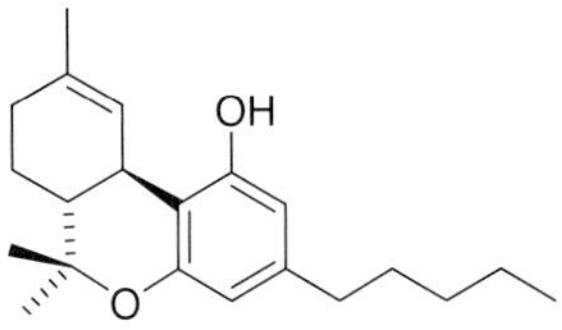

4.10 Conclusions

The seminal discovery that Δ^9-THC (**74**), a non nitrogenous lipophilic compound, binds to a specific receptor (CB1), set in motion what is now referred to as lipidomics [124]. Δ^9-THC was the first non-nitrogenous compound to show affinity for a receptor of the CPCR type [124], and, in the wake of this seminal discovery, interest has also been mounting in lipophilic compounds such as alkamides. Indeed, capsaicin has played a leading role in the emergence of lipidomics as an important branch of the neurosciences, and is now accepted as a standard pharmacological tool to investigate a host of physiological and pathological conditions. The discovery that capsaicin binds to a receptor of the TRP-type has also led to systematic investigations on this class of ion channels, validating the taste-active constituents of various spices as molecular probes to de-orphanize these sensors in terms of chemical ligands [125]. As a result of these studies, capsaicin has become a natural product whose ecological role is better understood, and the way it fits into the reproductive strategy of peppers testifies to the complexity of natural interactions. Finally, the molecular advances in our understanding of the biological properties of capsaicin can rationalize a host of curious culinary observations, vividly testifying how research can still thrive on curiosity and produce significant results.

4.11 Acknowledgments

This chapter is based on a series of seminars on spices I presented at the Università di Scienze Gastronomiche (USG, Pollenzo, Italy). I am grateful to Prof. Gabriella Morini and the teaching staff of USG for their invitation, and to their students for the enthusiasm and interest for what was basically an attempt to highlight the relevance of molecular sciences, and organic chemistry in particular, for their curricula. I am grateful to Arpad Szallasi, Vincenzo Di Marzo, and Eduardo Muñoz for stimulating me to moonlight from organic chemistry and explore the subtleties of vanilloid research, and to Renato Dominici (*La Carmagnole*, Carmagnola, Italy) for fostering my interest in the chemical aspects of cuisine with his savory creations.

Finally, I would like to thank my wife Enrica and my daughter Silvia for their constant support and love, and all my collaborators from the Università del Piemonte Orientale for their commitment to research and for showing me, every day, how important is "to see what everyone has seen and think what no one has thought" (A. Szent-Györgyi).

References

1 As at March 3, 2006, the SciFinder database shows 16 012 entries for capsaicin. For comparison, taxol and phorbol had 11 608 and 66 862 entries, respectively. A search in the same database produced 904 patents related to capsaicin.

2 See http://www.m-w.com/dictionary/capsaicin. There is considerable uncertainty regarding the correct pronunciation of capsaicin, even within scientists whose native language is English. According to the Webster Dictionary, capsaicin should be pronounced kap-' sA-&-s&n

3 A search on "capsaicin reviews" in the PubMed database gave 618 entries as at July 1, 2007. Of these reviews, 84 deal specifically with capsaicin, its receptor, or its ligands.

4 Suzuki, T. and Iwai, K. (1984) Constituents of red pepper species. Chemistry, biochemistry, pharmacology, and food science of the pungent principle of capsicum species, Brossi, A. (Ed.) *The Alkaloids*, Vol. **23**, Academic Press, Orlando, FL, 227–299.

5 Perkins, B., Bushway, R., Guthire, K., Fan, T., Stewart, B., Price, A., Williams, M. (2002) *Journal of AOAC International*, **85**, 82.

6 For a compelling book on the history and biology of peppers, see: Anderson, J. (1984) *Peppers. The Domesticated Capsicum*, University of Texas Press, Austin, TX.

7 For a definition of molecular gastronomy, see: This, H. (2002) *Angewandte Chemie International Edition England*, **41**, 83–88.

8 Pelletier, S. W. (1983) The nature and definition of an alkaloid, In Pelletier S. W. (Ed.) *Alkaloids: Chemical and Biological Perspectives*, Vol. **1**, Wiley, New York, 1–31. Qu, S.-J., Liu, Q.-W., Tan, C.-H., Jiang, S.-H., Zhu, D.-Y. (2006) Bufotenine is common in rutaceous plants. *Planta Medica*, **72**, 264–266.

9 For an account on the life of C. F. W. Meissner, see: Friedrich, C. and Von Domarus, C. (1998) *Pharmazie*, **53**, 67–73.

10 For a discussion of these early attempts to define the term alkaloid, see: Hesse, M. (2002) *Alkaloids. Nature's Curse or Blessing?* Wiley-VCH, Weinheim, pp. 1–5 (the term acidoids to refer to organic acids was seemingly never used in the chemical literature).

11 Winterstein, E. and Trier, G. (1910) *Die Alkaloide, eine Monographie der natürlichen Basen*, Bornträger, Berlin.

12 Hegnauer, R. (1963) The taxonomic significance of alkaloids, In Swain T. (Ed.) *Chemical Plant Taxonomy*, Academic Press, New York, 389–399.

13 Meissner, W. (1819) *Journal of Chemical Physics (Schweiger)*, **25**, 379.

14 Ladenburg, A. (1886) *Berichte der Deutschen Chemischen Gesellschaft*, **19**, 439–457.

15 Appendino, G., Prosperini, S., Valdivia, C., Ballero, M., Colombano, G., Billington, R.A., Genazzani, A.A., Sterner, O. (2005) *Journal of Natural Products*, **68**, 1213–1217.

16 For a general treatise on alkaloids biosynthesis, see: Cordell, G. A. (1981) *Introduction to Alkaloids. A Biogenetic Approach*, Wiley, New York.

17 Long, R. M. and Croteau, R. (2005) *Biochemical and Biophysical Research Communications*, **338**, 410–417.

18 Hesse, M. and Schmidt, H. (1977) Macrocyclic spermidine and spermine alkaloids, In Wiesner K. (Ed.) *MTP International Revies of Science, Series 2,* Vol. **9**, Butterworths, London, 265–307.
19 Lobata, K., Todo, T., Yazawa, S., Iwai, K., Watanabe, T. (1998) *Journal of Agricultural and Food Chemistry*, **46**, 1695–1697.
20 Bosland, P. W. (1994) Chiles: history, cultivation, and uses, In Charalambous G. (Ed.) *Spices, Herbs, and Edible Fungi*, Elsevier Science, Amsterdan, 347–366.
21 http://www.chilepepperinstitute.org. The five domesticated peppers are *C. annuum, C. fructescens, C. chinense, C. pendulum*, and *C. pubescens*. All non pungent peppers belong to *C. annuum* var. *annuum*.
22 Zewdie, Y. and Bosland, P. W. (2001) *Biochemical Systematics and Ecology*, **29**, 161–169.
23 Jurenitsch, J., David, M., Heresch, F., Kubelka, W. (1979) *Planta Medica*, **36**, 61–67.
24 Ochi, T., Takaishi, Y., Kogure, K., Yamauti, I. (2003) *Journal of Natural Products*, **66**, 1094–1096.
25 Nakamura, H., Matsuyama, K., Yoshitani, K., Tamura, W., Kagami, Y., Satoshi, Y. (2005) Isolation of Capsaicin Glycosides from Red Pepper. JP 2005 298,360.
26 Constant, H., Cordell, G. A., West, D. P., Johnson, J. H. (1995) *Journal of Natural Products*, **58**, 1925–1928.
27 Prasad, B. V. N., Gururaj, H. B., Kumar, V., Giridhar, P., Parimalan, R., Sharma, A., Ravishankar, G. A. (2006) *Journal of Agricultural and Food Chemistry*, **54**, 1854–1859.
28 Steward, C., Jr.,Kang, B.-C., Liu, K., Mazourek, M., Moore, S. L., Yoo, E. Y., Kim, B.-D., Paran, I., Jahn, M. M. (2005) *The Plant Journal*, **42**, 675–688.
29 Heiser, C. B. and Smith, P. G. (1953) *Economic Botany*, **7**, 214–227.
30 For recent studies, see: Kozukue, N., Han, J.-S., Kozukue, E., Lee, S.-J., Kim, J.-A., Lee, K. R., Levin, C. E., Friedman, M. (2005) *Journal of Agricultural and Food Chemistry*, **53**, 9172–9181.
31 Kopp, B. and Jurenitsch, J. (1981) *Planta Medica*, **43**, 272–279 (Since valine serves as a precursor of leucine, capsaicinoids from the iso-series are derived (or derivable) from valine, and those of the anteiso series form isoleucine. In the light of its biogenetic derivation, the configuration of the stereogenic centre of homocapsaicin II should be (S).
32 Lee, B.-A., Choi, D., Lee, K.-W.J. (1998) *Plant Biology*, **41**, 9–15.
33 Also the peroxidase involved in the biosynthesis of dimeric indole alkaloids is localized in vacuoles. For a discussion, see: Sottomayor, M., de Pinto, M. C., Salema, R., Di Cosmo, F., Pedreño, M. A., Ros Barcel, A. (1996) *Plant Cell and Environment*, **19**, 761–767.
34 Bernal, M. A. and Barcel, A. R. (1996) *Journal of Agricultural and Food Chemistry*, **44**, 3085–3089.
35 (a) Lawson, T. and Gannett, P. (1989) *Cancer Letters*, **48**, 109–113. (b) Nakamura, T., Ooi, T., Kogure, K., Nishimura, M., Terada, H., Kusumi, T. (2002) *Tetrahedron Letters*, **43**, 8181–8183.
36 Glinsukon, T., Stitmunnaithum, V., Toskulkao, T., Burannuti, T., Tangkrisanavinont, V. (1980) *Toxicon*, **18**, 215–221.
37 Olajos, E. J. and Salem, H. (2001) *Journal of Applied Toxicology*, **21**, 355–391.
38 Reilly, C. A., Ehlardt, W. J., Jackson, D. A., Kulanthaivel, P., Mutlib, A. E., Espina, R. J., Moody, D. E., Crouch, D. J., Yost, G. S. (2003) *Chemical Research in Toxicology*, **16**, 336–349.
39 (a) Surh, Y.-J., Ahn, S. H., Kim, K.-C., Park, J. B., Sohn, Y. W., Lee, S. (1995) *Life Science*, **56**, 305–311. (b) Lee, S. S., Beak, Y. H., Ko, S. Y., Kumar, S. (1981) *Proceedings 1st International Conference on Chemistry and Biotechnology of Biologically Active Natural Products*, **3**, 189–193.
40 Appendino, G., Daddario, N., Minassi, A., Schiano Morello, A., DePetrocellis, L., Di Marzo, V. (2005) *Journal of Medicinal Chemistry*, **48**, 4663–4669.

41 Markai, S., Marchand, P. A., Mabon, F., Baguet, E., Billaut, I., Robbins, R. J. (2002) *Chemistry and Biochemistry*, **3**, 212–218.

42 *Spices and Condiments: Chillies; Determination of Scoville Index.* Technical Committee ISQ/TC 34: Agricultural Good Products. ISO-3513. International Organization for Standardization, Geneva, 1977.

43 Krajewska, A. M. and Powers, J. J. (1988) *Journal Food Science*, **53**, 902–905.

44 For recent articles discussing the analytical separation of capsaicinoids, see: Thomas, B. V., Schreiber, A. A., Weisskopf, C. P. (1998) *Journal of Agricultural and Food Chemistry*, **46**, 2655–2663 (capillary GC). Thompson, R. Q., Phinney, K. W., Sander, L. C., Welch, M. J. (2005) *Analytical and Bioanalytical Chemistry*, **381**, 1432–1440.

45 *United States Pharmacopoeia* 2005; **28**: 337.

46 Constant, H. L. and Cordell, G. A. (1995) *Journal of Natural Products*, **58**, 1923–1928.

47 Balbaa, S. I., Karawya, S., Girgis, A. N. (1968) *Lloydia*, **31**, 272–274. (a) For recent studies on the distribution of capsaicinoids in pepper plants, see: Kirshbaum-Titze, P., Mueller-Seitz, E., Petz, M. (2002) *Journal of Agricultural and Food Chemistry*, **50**, 1260–1263. (b) Kirshbaum-Titze, P., Mueller-Seitz, E., Petz, M. (2002) *Journal of Agricultural and Food Chemistry*, **50**, 1264–1266.

48 Kozukue, N., Han, J.-S., Kozukue, E., Lee, S.-J., Kim, J.-A., Lee, K.-R., Levin, C. E., Friedman, M. (2005) *Journal of Agricultural and Food Chemistry*, **53**, 9172–9181.

49 Estrada, B., Bernal, M. A., Diaz, J., Pomin, F., Merino, F. (2002) *Journal of Agricultural and Food Chemistry*, **50**, 1188–1191.

50 Contreras-Padilla, M. and Yahiam, E. M. (1998) *Journal of Agricultural and Food Chemistry*, **46**, 2075–2079.

51 For an investigation on the molecular basis of the effect of water stress on the production of capsaicinoids, see: Sung, Y., Chang, Y.-Y., Ting, N.-L. (2005) *Botanical Bulletin of Academia Sinica*, **46**, 35–42.

52 Nelson, E. K. (1910) *Industrial and Engineering Chemistry*, **2**, 419.

53 Sass, N. L., Rounsavill, M., Combs, H. (1977) *Journal of Agricultural and Food Chemistry*, **25**, 1419–1420.

54 Kobata, K., Kobayashi, M., Kinpara, S., Watanabe, T. (2003) *Biotechnology Letters*, **25**, 1575–1578.

55 Yang, D., Wang, D., Pang, Z., Wang, F. (2002) Method for Enriching and Purifying Capsaicin by Using Macroporous Adsorption Resin. CN 2002-134,384.

56 Segi, H., Yamada, S., Kato, S., Murasugi, S. (1997) Chrmatography-Free Method for the Industrial-Scale Purification of Capsaicin from Capsinoid (sic) Mixtures using its Precipitable Silver Complex. JP 97 193760 19,970,718.

57 Späth, E. and Darling, S. F. (1930) *Berichte der Deutschen Chemischen Gesellschaft*, **110**, 737–743.

58 Crombie, L., Dandegaonker, S. H., Simpson, K. S. (1955) *Journal of the Chemical Society*, 1025–1027.

59 Gannet, P. M., Nagel, D. L., Reilly, P. J., Lawson, T., Sharpe, J., Toth, B. (1988) *Journal of Organic Chemistry*, **53**, 1064–1071.

60 Kaga, H., Miura, M., Orito, K. (1989) *Journal of Organic Chemistry*, **54**, 3477–3478.

61 Kaga, H., Goto, K., Fukuda, T., Orito, K. (1992) *Bioscience Biotechnology and Biochemistry*, **56**, 946–948.

62 Szallasi, A., Cortright, D.N., Blum, C.A., Eid, S.R., (2007) *Nature Reviews Drug Discovery*, **6**, 357–372.

63 Kaga, H., Goto, K., Takahashi, T., Hino, M., Tokuhashi, T., Orito, K. (1996) *Tetrahedron*, **52**, 8451– 8470.

64 Vig, O. P., Aggarwal, R. C., Sharma, M. L., Sharma, S. D. (1979) *Indian Journal of Chemistry Section B*, **17B**, 558–559.

65 McIlvain, S., Chen, W., Mamiya, P. H., Burch, R., Carter, R. B., Anderson, T. A. (2004) Preparation and Purification of Synthetic Capsaicin. WO 2004-US 10745 20,040,408

66 Appendino, G., Minassi, A., Morello, A. S., DePetrocellis, L., Di Marzo, V. (2002) *Journal of Medicinal Chemistry*, **45**, 3739–3745 and references therein.

67 Janusz, M. J., Buckwalter, B. L., Young, P. A., Lahann, T. R., Farmer, R. W., Kasting, G. B., Loomans, M. E., Kerckaert, G. A., Maddin, C. S., Berman, E. F., Bohne, R. L., Cupps, T. L., Milstein, J. R. (1993) *Journal of Medicinal Chemistry*, **36**, 2595–2604.

68 Walpole, C. S. J., Wrigglesworth, R., Bevan, S., Campbell, E. A., Dray, A., James, I. P., Masdin, K. J., Perkins, M. N., Winter, J. (1993) *Journal of Medicinal Chemistry*, **36**, 2381–2389.

69 Kobata, K., Toyoshima, M., Kawamura, M., Watanabe, T. (1998) *Biotechnology Letters*, **20**, 781–783.

70 vanden Heuvel, R. H. H., Fraaije, M. W., Laane, C., van Berkel, W. J. H. (2001) *Journal of Agricultural and Food Chemistry*, **49**, 2954–2958.

71 For a historical account on the discovery of cannabinoids and endocannabinoids, see: Mechoulam, R. and Ben-Shabat, S. (1999) *Natural Product Reports*, **16**, 131–143.

72 For a review on the early Hungarian work on the existence of a specific receptor for capsaicin, see: Szolcsányi, J. (2004) *Neuropeptides*, **38**, 377–384.

73 Szallasi, A. and Blumberg, P. M. (1989) *Neuroscience*, **30**, 515–520.

74 Caterina, M. J., Schumacher, M. A., Tominaga, M., Rosen, F. T. A., Levine, J. D., Julius, D. (1997) *Nature*, **389**, 816–824.

75 For a general review on the pharmacology of capsaicin, see: Szallasi, A. and Blumberg, P. M. (1999) *Pharmacological Reviews*, **51**, 159–211. For subsequent developments, see: Szallasi, A. and Appendino, G.*Journal of Medicinal Chemistry*, **47**, (2004) 2717–2723.

76 For a review article on the regulation of TRPV1, see: Cortright, D. N. and Szallasi, A. (2004) *European Journal of Biochemistry*, **271**, 814–819.

77 Zhang, X., Huang, X., Huang, J., McNauthton, P. A. (2005) *EMBO Journal*, **24**, 211–4223.

78 vander Stelt, M. and Di Marzo, V. (2004) *European Journal of Biochemistry*, **271**, 1827–1834.

79 (a) Human: Hayes, P., Meadows, H. J., Gunthorpe, M. J., Harries, M. H., Duckworth, D. M., Cairns, W., Harrison, D. C., Clarke, C. E., Ellington, K., Prinjha, R. K., Barton, A. J., Medhurst, A. D., Smith, G. D., Topp, S., Murdock, P., Sanger, G. J., Terrett, J., Jenkins, O., Benham, C. D., Randall, A. D., Gloger, I. S., Davis, J. B. (2000) *Pain*, **88**, 205–218. McIntyre, P., McLatchie, L. M., Chambers, A., Phillips, E., Clarke, M., Savidge, J., Toms, C., Peacock, M., Shah, K., Winter, J., Weerasakera, N., Webb, M., Rang, H. P., Bevan, S., James, I. F.*British Journal of Pharmacology*, **132**, (2001) 1084–2001. (b) Guinea pig: Savidge, J., Davis, C., Shah, K., Colley, S., Phillips, E., Ranasinghe, S., Winter, J., Kotsonis, P., Rang, H. P., McIntyre, P.*Neuropharmacology*, **43**, (2002) 459–465. (c) Rabbit: Gavva, N. R., Klionsky, L., Qu, Y., Shi, L., Tamir, R., Edenson, S., Zhang, T. J., Viswanadhan, V. N., Toth, A., Pearce, L. V., Vanderah, T. W., Porreca, F., Blumberg, P. M., Lile, J., Sun, Y., Wild, K., Louis, J. C., Treanor, J. J. (2004) *Journal of Biological Chemistry*, **279**, 20283–20295. (d) Chicken: Jordt, S. E. and Julius, D.*Cell*, **108**, (2002) 421–430. (e) Mouse: Correll, C. C., Phelps, P. T., Anthes, J. C., Umland, S., Greenfeder, S. (2004) *Neuroscience Letters*, **370**, 55–60. (f) Dog: Phelp, P. T., Anthes, J. C., Correl, C. E. (2005) *European Journal of Pharmacology*, **513**, 77–66.

80 Tewksbury, J. J. and Nabhan, G. P. (2001) *Nature*, **412**, 403–404.

81 McGee, H. (2004) *On Food and Cooking. The Science and Lore of the Kitchen*, Scribner, New York, p. 418.

82 Careaga, M., Fernández, E., Dorantes, L., Mota, L., Jaramillo, M. E., Hernandez-Sanchez, H. (2003) *International Journal of Food Microbiology*, **83**, 331–335.

83 (a) Appendino, G., Muñoz, E., Fiebich, B. (2003) *Expert Opinion on Therapeutic*

Patents, **13**, 1825–1837. (b) Calixto, J. B., Kassuya, C. A., Andre, E., Ferreira, J. (2005) *Pharmacology and Therapeutics*, **106**, 179–208.

84 Szallasi, A. (2005) *Trends in Pharmacological Sciences*, **26**, 437–439.

85 Okumura, R., Shima, K., Muramatsu, T., Nakagawa, K., Shimono, M., Suzuki, T., Magloire, H., Shibukawa, Y. (2005) *Archives of Histology and Cytology*, **68**, 251–257.

86 For a review see: Sterner, O. and Szallasi, A. (1999) *Trends in Pharmacological Sciences*, **20**, 431–439.

87 (a) Macpherson, L. J., Geierstanger, B. H., Viswanath, V., Bandell, M., Eid, S. R., Hwang, S., Patapoutian, A. (2005) *Current Biology*, 929–934. (b) Bautista, D. M., Movahed, P., Hinman, A., Axelsson, H. E., Sterner, O., Hogestatt, E. D., Julius, D., Jordt, S. E., Zygmunt, P. M. (2005) *Proceedings of the National Academy of Sciences of the United States America*, **102**, 12248–12252.

88 Block, E. (1992) *Angewandte Chemie*, **31**, 1135–1178.

89 Toth, A., Kedei, N., Szabo, T., Wang, Y., Blumberg, P. M. (2002) *Biochemical and Biophysical Research Communications*, **293**, 777–782.

90 Kitaguchi, T. and Swartz, K. J. (2005) *Biochemistry*, **44**, 15544–15549.

91 For a review on endovanilloids, see: Di Marzo, V., Blumberg, P. M., Szallasi, A. (2002) *Current Opinion in Neurobiology*, **12**, 372–379.

92 Ahern, G. P., Wang, X., Miyares, R. L. (2006) *Journal of Biological Chemistry* **281**, 8991–8995.

93 Trevisani, M., Patacchini, R., Nicoletti, P., Gatti, R., Gazzieri, D., Lissi, N., Zagli, G., Creminon, C., Geppetti, P., Harrison, S. (2005) *British Journal of Pharmacology*, **145**, 1123–1131. (a) For leading studies on the relationships between structure and pungency of capsaicinoids, see: Nelson, E. K. (1919) *Journal of the American Chemical Society*, **41**, 2121–2122. (b) Szolcsányi, J. and Jancsò-Gábor, A. (1975) *Arzneimittel-Forschung*, **25**, 1877–1881.

94 Lazar, J., Braun, D. C., Toth, A., Wang, Y., Pearce, L. V., Pavlyukovets, V. A., Blumberg, P. M., Garfield, S. H., Wincovitch, S., Choi, H. K., Lee. (2006) *Journal of Molecular Pharmacology*, **69**, 1166–1173.

95 Appendino, G., DePetrocellis, L., Trevisani, M., Minassi, A., Daddario, N., Schiano Morello, A., Mazzieri, D., Ligresti, A., Campi, B., Fontana, G., Pinna, C., Geppetti, P., Di Marzo, V. (2005) *The Journal of Pharmacology and Experimental Therapeutics*, **312**, 561–570.

96 Appendino, G., Cascio, M. G., Bacchiega, S., Morello, A. S., Minassi, A., Thomas, A., Ross, R., Pertwee, R., DePetrocellis, L., Di Marzo, V. (2005) *FEBS Letters*, **580**, 568–574.

97 Anonymous (2002) *Opinion of the Scientific Committee of Food on Capsaicin*, European Commission, Health & Consumer Protection Directorate General, Brussels, Belgium, Available at http://europa.eu.int/comm/food/fs/sc/scf/index_en.html.

98 Wang, H., Yu, M., Ochani, M., Amella, C. A., Tanovic, M., Susarla, S., Li, J. H., Wang, H., Yang, H., Ulloa, L., Al-Abed, Y., Czura, C. J., Tracey, K. J. (2003) *Nature*, **421**, 384–388.

99 Lyall, V., Heck, G. L., Vinnikova, A. K., Ghosh, S., Phan, T.-H.T., Alam, R. I., Russel, O. F., Malik, S. A., Bigbee, J. W., DeSimone, J. A. (2004) *Journal of Physiology*, **558**, 147–159.

100 Talavera, K., Yasumatsu, K., Voets, T., Droogmans, G., Shigemura, N., Ninomiya, Y., Margolskee, R. F., Nilius, B. (2005) *Nature*, **438**, 1022–1025.

101 Haxton, H. A. (1948) *Brain*, **71**, 16–25.

102 Fujii, T., Ohbuchi, Y., Takahashi, S., Sakurada, T., Sakurada, S., Ando, R., Risara, K. (1986) *Archives Internationales de Pharmacodynamie et de Therapie*, **280**, 165–176.

103 Jordt, S. E., Bautista, D. M., Chuang, H. H., McKemy, D. D., Zygmunt, P. M., Hogestatt, E. D., Meng, I. D., Julius, D. (2004) *Nature*, **427**, 260–265.

104 Szolcsányi, J. (1977) *Journal of Physiology (Paris)*, **73**, 251–259.

105 For a review, see: Ome, A.-S., Szolcsányi, J., Mozsik, G. y. (1997) *Journal of Physiology (Paris)*, **91**,

151–171 (It is interesting to remark that the first medicinal indication for hot pepper was as a digestion aid, as quoted by the herbalist John Gerard in his Herbal, a book published in 1597. According to Gerard, hot pepper *"helpeth greatly the digestion of meates"*.)

106 Jones, N. L., Shabib, S., Sherman, P. M. (1997) *FEMS Microbiology Letters,* **146**, 223–227.

107 Massa, F., Sibaev, A., Marsicano, G., Blaudzun, H., Storr, M., Lutz, B. (2006) *Journal of Molecular Medicine,* **84**, 142–146.

108 Lopez-Carrillo, L., Lopez-Cervantes, M., Robles-Diaz, G., Ramirez-Espitia, A., Mohar-Betancourt, A., Meneses-Garcia, A., Lopez-Vidal, Y., Blair, A. (2003) *International Journal of Cancer,* **106**, 277–282.

109 For reviews on this debated issue, see:(a) Surh, Y. J. and Lee, S. S. (1996) *Food and Chemical Toxicology,* **34**, 313–316. (b) Archer, V. E. and Jones, D. W.*Medical Hypotheses,* **59**, (2002) 450–457.

110 Surh, Y.-J. (2002) *Journal of the National Cancer Institute,* **94**, 1263–1265. For recent results on the potential of capsaicin against prostate cancer, see: Mori, A., Lehmann, S., O'Kelly, J., Kumagai, T., Desmond, J. C., Pervan, M., McBride, W. H., Kizaki, M., Koeffler, H. P.*Cancer Research,* **66**, (2006) 3222–3229.

111 Noguchi, M., Ikarashi, Y., Yuzurihara, M., Kase, Y., Takeda, S., Aburada, M. (2003) *Journal of Pharmacological Sciences,* **93**, 80–86.

112 Chen, S. C., Chang, T. J., Wu, F. S. (2004) *The Journal of Pharmacology and Experimental Therapeutics,* **311**, 529–536.

113 Wahlqvist, M. L. and Wattanapenpaiboon, N. (2001) *The Lancet,* **358**, 348–349.

114 Tokuyama, N.Kaju, T. (2000) Health Beverages Containing Capsaicin. JP 2000 189,121.

115 (a) Yoshioka, M., St-Pierre, S., Suzuki, M., Tremblay, A. (1998) *British Journal of Nutrition,* **80**, 503–510. (b) Westerterp-Plantenga, M. S., Smeets, A., Lejeune, M. P. G. (2005) *International Journal of Obesity,* **29**, 682–688.

116 (a) Yoshioka, M., St.-Pierre, S., Drapeau, V., Dionne, I., Doucet, E., Suzuki, M., Tremblay, A. (1999) *British Journal of Nutrition,* **82**, 115–123. (b) Yoshioka, M., Imanaga, M., Ueyama, H., Yamane, M., Kubo, Y., Boivin, A., St-Amand, J., Tanaka, H., Kiyonaga, A. (2004) *British Journal of Nutrition,* **91**, 991–995.

117 Ohnuki, K., Haramizu, S., Oki, K., Watanabe, T., Yazawa, S., Fushiki, T. (2001) *Bioscience Biotechnology and Biochemistry,* **65**, 2735–2740.

118 Hamada, H. and Takeda, T. (2002) *Food Style,* **21**, 69–71.

119 Okada, S., Kanbara, I., Yonetani, T., Tanimoto, S., Nishimura, T. (1993) Glycosides of Capsaicin and Dihydrocapsaicin Manufacture with Plant Tissue Culture. JP 93-254653.

120 Kobata, K., Kawaguchi, M., Watanabe, T. (2002) *Bioscience Biotechnology and Biochemistry,* **66**, 319–327.

121 Trevisani, M., Smart, D., Gunthorpe, M. J., Tognetto, M., Barbieri, M., Campi, B., Amadesi, S., Gray, J., Jerman, J. C., Brough, S. J., Owen, D., Smith, G. D., Randall, A. D., Harrison, S., Bianchi, A., Davis, J. B., Geppetti, P. (2002) *Nature Neuroscience,* **5**, 546–551.

122 Watanabe, T., Kobata, K., Morita, A., Iwasaki, Y. (2005) *Food and Food Ingredients,* **210**, 214–221.

123 Rozin, P. and Schiller, P. (1980) *Motivation and Emotion,* **4**, 77–101.

124 Matsuda, L. A., Lolait, S. J., Brownstein, A. C., Young, A. C., Bonner, I. I. (1990) *Nature,* **346**, 561–564.

125 For a general review on TRP receptors, see: Clapham, D. E. (2003) *Nature,* **426**, 517–524.

5
Glycosidase-Inhibiting Alkaloids: Isolation, Structure, and Application

Naoki Asano

5.1
Introduction

A large number of alkaloids mimicking the structures of monosaccharides or oligosaccharides have been isolated from plants and microorganisms [1,2]. Such alkaloids are easily soluble in water because of their polyhydroxylated structures and inhibit glycosidases because of a structural resemblance to the sugar moiety of the natural substrate. Glycosidases are involved in a wide range of important biological processes, such as intestinal digestion, posttranslational processing of the sugar chain of glycoproteins, quality-control systems in the endoplasmic reticulum (ER) and ER-associated degradation mechanism, and the lysosomal catabolism of glycoconjugates. Inhibition of these glycosidases can have profound effects on carbohydrate catabolism in the intestines, on the maturation, transport, and secretion of glycoproteins, and can alter cell–cell or cell–virus recognition processes. The realization that glycosidase inhibitors have enormous therapeutic potential in many diseases such as diabetes, viral infection, and lysosomal storage disorders has led to increasing interest in and demand for them [3,4]. In recent years, combinatorial methods and the rapid generation of large libraries of potential lead compounds have been favored for drug discovery. However, the investigation of natural products is still continuing to inspire the development of new drugs, because of their structural and biological diversity.

Glycosidase-inhibiting alkaloids from plants are classified into five structural classes: polyhydroxylated pyrrolidines, piperidines, indolizidines, pyrrolizidines, and nortropanes. Furthermore, they also occur as the glycosides. This review describes recent studies on isolation, characterization, glycosidase inhibitory activity, and therapeutic application of the sugar-mimicking alkaloids from plants.

5.2
Isolation and Structural Characterization

For a long time (and also currently), drug discovery programs have typically used organic solvents such as methanol, butanol, ethyl acetate, chloroform, or hexane for

Modern Alkaloids: Structure, Isolation, Synthesis and Biology. Edited by E. Fattorusso and O. Taglialatela-Scafati

ISBN: 978-3-527-31521-5

extraction. Hence, most of the water-soluble compounds have escaped detection and isolation. However, most preparations used in traditional medicines are formulated in water or hot water.

Polyhydroxylated alkaloids and N-containing sugar analogs are usually extracted with 50% aqueous MeOH or 50% aqueous EtOH and isolated by chromatography, using a variety of ion-exchange resins such as Amberlite IR-120B (H^+ form), Amberlite CG-50 (NH_4^+ form), CM-Sephadex C-25 (NH_4^+ form), and Dowex 1-X2 (OH^- form). Their structures are elucidated using various spectroscopic techniques including UV, IR, MS, one-dimensional NMR, and two-dimensional NMR such as COSY, HMBC, NOESY, and so on.

5.2.1 Deoxynojirimycin and Related Compounds

1-Deoxynojirimycin (DNJ, **1**) was originally prepared by catalytic hydrogenation of nojirimycin, which was discovered as the first glucose-mimicking antibiotic produced by *Streptomyces* spp., with a platinum catalyst or by chemical reduction with $NaBH_4$ [5,6]. Later it was isolated from the root bark of mulberry trees and called moranoline [7]. DNJ is also produced by many strains in the genera *Bacillus* and *Streptomyces* [8,9].

5.2.1.1 Isolation from *Morus* spp. (Moraceae)

Mulberry trees (*Morus* spp.) are cultivated in China, Korea, and Japan, and their leaves are used to feed silkworms (*Bombyx mori*). Mulberry leaves have been used traditionally in Chinese herbal medicine to cure and prevent "Xiao-ke" (diabetes). The root bark of mulberry trees has been used as a Chinese herbal medicine called "Sang-bai-pi" (Japanese name "Sohakuhi") for anti-inflammatory, diuretic, antitussive, and antipyretic purposes, while the fruits are used as both tonic and sedative. In 1994, the improvement of the purification procedures using a variety of ion-exchange resins described above led to the isolation of a number of water-soluble alkaloids from the genus *Morus* (Moraceae) [10,11]. Seven alkaloids were isolated from the leaves of *Morus bombycis* in Japan: DNJ (**1**), *N*-methyl-DNJ (**2**), and fagomine (1,2-dideoxynojirimycin) (**3**), 2-*O*-α-D-galactopyranosyl-DNJ (**4**), which are polyhydroxylated piperidine derivatives; 1,4-dideoxy-1,4-imino-D-arabinitol (DAB, **5**) which is a polyhydroxylated pyrrolidine and 2-*O*-β-D-glucopyranosyl-DAB (**6**) and calystegine B_2 (**7**) which are polyhydroxylated nortropane alkaloids [10]. Furthermore, eighteen alkaloids including the seven alkaloids described above were isolated from the root bark of *M. alba* in China [11]. Additional alkaloids are 3-*epi*-fagomine (**8**), 1,4-dideoxy-1,4-imino-D-ribitol (**9**), calystegine C_1 (**10**), and eight glycosides of DNJ: the 6-*O*-α-D-galactopyranosyl (**11**), 2-*O*-α-D-glucopyranosyl (**12**), 3-*O*-α-D-glucopyranosyl (**13**), 4-*O*-α-D-glucopyranosyl (**14**), 2-*O*-β-D-glucopyranosyl (**15**), 3-*O*-β-D-glucopyranosyl (**16**), 4-*O*-β-D-glucopyranosyl (**17**), and 6-*O*-β-D-glucopyranosyl (**18**). In 2001, calystegine B_1 (**19**), 4-*O*-β-D-glucopyranosyl fagomine (**20**), (2*R*, 3*R*, 4*R*)-2-hydroxymethyl-3,4-dihydroxypyrrolidine-*N*-propionamide (**21**) in addition to alkaloids **1–9**, **12**, and **16–18** were isolated from the root bark of *M. alba* from An-Huei-Shong in China, which has a very high

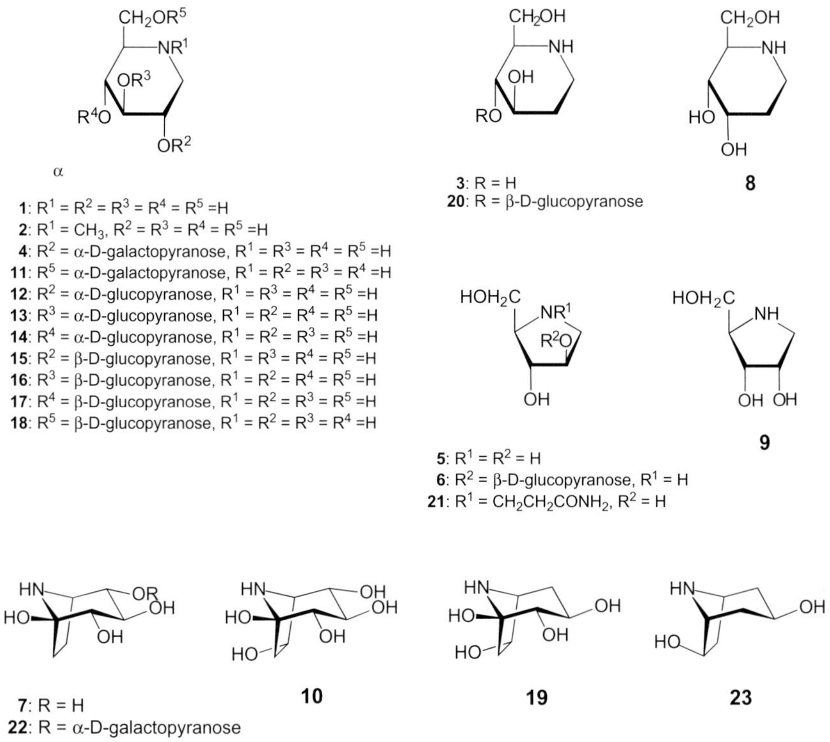

α

1: $R^1 = R^2 = R^3 = R^4 = R^5 = H$
2: $R^1 = CH_3$, $R^2 = R^3 = R^4 = R^5 = H$
4: R^2 = α-D-galactopyranose, $R^1 = R^3 = R^4 = R^5 = H$
11: R^5 = α-D-galactopyranose, $R^1 = R^2 = R^3 = R^4 = H$
12: R^2 = α-D-glucopyranose, $R^1 = R^3 = R^4 = R^5 = H$
13: R^3 = α-D-glucopyranose, $R^1 = R^2 = R^4 = R^5 = H$
14: R^4 = α-D-glucopyranose, $R^1 = R^2 = R^3 = R^5 = H$
15: R^2 = β-D-glucopyranose, $R^1 = R^3 = R^4 = R^5 = H$
16: R^3 = β-D-glucopyranose, $R^1 = R^2 = R^4 = R^5 = H$
17: R^4 = β-D-glucopyranose, $R^1 = R^2 = R^3 = R^5 = H$
18: R^5 = β-D-glucopyranose, $R^1 = R^2 = R^3 = R^4 = H$

3: R = H
20: R = β-D-glucopyranose

8

5: $R^1 = R^2 = H$
6: R^2 = β-D-glucopyranose, $R^1 = H$
21: $R^1 = CH_2CH_2CONH_2$, $R^2 = H$

9

7: R = H
22: R = α-D-galactopyranose

10

19

23

Fig. 5.1 Structures of N-containing sugars from mulberry trees (*Morus* spp.).

content (0.165% of dry weight) of DNJ [12]. The mulberry fruits have been commonly used as jam as well as a Chinese herbal medicine called "Sang-zi", and two new alkaloids.

4-*O*-α-D-galactopyranosylcalystegine B_2 (**22**) and 3β,6β-dihydroxynortropane (**23**), were isolated from the fruits of *M. alba* [12]. DNJ is present in high concentrations in all parts of the mulberry tree. Interestingly, silkworms feed exclusively on its leaves and appear to accumulate DNJ in their bodies since the DNJ content in silkworms is 2.7-fold higher than that in the leaves [12] (Figure 5.1).

5.2.1.2 Isolation from Thai Medicinal Plants "Thopthaep" and "Cha Em Thai"

1-Deoxymannojirimycin (DMJ, 1,5-dideoxy-1,5-imino-D-mannitol, **24**) was first isolated from the seeds of *Lonchocarpus sericeus* (Leguminosae), native to the West Indies and tropical America [13]. Later, it was also found in the other legumes *Angylocalyx* spp. [14] and *Derris malaccensis* [15], and the Euphorbiaceae family plants *Omphalea diandra* [16] and *Endospermum medullosum* [17]. DMJ is usually coproduced with the pyrrolidine alkaloid DMDP (2,5-dideoxy-2,5-imino-D-mannitol, **25**). This coproduction of DMJ and DMDP could arise from the six- or five-membered ring closure of a common precursor. Our group has isolated DMJ and DMDP in high yields of 0.05 and 0.04%, respectively, from "Thopthaep" [18], *Connarus ferrugineus* (Connaraceae) [18], which is used traditionally as ointment to treat scabies, and as an oral drug to treat stomach ache and constipation. This Thai medicinal plant produces DNJ, 1-deoxyallonijirimycin (**26**), 1-deoxyaltronojirimycin (**27**), 1,4-dideoxymanno-jirimycin (**28**), 1,4-dideoxyallonojirimycin (**29**), 1,4-dideoxyaltronojirimycin (**30**), 2,5-dideoxy-2,5-imino-D-glucitol (DIDG, **31**), 2-*O*-α-D-galactopyranosyl-DMJ (**32**), and 3-*O*-β-D-glucopyranosyl-DMJ (**33**), together with DMJ and DMDP (Figure 5.2).

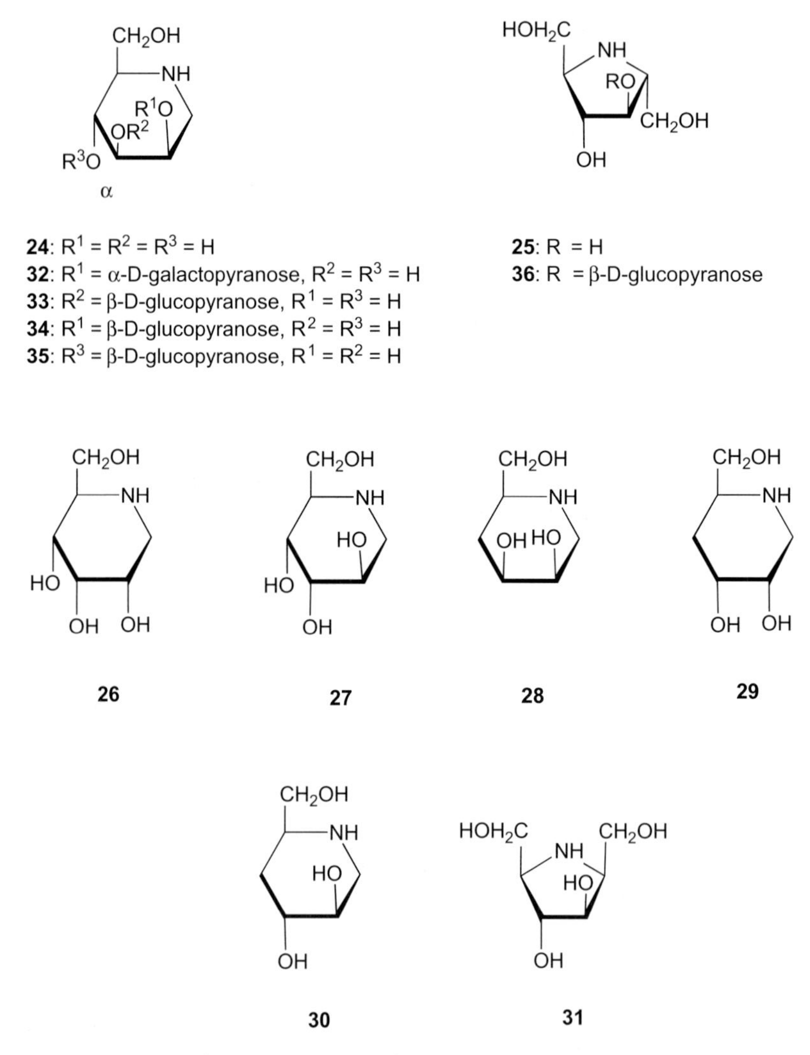

Fig. 5.2 Structures of N-containing sugars from Thai medicinal plants.

Recently, D- and L-enantiomers of DNJ and its six epimers other than 1-deoxytalonojirimycin have been synthesized enantiospecifically [19–21]. Consequently, the absolute configurations of natural DNJ, DMJ, 1-deoxyallonojirimycin, and 1-deoxyaltronojirimycin were determined to be D from the value and sign of the optical rotation [21]. DIDG has been previously prepared by chemoenzymatic approaches and the optical rotation of the natural product was in agreement with that of the synthetic compound [22–26].

A Thai traditional crude drug "Cha em thai," obtained from *Albizia myriophylla* (Leguminosae), is used to relieve thirst and sore throats and as a substitute for licorice owing to its sweet taste [27]. Several triterpene saponins (albiziasaponins A–E) have been characterized as sweet-tasting substances [28]. The wood of this plant was found to have the high DMJ content of 0.168% (dry weight) [18]. The ion-exchange resin chromatography of the 50% aqueous MeOH extract led to the isolation of **24**, **25**, **31**, 2-*O*-β-D-glucopyranosyl-DMJ (**34**), 4-*O*-β-D-glucopyranosyl-DMJ (**35**), and 3-*O*-β-D-glucopyranosyl-DMDP (**36**). The isolation of **36** from this plant is the first report of a glycoside of DMDP.

5.2.2
α-Homonojirimycin and Related Compounds

In 1988, α-homonojirimycin (α-HNJ, **37**) was isolated from the neotropical liana, *Omphalea diandra* (Euphorbiaceae), as the first example of a naturally occurring DNJ derivative with a carbon substituent at C-1 [16]. However, before the isolation of the natural product, the 7-*O*-β-D-glucopyranosyl-α-HNJ (**38**) had been designed and prepared as a potential drug for the treatment of diabetes [29,30]. α-HNJ has been detected in adults, pupae, and eggs of the neotropical moth, *Urania fulgens*, whose larvae feed on *O. diandra*, and the level (percentage dry weight) of α-HNJ in pupae was about 0.5% [17]. Until 1990, the known natural occurrence of α-HNJ had been strictly limited to the Euphorbiaceae family plants.

5.2.2.1 Isolation from Garden Plants

Aglaonema treubii (Araceae) is a very common indoor foliage plant and a native of the tropical rainforests of South-East Asia. In 1997, a 50% aqueous EtOH extract of *A. treubii* was found to potently inhibit α-glucosidase and subjected to various ion-exchange column chromatographic steps to give **37** (0.01% fresh weight), **38**, 5-*O*-α-D-galactopyranosyl-α-HNJ (**39**), β-homonojirimycin (**40**), α-homomannojirimycin (**41**), β-homomannojirimycin (**42**), α-homoallonojirimycin (**43**), and β-homoaltronojirimycin (**44**), together with **25** [31]. Although the structure of **43** was originally reported to be α-3,4-di*epi*-homonojirimycin, it was later revised to be α-4-*epi*-homonojirimycin on the basis of NMR analysis and synthetic studies [32]. Alkaloids **40**, **41**, and **42** were chemically synthesized before their isolation as natural products [33–36] (Figure 5.3).

Hyacinthus orientalis (Hyacinthaceae) is a plant native to North Africa and Eurasia, and commonly known as hyacinth. A search for polyhydroxylated alkaloids in species of the Hyacinthaceae family by GC–MS led to isolation and characterization of eleven

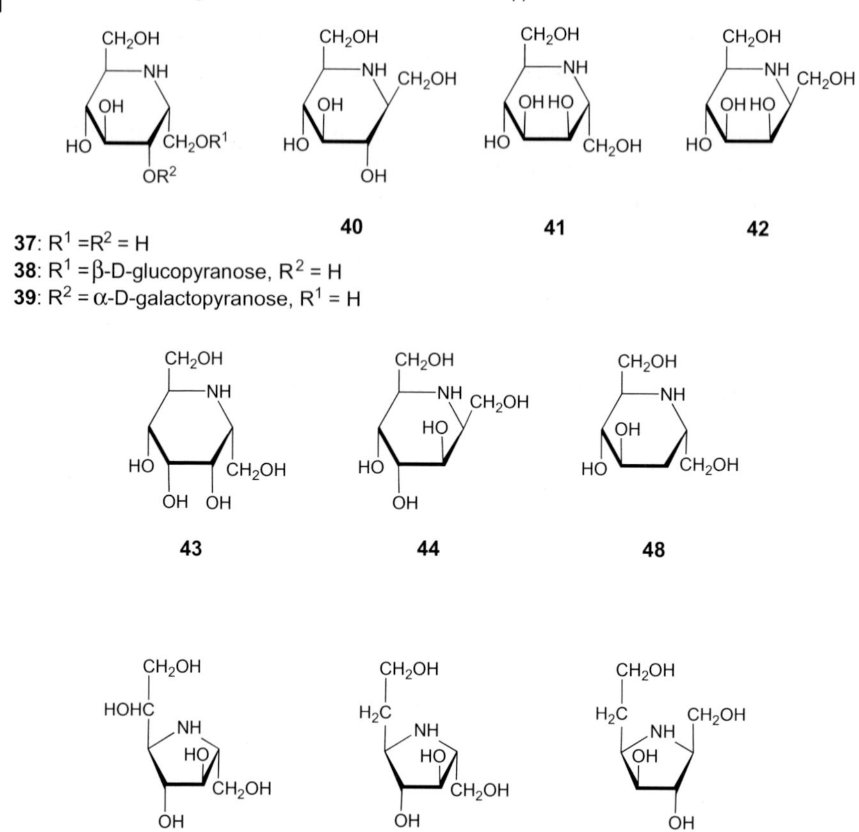

Fig. 5.3 Structures of N-containing sugars from α-homonojirimycin-producing plants.

alkaloids including α-HNJ [37]. A 50% aqueous MeOH extract from the bulbs of *H. orientalis* was subjected to ion-exchange chromatographies to give **37** (0.032% fresh weight), **38**, **40**, **41**, and **42**, together with **1**, **24**, **25**, homoDMDP (**45**), 6-deoxy-homoDMDP (**46**), and 2,5-imino-2,5,6-trideoxy-D-*gulo*-heptitol (**47**). The structure of homoDMDP has been assigned as 2,5-dideoxy-2,5-imino-*glycero*-D-*manno*-heptitol or its enantiomer from its NMR spectroscopic data. However, the relative configuration at C-6 was not determined since it could not be deduced from NMR data. In the course of the synthesis of a series of polyhydroxylated pyrrolidine alkaloids, Takebayashi *et al.* reported the enantiospecific synthesis of (1′*S*,2*R*,3*R*,4*R*,5*R*)-3,4-dihydroxy-2-(1,2-hydroxyethyl)pyrrolidine [38], and this compound and homoDMDP were found to be identical from comparison of their ^{1}H NMR and ^{13}C NMR spectroscopic data and optical rotation values. Hence, the structure of homoDMDP was determined to be 2,5-dideoxy-2,5-imino-D-*glycero*-D-*manno*-heptitol. α-HNJ and homoDMDP could be synthesized from a common precursor in the biosynthetic pathway.

5.2.2.2 Isolation from the Thai Medicinal Plant "Non Tai Yak"

In the course of a search for α-glucosidase inhibitors, Kitaoka *et al.* found that such inhibitors are present in some Thai traditional crude drugs [39]. For example, α-HNJ occurs in "Non tai yak" at a level of 0.1% (dry weight). The "Non tai yak" sample is known to be *Stemona tuberosa* (Stemonaceae), which has been used in China and Japan for various medicinal purposes. In particular, an extract from the fresh tuberous roots of *S. tuberosa* is used to treat respiratory disorders, including pulmonary tuberculosis and bronchitis, and is also recommended as an insecticide [27,40,41]. In 2005, re-examination of polyhydroxylated alkaloids in *S. tuberosa* led to the isolation of thirteen alkaloids, **1**, **24**, **25**, **31**, **36**, **37** (0.1% of dry weight), **38**, **40**, **41**, **42**, **43**, **44**, and α-5-deoxy-HNJ (α-1-*C*-hydroxymethylfagomine, **48**) [18]. The ^{1}H NMR, ^{13}C NMR, and optical rotation spectroscopic data of α-7-deoxy-HNJ were superimposable with those of the synthetic 2,3,6-trideoxy-2,6-imino-D-*manno*-heptitol [42].

5.2.2.3 Isolation from *Adenophora* spp. (Campanulaceae)

In 2000, adenophorine (**49**), a rare example of an α-HNJ homolog with a hydrophobic alkyl substituent at the pseudoanomeric position, was isolated from a commercially available Chinese crude drug "Sha-sheng," the roots of *Adenophora* spp. (Campanulaceae) [43]. This crude drug additionally contains 1-deoxyadenophorine (**50**), 5-deoxyadenophorine (**51**), α-1-*C*-ethylfagomine (**52**), β-1-*C*-butyldeoxygalactonojirimycin (**53**), 1-*O*-β-D-glucopyranosyladenophorine (**54**), and 1-*O*-β-D-glucopyranosyl- 5-deoxyadenophorine (**55**), together with **24** and **25**. In 2003, Davies *et al.* reported the first synthesis of (−)-adenophorine and assigned the absolute configuration of the natural product (+)-adenophorine to the enantiomer of the synthetic compound [44] (Figure 5.4).

Adenophora triphylla var. *japonica*, which was grown at the medicinal plant garden in Japan, contains 6-*C*-butyl-DMDP (**56**) in addition to **1**, **5**, **24**, **25**, and **52** [45]. Interestingly, 6-*C*-butyl-DMDP and β-1-*C*-butyldeoxygalactonojirimycin possess the same carbon backbone. Hence, both alkaloids could be biosynthesized by the five- or six-membered ring closure of a common precursor.

5.2.3 Indolizidine and Pyrrolizidine Alkaloids

Certain poisonous plants often cause serious livestock losses. The Australian legume, *Swainsona*, is known as "poison peas," and sheep eating them develop a syndrome called "pea struck" [46,47]. Livestock is also poisoned by the closely related *Astragalus* and *Oxytropis* species, which are found throughout the world, and intoxication of livestock by some species known as locoweeds in the western

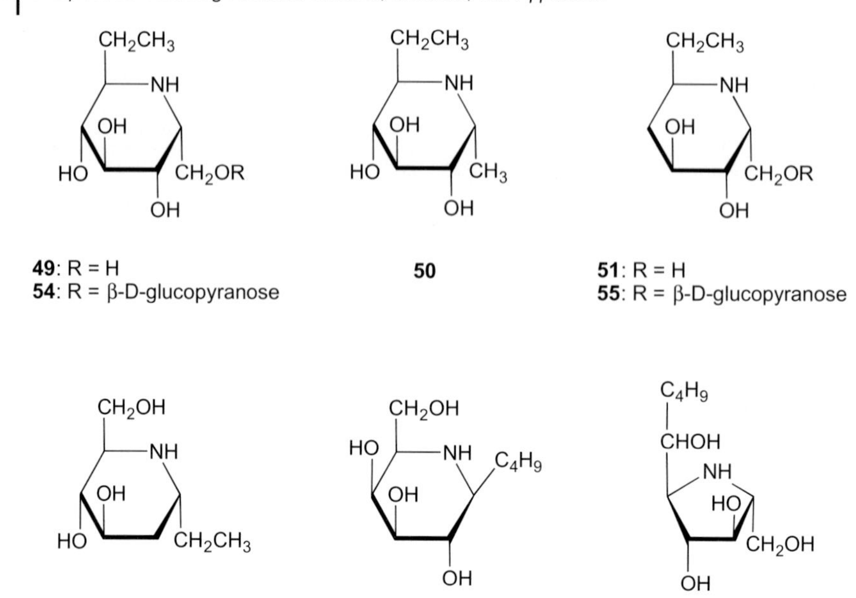

Fig. 5.4 Structures of α-homonojirimycin analogs from *Adenophora* spp. (Campanulaceae).

United States is called "locoism" [47,48]. The poisoning is characterized by cytoplasmic vacuolation of neuronal cells due to accumulation of mannose-rich oligosaccharides in lysosomes [49]. The trihydroxyindolizidine alkaloid swainsonine (**57**) occurs in these legumes and has been identified as a causative agent in locoism [48,50]. The toxicity of the other legume *Castanospermum australe* for livestock led to the isolation of the toxic principle castanospermine (**58**) [51] and these two alkaloids gave rise to a great impetus in research on N-containing sugars and their application.

5.2.3.1 **Isolation from the Leguminosae Family**

In 1981, castanospermine was first isolated from the immature seeds of *Castanospermum australe*, with the yield of 0.057% [51]. X-Ray crystallography showed that the stereogenic centers of the six-membered ring of castanospermine correspond to the *gluco* configuration [51], while 6-*epi*-castanospermine (**59**) isolated later from the seeds has the D-*manno* configuration in the piperidine ring [52]. *C. australe* coproduces 7-deoxy-6-*epi*-castanospermine (**60**) [53] and 6,7-di*epi*-castanospermine (**61**) [54]. We have isolated a new castanospermine isomer 6,8-di*epi*-castanospermine (**62**) from the leaves and twigs, which contain castanospermine at the high level of 1% (unpublished data). Lentiginosine (**63**) and 2-*epi*-lentiginosine

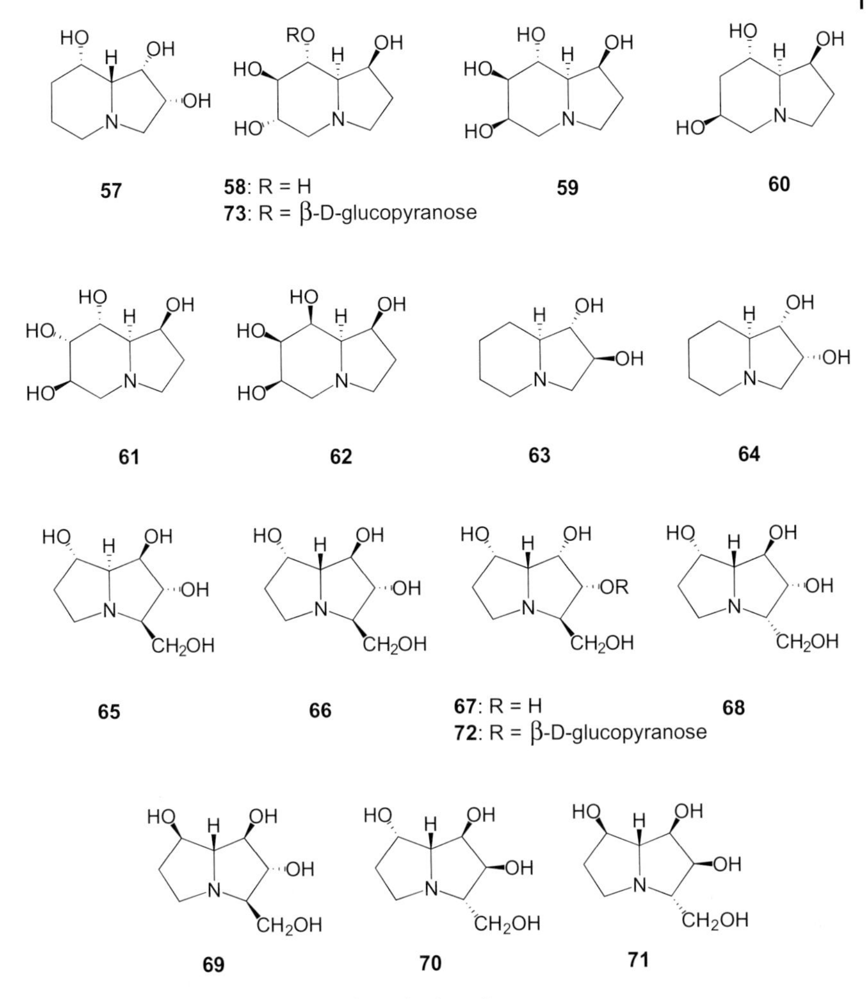

Fig. 5.5 Structures of indolizidines and pyrrolizidines from legumes.

(**64**) have been isolated from the leaves of *Astragalus lentiginosus* and these two dihydroxyindolizidines are probably biosynthesized from 1-hydroxyindolizidine by hydroxylation at C-2 [55] (Figure 5.5).

In 1988, alexine (**65**), a polyhydroxylated pyrrolizidine alkaloid, was isolated from the pods of legume *Alexa leiopetala* [56]. Although the broad class of pyrrolizidine alkaloids bear a carbon substituent at C-1 [57,58], alexine is the first example of a pyrrolizidine alkaloid with a carbon substituent at C-3. At about the same time, australine (**66**) was isolated from the seeds of *C. australe* and found to be 7a-*epi*-alexine by X-ray crystallographic analysis [59]. The isolation of 1-*epi*-australine (**67**),

3-*epi*-australine (**68**), and 7-*epi*-australine (**69**) from the same plant was later reported [60–62]. The structure of 1-*epi*-australine was firmly established by X-ray crystallographic analysis of the corresponding 1,7-isopropylidene derivative [61], and the absolute configurations of 3-*epi*-australine were also identified by X-ray crystal structure analysis [61]. Alkaloid **69** was tentatively assigned as 7-*epi*-australine, based on the difference between its NMR parameters and those reported for australine [62]. The unambiguous synthesis of australine [63] and 7-*epi*-australine [64,65] and extensive NMR studies on the natural and synthetic isomers of australine [65] by Denmark *et al.* elucidated that the natural product reported as 7-*epi*-australine is really australine. This means that 7-*epi*-austaline has not yet been found as a natural product. Reinvestigation of the natural occurrence of 7-*epi*-austaline in *C. australe* led to the isolation of new alkaloids, 2,3-di*epi*-australine (**70**), 2,3,7-tri*epi*-australine (**71**), 2-*O*-β-D-glucopyranosyl-1-*epi*-australine (**72**), and 8-*O*-β-D-glucopyranosylcastanospermine (**73**) [66]. Alkaloid **73** is the first naturally occurring glycoside of castanospermine.

5.2.3.2 Isolation from the Hyacinthaceae Family

Polyhydroxylated pyrrolizidine alkaloids with a hydroxymethyl substituent at C-3 have been thought to be of very restricted natural occurrence. The alexines and australines have been reported in only two small genera of the Leguminosae (*Castanospermum* and *Alexa*). However, a number of such pyrrolizidine alkaloids were found from the quite different family Hyacinthaceae. In 1999, new polyhydroxylated pyrrolizidines different from alexines and australines were isolated from Hyacinthaceae family plants and designated as hyacinthacines: hyacinthacines B_1 (**74**) and B_2 (**75**) from the immature fruits and stalks of *Hyacinthoides non-scripta* and hyacinthacine C_1 (**76**) from the bulbs of *Scilla campanulata* [67]. Shortly after their isolation, four new hyacinthacines A_1 (**77**), A_2 (**78**), A_3 (**79**), and B_3 (**80**) were isolated from the bulbs of *Muscari armeniacum* in addition to **75** [68]. In 2001, Martin and coworkers reported the first synthesis of (+)-hyacinthacine A_2 from commercially available 2,3,5-tri-*O*-benzyl-D-arabinofuranose, and confirmed the absolute configuration of natural (+)-hyacinthacine A_2 as (1*R*,2*R*,3*R*,7a*R*)-1,2-dihydroxy-3-hydroxymethylpyrrolizidine [69]. Subsequently, the natural (+)-hyacinthacine A_3 was enantiospecifically synthesized from an adequately protected DMDP (**25**) and its absolute configuration was determined to be (1*R*,2*R*,3*R*,5*R*,7a*R*)-1,2-dihydroxy-3-hydroxymethyl-5-methylpyrrolizidine [70]. More recently, the natural (+)-hyacinthacine A_1 has been determined to be (1*S*,2*R*,3*R*,7a*R*)-1,2-dihydroxy-3-hydroxymethylpyrrolizidine from total synthesis [71]. Many species of the genera *Muscari* and *Scilla* are very common as garden plants. The GC-MS analysis of the extract of commercially available bulbs of *S. sibirica* demonstrated the existence of many kinds of polyhydroxylated alkaloids, five pyrrolidines and two pyrrolidine glycosides, six piperidines and one piperidine glycoside, and eight pyrrolizidines [72]. Surprisingly, seven pyrrolizidines other than the known alkaloid **73** were new hyacinthacines. They are hyacinthacines A_4 (**81**), A_5 (**82**), A_6 (**83**), A_7 (**84**), B_4 (**85**), B_5 (**86**), and B_6 (**87**) (Figure 5.6).

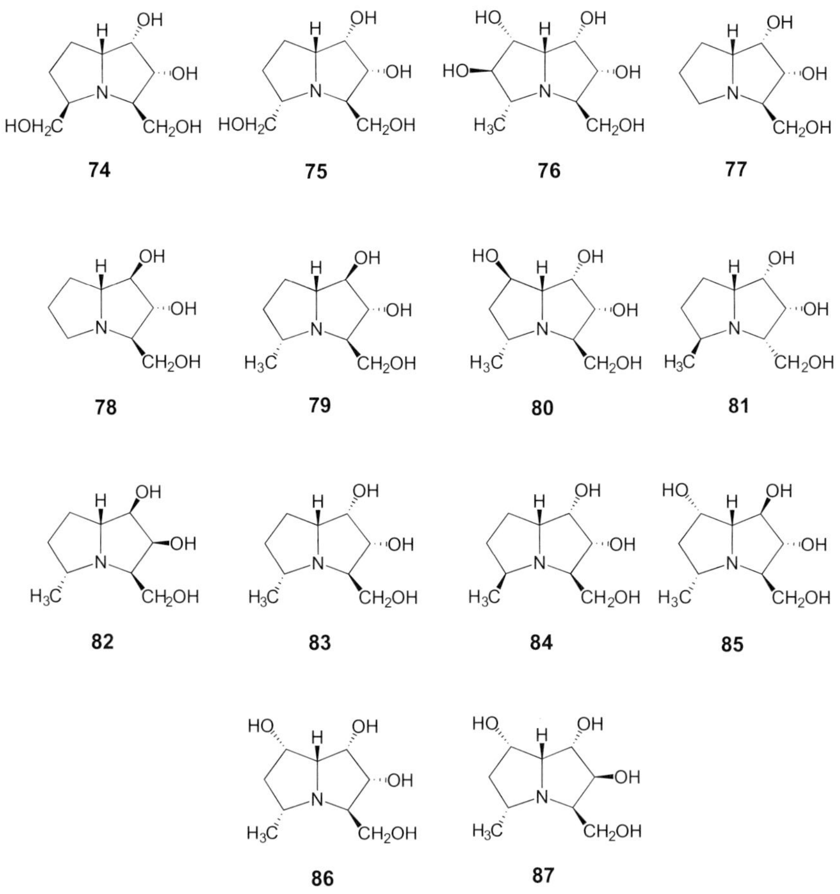

Fig. 5.6 Structures of pyrrolizidines from the Hyacinthaceae family.

In 1999, it was reported that *Broussonetia kajinoki* (Moraceae) produces a pyrrolizidine alkaloid with the C_{10} side chain at the C-5α position, which was designated as broussonetine N (**88**), together with pyrrolidine alkaloids with a long (C_{13}) side chain [73]. Broussonetine N can be regarded as the α-5-*C*-(1,19-dihydroxy-6-oxodecyl)-hyacinthacine A_2. In 2004, four new hyacinthacine A_1 and 7-*epi*-australine derivatives with the hydroxybutyl side chain at the C-5α position from the bulbs of *S. peruviana*, which also coproduces pyrrolidine alkaloids with a highly hydroxylated long side chain [74]. These four pyrrolizidine alkaloids were determined to be α-5-*C*-(3-hydroxybutyl)-7-*epi*-australine (**89**), α-5-*C*-(3-hydroxybutyl)- hyacinthacine A_1 (**90**), α-5-*C*-(1,3-dihydroxybutyl)-hyacinthacine A_1 (**91**), and α-5-*C*-(1,3,4-trihydroxybutyl)-hyacinthacine A_1 (**92**). NOE experiments on pyrrolizidines **91** and **92** suggested the configurations at C-1′ (Figure 5.7).

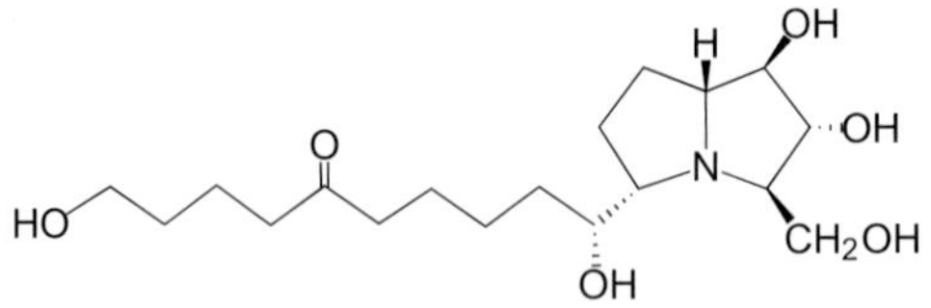

88

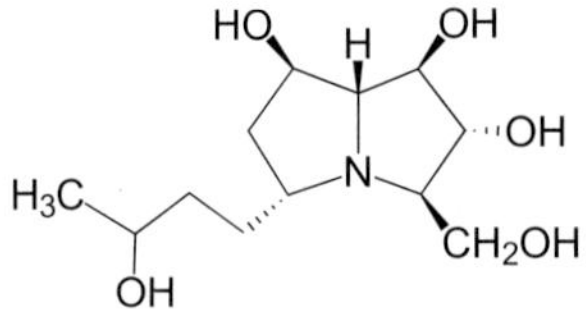

89

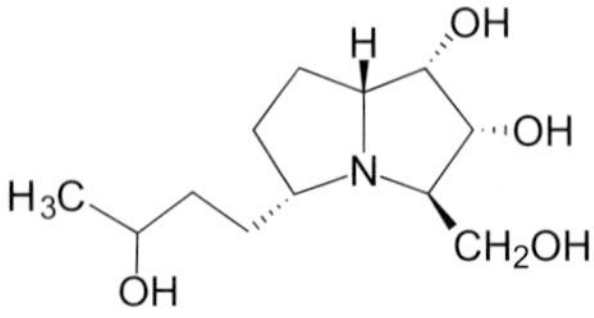

90

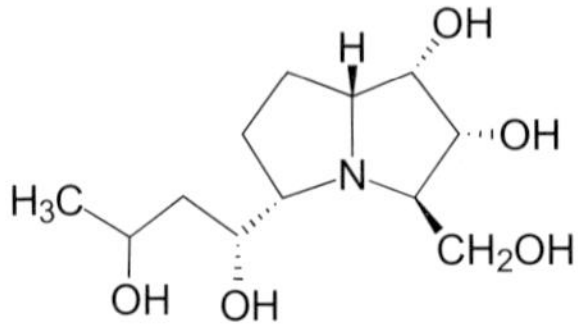

91

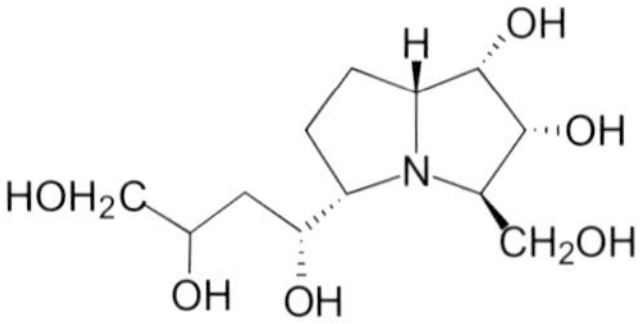

92

Fig. 5.7 Structures of pyrrolizidines with a long side chain.

5.2.4
Nortropane Alkaloids

Before the isolation of calystegines B_1 (**19**), B_2 (**7**), and C_1 (**10**) from *Morus* spp. [10–12], Tepfer *et al.* had reported the presence of a group of compounds in plants of the Convolvulaceae and Solanaceae familes and designated these compounds as calystegines [75]. Until the discovery of calystegines, four structural classes were encompassed as naturally occurring *N*-containing sugars: polyhydroxylated piperidines, pyrrolidines, indolizidines, and pyrrolizidines. Calystegines possess three structural features in common: a nortropane ring system; two to four secondary hydroxyl groups varying in position and stereochemistry; and a novel aminoketal functionality, which generates a tertiary hydroxyl group at the bicyclic ring bridgehead. The known calystegines have been subdivided into three groups on the basis of the number of hydroxyl groups present, namely calystegines A with three OH groups, B with four OH groups, and C with five OH groups. Tropane alkaloids bear a methyl substituent on the nitrogen atom, while nortropane alkaloids lack the

N-methyl group and occur occasionally as minor constituents in plants producing tropane alkaloids [76,77]. A recent survey of the occurrence of calystegines in Solanaceae and Convolvulaceae plants discovered that they are widely distributed in these families [1,2,78–80].

5.2.4.1 Isolation from the Solanaceae Family

The occurrence in Solanaceae is documented for 12 genera: *Atropa, Brunfelsia, Datura, Duboisia, Hyoscyamus, Lycium, Mandragora, Nicandra, Physalis, Scopolia, Solanum,* and *Withania* [2,79]. Our group isolated and characterized calystegines A_3 (**93**), A_5 (**94**), B_1, B_2, and B_3 (**95**) from the roots of *Physalis alkekengi* var. *francheti* [81], calystegines A_5, A_6 (**96**), B_1, B_2, B_3, and N_1 (**97**) from the whole plant of *Hyoscyamus niger* [82], calystegines A_3, A_5, B_1, B_2, B_3, B_4 (**98**), and C_1 from the roots of *Scopolia japonica* [83], and calystegines B_1, B_2, B_4, C_1, and C_2 (**99**) from the leaves and twigs of *Duboisia leichhardtii* [84]. An examination of the roots of *Lycium chinense* led to the discovery of two new calystegines A_7 (**100**) and B_5 (**101**), and two novel tropane alkaloids *N*-methylcalystegines B_2 (**102**) and C_1 (**103**), unlike the previously reported nortropane alkaloids [85]. Our recent work elucidated the presence of calystegine A_8 (**104**) in *H. niger,* and calystegine B_6 (**105**) in *S. japonica,* and calystegines C_3 (**106**) and N_2 (**107**) in *D. leichhardtii* (unpublished data). Calystegines N_1 and N_2 are assigned to an entirely new group of calystegines (the N series). FABMS analysis of calystegines N_1 and N_2 gives odd-numbered $[M+H]^+$ ions of m/z 175 and 191, owing to the replacement of an OH group by an NH_2 group relative to calystegines B_2 and C_1, respectively. The additional amino groups are located on C-1 in the parent alkaloids with the chemical shifts of the sole quaternary carbon at δ 78.3 (N_1) or 79.2 (N_2) in the ^{13}C NMR, in contrast to all other calystegines in which the hydroxyl-substituted quaternary carbon resonance occurs at an essentially invariant value of δ 93–94 ppm (Figure 5.8).

Besides free calystegines, calystegine B_1 occurs as the 3-*O*-β-D-glucoside in *Nicandra physalodes* fruits [86], and *Atropa belladonna* contains several glycosides including 3-*O*-β-D-glucopyranosyl-calystegine B_1 and 4-*O*-α-D-galactopyranosyl-calystegine B_2 [87]. The latter galactoside is also found in mulberry fruits, as previously described [12]. Microbial β-transglucosylation of calystegine B_1 or B_2 using whole cells of the yeast *Rhodotorula lactosa* gives 3-*O*-β-D-glucopyranosylcalystegine B_1 or 4-*O*-β-D-glucopyranosylcalystegine B_2, respectively [88]. Glucose transfer to calystegine B_1 by commercially available rice α-glucosidase provides 3-*O*-α-D-glucopyranosylcalystegine B_1, but this enzyme does not transfer D-glucose to calystegine B_2 [88]. The lack of α-glucosyl transfer to calystegine B_2 could be due to the inhibition of rice α-glucosidase by calystegine B_2.

Enantioselective syntheses of (+)- and (−)-calystegines B_2 have determined that (+)-calystegine, (1*R*,2*S*,3*R*,4*S*,5*R*)-1,2,3,4-tetrahydroxynortropane, is the natural molecule [89,90], while the absolute configuration of natural (−)-calystegine A_3 has been established as (1*R*,2*S*,3*R*,5*R*)-1,2,3-trihydroxynortropane by the syntheses

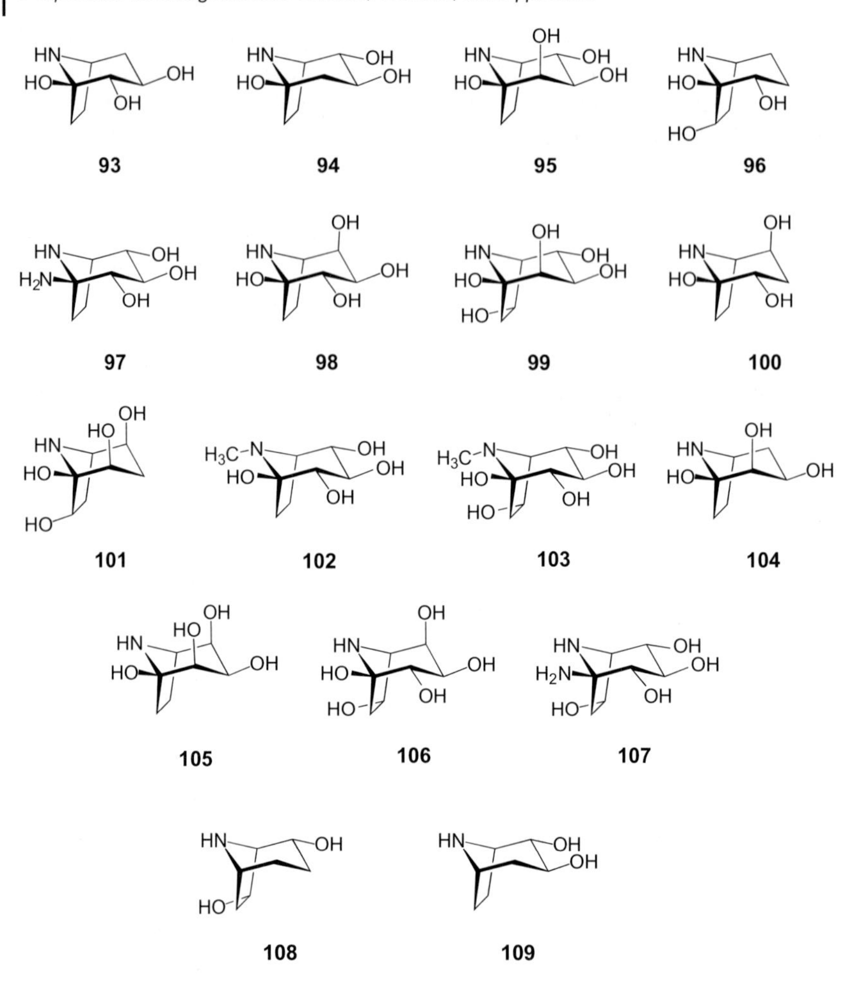

Fig. 5.8 Structures of nortropanes from the Solanaceae and Convolvulaceae families.

of both enantiomers [91]. Later, natural (+)-calystegine B_3 and (−)-calystegine B_4 were prepared from D-galactose and D-mannose, respectively, for the first time and their absolute configuration confirmed [92,93].

5.2.4.2 **Isolation from the Convolvulaceae Family**

Calystegines appear to be widely distributed in the Convolvulaceae family [78,80]. Eich and coworkers focused on the occurrence of seven calystegines (A_3, A_5, B_1, B_2, B_3, B_4, C_1) identified by GC-MS analysis with authentic samples as references and analyzed the extracts of 65 Convolvulaceae species from predominantly tropical habitats in all continents except Australia [78]. Consequently, they revealed their occurrence in 30 species belonging to 15 genera in this family. The qualitatively dominant alkaloid of the A series is calystegine A_3, detected in 43% of the 30 positive species followed by A_5 (20%). In the B series, calystegine B_2 exhibits the highest occurrence (86%) followed by B_1 (70%), B_3 (20%), and B_4 (13%). Calystegine C_1 is found in less than 1% of the species.

Convolvulaceae species produce polyhydroxylated alkaloids other than calystegines, as shown in Table 5.1. 2α,7β-Dihydroxynortropane (**108**) was isolated from seven species of Convolvulaceae, while only *Calystegia soldanella* contained both 2α,7β-dihydroxynortropane and 2α,3β-dihydroxynortropane (**109**) [80]. More interestingly, *Ipomoea carnea* produces the indolizidine alkaloids swainsonine (**57**) and 2-*epi*-lentiginosine (**64**) in addition to calystegines B_1, B_2, B_3, and C_1 [94]. *I. carnea* is a plant of tropical American origin but is now widely distributed in the tropical regions of the world. Chronic ingestion of *I. carnea* sometimes causes natural poisoning in livestock. Swainsonine is also found in other species of the Convolvulaceae family. *Ipomoea* sp. Q6 aff. *calobra* (Weir Vine) grows in a small area of southern Queensland in Australia and is reported to produce neurological disorders in livestock. The clinical symptoms are similar to those caused by swainsonine-containing legumes. Molyneux *et al.* detected swainsonine in the seeds, estimating the level as 0.058% by GC-MS, and also demonstrated the existence of swainsonine and calystegine B_2 in the seeds of *I. polpha* collected in the Northern Territory [95].

5.3 Biological Activities and Therapeutic Application

Glycosidase inhibitors are currently of great interest as potential therapeutic agents because they may be able to modify or block biological processes that can significantly affect carbohydrate anabolism and catabolism [96]. It has been increasingly realized that glycosidase inhibitors have enormous potential in many diseases such as diabetes, viral infection, and lysosomal storage disorders [1–3].

5.3.1 Antidiabetic Agents

5.3.1.1 α-Glucosidase Inhibitors

The intestinal oligo- and disaccharidases are fixed components of the cell membrane of the brush border region of the wall of the small intestine. These enzymes digest dietary carbohydrate to monosaccharides which are absorbed through the intestinal wall. They include sucrase, maltase, isomaltase, lactase, trehalase, and hetero-β-glucosidase. In the late 1970s, it was realized that inhibition of all or some of these activities could regulate the absorption of carbohydrate, and that these inhibitors could

Tab. 5.1 Distribution of polyhydroxylated nortropane and indolizidine alkaloids in the Convolvulaceae.

Plant	Organ[a]	108	109	A_3 (93)	A_5 (94)	B_1 (19)	B_2 (7)	B_3 (95)	B_4 (98)	C_1 (10)	Sw[b] *(57)*	El[c] *(64)*
Calystegia japonica Choisy	C	•	—[d]	•	—	•	•	—	—	—	—	—
Calystegia sepium (L.) R. Br.	D	•	—	•	—	•	•	—	—	—	—	—
Calystegia soldanella (L.) Roem. et Schult.	A	•	•	•	—	•	•	•	—	—	—	—
Ipomoea batatas Lam. var. *edulis* Makino	B	•	—	—	—	•	•	—	—	—	—	—
Ipomoea carnea Jacq. (Japan)	B	•	—	—	—	•	•	—	—	—	•	—
Ipomoea carnea Jacq. (Brazil)	B	—	—	—	—	•	•	•	—	•	•	•
Ipomoea nil (L.) Roth	A	—	—	—	—	—	—	—	—	—	—	—
Ipomoea obscura Ker.	A	—	—	—	—	•	•	•	•	•	—	—
Ipomoea pes-caprae (L.) Sweet	B	—	—	—	—	—	•	—	—	—	—	—
Ipomoea vitifolia Vanprul	B	•	—	—	—	—	•	•	•	•	—	—
Quamoclit angulata Bojer.	B	•	—	—	—	•	•	—	—	—	—	—

• present; — absent.
[a] A: whole parts; B: aerial parts; C: roots; D: root cultures.
[b] Swainsonine.
[c] 2-*epi*-Lentiginosine.
[d] Not determined.

be used therapeutically in the oral treatment of the noninsulin-dependent diabetes mellitus (NIDDM or type 2 diabetes) [97]. Acarbose (**110**) is a potent inhibitor of pig intestinal sucrase with an IC_{50} value of 0.5 μM and it was also effective in carbohydrate loading tests in rats and healthy volunteers, reducing postprandial blood glucose and increasing insulin secretion [98]. After intensive clinical development, acarbose (Glucobay) was first launched in Germany in 1990 and has been successfully marketed in Europe and Latin America. In 1996, it was introduced onto the market in the United States under the brand name Precose.

Valiolamine (**111**), an aminocyclitol produced by *Streptomyces hygroscopicus* var. *limoneus*, is a potent inhibitor of pig intestinal maltase and sucrase, with IC_{50} values of 2.2 and 0.049 μM, respectively [99]. Numerous *N*-substituted valiolamine derivatives were synthesized to enhance its α-glucosidase inhibitory activity *in vitro* and the very simple derivative voglibose (**112**) which is obtained by reductive amination of valiolamine with dihydroxyacetone, was selected as the potential oral antidiabetic agent [100]. Its IC_{50} values toward maltase and sucrase were 0.015 and 0.0046 μM, respectively. Voglibose (brand name Basen) has been commercially available for the treatment of type 2 diabetes in Japan since 1994 (Figure 5.9).

Mulberry leaves have traditionally been used to cure "Xiao-ke" (diabetes) in Chinese medicine. The strong inhibition of digestive α-glucosidases by DNJ attracted the interest of various research groups and a large number of *N*-substituted DNJ derivatives were prepared in the hope of increasing the *in vivo* activity. Miglitol (**113**) was identified as one of the most favorable candidates showing a desired glucosidase inhibitory profile [101]. Miglitol differs from acarbose in that it is almost completely

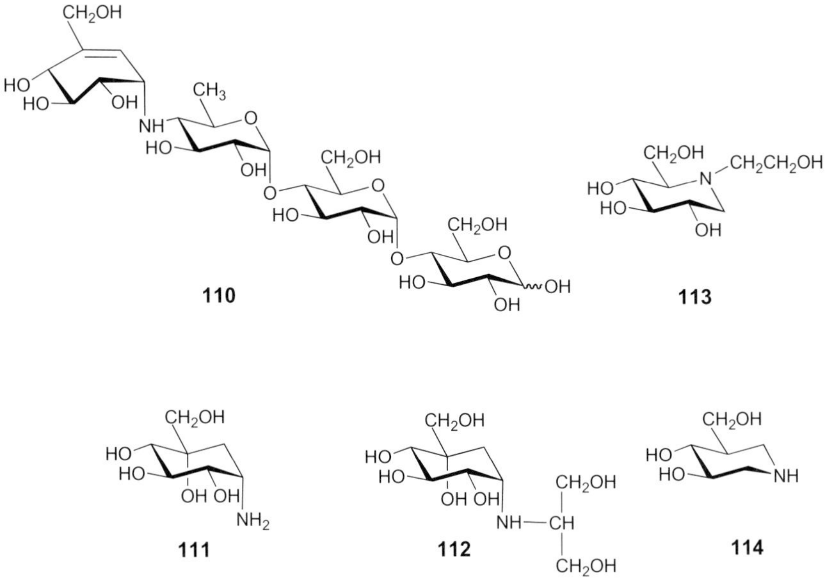

Fig. 5.9 Structures of α-glucosidase inhibitors in clinical

absorbed from the intestinal tract, and may possess systemic effects in addition to its effects in the intestinal border [102,103]. In 1996, Glyset (miglitol) tablets were granted market clearance by the US Food and Drug Administration (FDA) and introduced onto the market in 1999 as a more effective second-generation α-glucosidase inhibitor with fewer gastrointestinal side effects. In 2006, it was introduced onto the market in Japan under the brand name Seibule.

α-Glucosidase inhibitors are especially suitable for patients whose blood glucose levels are slightly above normal and can also be beneficial to those having high blood glucose immediately after eating, a condition known as postprandial hyperglycemia. These drugs slow the rate at which carbohydrates are broken down into monosaccharides in the digestive tract and therefore lengthen the digestive process. Other antidiabetic agents such as sulfonylureas and biguanides sometimes are prescribed in combination with α-glucosidase inhibitors to help increase the effectiveness of this therapy. The protective effects of α-glucosidase inhibitors have been reported for various diabetic complications. Interestingly, α-glucosidase inhibitors are also being studied as a possible treatment for heart disease, a common complication in diabetic patients. Although repetitive postprandial hyperglycemia increases ischemia/reperfusion injury, this effect can be prevented by treatment with α-glucosidase inhibitors [104].

5.3.1.2 Glycogen Phosphorylase Inhibitors

In type 2 diabetes, hepatic glucose production is increased [105]. A possible way to suppress hepatic glucose production and lower blood glucose in type 2 diabetes patients may be through inhibition of hepatic glycogen phosphorylase [106]. Fosgerau *et al.* reported that in enzyme assays DAB (**5**) is a potent inhibitor of hepatic glycogen phosphorylase [107]. Furthermore, in primary rat hepatocytes, DAB was shown to be the most potent inhibitor (IC_{50} 1 μM) of basal and glucagon-stimulated glycogenolysis ever reported [108]. Jakobsen *et al.* have reported that (3*R*,4*R*,5*R*)-5-hydroxymethylpiperidine-3,4-diol (isofagomine, **114**) synthesized chemically is a potent inhibitor of hepatic glycogen phosphorylase, with an IC_{50} value of 0.7 μM, and, furthermore, is able to prevent basal- and glucagon-stimulated glycogen degradation in cultured hepatocytes, with IC_{50} values of 2–3 μM [109]. However, its *N*-substitution always resulted in a loss of activity compared to the parent compound, and fagomine ((2*R*,3*R*,4*R*)-5-hydroxymethylpiperidine-3,4-diol) was a weak inhibitor of this enzyme, with an IC_{50} value of 200 μM [109]. Glycogen phosphorylase inhibitors would be a beneficial target to attack in the development of new anti-hyperglycemic agents.

5.3.1.3 Herbal Medicines

Current scientific evidence demonstrates that morbidity and mortality of diabetes can be eliminated by aggressive treatment with diet, exercise, and new pharmacological approaches to achieve better control of blood glucose levels. In recent years, the possibility of preventing the onset of diabetes using dietary supplements and/or herbal medicines has attracted increasing attention. Mulberry leaves have been shown to have some antidiabetic properties. It was found that mulberry leaf extract administered in a single dose of 200 mg/kg led to significant improvements in blood glucose levels in streptozotocin (STZ)-induced diabetic mice [110]. A study of 24 humans with type 2 diabetes found that patients treated with the mulberry agent showed significant improvement in blood glucose control compared to a group treated with glibenclamide [111]. It is known that DNJ, fagomine, and DAB are present in mulberry leaves [10]. Evaluation for antihyperglycemic effects in STZ-induced diabetic mice has been carried out with fagomine [112]. Fagomine significantly reduced the blood glucose level 2 h after intraperitoneal administration and its effect was sustained for 2–6 h after administration. The effect of fagomine on immunoreactive insulin (IRI) release was investigated with the perfused pancreas of normal rats. The 8.3 mM glucose-induced IRI release was increased in the presence of fagomine in a concentration-dependent manner. The antihyperglycemic effects of the mulberry leaf extract would appear to be a combination of α-glucosidase inhibition by DNJ and related compounds, the insulin-releasing effect of fagomine, and glycogen phosphorylase inhibition by DAB.

Thus, although traditional herbal medicines are candidates for diabetes prevention, it is very important to give scientific evidence for their antidiabetic effects. *Commelina communis* (Commelinaceae) has been used in traditional Chinese medicine as an antipyretic for noninfectious fever and to treat ascites, edema, and hordeolum [113], and is now very popular in Korea for the treatment of diabetes [114]. The MeOH extract of this plant shows strong inhibitory activity against porcine intestinal α-glucosidases and contains DNJ (**1**), DMJ (**24**), DMDP (**25**), α-HNJ (**37**), and 7-β-Glc-α-HNJ (**38**) [115]. Alkaloids **1**, **37**, and **38** are very potent inhibitors of digestive α-glucosidases, and DMDP as well as fagomine shows antihyperglycemic effects in STZ-induced diabetic mice [112]. These results support the pharmacological basis of this herb, which has been used as a folklore medicine for the treatment of diabetes.

5.3.2
Molecular Therapy for Lysosomal Storage Disorders

Experimental data show that some human genetic diseases are due to mutations in proteins that influence their folding and lead to retention of mutant proteins in the endoplasmic reticulum (ER) and successive degradation [116,117]. Lysosomes are membrane-bound cytoplasmic organelles that serve as a major degradative compartment in eukaryotic cells. The degradative function of lysosomes is carried out by more than 50 acid-dependent hydrolases contained within the lumen [118]. Glycosphingolipid (GSL) storage diseases are genetic disorders in which a mutation of one of the GSL glycohydrolases blocks GSL degradation, leading to lysosomal accumulation of undegraded GSL [119]. Possible strategies for the treatment of these lysosomal storage diseases include enzyme replacement therapy, gene therapy, substrate deprivation, and bone marrow transplantation. The successful treatment for such diseases to date is enzyme replacement therapy for patients with type 1 Gaucher disease and Fabry disease. However, this enzyme replacement therapy is only useful in diseases in the absence of neuropathology since enzymes do not cross the blood–brain barrier, and another problem in this therapy is the cost, which prevents many patients from obtaining the treatment. In recent years, remarkable progress has been made in developing a molecular therapy for GSL storage disorders [3,4,120,121]. There are two novel approaches in this field. One is substrate reduction therapy and another is pharmacological (or chemical) chaperone therapy.

5.3.2.1 **Substrate Reduction Therapy**

As long as the biosynthesis of substrate continues together with a decrease in the corresponding enzyme activity, the pathological accumulation of undegraded substrate in the lysosomes proceeds. The aim of substrate reduction therapy is to reduce GSL substrate influx into the lysosomes by inhibitors of GSL synthesis. *N*-Butyl-DNJ (miglustat, Zavesca, **115**) is an inhibitor of ceramide-specific glucosyltransferase [122]. Miglustat is the first orally active agent in the treatment of type 1 Gaucher disease. Gaucher disease is the most common lysosomal storage disorder caused by a deficiency of lysosomal β-glucosidase (known as β-glucocerebrosidase), resulting in the progressive accumulation of glucosylceramide. Type 1 Gaucher disease is nonneuronopathic and sometimes called the "adult" form. Ceredase in 1991 and its recombinant successor Cerezyme in 1994 were introduced as the enzyme replacement therapy of this type. In 2001, Genzyme released preclinical data supporting Genz-78132 (**116**) as the second-generation substrate reduction agent. The company has reported that Genz-78132 is 100–5000 times more potent *in vitro* for inhibition of cell surface ganglioside GM1, an indicator of glycosphingolipid synthesis, than the first-generation miglustat and *N*-(5-adamantane-1-yl-methoxypentyl)-DNJ (**117**). Furthermore, Genz-78132 is at least 20 times more potent *in vivo* than **117**, the most potent DNJ derivative [123]. Although there is no supporting clinical data, in preclinical studies Genz-78132 has a substantially greater therapeutic index than the first-generation inhibitors, which have shown limited efficacy and significant toxicity (Figure 5.10).

5.3.2.2 **Pharmacological Chaperone Therapy**

The concept of pharmacological chaperone therapy is that an intracellular activity of misfolded mutant enzymes can be restored by administering competitive inhibitors that serve as pharmacological chaperones. These inhibitors appear to act as a template that stabilizes the native folding state in the ER by occupying the active site of the mutant enzyme, thus allowing its maturation and trafficking to the lysosome [120]. This concept was first introduced with Fabry disease [124]. Residual α-galactosidase A (α-Gal A) activity in lymphoblasts derived from Fabry patients and in tissues of R301Q α-Gal A transgenic mice was enhanced by treatment with 1-deoxygalactonojirimycin (DGJ, **118**), a competitive inhibitor of α-Gal A with a K_i value of 40 nM. Yam *et al.* showed that DGJ induces trafficking of ER-retained

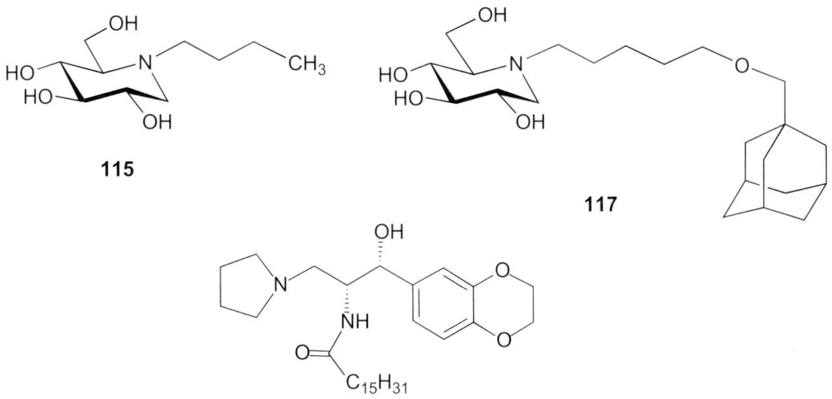

Fig. 5.10 Structures of ceramide-specific glucosyltransferase inhibitors.

R301Q α-Gal A to lysosomes of transgenic mouse fibroblasts, and DGJ treatment results in efficient clearance of the substrate, globotriaosylceramide (Gb3) [125]. By testing a series of α-Gal A inhibitors for both *in vitro* inhibitory and chaperoning activities in lymphoblasts from Fabry patients, it was demonstrated that a potent inhibitor shows an effective chaperoning activity, whereas less-potent inhibitors require higher concentrations to achieve the same effect [126]. DGJ, α-homo-galactonojirimycin (**119**), **42**, and **52** are inhibitors of α-Gal A with IC_{50} values of 0.04, 0.21, 4.3, and 16 μM, respectively, and the respective addition at 100 μM to culture medium of Fabry lymphoblasts increases the intracellular α-Gal A activity 14-, 5.2-, 2.4-, and 2.3-fold. Thus, potent and specific inhibitors of lysosomal glycosidases are expected to have therapeutic effects at lower concentrations (Figure 5.11).

Sawker *et al.* reported that *N*-nonyl-DNJ (**120**) is a potent inhibitor of lysosomal β-glucosidase, with an IC_{50} value of 1 μM, and the addition of a subinhibitory concentration (10 μM) of this compound to a fibroblast culture medium leads to a two-fold increase in the mutant (N370S) enzyme activity [127]. Examination of a series of DNJ analogs on the residual activities of various lysosomal β-glucosidase variants has revealed that the nature of the alkyl moiety greatly influences their chaperoning activity: *N*-Butyl-DNJ is inactive, the DNJ derivatives with *N*-nonyl and *N*-decyl chains are active, and *N*-dodecyl-DNJ is predominantly inhibitory [128]. However, it is also known that *N*-nonyl-DNJ is a potent inhibitor of ER processing α-glucosidases like *N*-butyl-DNJ, and hence has potential as an antiviral agent to inhibit folding and trafficking of viral envelope glycoproteins [129,130]. Inhibitors targeting a host function such as ER processing α-glucosidases must be carefully considered in terms of side effects since they may inhibit

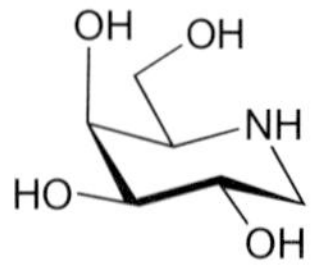

118 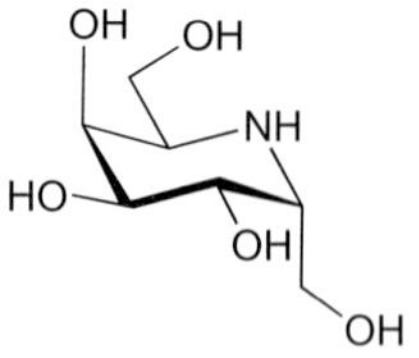 **119**

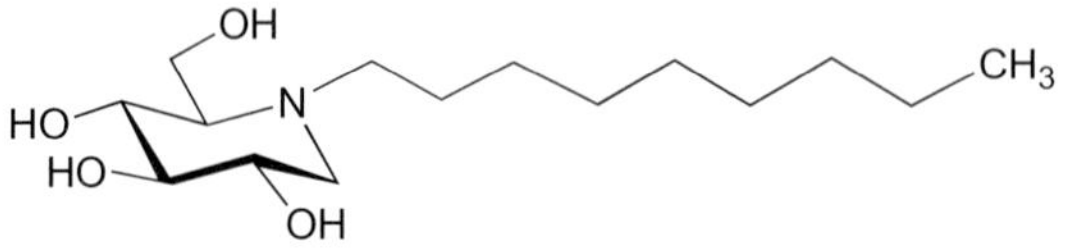

120

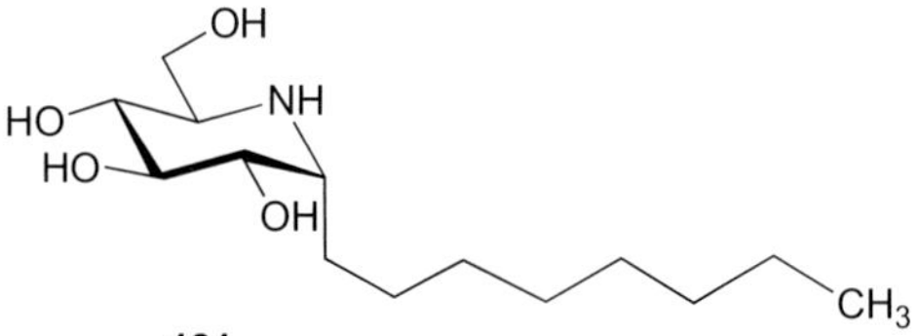

121

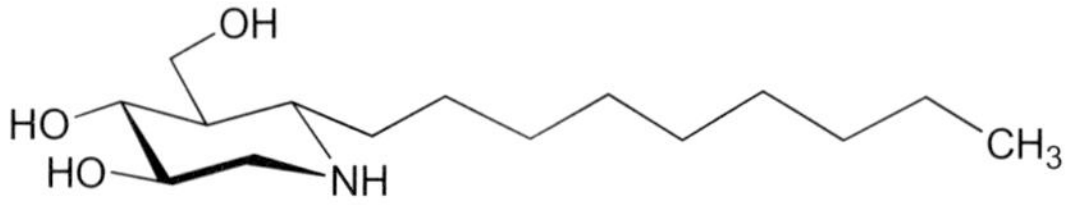

122

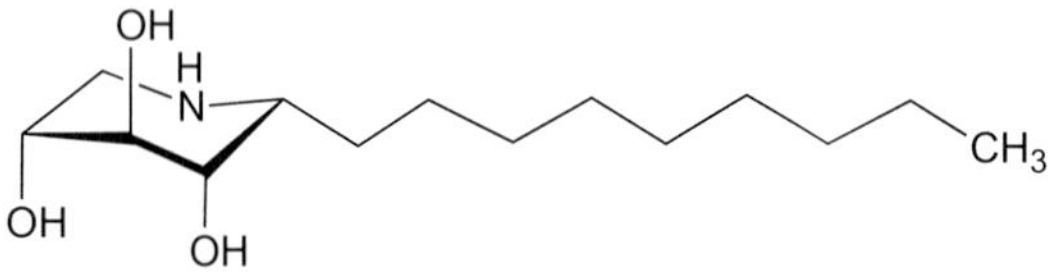

123

Fig. 5.11 Structures of pharmacological chaperones for lysosomal storage disorders.

folding, secretion, and trafficking of other glycoproteins in patient's cells or may inhibit directly lysosomal α-glucosidase after being taken up into cells. In fact, addition of *N*-nonyl-DNJ at 10 μM lowered the cellular lysosomal α-glucosidase activity by 50% throughout the assay period (10 days) in spite of its excellent chaperoning activity for the mutant β-glucosidase. The inhibition of lysosomal α-glucosidase as the side effect may induce storage of glycogen in the lysosomes, as observed in Pompe disease. On the other hand, α-1-*C*-octyl-DNJ (**121**), with a K_i value of 0.28 μM, showed a novel chaperoning activity for N370S Gaucher variants, minimizing the potential for undesirable side effects such as lysosomal α-glucosidase inhibition [131]. In addition, isofagomine [132], α-6-*C*-nonylisofagomine (**122**) [133], and α-1-*C*-nonyl-1,5-dideoxy-1,5-imino-D-xylitol (**123**) [134] are very potent inhibitors of lysosomal β-glucosidase and candidates as pharmacological chaperones for Gaucher disease.

5.4 Concluding Remarks and Future Outlook

From the success of α-glucosidase inhibitors as antidiabetic agents and neuraminidase inhibitors as anti-influenza drugs, the practical uses of glycosidase inhibitors appear to be limited to diabetes and viral infection. However, since glycosidases are involved in a wide range of anabolic and catabolic processes of carbohydrates, their inhibitors could have many kinds of beneficial effects as therapeutic agents. Pharmacological chaperone therapy for lysosomal storage disorders is a quite new application of glycosidase inhibitors. Although the safety and effectiveness of enzyme replacement therapy for such diseases were demonstrated in type 1 Gaucher disease and Fabry disease, the application is restricted to nonneuronopathic diseases since enzyme proteins do not cross the blood–brain barrier. Pharmacological chaperone therapy and substrate reduction therapy with small molecules are attracting considerable interest, particularly for neuronopathic lysosomal storage disorders. Many inhibitors of lysosomal glycosidases are waiting for the evaluation of pharmacological chaperone therapy for such diseases and some of them are in preclinical or phase I/II clinical trials (Amicus Therapeutics Inc., Cranbury, NJ).

References

1 Asano, N., Nash, R. J., Molyneux, R. J., Fleet, G. W. J. (2000) *Tetrahedron: Asymmetry*, **11**, 1–36.
2 Watson, A. A. Fleet, G. W. J., Asano, N., Molyneux, R. J., Nash, R. J. (2001) *Phytochemistry*, **56**, 265–295.
3 Asano, N. (2003) *Glycobiology*, **13**, 93R–104R.
4 Butters, T. D., Dwek, R. A., Platt, F. M. (2005) *Glycobiology*, **15**, 43R–52R.
5 Inoue, S., Tsuruoka, T., Niida, T. (1966) *Journal of Antibiotics*, **19**, 288–292.
6 Inoue, S., Tsuruoka, T., Ito, T., Niida, T. (1968) *Tetrahedron*, **24**, 2125–2144.
7 Yagi, M., Kouno, T., Aoyagi, Y., Murai, H. (1976) *Nippon Nogeikagaku Kaishi*, **50**, 571–572.
8 Schmidt, D. D., Frommer, W., Müller, L., Truscheit, E. (1979) *Naturwissenshaften*, **66**, 584–585.
9 Murao, S. and Miyata, S. (1980) *Agricultural and Biological Chemistry*, **44**, 219–221.

10 Asano, N., Tomioka, E., Kizu, H., Matsui, K. (1994) *Carbohydrate Research*, **253**, 235–245.

11 Asano, N., Oseki, K., Tomioka, E., Kizu, H., Matsui, K. (1994) *Carbohydrate Research*, **259**, 243–255.

12 Asano, N., Yamashita, T., Yasuda, K., Ikeda, K., Kizu, H., Kameda, Y., Kato, A., Nash, R. J., Lee, H.-S., Ryu, K.-S. (2001) *Journal of Agricultural and Food Chemistry*, **49**, 4208–4213.

13 Fellows, L. E., Bell, A., Lynn, D. G., Pilkiewicz, F., Miura, I., Nakanishi, K. (1979) *Journal of the Chemical Society, Chemical Communication*, 977–978.

14 Nash, R. J., Watson, A. A., Asano, N. (1996) *Alkaloids: Chemical and Biological Perspectives*, **11**, Elsevier Science, Oxford, UK, pp. 345–376.

15 Asano, N., Oseki, K., Kizu, H., Matsui, K. (1994) *Journal of Medicinal Chemistry*, **37**, 3701–3706.

16 Kite, G. C., Fellows, L. E., Fleet, G. W. J., Liu, P. S., Scofield, A. M., Smith, N. G. (1988) *Tetrahedron Letters*, **29**, 6483–6486.

17 Kite, G. C., Horn, J. M., Romeo, J. T., Fellows, L. E., Lees, D. C., Scofield, A. M., Smith, N. G. (1990) *Phytochemistry*, **29**, 103–105.

18 Asano, N., Yamauchi, T., Kagamifuchi, K., Shimizu, N., Takahashi, S., Takatsuka, H., Ikeda, K., Kizu, H., Chuakul, W., Kettawan, A., Okamoto, T. (2005) *Journal of Natural Products*, **68**, 1238–1242.

19 Takahata, H., Banba, Y., Sasatani, M., Nemoto, H., Kato, A., Adachi, I. (2004) *Tetrahedron*, **60**, 8199–8205.

20 Takahata, H., Banba, Y., Ouchi, H., Nemoto, H. (2003) *Organic Letters*, **5**, 2527–2529.

21 Kato, A., Kato, N., Kano, E., Adachi, I., Ikeda, K., Yu, L., Okamoto, T., Banba, Y., Takahata, H., Asano, N. (2005) *Journal of Medicinal Chemistry*, **48**, 2036–2044.

22 Reiz, A. and Baxter, E. W. (1990) *Tetrahedron Letters*, **31**, 6777–6780.

23 Liu, K.K.-C., Kajimoto, T., Zhong, L., Ichikawa, Y., Wong, C.-H. (1991) *Journal of Organic Chemistry*, **56**, 6280–6289.

24 Legler, G., Korth, A., Berger, A., Ekhart, C., Gradnig, G., Stütz, A. E. (1993) *Carbohydrate Research*, **250**, 67–77.

25 Baxter, E. W. and Reiz, A. B. (1994) *Journal of Organic Chemistry*, **59**, 3175–3185.

26 Takayama, S., Martin, R., Wu, J., Laslo, K., Siuzdak, G., Wong, C.-H. (1997) *Journal of American Chemical Society*, **119**, 8146–8151.

27 Saralamp, P.Chuakul, W.Temsiririrkkul, R.Clayton T. (1996) *Medicinal Plants in Thailand*, Vol. **1**, Mahidol University, Bangkok.

28 Yoshikawa, M., Morikawa, T., Nakano, K., Pongpiriyadacha, Y., Murakami, T., Matsuda, H. (2002) *Journal of Natural Products*, **65**, 1638–1642.

29 Liu, P. S. (1987) *Journal of Organic Chemistry*, **52**, 4717–4721.

30 Rhinehart, B. L., Robinson, K. M., Liu, P. S., Payne, A. J., Wheatley, M. E., Wagner, S. R. (1987) *Journal of Pharmacology and Experimental Therapeutics*, **241**, 915–920.

31 Asano, N., Nishida, M., Kizu, H., Matsui, K. (1997) *Journal of Natural Products*, **60**, 98–101.

32 Martin, O. R., Compain, P., Kizu, H., Asano, N. (1999) *Bioorganic and Medicinal Chemistry Letters*, **9**, 3171–3174.

33 Martin, O. R. and Saavedra, O. M. (1995) *Tetrahedron Letters*, **36**, 799–802.

34 Saavedra, O. M. and Martin, O. R. (1996) *Journal of Organic Chemistry*, **61**, 6987–6993.

35 Bruce, I., Fleet, G. W. J., Cenci di Bello, I., Winchester, B. (1992) *Tetrahedron*, **48**, 10191–10200.

36 Holt, K. E., Leeper, F. J., Handa, S. (1994) *Journal of Chemical Society Transactions*, **1**, 231–234.

37 Asano, N., Kato, A., Miyauchi, M., Kizu, H., Kameda, Y., Watson, A. A., Nash, R. J., Fleet, G. W. J. (1998) *Journal of Natural Products*, **61**, 625–628.

38 Takebayashi, M., Hiranuma, S., Kanie, Y., Kajimoto, T., Kanie, O., Wong, C. H. (1999) *Journal of Organic Chemistry*, **64**, 5280–5291.

39 Kitaoka, M., Ichikawa, K., Sakurai, Y., Matsushita, Y., Iijima, Y., Akiyama, T., Boriboon, M. (1993) *Annual Report of Sankyo Research Laboratories*, **45**, 99–104.
40 Terada, M., Sano, M., Ishii, A. I., Kino, H., Fukushima, S., Noro, T. (1982) *Nippon Yakurigaku Zasshi*, **79**, 93–103.
41 Sakata, K., Aoki, K., Chang, C. F., Sakurai, A., Murakoshi, J. (1978) *Agricultural and Biological Chemistry*, **42**, 457–463.
42 Goujon, J.-Y., Gueyrard, D., Compaine, P., Martin, O. R., Ikeda, K., Asano, N. (2005) *Bioorganic and Medicinal Chemistry*, **13**, 2313–2324.
43 Ikeda, K., Takahashi, M., Nishida, M., Miyauchi, M., Kizu, H., Kameda, Y., Arisawa, M., Watson, A. A., Nash, R. J., Fleet, G. W. J., Asano, N. (2000) *Carbohydrate Research*, **323**, 73–80.
44 Maughan, M. A. T. and Davies, I. G. (2003) *Angewandte Chemie-International Edition*, **42**, 3788–3792.
45 Asano, N., Nishida, M., Miyauchi, M., Ikeda, K., Yamamoto, M., Kizu, H., Kameda, Y., Watson, A. A., Nash, R. J., Fleet, G. W. J. (2000) *Phytochemistry*, **53**, 379–382.
46 James, L. F., VanKampen, K. R., Hartley, W. J. (1970) *Veterinary Pathology*, **7**, 116–125.
47 Hartley, W.J., Baker, D. C., James, L. F. (1989) In James, L. F.Elbein, A. D.Molyneux, R. J. Warren C. D. (Eds.) *Swainsonine and Related Glycosidase Inhibitors*, Iowa State University Press, Ames, IA, 50–56.
48 Molyneux, R. J. and James, L. F. (1982) *Science*, **216**, 190–191.
49 Dorling, P. R., Huxtable, C. R., Vogel, P. (1978) *Neuropathology and Applied Neurobiology*, **4**, 285–295.
50 Colegate, S. M., Dorling, P. R., Huxtable, C. R. (1979) *Australian Journal of Chemistry*, **32**, 2257–2264.
51 Hohenschutz, L. D., Bell, E. A., Jewess, P. J., Leworthy, D. P., Pryce, R. J., Arnold, E., Clardy, J. (1981) *Phytochemistry*, **20**, 811–814.
52 Molyneux, R. J., Roitman, J. N., Dunnheim, G., Szumilo, T., Elbein, A. D. (1986) *Archives of Biochemistry and Biophysics*, **251**, 450–457.
53 Molyneux, R. J., Tropea, J. E., Elbein, A. D. (1990) *Journal of Natural Products*, **53**, 609–614.
54 Molyneux, R. J., Pan, Y. T., Tropea, J. E., Benson, M., Kaushal, G. P., Elbein, A. D. (1991) *Biochemistry*, **30**, 9981–9987.
55 Pastuszak, I., Molyneux, R. J., James, L. F., Elbein, A. D. (1990) *Biochemistry*, **29**, 1886–1891.
56 Nash, R. J., Fellows, L. E., Dring, J. V., Fleet, G. W. J., Derome, A. E., Hamor, T. A., Scofield, A. M., Watkin, D. J. (1988) *Tetrahedron Letters*, **29**, 2487–2490.
57 Wrobel, J. T. (1985) In Brossi A. (Ed.) *The Alkaloids: Chemistry and Pharmacology*, Vol. **26**, Academic Press, New York, 327–385.
58 Robins, D. J. (1995) In Cordell G. A. (Ed.) *The Alkaloids: Chemistry and Pharmacology*, Vol. **46**, Academic Press, New York, 1–61.
59 Molyneux, R. J., Benson, M., Wong, R. Y., Tropea, J. E., Elbein, A. D. (1988) *Journal of Natural Products*, **51**, 1198–1206.
60 Harris, C. M., Harris, T. M., Molyneux, R. J., Tropea, J. E., Elbein, A. D. (1989) *Tetrahedron Letters*, **30**, 5685–5688.
61 Nash, R. J., Fellows, L. E., Dring, J. V., Fleet, G. W. J., Girdhar, A., Ramsden, N. G., Peach, J. M., Hegarty, M. P., Scofield, A. M. (1990) *Phytochemistry*, **29**, 111–114.
62 Nash, R. J., Fellows, L. E., Plant, A. C., Fleet, G. W. J., Derome, A. E., Baird, P. D., Hegarty, M. P., Scofield, A. M. (1988) *Tetrahedron*, **44**, 5959–5964.
63 Denmark, S. E. and Martinborough, E. A. (1999) *Journal of American Chemical Society*, **121**, 3046–3056.
64 Denmark, S. E. and Herbert, B. (1998) *Journal of American Chemical Society*, **120**, 7357–7358.
65 Denmark, S. E. and Herbert, B. (2000) *Journal of Organic Chemistry*, **65**, 2887–2896.
66 Kato, A., Kano, E., Adachi, I., Molyneux, R. J., Watson, A. A., Nash, R. J., Fleet, G. W. J., Wormald, M. R.,

Kizu, H., Ikeda, K., Asano, N. (2003) *Tetrahedron: Asymmetry*, **14**, 325–331.
67 Kato, A., Adachi, I., Miyauchi, M., Ikeda, K., Komae, T., Kizu, H., Kameda, Y., Watson, A. A., Nash, R. J., Wormald, M. R., Fleet, G. W. J., Asano, N. (1999) *Carbohydrate Research*, **316**, 95–103.
68 Asano, N., Kuroi, H., Ikeda, K., Kizu, H., Kameda, Y., Kato, A., Adachi, I., Watson, A. A., Nash, R. J., Fleet, G. W. J. (2000) *Tetrahedron: Asymmetry*, **11**, 1–8.
69 Rambaud, L., Compain, P., Martin, O. R. (2001) *Tetrahedron: Asymmetry*, **12**, 1807–1809.
70 Izquierdo, I., Plaza, M. T., Franco, F. (2002) *Tetrahedron: Asymmetry*, **13**, 1581–1585.
71 Chabaud, L., Landais, Y., Renaud, P. (2005) *Organic Letters*, **7**, 2587–2590.
72 Yamashita, T., Yasuda, K., Kizu, H., Kameda, Y., Watson, A. A., Nash, R. J., Fleet, G. W. J., Asano, N. (2002) *Journal of Natural Products*, **65**, 1875–1881.
73 Shibano, M., Tsukamoto, D., Kusano, G. (1999) *Chemical and Pharmaceutical Bulletin*, **47**, 907–908.
74 Asano, N., Ikeda, K., Kasahara, M., Arai, Y., Kizu, H. (2004) *Journal of Natural Products*, **67**, 846–850.
75 Tepfer, D. A., Goldman, A., Pamboukdjian, N., Maille, M., Lepingle, A., Chevalier, D., Denarie, J., Rosenberg, C. (1988) *Journal of Bacteriology*, **170**, 1153–1161.
76 Lounasmaa, M. (1988) In BrossiA. (Ed.) *The Alkaloids: Chemistry and Pharmacology*, Vol. **33**, Academic Press, New York, 1–81.
77 Lounasmaa, M. and Tamminen, T. (1993) In Cordell G. E. (Ed.) *The Alkaloids: Chemistry and Pharmacology*, Vol. **44**, Academic Press, New York, 1–115.
78 Schimming, T., Tofern, B., Mann, P., Richter, A., Jenett-Siems, K., Dräger, B., Asano, N., Cupta, M. P., Correa, M. D., Eich, E. (1998) *Phytochemistry*, **49**, 1989–1995.
79 Bekkouche, K., Daali, Y., Cherkaoui, S., Veuthey, J.-L., Christen, P. (2001) *Phytochemistry*, **58**, 455–462.
80 Asano, N., Yokoyama, K., Sakurai, M., Ikeda, K., Kizu, H., Kato, A., Arisawa, M., Höke, D., Dräger, B., Watson, A. A., Nash, R. J. (2001) *Phytochemistry*, **57**, 721–726.
81 Asano, N., Kato, A., Oseki, K., Kizu, H., Matsui, K. (1995) *European Journal of Biochemistry*, **229**, 369–376.
82 Asano, N., Kato, A., Yokoyama, Y., Miyauchi, M., Yamamoto, M., Kizu, H. (1995) *Carbohydrate Research*, **284**, 169–178.
83 Asano, N., Kato, A., Kizu, H., Matsui, K., Watson, A. A., Nash, R. J. (1996) *Carbohydrate Research*, **293**, 195–204.
84 Kato, A., Asano, N., Kizu, H., Matsui, K., Suzuki, S., Arisawa, M. (1997) *Phytochemistry*, **45**, 425–429.
85 Asano, N., Kato, A., Miyauchi, M., Kizu, H., Tomimori, T., Matsui, K., Nash, R. J., Molyneux, R. J. (1997) *European Journal of Biochemistry*, **248**, 296–303.
86 Griffiths, R. C., Watson, A. A., Kizu, H., Asano, N., Sharp, H. J., Jones, M. G., Wormald, M. R., Fleet, G. W. J., Nash, R. J. (1996) *Tetrahedron Letters*, **37**, 3207–3208.
87 Nash, R. J., Watson, A. A., Winters, A. L., Fleet, G. W. J., Wormald, M. R., Dealer, S., Lees, E., Asano, N., Molyneux, R. J. (1998) In Garland T. and BarrC. (Eds.) *Toxic Plants and Other Natural Toxicants*, CAB International, Wallingford, 276–284.
88 Asano, N., Kato, A., Kizu, H., Matsui, K., Griffiths, R. C., Jones, M. G., Watson, A. A., Nash, R. J. (1997) *Carbohydrrate Research*, **304**, 173–178.
89 Duclos, O., Mondange, M., Depezay, J. C. (1992) *Tetrahedron Letters*, **33**, 8061–8064.
90 Boyer, F. D. and Lallemand, J. Y. (1994) *Tetrahedron*, **50**, 10443–10458.
91 Johnson, C. R. and Bis, S. J. (1995) *Journal of Organic Chemistry*, **60**, 615–623.
92 Skaanderup, P. R. and Madsen, R. (2001) *Chemical Communications*, 1106–1107.
93 Skaanderup, P. R. and Madsen, R. (2003) *Journal of Organic Chemistry*, **68**, 2115–2122.

94 Haraguchi, M., Gorniak, S. L., Ikeda, K., Minami, Y., Kato, A., Watson, A. A., Nash, R. J., Molyneux, R. J., Asano, N. (2003) *Journal of Agricultural and Food Chemistry*, **51**, 4995–5000.

95 Molyneux, R. J., McKenzie, R. A., O'Sullivan, B. M., Elbein, A. D. (1995) *Journal of Natural Products*, **58**, 878–886.

96 Winchester, B. and Fleet, G. W. J. (1992) *Glycobiology*, **2**, 199–210.

97 Schmidt, D. D., Frommer, W., Müller, L., Truscheit, E. (1979) *Naturwissenschaften*, **66**, 584–585.

98 Puls, W., Keup, U., Krause, H. P., Thomas, G., Hoffmeister, F. (1977) *Naturwissenschaften*, **64**, 536–537.

99 Kameda, Y., Asano, N., Yoshikawa, M., Takeuchi, M., Yamaguchi, T., Matsui, K., Horii, S., Fukase, H. (1984) *Journal of Antibiotics*, **37**, 1301–1307.

100 Horii, S., Fukase, H., Matsuo, T., Kameda, Y., asano, N., Matsui, K. (1986) *Journal of Medicinal Chemistry*, **29**, 1038–1046.

101 Junge, B., Matzke, M., Stltefuss, J. (1996) In Kuhlmann J. and Plus W. (Eds.) Handbook of Experimental Pharmacology, Vol. **119**, Springer-Verlag, New York, 411–482.

102 Joubert, P. H., Foukaridis, G. N., Bopape, M. L. (1987) *European Journal of Clinical Pharmacology*, **31**, 723–724.

103 Joubert, P. H., Venter, H. L., Foukaridis, G. N. (1990) *British Journal of Clinical Pharmacology*, **30**, 391–396.

104 Franz, S., Calvillo, L., Tillmanns, J., Elbing, I., Dienesch, C., Bischoff, H., Ertl, G., Bauersachs, J. (2005) *FASEB Journal*, **19**, 591–593.

105 Defronzo, R. A., Bonadonna, R. C., Ferrannini, E. (1992) *Diabetes Care*, **15**, 318–368.

106 Martin, J. L., Veluraja, K., Ross, K., Johnson, L. N., Fleet, G. W. J., Ramsden, N. G., Bruce, I., Orchard, M. G., Oikonomakos, N. G., Papageorgiou, A. C., Leonidas, D. D., Tsitoura, H. S. (1991) *Biochemistry*, **30**, 10101–10116.

107 Fosgerau, K., Westergaard, N., Quistorff, B., Grunner, N., Kristiansen, M., Lundgren, K. (2000) *Archives of Biochemistry and Biophysics*, **380**, 274–284.

108 Andersen, B., Rassov, A., Westergaard, N., Lundgren, K. (1999) *Biochemical Journal*, **342**, 545–550.

109 Jakobsen, P., Lundbeck, J. M., Kristiansen, M., Breinholt, J., Demuth, H., Pawlas, J., Torres Candela, M. P., Andersen, B., Westergaard, N., Lundgren, K., Asano, N. (2001) *Bioorganic and Medicinal Chemistry*, **9**, 733–744.

110 Chen, F., Nakashima, N., Kimura, I., Kimura, M. (1995) *Yakugaku Zasshi*, **115**, 476–482.

111 Andallu, B., Suryakantham, V., Lakshmi Srikanthi, B., Reddy, G. K. (2001) *Clinica Chimica Acta*, **314**, 47–53.

112 Nojima, H., Kimura, I., Chen, F., Sugihara, Y., Haruno, M., Kato, A., Asano, N. (1998) *Journal of Natural Products*, **61**, 397–400.

113 Huang, K. C. (1993) *The Pharmacology of Chinese Herbs*, CRC Press, Boca Raton, FL, pp. 296–297.

114 Kim, O.K., Park, S. Y., Cho, K. H. (1991) *Korean Journal of Pharmacognosy*, **22**, 225–232.

115 Kim, H. S., Kim, Y. H., Hong, Y. S., Paek, N. S., Lee, H. S., Kim, T. H., Kim, K. W., Lee, J. J. (1999) *Planta Medica*, **65**, 437–439.

116 Bychkova, V. E. and Ptitsyn, O. B. (1995) *FEBS Letters*, **359**, 6–8.

117 Welch, W. J. and Brown, C. R. (1996) *Cell Stress and Chaperones*, **1**, 109–115.

118 deDuve, C (1963) In de Reuck A. V. S. and Cameron M. P. (Eds.) *Lysosomes*, Churchill, London, 1–35.

119 Kornfeld, S. and Mellman, I. (1989) *Annual Review of Cell Biology*, **5**, 483–525.

120 Fan, J.-Q. (2003) *Trends in Pharmacological Sciences*, **24**, 355–360.

121 Cohen, F. E. and Kelly, J. W. (2003) *Nature*, **426**, 905–909.

122 Platt, F. M., Neises, G. R., Dwek, R. A., Butters, T. D. (1994) *Journal of Biological Chemistry*, **269**, 8362–8365.

123 Overkleeft, H. S., Renkema, G. H., Neele, J., Vianello, P., Hung, I. O., Strijland, A., vander Burg, A. M., Koomen, G.-P., Pandit, U. K., Aerts,

J.M.F.G. (1998) *Journal of Biological Chemistry*, **41**, 26522–26527.

124 Fan, J.-Q., Ishii, S., Asano, N., Suzuki, Y. (1999) *Nature Medicine*, **5**, 112–115.

125 Yam, G.H.-F., Zuber, C., Roth, J. (2005) *FASEB Journal*, **19**, 12–18.

126 Asano, N., Ishii, S., Kizu, H., Ikeda, K., Yasuda, K., Kato, A., Martin, O. R., Fan, J.-Q. (2000) *European Journal of Biochemistry*, **267**, 4179–4186.

127 Sawker, A. R., Chen, W.-C., Beautler, E., Wong, C.-H., Baich, W. E., Kelly, J. W. (2002) *Proceedings of the National Academy of Sciences of the United States of America*, **99**, 15428–15433.

128 Sawker, A. R., Adamsky-Werner, S. L., Chen, W.-C., Wong, C.-H., Beautler, E., Zimmer, K.-P., Kelly, J. W. (2005) *Chemistry and Biology*, **12**, 1235–1244.

129 Block, T. M., Lu, X., Mehta, A. S., Blumberg, B. S., Tennant, B., Ebling, M., Korba, B., Lansky, D. M., Jacob, G. S., Dwek, R. A. (1998) *Nature Medicine*, **4**, 610–614.

130 Zitzmann, N., Mehta, A. N., Carrouée, S., Butters, T. D., Platt, F. M., McCauley, J., Blumberg, B. S., Dwek, R. A., Block, T. M. (1999) *Proceedings of the National Academy of Sciences of the United States of America*, **96**, 11878–11882.

131 Yu, L., Ikeda, K., Kato, A., Adachi, I., Godin, G., Compain, P., Martin, O. R., and Asano, N., submitted for publication.

132 Chang, H.-H., Asano, N., Ishii, S., Ichikawa, Y., and Fan, J.-Q., submitted for publication.

133 Zhu, X., Sheth, K. A., Li, S., Chang, H.-H., Fan, J.-Q. (2005) *Angewandte Chemie International Edition*, **44**, 7450–7453.

134 Compain, P., Martin, O. R., Boucheron, C., Godin, G., Yu, L., Ikeda, K., and Asano, N., submitted for publication.

6
Neurotoxic Alkaloids from Cyanobacteria

Rashel V. Grindberg, Cynthia F. Shuman, Carla M. Sorrels, Josh Wingerd, William H. Gerwick

6.1
Introduction

Toxic cyanobacterial blooms in brackish or freshwater environments have attracted the attention of both researchers and the general public for many years. George Francis of Adelaide, Australia, published the first scholarly description of a freshwater cyanobacterial bloom in 1878 [1]. His letter to *Nature* described a "thick scum" of what he believed to be *Nodularia* sp. in an estuary of the Murray River in Australia. Since that time an increase in harmful algal blooms, thought to be partially influenced by an increase in detergent and fertilizer runoff, has led to global concern over human health and environmental aspects [2]. There are an estimated 40 genera of cyanobacterial species that are responsible for the production of freshwater and marine cyanobacterial toxins. These toxins can be grouped according to their toxic mechanism in vertebrates as hepatotoxins (e.g. microcystin and nodularin), general cytotoxins (e.g. cylindrospermopsin), neurotoxins (e.g. anatoxins and saxitoxins), and irritant and dermatoxins (e.g. lipopolysaccharides and lyngbyatoxin). Of these, the microcystins, nodularins, anatoxins, saxitoxins, and cylindrospermopsin are currently recognized as potential health hazards that should be monitored in drinking and bathing water. As the demand for fresh drinking water increases, and methods for detection and characterization continue to improve, additional cyanobacterial toxins will undoubtedly be added to this list.

This chapter reviews the nitrogen-containing neurotoxic compounds produced by cyanobacteria, and follows an earlier review in this series which more broadly covered the alkaloid chemistry of these life forms from the marine environment [3]. A description of the discovery, isolation, structural elucidation, biosynthesis, mechanism of action, structure–activity relationship (SAR), and some aspects of chemical synthesis of cyanobacterial toxins is provided.

Historically, inquiries have been prompted by toxic cyanobacterial events, and the first section of this chapter will describe compounds discovered through such

Modern Alkaloids: Structure, Isolation, Synthesis and Biology. Edited by E. Fattorusso and O. Taglialatela-Scafati

ISBN: 978-3-527-31521-5

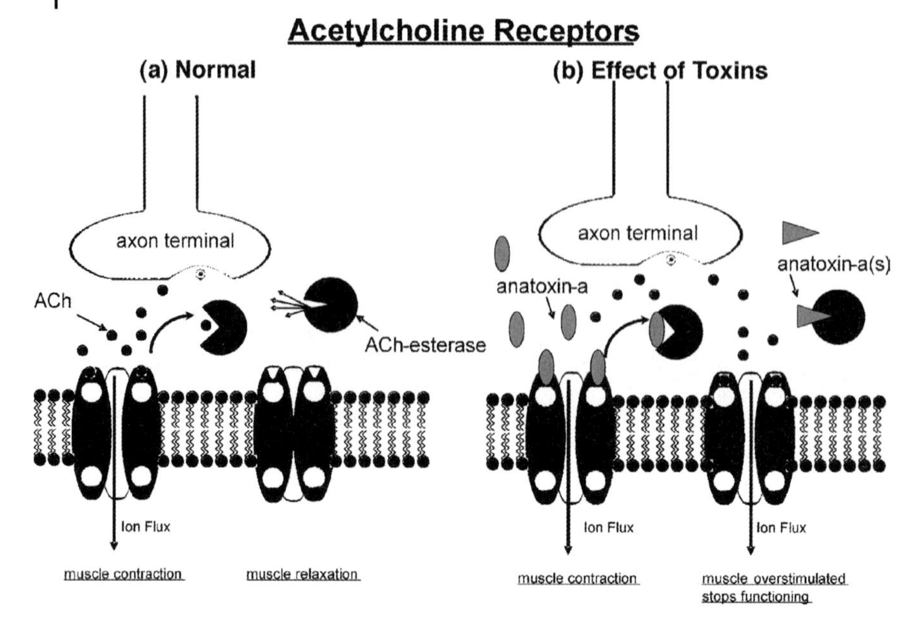

Fig. 6.1 (a) The neurotransmitter acetylcholine (ACh) (small circles) bind to the ACh receptor and stimulate muscle contraction. Acetylcholinesterase (black partial circle) degrades ACh allowing the muscle cells to return to a resting state. (b) Anatoxin-a, homoanatoxin-a and analogs (gray ovals) also bind to the ACh receptor, but are not degraded by acetylcholinesterase, resulting in overstimulation of the ACh receptors. Anatoxin-a(s) (gray triangle) inhibits acetylcholinesterase, preventing breakdown of ACh, and leading to muscle overstimulation.

investigations. Considerable research throughout the twentieth century has focused on toxic freshwater cyanobacterial blooms, detection of toxins, and preventative measures to protect both livestock and human populations. However, it was not until the mid-1970s that isolation and structural elucidation of the extremely potent cyanobacterial neurotoxins anatoxin-a and saxitoxin was achieved. Anatoxin-a is a nicotinic acetylcholine (nACh) receptor agonist (Figure 6.1), and saxitoxin selectively blocks sodium channels in excitable membranes (Figure 6.2). These neurotoxins, together with analogs, have proven to be invaluable biomedical tools for the characterization of neuronal nACh receptors and ion-gated channels.

Neurotoxic events linked to cyanobacteria are not limited to freshwater or marine conditions. For example, there is growing evidence that the neurological condition amyotrophic lateral sclerosis–parkinsonism-dementia complex (ALS-PD) is caused by consumption of β-methylaminoalanine (BMAA). BMAA is found in cyanobacteria which grow in the tissues of plants and is possibly biomagnified in the food chain in areas affected by ALS-PD.

Recognition that the secondary metabolites produced by cyanobacteria represent an untapped reservoir of chemically rich compounds prompted more focused efforts

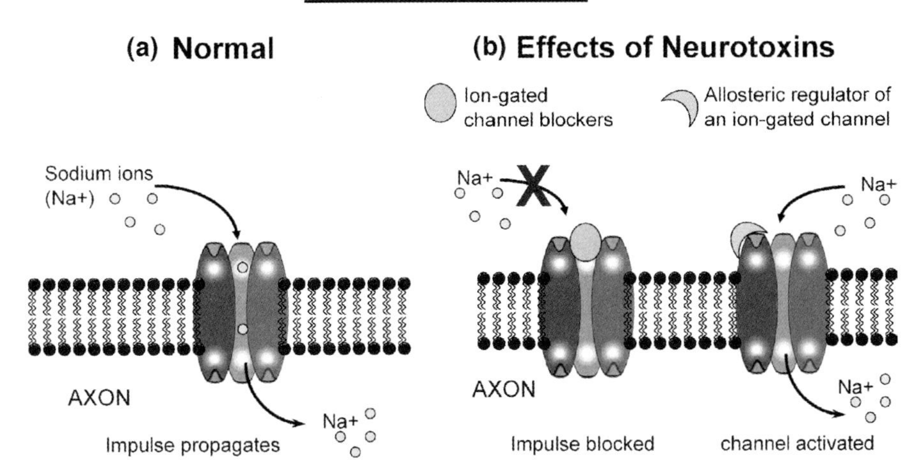

Fig. 6.2 (a) Normal propagation of nerve impulses requires influx of sodium ions (small circles) into cells via voltage-gated sodium channels. (b) Saxitoxin, neosaxitoxin, and kalkitoxin (large circle) block the channel and prevent influx of ions. Antillatoxin (crescent) acts by specifically activating voltage-gated sodium channels through binding to an allosteric site.

to discover unique compounds from these sources. This effort has led to the discovery of several important neurotoxic alkaloids from marine cyanobacteria: the antillatoxins, jamaicamides, and kalkitoxins, which will be the focus of the second section of this chapter. Antillatoxin, isolated from a shallow-water collection of *Lyngbya majuscula*, is an elegant example of bioassay-guided isolation of a potent neurotoxin from marine cyanobacteria. The jamaicamides are an extraordinary example of the focus on elucidating the biosynthetic pathways of cyanobacterial secondary metabolites. The phenomenal potential of such endeavors, and the unique enzymes they are revealing, is only just now being realized. Finally, kalkitoxin will also be an important biomedical tool because its interaction with neuronal voltage-gated sodium channels (VGSC) is distinct from that of other natural product derived VGSC mediating neurotoxins, such as antillatoxin.

6.2 Neurotoxic Alkaloids of Principally Freshwater and Terrestrial Cyanobacteria

6.2.1 Anatoxin-a, Homoanatoxin-a, Anatoxin-a(s), and Analogs

The anatoxins are neurotoxins produced by freshwater cyanobacteria that have caused periodic poisonings of wildlife, livestock, fowl, and fish in several countries.

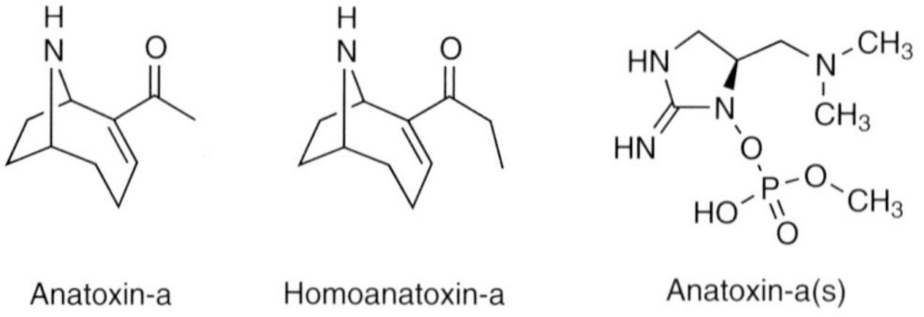

Anatoxin-a Homoanatoxin-a Anatoxin-a(s)

Fig. 6.3 Structures of anatoxin-a, homoanatoxin-a, and anatoxin-a(s).

These toxins are composed of the structurally related alkaloids anatoxin-a, homoanatoxin-a, and synthetic analogs, as well as the phosphate ester of a cyclic *N*-hydroxyguanidine, anatoxin-a(s) (Figure 6.3). The path to the isolation and characterization of these compounds began in the early 1960s when toxic strains were obtained from blooms of *Anabaena flos-aquae* (Lyngb.) de Bréb., which caused cattle poisoning at Burton Lake in Saskatchewan, Canada [4]. The toxic effect on mice was rapid (2–5 min) and the toxin, now known as anatoxin-a, was therefore named very fast death factor (VFDF) [4].

6.2.1.1 **Anatoxin-a**

Anatoxin-a is a naturally occurring homotropane alkaloid produced by freshwater cyanobacteria of the genera *Anabaena* (*A. flos-aquae* and *A. circinalis*), *Aphanizomenon, Cylindrospermum, Planktothrix, Microcystis aeruginosa* [5–7], and *Phormidium favosum* [8]. Fatal intoxications have typically included cattle and birds [9], and, more recently, dogs [8] and flamingos [10].

Anatoxin-a, the first highly potent cyanotoxin to have its structure and absolute stereochemistry elucidated, was originally isolated from a unialgal clone of *Anabaena flos-aquae* (NRC-44h) [5]. The structure was confirmed by X-ray crystallographic data for the *N*-acetyl derivative [11] and additional studies have since provided further proof for the structure and stereochemistry (for example, [12]). Anatoxin-a is an unsymmetrical bicyclic secondary amine, and was the first naturally occurring alkaloid discovered to contain a 9-azabicyclo[4,2,1]nonane (homotropane) skeleton. Homotropanes are one-carbon analogs of the tropanes and, as such, are structurally closely related to the well-known alkaloid cocaine.

The biosynthesis of anatoxin-a in *Anabaena flos-aquae* has been examined using feeding experiments with ^{13}C-labeled glutamate and acetate, indicating an intermediate formed by glutamic acid and three acetate units (Figure 6.4) [13,14]. This experiment showed that the entire C_5 carbon skeleton of [$^{13}C_5$]-(*S*)-glutamate was incorporated into anatoxin-a, and thus demonstrated that the biochemical mechanism of the assembly of the carbon skeleton of anatoxin-a and homoanatoxin-a differs from the structurally similar homotropane and tropane alkaloids, such as tropine and cocaine. This was in conflict with previous reports postulating incorporation of the C_4 diamine putrescine into anatoxin-a [15,16].

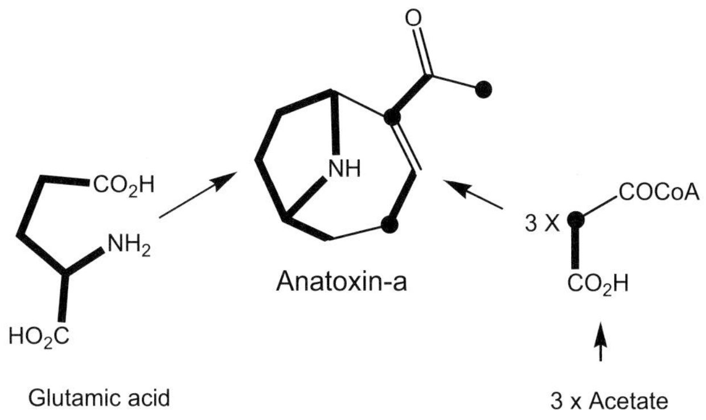

Fig. 6.4 Biosynthetic subunits to anatoxin-a from feeding various labeled substrates [13].

Anatoxin-a is an extremely potent nicotinic acetylcholine (nACh) receptor agonist, mimicking the neurotransmitter acetylcholine (ACh) as a depolarizing neuromuscular agent but with higher affinity than ACh itself [17]. Unlike ACh, anatoxin-a is not hydrolyzed by acetylcholinesterase. These two mechanisms lead to an over-stimulation of muscle tissue and resulting toxicity characterized by paralysis of the respiratory and peripheral skeletal muscles. In turn, this leads to tremors, convulsions, and eventually respiratory failure and death. The toxicity has been well characterized in a number of different biological systems (for example, refs [17–19] and references therein). Originally, the toxic effect of a crude toxin preparation was studied in calves, rats, ducks, and goldfish, as well as in isolated muscle such as rat, duck, and pheasant sciatic nerve-gastrocnemius, frog rectos abdomens, and guinea pig ileum by Carmichael and coworkers in the 1970s [20,21]. Anatoxin-a exerts its toxic effect primarily through binding muscle nACh receptors, leading to the observed muscle paralysis and respiratory failure.

Anatoxin-a has been a key biomolecular tool for exploring structural features of nACh receptors. Owing to the inability of acetylcholinesterase to degrade anatoxin-a and its analogs, these compounds can be used to mimic ACh to study mechanisms of binding and nACh receptor function. Recent research with this drug has focused on neuronal nACh receptors and developing a detailed understanding of its pharmacology and mechanism of action; these studies have been enhanced by the SAR studies discussed above. The natural enantiomer binds with high affinity to the $\alpha4\beta2$ nicotinic receptor ($K_i = 0.34$ nM) as compared to $\alpha3\beta4$ and $\alpha7$ nicotinic receptors ($K_i = 2.5$ and 91 nM, respectively) (see ref. [22] and references therein). The successful syntheses of analogs specific to neuronal subtypes could be of great value to the treatment and study of a wide range of neurological conditions.

The unique structure of anatoxin-a, and its potential as a pharmacological tool, has inspired many different chemical syntheses of anatoxin-a and analogs. The earliest synthesis, in 1977, used (−)-cocaine as starting material [23]. Subsequently, there have been numerous approaches to its synthesis from acyclic or commercially available cyclic starting materials. These have included ring expansion of tropanes, cyclization of cyclooctanes, cyclization of iminium salts, cycloaddition of nitrones, and electrophilic cyclization of allenes, all of which have been thoroughly reviewed up to 1996 [24]. Additional synthetic approaches since then have utilized palladium-mediated transannular cyclization [25], β-lactam ring opening [26,27], enyne metathesis [28–30], and diastereoselective cycloaddition of dichloroketene [31]. In 2000, (−)-cocaine was used to obtain enantiopure (+)-anatoxin-a in an approach that has since yielded a number of biologically active analogs [32]. Indeed, the variety of approaches employed for anatoxin-a synthesis has produced a wide range of structural analogs.

The SARs of anatoxin-a and analogs have been extensively studied [33–36], and only a few important findings dealing with chirality, conformation, and other structural features of anatoxin-a and homologs will be summarized here. A recent review that addresses the chemistry and pharmacology of anatoxin-a and analogs provides a more detailed treatment of this topic [37].

The naturally occurring enantiomer, (+)-anatoxin-a, is much more potent than (−)-anatoxin-a [38]. For example, using conditions that preferentially labeled rat brain α4β2 nACh receptors with [^{3}H]nicotine, the natural enantiomer was found to be 1000-fold more potent than (−)-anatoxin [39]. This enantiospecificity of anatoxin-a has provided an excellent tool for probing the stereospecificity of the ACh binding site on the nicotinic receptor.

As compared to the natural ligand ACh, the conformation of anatoxin is relatively rigid, with only one rotatable bond as compared to four in ACh. This simplifies the ability to correlate the structural conformation of the ligand to efficacy, and, consequently, attention has been directed toward determining the active conformation through the use of sterically constrained analogs of the s-*cis* and s-*trans* conformers of anatoxin-a [19,40–44]). Although current evidence leans toward s-*trans* as the active conformation, conclusive evidence, such as a crystal structure of (+)-anatoxin-a in complex with an nACh receptor, is still needed.

The synthesis of anatoxin-a analogs has primarily focused on modifications at N-9 or C-10 and C-11, as these are relatively accessible by synthesis and are important for bioactivity. N-Methylation at N-9 results in reductions in activity, as measured by toxicity, binding to neuronal nACh receptors, and activation of muscle nACh receptors [34,37,45]. A variety of modifications at C-10 also resulted in reduced activity [46]. Interestingly, two *N*-alkoxy amide variants retained more affinity for the neuronal α4β2 nACh receptors as compared to the neuronal α7 or muscle nACh receptors, opening the possibility that anatoxin-a homologs could demonstrate subtype specificity. Modifications at C-11 have been more successful. Extension of this position by one methylene unit resulted in homoanatoxin-a, a homolog with similar activity to anatoxin-a [33] and which was subsequently isolated from natural sources (see below). Ensuing studies have investigated the influence of steric bulk [26] and altered functionality [36] at C-11 on ligand–receptor interactions. In addition, hybrids of anatoxin-a and other nicotinic ligands have been useful both for probing the SAR of nACh receptor–ligand interactions and for generating drug leads. Kanne and coworkers synthesized structurally constrained anatoxin/nicotine variants in the late 1980s [40–42], a theme that has also been pursued by Seitz and coworkers [43]. Several generations of epibatadine/anatoxin-a homologs have been synthesized and used to probe the SAR of subtypes of neuronal $\alpha4\beta2$, $\alpha7$, and $\alpha3\beta4$ nACh receptors [37,47].

At present, there is no antidote for poisoning by anatoxin-a, making prevention and early detection of contaminated water essential. A wide variety of approaches have been utilized for detection of both anatoxin-a and homoanatoxin-a, including high performance liquid chromatography (HPLC), GC-MS, LC-MS, and capillary electrophoresis-based methods (see refs [48,49] and references therein). Anatoxin-a is sensitive to sunlight and high pH and the metabolites created by photochemical- or oxygen-mediated degradation, such as dihydroxyanatoxin-a, may not be toxic. These oxidative derivatives represent potentially important biomarkers of toxicity that may also be included in detection assays. For example, James and coworkers [50] described the simultaneous determination of anatoxin, homoanatoxin-a and their dihydro and epoxy degradation products by HPLC using derivatization with 4-fluoro-7-nitro-2,1,3-benzoxadiazole and fluorimetric detection.

6.2.1.2 **Homoanatoxin-a**

Homoanatoxin-a, obtained from various freshwater cyanobacteria, is a relatively rare natural analog of anatoxin-a for which the C-11 side chain is extended by one methylene unit (Figure 6.4). It was originally isolated from *Planktothrix* sp. (formerly *Oscillatoria*) in 1992 [51]. It has recently been isolated from *Raphidiopsis mediterranea* Skuja from Japan [52] and *Planktothrix* (formerly *Oscillatoria*) *formosa* blooms in Ireland [53].

The total synthesis of homoanatoxin-a, as an analog of anatoxin-a, was also accomplished in 1992 [33]. It is a potent nicotinic agonist active at the postsynaptic nicotinic ACh receptor channel complex [54]. Although not as potent as anatoxin-a, homoanatoxin-a has provided valuable insight for SAR studies of anatoxin-a and its homologs.

Biosynthesis of homoanatoxin-a was examined using *Oscillatoria formosa* and a mechanism similar to that for anatoxin-a was proposed [13,14]. The origin of the C-12 methyl group that distinguishes homoanatoxin-a from anatoxin-a, was shown through feeding experiments performed with L-[methyl-^{13}C]-methionine in the culture of *Raphidiopsis mediterranea* Skuja. It was proposed that the *S*-methyl of methionine is transferred to the toxin via *S*-adenosyl-L-methionine (SAM)-mediated methylation [55].

6.2.1.3 **Anatoxin-a(s)**

Anatoxin-a(s) is produced in both *Anabaena flos-aquae* [56] and *Anabaena lemmermannii* [57]. The symptoms of anatoxin-a(s) intoxication are similar to those of anatoxin-a but cause increased salivation in vertebrates; hence, the similarity in the names of these compounds with the (s) added to indicate "salivation" [18]. The structures, however, are quite different, as are the mechanisms of action.

Anatoxin-a(s) was originally isolated from *A. flos-aquae* in 1963 from Buffalo Pound Lake, Saskatchewan, Canada [56]. The structure was determined in 1989 by the Moore group [58]. Anatoxin-a(s) is a naturally occurring organophosphate that is similar in structure to synthetically produced organophosphate-based insecticides (Figure 6.4). Although organophosphate-based acetylcholinesterase inhibitors have been found in terrestrial bacteria, such as *Streptomyces antibioticus* [59], anatoxin-a(s) is the only known example produced by cyanobacteria.

Biosynthesis of anatoxin-a(s) has been investigated in *Anabaena flos-aquae* through feeding experiments with a series of labeled precursors. The amino acid L-arginine was identified as the source of the carbons of the triaminopropane backbone and the guanidine function [60], and (2*S*,4*S*)-4-hydroxyarginine was shown to be an intermediate in the biosynthetic pathway [13,14].

Anatoxin-a(s) inhibits acetylcholinesterase by acting as an irreversible active-site-directed inhibitor [61]. This prevents degradation of ACh and leads to over-stimulation of the muscle cells (Figure 6.1) [56,62]. Thus, although the mechanism of action of anatoxin-a(s) is quite different from that of anatoxin-a, the observed toxicity is similar. In addition, it was the first irreversible acetylcholinesterase inhibitor to be found in a cyanobacterium.

Carmichael speculates that anatoxin-a(s), being more water soluble than other insecticides and more biodegradable, could be a theoretical starting point for designing safer insecticides. This depends upon whether the analogs could penetrate the lipid-rich exoskeletons of insects. Thus, "by tinkering with the structure of anatoxin-a(s), investigators might be able to design a compound that would minimize accumulation in tissues of vertebrates but continue to kill agricultural pests" [18]. Appealing as this idea is, to date there are no reports of a successful reduction-to-practice of this concept.

6.2.2
β-Methylaminoalanine

Amyotrophic lateral sclerosis, parkinsonism, and dementia are neurodegenerative diseases commonly diagnosed throughout the world; however, in a few specific locations the clinical symptoms of these three diseases combine into a single fatal disorder known as amyotrophic lateral sclerosis–parkinsonism-dementia complex (ALS-PD) also known as "lytico-bodig." This rare disease is commonly seen in certain regions of the eastern Pacific including Irian Jaya in Indonesia, the Kii peninsula in Japan, and among the Chamorro population of Guam [63].

ALS-PD is a disease of the upper and lower motor neurons characterized by muscular atrophy, weakness, spasticity, slowed movements, tremors, and rigidity, as well as cognitive dysfunctions or dementia similar to that seen in patients diagnosed with Alzheimer's disease [64,65]. While the disease is found in all three regions mentioned above, most of the etiological studies have focused on the Chamorro population of Guam. In the 1950s, it was determined that the incidence rate of ALS-PD in Guam was 50–100 times that of more-developed countries [66]. Extensive studies of familial genealogy showed no indication of familial clustering or genetic inheritance. Other studies showed no apparent viral or transmissible factors common to the disease, leading to the proposal that an environmental agent was a contributing factor [64].

Field studies in each of the three high-incidence regions focused on commonalities in customary food and medicinal sources, and revealed that palm-like plants from the family Cycadaceae were commonly used in all three regions. Six of the nine genera of these primitive seed plants, mainly found in the tropics and subtropics, are implicated in initiating toxic symptoms in man or higher animals [67,68]. The Chamorro people used cycad plants as a food source, particularly in the preparation of flour, and recognized the toxicity of these seeds as indicated by their thorough washing prior to use [63,69].

Plants of the family Cycadaceae were first connected to neurological disease in 1966 when Mason and Whiting reported a disorder in cattle involving irreversible paralysis of the hindquarters after the ingestion of the leaves of four genera [67]. This type of paralysis was also common in Australia and the West Indies [70] and was attributed to being similar to classical lathyrism, a disease of the nervous system found in higher animals after the consumption of β-oxalylaminoalanine (BOAA) found in the chickling pea, *Lathyrus sativus* [64]. After careful analysis of *Cycas circinalis*, a member of the Cycadaceae family, Vega and Bell identified the structurally related amino acid β-methylaminoalanine (BMAA) as a possible culprit for the neurotoxicity of these plants (Figure 6.5) [67].

After failed attempts to induce neurological symptoms similar to ALS-PD experimentally by BMAA administration, investigations on cycad-derived BMAA as a neurological agent were abandoned until 1987 when Spencer and coworkers were able to induce a motor-system disorder with involvement of the upper and lower motor neurons in macaque primates (*Macaca fascicularis*) [66]. The renewed interest in BMAA as a neurotoxin led to the question of its connection to ALS-PD in the Chamorro people. Sieber and coworkers challenged the amount of cycad seed needed to be consumed to produce neurotoxic effects, after studies showed no neurological changes in primates fed large quantities of unprocessed cycad meal. Studies indicate that 100 g of seed kernels processed into flour contain between 64 and 143 mg of BMAA before processing, but after preparative washing, 80 % of the BMAA is washed away [63]. Therefore, on average a person consuming two cycad tortillas per day would ingest 1.1 mg/day BMAA, believed to be a nontoxic quantity [69]. The low

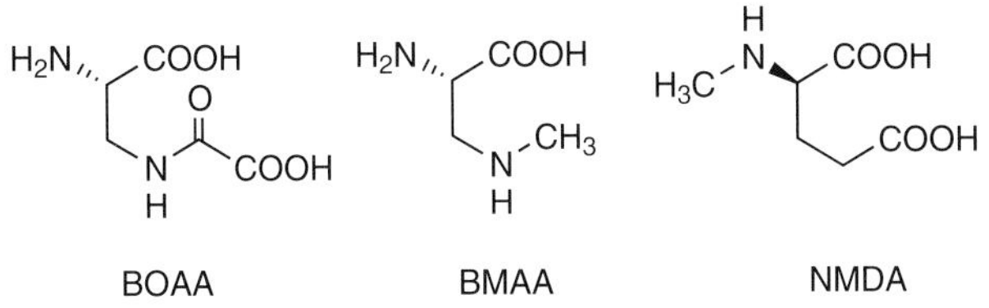

BOAA BMAA NMDA

Fig. 6.5 Structures of β-methylaminopropionic acid (BMAA) and two related compounds, BOAA and NMDA (see text for discussion).

concentrations consumed and the prevalence of the use of cycad seed in many other regions including Indochina, India, Fiji, and Australia for centuries without evidence of increased incidence of neurological disease, led to the examination of other sources for BMAA [71].

In 2003, Banack and Cox proposed that consumption of flying foxes during traditional feasts by the Chamorro people may provide the required higher dosage source of BMAA [72]. The abundant feeding by flying foxes on cycad seeds may provide a means for transfer of large amounts of BMAA up the food chain to humans [65]. The Chamorro people are known traditionally to celebrate during feasts by consuming entire flying foxes, sometimes several in a single week [73]. Because flying foxes are an endangered species in Guam, Banack, and Cox used skin-tissue samples from museum specimens of *Pteropus mariannus* taken from the Museum of Vertebrate Zoology (MVZ) at the University of California, Berkley to test for BMAA. When compared to cycads, these skins contained quantities of BMAA between 1287 μg/g and 7502 μg/g [72]. The idea that the ingestion of flying foxes leads to increased exposure to BMAA by humans correlates with a decline in the incidence of neurological disease among the Chamorro people some 10–20 years after the collapse of the flying fox population [73].

The notion that BMAA could be biomagnified through the Guam ecosystem led Cox, Banack, and Murch to reinvestigate the distribution pattern of BMAA in various tissues of *Cycas micronesica*. They found BMAA was concentrated in morphologically specialized "coralloid" roots but not in the unspecialized roots. These specialized roots were also found to contain a nitrogen-fixing cyanobacterium, identified as *Nostoc* sp., which produces 0.3 μg BMAA per gram of cyanobacterial tissue. It was subsequently shown that the BMAA concentration in these roots depended on the health of the cyanobacterial partner [73]. Testing of various cyanobacterial cultures obtained from the University of Dundee, Stockholm University, and the University of Hawaii, as well as natural bloom samples, showed that 73 % of *Nostoc* sp. strains that are normally in symbiotic relationships produce BMAA. This included *Nostoc* in symbiosis with both *Azolla filiculoides* (2 μg/g) and *Gunnera kaudiensis* (4 μg/g). Both free and protein-associated BMAA were quantified using fluorescent derivatization of amino acids coupled with HPLC, and led to the finding that BMAA is present in all five major taxonomic sections of cyanobacteria (Table 1). Cox and coworkers also found BMAA production in 95 % (20/21) of the cyanobacterial genera tested and 97 % (29/30) of the strains tested, including a marine *Trichodesmium* from a Hawaiian bloom [74].

The emerging hypothesis is that cyanobacteria produce BMAA, which is then biomagnified through first the cycad and then the flying fox trophic levels [73]. However, whether neurotoxicity results from the transfer of BMAA to humans by consumption of contaminated food still remains uncertain. Cox and coworkers examined the superior frontal gyrus from the deceased brains of six ALS-PD patients from Guam, two Alzheimer's patients from Canada, two asymptomatic Chamorros, and 13 individuals with no signs of neurodegeneration. They found BMAA in the brain tissues of the six Chamorro ALS-PD patients and in the two patients from Canada, but BMAA was absent in the brain tissues of the 13

Tab. 6.1 BMAA detected in both free and protein-associated forms in free-living cyanobacteria derived from.

Cyanobacterial species/strains	Section	Habitat	Origin	Free BMAA (μg/g)	Protein-associated BMAA (μg/g)
Microcystis PCC 7806	I	Freshwater	The Netherlands	4	6
Synechococcus PCC 6301	I	Freshwater	USA	25	ND
Myxosarcina burmensis GB-9-4	II	Marine coral	Marshall Islands	79	1,943
Lyngbya majuscula	III	Marine	Zanzibar	32	4
Symploca PCC 8002	III	Marine, intertidal	UK	3	262
Trichodesmium thiebautii	III	Marine	Caribbean	145	8
Anabaena variabilis ATCC 29413	IV	Freshwater	USA	35	ND
Cylindrospermopsis raciborskii CR3	IV	Freshwater	Australia	6,478	14
Nostoc sp. CMMED 01	IV	Marine	Hawaiian Islands	1,243	1,070
Fischerella PCC 7521	V	Yellowstone hot spring	USA	44	175
Scytonema PCC 7110	V	Limestone cave	Bermuda	ND	1,733

individuals without neurodegenerative disease [73]. That BMAA was found in the brain tissues of patients from Canada where cycads are not a part of the normal flora or diet suggests that cyanobacteria ingested from other foods may also be a source of BMAA [74].

Because BMAA may be involved in neurodegenerative diseases, understanding the mechanism of action of this nonprotein amino acid is important. A long latency period is known between exposure to BMAA and the occurrence of ALS-PD in the Chamorros; onset of disease occurs mostly in adults over 40 years [75]. As a result, Spencer suggested in 1991 that BMAA may act as a "slow toxin." If ALS-PD is caused by BMAA, then it seems necessary that a buildup of BMAA must develop. This is required because at present there are no known environmental neurotoxins which produce a significantly delayed onset of symptoms as well as a progressive neurological disease from a single exposure. The amino acid nature of BMAA would initially suggest that it is a poor candidate for biomagnifications, because it is not lipophilic and will not accumulate in fatty tissues as has been determined for other environmental agents. However, because it is an amino acid, it may be incorporated into proteins. Murch, Cox, and Banack explored this idea by looking for a "bound" or protein-associated form of BMAA. They found a 60–130-fold greater quantity of BMAA in the protein form than was recovered from the free amino acid pool throughout most of the trophic stages of the Guam ecosystem. Figure 6.6 shows a

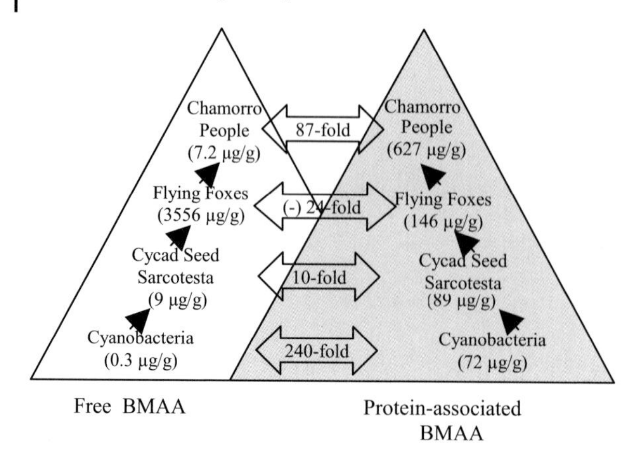

Fig. 6.6 Comparison of free and protein-associated BMAA biomagnified in the Guam ecosystem.
Source: Redrawn from ref. [76].

comparison of the biomagnification of both the free and the protein-associated form of BMAA in Guam.

It has been suggested that BMAA in protein may function as an endogenous neurotoxic reservoir in humans, and may slowly release the neurotoxin directly into the brain during protein metabolism. Incorporation of a nonproteinaceous amino acid into a protein may have other serious impacts, such as creating proteins of aberrant function or occurrence [76].

The involvement of BMAA in neurodegenerative disease is still a controversial topic. The debate centers around the lack of an animal model to demonstrate the direct onset of the symptoms of ALS-PD after exposure to BMAA [76]. However, *in vitro* studies on explanted fetal mouse spinal cord [77] indirectly indicate that BMAA may act on glutamate receptors in a manner similar to other excitatory amino acids, including BOAA [78]. Of the three subtypes of glutamate receptors, differentiated by their preferred activation by *N*-methyl-D-aspartate (NMDA), quisqualate, or kainate, BOAA appears to operate on those activated by kainate while BMAA is thought to operate on receptors activated by NMDA [77]. Specifically, BOAA was found to be blocked by a broad spectrum of glutamate antagonists but not by NMDA antagonists while the effects of BMAA are blocked by specific NMDA antagonists [64]. However, BMAA may not function solely as an agonist of NMDA receptors. Copani and coworkers showed that BMAA may also be an activator of "metabolotropic" glutamate receptors and Weiss and coworkers showed that the mechanism of action may be dependent on the concentration of BMAA present. In this latter study, it was found that BMAA is active on non-NMDA receptors (kainate and quisqualate receptors) at very low concentrations [79,64].

Chemically, BMAA agonism seems incompatible with activity at glutamate receptors, as it lacks the characteristic dicarboxylic acid structure of other excitotoxins [80]. However, the activity of BMAA is dependent on the presence of

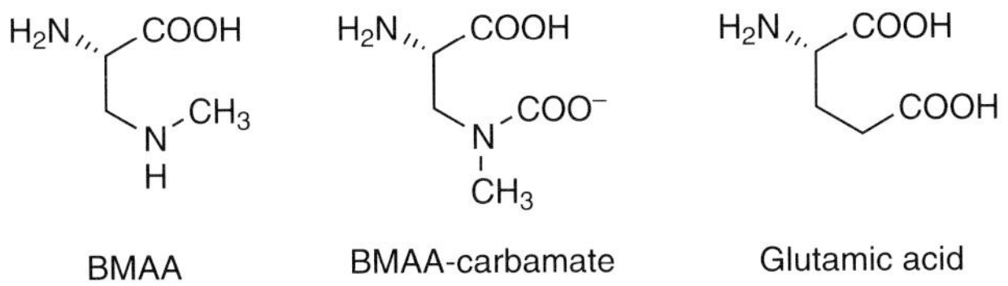

BMAA BMAA-carbamate Glutamic acid

Fig. 6.7 Structural comparison of BMAA, the carbamate of BMAA, and glutamic acid, three substrates for NMDA receptors [65].

extracellular bicarbonate [81]. Bicarbonate may interact with BMAA, either non-covalently or through forming a carbamate structure (Figure 6.7), which then possesses a terminal electronegative moiety that is more compatible with the receptor [64].

Although more research is needed to completely understand the mechanism of action of BMAA, Allen and coworkers suggest that the overall effects of BMAA may be to depolarize postsynaptic neurons, thus relieving the magnesium blockade of the calcium ion channel and producing a calcium influx. This in turn triggers post-synaptic swelling and neuronal degeneration [82]. This mechanism is similar to that of domoic acid, a neurotoxin produced by marine diatoms [65].

As noted above, many questions remain concerning the role of BMAA in causing ALS-PD. Repeated analysis of Guamanian patients diagnosed with neurological disease fail to show a buildup of free BMAA in brain tissue [83]. Chronic administration of large doses of BMAA produces no alteration in the levels of glutamate and aspartate in rodents; however, this may result from pharmacological attributes unique to rodents [78]. Nevertheless, the facts that a majority of cyanobacteria produce BMAA and that BMAA can biomagnify through trophic levels to potentially toxic levels in proteins of the human brain, leads to global concerns about this natural product. A clearer understanding of the etiology and mechanism of action of BMAA is needed to make a definitive prediction as to its true role in neurological disease. It remains possible that the high prevalence of ALS-PD in some human populations is caused by another environmental agent or by a combination of agents, such as cycasin and BMAA, which are both found in cycad plants [76,84]. The one certainty about BMAA is that many questions remain, despite its nearly 40-year scientific history.

6.2.3 Saxitoxin

Saxitoxin (STX) is one of the most potent neurotoxic alkaloids known, and is produced by a taxonomically diverse group of algae and cyanobacteria. It first gained notoriety for causing the intoxication associated with ingesting toxic bivalves such as clams, mussels, and scallops. This intoxication, commonly referred to as paralytic shellfish poisoning (PSP), exerts its effects on the neuromuscular system through a specific blockage of voltage-gated sodium channels

(VGSC). Saxitoxin poisoning was first observed in temperate waters along the coasts of North America, Japan, and Europe during dinoflagellate blooms called "red tides." It is now found widely distributed in both hemispheres in marine and freshwater environments, associated closely with cyanobacteria and dinoflagellate blooms. However, even one species of macroalgae, *Jania* sp., is reported to produce STX [85]. An estimated 40 genera of freshwater and marine cyanobacteria produce "cyanobacteria toxin poisons" (CTP); among these are several which produce STX, including *Anabaena*, *Aphanizomenon*, *Cylindrospermopsis*, *Oscillatoria* (*Planktothrix*), *Trichodesmium*, and *Lyngbya* [86–89].

There are more than 30 saxitoxin analogs that are naturally occurring. These include various patterns of ring hydroxylation and sulfation. Paralytic shellfish toxin (PST) is a term recently used to distinguish saxitoxin and its analogs from the syndrome they cause, PSP [90]. These analogs include the hydroxylated derivative neosaxitoxin (neoSTX) as well as decarbamoylated derivates known as the gonyautoxins (GTX) and "C toxins" [91–95] (Figure 6.8). There are many groups of animals that are able to harbor PSTs without ill effects, including bivalves, crustaceans, fish, gastropods, echinoderms, and even ascidians. Comprehensive tables of producers and carriers of PSTs can be found in a review by Llewellyn *et al.* [90].

One of the earliest studied events involving PST-producing cyanobacteria occurred in the US between 1966 and 1980 when a series of *Aphanizomenon flos-aquae* blooms afflicted rural New Hampshire. The first of these caused the death of more than six tons of fish in Kezar Lake [96]. A later bloom occurred in a small farm pond in Durham, New Hampshire. The active material first isolated and tested from the Kezar Lake location was called "aphantoxin" [97]. Four PST-like compounds were isolated from this outbreak, one of which was shown to be STX [98]. NeoSTX and STX were later identified and confirmed in the strain isolated from the latter location [99,100]. The extraction of aphantoxin was carried out by sonication of lyophilized cells followed by ultrafiltration through a 10 kDa membrane. Multiple fractions were collected by a combination of extraction and chromatography, and these were evaluated by the standard mouse bioassay along with a fluorescence assay based on the alkaline hydrogen peroxide oxidation of STX [99]. Further analysis included electrophysiological voltage clamp experiments on squid axons and the use of HPLC [101].

Although the New Hampshire blooms of *A. flos-aquae* occurred sporadically over a period of 14 years, they did not attract the widespread attention that a later Australian bloom caused. In 1991, one of the largest known cyanobacterial blooms to date occurred in the Barwon-Darling River, Australia, covering around 1000 km^2 and causing the death of huge numbers of fish and some livestock [102,103]. The dominant species responsible for the bloom was found to be *Anabaena circinalis* [104]. *A. circinalis* had previously been reported to be both neurotoxic and hepatotoxic in mouse bioassays [105,106]; however, PSTs were subsequently reported to be the primary, if not the only, toxin. Cyanobacterial samples were shown to contain STX, multiple gonyautoxins and at least two C-toxins [18,104]. The cause of the hepatotoxic symptoms has not yet been explained [102].

In the late 1950s, after several years' effort by various academic and government institutions, saxitoxin was finally isolated in pure form. Purification was accomplished by ion-exchange chromatography on carboxylic acid resins, followed by chromatography on acid-washed alumina in absolute ethanol [107,108]. Purified STX is a white, hygroscopic, water-soluble compound with no ultraviolet absorption above 210 nm. As the dihydrochloride salt it tends to be stable in acidic solution but loses activity around pH 7–9 [109,110]. STX possesses two pK_as, 8.22 and 11.28, which correspond to the 7,8,9- and 1,2,3-guanidium groups, respectively [90]. NeoSTX possesses an additional pK_a at 6.75 which is correlated with the N-1-hydroxyl group. In 1964, Rapoport *et al.* suggested saxitoxin possessed an unusually substituted tetrahydropurine structure and coined the name "saxitoxin," derived from *Saxidonus giganteus* (Alaskan Butter Clam), from which his material was obtained [111].

Determination of the molecular structure of STX was difficult owing to its noncrystalline, highly polar, and nonvolatile nature. Thus, it was not until 1975 when the Clardy and Schantz groups as well as the Rapoport group independently determined the structure of STX by X-ray crystallography [92,112]. Both groups synthesized analogs of STX, an ethyl hemiketal analog (Bordner *et al.* [92]) and a *p*-bromobenzenesulfonate analog (Schantz *et al.* [112]), which made crystallization possible. The definitive crystal structure of STX permitted its chemical synthesis, the elucidation of its biosynthetic pathway, and a physiological study of its mechanism of action.

In the 1980s, Shimizu *et al.* published a series of papers on the biosynthesis of STX and some of its analogs, including neoSTX. Feeding studies using the cyanobacterium *Aphanizomenon flos-aquae* and a variety of radiolabeled amino acids, as well as $^{13}CO_2$[113] revealed that STX is not formed from a purine derivative as one might first postulate (Figure 6.8). Instead, it is produced by a complicated biosynthetic pathway that involves a rare Claisen-type condensation of acetate or its derivative with arginine. This conclusion was based on results which showed two units of acetate are incorporated at C-5–C-6 and C-10–C-11. Labeling studies with [^{13}C–^{15}N]-arginine with NMR detection demonstrated that N-2 and C-2 of arginine are incorporated intact. The amino group is then converted to a guanidino group by the transfer of an amidine moiety from a second arginine unit. The side chain carbon C-13 was shown to be introduced by methylation of the ring system with *S*-adenosylmethionine (SAM). Incorporation of SAM was also based on the more efficient utilization of methionine over other one-carbon donors (e.g. methyl or methylene tetrahydrofolate). Overall, portions of three arginines, the *S*-methyl group of one methionine, and one acetate are needed for the biosynthesis of one STX molecule [114–118] (Figure 6.9).

One of the more prominent features of STX and its analogs is the presence of two positively charged guanidinium groups (Figure 6.8). Because guanidine is one of the few cations that can effectively act as a substitute for sodium in the generation of action potentials, a proposed mechanism of action for STX was developed. Kao and Nishiyama first suggested that the charged guanidinium group of STX enters the sodium channel like sodium or guanidine, but that the bulky toxin tightly binds to the

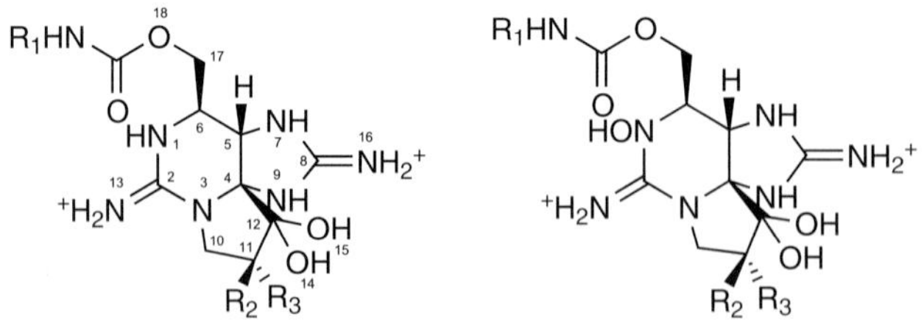

Saxitoxin (STX): R_1, R_2, R_3=H
Gonyautoxin-II (GTXII): R_1, R_2=H, R_3=OSO^-
Gonyautoxin-III (GTXIII): R_1, R_3=H, R_2=OSO_3^-
Gonyautoxin-V (GTXV or B1): R_1=SO_3^-, R_2, R_3=H
epi-Gonyautoxin-VIII (*epi*-GTXVIII or C1): R_1=SO_3^-, R_2=H, R_3=OSO_3^-
Gonyautoxin-VIII (GTXVIII or C2): R_1=SO_3^-, R_2=OSO_3^-, R_3=H

Neosaxitoxin (neoSTX): R_1, R_2, R_3=H
Gonyautoxin-IV (GTXIV): R_1, R_2=H, R_3=OSO^-
Gonyautoxin-I (GTXI): R_1, R_3=H, R_2=OSO_3^-
Gonyautoxin-VI (GTXVI or B2): R_1=SO_3^-, R_2, R_3=H
C3: R_1=SO_3^-, R_2=H, R_3=OSO_3^-
C4: R_1=SO_3^-, R_2=OSO_3^-, R_3=H

Fig. 6.8 Naturally occurring analogs of saxitoxin produced by both cyanobacteria and dinoflagellates. STX and neoSTX differ in the hydroxylation of N-1. DecarbamoylSTX (dcSTX) lacks the carbamoyl unit on O-18.
Source: Redrawn and expanded from ref. [104].

channel and effectively prevents sodium influx [119]. However, it was not until 1983 that this hypothesis was tested when Kao *et al.* identified the 7,8,9-guanidinium group as the biologically active feature (Figure 6.8) [120]. This was accomplished by comparing activities of the 1,2,3-guanidinium group of STX and neoSTX under differing pH conditions. The essential difference between these closely related analogs is the hydroxylated N-1 group on neoSTX (Figure 6.8). By adjusting pH conditions, the abundance of the protonated form of one of the guanidinium groups

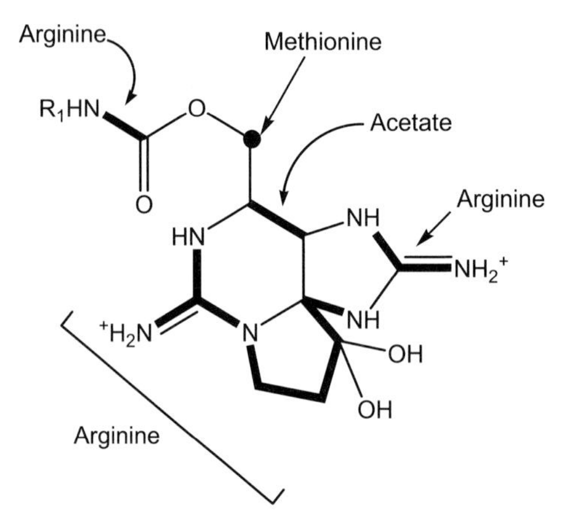

Fig. 6.9 Biosynthetic subunits of saxitoxin as determined by various precursor-feeding studies [104].

could be controlled, as well as the overall charge [120,121]. This manipulation of the two protonated guanidinium groups allowed a relative activity comparison to be made, leading to the discovery of the 7,8,9-guanidinium group as the key feature for activity.

STX is a heterocyclic guanidine, which selectively blocks VGSCs in excitable membranes [121,122]. Its use has led to breakthrough discoveries in the field of ion channel physiology pertaining to ion channel structure and function. Because STX blocks sodium channels at low concentrations, it was predicted that the toxin acts at a small number of discrete sites on excitable membranes. This was quantitatively shown using tritium-labeled STX preparations for membrane-binding studies [123–126]. Similar investigations based on this concept were used to demonstrate mechanisms of selectivity and estimate VGSC densities on various cells and tissue [124,125,127–130].

There is strong evidence that STX-induced lethality is caused by a combination of central and peripheral cardiorespiratory effects. Changes in cardiac output can be attributed to the effect of STX on fast sodium channels in contractile myocardium and Purkinje fibers. Cardiovascular shock resulting from high doses of STX is in response to a combination of vascular hypotension and reduced cardiac output, followed by a lack of venous return and finally hypoxia [131–134].

Once thought to be selective for Na^+ channels, new evidence suggests STX may interact with subtypes of both calcium (Ca^{2+}) and potassium (K^+) channels [135]. The "human ether-a-go-go-related gene" (hERG), a recently discovered K^+ channel subtype, plays a role in myocardial repolarization [136,137]. STX has been found to modify the voltage-sensing mechanism of this channel in a complex manner, not at all similar to the simple pore-blocking model for VGSCs [135]. Like the hERG channels, L-type calcium channels are cardiac related and act as the major pathway of calcium influx during excitation–contraction coupling in ventricular myocytes [138]. Voltage clamp data has shown that STX causes partial inhibition of L-type Ca^{2+} channels in a dose-dependent fashion [139].

Although research surrounding STX and its analogs has been extensive, many questions still remain. For example, Carmichael *et al.* have isolated a freshwater strain of cyanobacteria, *Lyngbya wollei* which was found to produce a decarbamoyl saxitoxin [140,141]. Even more intriguing are the variety of naturally occurring analogs available in the PST family. With over 30 analogs found in nature and an even larger number available through synthetic means, a wealth of structure–activity investigations are possible. Could any of these analogs serve a therapeutic purpose? For example, there are already several patent applications and research reports which demonstrate the analgesic effects of STX in combination with other known anesthetics. Combination dosing increases efficacy and potency without increased toxicity [142–145]. Furthermore, new STX targets such as hERG and L-type Ca^{2+} channels demonstrate that STX has many physiological interactions not well understood. Almost eight decades after its isolation, STX is still proving to be an intriguing natural product!

6.3
Neurotoxic Alkaloids of Marine Cyanobacteria

6.3.1
Antillatoxin A and B

A shallow-water Curacao collection of *Lyngbya majuscula* in 1991 yielded an extract that was toxic to several different classes of potential predators, including snails (*Biomphaleria glabrata*), arthropods (*Artemia salina*), and fish (*Carassius carassius*). Each of these model organisms was used to track the isolation of bioactive constituents, and led to the isolation of several classes of potently bioactive natural products. The fish toxicity assay was particularly interesting as enriched fractions were exceptionally potent and very fast acting, killing fish in a matter of seconds of exposure at less than part-per-million levels. The bioassay-guided isolation resulted in the recovery of 1.7 mg of a lipopeptide substance, termed antillatoxin, which was active in killing fish with an LD_{50} of approximately 50 ng/mL [146]. It is proposed that this potent ichthyotoxic property may be useful in defending the succulent strands of this cyanobacterium from fish predation.

The structure of this new ichthyotoxin was determined by interplay of 2D NMR and mass spectrometry, and involves a cyclic tripeptide composed of alanine, *N*-methyl valine, and glycine, linked together with an unsaturated lipid containing an exceptional number of methyl groups (seven of 17 carbon atoms are methyls or methyl group equivalents) (Figure 6.10). The absolute stereochemistry of the two chiral amino acids was determined by chiral HPLC of the acid hydrolysate. The C-4 and C-5 stereocenters were predicted based on additional NMR data, modeling, and CD spectra. Unfortunately, the stereochemistry at C-4 relative to C-5 was initially mis-assigned based on faulty 2D NMR data, and this was only clarified after chemical synthesis of all four possible C-4 and C-5 stereoisomers [147]. However, access to these four stereoisomers allowed a limited scope SAR evaluation, and showed that the natural stereoisomer (4*R*,5*R*) was at least 25-fold more active (ichthyotoxicity, microphysiometry, lactose dehydrogenase (LDH) efflux assay, neuro-2a cytotoxocitiy assay) than any of the other three configurational isomers (4*S*,5*R*; 4*R*,5*S*; 4*S*,5*S*) [148]. A detailed NMR analysis and molecular modeling of these four isomers showed that natural antillatoxin (ATX) has a distinct topology that is overall "L-shaped" with a preponderance of polar functional groups on the outer surface of the macrocycle.

Subsequently, shallow water collections of *Lyngbya majuscula* from Puerto Rico and the Dry Tortugas yielded additional supplies of ATX as well as a new congener termed "antillatoxin B" (Figure 6.10) [149]. The structure of the new metabolite was determined largely by comparison with the spectroscopic data set for ATX, and stereochemistry deduced by Marfey's analysis for L-alanine while the L-*N*-methyl homophenylalanine was proposed based on nuclear Overhauser effect (nOe) and bioassay results. Substitution of L-*N*-methyl homo-phenylalanine, an intriguing amino acid of quite rare occurrence in natural products, for L-*N*-methyl valine

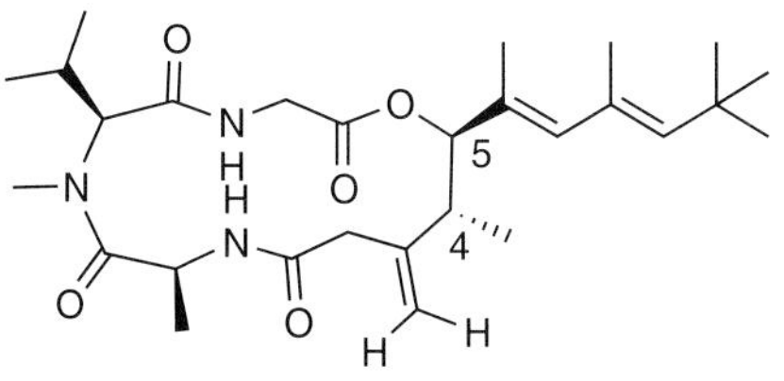

"Naturally Occurring" (4*R*,5*R*)-Antillatoxin

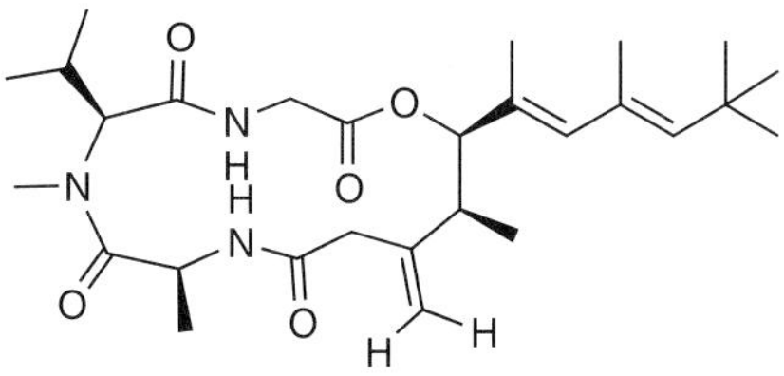

(4*S*,5*R*)-Antillatoxin

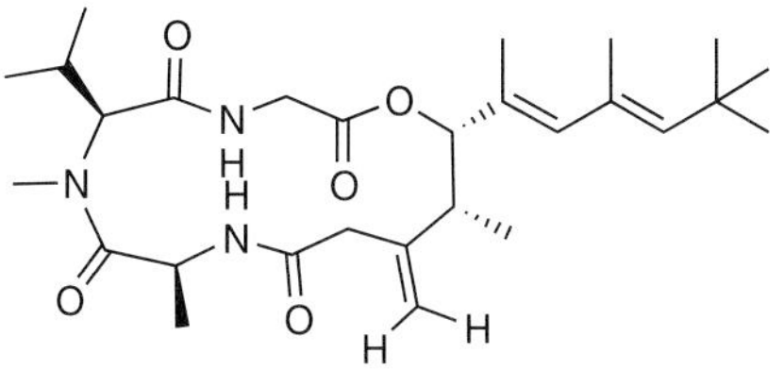

(4*R*,5*S*)-Antillatoxin

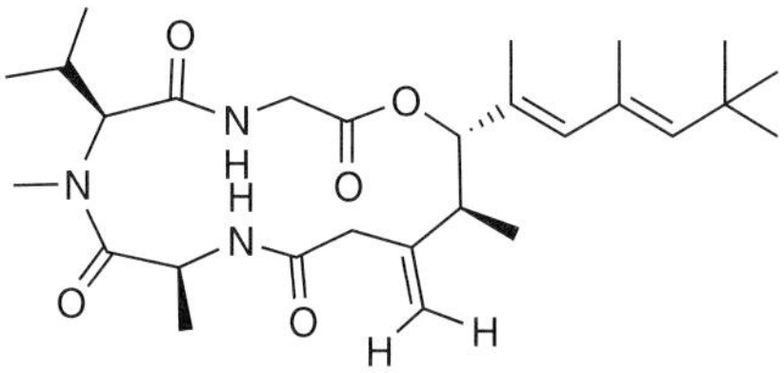

(4*S*,5*S*)-Antillatoxin

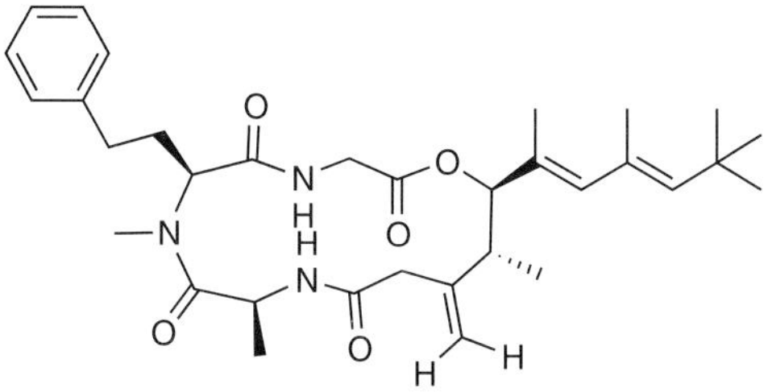

Antillatoxin B

Fig. 6.10 Structures of natural antillatoxin, three synthetic C-4 and C-5 stereoisomers, and antillatoxin B.

appears to decrease the potency of the antillatoxins by about 10-fold as measured by activity to neuro-2a cells and ichthyotoxicity.

ATX was shown to be a rapidly acting toxin to primary cultures of rat cerebellar granule neurons (CGNs), and its toxicity could be effectively blocked by co-exposure to two antagonists of the NMDA receptor (dextrorphan, MK-801) [150]. However, these antagonists were ineffective at protecting cells when applied subsequent to a 2-h ATX exposure. Specific activation of VGSCs was directly demonstrated in CGNs by measuring influx of $^{22}Na^+$ upon treatment with ATX; this response was completely blocked by co-treatment with tetrodotoxin (a site 1 antagonist) [151]. The putative binding site of ATX was examined by measuring Ca^{2+} mobilization in CGNs upon co-treatment with other VGSC ligands. An increase in Ca^{2+} mobilization was observed upon co-treatment with [^{3}H]batrachotoxin, a site 2 binding agent, and a synergistic effect on [^{3}H]batrachotoxin binding was observed upon co-treatment with brevetoxin, a site 5 agonist. However, no increase in the ATX-enhanced binding of [^{3}H]batrachotoxin was observed following application of either sea anemone toxin (a site 3 ligand) or a synthetic pyrethroid (RU39568, a site 7 ligand), thus distinguishing ATX's effect from that of brevetoxin. Thus, it seems likely that ATX binds to a distinct site on the α-subunit of the VGSC.

Based on experimental work on the biosynthesis of other cyanobacterial natural products [152,153], the biogenesis of ATX likely involves a cluster of genes with initiation occurring by a loading module that incorporates either *t*-butyl pentanoic acid or a less-methylated acid that is converted to *t*-butyl pentanoic acid by SAM methylation. This is followed by four rounds of acetate extension, the first three of which also are C-methylated at the corresponding C-2 positions of acetate by SAM. The penultimate acetate unit is also branched at the C-1 deriving carbon, and this likely involves addition of acetate to the carbonyl functionality by an HMGCoA synthase-like reaction, followed by decarboxylation and dehydration, as has been characterized in other marine cyanobacterial natural products [153]. The tripeptide section would logically involve three NRPS modules with specificity for alanine, valine, and glycine. A terminating thioesterase could catalyze both cleavage from the enzyme complex and macrocyclization.

6.3.2
Jamaicamide A, B, and C

A bioassay-guided fractionation of a laboratory-cultured strain of *Lyngbya majuscula* (strain JHB) collected from Hector's Bay, Jamaica led to the isolation of three new lipopeptides, jamaicamide A–C (Figure 6.11). A small tissue sample preserved at the time of collection was extracted and shown to possess potent brine shrimp toxicity. Laboratory pan cultures were grown and extracted by standard methods for lipid natural products. Evaluation of the extract and its vacuum liquid chromatographic (VLC) fractions identified a mid-polarity fraction with activity in a cellular assay designed to detect mammalian VGSC-blocking substances [154]. This fraction was also identified as possessing potentially novel substances by thin layer chromatography (TLC) and ^{1}H NMR analysis. Subsequent HPLC of the fraction led to the isolation of jamaicamides A, B, and C. These three new compounds exhibited low micromolar levels of channel-blocking activity in neuro-2a cells. Partial structures of jamaicamide A were constructed and connected using various 2D NMR spectroscopic methods and mass spectrometry (Figure 6.11) [155].

Jamaicamides A, B, and C exhibited cytotoxicity to both the H-460 human lung carcinoma and neuro-2a mouse neuroblastoma cell lines. The LC_{50}s were approximately 15 μM for all three compounds to both cell lines. All three compounds also exhibited sodium channel-blocking activity at 5 μM concentration. In the goldfish toxicity assay, a system that has been useful for the detection of neurotoxic activity in crude extracts as well as purified compounds, jamaicamide B was the most active (100 % lethality at 5 ppm after 90 min), followed by jamaicamide C (100 % lethality at

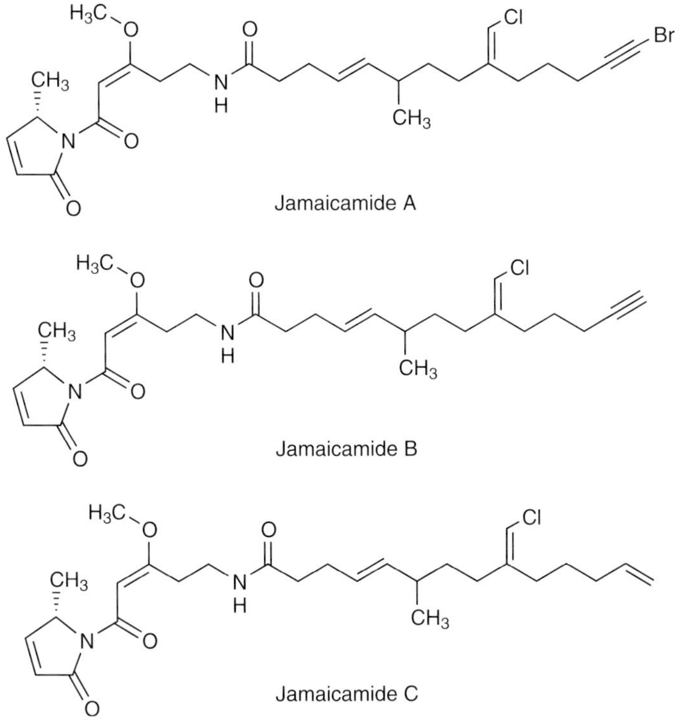

Jamaicamide A

Jamaicamide B

Jamaicamide C

Fig. 6.11 Structures of jamaicamides A, B, and C.

10 ppm after 90 min). Interestingly, jamaicamide A was the least active fish toxin (sublethal toxicity at 10 ppm after 90 min). Neither jamaicamide A nor B showed significant brine shrimp toxicity, while jamaicamide C was only modestly active at 10 ppm (25 % lethality) [156].

Dissection of the chemical structure of jamaicamides A–C led to the speculation that these metabolites derive from a mixture of polyketides (nine acetate units), amino acids (L-Ala and β-Ala), and the *S*-methyl group of methionine. To map out the biosynthetic subunits of these molecules, isotopically labeled precursors were supplied to *L. majuscula* JHB, and the labeling patterns discerned by NMR spectroscopy (Figure 6.12). From these experiments, insights were gained into the biochemical transformations that produce the jamaicamides, especially the mechanism of formation of the vinyl chloride group [157].

Analysis of jamaicamide A–C biosynthesis was further investigated by Edwards *et al.* using a variety of molecular genetic approaches [157]. Information gained from the biosynthetic feeding studies led to the rational design of a set of gene probes that were used to efficiently screen a total DNA library of *L. majuscula* JHB. These experiments identified several fosmids containing portions of the jamaicamide gene cluster. Sequencing of these and assembly of the pathway revealed a remarkably collinear set of genes that code for enzymes catalyzing jamaicamide biosynthesis.

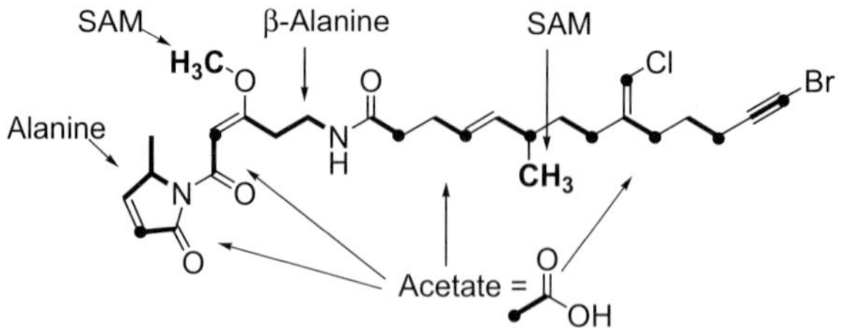

Fig. 6.12 Biosynthetic origins of jamaicamide A [157].

Several of these genes were deduced to program for polyketide synthase (PKS) and nonribosomal peptide synthetase (NRPS) enzymes as well as novel tailoring enzymes, including those that create the various unusual functional groups in jamaicamide A.

The complete Jam pathway contains six PKS modules [JamE, JamJ, JamK, JamL (module 4), JamM/N, JamP], two NRPS modules [Jam L (module 5), Jam O], and a host of tailoring enzymes [Jam A, JamG, JamH, JamI, JamQ] (Figure 6.13). JamA initiates Jam biosynthesis by the ATP-dependent activation of free 5-hexenoic or 5-hexynoic acid to the acyl-adenylate followed by subsequent loading to the ACP protein JamC [157]. The Walsh, Kelleher, and Gerwick groups subsequently utilized an FTMS-based assay to elucidate the timing of bromination [158]. Incubation of holo-JamC, JamA, 6-bromo-5-hexynoic acid, 5-hexenoic acid, and ATP resulted only in the activation and loading of hexenoic acid. These results suggest that bromination takes place either while the substrate is attached to JamC, while attached to another of the NRPS or PKS proteins found in the jamaicamide biosynthetic pathway, or after jamaicamide is fully assembled (e.g. on jamaicamide B) [158]. Next, a polyketide chain extension occurs generating an unreduced eight-carbon β-ketothioester inter-

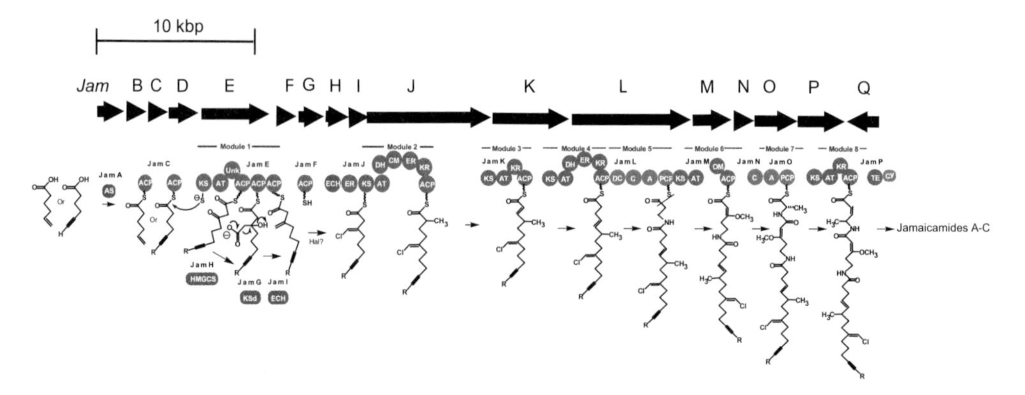

Fig. 6.13 Gene cluster and predicted biosynthetic assembly of jamaicamide A [157].

mediate. This intermediate is then transformed, via an extensive array of enzymatic transformations, to generate a pendent vinyl chloride functionality. As noted above, feeding experiments established that C-27, the vinyl chloride-bearing carbon, is derived from C-2 of acetate, a labeling pattern consistent with a hydroxymethylglutaryl CoA synthase-like (HMGCS) addition of acetate to a β-keto-polyketide intermediate. The high degree of sequence identity between JamH and HMGCSs support this proposal, and suggests that this Jam enzyme initiates this process by adding acetate to generate a branched acyl–S–acyl carrier protein (ACP) intermediate. From parallel work with the curacin A HMGCS-gene motif [159], subsequent dehydration of the tertiary alcohol and decarboxylation is predicted to yield the pendent vinyl group. This is presumably next chlorinated by a radical-mediated halogenation [160,161], and then the nascent polyketide undergoes three more rounds of PKS chain elongation. The *S*-methyl of SAM is predicted to form the *C*-methyl at C26 and the *O*-methyl, and this was demonstrated experimentally through feeding experiments. Acetate unit extensions which interdigitate first with β-Ala and then D-Ala complete the primary jamaicamide skeleton. Chain termination appears to occur with an unusual cyclization of the pyrrolinone ring to yield jamaicamide A [157].

6.3.3 Kalkitoxin

Isolated in 1996 from a strain of *L. majuscula* collected at Playa Kalki, Curacao, kalkitoxin (KTX) exhibited strong ichthyotoxic activity (LC_{50} 700 nM) and highly potent brine shrimp toxicity (LC_{50} 170 nM) [162]. KTX is a lipoamide with five stereogenic centers, four methyl groups on the carbon chain, an *N*-methylamide, and a thiazoline ring (Figure 6.14). The challenges to synthetic chemists presented by this intriguing stereochemical arrangement, as well as its potent biological properties, spurred several research groups to investigate different routes toward the total synthesis of KTX [162–164]. The first total synthesis of KTX was reported along with the original structure elucidation and was crucial to determination of the absolute stereochemistry. Synthetic analogs having all possible configurations were synthesized and the ^{13}C NMR spectrum of each compound was compared to the natural substance. Synthetic (3*R*, 7*R*, 8*S*, 10*S*, 2′*R*)-kalkitoxin was found to be identical to the natural product [162]. Moreover, the efficient synthesis of KTX and its analogs provided the means for further investigation into its remarkable biological properties.

The neurotoxicity of KTX was assayed using rat MCGNs and exhibited potent concentration-dependent toxicity (LD_{50} value of 3.9 ± 1.9); however, this toxicity showed a unique time delay of 22 h post exposure. Furthermore, KTX toxicity was prevented by co-treatment with NMDA receptor channel antagonists (dextrorphan and MK-801), indicating that KTX-induced neuronal death (morphology and LDH efflux assay) is mediated through an NMDA receptor-dependent mechanism [150].

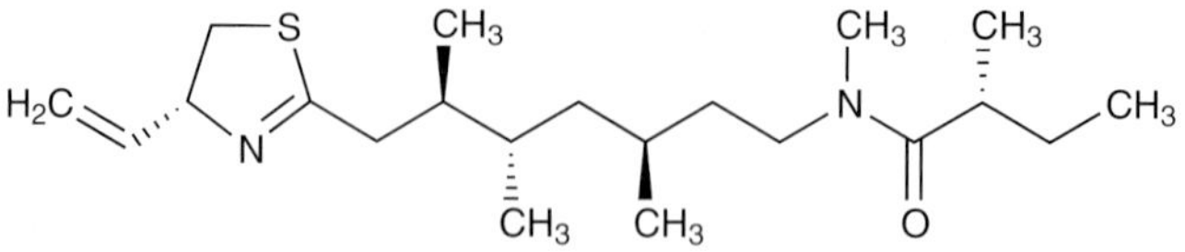

Kalkitoxin

Fig. 6.14 Structure of kalkitoxin, a potent voltage-gated sodium channel (VGSC) blocker.

The receptor site of KTX was probed via a series of competitive binding experiments [165] Veratridine (a site 2 ligand on the VGSC)] [166] activation of the VGSC with consequent rapid Ca^{2+} influx into the cell is potently inhibited by KTX; however, this is not definitive for direct action on site 2. Co-incubation of KTX and tritiated-batrachotoxin ([^{3}H]BTX) (site 2 ligand) did not affect binding of the radiolabeled probe to the tetrodotoxin-sensitive VGSC. More revealing was the finding that KTX inhibits the binding of [^{3}H]BTX in the presence of deltamethrin, a site 7 ligand and an allosteric agonist of BTX (site 7 shows allosteric coupling to site 2 of the VGSC). Based on these experiments and by deduction, it can be concluded that KTX is not a ligand of sites 1 or 2 because it does not inhibit [^{3}H]BTX binding, and thus may interact at a unique high affinity site on VGSCs.

6.4 Conclusion

The intent of this chapter has been to summarize key features of the occurrence, chemistry, and pharmacology of neurotoxic cyanobacterial alkaloids from terrestrial, freshwater, and marine habitats. In doing so, insights are gained into several aspects of their properties. In freshwater systems, cyanobacteria are common contributors to harmful algal blooms (HABs), and produce a chemically and pharmacologically diverse array of toxins. However, only a few of these are neurotoxic, as reviewed above for saxitoxin (although a few marine occurrences are reported) and the anatoxins, and most are either hepatotoxic or skin irritants. Recent increases in the occurrence of these pond, lake, river, and esturarine HABs are likely due to enrichments from nutrient-rich agricultural and sewage runoff.

In the marine environment, cyanobacteria are emerging as contributors to HABs, especially in tropical environments; however, their impact is likely more pronounced on the ecological structure of shallow-water reef systems than on human populations. But in some locations, such as the East Coast of Australia, extensive blooms of *Lyngbya majuscula*, which produce the hepatotoxic lyngbyatoxins, have influenced human health. Nevertheless, some populations of this same species from elsewhere in the world do produce neurotoxic natural products, as reviewed in the latter part of this chapter, and the potential exists for these neurotoxin-producing strains to contribute to HABs in the future.

It is intriguing to note the two dominant pharmacological trends in the neurotoxic alkaloids of cyanobacteria (Figures 6.1 and 6.2). On the one hand, the anatoxins have powerful effects on two sites of action within the acetylcholine neurotransmission system. Anatoxin-a and homoanatoxin-a both act as agonists at nACh receptors causing neuromuscular depolarization, an effect which is compounded by the inability of the ACh-esterase to degrade these toxins. Anatoxin-a(s) has the same effect, but exerts this activity through competitive inhibition of the ACh-esterase. The remainder of the cyanobacterial neurotoxins impact the VGSC involving sodium ion-based membrane depolarization and downstream NMDA receptor activation. BMAA, initially found in cyanobacteria living in association with cycads, but now recognized to be broadly prevalent in these prokaryotes, directly acts on the NMDA receptor to induce calcium influx and ensuing neuronal death. The marine toxins described in this chapter, all of which were isolated within a program focused on new compound discovery, act in different ways to modulate the function of the VGSC. For example, kalkitoxin and jamaicamide decrease sodium passage through the channel whereas the antillatoxins have the opposite effect of increasing sodium influx.

The natural function of the cyanobacterial neurotoxins is poorly understood at present, but represents a key aspect to understanding the incidence and severity of neurotoxic HABs. In the saxitoxin case, there are a few studies that alternately suggest a physiological role in osmoregulation [167] or a defensive role against competitive and potential predatory species [168]. A defensive role for the anatoxins has also been proposed [169]. The function of BMAA in cyanobacteria is completely unknown and at present has not been addressed in any regard. In the case of marine cyanobacterial toxins in general, there has been limited experimental investigation of their chemical ecology, and a defensive role has been assigned [170]. Unfortunately, investigation of the chemical ecology or other potential natural functions of the specific marine neurotoxins presented in this chapter have not occurred to date, and this represents a rich opportunity for further examination.

It is intriguing to consider that the biological activities of natural products from cyanobacteria fall into only a few distinct pharmacological classes. These include a robust number of metabolites, mainly lipopeptides, which interfere with protein polymerization processes involved in actin and tubulin assembly [3]. Indeed, there are approximately 10 metabolite classes which have as their site of action each of these protein targets, and all of these show profound antiproliferative and cytotoxic properties. Presumably, convergent evolutionary pressures have operated to produce an impressive number of structural scaffolds that target these key metabolic processes in eukaryotes, and, as such, they represent ideal targets of cyanobacterial defensive chemistry because of the inherent lack of toxicity to the prokaryotic producers. Joining these as a newer pharmacological class are several of the neurotoxins described in this review, which target the critical ion transport processes in the neurosystems of higher organisms. In parallel to the findings with diverse natural product structures with antitubulin or antiactin activity, it can be anticipated that cyanobacteria will be a rich source of additional structurally diverse neurotoxins. Moreover, this suggests a generalized strategy by which to utilize the ecology and biology of various life forms to search for specific classes of pharmacological agents. From the precedent set here with cyanobacteria, it seems a reasonable strategy that the defensive chemistry of prokaryotes will specifically target the biochemical systems distinctive to eukaryotes, and, by doing so, will suffer few ill effects from harboring large quantities of potent adaptive natural products.

As reflected throughout this chapter on the neurotoxic alkaloids of cyanobacteria, many questions remain on this relatively little explored facet of natural products chemistry. Results from screening efforts indicate that marine cyanobacteria, and likely freshwater species as well, are very rich in neurotoxic natural products, and this will be a productive area for further investigation. While in some cases the neurochemical targets of cyanobacterial toxins are known, in general this is at a very low level of resolution, and detailed molecular pharmacological studies will be highly productive in revealing the subtleties of toxin as well as drug action. Toxins have played a historically important role in revealing neurochemical receptors and pathways of importance to health as well as disease, and are thus important tools for molecular pharmacologists; marine cyanobacterial toxins have and will continue to contribute to this toolbox. The biosyntheses of marine neurotoxins have been studied to some level in a few cases, such as for saxitoxin and jamaicamide, but for most this knowledge is absent. Understanding the pathways, and perhaps more importantly their regulation and natural function, will be critical to predicting and perhaps controlling the production of harmful natural toxins in our marine and freshwater systems.

References

1 Francis, G. (1878) *Nature*, **18**, 11.
2 Antoniou, M. G., de laCruz, A. A., Dionysiou, D. D. (2005) *Journal of Environmental Engineering (ASCE)*, **131**, 1239.
3 Gerwick, W. H., Tan, L. T., Sitachitta, N. (2001) *The Alkaloids, Chemistry and Biology*, **57**, 75.
4 Gorham, P. R., McLachlan, J., Hammar, U. T., Kim, W. K. (1964) *Verh Internat Verein Limnol*, 796.
5 Devlin, J. P., Edwards, O. E., Gorham, P. R., Hunter, N. R., Pike, R. K., Stavric, B. (1977) *Canadian Journal of Chemistry*, **55**, 1367.
6 Edwards, C., Beattie, K. A., Scrimgeour, C. M., Codd, G. A. (1992) *Toxicon*, **30**, 1165.
7 Sivonen, K., Kononen, K., Carmichael, W. W., Dahlem, A. M., Rinehart, K. L., Kiviranta, J., Niemela, S. I. (1989) *Applied and Environmental Microbiology*, **55**, 1990.
8 Gugger, M., Lenoir, S., Berger, C., Ledreux, A., Druart, J. C., Humbert, J. F., Guette, C., Bernard, C. (2005) *Toxicon*, **45**, 919.
9 Dittmann, E. and Wiegand, C. (2006) *Molecular Nutration and Food Research*, **50**, 7.

10 Krienitz, L., Ballot, A., Kotut, K., Wiegand, C., Putz, S., Metcalf, J. S., Codd, G. A., Pflugmacher, S. (2003) *FEMS Microbiology Ecology*, **43**, 141.
11 Huber, C. S. (1972) *Acta Crystallographica Section B: Structural Science*, **28**, 37.
12 Koskinen, A. M. P. and Rapoport, H. (1985) *Journal of Medicinal Chemistry*, **28**, 1301.
13 Hemscheidt, T., Rapala, J., Sivonen, K., Skulberg, O. M. (1995) *Journal of Chemical Society, Chemical Communications*, 1361.
14 Hemscheidt, T., Burgoyne, D. L., Moore, R. E. (1995) *Journal of Chemical Society, Chemical Communications*, 205.
15 Gallon, J. R., Kittakoop, P., Brown, E. G. (1994) *Phytochemistry*, **35**, 1195.
16 Gallon, J. R., Chit, K. N., Brown, E. G. (1990) *Phytochemistry*, **29**, 1107.
17 Carmichael, W. W., Biggs, D. F., Peterson, M. A. (1979) *Toxicon*, **17**, 229.
18 Carmichael, W. W. (1997) *Advances in Botanical Research*, **27**, 211.
19 Holladay, M. W., Dart, M. J., Lynch, J. K. (1997) *Journal of Medicinal Chemistry*, **40**, 4169.
20 Carmichael, W. W. and Biggs, D. F. (1978) *Canadian Journal of Zoology—Revue Canadienne de Zoologie*, **56**, 510.
21 Carmichael, W. W., Biggs, D. F., Gorham, P. R. (1975) *Science*, **187**, 542.
22 Daly, J. W. (2005) *Cellular and Molecular Neurobiology*, **25**, 513.
23 Campbell, H. F., Edwards, O. E., Kolt, R. (1977) *Canadian Journal of Chemistry—Revue Canadienne de Chimie*, **55**, 1372.
24 Mansell, H. L. (1996) *Tetrahedron*, **52**, 6025.
25 Oh, C. Y., Kim, K. S., Ham, W. H. (1998) *Tetrahedron Letters*, **39**, 2133.
26 Parsons, P. J., Camp, N. P., Edwards, N., Sumoreeah, L. R. (2000) *Tetrahedron*, **56**, 309.
27 Hemming, K., O'Gorman, P. A., Page, M. I. (2006) *Tetrahedron Letters*, **47**, 425.
28 Brenneman, J. B., Machauer, R., Martin, S. F. (2004) *Tetrahedron*, **60**, 7301.
29 Brenneman, J. B. and Martin, S. F. (2004) *Organic Letters*, **6**, 1329.
30 Mori, M., Tomita, T., Kita, Y., Kitamura, T. (2004) *Tetrahedron Letters*, **45**, 4397.
31 Muniz, M. N., Kanazawa, A., Greene, A. E. (2005) *Synlett*, 1328.
32 Wegge, T., Schwarz, S., Seitz, G. (2000) *Tetrahedron: Asymmetry*, **11**, 1405.
33 Wonnacott, S., Swanson, K. L., Albuquerque, E. X., Huby, N. J. S., Thompson, P., Gallagher, T. (1992) *Biochemical Pharmacology*, **43**, 419.
34 Wonnacott, S., Jackman, S., Swanson, K. L., Rapoport, H., Albuquerque, E. X. (1991) *The Journal of Pharmacology and Experimental Therapeutics*, **259**, 387.
35 Swanson, K. L., Aronstam, R. S., Wonnacott, S., Rapoport, H., Albuquerque, E. X. (1991) *The Journal of Pharmacology and Experimental Therapeutics*, **259**, 377.
36 Magnus, N. A., Ducry, L., Rolland, V., Wonnacott, S., Gallagher, T. (1997) *Journal of the Chemical Society, Perkin Transactions*, **1**, 2313.
37 Wonnacott, S. and Gallagher, T. (2006) *Marine drugs*, **4**, 228.
38 Swanson, K. L., Allen, C. N., Aronstam, R. S., Rapoport, H., Albuquerque, E. X. (1986) *Molecular Pharmacology*, **29**, 250.
39 Macallan, D. R. E., Lunt, G. G., Wonnacott, S., Swanson, K. L., Rapoport, H., Albuquerque, E. X. (1988) *FEBS Letters*, **226**, 357.
40 Kanne, D. B. (1988) *Journal of Medicinal Chemistry*, **31**, 2227.
41 Kanne, D. B. and Abood, L. G. (1988) *Journal of Medicinal Chemistry*, **31**, 506.
42 Kanne, D. B., Ashworth, D. J., Cheng, M. T., Mutter, L. C., Abood, L. G. (1986) *Journal of the American Chemical Society*, **108**, 7864.
43 Gundisch, D., Kampchen, T., Schwarz, S., Seitz, G., Siegl, J., Wegge, T. (2002) *Bioorganic and Medicinal Chemistry*, **10**, 1.

44 Hernandez, A. and Rapoport, H. (1994) *Journal of Organic Chemistry*, **59**, 1058.

45 Costa, A. C. S., Swanson, K. L., Aracava, Y., Aronstam, R. S., Albuquerque, E. X. (1990) *The Journal of Pharmacology and Experimental Therapeutics*, **252**, 507.

46 Howard, M. H., Sardina, F. J., Rapoport, H. (1990) *Journal of Organic Chemistry*, **55**, 2829.

47 Gohlke, H., Gundisch, D., Schwarz, S., Seitz, G., Tilotta, M. C., Wegge, T. (2002) *Journal of Medicinal Chemistry*, **45**, 1064.

48 Vasas, G., Gaspar, A., Pager, C., Suranyi, G., Mathe, C., Hamvas, M. M., Borbely, G. (2004) *Electrophoresis*, **25**, 108.

49 Furey, A., Crowley, J., Hamilton, B., Lehane, M., James, K. J. (2005) *Journal of Chromatography A*, **1082**, 91.

50 James, K. J., Furey, A., Sherlock, I. R., Stack, M. A., Twohig, M., Caudwell, F. B., Skulberg, O. M. (1998) *Journal of Chromatography A*, **798**, 147.

51 Skulberg, O. M., Carmichael, W. W., Andersen, R. A., Matsunaga, S., Moore, R. E., Skulberg, R. (1992) *Environmental Toxicology and Chemistry*, **11**, 321.

52 Namikoshi, M., Murakami, T., Watanabe, M. F., Oda, T., Yamada, J., Tsujimura, S., Nagai, H., Oishi, S. (2003) *Toxicon*, **42**, 533.

53 Furey, A., Crowley, J., Shuilleabhain, A. N., Skulberg, A. M., James, K. J. (2003) *Toxicon*, **41**, 297.

54 Lilleheil, G., Andersen, R. A., Skulberg, O. M., Alexander, J. (1997) *Toxicon*, **35**, 1275.

55 Namikoshi, M., Murakami, T., Fujiwara, T., Nagai, H., Niki, T., Harigaya, E., Watanabe, M. F., Oda, T., Yamada, J., Tsujimura, S. (2004) *Chemical Research in Toxicology*, **17**, 1692.

56 Mahmood, N. A. and Carmichael, W. W. (1986) *Toxicon*, **24**, 425.

57 Onodera, H., Oshima, Y., Henriksen, P., Yasumoto, T. (1997) *Toxicon*, **35**, 1645.

58 Matsunaga, S., Moore, R. E., Niemczura, W. P., Carmichael, W. W. (1989) *Journal of the American Chemical Society*, **111**, 8021.

59 Neumann, R. and Peter, H. H. (1987) *Experientia*, **43**, 1235.

60 Moore, B. S., Ohtani, I., Dekoning, C. B., Moore, R. E., Carmichael, W. W. (1992) *Tetrahedron Letters*, **33**, 6595.

61 Hyde, E. G. and Carmichael, W. W. (1991) *Journal of Biochemistry and Toxicology*, **6**, 195.

62 Mahmood, N. A. and Carmichael, W. W. (1987) *Toxicon*, **25**, 1221.

63 Duncan, M. W., Kopin, I. J., Garruto, R. M., Lavine, L., Markey, S. P. (1988) *The Lancet*, 631.

64 Weiss, J. H., Koh, J. Y., Choi, D. W. (1989) *Brain Research*, **497**, 64.

65 Brownson, D. M., Mabry, T. J., Leslie, S. W. (2002) *Journal of Ethnopharmacology*, **82**, 159.

66 Spencer, P. S., Nunn, P. B., Hugon, J., Ludolph, A. C., Ross, S. M., Roy, D. N., Robertson, R. C. (1987) *Science*, **237**, 517.

67 Vega, A. and Bell, E. A. (1967) *Phytochemisty*, **6**, 759.

68 Schneider, D., Wink, M., Sporer, F., Lounibos, P. (2002) *Naturwissenschaften*, **89**, 281.

69 Kisby, G. E., Ellison, M., Spencer, P. S. (1992) *Neurology*, **42**, 1336.

70 Seawright, A. A., Brown, A. W., Nolan, C. C., Cavanagh, J. B. (1990) *Neuropathology and Applied Neurobiology*, **16**, 153.

71 Duncan, M. W. (1992) *Annals of the New York Academy of Sciences*, 161.

72 Banack, S. A. and Cox, P. A. (2003) *Neurology*, **61**, 387.

73 Cox, P. A., Banack, S. A., Murch, S. J. (2003) *Proceedings of the National Academy of Sciences of the United States of America*, **100**, 13380.

74 Cox, P. A., Banack, S. A., Murch, S. J., Rasmussen, U., Tien, G., Bidigare, R. R., Metcalf, J. S., Morrison, L. F., Codd, G. A., Bergman, B. (2005) *Proceedings of the National Academy of Sciences of the United States of America*, **102**, 5074.

75 Armon, C. (2003) *Neurology*, **61**, 291.

76 Murch, S. J., Cox, P. A., Banack, S. A. (2004) *Proceedings of the National Academy of Sciences of the United States of America*, **101**, 12228.

77 Nunn, P. B., Seelig, M., Zagoren, J. C., Spencer, P. S. (1987) *Brain Research*, **410**, 375.

78 Perry, T. L., Bergeron, C., Biro, A. J., Hansen, S. (1989) *Journal of the Neurological Sciences*, **94**, 173.

79 Copani, A., Canonico, P. L., Nicoletti, F. (1990) *European Journal of Pharmacology*, **181**, 327.

80 Allen, C. N., Omelchenko, I., Ross, M., Spencer, P. (1995) *Neuropharmacology*, **34**, 651.

81 Copani, A., Canonico, P. L., Catania, M. V., Aronica, E., Bruno, V., Ratti, E., van Amsterdam, F. T. M., Gaviraghi, G., Nicoletti, F. (1991) *Brain Research*, **558**, 79.

82 Allen, C. N., Spencer, P. S., Carpenter, D. O. (1993) *Neuroscience*, **54**, 567.

83 Montine, T. J., Li, K., Perl, D. P., Galasko, D. (2005) *Neurology*, **65**, 768.

84 Tanner, C. M. and Kurland, L. T. (1993) *Parkinsonian Syndromes*, 279.

85 Kotaki, Y., Tajiri, M., Oshima, Y., Yasumoto, T. (1983) *Bulletin of the Japanese Society of Scientific Fisheries*, **49**, 283.

86 Carmichael, W. W. (2001) *Human and Ecological Risk Assessment*, **7**, 1393.

87 Pomati, F., Sacchi, S., Rossetti, C., Giovannardi, S., Onodera, H., Oshima, Y., Neilan, B. A. (2000) *Journal of Phycology*, **36**, 553.

88 Pomati, F., Moffitt, M. C., Cavaliere, R., Neilan, B. A. (2004) *Biochimica et Biophysica Acta*, **1674**, 60.

89 Pomati, F., Burns, B. P., Neilan, B. A. (2004) *Applied and Environmental Microbiology*, **70**, 4711.

90 Llewellyn, L. E. (2006) *Natural Product Reports*, **23**, 200.

91 Alam, M., Oshima, Y., Shimizu, Y. (1982) *Tetrahedron Letters*, **23**, 321.

92 Bordner, J., Thiessen, W. E., Bates, H. A., Rapoport, H. (1975) *Journal of the American Chemical Society*, **97**, 6008.

93 Koehn, F. E., Ghazarossian, V. E., Schantz, E. J., Schnoes, H. K., Strong, F. M. (1981) *Bioorganic Chemistry*, **10**, 412.

94 Schantz, E. J. (1986) *Annals of the New York Academy of Sciences*, **479**, 15.

95 Walkers, S. and Kao, C. Y. (1980) *Federation Proceedings*, **39**, 380.

96 Sawyer, P. J., Gentile, J. H., Sasner, J. J. (1968) *Canadian Journal of Microbiology*, **14**, 1199.

97 Mahmood, N. A. and Carmichael, W. W. (1986) *Toxicon*, **24**, 175.

98 Alam, M., Shimizu, Y., Ikawa, M., Sasner, J. J. (1978) *Journal of Environmental Science and Health, Part A*, **13**, 493.

99 Ikawa, M., Wegener, K., Foxall, T. L., Sasner, J. J. (1982) *Toxicon*, **20**, 747.

100 Sasner, J. J., Foxall, T. L., Ikawa, M. (1983) *Abstract Papers from the American Chemical Society*, **186**, 184.

101 Adelman, W. J., Fohlmeister, J. F., Sasner, J. J., Ikawa, M. (1982) *Toxicon*, **20**, 513.

102 Bowling, L. C. and Baker, P. D. (1996) *Marine and Freshwater Research*, **47**, 643.

103 Humpage, A. R., Rositano, J., Baker, P. D., Nicholson, B. C., Steffensen, D. A. (1993) *The Medical Journal of Australia*, **159**, 423.

104 Humpage, A. R., Rositano, J., Bretag, A. H., Brown, R., Baker, P. D., Nicholson, B. C., Steffensen, D. A. (1994) *Australian Journal of Marine and Freshwater Research*, **45**, 761.

105 May, V. and Mcbarron, E. J. (1973) *Journal of Australian Institute for Agricultural Science*, **39**, 264.

106 Mcbarron, E. J., Walker, R. I., Gardner, I., Walker, K. H. (1975) *Australian Veterinary Journal*, **51**, 587.

107 Mold, J. D., Bowden, J. P., Stanger, D. W., Maurer, J. E., Lynch, J. M., Wyler, R. S., Schantz, E. J., Riegel, B. (1957) *Journal of the American Chemical Society*, **79**, 5235.

108 Schantz, E. J., Mold, J. D., Stanger, D. W., Shavel, J., Riel, F. J., Bowden, J. P., Lynch, J. M., Wyler, R. S., Riegel, B., Sommer, H. (1957)

Journal of the American Chemical Society, **79**, 5230.

109 Schantz, E. J., Mcfarren, E. F., Schafer, M. L., Lewis, K. H. (1958) *Journal of the Association of Official Agricultural Chemists*, **41**, 160.

110 Schantz, E. J., Howard, W. L., Stanger, D. W., Lynch, J. M., Walters, D. R., Mold, J. D., Bowden, J. P., Winterst, O. P., Riegel, B., Dutcher, J. D. (1961) Canadian Journal of Chemistry—Revue Canadienne de Chimie 39, 2117.

111 Schantz, E. J. (1969) *Journal of Agricultural and Food Chemistry*, **17**, 413.

112 Schantz, E. J., Ghazarossian, V. E., Schnoes, H. K., Strong, F. M., Springer, J. P., Pezzanite, J. O., Clardy, J. (1975) *Journal of the American Chemical Society*, **97**, 1238.

113 Gupta, S., Norte, M., Shimizu, Y. (1989) *Journal of Chemical Society: Chemical Communications*, 1421.

114 Shimizu, Y., Hori, A., Kobayashi, M., Genenah, A. (1983) *Abstract Papers from the American Chemical Society*, **186**, 116.

115 Shimizu, Y. (1996) *Annual Review of Microbiology*, **50**, 431.

116 Shimizu, Y., Norte, M., Hori, A., Genenah, A., Kobayashi, M. (1984) *Journal of the American Chemical Society*, **106**, 6433.

117 Shimizu, Y. (1986) *Pure and Applied Chemistry*, **58**, 257.

118 Shimizu, Y., Gupta, S., Masuda, K., Maranda, L., Walker, C. K., Wang, R. H. (1989) *Pure and Applied Chemistry*, **61**, 513.

119 Kao, C. Y. and Nishiyama, A. (1965) *Journal of Physiology (London)*, **180**, 50.

120 Kao, P. N., Jameskracke, M. R., Kao, C. Y. (1983) *Pflugers Archiv—European Journal of Physiology*, **398**, 199.

121 Catterall, W. A. (1980) *Annual Review of Pharmacology and Toxicology*, **20**, 15.

122 Kao, C. Y. (1966) *Pharmacological Reviews*, **18**, 997.

123 Henderson, R., Ritchie, J. M., Strichartz, G. (1973) *Journal of Physiology (London)*, **235**, 783.

124 Ritchie, J. M. (1975) *Philosophical Transactions of the Royal Society of London Series B—Biological Sciences*, **270**, 319.

125 Ritchie, J. M., Rogart, R. B., Strichartz, G. R. (1976) *Journal of Physiology (London)*, **261**, 477.

126 Ritchie, J. M., Rogart, R. B., Strichartz, G. (1976) *Journal of Physiology (London)*, **258**, 99.

127 Keynes, R. D. and Ritchie, J. M. (1984) *Proceedings of the Royal Society of London Series B—Biological Sciences*, **222**, 147.

128 Oaklander, A. L., Pellegrino, R. G., Ritchie, J. M. (1984) *Brain Research*, **307**, 393.

129 Ritchie, J. M. and Rogart, R. B. (1976) *Journal of Physiology (London)*, **263**, 129.

130 Ritchie, J. M. and Rogart, R. B. (1977) *Journal of Physiology (London)*, **269**, 341.

131 Andrinolo, D., Michea, L. F., Lagos, N. (1999) *Toxicon*, **37**, 447.

132 Borison, H. L., Culp, W. J., Gonsalves, S. F., Mccarthy, L. E. (1980) *British Journal of Pharmacology*, **68**, 301.

133 Chang, F. C. T., Benton, B. J., Lenz, R. A., Capacio, B. R. (1993) *Toxicon*, **31**, 645.

134 Kao, C. Y., Suzuki, T., Kleinhaus, A. L., Siegman, M. J. (1967) *Archives Internationales de Pharmacodynamie et de Therapie*, **165**, 438.

135 Wang, J. X., Salata, J. J., Bennett, P. B. (2003) *The Journal of General Physiology*, **121**, 583.

136 Sanguinetti, M. C., Jiang, C. G., Curran, M. E., Keating, M. T. (1995) *Cell*, **81**, 299.

137 Warmke, J. W. and Ganetzky, B. (1994) *Proceedings of the National Academy of Sciences of the United States of America*, **91**, 3438.

138 AdachiAkahane, S., Lu, L. Y., Li, Z. P., Frank, J. S., Philipson, K. D., Morad, M. (1997) *The Journal of General Physiology*, **109**, 717.

139 Su, Z., Sheets, M., Ishida, H., Li, F. H., Barry, W. H. (2004) *The Journal of Pharmacology and Experimental Therapeutics*, **308**, 324.

140 Carmichael, W. W., Evans, W. R., Yin, Q. Q., Bell, P., Moczydlowski, E. (1997) *Applied and Environmental Microbiology*, **63**, 3104.

141 Onodera, H., Satake, M., Oshima, Y., Yasumoto, T., Carmichael, W. W. (1997) *Natural Toxins*, **5**, 146.

142 Adams, H. J., Blair, M. R., Takman, B. H. (1976) *Archives Internationales de Pharmacodynamie et de Therapie*, **224**, 275.

143 Herbert J. F. Adams, Bertil H Takman, Inc. (1977) Astra Pharmaceutical Products, 629–633, 1.

144 Barnet, C. S., Tse, J. Y., Kohane, D. S. (2004) *Pain*, **110**, 432.

145 Kohane, D. S., Lu, N. T., Gokgol-Kline, A. C., Shubina, M., Kuang, Y., Hall, S., Strichartz, G. R., Berde, C. B. (2000) *Regional Anesthesia and Pain Medicine*, **25**, 52.

146 Orjala, J., Nagle, D. G., Hsu, V., Gerwick, W. H. (1995) *Journal of the American Chemical Society*, **117**, 8281.

147 Yokokawa, F., Fujiwara, H., Shioiri, T. (2000) *Tetrahedron*, **56**, 1759.

148 Li, W. I., Marquez, B. L., Yokokawa, F., Shioiri, T., Gerwick, W. H., Murray, T. F. (2004) *Journal of Natural Products*, **67**, 559.

149 Nogle, L. M., Okino, T., Gerwick, W. H. (2001) *Journal of Natural Products*, **64**, 983.

150 Berman, F. W., Gerwick, W. H., Murray, T. F. (1999) *Toxicon*, **37**, 1645.

151 Li, W. I., Berman, F. W., Okino, T., Yokokawa, F., Shioiri, T., Gerwick, W. H., Murray, T. F. (2001) *Proceedings of the National Academy of Sciences of the United States of America*, **98**, 7599.

152 Chang, Z., Flatt, P., Gerwick, W. H., Nguyen, V. -A., Willis, C. L., Sherman, D. H. (2002) *Gene*, **296**, 235.

153 Chang, Z., Sitachitta, N., Rossi, J. V., Roberts, M. A., Flatt, P. M., Jia, J., Sherman, D. H., Gerwick, W. H. (2004) *Journal of Natural Products*, **67**, 1356.

154 Manger, R. L., Leja, L. S., Lee, S. Y., Hungerford, J. M., Hokama, Y., Dickey, R. W., Granade, H. R., Lewis, R., Yasumoto, T., Wekell, M. M. (1995) *Journal of AOAC International*, **78**, 521.

155 Williamson, R. T., Marquez, B. L., Gerwick, W. H., Koehn, F. E. (2001) *Magnetic Resonance Chemistry*, **39**, 544.

156 Meyer, B. N., Ferrigni, N. R., Putnam, J. E., Jacobsen, L. B., Nichols, D. E., McLaughlin, J. L. (1982) *Planta Medica*, **45**, 31.

157 Edwards, D. J., Marquez, B. L., Nogle, L. M., McPhail, K., Goeger, D. E., Roberts, M. A., Gerwick, W. H. (2004) *Chemistry and Biology*, **11**, 817.

158 Dorrestein, P. C., Blackhall, J., Straight, P. D., Fischbach, M. A., Garneau-Tsodikova, S., Edwards, D. J., McLaughlin, S., Lin, M., Gerwick, W. H., Kolter, R., Walsh, C. T., Kelleher, N. L. (2006) *Biochemistry*, **45**, 1537.

159 Chang, Z., Sitachitta, N., Rossi, J. V., Roberts, M. A., Flatt, P. M., Jia, J., Sherman, D. H., Gerwick, W. H. (2004) *Journal of Natural Products*, **67**, 1356.

160 Vaillancourt, F. H., Yeh, E., Vosburg, D. A., O'Connor, S. E., Walsh, C. T. (2005) *Nature*, **436**, 1191.

161 Flatt, P. M., O'Connell, S. J., McPhail, K. L., Zeller, G., Willis, C. L., Sherman, D. H., Gerwick, W. H. (2006) *Journal of Natural Products*, **69**, 938.

162 Wu, M., Okino, T., Nogle, L. M., Marquez, B. L., Williamson, R. T., Sitachitta, N., Berman, F. W., Murray, T. F., McGough, K., Jacobs, R., Colsen, K., Asano, T., Yokokawa, F., Shioiri, T., Gerwick, W. H. (2000) *Journal of the American Chemical Society*, **122**, 12041.

163 White, J. D., Xu, Q., Lee, C. S., Valeriote, F. A. (2004) *Organic and Biomolecular Chemistry*, **2**, 2092.

164 Asano, F., Okino, T., Gerwick, T., Shioiria, W. H., Yokokawa, T. (2004) *Tetrahedron*, **60**, 6859.

165 LePage, K. T., Goeger, D., Yokokawa, F., Asano, T., Shioiri, T., Gerwick, W. H., Murray, T. F. (2005) *Toxicology Letters*, **158**, 133.

166 Cestele, S. and Catterall, W. A. (2000) *Biochimie*, **82**, 883.

167 Pomati, F., Rossetti, C., Manarolla, G., Burns, B. P., Neilan, B. A. (2004) *Microbiology*, **150**, 455.
168 Frangopulos, M., Guisande, C., deBlas, E., Maneiro, I. (2004) *Harmful Algae*, **3**, 131.
169 Kearns, K. D. and Hunter, M. D. (2001) *Microbial Ecology*, **42**, 80.
170 Nagle, D. G. and Paul, V. J. (1999) *Journal of Phycology*, **35**, 1412.

7
Lamellarin Alkaloids: Structure and Pharmacological Properties

Jérôme Kluza, Philippe Marchetti, Christian Bailly

7.1
Introduction

Over 100 of the most important drugs in use today derive from terrestrial organisms, either bacteria, fungi, or plants [1]. By contrast, the chemical diversity of marine life, which is estimated to about two million species, remains largely underexplored [2]. This exceptional reservoir represents a vast chemical, structural, and biological diversity of molecules, often very distinct from those found in terrestrial natural compounds. The most frequent marine species used as sources of bioactive drugs are sponges, ascidians, mollusks, echinoderms, bryozoans, algae, and coelenterates (sea whips, sea fans, and coral) [3]. One of the first key bioactive molecules of marine origin was spongothymidine, initially extracted from the sponge *Cryptotethia crypta* and identified by Werner Bergmann, an organic chemist in the 1950s [4–6]. This molecule is a nucleoside with a modified sugar, which interferes with DNA replication and inhibits the growth of some bacteria, viruses, and cancer cells. A few years later, Evans and coworkers synthesized other "false" nucleosides, including arabinosylcytosine (Cytarabine or ara-C) used mainly for the treatment of acute leukemia and non-Hodgkin's lymphoma [7]. Cytarabine also displays antiviral activities, of potential use for the treatment of herpes virus infections. But during the following two decades, this field of research implicating molecules of marine origin evolved relatively slowly, at least compared to the huge efforts devoted to the identification of new molecules from plants and microorganisms. The reason is that the marine material was generally collected by hand, and in most cases only small quantities of the living material could be obtained. It was (and remains) common to isolate less than one milligram of a bioactive substance from one kilogram (when available) of the marine organism. But today, extremely sensitive methods of NMR spectroscopy and mass spectrometry combined with liquid chromatography (HPLC, UPLC) are routinely used for the characterization of new drug structures [8]. Complex molecular structures can be solved with much less that 1 mg of compounds, without too much difficulty in most cases (although there are notorious exceptions). A few drugs derived from the sea are being developed for their biological activity in the field of inflammation (manoalide, pseudopterosins) and Alzheimer's disease (GTS-21), to

Modern Alkaloids: Structure, Isolation, Synthesis and Biology. Edited by E. Fattorusso and O. Taglialatela-Scafati

ISBN: 978-3-527-31521-5

cite only two examples [9–12]. The first totally marine-derived drug approved by the Food and Drug Administration was zicotinide (Prialt, developed by Elan Pharmaceuticals) in December 2004, for the treatment of chronic and severe pain [13]. The field of cancer chemotherapy has been also widely investigated using marine-derived molecules. Many natural products of marine origin are toxins and cytotoxic agents, potentially useful to inhibit cancer cell proliferation. A handful of marine compounds has reached phases II or III of clinical trials in oncology. This is the case for (i) certain dolastatin derivatives (soblidotin, tasitodin) as mitotic inhibitors [14,15], (ii) bryostatin 1, an activator of protein kinase C [16], (iii) ecteinascidin-743, a powerful DNA alkylating drug [17], (iv) neovastat, a pleiotropic antiangiogenic agent, which prevents the binding of the vascular endothelial growth factor (VEGF) to its receptors, inhibits gelatinolytic and elastinolytic activities of matrix metalloproteinase MMP-2, MMP-9, and MMP-12, and promotes the activity of tissue-type plasminogen activator (t-PA) [18], (v) squalamine, another antiangiogenic factor [20], and (vi) kahalalide F, which alters the function of the lysosomal membrane [20]. These are just a few examples, there are many more marine-based molecules of therapeutic interest in the field of cancer chemotherapy (see ref. [21] for a comprehensive survey). Ecteinascidin-743 (ET-743, Trabectidin, Yondelis®) is the most advanced marine natural product in clinical development, with a clinical efficacy established for the treatment of soft-tissue sarcoma [22]. This drug was originally developed by PharmaMar (Madrid, Spain), now in codevelopment with Johnson & Johnson (Ortho Biotech Products, NJ, USA). For more than 10 years, chemists and pharmacologists at PharmaMar have focused on the development of anticancer agents of marine origin (www.pharmamar.com). In addition to ET-743, the PharmaMar portfolio includes, at the clinical level, (i) aplidin, an apoptotic inducer originally isolated from the tunicate *Aplidium albicans*, (ii) kahalalide F, a depsipeptide isolated from the sea slug *Elysia rufescens* (iii) ES-285, an aminoalcohol isolated from the clam *Mactromeris polynyma*, (iv) Zalypsis, an analog of ET-743 structurally related to the marine natural compound jorumycin obtained from mollusks and, at the preclinical level, (v) thiocoraline, variolins, and the lamellarins, which represent original families of cytotoxic agents. These last molecules, the lamellarins, have attracted our interest owing to their complex mechanism of action and structural originality (Figure 7.1). This review is focused on the lamellarin family and summarizes the most recent discoveries concerning the mechanism of action of the lead compound lamellarin D (Lam-D) and its numerous naturally occurring and synthetic analogs.

7.2 The Discovery of Lamellarins

Lamellarins were originally extracted from a marine prosobranch mollusk *Lamellaria* sp. and subsequently from primitive chordate ascidians (tunicates) [23]. These ascidian species, known to produce many bioactive metabolites, likely represent the original producer of lamellarins because these organisms are presumed to be the dietary source of the *Lamellaria* mollusks. Lamellarins have been isolated from different tunicates, including recently from the Indian ascidian *Didemnum obscurum*

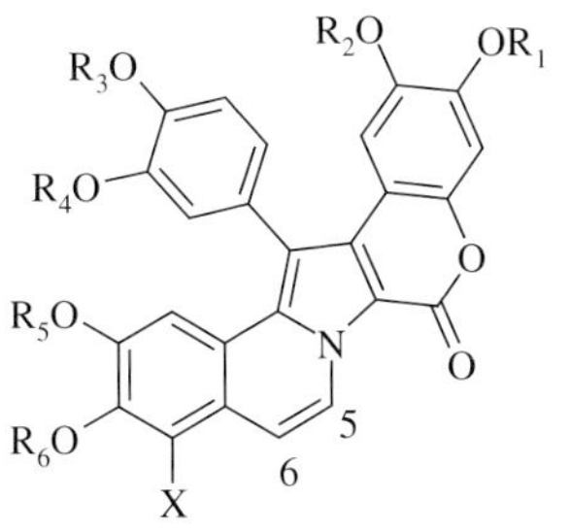

Lamellarin	R1	R2	R3	R4	R5	R6	X
D	**H**	**CH_3**	**H**	**CH_3**	**CH_3**	**H**	**H**
H	H	H	H	H	H	H	H
M	H	CH_3	CH_3	CH_3	CH_3	CH_3	OH
N	H	CH_3	CH_3	H	CH_3	H	H
20 α sulfate	SO_3Na	CH_3	CH_3	H	CH_3	CH_3	H
α	H	CH_3	CH_3	H	CH_3	CH_3	H
ζ	H	CH_3	CH_3	CH_3	CH_3	CH_3	OCH_3
η	H	CH_3	CH_3	CH_3	CH_3	CH_3	H
ϕ tri-acetate	Ac	CH_3	CH_3	CH_3	Ac	CH_3	OCH_3

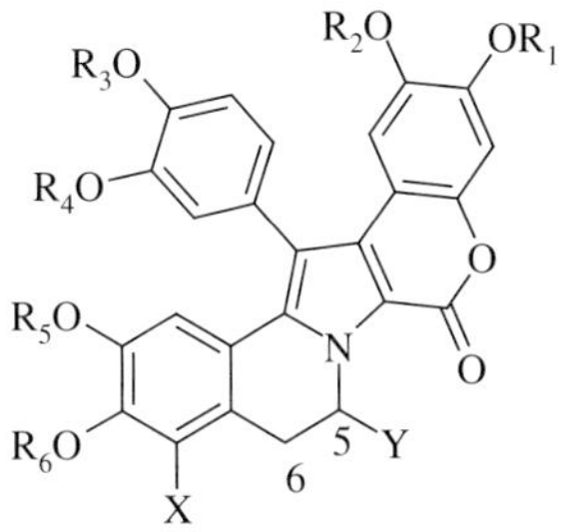

Lamellarin	R1	R2	R3	R4	R5	R6	X	Y
C	H	CH_3	H	CH_3	CH_3	CH_3	OCH_3	H
I	H	CH_3	CH_3	CH_3	CH_3	CH_3	OCH_3	H
K	H	CH_3	H	CH_3	CH_3	CH_3	OH	H
T	H	CH_3	CH_3	H	CH_3	CH_3	OCH_3	H
U	H	CH_3	CH_3	H	CH_3	CH_3	H	H
V	H	CH_3	CH_3	H	CH_3	CH_3	OCH_3	OH
γ	OH	OCH_3	OCH_3	OCH_3	OCH_3	OCH_3	OH	H
χ-tri acetate	Ac	CH_3	Ac	CH_3	CH_3	Ac	H	H

Lamellarin O P Q R

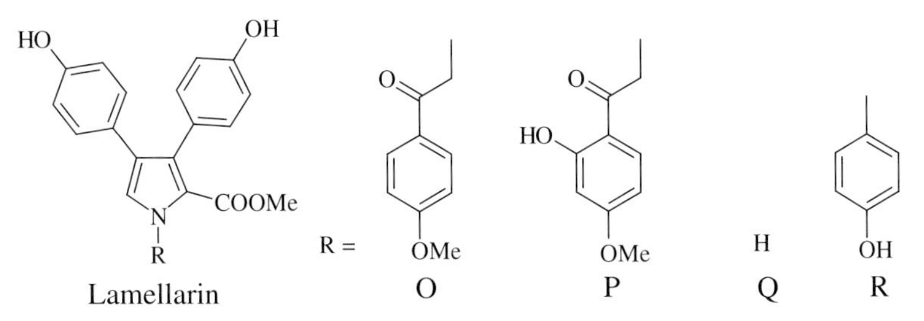

Fig. 7.1 Structures of selected lamellarins.

[24]. Ascidians are rich sources of cytotoxic compounds such as didemnin-B, eudistomin-C, lissoclinamides, ascididemnin, eilatin and segolins, lukianols, polycitrins, and ningalins [25–31]. However, it may also be postulated that lamellarins are produced by microorganisms symbiotic to the ascidians, since numerous natural products from marine invertebrates have been demonstrated to originate from symbiotic organisms, cyanobacteria in particular [32,33]. To date, 42 lamellarins (A–Z and α–ζ, including acetate and sulfate derivatives) have been isolated. Lamellarins ζ, η, ϕ, and χ are the most recent members of the family, having been isolated from the red colonial ascidian *Didemnum obscurum* [24,34]. They can be divided into three groups characterized by (i) the central ring pyrrole fused to adjacent rings having a single bond in the isoquinoline moiety at position 5 and 6, or (ii) a double bond, or (iii) the central ring pyrrole unfused to adjacent rings. Each series includes derivatives in which the phenolic hydroxyl groups are substituted by methoxy, sulfate, or acetate groups. Although, the exact "natural" role of lamellarins in the ascidians is unknown, we can hypothesize that these molecules participate in some forms of chemical communication or defense mechanism (against predation or overgrowth by competing species), as is the case for many other secondary metabolites [1,2]. In the following sections, the different biological properties of lamellarins are briefly summarized, including MDR modulation activities, antioxidant properties, HIV integrase inhibition and cytotoxicity against tumor cells (Figure 7.2).

7.3 Modulation of Multidrug Resistance

Multidrug resistance (MDR) is a term used to characterize the ability of tumors to exhibit simultaneous resistance to various chemotherapeutic agents [35]. Different mechanisms can explain this behavior including alteration of the activity of target enzymes (such as glutathione-S transferase and topoisomerases), changes in apoptotic processes, and/or, more frequently, modification of drug efflux, implicating ABC transporters such as the multidrug resistance-associated protein (MRP) or P-glycoprotein (Pgp). Pgp proteins have been implicated in resistance to many chemotherapeutic drugs such as doxorubicin, vincristine, etoposide, or taxol [36,37]. In fact, in tumor cells expressing Pgp, the efflux of a given anticancer drug increases across the plasma membrane, thereby reducing the intracellular drug concentration and hence its cytotoxicity. But Pgp activity can be reversed by a few specific molecules, referred to as MDR modulators [35]. For example, the calcium channel blocker verapamil possesses this reversing capacity at a concentration range from 5 μM to 50 μM [38]. Lamellarin I is also an MDR modulator, directly inhibiting P-glycoprotein-mediated drug efflux at nonlethal doses [39]. In P388/schabel cells, lamellarin I is able to reverse doxorubicin resistance at a concentration one-tenth of that necessary for verapamil. Other "sea world" molecules are endowed with this property to modulate MDR: discodermolide [40], pattelamide D [41], irciniasulphonic acid [42], agosterol A [43], and ningalin B [44]. This last compound bears a close structural analogy with lamellarin T, which is also a

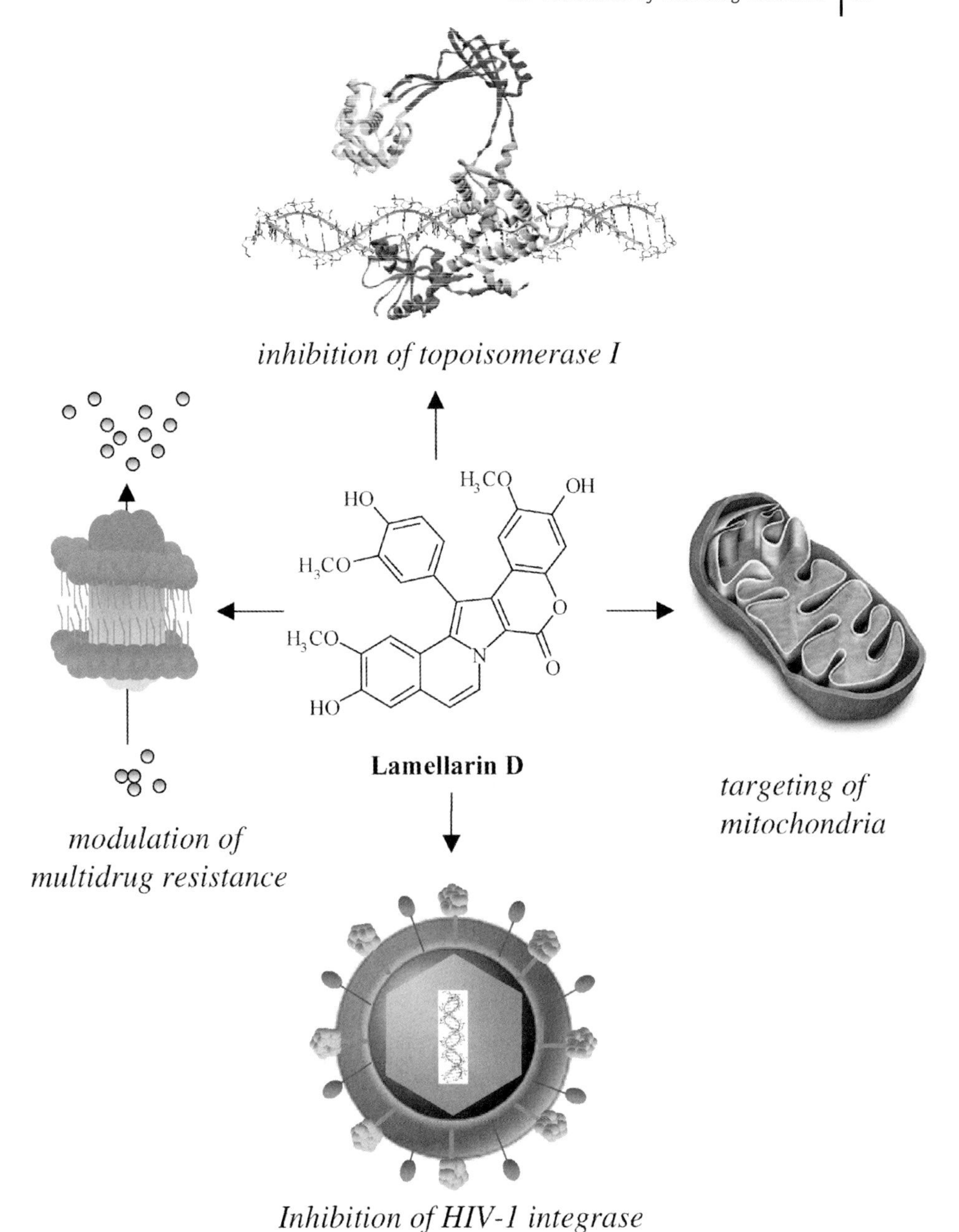

Fig. 7.2 Schematic illustration of the targets and mode of action of lamellarins.

modulator of multidrug resistance in addition to its cytotoxic and antibacterial properties [34]. Ningalin B was recently incorporated with combretastatin A4-like structures to design new antimitotic agents [45]. Combretastatin A-4 can easily isomerize to the thermodynamically more stable trans isomer, with reduced antimitotic activity. Its condensation with a lamellarin-type moiety provides an original approach to maintain the more potent cis configuration.

7.4
Antioxidant Properties

Oxidant products are reactive species usually toxic to cells through lesions caused to lipids, proteins, and DNA in particular. In cells, mitochondria are the major source of oxidants. In fact, using oxygen to generate energy, mitochondria naturally produce reactive oxygen species (ROS). To avoid the toxic effects, cells have developed complex chemical and/or enzymatic processes to prevent damage, including thioredoxin, the glutathione system, superoxide dismutase, and catalase [46,47]. When these protective systems are deficient, an abnormal ROS production arises and produces irreversible lesions potentially implicated in human pathologies such as cancer, artheriosclerosis, or neurodegeneration [48]. Antioxidants reduce the rate of particular oxidation reactions and can help to protect against lethal damage induced by ROS. Certain lamellarins such as lamellarins γ, K, U, I, and C-diacetate have revealed mild antioxidant properties [49]. Their effects are fairly modest (with IC_{50} above 2 mM) compared to reference drugs such as Trolox and α-tocopherol (IC_{50} around 50 μM) but this property may be enhanced upon adequate substitutions of the molecule. However, this is certainly not the most advantageous property of the lamellarins, given their significant cytotoxicity.

7.5
Inhibition of HIV-1 Integrase

AIDS remains a devastating disease, responsible for the death of millions of humans, principally in central Africa. The prevalence of AIDS remains extremely high and, despite the efficacy of the highly active antiretroviral therapy (HAART) and the continuous development of chemotherapeutic agents targeting HIV-1 reverse transcriptase and protease, newer effective drugs are still eagerly awaited [50]. HIV-1 integrase promotes the integration of proviral DNA into the host cell chromosomal DNA. It constitutes an attractive target because no human cellular homolog for this enzyme exists. Two major classes of integrase inhibitors have been described: the catechol-containing hydroxylated aromatics (L-choric acid) and the diketo acid-containing aromatics (5CITEP, S-1360) [51]. A few members of the lamellarin family revealed integrase inhibition activities. Lamellarin α 20-sulfate *in vitro* showed potent inhibition of HIV-1 integrase [52]. This compound, for which an efficient total synthesis has been reported [53], is able to act at two different steps of the catalytic cycle, terminal cleavage and strand transfer. Moreover, this compound inhibits viral replication in cultured cells at nontoxic doses. Its sulfate group plays a critical role because the analog lamellarin α is inactive against HIV-1 integrase. The nonsulfated analog lamellarin H may also be of interest as an HIV-1 integrase inhibitor ($IC_{50} = 8\ \mu$M) but unfortunately this compound is markedly toxic to cells (LD_{50} of

5.7 μM measured with Hela cells). Nevertheless, the lamellarins have an intrinsic potential for the inhibition of HIV-1 integrase. But so far this pharmacological activity has not been thoroughly explored.

7.6 Cytotoxicity

The most common properties of the lamellarins are their cytotoxic activities and in particular their capacity to inhibit the proliferation of cancer cells. This activity is usually easy to measure and a large panel of tumor cells can be cultivated *in vitro* without much difficulty. Different levels of activity corresponding to a lethal or a growth inhibition activity (cytotoxic versus cytostatic) have been reported in the literature with different cell lines. A representative selection of these values is collated in Table 7.1. Although comparison cannot strictly be made, because of distinct experimental conditions from one study to another, it is nevertheless obvious that the majority of lamellarins are considerably cytotoxic, with IC_{50} (or LD_{50}) values in the nanomolar to micromolar range, depending on the experimental conditions and

Tab. 7.1 Cytotoxic potential of lamellarins.

Molecules	Cell lines	Time incubation	Test[a]	IC_{50}[b]	Reference
Lamellarin-ξ	COLO-205	16 h	MTT	5.6 nM	[24]
Lamellarin-η	COLO-205	16 h	MTT	178 nM	[24]
Lamellarin-φ triacetate	COLO-205	16 h	MTT	56 nM	[24]
Lamellarin-χ triacetate	COLO-205	16 h	MTT	0.2 nM	[24]
Lamellarin α	Hela	—	MTT	5 μM	[54]
Lamellarin α 20-sulfate	Hela	3 d	MTT	274 μM	[55]
LAM-D	MDCK	5–8 d	ICF	22 nM	[56]
	P388	3 d	MTS	136 nM	[57]
	CEM	3 d	MTS	14 nM	[57]
	Hela	5–8 d	ICF	10 nM	[56]
	XC	5–8 d	ICF	12 nM	[56]
	Vero	5–8 d	ICF	10 nM	[56]
Lamellarin F	COLO-205	1–6 h	MTT	9 nM	[24]
Lamellarin H	Hela	ND	MTT	5.7 μM	[54]
Lamellarin I	COLO-205	16 h	MTT	25 nM	[24]
Lamellarin J	COLO-205	16 h	MTT	50 nM	[24]
Lamellearin K triacetate	COLO-205	16 h	MTT	700 nM	[24]
Lamellarin L triacetate	COLO-205	16 h	MTT	0.25 nM	[24]
Lamellarin N	SK-MEL-5	—	—	187 nM	[55]
Lamellarin U 20 sulfate	Hela	3 d	MTT	145 μM	[55]
Lamellarin V 20 sulfate	Hela	3 d	MTT	130 μM	[55]
Lamellarin T	Hela	3 d	MTT	27 μM	[55]
Lamellarin T diacetate	COLO-205	16 h	MTT	180 nM	[24]
Lamellarin W	Hela	3 d	MTT	28 μM	[55]

[a] MTT and MTS are conventional tetrazolium dyes used to evaluate cell proliferation.
[b] IC_{50}, drug concentration to inhibit cell proliferation by 50 % (in some cases, the numbers refer to LD_{50}, dose to induce cell death, depending on the assay).

the nature of the compounds. A noticeable exception is that of the sulfated lamellarins (α, U, and V) which are not cytotoxic, presumably owing to reduced cell uptake.

Lamellarin D (Lam-D) is one of the most cytotoxic compounds of the family. Its broad spectrum of toxicity to numerous tumor cell lines (Table 7.1) makes it a solid lead for the design of synthetic analogs [57]. Different procedures have been reported for the total synthesis of Lam-D and related lamellarins, thus opening a route to the rational design of a variety of analogs [56,58–62]. In the Lam-D series, precise structure–activity relationships have been delineated [63]. Compounds possessing hydroxyl groups at both the C-8 and C-20 positions are more cytotoxic than other Lam-D analogs, whereas the OH at C-14 and methoxy groups at C-13 and C-21 seem to be less important to maintaining the cytotoxic potential [64]. These structure–activity relationships (Figure 7.3) have been extended to delineate the importance of the lactone of Lam-D in the cytotoxicity [65]. Most of the Lam-D derivatives with an open lactone ring were found to be considerably less cytotoxic than Lam-D, except when a lactonization potential is preserved. In this case, a marked toxicity toward A-549 lung carcinoma, HT-29 colon carcinoma, and MDA-MD-231 breast adenocarcinoma cells was maintained [65]. The mode of action of these lactone-free derivatives of Lam-D is, however, unknown at present. It remains to be determined whether these cytotoxic derivatives can preserve an inhibitory activity against topoisomerase I, which is considered the privileged (but not unique) target of Lam-D. It is only since 2002 that the molecular mechanism of action of cytotoxic lamellarins, in particular Lam-D, has been investigated, essentially in our laboratory.

7.7 Topoisomerase I Inhibition

Topoisomerases are fundamental enzymes that ensure DNA replication, transcription, or recombination [66]. Both type I and type II topoisomerases act by changing

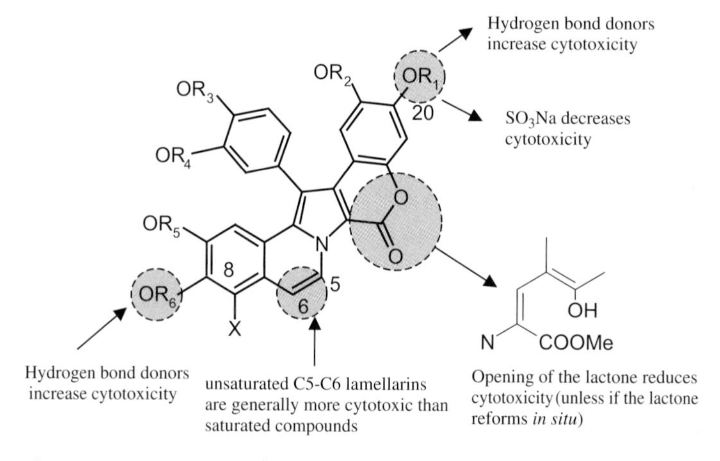

Fig. 7.3 Structure–activity relationships in the lamellarin series.

the supercoiling of DNA by nicking one (type I) or two (type II) of the DNA strands. But these vital enzymes can become lethal weapons against cancer cells. A significant number of chemotherapeutic drugs work by interfering with topoisomerases in cancer cells [67,68]. Topoisomerase II poisons were first used in cancer chemotherapy more than 50 years ago and different drugs in this category remain in wide clinical use, including certain anthracyclines such as doxorubicin and daunorubicin, the anthracenedione mitoxantrone, as well as the epipodophyllotoxin etoposide [69]. Topoisomerase I poisons were characterized in the mid-1980s, and two molecules, topotecan and irinotecan, both derived from the plant alkaloid camptothecin (CPT), have been approved for the treatment of cancers, essentially ovarian and colon cancers [70,71]. CPT stabilizes topoisomerase I–DNA complexes in the form of a covalent intermediate whereby the enzyme is linked to one strand of the DNA so as to promote the accumulation of DNA single-strand breaks [72]. The search for efficient topoisomerase I inhibitors has been an area of intense research since the late 1980s, but apart from the camptothecins, which have been extensively investigated, only a limited number of potent topoisomerase I poisons capable of stabilizing topoisomerase I–DNA covalent complexes has been identified [73,74]. The non-CPT topoisomerase I poisons with potent activities include different indenoisoquinolines such as MJ-III-65 [75,76], certain glycosyl indolocarbazoles such as NB-506 and edotecarin (formerly J-107088 and ED-749) [57] and lamellarin D which we identified as a potent topoisomerase I poison in 2003 [57]. Lam-D binds relatively weakly to DNA, presumably via the insertion of its planar pentacyclic chromophore between DNA base pairs. Intercalation of the flat chromophore between two adjacent base pairs exposes the perpendicular methoxyphenol moiety toward the major groove of DNA where the enzyme can be trapped [64]. This DNA interaction, although relatively weak, provides the necessary anchorage for stabilization of the enzyme–DNA complex. Inhibition of topoisomerase I by Lam-D has been demonstrated by a variety of approaches, both *in vitro* using recombinant enzymes and model DNA substrates and in a cellular context using immunoblot assays to detect the drug-stabilized topoisomerase I–DNA complexes in cells. The key discovery that topoisomerase I was a major target for Lam-D has opened the door to the determination of structure–function relationships and the rational design of lamellarin analogs of pharmaceutical interest. The double bond between carbons 5 and 6 in the quinoline B-ring is a crucial element for topoisomerase I inhibition [63]. This C-5=C-6 double bond confers a planar structure on the molecule and when this double bond is lacking (as in Lam-K, for example) the planar conformation no longer exists and the drug loses its capacity to interfere with topoisomerase I. This was clearly demonstrated using the synthetic compound Lam-501, which only differs from Lam-D in the absence of the C-5=C-6 double bond. This compound showed no inhibition of topoisomerase I, and its cytotoxicity was considerably reduced [57]. Additional structure–function relationships have been delineated, thanks to the synthesis of different series of Lam-D derivatives. In particular, using Lam-D analogs with distinct OMe/OH substitution on the core structure, it was demonstrated that the 8-OH and the 20-OH are crucial for topoisomerase I inhibition and in most cases a direct link could be established between enzyme inhibition and cytotoxicity [64]. The use of cancer cell lines with topoisomerase I mutated genes has also contributed to the characterization of topoisomerase I as the main target for Lam-D and related analogs. It is thus not surprising to understand why Lam-H, which is structurally close to Lam-D, has also been identified as a topoisomerase inhibitor. Lam-H was previously shown to inhibit topoisomerase I from the *Molluscum contagiosum* virus (MCV) [54].

In terms of efficacy, Lam-D is less efficient than CPT in inhibiting topoisomerase I (IC_{50} of 0.42 versus 0.087 μM, respectively) [57]. But the sequencing of DNA cleavage products revealed that the cleavage profiles obtained with Lam-D were distinct from those seen with CPT. Both molecules intercalate at the sites of DNA cleavage corresponding to $T^{\downarrow}G$ sites but in addition Lam-D can induce cleavage at certain $C^{\downarrow}G$ sites. Lam-D and CPT share a common mechanism of action at the topoisomerase I–DNA complex level but the two drugs are positioned in a slightly different manner at the DNA–enzyme interface, thus providing distinct molecular contacts at the origin of the nonidentical cleavage profiles. Molecular modeling studies have greatly contributed to the explanation of how Lam-D fits into the topoisomerase I–DNA active site [64].

But the story of Lam-D is not so simple. If topoisomerase I can be considered as the unique target of CPT, the situation is clearly more complicated for Lam-D because its cytotoxicity cannot be explained uniquely on the basis of the topoisomerase I targeting. The cytotoxicity of Lam-D has been evaluated using the mutant cell lines P388CPT5 and CEM-C2, which both express a mutant topoisomerase I conferring a high resistance to CPT and other CPT-derived topoisomerase I poisons [57,78]. The two cells lines are more resistant to CPT and Lam-D than the corresponding wild-type P388 and CEM cell lines. This observation confirmed the direct implication of topoisomerase I in the cytotoxicity of Lam-D. But the resistance indexes (i.e., the ratio of IC_{50} values for a pair of sensitive and resistant cell lines) was clearly different for Lam-D and CPT. For example, the *top1* mutated gene in CEMC2 cells confers a huge resistance of these leukemia cells to CPT (RI $>$ 2000) whereas the RI was about 30 times lower, around 70, with Lam-D. Moreover, we discovered that Lam-D, but not CPT, was able to induce apoptosis of the P388CPT5-resistant cell line. These observation strongly suggested the existence of alternative target(s) for Lam-D in cancer cells. This hypothesis was rapidly validated with the discovery of a nuclear topoisomerase I-independent, direct effect of Lam-D on mitochondria.

7.8
Targeting of Mitochondria and Proapoptotic Activities

Mitochondria are the well-characterized intracellular organelles needed for the production of ATP for all cellular processes. They also play a major role in the regulation of calcium flux and redox state. Moreover, since the mid-1990s, mitochondria have been considered as central effectors for the regulation of cell apoptosis [79]. Apoptosis is a cell death characterized by a number of distinctive biochemical and morphological changes occurring in cells, such as DNA fragmentation, chromatin condensation, and loss of phospholipid asymmetry in the plasma membrane and membrane budding. During apoptosis, a family of proteases known as caspases is generally activated. These proteins cleave key cellular substrates required for normal cellular functions, including structural proteins of the cytoskeleton and nuclear proteins such as DNA repair enzymes [80]. It is now well established that during apoptosis major changes occur at the level of mitochondria. A reduction of the mitochondrial membrane potential ($\Delta\Psi m$) is generally observed during the first steps of apoptosis, associated with the release into the cytosol of various proteins localized in the intermembrane space, such as cytochrome c, AIF, Smac/DIABLO and/or endoG [81]. Our initial study on the apoptotic pathway induced by Lam-D revealed a very early mitochondrial dysfunction, including a reduction of $\Delta\Psi m$ (Figure 7.4) and the release of cytochrome c and AIF from the mitochondria to the cytosol [82]. It was then observed that Lam-D could induce these alterations directly on isolated mitochondria. Both functional assays (Figure 7.4) and direct observations by electron microscopy indicated that mitochondria represented targets for Lam-D, and similarly for Lam-M (Figure 7.5). Although the exact mechanism implicated in the targeting of mitochondria is not fully understood, we have accumulated evidence that an opening of the mitochondrial permeability transition pore (mPTP) is induced by Lam-D. The drug induces a swelling of isolated mitochondria and this effect can be prevented by a preincubation with the PTP inhibitor cyclosporin A. Each protein constituting the mPTP can be considered as a molecular target for Lam-D, including the adenine nucleotide translocator (ANT), voltage-dependent anion channel (VDAC), peripheral benzodiazepine receptor (PBR), hexokinase (HK), Bcl-2 family members (including Bax and Bcl-2), mitochondrial creatine kinase (CK), and cyclophilin D [83]. These proteins and receptors represent potential drug targets to control tumor cell growth [84]. But we also cannot exclude the possibility that Lam-D induces an indirect opening of the mPTP, forming channels or pores through phospholipids from the outer or inner membranes or acting with specific complexes of the respiratory chain. The detailed mechanism is currently being investigated further.

Based on our previous studies, we are confronted with an uncommon situation where Lam-D seems to reside in two preferred sites within cells: (i) the nucleus, where the drug forms tight and stable complexes with DNA and topoisomerase I; (ii) mitochondria, where Lam-D interacts with an as yet unknown molecular target, potentially associated with the mPTP. Both sites and their associated mechanisms can contribute to cell death induced by Lam-D but which is the major contributor, the nucleus or the mitochondrion? The question remains unresolved at this stage but our recent data suggest that topoisomerase I targeting may be the main mechanism responsible for the antitumor effects of Lam-D. Recent results (unpublished) indicate that at low concentrations ($<1\,\mu M$), Lam-D affects preferentially the topoisomerase I pathway, without any direct interference with mitochondria. At higher concentrations (in particular for concentrations $>5\,\mu M$), Lam-D affects both the nucleus and mitochondria, with a dual action leading to massive and

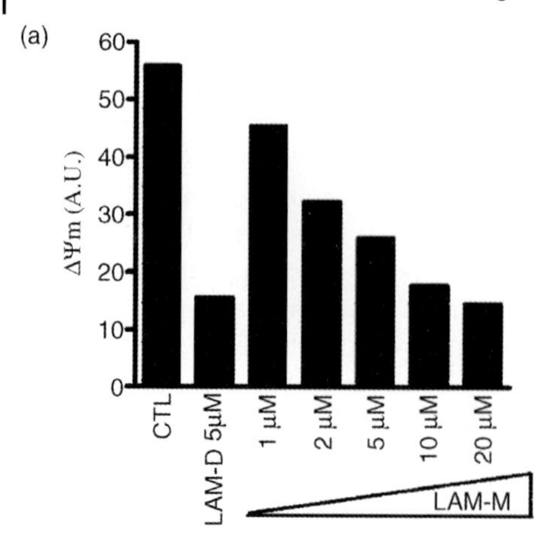

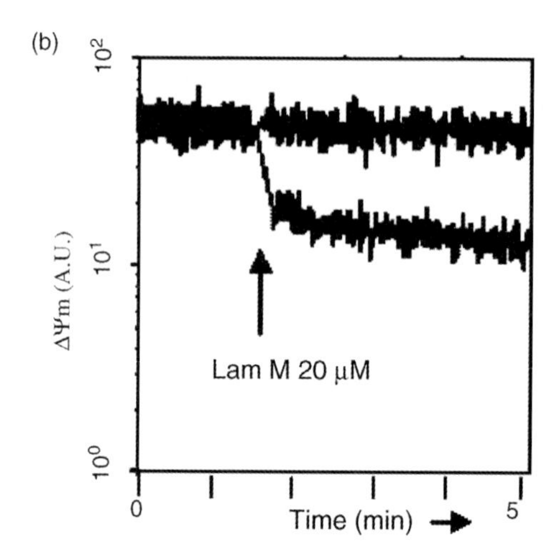

Fig. 7.4 (a) Dose–response of the lamellarin M-induced mitochondrial depolarization in P388 cells. (b) Monitoring of the mitochondrial membrane potential (ΔΨm) by real-time flow cytometry, using functional mitochondria isolated from P388 cells and the fluorescent probe JC-1.

rapid cell death. We must also consider the hypothesis, as yet purely conjectural, that the drug can target simultaneously both the nuclear and the mitochondrial topoisomerase I. The topology of the mitochondrial genome – a closed circular mtDNA – is regulated during replication by a protein encoded by the gene mtDNA topoisomerase I (*top1*mt) localized on human chromosome 8q24. This gene is highly homologous to the nuclear *top1* gene and the mitochondrial topoisomerase I enzyme is fully inhibited by CPT [85,86]. The close functional homology between CPT and Lam-D suggests that the mitochondrial enzyme can

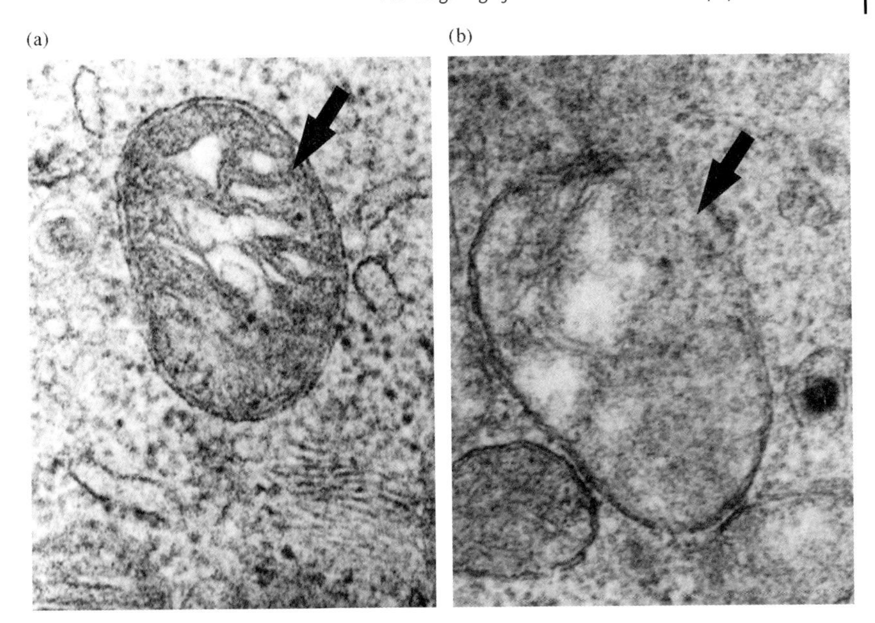

Fig. 7.5 Transmission electron microscopic images of a mitochondrion in P388 cells; (a) control, (b) after treatment for 15 min with 5 μM lamellarin D. Magnification, ×20 000.

also be targeted by Lam-D. It will be interesting to see if Lam-D does interfere with mitochondrial topoisomerase I and if this molecular effect is at the origin of the mitochondrial dysfunctions (reduction of $\Delta\Psi$m and mPTP opening) observed in cells upon treatment with Lam-D.

Lamellarin D is not the only marine compounds capable of altering mitochondrial functions. This is also the case for stolonoxides, which are oxidized fatty acid metabolites isolated from the Mediterranean tunicate *Stolonica socialis*. These compounds are able to inhibit mitochondrial respiration; more precisely they are specific inhibitors of complexes II and III of the respiratory chain [87]. Another example is that of cephalostatin 1, a bis-steroidal compound isolated from the marine worm *Cephalodiscus gilchristi*. Like Lam-D, cephalostatin 1 induces a reduction of the $\Delta\Psi$m, but interestingly this alteration is not associated with the release of cytochrome c or the apoptosis-inducing factor, but only with the release of the Smac/DIABLO protein. Drugs that induce apoptosis by directly disrupting mitochondrial functions or membrane integrity seem to have an attractive potential for preventing tumor growth and eliminating tumor cells [84]. But the next challenge will be to identify or to rationally design a mitochondriotoxic drug able to kill selectively cancer cells without affecting normal cells. At present this selectivity issue for cancer cells versus nontumor cells is a major obstacle to the development of anticancer drugs targeting mitochondria.

7.9 Conclusion

Studies on lamellarins have demonstrated that these marine molecules represent an original source of structures for the identification and design of drugs active against severe human pathologies such as AIDS and cancer. Only a very few aspects of their mechanisms of action are known, in particular their capacities to interfere with topoisomerase I and mitochondria, both contributing to their potent cytotoxicity. The recent discovery that Lam-D behaves as a CPT-like topoisomerase I poison has encouraged the synthesis of analogs with improved pharmaceutical profiles. Lam-D was first identified more than 20 years ago (1985) but its pharmacological interest has only recently been raised, essentially since 2003. More than 300 derivatives have been synthesized by chemists at PharmaMar (Spain) and drug candidates have been selected for the extended preclinical and toxicological studies required prior to human clinical trials. It took around 30 years from the first extraction of CPT from the Chinese tree *Campthoteca acuminata* (1966), to the identification of its topoisomerase I target (1985) and the FDA-approval of the first CPT derivative (early 1990s). Even in the best scenario, it will take even longer from the first identification of Lam-D (1985) to the characterization of topoisomerase I as its major target (2003) and hopefully the clinical development of a tumor-active derivative. However, even if Lam-D is only a laboratory tool at present, it represents a very interesting lead for the development of marine-derived anticancer agents.

References

1 Kelecom, A. (2002) *Anais da Academia Brasileira de Ciencias*, **74**, 151–170.

2 Simmons, T. L., Andrianasolo, E., McPhail, K., Flatt, P., Gerwick, W. H. (2005) *Molecular Cancer Therapeutics*, **4**, 333–342.

3 Kijjoa, A. and Sawangwong, P. (2004) *Marine Drugs*, **2**, 73–82.

4 Bergmann, W. and Feeney, R. J. (1950) *Journal of the American Chemical Society*, **72**, 2809–2812.

5 Bergmann, W. and Feeney, R. J. (1951) *Journal of Organic Chemistry*, **16**, 981–989.

6 Bergmann, W. and Burke, D. C. (1955) *Journal of Organic Chemistry*, **20**, 1501–1511.

7 Evans, J. S., Musser, E. A., Mengel, G. D., Forsblad, K. R., Hunter, J. H. (1961) *Proceedings of the Society for Experimental Biology and Medicine*, **2**, 350–353.

8 Newman, D. J. and Cragg, G. M. (2004) *Journal of Natural Products*, **67**, 1216–1238.

9 Lombardo, D. and Dennis, E. A. (1985) *The Journal of Biolgical Chemistry*, **260**, 7234–7240.

10 Look, S. A., Fenical, W., Jacobs, R. S., Clardy, J. (1986) *Proceedings of the National Academy of Sciences of the United States of America*, **83**, 6238–6240.

11 Burns, L. H., Jin, Z., Bowersox, S. S. (1999) *Journal of Vascular Surgery*, **30**, 334–343.

12 Woodruff-Pak, D. S., Li, Y. T., Kem, W. R. (1994) *Brain Research*, **645**, 309–317.

13 O'Hanlon, L. H. (2006) *Journal of the National Cancer Institute*, **98**, 662–663.

14 Watanabe, J., Minami, M., Kobayashi, M. (2006) *Anticancer Research*, **26**, 1973–1981.

15 Cunningham, C., Appleman, L. J., Kirvan-Visovatti, M., Ryan, D. P., Regan, E., Vukelja, S., Bonate, P. L., Ruvuna, F., Fram, R. J., Jekunen, A., Weitman, S., Hammond, L. A., Eder, J. P. (2005) *Clinical Cancer Research*, **11**, 7825–7833.

16 Ajani, J. A., Jiang, Y., Faust, J., Chang, B. B., Ho, L., Yao, J. C., Rousey, S., Dakhil, S., Cherny, R. C., Craig, C., Bleyer, A. (2006) *Investigational New Drugs*, **24**, 353–357.

17 Fayette, J., Coquard, I. R., Alberti, L., Ranchere, D., Boyle, H. (2005) *Oncologist*, **10**, 827–832.

18 Latreille, J., Batist, G., Laberge, F., Champagne, P., Croteau, D., Falardeau, P., Levinton, C., Hariton, C., Evans, W. K., Dupont, E. (2004) *Clinical Lung Cancer*, **4**, 231–236.

19 Herbst, R. S., Hammond, L. A., Carbone, D. P., Tran, H. T., Holroyd, K. J., Desai, A., Williams, J. I., Bekele, B. N., Hait, H., Allgood, V., Solomon, S., Schiller, J. H. (2003) *Clinical Cancer Research*, **9**, 4108–4115.

20 Rawat, D. S., Joshi, M. C., Joshi, P., Atheaya, H. (2006) *Anticancer Agents in Medicinal Chemistry*, **6**, 33–40.

21 Kornprobst, J. M. (2005) *Substances Naturelles d'Origine Marine*, Vols. 1 and 2, TEC & DOC.

22 Fayette, J., Coquard, I. R., Alberti, L., Boyle, H., Meeus, P., Decouvelaere, A. V., Thiesse, P., Sunyach, M. P., Ranchere, D., Blay, J. Y. (2006) *Current Opinion in Oncology*, **18**, 347–353.

23 Andersen, R. J., Faulkner, D., He, C. H., VanDuyne, G. D., Clardy, J. (1985) *Journal of the American Chemical Society*, **107**, 5492–5495.

24 Malla Reddy, S., Srinivasulu, M., Satyanarayana, N., Kondapi, A. K., Venkateswarlu, Y. (2005) *Tetrahedron*, **61**, 9242–9247.

25 Rinehart, K. L., Gloer, J. B., Hughes, R. G., Renis, H. E., McGovren, J. P., Swynenberg, E. B., Rinehart Stringfellow, D. A., Kuentzel, S. L., Li, L. H. (1981) *Science*, **212**, 933–935.

26 Schmitz, F. J., Ksebati, M. B., Chang, J. S., Wang, J. H., Hossain, M. B., Vander Helm, D. (1989) *Journal of Organic Chemistry*, **54**, 3463–3472.

27 Kobayashi, J., Chang, J., Walchli, M. R., Nakamura, H., Yoshimasa, H., Takuma, S., Ohizumi, T. (1988) *Journal of Organic Chemistry*, **53**, 1800–1804.

28 Rudi, A. and Kashman, Y. (1989) *Journal of Organic Chemistry*, **54**, 5331–5337.

29 Yoshida, W. Y., Lee, K. K., Carrol, A. R., Scheuer, P. J. (1992) *Helvetica Chimica Acta*, **75**, 1721–1725.

30 Rudi, A., Goldberg, I., Stein, Z., Frolow, F., Benayahu, Y., Schleyer, M., Kashman, Y. (1994) *Journal of Organic Chemistry*, **59**, 999–1003.

31 Kang, H. and Fenical, W. (1997) *Journal of Organic Chemistry*, **62**, 3254–3262.

32 Proksch, P., Adrada, R. A., Ebel, R. (2002) *Applied Microbiology and Biotechnology*, **59**, 125–134.

33 Faulkner, D. J. (2001) *Natural Product Reports*, **18**, 1–49.

34 Bailly, C. (2004) *Current Medicinal Chemistry—Anti-Cancer Agents*, **4**, 363–378.

35 Krishna, R. and Mayer, L. D. (2001) *Current Medicinal Chemistry—Anti-Cancer Agents*, **1**, 163–174.

36 Gottesman, M. M. and Pastan, I. (1993) *Annual Review of Biochemistry*, **62**, 385–427.

37 Debenham, P. G., Kartner, N., Siminovitch, L., Riordan, J. R., Ling, V. (1982) *Molecular and Cellular Biology*, **2**, 881–889.

38 Mickley, L. A., Bates, S. E., Richert, N. D., Currier, S., Tanaka, S., Foss, F., Rosen, N., Fojo, A. T. (1989) *The Journal of Biolgical Chemistry*, **264**, 18031–18040.

39 Quesada, A. R., Garcia Gravalos, M. D., Fernandez Puentes, J. L. (1996) *British Journal of Cancer*, **74**, 677–682.

40 Kowalski, R. J., Giannakakou, P., Gunasekera, S. P., Longley, R. E., Day, B. W., Hamel, E. (1997) *Molecular Pharmacology*, **52**, 613–622.

41 Williams, A. B. and Jacobs, R. S. (1993) *Cancer Letters*, **71**, 97–102.

42 Kawakami, A., Miyamoto, T., Higuchi, R., Uchiumi, T., Kuwano, M., Van Soest, R. W. M. (2001) *Tetrahedron Letters*, **42**, 3335–3337.

43 Aoki, S., Setiawan, A., Yoshioka, Y., Higuchi, K., Fudetani, R., Chen, Z. S., Sumizawa, T., Akiyama, S., Kobayashi, M. (1999) *Tetrahedron*, **55**, 13965–13972.

44 Tao, H., Hwang, I., Boger, D. L. (2004) *Bioorganic and Medicinal Chemistry Letters*, **14**, 5979–5981.

45 Banwell, M. G., Hamel, E., Hockless, D. C., Verdier-Pinard, P., Willis, A. C., Wong, D. J. (2006) *Bioorganic and Medicinal Chemistry*, **14**, 4627–4638.

46 Turrens, J. F. (2003) *The Journal of Physiology*, **552**, 335–344.

47 Ercal, N., Gurer-Orhan, H., Aykin-Burns, N. (2001) *Current Topics in Medicinal Chemistry*, **1**, 529–539.

48 Kang, D. and Hamasaki, N. (2005) *Current Medicinal Chemistry*, **12**, 429–441.

49 Krishnaiah, P., Reddy, V. L., Venkataramana, G., Ravinder, K., Srinivasulu, M., Raju, T. V., Kobayashi, J., Harbour, G. C., Hughes, R. G., Mizsak, S. A., Scahill, T. A. (1984) *Journal of the American Chemical Society*, **4**, 1524–1526.

50 Imamichi, T. (2004) *Current Pharmaceutical Design*, **10**, 4039–4053.

51 De Clercq, E. (2004) *The International Journal of Biochemistry and Cell Biology*, **36**, 1800–1822.

52 Reddy, M. V., Rao, M. R., Rhodes, D., Hansen, M. S., Rubins, K., Bushman, F. D., Venkateswarlu, Y., Faulkner, D. J. (1999) *Journal of Medicinal Chemistry*, **42**, 1901–1907.

53 Yamaguchi, T., Fukuda, T., Ishibashi, F., Iwao, M. (2006) *Tetrahedron Letters*, **47**, 3755–3757.

54 Ridley, C. P., Reddy, M. V., Rocha, G., Bushman, F. D., Faulkner, D. J. (2002) *Bioorganic and Medicinal Chemistry*, **10**, 3285–3290.

55 Reddy, V. R. and Faulkner, D. J. (1997) *Tetrahedron*, **53**, 3457–3466.

56 Ishibashi, F., Tanabe, S., Oda, T., Iwao, M. (2002) *Journal of Natural Products*, **65**, 500–504.

57 Facompré, M., Tardy, C., Bal-Mahieu, C., Colson, P., Perez, C., Manzanares, I., Cuevas, C., Bailly, C. (2003) *Cancer Research*, **63**, 7392–7399.

58 Ploypradith, P., Jinaglueng, W., Pavaro, C., Ruchirawat, S. (2003) *Tetrahedron Letters*, **44**, 1363–1366.

59 Cironi, P., Manzanares, I., Albericio, F., Alvarez, M. (2003) *Organic Letters*, **5**, 2959–2962.

60 Marfil, M., Albericio, F., Alvarez, M. (2004) *Tetrahedron*, **60**, 8659–8668.

61 Pla, D., Marchal, A., Olsen, C. A., Albericio, F., Alvarez, M. (2005) *Journal of Organic Chemistry*, **70**, 8231–8234.

62 Fujikawa, N., Ohta, T., Yamaguchi, T., Fukuda, T., Ishibashi, F., Iwao, M. (2006) *Tetrahedron*, **62**, 594–604.

63 Tardy, C., Facompré, M., Laine, W., Baldeyrou, B., Garcia-Gravalos, D., Francesch, A., Mateo, C., Pastor, A., Jimenez, J. A., Manzanares, I., Cuevas, C., Bailly, C. (2004) *Bioorganic and Medicinal Chemistry*, **12**, 1697–1712.

64 Marco, E., Laine, W., Tardy, C., Lansiaux, A., Iwao, M., Ishibashi, F., Bailly, C., Gago, F. (2005) *Journal of Medicinal Chemistry*, **48**, 3796–3807.

65 Pla, D., Marchal, A., Olsen, C. A., Francesch, A., Cuevas, C., Albericio, F., Alvarez, M. (2006) *Journal of Medicinal Chemistry*, **49**, 3257–3268.

66 Wang, J. C. (1996) *Annual Review of Biochemistry*, **65**, 635–691.

67 Nitiss, J. L. (2002) *Current Opinion in Investigational Drugs*, **3**, 1512–1516.

68 Giles, G. I. and Sharma, R. P. (2005) *Medicinal Chemistry*, **1**, 383–394.

69 Fortune, J. M. and Osheroff, N. (2000) *Progress in Nucleic Acid Research and Molecular Biology*, **64**, 221–253.

70 Garcia-Carbonero, R. and Supko, J. G. (2002) *Clinical Cancer Research*, **8**, 641–661.

71 Zunino, F., Dallavalleb, S., Laccabuea, D., Berettaa, G., Merlinib, L., Pratesi, G. (2002) *Current Pharmaceutical Design*, **8**, 2505–2520.

72 Bailly, C. (2000) *Current Medicinal Chemistry*, **7**, 39–58.

73 Li, Q. Y., Zu, Y. G., Shi, R. Z., Yao, L. P. (2006) *Current Medicinal Chemistry*, **13**, 2021–2039.

74 Meng, L. H., Liao, Z. Y., Pommier, Y. (2003) *Current Topics in Medicinal Chemistry*, **3**, 305–320.

75 Antony, S., Jayaraman, M., Laco, G., Kohlhagen, G., Kohn, K. W., Cushman, M., Pommier, Y. (2003) *Cancer Research*, **63**, 7428–7435.

76 Marchand, C., Antony, S., Kohn, K. W., Cushman, M., Ioanoviciu, A., Staker, B. L., Burgin, A. B., Stewart, L., Pommier, Y. (2006) *Molecular Cancer Therapeutics*, **5**, 287–295.

77 Saif, M. W. and Diasio, R. B. (2005) *Clinical Colorectal Cancer*, **5**, 27–36.

78 Vanhuyse, M., Kluza, J., Tardy, C., Otero, G., Cuevas, C., Bailly, C., Lansiaux, A. (2005) *Cancer Letters*, **221**, 165–175.

79 Green, D. R. and Kroemer, G. (2004) *Science*, **305**, 626–629.

80 Zhivotovsky, B. (2003) *Essays in Biochemistry*, **39**, 25–40.

81 Ly, J. D., Grubb, D. R., Lawen, A. (2003) *Apoptosis*, **8**, 115–128.

82 Kluza, J., Gallego, M. A., Loyens, A., Beauvillain, J. C., Sousa-Faro, J. M., Cuevas, C., Marchetti, P., Bailly, C. (2006) *Cancer Research*, **66**, 3177–3187.

83 Halestrap, A. P., McStay, G. P., Clarke, S. J. (2002) *Biochimie*, **84**, 153–166.

84 Dias, N. and Bailly, C. (2005) *Biochemical Pharmacology*, **70**, 1–12.

85 Zhang, H., Barcelo, J. M., Lee, B., Kohlhagen, G., Zimonjic, D. B., Popescu, N. C., Pommier, Y. (2001) *Proceedings of the National Academy of Sciences of the United States of America*, **98**, 10608–10613.

86 Zhang, H., Meng, L. H., Zimonjic, D. B., Popescu, N. C., Pommier, Y. (2004) *Nucleic Acids Research*, **32**, 2087–2092.

87 Fontana, A., Cimino, G., Gavagnin, M., Gonzalez, M. C., Estornell, E. (2001) *Journal of Medicinal Chemistry*, **44**, 2362–2365.

8
Manzamine Alkaloids

Jiangnan Peng, Karumanchi V. Rao, Yeun-Mun Choo, Mark T. Hamann

8.1
Introduction

Approximately 30 % of the drugs in the market worldwide are natural products or their derivatives. Natural products show a diversity of chemical structures wider than that accessible by even the most sophisticated synthetic methods [1]. Moreover, natural products have often opened up completely new therapeutic approaches. Work on marine natural products started in the early 1950s, when Bergman discovered the novel bioactive arabino-nucleoside from the marine sponge *Cryptotethya crypta* [2]. According to the numerous marine natural product reviews published since then, such as those by Faulkner, Blunt, Gribble, Tolvanen, and Lounasmaa [3], marine natural product evaluations were primarily focused on anticancer and anti-inflammatory activity. Marine alkaloids extracted from sponges have become increasingly important since the late 1990s, representing approximately 13.5 % of all reported marine natural products [4]. Their unique structural properties and potent biological activities have made them attractive in terms of natural product synthesis and pharmaceutical research. However, one class of compounds that has become especially interesting to both synthetic chemists and biochemists are the manzamines, a rapidly growing family of novel marine alkaloids.

Manzamine alkaloids are characterized by a unique polycyclic ring system, which may biogenetically derive from ammonia, C_{10} and C_3 units, and tryptamine [5]. In the majority of the manzamine alkaloids, the multicyclic units are condensed with tryptamine to form β-carboline. The first representative of this class of alkaloids, manzamine A (**1**) (Figure 8.1), was isolated as the hydrochloride salt from an Okinawan sponge *Haliclona* sp. by Higa's group in 1986 [6]. X-Ray diffraction crystallographic analysis revealed an unprecedented structure including absolute configuration, which consisted of a β-carboline and complicated 5-, 6-, 6-, 8-, and 13-membered rings. The piperidine and cyclohexene ring systems adopt chair and boat conformations, respectively, and are bridged by an eight-carbon chain creating a 13-membered macrocycle. The conformation of the 8-membered ring is in an envelope-boat, with a mirror plane passing through C-32 and C-28. The chloride ion is held by hydrogen bonding with two NH and one OH groups. The positive charge on the

Modern Alkaloids: Structure, Isolation, Synthesis and Biology. Edited by E. Fattorusso and O. Taglialatela-Scafati

ISBN: 978-3-527-31521-5

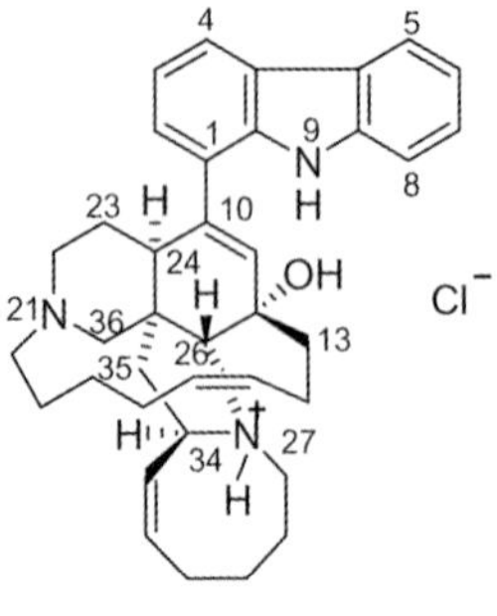

manzamine A (1)

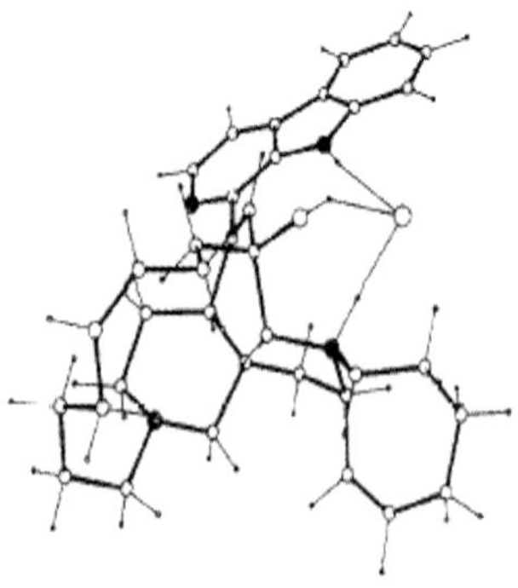

Fig. 8.1 X-ray crystal structure of manzamine A hydrochloride.

pyrrolidinium nitrogen atom was indicated by the longer than usual bond lengths of the attached α-carbon atoms [6].

Manzamines have shown a variety of bioactivities including cytotoxicity [6,7]; they have antimicrobial [8], pesticidal [9–11], and anti-inflammatory [12] properties, and have an effect on HIV and AIDS opportunistic infections [13,14]. To date, the greatest potential for the manzamine alkaloids appears to be against malaria with manzamine A and 8-hydroxymanzamine A (**2**) showing improved activity over the clinically used drugs chloroquine and artemisinin both *in vitro* and *in vivo* [15]. Recently, the manzamines have been patented for their immune depressing activity, which may be used in organ transplant or autoimmune disease [16]. The diversity of biological activity for this class of compounds suggests that they may act on multiple targets, which could be controlled through synthetic modification.

To date, there are over 80 manzamine-related alkaloids reported from more than 16 species of marine sponges belonging to five families distributed from the Red Sea to Indonesia. Wide structural variation has been reported, including the complex manzamine dimer kauluamine [17] and *neo*-kauluamine [15a], manadomanzamines A and B with unprecedented ring arrangements [13], and ircinal-related alkaloids such as keramaphidins [18–20], ingamines [21], ingenamines [22,23], and madangamines [24–26]. The major manzamine-producing sponges are in genus *Amphimedon* [27,20], *Acanthostrongylophora* [28,13], *Haliclona* [6], *Xestospongia* [29], and *Ircinia* [27,30].

The unique structure and extraordinary bioactivity of manzamine alkaloids have attracted great interest from synthetic chemists as one of the most challenging natural product targets for total synthesis. Great efforts have been made to achieve total synthesis of manzamine A and related alkaloids. Methodological studies towards the synthesis of manzamine structural units have also been reported [31–33].

There are several reviews referring to manzamine alkaloids [7,31–35]. In this chapter, we will review the structure and source of manzamine alkaloids including the large-scale preparation of manzamine alkaloids for preclinical evaluation. We also discuss the detailed synthesis and bioactivities of manzamine alkaloids.

8.2
Manzamine Alkaloids from Marine Sponges

8.2.1
β-Carboline-containing Manzamine Alkaloids

8.2.1.1 Manzamine A Type

This type of manzamine has a closed eight-membered bottom ring and a double bond between C-10 and C-11. In addition to manzamine A, this group of manzamines includes: manzamines D (**14**)[36,37], E (**7**)[38], F (**9**)[38], G (**2**)[20,39], M (**12**)[40], X (**10**)[41], and Y (**3**)[20,42], *ent*-8-hydroxymanzamine A (**2a**) [15a], *ent*-manzamine F (**9a**) [15a], 3,4-dihydromanzamine A (**4**)[42], 8-hydroxymanzamine D (**15**)[43], 8-hydroxy-2-*N*-methylmanzamine D (**16**) (43), 3,4-dihydro-6-hydroxymanzamine A (**13**)[40], 6-hydroxymanzamine E (**8**), [44], 6-deoxymanzamine X (**11**)[45], manzamine A *N*-oxide (**5**)[10], 3,4-dihydromanzamine A *N*-oxide [10], *epi*-manzamine D (**17**)[46], *N*-methyl-*epi*-manzamine D (**18**)[46], *ent*-12,34-oxamanzamines E (**19**) and F (**20**) [28a], 12,34-oxamanzamine A (**21**)[28a], 12,28-oxamanzamine A (**22**), 12,28-oxa-8-hydroxymanzamine A (**23**), 31-keto-12,34-oxa-32,33-dihydroircinal A (**61**)[47], 12,34-oxamanzamine E (**24**)[44], 6-hydroxy-12,34-oxamanzamine E (**25**), 12,28-oxamanzamine E (**26**)[48], 32,33-dihydro-31-hydroxymanzamine A (**27**), 32,33-dihydro-6,31-dihydroxymanzamine A (**28**), and 32,33-dihydro-6-hydroxymanzamine A-35-one (**29**)[49].

8-Hydroxymanzamine A was isolated from the Indonesian sponge *Pachypellina* sp. with moderate antitumor and anti-HSV-II activity, in 1994 [39]. It was named as manzamine G in some literature [20]. 6-Hydroxymanzamine A (called manzamine Y in some references [20,41]) and 3,4-dihydromanzamine A were isolated from the Okinawan marine sponge *Amphimedon* sp. The Philippine sponge *Xestospongia (=Acanthostrongylophora) ashmorica* Hooper yielded manzamine A *N*-oxide, 3,4-dihydromanzamine A *N*-oxide, and 6-deoxymanzamine X [10].

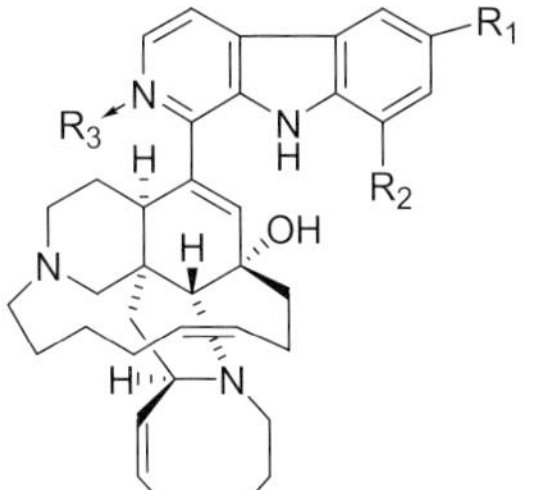

manzamine G (**2**): R_1 = H, R_2 = OH, no R_3
manzamine Y (**3**): R_1 = OH, R_2 = H, no R_3,
manzamine A *N*-oxide (**5**): R_1 = R_2 = H, R_3 = O

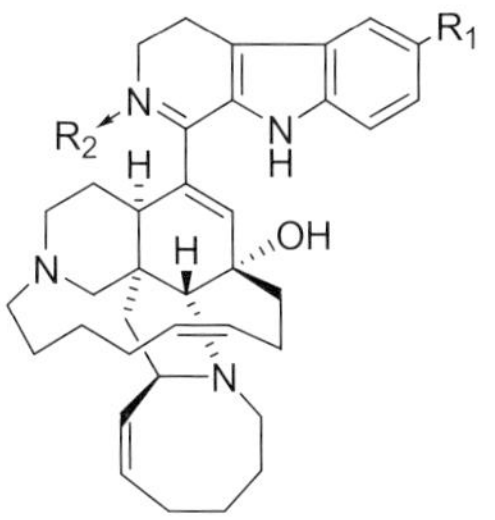

3,4-dihydromanzamine A (**4**): R_1 = H, no R_2
3,4-dihydromanzamine A *N*-oxide (**6**): R_1 = H, R_2 = O
3,4-dihydro-6-hydroxymanzamine A (**13**): R_1 = OH, no R_2

Manzamines E and F, with a ketonic carbonyl group in the eight-membered ring portion of the molecule, were isolated from an Okinawan *Xestospongia* sp. and patented as antitumor agents against P388 mouse leukemia cells, *in vitro* [38].

Manzamine F was proved to be identical with keramamine B, isolated earlier from a sponge *Pellina* sp. and assigned an incorrect structure [8]. 6-Hydroxymanzamine E was isolated from the sponge *Acanthostrongylophora* sp. collected from Manado, Indonesia [44]. Two manzamine enantiomers, *ent*-8-hydroxymanzamine A (**2a**) and *ent*-manzamine F (**9a**) were isolated from an Indonesian sponge originally identified as an undescribed *Petrosid* genus [15a].

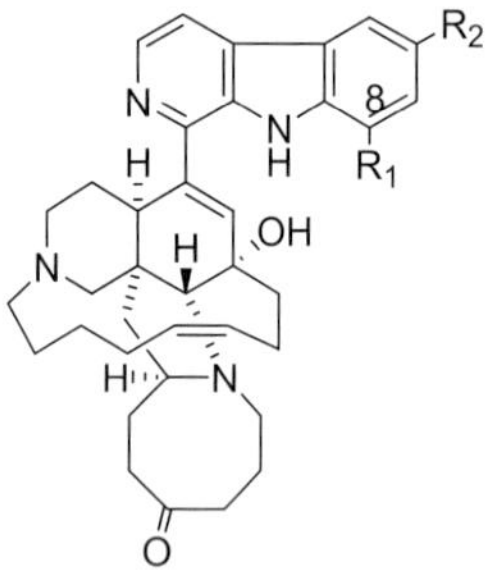

manzamine E (**7**): $R_1 = R_2 = H$
6-hydroxymanzamine E (**8**): $R_1 = H$, $R_2 = OH$
manzamine F (**9**): $R_1 = OH$, $R_2 = H$

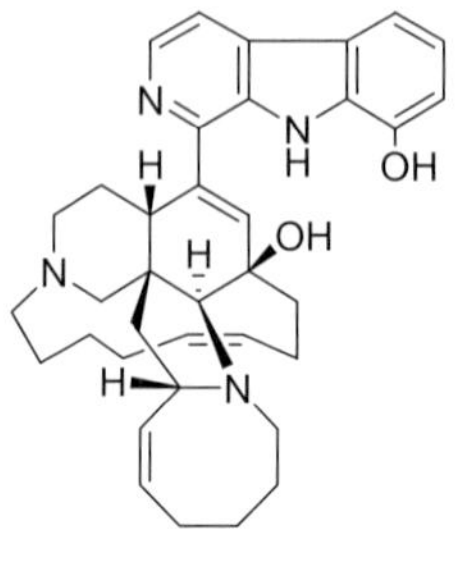

ent-8-hydroxymanzamine A (**2a**)

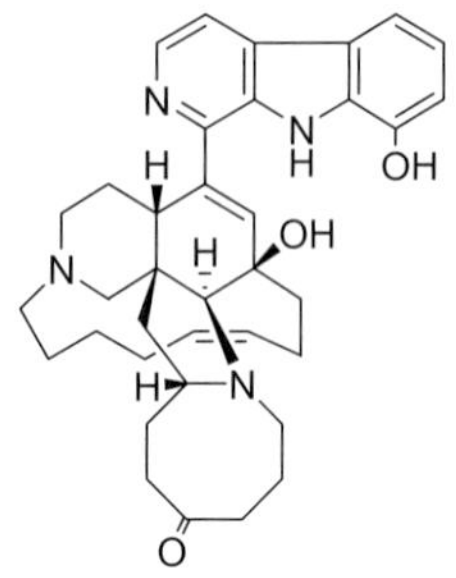

ent-manzamine F (**9a**)

Manzamine X was obtained from an Okinawan marine sponge *Xestospongia* sp. [41]. Its structure was confirmed by X-ray crystallographic analysis as including an inserted tetrahydrofuran ring in the lower part of the structure. Manzamine M and 3,4-dihydro-6-hydroxymanzamine A were isolated from the sponge *Amphimedon* sp. collected off the Kerama Islands [40]. Manzamine M is the only manzamine congener with a hydroxyl group on the C-13–C-20 chain and the absolute configuration of this hydroxyl group was determined by a modified Mosher's method.

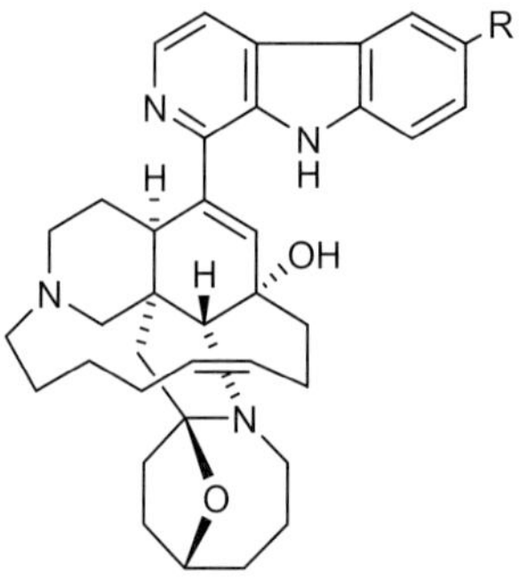

manzamine X (**10**): R = OH
6-deoxy-manzamine X (**11**): R = H

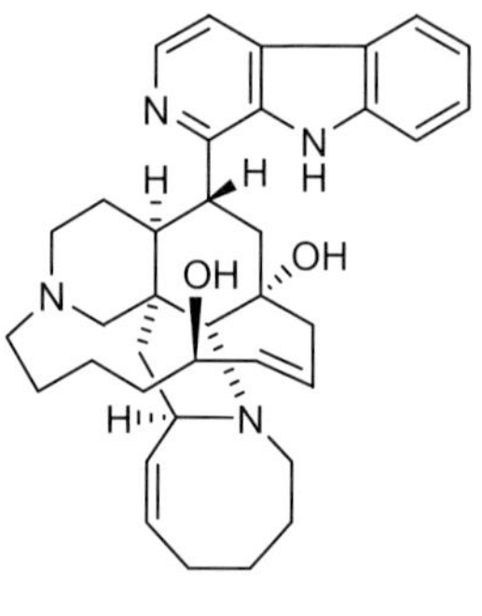

manzamine M (**12**)

Manzamine D was isolated as a minor component from the species that produce manzamine A in subsequent studies [36]. 8-Hydroxymanzamine D and its *N*-methylated derivative 8-hydroxy-2-*N*-methylmanzamine D were obtained from the Papua New Guinea sponges *Petrosia contignata* and *Cribrochalina* sp., which are in different families of the order Haplosclerida [43]. *epi*-Manzamine D and 2-*N*-methyl-*epi*-manzamine D were isolated from a Palaun sponge and the structure of **18** was confirmed by X-ray analysis [46]. Both of them were cytotoxic to HeLa and B16F10 mammalian cells, with **18** showing strong activity against the B16F10 cell line [46].

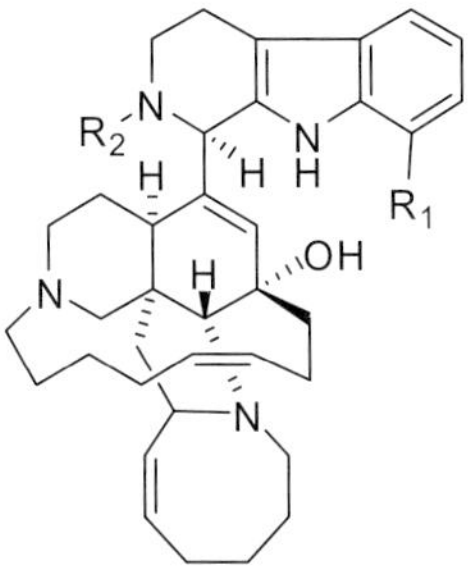

manzamine D (**14**): $R_1 = R_2 = H$
8-hydroxymanzamine D(**15**): $R_1 = OH$, $R_2 = H$
8-hydroxy-2-*N*-methyl-manzamine D(**16**): $R_1 = OH$, $R_2 = Me$

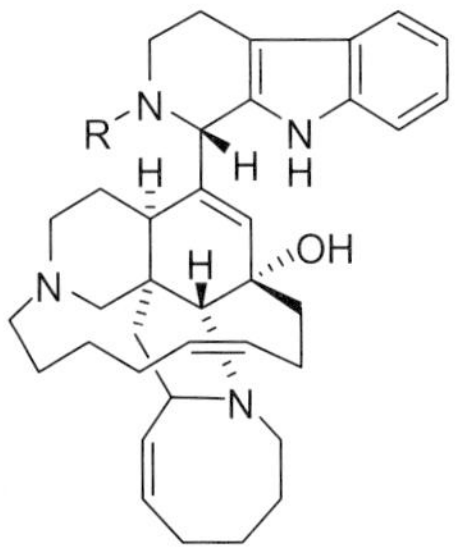

epi-manzamine D (**17**): R = H
N-methyl-*epi*-manzamine D (**18**): R = Me

A class of manzamine with an ether bridge between carbons 12 and 28 or between carbons 12 and 34 has been reported from Indo-Pacific sponges. *ent*-12,34-Oxamanzamines E and F, as well as 12,34-oxamanzamine A were isolated from three undescribed species of Indonesian sponge in the *Petrosiidae* genus. The biocatalytic transformation of *ent*-8-hydroxymanzamine A (**2a**) to **20**, using *Nocardia* sp. ATCC 21145 and *Fusarium oxysporum* ATCC 7601, has also been achieved [28a]. Three additional manzamines with an ether bridge, namely 12,28-oxamanzamine A, 12,28-oxa-8-hydroxymanzamine A, and 31-keto-12,34-oxa-32,33-dihydroircinal A (**61**), were obtained from a re-collection of the sponge [47].

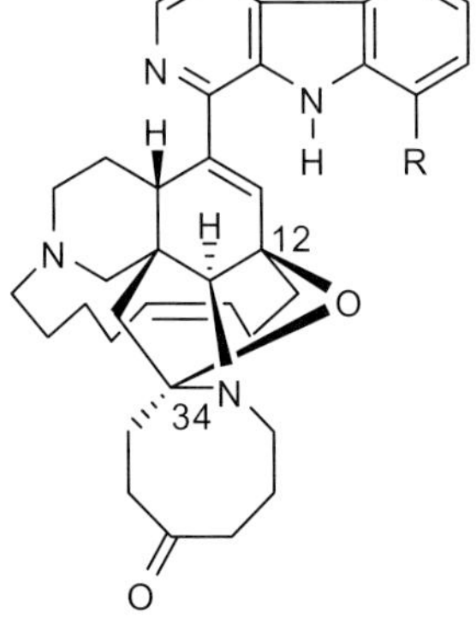

ent -12,34-oxamanzamine E (**19**): R = H
ent -12,34-oxamanzamine F (**20**): R = OH

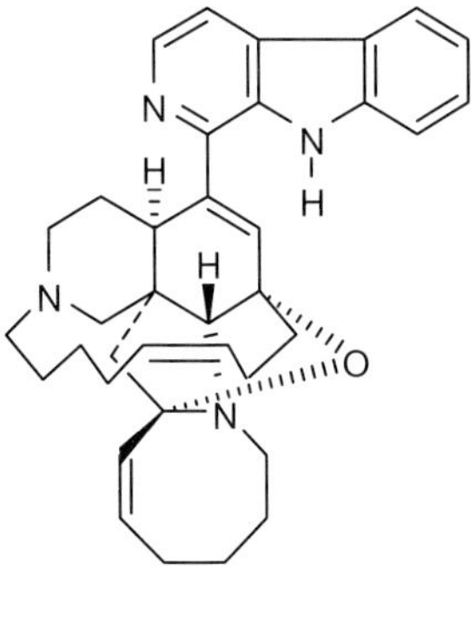

12,34-oxamanzamine A (**21**)

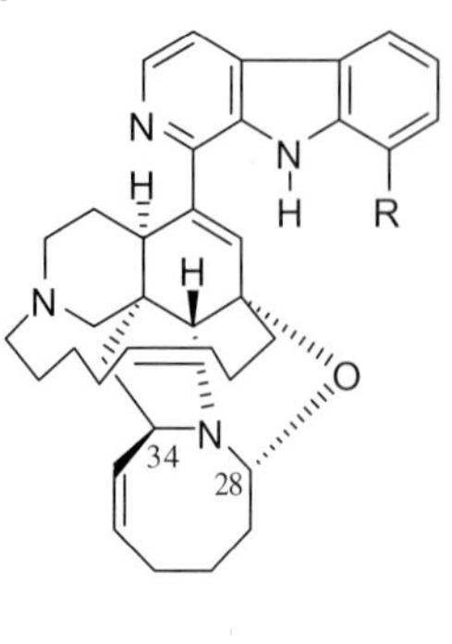

12,28-oxamanzamine A (**22**): R = H
12,28-oxa-8-hydroxymanzamine A (**23**): R = OH

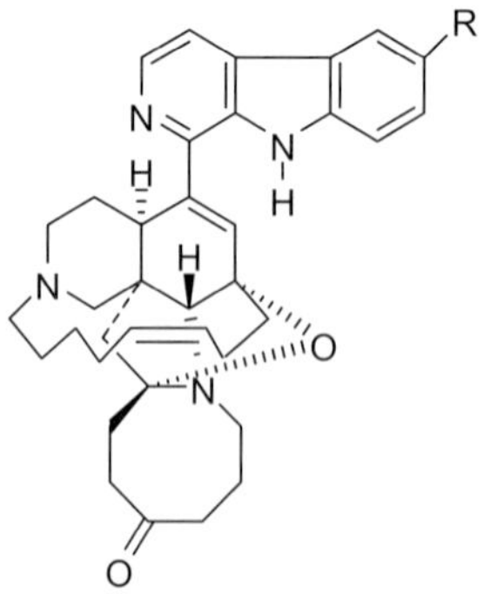

12,34-oxamanzamine E (**24**): R = H
6-hydroxy-12,34-oxamanzamine E (**25**): R = OH

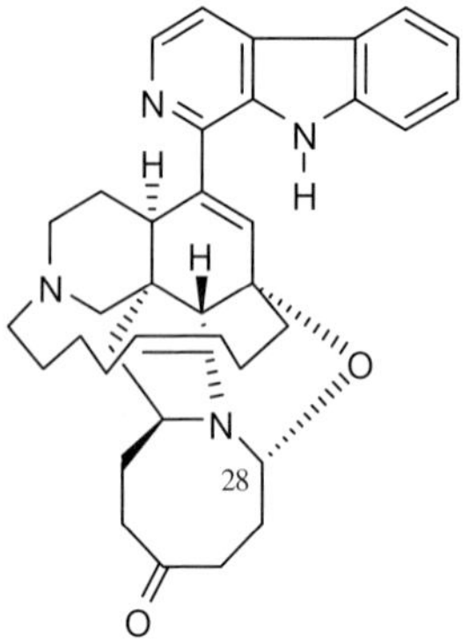

12,28-oxamanzamine E (**26**)

12,34-Oxamanzamine E [44], 6-hydroxy-12,34-oxamanzamine E and 12,28-oxamanzamine E [48] were isolated from a re-collection of the sponge *Acanthostrongylophora* sp. from Manado, Indonesia. In addition, 32,33-dihydro-31-hydroxymanzamine A, 32,33-dihydro-6,31-dihydroxymanzamine A, and 32,33-dihydro-6-hydroxymanzamine A-35-one have been reported from another collection of the same Indonesian sponge [49]. Compounds **27** and **28** are likely the reduced derivatives of manzamines E and F. Alkaloid **29** is unique in that it possesses a ketone moiety at C-35, instead of a typical C-31 ketone as seen in manzamines E and F.

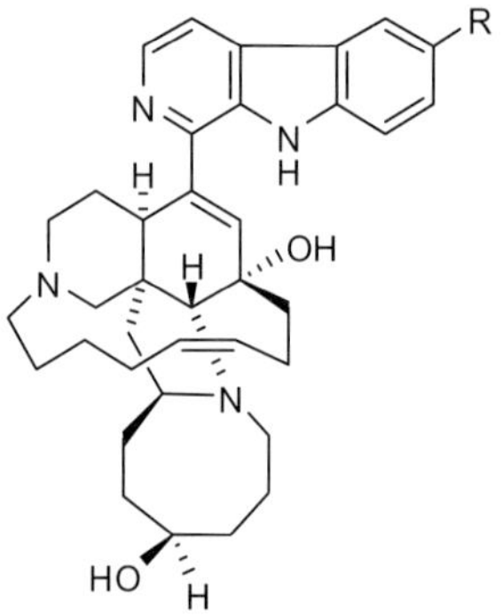

32,33-dihydro-31-hydroxymanzamine A (**27**): R = H
32,33-dihydro-6,31-dihydroxymanzamine A (**28**): R = OH

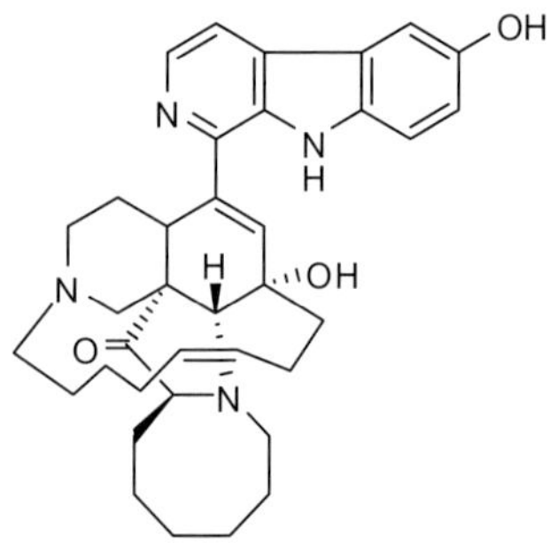

32,33-dihydro-6-hydroxy-
manzamine A-35-one (**29**)

8.2.1.2 Manzamine B Type

Higa's group reported mazamine B (**30**) in subsequent studies of an Okinawan sponge [36,37a] and its structure was characterized by X-ray analysis as having the bottom eight-membered ring opened. Recently its hydroxylated analog called 8-hydroxymanzamine B (**31**) has been isolated from an Indonesian sponge [44]. Manzamines H (**32**) and J (**33**) were first isolated from an Okinawan sponge *Ircinia* sp. [50]. Later, manzamine H and its C-1 epimer manzamine L (**34**) were isolated from the sponge *Amphimedon* sp. [20]. The absolute configuration of C-1 of manzamine L was deduced to be 1*S* from a negative Cotton effect, while **32** showed the opposite sign, implying the 1*R*-configuration [20].

manzamine B (**30**): R = H
8-hydroxymanzamine B (**31**) R = OH

manzamine H (**32**): 1*R*
manzamine L (**34**): 1*S*

In a continued search for new manzamine alkaloids, 1,2,3,4-tetrahydromanzamine B (**35**) and ma'eganedin A (**36**) has been isolated from the same species of the Okinawan marine sponge *Amphimedon* sp. [45,51]. Ma'eganedin A has a unique methylene carbon bridge between *N*-2 and *N*-27. Their structures, including the absolute configuration, were elucidated from spectroscopic data.

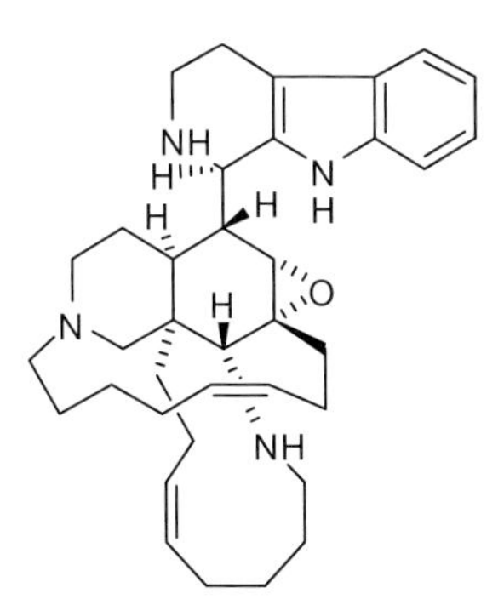

1,2,3,4-tetrahydromanzamine B(**35**)

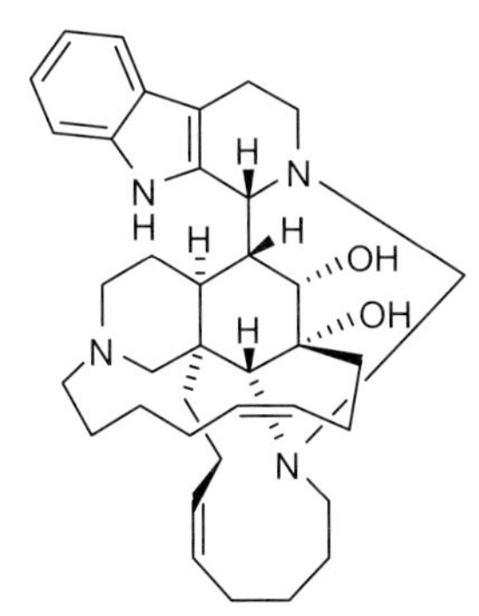

ma'eganedin A(**36**)

3,4-Dihydromanzamine J (**37**) was isolated from the Okinawan marine sponge *Amphimedon* sp. along with manzamine M and 3,4-dihydro-6-hydroxymanzamine A [40]. Manzamine J *N*-oxide was obtained from the Philippine sponge *Xestospongia* *ashmorica* [10] and 8-hydroxymanzamine J (**39**) was isolated from a species belongs to genus *Acanthostrongilophora* [44].

manzamine J (**33**): R_1 = H, R_2 = no
manzamine J *N*-oxide (**38**): R_1 = H, R_2 = O
8-hydroxymanzamine J (**39**): R_1 = OH, R_2 = no

3,4-dihydromanzamine J (**37**)

8.2.1.3 **Manzamine C Type**

Manzamine C (**40**) is characterized by a 2-ethyl-*N*-azacycloundec-6-ene connected to C-1 of a β-carboline. To date, only two analogs of this type of manzamine have been reported. Manzamine C was obtained from the same species *Haliclona* sp. that yields manzamine A [37a]. Keramamine C (**41**) was isolated from an Okinawan marine sponge *Amphimedon* sp. [18].

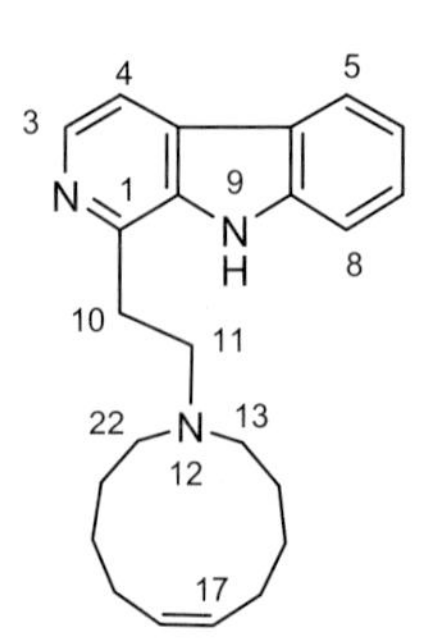

manzamine C (**40**)

keramamine C (**41**)

8.2.1.4 **Other β-Carboline-containing Manzamines**

Two manzamine dimers have been reported from two different Indonesian sponges. Kauluamine (**42**) was isolated by Scheuer's group from an Indonesian sponge originally identified as *Prianos* sp. [17]. This sponge was also recently identified as a species of *Acanthostrongylophora* (Hooper). Its structure was determined as a manzamine dimer by 2D NMR experiments, including HOHAHA and HMQC-HOHAHA techniques, adding yet another level of complexity to this fascinating group of alkaloids. Kauluamine showed moderate immunosuppressive activity in mixed lymphocyte reaction and did not show cytotoxicity. The second manzamine dimer, *neo*-kauluamine, was isolated from an Indonesian sponge originally identified as an undescribed petrosid genus, along with two manzamine enantiomers, *ent*-8-hydroxymanzamine A (**2a**) and *ent*-manzamine F (**9a**) [15a]. This sponge has now been shown to be of the genus *Acanthostrongylophora* [52]. The relative configuration of the nearly symmetric manzamine dimer *neo*-kauluamine was established through a detailed analysis of the NOE-correlations combined with molecular modeling [15a].

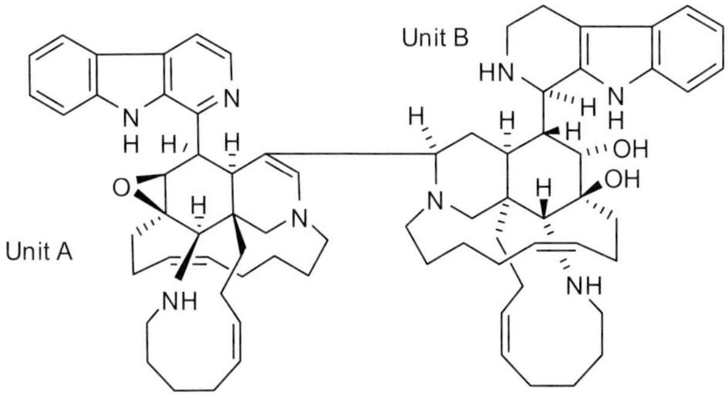

kauluamine(**42**)

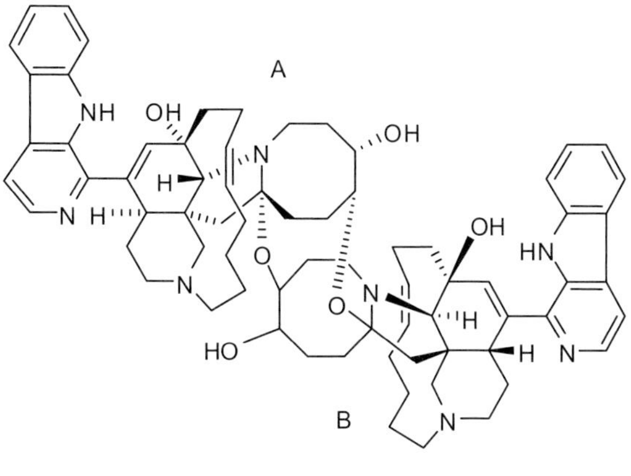

neo-kauluamine(**43**)

Recently, manadomanzamines A (**44**) and B (**45**), two novel alkaloids with unprecedented rearrangement of the manzamine skeleton, were isolated from the Indonesian sponge *Acanthostrongylophora* sp. Their structures were elucidated based on detailed 2D NMR analysis and shown to be a novel organic skeleton related to the manzamine-type alkaloids. Their absolute configurations and conformations were determined by CD, NOESY, and molecular modeling analysis [13]. Both of them exhibited activity against *Mycobacterium tuberculosis* (Mtb), HIV-1, and AIDS opportunistic fungal infections.

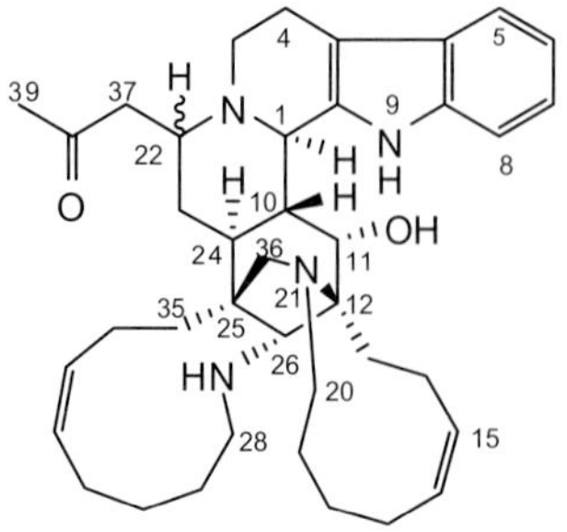

manadomanzamine A (**44**): 22 β-H
manadomanzamine B (**45**): 22 α-H

Xestomanzamine A (**46**) and B (**47**) were isolated from the Okinawan marine sponge *Xestospongia* sp. with manzamine X [41]. des-*N*-Methylxestomanzamine A (**48**) was reported from an undescribed Indonesian sponge [49]. Hyrtiomanzamine (**49**) is structurally similar to xestomanzamine A, and was isolated from the phylogenetically distant Red Sea sponge *Hyrtios erecta* [53].

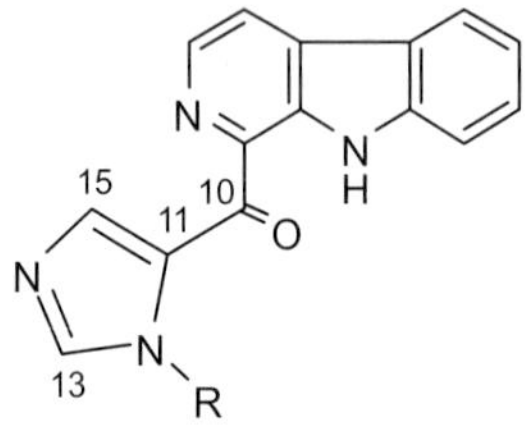

xestomanzamine A (**46**) R=CH_3
des-*N*-methylxestomanzamineA(**48**)R=H

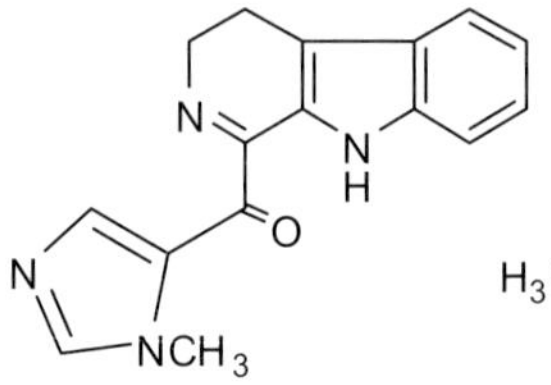

xestomanzamineB(**47**)

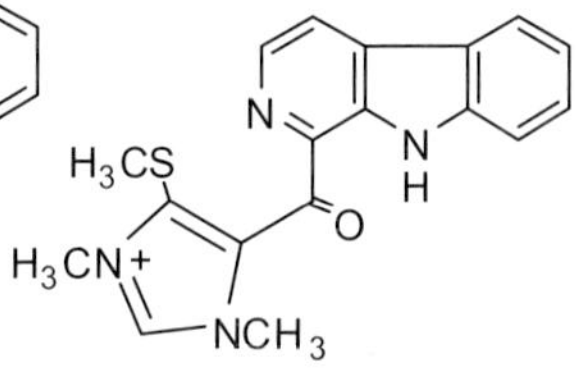

hyrtiomanzamine(**49**)

8.2.2
Ircinal-related Alkaloids

Keramaphidin C (**50**) was isolated from sponge *Amphimedon* sp. [18]. Haliclorensin (**51**) was isolated from *Haliclona* sp. and its structure was first proposed as **51** [54]. However, the synthetic (±)-**51**, (*R*)- and (*S*)-**51** showed significant differences in NMR and mass spectra from those of natural halicloresin [55]. The structure of haliclorensin was revised to (*S*)-7-methyl-1,5-diazacyclo-tetradecane (**51a**) recently [56].

keramaphidin C (**50**)

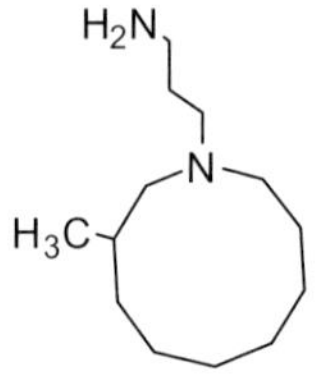

haliclorensin (**51**)

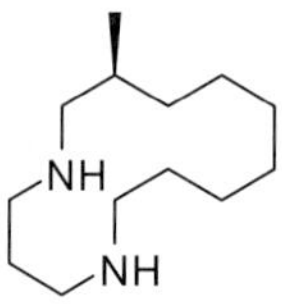

haliclorensin (**51a**)

Motuporamines A (**52**), B (**53**), and C (**54**), which contain a spermidine-like substructure, represent a new family of cytotoxic sponge alkaloids, were isolated as an inseparable mixture from *Xestospongia exigua* collected in Papua New Guinea [29].

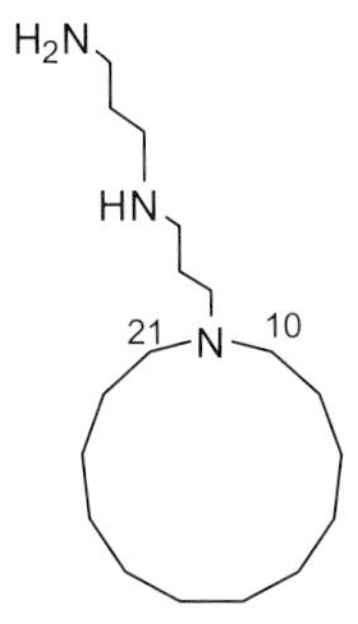

motuporamine A (**52**) 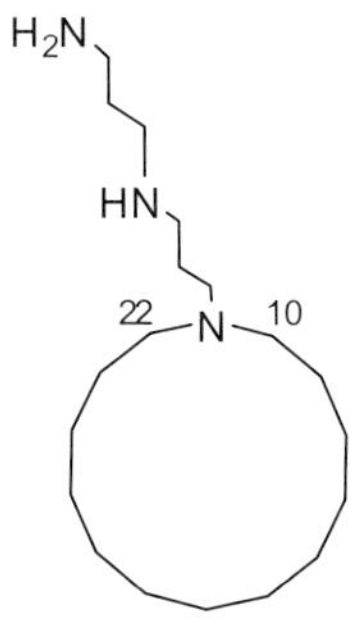 motuporamine B (**53**) 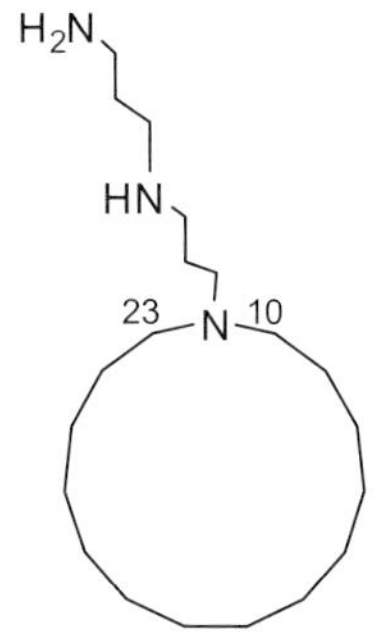 motuporamine C (**54**)

Two plausible biogenetic precursors of the manzamine alkaloids, ircinals A (**55**) and B (**56**), were isolated from the Okinawan sponge *Ircinia* sp. [50]. Ircinols A (**57**) and B (**58**), the antipodes of the manzamine-related alkaloids, were obtained from another Okinawan sponge *Amphimedon* sp., together with keramaphidins B (**59**) and C (**50**)[18–20,56]. Ircinols A and B were determined to be enantiomers of the C-1 alcoholic forms of ircinals A and B, respectively. Ircinols A and B represent the first occurrence of compounds which display the opposite absolute configurations to those of the manzamine alkaloids, and are a rare example of both enantiomeric forms being isolated from the same organism.

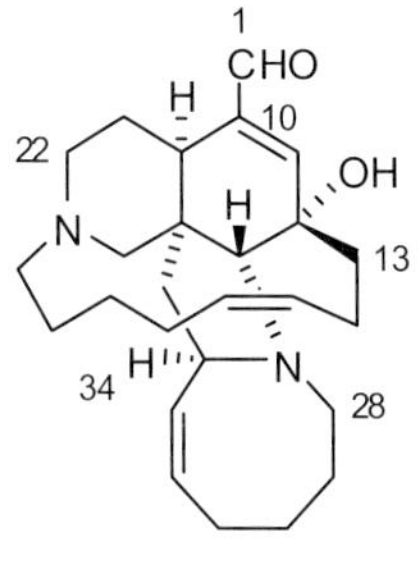

ircinal A(**55**) 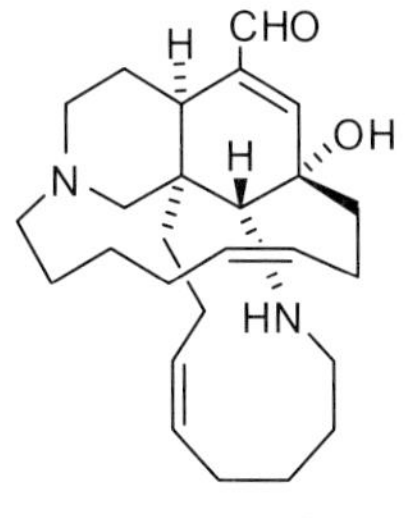 ircinal B(**56**)

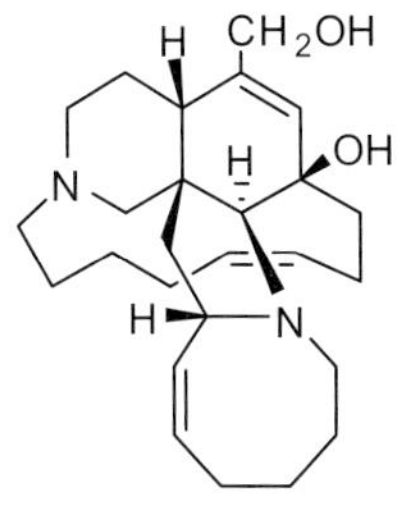

ircinol A(**57**) 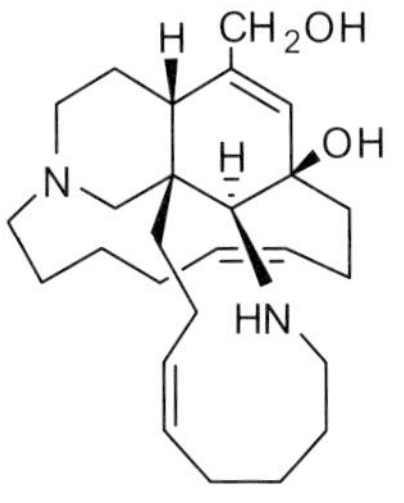 ircinol B(**58**)

12,28-Oxaircinal A (**60**), an usual ircinal A analog possessing a unique aminal ring system generated through an ether linkage between carbons 12 and 28, was isolated from *Acanthostrongylophora* sp. [48]. 31-Keto-12,34-oxa-32,33-dihydroircinal A possessing a unique aminal ring system generated through an ether linkage between carbons 12 and 34 was isolated from two collections of an Indo-Pacific sponge belongs to an undescribed species of the genus *Acanthostrongylophora* [47].

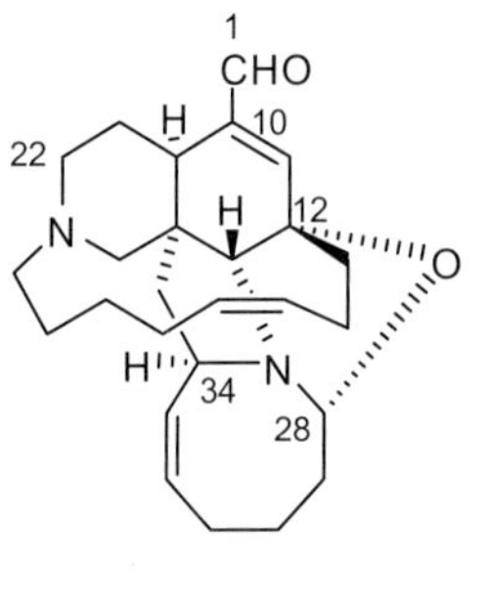

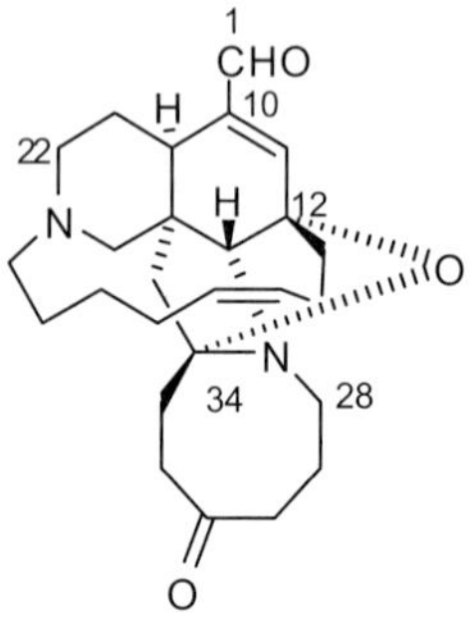

12,28-oxaircinal A (**60**) 31-keto-12,34-oxa-32,33dihydroircinal A (**61**)

Xestocyclamine A (**62**) was first isolated from the Papua New Guinea marine sponge *Xestospongia (=Acanthostrongylophora) ingens*, and its structure was revised in the following year with the isolation of xestocyclamine B (**63**)[57].

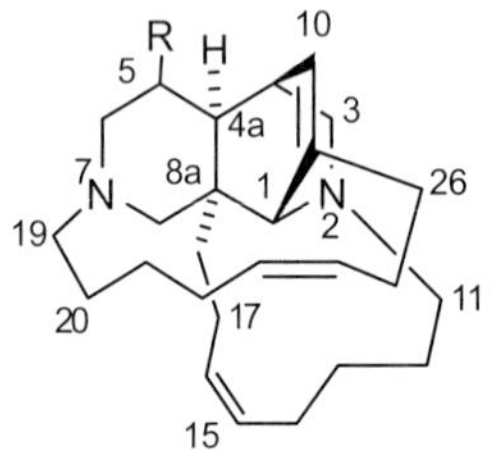

keramaphidin B(**59**): R = H
xestocyclamine B(**63**): R = −OH
ingenamine (**66**): R = −OH

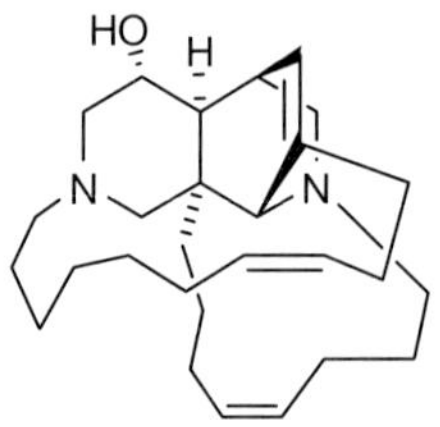

xestocyclamineA(**62**)

Ingamines A (**64**) and B (**65**)[21], ingenamine (**66**)[22], ingenamines B (**67**), C (**68**), D (**69**), E (**70**), and F (**71**)[23] were also isolated from the Papua New Guinea marine sponge *Xestospongia (=A.) ingens*.

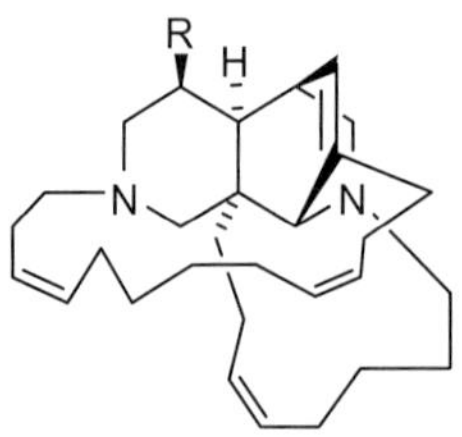

ingamine A (**64**): R = OH
ingamine B (**65**): R = H

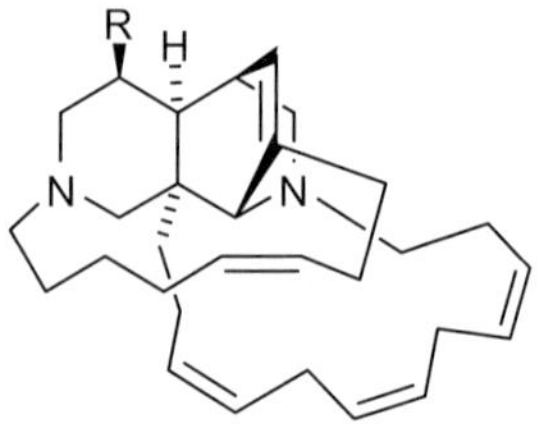

ingenamine E (**70**): R = OH
ingenamine F (**71**): R = H

R H 9 7N N X

ingenamineB(**67**): R = OH, X = N 7 ... C 9

ingenamineC(**68**): R = OAc, X = N 7 ... C 9

ingenamineD(**69**): R = OH, X = N 7 ... C 9

Ingenamine G (**72**) was isolated from *Pachychalins* sp. collected from Father's Island, Rio de Janerio, Brazil [58]

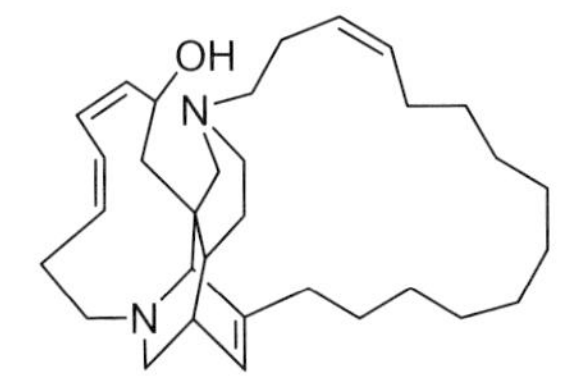

Ingenamine G (**72**)

Madangamine A (**73**), the first example of a new class of pentacyclic alkaloids was isolated from *Xestospongia* sp. collected off Madang, Papua New Guinea [24,25], and later madangamines B (**74**), C (**75**), D (**76**), and E (**77**) were also obtained from this same sponge [26]. They are extremely nonpolar compounds compared to the closely related ingenamine alkaloids, and it has been assumed that the madangamine skeleton arises via rearrangement of an ingenamine precursor.

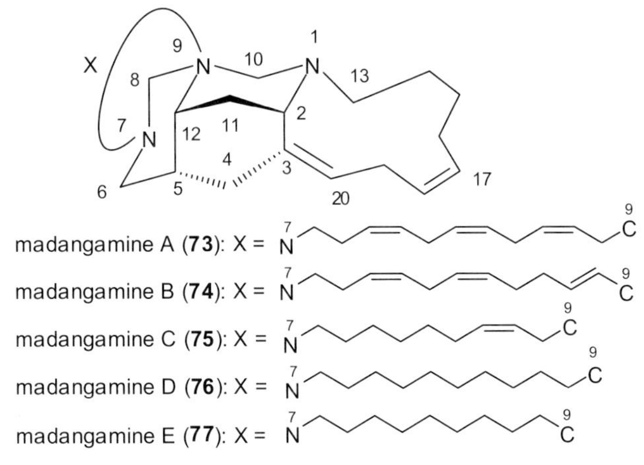

Misenine (**78**), which possesses an unprecedented tetracyclic 'cage-like' core, was isolated from an unidentified Mediterranean species *Reniera* sp. [59]. The spatial proximity between the carbonyl and the tertiary amine induces an interaction, or "proximity effect." The ^{1}H-NMR spectrum of this alkaloid showed significant variations with pH and it was concluded that **78a** is favored in neutral or weakly basic conditions whereas **78b** was preferred under acidic conditions. A similar transannular N/C=O "proximity effect" had previously been observed in saraine A (**79**) although, in this case, a lowering of pH enhanced the C–N linkage [60]. Nakadomarin A (**80**) was isolated from an *Amphimedon* sp., and its structure was reported to contain an unprecedented 8/5/5/5/15/6 ring system [61].

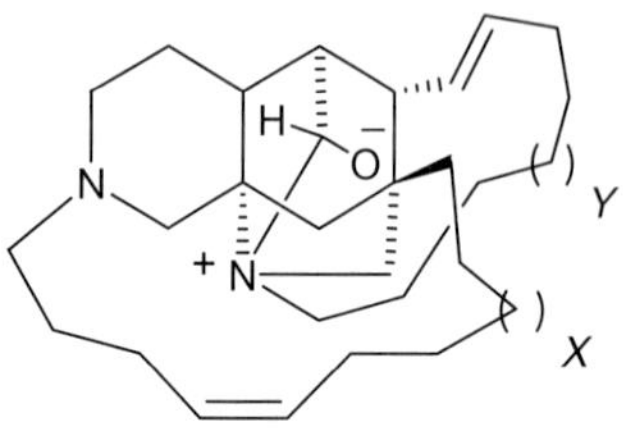

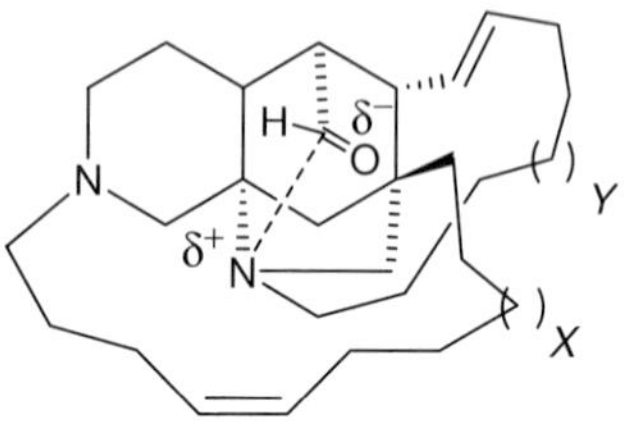

78a misenine (**78**): $(X + Y) = 6$ **78b**

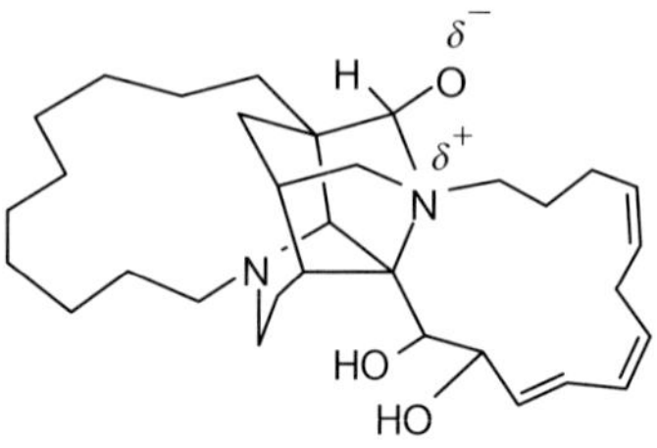

saraine A(**79**)

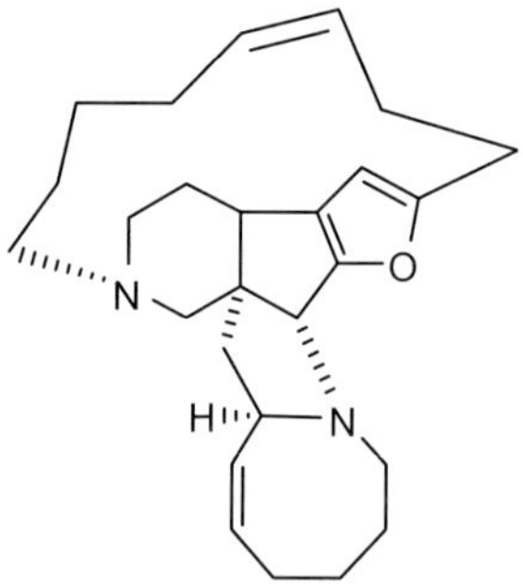

nakadomarin A(**80**)

8.3
Source and Large-scale Preparation of Manzamine Alkaloids

8.3.1
Source of Manzamine Alkaloids

Mazamine alkaloids have been reported from a variety of sponges. To date, there are 16 or more species belonging to five families and two orders of sponges that have been reported to yield the manzamine alkaloids (Table 8.1). These sponges have been collected from Okinawa, the Philippines, Indonesia, the Red Sea, Italy, South Africa, and Papua New Guinea. Most species produced a number of manzamine alkaloids and some of them give very high yields of β-carboline-containing manzamines. The most productive species are those in the genera *Amphimedon* sp. [20,27], and *Acanthostrongylophora* ([13,28]; see Table 8.1), which to date has yielded the greatest number of β-carboline-containing manzamine

Tab. 8.1 Marine sponges yielding manzamines and related alkaloids.

Taxonomy	Collection localities	Alkaloids	References
Order HAPLOSCLERIDA Topsent Family CHALINIDAE Gray	Manzamo and Amitori Bay (Iriomote Island), Okinawa	**1**, **30**, **40**, **14**, **3**	[6,36,37,41]
Genus *Haliclona* Grant *Haliclona* spp.			
Haliclona tulearensis VV&L[a]	Sodwana Bay, South Africa	**51**	[54]
Genus *Reniera* Nardo			
Reniera sp.	Capo Miseno,	**78**	[60]
Reniera sarai Pulitzeri-Finali	Naples, Italy; Bay of Naples	**79**	[61]
Family NIPHATIDAE van Soest Genus *Amphimedon* D&M[a] *Amphimedon* spp.	Kerama Islands, Okinawa	**1**, **30**, **40**, **41**, **14**, **2**, **32**, **34**, **12**, **3**, **4**, **13**, **37**, **35**, **36**, **50**, **55**, **56**, **57**, **58**, **59**, **80**	[18,27,19,20, 42,40,45,62]
Genus *Cribrochalina* Schmidt *Cribochalina* sp.	Madang, Papua New Guinea	**16**	[43]
Family PETROSIIDAE van Soest Genus *Acanthostrongylophora* Hooper[b] *Xestospongia* spp.	Miyako Island and Amitori Bay (Iriomote Island), Okinawa	**1**, **30**, **40**, **14**, **7**, **9**, **10**, **46**, **47**, **62**, **63**	[38,41,57,58]
Acanthostrongylophora sp.	Manado Bay, Indonesia	**8**, **60**, **24**, **25**, **26**, **27**, **28**, **29**, **31**, **39**, **48**	[44,48,49]
[*Prianos* sp.]	Manado Bay, Indonesia	**42**	[17]
[*Xestospongia ashmorica* Hooper]	Mindoro Island, Philippines	**1**, **7**, **9**, **33**, **11**, **5**, **38**, **6**	[10]
[*Xestospongia ingens* (Thiele)]	Papua New Guinea	**64**, **65**, **66**, **67**, **68**, **69**, **70**, **71**, **73**, **74**, **75**, **76**, **77**	[21–24,26]
[*Petrosiidae* n.g.] n. sp.	Manado Bay, Indonesia	**9a**, **2a**, **43**, **19**, **20**, **21**, **22**, **23**, **44**, **45**, **61**	[13,15,28,47]
[*Pachypellina* sp.]	Manado Bay, Indonesia	**1**, **2**	[39]

Tab. 8.1 (*Continued*)

Taxonomy	Collection localities	Alkaloids	References
Pellina sp.	Kerama islands, Okinawa	**1**, **9**	[8]
Genus *Xestospongia* de Laubenfels	Papua New Guinea	**52**, **53**, **54**	[29]
Xestospongia exigua (Kirkpatrick)			
Genus *Petrosia* Vosmaer	Milne Bay, Papua New Guinea	**15**, **16**	[43]
Petrosia contignata Thiele			
Order DICTYOCERATIDA Minchin	Red Sea	**49**	[53]
Family THORECTIDAE Bergquist			
Genus *Hyrtios* D&M[a]			
Hyrtios erecta (Keller)			
Family IRCINIIDAE Gray	Kise, Okinawa	**1**, **30**, **14**, **7**, **32**, **33**, **55**, **56**	[27,50]
Genus *Ircinia* Nardo			
Ircinia sp.			
Undetermined	Palau	**17**, **18**	[46]

[a]Taxonomic authorities: VV&L = Vacelet, Vasseur & Lévi; D&M = Duchassaing & Michelotti.
[b]The genus *Acanthostrongylophora* Hooper has been recently confirmed as the appropriate genus name for the group of sponges listed in square brackets above [52]. This table is an update of one previously published in ref. [7a].

and manzamine-related alkaloids. The wide distribution of manzamine alkaloids in taxonomically unrelated sponges strongly implies that microorganism(s) present as symbionts in these sponges may be the real producer of manzamine alkaloids.

8.3.2
Large-scale Preparation of Manzamines

Owing to the promising antimalarial and other activities, kilogram isolation of manzamine A is required for further preclinical evaluation, which include efficacy in different animal models, toxicological and pharmacokinetic studies, and so on. In addition, large amounts of manzamine A and its analogs such as 8-hydroxymanzamine A, ircinal A, and manzamine F are required for lead optimization and structure–activity relationship study. Thus, a method for kilogram-scale preparation of manzamines has been developed (see Figure 8.2) [62]. The fresh sponge was extracted with acetone, then an acid–base procedure was used to obtain a total alkaloid fraction. This mixture was subjected to one simple silica gel vacuum liquid chromatography procedure and pure (>90 %) manzamine A, 8-hydroxymanzamine A, and manzamine F were obtained as

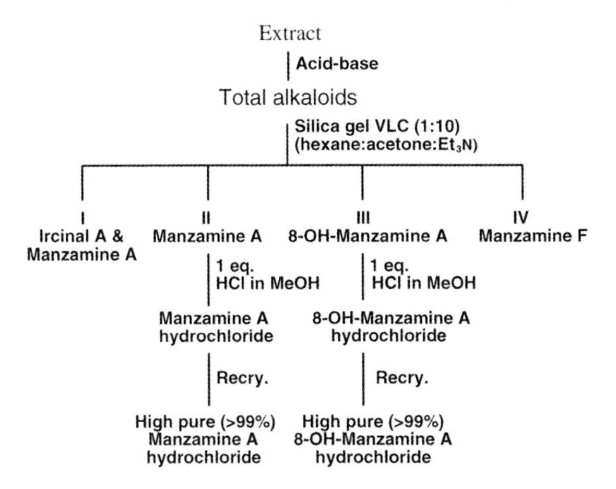

Fig. 8.2 Large-scale preparation of manzamine alkaloids.

free bases. Pure ircinal A was obtained after one additional chromatography step. High purity (>99 %) manzamine A and 8-hydroxymanzamine A can be obtained by converting the free base into the hydrochloride, followed by crystallization.

8.3.3 Supercritical Fluid Chromatography Separation of Manzamine Alkaloids

Although the above procedure for large-scale preparation of manzamine alkaloids is very simple and efficient, a deficiency of this method, like most chemical procedures, is the use of a large amount of organic solvent, which is expensive, toxic, and flammable. Supercritical carbon dioxide is widely recognized today as a "Green Chemistry" solvent. It is nontoxic, nonflammable, odorless, tasteless, inert, inexpensive, and environmentally benign. The unique physical and chemical properties of carbon dioxide in its supercritical state, namely the absence of surface tension, low viscosity, high diffusivity, and easily tunable solvent strength and density make it a useful solvent for extraction and chromatography.

A model large-scale supercritical fluid chromatography (SFC) system for producing manzamine alkaloids was established by adapting a supercritical fluid extraction unit SFT-150 [62]. The extraction chamber was used as the chromatography column, and 40 g of dry silica gel was placed in the chamber. A test preparative separation of 0.5 g of manzamine alkaloids was conducted using this system. A gradient elution was performed by increasing the amount of cosolvent (acetone, methanol) using an HPLC dual-pump unit. The fractions were evaluated by thin layer chromatography (TLC) and the results are shown in Figure 8.3. Fraction 8 only contained one spot, which corresponded to manzamine A. Fractions 6 and 13 each consist of a major spot and a minor one. This separation is comparable to the gravitational column separation of the manzamine alkaloids. The total elution time is less than 2 h, which is much less than gravitational column chromatography. Since the carbon dioxide was

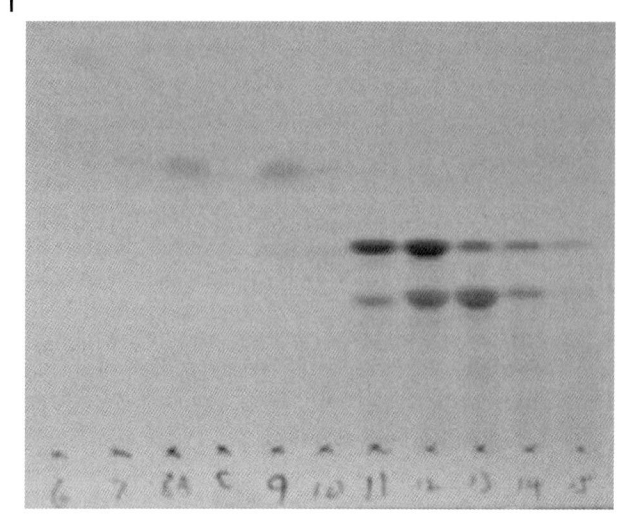

Fig. 8.3 TLC of fractions from SFC of manzamine alkaloids.

evaporated at room temperature, fractions collected are either dry powders or solutions in a minimal amount of cosolvent, and evaporation of a solvent is not required. Because the extraction chamber was utilized as the column, no expensive preparative SFC column is required. This process can be easily scaled to a 2-L column with a 2-L volume extraction chamber.

8.4
Synthesis of Manzamine Alkaloids

8.4.1
Total Synthesis of Manzamine A and Related Alkaloids

There are to date only two reports on total synthesis of manzamine A. The first total synthesis of manzamine A was reported in 1998 by Winkler *et al.*, along with those of manzamine D, ircinal A, and ircinol A [63]. The second total synthesis was reported a year later by Martin *et al.* [64] using a different approach from Winkler's.

The total synthesis of manzamine A reported by Winkler *et al.* involved a highly stereoselective intramolecular vinylogous amide photoaddition, retro-Mannich fragmentation, and a Mannich closure sequence (Scheme 8.1) in 17 steps starting from the bicyclic precursor **81**[63,65–67]. Reaction of the secondary amine **81** with ketone **82** gave the key substrate, vinylogous amide **83**, in high yield. Photoaddition and retro-Mannich fragmentation of amide **83** followed by Mannich closure gave the tetracyclic **88**. The required functionalities in ring B were achieved by carboxylation of C-10, reduction of C-11 ketone, followed by selenation, and finally the oxidation of selenide **93** to give the desired C-12α alcohol **94**. The formation of

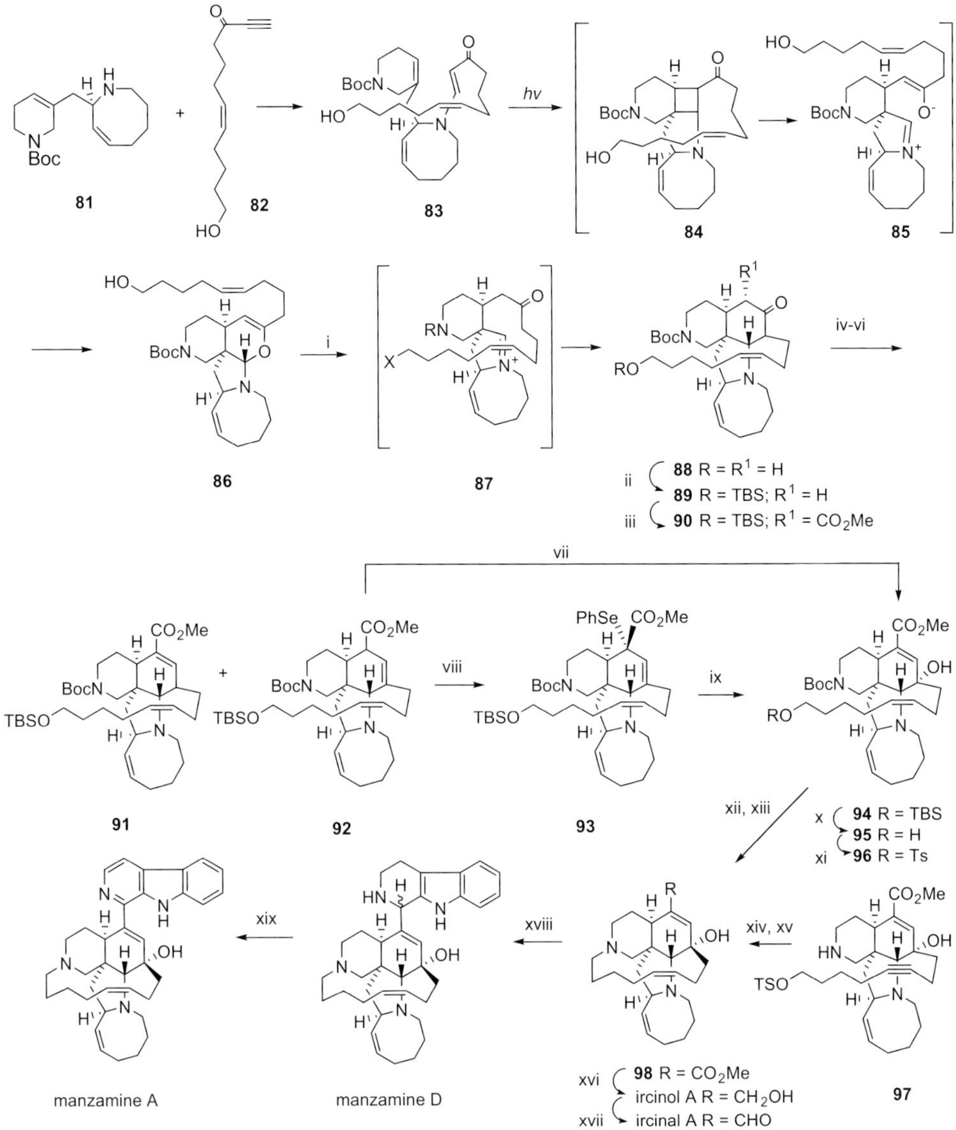

Scheme 8.1 Reagents: i, C_5H_5N, AcOH; ii, TBSCl; iii, LHMDS, MeOCOCN; iv, $NaBH_4$; v, MsCl, TEA; vi, DBU, benzene; vii, *m*-CPBA, NaOMe; viii, LiTMP, PhSeCl; ix, H_2O_2, pyridine; x, TBAF; xi, TsCl, TEA; xii, TFA; xiii, $(i\text{Pr})_2$NEt; xiv, $(i\text{Pr})_2$NEt; xv, Lindlar; xvi, DIBAL-H; xvii, Dess–Martin periodinane; xviii, tryptamine, TFA; xiv, DDQ.

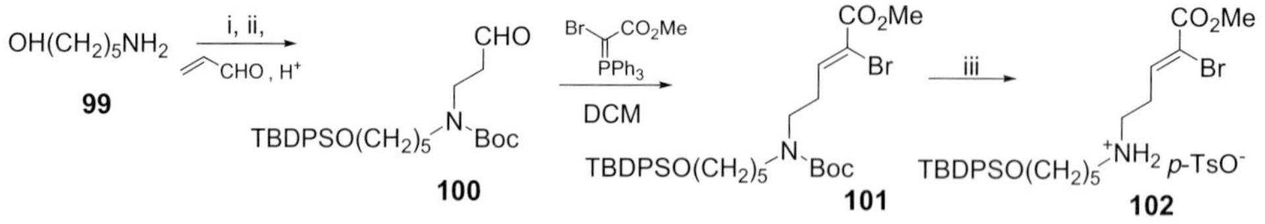

Scheme 8.2 Reagents: i, $(Boc)_2O$; ii, TBDPSCl; iii, TMS-OTf, 2,6-lutidine, *p*-TsOH.

the macrocyclic 13-membered ring E was achieved by exposing the secondary amine **96** to Hünig base at high dilution. Reduction of methyl ircinate **98** gave ircinol A, which was in turn oxidized to ircinal A. Pictet–Spengler reaction between **20** and tryptamine provided manzamine D, which on DDQ oxidation gave manzamine A.

The enantioselective total synthesis of manzamine A reported by Martin *et al.* utilizes a totally different approach from that of Winkler. A novel domino Stille/ Diels–Alder reaction was employed to construct the ABC tricyclic manzamine core and was followed by two sequential ring-closing metathesis (RCM) reactions to form the 8- and 13-membered rings D and E, respectively (Scheme 8.3) [64,68–71]. The key diene precursor **102**, which possessed the required functionalities for the eventual construction of the 13-membered ring E, was prepared in three steps (Scheme 8.2). Sequential reaction of **104** with oxalyl chloride and **102** afforded **105**, which then underwent the critical domino Still/Diels–Alder reactions. Reaction of **105** with vinyl tributylstannane in the presence of Pd(0) gave triene **106**, which subsequently underwent an intramolecular [4 + 2] cycloaddition to give the ABC tricyclic manzamine core **107**. Construction of the 13-membered ring E proceeded via a Wittig reaction, and stereoselective 1,2-addition of 4-butenyllithium to the tricyclic subunit gave diene **111** with the tertiary alcohol group internally protected as a cyclic carbamate. Treatment of diene **111** with Grubb's ruthenium catalyst afforded the RCM product **112**. Hydrolysis of the cyclic carbamate of **112** followed by N-acylation gave diene **113**. RCM cyclization of **113** with excess ruthenium furnished the pentacyclic **114**. Reduction of **38** with DIBAL-H gave ircinol A, which was in turn oxidized to ircinal A. The synthesis of ircinal A required a total of 24 steps from commercially available starting materials. Ircinal A was subsequently converted to manzamine D and manzamine A following the standard protocol [72].

8.4.2
Total Synthesis of Manzamine C

Manzamine C is the one of the simplest manzamine alkaloids in which an azacycloundecene ring is attached to a β-carboline moiety. Four total syntheses of this alkaloid employing different strategies have been reported to date [73–81]. Nakagawa *et al.* completed the first total synthesis of manzamine C in 1991 (Schemes 8.4 and 8.5) [73–74]. The β-carboline moiety was prepared utilizing the Bischler–Napieralsky protocol (Scheme 8.3) and the 11-membered azacycloundecene ring was constructed from the reaction between silyloxyacetylene **119** and

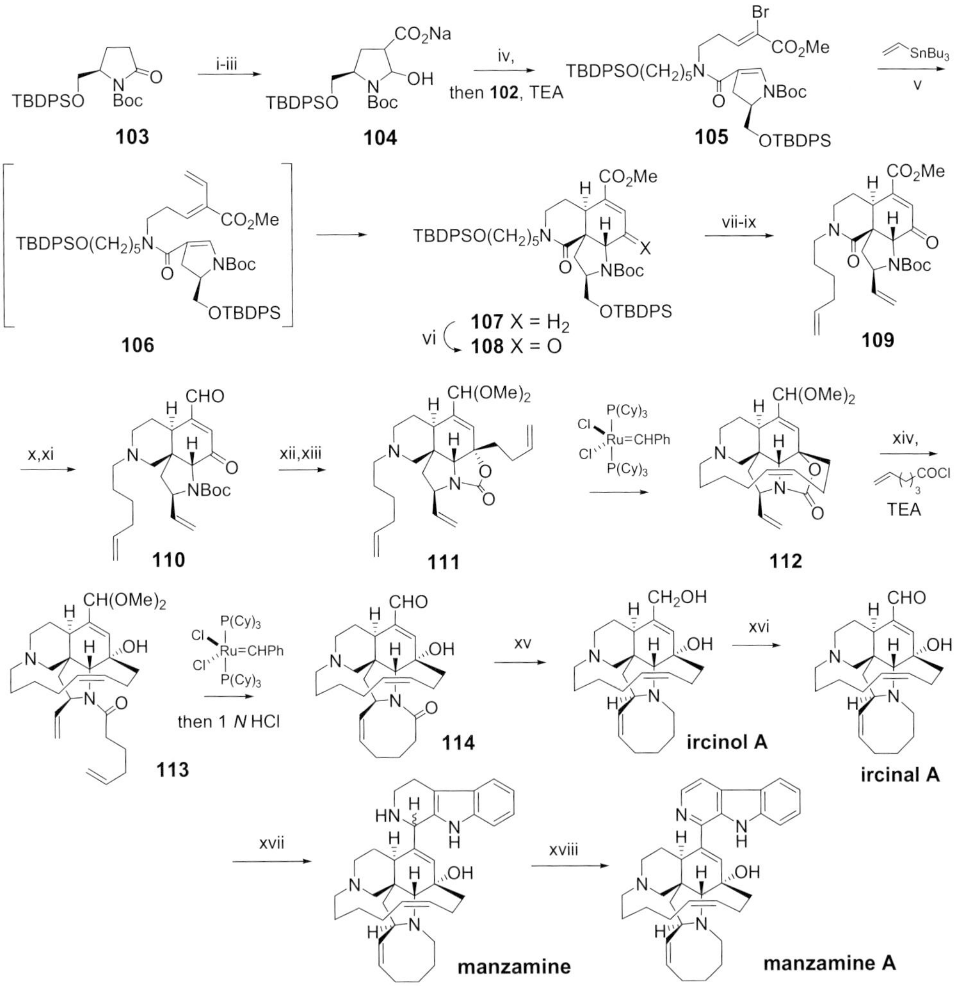

Scheme 8.3 Reagents: i, LHMDS, CO_2; ii, $NaBH_4$, EtOH; iii, Na_2CO_3; iv, $(COCl)_2$; v, $(Ph_3P)_4Pd$, toluene, Δ; vi, CrO_3, 3,5-$Me_2C_3H_2N_2$, DCM; vii, HCl, MeOH; viii, $(COCl)_2$, DMSO,TEA; ix, $Ph_3P{=}CH_2$; x, DIBAL-H; xi, Dess–Martin periodinane; xii, $HC(OMe)_3$, MeOH, HCl; xiii, $CH_2{=}CHCH_2CH_2Li$; xiv, KOH, MeOH; xv, DIBAL-H; xvi, Dess–Martin periodinane; xvii, tryptamine, TFA; xviii, DDQ, TEA.

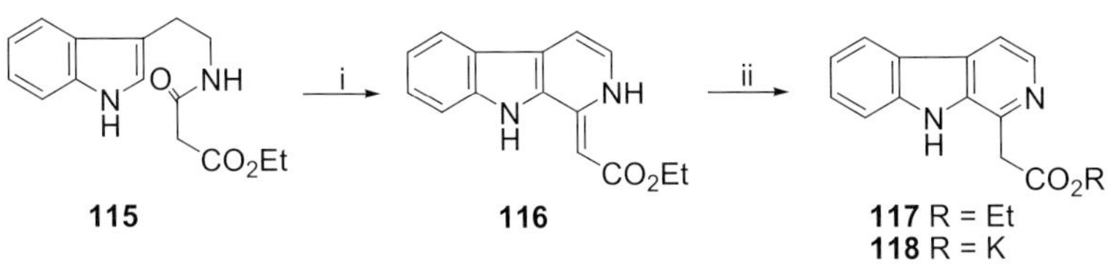

Scheme 8.4 Reagents: i, $POCl_3$; ii, Pd-C, *p*-cymene.

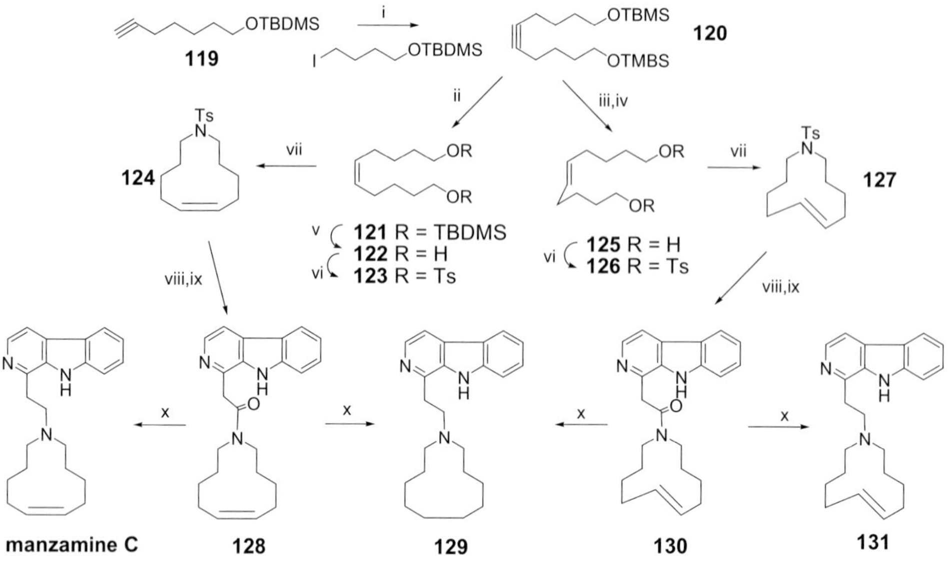

Scheme 8.5 i, *n*BuLi; ii, Pd-$CaCO_3$, quinoline; iii, Bu_4NF, THF; iv, $LiAlH_4$, diglyme; v, amberlyst 15, MeOH; vi, TsCl, pyridine; vii, $TsNH_2$, NBu_4NI, NaOH, PhH-H_2O; viii, Na, naphthalene; ix, 118, DPPA, DMF; x, $LiAlH_4$, THF.

1-iodo-4-*tert*-butyldimethylsilyloxybutane, which afforded the protected diol **120**. Hydrogenation of **120** in the presence of Lindlar catalyst gave stereoselectively the *Z*-alkene **121**, which was subsequently deprotected and tosylated, and treatment with tosylamine provided the *N*-tosyl-6-(*Z*)-azacycloundecene **124** (Scheme 8.4). Coupling of **124** with the potassium salt **118** in the presence of diphenylphosphoryl azide (DPPA) afforded amide **128**, which was then reduced to furnish manzamine C. On the other hand, the *E*-diol **125** was obtained after the desilylation of **120** and reduction. Using similar reaction conditions, **125** was converted to the *E*-isomer **131** of manzamine C and dihydromanzamine C **129**. The same group reported in 2000 a convenient method for the preparation of the 11-membered azacycloundecene ring using RCM (Scheme 8.6) [78].

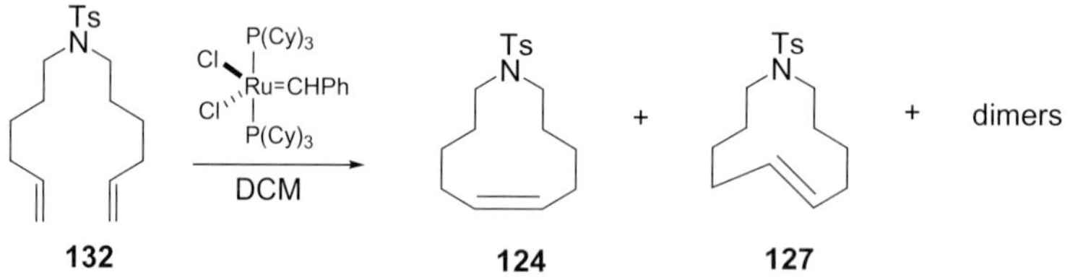

Scheme 8.6 Reagents: i, LHMDS, CO_2; ii, $NaBH_4$, EtOH; iii, Na_2CO_3; iv, $(COCl)_2$; v, $(Ph_3P)_4Pd$, toluene, Δ; vi, CrO_3, 3,5-$Me_2C_3H_2N_2$, DCM; vii, HCl, MeOH; viii, $(COCl)_2$, DMSO,TEA;ix, $Ph_3P{=}CH_2$; x, DIBAL-H; xi, Dess–Martin periodinane; xii, $HC(OMe)_3$, MeOH, HCl; xiii, $CH_2{=}CHCH_2CH_2Li$; xiv, KOH, MeOH; xv, DIBAL-H; xvi, Dess–Martin periodinane; xvii, tryptamine, TFA; xviii, DDQ, TEA.

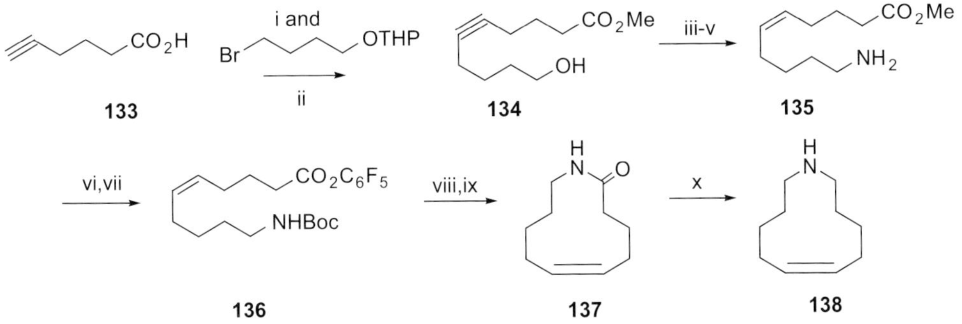

Fig. 8.7 Reagents: i, *n*BuLi; ii, H_2SO_4, MeOH; iii, TsCl, pyridine; iv, NaN_3, DMSO; v, Lindlar, quinoline, cyclohexane; vi, $(Boc)_2O$, KOH; vii, $Me_2N(CH_2)_3NCNEt$, HCl, C_6F_5OH; viii, TFA; xi, DMAP, THF; x, $LiAlH_4$, THF.

Gerlach *et al.* reported the second total synthesis of manzamine C in 1993 using a different synthetic approach [79]. The 11-membered azacycloundecene ring **138** was prepared starting from 5-hexynoic acid **133** in 10 steps, in which the macrolactamization was performed in the presence of 4-dimethylaminopyridine (DMAP) at high dilution (Scheme 8.7). The β-carboline moiety was prepared using a Pictet–Spengler condensation, and subsequent aromatization and methanolysis furnished compound **142**. Reaction between compound **142** and the 11-membered ring **138** in the presence of DMAP gave amide **128**, which was subsequently reduced to manzamine C (Scheme 8.8).

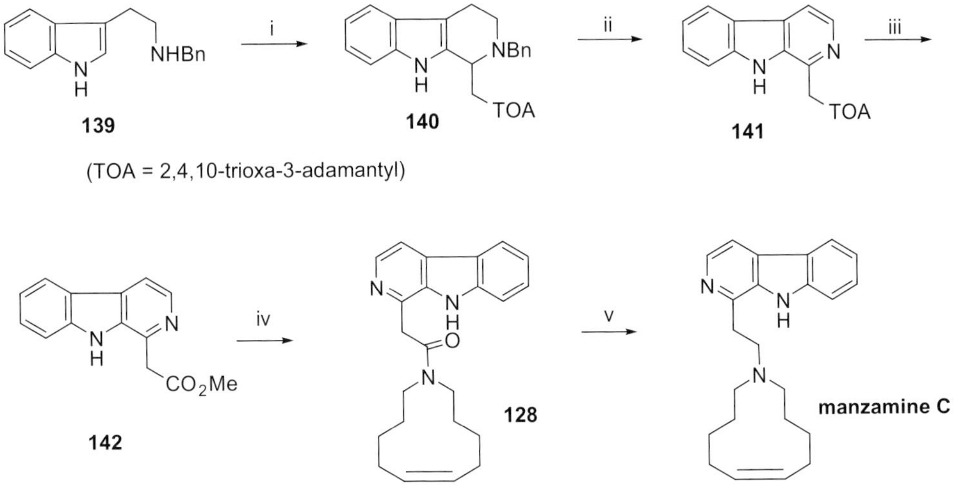

(TOA = 2,4,10-trioxa-3-adamantyl)

Scheme 8.8 Reagents: i, $TOACH_2CHO$, PhH; ii, Pd-C, mesitylene; iii, H_2SO_4, MeOH, dioxane; iv, 138, DMAP, mesitylene; v, $LiAlH_4$, THF.

In 1998, Langlois *et al.* completed the third synthesis of manzamine C using a strategy based on Sila–Cope elimination (Scheme 8.9) [80]. In this approach, the key intermediate, piperidine derivative **146**, was prepared in six steps from 2-methylpyridine **143**. Oxidation of **146** afforded a mixture of diastereomeric *N*-oxides **147** and **148**. Sila–Cope elimination of *N*-oxide **147** led to compound **149**, while Cope elimination of *N*-oxide **148** led to compound **150**. Oxidative cleavage of the N–O bond in **149** followed by treatment with *N*-chlorosuccinimide provided the chloroamine **153**. Elimination, hydrolysis, ditosylation, and finally basic treatment of the ditosylate intermediate afforded the cyclic sulfonamide **124**, which is the direct synthetic precursor of manzamine C [73].

The fourth approach to the synthesis of manzamine C involved a Ramberg–Bäcklund rearrangement in the key step for the preparation of the azacycloundecene ring [81]. Synthesis of this ring began with the condensation of 5-amino-1-pentanol **156**

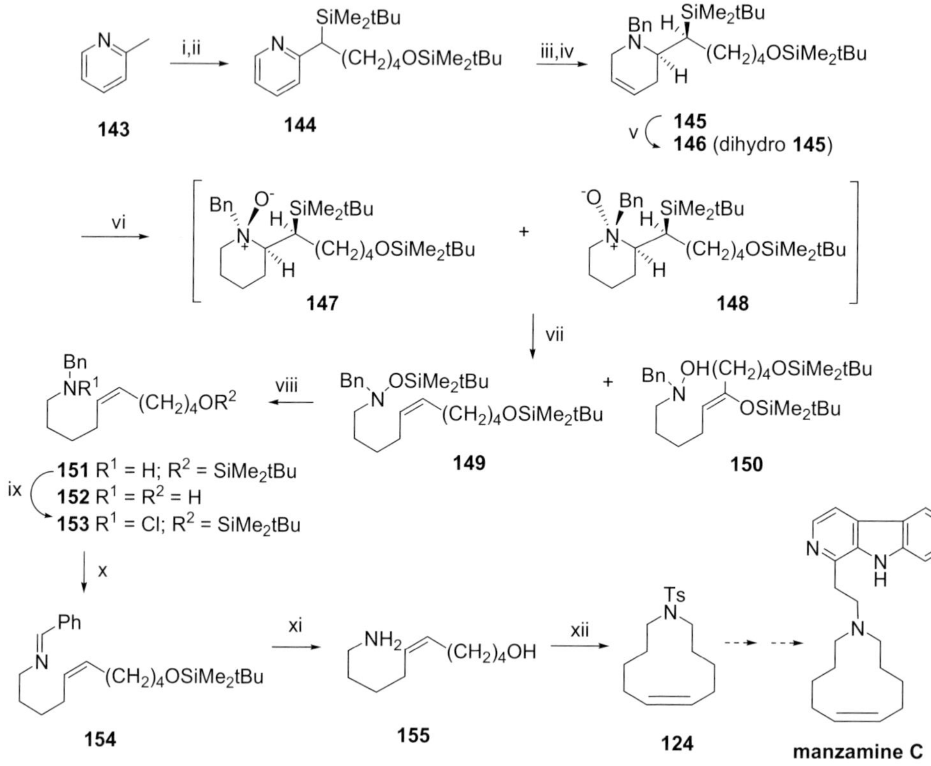

Scheme 8.9 Reagents: i, LDA, THF, then $ClSiMe_2tBu$; ii, LDA, THF, then $I(CH_2)_4OSiMe_2tBu$; iii, BnBr, MeCN; iv, $NaBH_4$, MeOH; v, H_2, PtO_2, EtOH; vi, *m*-CPBA, DCM; vii, MeCN; viii, Na-naphthalene, THF; ix, NCS, DCM; x, *t*BuOK, THF; xi, HCl, MeOH; xii, TsCl, pyridine, then Bu_4NI, NaOH, toluene.

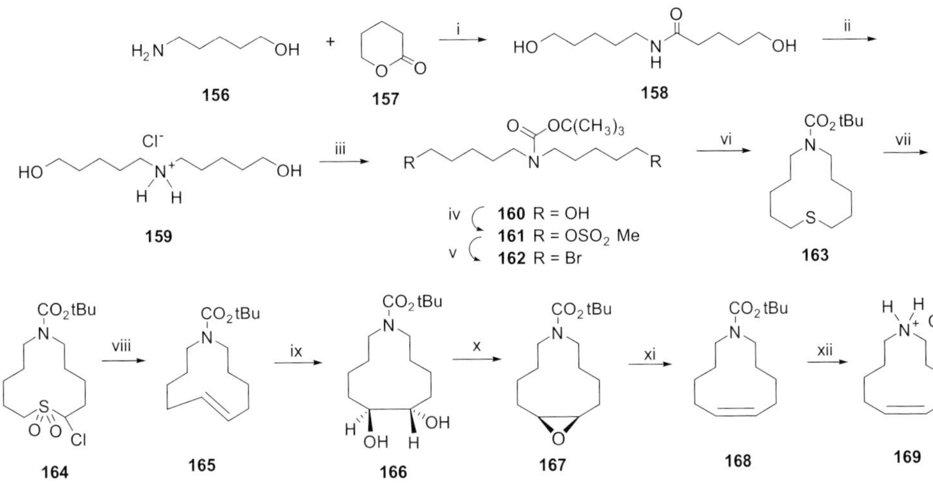

Scheme 8.10 Reagents: i, xylenes; ii, (a) BH_3.DMS,THF, (b) MeOH, anhyd.HCl; iii, Boc_2O, TEA, MeOH; iv, methanesulfonyl chloride, pyridine, DCM; v, LiBr, acetone; vi, $Na_2S{\cdot}9H_2O$, EtOH; vii, (a) NCS,CCl_4, (b) *m*-CPBA, DCM; viii, KO*t*Bu, DMSO; ix, OsO_4, NMO, THF-*t*BuOH-H_2O; x, (a) KO*t*Bu, THF, (b) 2,4,6-triisopropylbenzenesulfonyl chloride, (c) KO*t*Bu; xi, Rh(II) octanoate dimer, dimethyl diazomalonate, benzene; xii, HCl (g), EtOAc.

with lactone **157** to give amide-diol **158**. Reduction of amide **158** and subsequent amine protection gave diol **160**. Conversion of diol **160** to the dibromide **162** via the dimesylate **161**, and refluxing of dibromide **162** with $Na_2S{\cdot}9H_2O$ in ethanol at high dilution gave sulfide **163**, which was then converted to α-chlorosulfone **164**. Treatment of **164** with potassium *t*-butoxide produced the *Z*-azacycloundecene **165** in a stereospecific manner via a Ramberg–Bäcklund rearrangement followed by olefin inversion. Oxidation of olefin **165** to *trans*-diol **166** and dehydrolysis gave *cis*-epoxide **167**. Finally, deoxygenation and deprotection of the amine group provided compound **169**, the precursor for manzamine C (Scheme 8.10). Coupling of **169** with β-carboline **118** in the presence of DPPA and subsequent reduction of the resulting amide **128** furnished manzamine C (Scheme 8.11).

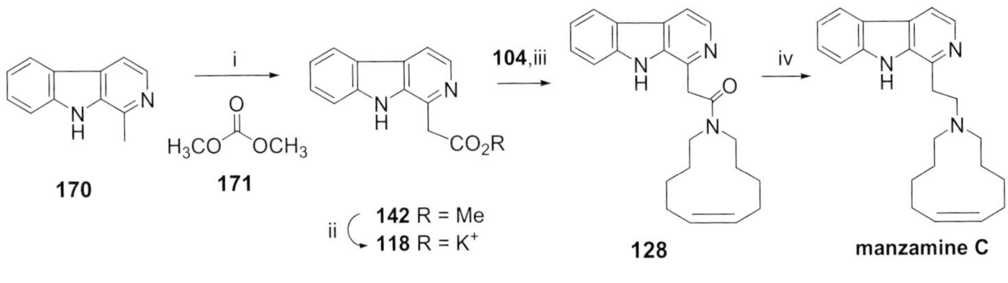

Scheme 8.11 Reagents: i, LDA, THF; ii, KOH, EtOH; iii, DPPA, TEA; iv, $LiAlH_4$, THF.

8.4.3
Total Synthesis of Nakadomarin A

The asymmetric total synthesis of the natural enantiomer (−)-nakadomarin A was completed by Nishida *et al.* in 2004 (Scheme 8.12) [82]. Diels–Alder reaction between siloxydiene **173** and chiral dienophile **172** (prepared from L-serine in 10 steps [83]) gave the highly functionalized key intermediate hydroisoquinoline **174**, which was subjected to Luche reduction, cyclization, and HCl treatment to furnish the tricyclic intermediate **175**. Compound **175** was converted to **177** via ozonolysis cleavage of ring B followed by recyclization of the unstable bisaldehyde to a five-membered ring by aldol condensation. The *Z*-olefin **178** was obtained from Wittig reaction of **177**, and was further converted to furan **180** via peroxide **179**. The

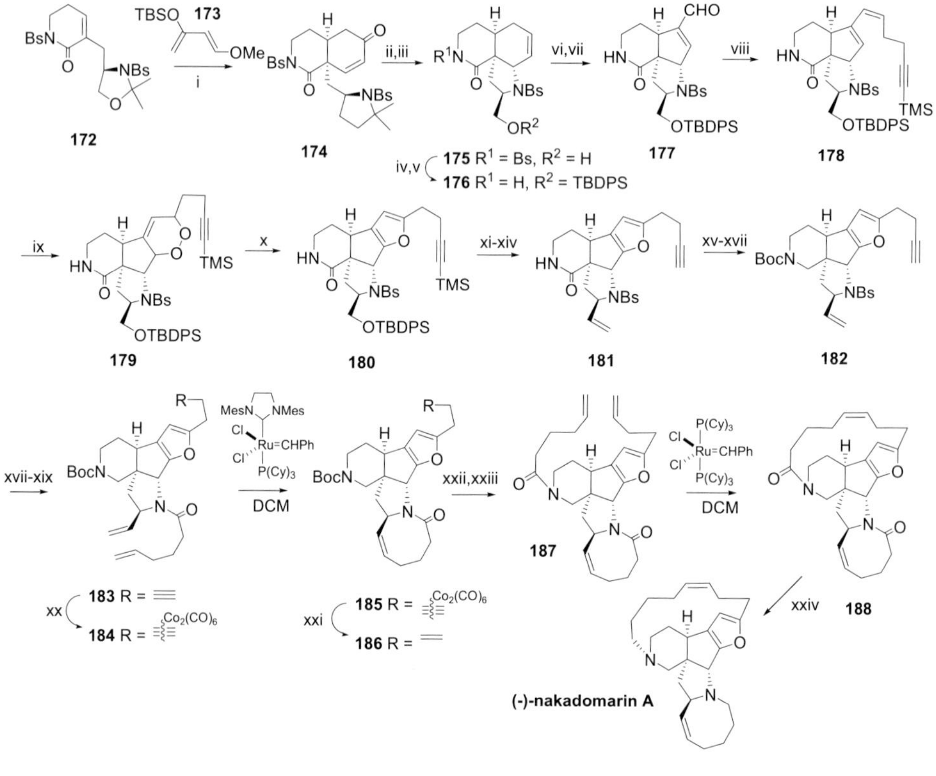

Scheme 8.12 Reagents: i, TFA, DCM; ii, $NaBH_4$,$CeCl_3.7H_2O$, DCM/MeOH; iii, HCl, benzene; iv, TBDPSCl, imidazole; v, Na/anthracene, DME; vi, O_3, DCM, Me_2S; vii, *N*-methylanilinium trifluoroacetate, THF; viii, $IPh_3PCH_2CH_2CH_2CCTMS$, NaH, THF; ix, O_2, halogen lamp, rose bengal, DCM/MeOH; x, *t*BuOK, THF, then HCl; xi, Dess–Martin periodinane; xii, $TMSCH_2MgCl$, Et_2O; xiii, $BF_3.Et_2O$, DCM; xiv, K_2CO_3, MeOH; xv, Boc_2O, DMAP, TEA, DCM; xvi, DIBAL-H, toluene; xvii, Et_3SiH, $BF_3.Et_2O$, DCM; xviii, Na/naphthalene, DME; xix, 5-hexenoyl chloride, TEA, DCM; xx, $Co_2(CO)_8$, DCM; xxi, $n$$Bu_3SnH$, benzene; xxii, TFA, DCM; xxiii, 5-hexenoyl chloride, TEA, DCM; xxiv, Red-Al, toluene.

tetracyclic furan **180** possesses the chiral ABCD-ring system of (−)-nakadomarin A. The formation of 8- and 15-membered rings was achieved by sequential RCM on compounds **184** and **187** utilizing Grubb's catalyst, which provided bislactam **188**. Reduction of bislactam **188** with Red-Al completed the total synthesis of (−)-nakadomarin A.

The same group has also developed a different synthetic route to the synthesis of the unnatural enantiomer (+)-nakadomarin A (**209**) (Scheme 8.13) [84]. The preparation of the tetracyclic **202** began from the condensation of (*R*)-(−)-**189** (prepared from commercially available methyl-4-oxopiperidinecarboxylate and

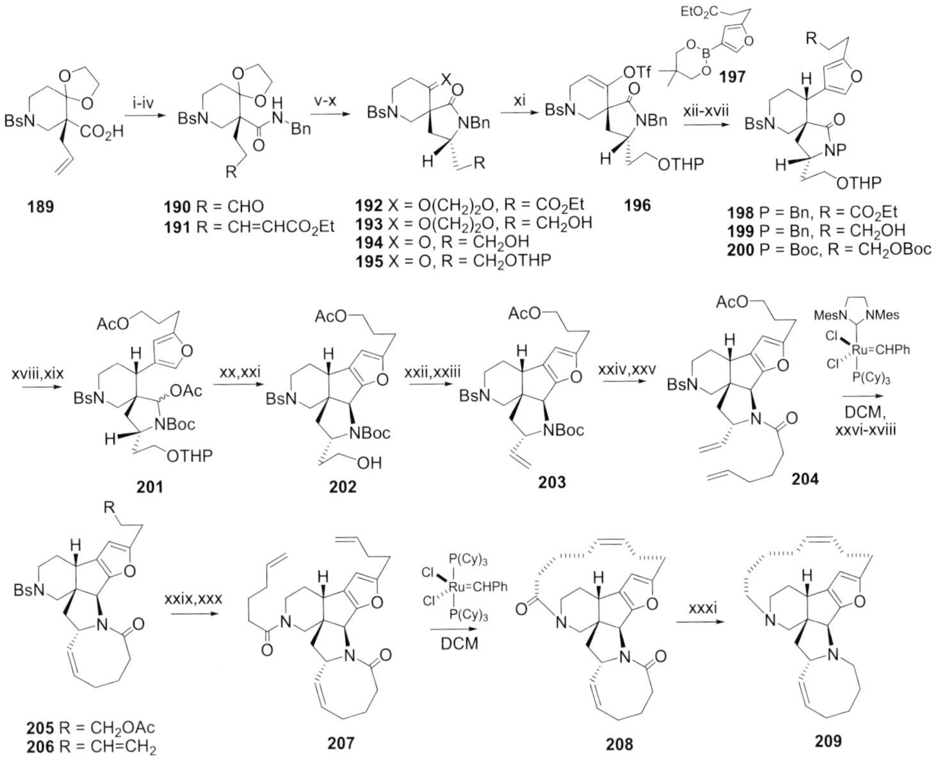

189
190 R = CHO
191 R = CH=CHCO2Et
192 X = O(CH2)2O, R = CO2Et
193 X = O(CH2)2O, R = CH2OH
194 X = O, R = CH2OH
195 X = O, R = CH2OTHP
196
197
198 P = Bn, R = CO2Et
199 P = Bn, R = CH2OH
200 P = Boc, R = CH2OBoc
201
202
203
204
205 R = CH2OAc
206 R = CH=CH2
207
208
209
i-iv
v-x
xi
xii-xvii
xviii,xix
xx,xxi
xxii,xxiii
xxiv,xxv
DCM,
xxvi-xviii
xxix,xxx
DCM
xxxi

Scheme 8.13 Reagents: i, $BnNH_2$, WSC.HCl, HOBt, DMF;ii,OsO_4, NMO, THF; iii, $NaIO_4$, DCM/H_2O; iv, $Ph_3P{=}CHCO_2Et$, DCM; v, DBU, EtOH; vi, NaOH, MeOH; vii, AcCl, MeOH; viii, $LiBH_4$, MeOH, THF; ix, $HClO_4$, DCM; x, DHP, CSA; xi, $LiN(TMS)_2$, THF, then $PhNTf_2$; xii, **197**, $PdCl_2$(dppf), K_3PO_4; xiii, H_2, Pd-C, MeOH, then PPTS, EtOH, then DHP, CSA; xiv, $LiBH_4$, MeOH,THF; xv, Li, liq.NH_3; xvi, $PhSO_2Cl$, aq. $NaHCO_3$; xvii, $(Boc)_2O$, TEA, DMAP; xviii, DIBAL-H, DCM, toluene; xix, Ac_2O, pyridine; xx, *p*-TsOH, DCM; xxi, HCl, THF; xxii, 2-nitrophenylselenocyanate, *n*-Bu_3P; xxiii, *m*-CPBA, aq. K_2HPO_4; xxiv, TFA, DCM; xxv, 5-hexenoic acid, WCS.HCl, HOBt; xxvi, NaOH, MeOH; xxvii, Dess–Martin periodinane; xxviii, $Ph_3P{=}CH_2$; xxix, Na, naphthalene; xxx, 5-hexenoic acid, WSC.HCl, HOBt; xxxi, Red-Al, toluene.

the absolute stereochemistry determined by X-ray analysis of its cinchoninium salt [84,85]) with benzylamine, followed by catalytic dihydroxylation and oxidative cleavage of the 1,2-diol to give aldehyde **190**, which was immediately converted to α,β-unsaturated ester **191** by Wittig olefination. The desired spirolactam **192** was obtained upon treatment of **191** with DBU/EtOH via intramolecular Michael addition. Suzuki–Miyaura coupling of enol triflate **196** with furan-3-boronic ester **197** under strongly basic conditions provided the coupling product **198** in high yield. The cyclization of ring B proceeded via treatment of **201** with *p*-TsOH and finally deprotection of the THP ether gave the desired tetracyclic **202**. The construction of the 8- and 15-membered rings proceeded through sequential RCM on compounds **204** and **207**, and reduction of bislactam **208** provided the unnatural (+)-nakadomarin A **209**.

8.4.4
Synthetic Studies of Manzamine Alkaloids

Manzamine alkaloids have been the subject of numerous synthetic studies owing to their novel molecular structure coupled with promising biological properties. Most of the synthetic studies on manzamine alkaloids focused on the construction of the ABCDE pentacyclic lower half of the manzamine structure.

Pandit *et al.* reported in 1996 the first synthesis of the pentacyclic core reminiscent of the ircinal and ircinol structures [86,87]. The tricyclic key intermediate **216** was prepared by intramolecular Diels–Alder reaction of **214**, which was in turn prepared from L-serine **210** in 10 steps (Scheme 8.14) [77,88–93]. The second phase of the synthesis involved the construction of the 8- and 13-membered rings, which was realized by sequential RCM on compounds **220** and **224** employing Grubb's catalyst, which therefore completed the first construction of the ABCDE pentacyclic core (Scheme 8.15). The application of the RCM methodology has been

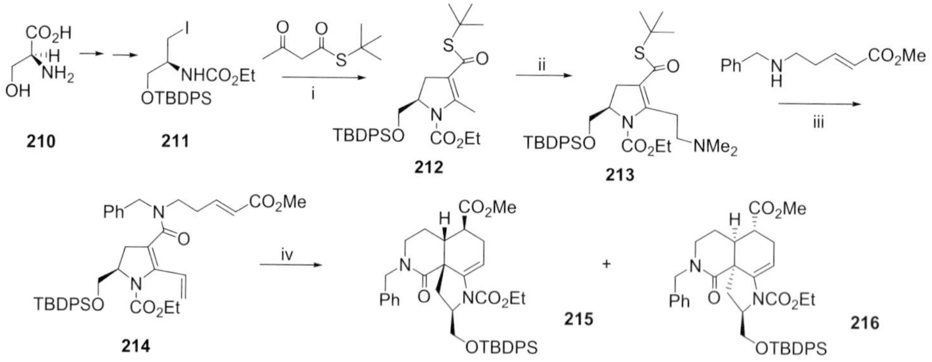

Scheme 8.14 Reagents: i, (a) NaH, DME, (b) TsOH, PhMe; ii, TMSOTf, TEA, Eschenmoser's salt; iii, (a) MeI, MeCN, (b) AgOTf, DIPEA, MeCN; iv, xylenes.

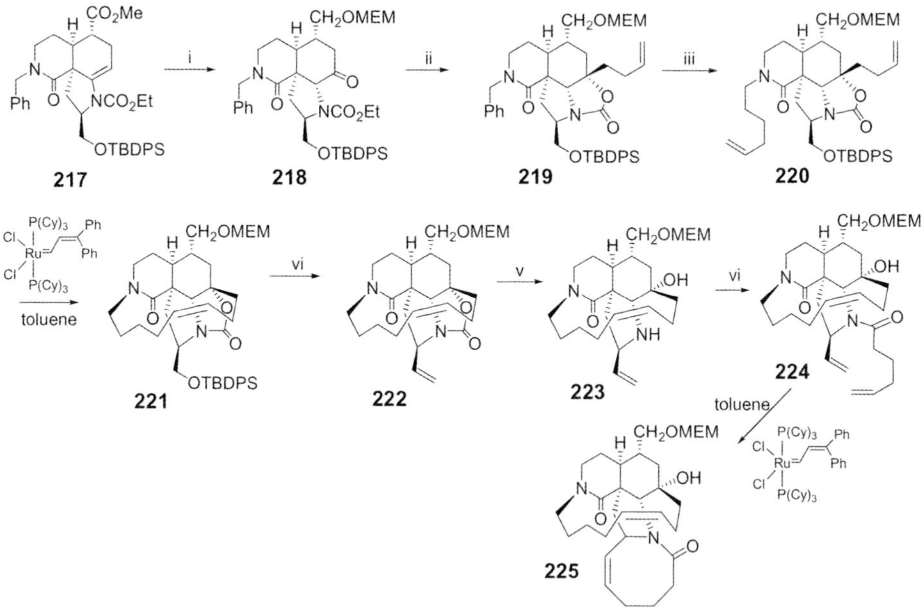

Scheme 8.15 Reagents: i, $LiBH_4$, THF, then MEMCl, NaH, DMSO, then OsO_4, pyridine, then H^+; ii, homoallyl MgBr, THF, then NaH, THF; iii, Li/NH_3, then Bn_2O, then $I(CH_2)_4CH{=}CH_2$, KOH, DMSO; iv, TBAF, THF, then Dess–Martin periodinane, then $(Ph_3)PCH_3Br$, BuLi, THF; v, 40% KOH/MeOH; vi, $CH_2CH{=}CH(CH_2)_3CO_2H$, EDC, DCM.

widely utilized in other synthetic studies of manzamine alkaloids since the publication of this report by Pandit *et al.* [64,78,82,84]. Other reports involving different synthetic approach toward the construction of ABCDE pentacyclic core are summarized in Table 8.2.

8.4.5 Studies on Biomimetic Synthesis

The biogenesis pathway for the manzamine alkaloids was proposed by Baldwin and Whitehead in 1992 (Scheme 8.16) [5], in which the complex manzamine structure could be reduced to four simple building blocks consisting of tryptophan, ammonia, a C_{10}, and a C_3 unit. Since the publication of this hypothesis, many natural products bearing striking resemblances to the intermediates proposed in the hypothesis have been isolated, such as ircinal A and keramaphidin B. The key step in the Baldwin–Whitehead hypothesis is the intramolecular Diels–Alder cyclization of the bisdihydropyridine **230**. To verify the feasibility of the proposed biosynthetic pathway, a biomimetic synthesis of keramaphidin B was successfully demonstrated by the same

Tab. 8.2 Synthetic approaches to the construction of the manzamine pentacyclic core.

Ring system	Synthetic approach	Author and references
AB	Intermolecular Diels–Alder	Nakagawa *et al.* [82,94,95]
ABC	Intermolecular Diels–Alder	Nakagawa *et al.* [96,97]
ABCD	Intermolecular Diels–Alder	Nakagawa *et al.* [75,98–101]
ABC	Intermolecular Diels–Alder	Langlois *et al.* [102]
ABE	Intermolecular Diels–Alder	Langlois *et al.* [103]
ABE	Intermolecular Diels–Alder	Simpkins *et al.* [104]
ABC	Intermolecular Diels–Alder	Marazano *et al.* [105]
ABC	Intramolecular Diels–Alder	Coldham *et al.* [106,107]
ABCD	Intramolecular Diels–Alder	Coldham *et al.* [108]
AB	Intramolecular Diels–Alder	Marko *et al.* [109]
CD	Intramolecular Diels–Alder	Marko *et al.* [110]
ABC	Intramolecular Diels–Alder	Marko *et al.* [111,112]
AB	Intramolecular Diels–Alder	Clark *et al.* [113]
AB	Intramolecular Diels–Alder	Leonard *et al.* [114]
ABC	Intramolecular Diels–Alder	Leonard *et al.* [115]
AD	Ionic cyclization	Yamamura *et al.* [116]
ABC	Ionic cyclization	Yamamura *et al.* [117,118]
ABCE	Ionic cyclization	Yamamura *et al.* [119–121]
CD	Ionic cyclization	Clark *et al.* [122,123]
BC	Ionic cyclization	Hart *et al.* [124]
BCD	Ionic cyclization	Hart *et al.* [125]
ABC	Ionic cyclization	Overman *et al.* [126]
ABC	Ionic cyclization	Brands *et al.* [127]
ABC	Ionic cyclization	Magnus *et al.* [128]
AB	Free radical cyclization	Hart *et al.* [129]
ABCD	Free radical cyclization	Hart *et al.* [130]
AD	Aldol-type coupling	Nakagawa *et al.* [131]

group (Scheme 8.17) [132,133]. In this study, the key bisiminium **231** was prepared from tolsylate **236** via a Finkelstein/dimerization/macrocyclization reaction followed by reduction, and finally a Polonovski–Potier reaction. The proposed Diels–Alder reaction was achieved by treatment of **231** in methanol/buffer followed by reduction to give a small amount of keramaphidin B (0.2–0.3 %) with recovery of bistetrahydropyridine **238** (60–85 %). The low yield of keramaphidin B was rationalized by the kinetic preference of **231** to disproportionate and an *in vivo* Diels–Alderase that would mediate the conversion of bisiminium **231** to keramaphidin B was envisaged [132,133].

A semibiomimetic synthesis of keramaphidin B was also reported utilizing RCM methods (Scheme 8.18) [133]. The keramaphidin B precursor **240** (prepared from dihydropyridinium salt **239** [134,135]) was treated with Grubb's catalyst to give the monocyclized **241** and keramaphidin B.

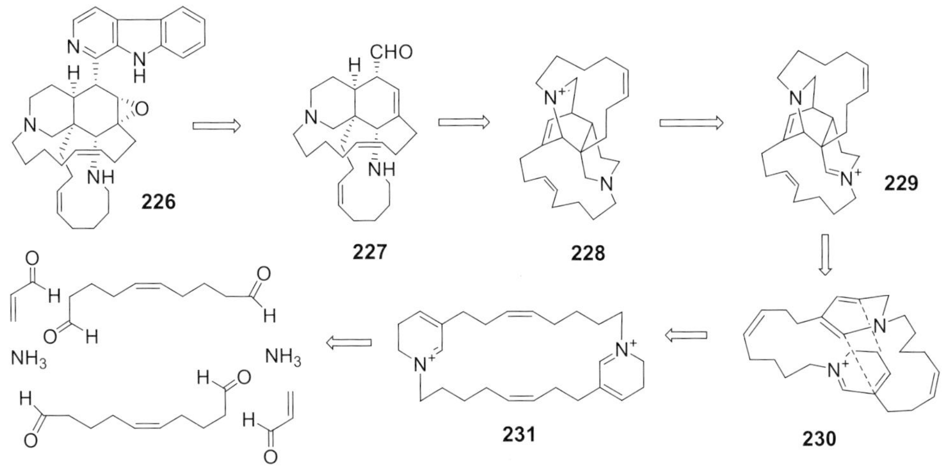
226
227
228
229
230
231
CHO
NH
NH_3

Scheme 8.16

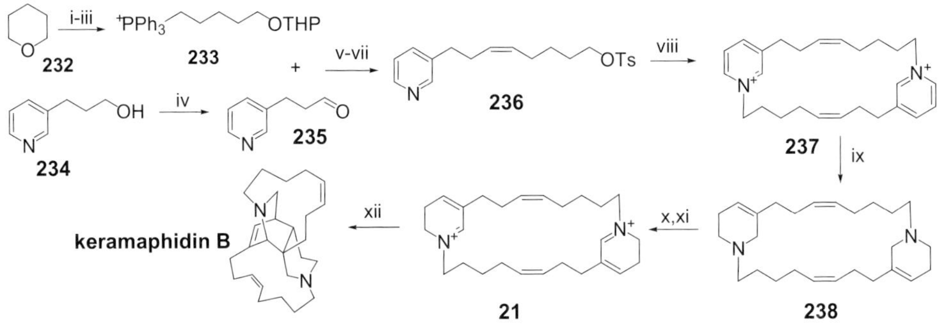
i-iii
232
233
234
235
236
237
238
21
keramaphidin B
OTHP
OTs
OH
iv
v-vii
viii
ix
x,xi
xii

Fig. 8.17 Reagents: i, AcBr, Zn; ii, Ph_3P, then K_2CO_3, H_2O, MeOH; iii, 3,4-dihydro-2*H*-pyran, PPT, DCM; iv, $(COCl)_2$, DMSO, TEA; v, KHMDS, THF; vi, HCl, MeOH; vii, TsCl, TEA, DCM; viii, NaI, butan-2-one; ix, $NaBH_4$, MeOH; x, *m*-CPBA, DCM; xi, TFAA, DCM; xii, MeOH/buffer (pH 7.3), then $NaBH_4$, MeOH.

8.4.6
Synthesis of Manzamine Analogs

With the completion of the total synthesis of manzamine C by Nakagawa *et al.* in 1991 [73], the synthesis and biological evaluation of manzamine C analogs were subsequently reported by the same group [136]. In this study, the azacyclic ring was modified from 11-membered to 5-, 6-, 7-, and 8-membered rings (Scheme 8.19).

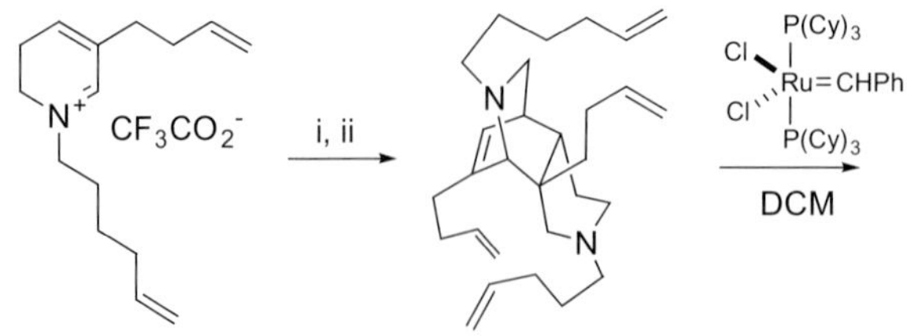

239 → (i, ii) → 240 → ($Cl_2Ru(=CHPh)(P(Cy)_3)_2$, DCM) → 241 + keramaphidin B

Scheme 8.18 Reagents: i, TRIS/HCl (pH 8.3); ii, NaBH$_4$.

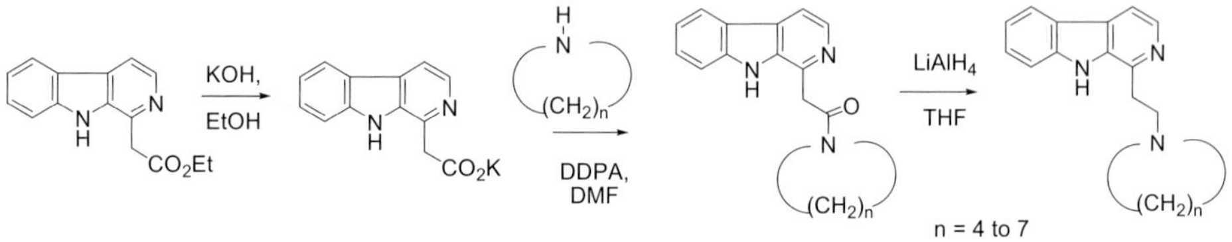

KOH, EtOH; DDPA, DMF; LiAlH$_4$, THF; n = 4 to 7

Scheme 8.19

8.5 Biological Activities of Manzamines

8.5.1 Anticancer Activity

Reports of effective cell proliferation inhibition by manzamines against various tumor cell lines have been published. Manzamine A showed the most pronounced biological activity of all the reported manzamines. Manzamine A displayed potent cytotoxicity against human colon tumor cells, lung carcinoma cells, and breast cancer cells at a concentration of 0.5 μg/mL [137]. It also showed significant *in vitro* activity against KB cell lines with an IC_{50} of 0.05 μg/mL, LoVo cell lines with an IC_{50} of 0.15 μg/mL, and HSV-II cell lines with an MIC of 0.05 μg/mL [39]. Manzamine A hydrochloride inhibited the growth of P388 mouse leukemia cells at an IC_{50} of 0.07 μg/mL [6]. 8-Hydroxymanzaine A also known as manzamine G or manzamine K inhibited the KB human epidermoid carcinoma cell line with an IC_{50} value of 0.03 μg/mL and was relatively active in the LoVo and HSV-II assays with an IC_{50} of 0.26 μg/mL and an MIC of 0.1 μg/mL, respectively [39]. Its enantiomer (**2a**) exhibits better activity against P388 with an IC_{50} of 0.25 μg/mL [15]. The cytotoxicity of 6-hydroxymanzamine A or manzamine Y, and 3,4-dihydromanzamine A were found to be active against L1210 murine leukemia cells with IC_{50} values of 1.5 and 0.48 μg/mL, respectively. Cytotoxicity against KB human epidermoid carcinoma cells was found with IC_{50} values of 12.5 and 0.61 μg/mL, respectively [42]. Manzamines B, C, D, E, and F exhibited antitumor activity against P388 with IC_{50} values of 6.0, 3.0, 0.5, 5.0, and 5.0 μg/mL, respectively [38]. Manzamine C, which is much simpler than the other manzamines, consisting of only a β-carboline unit and a symmetrical azacycloundecene ring, is just as potent against P388 leukemia cells ($IC_{50} = 3.0$ μg/mL) as the other manzamines, such as manzamine E, which exhibited an IC_{50} of 5.0 μg/mL against P388 leukemia cells. Manzamine F also showed cytotoxicity against L5178 mouse lymphoma cells with an ED_{50} of 2.3 μg/mL. 2-*N*-Methyl-8-hydroxymanzamine D showed an ED_{50} of 0.8 μg/mL against P388 [43]. Manzamine M, 3,4-dihydromanzamine J, 3,4-dihydro-6-hydroxymanzamine A, and 1,2,3,4-tetrahydromanzamine B inhibited L1210 murine leukemia cells effectively; the 50 % inhibition concentrations were 1.4, 0.5, 0.3, and 0.3 μg/mL, respectively [40,45]. Manzamine L exhibited cytotoxicity against murine lymphoma L1210 cells and human epidermoid carcinoma KB cells with an IC_{50} of 3.7 and 11.8 μg/mL, respectively [20]. The cytotoxicity exhibited by 1,2,3,4-tetrahydromanzamine B against KB human epidermoid carcinoma cells was moderate with an IC_{50} value of 1.2 μg/mL. Madangamine A showed significant cytotoxic activity against murine leukemia P388 (ED_{50} 0.93 μg/mL) and human lung A549 (ED_{50} 14 μg/mL), brain U373 (ED_{50} 5.1 μg/mL), and breast MCF-7 (ED_{50} 5.7 μg/mL) cancer cell lines [24]. Against L1210 murine leukemia cells, IC_{50} concentrations of ircinals A at 1.4 μg/mL, and B at 1.9 μg/mL, and manzamines H at 1.3 μg/mL and J 2.6 μg/mL were exhibited. The same compounds displayed cytotoxicity against KB human epidermoid carcinoma cells, with IC_{50} values of 4.8, 3.5, 4.6, and >10 μg/mL, respectively [50]. Keramaphidin B inhibited P388 murine leukemia and KB human epidermoid carcinoma cells, with IC_{50} values of 0.28 and 0.3 μg/mL, respectively [19].

The cytotoxic activities of manzamines X and Y, and xestomanzamine B against KB human epidermoid carcinoma cells were weak with IC_{50} values at 7.9, 7.3, and 14.0 μg/mL, respectively [41]. 6-deoxymanzamine X, the *N*-oxide of manzamine J, 3,4-dihydromanzamine A, and manzamine A exhibited cytotoxicity against the L5178 murine lymphoma cell line, with ED_{50} values of 1.8, 3.2, 1.6, and 1.6 μg/mL, respectively [10]. Although xestomanzamines A and B contain the substructures of β-carboline and imidazole, their biological activities have not matched those of manzamine A and some of the other imidazole-containing marine natural products. Xestomanzomine B, however, was shown to exhibit weak cytotoxicity against KB human epidermoid carcinoma cells with an IC_{50} of 14.0 μg/mL [77].

Kauluamine was inactive against tumor cell lines, but *neo*-kauluamine possesses cytotoxicity against human lung and colon carcinoma cells with an IC_{50} of 1.0 μg/mL [15]. The manadomanzamines **44** and **45** exhibit modest cytotoxic activity against human tumor cells and **44** is active against human lung carcinoma A-549 (IC_{50} 2.5 μg/mL) and human colon carcinoma H-116 (IC_{50} 5.0 μg/mL), while **45** is only active against H-116 with an IC_{50} value of 5.0 μg/mL [13].

Ircinols A and B exhibit cytotoxicity against L1210 murine leukemia cells, with IC_{50} values of 2.4 and 7.7 μg/mL, respectively; KB human epidermoid carcinoma cells were inhibited effectively with IC_{50} values of 6.1 and 9.4 μg/mL, respectively [56]. Xestocyclamine A exhibited moderate cytotoxicity against Protein Kinase C (PKC), with a 50 % inhibition concentration of 4.0 μg/mL, and also showed activity in a whole-cell IL-1 release assay with an IC_{50} of 1.0 μM. This action appeared to be selective, as compound **62** was inactive against other cancer-relevant targets, including PTK and

IMPDH. At doses as high as 100 μM, **62** did not show *in vitro* growth inhibition effects against cancer cells in the NCI's disease-oriented screening program [58].

Ingamines A and B both showed *in vitro* cytotoxicity against murine leukemia P388 with an ED_{50} of 1.5 μg/mL [21] and ingenamine showed cytotoxicity against murine leukemia P388 with an ED_{50} 1.0 μg/mL [22]. Ingenamine G displayed cytotoxic activity against HCT-8 (colon), B16 (leukemia), and MCF-7 (breast) cancer cell lines at the level of 8.6, 9.8, and 11.3 μg/mL, respectively [58]. Saraine A generally displays significant biological properties including vasodilative, antineoplastic, and cytotoxic activities [61]. Nakadomarin A showed inhibitory activity against cyclin-dependent kinase 4 with an IC_{50} of 9.9 μg/mL and cytotoxicity against murine lymphoma L1210 cells with an IC_{50} of 1.3 μg/mL [62].

8.5.2
Antimalarial Activity

Manzamine A demonstrated a potent *in vitro* bioactivity against chloroquine-sensitive (D6, Sierra Leone) and -resistant (W2, Indo-China) strains of *Plasmodium falciparum* respectively with respective IC_{50} values of 4.5 and 8.0 ng/mL. Its corresponding 8-hydroxyderivative exhibited almost the same bioactivity against the same strains at the same concentrations. Manzamines E, F, J, X, and Y, and 6-deoxymanzamine X exhibited antimalarial activity against chloroquine-sensitive strains of *P. falciparum* with IC_{50} values of 3400, 780, 1300, 950, 420, and 780 ng/mL and their IC_{50} values against chloroquine-resistant strains of *P. falciparum* respectively are 4760, 1700, 750, 2000, 850, and 1400 ng/mL. 6-Hydroxymanzamine E showed activity against chloroquine-sensitive (D6, Sierra Leone) and -resistant strains of *P. falciparum* respectively with IC_{50} values of 780 and 870 ng/mL [44]. *neo*-Kauluamine also displayed significant antimalarial activity *in vivo*, which was assayed against *P. berghei* with a single intraperitoneal (i.p.) dose of 100 μM/kg and no apparent toxicity. Ircinol A exhibited antimalarial activity against both chloroquine-sensitive and -resistant strains of *P. falciparum* with IC_{50} values of 2400 and 3100 ng/mL respectively.

12,34-Oxamanzamine A and E, 12,28-oxamanzamine A, 12,28-oxa-8-hydroxymanzamine A, 12,28-oxamanzamine E, 12,28-oxaircinal A, *ent*-12,34-oxamanzamine E, and *ent*-12,34-oxamanzamine F did not exhibit antimalarial activity against chloroquine-sensitive and -resistant strains of *P. falciparum*. The significant difference in biological activities of manzamines A, E, and F and 8-hydroxymanzamine A and the corresponding oxa-derivatives indicate that the C-12 hydroxy, C-34 methine, or the conformation of the lower aliphatic rings play a key role in the antimalarial activity, and this provides valuable insight into the structural moieties required with this class of compound for activity against malarial parasites.

The activity of manzamine Y against both the D6 clone and W2 clone of the malarial parasite *P. falciparum* (IC_{50} 420 and 850 ng/mL, respectively) is significantly lower than that of 8-hydroxymanzamine A (IC_{50} 6.0 and 8.0 ng/mL, respectively). This difference indicates that the change of the hydroxyl substitution from the C-8 position of the β-carboline moiety to the C-6 position decreases the antimalarial activity dramatically.

The ability of manzamine A to inhibit the growth of the rodent malaria parasite *P. bergheiin vivo* is the most promising activity. A single i.p. injection of manzamine A into infected mice inhibited more than 90 % of the asexual erythrocytic stages of *P. berghei*. The interesting aspect of manzamine A treatment is its ability to prolong the survival of highly parasitemic mice, with 40 % recovery 60 days after a single injection. A significant reduction (90 %) in parasitemia was observed by oral administration of an oil suspension of manzamine A or (−)-8-hydroxymanzamine A at a dose of 2×100 μM/kg. Morphological changes of *P. berghei* were observed 1 h after manzamine A treatment of infected mice, and plasma manzamine A concentration peaked 4 h after injection and remained high even at 48 h.

Initial *in vivo* studies of manzamine A and 8-hydroxymanzamine A against *P. berghei* at dosages of 50 and 100 μmol/kg (i.p.) showed significant improvements in survival times over mice treated with either chloroquine or artemisinin [138]. In addition it was observed that manzamine A possessed a rapid onset of action (1–2 h) against malaria in mice and provided a continuous and sustained level of the drug in plasma when measured as long as 48 h after administration. At an i.p. dose of 500 μmol/kg, both manzamine A and chloroquine were shown to be toxic to mice; however, the toxicity of manzamine A is slower acting than chloroquine indicating that the *in vivo* toxicity of manzamine A may be associated with its cytotoxicity. The fact that mice treated with a single 100 μmol/kg dose of manzamine A could survive longer carrying fulminating recurrent parasitemia and could in some cases clear the parasite led to speculation that manzamine A induced an immunostimulatory effect.

The serum concentrations of immunoglobulin G (IgG), interferon-α (IFN-α), interlukin-10 (IL-10) and tumor necrosis factor-α (TNF-α) of mice infected with *P. berghei* were evaluated to study the effect of manzamine A on the immune system of *P. berghei* infected mice [15c]. It was observed that Th1-mediated immunity was suppressed in *P. berghei* infected mice treated with manzamine A while Th2-mediated immunity was found to be upregulated. The concentrations of IL-10 and IgG did not increase with manzamine A alone, indicating that the immune-mediated clearance of malaria in mice may be a product of the long half-life of manzamine A resulting in a delayed rise of the parasitemia. This delay in parasitemia rise may provide the time necessary to upregulate a Th2-mediated response in the infected mice. In addition, the possibility that manzamine A and the other active manzamines form a conjugate with a malaria protein may also help to explain the Th2-mediated immune response.

The *in vivo* assay results of manzamine A against *Toxoplasma gondii* at a daily i.p. dose of 8 mg/kg, for eight consecutive days, beginning on day 1 following the infection prolonged the survival of Swiss Webster (SW) mice to 20 days, as compared with 16 days for the untreated control, indicate that **1** may be a valuable candidate for further investigation and development against several serious infectious diseases, and in particular malaria [15,138]. The promising antimalarial activity of manzamine A in laboratory mice, as well as its activity against tuberculosis *in vitro* indicates that it could have an outstanding impact on infectious diseases in developing countries. The diversity of biological activity associated with manzamine A further strengthens the growing possibility that these manzamines are broad-spectrum antiparasitic–antibiotics generated ultimately by the sponge-associated microbial communities.

(−)-8-Hydroxymanzamine A and *neo*-kauluamine displayed *in vivo* antimalarial activity, against *P. berghei* with a single i.p. dose of 100 μM/kg with no apparent toxicity. They efficiently reduced parasitemia with an increase in the average survival days of *P. berghei*-infected mice (9–12 days), as compared with untreated controls (2–3 days). Three 50 μmol/kg i.p. doses totally cleared the parasite and were found to be curative. Two oral doses at a rate of 100 μmoles/kg, provided a notable reduction of parasitemia.

8.5.3
Antimicrobial and Antituberculosis Activity

Keramamine-A (manzamine A, **1**) exhibited antimicrobial activity with a minimum inhibitory concentration (MIC) of 6.3 μg/mL, against *Staphylococcus aureus* [8]. Manzamine A, (−)-8-hydroxymanzamine A (**2a**), and *ent*-manzamine F (**9a**) induced 98–99 % inhibition of *Mycobacterium tuberculosis* (H37Rv) with an MIC < 12.5 μg/mL, and **1** and **2a** exhibit an MIC endpoint of 1.56 μg/mL and 3.13 μg/mL (**2a**), respectively [15,138]. 6-Hydroxymanzamine A and 3,4-dihydromanzamine A showed antibacterial activity against a gram-positive bacterium, *Sarcina lutea* (MIC values 1.25 and 4.0 μg/mL, respectively). Manzamine Y (6-hydroxymanzamine A, **3**) and 3,4-dihydromanzamine A also showed antibacterial activity against *S. lutea*, with a MIC values of 1.25 and 4.0 μg/mL, respectively. Manzamine F (**9**, keramamine-B) showed antimicrobial activity (MIC 25 μg/mL) against *Staphylococcus aureus* [8]. Manzamine L exhibited antibacterial activity against *Sarcina lutea*, *Staphylococcus aureus*, *Bacillus subtilis*, and *Mycobacterium* 607 with MIC values of 10, 10, 10, and 5 μg/mL, respectively [20]. Nakadomarin A exhibited antimicrobial activity against a fungus *Trichophyton mentagrophytes*, with an MIC of 23 μg/mL and a gram-positive bacterium *Corynebacterium xerosis* with an MIC of 11 mg/mL [62].

Most manzamines were active against *Mycobacterium tuberculosis* with MICs < 12.5 μg/mL. (+)-8-Hydroxymanzamine A had an MIC of 0.91 μg/mL, indicating improved activity for the (+) over the (−) enantiomer. The little difference in the *M. tuberculosis* activity may indicate that the structure–activity relationship for *M. tuberculosis* and *P. falciparum* are different. Comparison of the *M. tuberculosis* and *P. falciparum* activities of manzamine E and its hydroxy derivatives (**8** and **9**) indicates that hydroxyl functionality and its position on the β-carboline moiety may play a role in biological activity. These results may provide valuable information regarding the structural moieties required for activity against malaria. Manadomanzamine B and xestomanzamine A are active against the fungus *Cryptococcus neoformans* with IC_{50} values of 3.5 and 6.0 μg/mL. Manadomanzamine A was active against the fungus *Candida albicans* with an IC_{50} of 20 μg/mL. Both **44** and **45** showed strong activity against Mtb with MIC values of 1.86 and 1.53 μg/mL, indicating that the manadomanzamines may be a new class of antituberculosis leads.

Manzamine M, 3,4-dihydromanzamine J, and 3,4-dihydro-6-hydroxy-manzamine A showed antibacterial activity against *Sarcina lutea* with MIC values of 2.3, 12.5, and 6.3 mg/mL, respectively, and against *Corynebacterium xerosis* with MIC values of 5.7,

12.5, and 3.1 mg/mL, respectively. 32,33-Dihydro-31-hydroxymanzamine A, 32,33-dihydro-6,31-dihydroxymanzamine A, and 32,33-dihydro-6-hydroxymanzamine A-35-one were not active against chloroquine-sensitive and -resistant strains of *P. falciparum* [49]. Ingenamine G exhibited antibacterial activity against *S. aureus* at 105 μg/mL, *Escherichia coli* at 75 μg/mL, and four oxacillin-resistant *S. aureus* strains, two at concentrations between 10 and 50 μg/mL, as well as antimycobacterial activity against *M. tuberculosis* H37Rv at 8.0 μg/mL [58].

From the antimicrobial and HIV-1 activities of the manzamines and the corresponding oxa-derivatives, it is evident that manzamine A and 8-hydroxymanzamine A are more potent than manzamines E, F, and Y and these results provide valuable data to show that the nature of the eight-membered ring, and the hydroxyl functionality position on the β-carboline moiety are essential for antimicrobial and HIV-1 activity. This observation also suggests that reduction of the C-32–C-33 olefin and oxidation of C-31 to the ketone reduced the antimicrobial activity for the manzamine alkaloids. Significant differences in biological activities of manzamine A and 8-hydroxymanzamine A and the corresponding oxa-derivatives further indicate that the C-12 hydroxy, C-34 methine, or the conformation of the lower aliphatic rings play a key role in the antimicrobial and HIV-1 activity and provides valuable insight into the structural moieties required for activity.

8.5.4 Miscellaneous Biological Activities

Manzamine A was able to block the cytosolic calcium increase induced by KCl depolarization on human neuroblastoma SH-SY5Y cells. The maximum effect was observed at 1.0 μM, inhibiting by 31.5 %. 8-Hydroxymanzamine A also showed a similar effect whereas manzamines E, F, and Y, ircinal A and *neo*-kauluamine did not show any significant effect [44]. In the B lymphocytes reaction assay hyrtiomanzamine showed immunosuppressive activity with an EC_{50} of 2.0 μg/mL, but no activity was displayed against the KB human epidermoid carcinoma cell line [53].

Interestingly, when tested in an *in vitro* enzymatic assay, manzamines A (**1**, 73.2 %), E (**7**, 53.6 %), F (**9**, 29.9 %), and Y (**3**, 74.29 %), 8-hydroxymanzamine A (**2**, 86.7 %) and *neo*-kauluamine (**43**, 82.0 %) did show a moderate, although significant, effect in inhibiting human GSK3β activity, (shown in parentheses). Furthermore, when tested in a cell-based assay that measures GSK3β-dependent tau phosphorylation, manzamine A and 8-hydroxymanzamine A, showed a strong ability to inhibit tau phosphorylation within cells at concentrations as low as 5 μM without any cytotoxicity. Together, these data suggest that the manzamines may be interesting prototypes for the potential development of novel drugs for the treatment of Alzheimer's Disease (AD) and other tauopathies. Manzamine F at a dose of 132 ppm, inhibited 50.6 % growth of the insect *Spodoptera littoralis* larvae [44].

Manzamine A, *neo*-kauluamine, and chloroquine were evaluated against both medaka fry and eggs and were found to be more toxic than ethanol alone in both cases (in control groups, percentage fry survival and eggs hatching were over 94 % and 90 %, respectively). Medaka fry were 2.3- and 3.0-times more sensitive than were eggs when exposed to manzamine A and *neo*-kauluamine, respectively. Manzamine A was about 11 times more toxic than *neo*-kauluamine to medaka fry ($p = 0.0017$, *t*-test), and about 14.4 times more toxic (unsuccessful hatch) to eggs ($p = 0.012$, *t*-test) than *neo*-kauluamine. Chloroquine was approximately 150- and 1600-times less toxic than *neo*-kauluamine and manzamine A, respectively. Up to a 10.6 μM concentration of chloroquine did not affect medaka hatch relative to a control [44].

Manzamines A, E, F, and Y, 8-hydroxymanzamine A and *neo*-kauluamine did not show any effect on acetylcholinesterase (AChE) or α-amyloid cleaving enzyme (α-secretase, BACE1) using *in vitro* enzymatic assays. Likewise these compounds did not exhibit any significant ability to protect human neuroblastoma SH-SY5Y cells against oxidative stress-induced cell death [44].

Manzamine A at 132 ppm dose can induce 80 % growth inhibition of the insect *S. littoralis* larvae [10]. It also exhibited insecticidal activity toward neonate larvae of *S. littoralis*, the polyphagous pest insect with an ED_{50} of 35 ppm, when incorporated into an artificial diet and offered to larvae in a chronic feeding bioassay [44].

Mayer discovered that manzamine A, inhibits mediator formation in microglia isolated from newborn rats without killing healthy cells. Kauluamine showed moderate immunosuppressive activity. Manzamine F exhibited 50.6 % growth inhibition of the larvae of *S. littoralis* at 132 ppm dose [12].

Manzamines A, E, F, J, X, and Y, 8-hydroxymanzamine A, and 6-deoxymanzamine X exhibited leishmanicidal activity against *Leishmania donovani* with respective IC_{50} values of 0.9, 3.8, 4.2, 25, 5.7, 9.8, 6.2, and 3.2 μg/mL and their IC_{90} values respectively are 1.8, 6.8, 7.0, 45, 11, 14, 11 and 11 μg/mL. The oxa-derivatives of the corresponding manzamines did not exhibit any leishmanicidal activity. *neo*-Kauluamine and ircinal A exhibited leishmanicidal activity, with IC_{50} values of 4.2 and 4.6 μg/mL, respectively; their corresponding IC_{90} values are 8.2 and 5.5 μg/mL. The significant leishmanicidal activity of ircinol A (**57**, IC_{50} 0.9 μg/mL and IC_{90} 1.7 μg/mL) indicates that the β-carboline moiety may not be essential for activity against the leishmania parasite *in vitro* which is significantly different from the malaria SAR.

8.6 Concluding Remarks

The manzamine alkaloids are unique leads for the possible treatment of many diseases, including cancer and infectious diseases. There is considerable structural diversity in the manzamine class and, although many manzamines have been isolated and their biological activities evaluated, little is known about the SAR necessary for the potent biological activity observed. Since the mechanism of action of antimalarial activity of manzamine A remains unknown, the structural diversity may be a reflection of the variety of putative targets that exists for *Plasmodium*, an organism of considerable complexity. Although further investigations are required to completely understand the SAR for this class of compounds, to date the greatest potential for the manzamine alkaloids appears to be viable leads for the treatment of malaria, as well as other infectious or tropical parasitic diseases, based on their significant activity in animal models. In particular manzamine A and 8-hydroxymanzamine A show improved antimalarial activity over the clinically used drugs chloroquine and artemisinin in animal models. The effectiveness of manzamine A against malaria in laboratory mice, as well as against tuberculosis *in vitro* suggests that they could have an extraordinary impact on infectious diseases in developing countries. In addition, the diversity of biological activity associated with this molecule further supports the growing possibility that these alkaloids are broad-spectrum antiparasitic–antibiotics generated ultimately by sponge-associated microbial communities. The sequencing of the *P. falciparum* genome has raised hopes that valuable information with respect to drug targets will be unraveled to defeat a disease rivaled only by AIDS and tuberculosis as the world's most pressing health problem. If supported by adequate funding, this would give antimalarial drug discovery and vaccine development a decisive boost and optimistically, signal the demise of a scourge that has plagued humankind with such impunity. We strongly advocate expanding, not decreasing, the exploration of nature as a source of novel active agents, which may serve as the leads and scaffolds for elaboration into desperately needed efficacious drugs for a multitude of diseases.

References

1 Grabley, S. and Sattler, I. (2003) Natural products for lead identification: nature is a valuable resource for providing tools, Hillisch A. and Hilgenfeld R. *Modern Method of Drug Discovery*, 87–107.

2 Bergman, W. and Feeney, R. J. (1951) *Journal of Organic Chemistry*, **16**, 981–987.

3 (a) Blunt, J. W., Copp, B. R., Munro, M. H. G., Northcote, P. T., Prinsep, M. R. (2006) *Natural Products Reports*, **23**, 26–78. (b) MarinLit database, Department of Chemistry, University of Canterbury, http://www.chem.canterbury.ac.nz/marinlit/marinlit.shtml.

4 Magnier, E. and Langlois, Y. (1998) *Tetrahedron*, **54**, 6201–6258.

5 Baldwin, J. E. and Whitehead, R. C. (1992) *Tetrahedron Letters*, **33**, 2059–2062.

6 Sakai, R., Higa, T., Jefford, C. W., Bernardinelli, G. (1986) *Journal of the American Chemical Society*, **108**, 6404–6405.

7 (a) Hu, J.-F., Hamann, M. T., Hill, R., Kelly, M. (2003) Cordell G. A. *The Alkaloids: Biology and Chemistry*, Vol. **60** Elsevier, San Diego, CA, 207–285. (b) Urban, S., Hickford, A., Blunt, S. J. W., Munro, M. H. G. (2000) *Current Organic Chemistry*, **4**, 765–807.

8 Nakamura, H., Deng, S., Kobayashi, J., Ohizumi, Y., Tomotake, Y., Matsuzaki, T. (1987) *Tetrahedron Letters*, **28**, 621–624.

9 Peng, J., Shen, X., El Sayed, K. A., Dunbar, D. C., Perry, T. L., Wilkins, S. P., Hamann, M. T. (2003) *Journal of Agricultural and Food Chemistry*, **51**, 2246–2252.

10 Edrada, R. A., Proksch, P., Wray, V., Witte, L., Müeller, W. E. G., Van Soest, R. W. M. (1996) *Journal of Natural Products*, **59**, 1056–1060.

11 (a) El Sayed, K. A., Dunbar, D. C., Perry, T. L., Wilkins, S. P., Hamann, M. T., Greenplate, J. T., Wideman, M. A. (1997) *Journal of Agricultural and Food Chemistry*, **45**, 2735–2739. (b) Nugroho, B. W., Edrada, R. A., Bohnenstengel, F., Supriyono, A., Eder, C., Handayani, D., Proksch, P. (1997) *Spodoptera* littoralis as a test model for insecticidal bioassays,

Natural Product Analysis: Chromatography, Spectroscopy, Biological Testing (Symposium), Wuerzburg, Germany, 373–375.

12 Mayer, A. M. S., Gunasekera, S. P., Pomponi, S. A., Sennett S. H. (2000) PCT International Patent Application. CODEN: PIXXD2 WO 0056304 A2 20000928.

13 Peng, J., Hu, J.-F., Kazi, A. B., Franzblau, S. G., Zhang, F., Schinazi, R. F., Wirtz, S. S., Tharnish, P., Kelly, M., Hamann, M. T. (2003) *Journal of the American Chemical Society*, **125**, 13382–13386.

14 Hamann, M. T., Rao, K. V., and Peng, J. (2005) U.S. Patent Application Publications CODEN: USXXCO US 2005085554 A1 20050421, 21 pp.

15 (a) El Sayed, K. A., Kelly, M., Kara, U. A. K., Ang, K. K. H., Katsuyama, I., Dunbar, D. C., Khan, A. A., Hamann, M. T. (2001) *Journal of the American Chemical Society*, **123**, 1804–1808. (b) Hamann, M. T. and El-Sayed, K. A. (2002) PCT International Patent Application. CODEN: PIXXD2 WO 0217917 A1 200020307. (c) Ang, K. K. H., Holmes, M. J., Kara, U. A. (2001) *Parasitology Research*, **87**, 715–721. (d) Kara, A. U., Higa, T., Holmes, M., and Ang, K. H. (1999) PCT International Patent Application. CODEN: PIXXD2 WO 9959592 A1 19991125. US Patent 6143756. CA131: 346498.

16 Hiestand, P. C., Frank, P., Roggo, S., Hamann, M. T. (2006) PCT International Patent Application. CODEN: PIXXD2 WO 2006005620 A1 20060119, 24 pp.

17 (a) Ohtani, I. I., Ichiba, T., Isobe, M., Kelly-Borges, M., Scheuer, P. J. (1995) *Journal of the American Chemical Society*, **117**, 10743–10744. (b) Ohtani, I. I., Ichiba, T., Isobe, M., Kelly-Borges, M., Scheuer, P. J. (1995) *Tennen Yuki Kagobutsu Toronkai Koen Yoshishu 37th*, 236–241 CA124: 202710.

18 (a) Tsuda, M., Kawasaki, N., Kobayashi, J. (1994) *Tetrahedron Letters*, **35**, 4387–4388. (b) Tsuda, M., Kawasaki, N., Kobayashi, J. (1994) *Tennen Yuki Kagobutsu Toronkai Koen Yoshishu 36th*, 509–516 CA123: 193590.

19 (a) Kobayashi, J., Tsuda, M., Kawasaki, N., Matsumoto, K., Adachi, T. (1994) *Tetrahedron Letters*, **35**, 4383–4386. (b) Kobayashi, J., Kawasaki, N., Tsuda, M. (1996) *Tetrahedron Letters*, **37**, 8203–8204.

20 (a) Tsuda, M., Inaba, K., Kawasaki, N., Honma, K., Kobayashi, J. (1996) *Tetrahedron*, **52**, 2319–2324. (b) Tsuda, M., Kawasaki, N., Kobayashi, J. (1994) *Tetrahedron*, **50**, 7957–7960.

21 Kong, F., Andersen, R. J., Allen, T. M. (1994) *Tetrahedron*, **50**, 6137–6144.

22 Kong, F., Andersen, R. J., Allen, T. M. (1994) *Tetrahedron Letters*, **35**, 1643–1646.

23 Kong, F. and Andersen, R. J. (1995) *Tetrahedron*, **51**, 2895–2906.

24 Kong, F., Andersen, R. J., Allen, T. M. (1994) *Journal of the American Chemical Society*, **116**, 6007–6008.

25 Matzanke, N., Gregg, R. J., Weinreb, S. M., Parvez, M. (1997) *Journal of Organic Chemistry*, **62**, 1920–1921.

26 Kong, F., Graziani, E. I., Andersen, R. J. (1998) *Journal of Natural Products*, **61**, 267–271.

27 (a) Tsuda, M. and Kobayashi, J. (1997) *Heterocycles*, **46**, 765–794. (b) Baker, B. J. (1996) Pelletier W. *Alkaloids: Chemical and Biological Perspectives*, **10**, Pergamon, Oxford, 357–407. (c) Ihara, M. and Fukumoto, K. (1995) *Natural Products Reports*, **12**, 277–301.

28 (a) Yousaf, M., El Sayed, K. A., Rao, K. V., Lim, C. W., Hu, J.-F., Kelly, M., Franzblau, S. G., Zhang, F., Peraud, O., Hill, R. T., Hamann, M. T. (2002) *Tetrahedron*, **58**, 7397–7402. (b) Kasanah, N., Rao, K. V., Yousaf, M., Wedge, D. E., Hamann, M. T. (2003) *Tetrahedron Letters*, **44**, 1291–1293.

29 Williams, D. E., Lassota, P., Andersen, R. J. (1998) *Journal of Organic Chemistry*, **63**, 4838–4841.

30 Kondo, K., Shigemori, H., Kikuchi, Y., Ishibashi, M., Sasaki, T., Kobayashi, J. (1992) *Journal of Organic Chemistry*, **57**, 2480–2483.

31 Nakagawa, M., Nagata, T., Ono, K., Nishida, A. (2005) *Yuki Gosei Kagaku Kyokaishi*, **63**, 200–210.

32 Nakagawa, M., Nagata, T., Ono, K., Uchida, H., Watanabe, T., Hatakeyama, K., Akiba, M., Fuwa, M., Arisawa, M., Nishida, A. (2003) *Advances in Experimental Medicine and Biology*, **527**, 609–620 (Developments in Tryptophan and Serotonin Metabolism).

33 Rodriguez, J. (2000) *Studies in Natural Products Chemistry*, **24**, 573–681 [Bioactive Natural Products (Part E)].

34 Whitehead, R. (1999) *Annual Reports on the Progress of Chemistry Section B*, **95**, 183–205.

35 (a) Urban, S., Hickford, S. J. H., Blunt, J. W., Munro, M. H. G. (2000) *Current Organic Chemistry*, **4**, 765–807. (b) Higa, T., Ohtani, I. I., Tanaka, J. (2000) *ACS Symposium Series*, **745**, 12–21 (Natural and Selected Synthetic Toxins). (c) Higa, T., Tanaka, J., Ohtani, I. I., Musman, M., Roy, M. C., Kuroda, I. (2001) *Pure and Applied Chemistry*, **73**, 589–593. (d) Higa, T., Tanaka, J., Tan, L. T. (1998) Cytotoxic Macrocycles from Marine Sponges, Rahman A. U. and Choudhary M. I. *New Trends Nat. Prod. Chem. (Int. Symp. Nat. Prod. Chem., 6th, Meeting Data 1996)*, Harwood, Amsterdam, 109–120 CA129:185141. (e) Barriault, L. and Paquette, L. A. (1999) *Chemtracts*, **12**, 76–281 CA131:73818.

36 Higa, T., Sakai, R., Kohmoto, S., Lui, M. S. (22 June 1988) European Patent Application. EP 272056 (Cl. C07D471/04), US Appl. 943,609, 18 Dec 1986; 14 pp.

37 (a) Sakai, R., Kohmoto, S., Higa, T., Jefford, C. W., Bernardinelli, G. (1987) *Tetrahedron Letters*, **28**, 5493–5496. (b) Seki, H., Nakagawa, M., Hashimoto, A., Hino, T. (1993) *Chemical and Pharmaceutical Bulletin*, **41**, 1173–1176. (c) Seki, H., Hashimoto, A., Hino, T. (1993) *Chemical and Pharmaceutical Bulletin*, **41**, 1169–1172.

38 (a) Ichiba, T., Sakai, R., Kohmoto, S., Saucy, G., Higa, T. (1988) *Tetrahedron Letters*, **29**, 3083–3086. (b) Higa, T., Sakai, R., and Ichiba, T. (1990) U.S. Patent. CODEN: USXXAM US 4895852 A 19900123. (c) Kitagawa, I. and Kobayashi, M. (1991) *Yuki Gosei Kagaku Kyokaishi*, **49**, 1053–1061 CA116: 50671.

39 Ichiba, T., Corgiat, J. M., Scheuer, P. J., Kelly-Borges, M. (1994) *Journal of Natural Products*, **57**, 168–170.

40 Watanabe, D., Tsuda, M., Kobayashi, J. (1998) *Journal of Natural Products*, **61**, 689–692.

41 Kobayashi, M., Chen, Y. J., Aoki, S., In, Y., Ishida, T., Kitagawa, I. (1995) *Tetrahedron*, **51**, 3727–3736.

42 Kobayashi, J., Tsuda, M., Kawasaki, N., Sasaki, T., Mikami, Y. (1994) *Journal of Natural Products*, **57**, 1737–1740.

43 Crews, P., Cheng, X. C., Adamczeski, M., Rodriguez, J., Jaspars, M., Schmitz, F. J., Traeger, S. C., Pordesimo, E. O. (1994) *Tetrahedron*, **50**, 13567–13574.

44 Rao, K. V., Kasanah, N., Wahyuono, S., Tekwani, B. L., Schinazi, R. F., Hamann, M. T. (2004) *Journal of Natural Products*, **67**, 1314–1318.

45 Tsuda, M., Watanabe, D., Kobayashi, J. (1999) *Heterocycles*, **50**, 485–488.

46 Zhou, B.-N., Slebodnick, C., Johnson, R. K., Mattern, M. R., Kingston, D. G. I. (2000) *Tetrahedron*, **56**, 5781–5784.

47 Yousaf, M., Nicholas, H. L., Peng, J., Subagus, W., McIntosh, K. A., Charman, W. N., Mayer, A. M. S., Hamann, M. T. (2004) *Journal of Medicinal Chemistry*, **47**, 3512–3517.

48 Rao, K. V., Donia, M. S., Peng, J., Garcia-Palomero, E., Alonso, D., Martinez, A., Medina, M., Franzblau, S. G., Tekwani, B. L., Khan, S. I., Wahyuono, S., Willett, K., and Hamann, M. T. (2006) *Journal of Natural Products*, **69**, 1034–1040.

49 Rao, K. V., Santarsiero, B. D., Mesecar, A. D., Schinazi, R. F., Tekewani, B. L., Hamann, M. T. (2003) *Journal of Natural Products*, **66**, 823–828.

50 (a) Kondo, K., Shigemori, H., Kikuchi, Y., Ishibashi, M., Sasaki, T., Kobayashi, J. (1992) *Journal of Organic Chemistry*, **57**, 2480–2483. (b) Kondo, K., Shigemori, H., Kikuchi, Y., Ishibashi, M., Kobayashi, K., Sasaki, T. (1992) *Tennen Yuki Kagobutsu Toronkai 34th*, 463–469.

51 (a) Tsuda, M., Watanabe, D., Kobayashi, J. (1998) *Tetrahedron Letters*, **39**, 1207–1210. (b) Kobayashi, J., Tsuda, M., Ishibashi, M. (1999) *Pure and Applied Chemistry*, **71**, 1123–1126.

52 Desqueyroux-Faúndez, R. and Valentine, C. (2002) Family Petrosiidae Van Soest, 1980, Hooper, J. N. A. Soest, R. W. M.van Systema Porifera: A Guide to the Classification of Sponges, Kluwer Academi/Plenum Publishers, New York, 906–917.

53 Bourguet-Kondracki, M. L., Martin, M. T., Guyot, M. (1996) *Tetrahedron Letters*, **37**, 3457–3460.

54 Koren-Goldshlager, G., Kashman, Y., Schleyer, M. (1998) *Journal of Natural Products*, **61**, 282–284.

55 (a) Banwell, M. G., Bray, A. M., Edward, A. J., Wonga, D. J. (2001) *New Journal of Chemistry*, **25**, 1347–1350. (b) Heinrich, M. R. and Steglich, W. (2001) *Tetrahedron Letters*, **42**, 3287–3289.

56 Heinrich, M. R., Kashman, Y., Spiteller, P., Steglich, W. (2001) *Tetrahedron*, **57**, 9973–9978.

57 (a) Rodriguez, J. and Crews, P. (1994) *Tetrahedron Letters*, **35**, 4719–4722. (b) Rodriguez, J., Peters, B. M., Kurz, L., Schatzman, R. C., McCarley, D., Lou, L., Crews, P. (1993) *Journal of the American Chemical Society*, **115**, 10436–10437.

58 de Oliveira, J. H. H. L., Grube, A., Kock, M., Berlinck, R. G. S., Macedo, M. L., Ferreira, A. G., Hajdu, E. (2004) *Journal of Natural Products*, **67**, 1685–1689.

59 Guo, Y., Trivellone, E., Scognamiglio, G., Cimino, G. (1998) *Tetrahedron*, **54**, 541–550.

60 Cimino, G., Scognamiglio, G., Spinella, A., Trivellone, E. (1990) *Journal of Natural Products*, **53**, 1519–1525.

61 (a) Kobayashi, J., Watanabe, D., Kawasaki, N., Tsuda, M. (1997) *Journal of Organic Chemistry*, **62**, 9236–9239. (b) Tsuda, M., Watanabe, D., Kobayashi, J. (1998) *Tennen Yuki Kagobutsu Toronkai Koen Yoshishu 40th*, 467–472 CA131: 142191.

62 Peng, J., Diers, J., Hamann M. T. unpublished data from our laboratory.

63 Winkler, J. D. and Axten, J. M. (1998) *Journal of the American Chemical Society*, **120**, 6425–6426.

64 Martin, S. F., Humphrey, J. M., Ali, A., Hillier, M. C. (1999) *Journal of the American Chemical Society*, **121**, 866–867.

65 Winkler, J. D., Axten, J., Hammach, A. H., Kwak, Y. S., Lengweiler, U., Lucero, M. J., Houk, K. N. (1998) *Tetrahedron*, **54**, 7045–7056.

66 Winkler, J. D., Stelmach, J. E., Siegel, M. G., Haddad, N., Axten, J., Dailey, W. P. (1997) *Israel Journal of Chemistry*, **37**, 47–67.

67 Winkler, J. D., Siegel, M. G., Stamach, J. E. (1993) *Tetrahedron Letters*, **34**, 6509–6512.

68 Humphrey, J. M., Liao, Y., Ali, A., Rein, T., Wong, Y. L., Chen, H. J., Courtney, A. K., Martin, S. F. (2002) *Journal of the American Chemical Society*, **124**, 8584–8592.

69 Martin, S. F., Chen, H. J., Courtney, A. K., Liao, Y., Patzel, M., Ramser, M. N., Wagman, A. S. (1996) *Tetrahedron*, **52**, 7251–7264.

70 Martin, S. F., Liao, Y., Wong, Y. L., Rein, T. (1994) *Tetrahedron Letters*, **35**, 691–694.

71 Martin, S. F., Rein, T., Liao, Y. (1991) *Tetrahedron Letters*, **32**, 6481–6484.

72 Kondo, K., Shigemori, H., Kikuchi, Y., Ishibashi, M., Sasaki, T., Kobayashi, J. (1992) *Journal of Organic Chemistry*, **57**, 2480–2483.

73 Torisawa, Y., Hashimoto, A., Nakagawa, M., Seki, H., Hara, R., Hino, T. (1991) *Tetrahedron*, **47**, 8067–8078.

74 Torisawa, Y., Hashimoto, A., Nakagawa, M., Hino, T. (1989) *Tetrahedron Letters*, **30**, 6549–6550.

75 Hino, T. and Nakagawa, M. (1994) *Journal of Heterocyclic Chemistry*, **31**, 625–630.

76 Magnier, E. and Langlois, Y. (1998) *Tetrahedron*, **54**, 6201–6258.

77 Tsuda, M. and Kobayashi, J. (1997) *Heterocycles*, **46**, 765–794.

78 Arisawa, M., Kato, C., Kaneko, H., Nishida, A., Nakagawa, M. (2000) *Journal of the Chemical Society, Perkin Transactions*, **1**, 1873–1876.

79 Novak, W. and Gerlach, H. (1993) *Liebigs Annalen der Chemie*, 153–159.

80 Vidal, T., Magnier, E., Langlois, Y. (1998) *Tetrahedron*, **54**, 5959–5966.

81 MaGee, D. I. and Beck, E. J. (2000) *Canadian Journal of Chemistry*, **78**, 1060–1066.

82 Ono, K., Nakagawa, M., Nishida, A. (2004) *Angewandte Chemie International Edition*, **43**, 2020–2023.

83 Nakagawa, N., Uchida, H., Ono, K., Kimura, Y., Yamabe, M., Watanabe, T., Tsuji, R., Akiba, M., Terada, Y., Nagaki, D., Ban, S., Miyashita, N., Kano, T., Theeraladanon, C., Hatakeyama, K., Arisawa, M., Nishida, A. (2003) *Heterocycles*, **59**, 721–733.

84 Nagata, T., Nakagawa, M., Nishida, A. (2003) *Journal of the American Chemical Society*, **125**, 7484–7485.

85 Nagata, T., Nishida, A., Nakagawa, M. (2001) *Tetrahedron Letters*, **42**, 8345–8349.

86 Pandit, U. K., Borer, B. C., Bieraugel, H. (1996) *Pure and Applied Chemistry*, **68**, 659–662.

87 Pandit, U. K., Overkleeft, H. S., Corer, B. C., Bieraugel, H. (1999) *European Journal of Organic Chemistry*, **5**, 959–968.

88 Pandit, U. K. (1994) *Journal of Heterocyclic Chemistry*, **31**, 615–624.

89 Borer, B. C., Deerenberg, S., Bieraugel, H., Pandit, U. K. (1994) *Tetrahedron Letters*, **35**, 3191–3194.

90 Pandit, U. K., Borer, B. C., Bieraugel, H., Deerenberg, S. (1994) *Pure and Applied Chemistry*, **66**, 2131–2134.

91 Brands, K. M. J., Meekel, A. A. P., Pandit, U. K. (1991) *Tetrahedron*, **47**, 2005–2026.

92 Brands, K. M. J. and Pandit, U. K. (1990) *Heterocycles*, **30**, 257–261.

93 Brands, K. M. J. and Pandit, U. K. (1989) *Tetrahedron Letters*, **30**, 1423–1426.

94 Ma, J., Nakagawa, M., Torisawa, Y., Hino, T. (1994) *Heterocycles*, **38**, 1609–1618.

95 Nakagawa, M., Lai, Z., Torisawa, Y., Hino, T. (1990) *Heterocycles*, **31**, 999–1002.

96 Torisawa, Y., Nakagawa, M., Hosaka, T., Tanabe, K., Lai, Z., Ogata, K., Nakata, T., Oishi, T., Hino, T. (1992) *Journal of Organic Chemistry*, **57**, 5741–5747.

97 Torisawa, Y., Nakagawa, M., Arai, H., Lai, Z., Hino, T., Nakata, T., Oishi, T. (1990) *Tetrahedron Letters*, **31**, 3195–3198.

98 Nakagawa, M. (2000) *Journal of Heterocyclic Chemistry*, **37**, 567–581.

99 Uchida, H., Nishida, A., Nakagawa, M. (1999) *Tetrahedron Letters*, **40**, 113–116.

100 Torisawa, Y., Hosaka, T., Tanabe, K., Suzuki, N., Motohashi, Y., Hino, T., Nakagawa, M. (1996) *Tetrahedron*, **52**, 10597–10608.

101 Nakagawa, M., Torisawa, Y., Hosaka, T., Tanabe, K., Da-te, T., Okamura, K., Hino, T. (1993) *Tetrahedron Letters*, **34**, 4543–4546.

102 Magnier, E., Langlois, Y., Merienne, C. (1995) *Tetrahedron Letters*, **36**, 9475–9478.

103 Magnier, E. and Langlois, Y. (1998) *Tetrahedron Letters*, **39**, 837–840.

104 Imbroisi, D. D. O. and Simpkins, N. S. (1991) *Journal of the Chemical Society, Perkin Transactions*, **1**, 1815–1823.

105 Herdemann, M., Al-Mourabit, A., Martin, M. T., Marazano, C. (2002) *Journal of Organic Chemistry*, **67**, 1890–1897.

106 Coldham, I., Crapnell, K. M., Fernandez, J. C., Moseley, J. D., Rabot, R. (2002) *Journal of Organic Chemistry*, **67**, 6181–6187.

107 Coldham, I., Coles, S. J., Crapnell, K. M., Fernandez, J. C., Haxell, T. F. N., Hursthouse, M. B., Moseley, J. D., Treacy, A. B. (1999) *Chemical Communications*, **17**, 1757–1758.

108 Coldham, I., Pih, S. M., Rabot, R. (2005) *Synlett*, **11**, 1743–1745.
109 Marko, I. E. and Chesney, A. (1992) *Synlett*, **4**, 275–278.
110 Solberghe, G. F. and Marko, I. E. (2002) *Tetrahedron Letters*, **43**, 5061–5065.
111 Turet, L., Marko, I. E., Tinant, B., Declercq, J. P., Touillaux, R. (2002) *Tetrahedron Letters*, **43**, 6591–6595.
112 Marko, I. E., Southern, J. M., Adams, H. (1992) *Tetrahedron Letters*, **33**, 4657–4660.
113 Clark, J. S., Townsend, R. J., Blake, A. J., Teat, S. J., Johns, A. (2001) *Tetrahedron Letters*, **42**, 3235–3238.
114 Leonard, J., Fearnley, S. P., Hickey, D. M. B. (1992) *Synlett*, **4**, 272–274.
115 Leonard, J., Fearnley, S. P., Finlay, M. R., Knight, J. A., Wong, G. (1994) *Journal of the Chemical Society, Perkin Transactions*, **1**, 2359–2361.
116 Li, S., Kosemura, S., Yamamura, S. (1994) *Tetrahedron Letters*, **35**, 8217–8220.
117 Li, S., Kosemura, S., Yamamura, S. (1998) *Tetrahedron*, **54**, 6661–6676.
118 Li, S., Ohba, S., Kosemura, S., Yamamura, S. (1996) *Tetrahedron Letters*, **37**, 7365–7368.
119 Li, S. and Yamamura, S. (1998) *Tetrahedron*, **54**, 8691–8710.
120 Li, S. and Yamamura, S. (1998) *Tetrahedron Letters*, **39**, 2597–2600.
121 Li, S., Yamamura, S., Hosomi, H., Ohba, S. (1998) *Tetrahedron Letters*, **39**, 2601–2604.
122 Clark, J. S., Hodgson, P. B., Goldsmith, M. D., Blake, A. J., Cooke, P. A., Street, L. J. (2001) *Journal of the Chemical Society, Perkin Transactions*, **1**, 3325–3337.
123 Clark, J. S. and Hodgson, P. B. (1995) *Tetrahedron Letters*, **36**, 2519–2522.
124 Bland, D., Hart, D. J., Lacoutiere, S. (1997) *Tetrahedron*, **53**, 8871–8880.
125 Bland, D., Chambournier, G., Dragan, V., Hart, D. J. (1999) *Tetrahedron*, **55**, 8953–8966.
126 Kamenecka, T. M. and Overman, L. E. (1994) *Tetrahedron Letters*, **35**, 4279–4282.
127 Brands, K. M. J. DiMichele, L. M. (1998) *Tetrahedron Letters*, **39**, 1677–1680.
128 Magnus, P., Fielding, M. R., Wells, C., Lynch, V. (2002) *Tetrahedron Letters*, **43**, 947–950.
129 Hart, D. J. and McKinney, J. A. (1989) *Tetrahedron Letters*, **30**, 2611–2614.
130 Campbell, J. A. and Hart, D. J. (1992) *Tetrahedron Letters*, **33**, 6247–6250.
131 Torisawa, Y., Soe, T., Katoh, C., Motohashi, Y., Nishida, A., Hino, T., Nakagawa, M. (1998) *Heterocycles*, **47**, 655–659.
132 Baldwin, J. E., Claridge, T. D. W., Culshaw, A. J., Heupel, F. A., Lee, V., Spring, D. R., Whitehead, R. C., Boughtflower, R. J., Mutton, I. M., Upton, R. J. (1998) *Angewandte Chemie International Edition*, **37**, 2661–2663.
133 Baldwin, J. E., Claridge, T. D. W., Culshaw, A. J., Heupel, F. A., Lee, V., Spring, D. R., Whitehead, R. C. (1999) *Chemistry—A European Journal*, **5**, 3154–3161.
134 Baldwin, J. E., Bischoff, L., Claridge, T. D. W., Heupel, F. A., Spring, D. R., Whitehead, R. C. (1997) *Tetrahedron*, **53**, 2271–2290.
135 Baldwin, J. E., Claridge, T. D. W., Heupel, F. A., Whitehead, R. C. (1994) *Tetrahedron Letters*, **35**, 7829–7832.
136 Torisawa, Y., Hashimoto, A., Okouchi, M., Iimori, T., Nagasawa, M., Hino, T., Nakagawa, M. (1996) *Bioorganic and Medicinal Chemistry Letters*, **6**, 2565–2570.
137 Higa, T. and Sakai, R. (1988) PCT International Patent Application. CODEN: PIXXD2 WO 88,001,98 A1 19880114, 27 pp.
138 Ang, K. K. H., Holmes, M. J., Higa, T., Hamann, M. T., Kara, U. A. K. (2000) *Antimicrobial Agents and Chemotherapy*, **44**, 1645–1649.

9
Antiangiogenic Alkaloids from Marine Organisms

Ana R. Diaz-Marrero, Christopher A. Gray, Lianne McHardy, Kaoru Warabi, Michel Roberge, Raymond J. Andersen

9.1
Introduction

The three main treatment modalities currently available to cancer patients are surgery, radiotherapy, and chemotherapy [1]. Radiotherapy and chemotherapy non-selectively inhibit rapidly proliferating cells, including cancer cells, and the generality of the antiproliferative effect of these treatments leads to the severe dose-limiting toxicities experienced by cancer patients. In addition, the genetic instability of tumor cells facilitates the development of resistance to radiotherapy and cytotoxic chemotherapy that eventually causes these treatment approaches to fail in most patients with solid or metastatic tumors.

A fourth anticancer treatment modality, antiangiogenic therapy, was proposed in a seminal paper by Folkman published in 1971 [2]. He suggested that many cells capable of forming tumors appear in the body with regular frequency, but the majority never develop into detectable tumors. The growth of these "microtumors" is limited by inadequate blood supply to provide the required nutrients and oxygen. A small percentage of size-restricted "microtumors" eventually gain the ability to induce new blood vessel formation in the surrounding tissue (angiogenesis) and then they increase in size. Since the ability of tumors to grow and become lethal depends on angiogenesis, Folkman hypothesized that blocking angiogenesis would be an effective strategy for arresting tumor growth. This hypothesis has been extensively tested in animal models during the intervening period and is now widely accepted [3].

In more recent papers, Folkman has pointed out that an advantage of targeting the microvascular endothelium instead of tumor cells directly is that the genetic stability of endothelial cells might make them less susceptible to developing resistance. His argument for this contention comes from the observation that, unlike tumor cells, "normal" dividing cells of rapidly regenerating tissue in the gastrointestinal tract (gut mucosa), hair follicles, and bone marrow do not develop resistance to conventional cytotoxic anticancer chemotherapy [4]. Folkman also proposed that harmful side effects of antiangiogenic therapy would not be expected, provided the therapy is

Modern Alkaloids: Structure, Isolation, Synthesis and Biology. Edited by E. Fattorusso and O. Taglialatela-Scafati

ISBN: 978-3-527-31521-5

sufficiently specific, because angiogenesis does not occur in adults under normal physiological conditions (with the exceptions of wound healing and menstruation) [2].

Metastasis, the spread of cancer cells from a primary tumor to remote sites in the body, is frequently the ultimate cause of death from solid tumors. Current options for treatment or prevention of metastasis are very limited. Since vascularization of tumors (angiogenesis) provides conduits required for the spread of the tumor cells to other sites (metastasis), targeting tumor vasculature not only offers a noncytotoxic approach to eradicating primary solid tumors, it should also aid in preventing their spread via metastasis.

Folkman's early papers stimulated an intensive search for endogenous proangiogenic and antiangiogenic factors. It is now recognized that angiogenesis is tightly regulated by a fine balance between proangiogenic proteins (for example, vascular endothelial growth factor – VEGF) and antiangiogenic proteins (for example, angiostatin and endostatin). In tumors, the "angiogenic switch" is turned "on" when the effect of endogenous proangiogenic molecules exceeds that of the antiangiogenic molecules. There has been significant effort to utilize these endogenous regulatory proteins to develop "antiangiogenic" cancer treatments. For example, endostatin has entered into a number of Phase I clinical trials, and the most encouraging evidence to date for the efficacy of antiangiogenic therapy in human cancer comes from the recent FDA approval of Avastin, an antibody against VEGF that has shown significant responses in the treatment of metastatic colon cancer [5].

In parallel with investigations of the therapeutic potential of endogenous proteins that regulate angiogenesis, there has also been an active search for small-molecule antiangiogenic agents. Several antiangiogenic small molecules have also entered clinical trials and three of them, a combretastatin A-4 prodrug (**1**), the fumagillin analog TNP-470 (**2**), and squalamine (**3**) are either natural products or synthetic compounds derived from a natural product "lead structure".

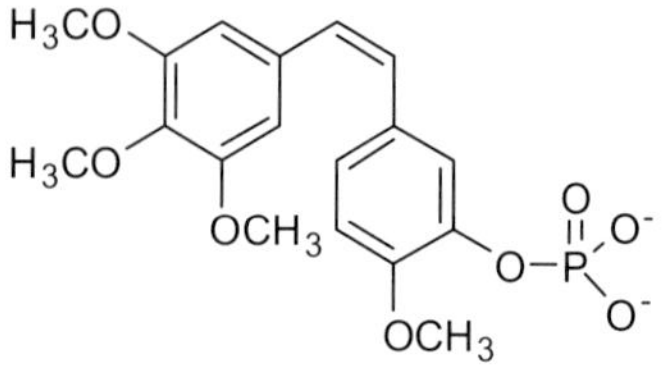

1

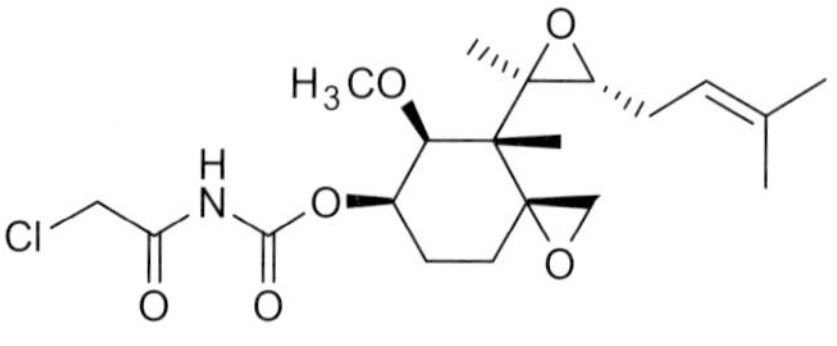

2

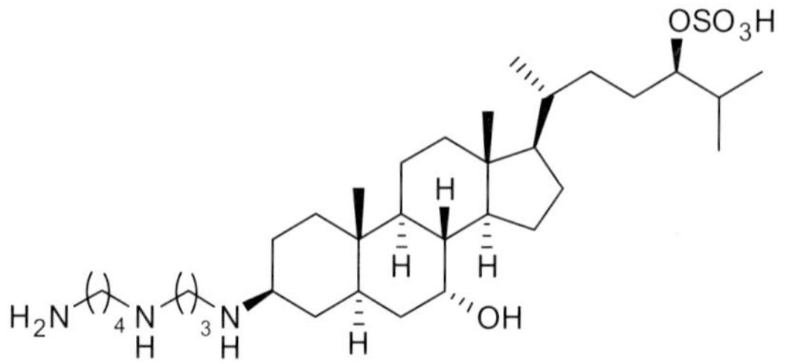

3

Marine organisms are an extremely rich source of novel bioactive secondary metabolites and several research groups have been exploring this pool of natural chemical diversity for the presence of promising antiangiogenic compounds. Many of the reported antiangiogenic marine natural products are alkaloids, and their structures and biological activity profiles are discussed below. The compounds have been grouped according to their putative biogenetic origins and we have adopted a very broad definition of an "alkaloid" that includes any marine natural product containing a basic nitrogen or clearly derived from an amino acid and still retaining its nitrogen atom, whether it is basic or not.

9.2 Purine Alkaloids

1,3-Dimethylisoguanine (**4**) and its conjugate acid 1,3-dimethylisoguaninium (**5**) have been found in several marine organisms as summarized in Table 9.1. These purine alkaloids were first identified in the marine sponge *Amphimedon viridis* [6,7] and later in *A. paraviridis* [8]. Copp and coworkers have also reported the isolation of 1,3-dimethylisoguanine from the ascidian *Cnemidocarpa bicornuta* [9].

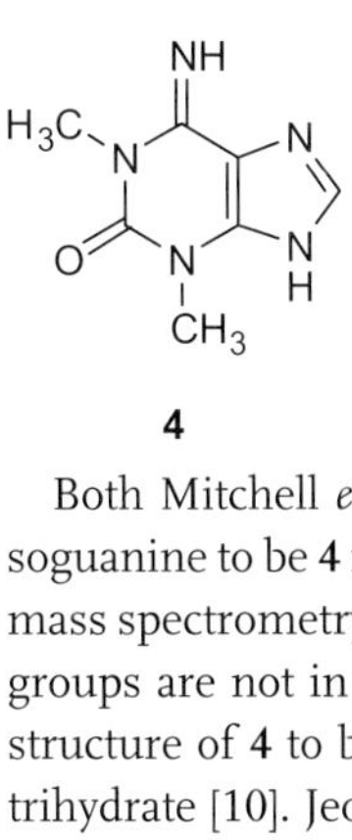

4 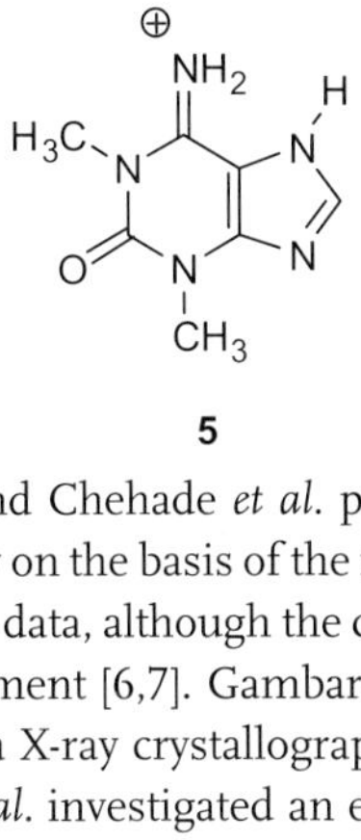 **5** 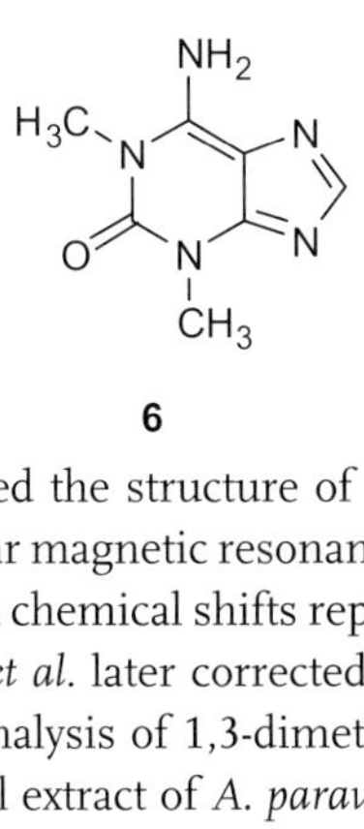 **6**

Both Mitchell *et al.* and Chehade *et al.* proposed the structure of 1,3-dimethylisoguanine to be **4** mainly on the basis of the nuclear magnetic resonance (NMR) and mass spectrometry (MS) data, although the carbon chemical shifts reported by these groups are not in agreement [6,7]. Gambardella *et al.* later corrected the proposed structure of **4** to be **6** via X-ray crystallographic analysis of 1,3-dimethylisoguanine trihydrate [10]. Jeong *et al.* investigated an ethanol extract of *A. paraviridis* showing inhibition of the proliferation of bovine aorta endothelial cells (BAECs). Assay-guided fractionation led to the isolation of a crystalline solid, which was determined to be the cationic form 1,3-dimethylisoguaninium by X-ray diffraction analysis, although no

Tab. 9.1 Marine organisms producing 1,3-dimethylisoguanine.

Marine organism	Taxonomic identification	Collection sites	Structure	References
Sponge	*Amphimedon viridis*	Bermuda	**44**	[6]
Sponge	*Amphimedon viridis*	Brazil	**66**	[7,10]
Ascidian	*Cnemidocarpa bicornuta*	New Zealand	**44**	[9]
Sponge	*Amphimedon paraviridis*	Japan	**55**	[8]
Sponge	*Xestospongia exigua*	Australia	**66**	[11]

charge-balancing counter ion was reported by this group [8]. Panthong *et al.* have subsequently confirmed the crystal structure of 1,3-dimethylisoguanine to be 6-amino-1,3-dimethyl-1*H*-purin-2(3*H*)-one (**6**) and suggest that the compound isolated from *Amphimedon paraviridis* was erroneously identified because of the pronounced hydrogen bonding between molecules in the crystal lattice [11]. The X-ray data reported by Panthong *et al.* are in agreement with the previous data for the trihydrate form published by Gambardella *et al.* and the NMR data obtained closely matched those published by Mitchell *et al.* and Jeong *et al.*

Low micromolar concentrations of **5** inhibited the proliferation of BAECs stimulated by serum or by basic fibroblast growth factor (bFGF). Compound **5** did not significantly inhibit the proliferation of human nasopharyngeal carcinoma KB cells, suggesting some selectivity toward vascular endothelial cells. At 0.1 μM, **5** also inhibited the ability of BAECs to form tubes resembling neovessels *in vitro* [8]. To our knowledge, no studies of the mechanism of action or *in vivo* antiangiogenic activity of **5** have been published.

9.3 Terpenoid Derivatives

9.3.1 Avinosol

Avinosol (**7**) was obtained from a methanolic extract of the marine sponge *Dysidea* sp. collected in Papua New Guinea that showed anti-invasion activity in a cell-based assay [12]. The structure of **7** was elucidated by analysis of spectroscopic data. Its molecular formula was established as $C_{31}H_{40}N_4O_6$ by high-resolution electrospray ionization mass spectrometry (HRESIMS), and analysis of the one-dimensional (1D) and two-dimensional (2D) NMR data allowed the identification of a sesquiterpenoid fragment linked to a hydroquinone ring bearing a 2′-deoxyinosine unit. Interpretation of 2D nuclear Overhauser effect spectroscopy (NOESY) correlations confirmed the relative configuration of the sesquiterpenoid fragment and the ribose moiety.

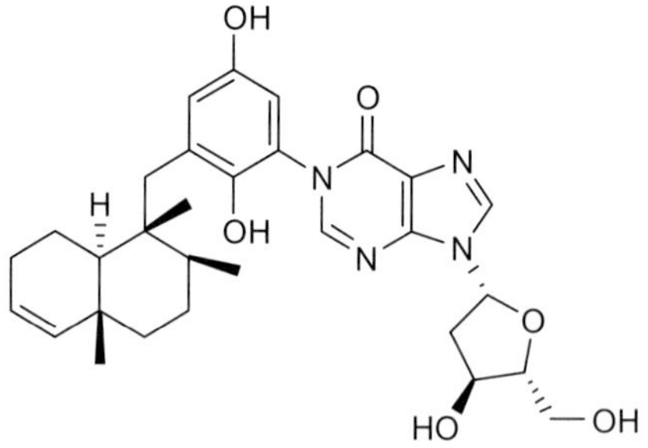

OH
O
N
N
H
OH
N
N
O
HO
OH

7

The anti-invasive activity of avinosol was examined using two human tumor cell lines that show distinct mechanisms of movement through the extracellular

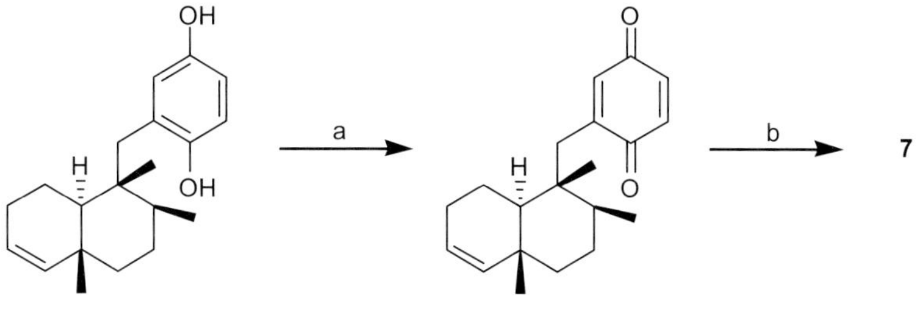

Reagents and Conditions: a) MnO_2, Et_2O, RT, 10 min; b) 2'-deoxyinosine, K_2CO_3, DMF, RT, 30 min

Scheme 9.1 The semisynthesis of avinosol (7) [12].

matrix: MDA-MB-231 breast cancer cells, which utilize a mesenchymal mode of invasion involving a path-generating process and LS174T colon carcinoma cells, which invade in an amoeboid manner. Avinosol showed an IC_{50} of ~50 μg/mL for both cell lines.

The putative biogenesis of avinosol involves terpenoid, acetate and/or shikimate, purine, and carbohydrate biosynthetic pathways. Diaz-Marrero *et al.* proposed that avinosol was constructed by nature in an efficient convergent manner by utilizing a common secondary metabolite pathway to form the meroterpenoid fragment and a primary metabolic pathway to form the nucleoside. The linkage of the two fragments to form avinosol would be favored by the intrinsic reactivity of the *p*-quinone subunit of the meroterpenoid and the purine ring in deoxyinosine.

The semisynthesis of avinosol was performed from avarone (**8**) and 2′-deoxyinosine to confirm the structure of the natural product. Avarone was prepared by oxidation of the naturally occurring avarol (**9**) obtained in the same extract (Scheme 9.1).

9.3.2
Cortistatins A–D

Cortistatins A–D (**10–13**) were isolated from the marine sponge *Corticium simplex* [13]. These steroidal alkaloids are unrelated to cortistatin, a neuropeptide structurally related to the hormone somatostatin which has antiangiogenic activity [14].

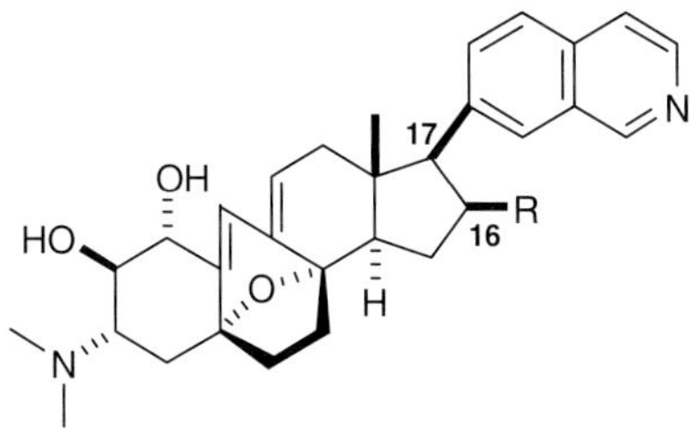

10 : R = H
11 : R = OH

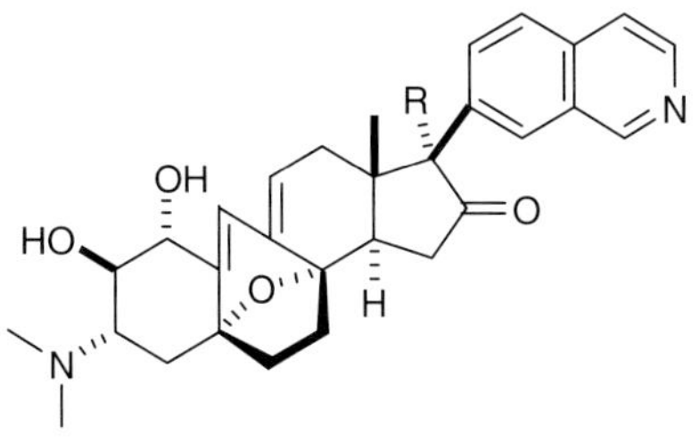

12 : R = H
13 : R = OH

The molecular formula of cortistatin A (**10**) was established as $C_{30}H_{36}N_2O_3$ by electrospray-ionization time-of-flight mass spectrometry (ESI-TOF MS) and NMR data. A planar structure for cortistatin A consisting of linked 9(10–19)-*abeo*-androstane and isoquinoline fragments was elucidated by analysis of COSY, HMQC, and HMBC data. Interpretation of the NOESY data as well as the proton coupling constant values revealed the relative configuration, which was confirmed by single-crystal X-ray diffraction analysis. The absolute configuration of **10** was determined via application of the circular dichroism (CD) exciton chirality method to the diene and isoquinoline chromophores [13].

Interpretation of 2D NMR data obtained for cortistatin B (**11**) showed that it had the same skeleton as **10**, but contained a hydroxyl group on C-16 with the β configuration. The carbon chemical shift value of δ 214.4 and the IR absorption at 1740 cm^{-1} indicated the existence of a ketone at C-16 in cortistatin C (**12**). Cortistatin D (**13**) was found to be a 17-hydroxy analog of cortistatin C, which was confirmed by ^{13}C NMR deuterium shift experiments as well as HMBC data. Analysis of NOESY data for **13** revealed the orientation of hydroxyl group at C-17 as α [13].

Cortistatin A exhibited antiproliferative activity against HUVECs at low nanomolar concentrations [13]. This potent activity appeared highly selective since the proliferation of four cancer cell lines was inhibited only at 3000-fold higher concentrations. Cortistatin C inhibited HUVEC proliferation 10-fold less potently than cortistatin A but it retained the remarkable selectivity of **10**. Cortistatins B and D were less potent still, but nevertheless exhibited good selectivity toward HUVECs. Cortistatin A was also shown to prevent bFGF- and VEGF-induced HUVEC migration and tube formation at low nanomolar concentrations [13].

9.3.3 Squalamine

Squalamine, a steroidal alkaloid, was isolated from the stomach tissues of the dogfish shark *Squalus acanthias*. This compound was first reported to exhibit potent bactericidal activity against both gram-positive and gram-negative bacteria. Squalamine also showed fungicidal activity and proved to induce osmotic lysis of protozoa [15]. The molecular formula of $C_{34}H_{66}O_5N_3S$ was determined by fast atom bombardment mass spectrometry (FABMS) data. Interpretation of 1D and 2D NMR data disclosed a steroid skeleton linked to a spermidine moiety, which was confirmed by analysis of the fragment peaks in the negative-mode FABMS spectrum. Observation of fragment ion peaks 80 mass units apart in the negative mode FABMS spectrum also revealed the existence of one sulfate group, which was assigned to C-24 by interpretation of the proton chemical shift value of H-24. The relative configurations in the steroid nucleus were deduced from NOE data as well as by analysis of proton coupling constant values [16].

Squalamine inhibits mitogen-induced endothelial migration and proliferation [17] and VEGF-induced capillary tube formation [18]. A large body of work has been carried out to understand the mechanism of action of squalamine and to develop the compound as a potential anticancer drug (for a review see ref. [19]). Squalamine interferes with important cell signaling pathways that affect vascular endothelial cell proliferation, adhesion, and mobility through blockage of the VEGF-stimulated

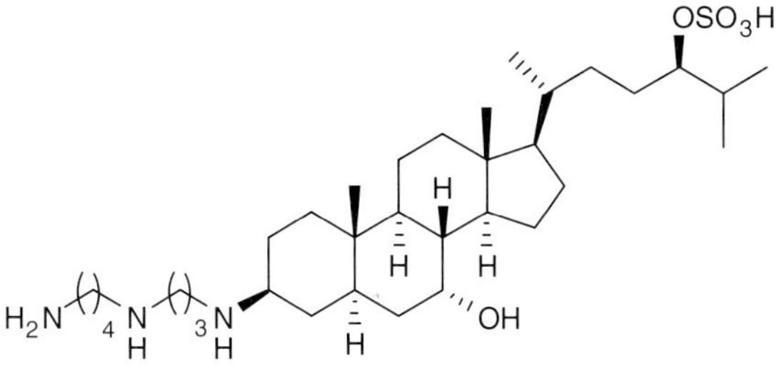

3

phosphorylation of p44/p42 MAP kinase, focal adhesion kinase (FAK), and stress-activated protein kinase-2/p38 (SAPK2) [18,19]. Squalamine-induced reduction of trans-membrane VE-cadherin-containing adhesions [20], its disruption of the actin cytoskeleton [20,21], and its ability to induce changes in cell shape [21] have been proposed to contribute to the antiangiogenic properties of the molecule [19]. The compound has also been shown to act as a calmodulin chaperone [21] and to inhibit the Na^+/H^+ exchanger NHE3 [22].

Squalamine has been tested in mammary, ovary, and lung cancer xenograft mouse models [20,23–26]. Results showed that squalamine alone had a modest effect on tumor growth delay and in some cases it resulted in decreased number of lung metastases. However, in most studies, squalamine treatment was more efficacious when combined with previously established anticancer agents such as cyclophosphamide, cisplatin, carboplatin, paclitaxel, 5-fluorouracil, or genestein, or with radiation therapy.

In phase I and phase II clinical trials, squalamine has been administered to patients with advanced nonleukemic cancers, advanced solid tumor malignancies, nonsmall-cell lung cancer, or stage III or IV human ovarian cancer [27–30]. When given alone, squalamine showed minimal toxic side effects, although some hepatoxicity was noted [27,28]. Transient tumor regression was also observed [27], but squalamine treatment alone did not result in any major tumor regression. When administered in combination with carboplatin or paclitaxel, results suggested that squalamine had increased clinical benefit. In patients with stage IIIB or IV nonsmall-cell lung cancer, 28 % showed partial clinical response and 19 % showed stable disease [29].

The promising biological activity of squalamine has stimulated significant interest in developing synthetic approaches to both the natural product and analogs. All of the routes published are, in principle, semisynthetic as they employ starting materials possessing the requisite steroid perhydrocyclopenta[*a*]phenanthrene ring system. The synthesis of **3** as an equal mixture of C-24 diastereomers was reported nearly concurrently by Moriarty *et al.* [31] and Frye and coworkers [32]. Both groups achieved the synthesis of **3** in 17 steps from 3β-acetoxy-5-cholenic acid and 3β-hydroxy-5-cholenic acid, respectively. The (24*R*) stereochemistry of squalamine was determined by Moriarty *et al.* through the stereoselective synthesis of both squalamine dessulfate epimers from stigmasterol and the two enantiomeric epoxides **14** and **15** derived from (*S*)-(+)-valine [33]. The key step in the synthesis was the condensation of either **14** or **15** with the C-22 anion obtained from the phenyl sulfone **16**, and this report constituted a 20-step formal stereoselective synthesis of **3**.

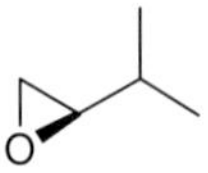

14 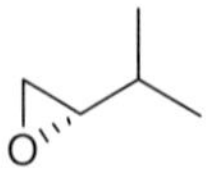 15 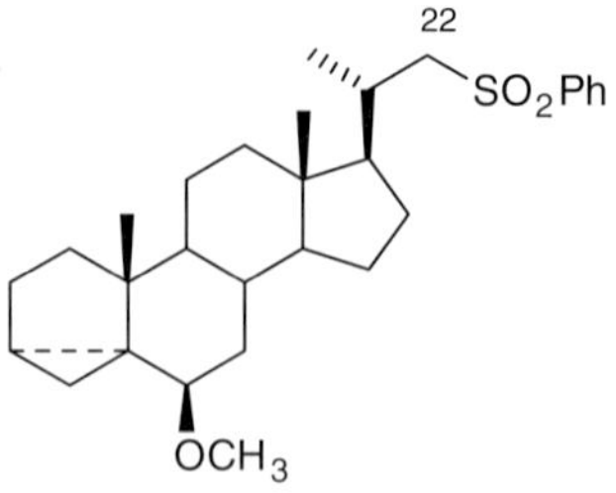 16

Subsequent efforts have focused upon improving the synthesis of squalamine in order to supply adequate material for clinical trials. A second synthesis from stigmasterol [34,35] and the development of efficient routes from 3-keto-5α-chenodeoxycholanate [36] and methyl chenodeoxycholanate [37,38] provided stereoselective access to **3** in fewer steps. The most significant and applicable advances in the synthetic strategy to squalamine, however, have involved developments in accessing adequate supplies of suitable, inexpensive steroid precursors. Formal syntheses employing the microbial 7α-oxidation of 3-keto-23,24-bisnorchol-4-en-22-ol [39] and desmosterol, a waste product obtained from the saponification of wool-grease [40], have illustrated the potential of alternative steroid sources to circumvent some of the more laborious synthetic steps and provide short and efficient routes to squalamine.

Numerous squalamine analogs have been prepared and examined for biological activity (Figure 9.1). The squalamine structure has been varied through the length of the polyamine chain [41–44]; the nature of the anionic functional group [41]; the position of the polyamine [43–45], sulfate [45,46], and free hydroxyl [41,43,44,47] on the steroid scaffold; the stereochemistry and substitution at C-3, C-7, and C-24 [41–44]; the length of the steroid side chain [41,45,46]; and the unsaturation of the steroid [43,44].

All of the analogs tested in antimicrobial assays [41–44,46] have exhibited broad-scale bioactivity comparable with that of squalamine, and a number of the amides **25–32** exhibited *in vitro* activity against the parasitic protozoa responsible for leishmaniasis, Chagas disease and African trypanosomiasis [45]. It is unfortunate that none of the squalamine analogs have been assessed for antiangiogenic activity, although **30**, **31**, and **34** were cytotoxic against the human nasopharynx carcinoma KB cell-line [45] and the unsaturated analogs **45–48** were toxic to a human nonsmall-cell bronchopulmonary carcinoma NSCLC-N6 cell-line [43,44].

9.4
Motuporamines

Motuporamines A–I **49–55** were found in the extract from the marine sponge *Xestospongia exigua* Kirkpatrick collected off Motupore Island, Papua New Guinea [48,49]. The structures of the motuporamines are composed of a 13-, 14-, or

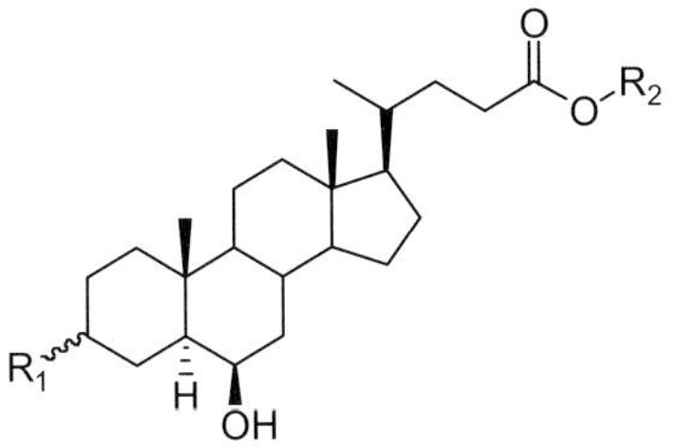

17: R_1 = α-spermidine, R_2 = Me
18: R_1 = β-spermidine, R_2 = Me
19: R_1 = α-spermidine, R_2 = H
20: R_1 = α-spermine, R_2 = Me
21: R_1 = β-spermine, R_2 = Me
22: R_1 = α-spermine, R_2 = H

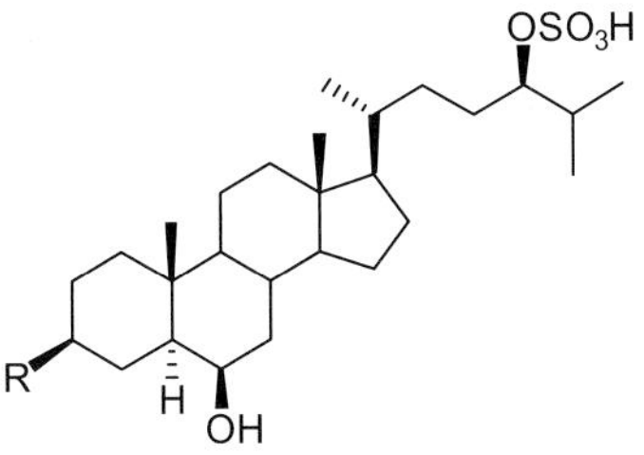

24: R = spermidine

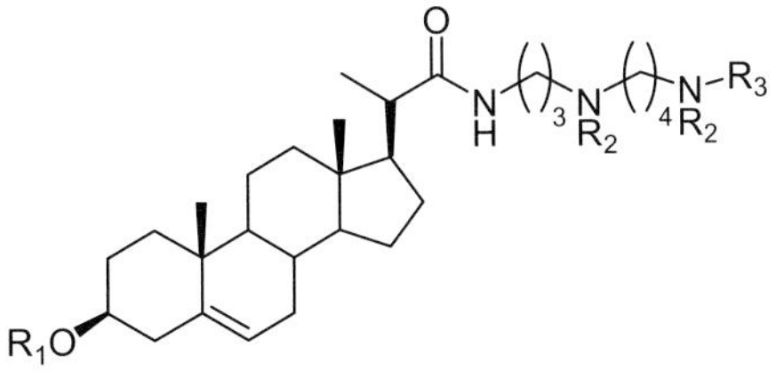

23: R = spermidine

25: R_1 = Ac, R_2 = BOC, R_3 = BOC
26: R_1 = Ac, R_2 = BOC, R_3 = $(CH_2)_3NHBOC$
27: R_1 = H, R_2 = BOC, R_3 = BOC
28: R_1 = H, R_2 = BOC, R_3 = $(CH_2)_3NHBOC$
29: R_1 = SO_3H, R_2 = BOC, R_3 = BOC
30: R_1 = SO_3H, R_2 = BOC, R_3 = $(CH_2)_3NHBOC$
31: R_1 = H, R_2 = H, R_3 = H
32: R_1 = H, R_2 = H, R_3 = $(CH_2)_3NH_2$
33: R_1 = SO_3H, R_2 = H, R_3 = H
34: R_1 = SO_3H, R_2 = H, R_3 = $(CH_2)_3NH_2$

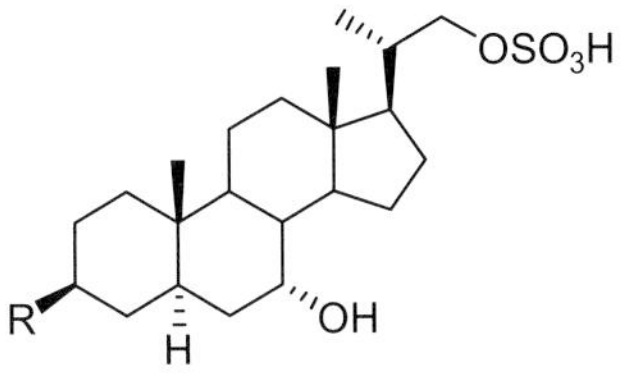

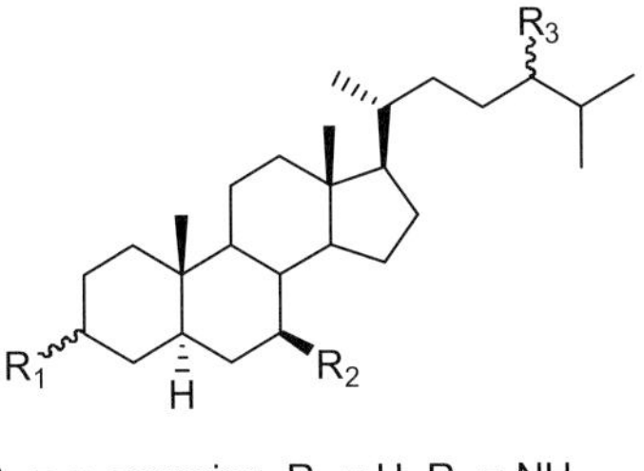

35: R_1 = α-spermine, R_2 = H, R_3 = NH_2
36: R_1 = β-spermine, R_2 = H, R_3 = NH_2
37: R_1 = α-spermine, R_2 = H, R_3 = OH
38: R_1 = β-spermine, R_2 = H, R_3 = OH
39: R_1 = α-spermine, R_2 = OH, R_3 = OH
40: R_1 = β-spermine, R_2 = OH, R_3 = OH
41: R_1 = α-spermine, R_2 = OH, R_3 = *R*-OH
42: R_1 = β-spermine, R_2 = OH, R_3 = *R*-OH
43: R_1 = α-spermine, R_2 = OH, R_3 = *R*-OSO_3H
44: R_1 = β-spermine, R_2 = OH, R_3 = *R*-OSO_3H

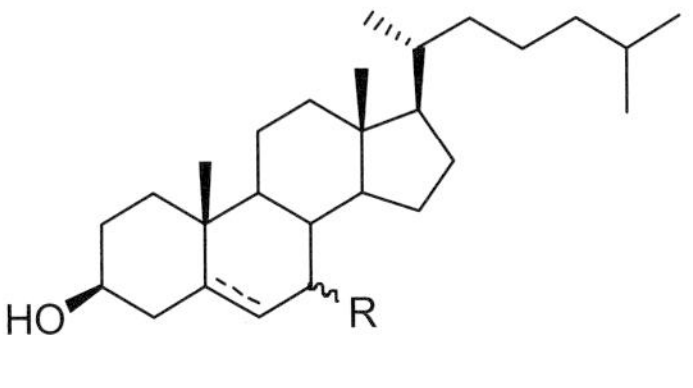

45: Δ^5, R = 7α-spermine
46: Δ^5, R = 7β-spermine
47: 5α-H, R = 7α-spermine
48: 5α-H, R = 7β-spermine

Fig. 9.1 Synthetic analogs of squalamine.

15-membered macrocyclic amine, linked to a diamino spermidine-like side chain. Motuporamine C (**51**) was the most abundant natural product, and its structure was elucidated by analysis of NMR and MS data. The position of the double bond in the macrocyclic amine moiety of **51** was confirmed via total synthesis [50]. The structures of the other motuporamines, except the methylated derivatives motuporamines G–I (**53–55**), were also characterized by spectroscopic analysis. Motuporamines G–I were obtained in such small quantities that the position of the methyl group in each macrocyclic ring could not be assigned through spectroscopy.

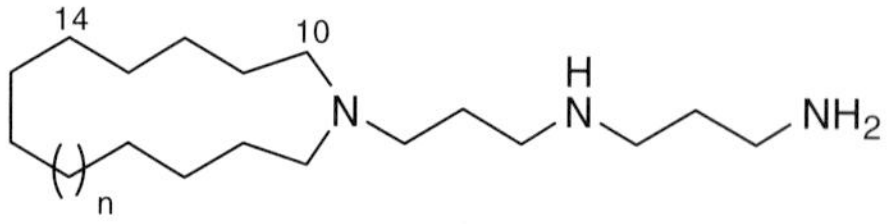

Motuporamine A (**49**): n=1
Motuporamine B (**50**): n=2
Motuporamine C (**51**): n=3; Δ^{14}
Motuporamine D (**52**): n=2; Δ^{13}

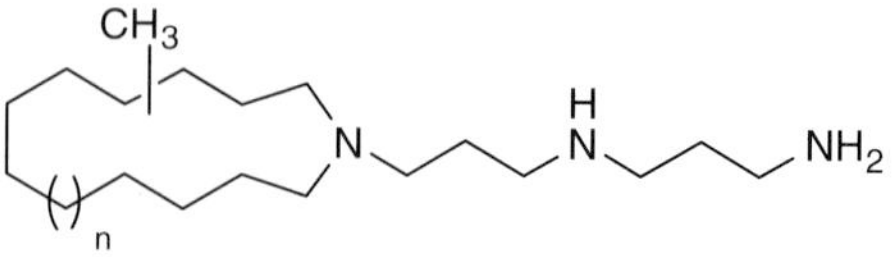

Motuporamines G, H and I (**53-55**): n=3

The motuporamines were first identified based on their modest cytotoxicity to human solid tumor cancer cell lines [48]. They were subsequently rediscovered in a screen for inhibitors of invasion of the tumor-derived basement membrane Matrigel by MDA-MB-231 breast carcinoma cells [51]. Subsequent studies revealed that motuporamine C inhibits *in vitro* MDA-MB-231 cell migration and *in vitro* angiogenesis in a human endothelial cell-sprouting assay. *In vivo* angiogenesis in the chick choriallantoic membrane assay was also inhibited [51]. The synthetic analog dihydromotuporamine C (**56**) activates RhoGTPase and remodels actin stress fibers and focal adhesions in a Rho-dependent manner. Motuporamines also stimulate Rho kinase-dependent sodium proton exchanger activity in mouse Swiss 3T3 fibroblasts [53], as well as Rho/Rho kinase-dependent stimulation of neuronal growth cone collapse [53]. Attenuation of Rho signaling with the Rho-specific inhibitor C3 exoenzyme re-established invasion in motuporamine-treated cells, thus implicating RhoGTPase as a key signaling component in the mechanism of tumor invasion inhibition by motuporamines [52].

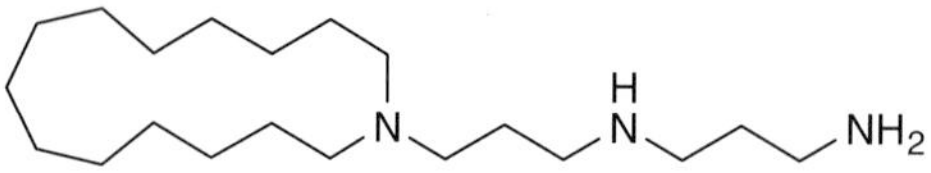

56

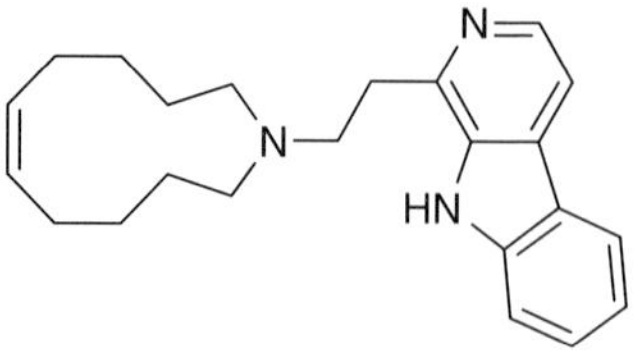

57

The motuporamines are considered as manzamine-related alkaloids owing to their structural relationship with manzamine C (**57**), the simplest alkaloid of this family of compounds. Baldwin and Whitehead have proposed a plausible biogenesis for these 3-alkylpiperidine alkaloids, suggesting that they could be formed from three basic building blocks: ammonia, a C_3 unit, and a C_{10} unit [54]. The motuporamines, which contain a spermidine-like side chain instead of the alkylated β-carboline substructure of manzamine C, appear to be biogenetically derived from the same precursors, ammonia, acrolein, and a long-chain dialdehyde involved in the Baldwin–Whitehead pathway (Scheme 9.2).

Syntheses of motuporamines A (**49**) and B (**50**) were simultaneously reported rapidly after their isolation by the groups of Baldwin [55] and Weiler [50]. Both syntheses employed lactams as precursors to the motuporamine macrocycles, with Baldwin choosing a reductive amination strategy to couple the macrocycle and nor-spermidine side chain while Weiler achieved the alkylation of the cyclic amine through Michael addition of methyl acrylate. Weiler also accomplished the synthesis of motuporamine C using ring-closing metathesis (RCM) to prepare the requisite 15-membered unsaturated lactam precursor (Scheme 9.3) and the preparation of diacetamide **61** allowed the unambiguous assignment of the site of unsaturation in the natural product **51**. The predominance of the undesired *E* isomer of **59**, an inherent problem with the preparation of macrocycles by RCM, prompted Fürstner and Rumbo to apply a stereoselective two-step alkyne metathesis/reduction strategy and develop an improved synthesis of **51** [56].

The strategies developed for the synthesis of motuporamines A–C have been applied to the preparation of a series of motuporamine analogs that were evaluated for anti-invasion activity against MDA-MB-231 cells [49]. Analysis of the structure–activity relationship (SAR) data obtained for these analogs and the motuporamine natural product series allowed the definition of a motuporamine pharmacophore model and the rational design of the simple carbazole derivative **62**, which displayed promising anti-invasive activity *in vitro*. Studies performed on the most active of the synthetic motuporamine series, dihydromotuporamine C (**56**), and its homospermidine analog **63** have indicated that these alkaloids do not gain cellular access via the polyamine transport system [57].

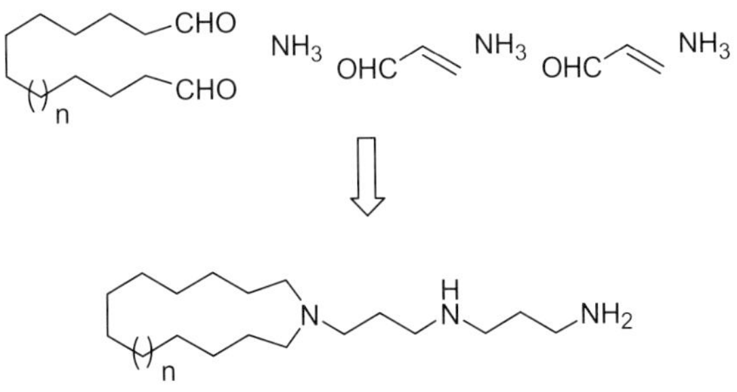

CHO CHO n NH3 OHC NH3 OHC NH3

N H N NH2 n

Scheme 9.2 Proposed biogenesis of motuporamines by Williams *et al.* [48].

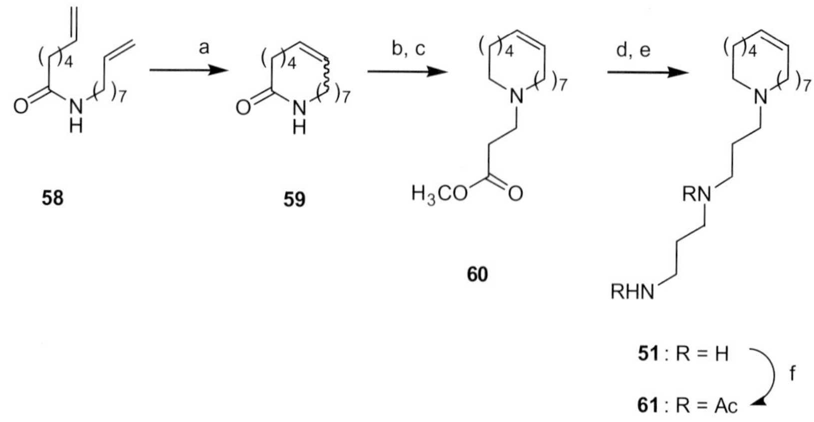

58 → (a) → 59 → (b, c) → 60 → (d, e) → 51

O, N, H, ( )4, ( )7 — 58

O, N, H, ( )4, ( )7 — 59

H_3CO, O, N, ( )4, ( )7 — 60

RN, RHN, N, ( )4, ( )7

51 : R = H

61 : R = Ac (f)

Reagents and Conditions : a) $Cl_2(PCy_3)_2R{=}CHPh$, CH_2Cl_2, 45°C, 4h, *E*:*Z* = 63:37; b) $LiAlH_4$, THF, 70°C, 19h; c) $MeO_2CCH{=}CH_2$, MeOH, RT, 29h; d) $H_2N(CH_2)_3NH_2$, MeOH, RT, 72h; e) $LiAlH_4$, THF, 70°C, 19h; f) Ac_2O, pyridine, RT, 71h.

Scheme 9.3 The synthesis of motuporamine C (**51**) [50].

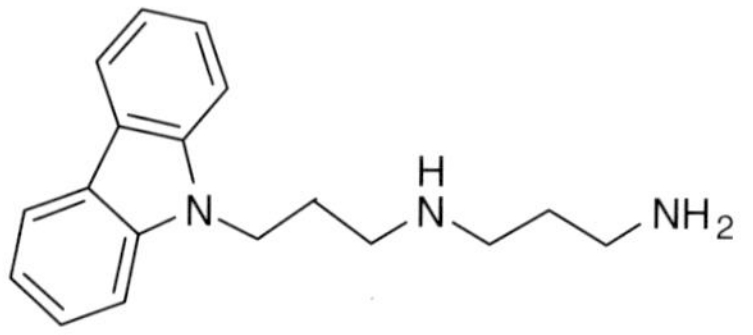

N, H, N, NH_2

62

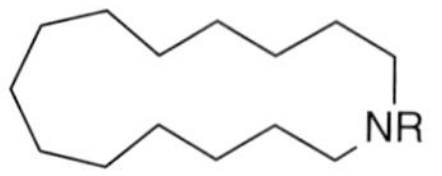

NR

56 : R = $(CH_2)_3NH(CH_2)_3NH_2$
63 : R = $(CH_2)_4NH(CH_2)_4NH_2$

9.5
Pyrrole-Imidazole Alkaloids: "Oroidin"-Related Alkaloids

This class of compounds is extensively described in Chapter 10. Oroidin (**64**) is considered the precursor in the elaboration of polycyclic $C_{11}N_5$ "oroidin" derivatives isolated from sponges belonging to the Axinellidae and Agelasidae families. These include the antiangiogenic compounds agelastatin A (**65**) and ageladine A (**70**).

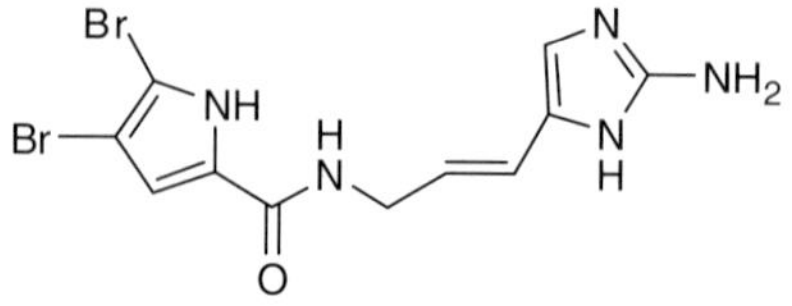

Br, Br, NH, H, N, O, N, N, H, NH_2

64

9.5.1
Agelastatin A

Agelastatin A has been found in extracts of sponges in the order Axinellida, such as the New Caledonian sponge *Agelas dendromorpha* [58] and the Indian Ocean sponge *Cymbastela* sp. [59]. It was first isolated as a cytotoxic agent active below 1 μg/mL toward KB cells [58]. In 1996, agelastatin A was reported to inhibit the proliferation of the human KB nasopharyngeal and murine L1210 tumor cell lines and to prolong the life expectancy of mice with L1210 leukemia [60]. Agelastatin A reduced vascular capillary tube formation *in vitro* [61]. The compound also showed insecticidal activity [59] as well as inhibition of the protein kinase GSK-3β activity with an IC_{50} of 12 μM [62].

The molecular formula of the methylated agelastatin A derivative (**66**) was determined as $C_{15}H_{19}BrN_4O_3$ by analysis of NMR and EIMS data. Interpretation of 1D and 2D NMR data for **66** led to the planar structure, and NOE correlations observed between H-4 and H-18 and between H-7 and H-8 revealed a cis configuration between H-4 and the methoxy group at C-5 and between H-7 and H-8, respectively [58]. The trans configuration between H-4 and H-8 was established by the small coupling constant value ($J < 0.5$ Hz) between those two proton signals. SAR studies suggested that the trans relationship between B and D rings, the C-5 hydroxyl group and the secondary amides at positions 3 and 9 are necessary for biological activity of **65** [60]. The absolute configuration of agelastatin A was established on the basis of molecular modeling studies and CD exciton-coupling measurements of two analogs of agelastatin [63]. These results have subsequently been confirmed by single-crystal X-ray diffraction of the natural product [61].

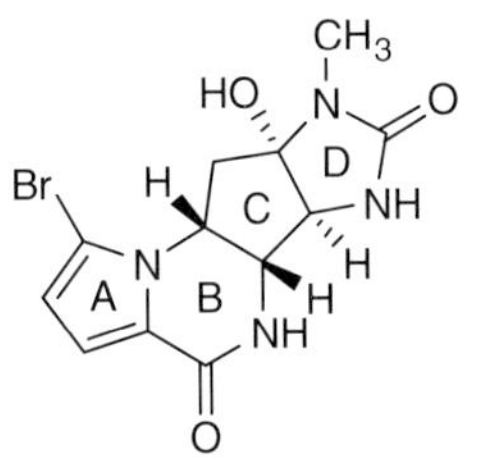

65 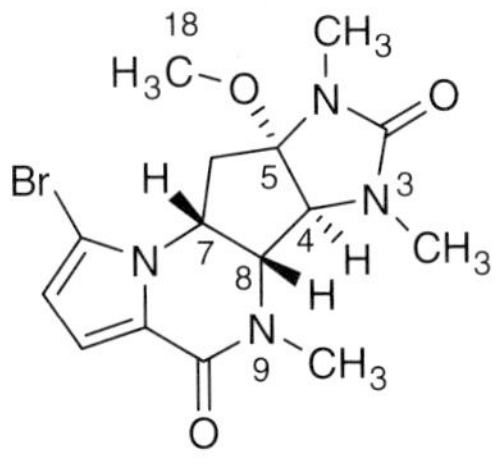 **66**

Kerr and coworkers identified proline/ornithine and histidine as building blocks for the biogenesis of the oroidin analog stevensine, a $C_{11}N_5$ alkaloid isolated from the sponge *Teichaxinella morchella* (Axinellidae), through feeding experiments using radiolabeled amino acids [64]. Based on these results, Al Mourabit and Potier proposed a biogenetic mechanism in which the enzymatic condensation of tautomeric forms of 3-amino-1-(2-aminoimidazolyl)-1-propene (an intermediate derived from histidine) with pyrrole-2-carboxylic acid or its 4- or 5-brominated derivatives (derived from proline) can account for the formation of over sixty marine alkaloids in the oroidin series [65]. Agelastatin A is proposed to be formed by initial oxidation of

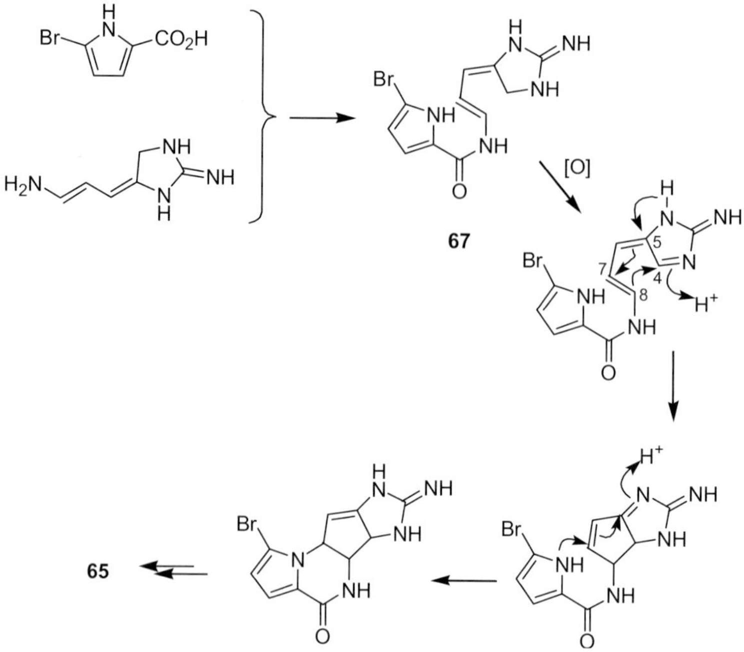

Scheme 9.4 Proposed biogenesis of agelastatin A (**65**) and its derivatives by Al Mourabit and Potier [65].

the corresponding pyrrole-2-carboxylic acid and aminoimidazol-propadiene adduct **67** followed by C-8–C-4 intramolecular cyclization and attack of the pyrrolic nitrogen at C-7 to form the agelastatin carbon skeleton. A sequence of enzymatic transformations will then lead to agelastatin A and its derivatives (Scheme 9.4) [65].

The unusual heterocyclic framework of the agelastatins in conjunction with the potent bioactivity of agelastatin A and the problems associated with isolating sufficient material from natural sources has made **65** an attractive target for total synthesis. The racemic synthesis of agelastatin A was achieved by Weinreb and coworkers in 1999 via an elegant 14 step route giving (±)-**65** in 7 % overall yield [66,67]. Key transformations employed in the construction of the agelastatin tetracyclic core in this case included a *N*-sulfinyl dienophile hetero Diels–Alder reaction to incorporate the five-membered carbocyclic ring-C; two [2,3]-sigmatropic rearrangements to situate functionality in the desired positions for elaboration of the cyclic core and addition of ring-A; an intramolecular Michael addition to form ring-B; and finally the annulation of the D-ring employing the addition of methyl isocyanate.

Asymmetric syntheses of agelastatin A rapidly followed Weinreb's racemic synthesis, with Feldman and coworkers publishing the first stereospecific approach to the natural enantiomer [68,69]. This route employed a chiral alkynyliodonium triflate–alkylidenecarbene–cyclopentene transformation to form the key intermediate **69** from which (−)-**65** could be efficiently assembled. Hale *et al.* initially reported a chiral formal synthesis of (−)-agelastatin A through the enantiospecific preparation of an advanced intermediate from Weinreb's racemic route [70] and later described a total

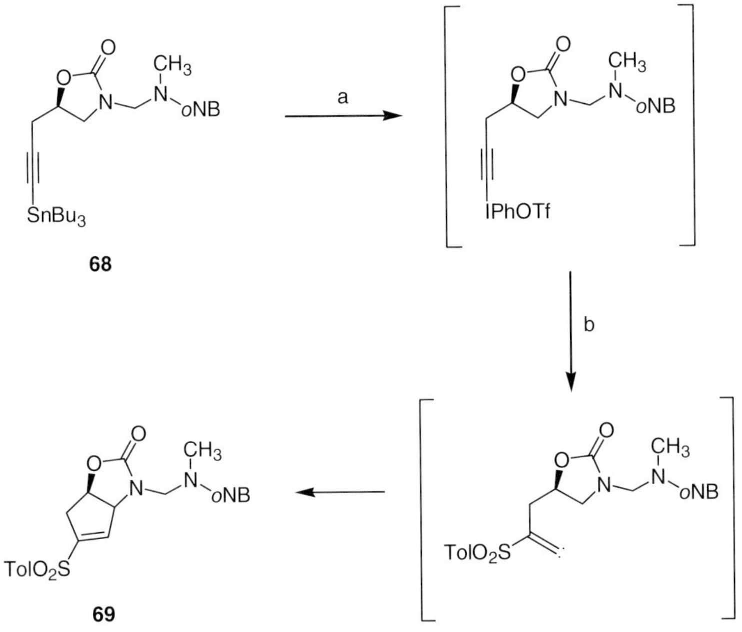

Reagents and Conditions: a) PhI(CN)OTf, CH_2Cl_2, -42 °C, 1h; then b) $TolO_2SNa$, DME, Δ, 5 min

Scheme 9.5 The key step in the synthesis of agelastatin A by Feldman *et al.* [68,69].

synthesis modified to avoid the use of methyl isocyanate [71]. More recently, the natural enantiomer has been prepared using a chiral sulfinimine [72] and (+)-**65** synthesized for the first time via palladium-catalyzed asymmetric allylic alkylation of methyl 5-bromopyrrole-2-carboxylate [73] (Scheme 9.5).

9.5.2
Ageladine A

A hydrophilic extract of the Japanese marine sponge *Agelas nakamurai* [74] showed inhibitory activity against MMP-2, a matrix metalloproteinase present at the surface of endothelial cells that degrades components of the extracellular matrix and facilitates angiogenesis. Bioassay-guided fractionation led to the identification of ageladine A (**70**), composed of a dibromopyrrole and an imidazopyridine moiety. Ageladine A is a broad-spectrum inhibitor of matrix metalloproteinases at low microgram per milliliter concentrations, which also inhibits the migration of BAECs and shows inhibitory activity in an *in vitro* vascular organization model [74].

Structure elucidation of **70** involved analysis of spectroscopic data and chemical degradation experiments. Although the structure elucidation via interpretation of the 2D NMR data for **70** was limited to partial structures, analysis of the 2D NMR data for the methylated derivatives revealed the connectivity between those partial units [74].

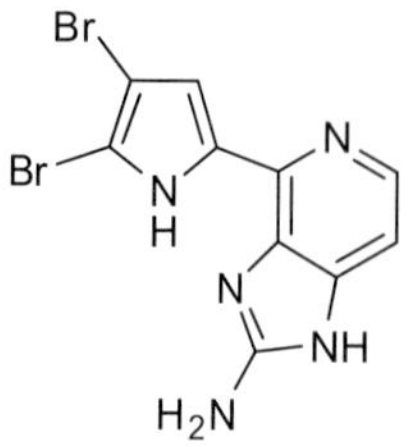

70

The typical 4,5-dibromo-pyrrole and 2-aminoimidazole fragments of the oroidin-derived alkaloids can be identified in the chemical structure of ageladine A. However, the biogenesis of this compound, which contains 10 carbon and 5 nitrogen atoms in its structure, cannot be postulated to occur from the usual $C_{11}N_5$ "oroidin"-building block. The possible biogenesis proposed by Fusetani and coworkers [74] involves a $C_{10}N_5$ precursor, derived from the amino acids proline and histidine, which undergoes intramolecular cyclization and dehydrogenation to afford ageladine A (Scheme 9.6).

Despite their relatively simple structure, the preparation of achiral marine alkaloids via classical synthetic routes and methodology can be extremely challenging. Ageladine A (**70**) provides a good example of such a case, where unpredictably inert intermediates and problematic protecting-group chemistry resulted in a demanding total synthesis finally achieved in 12 steps and 5% overall yield by Meketa and Weinreb [75] (Scheme 9.7). Sequential metallation and introduction of thiomethyl and formamide groups to benzyloxymethyl-protected tribromoimidazole (**71**) allowed the preparation of vinyl imidazole **73** through Wittig olefination and lithiation at the remaining brominated position on the imidazole ring. Conversion of **73** to the key intermediate *O*-methyloxime **74** allowed the formation of the imidazopyridine ring system through a 6π-azaelectrocyclization. A series of functional group interconversions were then necessary before chloropyridine **76** could be subjected to the second key step in the synthesis, Suzuki–Miyaura coupling with *N*-*t*-butoxycarbonyl-2-pyrryl boronic acid. Removal of the *t*-butoxycarbonyl and benzyloxymethyl protecting groups followed by bromination of the pyrrole ring gave ageladine A along with the monobromo- and tribromo-pyrroles **78** and **79**.

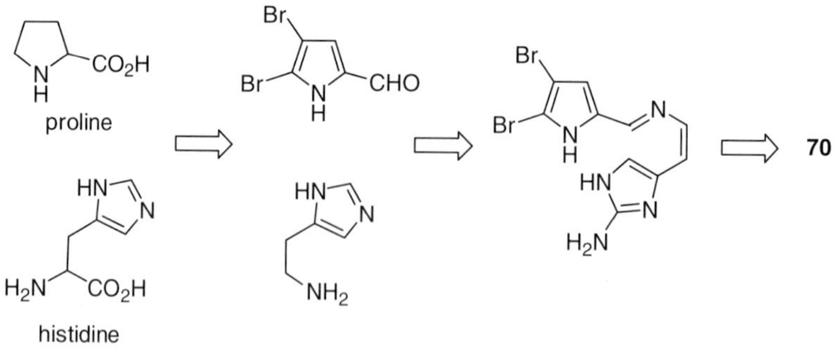

proline

histidine

70

Scheme 9.6 Proposed biogenesis of ageladine A (**70**) by Fusetani and coworkers [74].

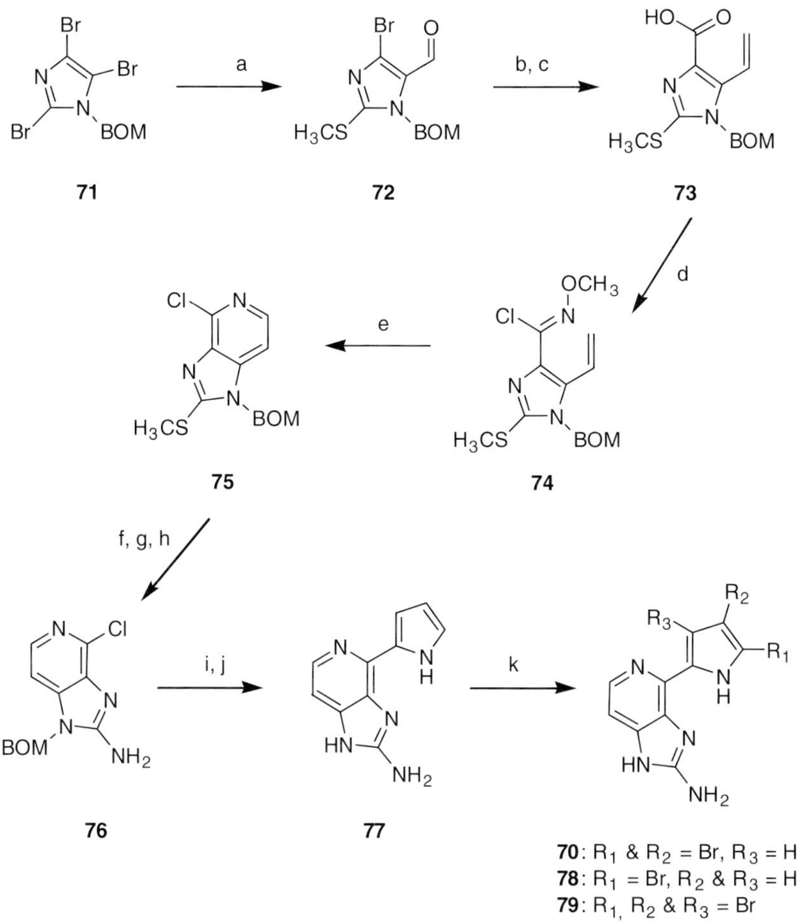
71
72
73
74
75
76
77
a
b, c
d
e
f, g, h
i, j
k
70: R1 & R2 = Br, R3 = H
78: R1 = Br, R2 & R3 = H
79: R1, R2 & R3 = Br

Reagents and Conditions: a) *n*-BuLi, Me_2S_2, THF, –78°C, then *n*-BuLi, DMF, THF, –78°C; b) Ph_3PMeBr, KO*t*-Bu, THF, RT; c) *n*-BuLi, CO_2, THF, RT; d) $MeONH_2$-HCl, CCl_4, PPh_3, pyridine, CH_3CN, RT; e) o-xylene, 150°C; f) Oxone, MeOH, H_2O, RT; g) NaN_3, DMF, RT; h) H_2, 10% Pd/C, EtOH; i) $Pd_2(dba)_3$, *N*-Boc-2-pyrryl boronic acid, biphenyl-PCy_2, K_3PO_4, 1,4-dioxane, RT; j) 6N HCl, EtOH, Δ; k) Br_2, HOAc, MeOH, 0°C.

Scheme 9.7 Meketa and Weinreb's synthesis of ageladine A (**70**) [75].

In stark contrast, a biomimetic approach [76] that was reported shortly after Weinreb's classical synthesis gave access to **70** in three steps (Scheme 9.8) from the commercially available 4,5-dibromo-2-formylpyrrole **80** and 2-aminohistidine **81**. Protection of the pyrrole **80** followed by Lewis acid-catalyzed Pictet–Spengler condensation with **81** gave tetrahydroageladine A **83**. Treatment of **83** with chloranil in refluxing chloroform effected aromatization of the imidazopyridine ring system

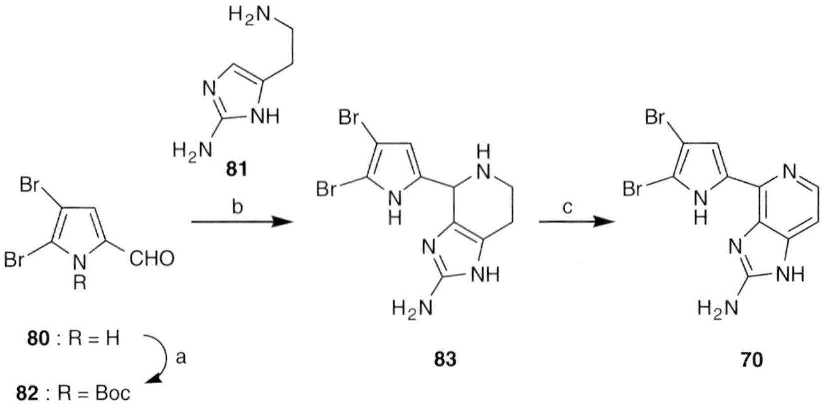

Reagents and Conditions: a) Boc_2O, TBAI, K_2CO_3, DMF; b) $Sc(OTf)_3$, EtOH, rt, 5h; c) chloranil, $CHCl_3$, Δ, 8h.

Scheme 9.8 Shengule and Karuso's biomimetic synthesis of ageladine A [76].

and deprotection of the dibromopyrrole in one pot to give **70** in an overall yield of about 28 %. The brevity of Karuso's biomimetic synthesis makes this approach an attractive and applicable route for supplying sufficient quantities of ageladine A for further biological testing or clinical trials as well as providing access to a range of analogs for SAR studies [76].

9.6 Tyrosine-derived Alkaloids

9.6.1 Aeroplysinin-1

Aeroplysinin-1 (**84**) has been obtained from marine sponges belonging to the order Verongida. The first isolation of (+)-aeroplysinin-1 from *Verongia aerophoba* collected in the Bay of Naples (Italy) was reported by Fattorusso *et al.* in 1970 [77]. Since then, (+)-aeroplysinin-1 has been reported from a number of marine sponges from various geographic locations (see Table 9.2) and the (−)-isomer has been isolated from *Ianthella ardis* [78].

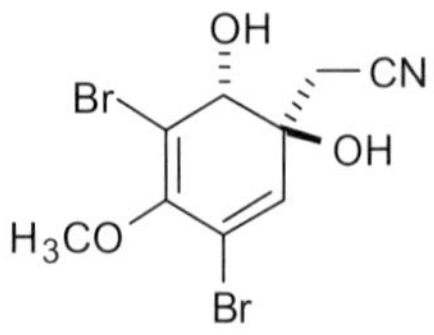

84 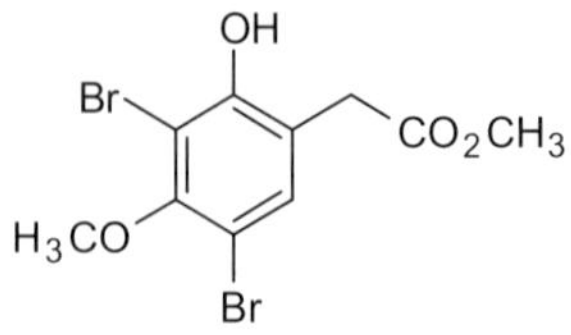 **85**

Tab. 9.2 Marine sponges producing aeroplysinin-1 (**84**).

Taxonomic identification	Collection sites	Enantiomer	References
Verongia aerophoba	Italy	(+)	[77,79]
Verongia aerophoba	Yugoslavia	(+)	[80]
Verongia aerophoba	Canary Islands	nd[a]	[81]
Psammaplysilla purpurea	Marshall Islands	(+)	[82]
Aplysia caissara	Brazil	(+)	[83]
Aplysina laevis	Australia	(+)	[84]
Inthella ardis	Caribbean Sea	(−)	[78,85]

[a]Not determined.

The structure of (+)-aeroplysinin-1 was deduced from analysis of its ultraviolet (UV), infrared (IR), ^{1}H and ^{13}C NMR, and MS data [77]. The proposed structure was confirmed by chemical degradation experiments in which the degraded material was found to be identical with methyl 3,5-dibromo-2-hydroxy-4-methoxyphenylacetate (**85**). Cosulich *et al.* confirmed the structure and relative configuration of the (−)-isomer by X-ray crystallographic analysis [85] while Fulmor *et al.* determined the absolute configuration of (−)-**84** on the basis of circular dichroism [78].

Low micromolar doses of aeroplysinin-1 induced cytotoxicity to various tumor cell lines including HeLa, Ehrlich ascites tumor cells, L5178y mouse lymphoma cells, human mammary carcinoma, and human colon carcinoma cells [80,81,86]. A further study by Rodriguez-Nieto *et al.* showed that aeroplysinin-1 inhibited the growth of BAECs and induced apoptotic cell death [87]. The presence of matrix-metalloproteinase 2 and urokinase in endothelial cell conditioned medium was reduced by aeroplysinin-1. The compound also inhibited endothelial cell migration and capillary tube formation in matrigel. In the same study, *in vivo* inhibition of angiogenesis was demonstrated in the CAM and matrigel plug assays. In a modified CAM assay using quail embryos, Gonzalez-Iriarte *et al.* showed that aeroplysinin-1 induced apoptotic death was preferentially directed toward endothelial cells [88]. Preferential cytotoxicity toward L5178y mouse lymphoma cells compared to murine spleen lymphocytes has also been shown by assaying incorporation rates of ^{3}H-thymidine [80]. *In vivo* studies by the same group demonstrated an antileukemic activity of aeroplysinin-1 in an L5178y cell/NMRI mouse system [80]. These results indicate that aeroplysinin-1 may inhibit both angiogenesis and the growth of tumor cells. Aeroplysinin-1 inhibited the *in vitro* kinase activity of epidermal growth factor receptor (EGFR) and blocked ligand-induced endocytosis of the EGF receptor *in vitro* [89]. Thus, an EGFR-dependent mechanism for activity is indicated, although no significant evidence to further support this mechanism has so far been reported in the literature.

Feeding experiments carried out with the sponge *Aplysina* (formerly *Verongia*) *fistularis* have indicated that labeled [*U*-^{14}C]-L-tyrosine, [*U*-^{14}C]-L-3-bromotyrosine, and [*U*-^{14}C]-L-3,5-dibromotyrosine were incorporated into aeroplysinin-1 [90]. [*Methyl*-^{14}C]methionine was found to be incorporated specifically into the *O*-methyl group [90]. Various ecological studies of *Aplysina* sponges [81,91–94] suggest that aeroplysinin-1 could be biosynthesized by the sponge as a chemical defense mechanism. These studies conclude that sponge tissue damage induces the enzymatically

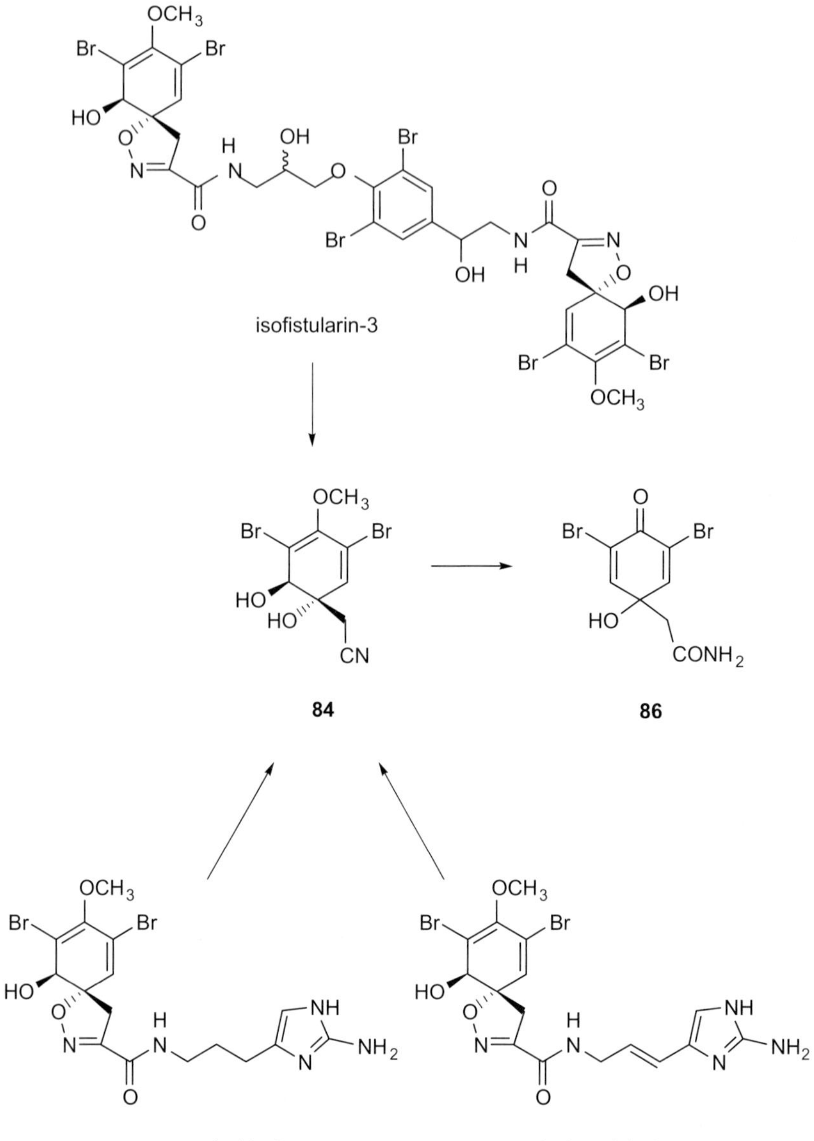

isofistularin-3

84

86

aerophobin-2

aplysinamisin-1

Scheme 9.9 Enzymatic transformation of isoxazoline alkaloids in *Aplysina* sponges into aeroplysinin-1 (**84**) and a dienone analog (**86**).

catalyzed transformation of more complex brominated isoxazoline alkaloids into aeroplysinin-1 and a dienone analog (**86**; see Scheme 9.9). The substitution pattern in the cyclohexadiene moieties of aeroplysinin-1 and the isoxazoline alkaloids supports the proposal that arene oxides are likely biogenetic precursors to this group of metabolites [95] (±)-Aeroplysinin-1 was first synthesized by Andersen and Faulkner

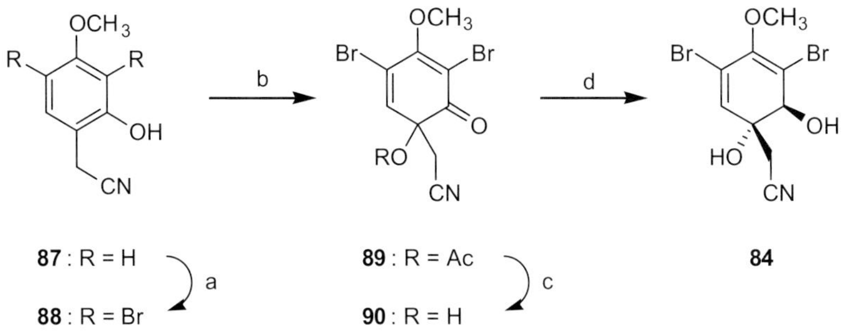

87 : R = H
88 : R = Br
89 : R = Ac
90 : R = H
84

Reagents and Conditions: a) Pyridinium tribromide (2 eq), pyridine; b) $Pb(O_4,Ac)$ HOAc, RT, 18h; c) TsOH, MeOH; d) $NaBH_4$, EtOH, 0°C, 10 min.

Scheme 9.10 Andersen and Faulkner's racemic synthesis of aeroplysinin-1 (**84**) [96].

[9] in 9 % overall yield and five steps from 2-hydroxy-4-methyoxyphenylacetonitrile (**87**) via the stereospecific oxygen-directed borohydride reduction of the α′-hydroxy α, β-unsaturated ketone **90** (Scheme 9.10). This provided a novel route to the synthesis of arene glycols that allows the preparation of both *trans*- and, when the acetylated intermediate **89** is reduced directly, *cis*- glycols.

An improved strategy employing the formation and subsequent ring opening of the spiroisoxazoline intermediate **91** has been published [97] (Scheme 9.11). Anodic two-electron oxidation of oxime **92** allowed the formation of the spiroisoxazoline moiety, which was reduced with zinc borohydride to give the isomeric alcohols **93** and **94**. Silylation and saponification of **93** followed by ring opening via thermal decarboxylation gave the protected alcohol **97**. Cleavage of the silyl ether gave (±)-aeroplysinin-1 in six steps and 22 % yield. The enantiospecific synthesis of either (+)- or (−)-aeroplysinin-1 is yet to be accomplished.

A series of aeroplysinin-1 analogs have been synthesized and evaluated as receptor tyrosine kinase inhibitors [98]. The natural product had shown inhibitory activity against epidermal growth factor receptor tyrosine kinase in isolated enzyme assays, but was inactive when tested in a whole-cell assay system [89]. Analogs were designed in an attempt to enhance membrane permeability in cell-based assays without loss of potential for nucleophilic covalent binding to the enzyme-binding site. Although four of the synthetic analogs (**98**–**101**) exhibited promising inhibitory activity against epidermal growth factor and platelet-derived growth factor receptor tyrosine kinases in cell-based assays, none of the analogs have been evaluated for antiangiogenic activity.

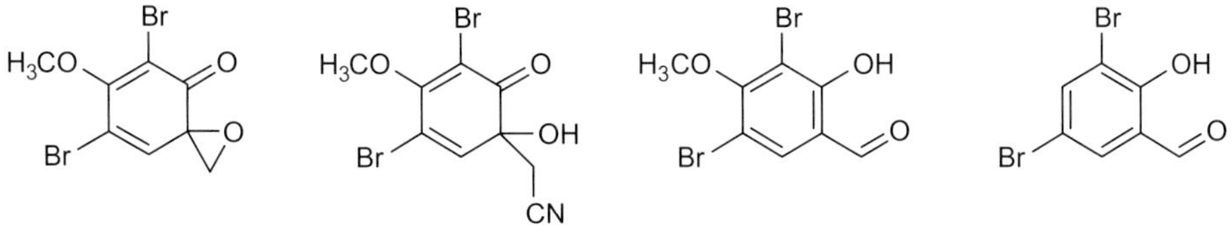

98 **99** **100** **101**

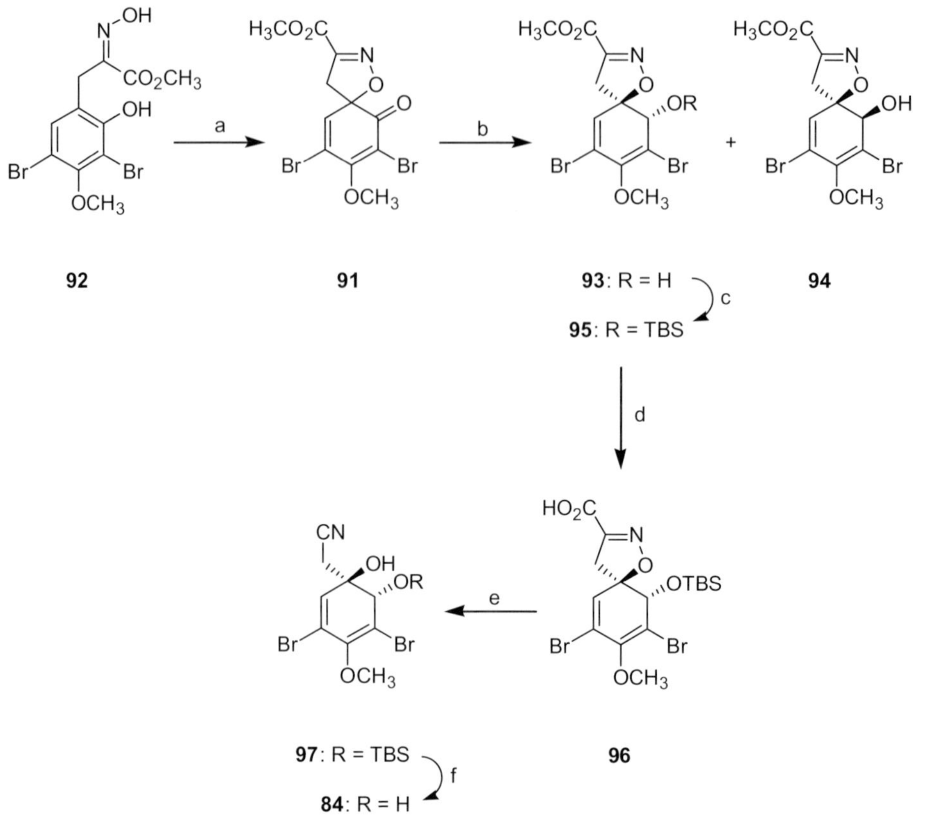

Reagents and Conditions : a) Constant potential electrolysis, $n\mathrm{Bu_4NClO_4}$, CH_3CN, Pt wire (cathode)-glassy carbon beaker (anode); b) $Zn(BH)_4$, CH_2Cl_2, RT, 10 min; c) TBSOTf, 2,6-lutidine, CH_2Cl_2, 0° C, overnight; d) NaOH, aq MeOH, RT, 30 min; e) DMF, 60 ^{o}C, 2h; f) TBAF, THF, 0 ^{o}C, 1h.

Scheme 9.11 Ogamino and Nishiyama's racemic synthesis of aeroplysinin-1 (**84**) [97].

9.6.2 Psammaplin A

Psammaplin A was simultaneously reported by Rodríguez *et al.* [99] and Quiñoá and Crews [100] from the sponges *Thorectopsamma xana* and *Psammaplysilla purpurea*, belonging to the order Verongida, and later by Arabshahi and Schmitz from an unidentified sponge [101]. Psammaplin A has also been isolated from other *Psammaplysilla purpurea* samples [102], *Aplysinella rhax* [103–105], and *Pseudoceratina purpurea* [106]. The two non-Verongid sponges *Jaspis wondoensis* and *Poecillastra wondoensis* were also reported to contain psammaplin A [107]. All known sponge sources for psammaplin A are summarized in Table 9.3.

Tab. 9.3 The sponge sources for psammaplin A.

Taxonomic identification	Collection sites	References
Unidentified	Guam	[101]
Thorectopsamma xana	Guam	[99]
Psammaplysilla purpurea	Tonga	[100]
Psammaplysilla purpurea	Tonga	[102]
Aplysinella rhax	Australia	[103]
Aplysinella rhax	Pohnpei	[104]
Aplysinella rhax	Fiji	[105]
Jaspis wondoensis	Korea	[107]
Poecillastra wondoensis	Korea	[107]
Pseudoceratina purpurea	Papua New Guinea	[106]

Psammaplin A is a symmetrical brominated tyrosine metabolite containing a disulfide linkage that exists as an interconverting mixture of isomers, **102** and **103**, that differ in the configuration of their oxime groups. The (*E,E*)-isomer (**102**) was obtained in greater abundance than the (*E,Z*)-isomer (**103**); however, it was observed that the (*E,Z*)-isomer was transformed into the (*E,E*)-isomer in solution. Thus, it is likely that the (*E,Z*)-isomer is the natural metabolite and that isomerization may have occurred during the isolation procedure. The symmetrical structure of the (*E,E*)-psammaplin A isomer (**102**) was deduced from the evidence that the ^{13}C NMR spectrum showed only eleven signals despite FABMS evidence for a molecular formula of $C_{26}H_{32}Br_2N_4O_6S_2$. Analysis of the NMR data revealed the structure of the monomeric unit, which was confirmed by data obtained for the triacetate (**104**) [101].

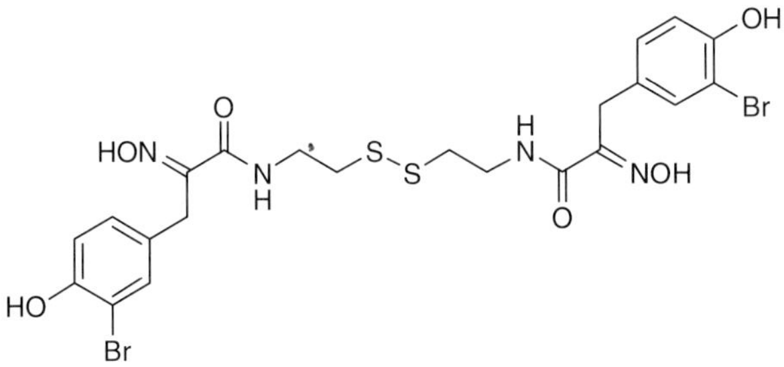

102: *(E,E)*-isomer
103: *(E,Z)*-isomer

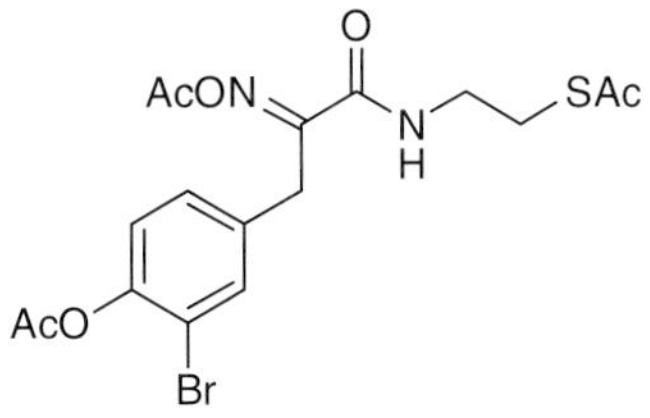

104

The isolation of psammaplin derivatives from the sponge *Psammaplysilla purpurea*, in particular the only nonhalogenated analog, prepsammaplin A (**105**), allowed Jiménez and Crews to propose a biogenesis for this family of amino acid derivatives in which psammaplin A is formed by the linear combination of two bromotyrosine oxime derivatives and two rearranged cysteines [102].

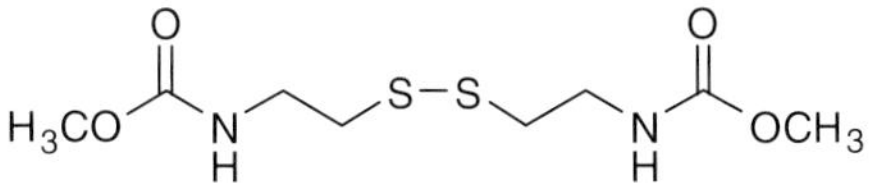

105

The cytotoxicity of psammaplin A toward a wide range of human cancer cells has been well documented [104,107–110]. The compound also inhibits the proliferation of BAECs [110]. Psammaplin A has been reported to affect several biochemical activities, including inhibition of topoisomerase II [108], inhibition of DNA synthesis and DNA gyrase [111], inhibition of farnesyl protein transferase and leucine aminopeptidase [104], inhibition of chitinase B from *Serratia mercescens* [105], inhibition of SV40 DNA replication *in vitro* through targeting of α-primase [109] and activation of peroxisome proliferator-activated receptor gamma [112]. Psammaplin A also inhibits DNA methyltransferase activity *in vitro* [113], although the lack of *in vivo* inhibition of DNA methylation suggests that this enzyme is not a major cellular target. In addition, psammaplin A also causes A549 cells to arrest cell cycle progression at the G_2/M phase and shows weak antioxidant activity [114]. The compound also inhibits aminopeptidase N (APN), a Zn-dependent metalloproteinase that has been implicated in tumor invasion and angiogenesis [110]. Psammaplin A suppressed tumor invasion and bFGF-induced endothelial cell tube formation in tissue culture models [110].

Psammaplin A has been efficiently synthesized in four steps and 23 % overall yield by Hoshino *et al.* through conversion of 3-bromotyrosine to the corresponding α-hydroxyimino acid followed by dimerization through coupling with cystamine [115]. A similar three-step synthesis from 4-hydroxyphenylpyruvic acid that yielded **102** in 43% yield has also been reported by Godert *et al.* [113]. Nicolaou *et al.* modified Hoshino's synthesis to gain access to **104** and homodimeric analogs that were used to construct a psammaplin A inspired combinatorial library of disulfide heterodimers [116]. Both the homodimers and heterodimers were screened for activity against methicillin-resistant *Staphylococcus aureus* and structure refinement of promising lead compounds through parallel synthesis afforded a series of antibacterial agents significantly more active than the natural product [117]. Heterodimer **106** in particular exhibited *in vitro* activities similar to several antibiotics currently in clinical use. However, no antiangiogenic assays were run on the synthetic analogs.

9.6.3 Bastadins

Bastadins have been isolated from the Indo-Pacific Verongid sponges *Ianthella basta* [118–129], *Ianthella quadrangulata* [130], *Ianthella* sp. [131] and *Psammaplysilla purpurea* [132,133] as well as the Dendroceratid sponge *Dendrilla cactos* [134]. To date, a total of 23 bastadin analogs have been reported [118–127,130–134], all of which are heterodimers derived biogenetically from the oxidative coupling of two brominated tyrosine–tyramine amides. The bastadins are structurally classified into three groups depending upon the degree and position of the phenolic couplings linking the monomeric, or hemibastadin [122,135], units. The linear (or "acyclic") bastadins contain a single ether or biaryl linkage (for example, bastadin 1, **112**, and bastadin 3, **113**), while the macrocyclic members of the series possess either the bastarane skeleton (for example, bastadin 6, **114**), in which the hemibastadin units are linked by phenolic ethers from C-10 to C-14 and from C-29 to C-33, or the isobastarane skeleton (for example, bastadin 13, **115**), in which ethers link C-9 to

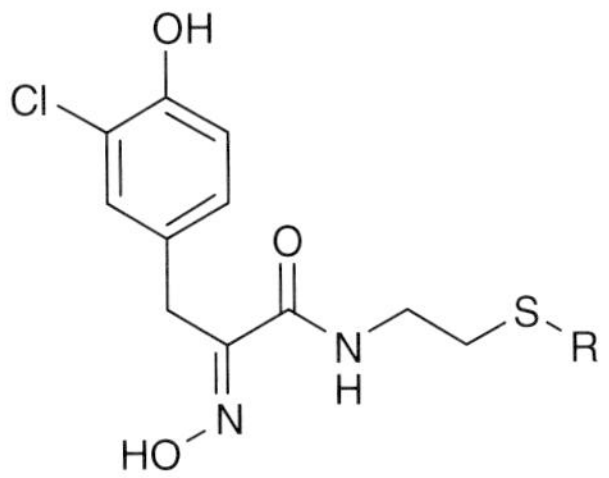

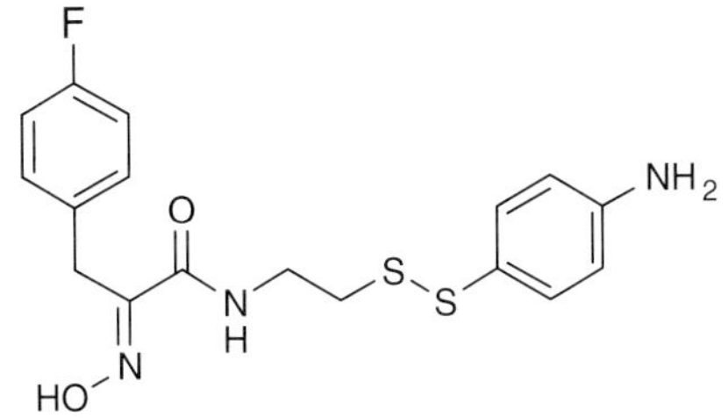

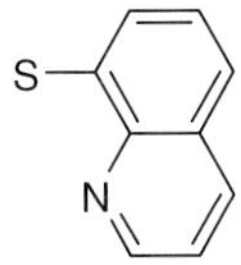

C-14 and C-29 to C-33. The structures of the bastadins have in most cases been elucidated through the application of standard spectroscopic techniques, although those of bastadin 4 and the tetramethyl ether of bastadin 5 have been established through X-ray crystallography [119].

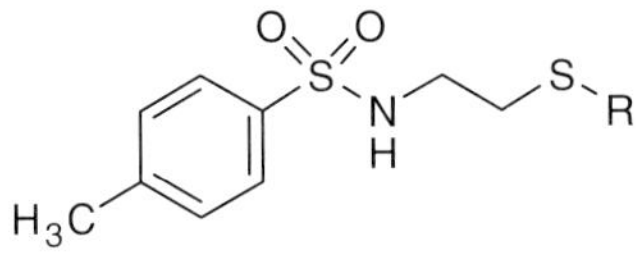

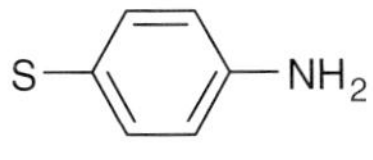

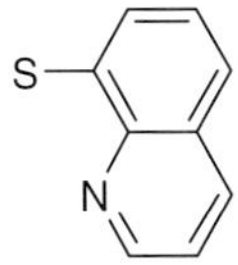

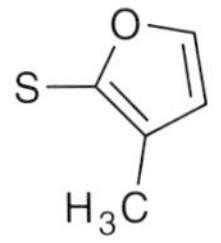

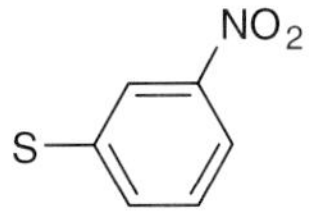

The absolute stereochemistry of the 6-hydroxybastadins 8, 10, and 12 has been found to be 6*S* through application of the Mosher–Trost method [136]. The first members of the bastadin series were isolated from the methanol extract of an Australian sample of *Ianthella basta* because of its potent *in vitro* antimicrobial activity against gram-positive bacteria [118,119]. Since then, bastadins have been shown to exhibit a variety of biological activities including cytotoxicity [120,121,132], anti-inflammatory activity [120], inhibition of the enzymes topoisomerase II, dehydrofolate reductase, inosine 5′-phosphate dehydrogenase, 12-human lipoxygenase, and 15-human lipoxygenase [123,132,137], agonistic activity toward the sarcoplastic reticulum Ca^{2+} channel through modulation of the Ry_1R FKBP12 receptor complex [127,138,139], and antiproliferative effects against a number of cancer cell lines [134,136]. The antiangiogenic activity of the bastadins, however, has only been reported by Kobayashi *et al.* [128,129].

Bioassay guided fractionation of the methanol extract of an Indonesian sample of *I. basta* resulted in the isolation of a total of eight bastadins representing each of the three structural motifs found in the series [128,129]. Bastadin 6 (**114**) was isolated as the major bioactive constituent of the sponge and inhibited the VEGF- and

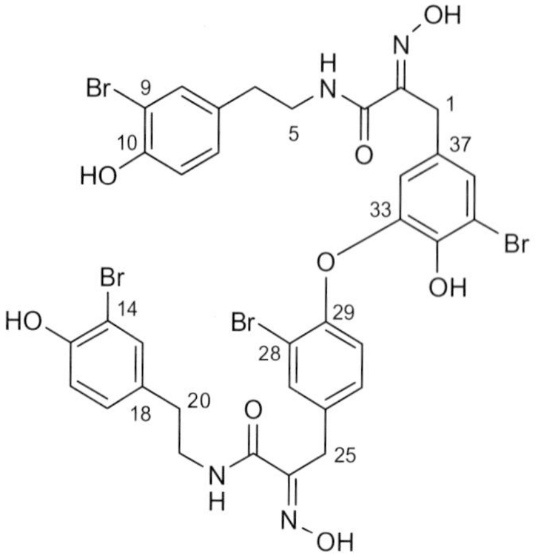

112

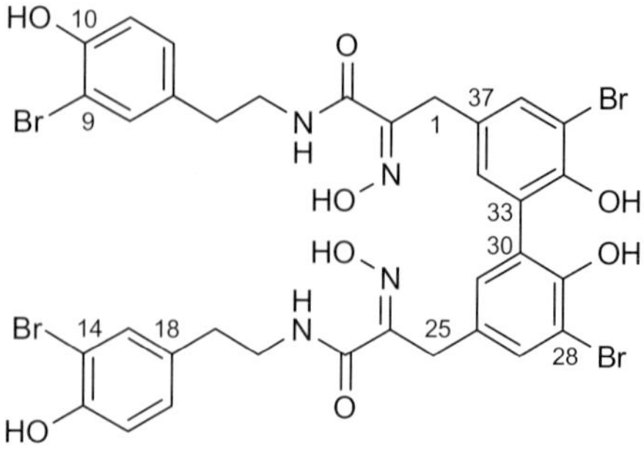

113

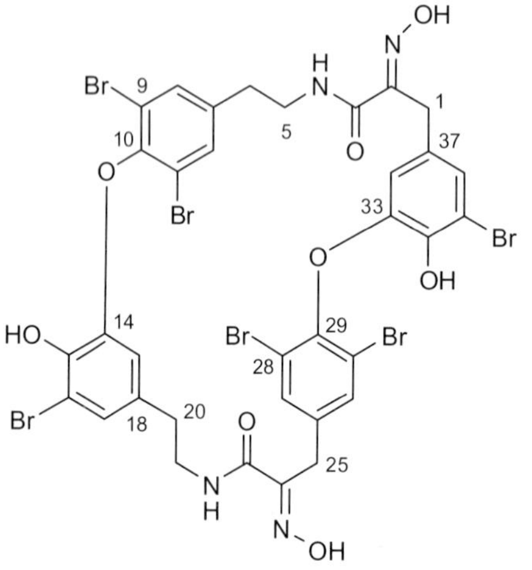

114

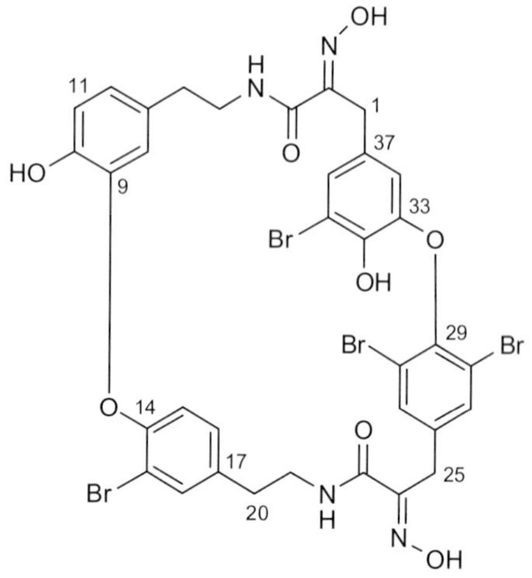

115

bFGF-dependent proliferation of HUVECs at sub-micromolar concentrations. The activity of bastadin 6 was found to be selective toward HUVECs when compared with normal fibroblasts (3Y1) and several tumor cell lines (KB3-1, K562, and Neuro2A) [129]. Bastadin 6 also inhibited the VEGF- or bFGF-induced capillary tube formation and migration of HUVECs *in vitro* and blocked VEGF- or bFGF-induced neovascularization *in vivo* in the mouse corneal assay. When assessed in the nude mouse A431 solid tumor xenograft model, bastadin 6 exhibited significant inhibitory activity against tumor growth without displaying acute toxicity [129]. A SAR study of the antiproliferative activity of the eight *I. basta* alkaloids against HUVECs indicated that a macrocyclic structure is critical for antiangiogenic activity [128]. Bastadins possessing the bastarane skeleton exhibited greater selectivity against endothelial cell proliferation than did those with the isobastarane skeleton, and a molecular-mechanics-based conformational analysis of the bastadins suggested that this was linked to the greater rigidity of the bastarane macrocycle [128].

Bastadin 6 was first synthesized by Nishiyama *et al.* in 1982 via a biomimetic strategy that employed thallium trinitrate to achieve the key oxidative macrocyclization step [140–142]. More recently, a chemoenzymatic approach has been reported in which the oxidative phenolic coupling reactions were accomplished using horseradish and soybean peroxidases [143]. In order to continue with their mechanistic and SAR studies of **114** and facilitate the development of a new antiangiogenic drug candidate, Kobayashi and coworkers have developed an efficient convergent synthesis that gave bastadin 6 in an overall yield of 26 % with a longest linear sequence of nine steps [144]. Kobayashi's synthesis was made possible through the development of a novel Ce(IV)-mediated oxidative coupling reaction that allowed the preparation of the key diaryl ether fragments **116** and **117** in reasonable yield.

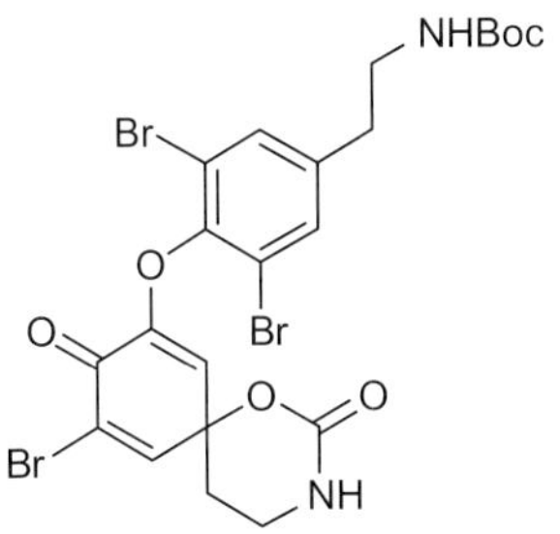

116

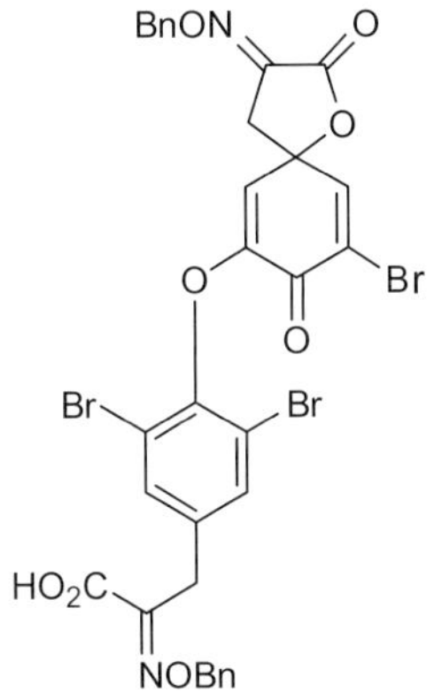

117

Reduction of the cyclic carbamate **116** to the primary amine **118** allowed the two key intermediates to be coupled using 1-(3-dimethylaminopropyl)-1-ethylcarbodiimide hydrochloride in the presence of 1-hydroxybenzotriazole hydrate (Scheme 9.12). The resulting amide **119** was then prepared for the macrocyclizaton step through reduction of the spirodienone moiety to yield acid **120** and removal of the Boc protecting group to give amine **121** as the HCl salt. Formation of the bastarane ring system was accomplished under similar condensation conditions as those employed previously in the synthesis and bastadin 6 was obtained in good yield following selective deprotection of the two oxime groups using boron trichloride–dimethyl sulfide.

9.7 Tryptophan-derived Alkaloids

Scytonemin (**123**) is a yellow-brown pigment of cyanobacterial sheaths. While a plethora of cyanobacterial species with yellow-brown sheaths are described in the

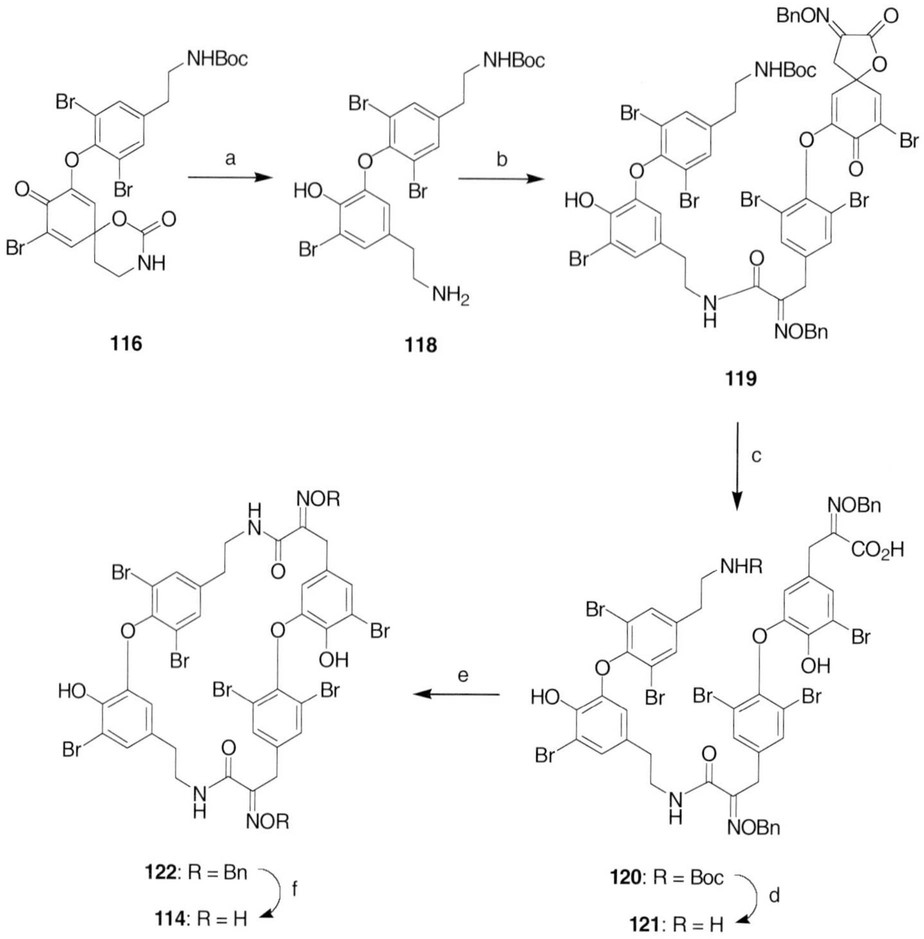

Reagents and Conditions: a) $Na_2S_2O_4$, THF/H_2O, RT, 10min; b) **117**, EDCI, HOBt, THF, 0°C, 6h; c) $Na_2S_2O_4$, CH_3CN/H_2O, RT, 30min; d) TFA, CH_2Cl_2, RT, 10min then HCl/Et_2O; e) EDCI, HOBt, THF, RT, 3h; f) BCl_3-SMe_2, CH_2Cl_2, RT, 3h.

Scheme 9.12 Kobayashi and coworkers' total synthesis of bastadin 6 [144].

literature, scytonemin has been detected in only about 30 species [145]. In order to elucidate its chemical structure, scytonemin was isolated from the sheaths of several cyanobacteria such as *Stigonema* sp., *Scytonema* sp., and *Lyngbya* sp. [146]. The molecular formula of scytonemin (**123**) was determined to be $C_{36}H_{22}N_2O_4$ from FABMS data and a UV absorption band at 370 nm suggested extended conjugation. The ^{13}C NMR spectrum showed only 16 signals, indicating that **123** had a symmetrical structure. Analysis of 1D and 2D NMR data for reduced

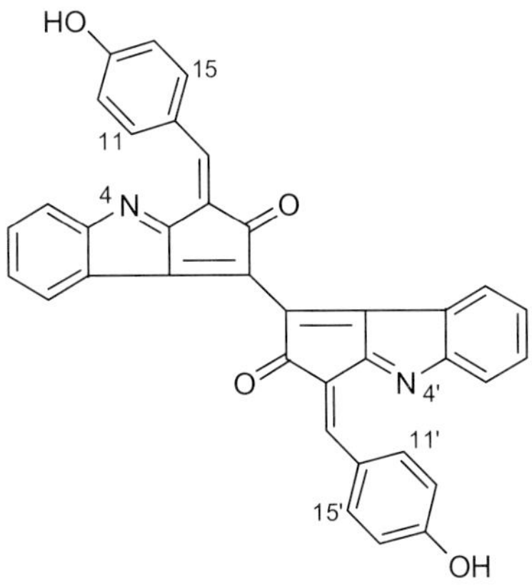

123 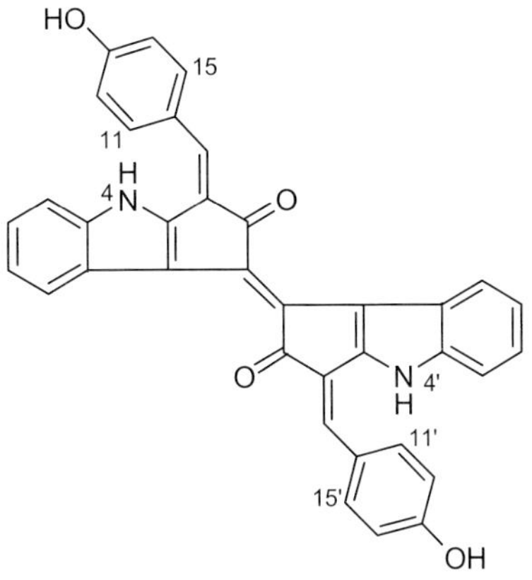 **124**

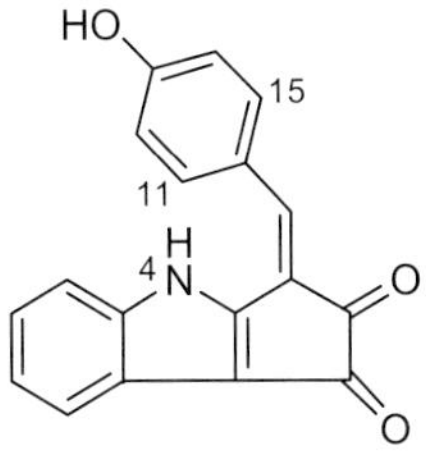

125

scytonemin (**124**) identified the monomeric unit of the molecule. Ozonolysis of **124** afforded **125**, indicating that the half units are connected to each other via a double bond. The NOE correlations between 4-NH and H-11 (H-15) in **125** revealed an *E*-configuration of the trisubstituted olefin [146].

According to Proteau *et al.*, condensation of two subunits of tryptophan and two phenylpropanoid-derived units would explain the biogenetic pathway to form the symmetrical structure of scytonemin (**123**) [146]. No syntheses of **123** have been reported.

Scytonemin inhibited the proliferation of HUVECs in the low micromolar concentration range [147]. Scytonemin was the first identified small-molecule inhibitor of polo-like kinase 1 (PLK1). The compound also inhibited the cell cycle regulatory kinases Myt1, checkpoint kinase 1, cyclin-dependent kinase 1/cyclin B and protein kinase Cβ2 at similar low micromolar concentrations, indicating that it is a relatively broad specificity protein kinase inhibitor [148]. Inhibition of HUVEC proliferation is expected to result in antiangiogenic activity. Use of scytonemin as an antiangiogenic agent has been proposed in the patent literature but we are unaware of scientific reports of *in vitro* or *in vivo* antiangiogenic activity for this compound.

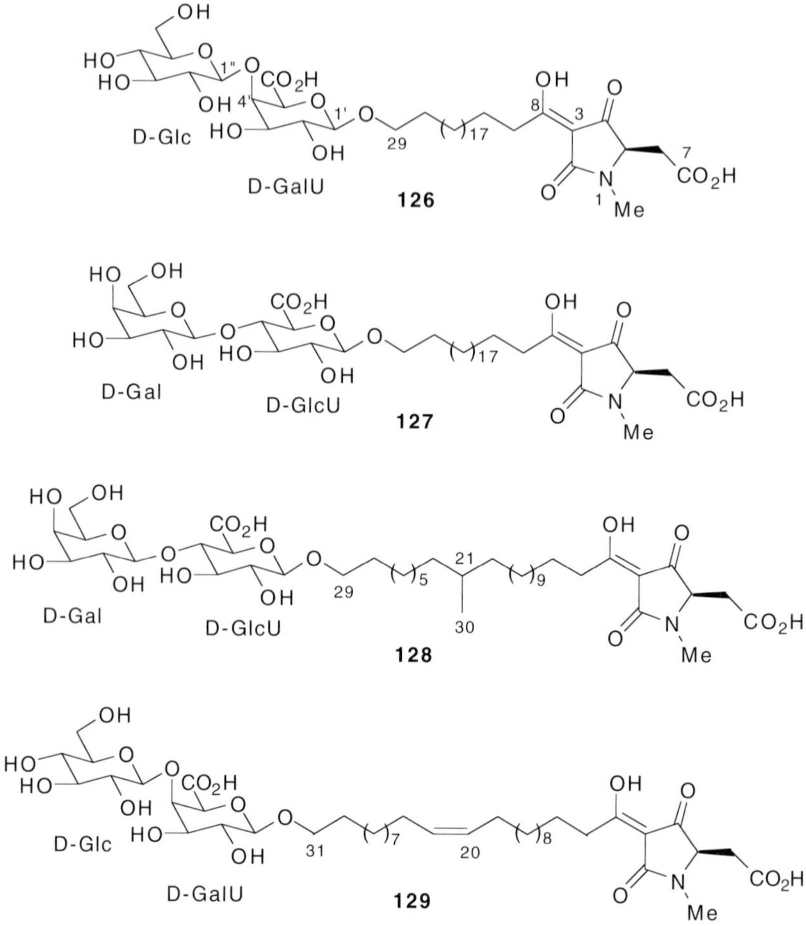

Fig. 9.2 Ancorinosides A–D.

9.8
Ancorinosides

Ancorinosides A–D are tetramic acid glycosides isolated from specimens of the Japanese sponges *Ancorina* sp. [149,150] and *Penares sollasi* [151]. Ancorinoside A (**126**) was originally isolated due to its ability to inhibit blastulation of starfish embryos both as the free acid [149] and as a magnesium salt [150]; however, Fusetani and coworkers subsequently isolated **126** and three new analogs, ancorinosides B–D (**127–129**), as inhibitors of membrane type 1 matrix metalloproteinase (MT1-MMP) [151]. MT1-MMP is one of the enzymes responsible for the conversion of progelatinase A into MMP-2, a metalloprotein particularly implicated in tumor progression and angiogenesis. The planar structures of the ancorinosides were elucidated from detailed analysis of 1D and 2D NMR data in conjunction with HRFABMS data for ancorinoside A [149] and HRFABMS supplemented with FAB-MS/MS data for ancorinosides B–D [151]. All of the ancorinosides consist of a tetramic acid unit linked through a long-chain fatty alcohol to a disaccharide moiety derived from glucose and galactose (Figure 9.2). Structural differences between analogs are introduced by virtue of the constitution of the disaccharide unit and the length, functionality, and branching of the fatty alcohol. The absolute stereochemistries of anchorinosides A–D have been determined through chemical degradation and derivatization, and it was found that in this series of natural products all the sugar residues possess the D-configuration and the tetramic acid moiety is derived from D-aspartic acid [149,151].

Ancorinosides A–D inhibited MT1-MMP, with median inhibitory concentrations of 440, 500, 370, and 180 μg/mL respectively, and are an order of magnitude less potent than the known synthetic MMP inhibitor FN-439 (**130**)[151,152]. A limited SAR study of the natural products and the aglycon of ancorinoside B indicated that it is the tetramic acid moiety that is responsible for the inhibitory activity of the ancorinosides. Indeed, tenuazonic acid (**131**), a natural product isolated from the fungus *Alternaria tenuis* [153] that has the tetramic acid unit as its only functionality, exhibited inhibitory activity against MT1-MMP and MMP2 comparable to FN-439 [151].

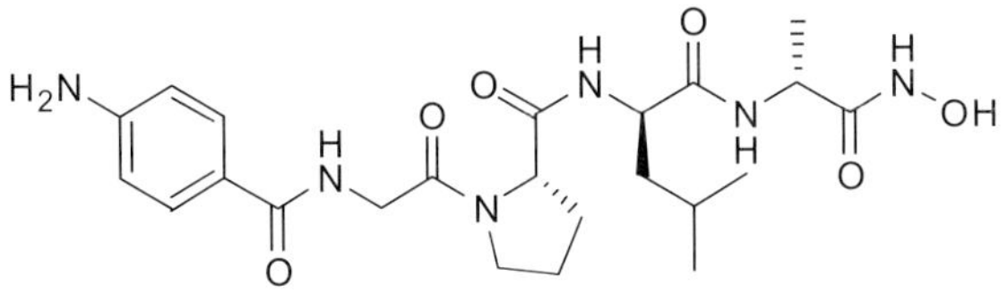

H_2N H N O O N O H N O N H O H N OH HN O OH O

130 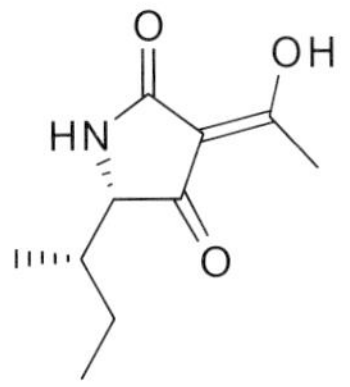 **131**

9.9 Concluding Remarks

Marine natural products provide a rich diversity of unprecedented chemical structures that can act as lead compounds for drug development and/or tools for basic cell biology investigations of diseases. Screening marine natural product crude extracts and pure compounds using sophisticated assay systems has resulted in the identification of a number of marine alkaloids with antiangiogenic activity that possess genuine potential to become or inspire future therapeutic agents. Squalamine is currently being developed as an antiangiogenic drug under license to Genaera and has progressed to phase II clinical trials as a combination therapy with standard agents for nonresponding solid tumors and as a primary treatment for advanced ovarian cancer [154]. It is also present in AE-941 (Neovastat), a standardized shark cartilage extract being developed by Æterna Zentaris that has been in phase III trials for nonsmall-cell lung cancer as well as renal carcinoma [154,155]. The motuporamines (**49–55**) have demonstrated their usefulness as biological tools and given significant insights into important aspects of the angiogenic process at the cellular level [52]. Their drug potential is still undergoing preclinical evaluation. The bastadins, particularly bastadin 6 (**114**), promise to be an important suite of compounds in the development of natural-product-based antiangiogenic drug candidates, and scytonemin (**123**), the first small-molecule inhibitor of PLK1, represents a new pharmacophore that may be of enormous value both therapeutically and as a biological tool [148]. We have only begun to evaluate the drug potential of the natural products diversity present in the world's oceans and we can expect that many marine natural products possessing useful antiangiogenic and antimetastatic activities will be discovered in the future.

References

1 Abdollahi, A., Hlatky, L., Huber, P. E. (2005) *Drug Resistance Update*, **8**, 59– 74.

2 Folkman, J. (1971) *The New England Journal of Medicine*, **285**, 1182–1186.

3 Carmeliet, P. and Jain, R. K. (2000) *Nature*, **407**, 249–257.

4 Boehm, T., Folkman, J., Browder, T., O'Reilly, M. S. (1997) *Nature*, **390**, 404–407.

5 Hurwitz, H., Fehrenbacher, L., Novotny, W., Cartwright, T., Hainsworth, J., Heim, W., Berlin, J., Baron, A., Griffing, S., Holmgren, E., Ferrara, N., Fyfe, G., Rogers, B., Ross, R., Kabbinavar, F. (2004) *The New England Journal of Medicine*, **350**, 2335–2342.

6 Mitchell, S. S., Whitehill, A. B., Trapido-Rosenthal, H. G., Ireland, C. M. (1997) *Journal of Natural Products*, **60**, 727–728.

7 Chehade, C. C., Dias, R. L. A., Berlinck, R. G. S., Ferreira, A. G., Costa, L. V., Rangel, M., Malpezzi, E. L. A., de Freitas, J. C., Hajdu, E. (1997) *Journal of Natural Products*, **60**, 729–731.

8 Jeong, S. -J., Inagaki, M., Higuchi, R., Miyamoto, T., Ono, M., Kuwano, M., Van Soest, R. W. M. (2003) *Chemical and Pharmaceutical Bulletin*, **51**, 731–733.

9 Lindsay, B. S., Battershill, C. N., Copp, B. R. (1998) *Journal of Natural Products*, **61**, 857–858.

10 Gambardella, M. T. D. P., Dias, R. L. A., Chehade, C. C., Berlinck, R. G. D. S. (1999) *Acta Crystallographica Section C*, **55**, 1585–1587.

11 Panthong, K., Garson, M. J., Bernhardt, P. V. (2006) *Acta Crystallographica Section C*, **62**, 193–195.

12 Diaz-Marrero, A. R., Austin, P., Van Soest, R., Matainaho, T., Roskelley, C. D., Roberge, M., Andersen, R. J. (2006) *Organic Letters*, **8**, 3749–3752.

13 Aoki, S., Watanabe, Y., Sanagawa, M., Setiawan, A., Kotoku, N., Kobayashi, M. (2006) *Journal of American Chemical Society*, **128**, 3148–3149.

14 Dasgupta, P. (2004) *Pharmacology and Therapeutics*, **102**, 61–85.

15 Moore, K. S., Wehrli, S., Roder, H., Rogers, M., Forrest, J. N., McCrimmon, D., Zasloff, M. (1993) *Proceedings of the National Academy of Sciences of the United States of America*, **90**, 1354–1358.

16 Wehrli, S. L., Moore, K. S., Roder, H., Durell, S., Zasloff, M. (1993) *Steroids*, **58**, 370–378.

17 Sills, A. K. Jr.,Williams, J. I., Tyler, B. M., Epstein, D. S., Sipos, E. P., Davis, J. D., McLane, M. P., Pitchford, S., Cheshire, K., Gannon, F. H., Kinney, W. A., Chao, T. L., Donowitz, M., Laterra, J., Zasloff, M., Brem, H. (1998) *Cancer Research*, **58**, 2784–2792.

18 Pietras, R. J., Gorrin-Rivas, M., Chen, H. -W. (2003) *International Journal of*

Gynecology and Obstetrics, **83** (suppl 3), 59–60.

19 Pietras, R. J. and Weinberg, O. K. (2005) *Evidence-Based Complementary and Alternative Medicine*, **2**, 49–57.

20 Williams, J. I., Weitman, S., Gonzalez, C. M., Jundt, C. H., Marty, J., Stringer, S. D., Holroyd, K. J., McLane, M. P., Chen, Q., Zasloff, M., Von Hoff, D. D. (2001) *Clinical Cancer Research*, **7**, 724–733.

21 Chen, Q., William, J. I., Anderson, M., Kinney, W. K., Zasloff, M. (1999) *Clinical Cancer Research*, **5** (Suppl.), 3768s.

22 Akhter, S., Nath, S. K., Tse, C. M., Williams, J., Zasloff, M., Donowitz, M. (1999) *American Journal of Physiology*, **276**, C136–C144.

23 Teicher, B. A., Williams, J. I., Takeuchi, H., Ara, G., Herbst, R. S., Buxton, D. (1998) *Anticancer Research*, **18**, 2567–2573.

24 Schiller, J. H. and Bittner, G. (1999) *Clinical Cancer Research*, **5**, 4287–4294.

25 Williams, J. I., Mangold, G., Zasloff, M., Von Hoff, D. D. (1997) *Breast Cancer Research and Treatment*, **46**, 723–733.

26 Li, D., Williams, J. I., Pietras, R. J. (2002) *Oncogene*, **21**, 2805–2814.

27 Bhargava, P., Marshall, J. L., Dahut, W., Rizvi, N., Trocky, N., Williams, J. I., Hait, H., Song, S., Holroyd, K. J., Hawkins, M. J. (2001) *Clinical Cancer Research*, **7**, 3912–3919.

28 Hao, D., Hammond, L. A., Eckhardt, S. G., Patnaik, A., Takimoto, C. H., Schwartz, G. H., Goetz, A. D., Tolcher, A. W., McCreery, H. A., Mamun, K., Williams, J. I., Holroyd, K. J., Rowinsky, E. K. (2003) *Clinical Cancer Research*, **9**, 2465–2471.

29 Herbst, R. S., Hammond, L. A., Carbone, D. P., Tran, H. T., Holroyd, K. J., Desai, A., Williams, J. I., Bekele, B. N., Hait, H., Allgood, V., Solomon, S., Schiller, J. H. (2003) *Clinical Cancer Research*, **9**, 4108–4115.

30 Davidson, S. A., Chap, L., Pietras, R., Astrow, A., Gajewski, W., Brader, K., Petrone, M., Desai, A., Solomon, S., Holroyd, K., Major, F., Adler, L., Cohn, A. (2002) *Proceedings of the American Society of Clinical Oncology*, **9**, 2465–2471.

31 Moriarty, R. M., Tuladhar, S. M., Guo, L., Wehrli, S. (1994) *Tetrahedron Letters*, **35**, 8103–8106.

32 Pechulis, A. D., Bellevue, F. H.III., Cioffi, C. L., Trapp, S. G., Fojtik, J. P., McKitty, A. A., Kinney, W. A., Frye, L. L. (1995) *The Journal of Organic Chemistry*, **60**, 5121–5126.

33 Moriarty, R. M., Enache, L. A., Kinney, W. A., Allen, C. S., Canary, J. W., Tuladhar, S. M., Guo, L. (1995) *Tetrahedron Letters*, **36**, 5139–5142.

34 Jones, S. R., Selinsky, B. S., Rao, M. N., Zhang, X., Kinney, W. A., Tham, F. S. (1998) *The Journal of Organic Chemistry*, **63**, 3786–3789.

35 Zhang, X., Rao, M. N., Jones, S. R., Shao, B., Feibush, P., McGuigan, M., Tzodikov, N., Feibush, B., Sharkansky, I., Snyder, B., Mallis, L. M., Sarkahian, A., Wilder, S., Turse, J. E., Kinney, W. A., Kjrsgaard, H. J., Michalak, R. S. (1998) *The Journal of Organic Chemistry*, **63**, 8599–8603.

36 Zhou, X.-D., Cai, F., Zhou, W.-S. (2002) *Tetrahedron*, **58**, 10293–10299.

37 Zhang, D.-H., Cai, F., Zhou, X.-D., Zhou, W.-S. (2003) *Organic Letters*, **5**, 3257–3259.

38 Zhang, D. H., Cai, F., Zhou, X. D., Zhou, W. S. (2005) *Chinese Journal of Chemistry*, **23**, 176–181.

39 Kinney, W. A., Zhang, X., Williams, J. I., Johnston, S., Michalak, R. S., Deshpande, M., Dostal, L., Rosazza, J. P. N. (2000) *Organic Letters*, **2**, 2921–2922.

40 Okumura, K., Nakamura, Y., Takeuchi, S., Kato, I., Fujimoto, Y., Ikekawa, N. (2003) *Chemical and Pharmaceutical Bulletin*, **51**, 1177–1182.

41 Jones, S. R., Kinney, W. A., Zhang, X., Jones, L. M., Selinsky, B. S. (1996) *Steroids*, **61**, 565–571.

42 Shu, Y., Jones, S. R., Kinney, W. A., Selinsky, B. S. (2002) *Steroids*, **67**, 291–304.

43 Choucair, B., Dherbomez, M., Roussakis, C., El Kihel, L. (2004) *Bioorganic and Medicinal Chemistry Letters*, **14**, 4213–4216.

44 Choucair, B., Dherbomez, M., Roussakis, C., El Kihel, L. (2004) *Tetrahedron*, **60**, 11477–11486.
45 Khabnadideh, S., Tan, C. L., Croft, S. L., Kendrick, H., Yardley, V., Gilbert, I. H. (2000) *Bioorganic and Medicinal Chemistry Letters*, **10**, 1237–1239.
46 Kim, H. S., Choi, B. S., Kwon, K. C., Lee, S. O., Kwak, H. J., Lee, C. H. (2000) *Bioorganic and Medicinal Chemistry*, **8**, 2059–2065.
47 Cai, F. and Zhou, W. S. (2004) *Chinese Journal of Chemistry*, **22**, 1019–1021.
48 Williams, D. E., Lassota, P., Andersen, R. J. (1998) *The Journal of Organic Chemistry*, **63**, 4838–4841.
49 Williams, D. E., Craig, K. S., Patrick, B., McHardy, L. M., Van Soest, R., Roberge, M., Andersen, R. J. (2002) *The Journal of Organic Chemistry*, **67**, 245–258.
50 Goldring, W. P. D. and Weiler, L. (1999) *Organic Letters*, **1**, 1471–1473.
51 Roskelley, C. D., Williams, D. E., McHardy, L. M., Leong, K. G., Troussard, A., Karsan, A., Andersen, R. J., Dedhar, S., Roberge, M. (2001) *Cancer Research*, **61**, 6788–6794.
52 McHardy, L. M., Sinotte, R., Troussard, A., Sheldon, C., Church, J., Williams, D. E., Andersen, R. J., Dedhar, S., Roberge, M., Roskelley, C. D. (2004) *Cancer Research*, **64**, 1468–1474.
53 To, K. C. W., Loh, K. T., Roskelley, C. D., Andersen, R. J., O'Connor, T. P. (2006) *Neuroscience*, **139**, 1263–1274.
54 Baldwin, J. E. and Whitehead, R. C. (1992) *Tetrahedron Letters*, **33**, 2059–2062.
55 Baldwin, J. E. and Vollmer, H. R.Lee, V. (1999) *Tetrahedron Letters*, **40**, 5401–5404.
56 Fürstner, A. and Rumbo, A. (2000) *The Journal of Organic Chemistry*, **65**, 2608–2611.
57 Kaur, N., Delcros, J.-G., Martin, B., Phanstiel, O. (2005) *Journal of Medicinal Chemistry*, **48**, 3832–3839.
58 D'Ambrosio, M., Guerriero, A., Debitus, C., Ribes, O., Pusset, J., Leroy, S., Pietra, F. (1993) *Chemical Communications*, 1305–1306.
59 Hong, T. W., Jimenez, D. R., Molinski, T. F. (1998) *Journal of Natural Products*, **61**, 158–161.
60 D'Ambrosio, M., Guerriero, A., Ripamonti, M., Debitus, C., Waikedre, J., Pietra, F. (1996) *Helvetica Chimica Acta*, **79**, 727–735.
61 Pettit, G. R., Ducki, S., Herald, D. L., Doubek, D. L., Schmidt, J. M., Chapuis, J.-C. (2005) *Oncology Research*, **15**, 11–20.
62 Meijer, L., Thunnissen, A. M. W. H., White, A. W., Garnier, M., Nikolic, M., Tsai, L. H., Walter, J., Cleverley, K. E., Salinas, P. C., Wu, Y. Z., Biernat, J., Mandelkow, E. M., Kim, S. H., Pettit, G. R. (2000) *Chemistry and Biology*, **7**, 51–63.
63 D'Ambrosio, M., Guerriero, A., Chiasera, G., Pietra, F. (1994) *Helvetica Chimica Acta*, **77**, 1895–1902.
64 Andrade, P., Willoughby, R., Pomponi, S. A., Kerr, R. G. (1999) *Tetrahedron Letters*, **40**, 4775–4778.
65 Al Mourabit, A. and Potier, P. (2001) *European Journal of Organic Chemistry*, **2001**, 237–243.
66 Anderson, G. T., Chase, C. E., Koh, Y., Stien, D., Weinreb, S. M., Shang, M. (1998) *The Journal of Organic Chemistry*, **63**, 7594–7595.
67 Stien, D., Anderson, G. T., Chase, C. E., Koh, Y., Weinreb, S. M. (1999) *Journal of American Chemical Society*, **121**, 9574–9579.
68 Feldman, K. S. and Saunders, J. C. (2002) *Journal of American Chemical Society*, **124**, 9060–9061.
69 Feldman, K. S., Saunders, J. C., Wrobleski, M. L. (2002) *The Journal of Organic Chemistry*, **67**, 7096–7109.
70 Hale, K. J., Domostoj, M. M., Tocher, D. A., Irving, E., Scheinmann, F. (2003) *Organic Letters*, **5**, 2927–2930.
71 Domostoj, M. M., Irving, E., Scheinmann, F., Hale, K. J. (2004) *Organic Letters*, **6**, 2615–2618.
72 Davis, F. A. and Deng, J. (2005) *Organic Letters*, **7**, 621–623.
73 Trost, B. M. and Dong, G. (2006) *Journal of American Chemical Society*, **128**, 6054–6055.
74 Fujita, M., Nakao, Y., Matsunaga, S., Seiki, M., Itoh, Y., Yamashita, J., vanSoest, R. W. M., Fusetani, N. (2003) *Journal of American Chemical Society*, **125**, 15700–15701.

75 Meketa, M. L. and Weinreb, S. M. (2006) *Organic Letters*, **8**, 1443–1446.

76 Shengule, S. R. and Karuso, P. (2006) *Organic Letters*, **8**, 4083–4084.

77 Fattorusso, E., Minale, L., Sodano, G. (1970) *Chemical Communications*, 751–752.

78 Fulmor, W., Van Lear, G. E., Morton, G. O., Mills, R. D. (1970) *Tetrahedron Letters*, **11**, 4551–4552.

79 Fattorusso, E., Minale, L., Sodano, G. (1972) *Journal of the Chemical Society, Perkin Transactions*, **1**, 16–18.

80 Kreuter, M. H., Bernd, A., Holzmann, H., Muller-Klieser, W., Maidhof, A., Weissmann, N., Kljajic, Z., Batel, R., Schroder, H. C., Muller, W. E. (1989) *Zeitschrift Fur Naturforschung C: A Journal of Biosciences*, **44**, 680–688.

81 Teeyapant, R., Woerdenbag, H. J., Kreis, P., Hacker, J., Wray, V., Witte, L., Proksch, P. (1993) *Zeitschrift Fur Naturforschung C: A Journal of Biosciences*, **48**, 939–945.

82 Chang, C. W. J. and Weinheimer, A. J. (1977) *Tetrahedron Letters*, 4005–4008.

83 Saeki, B. M., Granato, A. C., Berlinck, R. G. S., Magalhaes, A., Schefer, A. B., Ferreira, A. G., Pinheiro, U. S., Hajdu Eduardo (2002) *Journal of Natural Products*, **65**, 796–799.

84 Capon, R. J. and MacLeod, J. K. (1987) *Australian Journal of Chemistry*, **40**, 341–346.

85 Cosulich, D. B. and Lovell, F. M. (1971) *Chemical Communications*, 397–398.

86 Koulman, A., Proksch, P., Ebel, R., Beekman, A. C., van Uden, W., Konings, A. W., Pedersen, J. A., Pras, N., Woerdenbag, H. J. (1996) *Journal of Natural Products*, **59**, 591–594.

87 Rodriguez-Nieto, S., Gonzalez-Iriarte, M., Carmona, R., Munoz-Chapuli, R., Medina, M. A., Quesada, A. R. (2002) *FASEB Journal*, **16**, 261–263.

88 Gonzalez-Iriarte, M., Carmona, R., Perez-Pomares, J. M., Macias, D., Medina, M. A., Quesada, A. R., Munoz-Chapuli, R. (2003) *Angiogenesis*, **6**, 251–254.

89 Kreuter, M. H., Leake, R. E., Rinaldi, F., Muller-Klieser, W., Maidhof, A., Muller, W. E., Schroder, H. C. (1990) *Comparative Biochemistry and Physiology: Part B*, **97**, 151–158.

90 Carney, J. R. and Rinehart, K. L. (1995) *Journal of Natural Products*, **58**, 971–985.

91 Ebel, R., Brenzinger, M., Kunze, A., Gross, H. J., Proksch, P. (1997) *Journal of Chemical Ecology*, **23**, 1451–1462.

92 Teeyapant, R. and Proksch, P. (1993) *Naturwissenschaften*, **80**, 369–370.

93 Thoms, C., Ebel, R., Proksch, P. (2006) *Journal of Chemical Ecology*, **32**, 97–123.

94 Weiss, B., Ebel, R., Elbrachter, M., Kirchner, M., Proksch, P. (1996) *Biochemical Systematics and Ecology*, **24**, 1–7.

95 Minale, L., Sodana, G., Chan, W. R., Chen, A. M. (1972) *Chemical Communications*, 674–675.

96 Andersen, R. J. and Faulkner, D. J. (1975) *Journal of American Chemical Society*, **97**, 936–937.

97 Ogamino, T. and Nishiyama, S. (2003) *Tetrahedron*, **59**, 9419–9423.

98 Hinterding, K., Knebel, A., Herrlich, P., Waldmann, H. (1998) *Bioorganic and Medicinal Chemistry*, **6**, 1153–1162.

99 Rodriguez, A. D., Akee, R. K., Scheuer, P. J. (1987) *Tetrahedron Letters*, **28**, 4989–4992.

100 Quinoa, E. and Crews, P. (1987) *Tetrahedron Letters*, **28**, 3229–3232.

101 Arabshahi, L. and Schmitz, F. J. (1987) *The Journal of Organic Chemistry*, **52**, 3584–3586.

102 Jimenez, C. and Crews, P. (1991) *Tetrahedron*, **47**, 2097–2102.

103 Pham, N. B., Butler, M. S., Quinn, R. J. (2000) *Journal of Natural Products*, **63**, 393–395.

104 Shin, J., Lee, H. S., Seo, Y., Rho, J. R., Cho, K. W., Paul, V. J. (2000) *Tetrahedron*, **56**, 9071–9077.

105 Tabudravu, J. N., Eijsink, V. G. H., Gooday, G. W., Jaspars, M., Komander, D., Legg, M., Synstad, B., van Aalten, D. M. F. (2002) *Bioorganic and Medicinal Chemistry*, **10**, 1123–1128.

106 Pina, I. C., Gautschi, J. T., Wang, G.-Y.-S., Sanders, M. L., Schmitz, F. J., France, D., Cornell-Kennon, S., Sambucetti, L. C., Remiszewski, S. W.,

Perez, L. B., Bair, K. W., Crews, P. (2003) *The Journal of Organic Chemistry*, **68**, 3866–3873.

107 Park, Y., Liu, Y., Hong, J., Lee, C. O., Cho, H., Kim, D.-K., Im, K. S., Jung, J. H. (2003) *Journal of Natural Products*, **66**, 1495–1498.

108 Kim, D., Lee, I. S., Jung, J. H., Lee, C. O., Choi, S.-U. (1999) *Anticancer Research*, **19**, 4085–4090.

109 Jiang, Y., Ahn, E.-Y., Ryu, S. H., Kim, D.-K., Park, J.-S., Yoon, H. J., You, S., Lee, B.-J., Lee, D. S., Jung, J. H. (2004) *BMC Cancer*, **4**, 70.

110 Shim, J. S., Lee, H.-S., Shin, J., Kwon, H. J. (2004) *Cancer Letters*, **203**, 163–169.

111 Kim, D., Lee, I. S., Jung, J. H., Yang, S.-I. (1999) *Archives of Pharmacal Research*, **22**, 25–29.

112 Mora, F. D., Jones, D. K., Desai, P. V., Patny, A., Avery, M. A., Feller, D. R., Smillie, T., Zhou, Y. D., Nagle, D. G. (2006) *Journal of Natural Products*, **69**, 547–552.

113 Godert, A. M., Angelino, N., Woloszynska-Read, A., Morey, S. R., James, S. R., Karpf, A. R., Sufrin, J. R. (2006) *Bioorganic and Medicinal Chemistry Letters*, **16**, 3330–3333.

114 Jiang, Y., Ryu, S.-H., Ahn, E.-Y., You, S., Lee, B.-J., Jung, J. H., Kim, D.-K. (2004) *Natural Product Sciences*, **10**, 277–283.

115 Hoshino, O., Murakata, M., Yamada, K. (1992) *Bioorganic and Medicinal Chemistry Letters*, **2**, 1561–1562.

116 Nicolaou, K. C., Hughes, R., Pfefferkorn, J. A., Barluenga, S., Roecker, A. J. (2001) Chemistry—A European Journal 7, 4280–4295.

117 Nicolaou, K. C., Hughes, R., Pfefferkorn, J. A., Barluenga, S. (2001) Chemistry—A European Journal **7**, 4296–4310.

118 Kazlauskas, R., Lidgard, R. O., Murphy, P. T., Wells, R. J. (1980) *Tetrahedron Letters*, **21**, 2277–2280.

119 Kazlauskas, R., Lidgard, R. O., Murphy, P. T., Wells, R. J., Blount, J. F. (1981) *Australian Journal of Chemistry*, **34**, 765–786.

120 Pordesimo, E. O. and Schmitz, F. J. (1990) *The Journal of Organic Chemistry*, **55**, 4704–4709.

121 Miao, S., Andersen, R. J., Allen, T. M. (1990) *Journal of Natural Products*, **53**, 1441–1446.

122 Butler, M. S., Lim, T. K., Capon, R. J., Hammond, L. S. (1991) *Australian Journal of Chemistry*, **44**, 287–296.

123 Jaspars, M., Rali, T., Laney, M., Schatzman, R. C., Diaz, M. C., Schmitz, F. J., Pordesimo, E. O., Crews, P. (1994) *Tetrahedron*, **50**, 7367–7374.

124 Park, S. K., Jurek, J., Carney, J. R., Scheuer, P. J. (1994) *Journal of Natural Products*, **57**, 407–410.

125 Park, S. K., Park, H., Scheuer, P. J. (1994) *Bulletin of the Korean Chemical Society*, **15**, 534–537.

126 Park, S. K., Ryu, J. K., Scheuer, P. J. (1995) *Bulletin of the Korean Chemical Society*, **16**, 677–679.

127 Franklin, M. A., Penn, S. G., Lebrilla, C. B., Lam, T. H., Pessah, I. N., Molinski, T. F. (1996) *Journal of Natural Products*, **59**, 1121–1127.

128 Aoki, S., Cho, S. h., Hiramatsu, A., Kotoku, N., Kobayashi, M. (2006) *Journal of Natural Medicines*, **60**, 231–235.

129 Aoki, S., Cho, S. h., Ono, M., Kuwano, T., Nakao, S., Kuwano, M., Nakagawa, S., Gao, J. Q., Mayumi, T., Shibuya, M., Kobayashi, M. (2006) *Anti-Cancer Drugs*, **17**, 269–278.

130 Coll, J. C., Kearns, P. S., Rideout, J. A., Sankar, V. (2002) *Journal of Natural Products*, **65**, 753–756.

131 Dexter, A. F., Garson, M. J., Hemling, M. E. (1993) *Journal of Natural Products*, **56**, 782–786.

132 Carney, J. R., Scheuer, P. J., Kelly-Borges, M. (1993) *Journal of Natural Products*, **56**, 153–157.

133 Venkateswarlu, Y., Venkatesham, U., Rao, M. R. (1999) *Journal of Natural Products*, **62**, 893–894.

134 Reddy, A. V., Ravinder, K., Narasimhulu, M., Sridevi, A., Satyanarayana, N., Kondapi, A. K., Venkateswarlu, Y. (2006) *Bioorganic and Medicinal Chemistry*, **14**, 4452–4457.

135 Pettit, G. R., Butler, M. S., Williams, M. D., Filiatrault, M. J., Pettit, R. K. (1996) *Journal of Natural Products*, **59**, 927–934.

136 Pettit, G. R., Butler, M. S., Bass, C. G., Doubek, D. L., Williams, M. D., Schmidt, J. M., Pettit, R. K., Hooper, J. N. A., Tackett, L. P., Filliatrault, M. J. (1995) *Journal of Natural Products*, **58**, 680–688.

137 Segraves, E. N., Shah, R. R., Segraves, N. L., Johnson, T. A., Whitman, S., Sui, J. K., Kenyon, V. A., Cichewicz, R. H., Crews, P., Holman, T. R. (2004) *Journal of Medicinal Chemistry*, **47**, 4060–4065.

138 Mack, M. M., Molinski, T. F., Buck, E. D., Pessah, I. N. (1994) *The Journal of Biological Chemistry*, **269**, 23236–23249.

139 Chen, L., Molinski, T. F., Pessah, I. N. (1999) *The Journal of Biological Chemistry*, **274**, 32603–32612.

140 Nishiyama, S. and Yamamura, S. (1982) *Tetrahedron Letters*, **23**, 1281–1284.

141 Nishiyama, S., Suzuki, T., Yamamura, S. (1982) *Tetrahedron Letters*, **23**, 3699–3702.

142 Nishiyama, S., Suzuki, T., Yamamura, S. (1982) *Chemistry Letters*, 1851–1852.

143 Guo, Z. W., Machiya, K., Salamonczyk, G. M., Sih, C. J. (1998) *The Journal of Organic Chemistry*, **63**, 4269–4276.

144 Kotoku, N., Tsujita, H., Hiramatsu, A., Mori, C., Koizumi, N., Kobayashi, M. (2005) *Tetrahedron*, **61**, 7211–7218.

145 Garcia-Pichel, F. and Castenholz, R. W. (1991) *Journal of Phycology*, **27**, 395–409.

146 Proteau, P. J., Gerwick, W. H., Garcia-Pichel, F., Castenholz, R. (1993) *Experientia*, **49**, 825–829.

147 Stevenson, C. S., Capper, E. A., Roshak, A. K., Marquez, B., Grace, K., Gerwick, W. H., Jacobs, R. S., Marshall, L. A. (2002) *Inflammation Research*, **51**, 112–114.

148 Stevenson, C. S., Capper, E. A., Roshak, A. K., Marquez, B., Eichman, C., Jackson, J. R., Mattern, M., Gerwick, W. H., Jacobs, R. S., Marshall, L. A. (2002) *The Journal of Pharmacology and Experimental Therapeutics*, **303**, 858–866.

149 Ohta, S., Ohta, E., Ikegami, S. (1997) *The Journal of Organic Chemistry*, **62**, 6452–6453.

150 Ohta, E., Ohta, S., Ikegami, S. (2001) *Tetrahedron*, **57**, 4699–4703.

151 Fujita, M., Nakao, Y., Matsunaga, S., Seiki, M., Itoh, Y., van Soest, R. W. M., Fusetani, N. (2001) *Tetrahedron*, **57**, 1229–1234.

152 Odake, S., Morita, Y., Morikawa, T., Yoshida, N., Hori, H., Nagai, Y. (1994) *Biochemical and Biophysical Research Communications*, **199**, 1442–1446.

153 Rosett, T., Sankhala, R. H., Stickings, C. E., Taylor, M. E. U., Thomas, R. (1957) *Biochemical Journal*, **67**, 390–400.

154 Newman, D. J. and Cragg, G. M. (2004) *Journal of Natural Products*, **67**, 1216–1238.

155 Butler, M. S. (2005) *Natural Product Report*, **22**, 162–195.

10
A Typical Class of Marine Alkaloids: Bromopyrroles

Anna Aiello, Ernesto Fattorusso, Marialuisa Menna, Orazio Taglialatela-Scafati

10.1
Introduction

Bromopyrrole alkaloids constitute a family of exclusively marine alkaloids and represent a fascinating example of the large variety of secondary metabolites elaborated by marine sponges. The first member of this group to be isolated was oroidin (**1**), initially from the sponge *Agelas oroides* in 1971 and then from several other sponges [1,2]. Oroidin is considered the key metabolite of this family of alkaloids, since many bromopyrrole alkaloids possessing a pyrrole–imidazole structure can be conceived as derivatives of the $C_{11}N_5$ skeleton of oroidin. These can vary with regard to (a) oxidation, reduction, or hydration of the 2-amino-4(5)-vinylimidazole unit, (b) dimerization, and (c) cyclization. The pyrrole-2-carboxamide moiety can be non-, mono-, or dibrominated exclusively in the 2- and 3-positions (numbering according to Figure 10.1). Bromination of the pyrrole 4-position or of the imidazole part has never been observed. Actually, small chemical changes within the building blocks, such as bromination of the pyrrole-2-carboxamide moiety or tautomeric equilibrium and the consequent bivalent reactivity of the 2-aminoimidazole moiety, are mostly responsible for the great molecular diversity observed in this group of alkaloids.

Since the mid-1970s, about 140 derivatives, with various structures and interesting biological activities, have been isolated from more than 20 different sponges of various genera, essentially (but not exclusively) belonging to the Agelasidae, Axinellidae, and Halichondridae families. It is currently believed that these alkaloids are taxon-specific of at least the Agelasida order and can be used as chemical markers of these phylogenetically related sponges [3].

From the ecological point of view, the antipredatory role of these alkaloids can be considered their most important biological function. It has been determined that Caribbean reef sponges of the genus *Agelas* are chemically defended from fish predation by bromopyrrole alkaloids [4–6]. The relative activity of the naturally occurring compounds and the chemical functionalities necessary and sufficient for this activity have also been determined, thus providing the current knowledge on

Modern Alkaloids: Structure, Isolation, Synthesis and Biology. Edited by E. Fattorusso and O. Taglialatela-Scafati

ISBN: 978-3-527-31521-5

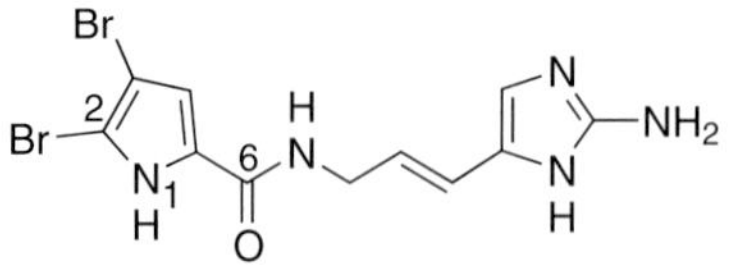

Fig. 10.1 Oroidin, the postulated key precursor of bromopyrrole alkaloids.

the relationships between molecular structure and fish feeding deterrence [7]. The role of these alkaloids in defense mechanisms is also supported by a study demonstrating that they have a sponge cellular origin and have not been detected in the sponge-associated microorganisms [8].

Bromopyrrole alkaloids are important not only for their ecological role and for chemotaxonomic considerations, but also for the number of interesting pharmacological activities they have been shown to possess. Among them, the cytotoxicity of agelastatin (see below) and the immunosuppressive activity of palau'amine (see below) are remarkable. This is probably why the interest of organic chemists in total syntheses of bromopyrrole alkaloids has strongly increased since 2000.

Until now, only one review article on bromopyrrole alkaloids has been published, covering the literature appearing up to the end of 2002 [9]; this article was mainly focused on the chemistry of this alkaloid family, and a number of biomimetic and nonbiomimetic total syntheses of bromopyrrole alkaloids were discussed comprehensively.

In this chapter we intend to provide a complete picture of the structures of bromopyrrole alkaloids updated to the first half of 2006. The biological activities of these compounds are described in outline, together with the plausible biogenetic pathways that have been proposed so far. An account of the synthetic approaches reported for preparation of some alkaloids is also given, even if it is not intended to be a systematic picture but only a few scattered notes.

These alkaloids can be classified from a number of points of view but, in the present chapter, we have divided them into four groups with respect to their chemical architecture:

- *oroidin-like linear monomers*, whose structures contain the skeleton of oroidin without any further C–C or C–N bond formation;
- *polycyclic oroidin-derivatives*, whose structures can be rationalized by one of the many intramolecular cyclizations of oroidin or oroidin-like linear monomers. In fact, six modes of cyclization of oroidin have been found in Nature, which will be classified according to the oroidin atoms involved in the linkage formation;
- *simple or cyclized oroidin-like dimers*;
- *other bromopyrrole alkaloids*, which cannot be included in the previous groups, basically because they do not possess the pyrrole–imidazole moieties.

Some nonbrominated pyrrole–imidazole alkaloids are included in this chapter, but only when they are close analogs of the corresponding brominated compounds.

10.2 Oroidin-like Linear Monomers

Oroidin was first isolated in 1971 [1] from the sponges *Agelas oroides*, but its structure was definitely assigned in 1973 by synthesis of its monoacetylated derivative dihydrooroidin [2]. Subsequently, oroidin has been found in several different sponges, co-occurring with cyclized and/or dimeric related compounds. In addition to the above cited feeding deterrent activity against the reef fish *Thalassomia bifasciatum* [4–7], this molecule is also able to inhibit larval metamorphosis of the barnacle *Balanus amphirite* [10]. The first synthesis of oroidin was performed in 1986 [11]; since then, several different approaches have been proposed [12–17]. The key step of most of these synthetic strategies is a Wittig–Schweizer reaction employing a phosphonium salt prepared *in situ* via nucleophilic attack of either phthalimide or trichloroacetamide to tributylvinylphosphonium bromide [13].

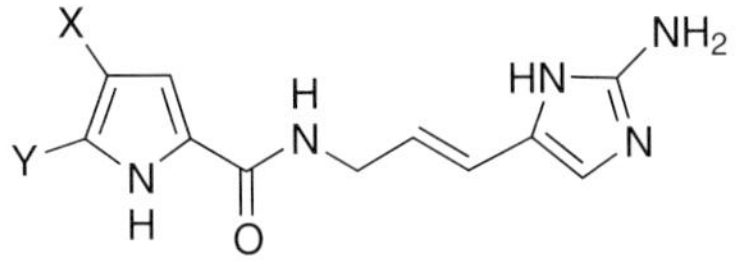

1 X = Y = Br
2 X = Br Y = H
3 X = Y = H

Hymenidin (**2**), isolated from the Okinawan sponge *Hymeniacidon* sp. [18], is the 2-debromo derivative of oroidin, while clathrodin (**3**), isolated from the Caribbean sea sponge *Agelas clathrodes* [19], is its 2,3-debromo derivative. The degree of bromination of the pyrrole moiety has been shown to affect the biological properties of these compounds. Recently, oroidin and hymenidin have been found to reduce voltage-dependent calcium elevation in PC12 cells, and the potency of the tested alkaloids increases with the number of bromine atoms associated with the pyrrole ring [20]. The degree of bromination is also important for the feeding deterrent properties; as expected on the basis of the experiments performed on pyrrole-2-carboxilic acid derivatives [7], hymenidin showed a smaller feeding deterrence than oroidin [21]. Oroidin, hymenidin, and clathrodin showed marked antiserotonergic and anticholinergic activities [18,22]. Clathrodin and oroidin have been tested for their effect on membrane sodium currents; these experiments demonstrated that clathrodin is a sodium channel neurotoxin, which acts by influencing channel ionic conductance [23].

The same synthetic approach starting from imidazolemethanol and via Wittig–Schweizer olefination, which provided an improved synthesis of oroidin, was used for the preparation of hymenidin [13] and clathrodin [24]. A further total synthesis of clathrodin has been reported [14].

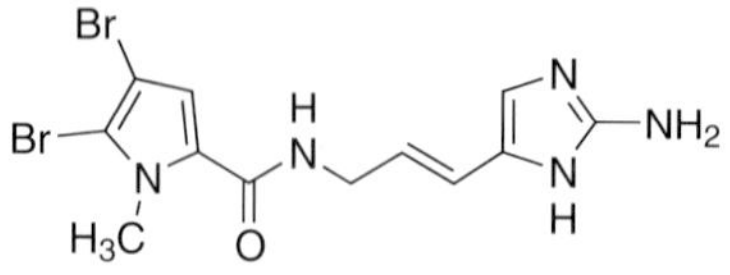

4

Sventrin (**4**) is the pyrrole N-methylated analog of oroidin and it has been isolated from the Caribbean sponge *Agelas sventres* [22]. Sventrin was shown to be a feeding deterrent compound against the reef fish *Thalassoma bifasciatum* in aquarium assays, and it has been observed that N-methylation of the pyrrole nitrogen reduces the feeding deterrent activity [22]. Since attempts to synthesize **4** by regioselective methylation of oroidin failed, it was obtained by stereoselective reduction of an alkyne intermediate [25].

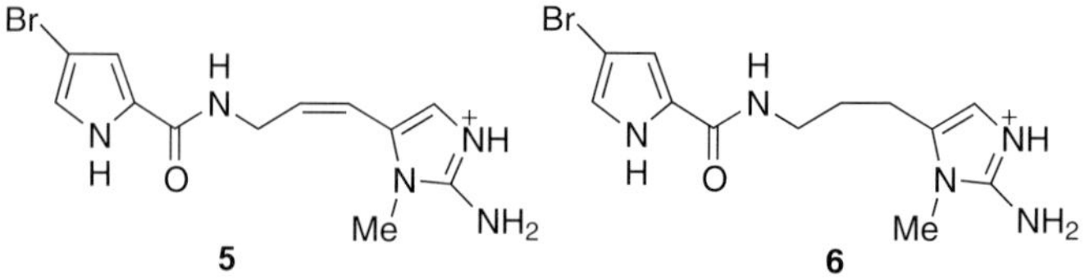

5 **6**

Keramadine (**5**) has been isolated from an Okinawan *Agelas* sp. and contains an N-methylated 2-aminoimidazole ring and a (*Z*)-double bond [26]. Its 9,10-dihydro analog, dihydrokeramadine (**6**) has been isolated, again from an *Agelas* sp. [27]. Total synthesis of keramadine has been performed both via Wittig–Schweizer olefination [13,24] and via an alkyne pathway [28,15]. In the first case, a problem was the partial isomerization of the (*Z*)-double bond, which was avoided in the second case by masking the unsaturation as an alkyne throughout the synthetic sequence. Like compounds **1–3**, keramadine showed antagonistic activity against serotonergic and cholinergic receptors with an $ED_{50} = 1.5 \times 10^{-5}$ M [22].

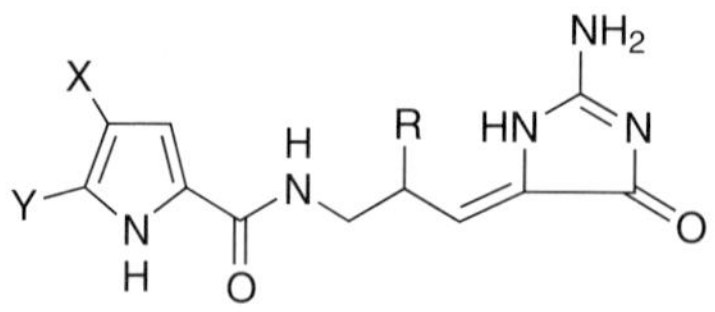

7 X = Y = Br, R=H
8 X = Br Y = H, R=H
9 X = Y = Br, R=OH
10 X = Br Y = H, R=OH (dispacamide D/mukanadin A)
11 X = Br, Y = H , R = OMe

In dispacamides A–D (**7–10**), isolated from four Caribbean *Agelas* sponges (*A. conifera*, *A. longissima*, *A. clathrodes*, *A. dispar*) [29,30], the 2-aminoimidazole moiety is oxidized to an alkylidene glycocyamidine. Dispacamide A (**7**) and its monobromo derivative dispacamide B (**8**) differ from oroidin and hymenidin, respectively, both in an isomerization of the double-bond position and in the presence of the aminoimidazolone moiety. Dispacamides C (**9**) and D (**10**) have been isolated as racemates and their structures have been found to be 9-hydroxylated forms of dispacamides A and B, respectively [30]. Interestingly, despite the plain structural resemblance with alkaloids such as keramadine or clathrodin, dispacamides A and B were found fully inactive both as anticholinergic and antiserotonergic agents, even at millimolar concentrations. On the other hand, all dispacamides exhibited a remarkable antihistaminic activity on the guinea pig ileum [29]. They were shown to produce a reversible noncompetitive antagonistic effect, specific toward histamine receptors; dispacamide A was the most potent compound in the series, while the activity of dispacamides C and D was mild, when compared to dispacamides A and B. Thus, the insertion of a hydroxyl group in the alkyl chain causes a marked reduction of the antihistaminic activity, indicating the importance of the central segment for the pharmacological activity. In has been hypothesized that this portion of the molecule could interact with hydrophobic groups in the receptor zone, so that the insertion of a polar group achieves the effect of making this interaction less marked [30].

Total synthesis of dispacamide A has been performed through condensation of 2-thiohydantoin with α-pyrrolecarboxamidoaldehyde, followed by a one-step treatment with aqueous ammonia in the presence of *tert*-butylhydroperoxide to give the glycocyamidine ring [31]. An alternative synthesis of dispacamide A, in which assembly of the imidazole ring occurs during the synthetic sequence, has been developed. It starts from ornithine methyl ester and leads to dihydrooroidin, which is successively oxidized to either oroidin or dispacamide A [14]. Compound **11**, isolated as a racemate from the Mediterranean sponge *Axinella verrucosa* [32], has been shown to be the 9-methoxy derivative of dispacamide B. Compound **11** was found to display neuroprotective activity against the agonist serotonin *in vitro*, with a potential to treat psychosis, different phobias, and mood fluctuation disorders. It acts by reducing the exitotoxic effect related to the massive entry of calcium ions into the cells as a consequence of activation of serotonin receptors.

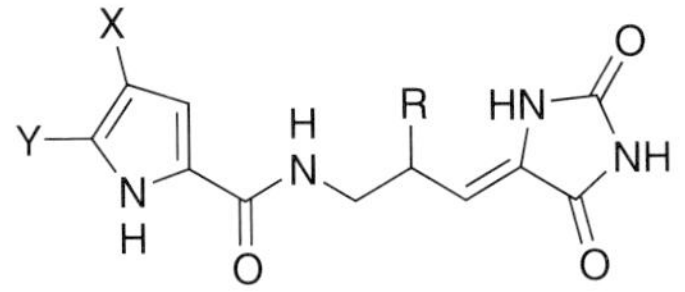

12 X = Br, Y = R = H
13 X = Y = Br, R = H
14 X = Br, Y = H, R = OH

Mukanadins A (**10**) and B (**12**) have been isolated from the Okinawan sponge *Agelas nakamurai* [33]; actually, the structure of mukanadin A corresponds to that of dispacamide D. Mukanadin D (**13**) has been found in the acetone extract of the Jamaican sponge *Didiscus oleata* [34]. In mukanadins B and D the 2-aminoimidazole ring of dispacamides is replaced by a hydantoin moiety; the same feature is present in compound **14**, isolated from *Axinella verrucosa* [32]. Like compound **11**, compound **14** displayed neuroprotective activity but it has been demonstrated that it acts as a potent glutamate antagonist [32].

The racemic midpacamide (**15**), possessing both pyrrole-2-carboxamide and N-methylated hydantoin moieties, has been isolated from an unidentified marine sponge collected in the Marshall Islands [35] and, subsequently, from the sponge *Agelas mauritiana* [36]. Two different total syntheses have been reported for this compound [37,38].

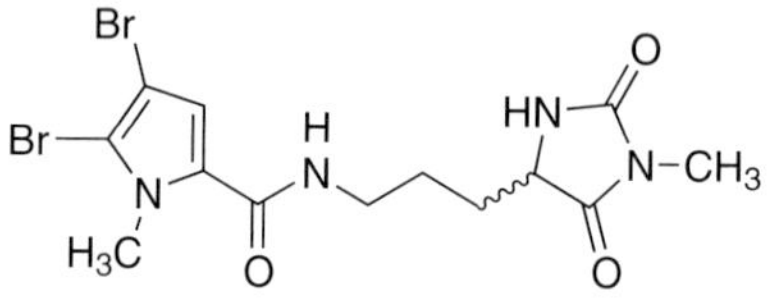

15

Mauritiamide A (**16**), isolated from the Fijian sponge *Agelas mauritiana*, contains the same *N*-methylpyrrole-2-carboxamide partial structure of midpacamide, but the second heterocycle is an aminoimidazolone ring [36].

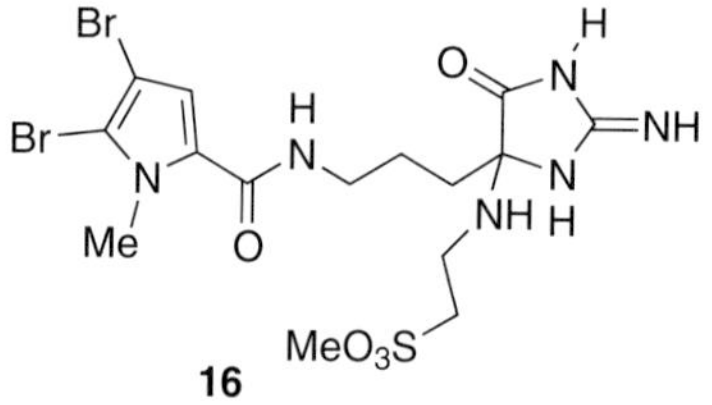

16

Mauritiamide A was the first member of the bromopyrrole alkaloids class to include a taurine moiety. This uncommon feature has been found successively in tauroacidin A (**17**) and its debromoderivative tauroacidin B (**18**), isolated from an Okinawan *Hymeniacidon* sp. [39], in taurodispacamide (**19**), isolated from the Mediterranean sponge *Agelas oroides* [40], and its debromoderivative **20**, isolated from *Axinella verrucosa* [32]. All the above-mentioned compounds show the taurine residue attached to a 2-aminoimidazole ring. Tauroacidins A and B are analogs of taurodispacamide and of compound **20**, respectively, bearing a hydroxyl group at position 9. Different pharmacological activities have been evidenced for these four compounds. Tauroacidins A and B exhibited inhibitory activity against EGF receptor kinase and *c-erbB*-2 kinase, with IC_{50} of 20 μg/mL. Taurodispacamide (**19**) was shown to have a good antihistaminic activity; in particular, the 0.1 μM response of histamine was almost completely abolished, in a reversible manner, by a 10 μM solution of **19**. Compound **20** was shown to exert a very potent neuroprotective effect, since it acts as a glutamate, as well as a serotonin, antagonist [32]. It should be noted that noncyclized pyrrole–imidazole alkaloids with taurine side chains (compounds **16–20**) have not yet been synthesized.

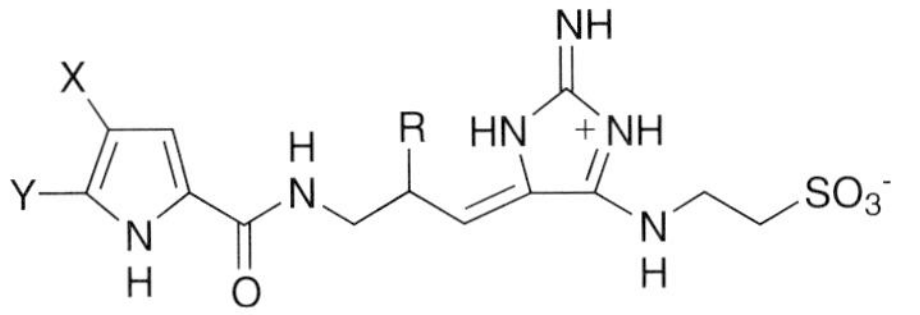

17 X = Y = Br, R = OH (9S/9R = 6:4)
18 X = Br, Y = H, R = OH (9S/9R = 1:1)
19 X = Y = Br, R = H
20 X = Br, Y = R = H

Slagenins A–C (**21–23**), isolated from the Okinawan sponge *Agelas nakamurai* [41], possess a unique tetrahydrofuro[2,3-*d*]imidazolidin-2-one moiety; their structures were determined devoid of absolute configurations. Although they are formally tricyclic, these alkaloids are included in this section because the third ring is not formed by an intramolecular C–C or C–N cyclization of the key oroidin building block but through the participation of a hydroxy group at C-9. Slagenins B and C exhibited cytotoxicity against murine leukemia L1210 cells *in vitro* with IC_{50} values of 7.5 and 7.0 μg/mL, respectively, while slagenin A was inactive [41].

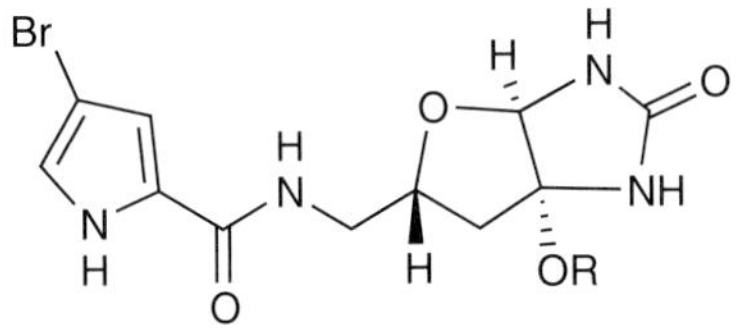

21 R = H
22 R = CH_3

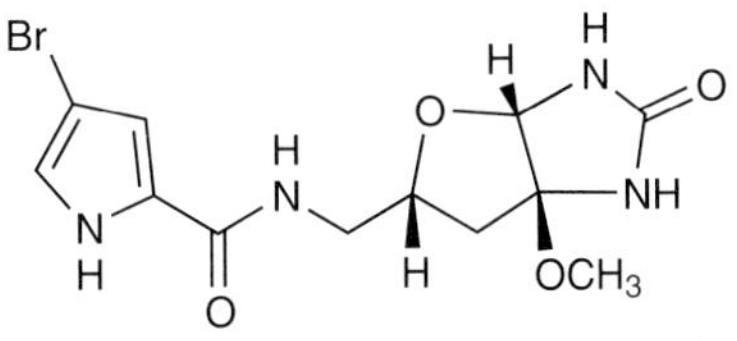

23

The first total synthesis of slagenins A–C has been accomplished starting from ornithine [42], but it produced racemic compounds; thus, the absolute configurations of the natural compounds remained undetermined. Subsequently, a short synthesis of the (−)-antipode of slagenin B and the (+)-antipode of slagenin C has been performed and the absolute configurations (9*R*, 11*R*, 15*R*) and (9*R*, 11*S*, 15*S*) for the natural compounds **22** and **23**, respectively, have been indicated [43]. Confirmation of these configurations has been achieved through the first enantioselective synthesis of slagenins B and C, performed starting from L-arabinose [44]. Other enantioselective syntheses of slagenins A–C have been proposed [45,46], which allowed the absolute configuration of slagenin A to be established as (9*R*, 11*R*, 15*R*).

10.3
Polycyclic Oroidin Derivatives

This class of bromopyrrole alkaloids includes all those molecules whose cyclized skeleton can be formally derived through the formation of one (or more) C–C or C–N bonds within the oroidin framework. Consequently, we have found it very useful to classify these molecules according to the oroidin atoms involved in the formal cyclization. More than 40 molecules have been thus classified into six different groups.

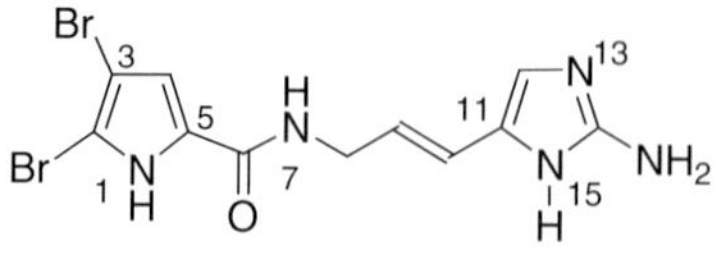

Br 3 5 H N 7 11 N 13 NH$_2$ Br 1 N H O N 15 H

1

Several alkaloids belonging to this class have been shown to possess interesting pharmacological activity, particularly as antitumor agents (e.g. agelastatins and palau'amine) or inhibitors of pro-inflammatory cytokines (e.g. hymenialdisins). These activities and their intriguing chemical structures continue to stimulate a series of elegant synthetic strategies, whose detailed discussion is beyond the scope of the present chapter.

10.3.1
C-4/C-10 Derivatives

Hymenin (**24**) has been isolated as an α-adrenoceptor blocking agent, tested on the isolated rabbit aorta, from the sponge *Hymeniacidon* sp. [47,48]. In addition, **24** also exhibited good antibacterial activity against *Bacillus subtilis* and *Escherichia coli* [47]. The absolute configuration at the single asymmetric carbon of hymenin (**24**) has not been assigned.

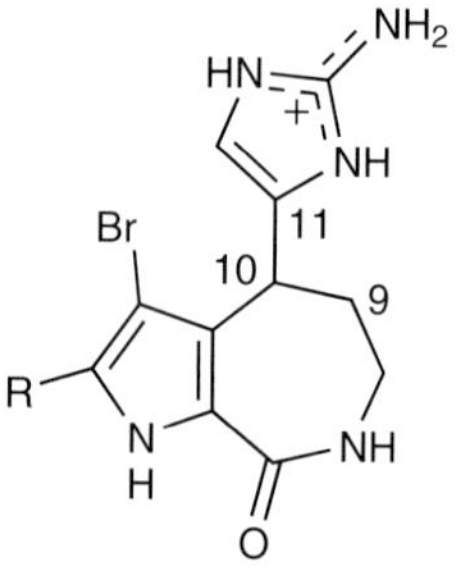

NH$_2$ HN + NH Br 11 10 9 R N H NH O

24 R = Br
25 R = H

A 2-debrominated hymenin (**25**) has been isolated from the sponge *Stylissa* (= *Axinella*) *carteri* [49]. A series of alkaloids of the hymenin family differ from this parent compound only in the number and/or the position of the double bonds. Stevensine (= odiline) (**26**) was isolated by Faulkner *et al.* from an unidentified marine sponge [50], but subsequently it has been re-obtained from *Pseudaxynissa cantharella* [51], *Axinella verrucosa* [32], and *Stylissa carteri* [49]. 2-Debromostevensine (**27**) has been isolated from *Stylissa carteri* [49]. Stevensine has been demonstrated to be one of the main factors in the chemical defense of the reef sponge *Axinella corrugata* against predatory fish [52].

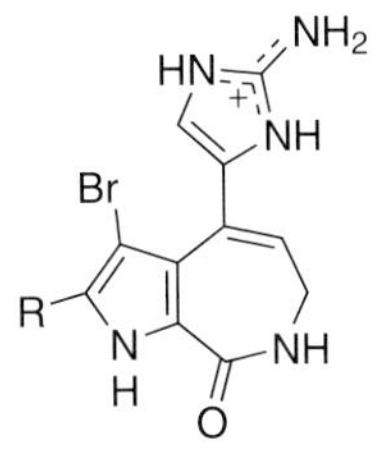

26 R = Br
27 R = H

Hymenin analogs possessing a double bond $\Delta^{10,11}$ and a carbonyl group at C-12 are highly bioactive compounds called hymenialdisins. Both (*E*) (**28–30**) and (*Z*) (**31–33**) configurations of the double bond $\Delta^{10,11}$, connecting the azepine and the imidazole rings, have been found in the hymenialdisin class.

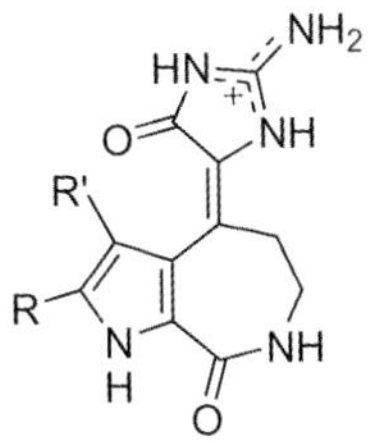

28 R = R' = H
29 R = Br R' = H
30 R = R' = Br

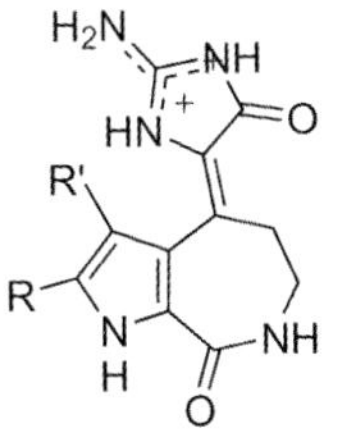

31 R = R' = H
32 R = Br R' = H
33 R = R' = Br

These molecules, initially called "yellow compounds," have been isolated from *Phakellia flabellata* [53], *Axinella verrucosa* [54], *Acanthella aurantiaca* [54], *Hymeniacidon aldis* [55], *Stylissa carteri* [49], and *Pseudaxynissa cantharella* [51]. The observation that, in solvents such as DMSO, (*E*)-hymenialdisins slowly converted into their corresponding (*Z*)-isomers, as shown below, provided a reasonable explanation for the finding of both geometrical isomers in the hymenialdisin series (Scheme 10.1).

Hymenialdisins proved to be nanomolar inhibitors of G_2 DNA damage checkpoint and of the protein kinases Chk1 and Chk2 [56], mitogen-activated protein kinase 1 (MEK-1) [57], and of other kinases [58]; therefore, they could be valuable agents in cancer therapy. In addition, hymenialdisins have been shown to inhibit

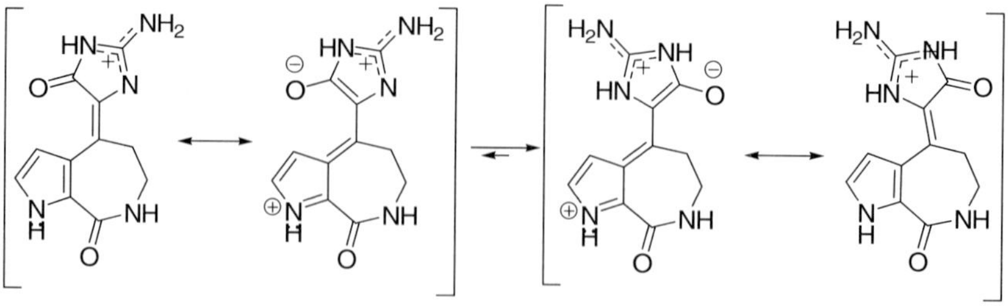

Scheme 10.1 Interconversion of double bond geometry in the hymenialdisin series.

several pro-inflammatory cytokines (interleukin-2, IL-6, IL-8, nitric oxide) through inhibition of the NF-*k*B signaling pathway [59,60]. This activity is potentially useful in the treatment of rheumatoid arthritis and osteoarthritis, a pathology for which specific pharmaceutical agents are badly needed. It is interesting to note that several hymenialdisin analogs have been tested for MEK-1 inhibitory activity [56] and the following structure–activity relationships established: (i) a change in the geometry of the double bond is not important; (ii) the activity is strictly dependent on the bromination of the pyrrole ring; (iii) the presence of an aminoimidazolone ring is essential for the activity, indeed hymenin (**24**) proved to be much less active.

The deaminated analogs of hymenialdisins have also been found. Compounds with the *E*-geometry of the double bond are called axinohydantoins (**34**–**35**) and these have been isolated from *Stylotella aurantium* [61]; compounds with the *Z*-geometry of the double bond are called spongiacidins (**36**–**37**), isolated from *Hymeniacidon* sp. [62] Interestingly, axinohydantoins and spongiacidins do not share the interesting bioactivities shown by hymenialdisins, indicating the key role played by the guanidinium group in the interaction with the targets.

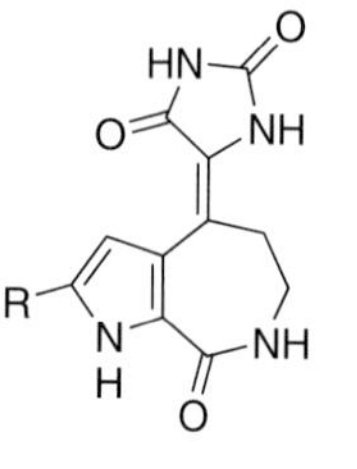

34 R = H
35 R = Br

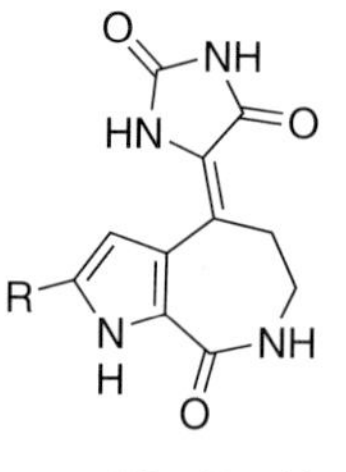

36 R = H
37 R = Br

The synthesis of C-4/C-10 cyclic bromopyrrole alkaloids has been intensively explored, and several approaches to their pyrrolo[2,3-*c*]azepin-8-one ring system (bearing an imidazole-derived ring at position 4) have been proposed [63–65].

10.3.2
N-1/C-9 Derivatives

The parent compound of this class is cyclooroidin (**38**), isolated from the Mediterranean sponge *Agelas oroides* [40]. Two syntheses of this alkaloid have been proposed [66,67], and one of them, obtaining *rac*-cyclooroidin by intramolecular cyclization of oroidin formate at 95 °C in protic solvents, could have biomimetic significance [67].

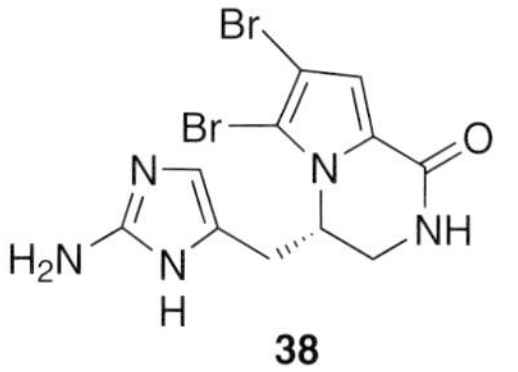

Br Br N N NH O H₂N N H **38**

It should be noted that cyclooroidin (**38**) could be envisaged as the precursor of the nonimidazole bromopyrrole alkaloids longamides (see below). Two cyclooroidin analogs have been isolated: agesamide (**39**) from *Agelas* sp. [68] and oxocyclostylidol (**40**) from *Stylissa caribica* [69]. This latter compound represents the first pyrrole–imidazole alkaloid that includes an oxidized pyrrole moiety.

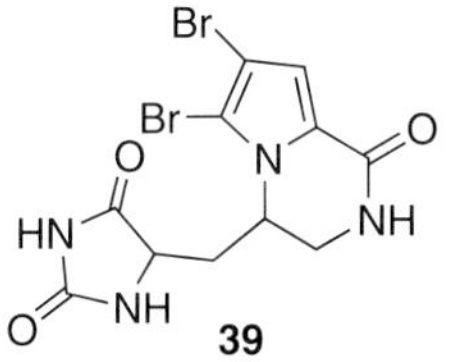

Br Br O N O HN NH O NH **39**

Br OH O N O N NH H₂N N H **40**

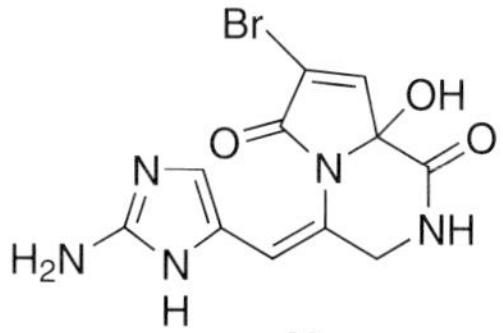

10.3.3
N-7/C-11 + N-1/C-12 Derivatives

Phakellins (**41**, **42**), tetracyclic derivatives in which both the pyrrole and the amidic nitrogen atoms are involved in the formation of a linkage with carbon atoms of the imidazole ring, represented the first members of the family of oroidin-related cyclic bromopyrrole alkaloids to be isolated. They were found in 1971 by Sharma *et al.* in the marine sponge *Phakellia flabellata*; their structure was confirmed by X-ray diffraction analysis of a single crystal of the monoacetyl derivative of **41** [70]. Complete spectral data were provided by the same authors a few years later [71].

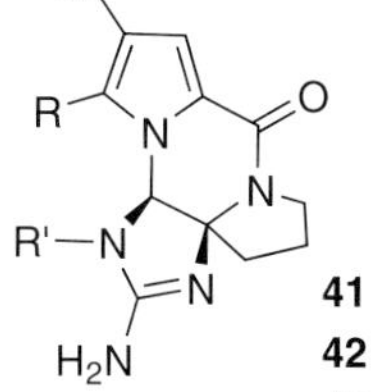

41 R = H R' = H
42 R = Br R' = H
45 R = H R' = CH_3
46 R = Br R' = CH_3

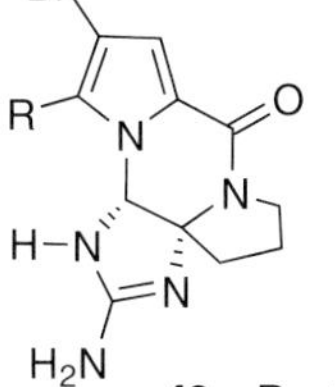

43 R = H
44 R = Br

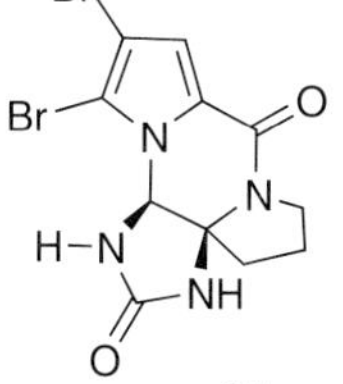

47

More recently, some closely related analogs of phakellins have been obtained. Firstly, the phakellins **43** and **44**, enantiomeric to **41** and **42**, respectively, were obtained from *Pseudaxynissa cantharella* [51]; later, the 7-*N*-methylphakellins **45** and **46** were isolated from an *Agelas* sp. [72], and phakellistatin (**47**) from *Phakellia mauritiana* [73]. This last compound was reported to exhibit potent cell growth inhibitory activity against a variety of human cancer cell lines (ED_{50} from 0.3 to 4.0 μM) [73].

Phakellins represent an attractive synthetic target, owing to their compact and heteroatom-dense structure, including two vicinal, stereogenic aminal centers. A concise and elegant biomimetic synthesis of racemic dibromophakellin was proposed by Büchi in 1982 [74], while an enantioselective strategy toward phakellistatin and phakellin, through highly diastereoselective desymmetrization of the diketopiperazine cyclo(Pro,Pro), has been proposed more recently [75].

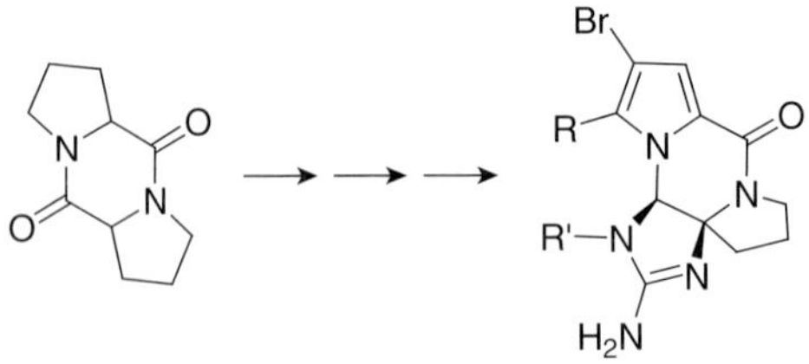

Br
R
N
O
O
N
N
O
N
R'—N
N
H2N

Interestingly, a series of similar cyclo(Pro,Pro) diketopiperazines, named verpacamides (e.g. **48**), have been isolated from the marine sponge *Axinella vaceleti* [76]. It was proposed that these molecules could be the biogenetic precursors of the $C_{11}N_5$ skeleton, common to all the alkaloids of this class, through an intramolecular oxidative rearrangement, via a dioxetanone intermediate.

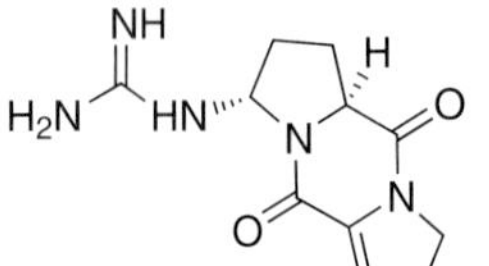

NH
H
H2N
HN
N
O
O
N
48

The class of N-7/C-11 + N-1/C-12 derivatives also includes some complex metabolites possessing a phakellin skeleton conjugated to another heterocyclic moiety: the bisguanidine derivatives called palau'amines. Palau'amine (**49**) was first isolated by Scheuer *et al.* as a strongly cytotoxic (IC_{50} about 0.1 μg/mL against several cancer cell lines) and immunosuppressive agent from the sponge *Stylotella aurantium* [77]. Good antibacterial activity against *Staphylococcus aureus* and *Bacillus subtilis* was also reported [77]. The monobrominated (**50**) and the dibrominated (**51**) analogs have been subsequently reported by the same group [78]. Interestingly, they proved to be much less bioactive than palau'amine.

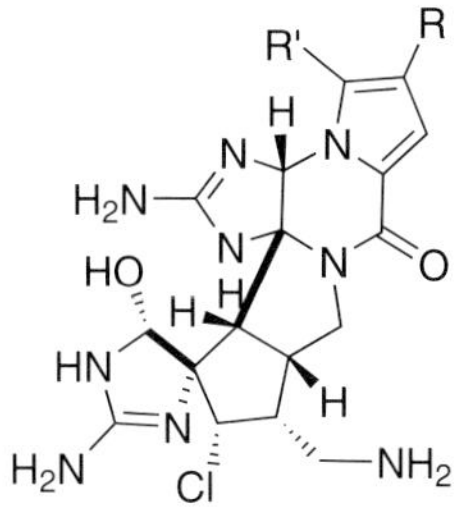

49 R = R' = H
50 R = Br R' = H
51 R = R' = Br

The authors hypothesized the following biogenetic origin for palau'amines (Scheme 10.2).

According to this hypothesis, a phakellin derivative would give rise to a Diels–Alder reaction with 3-amino-1-(2-aminoimidazolyl)-1-propene, a truncated oroidin which is clearly related to the alkaloid girolline, also found in the sponge. The product formed would undergo a chloroperoxidase-initiated chlorination, 1,2-shift/ring contraction, and reaction with water, as shown in the following scheme. This hypothesis has been recently refined by Romo *et al.* [79]. Alternative biosynthetic proposals have been formulated by Al Mourabit and Potier [80], and, more recently, by Baran *et al.* [81]. These latter authors proposed that palau'amine biosynthesis could start with ring expansion of the dimeric alkaloid sceptrin, and could end with the formation of the dimeric konbu'acidin (the structures of these dimeric alkaloids are reported in the next section). Palau'amine would derive from konbu'acidin by loss of one of the pyrrole units [81].

As might be expected, the complex bisguanidine-containing hexacyclic skeleton of palau'amines stimulated a series of creative synthetic efforts, but, while some moieties of the palau'amine skeleton have been prepared, a total synthesis is still lacking [82–84].

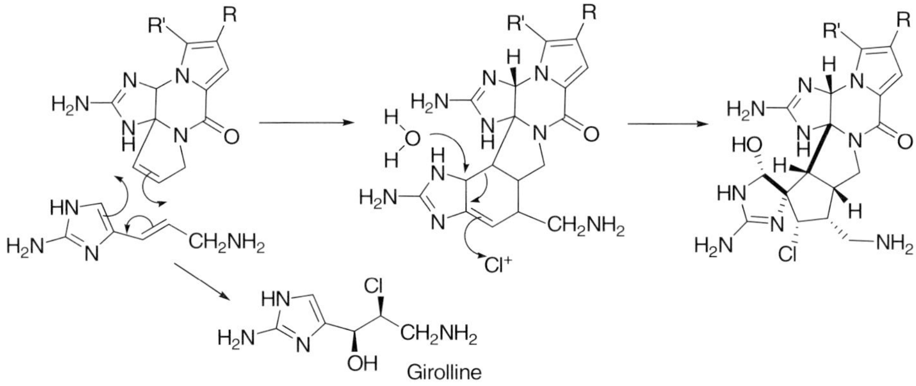

Girolline

Scheme 10.2 Postulated biogenetic origin of palau'amines.

10.3.4
N-7/C-11 + C-4/C-12 Derivatives

Molecules of this class are isomeric with those of the previous class: the only difference being the linkage of the imidazole carbon C-12 with C-4 in place of N-1. The parent compounds are called isophakellins (**52** and **53**), isolated from *Acanthella carteri* [85] and *Agelas* sp. [86], respectively. *N*-Methyldibromophakellin (**54**) has been obtained from *Stylissa caribica* [87]. Another member of the isophakellin family (**55**) has been obtained from *Axinella brevistyla* [88]; interestingly, this molecule possesses the chlorohydrin functionality already encountered in girolline (see Scheme 10.2), and a biogenetic relationship between these two molecules is hypothesized. Also in the case of isophakellins, the sponge *Pseudaxynissa cantharella* is able to elaborate a molecule which is completely enantiomeric to dibromoisophakellin: this metabolite has been called dibromocantharelline (**56**) [51]. Finally, an achiral *seco*-dibromoisophakellin, called ugibohlin (**57**), has been isolated from *Axinella carteri* [89].

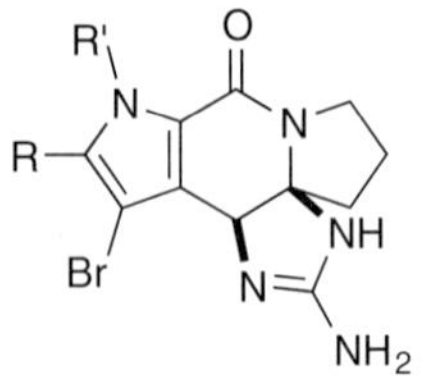

52 R = Br R' = H
53 R = H R' = H
54 R = Br R' = CH_3

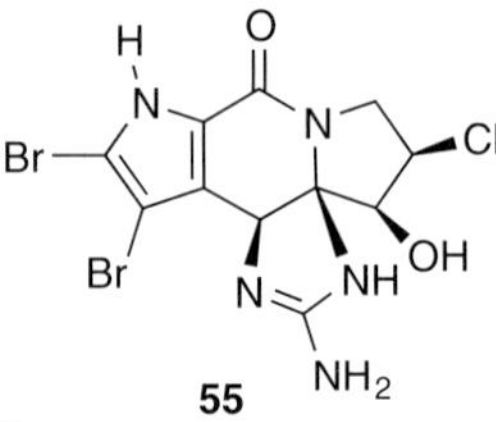

55

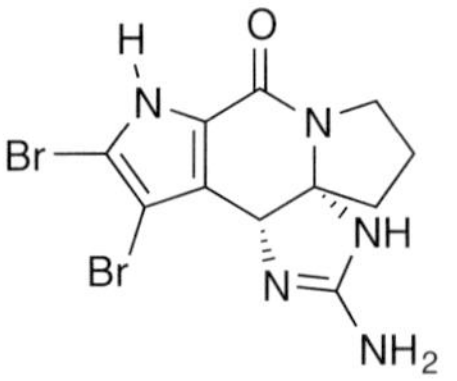

56

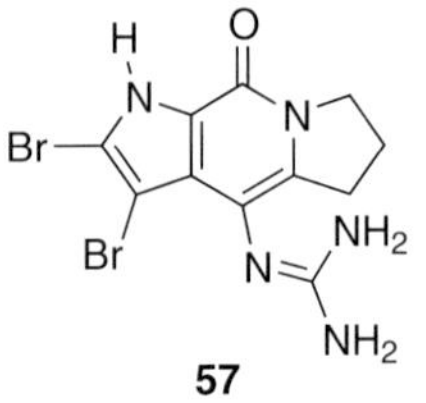

57

This class of cyclic oroidin derivatives also parallels the above N-7/C-11 + N-1/C-12 class in the presence of complex derivatives in which the isophakellin skeleton appears to be conjugated with an aminoimidazolyl propene unit. These isomers of palau'amines are called styloguanidines (**58–60**); they have been isolated from the same sponge that contains palau'amines, *Stylotella aurantium* [90].

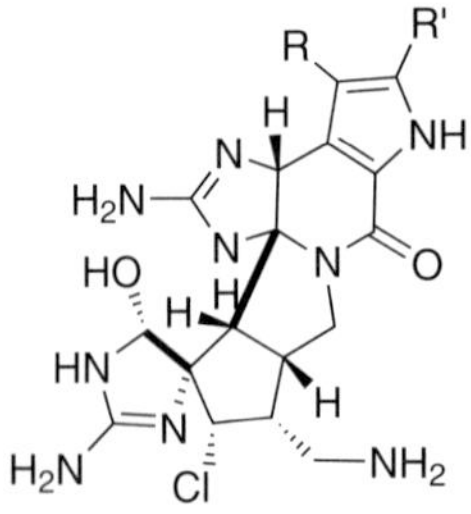

58 R = R' = H
59 R = Br R' = H
60 R = R' = Br

These molecules were shown to be potent inhibitors of chitinase, an enzyme involved in the ecdysis of many insects and crustaceans, whose inhibitors are supposed to control the settlement of barnacles and, therefore, could have a potential application as antifouling agents. Unfortunately, no data are available in the literature about cytotoxic, immunosuppressive, or antibacterial activities of styloguanidines and, thus, a comparison with the potently bioactive congeners palau'amines is not possible.

10.3.5 N-1/C-12 + N-7/C-12 Derivatives

The single member of this class is dibromoagelaspongin (**61**), isolated from an *Agelas* sp. collected along the Tanzanian coasts [91]. This molecule is closely related to dibromophakellin (**42**), but in this case the nitrogen atoms N-1 and N-7 link the same carbon of the imidazole ring, namely C-12, and, consequently, the pyrrole-condensed ring is five-membered and not six-membered. A biogenetic relationship has been proposed between the two molecules. The absolute configuration at the two chiral carbons of dibromoagelaspongin has not been determined.

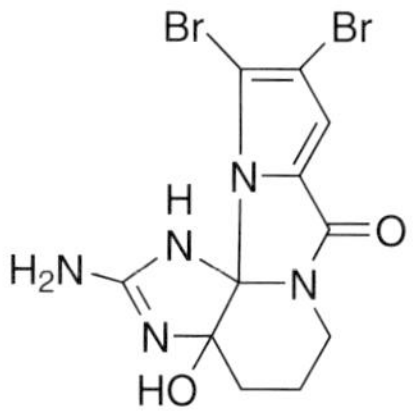

61

10.3.6 N-1/C-9 + C-8/C-12 Derivatives

This class includes only agelastatins A–D (**60–65**), a family of tetracyclic oroidin derivatives isolated from *Agelas dendromorpha* [92,93] and *Cymbastela* sp. [94]. The absolute configuration of these molecules was deduced through the application of exciton-coupling techniques [93].

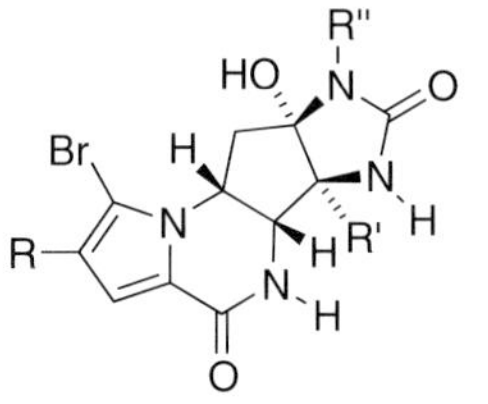

62 R = H R' = H R'' = CH_3
63 R = Br R' = H R'' = CH_3
64 R = H R' = OH R'' = CH_3
65 R = R' = R'' = H

Agelastatins are highly bioactive derivatives, possessing activity against several cancer cell lines at nanomolar concentrations, although the mechanism of this potent action has not yet been elucidated [93]. In addition, agelastatin A (**62**) inhibits glycogen synthase kinase-3β (GSK-3β), an activity that could be useful in the treatment of Alzheimer's disease [95], but, since GSK-3β is also related to the production of tumorigenic promoters involved in melanoma and colon cancer, its inhibition may also be significant in cancer prevention and therapy [95].

The above considerations make agelastatin A (**62**) a molecule of enormous biological interest and, consequently, it has been the object of intense synthetic activity. Indeed, since 2002, four total syntheses of agelastatin A have been reported [96–99]. The availability of multigram amounts of this alkaloid would be very useful for deep investigation of its very promising pharmacological potential.

10.4
Simple or Cyclized Oroidin-like Dimers

Dimeric pyrrole–imidazole alkaloids have been found almost exclusively as secondary metabolites of sponges belonging to the genus *Agelas*. In 1981, sceptrin (**66**) was isolated as the major antimicrobial constituent of *Agelas sceptrum* [100]. Analysis of spectral data and an X-ray study demonstrated that **66** is a symmetrical dimer of the 2-debromo derivative of oroidin. Sceptrin exhibits a broad range of biological activities. First, sceptrin showed an antimicrobial activity against *Staphylococcus aureus, Bacillus subtilis, Candida albicans, Pseudomonas aeruginosa, Alternaria* sp., and *Cladosporium cucumerinium* considerably greater than that recorded for oroidin [100,101]. In a study on the mechanism of action, sceptrin demonstrated a bacteriostatic rather than a bactericidial effect on exponentially growing *Escherichia coli* cells [102].

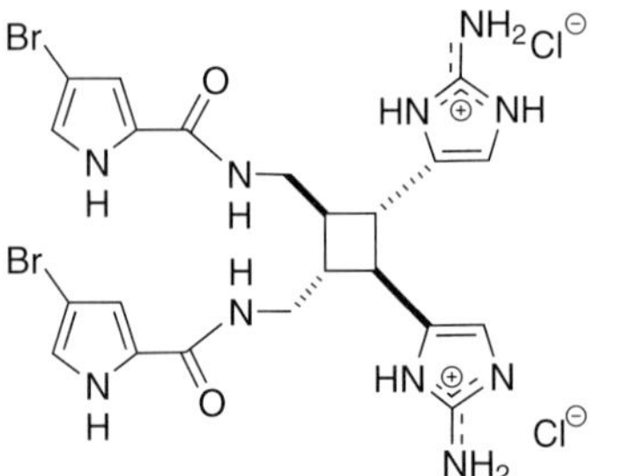

66

The fungicidal activity of sceptrin at 10 ppm was found to be >65 % inhibition of *Phytophthora infestans* (potato late blight) [103]. In addition to its antimicrobial, antiviral [104], antimuscarinic [22], and antihistaminic [30] properties, sceptrin was shown to be the first and most potent nonpeptide somatostatin inhibitor in the submicromolar range [105].

Despite its potent biological activity, sceptrin remained a prominent unanswered synthetic challenge until 2004. Although upon cursory inspection it appears to be a photodimer of hymenidin, synthetic strategies involving the dimerization of **2** have been futile. The first short total synthesis of sceptrin in its racemic form was reported by Baran *et al.* [106], but other approaches, including a short enantioselective total synthesis by programmed oxaquadricyclane fragmentation, have been proposed [107,108].

Sceptrin has also been found in several different sponges, co-occurring with related compounds. Oxysceptrin (**67**), previously found by Rinehart *et al.* in a sample of *Agelas conifera* [109], was re-isolated by Kobayashi *et al.* together with sceptrin from the Okinawan sponge *Agelas* cf. *nemoechinata* [110]. Its activity as an actimyosin-ATPase activator has been reported [110].

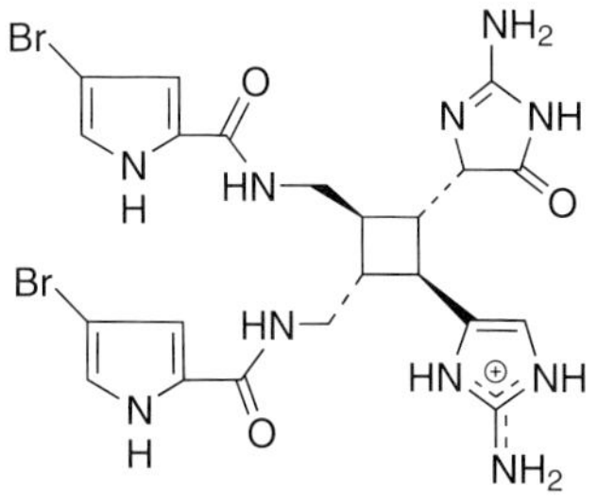

Numerous antiviral and antibacterial sceptrin/oxysceptrin derivatives (**68–72**), along with the diacetate salts of sceptrin and oxysceptrin, were isolated from *Agelas conifera* [104,111,112].

Dibromosceptrin (**69**), tested together with other brominated pyrrole alkaloids for interaction with cellular calcium homeostatis, was shown to reduce voltage-dependent calcium elevation in PC12 cells [113].

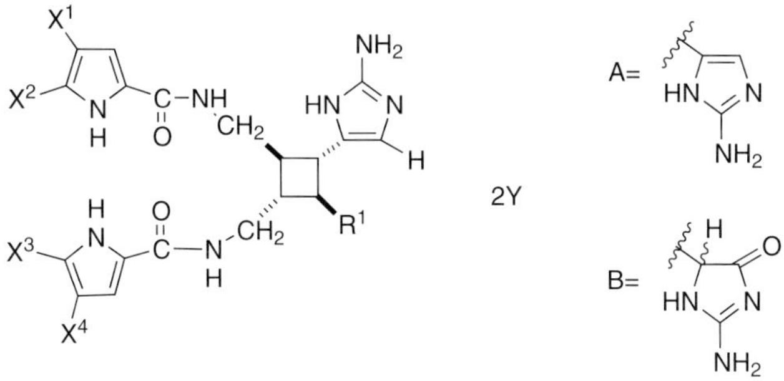

	X^1	X^2	X^3	X^4	R^1	Y
68	Br	H	H	H	A	HOAc
69	Br	Br	Br	Br	A	HOAc
70	H	H	H	Br	B	HOAc
71	Br	Br	H	Br	A	HOAc
72	H	H	H	H	A	HCl

Two sceptrin-related compounds, nakamuranic acid (**73**) and its corresponding methyl ester (**74**), have been isolated from the Indo-Pacific sponge *Agelas nakamurai* [114,115].

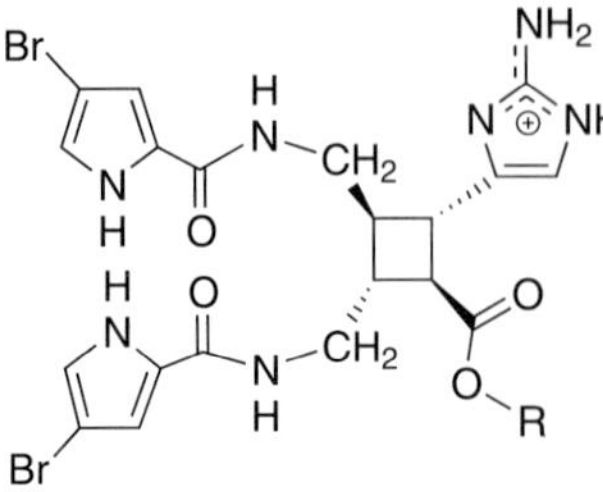

$CF_3COO^{\ominus}$

73 R=H
74 R=CH_3

Ageliferin (**75**), bromoageliferin (**76**), and dibromoageliferin (**77**) were first isolated in 1986 by Rinehart from the sponges *Agelas conifera* and *A.* cf. *mauritiana* [116] but detailed structure elucidation including stereochemistry was reported in 1990 by Kobayashi *et al.* [117]. Subsequently, (**75–77**) were found in the Caribbean sponge *Agelas conifera* [104] as well as in *Stylissa caribica* [118], a sponge closely related to *Axinella corrugata*. Both bromoageliferin and dibromoageliferin reduce voltage-dependent calcium entry in PC12 cells, but not store-operated calcium entry [119]. The first synthesis of ageliferin, accomplished by Baran *et al.*, was based on the thermal conversion of sceptrin into ageliferin and included the first vinyl cyclobutane rearrangement of a natural product [120].

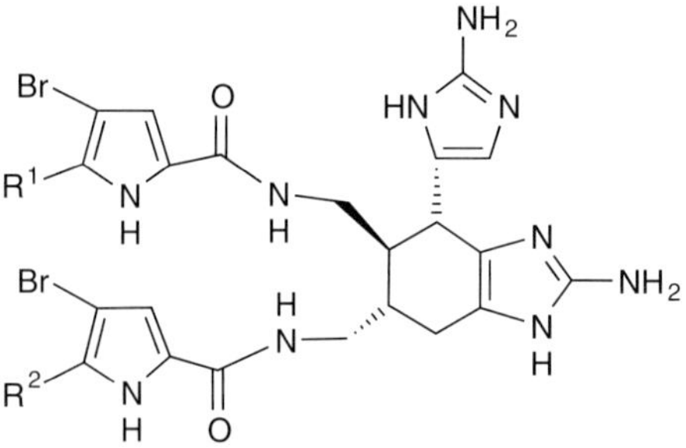

75	R^1=H	R^2=H
76	R^1=Br	R^2=H
77	R^1=Br	R^2=Br

Seven new ageliferin derivatives (**78–84**), characterized by methylation on one or both of the pyrrole nitrogens, together with previously isolated bromoageliferin and dibromoageliferin, were found in *Astrosclera willeyana*, a calcareous sponge coming from Pohnpei, Micronesia [121].

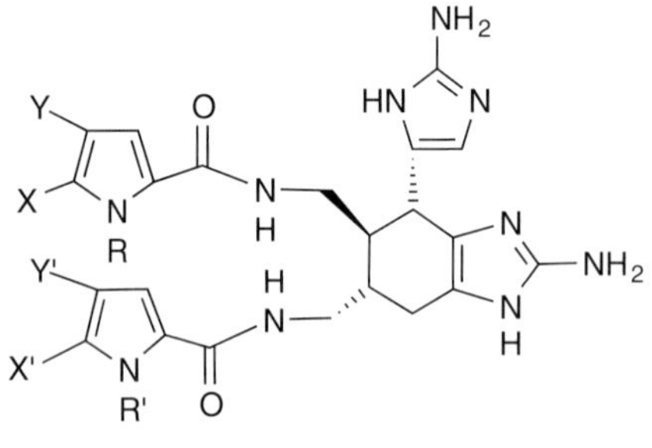

78	R=X=X'=H	R'=C H_3	Y=Y'=Br
79	R=R'=C H_3	X=X'=H	Y=Y'=Br
80	R=X=Y'=H	R'=C H_3	X'=Y=Br
81	R=R'=C H_3	X=Y'=H	X'=Y=Br
82	R=X'=H	R'=C H_3	X=Y =Y'=Br
83	R=X=H	R'=C H_3	X'=Y=Y'=Br
84	R=H	R'=C H_3	X=X'=Y=Y'=Br

Sclerosponges such as *A. willeyana* have been difficult to classify. However, the discovery of metabolites of the oroidin class lend support to the assignment of *A. willeyana* to the order Agelasida. Bioassay-guided fractionation of the methanol extract of *Agelas mauritiana* led to the isolation of a new antifouling oroidin dimer named mauritiamine (**85**) [122]. Its synthesis has been reported [123].

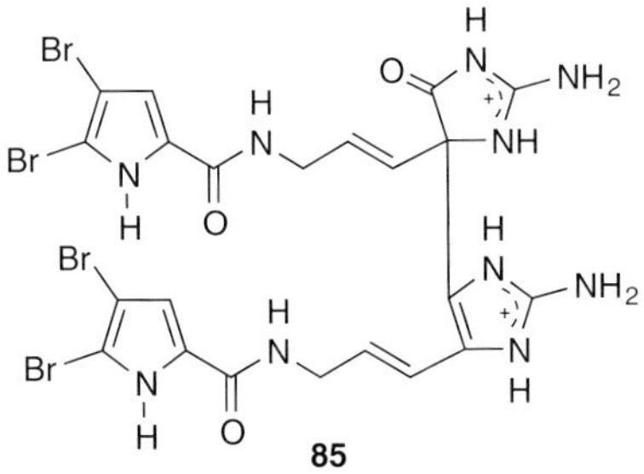

85

Konbu'acidin A (**86**), a dimeric pyrrole–imidazole alkaloid with a fused-hexacyclic skeleton containing two guanidine units, related to palau'amine (**49**) and styloguanidines (**58**–**60**), was isolated from an Okinawan *Hymeniacidon* sp. [124]. Konbu'acidin A (**86**) exhibited inhibitory activity against cyclin-dependent kinase 4 (cdk4) [124].

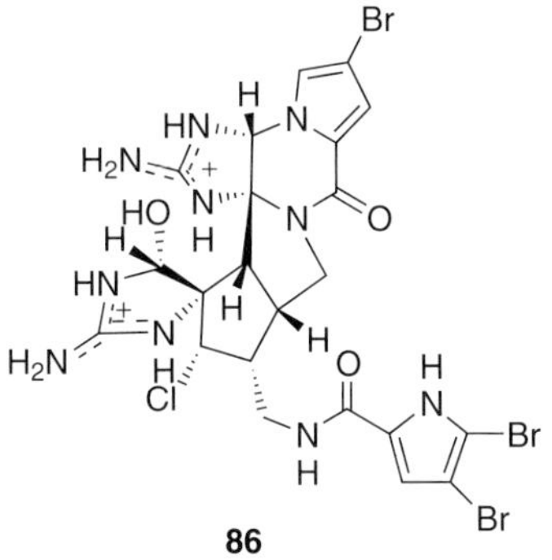

86

Four imidazo-azolo-imidazole alkaloids, axinellamines A–D (**87**–**90**), were isolated from an Australian *Axinella* sp.; compounds **88**–**90** had bactericidal activity against *Helicobacter pylori*, a Gram-negative bacterium associated with pepticular and gastric cancer [125].

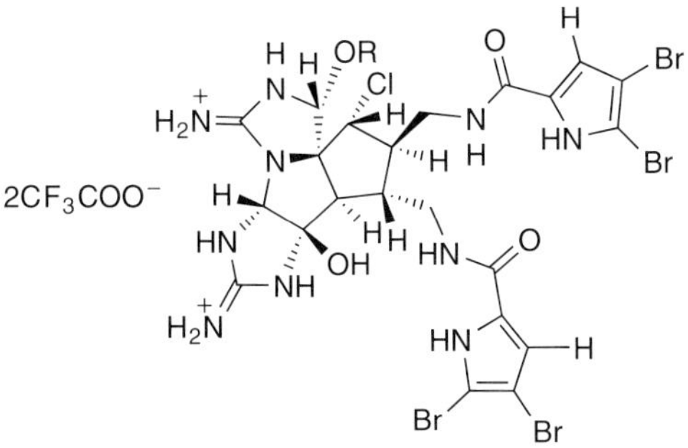

87 R=H
89 R=CH_3

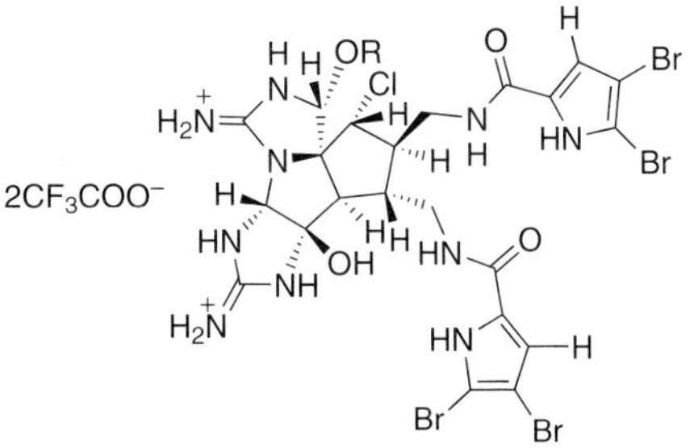

88 R=H
90 R=CH_3

Eight dimeric bromopyrrole alkaloids, nagelamides A–H (**91–98**) were isolated from *Agelas* spp. [126]. Nagelamides A–D (**91–94**) possess a connection between C-10 and C-15′, while nagelamide H (**98**), like mauritiamine (**85**), possesses a C-11–C-15′ bond. The elucidation of the relative stereochemistry of each cyclohexene ring in nagelamides E–G (**95–97**), was carried out by analysis of ROESY spectra, and these molecules were thus proved to be diastereomers of ageliferins.

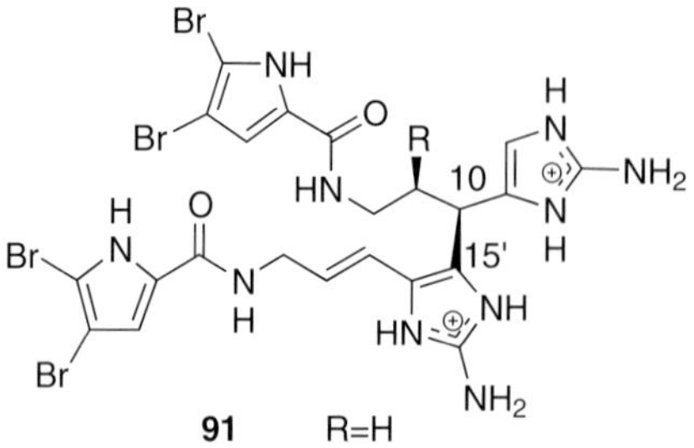

91 R=H
92 R=OH

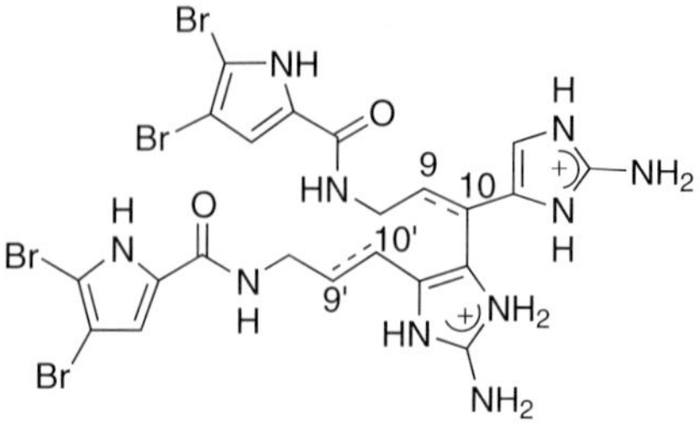

93 $\Delta^{9(10),9'(10')}$
94 9,9′,10,10′-tetrahydro

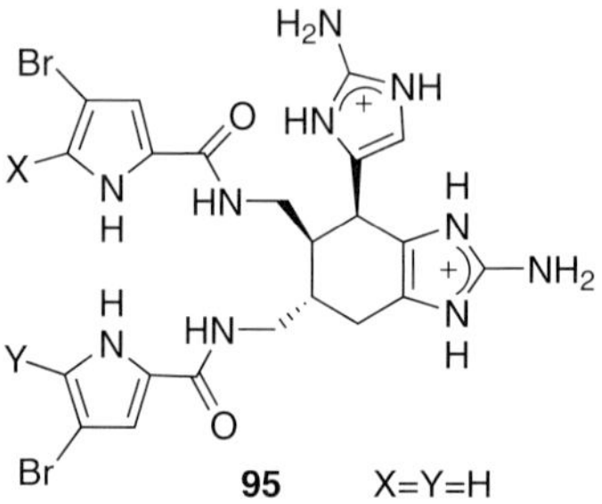

95 X=Y=H
96 X= Br Y=H
97 X= Y= Br

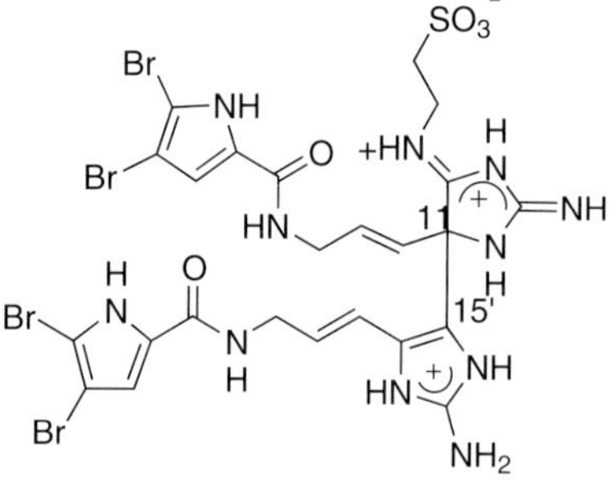

98

In 2003, massadine (**99**), a highly oxygenated congener of dimeric oroidin derivatives, was isolated from the marine sponge *Stylissa* aff. *massa* [127]. In the proposed structure two cyclic guanidines face each other in the endo positions of ring B, which adopts a boat conformation. The positive exciton split present in the CD spectrum of massadine due to 4,5-dibromopyrrole-2-carboxyamide side chains allowed the *1R,15S* configuration to be assigned.

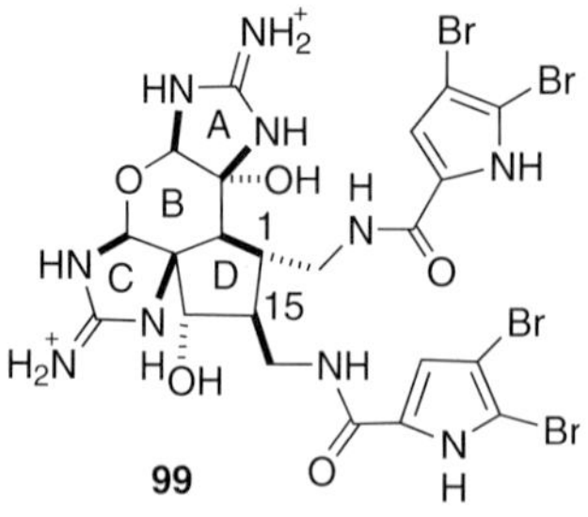

99

The first examples of tetrameric pyrrole imidazole alkaloids, stylissadine A (**100**) and B, were isolated from the Caribbean sponge *Stylissa caribica* in 2006 [128]. They possess a symmetric dimeric structure derived from condensation of two massadine units, and differ in the configuration at the center C-2′. Stylissadines represent the largest pyrrole-imidazole alkaloids isolated so far, and, with their 16 stereogenic centers, they are the most complex structures known within the oroidin alkaloid family.

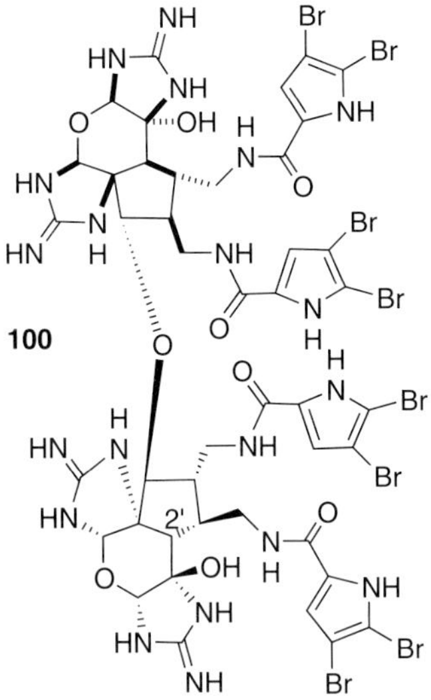

100

10.5 Other Bromopyrrole Alkaloids

Several simple bromopyrroles (**101–109**) have been isolated from *Agelas oroides* (**101**, **105**) [129,1], *Agelas flabelliformis* (**101**, **102**) [130], *Acanthella carteri* (**103**, **104**) [131], *Agelas nakamurai* (**106**, **107**) [132], *Homoaxinella* sp. (**108**) [133], and *Axinella verrucosa* (**109**)[32].

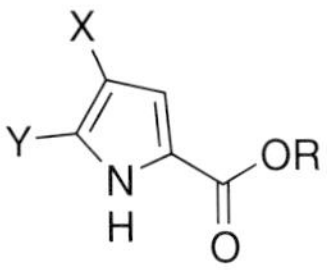

	X	Y	R
101	Br	Br	H
102	Br	Br	CH_3
103	Br	Br	NH_2
104	Br	H	NH_2
105	H	H	NHCHO

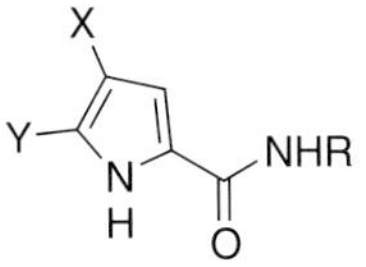

	X	Y	R
106	H	Br	H
107	H	Br	CH_2OCH_3
108	Br	Br	CH_2OCH_3
109	Br	H	CH_2OCH_3

Compounds **101** and **103** deterred fish feeding at 10 mg/mL [7]. 4,5-Dibromopyrrole-2-carboxylic acid (**101**) was investigated for its effects on cellular calcium homeostatis in PC12 cells. Its unpalatability for predatory reef fish is not only transduced by specific membrane receptors present on sensory nerve cells but additionally through a more general pharmacological effect on cellular calcium homeostatis, even in a rat phaeochromocytoma cell line [134].

A novel alkaloid with the unusual pyrrolopiperazine nucleus, (9*S*)-longamide (**110**) was obtained from the Caribbean sponge *Agelas longissima* [135], and more recently from a Japanese *Homoaxinella* sp. [133]. The 2-debromo derivative of **110** was isolated from the sponge *Axinella carteri* (**111**) [136] together with bromoaldisine (**112**) and aldisine (**113**), previously found in the sponge *Hymenaicidon aldis* [137].

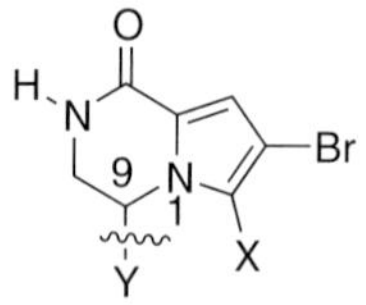

110 X=Br Y= —OH
111 X =H Y= —OH

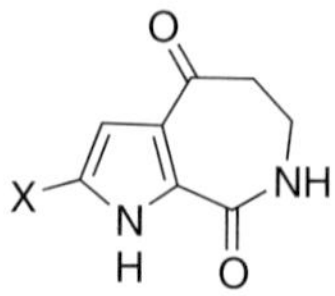

112 X=Br
113 X=H

Longamide B (**114**) has been isolated, for the first time in racemic form, from the Caribbean sponge *Agelas dispar* [138]. The methyl ester of longamide B (**115**) was found, in racemic form, in the Japanese *Homoaxinella* sp. [133], while the (9*S*)-enantiomer (**116**) was found in the sponge *Agelas ceylonica* [139]. The 3-debromo-derivative (**117**) of **115** was found in the sponge *Axinella tenuidigitata* [140]. Hanishin (**118**), longamide B ethyl ester, was isolated from a collection coming from the Hanish Islands (Red Sea) of the highly polymorphic sponge *Acanthella carteri* [131].

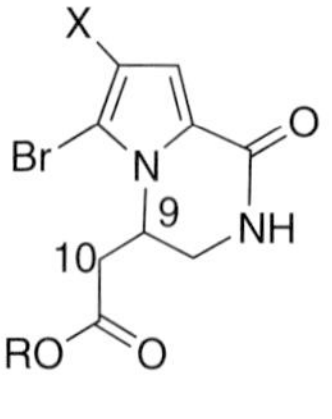

(±)**114** X= Br R=H
(±)**115** X= Br R=CH_3
(9*S*)-(+)**116** X= Br R=CH_3
(±)**117** X= H R=CH_3
(±)**118** X= Br R=CH_2CH_3

Racemic syntheses of the pyrroloketopiperazine marine natural products (±)-longamide (**111**) [141,142], (±)-longamide B (**114**) [142], (±)-longamide B methyl ester (**115**) [142], and (±)-hanishin (**118**)[142] have been performed in a concise fashion starting from pyrrole. The first syntheses of enantiopure (*S*)-(−) longamide B, (*S*)-(−)-hanishin, and (*S*)-(−) longamide B methyl ester (**116**) were described in 2005 [143]. In a study of the cyclization of 2-pyrrolecarboxamido acetals, the structure reported as homolongamide **119a**, a compound obtained by the quantitative cyclization of 4,5-dibromo-*N*-(2-formylethyl)-1*H*-pyrrole-2-carboxamide [141], was found to be in better agreement with hemiacetal **119b** [144].

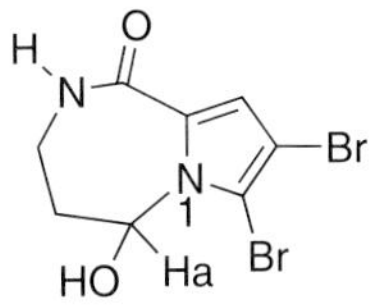

119a (CD_3OD) 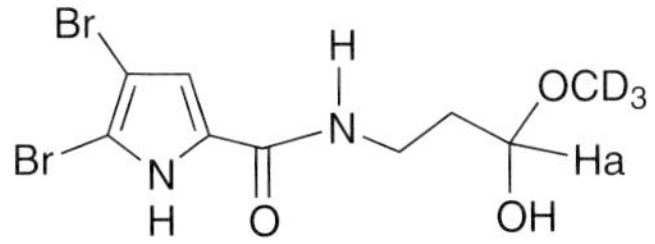 **119b** (CD_3OD)

A spermidine derivative with two 4,5-dibromopyrrole-2-carbamyl units, pseudoceratidine (**120**), has been isolated from the sponge *Pseudoceratina purpurea* as an antifouling compound [145].

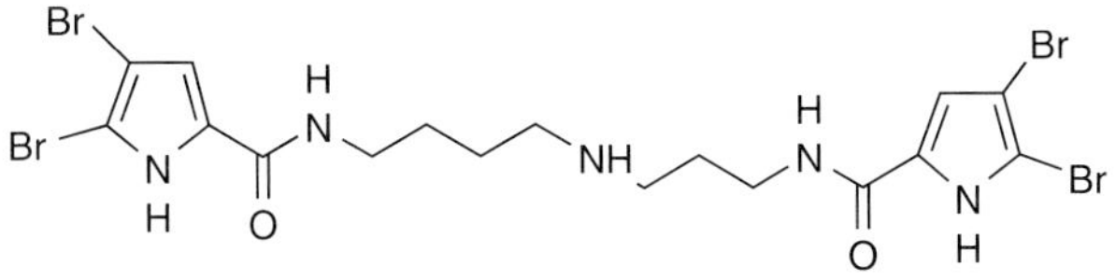

120

Two syntheses of pseudoceratidine (**120**), which was reported to inhibit metamorphosis of the barnacle *Balanus amphitrite,* have been described [146,147]. To delineate the structural requirements for this bioactivity, the corresponding N-1 and N-8 mono-(4,5-dibromo-2-pyrrolyl) amides of spermidine were also synthesized, as well as the closely related mono- and bis-(4,5-dibromo-2-pyrrolecarboxamido) residues to test the importance of the pyrrole substituents and the positively charged secondary amine group in **120** [148].

Agelongine (**121**), an alkaloid containing a pyridinium ring in place of the typical imidazole nucleus, has been isolated from *Agelas longissima*; its antiserotoninergic activity tested *in vitro* on the rat stomach fundus trip was also reported [149]. The bromine-free analog of **121**, daminin (**122**), was isolated from the sponge *Axinella damicornis* and its total synthesis was developed. *In vitro* tests on rat cortical cell cultures showed that daminin (**122**) might represent a new therapeutic tool for the treatment of CNS diseases such as Parkinson's and Alzheimer's diseases [150].

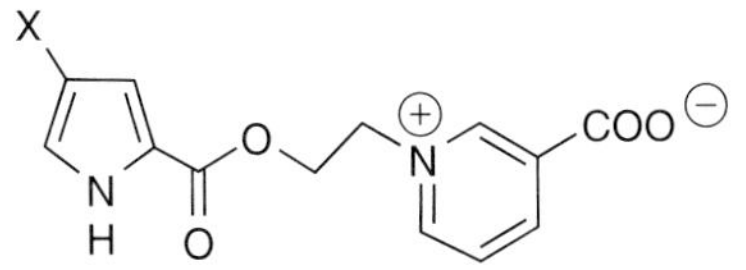

121 X=Br
122 X=H

Clathramides A–D (**123**–**126**) are imidazole-containing bromopyrrole alkaloids but, since in this case the imidazole ring is part of a histidine-like moiety, their skeleton appears to be different from that of oroidin. They were isolated from *Agelas clathrodes* [151] and *A. dispar* [138] as mild antifungal agents tested in agar disk assays on *Aspergillus niger.*

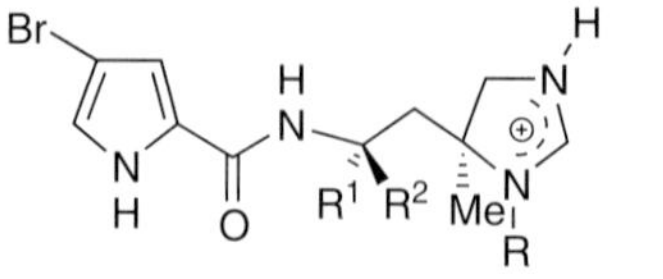

	R	R^1	R^2
123	CH_3	H	COO^-
124	CH_3	COO^-	H
125	H	H	COO^-
126	H	COO^-	H

Assmann *et al.* described, in addition to known alkaloids, the isolation of the first pyrrole–imidazole alkaloid with a guanidine function instead of the aminoimidazole (**127**) [152] from the sponge *Agelas wiedenmayeri*, while its decarboxylated derivate laughine (**128**) was isolated from the sponge *Eurypon laughlini* [153].

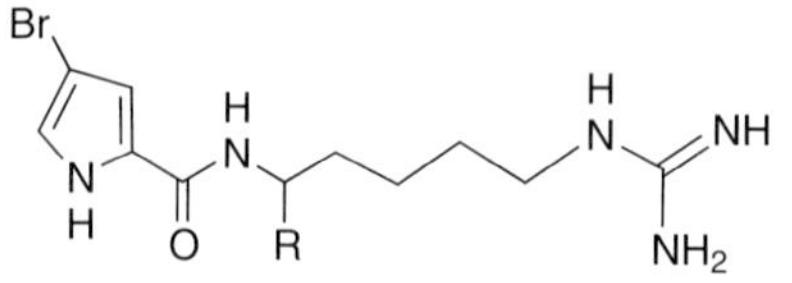

127 R=COOH
128 R=H

The hypothetical biosynthetic pathway of **127**, which does not correspond to the proposed biosynthesis of the oroidin-like alkaloids, is based on the formation of an amide bond between a pyrrole-2-carboxylic acid precursor and an aminopropylimidazole moiety, which are both derived from ornithine [3]. Therefore, the authors formulated the hypothesis that compound **127** may be alternatively a biosynthetic precursor of hymenidin/oroidin-related alkaloids in sponges of the genus *Agelas*. 4-Bromopyrrole-carboxyarginine (**129**) and 4-bromopyrrole-2-carboxy-*N*(ε)-lysine (**130**) were isolated from the sponge *Stylissa caribica* [154]. They differ from **127** only by replacement of homoarginine with arginine in **129** and with lysine in **130**.

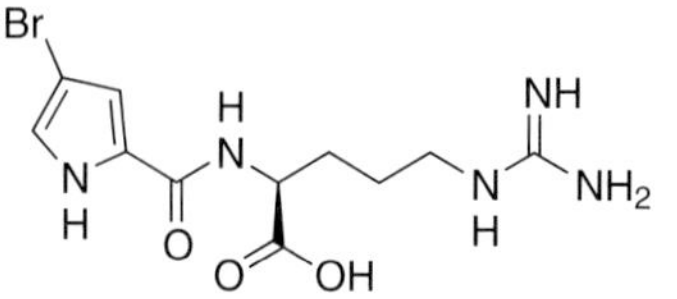

129 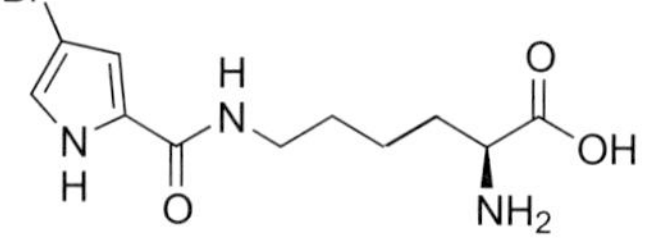 **130**

The first bromopyrrole alkaloids with a tetrahydropyrimidine ring attached through an ester linkage were manzacidins A–C (**131–133**), isolated from the sponge *Hymeniacidon* sp. [155]. By employing a new method for saturated C–H bond functionalization that makes possible the stereospecific installation of tetrasubstituted carbinolamines, a rapid, enantioselective, and high-throughput synthesis of both manzacidines A and C has been achieved [156]. More recently, a catalytic tandem asymmetric conjugate addition–protonation reaction, with C-6′-cinchona alkaloid derivatives as dual-function chiral catalysts, has been developed; this approach was employed in the development of a concise and highly stereoselective route to (−)manzacidine A [157].

The family of manzacidines was expanded with the isolation of manzacidine D (**134**) from a living fossil sponge *Astrosclera willeyana* [158] and, more recently, of compound **135**, *N*-methyl manzacidine C, from the marine sponge *Axinella brevistyla* [159]. The first total synthesis of manzacidin D, in 11 steps and 16 % overall yield, from commercially available glycine *tert*-butyl ester hydrochloride was reported in 2004 [160].

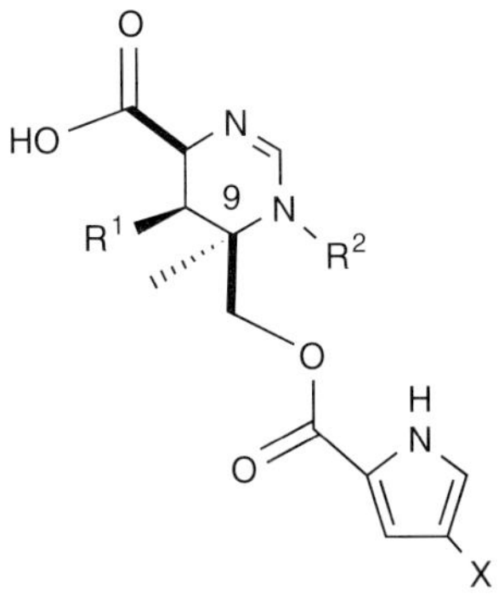

	R^1	R^2	X
131	H	H	Br
132	OH	H	Br
133	H	H	Br (9-*epi*)
134	H	CH_3	H
135	H	CH_3	Br (9-*epi*)

Ageladine A (**136**), an antiangiogenic matrix metalloproteinase inhibitor was isolated from the marine sponge *Agelas nakamurai* [161]. The antiangiogenic effect of **136** was evaluated by the *in vitro* vascular organization model using mouse ES cells. A 12-step total synthesis of the tricyclic heteroaromatic metabolite **136** has been achieved using a 6π-azaelectrocyclization and a Suzuki–Miyaura coupling of *N*-Boc-pyrrole-2-boronic acid with a chloropyridine as key step [162].

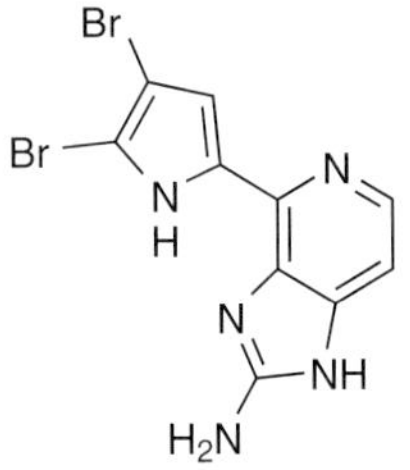

136

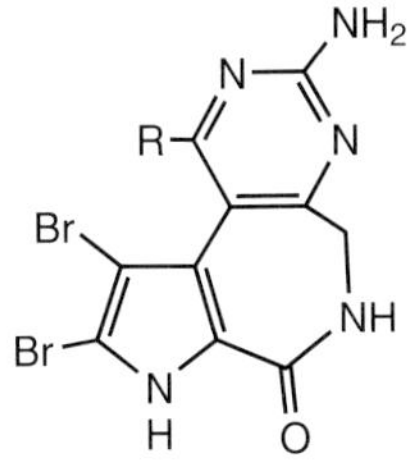

137 R=H
138 R=COOH

Latonduines A (**137**) and B (**138**), have been isolated from the Indonesian sponge *Stylissa carteri* [163]. Their skeleton cannot be derived from the $C_{11}N_5$ building block of the oroidins; it has been proposed that ornithine is the biogenetic precursor to the aminopyrimidine fragment of the latonduines. Agelasine G (**139**), a new antileukemic alkaloid, containing 9-methyladenine and a diterpene moiety, was isolated from an Okinawan *Agelas* sp. [164].

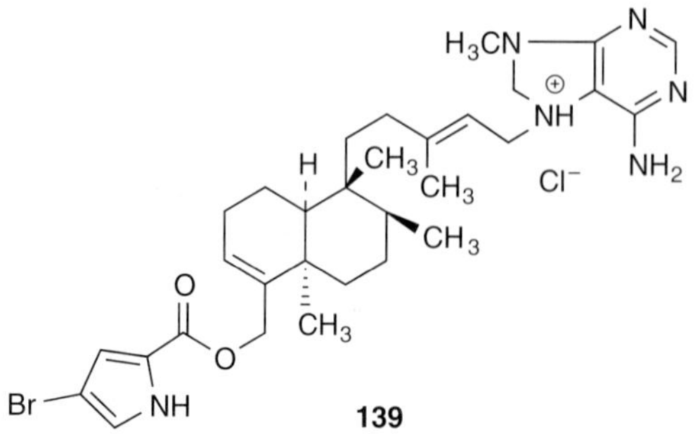

10.6 Conclusions

The structural complexity and diversity of bromopyrrole alkaloids continue to challenge organic chemists, as indicated by the high number of contributions appearing since 2000.

Their interesting pharmacological activities, outlined in Table 10.1, have inspired the development of several synthetic approaches to these alkaloids for the purpose of further biological evaluation. However, current interest in this group of marine compounds has been concerned not only with preparative synthesis and biological activity but also with their biogenetic chemical reactions. In the group of bromopyrrole alkaloids, all compounds can be considered closely related; they seem to share a common biogenetic pathway even if a plausible biosynthetic mechanism leading to these marine alkaloids is still unproven. The first (and only) biosynthetic study performed in cell cultures of the sponge *Teichaxinella morchella* using [^{14}C]-labeled amino acids revealed that histidine and proline/ornithine are precursors of stevensine (**26**) [165].

In the light of these studies, it has been proposed that both proline and ornithine can be converted to pyrrole-2-carboxylic acid prior to halogenation; subequently amide formation by reaction with 3-amino-1-(2-aminoimidazolyl)-prop-1-ene, derived from histidine, generates oroidin, followed by cyclization to stevensine (Scheme 10.3).

Different molecular mechanisms have been separately postulated for dibromophakellin [74], dibromoagelaspongin [91], agelastatin [92], mauritiamine [88], and palau'amine [78]. Al Mourabit and Potier proposed a universal chemical pathway, starting from the simple precursors **101** and **140** and leading to over 60 pyrrole–imidazole alkaloids [80]. A new biomimetic spontaneous conversion of proline to 2-aminoimidazolinone derivatives using a self-catalyzed intramolecular transamination reaction together with peroxide dismutation as key step has been described [166]. This work has pointed to dispacamide A as the forerunner of oroidin and compounds **101** and **140** as probable hydrolysis products of oroidin and not the precursors. In this

Tab. 10.1 Selected bioactivities of bromopyrrole alkaloids.

Biological activity	Compounds
Feeding deterrence[a]	Oroidin (**1**) [4–7], Sventrin (**4**) [22], Stevensine (**26**) [52], Sceptrin (**66**) [113]
Interaction with calcium and sodium homeostasis[a]	Oroidin (**1**) [20], Hymenidin (**2**) [20], Clathrodin (**3**) [23], Dibromosceptrin (**69**) [114], Compound **101** [134]
Antifouling[c]	Mauritiamine (**85**) [122], Pseudoceratidine (**120**) [145]
Antiserotonergic	Oroidin (**1**) [18,22], Hymenidin (**2**) [18,22], Clathrodin (**3**) [18,22], Keramadine (**5**) [22], Agelongine (**121**) [149]
Antimuscarinic	Oroidin (**1**) [18,22], Hymenidin (**2**) [18,22], Clathrodin (**3**) [18,22], Keramadine (**5**) [22], Sceptrin (**66**) [22]
Antihistaminic	Dispacamide A–D (**7–10**) [29], Taurodispacamide (**19**) [40], Sceptrin (**66**) [30]
Anti-α-adrenergic	Hymenin (**24**) [47]
Neuroprotective	Compound **11** [32], Compound **14** [32], Compound **20** [32]
Antibacterial	Hymenin (**24**) [47], Palau'amine (**49**) [77], Sceptrin (**66**) [100], Sceptrins (**68–72**) [104,111,112], Axinellamides B–D (**88–90**) [125]
Antifungal	Sceptrin (**66**) [103], Clathramides A–D (**123–126**) [151]
Antiviral	Sceptrin (**66**) [104], Sceptrins (**68–72**) [104,111,112]
Cytotoxic	Slagenins B–C (**22–23**) [41], Phakellstatin (**47**) [73], Palau'amine (**49**) [77], Agelastatins A–D (**62–65**) [93], Agelasine G (**139**) [164]
Antiangiogenic[b]	Ageladine A (**136**) [161]
Immunosuppressive	Palau'amine (**49**) [77]
Inhibition of mitogen-activated kinase 1[b]	Hymenialdisins (**28–33**) [57]
Inhibition of EGF kinase[b]	Tauroacidins A–B (**17–18**) [39]
Inhibition of cyclin-dependent kinase 4[b]	Konbu'acidin (**86**) [124]
Inhibition of glycogen synthase kinase-3β[b]	Agelastatins A–D (**62–65**) [95]
Inhibition of G_2 DNA damage checkpoint[b]	Hymenialdisins (**28–33**) [56]
Inhibition of pro-inflammatory cytokines	Hymenialdisins (**28–33**) [59,60]
Inhibition of chitinase[c]	Styloguanidines (**58–60**) [90]
Inhibition of somatostatin	Sceptrin (**66**) [105]
Actimyosin ATPase activation	Oxysceptrin (**67**) [110]

[a] These two activities could be related.
[b] These activities are potentially beneficial in the cancer therapy.
[c] These two activities are related.

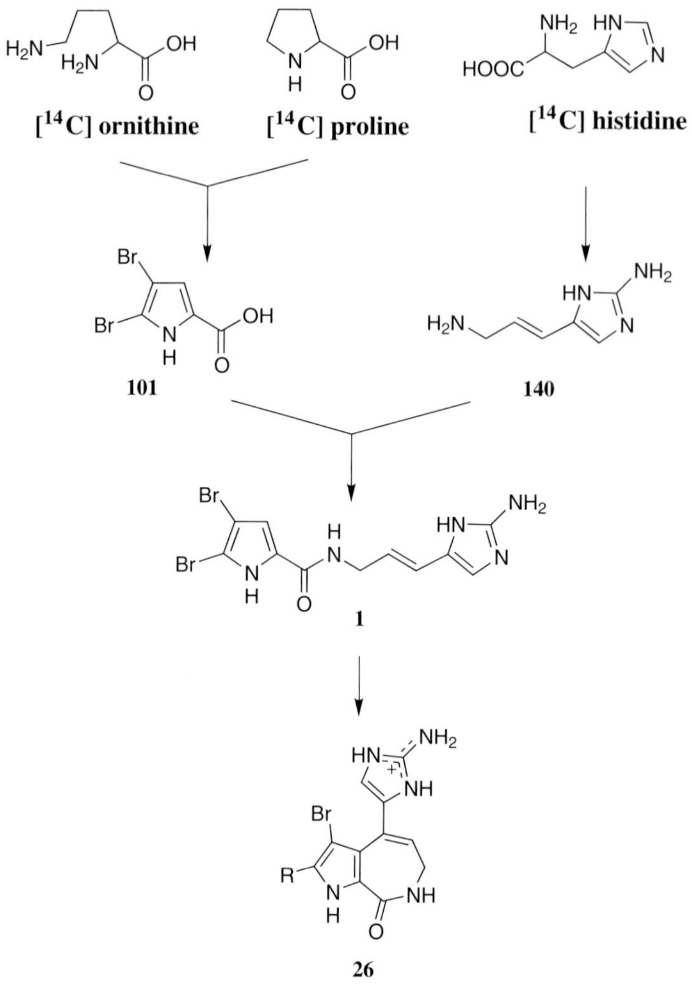

[^{14}C] ornithine [^{14}C] proline [^{14}C] histidine

101 140

1

26

Scheme 10.3 The biogenetic origin of stevensine from the sponge *T. morchella*.

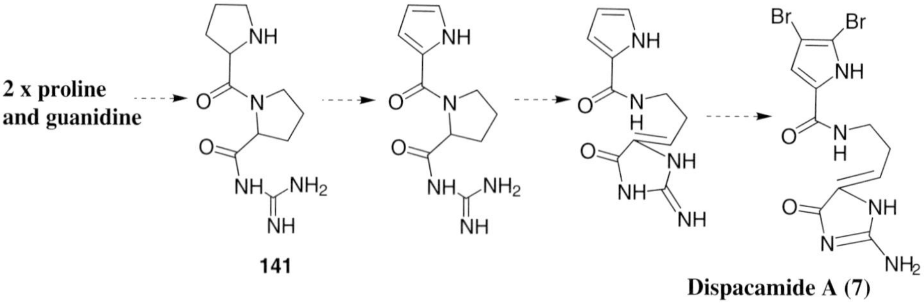

2 x proline and guanidine

141

Dispacamide A (7)

Scheme 10.4 Postulated formation of pyrrole and 2-aminoimidazolinone moieties.

view, the pseudo-dipeptide pyrrole–proline–guanidine **141** has been considered as a likely precursor leading to the amide-connected $C_{11}N_5$ pyrrole and 2-aminoimidazolinone moieties (Scheme 10.4) [166].

In conclusion, research on bromopyrrole alkaloids represents a striking example in which the combination of isolation of natural products and the study of biomimetic chemical reactivity can provide new ideas for both biogenetically inspired syntheses [167] and biogenetic studies.

References

1 Forenza, S., Minale, L., Riccio, R., Fattorusso, E. (1971) *Journal of the Chemical Society, D Chemical Communications*, 1129–1130.

2 Garcia, E. E., Benjamin, L. E., Fryer, R. I. (1973) *Journal of the Chemical Society, D Chemical Communications*, 78–79.

3 Braekman, J. C., Daloze, Q., Stoller, C., van Soest, R. W. (1992) *Biochemical Systemetics and Ecology*, **20**, 417–431.

4 Chanas, B., Pawlik, J. R., Lindel, T., Fenical, W. (1997) *Journal of Experimental Marine Biology and Ecology*, **208**, 185–196.

5 Wilson, D. M., Puyama, M., Fenical, W., Pawlik, J. R. (1999) *Journal of Chemical Ecology*, **25**, 2811–2823.

6 Assmann, M., Licthe, E., Pawlik, J. R., Kock, M. (2000) *Marine Ecology—Progress Series*, **207**, 255–262.

7 Lindel, T., Hoffmann, H., Hochgurtel, M., Pawlik, J. R. (2000) *Journal of Chemical Ecology*, **26**, 1477–1496.

8 Richelle-Maurer, E., De Kluijver, M. J., Feio, S., Gaudencio, S., Gaspar, H., Gomez, R., Tavares, R., Van de Vyver, G., Van Soest, R. W. M. (2003) *Biochemical Systemetics and Ecology*, **31**, 1073–1091.

9 Hoffmann, H.Lindel, T. (2003) *Synthesis*, 1753–1783.

10 Tsukamoto, S., Kato, H., Hirota, H., Fusetani, N. (1996) *Journal of Natural Products*, **59**, 501–503.

11 De Nanteuil, G., Ahond, A., Poupat, C., Thoison, O., Potier, P. (1986) *Bulletin de la Societe Chimique de France*, **5**, 813–816.

12 Little, T. L.Webber, S. E. (1994) *Journal of Organic Chemistry*, **59**, 7299–7305.

13 Daninos-Zeghal, S., Al Mourabit, A., Ahond, A., Poupat, C., Potier, P. (1997) *Tetrahedron*, **53**, 7605–7614.

14 Olofson, A., Yakushijin, K., Horne, D. A. (1998) *Journal of Organic Chemistry*, **63**, 1248–1253.

15 Lindel, T.Hochguertel, M. (2000) *Journal of Organic Chemistry*, **65**, 2806–2809.

16 Berree, F., Girard-Le Bleis, P., Carboni, B. (2002) *Tetrahedron Letters*, **43**, 4935–4938.

17 Poeverlein, C., Breckle, G., Lindel, T. (2006) *Organic Letters*, **8**, 819–821.

18 Kobayashi, J., Ohizumi, Y., Nnakamura, H., Hirata, Y. (1986) *Experientia*, **42**, 1176–1177.

19 Morales, J. J. and Rodriguez, A. D. (1991) *Journal of Natural Products*, **54**, 629–631.

20 Bickmeyer, U., Drechsler, C., Kock, M., Assmann, M. (2004) *Toxicon*, **44**, 45–51.

21 Assmann, M., Zea, S., Koeck, M. (2001) *Journal of Natural Products*, **64**, 1593–1595.

22 Rosa, R., Silva, W., Escalona de Motta, G., Rodriguez, A. D., Morales, J. J., Ortiz, M. (1992) *Experientia*, **48**, 885–887.

23 Rivera Rentas, A. L., Rosa, R., Rodriguez, A. D., Escalona de Motta, G., (1995) *Toxicon*, **33**, 491–497.

24 Daninos, S., Al Mourabit, A., Ahond, A., Zurita, M. B., Poupat, C., Potier, P. (1994) *Bulletin de la Societe Chimique de France*, **131**, 590–599.

25 Breckle, G., Polborn, K., Lindel, T. (2003) *Zeitschrift fur Naturforschung B: Journal of Chemical Sciences*, **58**, 451–456.

26 Nakamura, H., Ohizumi, Y., Kobayashi, J., Hirata, Y. (1984) *Tetrahedron Letters*, **25**, 2475–2478.

27 Endo, T., Tsuda, M., Okada, T., Mitsuhashi, S., Shima, H., Kikuchi, K., Mikami, Y., Fromont, J., Kobayashi, J. (2004) *Journal of Natural Products*, **67**, 1262–1267.

28 Lindel, T. and Hochgurtel, M. (1998) *Tetrahedron Letters*, **39**, 2541–2544.

29 Cafieri, F., Fattorusso, E., Mangoni, A., Taglialatela-Scafati, O. (1996) *Tetrahedron Letters*, **7**, 3587–3590.

30 Cafieri, F., Carnuccio, R., Fattorusso, E., Taglialatela-Scafati, O., Vallefuoco, T. (1997) *Bioorganic and Medicinal Chemistry Letters*, **7**, 2283–2288.

31 Lindel, T. and Hoffmann, H. (1997) *Tetrahedron Letters*, **38**, 8935–8938.

32 Aiello, A., D'Esposito, M., Fattorusso, E., Menna, M., Mueller, W. E. G., Perovic-Ottstadt, S., Schroeder, H. C. (2006) *Bioorganic and Medicinal Chemistry*, **14**, 17–24.

33 Uemoto, H., Tsuda, M., Kobayashi, J. (1999) *Journal of Natural Products*, **62**, 1581–1583.

34 Hu, J.-F., Peng, J., Kazi, A. B., Kelly, M., Hamann, M. T. (2005) *Journal of Chemical Research*, **7**, 427–428.

35 Chevolot, L., Padua, S., Ravi, B. N., Blyth, P. C., Scheuer, P. J. (1977) *Heterocycles*, **7**, 891–894.

36 Jimenez, C. and Crews, P. (1994) *Tetrahedron Letters*, **35**, 1375–1378.

37 Lindel, T. and Hoffmann, H. (1997) *Liebigs Annalen*, **7**, 1525–1528.

38 Fresneda, P. M., Molina, P., Sanz, M. A. (2001) *Tetrahedron Letters*, **42**, 851–854.

39 Kobayashi, J., Inaba, K., Tsuda, M. (1997) *Tetrahedron*, **53**, 16679–16682.

40 Fattorusso, E. and Taglialatela-Scafati, O. (2000) *Tetrahedron Letters*, **41**, 9917–9922.

41 Tsuda, M., Uemoto, H., Kobayashi, J. (1999) *Tetrahedron Letters*, **40**, 5709–5712.

42 Barrios Sosa, A. C., Yakushijin, K., Horne, D. A. (2000) *Organic Letters*, **2** (22), 3443–3444.

43 Jiang, B., Liu, J., Zhao, S. (2001) *Organic Letters*, **3** (25), 4011–4013.

44 Gurjar, M. K. and Bera, S. (2002) *Organic Letters*, **4** (21), 3569–3570.

45 Jiang, B., Liu, J., Zhao, S. (2002) *Organic Letters*, **4** (22), 3951–3953.

46 Jiang, B., Liu, J., Zhao, S. (2003) *Journal of Organic Chemistry*, **68** (6), 2376–2384.

47 Kobayashi, J., Ohizumi, Y., Nakamura, H., Hirata, Y., Wakamatsu, K., Miyazawa, T. (1986) *Experientia*, **42**, 1064-L 1065.

48 Kobayashi, J., Nakamura, H., Ohizumi, Y. (1988) *Experientia*, **44**, 86–87.

49 Eder, C., Proksch, P., Wray, V., Steube, K., Bringmann, G., Van Soest, R. W. M., SudarsonoFerdinandus, E., Pattisina, L. A., Wiryowidagdo, S., Moka, W. (1999) *Journal of Natural Products*, **62**, 184–187.

50 Albizati, K. F. and Faulkner, D. J. (1985) *Journal of Organic Chemistry*, **50**, 4163–4164.

51 De Nanteuil, G., Ahond, A., Guilhem, J., Poupat, C., Tran Huu Dau, E., Potier, P., Pusset, M., Pusset, J., Laboute, P. (1985) *Tetrahedron*, **41**, 6019–33.

52 Wilson, D. M., Puyana, M., Fenical, W., Pawlik, J. R. (1999) *Journal of Chemical Ecology*, **25**, 2811–2823.

53 Sharma, G. M., Buyer, J. S., Pomerantz, M. W. (1980) *Journal of the Chemical Society, Chemical Communications*, 435–436.

54 Cimino, G., De Rosa, S., De Stefano, S., Mozzarella, L., Puliti, R., Sodano, G. (1982) *Tetrahedron Letters*, **23**, 767–768.

55 Kitagawa, I., Kobayashi, M., Kitanaka, K., Kido, M., Kyogoku, Y. (1983) *Chemical and Pharmaceutical Bulletin*, **31**, 2321–2328.

56 Curman, D., Cinel, B., Williams, D. E., Rundle, N., Block, W. D., Goodarzi, A. A., Hutchins, J. R., Clarke, P. R., Zhou, B., Lees-Miller, S. P., Andersen, R. J., Roberge, M. (2001) *The Journal of the Biological Chemistry*, **276**, 17914–17919.

57 Tasdemir, D., Mallon, R., Greenstein, M., Feldberg, L. R., Kim, S. C., Collins, K., Wojciechowicz, D., Mangalindan, G. C., Concepcion, G. P., Harper, M. K., Ireland, C. M. (2002) *Journal of Medicinal Chemistry*, **45**, 529–532.

58 Wan, Y., Hur, W., Cho, C. Y., Liu, Y., Adrian, F. J., Lozach, O., Bach, S., Mayer, T., Fabbro, D., Meijer, L., Gray, N. (2004) *Chemistry and Biology*, **11**, 247–259.
59 Badger, A. M., Cook, M. N., Swift, B. A., Newman-Tarr, T. M., Gowen, M., Lark, M. (1999) *The Journal of Pharmacology and Experimental Therapeutics*, **290**, 587–593.
60 Sharma, V., Lansdell, T. A., Jin, G., Tepe, J. J. (2004) *Journal of Medicinal Chemistry*, **47**, 3700–3703.
61 Patil, A. D., Freyer, A. J., Killmer, L., Hofmann, G., Johnson, R. K. (1997) *Natural Product Letters*, **9**, 201–207.
62 Inaba, K., Sato, H., Tsuda, M., Kobayashi, J. (1998) *Journal of Natural Products*, **61**, 693–695.
63 Barrios Sosa, A. C., Yakushijin, K., Horne, D. A. (2000) *Journal of Organic Chemistry*, **65**, 610–611.
64 Portevin, B., Golsteyn, R. M., Pierre, A., De Nanteuil, G. (2003) *Tetrahedron Letters*, **44**, 9263–9265.
65 Papeo, G., Posteri, H., Borghi, D., Varasi, M. (2005) *Organic Letters*, **7**, 5641–5644.
66 Papeo, G., Gomez-Zurita, M. A., Borghi, D., Varasi, M. (2005) *Tetrahedron Letters*, **46**, 8635–8638.
67 Poeverlein, C., Breckle, G., Lindel, T. (2006) *Organic Letters*, **8**, 819–821.
68 Tsuda, M., Yasuda, T., Fukushi, E., Kawabata, J., Sekiguchi, M., Fromont, J., Kobayashi, J. (2006) *Organic Letters*, **8**, 4235–4238.
69 Grube, A. and Köck, M. (2006) *Journal of Natural Products*, **69**, 1212–1214.
70 Sharma, G. M. and Burkholder, P. R. (1971) *Journal of the Chemical Society, Chemical Communications*, 151–152.
71 Sharma, G. and Magdoff-Fairchild, B. (1977) *Journal of Organic Chemistry*, **42**, 4118–4124.
72 Gautschi, J. T., Whitman, S., Holman, T. R., Crews, P. (2004) *Journal of Natural Products*, **67**, 1256–1261.
73 Boyd, M. R., Pettit, G. R., McNulty, J., Herald, D. L., Doubek, D. L., Chapuis, J., Schmidt, J. M., Tackett, L. P. (1997) *Journal of Natural Products*, **60**, 180–183.
74 Foley, L. H. and Büchi, G. (1982) *Journal of the American Chemical Society*, **104**, 1776-L 1777.
75 Poullennec, K. G., Kelly, A. T., Romo, D. (2002) *Organic Letters*, **4**, 2645–2648.
76 Vergne, C., Boury-Esnault, N., Perez, T., Martin, M., Adeline, M., Dau, E. T. H., Al Mourabit, A. (2006) *Organic Letters*, **8**, 2421–2424.
77 Kinnel, R. B., Gehrken, H. P., Scheuer, P. J. (1993) *Journal of the American Chemical Society*, **115**, 3376–3377.
78 Kinnel, R. B., Gehrken, H. P., Swali, R., Skoropowski, G., Scheuer, P. J. (1998) *Journal of Organic Chemistry*, **63**, 3281–3286.
79 Dransfield, P. J., Dilley, A. S., Wang, S., Romo, D. (2006) *Tetrahedron*, **62**, 5223–5247.
80 Al Mourabit, A. and Potier, P. (2001) *European Journal of Organic Chemistry*, 237–243.
81 Baran, P. S., O'Malley, D. P., Zografos, A. L. (2004) *Angewandte Chemie International Edition*, **43**, 2674–2677.
82 Jacquot, D. E. N.Lindel, T. (2005) *Current Organic Chemistry*, **9**, 1551–1565.
83 Garrido-Hernandez, H., Nakadai, M., Vimolratana, M., Li, Q., Doundoulakis, T., Harran, P. G. (2005) *Angewandte Chemie International Edition*, **44**, 765–769.
84 Overman, L. E., Rogers, B. N., Tellew, J. E., Trenkle, W. C. (1997) *Journal of the American Chemical Society*, **119**, 7159–7160.
85 Fedoreev, S. A., Utkina, N. K., Il'in, S. G., Reshetnyak, M. V., Maksimov, O. B. (1986) *Tetrahedron Letters*, **27**, 3177–3180.
86 Assmann, M. and Kock, M. (2002) *Zeitschrift fuer Naturforschung C: Journal of Biosciences*, **57**, 153–156.
87 Assmann, M., Van Soest, R. W. M., Kock, M. (2001) *Journal of Natural Products*, **64**, 1345–1347.
88 Tsukamoto, S., Tane, K., Ohta, T., Matsunaga, S., Fusetani, N., Van Soest, R. W. M. (2001) *Journal of Natural Products*, **64**, 1576–1578.
89 Goetz, G. H., Harrigan, G. G., Likos, J. (2001) *Journal of Natural Products*, **64**, 1581–1582.

90 Kato, T., Shizuri, Y., Izumida, H., Yokoyama, A., Endo, M. (1995) *Tetrahedron Letters*, **36**, 2133–2136.

91 Fedoreev, S. A., Il'in, S. G., Utkina, N. K., Maksimov, O. B., Reshetnyak, M. V., Antipin, M. Y., Struchkov, Y. T. (1989) *Tetrahedron*, **45**, 3487–3492.

92 D'Ambrosio, M., Guerriero, A., Debitus, C., Ribes, O., Pusset, J., Leroy, S., Pietra, F. (1993) *Journal of the Chemical Society, Chemical Communications*, **16**, 1305–6.

93 D'Ambrosio, M., Guerriero, A., Ripamonti, M., Debitus, C., Waikedre, J., Pietra, F. (1996) *Helvetica Chimica Acta*, **79**, 727–735.

94 Hong, T. W., Jimenez, D. R., Molinski, T. F. (1998) *Journal of Natural Products*, **61**, 158–161.

95 Meijer, L., Thunnissen, A.-M. W. H., White, A. W., Garnier, M., Nikolic, M., Tsai, L.-H., Walter, J., Cleverley, K. E., Salinas, P. C., Wu, Y.-Z., Biernat, J., Mandelkow, E.-M., Kim, S.-H., Pettit, G. R. (2000) *Chemistry and Biology*, **7**, 51–63.

96 Trost, B. M. and Dong, G. (2006) *Journal of the American Chemical Society*, **128**, 6054–6055.

97 Davis, F. A. and Deng, J. (2005) *Organic Letters*, **7**, 621–623.

98 Domostoj, M. M., Irving, E., Scheinmann, F., Hale, K. J. (2004) *Organic Letters*, **6**, 2615–2618.

99 Hale, K. J., Domostoj, M. M., Tocher, D. A., Irving, E., Scheinmann, F. (2003) *Organic Letters*, **5**, 2927–2930.

100 Walker, R. P., Faulkner, D. J., Van Engen, D., Clardy, J. (1981) *Journal of the American Chemical Society*, **103** (22), 6772–6773.

101 Faulkner, D. J. (1983) PATENT U.S., 3 pp. [CODEN: USXXAM US 4,370,484 Appl. No.: 242728].

102 Bernan, V. S., Roll, D. M., Ireland, Chris M., Greenstein, M., Maiese, William M., Steinberg, D. A. (1993) *The Journal of Antimicrobial Chemotherapy*, **32** (4), 539–50.

103 Peng, J., Shen, X., El Sayed, K. A., Dunbar, D. C., Perry, T. L., Wilkins, S. P., Hamann, M. T., Bobzin, S., Huesing, J., Camp, R., Prinsen, M., Krupa, D., Wideman, M. A. (2003) *Journal of Agricultural and Food Chemistry*, **51** (8), 2246–2252.

104 Keifer, P. A., Schwartz, R. E., Koker, M. E. S., Hughes, R. G., Jr., Rittschof, D., Rinehart, K. L. (1991) *Journal of Organic Chemistry*, **56** (9), 2965–75.

105 Vassas, A., Bourdy, G., Paillard, J. J., Lavayre, J., Pais, M., Quirion, J. C., Debitus, C. (1996) *Planta Medica*, **62** (1), 28–30.

106 Baran, P. S., Zografos, A. L., O'Malley, P. (2004) *Journal of the American Chemical Society*, **126** (12), 3726–3727.

107 Birman, Vladimir B. and Jiang, Xun-Tian (2004) *Organic Letters*, **6** (14), 2369–2371.

108 Baran, P. S., Li, K., O'Malley, D. P., Mitsos, C. (2006) *Angewandte Chemie International Edition*, **45**, 249–252.

109 Rinehart, K. L., Jr. (1988) PATENT U.S., 15 pp. [CODEN: USXXAM US 4,737,510 Appl. No.: 913819].

110 Kobayashi, J., Tsuda, M., Ohizumi, Y. (1991) *Experientia*, **47** (3), 301–304.

111 Shen, X., Perry, T. L., Dunbar, C. D., Kelly-Borges, M., Hamann, M. T. (1998) *Journal of Natural Products*, **61** (10), 1302–1303.

112 Assmann, M. and Kock, M. (2002) *Zeitschrift fuer Naturforschung C: Journal of Biosciences*, **57** (1/2), 157–160.

113 Bickmeyer, U., Drechsler, C., Kock, M., Assmann, M. (2004) *Toxicon*, **44** (1), 45–51.

114 Ederm, C., Proksch, P., Wray, V., Van Soest, R. W. M., Ferdinandus, E., Pattisina, L. A., Sudarsono (1999) *Journal of Natural Products*, **62** (9), 1295–1297.

115 Hao, E., Fromont, J., Jardine, D., Karuso, P. (2001) *Molecules*, **6**, 130–141.

116 Rinehart, K. L. (1989) *Pure and Applied Chemistry*, **61**, 525–528.

117 Kobayashi, J., Tsuda, M., Murayama, T., Nakamura, H., Ohizumi, Y., Ishibashi, M., Iwamura, M., Ohta, T., Nozoe, S. (1990) *Tetrahedron*, 5579–5586.

118 Assmann, M., Van Soest, R. W. M., Koeck, M. (2001) *Journal of Natural Products*, **64** (10), 1345–1347.

119 Bickmeyer, U. (2005) *Toxicon*, **45**, 627–632.
120 Baran, P. S., O'Malley, D. P., Zografos, A. L. (2004) *Angewandte Chemie International Edition*, **43**, 2674–2677.
121 Williams, D. H. and Faulkner, D. J. (1996) *Tetrahedron*, **52**, 5381–5390.
122 Tsukamoto, S., Kato, H., Hirota, H., Fusetani, N. (1996) *Journal of Natural Products*, **59** (5), 501–503.
123 Olofson, A., Yakushijin, K., Horne, D. A. (1997) *Journal of Organic Chemistry*, **62** (23), 7918–7919.
124 Kobayashi, J., Suzuki, M., Tsuda, M. (1997) *Tetrahedron*, **53**, 15681–15684.
125 Urban, S., De Almeida Leone, P., Carroll, A. R., Fechner, G. A., Smith, J., Hooper, J. N. A., Quinn, R. J. (1999) *Journal of Organic Chemistry*, **64**, 731–735.
126 Endo, T., Tsuda, M., Okada, T., Mitsuhashi, S., Shima, H., Kikuchi, K., Mikami, Y., Fromont, J., Kobayashi, J. (2004) *Journal of Natural Products*, **67**, 1262–1267.
127 Nishimura, S., Matsunaga, S., Shibazaki, M., Suzuki, K., Furihata, K., Van Soest, R. W. M., Fusetani, N. (2003) *Organic Letters*, **5**, 2255–2257.
128 Grube, A. and Köck, M. (2006) *Organic Letters*, **8**, 4675–4678.
129 Koenig, G. M. and Wright, A. D. (1994) *Natural Product Letters*, **5**, 141–146.
130 Gunasekera, S. P., Cranick, S., Longley, R. E. (1989) *Journal of Natural Products*, **52**, 757–761.
131 Mancini, I., Guella, G., Amade, P., Roussakis, C., Pietra, F. (1997) *Tetrahedron Letters*, **38**, 6271–6274.
132 Tetsuo, I., Kaneko, M., Okamura, H., Nakatani, M., Van Soest, R. W. M. (1998) *Journal of Natural Products*, **61**, 1310–1312.
133 Umeyama, A., Ito, S., Yuasa, E., Arihara, S., Yamada, T. (1998) *Journal of Natural Products*, **61**, 1433–1434.
134 Bickmeyer, U., Assmann, M., Koeck, M., Schuett, C. (2005) *Environmental Toxicoloy and Pharmacology*, **19**, 423–427.
135 Cafieri, F., Fattorusso, E., Mangoni, A., Taglialetela-Scafati, O. (1995) *Tetrahedron Letters*, **36**, 7893–7896.
136 Li, C.-J., Schmitz, F. J., Kelly-Borges, M. (1998) *Journal of Natural Products*, **61**, 387–389.
137 Schmitz, F. J., Gunasekera, S. P., Lakshmi, V., Tillekeratne, L. M. V. (1985) *Journal of Natural Products*, **48**, 47–53.
138 Cafieri, F., Fattorusso, E., Taglialatela-Scafati, O. (1998) *Journal of Natural Products*, **61**, 122–125.
139 Srinivasa, R. N. and Venkateswarlu, Y. (2000) *Biochemical Systemetics and Ecology*, **28**, 1035–1037.
140 Srinivasa, R. N. and Venkateswarlu, Y. (2000) *Indian Journal of Chemistry, Section B*, **39B**, 971–972.
141 Marchais, S., Al Mourabit, A., Ahond, A., Poupat, C., Potier, P. (1999) *Tetrahedron Letters*, **40**, 5519–5522.
142 Banwell, M. G., Bray, A. M., Willis, A. C., Wong, D. J. (1999) *New Journal of Chemistry*, **23**, 687–690.
143 Patel, J., Pelloux-Leon, N., Minassian, F., Vallee, Y. (2005) *Journal of Organic Chemistry*, **70**, 9081–9084.
144 Barrios Sosa, A. C., Yakushijin, K., Horne, D. A. (2000) *Tetrahedron Letters*, **41**, 4295–4299.
145 Tsukamoto, S., Kato, H., Hirota, H., Fusetani, N. (1996) *Tetrahedron Letters*, **37**, 1439–1440.
146 Ponasik, J. A., Kassab, D. J., Ganem, B. (1996) *Tetrahedron Letters*, **37**, 6041–6044.
147 Behrens, C., Christoffersen, M. W., Gram, L., Nielsen, P. H. (1997) *Bioorganic and Medicinal Chemistry Letters*, **7**, 321–326.
148 Ponasik, J. A., Conova, S., Kinghorn, D., Kinney, W. A., Rittschof, D., Ganem, B. (1998) *Tetrahedron*, **54**, 6977–6986.
149 Cafieri, F., Fattorusso, E., Mangoni, A., Taglialatela-Scafati, O. (1995) *Bioorganic and Medicinal Chemistry Letters*, **5**, 799–804.
150 (a) Aiello, A., D'Esposito, M., Fattorusso, E., Menna, M., Mueller, W. E. G., Perovic-Ottstadt, S., Tsuruta, H., Gulder, T. A. M., Bringmann, G. (2005) *Tetrahedron*, **61**, 7266–7270.
(b) Bringmann, G., Lang, G., Tsuruta, H., Fattorusso, E., Aiello, A., D'Esposito, M., Menna, M., Mueller,

W. E. G., Schroeder, H. C., Petrovic-Ottstadt (2005) PATENT DE 102004002885, 18 pp. [CODEN: GWXXBX A1 20050818S].

151 Cafieri, F., Fattorusso, E., Mangoni, A., Taglialatela-Scafati, O. (1996) *Tetrahedron*, **52**, 13713–13720.

152 Assmann, M., Lichte, E., Van Soest, R. W. M., Koeck, M. (1999) *Organic Letters*, **1**, 455–457.

153 Williams, D. E., Patrick, B. O., Behrisch, H. W., Van Soest, R., Roberge, M., Andersen, R. J. (2005) *Journal of Natural Products*, **68**, 327–330.

154 Grube, A., Lichte, E., Kock, M. (2006) *Journal of Natural Products*, **69**, 125–1125.

155 Kobayashi, J., Kanda, F., Ishibashi, M., Shigemori, H. (1991) *Journal of Organic Chemistry*, **56**, 4574–4576.

156 When, Paul, M., DuBois, J. (2002) *Journal of the American Chemical Society*, **124**, 12950–12951.

157 Wang, Y., Liu, X., Deng, L. (2006) *Journal of the American Chemical Society*, **128**, 3928–3930.

158 Jahn, T., Koning, G. M., Wright, A. D. (1997) *Tetrahedron Letters*, **38**, 3883–3884.

159 Tsukamoto, S., Tane, K., Ohta, T., Matsunaga, S., Fusetani, N., Van Soest, R. W. M. (2001) *Journal of Natural Products*, **64**, 1576–1578.

160 Drouin, C., Woo, J. C. S., MacKay, D. B., Lavigne, R. M. A. (2004) *Tetrahedron Letters*, **45**, 7197–7199.

161 Fujita, M., Nakao, Y., Matsunaga, S., Seiki, M., Itoh, Y., Yamashita, J., Van Soest, R. W. M., Fusetani, N. (2003) *Journal of the American Chemical Society*, **125**, 15700–15701.

162 Meketa, M. L. and Weinreb, S. M. (2006) *Organic Letters*, **8**, 1443–1446.

163 Linington, R. G., Williams, D. E., Tahir, A., Van Soest, R. W., Andersen, R. J. (2003) *Organic Letters*, **5**, 2735–2738.

164 Ishida, K., Ishibashi, M., Shigemori, H., Sasaki, T., Kobayashi, J. (1992) *Chemical and Pharmaceutical Bulletin*, **40**, 766–767.

165 Andrade, P., Willoughby, R., Pomponi, S. A., Kerr, R. (1999) *Tetrahedron Letters*, **40**, 4775–4778.

166 Travert, N. and Al Mourabit, A. (2004) *Journal of the American Chemical Society*, **126**, 10252–10253.

167 Abou-Jneid, R., Ghoulami, S., Martin, M., Tran Huu Dau, E., Travert, N., Al Mourabit, A. (2004) *Organic Letters*, **6**, 3922–6.

11
Guanidine Alkaloids from Marine Invertebrates

Roberto G.S. Berlinck, Miriam H. Kossuga

11.1
Introduction

Guanidine alkaloids from marine invertebrates were first reviewed by Chevolot [1] and more recently in a series of reviews that included their occurrence in all natural sources [2–6]. This chapter includes only those guanidine alkaloids isolated from marine invertebrates. It does not include simple, primary metabolism related, alkaloid derivatives, which were reviewed by Chevolot [1], but only alkaloids and modified peptides derived from nonribosomal peptide synthases. Cyanobacteria guanidine metabolites such as cyclindrospermopsins, anatoxin-a(s), microcystins, and many other peptides, as well as guanidine alkaloids bearing a bromopyrrole moiety isolated from sponges will not be discussed here, since Chapters 6 and 10 of this volume review these guanidine-containing metabolites. This chapter attempts to provide an overview of marine invertebrate guanidine alkaloids. Considering the number of such compounds, only selected examples of the different classes have been included. The topics are presented by biogenetic relatedness.

Owing to their intrinsic basicity, marine invertebrate guanidine alkaloids have a rather polar behavior and their isolation from complex mixtures may be difficult. Isolation procedures frequently include chromatography on lipophilic Sephadex LH20, on reversed phase silica gel (such as C_{18} bonded, aminopropyl bonded, or cyanopropyl bonded), or even on ion-exchange resins. HPLC purification using acidic (TFA) or buffered eluents have frequently been employed.

11.2
Modified Creatinine Guanidine Derivatives

Aplysinopsin (**1**) and many of its derivatives have been isolated from several species of marine sponges, including *Thorecta* sp. [7], *Verongia* (= *Aplysina*) *spengelli* [8], *Dercitus* sp. [9], *Smenospongia aurea* [10], *Aplysina* sp. [11a], *Thorectandra* sp., and *Smenospongia* sp. [11b], from the anthozoan *Astroides calycularis*

Modern Alkaloids: Structure, Isolation, Synthesis and Biology. Edited by E. Fattorusso and O. Taglialatela-Scafati

ISBN: 978-3-527-31521-5

[12], from the scleratinian coral *Tubastrea aurea* [12], from *T. coccinea* and its nudibranch predator *Prestilla melanobranchia* [14], and from the scleratinian coral *Dendrophyllia* sp. [15]. Aplysinopsin was originally identified by analysis of spectroscopic data and total synthesis from indole-3-aldehyde and N^3-methylcreatinine [7]. The geometry of the creatinine substituted double bond was assigned as *E* by NOE NMR experiments [8]. Aplysinopsin derivatives isolated from *Dendrophyllia* sp. were obtained as mixtures of *E* and *Z* geometrical isomers, which could be distinguished by measuring ^{1}H–^{13}C heteronuclear coupling constants [15]. 6-Bromo-2′-*N*-demethyl-3′-*N*-methylaplysinopsin (**2**) isolated as a 85 : 15 *Z*/*E* mixture undergoes facile photoisomerization to give a mixture richer in the *E* isomer. Biological activities of aplysinopsin and naturally occurring derivatives include *in vivo* antineoplastic activity in mice with P388 leukemia cells [8], cytotoxic activity against murine lymphoma L-2110 and epidermoid carcinoma KB cells [11a], and inhibition of neuronal nitric oxide synthase [16]. Interestingly, aplysinopsin isolated from the sponge *Aplysinopsis reticulata* was considered one of the most promising hits obtained during the first program of a wide screening of marine invertebrate crude extracts, developed in Australia over seven years at the Roche Research Institute of Marine Pharmacology. It has been reported that aplysinopsin was isolated as a potent reversible inhibitor of monoamine oxidase. Although 36 aplysinopsin derivatives were synthesized and tested as protectors of tetrabenazine-induced ptosis, none displayed better activity than the natural product (**1**). Unfortunately, aplysinopsin was also hepatotoxic, and the project was closed [17,18]. Many syntheses of aplysinopsin and various derivatives have been developed [7,9,15,17–21].

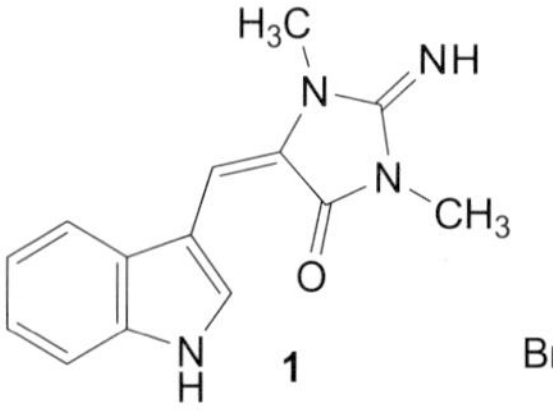

H_3C N NH N CH_3 O N H **1**

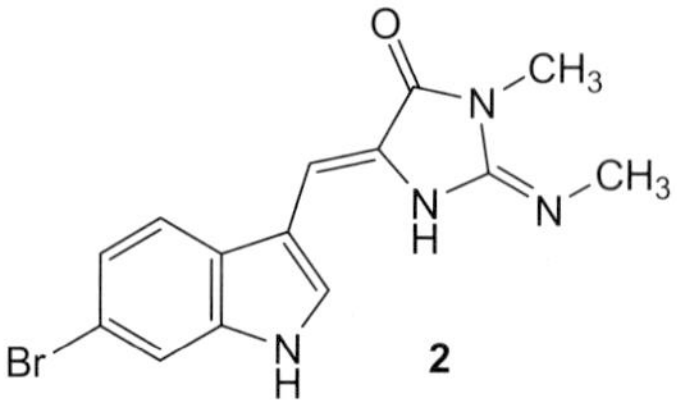

O N CH_3 N H N CH_3 Br N H **2**

Interesting "aplysinopsin-like dimers," tubastrindoles A–C (**3–5**), have been isolated from a Japanese *Tubastrea* sp. [22]. Several related modified creatinine metabolites have been isolated from sponges. Corallistine (**6**), was isolated from the sponge *Corallistes fulvodesmus* [23] and identified by analysis of spectroscopic data and X-ray diffraction analysis of an isopentyl acid derivative. Leucettamine B (**7**) was isolated from *Leucetta microraphis* and was shown to be inactive in a membrane receptor leukotriene, LTB4, binding assay [24]. Leucettamine B has been synthesized in four steps and 49 % overall yield [25]. Calcaridine A (**8**), (−)-spirocalcaridine A (**9**), (−)-spirocalcaridine C (**10**), (−)-spiroleucettadine (**11**), and leucettamine (**12**) have all been isolated from *Leucetta* spp. Compounds **11** and **12** displayed antibacterial activity against a series of human pathogenic bacteria [26–28].

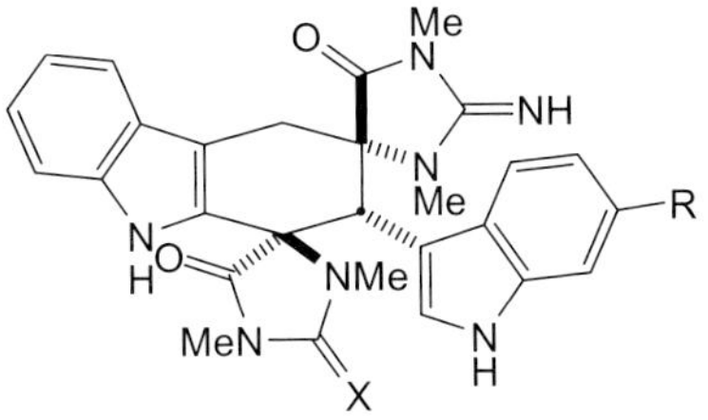

3 X = NH, R = Br
4 X = NH, R = H
5 X = O, R = H

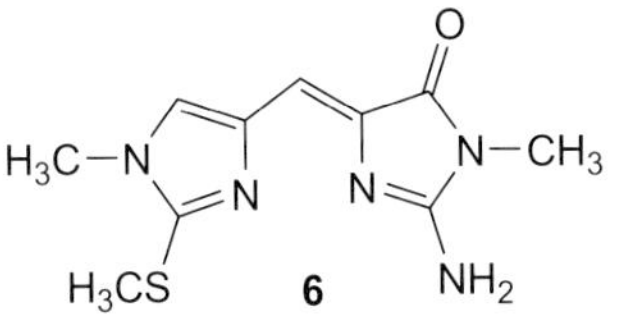

6 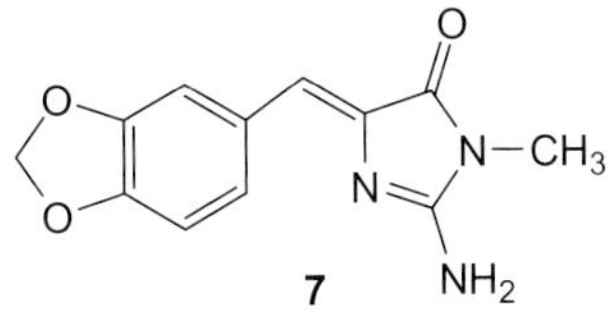 **7** 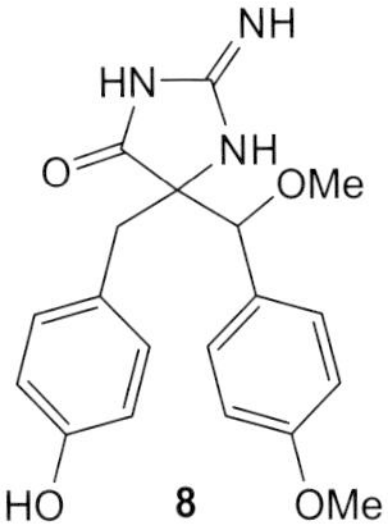 **8**

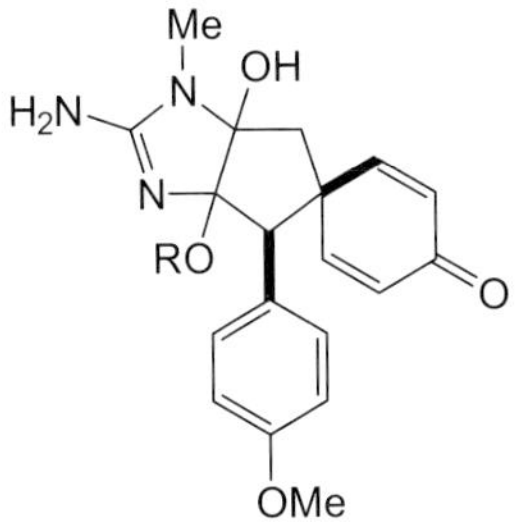

9 R = H
10 R = Me

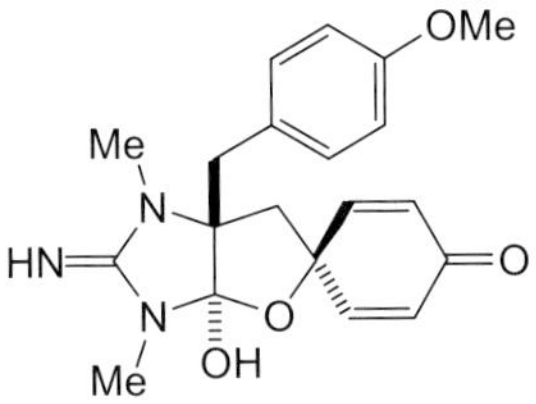

11 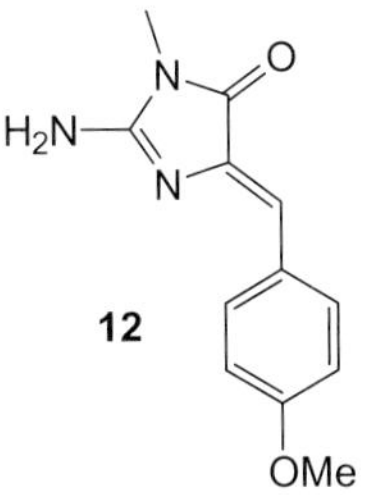 **12**

11.3
Aromatic Guanidine Alkaloids

The pyridopyrrolopyrimidine alkaloid 3′-methyltetrahydrovariolin (**13**) has been isolated from the sponge *Kirkpatrickia varialosa* along with other variolins, and displayed moderate cytotoxic and antifungal activities [29,30]. Tubastrine (**14**) has been isolated from the coral *Tubastrea aurea* and displayed antiviral activity against *Herpes simplex* virus [31]. It has been also recently isolated from the ascidian *Dendrodoa grossularia* as an inhibitor of epidermal growth factor receptor protein kinase [32]. Related antibacterial tubastrine derivatives have been isolated from the sponges *Spongosorites* sp. [33] and *Petrosia* cf. *contignata* [34]. The simple acetophenone derivative *N*-(methoxybenzoyl)-*N*′-methylguanidine (**15**) was isolated from the ascidian *Polycarpa aurata* [35] while the pyridine derivative pyraxinine (**16**) was isolated from the sponge *Cymbastela cantharella* and inhibited macrophagic NO synthase at 100 μM concentration [36]. The indole alkaloid discodermindole (**17**) has been isolated from *Discodermia polydiscus* [37] and exhibited *in vitro* cytotoxic activity against murine leukemia P388 (IC_{50} = 1.8 μg/mL), human lung A-549 (IC_{50} = 4.6 μg/mL), and human colon HT-29 (IC_{50} = 12 μg/mL) cancer cell lines, while the tetrahydroisoquinoline alkaloid fuscusine (**18**) has been isolated from the sea star *Perknaster fuscus antarcticus* [38]. Additional indole-derived guanidine alkaloids include compound **19** from the ascidian *Dendrodoa grossularia* [39] and the β-carboline alkaloids trypargine (**20**), tryparginine (**21**), and 1-carboxy-trypargine (**22**) isolated from the ascidian *Eudistoma* sp. [40].

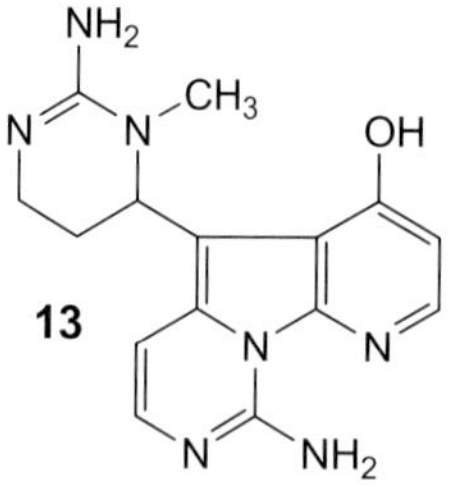
13

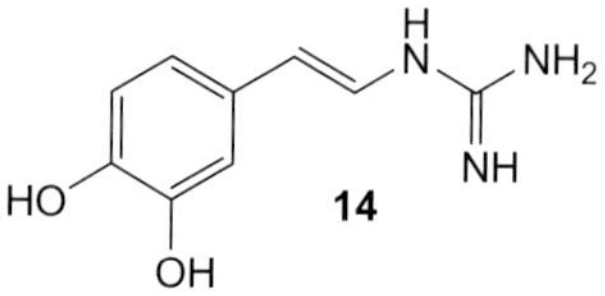
14

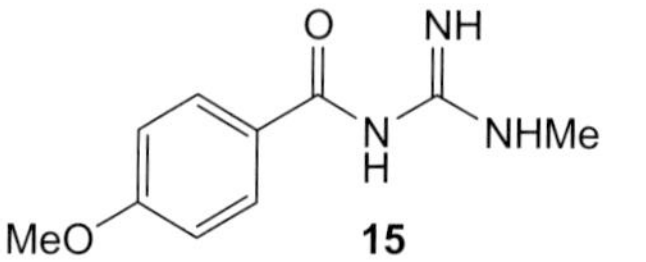
15

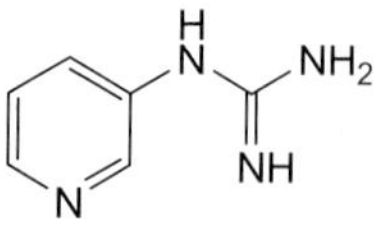
16

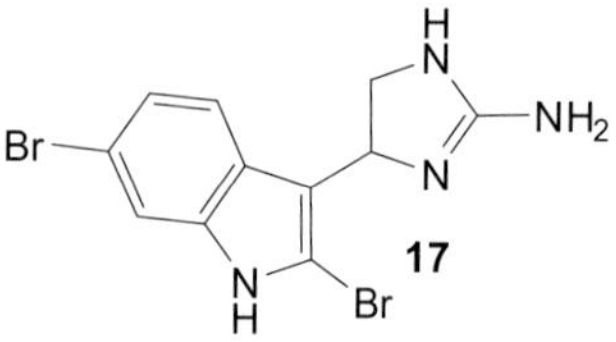
17

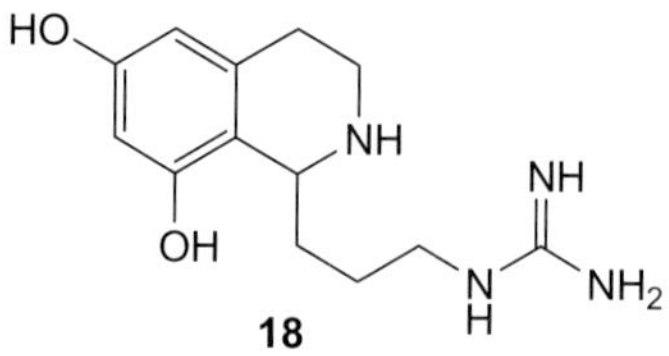
18

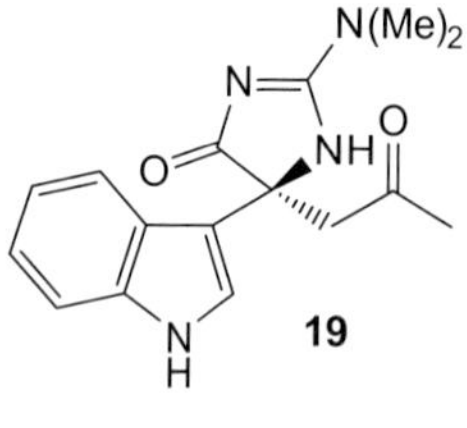
19

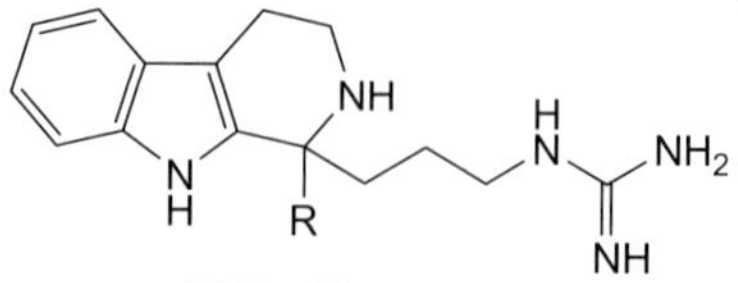
20 R = H
22 R = CO_2H

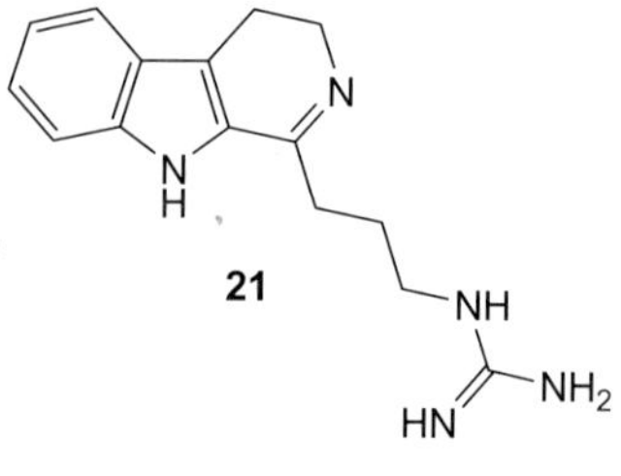
21

The bioluminescence of crustaceans belonging to genera *Cypridina* is due to the reaction of *Cypridina* luciferin (**23**) with luciferase in the presence of oxygen [41,42]. The structure of *Cypridina* luciferin was established by analysis of spectroscopic data, chemical degradations, extensive MS analysis, and total synthesis [41–44]. *Cypridina* luciferin is biosynthesized from L-tryptophan, L-arginine, and L-isoleucine [45,46]. Distomadines A (**24**) and B (**25**) have been isolated from the ascidian *Pseudodistoma aureum* and were not active in antifungal, antiviral, anti-inflammatory, and anti-mycobacterial bioassays [47].

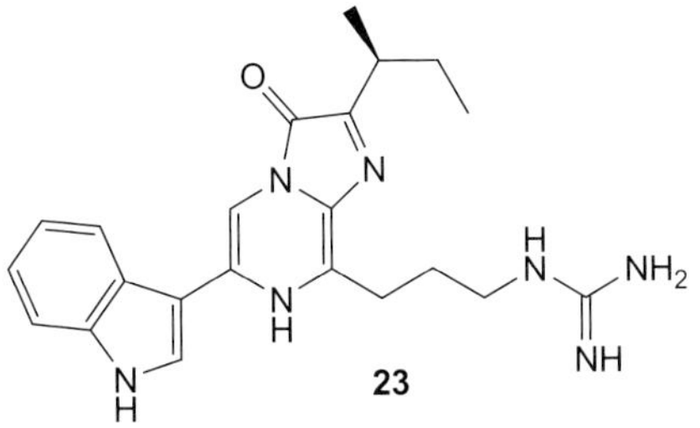

23

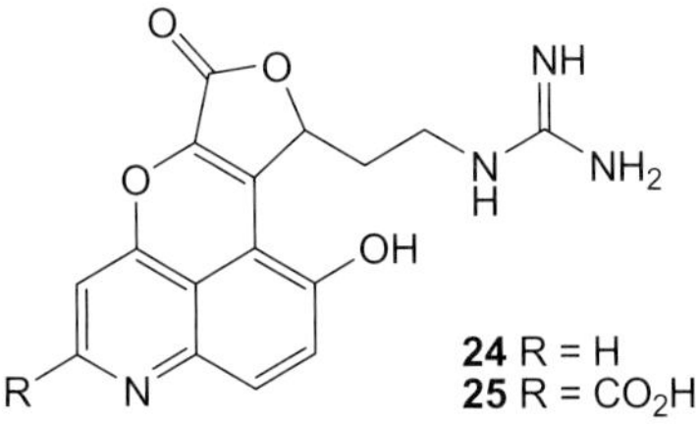

24 R = H
25 R = CO_2H

11.4 Bromotyrosine Derivatives

Aplysinamisine II (**26**) was isolated from the sponge *Aplysina cauliformis* and exhibited weak antimicrobial (*Staphylococcus aureus, Pseudomonas aeruginosa*, and *Escherichia coli*) and cytotoxic activity against human breast and leukemia cell lines [48]. Caissarine A (**27**) has been isolated from the marine sponge *Aplysina caissara* [49]. The *Mycobacterium tuberculosis* mycothiol S-conjugate amidase inhibitor **28** (unnamed) has been isolated from the sponge *Oceanapia* sp. [50] and subsequently synthesized through three distinct approaches [51–53].

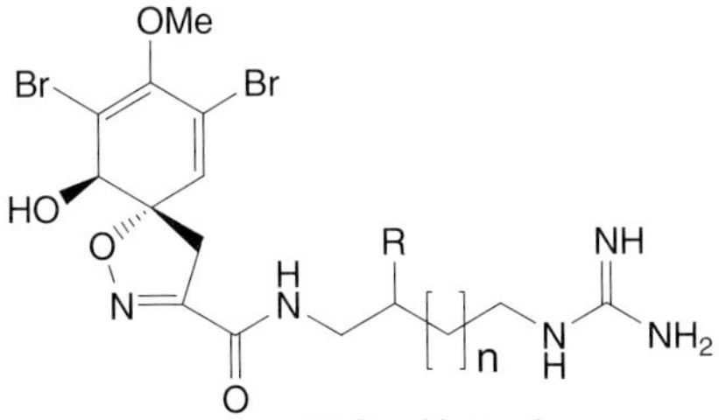

26 R = H, n = 2
27 R = OH, n = 1

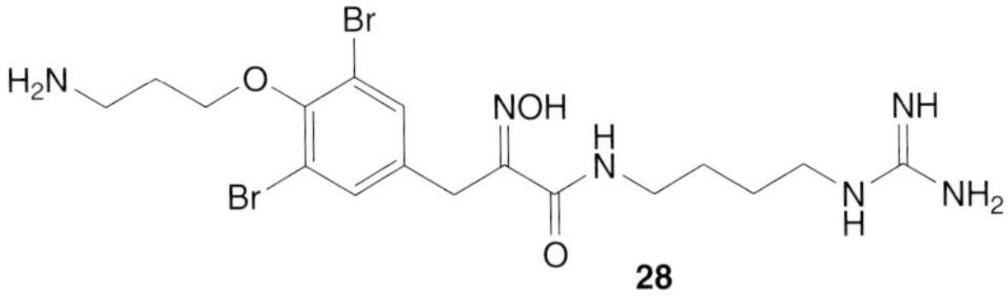

28

11.5
Amino Acid and Peptide Guanidines

Marine invertebrates are a rich source of guanidine-bearing amino acid and peptide derivatives. Simple amino acid derivatives include the diketopiperazines cyclo(-L-Arg-dehydrotyrosine) (**29**) isolated from the sponge *Anthosigmella* aff. *raromicrosclera* [54], the linear dipeptides *N*-(*p*-hydroxybenzoyl)-L-arginine (**30**) *N*-(1*H*-indol-3-ylcarbonyl)-D-arginine (**31**), *N*-(6-bromo-1*H*-indolyl-3-carbonyl)-L-arginine (**32**), *N*-(6-bromo-1*H*-indol-3-ylcarbonyl)-L-enduracididine (**33**) [55], as well as barettin (**34**), isolated as a 87 : 13 mixture of *E* and *Z* isomers, and dihydrobarettin (**35**), from the sponge *Geodia baretti*. Both **34** and **35** displayed antifouling activity against larvae of the barnacle *Balanus improvisus* at 0.9 and 7.9 mM concentrations, respectively [56,57]. Tokaramide A (**36**) is a modified tripeptide from the sponge *Theonella* aff. *mirabilis*, isolated as a cathepsin B inhibitor [58]. Eurypamide A (**37**) has been isolated from the sponge *Microciona eurypa*, along with related eurypamides devoid of the (2*S*, 3*S*, 4*R*)-3,4-dihydroxyarginine residue. None of the eurypamides displayed cytotoxic or anti-inflammatory activities [59a].

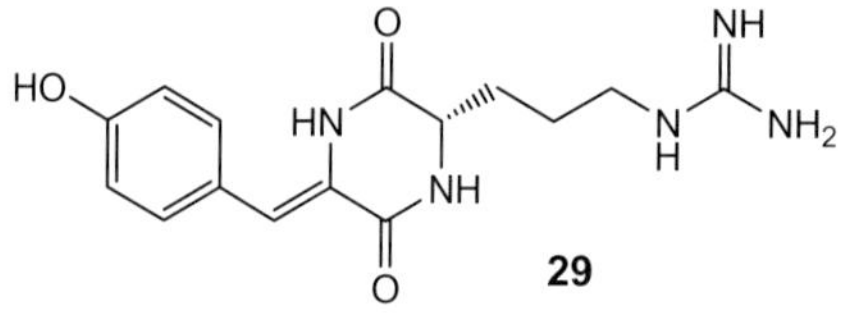

29 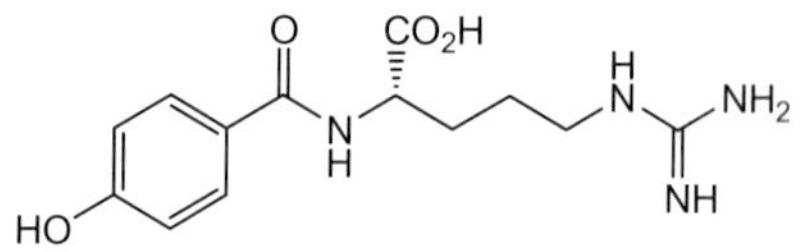 **30** 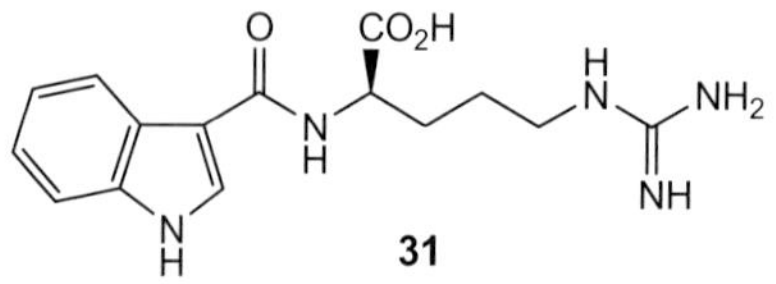 **31** 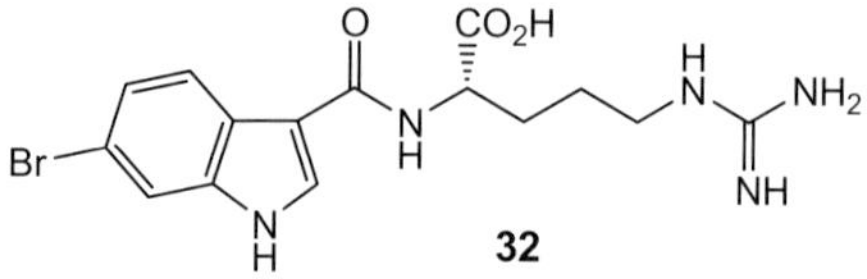 **32** 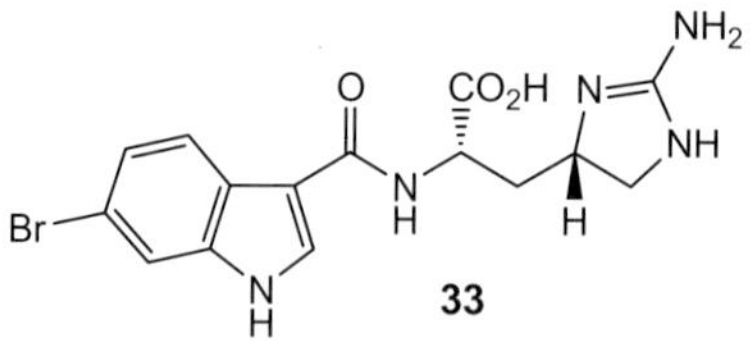 **33** 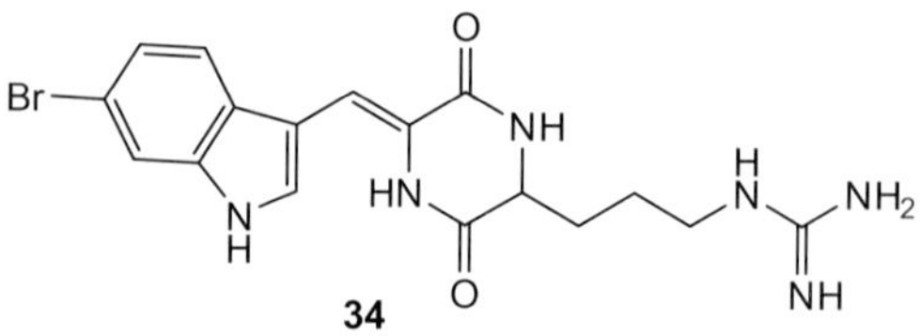 **34** 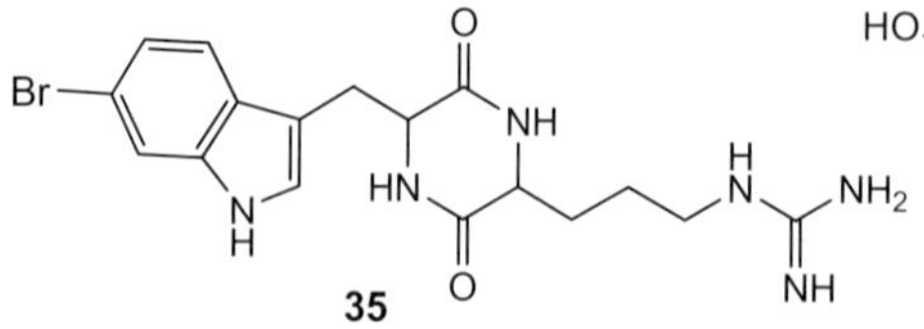 **35** 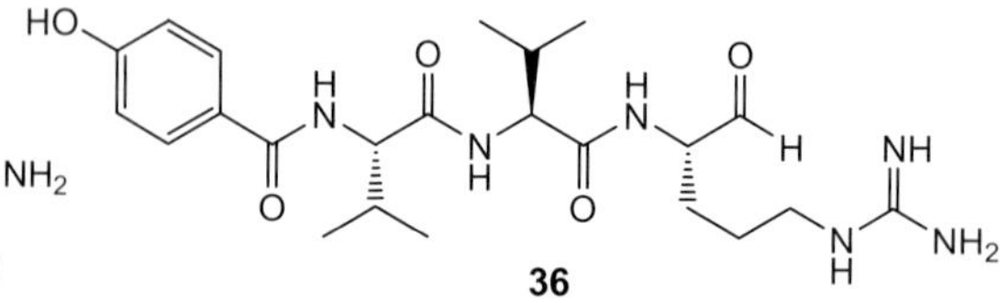 **36**

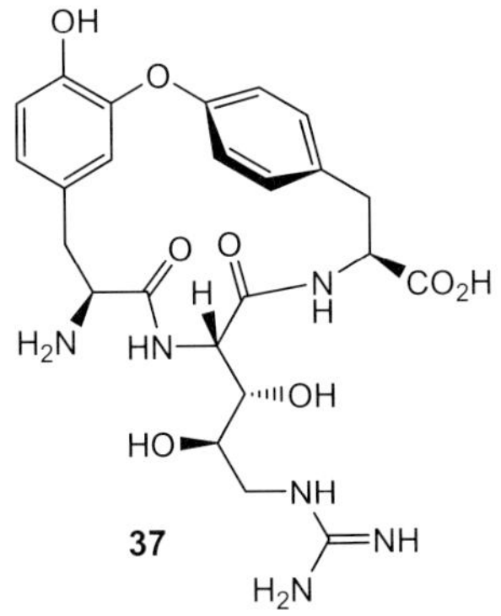

37

Dysinosins are alkaloidal tripeptides with very unusual α-substituted acid residues, including 5,6-dihydroxyoctahydroindole-2-carboxylic acid and 3-sulfonic-2-methoxy propionic acid, as well as the basic moiety 2-aminoethyl-(1-*N*-amidino-Δ-3-pyrroline). The first member of this series, dysidosin A (**38**), was isolated from an unidentified sponge belonging to the family Dysideidae, and was identified by analysis of spectroscopic data and X-ray diffraction analysis of the dysinosin A–thrombin–hirugen complex [59b]. Dysinosin A inhibited factor VIIa with a K_i of 108 nM and thrombin with a K_i of 452 nM [59b]. Further dysinosins have been isolated from the sponge *Lamellodysia chlorea*, and also displayed potent inhibition of factor VIIa and thrombin [59c]. Dysinosin A has been synthesized via a convergent route [59d].

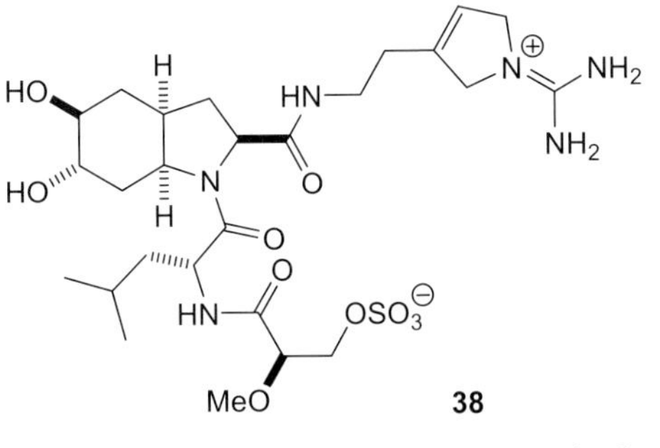

38

Several guanidine-modified tetrapeptides have been isolated from marine invertebrates, in particular marine sponges. Nazumamide A (**39**) has been isolated from the sponge *Theonella* sp. and identified by analysis of spectroscopic data [60] as well as by X-ray diffraction analysis of a nazumamide A–human thrombin complex [61]. Nazumamide A has been synthesized by conventional peptide synthesis [62]. A series of nazumamide derivatives have been prepared via combinatorial synthesis [63].

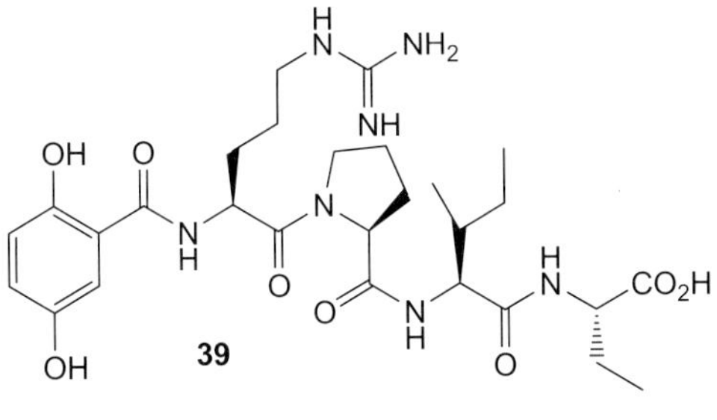

39

Criamides A and B (**40**) were isolated from the sponge *Cymbastela* sp. Criamide B displayed potent cytotoxic activity against human cancer cell lines such as murine leukemia P388 (ED_{50} = 7.3 ng/mL), breast cancer (MCF7 (ED_{50} = 6.8 μg/mL), glioblastome/astrocytoma U373 (ED_{50} = 0.27 μg/mL), ovarian carcinoma HEY (ED_{50} = 0.19 μg/mL), and human lung A549 (ED_{50} = 0.29 μg/mL) [64].

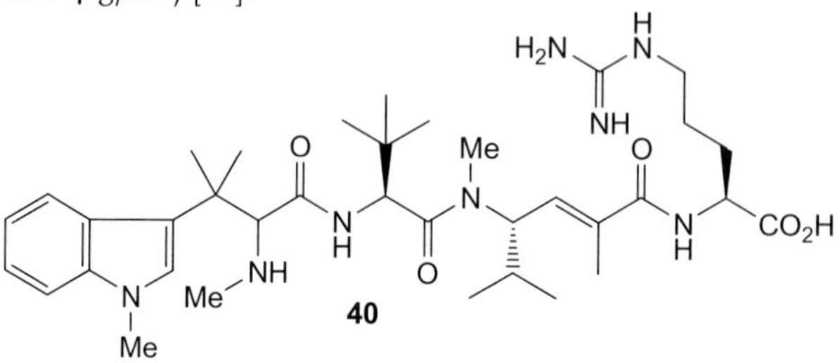

40

Kalahalide D (**41**) has been isolated from the marine mollusk *Elysia rufescens* and from the alga *Bryopsis* sp. and identified by analysis of spectroscopic data, chemical degradations, and chiral HPLC, GC, and TLC [65].

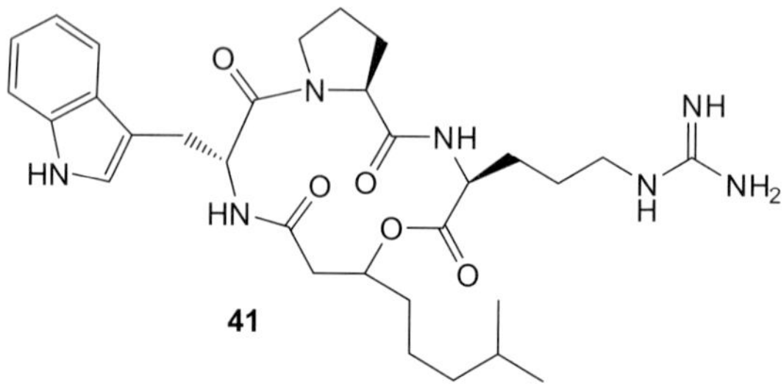

41

Eusynstyelamide (**42**) was obtained from the ascidian *Eusynstyela misakiensis* [66]. Minalemines (**43–47**) are a group of modified tripeptides bearing two homoagmatine moieties which have been isolated from the ascidian *Didemnum rodriguesi* [67]. The structures of these bis-guanidine derivatives have been established by analysis of spectroscopic data and total synthesis of minalemine A [68]. Very unusual modified tetrapeptides have been isolated from the sponge *Anchinoe tenacior*, among them anchinopeptolide A (**48**) and cycloanchinopeptolide C (**49**) [69,70]. The total synthesis of racemic anchinopeptolide D has been reported [71].

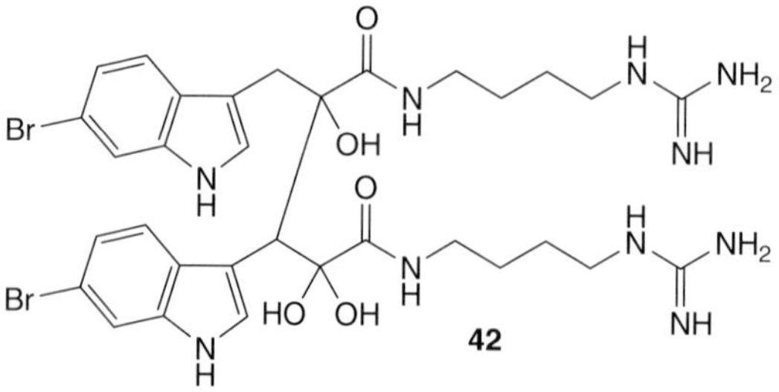

42

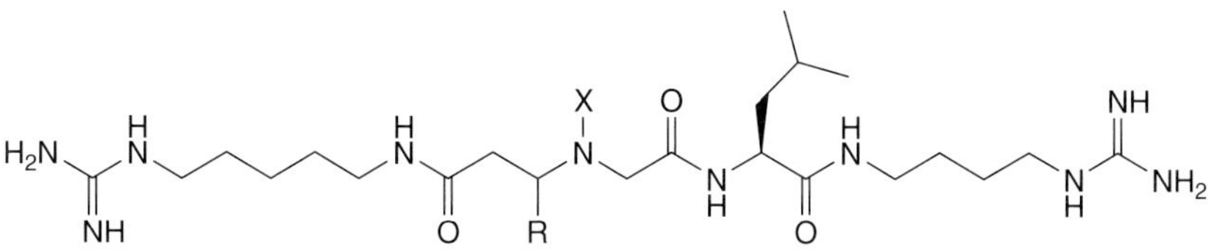

43 X=H, R=C_7H_{15}
44 X=H, R=C_8H_{17}
45 X=H, R=C_9H_{19}
46 X=SO_3H, R=C_7H_{15}
47 X=SO_3H, R=C_8H_{17}

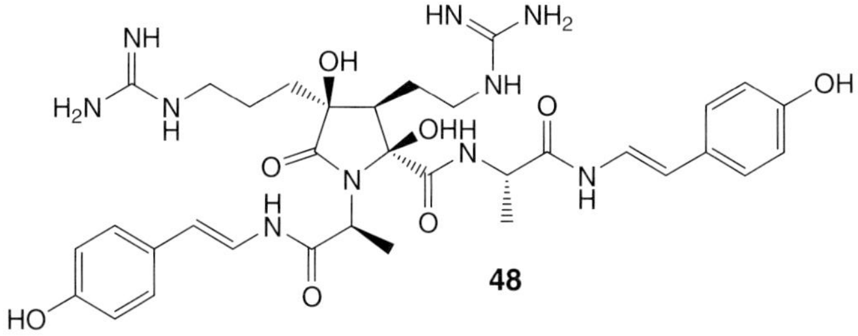

48

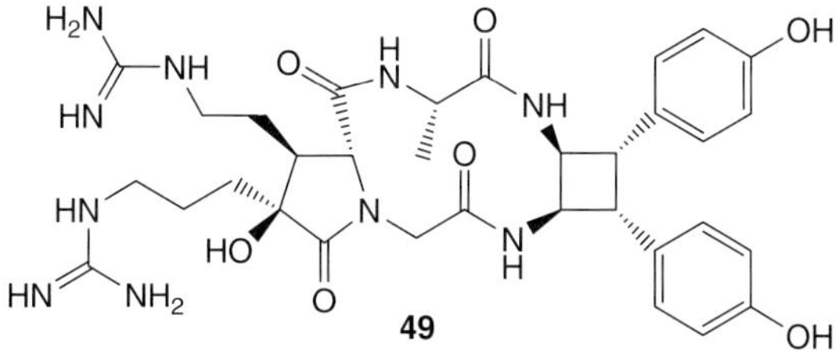

49

The cyclotheonamides are unusual pentapeptides isolated from sponges of the genus *Theonella*. The first two members of this series, cyclotheonamides A (**50**) and B (**51**), were isolated from the sponge *Theonella* sp. as thrombin inhibitors, and presented two new amino acids, β-ketohomoarginine and (*E*)-4-amino-5-(4-hydroxyphenyl)pent-2-enoic acid [72]. Subsequently, cyclotheonamides C–E have been isolated from *Theonella swinhoei*, also as thrombin inhibitors [73]. The bioassay results indicated that cyclotheonamides A, C, and D were the most potent thrombin inhibitors among the five peptides. Further cyclotheonamides were isolated from *Theonella* spp. as serine protease inhibitors. These include cyclotheonamides E2 and E3, which inhibited trypsin at the nanomolar concentration level [74]. Pseudotheonamides A1, A2, B2, C, and D and dihydrocyclotheonamide A have been isolated from *T. swinhoei*, and inhibited thrombin with IC_{50} values of 1.0, 3.0, 1.3, 0.19, 1.4, and 0.33 μM, respectively, while trypsin was inhibited with IC_{50} values of 4.5, >10, 6.2, 3.8, >10, and 6.7 μM, respectively. The inhibition of serine proteases by the cyclotheonamides is due to the presence of the β-keto group in the β-ketohomoarginine residue [75]. Cyclotheonamides E4 and E5 have been isolated from a sponge belonging to the genus *Ircinia*. Cyclotheonamide E4 has proven to be a more potent inhibitor of human tryptase ($IC_{50} = 5.1$ nM) than cyclotheonamide E5 ($IC_{50} = 4.7$ nM) [76]. Owing to their potent enzyme inhibitory activities [77–84], which were investigated at the atomic level by X-ray crystallographic analysis [85,86] and by NMR conformational studies [87], related to the unique peptide skeleton of these compounds, the cyclotheonamides have been widely explored as target of many synthetic approaches, including several total syntheses [88–96].

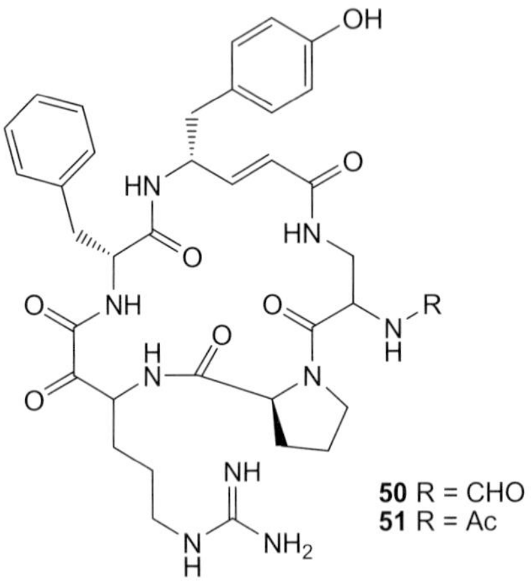

50 R = CHO
51 R = Ac

Hymenamide A (**52**) has been isolated from the sponge *Hymeniacidon* sp., and displayed antifungal activity against *Candida albicans* with MIC at 66 μg/mL as well as against *Cryptococcus neoformans* with MIC > 133 μg/mL. Hymenamide A did not present cytotoxic activity [97]. Hymenamide F (**53**) was also isolated from *Hymeniacidon* sp. in an *S*-oxide form [98]. Cupolamide A (**54**) has been isolated from the sponge *Theonella cupola* and displayed cytotoxic activity against P388 murine leukemia cells at $IC_{50} = 7.5$ μg/mL [99]. Kalahalide C (**55**), isolated from the sagoglossan marine mollusc *Elysia rufescens* and from the alga on which the mollusc feeds, *Bryopsis* sp. has an arginine residue and an *N*-phenylalanine terminus linked to butyric acid [65].

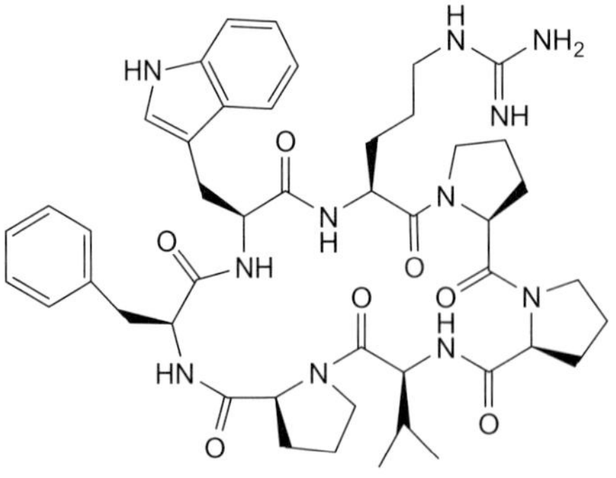

52 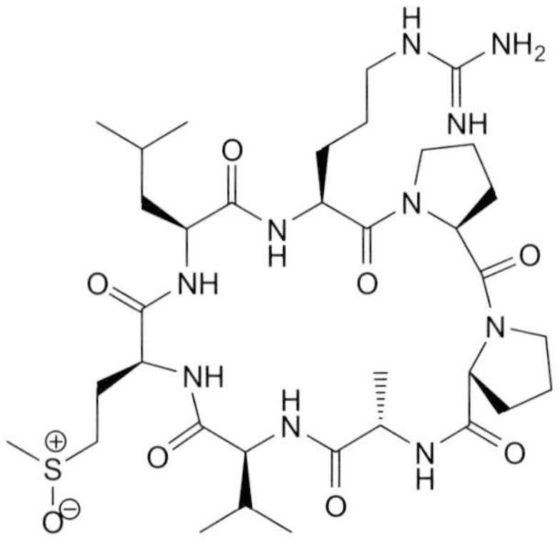 **53**

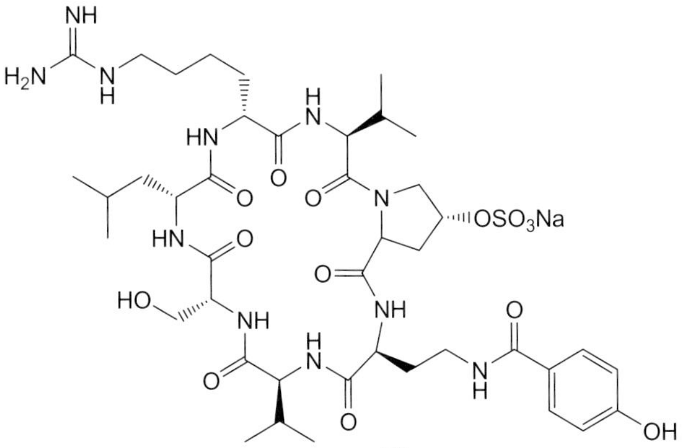

54

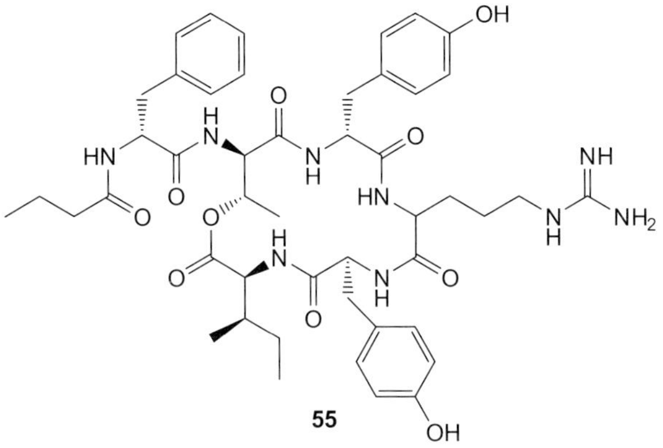

55

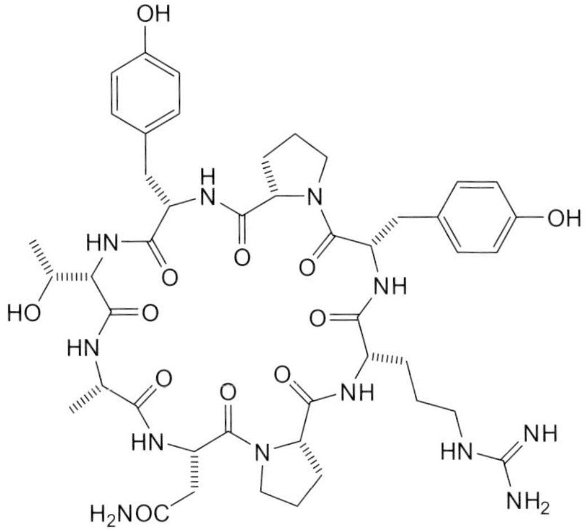

56

Cyclonellin, a new cyclooctapeptide, was isolated from the sponge *Axinella carteri* [100], and was shown to be inactive against cancer cell lines.

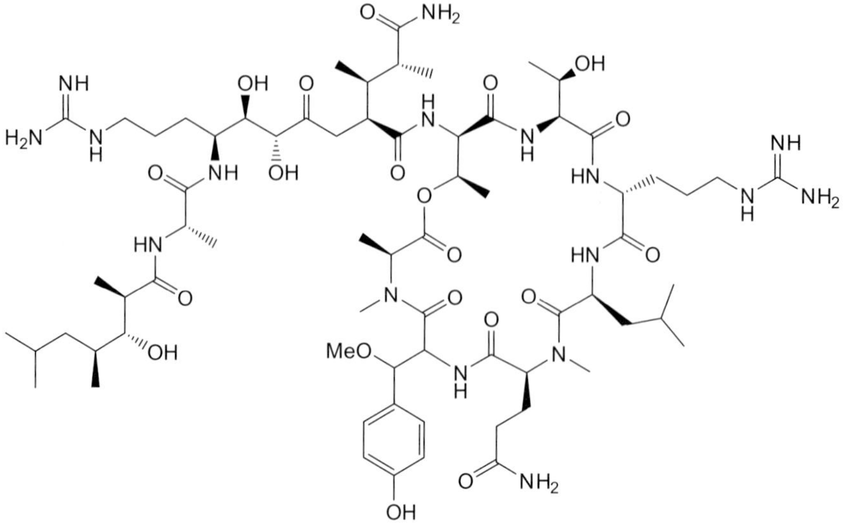

57

Callipeltins A (**57**), B (**58**), and C (the linear peptide corresponding to callipeltin A) are complex polypeptides, originally isolated from the sponge *Callipelta* sp. [101,102]. The structures of callipeltins A and C include a new guanidine amino acid, (2*R*,3*R*,4*S*)-4-amino-7-guanidino-2,3-dihydroxyheptanoic acid (AGDHE). The stereochemistry of the 3-hydroxy-2,4,6-trimethylheptanoic acid moiety in callipeltin A was subsequently defined by enantioselective synthesis [103].

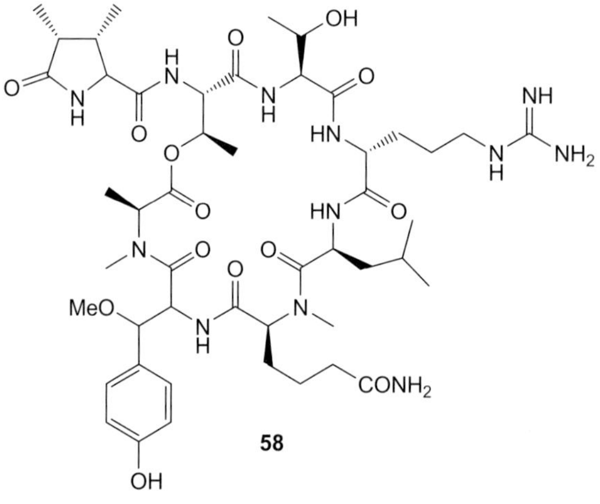

58

The stereochemistry of the *N*-methyl-alanine residue connected to the end of the 3-hydroxy-2,4,6-trimethylheptanoic acid moiety as well as of the allo-threonin placed between the L-methylglutamine and L-threonin residues have been revised to the correct actual stereostructure of callipeltin A (**57**) [104]. Two additional members of this class, named callipeltins D and E, have been isolated from

Latrunculia sp. [104], and represent fragments of callipeltin A. A new series of callipeltins have been isolated from the same extracts, and four additional peptides, callipeltins F–I have been isolated [105]. Callipeltins F–I are also fragments of the original peptide callipeltin A, in the case of callipeltins H and I with (*Z*)-2-amino-2-butenoic acid replacing the D-allo-threonin residue linked to the L-methylglutamine residue. The relative configuration of the *para*-methoxytyrosine residue and the absolute configuration of the Thr-2 unit in callipeltin A have been recently revised using an integrated NMR–quantum mechanical approach, relying on the comparison between calculated and experimental *J*-values [106]. Callipeltins A-C displayed *in vitro* cytotoxic activity against several human carcinoma cells, callipeltin A being the most active [101]. Callipeltins A and C also exhibited antifungal activity against *Candida albicans*, while callipeltin A has been shown to possess cytotoxic activity on CEM4 lymphocytic cell lines infected with HIV-1 [101]. Callipeltin A did not inhibit the dengue virus [107]. In cardiac sarcolemmal vesicles, callipeltin A induces a powerful and selective inhibition of the Na^+/Ca^{2+} exchanger, with IC_{50} at 0.85 μM. In electrically driven guinea-pig atria, callipeltin A induces a positive inotropic effect at concentrations ranging between 0.7 and 2.5 μM [108,109]. Callipeltin A appears to display an Na-ionophore action, since resting aorta responded to callipeltin A in a dose-dependent manner, with EC_{50} at 0.44 μM, which was not inhibited by common calcium channel blockers. Callipeltin A also increased Na^+ efflux of Na-loaded erythrocytes, with EC_{50} at 0.51 μM [110].

Two sponge-derived arginine-bearing peptides displayed inhibition of the HIV-1 cytopathic effect in CEM-SS target cells. These include neamphamide A (**59**), isolated from the sponge *Neamphius huxleyi*, with an EC_{50} of 28 nM [111] and microspinosamide (**60**) isolated from the sponge *Sidonops microspinosa* with an EC_{50} of 0.2 μg/mL [112]. Recently the stereochemistry of neamphamide was completely solved by synthesis of the unusual aminoacids present in the peptide structure and comparison with the hydrolyzed products by HPLC and NMR analyses [113].

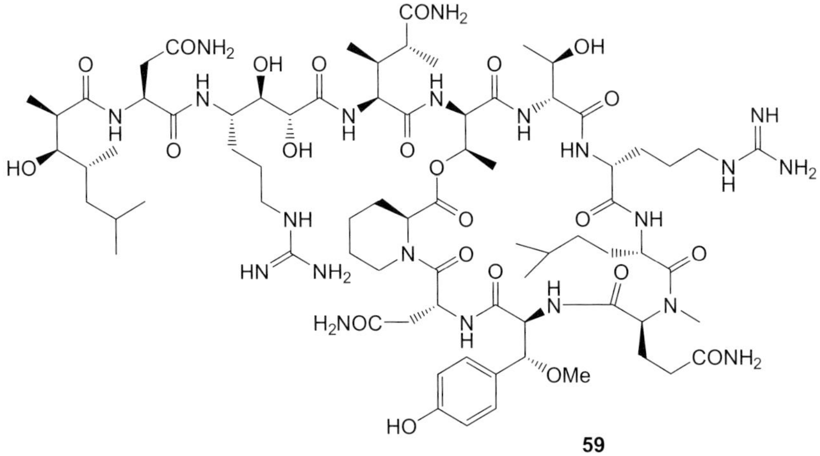

59

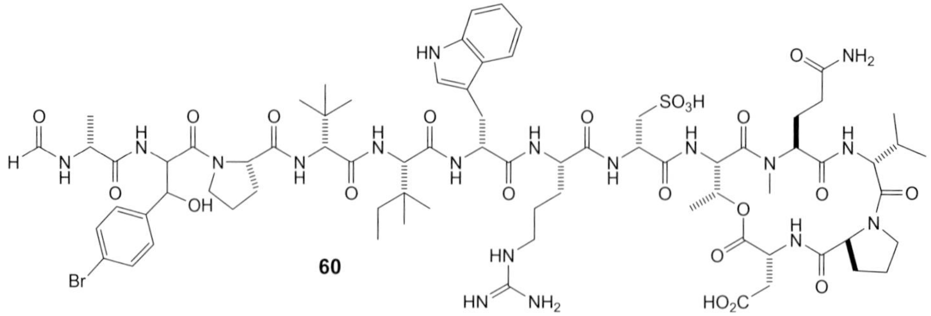

60

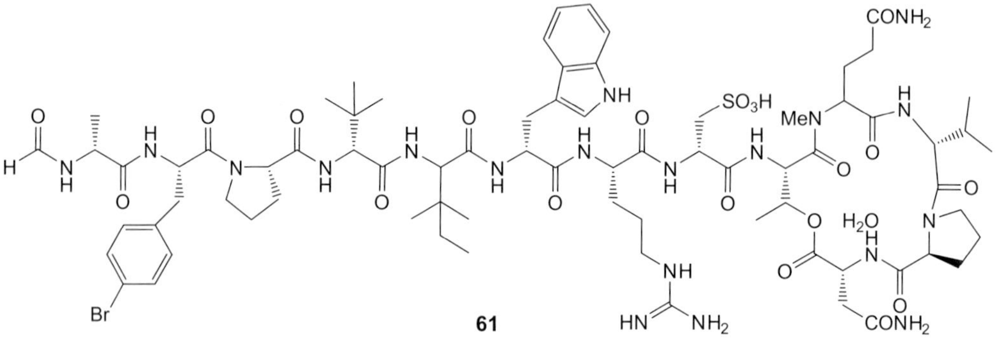

61

Polydiscamide A (**61**) is a depsitetradecapeptide which A has been isolated from the sponge *Discodermia* sp. [113a]. Polydiscamide A inhibited the growth of *Bacillus subtilis* (MIC = 3.1 μg/mL) and the proliferation of the cultured human lung cancer A549 cell line12 (IC_{50} = 0.7 μg/mL) *in vitro*. Related polypeptides, halicylindramides A–E, have been isolated from the sponge *Halichondria cylindrata* [114,115]. All halicylindramides displayed antifungal activity against *Mortierella ramanniana* and cytotoxic activity against P-388 murine leukemia cells. Additional similar peptides, the discodermins, were the first polypeptides to have been isolated from the marine sponge, *Discodermia kiiensis*. The discodermins presented a broad range of antimicrobial activity. Discodermin A (**62**) promotes permeabilization of the plasma membrane of A10 cells to the nonpermeable fluorescent probes ethidium homodimer-1 (MW = 857), calcein (MW = 623), as well as permeabilization of vascular tissue cells to Ca^{2+} and ATP [116–121].

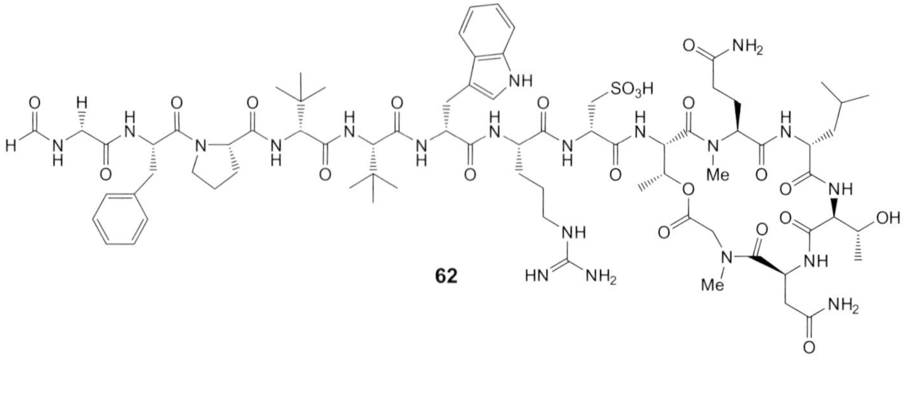

62

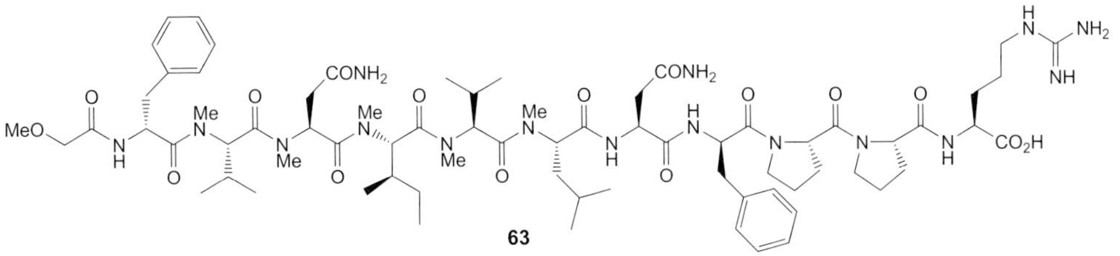

63

Koshikamide A2 (**63**) has been recently isolated from a sponge of the genus *Theonella*, and displayed moderate cytotoxic activity against P388 murine cancer cells, with IC_{50} at 6.7 μg/mL [122]. Neopetrosiamides A (**64**) and B are rather complex linear polypeptides, differing from each other by the stereochemistry of the sulfoxide group. Both peptides have been isolated from a sponge of the genus *Neopetrosia* sp., and were active in an amoeboid invasion assay at 6 μg/mL [123].

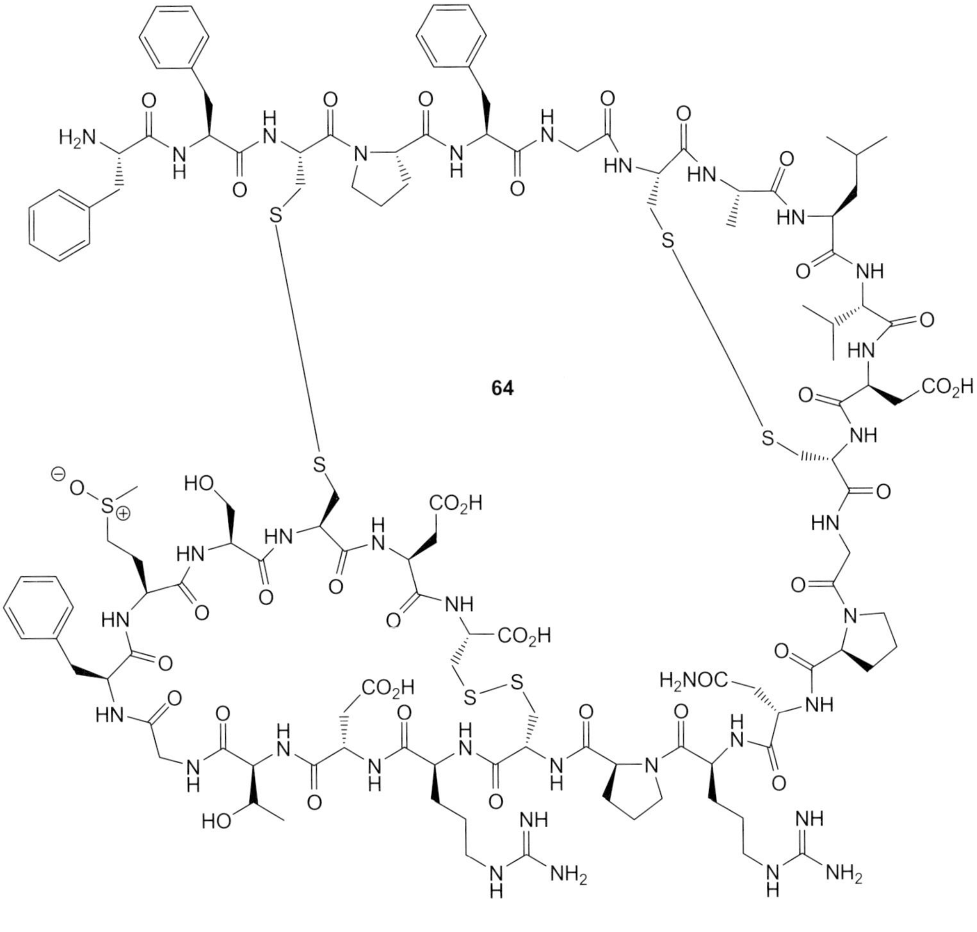

64

11.6
Terpenic Guanidines

Siphonodictidine, a guanidine sesquiterpene (**65**), was found in the mucus secreted by the sponge *Siphonodictyon* sp. and inhibits the coral growth in the vicinity of the sponge [124]. Two approaches for the synthesis of **65** have been developed [125,126]. The structure **66** initially proposed for suvanine isolated from the sponge *Ircinia* sp. [127] was later revised to **67** by ion-exchange chromatography, chemical degradations, and X-ray diffraction analysis, when it was also isolated from the sponge *Coscinoderma matthewsi* [128]. It must be noted that the correct sulfate group was incorrectly drawn as a sulfonic group. A related metabolite, sulfircin (**68**), was isolated from the sponge *Ircinia* sp. [129]. Several analogs of sulfircin have been synthesized [130].

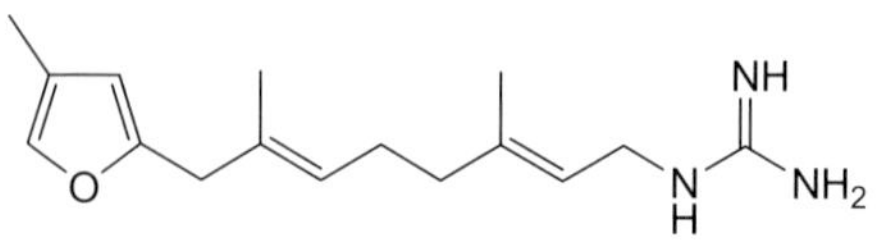

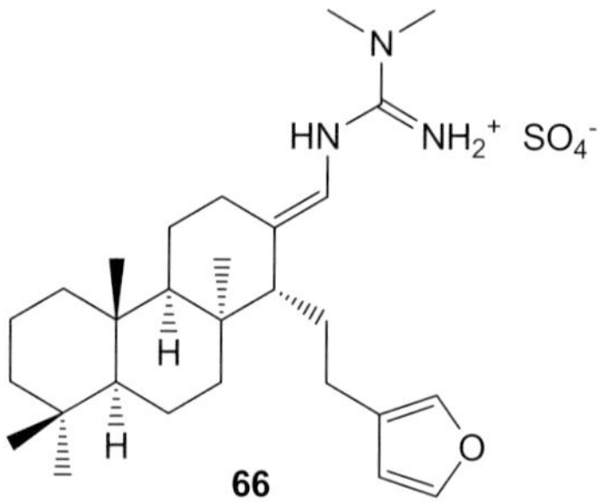

Agelasidines are diterpene guanidines which have been isolated from sponges of the genus *Agelas*. The first members of this series, agelasidines A–C (**69**–**71**), have been isolated from *Agelas nakamurai* [131], while (−)-agelasidine C (**72**) and agelasidine D (**73**) have been isolated from *Agelas clathrodes* [132]. The structures of agelasidines A–C were established not only by analysis of spectroscopic data but also by chemical correlations. Agelasidines A–C inhibited N^+,K^+-ATPase [133].

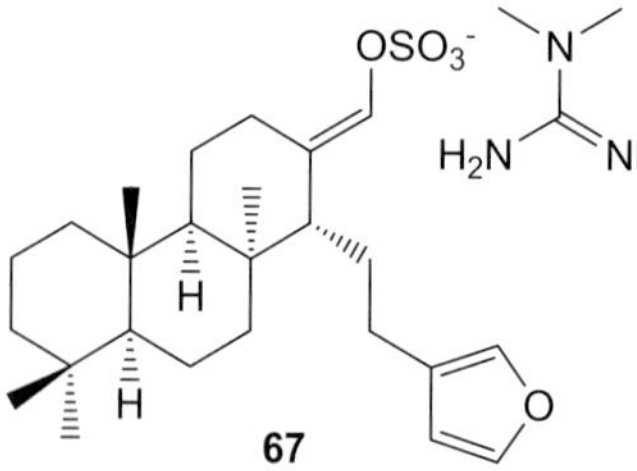

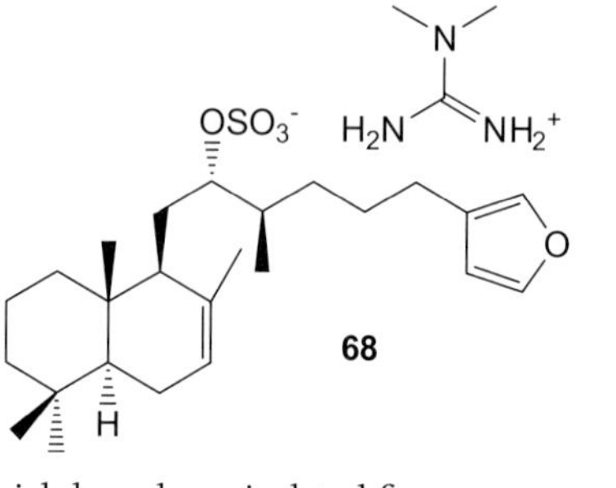

Stellettazoles A (**74**) and B (**75**) have been isolated from a *Stelleta* sp. marine sponge [134,135]. Stellettazole A (**74**) exhibited potent antibacterial activity against *Escherichia coli* and inhibitory activity against Ca^{2+}/calmodulin-dependent phosphodiesterase. Stellettazole B (**75**) displayed only marginal antibacterial activity against *E. coli* [135]. Bistellettadines A (**76**) and B (**77**) were also isolated from the same sponge, as moderate inhibitors of Ca^{2+}/calmodulin-dependent phosphodiesterase and potent antibacterial agents against *E. coli* (10 μg per disk) [136].

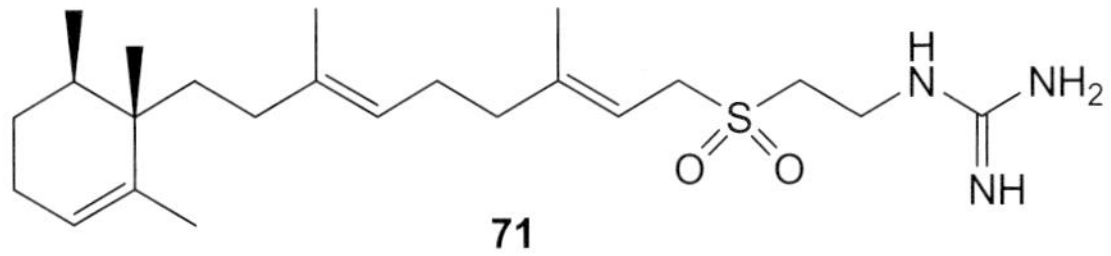

71

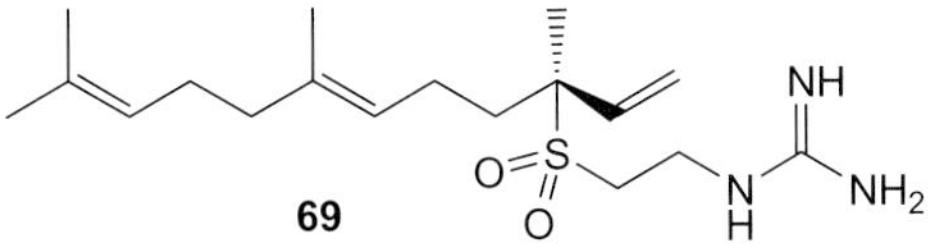

69

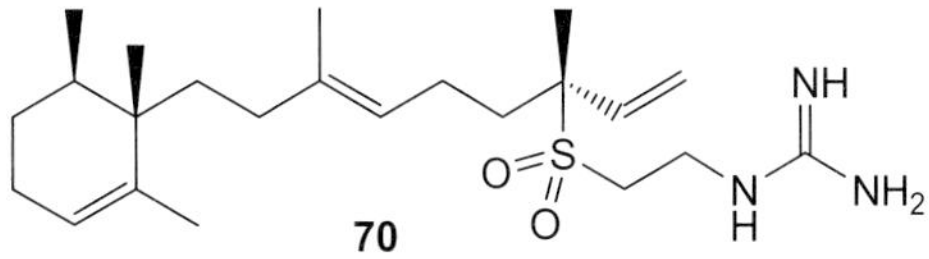

70

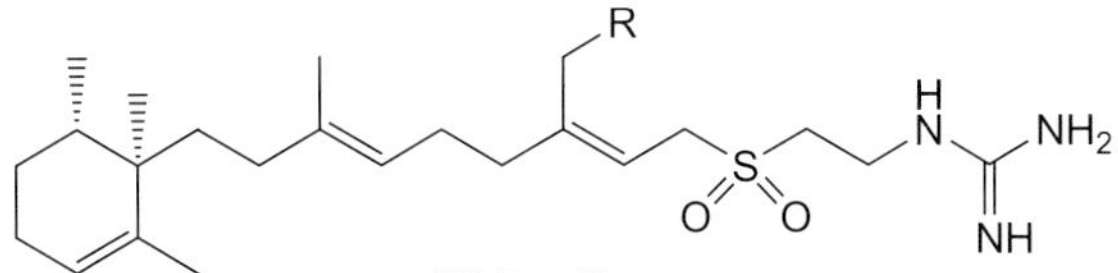

72 R = H
73 R = OH

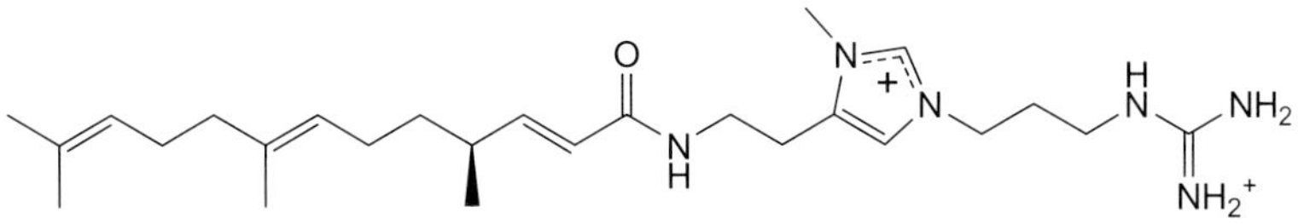

74

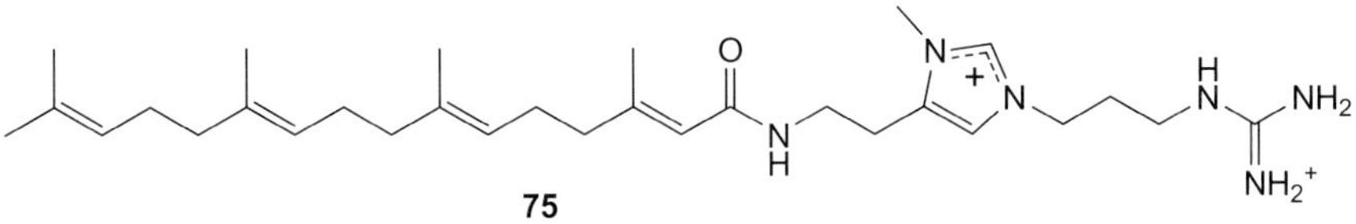

75

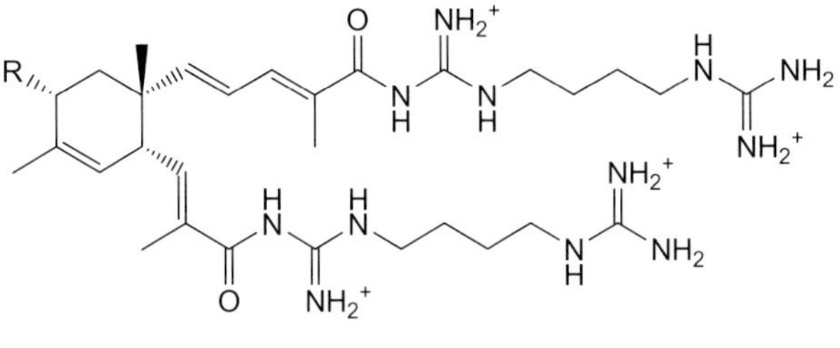

76 R=H 77 R=

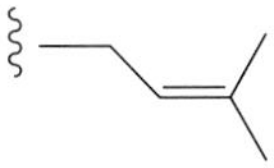

11.7 Polyketide-derived Guanidines

Polyketide guanidines comprise the largest and structurally more complex group of marine guanidine alkaloids. The antiviral and antimicrobial acarnidines **78–80** have been isolated from the sponge *Acarnus erithacus* [137], identified by analysis of spectroscopic data and later synthesized using different approaches [138–140]. Aplysillamides A (**81**) and B (**82**) have been isolated from the marine sponge

Psammaplysilla purea [141]. Aplysillamide B (**82**) has been synthesized in order to define the absolute stereochemistry of its stereogenic center [141,142]. Polyandrocarpidines A–D have been isolated from the ascidian *Polyandrocarpa* sp. and displayed antimicrobial and cytotoxic activities [143]. The structures initially proposed [143] were later revised to **83–86**, respectively [144]. The hexadyro derivative of polyandrocarpidine A has been synthesized in order to confirm the structure reassignment [145]. A series of antibacterial and moderately cytotoxic phloeodictynes Al–A7 (**87–93**) and Cl–C2 (**94–95**) have been isolated as mixtures of homologs from the sponge *Phloeodictyon* sp. [146]. The structure determination of compounds **87–95** was performed by NMR and MS analysis using B/E collisionally activated dissociation (CAD) FAB mass spectrometry [146]. Recently a series of 17 new phloeodictynes has been isolated from the sponge *Oceanapia fistulosa* from New Caledonia, and identified by LC-ESI-MS analysis [147]. Erylusidine (**95**) was isolated from the sponge *Erylus* sp. together with analogs containing *N,N*-dimethylputrescine and *N,N,N*-trimethylspermidine replacing the agmatine moiety [148]. Triophamine (**96**) was isolated from the dorid nudibranch *Triopha catalinae* [149] and *Polycera tricolor* [150]. The stereochemistry of the double bonds was assigned as *E* after total synthesis of **96** [151]. The biosynthesis of **96** proceeds from acetate through butyrate [152,153]. A related acylguanidine, limaciamine (**97**), was isolated from the North Sea nudibranch *Limacia clavigera* [154–155].

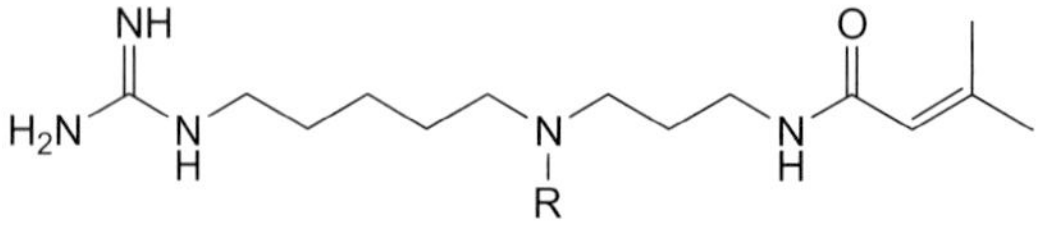

78 R = $CO(CH_2)_{10}CH_3$

79 R = $CO(CH_2)_3CH{=}CH(CH_2)_5CH_3$

80 R = $CO(CH_2)_{12}CH_3$

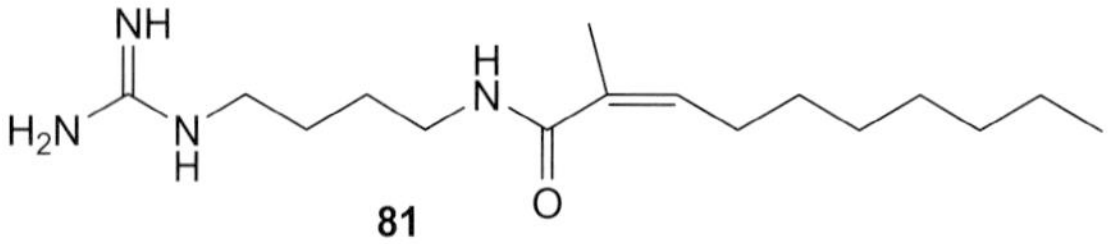

81

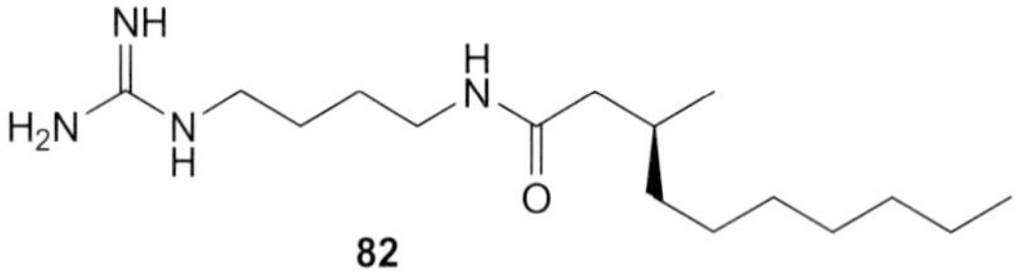

82

83 n = 2
84 n = 1

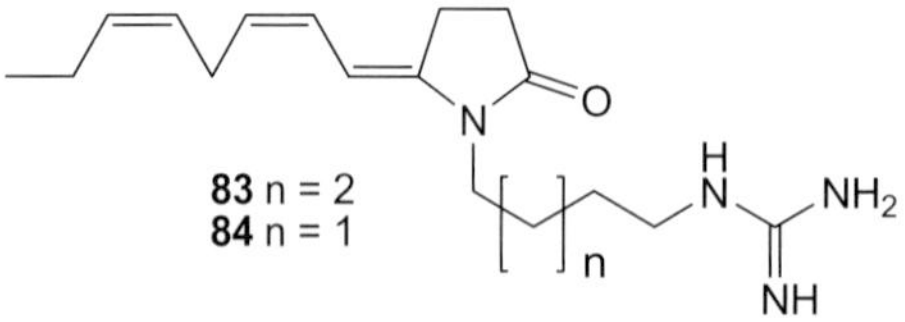

85 n = 2
86 n = 1

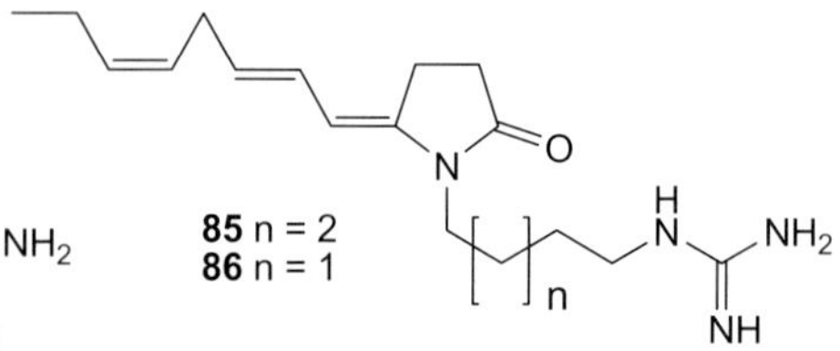

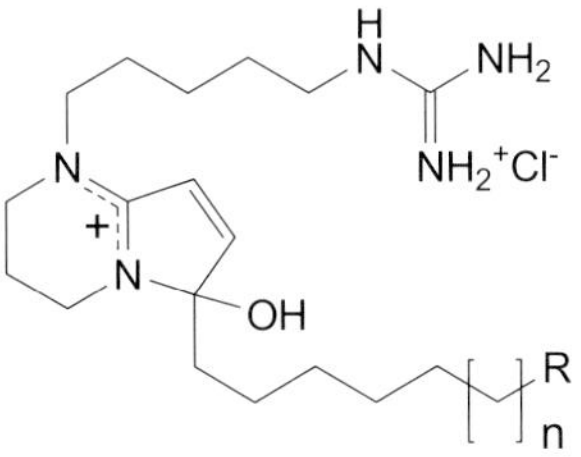

87 n = 4, R = CH_2 $CH=CH_2$
88 n = 2, R = CH_2 $CH=CH_2$
89 n =1 , R = CH_2 $CH=CH_2$
90 n = 5, R = $CHMe_2$

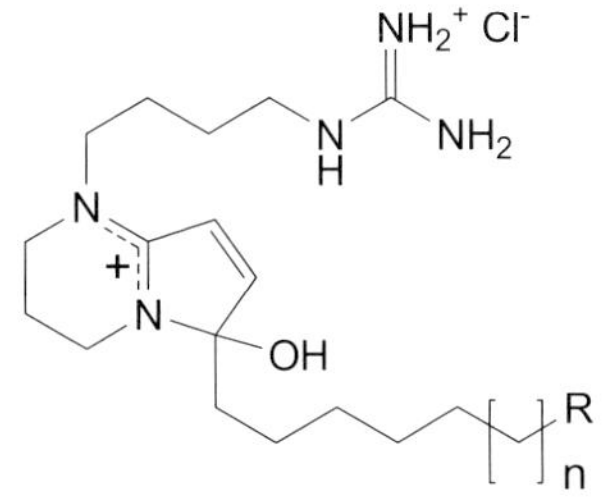

90 n = 4, R = CH_2 $CH=CH_2$
91 n = 2, R = CH_2 $CH=CH_2$
92 n = 5, R = $CHMe_2$

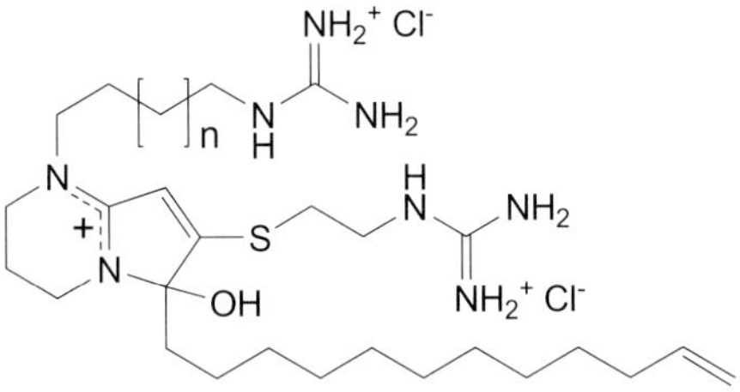

93 n = 2
94 n = 1

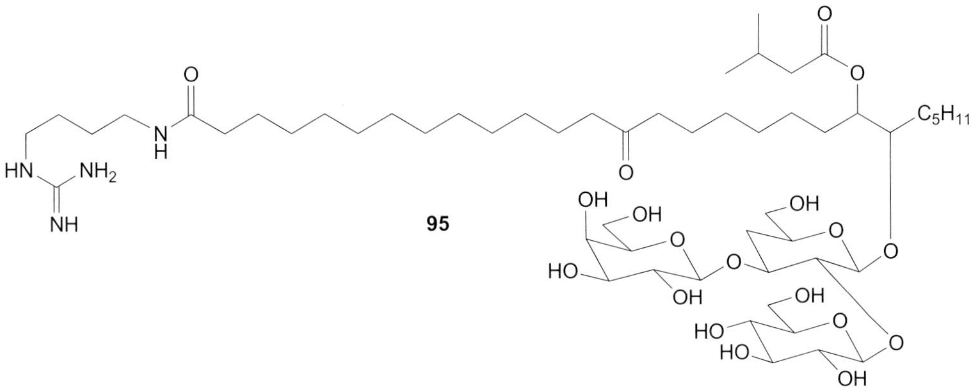

95

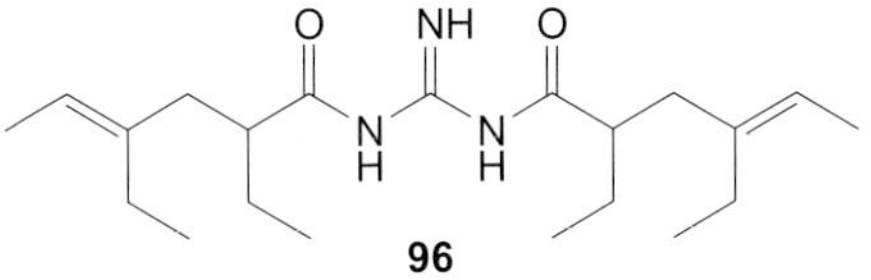

96

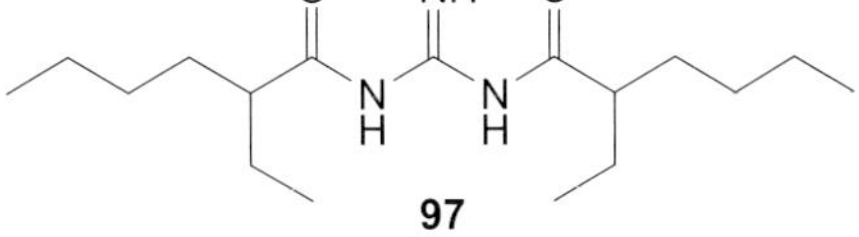

97

Onnamide A (**98**) and several of its related derivatives have been isolated from marine sponges of the genus *Theonella*. Onnamide A itself was first isolated from *Theonella* sp. [156] and displayed potent cytotoxic activity [157]. Subsequently, a series of related derivatives have been isolated from different species of marine sponges belonging to the genus *Theonella* [158,159]. The structure of onnamide A was confirmed by synthesis [160]. The mechanism-of-action of onnamide A cytotoxicity is related to the inhibition of protein synthesis and the induced activation of p38 mitogen-activated protein kinase and c-Jun NH_2-terminal protein kinase (JNK) [161]. Interestingly, the gene cluster that is probably responsible for the biosynthesis of onnamide A has been isolated and compared to the *Paederus fuscipes* beetle *Pseudomonas* symbiont gene cluster responsible for the biosynthesis of the related metabolite pederin [162]. It appears that the gene cluster isolated from *Theonella swinhoei* includes a variety of subclusters responsible for the biosynthesis of distinct synthases, including polyketide synthases, a 3-hydroxy-3-methylglutaryl-CoA synthase (HMGS), oxygenases, methyl transferases, a nonribosomal peptide synthase (NRPS), as well as additional genes with unidentified functions. These investigations have been reviewed [163].

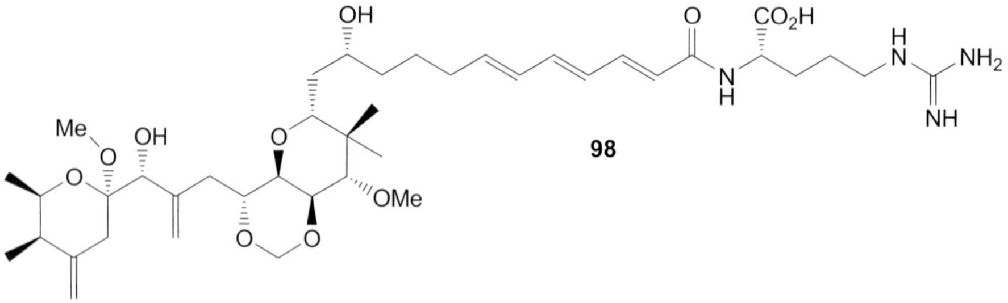

98

Several structurally complex polyketide-derived polycyclic guanidine alkaloids have been isolated from Poescilosclerida sponges belonging to genera *Ptilocaulis, Crambe,* and *Monanchora*. The first members of this class were ptilocaulin (**99**) and isoptilocaulin (**100**) isolated from the Caribbean sponge *Ptilocaulis* aff. *spiculifer.* Ptilocaulin (**99**) displayed cytotoxic activity against L1210 leukemia cells with ID_{50} at 0.39 μg/mL as well as antimicrobial minimum inhibitory concentrations (μg/mL) against the following strains: *Streptococcus pyogenes,* 3.9; *S. pneumoniae,* 15.6; *S. faecalis, Staphylococcus aureus,* and *Escherichia coli,* all at 62.5. Isoptilocaulin displayed cytotoxic activity against L1210 leukemia cells with ID_{50} 1.4 μg/mL and antimicrobial MIC's (μg/mL) against: *S. pyogenes,* 25; *S. pneumoniae* and *S. aureus,* 100; *S. faecalis* and *E. coli,* >100 [164]. Further cytotoxicity studies were performed with ptilocaulin [165,166]. Owing to their unique carbon structure framework and biological activities, these compounds were selected targets for several total syntheses [167–177]. Related ptilocaulin-like alkaloids have been isolated from sponges of the genera *Monanchora* [178,179], *Arenochalina* [180,181], and *Batzella* [182].

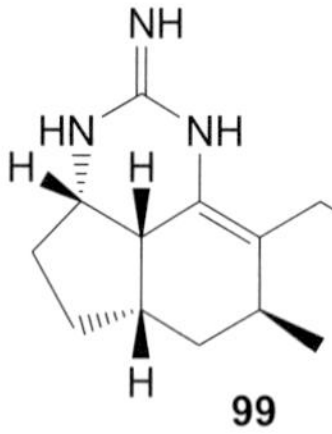

99 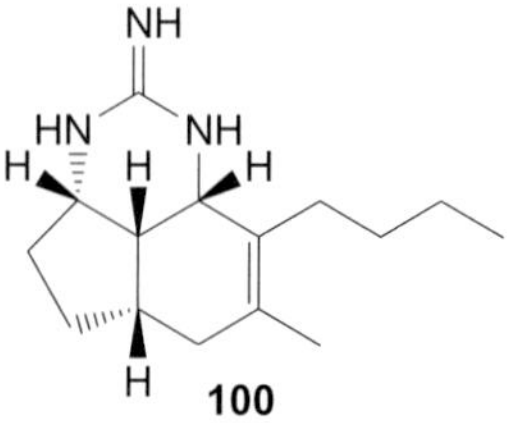 **100**

The isolation of ptilomycalin A (**101**) from the Caribbean sponge *Ptilocaulis spiculifer* and from the Red Sea *Hemimycale* sp. was published in 1989 [183–185]. Ptilomycalin A possesses a complex pentacyclic guanidine/fatty acid/polyamine skeleton and shows cytotoxicity against P388 cancer cell line ($IC_{50} = 0.1$ μg/mL), antifungal activity against *Candida albicans* (MIC = 0.8 μg/mL) as well as antiviral activity against *Herpes simplex* virus at 0.2 μg/mL. Ptilomycalin A was thus of special interest in organic synthesis [186–198] and further bioactivity studies, which showed that ptilomycalin A (**101**) inhibits Na^+, K^+ or Ca^{2+} ATPases [197, 199].

Subsequently, related ptilomycalin A derivatives were isolated from the marine sponge *Crambe crambe*, namely crambescidins 800 (**102**), 816 (**103**), 830 (**104**), and 844 (**105**), as well as isocrambescidin 800 (**106**) [200–202]. The absolute stereochemistries of compounds **102–106** were established by chemical degradations and chiral gas chromatography analysis [202]. Crambescidin 816 (**103**) was active against HCT-16 human colon carcinoma cells ($IC_{50} = 0.24$ μg/ml). The activity of crambescidin 816 on voltage-sensitive Ca^{2+} channels was also tested in a neuroblastoma X glioma cell line (NG 108-1 5) and acetylcholine-induced contractions of isolated guinea pig ileum were also evaluated [201]. Crambescidin 816 exerts a potent Ca^{2+} antagonist activity ($IC_{50} = 1.5 \times 10^{-4}$ μM), more potent than nifedipine ($IC_{50} = 1.2$ μM), a known selective blocker of L-type Ca^{2+} channels. Crambescidin 816 also inhibited acetylcholine-induced contraction of guinea pig ileum at very low concentrations. Apparently crambescidin 816 operates through a reversible blockage of Ca^{2+} channels [201]. Crambescidins 816 (**103**), 844 (**105**), and 800 (**102**) displayed antiviral activity against HSV-1, with diffuse cytotoxicity, at 1.25 μg/well and displayed 98 % cytotoxic activity against L1210 cell growth at 0.1 μg/mL [200]. Ptilomycalin A and crambescidin 800 were also isolated from the starfish *Celerina heffermani* and *Fromia monilis*, along with the novel alkaloids celeromycalin (**107**) and fromiamycalin (**108**) [203]. Compounds **107** and **108** displayed anti-HIV activity on CEM 4 cells infected by HIV-1, at concentrations of 0.32 μg/mL and 0.11 μg/mL, respectively. As in the case of ptilomycalin A, the crambescidin alkaloids have been a target of choice in organic synthesis [204–207].

101 n = 12, $R_1 = R_2 = H$
102 n = 12, $R_1 = H$, $R_2 = OH$
103 n = 12, $R_1 = R_2 = OH$
104 n = 13, $R_1 = R_2 = OH$
105 n = 14, $R_1 = R_2 = OH$

Additional but simpler members of this class of guanidine alkaloids are the icthyotoxic crambescins (formerly crambines) A (**109**), B (**110**), and C1 (**111**) isolated

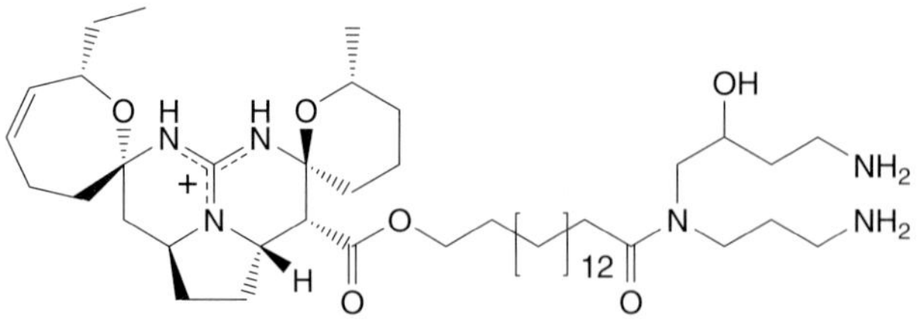

106

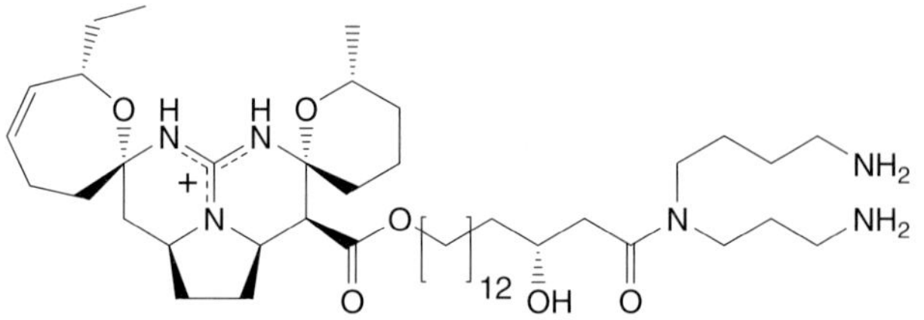

107

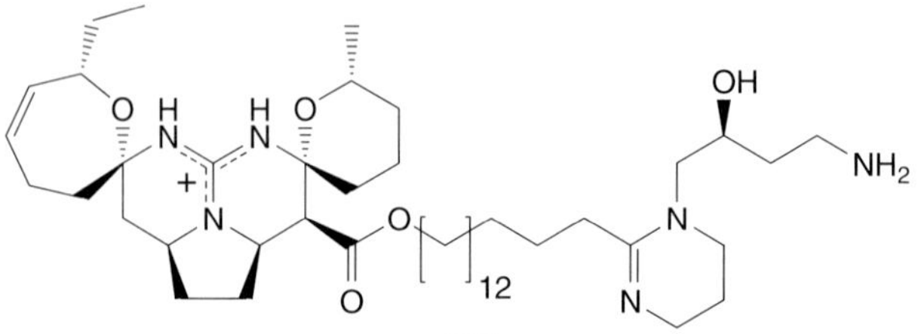

108

from the sponge *Crambe crambe* as mixtures of homologues difficult to separate [208,209]. The structure of crambescin C1 [209] has been revised to **111** by tandem mass spectrometric analysis [209], while the correct stereochemistry of crambescin B (**110**) was established after its total synthesis [210–212]. While crambescidin 800 (**102**) has also been re-isolated from a Brazilian specimen of the sponge *Monanchora arbuscula* [213], related members of this series, crambescidin 826 and dehydrocrambine A, have been isolated from *Monanchora* sp. along with the known compounds crambescidin 800 and fromiamycalin [214], and also from *Ptilocaulis spiculifer* [215]. The simplest members of the crambescidins, crambescidin 359 (**113**) and 431 (**113**) have been isolated from the sponge *Monanchora unguiculata* [216], while crambescidic acid (**114**) has been isolated from *M. unguifera* [217]. Less well-characterized crambescidin-related pentacyclic alkaloids have been isolated as mixtures from the sponge *Neofolitispa dianchora* [218]. Different syntheses of crambescidin 359 (**112**) have been reported [219–221].

Biogenetically related batzelladines have been isolated from several marine sponges belonging to the order Poescilosclerida. The first series of such compounds were batzelladines A (**115**, major homolog), B, C (**116**, major homolog), D, and E, isolated from a marine sponge of the genus *Batzella* as mixtures of homologs [222].

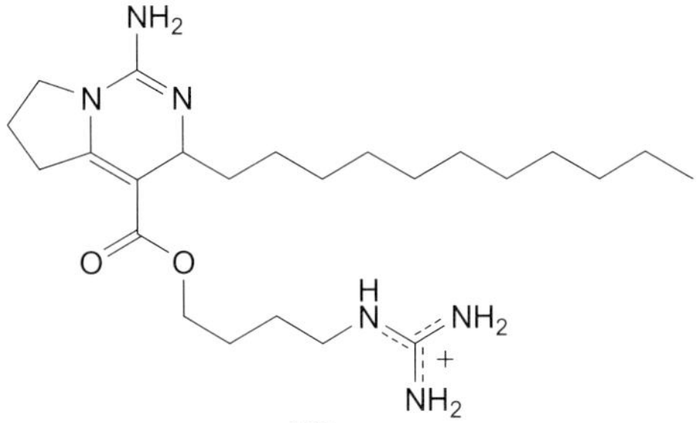

109 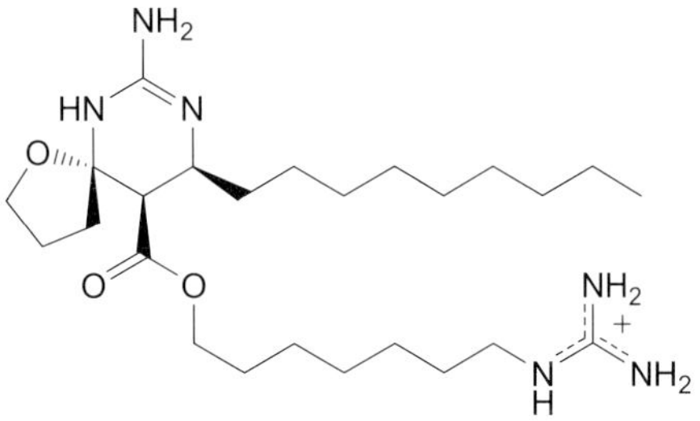 110

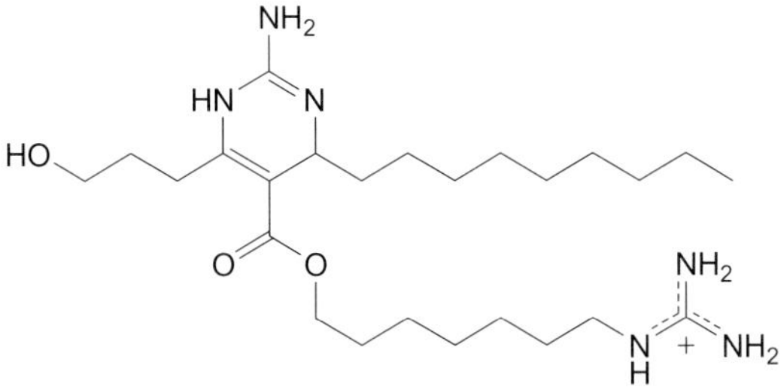

111

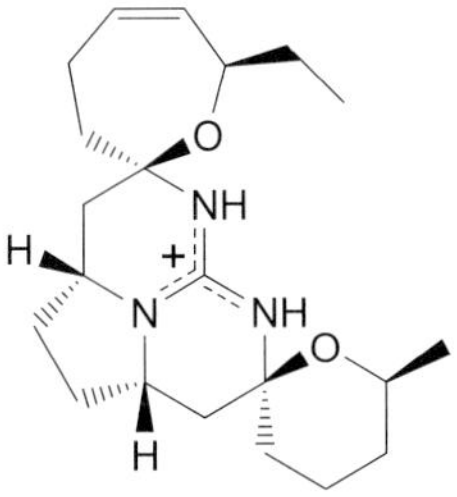

112 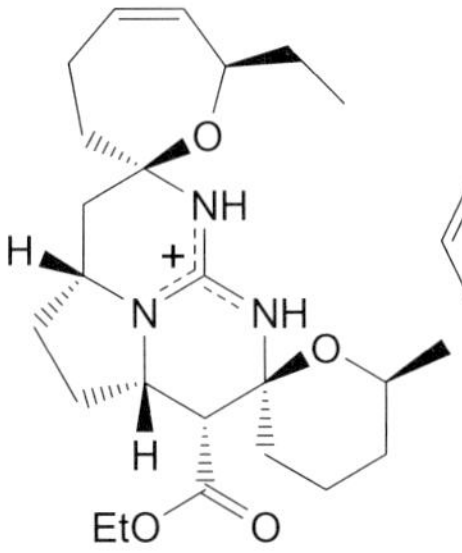 113 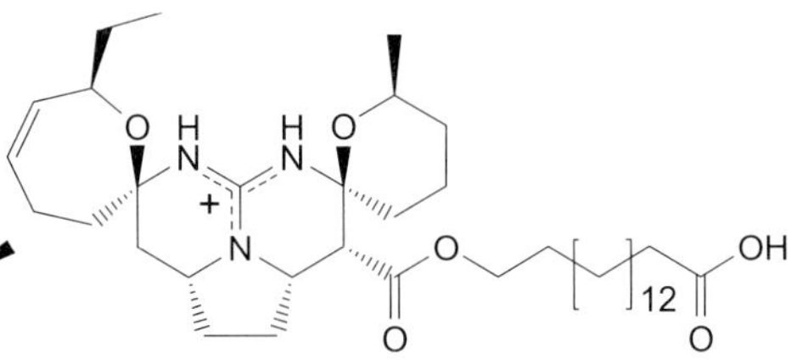 114

Batzelladine A and the structurally closely related batzelladine B inhibited the binding of the gp120 domain of HIV-envelope gp160 glycoprotein to the CD4 receptor on the surface of the human T cell [223]. A second series of batzelladine alkaloids isolated from another *Batzella* sp. sponge include the F (**117**) and G–I members, all of which are closely related [223]. Batzelladine F induced a 100 % dissociation of a p56lck-CD4 binding complex, while batzelladine G and an inseparable mixture of batzelladines H and I induced dissociation of the p56lck-CD4 complex at concentrations of 24 and 7.1 μM, respectively [223]. Further members of the batzelladine alkaloids are dehydrobatzelladine C (**118**) isolated from *Monanchora arbuscula* [216] and batzelladine J (**119**) from *M. unguifera* [217]. There have been several chemical and biological studies aimed at the development of the total synthesis of several batzelladine alkaloids [224–235] as well as investigations of their mode-of-action and the development of more-potent and selective antiviral and antiproliferative alkaloids [236–237].

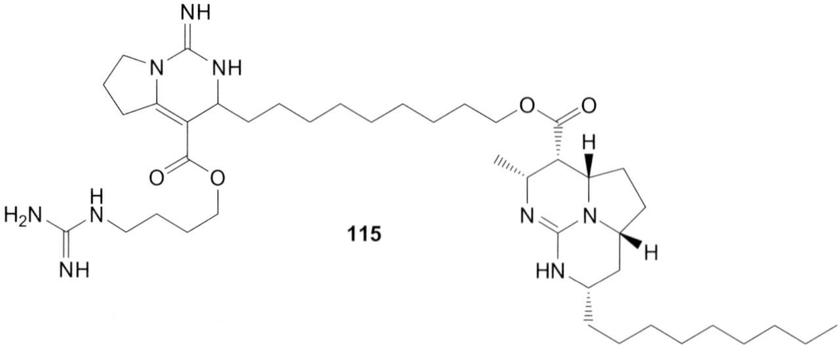

115

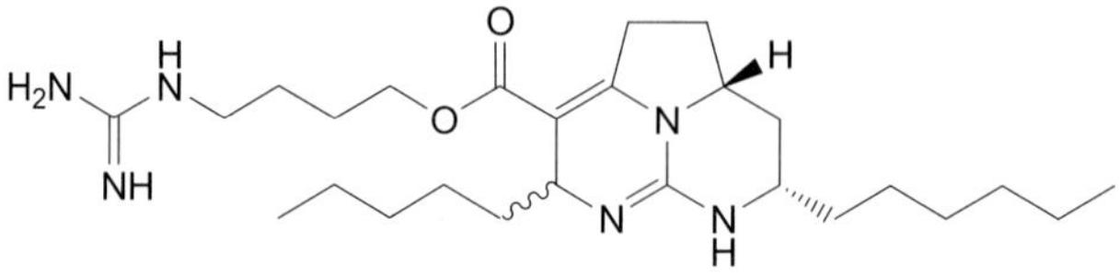

116

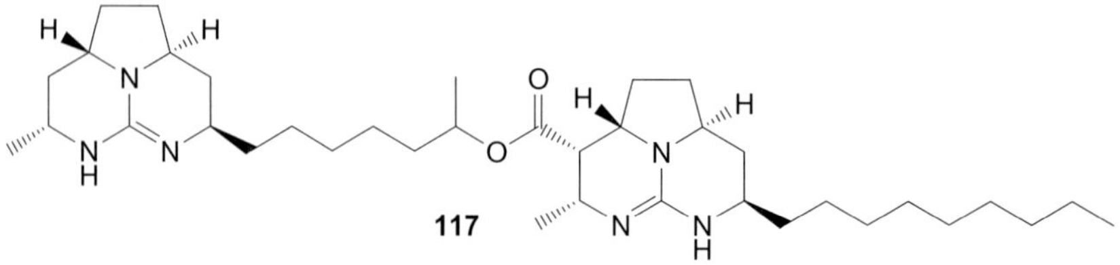

117

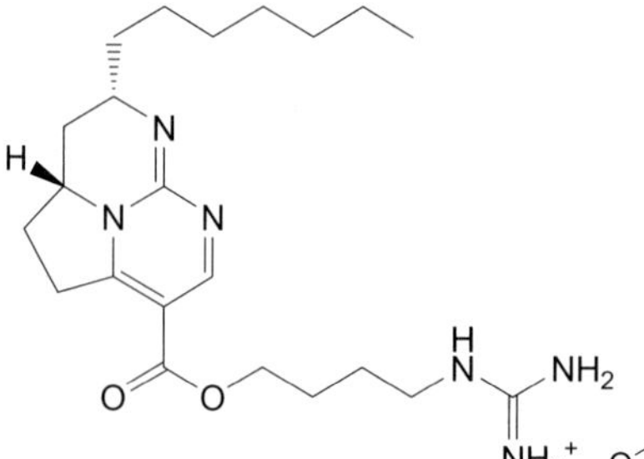

118

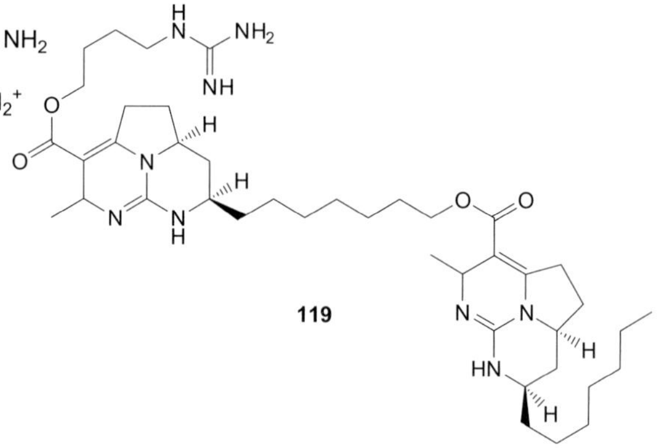

119

Monanchorin (**120**), which has been isolated from the sponge *Monanchora unguiculata* [238], is the last example of polyketide-derived guanidines.

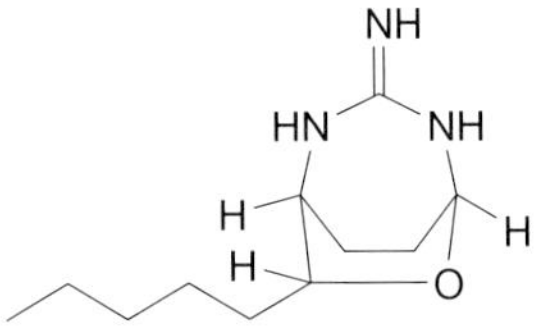

120

The chemistry (isolation, structures, and synthesis) and biological activities of this large class of guanidine alkaloids have been the subject of several reviews [239–243].

References

1 Chevolot, L. (1981) Guanidine derivatives, In Scheuer P. J. (Ed.) *Marine Natural Products: Chemical and Biological Perspectives*, Vol. 4, Academic Press, London, 53–91.

2 Berlinck, R. G. S. (1995) *Fortschritte der Chemie Organischer Naturstoffe (Progress in the Chemistry of Organic Natural Products)*, **66**, 119–295.

3 Berlinck, R. G. S. (1996) *Natural Product Reports*, **13**, 377.

4 Berlinck, R. G. S. (1999) *Natural Product Reports*, **16**, 339.

5 Berlinck, R. G. S. (2002) *Natural Product Reports*, **19**, 617.

6 Berlinck, R. G. S. and Kossuga, M. H. (2005) *Natural Product Reports*, **22**, 516.

7 Kazlauskas, R., Murphy, P. T., Quinn, R. J., Wells, R. J. (1977) *Tetrahedron Letters*, 61.

8 Hollenbeak, K. H. and Schmitz, F. J. (1977) *Journal of Nature Products*, **40**, 479.

9 Djura, P. and Faulkner, D. J. (1980) *The Journal of Organic Chemistry*, **45**, 735.

10 Tymiak, A. A., Rinehart, K. L. Jr., Bakus, G. J. (1985) *Tetrahedron*, **41**, 1039.

11 (a) Kondo, K., Nishi, J., Ishibashi, M., Kobayashi, J. (1994) *Journal of Natural Products*, **57**, 1008.
(b) Seagraves, N. L. and Crews, P. (2005) *Journal of Natural Products*, **68**, 1484–1488.

12 Fattorusso, E., Lanzotti, V., Magno, S., Novellino, E. (1985) *Journal of Natural Products*, **48**, 924.

13 Fusetani, N., Asano,Matsunaga, S., Hashimoto, K. (1986) *Comparative Biochemistry and Physiology*, **85B**, 845.

14 Okuda, R. K., Klein, D., Kinnel, R. B., Li, M., Scheuer, P. J. (1982) *Pure and Applied Chemistry*, **54**, 1907.

15 Guella, G., Mancini, I., Zibrowius, H., Pietra, F. (1989) *Helvetica Chimica Acta*, **72**, 1444.

16 Aoki, S., Ye, Y., Higuchi, K., Takashima, A., Tanaka, Y., Kitagawa, I., Kobayashi, M. (2001) *Chemical and Pharmaceutical Bulletin*, **49**, 1372.

17 Baker, J. T. and Wells, R. J. (1982) Biologically active substances from Australian marine organisms, In Beal J. L. and Reinhard E. (Eds.) *Natural Products as Medicinal Agents*, Hippokrates, Stuttgard, 281–318.

18 Baker, J. T. (1984) Modern drug research: the potential and the problems of marine natural products, In Krogsgaard-Larsen, P. Brøgger Christersen, S. Kofod H. (Eds.) *Natural Products and Drug Development*, Munksgaard, Copenhagen, 145–163.

19 Gulati, D., Chauhan, P. M. S., Pratap, R., Bhakuni, D. S. (1994) *Indian Journal of Chemistry Section B*, **33**, 4.

20 Selic, L., Jakse, R., Lampic, K., Golic, L., Golic-Grdadolnik, S., Stanovnik, B.

(2000) *Helvetica Chimica Acta*, **83**, 2802.

21 Jakse, R., Recnik, S., Svete, J., Golobic, A., Golic, L., Stanovnik, B. (2001) *Tetrahedron*, **57**, 8395.

22 Iwagawa, T., Miyazaki, M., Okamura, H., Nakatani, M., Doe, M., Takemura, K. (2003) *Tetrahedron Letters*, **44**, 2533.

23 Debitus, C., Cesario, M., Gilhem, J., Pascard, C., Pais, M. (1989) *Tetrahedron Letters*, **30**, 1535.

24 Chan, G. W., Mong, S., Hemling, M. E., Freyer, A. J., Offen, P. H., Debrosse, C. W., Sarau, H. M., Westley, J. W. (1993) *Journal of Natural Products*, **56**, 116.

25 Molina, P., Almenderos, P., Fresneda, P. M. (1994) *Tetrahedron Letters*, **35**, 2235.

26 Edrada, R. A., Stessman, C. C., Crews, P. (2003) *Journal of Natural Products*, **66**, 939.

27 Ralifo, P. and Crews, P. (2004) *The Journal of Organic Chemistry*, **69**, 9025.

28 Crews, P., Clark, D. P., Tenney, K. (2003) *Journal of Natural Products*, **66**, 177.

29 Perry, N. B., Ettouati, L., Litaudon, M., Blunt, J. W., Munro, M. H. G., Parkin, S., Hope, H. (1994) *Tetrahedron*, **50**, 3987.

30 Trimurtulu, G., Faulkner, D. J., Perry, N. B., Ettouati, L., Litaudon, M., Blunt, J. W., Munro, M. H. G., Jameson, G. B. (1994) *Tetrahedron*, **50**, 3993.

31 Sakai, R. and Higa, T. (1987) *Chemical Letters*, 127.

32 Barenbrock, J. S. and Köck, M. (2005) *Journal of Biotechnology*, **117**, 225.

33 Urban, S., Capon, R. J., Hooper, J. N. A. (1994) *Australian Journal of Chemistry*, **47**, 2279.

34 Sperry, S. and Crews, P. (1998) *Journal of Natural Products*, **61**, 859.

35 Wessels, M., König, G. M., Wright, A. D. (2001) *Journal of Natural Products*, **64**, 1556.

36 Mourabit, A. A., Pusset, M., Chtourou, M., Gaigne, C., Ahond, A., Poupat, C., Potier, P. (1997) *Journal of Natural Products*, **60**, 290.

37 Sun, H. H. and Sakemi, S. (1991) *The Journal of Organic Chemistry*, **56**, 4307.

38 Kong, F., Harper, M. K., Faulkner, D. J. (1992) *Natural Product Letters*, **1**, 71.

39 Loukaci, A., Guyot, M., Chiaroni, A., Riche, C. (1998) *Journal of Natural Products*, **61**, 519.

40 Van Wagoner, R. M., Jompa, J., Tahir, R., Ireland, C. M. (1999) *Journal of Natural Products*, **62**, 794.

41 Shimomura, O., Goto, T., Hirata, Y. (1957) *Bulletin of the Chemical Society of Japan*, **30**, 929.

42 Kishi, Y., Goto, T., Hirata, Y., Shimomura, O., Johnson, F. H. (1966) *Tetrahedron Letters*, 3427.

43 Kishi, Y., Goto, T., Eguchi, S., Hirata, Y., Watanabe, E., Aoyama, T. (1966) *Tetrahedron Letters*, 3437.

44 Kishi, Y., Goto, T., Inoue, S., Sugiura, S., Kishimoto, H. (1966) *Tetrahedron Letters*, 3445.

45 Oba, Y., Kato, S.-I., Ojika, M., Inouye, S. (2002) *Tetrahedron Letters*, **43**, 2389.

46 Kato, S.-I., Oba, Y., Ojika, M., Inouye, S. (2004) *Tetrahedron*, **60**, 11427.

47 Pearce, A. N., Appleton, D. R., Babcock, R. C., Copp, B. R. (2003) *Tetrahedron Letters*, **44**, 3897.

48 Rodriguez, A. D. and Piña, I. C. (1993) *Journal of Natural Products*, **56**, 907.

49 (a) Saeki, B. M., Granato, A. C., Berlinck, R. G. S., Magalhães, A., Schefer, A. B., Ferreira, A. G., Pinheiro, U. S., Hajdu, E. (2002) *Journal of Natural Products*, **65**, 796. (b) Saeki, B. M., Granato, A. C., Berlinck, R. G. S., Magalhães, A., Schefer, A. B., Ferreira, A. G., Pinheiro, U. S., Hajdu, E. (2003) *Journal of Natural Products*, **66**, 1038.

50 Nicholas, G. M., Newton, G. L., Fahey, R. C., Bewley, C. A. (2001) *Organic Letters*, **3**, 1543.

51 Fetterolf, B. and Bewley, C. A. (2004) *Bioorganic and Medicinal Chemistry Letters*, **14**, 3785.

52 Kende, A. S., Lan, J., Fan, J. (2004) *Tetrahedron Letters*, **45**, 133.

53 Chanda, B. M. and Sulake, R. S. (2005) *Tetrahedron Letters*, **46**, 6461.

54 Tsukamoto, S., Kato, H., Hirota, H., Fusetani, N. (1995) *Tetrahedron*, **51**, 6687.

55 García, A., Vásquez, M. J., Quiñoá, E., Riguera, R., Débitus, C. (1996) *Journal of Natural Products*, **59**, 782.
56 Sölter, S., Dieckmann, R., Blumenberg, M., Francke, W. (2002) *Tetrahedron Letters*, **43**, 3385.
57 Sjörgren, M., Göransson, U., Johnson, A.-L., Dahlström, M., Andersson, R., Bergmann, J., Jonsson, P. R., Bohlin, L. (2004) *Journal of Natural Products*, **67**, 368.
58 Fusetani, N., Fujita, M., Nakao, Y., Matsunaga, S., van Soest, R. W. M. (1999) *Bioorganic and Medicinal Chemistry Letters*, **9**, 3397.
59 (a) Reddy, M. V. R., Harper, M. K., Faulkner, D. J. (1998) *Tetrahedron*, **54**, 10649.
(b) Carroll, A. R., Pierens, G. K., Fechner, G., Leone, P. A., Ngo, A., Simpson, M., Hyde, E., Hooper, J. N. A., Boström, S. -L., Musil, D., Quinn, R. J. (2002) *Journal of the American Chemical Society*, **124**, 13340.
(c) Carroll, A. C., Buchanan, M. S., Edser, A., Hyde, E., Simpson, N., Quinn, R. J. (2004) *Journal of Natural Products*, **67**, 1291.
(d) Hanessian, S., Margarita, R., Hall, A., Johnstone, S., Tremblay, M., Parlanti, L. (2002) *Journal of the American Chemical Society*, **124**, 13342–13343.
60 Fusetani, N., Kakao, Y., Matsunaga, S. (1991) *Tetrahedron Letters*, **32**, 7073.
61 Nienaber, V. L. and Amparo, E. C. (1996) *Journal of the American Chemical Society*, **118**, 6807.
62 Hayashi, K., Hamada, Y., Shioiri, T. (1992) *Tetrahedron Letters*, **33**, 5075.
63 Kundu, B., Bauser, M., Betschinger, J., Kraas, W., Jung, G. (1998) *Bioorganic and Medicinal Chemistry Letters*, **8**, 1669.
64 Coleman, J. E., de Silva, E. D., Kong, F., Andersen, R. J., Allen, T. M. (1995) *Tetrahedron*, **51**, 10653.
65 Hamann, M. T., Otto, C. S., Scheuer, P. J., Dunbar, D. C. (1996) *The Journal of Organic Chemistry*, **61**, 6594.
66 Swersey, J. C., Ireland, C. M., Cornell, L. M., Peterson, R. W. (1994) *Journal of Natural Products*, **57**, 842.
67 Expósito, M. A., López, B., Fernández, R., Vásquez, M. J., Débitus, C., Iglesias, T., Jiménez, C., Quiñoá, E., Riguera, R. (1998) *Tetrahedron*, **54**, 7539.
68 Expósito, A., Fernández-Suárez, M., Fernández, R., Iglesias, T., Muñoz, L., Riguera, R. (2001) *The Journal of Organic Chemistry*, **66**, 4206.
69 Casapullo, A., Finamore, E., Minale, L., Zollo, F. (1993) *Tetrahedron Letters*, **34**, 6297.
70 Casapullo, A., Minale, L., Zollo, F., Lavayre, J. (1994) *Journal of Natural Products*, **57**, 1227.
71 Snider, B. B., Song, F. B., Foxman, B. M. (2000) *The Journal of Organic Chemistry*, **65**, 793.
72 Fusetani, N., Matsunaga, S., Matsumoto, H., Takebayashi, Y. (1990) *Journal of the American Chemical Society*, **112**, 7053–7054.
73 Nakao, Y., Matsunaga, S., Fusetani, N. (1995) *Bioorganic and Medicinal Chemistry*, **3**, 1115.
74 Nakao, Y., Oku, N., Matsunaga, S., Fusetani, N. (1998) *Journal of Natural Products*, **61**, 667.
75 Nakao, Y., Masuda, A., Matsunaga, S., Fusetani, N. (1999) *Journal of the American Chemical Society*, **121**, 2425.
76 Murakami, Y., Takei, M., Shindo, K., Kitazume, C., Tanaka, J., Higa, T., Fukamachi, H. (2002) *Journal of Natural Products*, **65**, 259.
77 Lewis, S. D., Ng, A. S., Baldwin, J. J., Fusetani, N., Naylor, A. M., Shafer, J. A. (1993) *Thrombosis Research*, **70**, 173–190.
78 Lewis, A. S. D., Ng, A. S., Baldwin, J. J., Fusetani, N., Naylor, A. M., Shafer, J. A. (1993) *Thrombosis and Haemostasis*, **69**, 668–1668.
79 (a) Maryanoff, B. E., Qiu, X. Y., Padmanabham, K. P., Tulinsky, A., Almond, H. R., Andradegordon, P., Greco, M. N., Kauffman, J. A., Nicolaou, K. C., Liu, A. J., Brungs, P. H., Fusetani, N. (1993) *Proceedings of the National Academy of Sciences of the United States of America*, **90**, 8048–8052.
(b) Maryanoff, B. E., Qiu, X. Y., Padmanabham, K. P., Tulinsky, A.,

Almond, H. R., Andradegordon, P., Greco, M. N., Kauffman, J. A., Nicolaou, K. C., Liu, A. J., Brungs, P. H., Fusetani, N. (1997) *Biopolymers*, **41**, 349–358.

80 Jones, D. M., Atrash, B., Ryder, H., Tegernilsson, A. C., Gyzander, E., Szelke, M. (1995) *Journal of Enzyme Inhibition*, **9**, 43–60.

81 Brady, S. F., Sisko, J. T., Stauffer, K. J., Colton, C. D., Qiu, H., Lewis, S. D., Ng, A. S., Shafer, J. A., Bogusky, M. J., Veber, D. F., Nutt, R. F. (1995) *Bioorganic and Medicinal Chemistry*, **3**, 1063–1078.

82 Maryanoff, B. E., Zhang, H. C., Greco, M. N., Glover, K. A., Kauffman, J. A., Andradegordon, P. (1995) *Bioorganic and Medicinal Chemistry*, **3**, 1025–1038.

83 Lewis, S. D., Ng, A. S., Lyle, E. A., Mellott, M. J., Appleby, S. D., Brady, S. F., Stauffer, K. J., Sisko, J. T., Mao, S. S., Veber, D. F., Nutt, R. F., Lyngh, J. J., Cook, J. J., Gardell, S. J., Shafer, J. A. (1995) *Thrombosis and Haemostasis*, **74**, 1107–1112.

84 Wiley, M. R. and Fisher, M. J. (1997) *Expert Opinion on Therapeutic Patents*, **7**, 1265–1282.

85 Lee, A. Y., Hagihara, M., Krmacharya, R., Albers, M. W., Schreiber, S. L., Clardy, J. (1993) *Journal of the American Chemical Society*, **115**, 12619–12620.

86 Ganesh, V., Lee, A. Y., Clardy, J., Tulinsky, A. (1996) *Protein Science*, **5**, 825–835.

87 McDonnell, P. A., Caldwell, G. W., Leo, G. C., Podlogar, B. L., Maryanoff, B. E., (1997) *Biopolymers*, **41**, 349–358.

88 Hagihara, M. and Schreiber, S. L. (1992) *Journal of the American Chemical Society*, **114**, 6570–6571.

89 (a) Wipf, P. and Kim, H. Y. (1993) *The Journal of Organic Chemistry*, **58**, 5592–5594.
(b) Wipf, P. and Kim, H. Y. (1994) *The Journal of Organic Chemistry*, **59**, 2914.

90 Deng, J. G., Hamada, Y., Shioiri, T., Matsunaga, S., Fusetani, N. (1994) *Angewandte Chemie (International Edition in English)*, **33**, 1729–1731.

91 Maryanoff, B. E., Greco, M. N., Zhang, H. C., Andradegordon, P., Kauffman, J. A., Nicolaou, K. C., Liu, A. J., Brungs, P. H. (1995) *Journal of the American Chemical Society*, **117**, 1225–1239.

92 Bastiaans, H. M. M., Vanderbaan, J. L., Ottenheijm, H. C. J. (1995) *Tetrahedron Letters*, **36**, 5963–5966.

93 Wipf, P. (1995) *Chemical Reviews*, **95**, 2115–2134.

94 Deng, J. G., Hamada, Y., Shioiri, T. (1996) *Tetrahedron Letters*, **37**, 2261–2264.

95 Bastiaans, H. M. M., vanderBaan, J. L., Ottenheijm, H. C. J. (1997) *The Journal of Organic Chemistry*, **62**, 3880–3889.

96 Humphrey, J. M. and Chamberlin, A. R. (1997) *Chemical Reviews*, **97**, 2243–2266.

97 Kobayashi, J., Tsuda, M., Nakamura, T., Mikami, Y., Shigemori, H. (1993) *Tetrahedron*, **49**, 2391.

98 Kobayashi, J., Nakamura, T., Tsuda, M. (1996) *Tetrahedron*, **52**, 6355–6360.

99 Bonnington, L. S., Tanaka, J., Higa, T., Kimura, J., Yoshimura, Y., Nakao, Y., Yoshida, W. Y., Scheuer, P. J. (1997) *The Journal of Organic Chemistry*, **62**, 7765.

100 Milanowski, D. J., Rashid, M. A., Gustafson, K. R., O'Keefe, B. R., Nawrocki, J. P., Pannell, L. K., Boyd, M. R. (2004) *Journal of Natural Products*, **67**, 441.

101 Zampella, A., D'Auria, M. V., Paloma, L. G., Casapullo, A., Minale, L., Débitus, C., Henin, Y. (1996) *Journal of the American Chemical Society*, **118**, 6202.

102 D'Auria, M. V., Zampella, A., Paloma, L. G., Minale, L., Débitus, C., Roussakis, C., Le Bert, V. (1996) *Tetrahedron*, **52**, 9589.

103 Zampella, A. and D'Auria, M. V. (2002) *Tetrahedron: Asymmetry*, **13**, 1237–1239.

104 Zampella, A., Randazzo, A., Borbone, N., Luciani, S., Trevisi, L., Debitus, U., D'Auria, M. V. (2002) *Tetrahedron Letters*, **43**, 6163–6166.

105 Sepe, V., D'Orsi, R., Borbone, N., D'Auria, M. V., Giuseppe, B. B. A., Monti, M. C., Catania, A., Zampella, A. (2006) *Tetrahedron*, **62**, 833–840.

106 Bassarello, C., Zampella, A., Monti, M. C., Gomez-Paloma, L., D'Auria, M. V., Riccio, R., Bifulco, G. (2006) *European Journal of Organic Chemistry*, 604–609.

107 Laille, M., Gerald, F., Debitus, C. (1998) *Cellular and Molecular Life Sciences*, **54**, 167–170.

108 Trevisi, L., Bova, S., Cargnelli, G., Danieli-Betto, D., Floreani, M., Germinario, E., D'Auria, M. V., Luciani, S. (2000) *Biochemical and Biophysical Research Communications*, **279**, 219–222.

109 Trevisi, L., Bova, S., Cargnelli, G., Danieli-Betto, D., Floreani, M., Germinario, E., D'Auria, M. V., Luciani, S. (2001) *British Journal of Pharmacology*, **133** (Suppl.), 128P.

110 Trevisi, L., Cargnelli, G., Ceolotto, G., Papparella, I., Semplicini, A., Zampella, A., D'Auria, M. V., Luciani, S. (2004) *Biochemical Pharmacology*, **68**, 1331–1338.

111 Oku, N., Gustafson, K. R., Cartner, L. K., Wilson, J. A., Shigematsu, N., Hess, S., Pannell, L., Boyd, K., McMahon, M. R., J. B. (2004) *Journal of Natural Products*, **67**, 1407.

112 Rashid, M. A., Gustafson, K. R., Cartner, L. K., Shigematsu, N., Pannell, L. K., Boyd, M. R. (2001) *Journal of Natural Products*, **64**, 117.

113 Oku, N., Krishnamoorthy, R., Benson, A. G., Ferguson, R. L., Lipton, M. A., Phillips, L. R., Gustafson, K. R., McMahon, J. B. (2005) *The Journal of Organic Chemistry*, **70**, 6842–6847.

113a Gulavita, N.K., Gunasekera, S.P., Pomponi, S.A., Robinson, E.V. (1992) *The Journal of Organic Chemistry*, **57**, 1767.

114 Li, H.-Y., Matsunaga, S., Fusetani, N. (1995) *Journal of Medicinal Chemistry*, **38**, 338.

115 Li, H.-Y., Matsunaga, S., Fusetani, N. (1996) *Journal of Natural Products*, **59**, 163.

116 Matsunaga, S., Fusetani, N., Konosu, S. (1985) *Journal of Natural Products*, **48**, 236.

117 Matsunaga, S., Fusetani, N., Konosu, S. (1984) *Tetrahedron Letters*, **25**, 5165.

118 Matsunaga, S., Fusetani, N., Konosu, S. (1985) *Tetrahedron Letters*, **26**, 855.

119 Ryu, G., Matsunaga, S., Fusetani, N. (1994) *Tetrahedron Letters*, **35**, 8251.

120 Ryu, G., Matsunaga, S., Fusetani, N. (1994) *Tetrahedron*, **50** (13), 409.

121 Sato, K., Horibe, K., Amano, K., Mitusi-Saito, M., Hori, M., Matsunaga, S., Fusetani, N., Ozaki, H., Karaki, H. (2001) *Toxicon*, **39**, 259–264.

122 Araki, T., Matsunaga, S., Fusetani, N. (2005) *Bioscience, Biotechnology, and Biochemistry*, **69**, 1318–1322.

123 Williams, D. E., Austin, P., Diaz-Marrero, A. R., Van Soest, R., Matainaho, T., Roskelley, C. D., Roberge, M., Andersen, R. J. (2005) *Organic Letters*, **7**, 4173–4176.

124 Sullivan, B., Faulkner, D. J., Webb, L. (1983) *Science*, **221**, 1175–1176.

125 Jefford, C. W., Huang, P. Z., Rossier, J. C., Sledeski, A. W., Boukouvalas, J. (1990) *Synlett*, 745–746.

126 Jefford, C. W., Rossier, J. C., Boukouvalas, J., Sledeski, A. W., Huang, P. Z. (2004) *Journal of Natural Products*, **67**, 1383–1386.

127 Manes, L. V., Naylor, S., Crews, P., Bakus, G. J. (1985) *The Journal of Organic Chemistry*, **50**, 284–286.

128 Manes, L. V., Crews, P., Kernan, M. R., Faulkner, D. J., Froczek, F. R., Gandour, R. D. (1988) *The Journal of Organic Chemistry*, **53**, 570–575.

129 Wright, A. E., McCarthy, P. J., Schulte, G. K. (1989) *The Journal of Organic Chemistry*, **54**, 3472–3474.

130 Cebula, R. E., Blanchard, J. L., Boisclair, M. D., Pal, K., Bockovich, N. J. (1997) *Bioorganic and Medicinal Chemistry Letters*, **7**, 2015–2020.

131 Nakamura, H., Wu, H., Kobayashi, J., Kobayashi, M., Ohizumi, Y., Hirata, Y. (1985) *The Journal of Organic Chemistry*, **50**, 2494–2497.

132 Morales, J. J. and Rodriguez, A. D. (1992) *Journal of Natural Products*, **55**, 389–394.

133 Kobayashi, M., Nakamura, H., Wu, H. M., Kobayashi, J., Ohizumi, Y. (1987) *Archives of Biochemistry and Biophysics*, **259**, 179–184.

134 Tsukamoto, S., Yamashita, T., Matsunaga, S., Fusetani, N. (1999) *Tetrahedron Letters*, **40**, 737.

135 Matsunaga, S., Yamashita, T., Tsukamoto, S., Fusetani, N. (1999) *Journal of Natural Products*, **62**, 1202.

136 Tsukamoto, S., Yamashita, T., Matsunaga, S., Fusetani, N. (1999) *The Journal of Organic Chemistry*, **64**, 3794.

137 Carter, G. T. and Rinehart, K. L. (1978) *Journal of the American Chemical Society*, **100**, 4302–4304.

138 Blunt, J. W., Munro, M. H. G., Yorke, S. C. (1982) *Tetrahedron Letters*, **23**, 2793–2796.

139 Yorke, S. C., Blunt, J. W., Munro, M. H. G., Cook, J. C., Rinehart, K. L. (1986) *Australian Journal of Chemistry*, **39**, 447–455.

140 Boukouvalas, J., Golding, B. T., McCabe, R. W., Slaich, P. K. (1983) *Angewandte Chemie (International Edition in English)*, **22**, 618–619.

141 Honma, K., Tsuda, M., Mikami, Y., Kobayashi, J. (1995) *Tetrahedron*, **51**, 3745.

142 Davies, S. G., Sanganee, H. J., Szolcsanyi, P. (1999) *Tetrahedron*, **55**, 3337.

143 Cheng, M. T. and Rinehart, K. L. (1978) *Journal of the American Chemical Society*, **100**, 7409–7411.

144 Carté, B. and Faulkner, D. J. (1982) *Tetrahedron Letters*, **23**, 3863–3866.

145 Rinehart, K. L., Harbour, G. C., Graves, M. D., Cheng, M. T. (1983) *Tetrahedron Letters*, **24**, 1593–1596.

146 Kourany-Lefoll, E., Laprevote, O., Sevenet, T., Montagnac, A., Païs, M., Debitus, C. (1994) *Tetrahedron*, **50**, 3415.

147 Mancini, I., Guella, G., Sauvain, M., Débitus, C., Duigou, A.-G., Ausseil, F., Menou, J. L., Pietra, F. (2004) *Organic and Biomolecular Chemistry*, **2**, 783.

148 Goobes, R., Rudi, A., Kashman, Y., Ilan, M., Loya, Y. (1996) *Tetrahedron*, **52**, 7921.

149 Gustafson, K. and Andersen, R. J. (1982) *The Journal of Organic Chemistry*, **47**, 2167.

150 Gustafson, K. (1985) Andersen, R. J. *Tetrahedron*, **41**, 1101.

151 Piers, E., Chong, J. M., Gustafson, K., Andersen, R. J. (1984) *Canadian Journal of Chemistry*, **62**, 1.

152 Graziani, E. I. and Andersen, R. J. (1996) *Chemical Communications*, 2377.

153 Kubanek, J. and Andersen, R. J. (1997) *Tetrahedron Letters*, **38**, 6327.

154 Graziani, E. I., Andersen, R. J. (1998) *Journal of Natural Products*, **61**, 285.

155 Cimino, G., Ghiselin, M. T. (1999) *Chemoecology*, **9**, 187–207.

156 Sakemi, S., Ichiba, T., Kohmoto, S., Saucy, G., Higa, T. (1988) *Journal of the American Chemical Society*, **110**, 4851–4853.

157 Burres, N. S. and Clement, J. J. (1989) *Cancer Research*, **49**, 2935–2940.

158 Matsunaga, S., Fusetani, N., Nakao, Y. (1992) *Tetrahedron*, **48**, 8369–8376.

159 Kobayashi, J., Itagaki, F., Shigemori, H., Sasaki, T. (1993) *Journal of Natural Products*, **56**, 976–981.

160 Hong, C. Y. and Kishi, Y. (1991) *Journal of the American Chemical Society*, **113**, 9693–9694.

161 Lee, K. H., Nishimura, S., Matsunaga, S., Fusetani, N., Horinouchi, S., Yoshida, M. (2005) *Cancer Science*, **96**, 357–364.

162 Piel, J., Hui, D., Wen, G., Butzke, D., Platzer, M., Fusetani, N., Matsunaga, S. (2004) *Proceedings of the National Academy of Sciences of the United States of America*, **101**, 16222.

163 Piel, J., Butzke, D., Fusetani, N., Hui, D. Q., Platzer, M., Wen, G. P., Matsunaga, S. (2005) *Journal of Natural Products*, **68**, 472–479.

164 Harbour, Gary C., Tymiak, Adrienne A., Rinehart, Kenneth L., Shaw, Paul D., Robert, Hughes Mizsak Stephen A., Coats, John H., Zurenko, Gary E., Li, Li H., Kuentzel, Sandra L. (1981) *Journal of the American Chemical Society*, **103** (18), 5604–5606.

165 Ruben, R. L. and Snider, B. B. (1988) *Proceedings of the American Association for Cancer Research*, **29**, 320.

166 Ruben, R. L., Snider, B. B., Hobbs, F. W., Confalone, P. N., Dusak, B. A. (1989) *Investigational New Drugs*, **7**, 147–154.

167 Snider, B. B. and Faith, W. C. (1983) *Tetrahedron Letters*, **24**, 861–864.

168 Roush, W. R. and Walts, A. E. (1984) *Journal of the American Chemical Society*, **106**, 721–723.

169 Snider, B. B. and Faith, W. C. (1984) *Journal of the American Chemical Society*, **106**, 1443–1445.
170 Walts, A. E. and Roush, W. R. (1985) *Tetrahedron*, **41**, 3463–3478.
171 Hassner, A. and Murthy, K. S. K. (1986) *Tetrahedron Letters*, **27**, 1407–1410.
172 Uyehara, T., Furuta, T., Kabasawa, Y., Yamada, J. I., Kato, T. (1986) *Chemical Communications*, 539–540.
173 Uyehara, T., Furuta, T., Kabawawa, Y., Yamada, J., Kato, T., Yamamoto, Y. (1988) *The Journal of Organic Chemistry*, **53**, 3669–3673.
174 Asaoka, M., Sakurai, M., Takei, H. (1990) *Tetrahedron Letters*, **31**, 4759–4760.
175 Murthy, K. S. K. and Hassner, A. (1991) *Israel Journal of Chemistry*, **31**, 239–246.
176 Cossy, J. and BouzBouz, S. (1996) *Tetrahedron Letters*, **37**, 5091–5094.
177 Schellhaas, K., Schmalz, H. G., Bats, J. W. (1998) *Chemistry—A European Journal*, **4**, 57–66.
178 Tavares, R., Daloze, ., Braekman, D., Hajdu, J. C., Van Soest, E., R.W.M. (1995) *Journal of Natural Products*, **58**, 1139–1142.
179 Hua, H. M., Peng, J., Fronczek, F. R., Kelly, M., Hamann, M. T. (2004) *Bioorganic and Medicinal Chemistry*, **12**, 6461–6464.
180 Barrow, R. A., Murray, L. M., Lim, T. K., Capon, R. J. (1996) *Australian Journal of Chemistry*, **49**, 767–773.
181 Capon, R. J., Miller, M., Rooney, F. (2001) *Journal of Natural Products*, **64**, 643–644.
182 Patil, A. D., Freyer, A. J., Offen, P., Bean, M. F., Johnson, R. K. (1997) *Journal of Natural Products*, **60**, 704–707.
183 Kashman, Y., Hirsh, S., McConnell, O. J., Ohtani, I., Kusumi, T., Kakisawa, H. (1989) *Journal of the American Chemical Society*, **111**, 8925–8926.
184 Ohtani, I., Kusumi, T., Kakisawa, H. (1992) *Tetrahedron Letters*, **33**, 2525–2528.
185 Ohtani, I., Kusumi, T., Kakisawa, H., Kashman, Y., Hirsh, S. (1992) *Journal of the American Chemical Society*, **114**, 8472–8479.
186 Snider, B. B. and Shi, Z. P. (1993) *Tetrahedron Letters*, **34**, 2099–2102.
187 Overman, L. E. and Rabinowitz, M. H. (1993) *The Journal of Organic Chemistry*, **58**, 3235–3237.
188 Snider, B. B. and Shi, Z. P. (1994) *Journal of the American Chemical Society*, **116**, 549–557.
189 Murphy, P. J., Williams, H. L., Hursthouse, M. B., Malik, K. M. A. (1994) *Chemical Communications*, 119–120.
190 Murphy, P. J. and Williams, H. L. (1994) *Chemical Communications*, 819–820.
191 Overman, L. E., Rabinowitz, M. H., Renhowe, P. A. (1995) *Journal of the American Chemical Society*, **117**, 2657–2658.
192 Grillot, A. L. and Hart, D. J. (1995) *Tetrahedron*, **51**, 11377–11392.
193 Anderson, G. T., Alexander, M. D., Taylor, S. D., Smithrud, D. B., Benkovic, S. J., Weinreb, S. M. (1996) *The Journal of Organic Chemistry*, **61**, 125–132.
194 Louwrier, S., Ostendorf, M., Boom, A., Hiemstra, H., Speckamp, W. N. (1996) *Tetrahedron*, **52**, 2603–2628.
195 Murphy, P. J., Williams, H. L., Hibbs, D. E., Hursthouse, M. B., Malik, K. M. A. (1996) *Tetrahedron*, **52**, 8315–8332.
196 (a) Nagasawa, K., Georgieva, A., Nakata, T. (2000) *Tetrahedron*, **56**, 187–192.
(b) Erratum: Nagasawa, K., Georgieva, A., Nakata, T. (2001) *Tetrahedron*, **57**, 4057.
197 Black, G. P., Coles, S. J., Hizi, A., Howard-Jones, A. G., Hursthouse, M. B., McGown, A. T., Loya, S., Moore, C. G., Murphy, P. J., Smith, N. K., Walshe, N. D. A. (2001) *Tetrahedron Letters*, **42**, 3377–3381.
198 Georgieva, A., Hirai, M., Hashimoto, Y., Nakata, T., Ohizumi, Y., Nagasawa, K. (2003) *Synthesis (Stuttgart)*, 1427–1432.
199 Ohizumi, Y., Sasaki, S., Kusumi, T., Ohtani, I. (1996) *European Journal of Pharmacology*, **310**, 95–98.
200 Jares-Erijman, E. A., Sakai, R., Rinehart, K. L. (1991) *The Journal of Organic Chemistry*, **56**, 5712–5715.

201 Berlinck, R. G. S., Braekman, J. C., Daloze, D., Bruno, I., Riccio, R., Ferri, S., Spampinato, S., Speroni, E. (1993) *Journal of Natural Products*, **56**, 1007–1015.

202 Jares-Erijman, E. A., Ingrum, A. L., Carney, J. R., Rinehart, K. L., Sakai, R. (1993) *The Journal of Organic Chemistry*, **58**, 4805–4808.

203 Palagiano, E., De Marino, S., Minale, L., Riccio, R., Zollo, F., Iorizzi, M., Carré, J. B., Debitus, C., Lucarain, L., Provost, J. (1995) *Tetrahedron*, **51**, 3675.

204 Coffey, D. S., McDonald, A. I., Overman, L. E., Stappenbeck, F. (1999) *Journal of the American Chemical Society*, **121**, 6944–6945.

205 Coffey, D. S., Overman, L. E., Stappenbeck, F. (2000) *Journal of the American Chemical Society*, **122**, 4904–4914.

206 Coffey, D. S., McDonald, A. I., Overman, L. E., Rabinowitz, M. H., Renhowe, P. A. (2000) *Journal of the American Chemical Society*, **122**, 4893–4903.

207 (a) Aron, Z. D., Pietraszkiewicz, H., Overman, L. E., Valeriote, F., Cuevas, C. (2004) *Bioorganic and Medicinal Chemistry Letters*, **14**, 3445–3449.
(b) Overman, L. E. and Rhee, Y. H. (2005) *Journal of the American Chemical Society*, **127**, 15652–15658.

208 Berlinck, R. G. S., Braekman, J. C., Daloze, D., Hallenga, K., Ottinger, R., Bruno, I., Riccio, R. (1990) *Tetrahedron Letters*, **31**, 6531–6534.

209 Berlinck, R. G. S., Braekman, J. C., Daloze, D., Bruno, I., Riccio, R., Rogeau, D., Amade, P. (1992) *Journal of Natural Products*, **55**, 528–532.

210 Jares-Erijman, E. A., Ingrum, A. A., Sun, F., Rinehart, K. L. (1993) *Journal of Natural Products*, **56**, 2186.

211 Snider, B. B. and Shi, Z. P. (1992) *The Journal of Organic Chemistry*, **57**, 2526.

212 Snider, B. B. and Shi, Z. (1993) *The Journal of Organic Chemistry*, **58**, 3828.

213 Tavares, R., Daloze, D., Braekman, J. C., Hajdu, E., Muricy, G., van Soest, R. W. M. (1994) *Biochemical Systematics and Ecology*, **22**, 645–646.

214 Chang, L. C., Whittaker, N. F., Bewley, C. A. (2003) *Journal of Natural Products*, **66**, 1490–1494.

215 Yang, S. W., Chan, T. M., Pomponi, S. A., Chen, G. D., Wright, A. E., Patel, M., Gullo, V., Pramanik, B., Chu, M. (2003) *The Journal of Antibiotics*, **56**, 970–972.

216 Braekman, J. C., Daloze, D., Tavares, R., Hajdu, E., Van Soest, R. W. M. (2000) *Journal of Natural Products*, **63**, 193–196.

217 Gallimore, W. A., Kelly, M., Scheuer, P. J. (2005) *Journal of Natural Products*, **68**, 1420–1423.

218 Venkateswarlu, Y., Reddy, M. V. R., Ramesh, P., Rao, J. V. (1999) *Indian Journal of Chemistry, Section B*, **38**, 254–256.

219 Nagasawa, K., Georgieva, A., Koshino, H., Nakata, T., Kita, T., Hashimoto, Y. (2002) *Organic Letters*, **4**, 177–180.

220 Moore, C. G., Murphy, P. J., Williams, H. L., McGown, A. T., Smith, N. K. (2003) *Tetrahedron Letters*, **44**, 251–254.

221 Aron, Z. D. and Overman, L. E. (2005) *Journal of the American Chemical Society*, **127**, 3380–3390.

222 Patil, A. D., Kumar, N. V., Kokke, W. C., Bean, M. F., Freyer, A. J., Debrosse, C., Mai, S., Truneh, A., Faulkner, D. J., Carté, B., Breen, A. L., Hertzberg, R. P., Johnson, R. K., Westley, J. W., Potts, B. C. M. (1995) *The Journal of Organic Chemistry*, **60**, 1182–1188.

223 Patil, A. D., Freyer, A. J., Taylor, P. B., Carte, B., Zuber, G., Johnson, R. K., Faulkner, D. J. (1997) *The Journal of Organic Chemistry*, **62**, 1814–1819.

224 Black, G. P., Murphy, P. J., Walshe, N. D. A., Hibbs, D. E., Hursthouse, M. B., Malik, K. M. A. (1996) *Tetrahedron Letters*, **37**, 6943–6946.

225 Snider, B. B., Chen, J. S., Patil, A. D., Freyer, A. J. (1996) *Tetrahedron Letters*, **37**, 6977–6980.

226 Black, G. P., Murphy, P. J., Walshe, N. D. A. (1998) *Tetrahedron*, **54**, 9481–9948.

227 Snider, B. B. and Chen, J. S. (1998) *Tetrahedron Letters*, **39**, 5697–5700.

228 Black, G. P., Murphy, P. J., Thornhill, A. J., Walshe, N. D. A., Zanetti, C. (1999) *Tetrahedron*, **55**, 6547–6655.

229 Snider, B. B. and Busuyek, M. V. (1999) *Journal of Natural Products*, **62**, 1707–1711.

230 (a) Duron, S. G. and Gin, D. Y. (2001) *Organic Letters*, **3**, 1551–2155. (b) Nagasawa, K., Koshino, H., Nakata, T. (2001) *Tetrahedron Letters*, **42**, 4155–4158.

231 Elliott, M. C. and Long, M. S. (2002) *Tetrahedron Letters*, **43**, 9191–9919.

232 Cohen, F., Collins, S. K., Overman, L. E. (2003) *Organic Letters*, **5**, 4485–4488.

233 Collins, S. K., McDonald, A. I., Overman, L. E., Rhee, Y. H. (2004) *Organic Letters*, **6**, 1253–1255.

234 Cohen, F. and Overman, L. E. (2006) *Journal of the American Chemical Society*, **128**, 2594–2603.

235 Rao, A. V. R., Gurjar, M. K., Vasudevan, J. (1995) *Chemical Communications*, 1369–1370.

236 Bewley, C. A., Ray, S., Cohen, F., Collins, S. K., Overman, L. E. (2004) *Journal of Natural Products*, **67**, 1319–1324.

237 Olszewski, A., Sato, K., Aron, Z. D., Cohen, F., Harris, A., McDougall, B. R., Robinson, W. E., Overman, L. E., Weiss, G. A. (2004) *Proceedings of the National Academy of Sciences of the United States of America*, **101**, 14079–14084.

238 Meragelman, K. M., McKee, T. C., McMahon, J. B. (2004) *Journal of Natural Products*, **67**, 1165–1167.

239 Laille, M., Gerald, F., Debitus, C. (1998) *Cellular and Molecular Life Sciences*, **54**, 167–170.

240 Heys, L., Moore, C. G., Murphy, P. J. (2000) *Chemical Society Reviews*, **29**, 57–67.

241 Nagasawa, K. (2003) *Journal of the Pharmaceutical Society of Japan*, **123**, 387–398.

242 Nagasawa, K. and Hashimoto, Y. (2003) *Chemical Record*, **3**, 201–211.

243 Aron, Z. D. and Overman, L. E. (2004) *Chemical Communications*, 253–265.

II
New Trends in Alkaloid Isolation and Structure Elucidation

Modern Alkaloids: Structure, Isolation, Synthesis and Biology. Edited by E. Fattorusso and O. Taglialatela-Scafati

ISBN: 978-3-527-31521-5

12
Analysis of Tropane Alkaloids in Biological Matrices

Philippe Christen, Stefan Bieri, Jean-Luc Veuthey

12.1
Introduction

Tropane alkaloids are an important class of structurally related natural products, having in common the azabicyclo[3.2.1]octane skeleton. It is now well established that the tropane ring system derives its pyrrolidine ring from ornithine and/or arginine [1] and that *N*-methyl-Δ^1-pyrrolinium salt is the common intermediate (Figure 12.1). The majority of these alkaloids are mono-, di-, and tri-esters of hydroxytropanes with various organic acids, such as tropic, atropic, cinnamic, angelic, tiglic, senecioic, isovaleric, and truxillic acids. Hyoscyamine and scopolamine (hyoscine) are important representatives of this class of alkaloids. They are ester derivatives of tropane-3α-ol and its 6–7 epoxide with tropic acid, respectively (Figure 12.1).

Tropane alkaloids are commonly found in genera belonging to three families: Solanaceae, Erythroxylaceae, and Convolvulaceae, but they occur also sporadically in a number of other families including Proteaceae and Rhizophoraceae [2]. To date, more than 200 tropane alkaloids have been isolated from biological material [1]. The natural form of hyoscyamine is the (−)-form. (−)-Hyoscyamine is easily racemized, yielding atropine. Dimeric (e.g. schizanthines) and trimeric (grahamine) tropane alkaloids have also been found [3]. There is considerable structural diversity within this class of compounds. Another important tropane alkaloid, cocaine, occurs in the genus *Erythroxylum*. It is an ester of ecgonine, a derivative of tropane-3β-ol carrying a carboxyl goup at C-2 (Figure 12.1). Recently, the calystegines, a new type of *nor*tropane alkaloids, have been structurally characterized. A typical feature of calystegines is a hydroxyl group on the bridgehead carbon 1 (Figure 12.1). According to the number of hydroxyl groups on the *nor*tropane skeleton (3, 4, or 5), they belong to the series A, B, or C, respectively. To date, 15 calystegines have been identified and, in contrast to most other tropane alkaloids, they are not esterified. They were first identified in *Calystegia sepium*, Convolvulaceae [4], and later detected in numerous other Convolvulaceae [5], Solanaceae, Moraceae [6], Erythroxylaceae [7], and Brassicaceae [8]. Particularly high concentrations are found in potato skins and eggplants.

Modern Alkaloids: Structure, Isolation, Synthesis and Biology. Edited by E. Fattorusso and O. Taglialatela-Scafati

ISBN: 978-3-527-31521-5

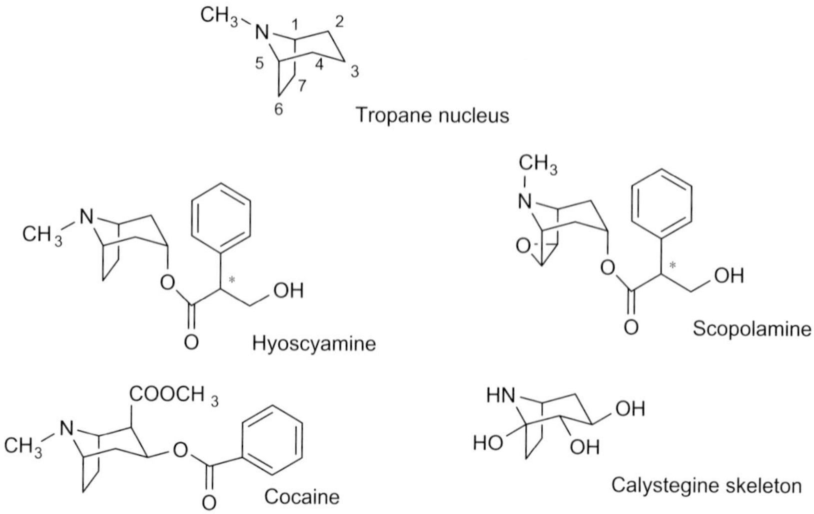

Fig. 12.1 Tropane alkaloids of pharmaceutical interest.

From a pharmacological point of view, tropane alkaloids show antimuscarinic activity (parasympathetic inhibition), which at the peripheral level translates into antispasmodic effects on the gastrointestinal and genitourinary systems. Atropinic drugs dilate the pupil (mydriasis) and result in a loss of accommodation (cycloplegia). In terms of its effects on the central nervous system, atropine can prevent vagal reflexes induced by the manipulation of visceral organs. Scopolamine is used to prevent nausea and vomiting caused by motion sickness. Cocaine is the parent compound of local anesthetic; however, its use is limited because cocaine has stimulatory effects on the central nervous system. To achieve the latter effect, leaves can be chewed or pure cocaine sniffed or injected.

Semisynthetic derivatives have also been commercialized. These compounds are quaternary derivatives that have no side effects on the central nervous system. They are used against spasm of the bladder or of the intestine and to treat bronchospasm associated with chronic obstructive pulmonary disease.

Owing to their structural similarity with monosaccharides, calystegines exhibit potent inhibitory activities against glycosidases. Thus, they are of considerable interest as potential antiviral, anticancer, and antidiabetic agents. The pharmacological properties of these compounds have been comprehensively reviewed [9,6].

Among the main toxic plants responsible for human deaths throughout the world, those producing alkaloids and, in particular, containing tropane alkaloids (e.g. *Atropa belladonna*, *Datura* sp., and *Hyoscyamus niger*) are of significant importance [10]. Therefore, it remains critical to have rapid and accurate analytical methods for the identification and quantification of tropane alkaloids in plants or in biological fluids.

In this review, plant material and numerous biological fluids, mainly blood plasma and serum, urine, and saliva, as well as some alternative matrices such as bile, amniotic fluid, cord blood, meconium, and maternal hair among others will be considered. The tropane alkaloids of interest reported in this review are cocaine and its metabolites, hyoscyamine, as well as its racemate atropine, scopolamine and their derivatives. Calystegines and other polyhydroxylated compounds will not be further discussed in this chapter.

12.2 Extraction

12.2.1 Plant Material

The classical method for the extraction of tropane alkaloids is based on the Stas–Otto procedure. In this protocol, the alkaloids are transferred to either aqueous or organic layers simply by changing the pH of the two-phase system: the salts are soluble in water whereas the free bases are soluble in organic solvents. Chloroform is frequently used in tropane alkaloid extraction and analysis. El Jaber-Vazdekis *et al.* [11] have demonstrated that dichloromethane, a solvent less hazardous than chloroform, can advantageously replace the latter for the extraction of tropane alkaloids with a similar eluotropic value. Classically, extraction of tropane alkaloids is carried out using percolation, maceration, digestion, decoction, and extraction using a Soxhlet apparatus [12]. These techniques have been used for many decades. However, they are very time-consuming and require relatively large quantities of polluting solvents. Furthermore, some problems may arise using some of these procedures or solvents. Prolonged treatment with strong alkali leads to hydrolysis or racemization of hyoscyamine to atropine. Diethyl ether should be avoided not only because of its volatility and flammability but also because of its tendency to form alkaloid *N*-oxides in the presence of peroxides [13].

Another major drawback of classical extractions is that additional clean-up procedures are frequently required before chromatographic analyses. Solid phase extraction (SPE) avoids the emulsion problems often encountered in liquid–liquid extraction. A wide range of adsorbents are commercially available and may be divided into three classes: polar, ion-exchange, and nonpolar adsorbents. Solid-supported liquid-liquid extraction on Extrelut columns is frequently reported for efficient clean-up of crude tropane alkaloid mixtures. Basified aqueous solutions of alkaloids may be transferred to Extrelut columns and the bases recovered in dichloromethane–isopropanol mixture [13].

12.2.2 Supercritical Fluid Extraction

Supercritical fluid extraction (SFE) has become a method of choice for the extraction of plant material [14]. It represents an interesting alternative technique compared to conventional liquid–solid extraction, with lower solvent consumption and working temperature. The free bases of hyoscyamine and scopolamine are extractable with pure supercritical CO_2, but the addition of alkaline modifiers such as methanol basified with diethylamine is necessary to extract the salts of hyoscyamine and scopolamine from plant material [15]. Chemometry was used to optimize supercritical fluid extraction of hyoscyamine and scopolamine from hairy root culture of *Datura candida* × *D. aurea* [16]. A polar modifier was required in order to extract hyoscyamine and scopolamine quantitatively. The optimal conditions were 20 % methanol in CO_2 at a pressure of 15 MPa and 85 °C. Extracted amounts were in good agreement with those obtained by traditional liquid–solid extraction.

Cocaine has been extracted from coca leaves and the optimization procedure was investigated by means of a central composite design [17]. Pressure, temperature, nature, and percentage of polar modifier were studied. A rate of 2 mL/min CO_2 modified by the addition of 29 % water in methanol at 20 MPa for 10 min allowed the quantitative extraction of cocaine. The robustness of the method was evaluated by drawing response surfaces. The same compound has also been extracted by SFE from hair samples [18–20].

12.2.3 Microwave-assisted Extraction

The use of microwave energy as a heating source in analytical laboratories started in the 1970s and was applied to acid digestions [21]. Later, the use of microwave-assisted extraction was reported by Ganzler [22,23]. The fundamental principles of microwave energy for digestion, extraction, and desorption have been reviewed by Zlotorzynski [24]. With the wide availability of microwaves ovens, this type of heating system has been introduced into many analytical laboratories [25]. Microwave heating depends on the presence of polar molecules or ionic species. Furthermore, the chosen solvent should absorb the microwaves without leading to strong heating so as to avoid degradation of the compounds of interest. Disruption of hydrogen bonds, resulting from dipole rotation of molecules, and migration of dissolved ions facilitate the penetration of solvent molecules into the matrix and allow the solvation of extracted components [26]. Heating with microwaves is instantaneous and occurs in the center of the sample, leading to homogenous, very fast extractions. Two technologies are available: either closed vessels, under controlled pressure and temperature, also called pressurized microwave-assisted extraction (PMAE) or open vessels, also called focused microwave-assisted extraction (FMAE) at atmospheric pressure [27]. Most of the papers dealing with microwave energy have been dedicated to the extraction of pesticides and organometallic compounds from environmental matrices, and very few applications have been published which concern the extraction of alkaloids. In particular, very little is reported on the microwave-assisted extraction of tropane alkaloids. Cocaine and benzoylecgonine have been extracted from coca leaves by FMAE [28]. Several parameters including the nature of the extracting solvent, the particle size distribution, the sample moisture, the applied microwave power, and the radiation time were studied by means of a central composite design. Bieri *et al.* [29] analyzed cocaine distribution in 51 wild *Erythroxylum* species. Extraction was performed using FMAE on 100 mg of hydrated plant material with 5 mL methanol for 30 s. After filtration, the samples were analyzed by GC-MS without further purification.

12.2.4 Pressurized Solvent Extraction

Pressurized solvent extraction (PSE), also called pressurized fluid extraction (PFE), accelerated solvent extraction (ASE®), pressurized liquid extraction (PLE), or enhanced solvent extraction (ESE), is a solid–liquid extraction that has been developed as an alternative to conventional extractions such as Soxhlet, maceration, percolation, or reflux. It uses organic solvents at high pressure and temperature to increase the efficiency of the extraction process. Increased temperature decreases the viscosity of the liquid solvent, enhances its diffusivity, and accelerates the extraction kinetics. High pressure keeps the solvent in its liquid state and thus facilitates its penetration into the matrix, resulting in increase extraction speed [30]. PSE was applied to the rapid extraction of cocaine and benzoylecgonine from coca leaves [31]. Several parameters including the nature of the extracting solvent, the pressure, temperature, extraction, addition of alkaline substances, and sample granulometry were investigated. Critical parameters were pressure, temperature, and extraction time. They were optimized by means of a central composite design. It was demonstrated that an extraction time of 10 min was sufficient to extract cocaine quantitatively at 80 °C and 20 MPa.

12.2.5 Solid-phase Microextraction

Solid-phase microextraction (SPME) was introduced by Arthur and Pawliszyn in 1990 [32]. It is a simple, solvent-free preparation method. It can be conducted as a direct extraction in which the coated fiber is immersed in the liquid sample or in a headspace configuration for sampling the volatiles from the headspace above the liquid sample placed in a vial. The SPME process consists of two steps: (a) the sorbent is exposed to the sample for a specified period of time; (b) the sorbent is transferred to a device that interfaces with an analytical instrument for thermal desorption using GC or for solvent desorption with HPLC. SPME has become a widely used technique in many areas of analytical chemistry, such as food analysis, environmental sampling, and biological analysis. The state-of-the-art of SPME including recent developments and future challenges has been comprehensively reviewed by O'Reilly *et al.* [33]. Reports on alkaloid analysis by SPME are still scarce and mainly concern cocaine in biological fluids [34,35] and human hairs [36]. SPME has been evaluated as an alternative injection technique in combination with fast GC to carry out a quantitative determination of cocaine first extracted from coca leaves by FMAE [37]. A 7 μm PDMS fiber allowed an extraction time of 2 min and a very short desorption time of 12 s for the compounds of interest (cocaine and cocaethylene as internal standard).

An effective combination of focused microwave-assisted extraction with solid-phase microextraction (FMAE-SPME) was carried out for the extraction of cocaine from coca leaves prior to GC analysis [38]. SPME was performed in the direct immersion mode with a 100 μm polydimethylsiloxane coated fiber. A significant gain in selectivity was obtained with the incorporation of SPME in the extraction procedure. Therefore, the analysis time was reduced to 6 min compared to 35 min with conventional GC. A comparison of extraction methods, namely ultrasonic bath, hot solvent under reflux, and pressurized liquid extraction applied to the extraction of hyoscyamine, scopolamine, and other related alkaloids has been published [39]. A mixed-mode reversed phase cation-exchange SPE was optimized for simultaneous recovery of (−)-hyoscyamine, scopolamine, and scopolamine *N*-oxide from various *Datura* species. The alkaloids were qualitatively and quantitatively analyzed by HPTLC-densitometry without derivatization and compared with reversed phase high-performance liquid chromatography with diode array detection (RP-HPLC-DAD). Another densitometric method for the analysis of hyoscyamine and scopolamine in different solanaceous plants and hairy roots has been reported by Berkov and Pavlov [40].

12.2.6
Biological Matrices

The analytical process can be divided into four major steps: sample preparation, separation, detection, and data treatment. For the analysis of drugs and metabolites in biological matrices, sample preparation remains the most challenging task since the compounds of interest are often present at trace level in a complex matrix containing a large number of biomolecules (e.g. proteins) and other substances, such as salts.

Different procedures can be used for the preparation of biological matrices prior to separation and detection techniques as a function of the selectivity and sensitivity of the latter as well as of the matrix complexity. However, for liquid samples, conventional methods are generally carried out: simple dilution/filtration and injection (dilute and shoot), protein precipitation (PP), liquid–liquid extraction (LLE), and solid phase extraction (SPE). These techniques present advantages and drawbacks; the final choice depending on several criteria, among them: analyte concentration, selected analytical method, nature of the matrix, number of samples, time delivery, ease of automation, and so on. They can be performed manually as well as automatically in conventional and high-throughput modes. Comprehensive textbooks have been published on this topic in bioanalysis [41,42]. For solid biological matrices such as tissues or hair, a preliminary extraction with an organic solvent is performed followed by a purification step (e.g. LLE and SPE), if needed.

Tropane alkaloids are basic compounds, generally water soluble at acidic pH, while their unionized form (basic pH) is more soluble in apolar organic solvents. Therefore, LLE has been largely used to extract these compounds from biological fluids after addition of sodium carbonate, sodium borate, or ammonia to attain a pH ≥ 10 [43,44]. SPE has also been used for extracting tropane alkaloids from biological matrices. Some advantages exist for SPE methods versus LLE, including the absence of emulsion, fewer tedious tasks, lower solvent consumption, and possible automation in both off-line and on-line modes. For this purpose, conventional cartridges packed with reversed phase materials (e.g. C8, C18, and other polymeric phases) have been extensively employed [45]. Other supports such as the hydrophilic–lipophilic water-wettable reversed phase sorbent Oasis HLB (Waters, Milford, MA, USA), commercialized in 1996, has gained considerable interest for the extraction of basic drugs, for example cocaine and metabolites. This kind of support presents a high retention capacity of polar analytes due to its "polar-hook." In order to enhance selectivity and sensitivity for basic compounds, a different water-wettable polymeric sorbent (Oasis MCX, Waters) was introduced in 1999. It provides a dual mode of retention with strong cation-exchange and reversed phase mechanisms. These sorbents are commercialized under different formats (syringe barrel cartridges, 96-well extraction plates, microelution plates, and on-line columns) permitting the selection of the method of choice as a function of the number of samples to be analyzed, the sample volume available, the sensitivity required, and so on. A great number of generic procedures have been developed for the analysis of cocaine and its metabolites in urine, serum, plasma, whole blood, among others, since cocaine abuse has increased dramatically during the last decades. Scopolamine, and its internal standard atropine, were also extracted from human serum with a simple and automated SPE on Oasis HLB before liquid chromatography tandem mass spectrometry analysis [46].

Another possibility for increasing selectivity during the extraction process is the use of immuno techniques based on molecular recognition. An excellent review was published on this subject in 2003 by Hennion and Pichon [47] and the reader is invited to study this comprehensive review, and references therein, for more information. Immunoaffinity extraction sorbents, also called immunosorbents (ISs), contain antibodies that can retain an antigen with high affinity and selectivity. Several ISs have been used for extracting drugs, hormones, peptides, and other large biomolecules in environmental and biological matrices. However, the major drawbacks of ISs are their long development time and their cost. Therefore, synthetic antibodies or plastic antibodies, called molecularly imprinted polymers (MIPs), have been described and their use as solid phase extraction materials reviewed [48]. In this case, the molecular imprinting is the synthesis of highly crossed-linked resins in the presence of a given molecule of interest. After washing, the molecule is eliminated and the polymer contains cavities with the appropriate size and shape. Thus, the MIP can be applied to selectively bind the molecule and analogs in different matrices. Using this strategy, Nakamura *et al.* [49] developed a uniformly sized MIP for atropine. This selective sorbent was applied for determining atropine and scopolamine in pharmaceutical preparations containing *Scopolia* extracts. A column-switching procedure was used with the MIP as a precolumn and a conventional cation exchanger as the analytical column. The automated method was validated and showed good performances in terms of linearity, recoveries, and precision.

Microdialysis has also been used as a sampling method for measuring the concentration of drugs in human subcutaneous tissues for pharmacokinetic studies [50]. The microdialysates are simpler than other biological fluids and do not contain proteins, permitting their direct injection for analysis by liquid chromatography.

However, the presence of nonvolatile salts in the microdialysate can cause some contamination (after several injections) of the mass spectrometer, for instance. Thus, the ionization source needs frequent cleaning. Microdialysis combined with LC-MS/MS quantitation has been reported for scopolamine [46] for concentrations ranging from 50 pg/mL to 10 ng/mL. Free cocaine and benzoylecgonine were also investigated in rat brain with *in vivo* microdialysis [51] with off-line LC-MS analysis. Monitoring cocaine and its metabolites is of prime importance for understanding its effects in the brain. Recently, the same kinds of methods were used to determine unbound cocaine in blood, brain, and bile of rats with LC-UV and LC-MS/MS [52].

As previously mentioned, besides conventional matrices such as urine and blood, alternative matrices have become of great interest in toxicology. Different reviews describe the analysis of drugs of abuse in saliva, sweat, and hair [53–56]. For conventional matrices, LLE and SPE are usually the methods of choice. However, for hair analysis, a more drastic extraction step is necessary initially, followed by a purification step [57].

12.3 Analysis of Plant Material and Biological Matrices

Current methods for tropane alkaloids analysis have been well covered in the literature. An excellent comprehensive review written by B. Dräger [45] appeared in 2002, describing the analysis of tropane and related alkaloids in plant material. Sample preparation procedures were reviewed, as well as the analytical methods used for performing the separation and detection of tropane alkaloids, such as gas chromatography (GC), liquid chromatography (LC), and capillary electrophoresis (CE). Therefore, this chapter will not describe in detail these well-known analytical methods but discuss some recently developed applications for the analysis of tropane alkaloids in plant material and biological matrices.

12.3.1 Gas Chromatography

The isolation of atropine, scopolamine, and cocaine occurred long before the development of modern analytical techniques. Gas chromatography was the first instrumental technique available in the field of separation science and thus it is not surprising that these alkaloids were firstly analyzed by GC despite their low volatility. With the advent of capillary columns and the proliferation of various sample introduction and detection methods, GC has evolved as the dominant analytical technique for screening, identification, and quantitation of tropane alkaloids of plant origin as well as in biological fluids. The state-of-the-art of GC analysis of tropane alkaloids has been the subject of two comprehensive reviews [45,58]. We shall therefore mainly focus on publications which have appeared since 2002.

Currently, rapid and unambiguous identification of tropane alkaloids is routinely accomplished by GC-MS by "fingerprint matching," comparing the MS spectra with those available for authentic compounds. The fragmentation of tropane alkaloids is very well known and may help in the tentative identification of unknown derivatives. A fundamental parameter for identification in GC is the reproducibility of retention data. Supporting this is the basic concept of the retention index system, introduced by Kováts for isothermal conditions and by Van den Dool and Kratz for linear programmed temperatures. Incorporation of retention indices in peak identification criteria strongly assists and complements MS matches or alternative confirmatory detection techniques such as Fourier transform infrared spectroscopy (FTIR) or element-specific detectors. Today, with more reproducible column manufacturing technologies and the development of electronic gas flow controllers, instead of relative retention-based identification, a new software called retention time locking (RTL) developed by Agilent Technologies allows reliable identification from absolute retention time values. This software enables the chromatographic system to be adjusted so that the retention times of analytes remain constant even after routine maintenance procedures, column replacement, column shortening, and method transference from instrument to instrument. The use of RTL has been evaluated by Savchuk *et al.* [59] during the development of a unified procedure for the detection of drugs (including cocaine) in biological fluids. The high precision obtained with RTL has also been pointed out by Rasanen *et al.* [60], who prefer to use absolute retention time instead of the more laborious retention index system during toxicological screening for drugs.

Finally, the development of fast GC [61–63] and comprehensive two-dimensional GC ($GC \times GC$) [64–66] address the continuous demand for increased speed and separation power in routine analysis. The former technique allows a dramatic reduction in analysis time without sacrificing resolution, while the latter offers a markedly increased separation power without altering the analysis time. A fast GC method for the analysis of cocaine and other drugs of forensic relevance has been published by Williams *et al.* [67]. They used a GC instrument in which the column was resistively heated at rates of up to $30^{\circ}/s$ which allowed separation of 19 compounds within 1.5 min. A $GC \times GC$ time-of-flight mass spectrometry (TOF-MS) method has been proposed by Song *et al.* [68] for the analysis of a mixture of 78 drugs of interest, including cocaine and benztropine.

As part of a comparative study, roots, leaves, and seeds of three varieties of *Datura stramonium* L., namely var. *stramonium*, *tatula*, and *godronii* were investigated by GC-MS for their tropane alkaloid pattern [69]. In total, 25 alkaloids were directly detected in these varieties. The identification was based upon electron impact ionization MS data from commercially available libraries or from literature. Hyoscyamine and scopolamine were detected in all plant organs of the three *Datura stramonium* varieties grown in Bulgaria but not in the roots, seeds or leaves from the Egyptian variety *stramonium*. Among the chromatographically separated tropane alkaloids, some showed superimposable mass spectra and were accordingly determined as isomeric series. The difficulty of unambiguously distinguishing tiglioyl from its isomeric angeloyl or senecioyl derivatives based upon mass spectral information only is worth emphasizing. Thus, identification by GC-MS of such tropane derivatives without referring to authentic compounds should be considered with care and regarded as tentative only.

Similarly, screening of 12 different species and their varieties belonging to the tribe Datureae has been reported [70]. GC-MS investigation permitted on-line identification of 66 tropane alkaloids from crude leaf and root extracts using no more than 300 mg of plant material. Many of the alkaloids were described for the first time in the corresponding species, and their identification was based on their fragmentation pathways. One new tropane alkaloid was reported as well; however, its stereochemistry could obviously not be assigned.

Kartal *et al.* have discussed the quantitative analysis of hyoscyamine in *Hyoscyamus reticulatus* L., a plant growing in east Anatolia [71].

GC-MS analysis of the tropane content of shoots from *in vitro* regenerated plantlets from *Schizanthus hookeri* [72] allowed the detection of ten alkaloids ranging from simple pyrrolidine derivatives to tropane esters derived from angelic, tiglic, senecioic, or methylmesaconic acids. One of them, 3α-methylmesaconyloxytropane, is a new alkaloid. Its structure was deduced by comparing mass spectral data and retention indices with those of a synthetic reference compound. To date, the fastest GC analysis of tropane alkaloids dealt with the separation of isomeric secondary metabolites from the stem-bark of *Schizanthus grahamii* [73]. This study presents a systematic investigation of very fast GC applied for the baseline separation of a series of four hydroxytropane esters. Theoretical and practical relationships were used in the optimization steps, including selection of stationary phase, temperature, internal column diameter, and optimal practical gas velocity. This work provided a challenging application for isothermal analysis in conjunction with very short, narrow-bore columns. The investigated approach allowed a baseline separation of the alkaloids of interest in less than 9 s (Figure 12.2).

Even though sample preparation is usually the most time-consuming step in natural product research, this particular case study demonstrated that phytochemical investigations can positively benefit from fast GC methods to reduce the overall analysis time.

The usefulness of GC-MS analysis for biosynthetic studies was demonstrated by Patterson and O'Hagan [74] in their investigation of the conversion of littorine to hyoscyamine after feeding transformed root cultures of *Datura stramonium* with deuterium-labeled phenyllactic acids. This study complements previous investigations on the biosynthesis of the tropate ester moiety of hyoscyamine and scopolamine [75], where GC-MS played a key role. It also has general relevance in the biosynthetic pathway of tropane alkaloids in the entire plant kingdom [76].

Gas chromatography of cocaine of plant origin has mainly involved the analysis of the coca plant [77–79]. Identification and quantitation GC methods of minor naturally occurring tropane alkaloids in illicit cocaine samples have also been reviewed [80]. Moore *et al.* presented an in-depth methodology for the analysis of the coca plant by GC-FID, GC-ECD, and GC-MS for the identification of alkaloids of unknown structure [81]. Recently, Casale *et al.* [82] have analyzed the seeds from *Erythroxylum coca* for their alkaloidal content. Several tropane alkaloids were detected and characterized and it appeared that methylecgonidine (MEG) was the primary constituent and not an analytical artifact.

Thus far, the genus *Erythroxylum* and, more particularly, the coca plant, represented by the cultivated species *Erythroxylum coca* and *Erythroxylum novogranatense,*

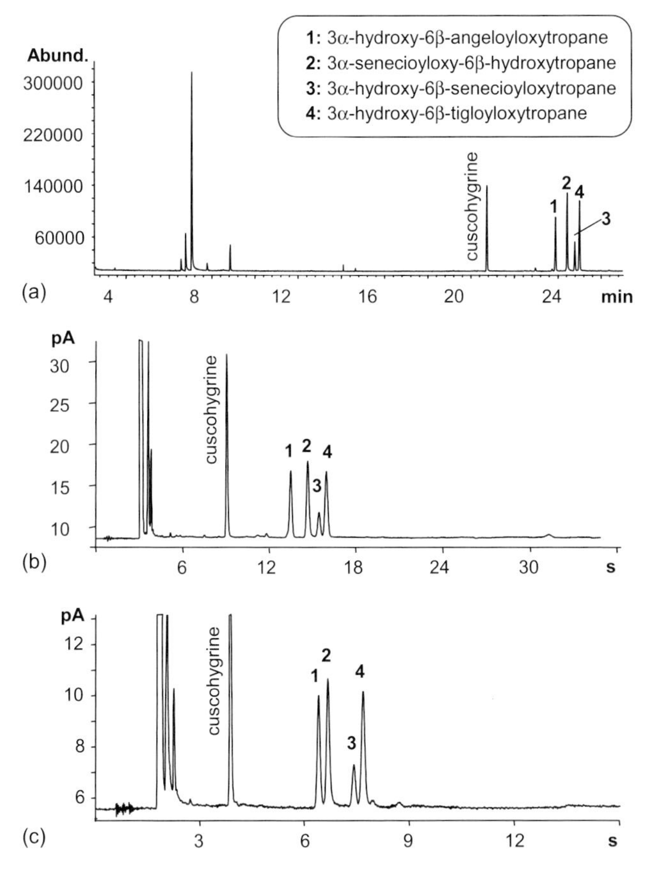

Fig. 12.2 Separation of four isomeric tropane alkaloids by GC: (a) Analysis in linear programmed temperature GC-MS using a conventional 30 m × 0.25 mm i.d. × 0.25 μm film thickness HP5-MS column. (b) Fast GC separation on a 3 m × 0.1 mm i.d. × 0.1 μm DB5 column. (c) 1.5 m × 0.05 mm i.d. × 0.05 μm microbore BGB-1701 column operating at an average linear gas velocity (H_2) of 150 cm/s.

is the only natural source of cocaine. However, little attention has been paid to noncultivated *Erythroxylum* species for the possible presence of cocaine.

During a screening of tropane alkaloids in *Erythroxylum* species from Southern Brazil, Zuanazzi *et al.* [83] identified a new alkaloid as 3β,6β-ditigloyloxynortropane. The five investigated species were also screened for MEG, tropacocaine, and cocaine. Tropacocaine and MEG were present in two plants but no cocaine was detected in any species.

Leaf samples of various *Erythroxylum* species have been investigated for their cocaine content by a fast GC-MS method [29]. Amongst the 51 analyzed species, 28 had not been examined previously and cocaine was detected in 23 wild species. Cocaine content was less than 0.001 % for all wild species, except for *Erythroxylum laetevirens* in which a 10-fold higher concentration was determined. The qualitative chromatographic profile of the latter species was very similar to that of cultivated coca species. Moreover, GC profiles and quantitative results showed that the so-called "Mate de coca," was mainly composed of unadulterated coca leaves. This result agrees with previous GC-MS investigations on the cocaine content in coca tea [84,85].

An effective combination of FMAE and SPME to enhance selectivity for the quantitative GC analysis of cocaine in leaves of *E. coca* has been proposed by Bieri *et al.* [38]. The dual extraction step greatly improved the selectivity, thus allowing much faster GC-FID analysis. Finally, by optimizing the desorption step after SPME sampling and by using a fast GC method, Ilias *et al.* [37] were able to complete a quantitative cocaine analysis from coca leaves (i.e. sample preparation, extraction, and chromatography) in less than 1 h (Figure 12.3).

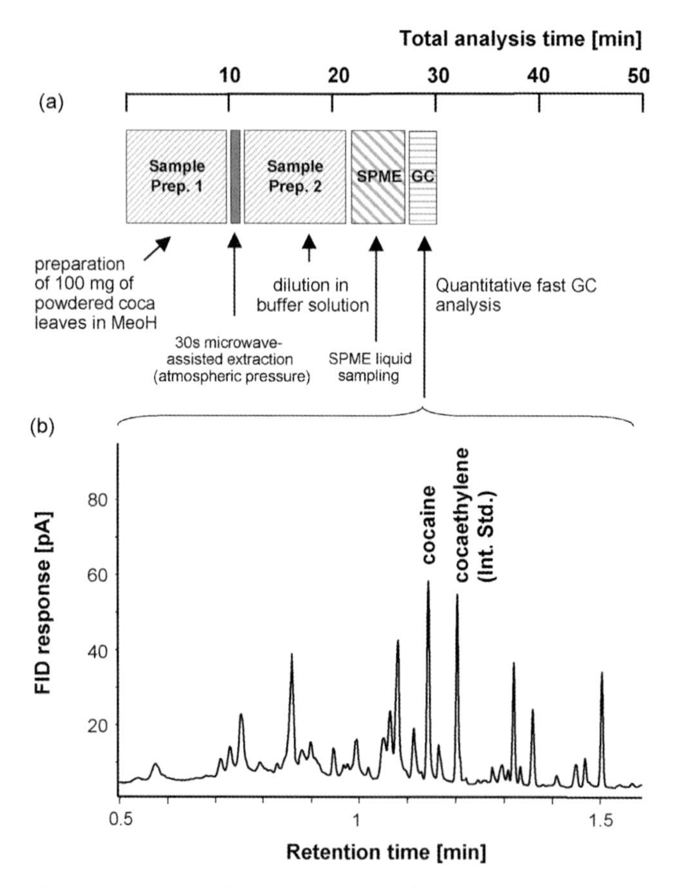

Fig. 12.3 Fast cocaine determination in coca leaves by GC according to Ilias *et al.* (adapted from [37]). (a) Schematic of the total analysis time. (b) Fast GC-FID chromatogram using a short 100 μm i.d. column and a fast oven temperature programming.

Gas chromatography is an established separation method of major importance in forensic sciences. In particular, GC-MS is one of the most frequently used techniques in toxicology. The increasing incidence and variety of "drugs of abuse" have resulted in a growing demand for rapid and universal screening methods for the analysis of biological matrices. In particular, identification and quantification of cocaine in biological fluids and tissues are of great importance. A large range of GC methods have been developed for the analysis of cocaine and its main metabolites in biological fluids. A critical review of chromatographic procedures for testing hair sample for drugs has been published by Sachs and Kintz [86]. An exhaustive review concerning GC-MS analysis of body fluids for drugs of abuse has been written by Cody and Foltz [87]. Owing to the limited sensitivity of the flame ionization detector (FID), some reports pointed out the effectiveness of the element-sensitive nitrogen–phosphorus detector (NPD) [88,89] or electron capture detector (ECD) [90].

A GC-MS method was developed for the determination of hyoscyamine and scopolamine in blood serum [91,92]. Extraction was carried out using aqueous basic solution followed by a purification step on an Extrelut column. Derivatization was done with *N,O*-bis(trimethylsilyl)trifluoroacetamide/trimethylchlorosilane (99 : 1). GC-MS was performed on a HP-5 MS column (30 m $\times$ 0.25 mm i.d. with a 0.25 μm film thickness). The linearity was good between 10 and 5000 ng/mL. The limit of detection (LOD) was 5 ng/mL for each compound.

SPE and LLE are considered the methods of choice for preparing biological samples before a GC analysis of cocaine and metabolites. Farina *et al.* [43] have developed a simple, rapid, and sensitive method for determining cocaine in urine with a single-step LLE and using GC with NPD. A mean extraction recovery of 74 % was reported and the limits of detection and quantitation were 5 and 20 ng/mL, respectively.

Stir-bar sorptive extraction (SBSE) [93] with polydimethylsiloxane sorbent followed by thermal desorption–capillary gas chromatography–mass spectrometry has been used for the detection of drugs of abuse in biological fluids [94]. The following biological fluids were investigated: urine, blood, bile, and stomach content. The method was developed for 34 drugs, including cocaine. The GC-MS data were plotted in a contour plot with locked retention time on the *x*-axis and ion traces on the *y*-axis. Target solutes were identified by a spot in specific positions in the plot and the color of the spots was related to peak abundances. Semiquantitative information was readily obtained from the contour plots while precise quantification required calibration procedures.

A rapid, automated procedure for single-step and simultaneous extraction of cocaine and 11 related compounds was developed by Lewis *et al.* [95]. Fluid and tissue specimens were extracted using an automated SPE system (Zymark Rapid-Trace) with a Bond Elute-Certify I cartridge. Samples were derivatized with pentafluoropropionic anhydride/2,2,3,3,3-pentafluoro-1-propanol prior to GC-MS analysis. The method allowed differentiation between smoking crack and intranasal/intravenous cocaine use and was able to elucidate whether ethanol and cocaine were used simultaneously. Another way to determine cocaine in biological matrices is to measure its metabolites and/or degradation products. MEG is produced when cocaine base is smoked and ecgonidine (EC) is a hydrolytic metabolite of MEG that has been identified in the urine of crack smokers. These compounds can be used as biomarkers to differentiate smoking from other routes of administration. A GC-MS method was developed by Scheidweiler *et al.* [96] to analyze MEG and EC in blood samples. The two compounds were extracted from plasma by SPE after methanol precipitation of proteins. The method was linear between 20 and 2500 μg/L for MEG and between 30 and 3000 μg/L for EC. The limit of detection was 10 μg/L.

Saliva is used increasingly as a matrix of choice for the detection of illicit drugs. The advantages of saliva over traditional fluids are that collection is almost noninvasive, easy to perform and can be achieved under close supervision to prevent adulteration or substitution of the samples. Saliva can be extracted and analyzed in the same manner as other biological fluids. In general, there is less interference from endogenous compounds than with blood or urine. The cocaine concentration in saliva, stored in a plastic container without the addition of citric acid or other stabilizers, remains unaltered at −4 °C for 1 week. Generally, cocaine and its metabolites can be extracted from saliva using a simple SPE procedure in acetate buffer at pH 4.0 [56]. Similarly, Campora *et al.* [97] developed an analytical method for the simultaneous determination of cocaine and its major metabolites, ecgonine methyl ester, and benzoylecgonine in the saliva of chronic cocaine users. The method involved LLE, derivatization with (99 : 1) BSTFA/TMCS, and GC-CIMS in positive ion mode using a single quadrupole detector with the appropriate deuterated standards. Chromatographic elution performed using a 12 m × 0.20 mm i.d. column, 0.33 μm film thickness of 5 % phenyl-methylsiloxane, and temperature programming. Selected ions were monitored, in SIM mode, for each compound studied. Samyn *et al.* [98] also used some alternative matrices, such as saliva and sweat for the detection of drugs of abuse in drivers. Cocaine was determined by GC-MS and the positive predictive values of saliva and sweat wipe were 92 % and 90 %, respectively.

Teske *et al.* [99] evaluated a programmed-temperature vaporizing injection for GC-MS determination of different drugs, among them cocaine, in biological fluids (blood, saliva, etc.). This method reduced the sample consumption (50 μL) and possessed good sensitivity.

Hair is also frequently a useful matrix for drug testing because drugs can be detected for a longer period than in blood or urine. A critical review of chromatographic procedures for drug testing of hair samples has been published by Sachs and Kintz [86]. Gruszecki *et al.* [100] reported the detection of cocaine and cocaethylene in postmortem biological specimens. Hair samples were washed successively at 37 °C with methanol and with phosphate buffer at pH 6.0 and dried. The powdered material was then refluxed in methanol. After filtration, the extract was evaporated to dryness, reconstituted in chloroform and analyzed by GC-MS on an HP I capillary column (50 m × 0.2 mm i.d., 0.1 μm film thickness). The MSD was operated in the SIM mode with a limit of detection for both cocaine and cocaethylene of 5 ng/mL.

It is also interesting to note that cocaine can be determined in amniotic fluid, cord blood, infant urine, meconium, and maternal hair to detect prenatal cocaine use [101]. GC-MS was performed after an appropriate extraction procedure and 51 of 115 subjects were positive for cocaine metabolites. Urine was most frequently positive in identified users, followed by hair.

12.3.2
High-performance Liquid Chromatography

The analysis of tropane alkaloids by liquid chromatography has been performed for over 30 years. The first high-performance liquid chromatography (HPLC) analysis of tropane alkaloids was published in 1973 and concerned the separation of atropine and scopolamine as well as homatropine and apoatropine [102]; the first analysis of these compounds in plant material appeared 12 years later [103].

Numerous HPLC methods for analyzing tropane alkaloids in plant material have been published and reversed phase (RP) columns appear to be the stationary phase of choice [45,58]. RP-18 is generally the preferred stationary phase, while others (RP-8, RP-6, and RP-cyano) are more rarely used. HPLC with UV detection is often employed for tropane alkaloids with UV-absorbing moieties. Gradient elution with a mixture of water–acetonitrile or water–methanol at acidic pH is generally used. A database of retention indices of 383 toxicologically relevant compounds, including some tropane alkaloids was published in 1994 [104] using gradient elution with a mixture of acetonitrile and phosphate buffer. The column was packed with a C18 stationary phase and the detection performed by a UV-diode array detector, permitting the on-line collection of UV spectra.

New generation columns made of high-purity silica with the absence or low content of metals can be endcapped to ensure low silanol activity [105] and thus prevent tailing of the alkaloids. Such columns (Luna 5 μm C-18, Phenomenex) coupled to a SPE unit (SupelClean LC-18) have been used for the simultaneous analysis of hyoscyamine, scopolamine, 6β-hydroxyhyoscyamine, and apoatropine in hairy roots of *Atropa belladonna* and of *Datura innoxia* [106]. The column was eluted with a mixture of acetonitrile, methanol, and 0.1 % TFA. Absolute LOD values were 0.6 and 0.8 ng for hyoscyamine and scopolamine, respectively. Another HPLC separation was developed to investigate the tropane alkaloid content in genetically transformed root cultures of *Atropa belladonna* [107]. After extraction with a mixture of chloroform–methanol and 25 % ammonia, the sample purification was carried out on a self-packed Extrelut column. Chromatography was performed using a Luna 5 μm C8 with an isocratic mixture of acetonitrile–phosphate buffer (pH 6.2)–methanol as eluent. The absolute LODs were 3 ng for hyoscyamine and 2.3 ng for scopolamine, with a corresponding signal-to-noise ratio of 3. A RP-HPLC-DAD method was developed and optimized for the analysis of hyoscyamine and scopolamine in 14 different leaf and seed samples of *Datura* sp. [39]. The separation was performed on an XTerra RP-18 column with gradient of acetonitrile in 15 mM ammonia solution. LODs were 0.25 and 0.29 ng/μL for hyoscyamine and scopolamine, respectively, with limits of quantification (LOQ) of 0.82 and 0.97 ng/μL. Kirchhoff *et al.* [108] developed an HPLC method without ion-pairing reagents to separate the degradation products and by-products of atropine. The separation was performed on a hydrophilic embedded RP-18 column (Thermo Hypersil Aquasil) characterized by hydrophilic endcapping. Acidic gradient elution with 20 mM phosphate buffer–acetonitrile mixture was used. The method was applied to the atropine assay for eye drops. The short retention time of atropine of about 2 min and the baseline separation of the main degradation products are very convenient for routine analysis.

Since 1990, with the emergence of high-throughput analyses, the approach in natural products discovery has changed considerably to direct analysis of crude plant extracts with minimal sample manipulation. This approach led to the dereplication method, avoiding the tedious isolation of known or undesirable compounds [109]. Conventional detection systems used with HPLC, such as UV or fluorescence spectroscopy, provide only limited information on the molecular structure of the separated compounds. The coupling of HPLC with mass spectrometry (LC-MS, LC-MS^n) resulted in a powerful analytical tool for qualitative and quantitative determination of drugs and their metabolites. In particular, LC-ion-trap multiple-stage mass spectrometry (LC-IT-MS^n) and LC-time-of-flight mass spectrometry (LC-TOF-MS) are used to study fragmentation patterns and determine elemental formulae of analyzed compounds. However, mass spectrometry cannot provide unequivocal structural determination, particularly when no reference material is available or in the case of isomeric compounds. The complementary use of LC-NMR [110] has therefore become extremely attractive as it offers unparalleled structure elucidation capabilities. In this respect, chemical screening strategies have been developed using hyphenated techniques LC-DAD-UV, LC-MS, and LC-NMR [111] for on-line identification of natural products in crude plant extracts. Using LC-DAD-UV, LC-APCI-MS^n, LC-APCI-TOF-MS, and LC-NMR experiments, Zanolari *et al.* [112] identified 24 tropane alkaloids from the bark of *Erythroxylum vacciniifolium*. Among them, six new compounds were characterized. Bieri *et al.* [113] have published two fully automated LC-NMR approaches, namely loop storage and trapping, using a LC-UV-MS/SPE-NMR set-up to deal with the limited NMR sensitivity and overcome the short acquisition times during on-flow measurements. Both approaches were applied to the separation of four isomeric tropane alkaloids isolated from the stem-bark of an endemic plant from Chile, *Schizanthus grahamii* (Solanaceae). The chromatographic separation was carried out on a Hypercarb porous graphitic carbon column (125 × 4.6 mm i.d., 5 μm particle size). In the loop storage approach, each peak was stored into a loop by means of a switching valve and after the separation of the individual compound, the content of each loop was transferred one at a time into the NMR flow cell and subjected to NMR spectroscopy without undesirable peak broadening (Figure 12.4a). In the second approach, LC was combined with parallel ion-trap MS detection for peak selection, and NMR spectroscopy using an SPE cartridge for postcolumn analyte trapping (Figure 12.4b). Nondeuterated solvents were used, reducing the cost and avoiding the interference in MS with the alcoholic exchangeable protons. After the trapping step, the cartridges were dried and analytes were sequentially flushed into the cryogenically cooled NMR cell with a deuterated solvent for measurements.

Analysis of tropane alkaloids in biological fluids has been developed mostly for cocaine and metabolites. Indeed, it is well recognized that cocaine remains one of the most widely consumed drugs of abuse worldwide. Generally, reversed phase liquid chromatography coupled with UV-VIS detection is employed for these analyses.

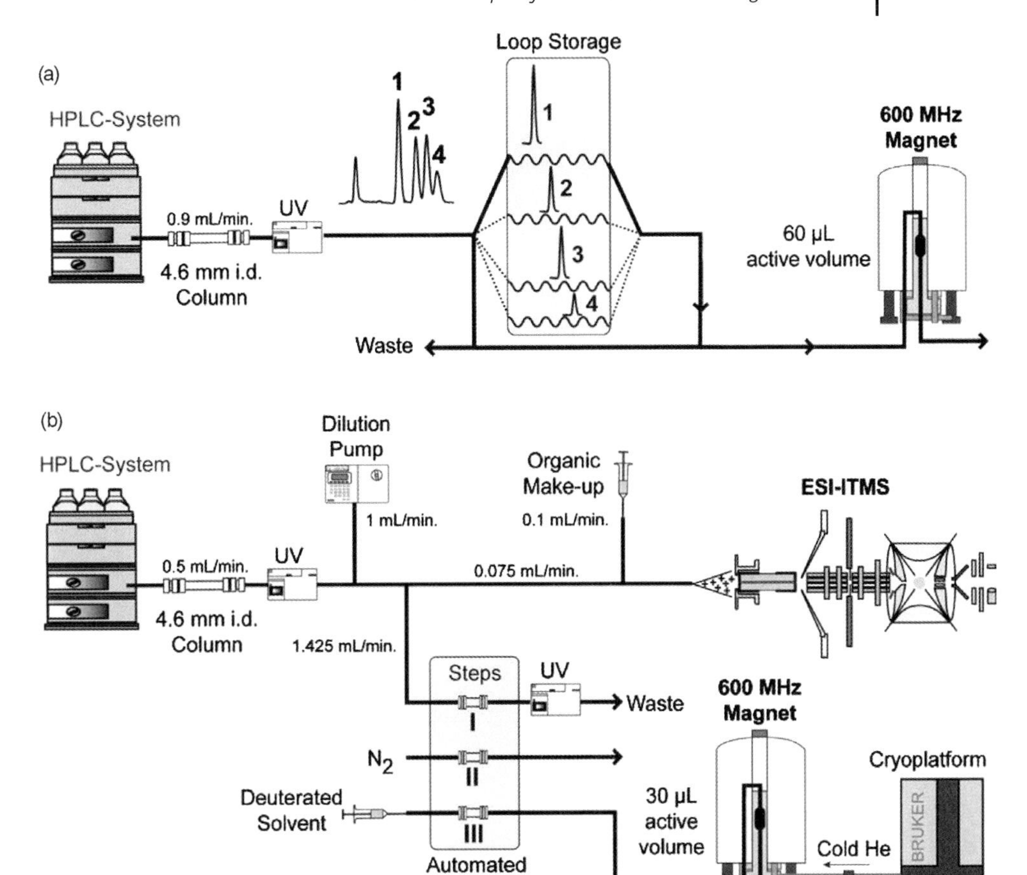

Fig. 12.4 Instrumental set-up for the HPLC-NMR hyphenated approaches. (a) Peak sampling unit using storage loops. (b) Peak trapping onto SPE cartridges with parallel MS and cryogenically cooled NMR detection [113]. *Source:* Reproduced with permission of Wiley.

In the case of cocaine, its persistence in the blood stream is short owing to hydrolysis of the ester linkage [114]. Therefore, sensitive methods are needed for measuring cocaine in biological fluids. With UV detection, Chen *et al.* [52] investigated the disposition of unbound cocaine in rat blood, brain, and bile after microdialysis. A C18 column was used with a mobile phase containing 25 % acetonitrile (with triethylamine) and 75 % of 10 mM phosphate buffer (pH 4.0). The UV detector was set at 235 nm and the LOQ was 0.05 μg/mL. The *in vivo* recoveries were lower than 50 % owing to different factors affecting the dialysis efficiency. Thus, the method was not sensitive enough to determine cocaine in the bile dialysate and a LC-tandem mass spectrometry assay was developed instead. Raje *et al.* [44] developed a sensitive LC-UV method for determining the benztropine analog AHN-1055 in rat plasma and brain. This compound, a novel cocaine abuse therapeutic, was first extracted by LLE and analyzed on a C18 column with a gradient profile. The mobile phase was a mixture of methanol and phosphate buffer set at pH 3.0 and detection was achieved at 220 nm. The sensitivity was sufficient (25 and 50 ppb in plasma and brain, respectively) for performance of a pharmacokinetic study. Harrison *et al.* [115] have developed a LC-UV method for determining the rapid hydrolysis of atropine by atropinesterase in plasma of dogs, goats, guinea pigs, humans, pigs, rabbits, and rhesus monkeys. The activity of this enzyme in rabbits and some plants is well described, while its presence in other animal species is very controversial. In this study, atropine and tropic acid were separated by reversed phase chromatography with a mixture of acetonitrile and phosphate buffer (pH 3.1) and detection was performed at 205 nm.

As mentioned above, UV detection is often not sensitive enough to determine low amounts of tropane alkaloids in biological fluids. Therefore, different LC methods have been published using MS and tandem MS. For this purpose an electrospray ionization (ESI) source is used, since tropane alkaloids are basic substances that can be easily protonated. Analysis of free cocaine and benzoylecgonine was performed by Fuh *et al.* [51] using a C18 column in the gradient mode with a mobile phase containing acetonitrile and acetic acid. With a single quadrupole, the authors investigated in-source collision-induced dissociation (CID) to promote fragmentation, with the skimmer voltage set at 125 V to obtain two major ions for each tested substance. Cocaine had characteristic ions at m/z 304 and m/z 182 and corresponding benzoylecgonine ions were seen at m/z 290 and m/z 168. The method was validated and applied to determine the free form of cocaine and its metabolite in rat brain following microdialysis. The same strategy was applied by Chen *et al.* [52] for the analysis of unbound cocaine in the bile fluid with a triple quadrupole MS. With the same ions (precursor at m/z 304 and product ion at m/z 182), the sensitivity was sufficient to determine that cocaine is excreted through the hepatobiliary system. Scopolamine and atropine, selected as internal standard, were also determined by LC-MS/MS in serum samples of volunteers after a simple and rapid SPE procedure and in microdialysates [46]. The sensitivity was excellent, with a limit of quantification of 20 pg/mL for scopolamine in serum, allowing the application of this method for conducting pharmacokinetic studies. Other authors used LC-MS/MS to analyze cocaine and its metabolite ecgonine methyl ester in human urine [116]. In this study, a nonconventional hydrophobic pentafluorophenylpropyl (PFPP) bonded silica stationary phase was used to strongly retain cocaine and its metabolite with good efficiency. A mobile phase with 90 % acetonitrile was then necessary to elute the compounds of interest, thus inducing a large enhancement of the ESI-MS signal. It is noteworthy that the MS signal of cocaine was 12 times greater with the PFPP column than with a conventional C18. Therefore, the urine was simply diluted (1 : 10) and injected directly in the LC-MS, saving time and money.

As already reported by Dräger in 2002 [45], the low UV absorption of tropane alkaloids means that other detection modes have been investigated. Besides mass spectrometry, electrochemical detection was tested by different authors, but sensitivity was not significantly improved since tropane alkaloids are only moderately responsive compounds. Fluorescence detection can also be used to improve selectivity and sensitivity. However, tropane alkaloids are not natively fluorescent and require a prederivatization step. With 1-anthroylnitrile (1-AN) used as reagent, atropine was determined by LC-fluorescence at a concentration of 10 ng/mL [117]. Jamdar *et al.* [118] modified an analytical procedure to achieve a rapid and sensitive quantitation of cocaine and its major metabolites in plasma and urine with two analytical columns packed in series with a C8 and a cyanopropyl phase. For determining ecgonine methylester (EME) and rendering this compound UV-detectable, a derivatization was performed with 4-fluorobenzoyl chloride. Under optimized conditions, quantitation limits were 25 ng/mL for cocaine, benzoylecgonine, and norcocaine and 50 ng/mL for EME. Therefore, this method was sensitive enough to perform pharmacokinetic studies.

12.3.3 Capillary Electrophoresis

Capillary electrophoresis (CE) coupled with UV or MS detection has been widely used for the determination of pharmaceutical compounds because of its efficiency, accuracy, and high resolution. This separation method appears particularly interesting for the analysis of alkaloids because these compounds are easily protonated if an appropriate acidic buffer is chosen. In 1998, Stöckigt *et al.* [119] published a review in which they summarized the use of CE for the analysis of various alkaloids.

Surprisingly, in comparison with HPLC or GC methods, CE was not frequently applied to the analysis of tropane alkaloids. A capillary zone electrophoresis (CZE) method in combination with an on-column diode array detection was developed and validated for the simultaneous determination of atropine, scopolamine, and derivatives in pharmaceutical preparations [120,121] or in plant material [122]. Enantioseparation of atropine using sulfated β-cyclodextrin has been optimized by means of a central composite design. The method has been developed for the enantiomeric purity evaluation of (−)-hyoscyamine as well as for the separation of littorine (a positional isomer of hyoscyamine) from atropine enantiomers in genetically transformed root cultures of *Hyoscyamus albus* [123,124]. In plant material, hyoscyamine and scopolamine are generally present together with other tropane alkaloids with similar structures and mass-to-charge ratios. Therefore, micellar electrokinetic capillary chromatography (MEKC) is frequently found more appropriate for analysis of the alkaloids in plant extracts. Using the same set-up as CZE, MEKC only requires the addition of a surfactant at a concentration above its critical micelle concentration to the running buffer. In particular, MEKC is able to separate neutral and ionic analytes in the same run. This method was applied to the simultaneous analysis of six tropane alkaloids, including hyoscyamine and scopolamine [125]. The optimized conditions have been applied to the analysis of tropane alkaloids found in hairy roots of *Datura candida* × *D. aurea*. MEKC was selected for the quantitative determination of hyoscyamine in *Belladonna* extract [126] and for the simultaneous quantitative determination of hyoscyamine, scopolamine, and littorine in different hairy root clones of *Hyoscyamus muticus* [127] and other solanaceous plant extracts [128].

The use of nonaqueous media has gained considerable importance in the analysis of pharmaceuticals by CE. In particular, very high efficiency and resolution, short analysis time, and the possibility of increasing the analyte solubility have been demonstrated. In addition, nonaqueous media are suitable for on-line coupling to mass spectrometry. Nonaqueous capillary electrophoresis (NACE) has been developed for the separation of eight tropane alkaloids [129]. The optimized method was applied to the analysis of hyoscyamine and scopolamine in genetically transformed roots of *Datura candida* × *D. aurea* and results have been compared with those obtained by MEKC. A simple NACE method has been described for the separation of several atropine- and scopolamine-related drugs. The robustness has been verified using a full factorial design and, after validation, the method was applied for the determination of *N*-butylscopolamine in different pharmaceutical preparations [130].

The on-line coupling of CE with electrospray ionization mass spectrometry (CE-ESI-MS) allows high separation efficiency together with high sensitivity and selectivity as well as molecular structural information. A CE-UV-ESI-MS method was developed for the analysis of hoscyamine, scopolamine, and other tropane derivatives [131]. The differentiation of hyoscyamine from littorine, commonly encountered in plant material, was demonstrated using in-source collision-induced dissociation. The developed method was applied to the analysis of these alkaloids in *Belladonna* leaf extract and in *Datura candida* × *D. aurea* hairy root extract. Recently, CE coupled with electrochemiluminescence detection has been used for the determination of atropine and scopolamine in *Flos daturae* [132].

Even if capillary electrophoresis is a very powerful technique and has gained considerable interest in pharmaceutical and biomedical analysis [133], its use in the determination of tropane alkaloids in biological matrices remains very restricted. Several reasons could explain this lack of interest in CE. First, sensitivity with UV detection, even at low wavelength, is often not sufficient to determine drugs and metabolites at sub-part-per-million levels. This lack of sensitivity is mainly due to the short optical path length afforded by the small internal diameter of the capillary and to the small injected volume. Second, the injection is matrix-dependent; therefore, a sample preparation is generally required, such as a liquid–liquid extraction or a solid phase extraction, for reducing the complexity of the biological matrix. Third, the coupling of MS with CE is not yet recognized as a completely routine technique. However, several strategies can be used to overcome these drawbacks and a large number of publications have appeared since the late 1990s dealing with the quantification of drugs and metabolites in blood, urine, and other biological matrices by CE [134–138]. Presently, tropane alkaloids are only rarely analyzed by CE. In 1998, Plaut and Staub developed a micellar electrokinetic chromatography (MEKC) procedure for the determination of atropine in the presence of strychnine and tetracaine in blood and gastric contents [139]. After sample preparation by LLE, atropine was detected at the part-per-million level by UV detection with scopolamine as internal standard. Tagliaro *et al.* [140–142] used capillary electrophoresis for analyzing illicit substances, such as cocaine, in hair samples. The sample preparation was conventional, with a washing procedure followed by extraction in acidic medium overnight and purification in ready-to-use Toxi-Tubes A. The analysis was conducted at pH 9.2, with UV detection. A field-amplified sample stacking technique was performed during the injection for improving sensitivity. The detection of concentrations of cocaine as low as 0.5 ng/mg have been reported.

CE also suffers from limited reproducibilities of compound migration times. The major reason is the variability of the electroosmotic flow (EOF). In order to overcome this drawback, capillaries can be permanently or dynamically coated. The latter is more advantageous, since the coating can be replaced and regenerated after each run. CElixir, also called CEofix, has been developed as a commercially available dynamic coating to stabilize the EOF [143] and Boone *et al.* [144] used this procedure for a systematic toxicological analysis. A set of 73 compounds was analyzed, among them atropine and cocaine. It was clearly demonstrated that a coated capillary gave better results for the identification of basic drugs in terms of higher reproducibility, identification power, and shorter analysis time than conventional CE. More recently, Alnajjar *et al.* [145] used CE with native fluorescence and laser-induced fluorescence for the separation and detection of multiple drugs of abuse in biological fluids. Cocaine was analyzed in urine after solid phase extraction and derivatization using fluorescein isothiocyanate isomer I. Before derivatization, cocaine was subjected to an N-demethylation reaction involving the use of 1-chloroethyl chloroformate. The sensitivity of the method was excellent (approximately 100 pg/mL) with excitation and emission wavelengths of 488 and 522 nm, respectively.

12.3.4 Desorption Electrospray Ionization Mass Spectrometry

Since 2000, the development of ambient desorption electrospray ionization (DESI) as an ion source for mass spectrometry has emerged as an interesting alternative in cases where the analyte is otherwise destroyed by sample preparation, as a simple preliminary screening test in emergency toxicology, or in high-throughput applications [146,147]. It allows for the rapid analysis of samples under ambient conditions and without any sample preparation. Typically, DESI is carried out by directing electrosprayed droplets and ions of solvent onto the surface of a complex sample of interest. The instrumentation, mechanisms, and applications of DESI in forensics, chemistry, and biology have been reviewed [148]. Another report presented the analysis of 21 commercial drugs as well as some illicit Ecstasy tablets [149]. DESI-MS has also been used for the screening of cannabis samples, resulting in the rapid detection of the major cannabinoids [150]. *In situ* detection of tropane alkaloids in *Datura stramonium* and *Atropa belladonna* has been investigated [151]. The effects on analytical performance of operating parameters, including the electrospray high voltage, heated capillary temperature, the solvent infusion rate, and the carrier gas pressure, were evaluated. Fifteen out of nineteen known alkaloids for *D. stramonium* and the principal alkaloids of *A. belladonna* were identified using DESI in combination with tandem mass spectrometry with methanol : water (1 : 1) as the spray solvent. Total analysis time was reduced to a minimum, as there is no sample preparation and no separation.

12.4 Conclusions

Tropane alkaloids are an important class of natural products possessing different and interesting pharmacological activities. Hyoscyamine (atropine in the racemate form), scopolamine, and cocaine are the major representatives of this class. They are commonly found in plant materials, mainly in genera belonging to three families: Solanaceae, Erythroxylaceae, and Convolvulaceae. The importance of these compounds requires that there are accurate analytical methods for their determination in plants and in biological matrices. This chapter describes the state-of-the-art of analytical procedures (extraction and analysis) for analyzing tropane alkaloids.

Extraction procedures of plant materials: classical percolation, maceration, digestion, decoction, and so on, as well as supercritical fluid extraction, microwave-assisted extraction, pressurized solvent extraction, and solid-phase microextraction are described. For biological matrices, liquid–liquid, and solid phase extractions are mainly used for different samples such as blood, urine, microdialysates, and saliva, among others.

Analyses of tropane alkaloids are mainly carried out by GC and HPLC and to a lesser extent by CE. This review describes recent applications developed for the analysis of this class of compounds in plant materials and biological matrices. Of course, mass spectrometry is generally used as the detection technique because of its high sensitivity and selectivity, but other techniques such as UV, fluorescence, flame ionization detection, nuclear magnetic resonance, among others have also been investigated. Finally, desorption electrospray ionization mass spectrometry is reported as a new interesting detection technique for the rapid analysis of samples without any sample preparation.

References

1 Lounasmaa, M. and Tamminen, T. (1993) The tropane alkaloids, In Cordell, G. A. (Ed.) *The Alkaloids*, Vol. 44, Academic Press, San Diego, CA.

2 Christen, P. (2000) Tropane alkaloids: old drugs used in modern medicine, In Atta-ur-Rahman (Ed.) *Studies in Natural Products Chemistry*, Vol. 22, Elsevier, Amsterdam, NL.

3 Griffin, W. J. and Lin, G. D. (2000) *Phytochemistry*, **53**, 623–637.

4 Goldmann, A., Milet, M. L., Ducrot, P. H., Lallemand, J. Y., Maille, M., Lepingle, A., Charpin, I., Tepfer, D. (1990) *Phytochemistry*, **29**, 2125–2128.

5 Schimming, T., Tofern, B., Mann, P., Richter, A., Jenett-Siems, K., Dräger, B., Asana, N., Gupta, M. P., Correa, M. D., Eich, E. (1998) *Phytochemistry*, **49**, 1989–1995.

6 Dräger, B. (2004) *Natural Products Report*, **21**, 211–223.

7 Brock, A., Bieri, S., Christen, P., Dräger, B. (2005) *Phytochemistry*, **66**, 1231–1240.

8 Brock, A., Herzfeld, T., Paschke, R., Koch, M., Dräger, B. (2006) *Phytochemistry*, **67**, 2050–2057.

9 Asano, N., Nash, R. J., Molyneux, R. J., Fleet, G. W. J. (2000) *Tetrahedron: Asymmetry*, **11**, 1645–1680.

10 Gaillard, Y. and Pepin, G. (1999) *Journal of Chromatography, B*, **733**, 181–229.

11 El Jaber-Vazdekis, N., Gutierrez-Nicolás, F., Ravelo, A. G., Zárate, R. (2006) *Phytochemical Analysis*, **17**, 107–113.

12 Cannell, R. J. P. (1998) Natural products isolation, *Methods in Biotechnology*, Vol. 4, Humana Press, Totowa, NJ.
13 Woolley, J. G. (1993) Tropane alkaloids, In Watermann, P. G. (Ed.) *Methods in Plant Biochemistry*, Vol. 8, Academic Press, London.
14 Castioni, P., Christen, P., Veuthey, J. -L. (1995) *Analusis*, **23**, 95–106.
15 Choi, Y. H., Chin, Y. -W., Kim, J., Jeon, S. H., Yoo, K. -P. (1999) *Journal of Chromatography, A*, **863**, 47–55.
16 Brachet, A., Mateus, L., Cherkaoui, S., Christen, P., Gauvrit, J. -Y., Lantéri, P., Veuthey, J. -L. (1999) *Analusis*, **27**, 772–778.
17 Brachet, A., Christen, P., Gauvrit, J. -Y., Longeray, R., Lantéri, P., Veuthey, J. -L. (2000) *Journal of Biochemical and Biophysical Methods*, **43**, 353–366.
18 Later, D. W., Richter, B. E., Knowles, D. E., Anderson, M. R. (1986) *Journal of Chromatographic Science*, **24**, 249–253.
19 Morrison, J. F., Chesler, S. N., Yoo, W. J., Selavka, C. M. (1998) *Analytical Chemistry*, **70**, 163–172.
20 Veuthey, J. -L., Edder, P., Staub, C. (1995) *Analusis*, **23**, 258–265.
21 Abu Samra, A., Morris, J. S., Koirtyohann, S. R. (1975) *Analytical Chemistry*, **47**, 1475–1477.
22 Ganzler, K., Salgò, A., Valko, K. (1986) *Journal of Chromatography*, **371**, 299–306.
23 Ganzler, K. and Salgò, A. (1987) *Zeitschrift fuer Lebensmittel-Untersuchung und -Forschung*, **184**, 274–276.
24 Zlotorzynski, A. (1995) *Critical Reviews in Analytical Chemistry*, **25**, 43–76.
25 Neas, E. D. and Collins, M. J. (1988) Microwave heating: theoretical concepts and equipment design, In Kingston, H. M. and Jassie, L. B. (Eds.) *Introduction to Microwave Sample Preparation. Theory and Practice*, American Chemical Society, Washington DC.
26 Ganzler, K., Szinai, I., Salgò, A. (1990) *Journal of Chromatography*, **520**, 257–262.
27 Camel, V. (2001) *Analyst*, **126**, 1182–1193.
28 Brachet, A., Christen, P., Veuthey, J. -L. (2002) *Phytochemical Analysis*, **13**, 162–169.
29 Bieri, S., Brachet, A., Veuthey, J. -L., Christen, P. (2006) *Journal of Ethnopharmacology*, **103**, 439–447.
30 Kaufmann, B. and Christen, P. (2002) *Phytochemical Analysis*, **13**, 105–113.
31 Brachet, A., Rudaz, S., Mateus, L., Christen, P., Veuthey, J. -L. (2001) *Journal of Separation Science*, **24**, 865–873.
32 Arthur, C. L. and Pawliszyn, J. (1990) *Analytical Chemistry*, **62**, 2145–2148.
33 O'Reilly, J., Wang, Q., Setkova, L., Hutchinson, J. P., Chen, Y., Lord, H. L., Linton, C. M., Pawliszyn, J. (2005) *Journal of Separation Science*, **28**, 2010–2022.
34 Gentili, S., Cornetta, M., Macchia, T. (2004) *Journal of Chromatography, B*, **801**, 289–296.
35 Follador, M. J. D., Yonamine, M., de Moraes Moreau, R. L., Silva, O. A. (2004) *Journal of Chromatography, B*, **811**, 37–40.
36 Bermejo, A. M., López, P., Álvarez, I., Tabernero, M. J., Fernandez, P. (2006) *Forensic Science International*, **156**, 2–8.
37 Ilias, Y., Bieri, S., Christen, P., Veuthey, J. -L. (2006) *Journal of Chromatographic Science*, **44**, 394–398.
38 Bieri, S., Ilias, Y., Bicchi, C., Veuthey, J. -L., Christen, P. (2006) *Journal of Chromatography, A*, **1112**, 127–132.
39 Mroczek, T., Glowniak, K., Kowalska, J. (2006) *Journal of Chromatography, A*, **1107**, 9–18.
40 Berkov, S. and Pavlov, A. (2004) *Phytochemical Analysis*, **15**, 141–145.
41 Wells, D. A. (2003) High throughput bioanalytical sample preparation, *Methods and Automation Strategies. Progress in Pharmaceutical and Biomedical Analysis*, Vol. 5, Elsevier, Amsterdam, NL.
42 Van Hout, M. W. J., Niederländer, H. A. G., de Zeeuw, R. A., de Jong, G. J. (2003) New developments in

integrated sample preparation for bioanalysis, In Smith, R. M. and Wilson, I. A. (Eds.) *Handbook of Analytical Separations: Bioanalytical Separations*, Vol. 4, Elsevier, Amsterdam, NL.

43 Farina, M., Yonamine, M., Silva, A. A. (2002) *Forensic Science International*, **127**, 204–207.

44 Raje, S., Dowling, T. C., Eddington, N. D. (2002) *Journal of Chromatography, B*, **768**, 305–313.

45 Dräger, B. (2002) *Journal of Chromatography, A*, **978**, 1–35.

46 Oertel, R. (2001) *Journal of Chromatography, B*, **750**, 121–128.

47 Hennion, M. -C. and Pichon, V. (2003) *Journal of Chromatography, A*, **1000**, 29–52.

48 Sellergren, B. (2001) Molecularly imprinted polymers in solid-phase extractions, In Lanza, F. (Ed.) *Molecularly Imprinted Polymers-Man-made Mimics of Antibodies and Their Applications in Analytical Chemistry: Techniques and Instrumentation in Analytical Chemistry*, Vol. 23, Elsevier, Amsterdam, NL.

49 Nakamura, M., Ono, M., Nakajima, T., Ito, Y., Aketo, T., Haginaka, J. (2005) *Journal of Pharmaceutical and Biomedical Analysis*, **37**, 231–237.

50 Müller, M., Schmid, R., Georgopoulos, A., Buxbaum, A., Wasicek, C., Eichler, H. -G. (1995) *Clinical Pharmacology and Therapeutics*, **57**, 371–380.

51 Fuh, M. -R., Tai, Y. -L., Pan, W. H. T. (2001) *Journal of Chromatography, B*, **752**, 107–114.

52 Chen, Y. -F., Chang, C. -H., Wang, S. -C., Tsai, T. -H. (2005) *Biomedical Chromatography*, **19**, 402–408.

53 Kidwell, D. A., Holland, J. C., Athanaselis, S. (1998) *Journal of Chromatography, B*, **713**, 111–136.

54 Samyn, N., Verstraete, A., van Haeren, C., Kintz, P. (1999) *Forensic Science Review*, **11**, 1–19.

55 Gaillard, Y. and Pépin, G. (1999) *Journal of Chromatography, B*, **733**, 231–246.

56 Kintz, P. and Samyn, N. (2002) *Therapeutic Drug Monitoring*, **24**, 239–246.

57 Kintz, P. (1996) *Drug Testing in Hair*, CRC Press, New York, USA.

58 Baerheim-Svendsen, A. and Verpoorte, R. (1984) Chromatography of alkaloids, *Part b: Gas–Liquid Chromatography, and High-Performance Liquid Chromatography*, Elsevier, Amsterdam, NL.

59 Savchuk, S. A., Simonov, E. A., Sorokin, V. I., Dorogokupets, O. B., Vedenin, A. N. (2004) *Journal of Analytical Chemistry*, **59**, 954–964.

60 Rasanen, I., Kontinen, I., Nokua, J., Ojanpera, I., Vuori, E. (2003) *Journal of Chromatography, B*, **788**, 243–250.

61 Cramers, C. A., Janssen, H. -G., van Deursen, M., Leclercq, P. A. (1999) *Journal of Chromatography, A*, **856**, 315–329.

62 Korytar, P., Janssen, H. -G., Matisova, E., Brinkman, U. A. T. (2002) *Trends in Analytical Chemistry*, **21**, 558–572.

63 Klee, M. S. and Blumberg, L. M. (2002) *Journal of Chromatographic Science*, **40**, 234–247.

64 Bertsch, W. (2000) *Journal of High Resolution Chromatography*, **23**, 167–181.

65 Marriott, P. and Shellie, R. (2002) *Trends in Analytical Chemistry*, **21**, 573–583.

66 Dalluge, J., Beens, J., Brinkman, U. A. T. (2003) *Journal of Chromatography, A*, **1000**, 69–108.

67 Williams, T. A., Riddle, M., Morgan, S. L., Brewer, W. E. (1999) *Journal of Chromatographic Science*, **37**, 210–214.

68 Song, S. M., Marriott, P., Kotsos, A., Drummer, O. H., Wynne, P. (2004) *Forensic Science International*, **143**, 87–101.

69 Berkov, S., Zayed, R., Doncheva, T. (2006) *Fitoterapia*, **77**, 179–182.

70 Doncheva, T., Berkov, S., Philipov, S. (2006) *Biochemical Systematics and Ecology*, **34**, 478–488.

71 Kartal, M., Kurucu, S., Altun, L., Ceyhan, T., Sayar, E., Cevheroglu, S., Yetkin, Y. (2003) *Turkish Journal Of Chemistry*, **27**, 565–569.

72 Jordan, M., Humam, M., Bieri, S., Christen, P., Poblete, E., Muñoz, O. (2006) *Phytochemistry*, **67**, 570–578.

73 Bieri, S., Muñoz, O., Veuthey, J. -L., Christen, P. (2006) *Journal of Separation Science*, **29**, 96–102.
74 Patterson, S. and O'Hagan, D. (2002) *Phytochemistry*, **61**, 323–329.
75 O'Hagan, D. and Robins, R. J. (1998) *Chemical Society Reviews*, **27**, 207–212.
76 Humphrey, A. J. and O'Hagan, D. (2001) *Natural Products Report*, **18**, 494–502.
77 Casale, J. F. and Moore, J. M. (1996) *Journal of Chromatography, A*, **749**, 173–180.
78 Casale, J. F. and Moore, J. M. (1996) *Journal of Chromatography, A*, **756**, 185–192.
79 Moore, J. M. and Casale, J. F. (1997) *Journal of Forensic Sciences*, **42**, 246–255.
80 Moore, J. M. and Casale, J. F. (1994) *Journal of Chromatography, A*, **674**, 165–205.
81 Moore, J. M., Casale, J. F., Klein, R. F., Cooper, D. A., Lydon, J. (1994) *Journal of Chromatography, A*, **659**, 163–175.
82 Casale, J. F., Toske, S. G., Colley, V. L. (2005) *Journal Of Forensic Sciences*, **50**, 1402–1406.
83 Zuanazzi, L. A. S., Tremea, V., Limberger, R. P., Sobral, M., Henriques, A. T. (2001) *A. Biochemical Systematics and Ecology*, **29**, 819–825.
84 Engelke, B. F. and Gentner, W. A. (1991) *Journal of Pharmaceutical Sciences*, **80**, 96.
85 Jenkins, A. J., Llosa, T., Montoya, I., Cone, E. J. (1996) *Forensic Science International*, **77**, 179–189.
86 Sachs, H. and Kintz, P. (1998) *Journal of Chromatography*, **713**, 147–161.
87 Cody, J. T. and Foltz, R. L. (1995) *Forensic Applications of Mass Spectrometry*, 1–59.
88 Hime, G. W., Hearn, W. L., Rose, S., Cofino, J. (1991) *Journal of Analytical Toxicology*, **15**, 241–245.
89 Chasin, A. A. M., De Lima, I. V., De Carvalho, D. G., Midio, A. F. (1997) *Acta Toxicológica Argentina*, **5**, 77–80.
90 Gunnar, T., Mykkanen, S., Ariniemi, K., Lillsunde, P. (2004) *Journal of Chromatography, B*, **806**, 205–219.
91 Namera, A., Yashiki, M., Hirose, Y., Yamaji, S., Tani, T., Kojima, T. (2002) *Forensic Science International*, **130**, 34–43.
92 Namera, A. (2005) Tropane alkaloids, In Suzuki, O. and Watanabe, K. (Eds.) *Drugs and Poisons in Human: A handbook of Practical Analysis*, Springer, Berlin.
93 Baltussen, E., Cramers, C. A., Sandra, P. J. F. (2002) *Analytical and Bioanalytical Chemistry*, **373**, 3–22.
94 Tienpont, B., David, F., Stopforth, A., Sandra, P. (2003) *LC-GC Europe*, **16** (12a), 5–13.
95 Lewis, R. J., Johnson, R. D., Angier, M. K., Ritter, R. M. (2004) *Journal of Chromatography, B*, **806**, 141–150.
96 Scheidweiler, K. B., Plessinger, M. A., Shojaie, J., Wood, R. W., Kwong, T. C. (2003) *Journal of Pharmacology and Experimental Therapeutics*, **307**, 1179–1187.
97 Campora, P., Bermejo, A. M., Tabernero, M. J., Fernandez, P. (2003) *Journal of Analytical Toxicology*, **27**, 270–274.
98 Samyn, N., DeBoeck, G., Verstraete, A. G. (2002) *Journal of Forensic Sciences*, **47**, 1380–1387.
99 Teske, J., Putzbach, K., Engewald, W., Kleemann, W. J., Müller, R. K. (2003) *Chromatographia*, **57** (Suppl.), S/271–S/273.
100 Gruszecki, A. C., Robinson, C. A., Jr., Embry, J. H., Davis, G. G. (2000) *American Journal of Forensic Medicine and Pathology*, **21**, 166–171.
101 Eyler, F. D., Behnke, M., Wobie, K., Garvan, C. W., Tebbett, I. (2005) *Neurotoxicology and Teratology*, **27**, 677–687.
102 Stutz, M. H. and Sass, S. (1973) *Analytical Chemistry*, **45**, 2134–2136.
103 Pekic, B., Slavica, B., Lepojevic, Z., Gorunovic, M. (1985) *Die Pharmazie*, **40**, 422–423.
104 Bogusz, M. and Erkens, M. (1994) *Journal of Chromatography, A*, **674**, 97–126.
105 Herrero-Martinez, J. M., Mendez, A., Bosch, E., Roses, M. (2004) *Journal of Chromatography, A*, **1060**, 135–145.

106 Kurzinski, L., Hank, H., László, I., Szöke, E. (2005) *Journal of Chromatography, A*, **1091**, 32–39.

107 Hank, H., Szöke, E., Tóth, K., László, I., Kursinszki, L. (2004) *Chromatographia*, **60**, S55–S59.

108 Kirchhoff, C., Bitar, Y., Ebel, S., Holzgrabe, U. (2004) *Journal of Chromatography, A*, **1046**, 115–120.

109 Wolfender, J. -L., Rodriguez, S., Hostettmann, K. (1998) *Journal of Chromatography, A*, **794**, 299–316.

110 Albert, K. (2002) *On-Line LC-NMR and Related Techniques*, John Wiley & Sons, Chichester, UK.

111 Wolfender, J. -L., Queiroz, E. F., Hostettmann, K. (2005) *Magnetic Resonance in Chemistry*, **43**, 697–709.

112 Zanolari, B., Wolfender, J. -L., Guilet, D., Marston, A., Queiroz, E. F., Paulo, M. Q., Hostettmann, K. (2003) *Journal of Chromatography, A*, **1020**, 75–89.

113 Bieri, S., Varesio, E., Veuthey, J. -L., Munoz, O., Tseng, L. -H., Braumann, U., Spraul, M., Christen, P. (2006) *Phytochemical Analysis*, **17**, 78–86.

114 Bowman, B. P., Vaughan, S. R., Walker, Q. D., Davis, S. L., Little, P. J., Scheffler, N. M., Thomas, B. F., Kuhn, C. M. (1999) *Journal of Pharmacology and Experimental Therapeutics*, **290**, 1316–1323.

115 Harrison, P. K., Tattersall, J. E. H., Gosden, E. (2006) *Naunyn Schmiedebergs Archives of Pharmacology*, **373**, 230–236.

116 Needham, S. R., Jeanville, P. M., Brown, P. R., Estape, E. S. (2000) *Journal of Chromatography, B*, **748**, 77–87.

117 Takahashi, M., Nagashima, M., Shigeoka, S., Nishijima, M., Kamata, K. (1997) *Journal of Chromatography, A*, **775**, 137–141.

118 Jamdar, S. C., Pantuck, C. B., Diaz, J., Mets, B. (2000) *Journal of Analytical Toxicology*, **24**, 438–441.

119 Stöckigt, J., Unger, M., Stöckigt, D., Belder, D. (1998) Analysis of alkaloids by capillary electrophoresis and capillary electrophoresis-electrospray mass spectrometry, In Pelletier, S. W. (Ed.) *Alkaloids: Chemical and Biological Perspectives*, Vol. 12, Elsevier, Oxford, UK.

120 Cherkaoui, S., Mateus, L., Christen, P., Veuthey, J. -L. (1997) *Journal of Chromatography, B*, **696**, 283–290.

121 Cherkaoui, S., Mateus, L., Christen, P., Veuthey, J. -L. (1998) *Journal of Pharmaceutical and Biomedical Analysis*, **17**, 1167–1176.

122 Eeva, M., Salo, J. -P., Oksman-Caldentey, K. -M. (1998) *Journal of Pharmaceutical and Biomedical Analysis*, **16**, 717–722.

123 Mateus, L., Cherkaoui, S., Christen, P., Veuthey, J. -L. (2000) *Journal of Chromatography, A*, **868**, 285–294.

124 Ye, N., Zhu, R., Gu, X., Zou, H. (2001) *Biomedical Chromatography*, **15**, 509–512.

125 Cherkaoui, S., Mateus, L., Christen, P., Veuthey, J. -L. (1997) *Chromatographia*, **46**, 351–357.

126 Mateus, L., Cherkaoui, S., Christen, P., Veuthey, J. -L. (1998) *Journal of Chromatography, A*, **829**, 317–325.

127 Mateus, L., Cherkaoui, S., Christen, P., Oksman-Caldentey, K. -M. (2000) *Phytochemistry*, **54**, 517–523.

128 Mateus, L., Cherkaoui, S., Christen, P., Veuthey, J. -L. (1999) *Current Topics in Phytochemistry*, **2**, 175–182.

129 Cherkaoui, S., Mateus, L., Christen, P., Veuthey, J. -L. (1999) *Chromatographia*, **49**, 54–60.

130 Cherkaoui, S., Mateus, L., Christen, P., Veuthey, J. -L. (1999) *Journal of Pharmaceutical and Biomedical Analysis*, **21**, 165–174.

131 Mateus, L., Cherkaoui, S., Christen, P., Veuthey, J. -L. (1999) *Electrophoresis*, **20**, 3402–3409.

132 Gao, Y., Tian, Y., Wang, E. (2005) *Analytica Chimica Acta*, **545**, 137–141.

133 Veuthey, J. -L. (2005) *Analytical and Bioanalytical Chemistry*, **381**, 93–95.

134 Boone, C. M., Waterval, J. C. M., Lingeman, H., Ensing, K., Underberg, W. J. M. (1999) *Journal of Pharmaceutical and Biomedical Analysis*, **20**, 831–863.

135 Petersen, J. R. and Mohammad, A. A. (2001) *Clinical and Forensic Applications of Capillary Electrophoresis*, Humana Press, Totowa, NJ.

136 Petersen, J. R., Okorodudu, A. O., Mohammad, A. A., Payne, D. A. (2003) *Clinica Chimica Acta*, **330**, 1–30.

137 Thormann, W. (2003) *Therapeutic Drug Monitoring*, **24**, 222–231.

138 Plaut, O. and Staub, C. (2002) *Chimia*, **56**, 96–100.

139 Plaut, O. and Staub, C. (1998) *Electrophoresis*, **19**, 3003–3007.

140 Tagliaro, F., Poiesi, C., Aiello, R., Dorizzi, R., Ghielmi, S., Marigo, M. (1993) *Journal of Chromatography*, **638**, 303–309.

141 Tagliaro, F., Manetto, G., Crivellente, F., Scarcella, D., Marigo, M. (1998) *Forensic Science International*, **92**, 259–268.

142 Tagliaro, F., Valentini, R., Manetto, G., Crivellente, F., Carli, G., Marigo, M. (2000) *Forensic Science International*, **107**, 121–128.

143 Chevigné, R. and Louis, P. (1997) US patent 5 611 903.

144 Boone, C. M., Jonkers, E. Z., Franke, J. P., de Zeeuw, R. A., Ensing, K. (2001) *Journal of Chromatography, A*, **927**, 203–210.

145 Alnajjar, A., Butcher, J. A., McCord, B. (2004) *Electrophoresis*, **25**, 1592–1600.

146 Takáts, Z., Wiseman, J. M., Gologan, B., Cooks, R. G. (2004) *Science*, **306**, 471–473.

147 Takáts, Z., Cotte-Rodriguez, J., Talaty, N., Chen, H., Cooks, R. G. (2005) *Chemical Communications*, 1950–1952.

148 Takáts, Z., Wiseman, J. M., Cooks, R. G. (2005) *Journal of Mass Spectrometry*, **40**, 1261–1275.

149 Leuthold, L. A., Mandscheff, J. -F., Fathi, M., Giroud, C., Augsburger, M., Varesio, E., Hopfgartner, G. (2006) *Rapid Communications in Mass Spectrometry*, **20**, 103–110.

150 Rodriguez-Cruz, S. E. (2006) *Rapid Communications in Mass Spectrometry*, **20**, 53–60.

151 Talaty, N., Takáts, Z., Cooks, R. G. (2005) *Analyst*, **130**, 1624–1633.

13
LC-MS of Alkaloids: Qualitative Profiling, Quantitative Analysis, and Structural Identification

Steven M. Colegate, Dale R. Gardner

13.1
Introduction

This chapter will assume a working knowledge of HPLC and mass spectrometry and the combination of both procedures in what is generically referred to as liquid chromatography–mass spectrometry (LC-MS). Since the mid-1990s, developments in LC-MS have resulted in an unprecedented availability of this analytical technology to researchers across a wide diversity of disciplines. Combined with improved and more efficient methods of extraction, the liquid chromatographic separation and mass spectrometric analysis of alkaloidal components of extracts has provided a frontline tool in the qualitative profiling and quantitative analysis of alkaloids. Analysis of mass spectral data, including tandem mass spectrometry in which selected ions are isolated and specifically fragmented in order to determine sequential relationships of ions, has been used to assist in the structural elucidation of unknown alkaloids. This latter application goes beyond simply obtaining low- and high-resolution mass spectrometric data for the molecular ion.

Whilst a review of the literature reveals an abundance of reports dealing with the LC-MS of alkaloids, this chapter will cover only some of those applications where they lend support to the aspects being described. In addition the chapter includes a more detailed description of research from the authors' laboratories on the extraction and LC-MS analysis of pyrrolizidine alkaloids and the alkaloids present in *Delphinium* spp. (larkspurs).

13.2
LC-MS Overview

There are a number of reviews of the application of various forms of LC-MS to the detection and identification and quantitation of specific analytes, whether they be natural products, synthetic drug leads, or metabolites. For example, Korfmacher [1] provides a good general introduction to LC-MS by describing its basic principles and how it facilitates new drug discovery. Prasain *et al.* [2] have reviewed mass

Modern Alkaloids: Structure, Isolation, Synthesis and Biology. Edited by E. Fattorusso and O. Taglialatela-Scafati

ISBN: 978-3-527-31521-5

spectrometric methods, including the ionization sources applicable to LC-MS, from the perspective of flavonoids in biological samples.

In its most basic form, although not strictly a liquid chromatographic technique, LC-MS involves simply the introduction of a solution of an analyte (or analytes) into the ion source of a mass spectrometer. This can be a usefully rapid analytical technique for the presence of known analytes (see Section 13.6.1). Involving both liquid chromatography and the mass spectrometric analysis of eluants, successful LC-MS of alkaloids depends heavily upon the optimization of both components. Whilst data management, deconvolution software can assist, the aim is to optimize the chromatographic resolution of components in an alkaloidal extract whilst optimizing the sensitivity of ion detection. Changes in the matrix used to effect separation, subtle changes in the constitution of the chromatographic mobile phases, and changes to various mass spectrometer parameters can all lead to more efficient and reliable LC-MS analysis. It is clear that the mobile phases used in the analyte separation procedure must be compatible with the mass spectrometric detection of the ions. It is the balanced effect of these optimization approaches that results in the final analytical method.

Derivatization of analytes and modification of the mobile phase can both contribute to markedly increased sensitivities for particular analytes [3]. It is probably the latter (mobile phase modification) that has the greater impact upon the LC-MS of alkaloids, since the need to derivatize to enhance ion formation is reduced by the presence of the ionizable nitrogen.

Tandem mass spectrometry is becoming the norm as the detector for HPLC effluents owing to its increased ability to provide structural information. In contrast to electron impact mass spectrometry, the atmospheric pressure modes of ionization (API) such as electrospray (ESI) and atmospheric pressure chemical ionization (APCI) result in strong molecular ion adduct peaks and very little molecular fragmentation. Techniques such as in-source collision-induced dissociation (CID) can help but the specificity offered by tandem MS experiments including MS/MS (MS^2), MS/MS/MS (MS^3) through to MS^n increases the value of these instruments to structure elucidation, qualitative profiling, and quantitative analysis. Shukla and Futrell [4] have described two basic configurations for tandem mass spectrometry. The "tandem-in-space" configuration relies upon multiple, sequential analyzers, such as the triple quadrupole and quadrupole-time-of-flight instruments whereas the "tandem-in-time" configuration refers to the selection of parent ions and their subsequent dissociation within the same analyzer, and is exemplified by the ion trap spectrometers that allow multiple tandem mass spectrometry (MS^n).

13.2.1 Optimization

13.2.1.1 Modification of Mobile Phases and Ionization Parameters

Providing rapid multiresidue analytical profiling is a first step toward effective quality control of herbal preparations. For example, Luo *et al.* [5] developed a method for the simultaneous analysis of protoberberine alkaloids, indolequinoline alkaloids, and quinolone alkaloids (Figure 13.1) extracted from the traditional Chinese medicinal

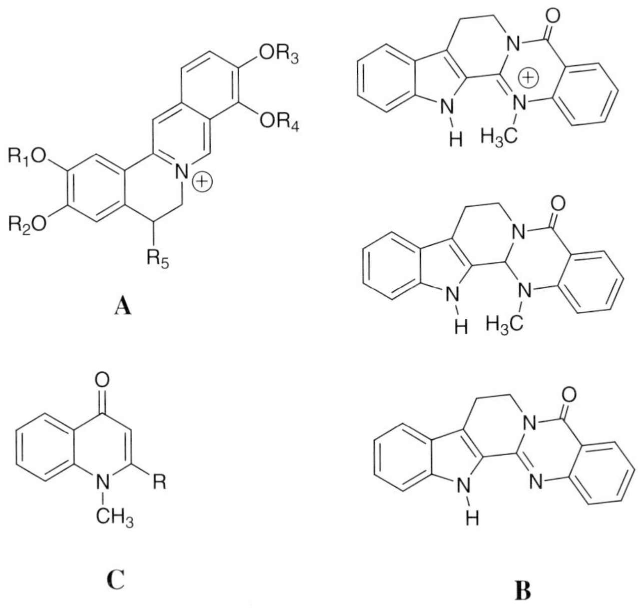

Fig. 13.1 Alkaloids extracted from the Chinese traditional medicinal herbs *Coptidis rhizoma* and *Evodiae fructus* and analyzed using C18 reversed phase HPLC-ESI-MS.
A: protoberberine alkaloids where $R_5 = H$ or OH and R_1/R_2 and R_3/R_4 are combinations of H, CH_3, or $-CH_2-$,
B: indolequinoline alkaloids, and **C**: quinolone alkaloids where R is one of various long-chain alkanes, alkenes, and dienes.

herbs *Coptidis rhizoma* and *Evodiae fructus* into 70% methanol in water (70% aqueous methanol). The filtered aqueous methanolic extract was applied to a C18 reversed phase (RP) chromatography column with the eluant directed into a diode array spectrophotometer and, subsequently, the ESI source of a quadrupole mass spectrometer. The concentration of an aqueous ammonium acetate and acetic acid buffer and its gradient mixing with various percentages of acetonitrile and methanol was optimized for the HPLC separation of the alkaloids. Once the tertiary mobile phase conditions had been optimized, the cone voltage of the ESI source was varied to maximize the intensity of the protonated molecular ion adducts (MH^+). The use of a tertiary mobile phase involved a gradient change of water, acetonitrile, and methanol and a consequent change in the ammonium acetate and acetic acid concentrations over the period of the gradient. There were no data to support the superiority of this tertiary gradient system over, for example, a binary system of water and acetonitrile both containing the ionic modifiers. However, the protoberberine alkaloids in *Coptidis chinensis* have been analyzed using a binary solvent system and HPLC-ESI-MS [6]. Similar to the previous example, the alkaloids were extracted from the plant samples into 75% aqueous methanol. Once again, the aqueous methanolic extracts were simply filtered and injected directly onto a RP C18 HPLC column

(250 mm × 4.6 mm, 5 μm particle size). Elution of the alkaloids was accomplished with a gradient flow of acetonitrile (with no additives) into water containing ammonium acetate (0.0034 mol/L) and acetic acid (2% v/v). Extensive definition of the MS/MS spectra for each of the standard alkaloids allowed unambiguous identification of those alkaloids in the extracts and also allowed a rational structural assignment of those for which authenticated standards were not available.

In a similar way, the use of formic acid/ammonium formate was shown to be superior to the use of acetic acid/ammonium acetate in the RP HPLC-ESI MS/MS analysis of a number of heterocyclic aromatic amines [7]. It is clear from numerous other examples in the literature that HPLC resolution and MS sensitivity can be dramatically influenced by the correct selection of mobile phases, organic modifiers, and ion-pairing additives.

13.2.1.2 **HPLC Versus UPLC**

The development of 1.7 μm silica-based columns in combination with solvent pumps capable of delivering the solutions at very high pressures (>10 000 psi) has enabled ultra-performance liquid chromatography (UPLC) [8]. Combined with the high-speed detection of eluants using a mass spectrometer, this offers potential improvements in the detection and routine analysis of alkaloids. A recent comparison of the HPLC and UPLC separations of four phenylamines from *Ephedra sinica*, directed the flows from an HPLC column (C18, 150 mm × 2 mm, 3 μm particle size) or a UPLC column (C18, 50 mm × 1 mm, 1.7 μm particle size) into the ESI source of a triple quadrupole mass spectrometer that was optimized for multiple reaction monitoring (MRM) analyte detection. The results with ephedrine (and its diastereomer pseudoephedrine), methylephidrine and *nor*-ephedrine clearly demonstrated the enhancement in overall performance (decreased retention times and increased sensitivity) resulting from UPLC-ESI-MS/MS [9].

13.2.1.3 **Fluorinated HPLC Solid Phases**

For the separation of alkaloids prior to MS analysis, perfluorinated stationary phases for HPLC columns can be a useful alternative to the reversed phase, alkyl-bonded silica-based stationary phases such as the C8 and C18 reversed phase columns [10].

Bell *et al.* [11] described the optimization of the separation of six ephedrine-related alkaloids, including diastereoisomers (Figure 13.2) on a pentafluorophenylpropyl-bonded silica column using high concentrations (85–90%) of organic phase (acetonitrile) in water as the mobile phase. The addition of ammonium acetate to the mobile phase and adjustment of the ambient temperature of the column were investigated for their effects on analyte retention, separation, and MS detection. Not only was the concentration of the ammonium acetate important for manipulating the resolution of the alkaloids, but, and in contrast to usual observations with C8 and C18 RP phases, the increase in ambient temperature of the column increased retention of the alkaloids. This is an important observation for the more polar alkaloids that usually require highly aqueous mobile phases to improve their retention on alkyl-bonded RP columns.

ephedrine : $R_1 = R_4 = H$, $R_2 = R_3 = CH_3$
methylephedrine : $R_1 = H$, $R_2 = R_3 = R_4 = CH_3$
norephedrine : $R_1 = R_4 = R_3 = H$, $R_2 = CH_3$
synephrine : $R_1 = OH$, $R_2 = R_3 = H$, $R_4 = CH_3$

pseudoephedrine : $R = CH_3$
norpseudoephedrine : $R = H$

Fig. 13.2 Ephedrine alkaloids separated using a pentafluorophenylpropyl-based HPLC column.

13.2.1.4 **Reduction of Ion Suppression**

Trifluoroacetic acid (TFA) has often been used as an additive in the HPLC of alkaloids since it improves peak shape by providing a controlled pH for the mobile phase and acting as an ion-pairing reagent. Owing to its volatility, TFA is also generally suitable as an acidic additive for LC-MS applications. However, it can suppress analyte ionization through the formation of ion pairs with the positively charged analyte ions in the gas phase.

One way of reducing this ion suppression effect is to infuse a solution of propionic acid in isopropanol into the postcolumn effluent [12]. This provides an alternative, excess source of ion-pairing agent for the TFA, thus freeing the analyte ions for subsequent MS analysis. A major disadvantage is the requirement for more hardware to effect the infusion, and the consequent dilution of the analyte concentrations. However, it has been shown [13] that reduction of TFA-related ion suppression can be achieved with the simple addition of a weak organic acid, such as acetic acid or propionic acid, directly to the TFA-containing mobile phases used then to elute analytes from the column. The addition of 0.5 % acetic acid or 1 % propionic acid to 0.025 % TFA in water and 0.025 % TFA in acetonitrile provided mobile phases that were effective in separating eight test alkaloids (Figure 13.3) and still allowing a two- to fivefold enhanced sensitivity compared to the absence of the acetic or propionic acids. The method was applied to the detection of sildenafil in human plasma. The plasma sample was applied to a properly conditioned bimodal (polar and strong cation exchange) solid phase extraction cartridge and washed with 5 % acetic acid in water followed by methanol. The alkaloids, in this case sildenafil, were eluted by washing the cartridge with 2 % ammonium hydroxide in acetonitrile. Subsequent evaporation of the elution solvent and reconstitution into 0.05 % TFA in acetonitrile provided the analytical sample ready for normal or reversed phase HPLC-ESI-MS/MS. It was particularly noted that the addition of formic acid, instead of acetic or propionic acids, to the TFA-containing mobile phase did not reduce the ion suppression but in fact

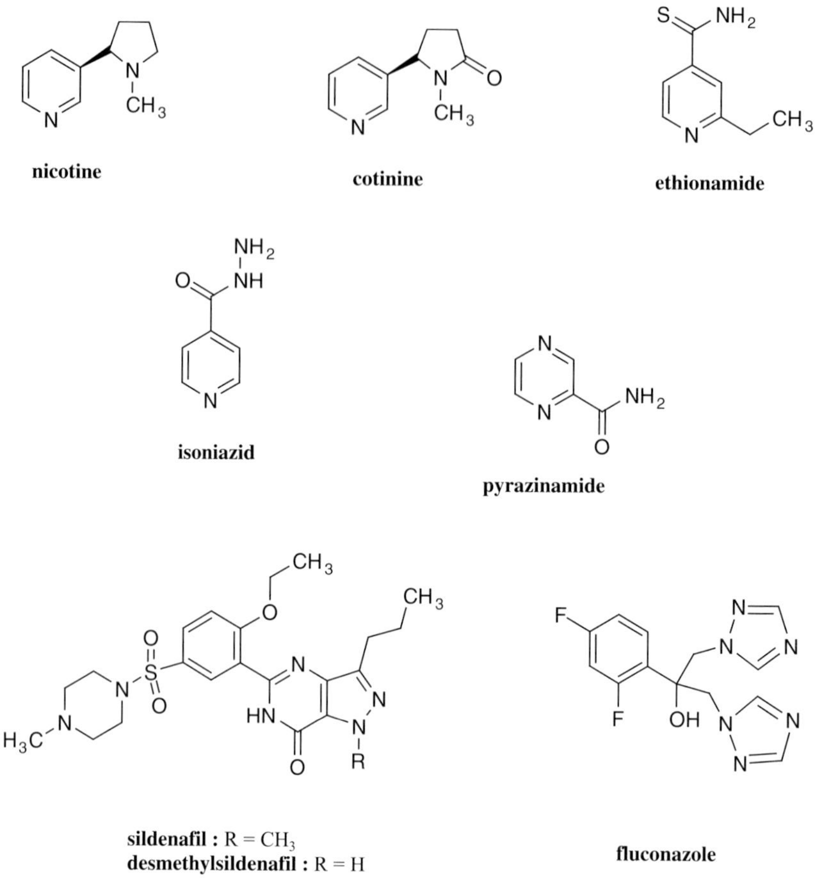

nicotine **cotinine** **ethionamide**

isoniazid **pyrazinamide**

sildenafil : R = CH_3
desmethylsildenafil : R = H

fluconazole

Fig. 13.3 Eight test alkaloids used to demonstrate the reduction of trifluoroacetic acid-related ion suppression.

enhanced it. However, in the application to the analysis of sildenafil in plasma it was shown that under reversed phase chromatography conditions, the use of mobile phases modified with formic acid alone were superior to the TFA and TFA/acetic acid modified mobile phases in terms of sensitivity, whilst retaining similar chromatography [13].

13.3
Clinical Chemistry and Forensic Applications

The use of alkaloids as therapeutic drugs, drugs of abuse and deliberate poisons has required the development of screening and confirmatory assays for these compounds. In addition, plant-based alkaloids have also required such assay development to allow or assist confirmation of a plant-related intoxication, including the overuse or misuse of traditional herbal remedies.

13.3.1 Extraction and Analytical Considerations

Maurer has reviewed the application of LC-MS and LC-MS/MS to the detection of alkaloids in human biofluids [14]. Extraction techniques include liquid–liquid extraction relying upon the ionization of alkaloids in aqueous acid, solid phase extraction (SPE) in which alkaloids are "cleaned up" and concentrated from the biomatrix by adsorption and subsequent elution from a small cartridge of solid phase adsorbent, and solid-phase microextraction (SPME), in which analytes are adsorbed directly from the matrix or the headspace above the heated matrix onto a fine fiber of adsorbent on fused silica. The latter process is more commonly used with GC-MS but is finding increasing use with LC-MS.

Maurer [14] highlights several aspects or considerations with respect to the application of LC-MS to the multianalyte screening of biological samples:

- Collision-induced dissociation (CID) or tandem MS techniques need to be developed to match the fragmentation information afforded by electron impact ionization fragmentation observed in GC-MS.
- Ion suppression effects, due to coeluting compounds or the mobile phase, need to be identified.
- The APCI mode may be preferable to ESI in terms of enhancing the sensitivity by reduction of ion suppression effects.
- Various factors require definition when using the LC-MS in a quantitative mode. These include stability testing of the analyte in the sample matrix, selectivity, sensitivity, or level of detection, the limit of quantitation, and recovery levels from the sample matrix.

13.3.2 Forensic Detection of Plant-derived Alkaloids

13.3.2.1 Plant-associated Intoxications

Gaillard and Pepin [15] have reviewed the application of LC-MS/MS to the multiresidue detection of alkaloids associated with those plants that have caused human intoxication. They developed two methods of C18 RP HPLC-ESI-MS analysis for the alkaloids, which were dependent of the pK_a values of the alkaloids. Thus, for those alkaloids with pK_a values between 6 and 9 the eluting mobile phase was adjusted to pH 8.2 with ammonium formate and ammonium hydroxide whereas in the case of alkaloids with pK_a values less than 6.5, the eluting mobile phase buffer was adjusted to pH 3 with ammonium formate and formic acid. To develop the method, the authors spiked blood samples with up to 14 different authenticated alkaloids. The whole blood was then treated with saturated ammonium chloride (pH 9.5) and the alkaloids extracted into chloroform–isopropanol. Subsequent standard acid/base alkaloid extraction provided the analytical samples. To achieve a multianalyte method, the MS conditions were varied throughout the chromatographic run, thereby providing nearer optimum conditions for individual alkaloids. Further confirmation of identity was achieved using a triple quadrupole mass spectrometer to obtain MS/MS data. In this latter instance, the collision energies were also tailored for individual alkaloids and programmed into the chromatographic run.

13.3.2.2 **Illicit Drug Use: Multiple Reaction Monitoring**

The determination of lysergic acid diethylamide (LSD), its isomer *iso*-LSD (which is a diastereomeric impurity indicative of illegally prepared LSD) and its major metabolite 2-*oxo*-3-hydroxyLSD using LC-MS/MS-based multiple reaction monitoring (MRM) in human blood illustrated the need for analyte separation prior to MS analysis, since LSD and *iso*-LSD resulted in the same fragmentation ions [16]. Thus, in a simple solvent–solvent extraction scheme, basified blood samples were extracted with butylacetate, which was subsequently collected by centrifugation and evaporated to dryness, and the residue reconstituted for LC-MS analysis. To improve the selectivity, and hence confidence of identification, of the analytical process, two MRMs for each compound were identified using the parent ion and two daughters.

13.3.2.3 **Quality Control of Herbal Preparations: APCI-MS**

The accidental or deliberate substitution of herbal plants with others that may be ineffective or toxic is a major issue facing the herbal medicine industry. One such example is the substitution of *Stephania tetrandra* with *Aristolochia fangchi*, the latter containing the nephrotoxic and carcinogenic nitrophenanthrene acid, aristocholic acid [17].

A reversed phase HPLC-APCI-MS/MS method was developed to analyze filtered methanolic extracts of plant samples for the presence of the tetrahydroisoquinoline alkaloids tetrandrine and fangchinoline (Figure 13.4). In this way it was clearly shown that 9 of 10 samples purported to be *S. tetrandra* were in fact the toxic substituent *Aristolochia fangchi* whilst the tenth sample was unidentified but also unrelated to *S. tetrandra* [17].

13.4 Metabolite Profiling and Structure Determination

Many alkaloids are used therapeutically, or are being developed for therapeutic use, for illness in humans. The specific and rapid analysis of such compounds and their metabolites in human plasma can be an important aspect of managing treatment regimes.

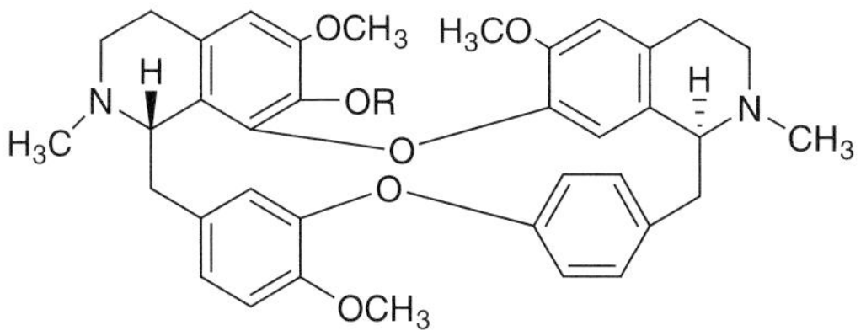

Tetrandrine R = CH_3
Fangchinoline R = H

Fig. 13.4 The tetrahydroisoquinoline alkaloids tetrandrine and fangchinoline extracted from *Stephania tetrandra* and analyzed using HPLC-APCI-MS/MS.

13.4.1
LC-MS/MS Approaches to the Identification/Structural Elucidation of Alkaloid Drug Metabolites

Even with the advent of powerful techniques such as HPLC-NMR [18], LC-MS techniques can offer significant insight into the structure of eluted compounds and can be used to search for compounds/metabolites that are very specifically related to the target compound.

Liu and Hop [19] have reviewed various LC-tandem MS strategies, with or without chemical modification or derivatization, which have been successfully applied in the identification and the rational, but tentative, structural determination of drug metabolites. These techniques are equally applicable to the analysis of known plant secondary metabolites or to the *de novo* structural identification of new compounds.

13.4.1.1 Tandem MS

The multiple MS stages provide a sequential mapping of associated fragments that can be used to help develop the structure of the analyte. MS/MS experiments can be performed with both "tandem-in-time" and "tandem-in-space" spectrometers [4] whereas the MS^n ($n > 2$) experiments can only be conveniently conducted with a "tandem-in-time" spectrometer. Other important tandem MS applications, that are, however, restricted to "tandem-in-space" instruments, include precursor ion and neutral loss experiments [19]. These two approaches are particularly valuable for actively searching for compounds that are closely related to the alkaloid (or other drug) of interest. In both cases, only those precursor ions that result in the generation of a common product or eliminate a common neutral product will be represented in the total ion chromatogram. This can help identify such compounds from a complex biological matrix (see examples in this Section). Tandem MS techniques are an invaluable tool to assist in the elucidation, or at least the tentative elucidation, of the structures of unknown alkaloids (Sections 13.5 and 13.6).

13.4.1.2 Accurate Mass Measurement

An important parameter to be determined for all novel alkaloids, the accurate high-resolution mass measurement is a more tolerant surrogate for combustion elemental analysis in the confirmation of proposed molecular formulae. It also has an important role in the structural determination of MS fragments, differentiating between molecular formula candidates that have the same integer (low-resolution) molecular weight. Confirming the molecular constituents of fragments then enables a more accurate piecing together of the fragments to deduce the structure of the parent compound.

For example, Liu and Hop [19] described the hepatocytic incubation of a drug candidate possessing a methoxy substituent (RCH_2OCH_3). The resultant observation of a compound isobaric (at low resolution) with the parent compound was resolved with high-resolution mass measurements determined using a quadrupole-time-of-flight mass spectrometer (Q-TOF). The accurate mass of the observed peak was 0.0362 amu lower than the parent compound and correlated to the formation of a carboxylic acid moiety (RCOOH) via demethylation and subsequent oxidation.

13.4.1.3 Chemical Modification

Microderivatization or chemical modification of analytical samples can be useful in enhancing the ion response, and hence the LC-MS sensitivity, for any particular analyte. In the case of alkaloids, which are already capable of ready ionization, chemical modification can be more valuable in elucidating structures by providing tandem MS evidence for positions of substitution in the parent molecule. Such derivatization or chemical modification for HPLC-MS can include H/D exchange by using deuterated mobile phases, Jones oxidation of aliphatic hydroxyls, selective acetylation of hydroxyl and amine groups, and *N*-oxide reduction [19,20].

To avoid the expensive use of deuterated solvents for H/D exchange experiments, Tolonen *et al.* [21] have described the postcolumn infusion of D_2O to facilitate the LC-MS detection and identification of labile protons in a column eluant. Whilst acknowledging the potential limitations with respect to a reduced level of exchange, and hence sensitivity, compared to the use of deuterated mobile-phase solvents, they optimized the column effluent flow rate (via a splitting connector) with the infused D_2O flow rate to enable the very useful determination of up to four labile protons. The method was exemplified by the differentiation of hydroxylated metabolites of the alkaloidal drugs imipramine and omeprazole (Figure 13.5) from the *N*-oxide and sulfone metabolites, respectively [21]. This was a differentiation that could not be achieved by high-resolution mass measurements.

13.4.2 Minimization of Sample Treatment

The piperidyl alkaloid propiverine is an anticholinergic drug that causes relaxation of the smooth muscle of the urinary bladder. Treatment of plasma samples with acetonitrile, in a ratio of 1 : 2 (plasma : acetonitrile) resulted in precipitation of proteins. Subsequent centrifugation provided a clear aqueous acetonitrile super-

omeprazole

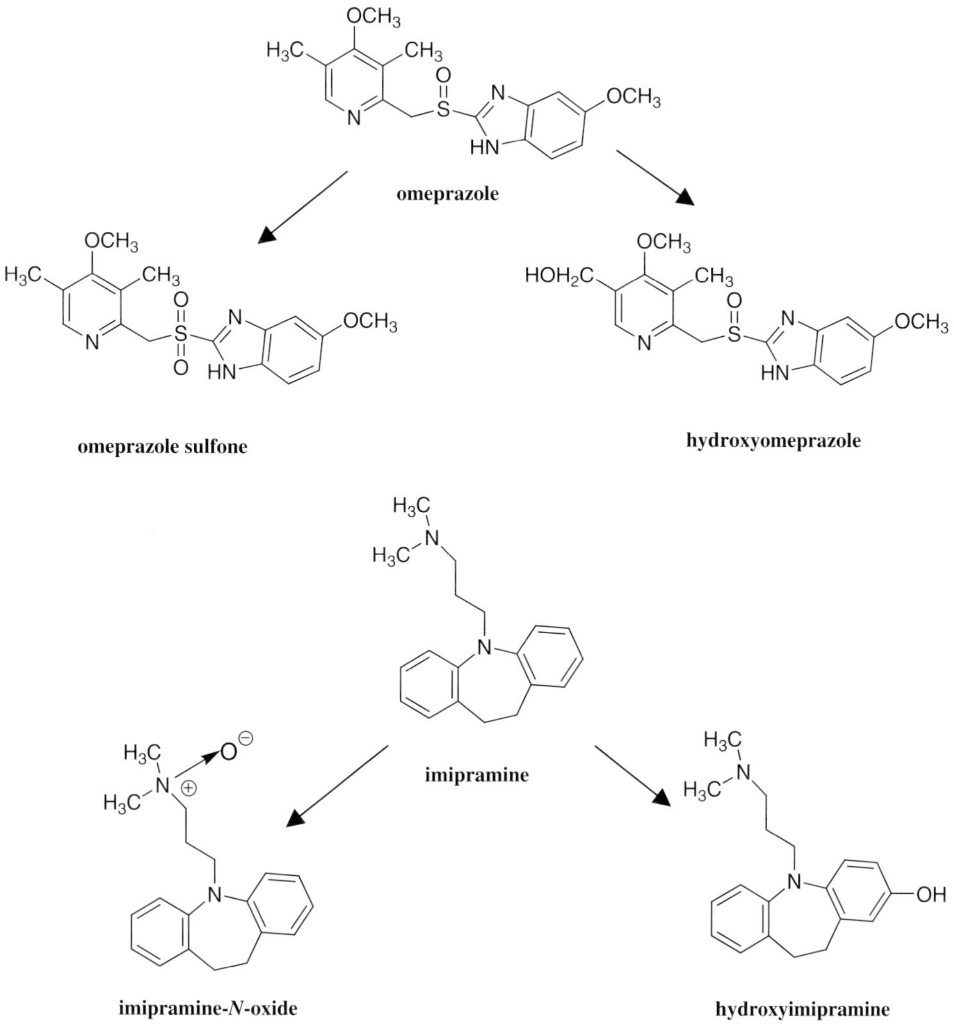

omeprazole sulfone

hydroxyomeprazole

imipramine

imipramine-*N*-oxide

hydroxyimipramine

Fig. 13.5 Hydrogen/deuterium exchange can be used to differentiate hydroxylated metabolites of omeprazole and imiprimine from their isobaric sulfone and *N*-oxide metabolites, respectively.

natant that was analyzed directly using HPLC-ESI-MS [22]. This approach provided obvious advantages in terms of decreased sample manipulation (as compared to a solvent extraction approach) and the consequent time required to complete the assay. The deproteinized aqueous acetonitrile solution was injected directly on to a silica-based C8 RP column and the sample components were subsequently isocratically eluted using water and acetonitrile (50 : 50) containing 0.2 % formic acid. The use of acetonitrile rather than methanol, and the concentrations of acetonitrile and formic acid in the water, were all variable factors that were assessed in the optimization of the

HPLC-MS analysis of propiverine and its primary metabolite, propiverine-*N*-oxide. To further enhance the specificity and sensitivity of the analytical procedure for propiverine and its *N*-oxide, an MS/MS method was developed using a triple quadrupole mass spectrometer. Thus, the MH^+ ions at m/z 368 and 384 for propiverine and its *N*-oxide respectively were isolated and specifically fragmented. Identification of the daughter ions allowed multiple reaction monitoring (MRM), which tolerated a less-rigorous sample preparation and a faster elution time owing to the higher specificity.

13.4.3
Structure Determination

NMR experiments (1D and 2D, homonuclear and heteronuclear) are the preeminent techniques for the determination of molecular structures. However, careful application and analysis of mass spectral data can provide sufficient information to postulate tentative structures. In this respect, the application of tandem MS experiments, sometimes in conjunction with selective derivatization of the unknown compound, can be very informative about the structure. The high-resolution mass spectral data are critical to the support of NMR-deduced structures by providing molecular formulae for unknowns.

13.4.3.1 Nudicaulins from *Papaver nudicaule*: High-resolution MS

Nudicaulin was a generic name given to the major pigment(s) in *Papaver nudicaule* (Iceland poppy) [23] but the structures have only recently been elucidated using NMR and high-resolution MS techniques [24].

Lyophilized plant material was extracted with aqueous methanol to provide an extract that was partitioned between hexane and water. The aqueous fraction contained the pigments, including the nudicaulins, which were separated using preparative reversed phase HPLC. Using ESI-Fourier transform ion cyclotron resonance mass spectrometry (ESI-FT-ICR-MS), the high-resolution mass measurements of the parent molecular ion adducts and some of their major MS fragments were used to structurally relate the eight isolated nudicaulins as glycosidic pentacyclic indole alkaloids (Figure 13.6).

13.4.3.2 Endophyte Alkaloids: An MS Fragment Marker

The intimate relationship of endophytes with plants can raise questions about the validity of assumptions that a secondary metabolite is a true phytochemical rather than biosynthesized by the endophyte(s) associated with the plant.

It has also been proposed that the presence of endophytes can alter the usual suite of secondary metabolites of plants. For example, whilst *Murraya* spp. have been reported to produce indole and carbazole alkaloids, the Brazilian *M. paniculata* did not [25]. An endophytic *Eupenicillium* sp. was isolated from surface-sterilized leaf material of the Brazilian *M. paniculata* and subsequently cultured on white corn. The *Eupenicillium* sp. produced hydrophobic spiroquinazoline alkaloids that were separated using silica gel column chromatography and preparative gel-filtration

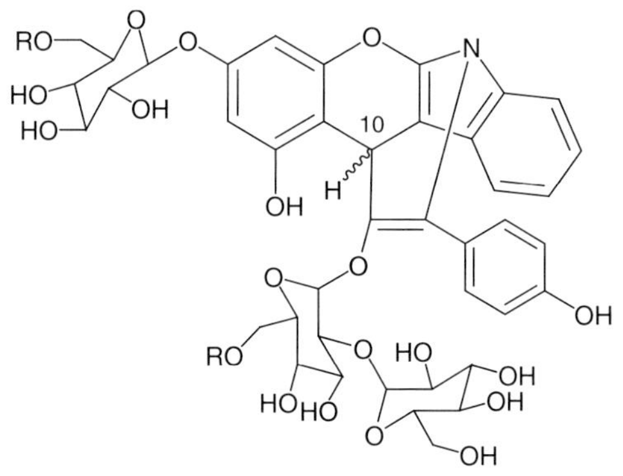

R = H, H : **nudicaulins I and II**
R = H, HO_2CCH_2CO : **nudicaulins III - VI**
R = HO_2CCH_2CO, HO_2CCH_2CO : **nudicaulins VII - VIII**

Fig. 13.6 Nudicaulin pigments isolated from *Papaver nudicaule* and analyzed using high-resolution ESI-Fourier transform ion cyclotron mass spectrometry.

HPLC. The structures were determined mainly by NMR investigations but a common APCI-MS fragment (*m*/*z* 226, Figure 13.7) was observed that could readily be used in tandem MS experiments to screen extracts for structurally related HPLC eluants.

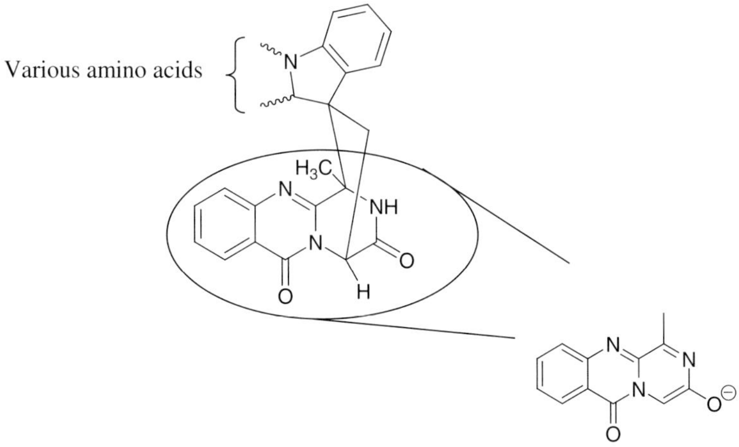

Various amino acids

m/z 226

Fig. 13.7 The common fragment ion indicative of the spiroquinazoline alkaloids isolated from a *Murraya paniculata* endophyte, *Eupenicillium* spp.

In a similar utilization of MS markers, the total ion chromatograms derived for extracts of *Delphinium* species were scanned for the presence of toxic alkaloids (Section 13.6.1).

13.5
Pyrrolizidine Alkaloids and their *N*-Oxides

Pyrrolizidine alkaloids or, more specifically, the mono or diesters of 1-hydroxymethyl-7-hydroxy-1,2-dehydropyrrolizine (Figure 13.8) are a diverse class of naturally occurring alkaloids.

Over 350 of this class of pyrrolizidine alkaloids have been identified in more than 6000 plant species belonging to the Boraginaceae, Leguminoseae, and Asteraceae families [26]. These alkaloids, which lead to metabolites that are hepatotoxic and can be pneumotoxic, genotoxic, and carcinogenic, occur as natural components of many herbal preparations, cooking spices, and honey, and can contaminate other food crops and animal-derived food destined for the human food supply [27]. Consequently, regulations governing human exposure to these toxic pyrrolizidine alkaloids have been instigated by several countries [28].

Whilst the N-oxidation of pyrrolizidine alkaloids is a recognized detoxifying mechanism in mammals [26], it has also been shown that ingested *N*-oxides are indeed toxic, presumably via *in vivo* reduction to the parent pyrrolizidine alkaloids

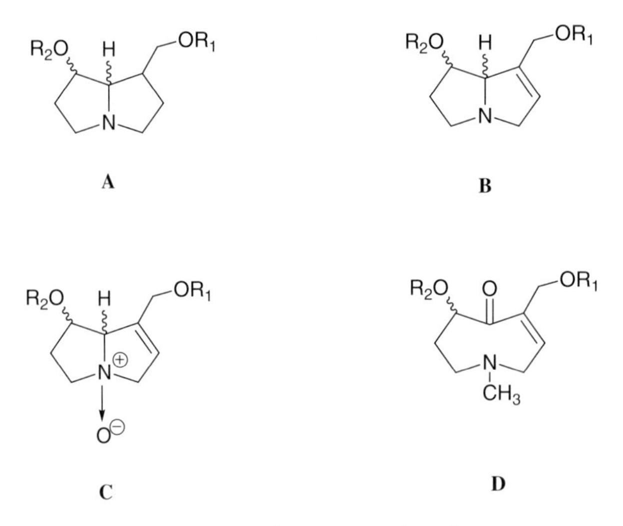

Fig. 13.8 General structures of some pyrrolizidine alkaloids. **A**: Pyrrolizidine alkaloid esters, **B**: 1,2-dehydro pyrrolizidine alkaloid esters, **C**: the *N*-oxide of 1,2-dehydro pyrrolizidine alkaloid esters, and **D**: an otonecine-based alkaloid. R_1 and R_2 can be H or esters of carboxylic acids that may or may not be cyclized forming a macrocyclic diester.

and subsequent hepatic oxidation to the toxic "pyrrolic" metabolites [29]. Since the pyrrolizidine alkaloids are biosynthesized in plants as their *N*-oxides [30], analysis of the *N*-oxide content is an important goal.

Gas chromatography coupled to mass spectrometry is a widely used analytical method [31] but does require prior reduction of the involatile *N*-oxides. Additionally, there are problems associated with some alkaloids that require prederivatization to enhance the GC characteristics. These requirements for prederivatization can adversely affect the GC-MS interpretation, especially at trace levels of alkaloids [32]. Combined SPE-LC-MS approaches have provided methods for qualitative profiling and quantitative analysis of pyrrolizidine alkaloids and their *N*-oxides.

13.5.1 Solid Phase Extraction

Extraction of the pyrrolizidine alkaloids and their *N*-oxides from plant sources is a critical first step in LC-MS analysis. Some methods of extraction already described (Section 13.2) consist simply of treatment with aqueous methanol and, following microfiltration (0.45 μm), immediate LC-MS analysis with no further treatment. However, the aqueous methanol treatment will extract more than just the alkaloids. This can result in a more complicated chromatogram, which may become a hindrance unless specific analytes are being searched for in the LC-MS total ion chromatogram. Insertion of a "clean-up" stage such as cation-exchange chromatography can simplify interpretation of the LC-MS ion chromatogram (Figure 13.9). The availability of solid phase extraction (SPE) cartridges facilitates the clean-up process allowing for high throughput of samples. The successful application of SPE to the capture and subsequent release of pyrrolizidine alkaloids and their *N*-oxides from aqueous acidic solutions depends upon the selection of the adsorbent phase [33]. Utilization of a strong cation-exchange (SCX), silica-based resin provides for both efficient capture of the pyrrolizidine alkaloids and their *N*-oxides and their subsequent elution using ammoniated methanol. The elution in a low volume of ammoniated methanol allows for facile evaporation of the solvent and subsequent reconstitution of the residue into methanol in preparation for LC-MS analysis. This is in contrast to the use of a polystyrene-based SCX resin that efficiently captures the pyrrolizidine alkaloids and *N*-oxides but requires the use of large volumes of strong aqueous acid for elution. This latter process then still requires reduction of the *N*-oxides and a traditional acid/base solvent extraction of the pyrrolizidine alkaloids for subsequent concentration and LC-MS analysis [34].

13.5.2 Qualitative Profiling

Plant samples (e.g. leaves, flowers, pollen) are first extracted into methanol or dilute aqueous sulfuric acid. In the case of the methanol extracts, the solvent is evaporated and the residue re-extracted with dilute aqueous sulfuric acid. The aqueous acid

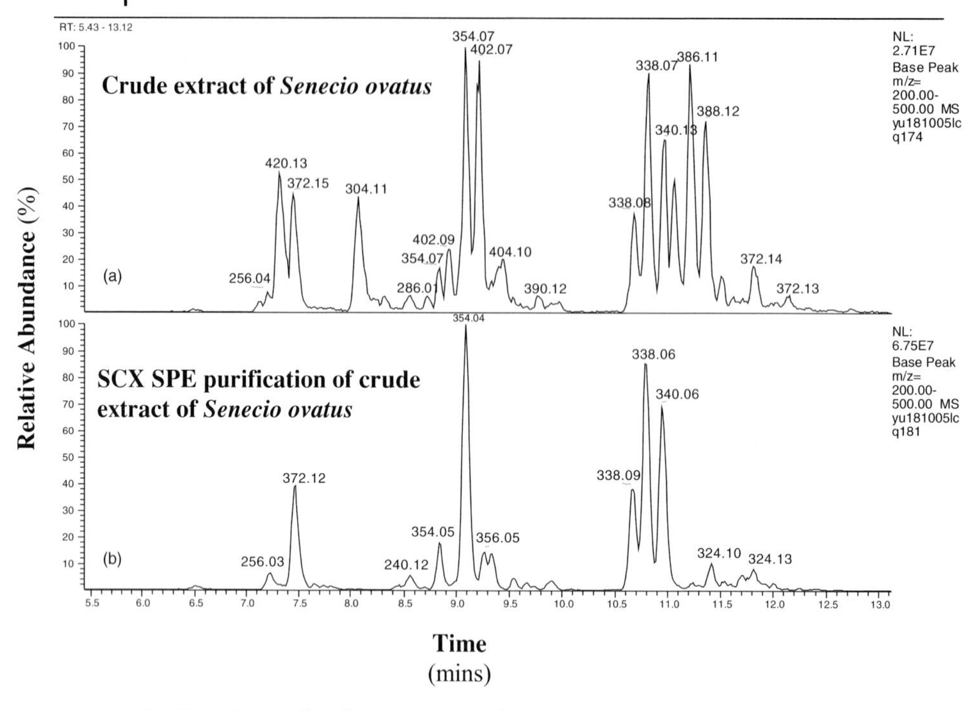

Fig. 13.9 The benefits of strong cation exchange (SCX), solid phase extraction (SPE) "clean-up" of samples. The reversed phase HPLC-ESI-MS base ion (*m*/*z* 200–500) chromatograms of a reduced (zinc/sulfuric acid) extract of *Senecio ovatus*. (a) Methanolic solubles of the reduced extract and, (b) the SCX SPE of the methanolic solubles of the reduced extract.

extracts, or aliquots of them, are then applied to the SCX SPE cartridges for capture of the pyrrolizidine alkaloids and their *N*-oxides. The combination of HPLC elution times and the mass spectral data, including the MS/MS spectra where the molecular ion adducts are trapped and selectively fragmented, helps identify individual alkaloids in cases where authenticated standards are available for comparison.

Initial investigations have revealed higher than previously reported (based upon differential quantitative analysis of the naturally occurring pyrrolizidine alkaloids in a sample and the total levels of pyrrolizidine alkaloids following a zinc/sulfuric acid reduction of the *N*-oxides present) of levels of *N*-oxides relative to their parent tertiary bases. During the profiling of the pyrrolizidine alkaloid and *N*-oxide content of plant-derived samples, several apparently undescribed alkaloids were identified and assigned tentative structures based upon their MS and MS^n data [33,35]. Since the new alkaloids were isolated from the plant samples in concert with several known alkaloids, a basic assumption of biogenetic comparability was made in assigning tentative structures. This was an important concession to the usual rigor of structural elucidation since much of the diversity in pyrrolizidine alkaloids involves isomeric changes not easily discernible using mass spectrometry.

Under the ESI-MS conditions used, the *N*-oxide character of eluted peaks was indicated by the appearance of a significant (5–10 % Relative Abundance) dimeric molecular ion adduct ($[2M + H]^+$) that was absent, or very low abundance, in the ESI mass spectrum of the parent tertiary pyrrolizidine alkaloids (Figure 13.10). The indicated *N*-oxide character was then confirmed by mixing a methanolic solution of the extracts with indigocarmine-based redox resin for about 2–4 h at 37 °C and then reanalyzing the solution using HPLC-ESI-MS. The formation of new peaks, usually eluting slightly earlier than the putative *N*-oxides, with molecular ion adducts 16 mu less than the *N*-oxides and no evidence of dimeric molecular ion adducts, strongly supported the pyrrolizidine alkaloid/*N*-oxide relationship (Figure 13.11).

All of the following examples involved HPLC separation of the alkaloids on a reversed phase (C18 or C8) column (150 × 2.1 mm i.d.) that was tolerant of high aqueous phases. The alkaloids were eluted using a combination of isocratic flow and gradient flow (200 μL/min) of 0.1 % formic acid in water (mobile phase A) and 0.1 % formic acid in acetonitrile (mobile phase B). The first isocratic stage (5 min) consisted of 93 % A and 7 % B. Then, over 15 min, the proportion of A : B was altered in a linear fashion to 30 : 70.

13.5.2.1 *Echium plantagineum* and *Echium vulgare*

Both *Echium plantagineum* and *E. vulgare* are of European origin but have become opportunistic weeds in other parts of the world. In particular, *E. plantagineum* is a major agricultural toxic weed in Australia whilst *E. vulgare* has infested large parts of New Zealand. Both have implications for livestock health, welfare, and productivity as well as human health implications via the presence of their alkaloids in honey and other food products [26–28].

The HPLC-ESI-ion trap MS base ion chromatograms for extracts of *E. plantagineum* and *E. vulgare* are shown in Figures 13.11 and 13.12. The profiles for both plants are very similar but with some obvious additional peaks in the chromatogram of the *E. vulgare* extract. Some of the peaks were identified as being the *N*-oxides of pyrrolizidine alkaloids previously isolated from these plants but a number of peaks were apparently undescribed alkaloids. Careful examination of the mass spectral data provided confident but tentative structural identifications [33,35]. Thus, a suite of new alkaloids was tentatively identified in *E. vulgare* in which the major *E. plantagineum* alkaloids were further esterified with angelic acid (or one of its configurational isomers) on the C-9 esterifying acid (Figure 13.12). These alkaloids were also identified in samples of honey produced from *E. plantagineum* and *E. vulgare*.

The selective application of MS/MS and MS/MS/MS experiments has been used to assist in the tentative identification of structures. For example, the ions with *m/z* 456, corresponding to acetylechimidine-*N*-oxide or acetylvulgarine-*N*-oxide (Figure 13.12) have been selected from the total ion spectrum derived from directly infusing an extract of *E. vulgare* into the ESI source (Figure 13.13). Without a chromatographic

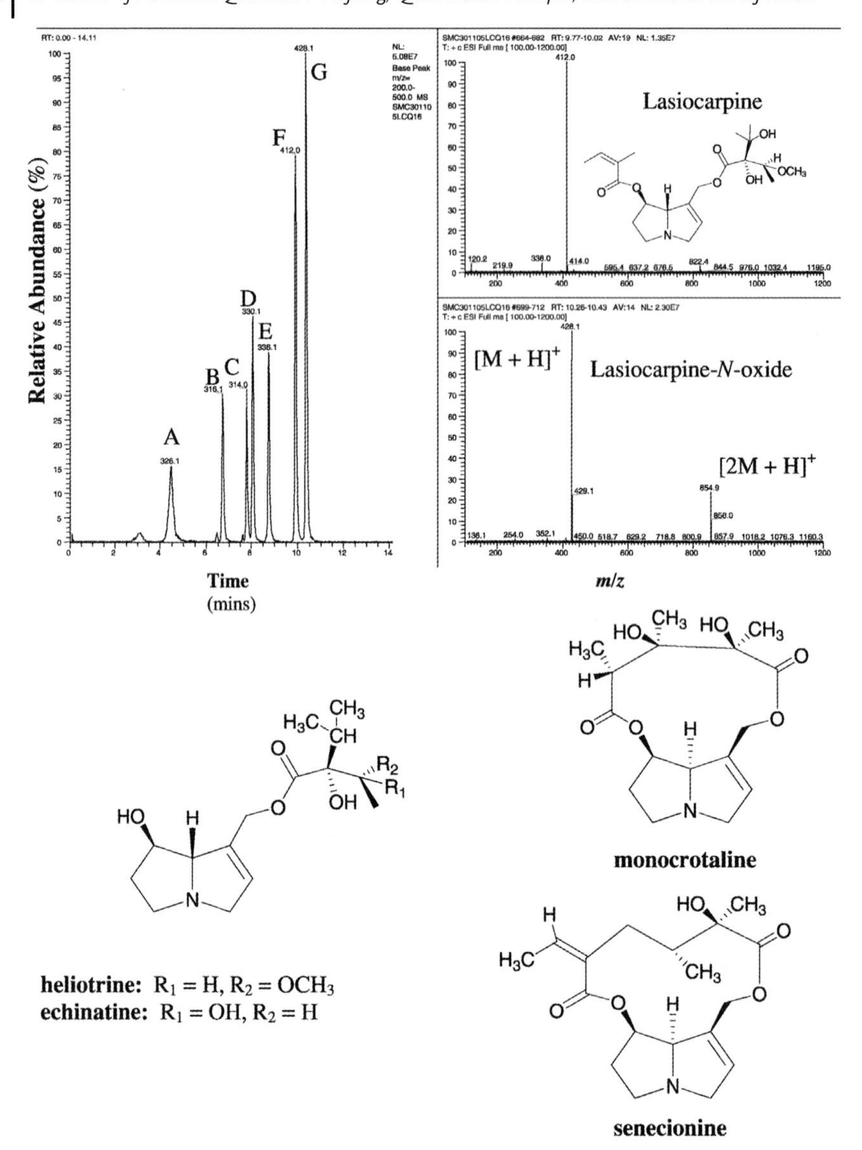

Fig. 13.10 The reversed phase HPLC-ESI-MS base ion (*m*/*z* 200–500) chromatogram of some authenticated standard pyrrolizidine alkaloids and *N*-oxides. **A**: Monocrotaline, **B**: echinatine-*N*-oxide, **C**: heliotrine, **D**: heliotrine-*N*-oxide, **E**: senecionine, **F**: lasiocarpine, and **G**: lasiocarpine-*N*-oxide. The mass spectra for lasiocarpine and lasiocarpine-*N*-oxide show the clear enhancement of the molecular ion dimer adduct.

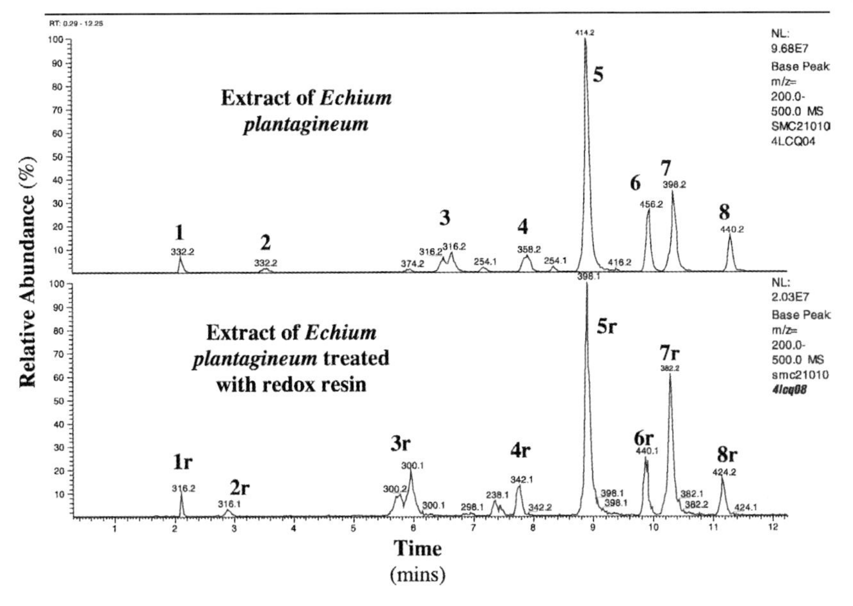

Fig. 13.11 Confirmation of *N*-oxide character. The reversed phase HPLC-ESI-MS base ion (*m*/*z* 200–500) chromatogram of an extract of *Echium plantagineum* petals shows the predominant presence of *N*-oxides (peaks **1**–**8**). Treatment of the same sample with an indigocarmine-based redox resin resulted in "mirrored" peaks (**1r**–**8r**) with $[M+H]^+$ 16 mass units less and eluting slightly earlier than their respective *N*-oxide peaks.

stage, there is no indication from Figure 13.13a whether the ions observed represent one or more isobaric compounds. However, careful analysis of the tandem MS data can identify multiple components or yield valuable structural information. For example, the MS/MS spectrum of *m*/*z* 456 (Figure 13.13b) shows five main ions (*m*/*z* 438, 396, 378, 352, 338, and 254). By then conducting MS/MS/MS experiments (Figure 13.13c–f) it can be deduced that the ion peak at *m*/*z* 456 represents two compounds and, further, that derivatization appears to occur at either of two positions on the C-9 esterifying acid [35].

13.5.2.2 *Senecio ovatus* and *Senecio jacobaea*

Senecio jacobaea, unlike *S. ovatus*, has been extensively studied around the world owing to its weedlike propensity and the toxic effect on livestock. The HPLC-ESI-ion trap MS profiles for extracts of both plants (Figure 13.14) showed the presence of pyrrolizidine-*N*-oxides (Colegate *et al.*, unpublished).

The *S. jacobaea* profiled (Figure 13.14) was evidently an erucifoline chemotype, in contrast to a jacobine chemotype, as described by Witte *et al.* [36] and Macel *et al.* [37] and as indicated by the mass spectral data of the major peak, which were consistent with erucifoline-*N*-oxide (MH^+, *m*/*z* 366; $[2M+H]^+$, *m*/*z* 731) (peak a, Figure 13.14). The *N*-oxides of seneciphylline (peak d, MH^+, *m*/*z* 350) and senecionine (peak e, MH^+, *m*/*z* 352) were readily identified by their mass spectra and comparison with

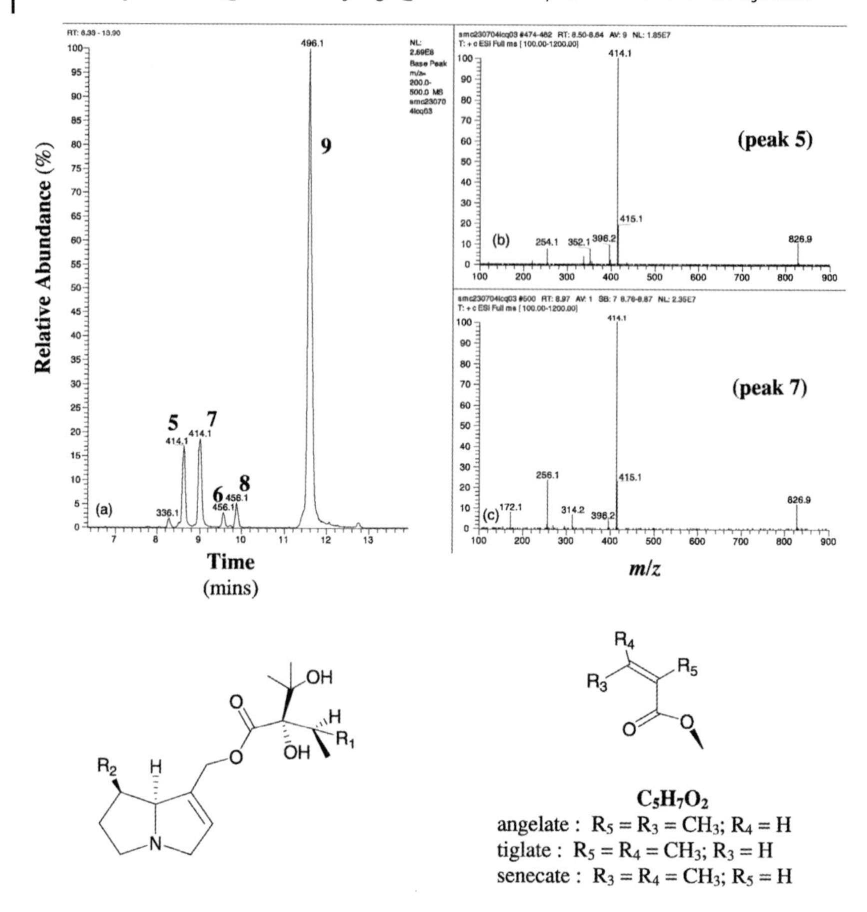

echimidine : $R_1 = OH$; $R_2 =$ angelate (peak 5)
acetylechimidine : $R_1 =$ acetate; $R_2 =$ angelate acid (peak 6)
echivulgarine *: $R_1 = C_5H_7O_2$; $R_2 =$ angelate (peak 9)
acetylvulgarine *: $R_1 = C_5H_7O_2$; $R_2 =$ acetate (peak 8)
vulgarine *: $R_1 = C_5H_7O_2$; $\mathbf{R_2} = OH$ (peak 7)

Fig. 13.12 A portion of the reversed phase HPLC-ESI-MS base ion (m/z 200–500) chromatogram for an extract of *Echium vulgare* pollen showing (a) the presence of a new suite of alkaloids (peaks 7, 8, and 9, structures below marked with an*) in addition to those alkaloids also found in *E. plantagineum* (peaks 5 and 6), (b) the mass spectrum for peak 5, common to both *E. vulgare* and *E. plantagineum* and shown to be echimidine-*N*-oxide, and (c) the mass spectrum for peak 7, a new alkaloid found in *E. vulgare*. The differences in fragmentation for peaks 5 and 7 are clear. Similar differences were observed for peaks 6 and 8.

authenticated standards. The other peaks have not been unequivocally identified but all have mass spectra indicative of pyrrolizidine-*N*-oxides (Colegate *et al.*, unpublished). For example, peaks b and c, both with an MH^+ at m/z 368 could correspond to the known alkaloids eruciflorine and any one of the hydroxylated senecionine-type

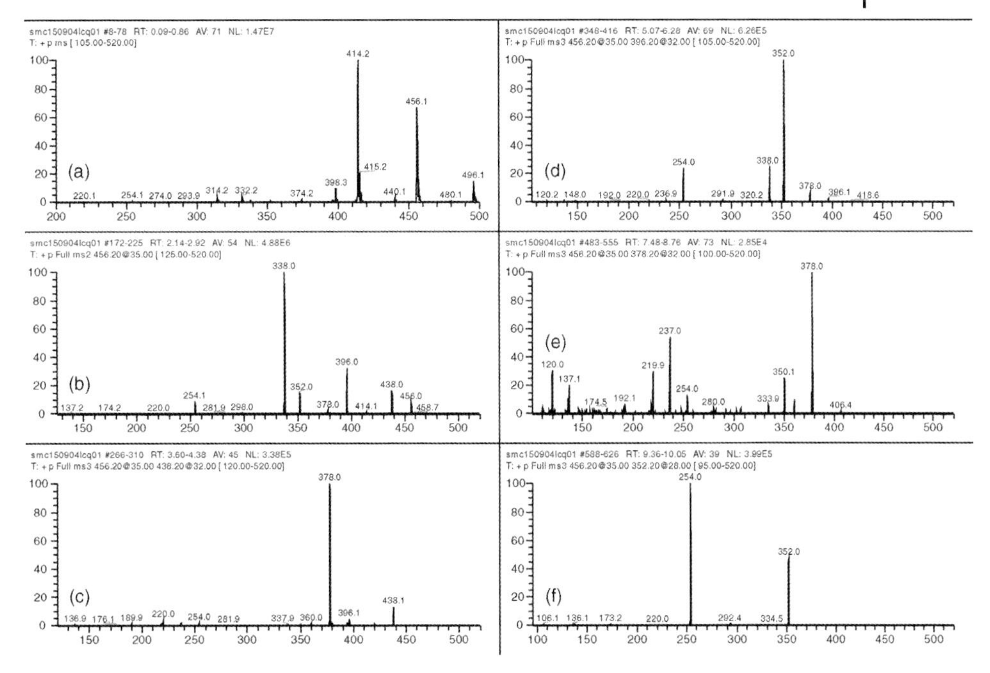

Fig. 13.13 Application of tandem ESI-MS experiments to structural elucidation of pyrrolizidine alkaloids. (a) The direct infusion ESI-MS of an extract of *Echium vulgare*, (b) the MS/MS of *m/z* 456, (c) MS/MS/MS *m/z* 456 → 438, (d) MS/MS/MS *m/z* 456 → 396, (e) MS/MS/MS *m/z* 456 → 378, and (f) MS/MS/MS *m/z* 456 → 352.

alkaloids. The ESI-MS and MS/MS data for the latest eluting peak (peak f, MH^+, *m/z* 408) strongly supported an acetylated derivative of erucifoline-*N*-oxide (peak a).

In contrast to *S. jacobaea*, very little research has been reported on *S. ovatus* (otherwise referred to as *S. fuchsii* or *S. nemorensis*). Consistent with the literature reports of pyrrolizidine alkaloids from *S. fuchsii* [38], the HPLC-ESI-MS and MS/MS analysis (Colegate *et al.*, unpublished) clearly identified the presence of 1,2-dehydropyrrolizidine alkaloids (fragment ions at *m/z* 120, 138, 220) and their 1,2 saturated pyrrolizidine analogs (fragment ions at *m/z* 122, 140, 222) (Figure 13.8). The MS/MS data were critical to the tentative identification of the minor peaks as pyrrolizidine alkaloids. Thus, MS/MS data for some low abundance peaks (Figure 13.14) were consistent with simple monoesters of platynecine and retronecine. In particular it appears from the MS/MS data that a series of a 1,2-unsaturated, a 1,2-dihydro, and a 1*H*-2-hydroxy monoesterified pyrrolizidine-*N*-oxide is present where the esterifying acid is angelic acid or one of its configurational isomers.

Careful analysis of MS and the MS/MS data for the major components of *S. ovatus* identified sarracine-related alkaloids. For example, the molecular ion data for peak 7 (Figure 13.14) (MH^+ *m/z* 354) potentially corresponded to the *N*-oxides of several 1,2-saturated, macrocyclic or open chain diester pyrrolizidine alkaloids previously

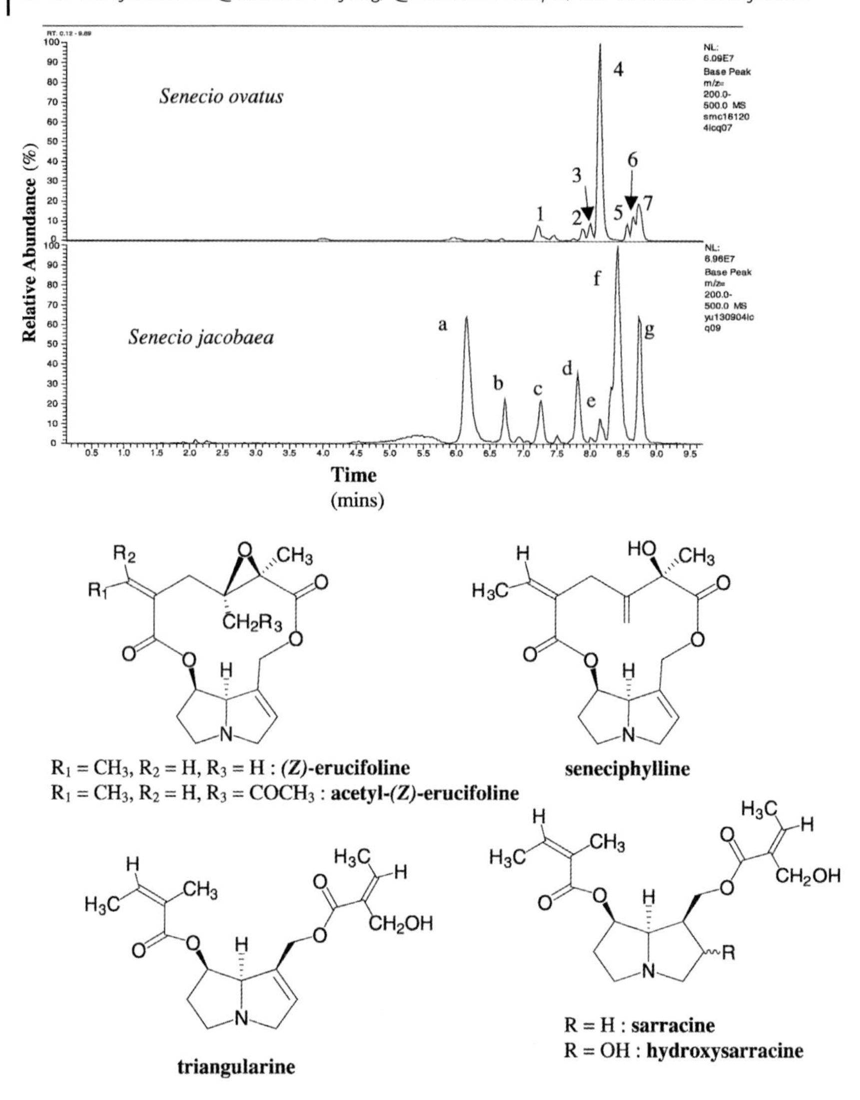

Fig. 13.14 The reversed phase HPLC-ESI-MS base ion (*m*/*z* 200–500) chromatograms for, and structures of some pyrrolizidine alkaloids identified in extracts of *Senecio jacobaea* and *S. ovatus*. The peak labels are referred to in the text.

isolated from *S. fuchsii* or *S. nemorensis* [38]. However, its identity as sarracine-*N*-oxide was indicated by coelution and MS/MS data comparison with an authenticated standard. In particular, the losses of 82 and 100 mu indicate an angelic acid substituent whilst the losses of 98 and 116 mu indicate a sarracinic acid substituent. In support of this assignment, the MS/MS data for peak 7 (and authentic sarracine-*N*-oxide) do not include a loss of 28 mu from the molecular ion adduct. This helps

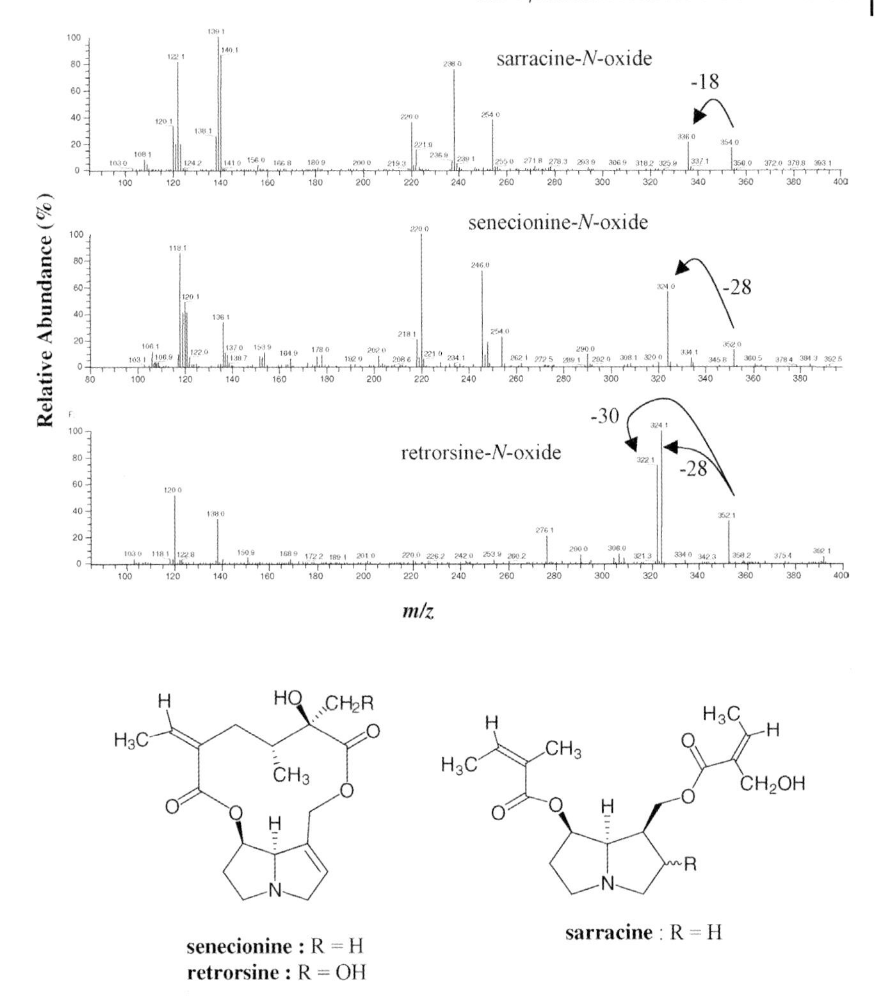

Fig. 13.15 HPLC-ESI-ion trap MS/MS spectra showing the diagnostic appearance of $[M + H]^+ - 28$ and 30 for pyrrolizidine alkaloid macrocyclic diesters.

differentiate sarracine-*N*-oxide from the isobaric macrocyclic diester isomers that can be expected to show $MH^+ - 28$ ions (Figure 13.15). The MS data for peak 6 corresponded to triangularine-*N*-oxide, the 1,2-unsaturated analog of sarracine-*N*-oxide. The major alkaloidal component of the *S. ovatus* (peak 4, Figure 13.14) was tentatively identified as the novel 2-hydroxysarracine (Figure 13.14) based upon the MS and MS/MS data. The MS/MS data for peak 4 were similar to data for peaks 6 and 7 and included losses of 82/100 and 98/116 mu that indicated an open chain diester with angelic and sarracinic acids, respectively. Unlike the rationale applied to EI-MS [39,40], the ESI-MS data cannot unambiguously differentiate the positions of esterification since loss of sarracinic acid, or angelic acid, from either the C-9 or

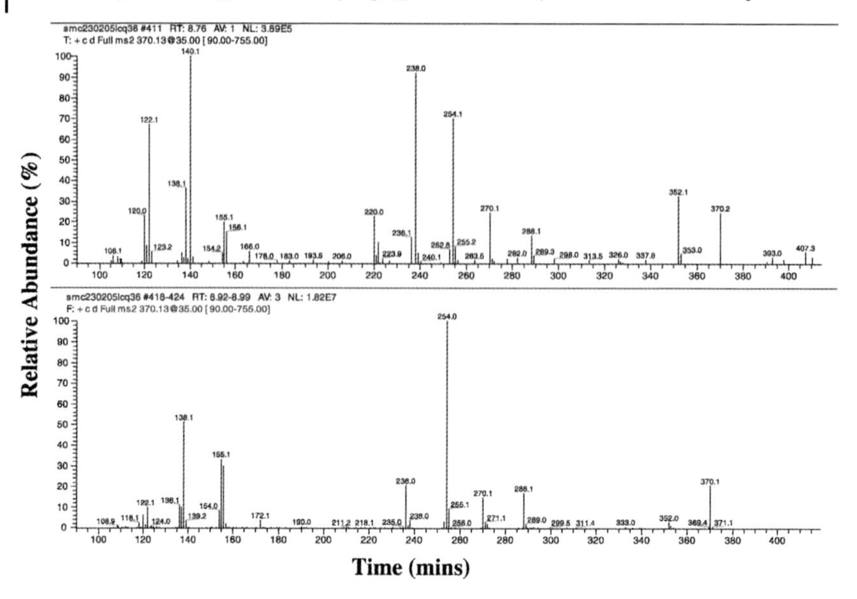

Fig. 13.16 HPLC-ESI-ion trap MS/MS spectra for peak 2 (top) and peak 4 (bottom) in the extract of *Senecio ovatus* (Figure 13.14). The differences in the fragmentation pattern support the extra hydroxylation of sarracine-*N*-oxide in the C-9 esterifying acid for peak 2 rather than in the necine base to yield 2-hydroxysarracine-*N*-oxide (peak 4, Figure 13.14).

C-7 position will result in ions with the same *m*/*z* ratio, one of which will be protonated under the ESI conditions. The difference in molecular ion data between peak 4 and peaks 6 and 7 is consistent with an additional hydroxylation relative to sarracine-*N*-oxide and it is tentatively suggested that this extra hydroxylation is a result of hydration of the 1,2 olefinic center of triangularine-*N*-oxide to yield 2-hydroxysarracine-*N*-oxide. In support of this, the abundant presence of an ion at *m*/*z* 138 in the ESI-MS/MS spectra of peak 4, and its redox resin reduction product, can be rationalized as the 1,2-hydrated analog of the *m*/*z* 120 ion characteristically observed in the ESI-MS spectra of 1,2-dehydropyrrolizidine alkaloids [33]. In contrast to the proposed structure for peak 4, the MS/MS data for the minor, but isobaric, peak 2 clearly indicated the structural difference between the two (Figure 13.16) and suggested the extra hydroxylation of the C-9 esterifying acid rather than in the necine base as with peak 4.

13.5.3 Quantitative Analysis

There is an inherent instability in the ion trap detector that makes inclusion of an internal standard mandatory for reliable quantitation. In this way, prior to estimating

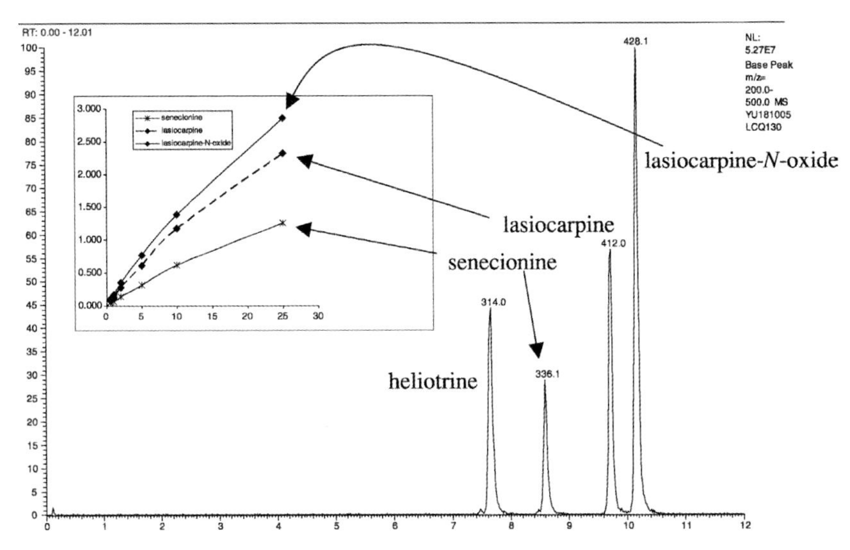

Fig. 13.17 Calibration curve generation. The main figure shows the reversed phase HPLC-ESI-MS base ion (*m/z* 200–500) chromatograms for authenticated standards of senecionine, lasiocarpine, and lasiocarpine-*N*-oxide. Heliotrine was included as an internal standard. The inset shows a calibration curve derived from serial dilutions of the standards.

quantities by comparison to a suitable calibration standard curve, the response of every peak is adjusted relative to the area response of the internal standard (Figure 13.17).

It has been demonstrated that, when many samples are to be analyzed in automatically injected sequences, the APCI mode is more stable than the ESI mode for quantitation [32]. However, the ESI mode is more sensitive for the detection of the pyrrolizidine-*N*-oxides.

The apparent sensitivity and resolution of the HPLC-ESI-ion trap MS ion chromatograms could be enhanced by software manipulation. Thus, a reconstructed ion chromatogram displaying only the ion of interest, usually the molecular ion adduct (MH^+), was capable of extracting quantifiable peaks from low abundance chromatograms or from superimposed peaks (Figure 13.18).

13.5.3.1 **Calibration Standards**

Because of the diversity of structures, including stereochemical orientations, in the pyrrolizidine alkaloids extracted from a plant, it is not usually possible to have authenticated standards for every alkaloid extracted. An alternative approach is to generate calibration curves using standard pyrrolizidine alkaloids that best approximate those extracted from the plant [35,41]. Thus, considerations such as N-oxidation, mono- or diesterification, open chain or cyclic diesterification, and the character of the esterifying acids are important in deciding which standards to use. The analytes are then described in terms of equivalents of the standard alkaloid. At a later stage, if it becomes important, specific authenticated alkaloids can be used

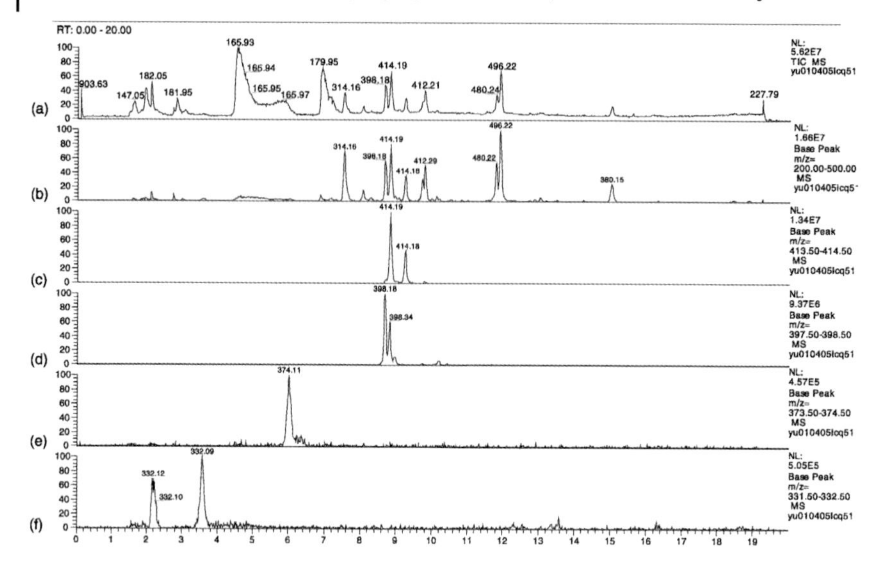

Fig. 13.18 Reversed phase HPLC-ESI-MS ion chromatograms of the strong cation exchange solid phase extract of an *Echium vulgare*-derived honey. (a) Total ion chromatogram, (b) base ion (*m/z* 200–500) chromatogram, and (c)–(f) reconstructed ion chromatograms displaying *m/z* 414, 398, 374, and 332, respectively.

to define response factors relative to the original calibration standard, and subsequent adjustments made to quantitation estimates.

13.5.3.2 **Honey**

The presence of toxic pyrrolizidine alkaloids in honey has been known for several decades [42]. The potential health concerns associated with pyrrolizidine alkaloids in food [27] and honey [28] in particular demand a rapid, sensitive method of detection in various matrices. The application of LC-MS methods to honey samples extracted using SCX SPE cartridges has facilitated the analysis of honey for the presence of pyrrolizidine alkaloids and their *N*-oxides [34,41]. The SPE and LC-MS analysis has shown that honey attributed to known pyrrolizidine alkaloid-producing sources can have levels in excess of 5000 μg/kg honey. Further to this, honey attributed to non-pyrrolizidine alkaloid-producing floral sources and unspecified blended honeys can also have significant amounts of alkaloids present.

Typical HPLC-ESI-ion trap MS ion chromatograms of a honey derived from *Echium vulgare* are shown in Figure 13.18, including the total ion chromatogram, the base ion (*m/z* 200–500) chromatogram and some representative reconstructed ion chromatograms that highlight the enhanced resolution and sensitivity capability of the software management of data. Individual alkaloids (including the *N*-oxides) were quantitated with respect to authenticated calibrations standards that were

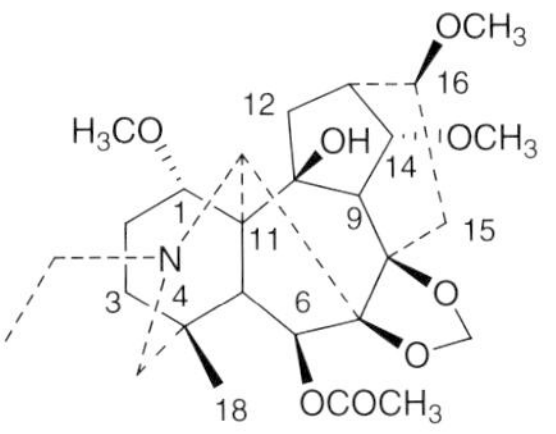

deltaline

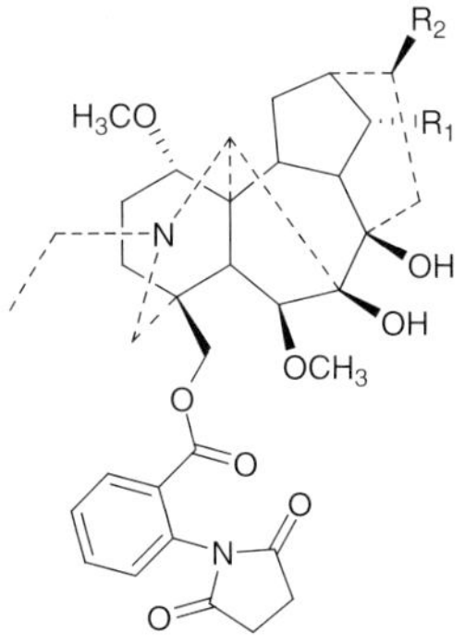

MSAL Alkaloids	R_1	R_2	Mol. Wt.
methyllycaconitine	OCH_3	OCH_3	682
14-deacetylnudicauline	OH	OCH_3	668
geyerline	OCH_3	$OCOCH_3$	710
nudicauline	O $COCH_3$	OCH_3	710
bearline	OH	O $COCH_3$	696
14-acetylbearline	O $COCH_3$	O $COCH_3$	738
16-deacetylgeyerline	OCH_3	OH	668

Fig. 13.19 Structures of norditerpene alkaloids from *Delphinium* spp.

judged to best mimic their specific response factors. An example of the levels of different alkaloids found in a commercial honey sample, as determined using HPLC-MS is shown in Table 13.1.

13.6 Alkaloids from *Delphinium* spp. (Larkspurs)

Diterpene (C_{20}) and norditerpene (C_{19}) alkaloids are typically found in species of the *Aconitum, Delphinium,* and *Consolida* genera [43–46]. The toxic effects of the *Delphinium* and *Aconitum* species can be attributed to their norditerpene alkaloid content. Thus, many of the examples presented in this chapter have been taken from the analysis of norditerpene alkaloids; however, similar experiments would be applicable for any of the diterpene alkaloids (Figure 13.19).

The biological activity of norditerpene alkaloids is a result of antagonism of nicotinic acetylcholine receptors resulting in neuromuscular paralysis [44,45, 47,48]. There is usually a large number of norditerpenoid alkaloids in a single plant species but they are not all equally toxic. The most toxic contain the *N*-methylsuccinimidoanthronyl (MSAL) ester functional group at C-18 [49–51], as represented by methyllycaconitine (Figure 13.19), and are therefore referred to as MSAL alkaloids.

Tab. 13.1 Quantities of pyrrolizidine alkaloids and their *N*-oxides extracted from *E. vulgare*-derived honey.

Retention time (min)	Molecular ion adduct ([MH]$^+$) (*m/z*)	Identity	Concentration (ppb)
11.94	496	Echivulgarine-*N*-oxide	558 ± 40
11.81	480	Echivulgarine	879 ± 35
10.12	456	7-*O*-Acetylvulgarine-*N*-oxide	Trace
9.97	440	7-*O*-Acetylvulgarine	Trace
9.80	456	Acetylechimidine-*N*-oxide	96 ± 4
9.72	440	Acetylechimidine	104 ± 7
9.27	414	Vulgarine-*N*-oxide	166 ± 3
9.01	398	Vulgarine	84 ± 7
8.90	414	Echimidine-*N*-oxide	291 ± 9
8.84	398	Echimidine	299 ± 25
8.73	398	Echiuplatine-*N*-oxide	140 ± 10
8.72	382	Echiuplatine	194 ± 13
6.07	374	Uplandicine-*N*-oxide	15 ± 1
3.51	332	Leptanthine-*N*-oxide	14 ± 1
2.10	332	Echimiplatine-*N*-oxide	10 ± 1

The total pyrrolizidine alkaloid content was 2850 ± 143 μg/kg honey (ppb) expressed as equivalents of lasiocarpine-*N*-oxide for the *N*-oxides present and equivalents of lasiocarpine for the tertiary base pyrrolizidine alkaloids present. The assignment of the minor alkaloids leptanthine-*N*-oxide and echimiplatine-*N*-oxide may be reversed

Whilst gas chromatography has been used for the analysis of many of the lycoctonine-based alkaloids [52], the larger, less volatile, and more thermally labile MSAL compounds require analytical procedures such as TLC and HPLC for separation and detection. For example, both normal phase liquid chromatography [53] and reversed phase liquid chromatography [54] with UV detection have been used for separation, detection, and quantitation of alkaloids from *Delphinium* species associated with livestock poisonings in the western US and Canada. The introduction of API techniques has allowed the analysis of all types of diterpene alkaloids by direct MS methods and with MS methods coupled to liquid chromatography.

13.6.1 Flow Injection (FI) Mass Spectrometry

One of the easiest MS techniques to use, without involving a chromatographic stage, is a single loop injection of a sample solution into a stream of solvent that flows directly into the API source. This is an alternative to infusing the sample solution directly into the API source via a syringe. The advantages of FI are the rapid analysis time, high-throughput capabilities, and the ability to obtain quantitative information for selected ions. Infusion of a sample via a syringe requires more time for system preparation and sample introduction but it does allow an increase in the number of different MS measurements that can be completed on such samples.

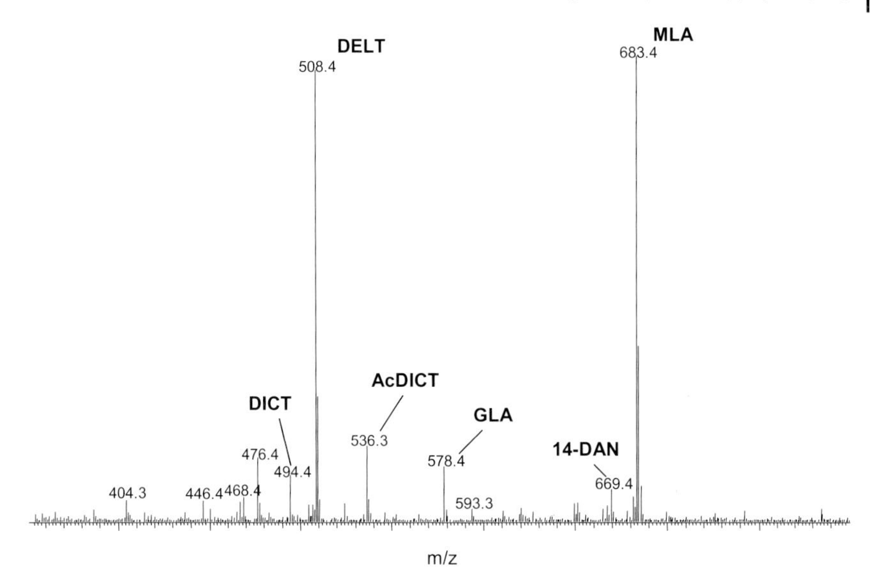

Fig. 13.20 Electrospray ionization mass spectrum of a crude methanol extract of *Delphinium barbeyi* plant material. The methanol extract was diluted with the electrospray flow solvent (50 % methanol in 1 % acetic acid) and a 20 μL loop injection made. Alkaloids identified based on their MH^+ are dictyocarpine (DICT), deltaline (DELT), 14-acetyldictyocarpine (AcDICT), glaucenine/barbisine (GLA), 14-deacetylnudicauline (14DAN), and methyllycaconitine (MLA).

13.6.1.1 Qualitative FI Analysis

Many of the earliest applications of API-MS used direct syringe infusion of crude alkaloidal extracts into the ESI source. In this manner, Marko and Stermitz [55] were able to detect the transfer of alkaloids from *Delphinium occidentale* to the root parasitic *Castilleja sulphurea*.

In more recent applications, FI-ESI-MS analysis of crude methanolic extracts, without any further purification, of *Delphinium* species (*D. barbeyi*, *D. geyeri*, and *D. nuttallianum*) have provided alkaloidal profiles of these plants [56]. Thus, a small amount (about 100 mg) of ground dry plant material was extracted with methanol (4 mL) with mechanical rotation at room temperature for several hours. An aliquot (about 20 μL) of the crude extract was diluted with the electrospray flow solvent (1 mL) and injected (20 μL) via a loop into the electrospray flow for subsequent ESI analysis. The resulting mass spectrum was a pattern of protonated molecular ion adducts (MH^+) from the alkaloids (Figure 13.20). The major alkaloids of *Delphinium barbeyi* are easily detected, that is, methyllycaconitine (MLA, $MH^+ m/z$ 683) and deltaline (DELT, $MH^+ m/z$ 508). Several other minor alkaloids, dictyocarpine (DICT, $MH^+ m/z$ 494), 14-acetyldictyocarpine (AcDICT, $MH^+ m/z$ 536), glaucenine/barbisine (GLA, $MH^+ m/z$ 578) and 14-deacetylnudicauline (14-DAN, $MH^+ m/z$ 669) were also readily identified. Additionally, this approach allows rapid screening for the presence of potentially toxic norditerpenoid alkaloids, specifically the MSAL alkaloids. Because of the higher molecular weights afforded by the MSAL esterification, selected ion scans of the region *m/z* 650–750 can quickly identify if such toxic alkaloids are present in the sample. It is to be noted that the confident identification of an individual alkaloid from such a basic mass spectrum, based on the presence of molecular ion data only, relies on a prior knowledge of the alkaloids present in the plant. Confirmation of MSAL-like structures of new compounds will require more rigorous structural information including MS fragmentation and NMR analysis.

Whilst both ESI and APCI can be used for analysis of most diterpene alkaloid-containing plant samples, ESI is recommended as the primary method for screening plant material, since the true molecular ion adduct is readily observed. With APCI, the lower molecular weight alkaloids (i.e. non-MSAL) produce good molecular ions with little fragmentation, but the larger esterified alkaloids (e.g. methyllycaconitine, MH^+ *m/z* 683) yield significant fragment ions resulting from loss of water from the MH^+ molecular ion adduct (i.e. *m/z* 665 for methyllycaconitine). Additionally, the loss of methanol and acetic acid observed in APCI spectra from other diterpene alkaloids, depending on the functional group at C-8 [57,58], could also complicate the interpretation of the APCI mass spectra from unknown samples.

13.6.1.2 **Quantitative FI Analyses**

By specifically and selectively reconstructing the total ion peak observed after a loop flow injection to display only the *m/z* values of interest, the area count for that ion can be extracted from the composite total ion peak. The measured area count can then be used to estimate the quantity of specific alkaloid by comparison to a calibration curve. Thus, component compounds in a mixture are separated by mass, as opposed to chromatography, for quantitation. To normalize the variability of the API response, an internal reference standard is added to the sample prior to loop injection. Alkaloids for which standards are not available are reported as equivalents of a closely related and available standard used to generate the calibration curve. For example, deltaline and methyllycaconitine have been used as calibration standards to represent the non-MSAL and MSAL types of alkaloids, respectively, in the plant material. Calibration curves for these two compounds were linear ($r^2 \geq 0.990$) and there appears to be no selective suppression of lower-level alkaloids (Figure 13.21). Multiple analyses of *Delphinium barbeyi* samples returned a level of precision that was less than 10 % (relative standard deviation) for all components [56].

Using the quantitative FI method a large number of samples can be extracted overnight, aliquots and dilutions made the next day, samples loaded into the autosampler, and run in sequence. The method was validated by comparison to the previously used HPLC method to measure toxic alkaloids [56]. The level of methyllycaconitine, 14-deacetylnudicauline, and barbinine, as determined from the peak areas for ion traces at *m/z* 683, 669, and 667, respectively, were summed to give total toxic alkaloids. Measured recoveries are slightly higher for the ESI-MS method since the HPLC method required further purification of the alkaloidal extract using an acid/base partition procedure, incurring some loss of alkaloids.

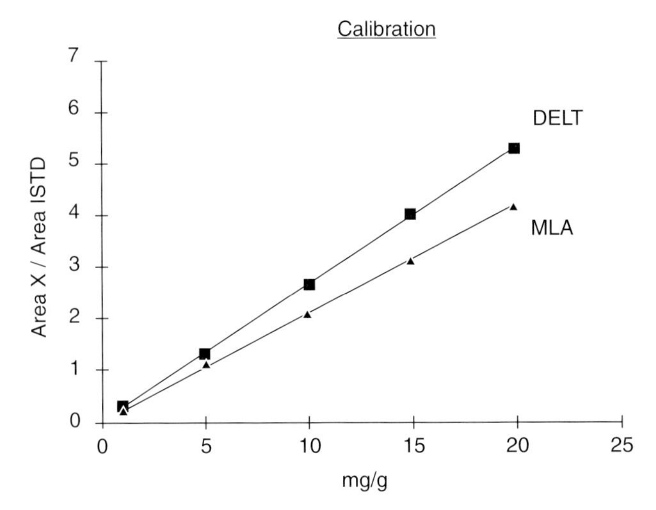

Fig. 13.21 Calibration curves for quantitative analysis of alkaloids in delphinium plant material. Standards (DELT, deltaline; MLA, methyllycaconitine) were prepared in chloroform and then diluted into electrospray flow solvent and analyzed by FI-ESI-MS. The area count of the alkaloid (Area X) is normalized by division by the area count of the internal standard (Area ISTD).

13.6.1.3 Chemotaxonomy of *Delphinium* Species

Flow injection ESI-MS has been applied to an examination of the chemotaxonomy of the major toxic *Delphinium* species [59]. Of the major toxic larkspur species found in the western US, *Delphinium glaucescens* is the most distinct taxonomically whereas the other three tall larkspurs, *D. barbeyi, D. glaucum,* and *D. occidentale,* are morphologically very similar with only small differences in the type and amount of pubescence on the inflorescence, and in the shape and arrangement of the flowers [60]. Those plants historically classified as *Delphinium occidentale* have recently been regrouped into *D. glaucum, D. barbeyi* has been restricted to a small geographically restricted population in central Utah, and the *D. occidentale* classification has been used to define those plants being a hybrid of *D. barbeyi* and *D. glaucum* [61].

Because of the difficulties associated with the *Delphinium* classification, the diterpene alkaloid content was assessed as an aid to determining the chemical taxonomic diversity of the toxic larkspur species [59]. Plant samples were collected from 18 different locations in five western states in the US. The crude methanolic extracts were analyzed for diterpene alkaloids using FI-ESI-MS. The data from the individual ESI mass spectra were statistically analyzed using canonical discriminant analysis and analysis of variance. In brief, the sample (100 mg) was extracted by mechanical shaking at room temperature with methanol (5 mL) for 16 h. Reserpine (500 μg) was added as an internal reference standard and the sample mixed for 5 min and then centrifuged. An aliquot (30 μL) of the supernatant was then diluted with of 1 : 1 methanol/1 % acetic acid (1.0 mL) and an aliquot (20 μL) injected for analysis. The mass spectral data was tabulated by recording the relative abundance for all ions above 0.1 %. The amount of each compound (as represented by a single mass unit) detected was calculated based on its abundance relative to reserpine (MH^+ m/z 609). The data was reduced to a final set of 33 individual masses by eliminating all components below 5 μg alkaloid/100 mg of plant and all obvious isotope peaks.

Based on the observed ESI-MS alkaloid profiles, those plants collected from the Sierra Nevada region and identified as *D. glaucum* using the morphological characteristics described by Ewan [60] were quite distinct from all other samples (Figure 13.22). The larkspur populations from the Central Rocky Mountain regions were not easily grouped by the simple presence or absence of particular alkaloids; however, statistical grouping analysis was able to separate the samples (Figure 13.23). Plants historically identified as *D. barbeyi* and *D. occidentale* were indeed found to be chemotaxonomically distinct groups, albeit somewhat closely related. Samples representing a putative hybrid between Ewan's [60] *D. barbeyi* and *D. occidentale* were found to be more closely related to *D. occidentale*, but were significantly different from all other groups. The chemotaxonomic data and geographic relationships are in agreement with those based on the historical morphological characteristics [60] and are contrary to the more recent relationships upon which it has been proposed to combine *D. occidentale* with *D. glaucum* into one species. In addition, enough of a difference between *D. occidentale* and *D. barbeyi* was observed that assigning to them subspecies status under another species is not warranted [62].

The classification of the larkspur species using the chemical alkaloid profiles was in complete agreement with the molecular genomic data [63].

13.6.2
LC-MS Analysis of Diterpene Alkaloids

Both normal phase [53] and reversed phase [54] HPLC methods have been used for the separation of diterpene alkaloids. Reversed phase HPLC coupled to APCI mass spectrometry has been used for the analysis of diterpene alkaloids of *Aconitum* spp. [64,65] and normal phase HPLC conditions [53] have been successfully used with APCI-MS for the detection of diterpene alkaloids in *Delphinium* species [56]. However, caution should be observed in the use of APCI sources with some normal phase HPLC solvents such as hexane, to ensure no oxygen is introduced into the system producing a possible explosive mixture in the API source.

13.6.2.1 Toxicokinetics and Clearance Times

Reversed phase HPLC conditions have been used with good success in the analysis of low levels of specific alkaloids. For example, the toxicokinetics of methyllycaconitine were determined by analyzing mouse sera and tissue samples (kidney, brain, liver, muscle) with detection down to one part per billion using selected ion monitoring MS/MS conditions [66]. Similar procedures are being used to measure alkaloid clearance times in sheep sera for methyllycaconitine and deltaline (Gardner, unpublished data).

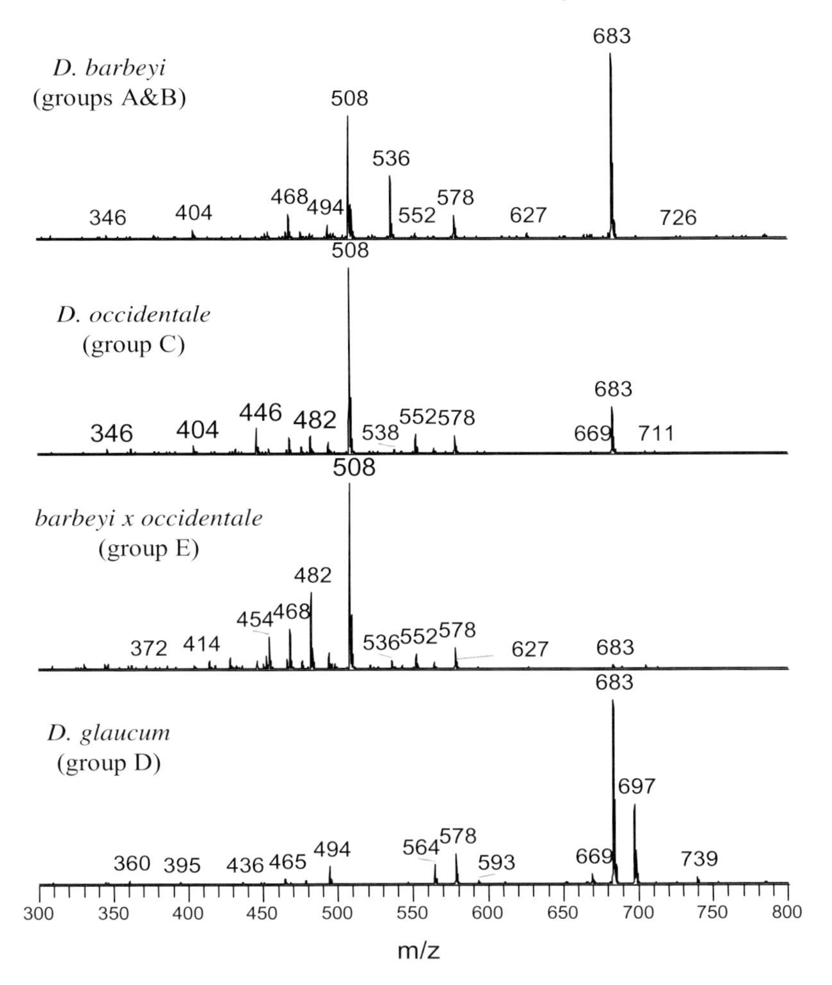

Fig. 13.22 ESI-MS spectra of composite plant samples (*Delphinium* spp.) used for chemotaxonomy (reserpine was not added for this qualitative comparison). Group D (*Delphinium glaucum*, from the Sierra Nevada region) was the most distinct group based on simple qualitative presence or absence of individual alkaloids. Interestingly, the proposed *barbeyi* × *occidentale* hybrid (Group E) plants contain very little toxic alkaloid ($m/z > 650$) in comparison to their proposed genetic parents. Plant identifications were based on Ewan's classification [59].

13.6.2.2 **Diagnosis of Poisoning**

For diagnostic purposes, extracted samples of rumen fluid, kidney, liver, and blood have all been analyzed using FI-ESI-MS and HPLC-ESI-MS to identify toxic alkaloids in suspected plant poisoning cases. From a recent submission of a suspected Death Camas (*Zigadenus* spp.)-related poisoning case, an ion was detected (m/z 536), using FI-ESI-MS, in extracted tissue samples and a blood sample that could correspond to the protonated molecular ion for the Death Camas alkaloid zygacine. However, the

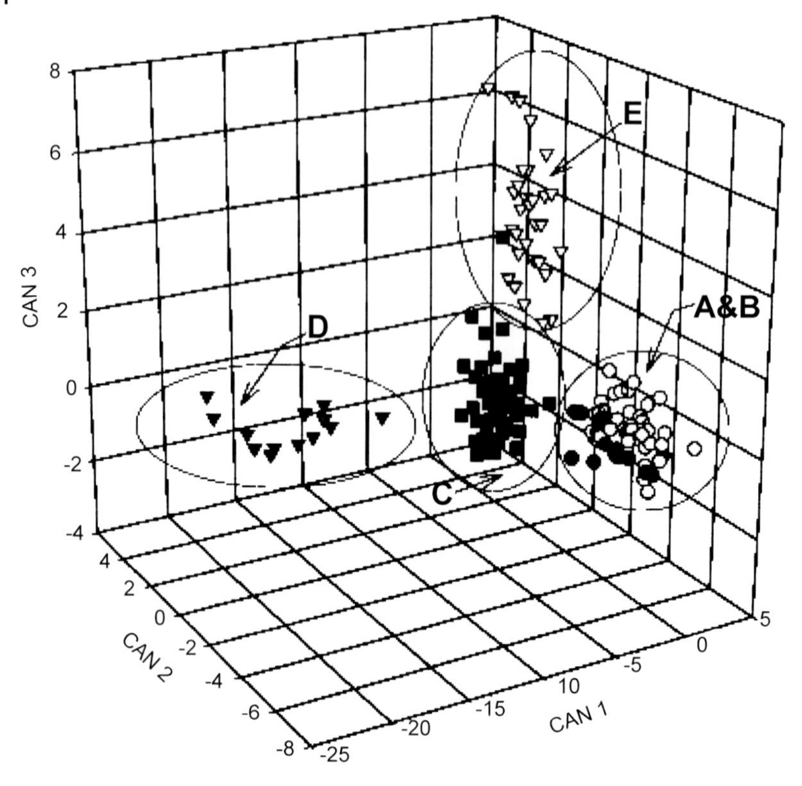

Fig. 13.23 Plot of the first three canonical variables from the discriminant analysis of the tabulated ESI-MS data for toxic *Delphinium* species according to Ewan [59]. Groups A and B are *Delphinium barbeyi*, group C is *D. occidentale*, group D is *D. glaucum*, and group E is a hybrid (*barb.* × *occi.*).

MS/MS data did not confirm the ion as being from zygacine. A number of other ions were observed indicating the presence of diterpene alkaloids, particularly the toxic MSAL alkaloids methyllycaconitine (m/z 683) and 14-deacetylnudicauline (m/z 669). The presence of methyllycaconitine was confirmed using HPLC-ESI-MS/MS analysis, providing evidence that the animal was poisoned by ingested larkspur and not Death Camas (Gardner *et al.*, unpublished).

13.6.3 Structural Elucidation of Norditerpenoid Alkaloids

13.6.3.1 Stereochemical Indications

APCI-MS has been used extensively for the structural analysis of diterpene alkaloids from various *Aconitum* species for both positional and stereoiosomeric determination [57,58]. By controlling the drift voltage between the first and second electrodes in the ion source, the configurations at C-1, C-6, or C-12 were determined for a series of isomeric compounds. The characteristic APCI fragment ion included loss of water, methanol, or acetic acid from the C-8 position and the relative abundance of these fragment ions was significantly higher for the C-1, C-6, or C-12 β-form alkaloids than for their corresponding α configurations.

13.6.3.2 Isomeric Differentiation Using Tandem Mass Spectrometry

Sequential tandem mass spectrometry experiments have also been useful for structural determination of a number of norditerpene alkaloids [56,67]. The most abundant fragment ions in MS^n-generated product ion scans occur from losses of oxygenated functional groups such as H_2O ($MH^+ - 18$), methanol ($MH^+ - 32$) and acetic acid ($MH^+ - 60$) [56,67]. For example, the MS/MS, MS/MS/MS, and MS^4 spectra for methyllycaconitine show three sequential losses of methanol ($MH^+ - 32$, $[MH^+ - 32] - 32$, and $\{[MH^+ - 32] - 32\} - 32$). For the MSAL-type compounds it was shown that the functional groups at C-16 are the most labile while those at C-14 are the most stable [56]. This preferential fragmentation can yield useful structural information. For example, the alkaloids nudicauline and geyerline cannot be distinguished based solely on their ESI-MS spectra because they have the same molecular weight; however, geyerline shows preferential loss of acetic acid ($MH^+ - 60$, m/z 651) while the major fragment loss from nudicauline is methanol ($MH^+ - 32$, m/z 679) (Figure 13.24). The stability of the functional group at C-14 is further supported by the lack of a sequential loss of acetic acid in the MS/MS

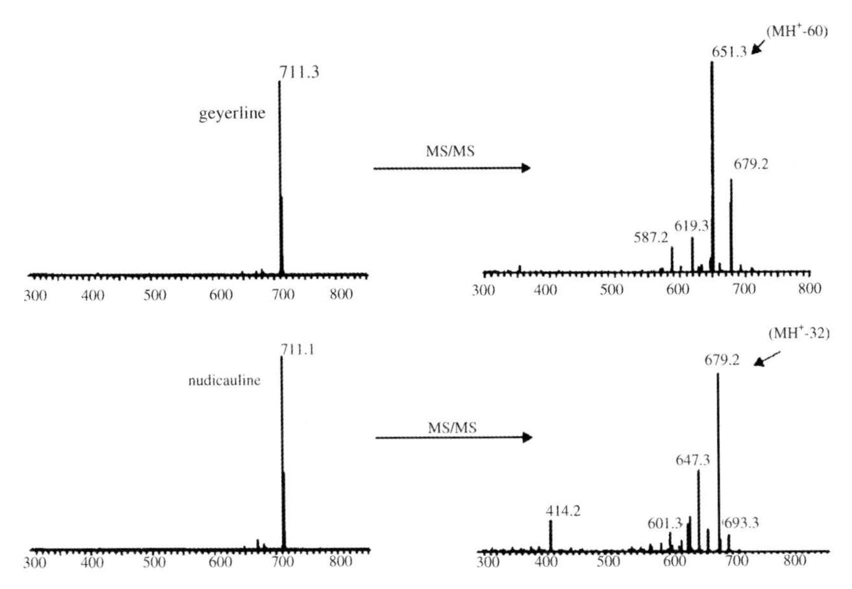

Fig. 13.24 ESI-MS and ESI-MS/MS analysis of the isomeric alkaloids geyerline (top) and nudicauline (bottom) which differ only in the methoxy and acetoxy substitution of C-14 and C-16 (Figure 13.19).

of nudicauline (i.e. $[MH^+ - 32] - 60$, m/z 619) or indeed in any of the sequential fragmentations up to MS^4.

It is important to note that the principal fragment ions generated during sequential MS experiments within an ion trap should not be confused with fragment ions generated in the ionization source. For example, whilst the functional groups at C-16 were the most labile under MS^n (ion trap) conditions, the principal fragmentation within the APCI source involves losses from the C-8 position [57,58].

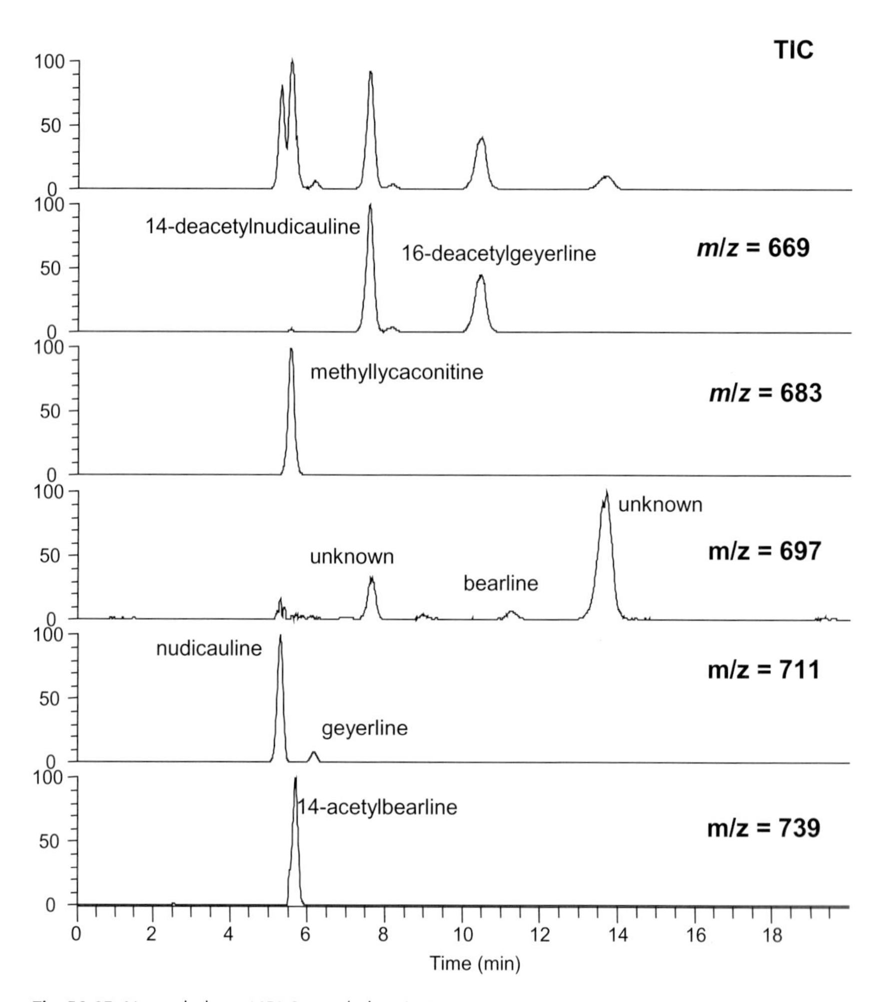

Fig. 13.25 Normal phase HPLC coupled to APCI-MS analysis of *Delphinium nuttallianum* from one geographical location. Individual alkaloids are identified from reconstructed ion chromatograms and their retention time. At least two unknown alkaloids, in addition to bearline, were observed in the reconstructed ion chromatogram displaying m/z 697.

13.6.3.3 Novel Diterpene Alkaloid Identification: Application of Tandem Mass Spectrometry

Delphinium nuttallianum and *D. andersonii* are amongst the "low larkspurs," a group of larkspurs from which ten different MSAL alkaloids have been detected using normal phase HPLC with APCI-MS detection. Not all of these alkaloids are present in every sample of *D. nuttallianum* or *D. andersonii* but many samples will contain up to seven to nine of the alkaloids as demonstrated by the HPLC-APCI-MS total ion chromatogram of the crude alkaloid fraction isolated from *D. nuttallianum* (Figure 13.25). Many of the major MSAL alkaloids present in extracts of *D. nuttallianum* samples were readily identified by comparison of their retention times and mass spectra with authenticated standards. However, some samples of *D. nuttallianum* indicated the presence of two unidentified alkaloids (MH^+ m/z 697 and m/z 739), sometimes in relatively high concentrations. Tandem MS^n experiments were conducted on both parent ions following direct infusion of the crude methanolic extract into the mass spectrometer under ESI conditions. The most abundant ion in the MS/MS product ion scan of m/z 697 was m/z 637 ($MH^+ - 60$) thereby indicating an acetate group at C-16, the most labile position of substitution. The MS/MS product ion spectrum of m/z 637 (i.e. the MS/MS/MS of m/z 697) and the subsequent MS/MS/MS product ion spectrum of m/z 637 (i.e., the MS^4 of m/z 697) indicated two additional methoxy groups at C-1 and C-6. The 1,6-di-methoxy, 14-hydroxy, 16-acetate compound was given the name bearline [56]. The unidentified alkaloid at m/z 739 corresponded to the 14-*O*-acetyl derivative of bearline. The MS/MS/MS product ion spectra (m/z 739 → m/z 679 → m/z 647 → m/z 615) were analogs to bearline. Subsequently, both compounds were isolated and their structural identities were confirmed using NMR spectroscopy [68].

13.7 Conclusions

The increasing accessibility of bench-top LC-MS systems to researchers of all disciplines, combined with the tandem and high-resolution mass spectrometry capabilities of such instruments, will only increase the number of applications to which LC-MS can be directed. The examples documented in this chapter illustrate some of the diversity and power of the techniques, including analytical applications for known analytes in various matrices, metabolomic analysis, the tentative structural identification of novel compounds, and the screening of extracts for minor, and perhaps novel, components of the alkaloidal profile of plants.

References

1 Korfmacher, W. A. (2005) *Drug Discovery Today*, **10**, 1357–1367.

2 Prasain, J. K., Wang, C. -C., Barnes, S. (2004) *Free Radical Biology and Medicine*, **37**, 1324–1350.

3 Gao, S., Zhang, Z. -P., Karnes, H. T. (2005) *Journal of Chromatography B*, **825**, 98–110.

4 Shukla, A. K. and Futrell, J. H. (2000) *Journal of Mass Spectrometry*, **35**, 1069–1090.

5 Luo, X., Chen, B., Yao, S. (2005) *Talanta*, **66**, 103–110.

6 Wu, W., Song, F., Yan, C., Liu, Z., Liu, S. (2005) *Journal of Pharmaceutical and Biomedical Analysis*, **37**, 437–446.

7 Bianchi, F., Careri, M., Corradini, C., Elviri, L., Mangia, A., Zagnoni, I. (2005) *Journal of Chromatography B*, **825**, 193–200.

8 Swartz, M. E. and Murphy, B. J. (2004) *LabPlus International*, June.

9 Churchwell, M. I., Twaddle, N. C., Meeker, L. R., Doerge, D. R. (2005) *Journal of Chromatography B*, **825**, 134–143.

10 Euerby, M. R., McKeown, A. P., Petersson, P. (2003) *Journal of Separation Science*, **26**, 295–306.

11 Bell, D. S., Cramer, H. M., Jones, A. D. (2005) *Journal of Chromatography A*, **1095**, 113–118.

12 Kuhlmann, F. E., Apffel, A., Fischer, S. M., Goldberg, G., Goodley, P. C. (1995) *Journal of the American Society for Mass Spectrometry*, **6**, 1221–1225.

13 Shou, W. Z. and Naidong, W. (2005) *Journal of Chromatography B*, **825**, 186–192.

14 Maurer, H. H. (2005) *Clinical Biochemistry*, **38**, 310–318.

15 Gaillard, Y. and Pepin, G. (1999) *Journal of Chromatography B*, **733**, 181–229.

16 Johansen, S. S. and Jensen, J. L. (2005) *Journal of Chromatography B*, **825**, 21–28.

17 Koh, H. L., Wang, H., Zhou, S., Chan, E., Woo, S. O. (2005) *Journal of Pharmaceutical and Biomedical Analysis*, **40**, 653–661.

18 Wolfender, J. -L., Queiroz, E. F., Hostettmann, K. (2005) *Magnetic Resonance in Chemistry*, **43**, 697–709.

19 Liu, D. Q. and Hop, C. E. C. A. (2005) *Journal of Pharmaceutical and Biomedical Analysis*, **37**, 1–18.

20 Lunn G. and Hellwig L. C. (1998) Handbook of Derivatization Reactions for HPLC, Wiley, New York.

21 Tolonen, A., Turpeinen, M., Uusitalo, J., Pelkonen, O. (2005) *European Journal of Pharmaceutical Sciences*, **25**, 155–162.

22 Yoon, K. -H., Lee, S. -Y., Jang, M., Ko, S. -H., Kim, W., Park, J. -s., Park, I., Kim, H. -J. (2005) *Talanta*, **66**, 831–836.

23 Price, J. R., Robinson, R., Scott-Moncrieff, R. (1939) *Journal of Chemical Society*, 1465–1468.

24 Schliemann, W., Schneider, B., Wray, V., Schmidt, J., Nimtz, M., Porzel, A., Böhm, H. (2006) *Phytochemistry*, **67**, 191–201.

25 Proenca Barros, F. A. and Rodrigues-Filho, E. (2005) *Biochemical Systemetics and Ecology*, **33**, 257–268.

26 Stegelmeier, B. L., Edgar, J. A., Colegate, S. M., Gardner, D. R., Schoch, T. K., Coulombe, R. A., Molyneux, R. J. (1999) *Journal of Natural Toxins*, **8**, 95–116.

27 Colegate, S. M., Edgar, J. A., Stegelmeier, B. L. (1998) Plant-associated toxins in the human food supply, Rose J. Environmental Toxicology: Current Developments, Gordon and Breach, Amsterdam, 317–344 (Chapter 15).

28 Edgar, J. A., Roeder, E., Molyneux, R. J. (2002) *Journal of Agricultural and Food Chemistry*, **50**, 2719–2730.

29 Chou, M. W., Wang, Y. -P., Yan, J., Yang, Y. -C., Beger, R. D., Williams, L. D., Doerge, D. R., Fu, P. P. (2003) *Toxicology Letters*, **145**, 239–247.

30 Hartmann, T. and Toppel, G. (1987) *Phytochemistry*, **26**, 1639–1643.

31 Asres, K., Sporer, F., Wink, M. (2004) *Biochemical Systemetics and Ecology*, **32**, 915–930.

32 Beales, K. A., Colegate, S. M., Edgar, J. A. (2003) Experiences with the trace analysis of pyrrolizidine alkaloids using GC-MS and LCMS, Acamovic, T. Stewart, C. S. Pennycott T. W. Poisonous Plants and Related Toxins, CABI Publishing, Wallingford, UK, 453–458.

33 Colegate, S. M., Edgar, J. A., Knill, A. M., Lee, S. T. (2005) *Phytochemical Analysis*, **16**, 108–119.

34 Beales, K. A., Betteridge, K., Colegate, S. M., Edgar, J. A. (2004) *Journal of Agricultural and Food Chemistry*, **52**, 6664–6672.

35 Boppré, M., Colegate, S. M., Edgar, J. A. (2005) *Journal of Agricultural and Food Chemistry*, **53**, 594–600.

36 Witte, L., Ernst, L., Adam, H., Hartmann, T. (1992) *Phytochemistry*, **31**, 559–565.

37 Macel, M., Vrieling, K., Klinkhamer, P. G. L. (2004) *Phytochemistry*, **65**, 865–873.

38 Röder, E. (1995) *Pharmazie*, **50**, 83–98.

39 Röeder, E., Wiedenfeld, H., Britz-Kirstgen, R. (1984) *Phytochemistry*, **23**, 1761–1763.

40 Roitman, J. N. (1983) *Australian Journal of Chemistry*, **36**, 1203–1213.

41 Betteridge, K., Cao, Y., Colegate, S. M. (2005) *Journal of Agricultural and Food Chemistry*, **53**, 1894–1902.

42 Culvenor, C. C. J., Edgar, J. A., Smith, L. W. (1981) *Journal of Agricultural and Food Chemistry*, **29**, 958–960.

43 Pelletier, S. W. (1970) *Chemistry of the Alkaloids*, Van Nostrand Reinhold Co., New York, pp. 549–590.

44 Jacyno, J. M. (1996) Chemistry and Toxicology of the Diterpenoid Alkaloids, Blum M. S. Chemistry and Toxicology of Diverse Classes of Alkaloids, Alken, Fort Collins, CO, 301–336.

45 Benn, M. H. and Jacyno, J. (1983) The Toxicology and Pharmacology of Diterpenoid Alkaloids, Pelletier S. W. Alkaloids: Chemical and Biological Perspectives, Vol. **1**, Wiley, New York, 153–210.

46 Joshi, B. S. and Pelletier, S. W. (1999) Recent Development in the Chemistry of Norditerpenoid an Diterpenoid Alkaloids, Pelletier S. W. Alkaloids: Chemical and Biological Perspectives, Vol. **13**, Pergamon, New York, 292–370.

47 Nation, P. N., Benn, M. H., Roth, S. H., Wilkens, J. L. (1982) *The Canadian Veterinary Journal*, **23**, 264–266.

48 Kukel, C. F. and Jennings, K. R. (1994) *Canadian Journal of Physiology and Pharmacology*, **72**, 104–107.

49 Turek, J. W., Kang, C. -H., Campbell, J. E., Arneric, S. P., Sullivan, J. P. (1995) *Journal of Neuroscience Methods*, **61**, 113–118.

50 Manners, G. D., Panter, K. E., Pelletier, S. W. (1995) *Journal of Natural Products*, **58**, 863–869.

51 Manners, G. D., Panter, K. E., Pfister, J. A., Ralphs, M. H., James, L. F. (1998) *Journal of Natural Products*, **61**, 1086–1089.

52 Manners, G. D. and Ralphs, M. H. (1989) *Journal of Chromatography*, **466**, 427–432.

53 Manners, G. D. and Pfister, J. A. (1993) *Phytochemical Analysis*, **4**, 14–18.

54 Majak, W., McDiarmid, R. E., Benn, M. H. (1987) *Journal of Agricultural and Food Chemistry*, **35**, 800–802.

55 Marko, M. D. and Stermitz, F. R. (1997) *Biochemical Systemetics and Ecology*, **25**, 279–285.

56 Gardner, D. R., Panter, K. E., Pfister, J. A., Knight, A. P. (1999) *Journal of Agricultural and Food Chemistry*, **47**, 5049–5058.

57 Wada, K., Mori, T., Kawahara, N. (2000) *Chemical and Pharmaceutical Bulletin*, **48**, 660–668.

58 Wada, K., Mori, T., Kawahara, N. (2000) *Journal of Mass Spectrometry*, **35**, 432–439.

59 Gardner, D. R., Ralphs, M. H., Turner, D. L., Welsh, S. L. (2002) *Biochemical Systemetics and Ecology*, **30**, 77–90.

60 Ewan, J. (1945) *A synopsis of the North American species of Delphinium*. Univ. Colorado Studies, Series D, **2**, 2, 55–244.

61 Warnock, M. J. (1995) *Physiologia*, **78**, 73–101.

62 Welsh, S. L., Atwood, N. D., Higgins, L. C., Goodrich, S. (1987) *Great Basin Naturalist Memoirs, No. 9*, Brigham Young University, pp. 508–509.

63 Li, X., Ralphs, M. H., Gardner, D. R., Wang, R. R. C. (2002) *Biochemical Systemetics and Ecology*, **30**, 91–102.

64 Wada, K., Bando, H., Kawahara, N. (1993) *Journal of Chromatography*, **644**, 43–48.

65 Wada, K., Bando, H., Kawahara, N., Mori, T., Murayama, M. (1994) *Biological Mass Spectrometry*, **23**, 97–102.

66 Stegelmeier, B. L., Hall, J. O., Gardner, D. R., Panter, K. E. (2003) *Journal of Animal Science*, **81**, 1237–1241.

67 Chen, Y., Koelliker, S., Oehme, M., Katz, A. (1999) *Journal of Natural Products*, **62**, 701–704.

68 Gardner, D. R., Manners, G. D., Panter, K. E., Lee, S. T., Pfister, J. A. (2000) *Journal of Natural Products*, **63**, 1127–1130.

14
Applications of ^{15}N NMR Spectroscopy in Alkaloid Chemistry

Gary E. Martin, Marina Solntseva, Antony J. Williams

14.1
Introduction

^{15}N has had an interesting role in the history of NMR spectroscopy. Following the appropriation of NMR techniques from the physicists by the chemical research community, ^{15}N studies were few in number owing to the intrinsic properties of this chemically important nuclide. Specifically, ^{15}N is a relatively low natural abundance nuclide, present in nature at only 0.13 %. The low relative abundance is further exacerbated by a low gyromagnetic ratio, γ_N (the observation frequency of ^{15}N is ~10 % that of ^1H), which served to make ^{15}N a difficult nuclide to observe directly. Nevertheless, the early literature still contains numerous citations that provide a wealth of information about ^{15}N chemical shift behavior. The development of pulsed Fourier transform NMR instruments in the early 1970s has ameliorated some of the difficulties associated with measuring ^{15}N spectra, as did the development of much stronger field magnets. Probably the biggest single factor in enhancing experimental access to ^{15}N data has been the development of indirect (sometimes referred to as "inverse") detection NMR pulse sequences [1,2] and probes whereby an insensitive nucleus such as ^{15}N can be detected via a much more sensitive, high abundance nuclide such as ^1H, ^{19}F, or ^{31}P [3–7]. Proton-detected measurement of ^{15}N spectra is 300 times more sensitive than direct ^{15}N excitation and observation [8]. Through the use of indirect detection techniques, it is possible to acquire direct (one-bond) or long-range $^nJ_{HN}$ $(n \geq 2)$ ^1H–^{15}N heteronuclear shift correlation information for even submilligram quantities of material in a few hours or overnight.

14.1.1
^{15}N Chemical Shift Referencing

^{15}N chemical shift referencing has changed over time. At present there are two conflicting referencing schemes in widespread use and it is important for investigators with an interest in ^{15}N data to be aware of both. Older work in the literature was most commonly referenced to nitromethane, which was assigned a chemical shift of 0 ppm. Nitrogen references downfield of nitromethane were assigned a positive (+)

Modern Alkaloids: Structure, Isolation, Synthesis and Biology. Edited by E. Fattorusso and O. Taglialatela-Scafati

ISBN: 978-3-527-31521-5

chemical shift while those upfield, which encompasses virtually all nitrogens commonly encountered in natural products, were assigned a negative (−) chemical shift. Ammonia, NH_3, the other ^{15}N chemical shift reference compound, had a chemical shift of ∼−379.5 ppm relative to nitromethane. More recently, the sign convention for nitromethane ^{15}N chemical shift referencing has been reversed. Now, ^{15}N resonances upfield of nitromethane are assigned a positive (+) chemical shift. Care must obviously be taken when delving into older literature to be certain of which sign convention is employed if the ^{15}N data are referenced to nitromethane.

To add still further confusion, the use of ammonia as a ^{15}N chemical shift reference has become increasingly prevalent as various heteronuclear 2D NMR studies of proteins have been reported. (NH_3 is assigned a chemical shift of 0 ppm on the NH_3 scale while nitromethane has a shift relative to liquid NH_3 that has been variously reported as +379.5, +380.2, and +381.7 ppm.) The authors of this chapter find it convenient to use the NH_3 chemical shift scale to maintain the "sense" of ^{15}N chemical shifts in parallel to those of ^{13}C and ^{1}H. Chemical shifts referenced to liquid ammonia in the ACD/Labs NNMR database are referenced relative to nitromethane at +380.2 ppm. Given what would be considered as normal F_1 digitization in long-range ^{1}H–^{15}N 2D NMR experiments, ^{15}N chemical shifts derived from these experiments are probably only accurate to ∼±1 ppm in any case. Hence, small deviations between what may be reported in the literature and what is contained in the ACD/Labs NNMR database are to be expected, since it would be very problematic to maintain several reference values for the ^{15}N chemical shift of nitromethane relative to liquid NH_3.

A number of other ^{15}N chemical shift referencing schemes have also appeared sporadically in the literature. The authors are aware that nitric acid, ammonium chloride, ammonium nitrate (variously using both of the ^{15}N resonances), and formamide have all been applied to natural product ^{15}N studies in general. Hence, care must be taken when using ^{15}N data from the literature to ensure that the chemical shift referencing scheme for the reported data is the same as that in use in an investigator's laboratory. Otherwise, reported ^{15}N chemical shift data must be converted to the referencing scheme used in the investigator's laboratory. The appropriate conversion factors are shown in Table 14.1.

Tab. 14.1 Inter-relation of ^{15}N chemical shifts in the various ^{15}N referencing schemes[a].

	^{15}N Chemical shift reference				
	Liq. NH_3	**CH_3NO_2**	**$^{15}NH_4NO_3$**	**$NH_4{}^{15}NO_3$**	**$^{15}NH_4Cl$**
Liq. NH_3	0.0	379.5	−19.9	383.5	−26.6
CH_3NO_2	379.5	0.0	359.6	4.0	352.9
$^{15}NH_4NO_3$	19.9	359.6	0.0	363.6	−6.7
$NH_4{}^{15}NO_3$	383.5	−4.0	363.6	0.0	356.9
$^{15}NH_4Cl$	26.6	352.9	6.7	356/0	0.0

[a] The standards shown have been the most commonly utilized although others such as nitric acid, formamide, and potassium nitrate have also been reported.

14.1.2 ^{15}N Chemical Shifts

The ^{15}N chemical shift range is very broad, encompassing approximately 900 ppm. For natural products, however, the majority of ^{15}N resonances of interest will be in the range from about 20 to 350 ppm, with the exception of a few functional groups such as nitro groups that resonate in the vicinity of ~380 ppm and methoxime (=N–OCH_3) groups that resonate at ~405 ppm. ^{15}N Chemical shifts of alkaloids range from those of amino groups, which are generally ~20–60 ppm down through thiazolyl and pyridyl nitrogens which resonate in the range of about 300–340 ppm (relative to NH_3).

Even with the restriction of the ^{15}N chemical shift range of interest to 20–350 ppm, this spectral window is still experimentally challenging from the standpoint of experiment parameterization. At the low observation frequency of ^{15}N (~60 MHz on a 600 MHz instrument), it is difficult to generate observed pulses short enough to effectively cover a 300+ ppm spectral window. For this reason it is preferable, when an investigator is working with a system that they understand reasonably well, to limit the F_1 spectral window to whatever extent possible. The use of ^{15}N chemical shift calculation algorithms such as Advanced Chemistry Development's ACD/NNMR software can help in this regard. Alternatively, as shown in the work of Hadden [9], it is also possible to employ adiabatic pulses to increase the ^{15}N excitation bandwidth in a uniform manner.

14.1.3 ^{15}N Reviews and Monographs

There are several reasonably extensive reviews of the reported applications of ^{15}N NMR. Unfortunately, monographs that have dealt with ^{15}N applications are becoming somewhat dated and, for the most part, do not discuss contemporary experimental methods applicable to ^{15}N. The first comprehensive review in the area of long-range ^{1}H–^{15}N NMR spectroscopy was that of Martin and Hadden published in 2000 [10]. Marek and Lycka again reviewed the growing body of long-range ^{1}H–^{15}N applications in 2002 [11]. The most recent review to appear was that of Martin and Williams published in mid-2005 [12]. While applications of long-range ^{1}H–^{15}N methods to alkaloids have been cited in the three published reviews [10–12] there has thus far been no comprehensive report that specifically deals with applications of these methods in alkaloid chemistry. A general review of applications of indirect detection methods in alkaloid chemistry has appeared [13], and a chapter has been published on the application of advanced multidimensional NMR methods to the structure determination of the *Amaryllidaceae* alkaloids [14]. Similarly, there has been no discussion of parameter considerations applicable to alkaloids in any of the work published to date.

Monographs dealing with ^{15}N include volumes by Levy and Lichter [15], Witanowski and Webb [16], Martin, Martin, and Gouesnard [17], and a volume dealing with the NMR of a number of nonmetallic elements by Berger, Braun, and

Kalinowski which includes ^{15}N [18]. There have also been a number of volumes devoted to ^{15}N NMR in the *Annual Reports on NMR Spectroscopy* series [19–24]. In spite of the age of these volumes, they still contain a wealth of valuable ^{15}N chemical shift information.

In addition to the published literature, a ^{15}N chemical shift database is being developed by Advanced Chemistry Development (ACD/Labs) that can be used interactively by an investigator both to predict ^{15}N chemical shifts for a molecule being investigated and to search the database by a multitude of parameters, including structure, substructure, and alphanumeric text values. This database is accessible in the NNMR software package offered by ACD/Labs and presently contains data on more than 8800 compounds with over 20 700 ^{15}N chemical shifts. Examples of the use of the NNMR database will be presented later in this chapter.

14.2 Indirect-Detection Methods Applicable to ^{15}N

The development of older heteronuclear 2D NMR experiments relied on the detection of the heteronuclide, typically ^{13}C. Quite early it was recognized that detection via a proton or other high sensitivity, high natural abundance nuclide, for example ^{19}F or ^{31}P, offered a considerable sensitivity advantage [2,3]. There were, however, experimental challenges to overcome, specifically in the case of ^{1}H–^{13}C shift correlation methods the need to suppress and reject ~99 % of the proton signal arising from ^{1}H–^{12}C species. In the case of ^{1}H–^{15}N experiments, the problem is magnified by another factor of 10 because of the difference in the natural abundance of ^{13}C versus ^{15}N. Nevertheless, investigators, instrument vendors, and probe developers rose to the inherent challenges of indirect detection methods. Bax, Griffey, and Hawkins [3] reported the application of ^{1}H–^{15}N indirect detection methods to the study of labeled proteins in 1983. By the mid-1980s, the HMQC (Heteronuclear Multiple Quantum Coherence) experiment pioneered by Bax and Subramanian [25] began to be applied in the structural characterization of small molecules. The HMQC experiment was followed in 1986 by the HMBC (Heteronuclear Multiple Bond Correlation) experiment of Bax and Summers [26] used for establishing long-range correlations between protons and a heteronuclide over two or more bonds. The HMBC experiment quickly supplanted the assortment of older, heteronucleus-detected long-range experiments that were reviewed by Martin and Zektzer [27].

Applications of indirect detection experiments to ^{1}H–^{15}N one bond (direct) and long-range (across two or more bonds) correlation initially differed in relative utility. While indirect-detection one-bond correlations work quite effectively in the authors' experience, the same could not always be said for ^{1}H–^{15}N HMBC experiments. While groups such as *N*-methyls could be readily observed, the observation of other long-range correlations to ^{15}N was challenging [28].

The development of gradient-enhanced heteronuclear shift correlation experiments in the early 1990s heralded a major improvement in the applicability of these experiments for ^{1}H–^{15}N direct and long-range heteronuclear shift correlation applications [29,30]. Based on these fundamental experimental developments, the first successful applications of long-range ^{1}H–^{15}N heteronuclear shift correlation experiments were reported by groups led by Koshino [31] and Martin [32–34].

14.2.1 Accordion-optimized Long-range ^{1}H–^{15}N Heteronuclear Shift Correlation Experiments

While ^{1}H–^{13}C heteronuclear long-range couplings are generally fairly uniform, the same cannot be said for the corresponding ^{1}H–^{15}N couplings. Rather, the orientation of the C–H bond vector of a proton that is long-range coupled to ^{15}N, relative to the orientation of the nitrogen lone pair of electrons, can have a significant impact on the size of the ^{1}H–^{15}N coupling constant. In cases where the C–H bond vector is synclinal to the orientation of the lone pair, couplings tend to be larger and are more readily exploited experimentally. In contrast, when the C–H bond vector is anticlinal, the corresponding couplings tend to be much smaller and are consequently more difficult to observe experimentally. The inherent variability of long-range ^{1}H–^{15}N couplings makes the optimization of the delay for the long-range transfer of magnetization between a proton and ^{15}N in an HMBC experiment a much more difficult undertaking for an investigator.

Unlike a ^{1}H–^{13}C long-range correlation experiment in which optimization of the long-range delay for an assumed coupling in the range of 6–10 Hz generally gives acceptable results, the same will not necessarily be true for a ^{1}H–^{15}N long-range correlation experiment. Consequently, early investigators either relied on the acquisition of several ^{1}H–^{15}N HMBC experiments, each recorded with a different optimization of the long-range coupling delay, or a single experiment with a long-range delay optimized for a small coupling, for example 2–2.5 Hz, that would refocus magnetization for several different larger long-range couplings of interest. Unfortunately, the downside of optimizing for a small delay are the signal losses during the very long delays. For example, 10 Hz optimization of an HMBC experiment corresponds to $[1/(2^{n}J_{CH})]$ to a 50 ms delay. In contrast, optimizing for a 2 Hz long-range coupling requires a 250 ms delay. Signal losses during a delay five times longer when the experiment is optimized for 2 rather than 10 Hz leads to an undesirable reduction in sensitivity.

A viable alternative to "static" optimization of the long-range delay in ^{1}H–^{15}N long-range heteronuclear shift correlation experiments is available through the use of accordion-optimized long-range experiments derived from the ACCORD-HMBC experiment described by Berger [35]. A modified accordion-optimized experiment, IMPEACH-MBC [36] has been applied to long-range ^{1}H–^{15}N heteronuclear shift correlation and offers a considerable advantage over the GHMBC experiment in a direct comparison [37]. More recently, Kline and Cheatham [38] reported a modification of the CIGAR-HMBC [39] experiment specifically intended for use in the acquisition of long-range ^{1}H–^{15}N heteronuclear shift correlation data with excellent results. It is useful to note that in the direct comparisons performed by Kline and Cheatham [38] some responses that were not observed at all in the HMBC experiment were in contrast observed with excellent intensity in the ^{15}N-optimized CIGAR-HMBC data.

Accordion-optimized heteronuclear 2D NMR experiments operate by the successive reoptimization of the long-range magnetization transfer delay in successive increments of the evolution time (t_1) in the experiment. The experiments are generally designed so that the longest magnetization transfer delay (the smallest coupling constant for which the experiment is optimized) occurs with the first increment of the evolution time. In this fashion, the duration of the long-range delay is successively decremented as the duration of the evolution time is incremented, thereby keeping the overall duration of the experiment as short as possible to avoid losses of magnetization arising from relaxation processes. Generally, in the authors' experience, the best optimization range, regardless of which of the accordion-optimized experiments is used for long-range ^1H–^{15}N heteronuclear shift correlation, is obtained with the optimization range set for 4–8 Hz.

14.2.2
Pulse Width and Gradient Optimization

Optimization of pulse widths and gradient ratios for long-range ^1H–^{15}N correlation experiments has been treated in several reviews [10,12] and in recent publications [9] to which the interested reader is referred.

14.2.3
Long-range Delay Optimization

The impact of delay optimization versus long-range correlation response intensity is graphically illustrated by the data presented in Figure 14.1, which were derived for

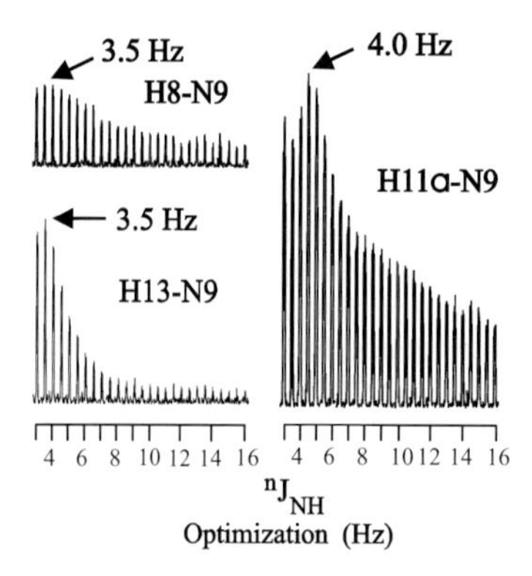

Fig. 14.1 Horizontal stack plots of a series of successively reoptimized selective 1D ^1H–^{15}N shift correlation experiments performed on the N9 amide nitrogen resonance of strychnine (**1**). The optimization of the long-range delay was varied in 0.5 Hz increments from 3 to 16 Hz. When strychnine was originally studied by one of the author's in 1995, a 10 Hz optimization of the long-range delay was utilized [33]. As can be readily seen from these comparative plots, the agreement between the delay optimization of 10 Hz (red peaks) and the actual coupling constants was rather poor, leading to minimal response intensity for the H13–N9 long-range correlation that could have been missed if working with a dilute sample.

strychnine (1) in a series of selective one-dimensional experiments in which the duration of the long-range delay was successively varied in 0.5 Hz steps over the range from 3 to 16 Hz. While it is certainly possible to cover the range of potential coupling constants by acquiring a series of GHMBC experiments with varied settings for the long-range delay, a more expeditious approach to ensuring that all long-range couplings are observed is to utilize one of the accordion-optimized long-range experiments described above.

A direct comparison of the results obtained using conventional GHMBC versus 3–8 Hz optimized IMPEACH-MBC experiments performed on strychnine is shown in Figure 14.2. Traces were extracted from an 8 Hz optimized GHMBC spectrum (Figure 14.2, bottom trace) and a 3–16 Hz optimized IMPEACH-MBC spectrum (Figure 14.2, top trace). While the signal-to-noise (s/n) ratio of the former was somewhat higher, that is solely due to the intensity of the correlation between the H11α and N9. In comparison, despite the slightly lower s/n in the IMPEACH-MBC data shown in Figure 14.2 (top trace), the response intensity for the H8–N9 and H13–N9 correlation responses is markedly better than was observed in the conventional 8 Hz optimized GHMBC experiment.

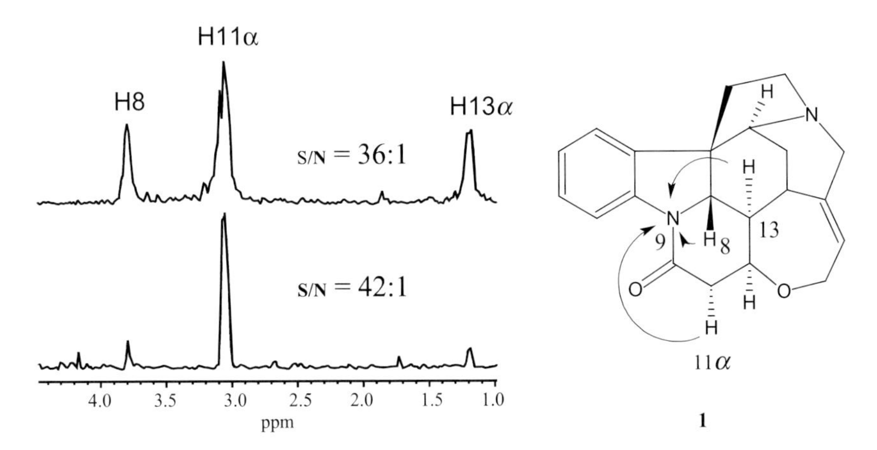

Fig. 14.2 Comparison traces for the N9 amide resonance of strychnine: top) trace extracted from a 3-16 Hz optimized IMPEACH-HMBC experiment; bottom) trace extracted from a conventional 8 Hz optimized GHMBC spectrum. Parameters were adjusted to give equivalent data matrices, the only difference being total acquisition time because of the difference between the statically optimized delay in the case of the GHMBC experiment and the accordion-optimized delay in the IMPEACH-MBC experiment [36]. While the overall s/n ratio of the IMPEACH-MBC spectrum is slightly lower than that of the GHMBC data (36 : 1 vs. 42 : 1), the response intensities for the H8 and H13 ^{2}J and ^{3}J correlations, respectively, is significantly improved in the IMPEACH-MBC data. The response intensity improvement can be attributed to the accordion-optimization range encompassing the actual long-range correlations of the H8–N9 and H13–N9 correlations (~3.5 Hz).

14.2.4
Establishing F_1 Spectral Windows

A fundamental consideration of any long-range ^{1}H–^{15}N heteronuclear shift correlation experiment is the determination of appropriate spectral windows in the two frequency domains. Setting the F_2 spectral window is straightforward, based on the acquisition of a proton spectrum, and does not merit any further discussion. The F_1 spectral window, in contrast, requires either some knowledge of the nature of the alkaloid structure in question and the ^{15}N chemical shifts characteristic of that alkaloid skeleton or a means of predicting the ^{15}N chemical shifts for the working structural hypothesis. Since the body of ^{15}N chemical shift information for alkaloids is still somewhat limited, coupled with a lack of familiarity with ^{15}N chemical shift ranges for most investigators, generally ^{15}N chemical shift prediction is the preferred approach for establishing F_1 spectral windows for the experiments to be performed.

One of the fundamental approaches to acquiring heteronuclear 2D NMR data for unknown molecules in many laboratories is to acquire data using "survey" spectrum conditions. For "survey" conditions, pre-established spectral windows are routinely used in both frequency domains. In the case of ^{1}H–^{13}C HSQC and HMBC experiments, using survey conditions works reasonably well since most investigators have a very good intuitive feel for where carbons in a molecule will resonate.

Further, the F_1 spectral windows generally used to acquire ^{1}H–^{13}C survey spectra are compatible with the probe performance, allowing good quality data to be acquired in most circumstances, assuming that parameters such as coupling constant optimization(s) are reasonable for the sample in question. In contrast, the possible F_1 spectral window for ^{1}H–^{15}N long-range heteronuclear shift correlation experiments is effectively twice the width of that used for ^{1}H–^{13}C long-range correlation survey spectra. In addition, on a 600 MHz NMR spectrometer, while ^{13}C resonates at $\sim$150 MHz, ^{15}N resonates in the vicinity of $\sim$60 MHz. When triple resonance inverse-detection gradient NMR probes are built, they are usually biased toward more efficient pulse capabilities for the ^{13}C end of the turning range of the doubly tuned X coil. At the other end of the range, ^{15}N pulses are generally less efficient, leading to a situation where the F_1 spectral width that can be effectively excited is more narrow for ^{15}N than for ^{13}C, which can be exasperating given that the ^{15}N long-range heteronuclear correlation spectral window can in practice be twice that required for the corresponding ^{13}C long-range correlation experiment.

Figure 14.3 shows what can happen when long-range ^{1}H–^{15}N 2D NMR data are acquired using survey conditions compared with acquiring these data following ^{15}N chemical shift calculations. The data shown in Figure 14.3a were acquired for the simple alkaloid harmaline (**2**) using the ^{1}H–^{15}N survey conditions established in the laboratory where these data were acquired. These survey conditions set the F_1 spectral window for the range from 50 to 200 ppm, which is perfectly adequate for indole alkaloids and related compounds. Unfortunately, the N1 resonance of harmaline involves a C=N double bond, which shifts the N1 resonance considerably downfield and outside of the F_1 spectral window used in Figure 14.3a, resulting in this resonance folding back into the observable spectral window. If an investigator were

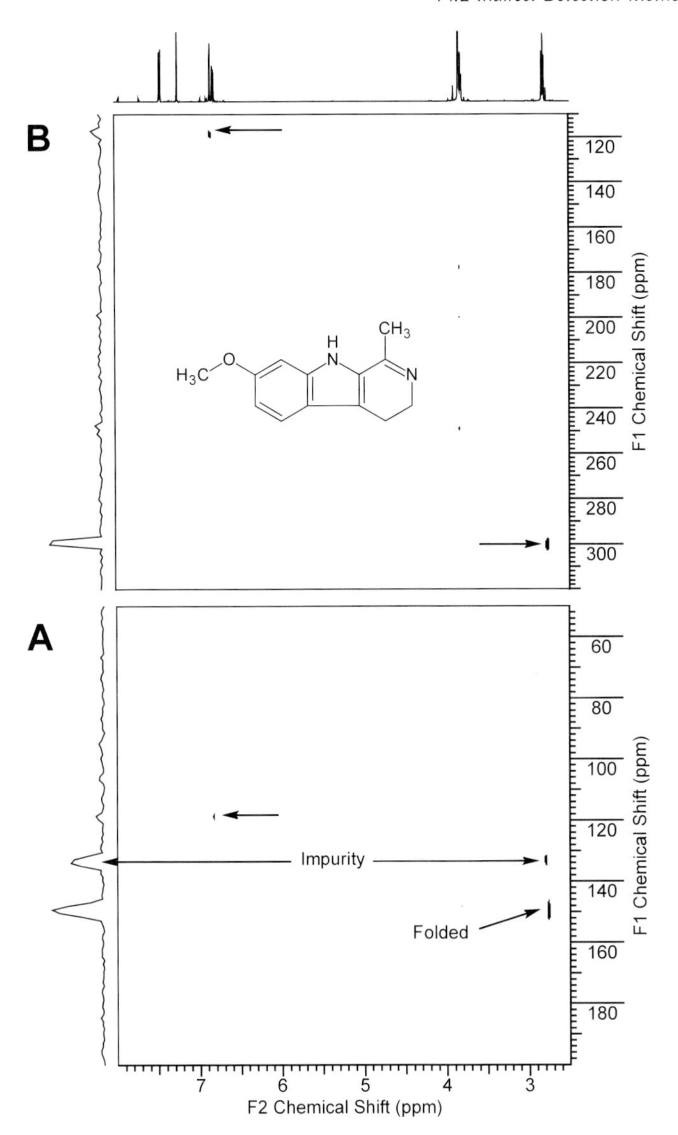

Fig. 14.3 Figure showing two ^1H–^{15}N IMPEACH-MBC spectra of the simple alkaloid harmaline (**2**, structure inset). The data were recorded using an accordion optimization range of 3–8 Hz. The spectrum shown in panel **A** was recorded using standard survey conditions routinely used in the laboratory in which these data were acquired, with the F_1 spectral window set from 50-200 ppm (relative to liq. $NH_3 = 0$ ppm). The data shown in panel **B** were acquired with the F_1 spectral window set from 100 to 320 ppm based on ^{15}N chemical shift calculations for harmaline performed using ACD/NNMR v 9.07 software. (See Figure 14.4).

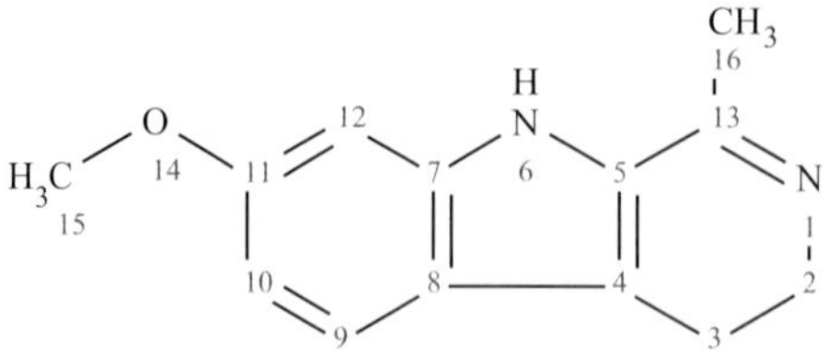

X	N1	Value(ppm)	Error
15N	6	131.6	6.8
15N	1	298.1	9.07

Fig. 14.4 ACD/NNMR chemical shift calculation for the simple alkaloid harmaline (**2**). The N6 resonance, as expected, has a ^{15}N chemical shift in a range typical for indoles while the N1 resonance exhibits a calculated ^{15}N chemical shift more typical of that of pyridine, which is well outside the F_1 window provided for the survey conditions shown in Figure 3A.

dealing with an unknown structure, folding of this type could suggest a completely different type of environment for the N1 nitrogen than it actually occupies, leading, in turn, to a potential misassignment of the structure in some cases. In contrast, when the ^{15}N chemical shifts of harmaline are calculated using ACD/NNMR software, the predicted ^{15}N shifts for N1 and N6 are 298.1 and 131.6 ppm, respectively. After the F_1 spectral window was adjusted from 50–200 to 100–320 ppm, the 3–8 Hz optimized IMPEACH-MBC spectrum of harmaline was reacquired, affording the results shown in Figure 14.1b. The N6 indole resonance again resonated in the vicinity of 120 ppm, as expected. More importantly, the N1 C=N resonance appeared at the correct position in the spectrum near 300 ppm as predicted. An example of the value of prediction prior to data acquisition will be described later in this chapter. Through the process of comparing experimental shifts with predicted chemical shifts we were able to identify potential issues with the experimental data, which were later confirmed to be erroneous (Figure 14.4).

14.3 ^{15}N Chemical Shift Calculation and Prediction

14.3.1 Structure Verification Using a ^{15}N Content Database

For the chemist attempting to elucidate a chemical structure, the ^{15}N chemical shift is a sensitive probe of the nitrogen environment. For the purpose of structure verification, a common approach is to review the literature for related species and use their chemical shifts and couplings as models to allow estimates of these properties for the new species. While there are a number of texts, reviews, and publications available that have brought together the spectral properties of tens to hundreds of molecules, these paper-based collections are cumbersome to use when it comes to searching for a particular chemical shift or a chemical structure or substructure. With both time and quality being of the essence for such searches of data, the most obvious approach is to compile an appropriate collection of data into an electronic database and enable the appropriate types of searches.

When a content database of chemical structures and associated spectral parameters is made available, this can greatly speed up the process of identifying the nature of a compound. Electronic content databases are available from a number of sources. The largest and most up-to-date source of ^{15}N data is that supplied by ACD/Labs. The content database is delivered with their ACD/NNMR Predictor program [40]. It can be searched by structure, substructure, similarity of structure, chemical shift or range of chemical shifts, as well as by coupling constants. Add to this the ability to search through the databases by formula and mass (nominal, average, or exact) and an NMR spectroscopist has immediate access to a warehouse of valuable information.

The ACD/NNMR v10 content database contains >8800 chemical structures (>21 000 ^{15}N chemical shifts). These data have been culled from the literature and checked for quality according to a number of stringent criteria prior to being added to the database. The chemical shift reference is homogenized during the process such that all shifts are relative to one reference (even though predictions can be referenced to four common standards: liq. NH_3, NH_4Cl, HNO_3, and CH_3NO_2). A single database record includes the chemical structure, the original literature reference, the ^{15}N chemical shift(s) and, where available, associated heteronuclear coupling constants.

The database is updated on an annual basis with new data extracted from the literature. This database is also the foundation of data supporting the prediction algorithms that are required to predict NMR spectral properties for chemical structures not contained within the database.

14.3.2
^{15}N NMR Prediction

NMR prediction brings the possibility of structure verification based on chemical shifts as well as offering the opportunity of using prediction to optimize experimental acquisition parameters and sweep widths for the acquisition of 2D spectra, a valuable facility considering the example of the survey conditions applied to harmaline as discussed earlier. ACD/Labs uses proprietary algorithms based on a modified form of HOSE-code technology [41]. In order to perform a prediction, the user simply sketches the chemical structure of interest using a structure editor. The calculation of the chemical shifts and coupling constants is performed in a matter of seconds. A resulting table of chemical shifts displays the number of the atom in the structure that gives rise to the predicted shift, the value of the predicted shift, and the uncertainty of the predicted shift, based on 95 % confidence limits for the structure fragment. The table also includes predicted coupling constants between pairs of atoms.

It is possible to determine how the chemical shifts were predicted, and the type of structural fragments used to derive the parameters though a *Calculation Protocol* window which shows a series of points, each representing an individual chemical structure and associated chemical shift that was used to influence the chemical shift prediction. If the prediction was performed on a compound not in the database, then a variety of different structures are shown in the *Calculation Protocol* and displayed as a histogram plot containing structures that are only fragmentally similar to the input structure. The general applicability and success of ^{15}N NMR prediction will be examined in further detail below.

14.3.3
Enhancing NMR Prediction With User-"trained" Databases

Since chemical structure diversity continues to shift and new compounds are synthesized and characterized almost daily, it is a significant effort to ensure that the content of the database and the associated improvement in prediction accuracy is maintained. ACD/Labs culls data on an ongoing basis with sources including the published literature, academic laboratories, and commercial projects in which the compounds are now of noncommercial interest or have been protected by patenting. The value of proprietary data generated in-house is limited if these data cannot be collected and incorporated into a database for the reasons cited above. The ideal of allowing these data to be searched together with the internal content database, as well as allowing them to contribute to the NMR predictions is possible by creating a *user database* of structures, associated shifts, and other data. User databases can be built to contain hundreds to thousands of chemical structures that can then serve to enhance chemical shift prediction for a given laboratory and to provide a legacy resource for any organization. For researchers studying specific alkaloid classes the ability to populate a database with experimentally determined values of both shifts and coupling constants offers significant benefits for future data-mining and for short-cutting the process to both structure identification and experimental condition optimization. Coupled with the ability to manage the collective wealth of related information, including thousands of data fields, the scientist can find such flexibility inherently beneficial to managing their laboratory operations.

14.3.4
Validating ^{15}N NMR Prediction

The validation of ^{15}N NMR prediction is best performed by comparing the predicted shifts for compounds not in the database with the experimental shifts available in the literature or measured directly. ACD/Labs have reported [42] a statistical analysis of their ^{15}N NMR prediction. Using a classical leave-one-out (LOO) approach they predicted the ^{15}N shifts for >8300 individual chemical structures contained within the ACD/NNMR v 8.08 NNMR program database. The resulting analysis gave a correlation coefficient of $R^2 = 0.97$ over 21 244 points. The distribution in deviations between the experimental values and the predicted values using this LOO approach is shown in Figure 14.5.

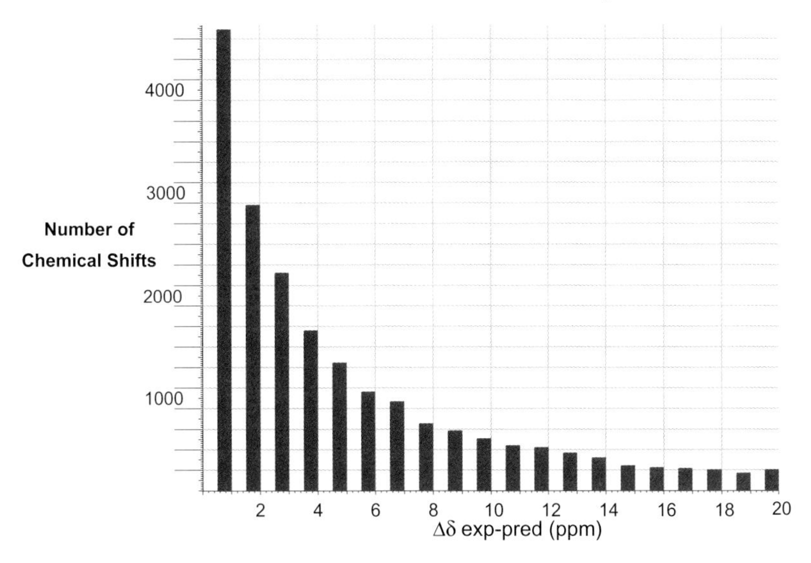

Fig. 14.5 LOO (leave-one-out) analysis of the ^{15}N NMR database contained in ACD/Labs NNMR chemical shift and coupling constant prediction package. The calculation was done by removing one ^{15}N shift and then calculating the ^{15}N shift in question using the remaining structures and data contained in the database. The resulting analysis gave a correlation coefficient of $R^2 = 0.97$ over 21 244 points. The presentation shows a plot of a number of chemical shifts vs. the difference between the experimental and predicted chemical shift values. Approximately 250 shifts out of >20,000 have a deviation of >20 ppm.

Selecting 24 random compounds (none from tables of related structures) from different sections of the authors' review [12] on long-range ^1H–^{15}N heteronuclear shift correlation that are not in the ACD/NNMR v8.08 database, the authors calculated the ^{15}N shifts for the 71 nitrogen resonances contained in these structures. Plotting calculated versus observed ^{15}N shifts for the 24 compounds used, we obtained the results shown in Figure 14.6. Regression analysis of these data gave $R^2 = 0.97$, with a standard error of 14.8 ppm. Based on the structural diversity of the 24 compounds used in this analysis, a standard error of $< \pm 15$ ppm is quite reasonable, and affords a basis for setting F_1 windows to acquire long-range ^1H–^{15}N data if one were dealing with an unknown molecule.

As described earlier the prediction algorithms are derived from a training set of over 21 000 chemical shifts. The training set is upgraded on an annual basis based on published literature data. For the chemical shifts reported in this chapter, almost 75 % of the data reported are contained within the database presently associated with the version 10 release. For the remaining 25 % of chemical shifts listed in this article a regression analysis was performed to compare experimental versus predicted chemical shifts. The results are represented in Figure 14.7. Regression delivers a value of $R^2 = 0.987$, an excellent correlation and demonstrative of the performance of the NMR prediction algorithms.

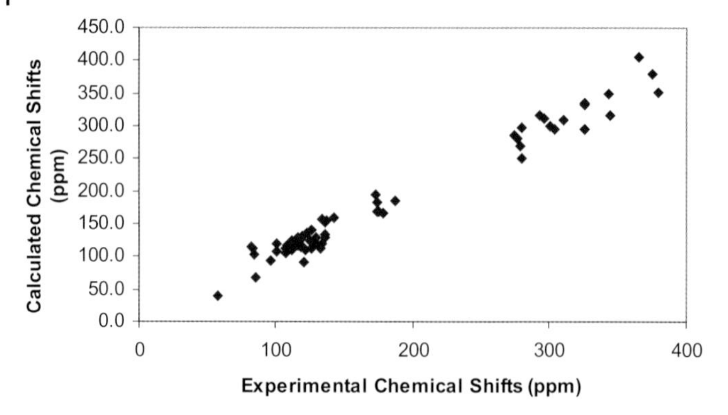

Fig. 14.6 Plot of observed versus calculated ^{15}N chemical shifts for 24 compounds from a previously published review (71 ^{15}N shifts) [12]. None of the 24 compounds selected were contained in the NNMR v8.08 database. When regression analysis was performed on these data the following results were obtained: $R^2 = 0.97$, standard error 14.8 ppm.

14.4 Computer-assisted Structure Elucidation (CASE) Applications Employing ^{15}N Chemical Shift Correlation Data

There have been very few reports of the combined use of computer-assisted structure elucidation methods that have employed any form of ^{15}N data, whether direct or long-range correlation. Two papers appeared in 1999, the first by Köck, Junker, and Lindel [43]. These authors compared the application of computer-assisted structure elucidation methods to the alkaloid oroidin (**3**) with and without the inclusion of ^{15}N chemical shift correlation data. The tabulated results obtained for oroidin and reported in the study are quite interesting in terms of the impact of the availability of ^{15}N chemical shift correlation data.

3

As is readily seen from the data contained in Table 14.2, when there were limited numbers of long-range ^{1}H–^{13}C HMBC correlations provided to the program with the COSY data, the number of possible structures generated was enormous. As the quality of the data set increased in terms of the numbers of ^{1}H–^{13}C HMBC correlations provided to the program, the number of structures generated dropped precipitously. However, when the ^{1}H–^{15}N HMBC data were included, even with an underdetermined ^{1}H–^{13}C HMBC data set, the program still

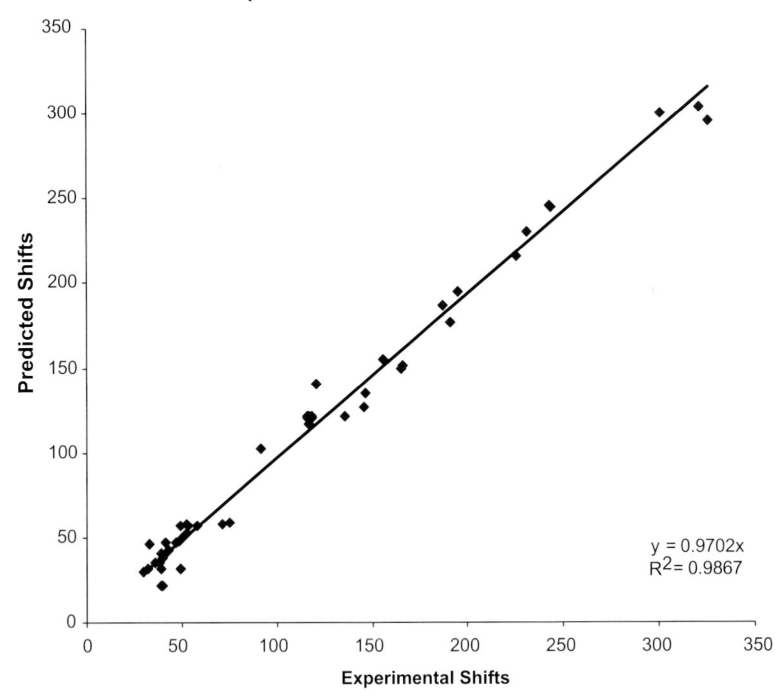

Fig. 14.7 Plot of observed vs. calculated ^{15}N chemical shifts for 49 ^{15}N chemical shifts from this chapter. These chemical shifts and associated compounds are not contained within the training set. Regression analysis delivered: $R^2 = 0.987$, standard error 14.8 ppm.

generated only a small number of structures that an investigator would have to sort through.

Nuzillard and coworkers [44] also reported the application of the LSD CASE program to the elucidation of the structure of acosmine acetate (**4**) in 1999. The

Tab. 14.2 Impact of information content on the application of the COCON CASE program for the elucidation of the structure of oroidin (**3**).

Oroidin datasets	#HMBC correlations	COSY, HMBC	COSY, HMBC, ^{15}N HMBC
A	9 (29 %)	234 336	10
B	18 (58 %)	27 142	10
C	23 (74 %)	690	10
D	26 (84 %)	60	6

The #HMBC column shows the number of correlations provided to the program as a percentage of the total number of possible correlations. The values in the two columns to the right are the number of structures generated from the input data.

available NMR data, when supplemented by the ^{1}H–^{15}N GHMBC data, led to the generation of a single structure by the program in 0.1 s.

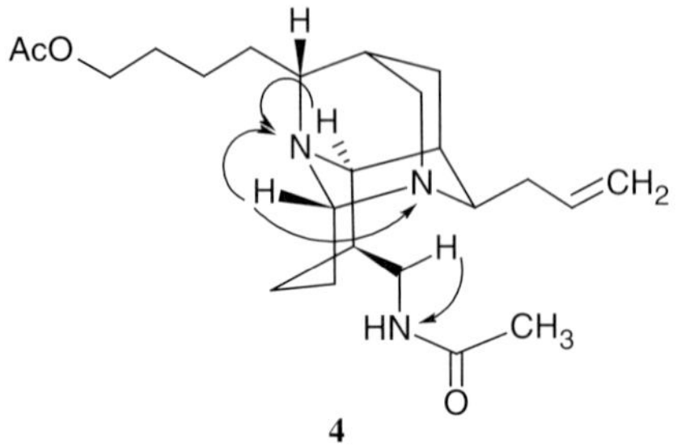

AcO H H N H N CH₂ H HN CH₃ O

4

The structure of a complex degradation product of cryptospirolepine, **5**, cryptoquinoline, **6**, was elucidated in parallel by both a spectroscopist and by colleagues using the Structure Elucidator CASE program [45]. After 10 years of storage in DMSO in a sealed NMR tube, a sample of cryptospirolepine (**5**) was totally degraded. Two of the major degradation products were isolated from the complex mixture. The more abundant degradant was readily identifiable as cryptolepinone, presumably derived by the oxidation at the spiro center of cryptospirolepine. The molecule gave a molecular ion ($MH^{+} = 479$) and HRMS data consistent with a molecular formula of $C_{34}H_{24}N_{4}O$. MS/MS fragment ions were observed at 464, 447, 435, 432, 247, 232, and 217 Da; the 232 daughter ion corresponds to a cryptolepine fragment minus a proton, suggesting that an 11-cryptolepinyl moiety (**7**) was an integral part of the degradant structure. The presence of an 11-cryptolepinyl fragment in the structure of the degradant was readily confirmed from the 2D NMR data available and the mass spectral data.

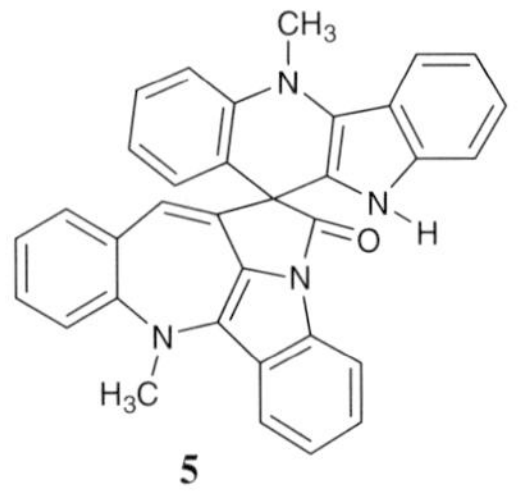

5 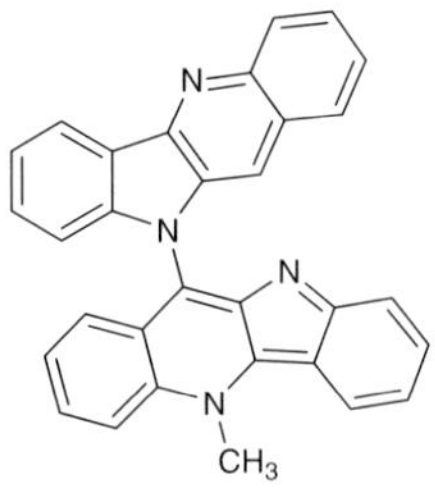 **6** 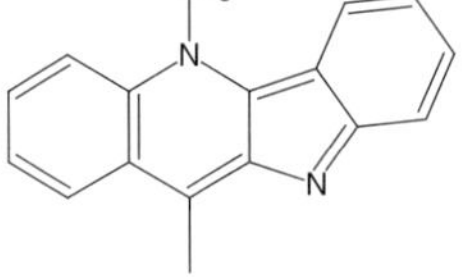 **7**

The 2D NMR data were interpreted to assemble the structural fragment **8**, which can be logically cyclized to afford **9**. Linking **7** and **9**, which is consistent with the fragmentation observed in the mass spectral data, affords the final structure of the degradant as cryptoquindolinone (**10**).

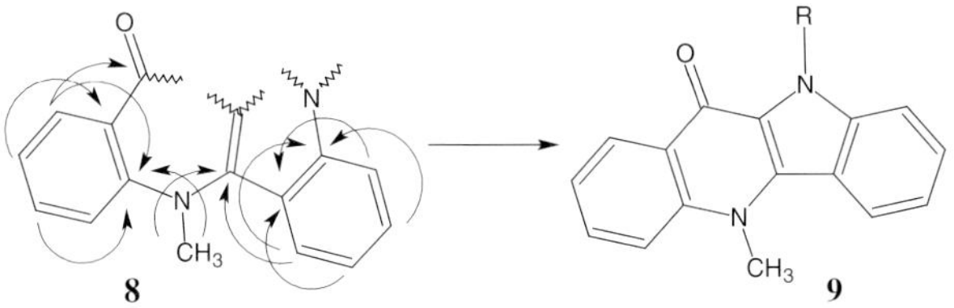

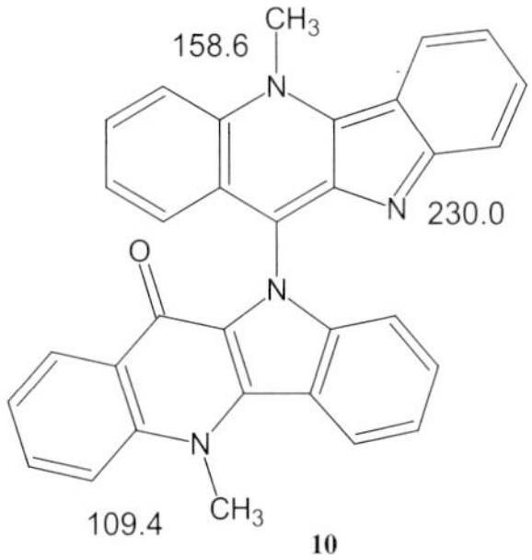

Long-range ^{1}H–^{15}N heteronuclear chemical shift correlations were observed for three of the four nitrogens in the structure of **10** in a 3–8 Hz optimized CIGAR-HMBC ^{1}H–^{15}N spectrum recorded for the ~100 μg sample using a 600 MHz Varian ColdProbe. As in the case of the report by Köck, Junker, and Lindel [43] discussed above, the ^{1}H–^{15}N data afforded a significant reduction in the computation time in the determination of the structure using the Structure Elucidator program. From the start of data interpretation to the completion of the structure elucidation, including the "quantum leap" provided by the recognition that the structure contained an 11-cryptolepinyl moiety based on the observed color and then confirmed by the data assignment, a highly competent spectroscopist required 72 h to assemble the structure. Using the same data, including the ^{1}H–^{15}N CIGAR-HMBC data, which afforded correlations to three of the four annular nitrogens in the structure, Structure Elucidator v7.0 consumed ~8 h of computer time, generating ~3300 structures in the process. After filtering, the output reduced to 355 structures, which after sorting, left the cryptoquindoline structure in the second position, with a deviation, $d = 3.947$.

Another example of the impact of ^{1}H–^{15}N chemical shift correlation was observed when the NMR data for the dimeric *Cryptolepis* alkaloid cryptomisrine (**11**) was used as input for the Structure Elucidator CASE program. Only ^{1}H–^{15}N HMQC direct correlation data were available. When this small piece of information was not provided as input, the program ran for 210 h, generating >75 million structures, of which >22 000 remained after filtering and the removal of duplicates. In contrast, when the ^{1}H–^{15}N HMQC data were provided as input, the program generated a total of only five structures in ~1 min [46].

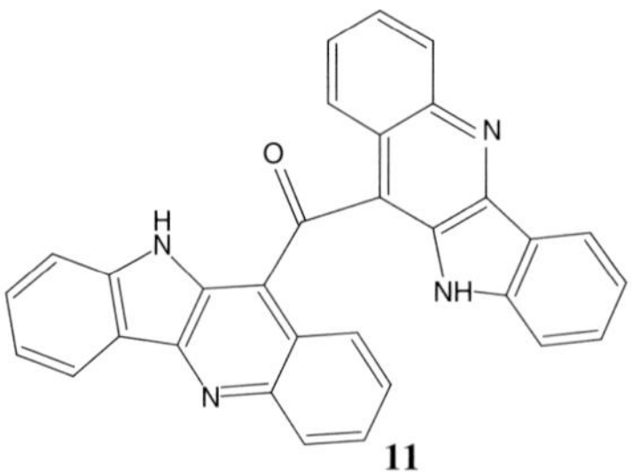

11

In another report utilizing CASE methods and long-range ^{1}H–^{15}N 2D NMR data, Grube and Köck [47] reported the characterization of oxocyclostylidol (**12**), an intramolecular cyclized oroidin derivative from the marine sponge, *Stylissa caribica*.

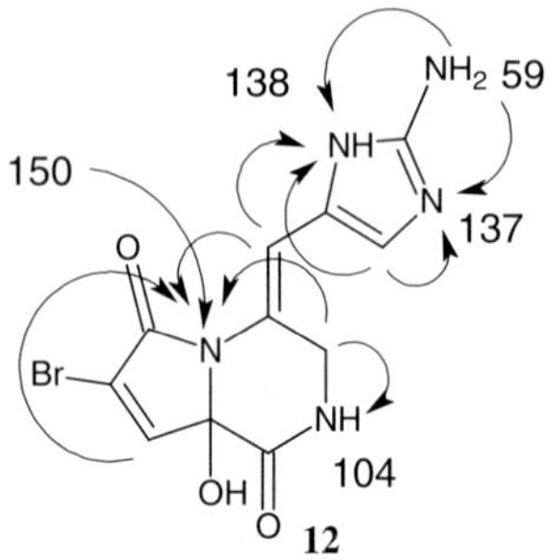

12

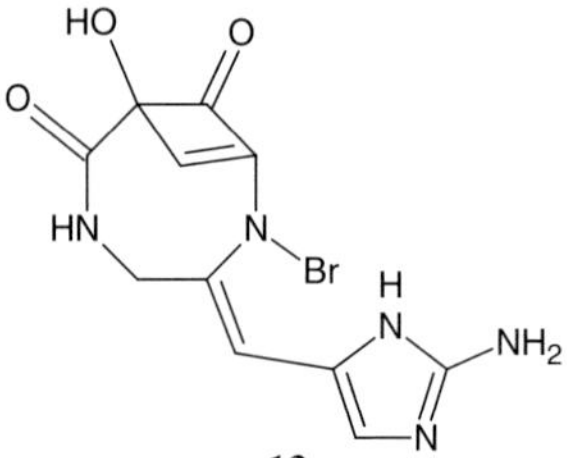

13

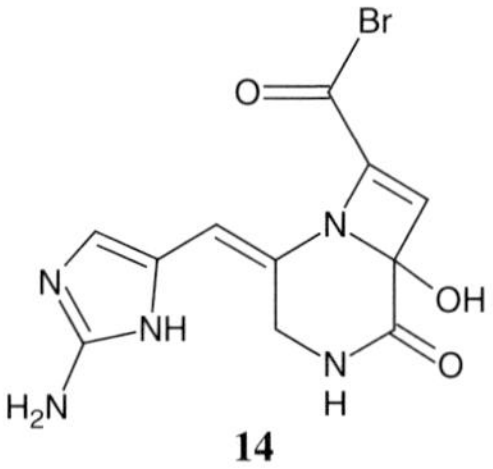

14

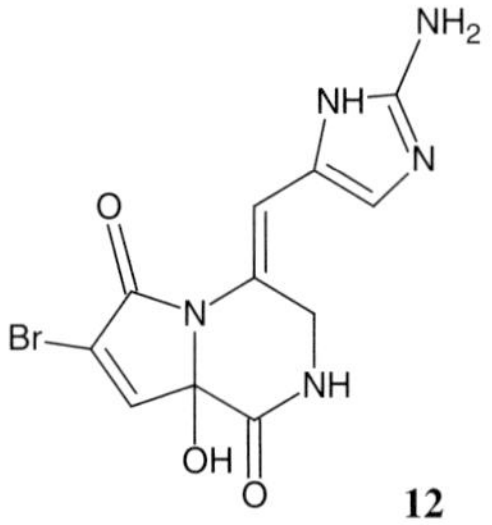

12

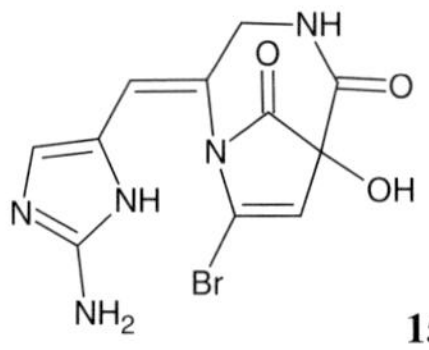

15

The authors used the COCON program to verify the assigned structure of oxocyclostylidol (**12**). The full complement of spectral data, including both long-range ^{1}H–^{15}N 2D and 1,1-ADEQUATE correlations were used as input. When the full NMR data set was used as input, four structures, **12–15**, were generated by the program. The calculation without the ^{1}H–^{15}N GHMBC data generated more than 50 000 structures; if the 1,1-ADEQUATE data were also excluded, the number of generated structures rose to more than 150 000. These observations clearly underscore the importance of long-range ^{1}H–^{15}N 2D NMR data in the computer-assisted structure elucidation of complex alkaloid structures.

A separate analysis of oxocyclostylidol has been performed recently by Elyashberg (private communication). Elyashberg, in collaboration with the authors of this chapter and other collaborators, previously reported on the applications of a CASE system, ACD/Structure Elucidator to complex molecules and specifically to natural products [45,46,48–58]. The details of the system and its associated algorithms will not be reviewed here and readers are referred to the previous articles for information.

In our earlier reports [56,59] we have already discussed the problems that arise when an expert system is used for molecular structure elucidation from 2D NMR data containing correlations assumed to correspond only to $^{2-3}J_{CH}$ and $^{2-3}J_{HH}$ coupling constants, respectively. We have defined such correlations as "standard" correlations [51,55,56]. For $^{n>3}J_{HH/CH}$ couplings we define these correlations as *nonstandard*. The origin and nature of nonstandard correlations (NSCs) has been discussed elsewhere in the literature [60]. It is generally believed by others in this field of study that correlations of nonstandard length are observed fairly rarely [61–63]. Results obtained in our work [59] contradict this opinion. Previously the solutions of more than 250 problems were investigated by applying ACD/Structure Elucidator [55,56] to the structural identification of complex natural products from 2D NMR and MS spectra. The studies indicated that almost half of the problems (45 %) contained nonstandard correlations in the 2D NMR data. Meanwhile, expert systems are usually optimized to structure elucidation assuming a set of correlations of common (standard) length. Prior to actually establishing the structure of the unknown molecule, the presence or absence of nonstandard correlations, as well their number and real lengths, remains unknown. Therefore, the problem adds up to molecular structure elucidation from spectrum-structural information that is not only fuzzy by nature ($^{2-3}J_{CH}$ in HMBC), but can also be both contradictory and uncertain (i.e. the number of nonstandard connectivities and their lengths are unknown).

In previous reports [56,58] we suggested approaches for solving problems in the presence of correlations of nonstandard lengths. The details will not be discussed in further detail here. Our experience has shown that the number of nonstandard correlations contained within the 2D NMR data associated with a molecule can be rather large – up to about 20 correlations. Elyashberg's review of the data presented by Grube and Köck [47] indicates that the lengths of the nonstandard correlations were likely determined post-elucidation by the authors. Six nonstandard correlations were identified, five within the HMBC data (unidirectional arrows) and one within the COSY (double-headed arrow) data. These are shown on structure **16**. The COCON program is unable to facilitate elucidation in the presence of such nonstandard correlations and would likely produce an empty structural file if the nonstandard correlations were included into the input data set.

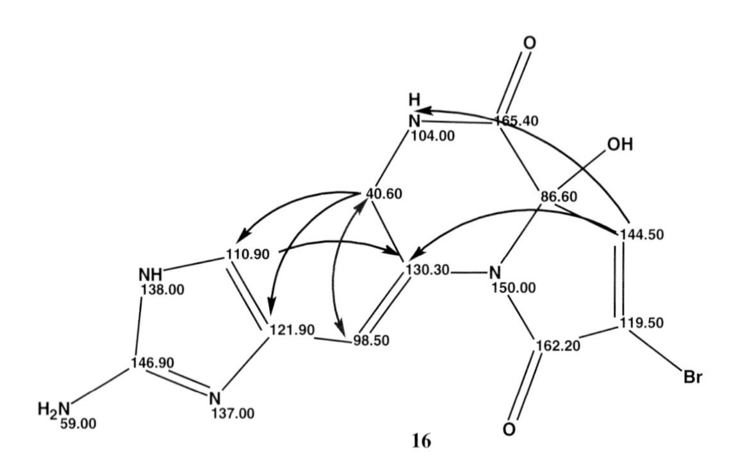

Elyashberg applied ACD/Structure Elucidator to the problem using the HMBC, COSY, ^{1}H–^{15}N HMBC, and 1,1-ADEQUATE data as inputs. A number of approaches to solving the problem were considered. The results demonstrated that the process of *fuzzy generation* [58] allowed elucidation of the structure with no assumptions being made regarding the number and lengths of the nonstandard correlations. The value of the ^{1}H–^{15}N HMBC data was demonstrated since when the data are excluded, the number of structures and generation time increase dramatically as previously demonstrated in applications of the COCON program [43,47]. The four structures generated as output by the Structure Elucidator program, ranked according to $d_A{}^{13}$C (accurately calculated ^{13}C shift) are shown in Figure 14.8.

14.5
Applications of ^{15}N Spectroscopy in Alkaloid Chemistry

Following the development and commercial availability of pulsed Fourier transform NMR instruments in the early 1970s there was considerable interest in the acquisition of ^{15}N chemical shift data. One of the first papers that the authors are aware of dealt with the assignment of the ^{13}C and ^{15}N shifts of the alkaloid physostigmine (**20**) [64]. While examining the literature for direct observation ^{15}N studies of alkaloids between the 1977 report by Steenberg [64] and the first reported applications of inverse-detected long-range ^{1}H–^{15}N 2D NMR studies at natural abundance by one of the authors in 1995 [32–34] is completely beyond the scope of this chapter, it is still

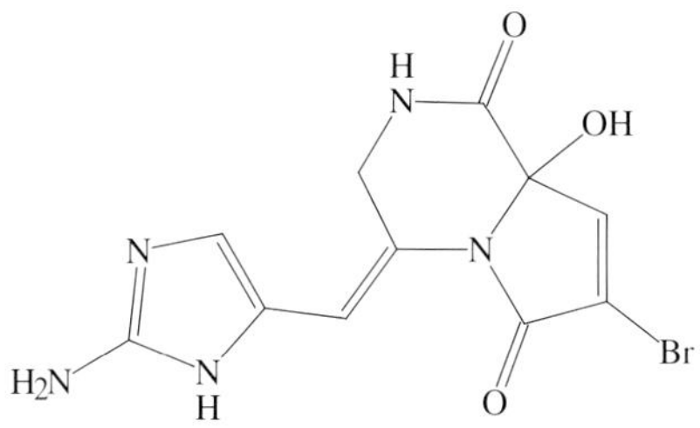

dA(13C): 4.392 (v.9.07)

12

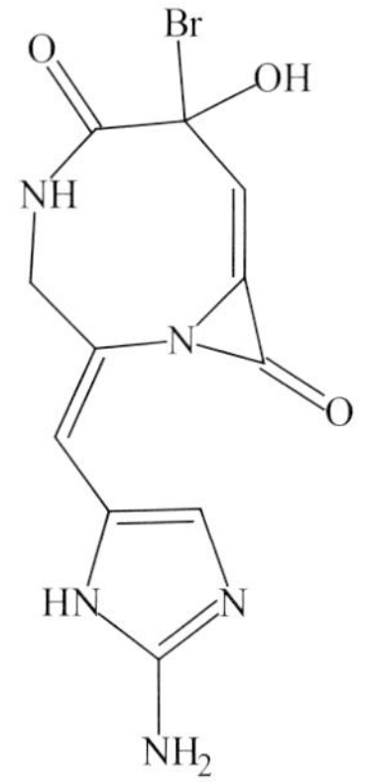

dA(13C): 8.155 (v.9.07)

17

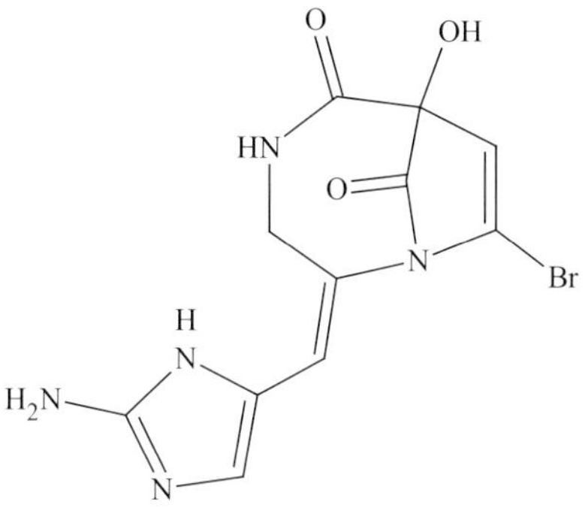

dA(13C): 8.205 (v.9.07)

18

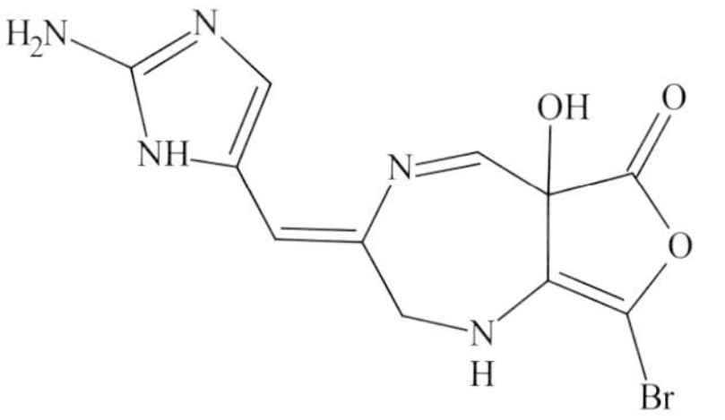

dA(13C): 9.345 (v.9.07)

19

Fig. 14.8 Structures generated by Structure Elucidator v9.07 for the data reported by Grube and Köck [47] when both the ^{1}H–^{15}N HMBC and 1,1-ADEQUATE data were included in the data input for the program. When the ^{1}H–^{15}N HMBC and 1,1-ADEQUATE were excluded, the number of structures generated by the program rose dramatically in a manner analogs to that reported by Grube and Köck [47] when the structure was solved using the COCON program. While the structures generated by Structure Elucidator are similar to the output from the COCON program, they are not identical, undoubtedly owing to differences in the fundamental algorithms of the two program packages.

interesting to note that interest in the application of ^{15}N NMR in the study of alkaloids dates back to the early history of FT NMR. For purposes of comparison, a 3–8 Hz optimized IMPEACH-MBC spectrum of physostigmine (**20**) was acquired and processed [36,37,65]. The chemical shifts were virtually identical to those reported from the ^{15}N directly observed spectrum. Long-range couplings, including a pair of weak $^5J_{NH}$ correlations (denoted by dotted arrows) from the aromatic protons flanking the point of attachment of the *N*-methyl carbamoyl group were observed in the spectrum as shown by **21**.

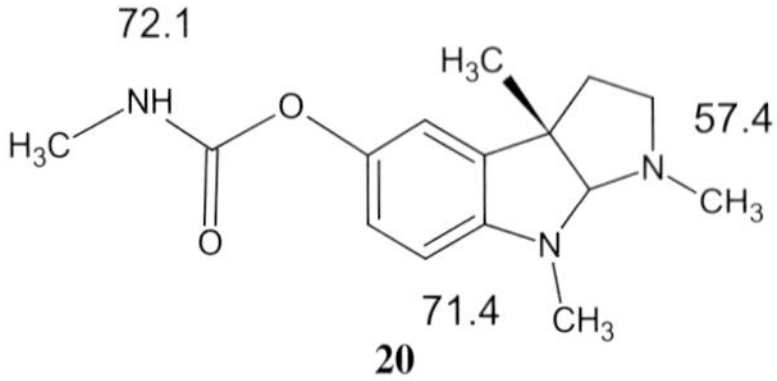

72.1 NH H_3C O H_3C 57.4 N CH_3 N 71.4 CH_3

20

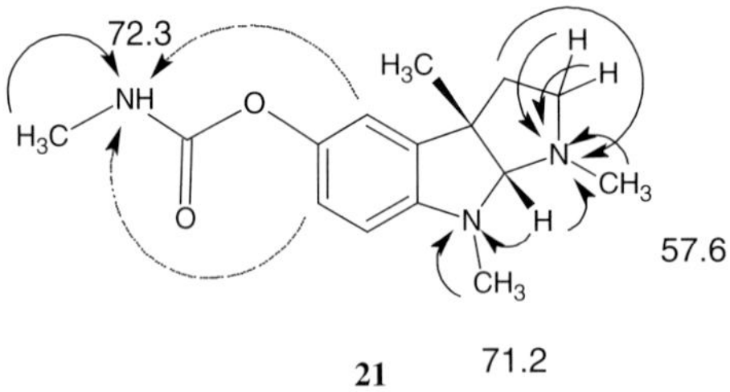

72.3 NH H_3C O O H_3C H H N CH_3 N H CH_3 57.6 71.2

21

Another useful collection of ^{15}N data on alkaloids is contained in the 1979 report of Fanso-Free and colleagues, who described a study of the *Rauwolfia* alkaloids and related model compounds [66]. Their results are summarized in Figure 14.9.

With any review of applications of a given technique, there is always the question of where to start and how to organize the applications. This chapter is no different in that regard. Applications are grouped in the following section on the basis of the nitrogen-containing species, for example, pyrrole, imidazole, and so on, and are arranged in order of increasing complexity. This grouping is completely arbitrary on the part of the authors.

14.6
Applications of Long-range ^1H–^{15}N 2D NMR

14.6.1
Five-membered Ring Alkaloids

The simplest alkaloid families to which ^1H–^{15}N chemical shift correlation methods have been applied are those containing a single five-membered ring with one or more

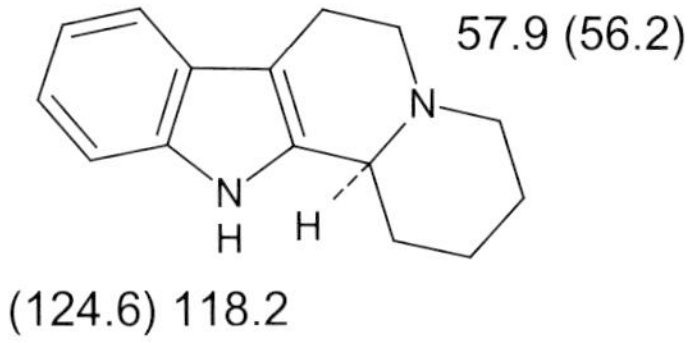

57.9 (56.2)
(124.6) 118.2
22

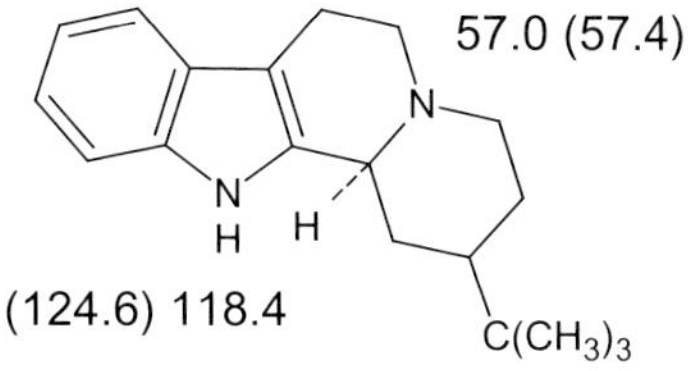

57.0 (57.4)
(124.6) 118.4
$C(CH_3)_3$
23

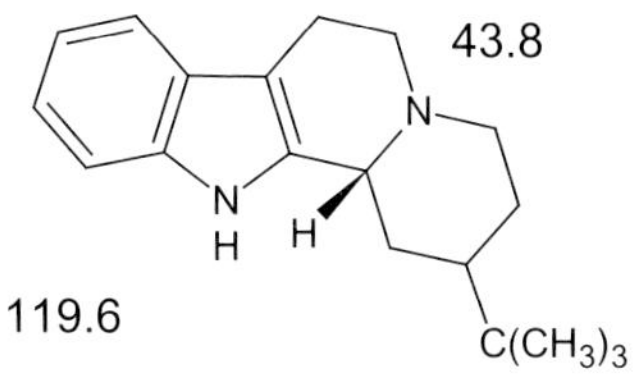

43.8
119.6
$C(CH_3)_3$
24

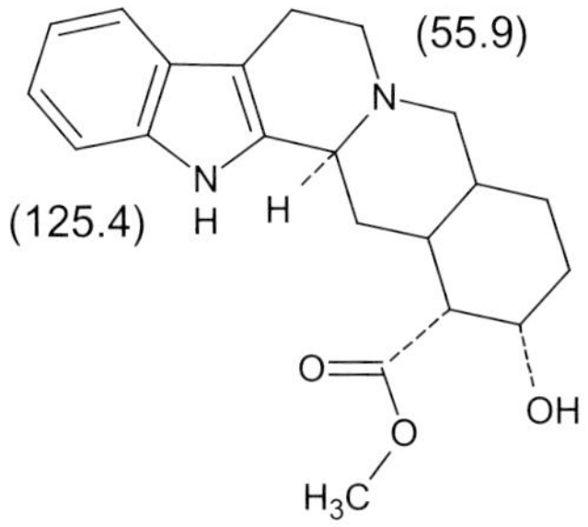

(55.9)
(125.4)
OH
H_3C
yohimbine (**25**)

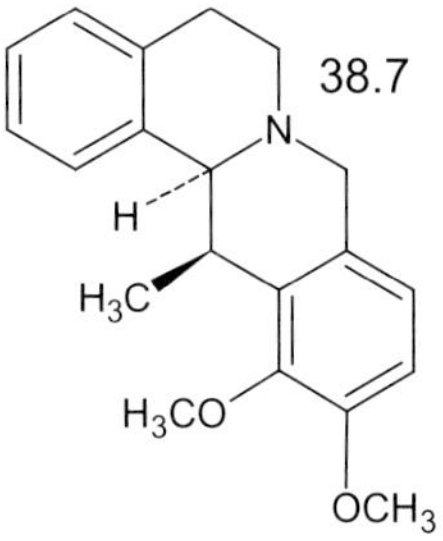

38.7
H_3C
H_3CO
OCH_3
corydaline (**26**)

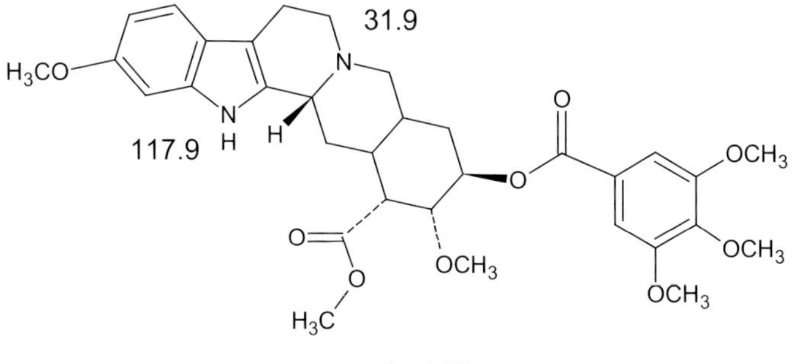

31.9
H_3CO
117.9
OCH_3
OCH_3
OCH_3
OCH_3
H_3C
reserpine (**27**)

Fig. 14.9 ^{15}N Chemical shifts of several Rauwolfia alkaloids and related model compounds. Chemical shifts are reported in chloroform or, parenthetically, in DMSO [66].

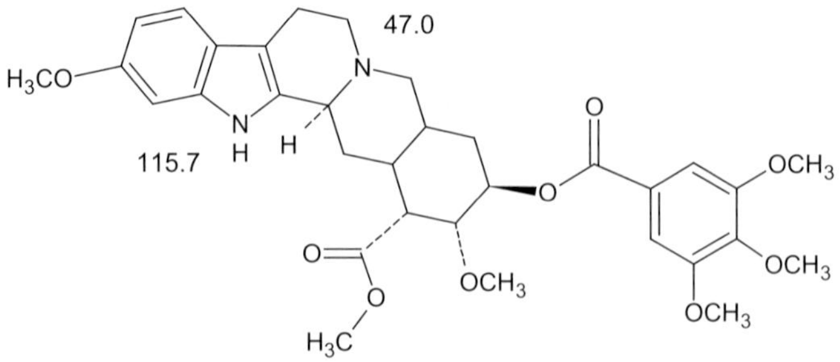

isoreserpine (**28**)

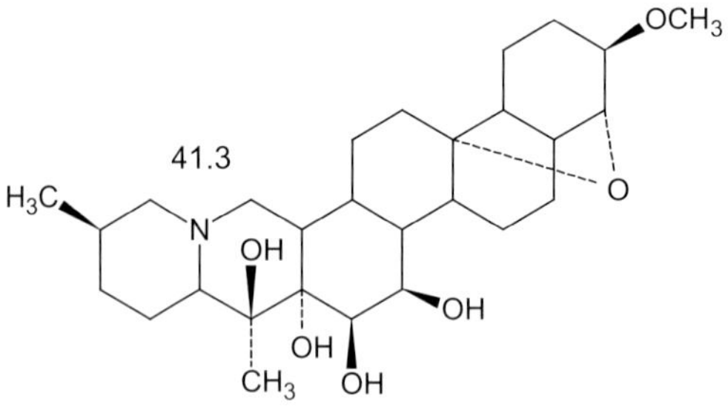

cevadine (**29**)

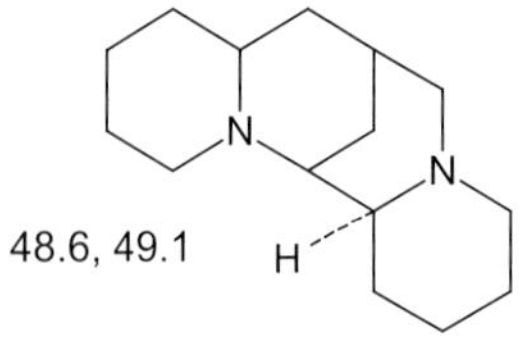

sparteine (**30**)

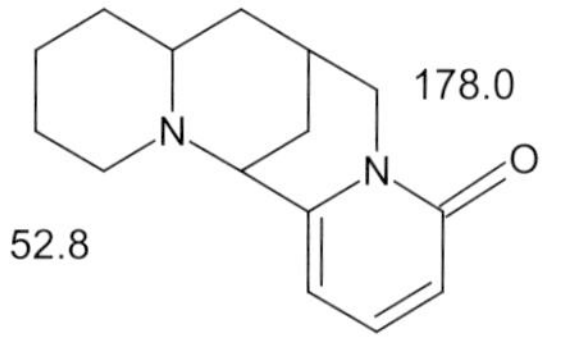

thermopsine (**31**)

Fig. 14.9 *(Continued)*

nitrogen atoms. In one of the initial reports on the application of ^{1}H–^{15}N long-range heteronuclear chemical shift correlation, Koshino and Uzawa reported the application of the technique to nicotine (**32**), in addition to other model compounds [67]

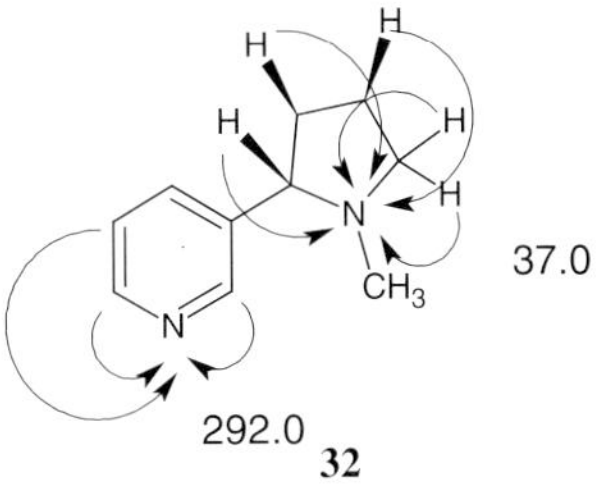

Kato *et al.* [68] reported the isolation and characterization of a novel imidazole antifungal alkaloid, fungerin (**33**) from the fungus *Fusarium* sp. The authors employed a ^1H–^{15}N GHMBC experiment optimized for 60 ms (8.3 Hz) to establish the substitution pattern of the *N*-methyl imidazole ring.

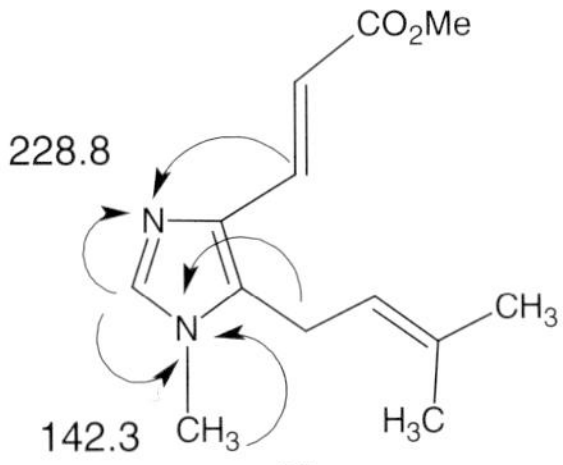

Fractionation of an alkaloid extract of *Psychotria colorata* flowers led to the isolation of six alkaloids, one of which was unknown. The new pyrrolidinoindoline alkaloid was, in part, characterized through the use of long-range ^1H–^{15}N GHMBC data as the highly symmetric structure **34** [69].

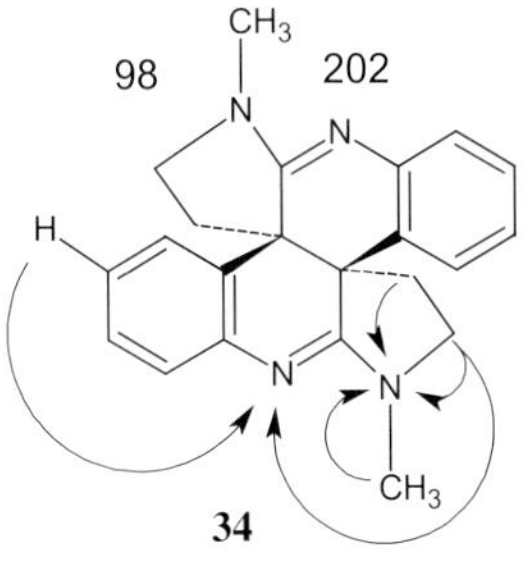

Later in 1998, two new pyrrole alkaloids were isolated and characterized from the berries of *Solanum sodameum* collected in the Libyan desert [70]. The structures were, in part, confirmed through the acquisition of ^1H–^{15}N GHMBC spectra. Long-range couplings observed and the ^{15}N chemical shifts of the alkaloid on which the proton–nitrogen experiments were performed are shown by **35**. The structure shown is reproduced as reported in the literature, although, in the opinion of the present authors, it is likely that the double bond between the two annular nitrogens within the imidazole ring probably favors the nitrogen resonating further downfield rather than as the structure is drawn because of the chemical shift.

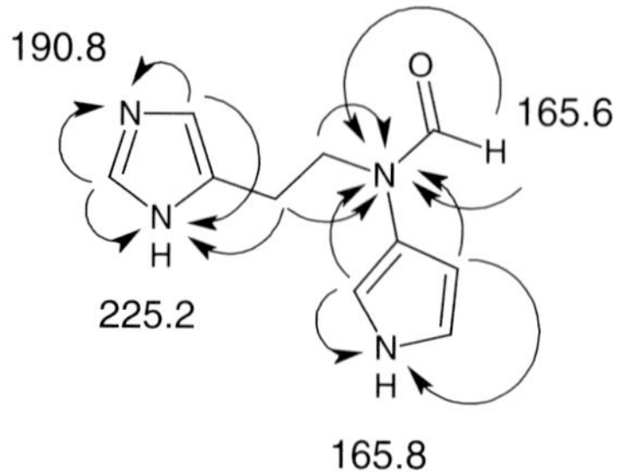

190.8 165.6 225.2 165.8 **35**

The next isolation and characterization of an imidazole alkaloid in which ^{15}N data were employed during the characterization is found in the work of Dunbar *et al.* [71] who reported a novel imidazole alkaloid, naamine-A **36**, from the calcareous sponge *Leucetta chagosensis.* The authors did not report the optimization used to acquire the data.

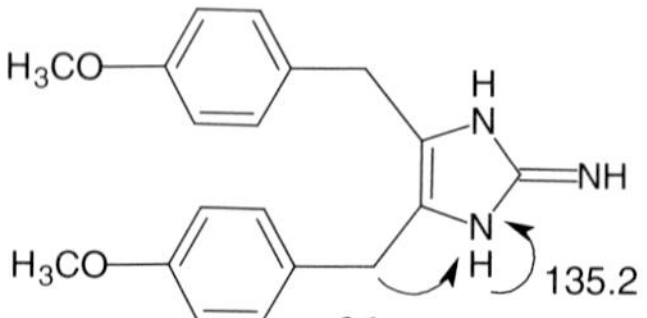

H_3CO H_3CO NH 135.2 **36**

Ford and Capon [72] used long-range ^{1}H–^{15}N 2D NMR data in the process of characterizing the complex pyrroloiminoquinone alkaloid discorhabdin-R (**37**) from two latrunculiid marine sponges *Latrunculia* sp. and *Negombata* sp. Unfortunately, the authors only reported the ^{15}N chemical shift of one of the three nitrogen resonances in the structure; they did not report any experimental details.

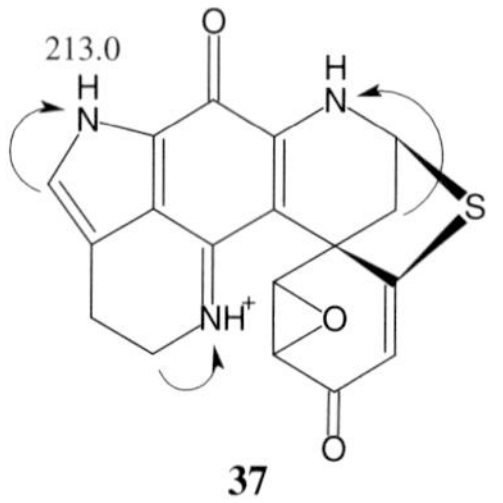

213.0 **37**

In a very interesting report from 2001, Ciminiello and coworkers [73] reported the isolation and structural characterization of the complex bromotyrosine-derived imidazole alkaloid archerine (**38**) that was isolated from the Caribbean marine sponge *Aplysina archeri*. The authors used long-range ^{1}H–^{15}N correlation data to assemble four substructural fragments, **39–42**. ^{15}N Chemical shift data reported in conjunction with these fragments are shown with the individual fragments. The final structure was assembled from the substructural components through the use of long-range ^{1}H–^{13}C correlation data. This study demonstrates quite clearly how long-range ^{1}H–^{15}N heteronuclear shift correlation can be used strategically in the assembly of a complex alkaloid structure.

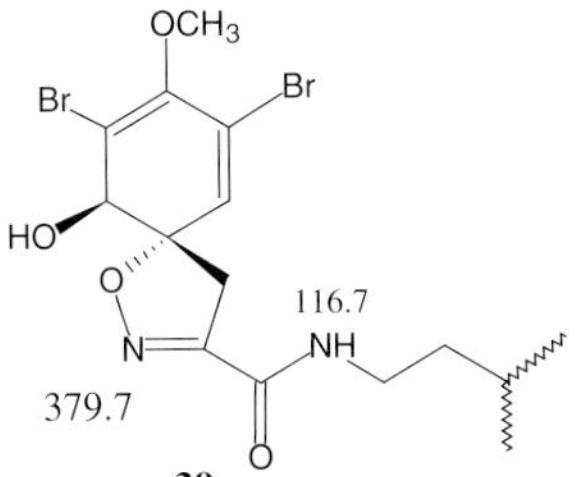

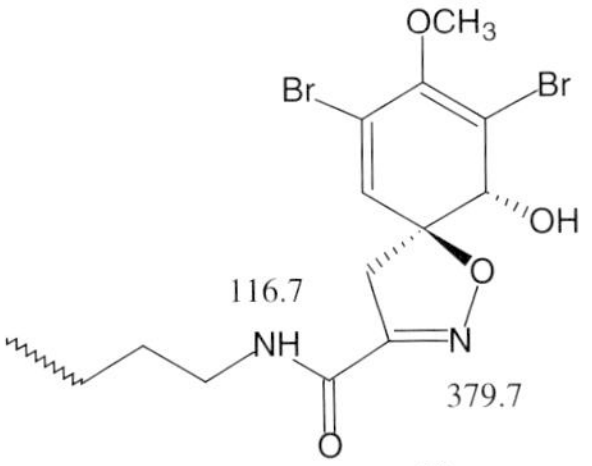

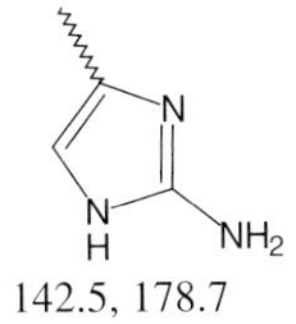

More recently, Kretsi and coworkers [74] reported the isolation and chemical structure characterization of several hepatotoxic pyrrazolidine alkaloids from *Onosma leptantha*. The structure of leptanthine (**43**) and leptanthine *N*-oxide (**44**) are shown. As would be expected following N-oxidation, the ^{15}N resonance of the latter is shifted downfield by +80.9 ppm relative to the parent compound. This behavior is consistent with the authors' experience, which has shown that N-oxidized nitrogens are shifted downfield from +65 to approximately +105 ppm [75].

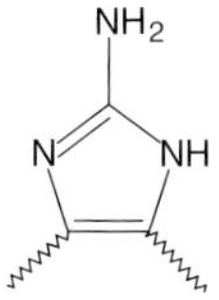

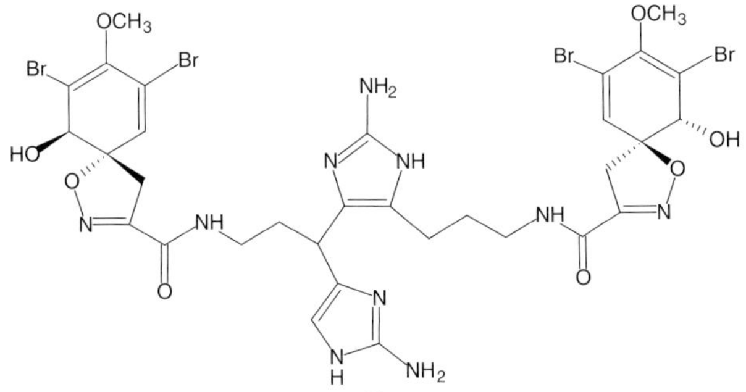

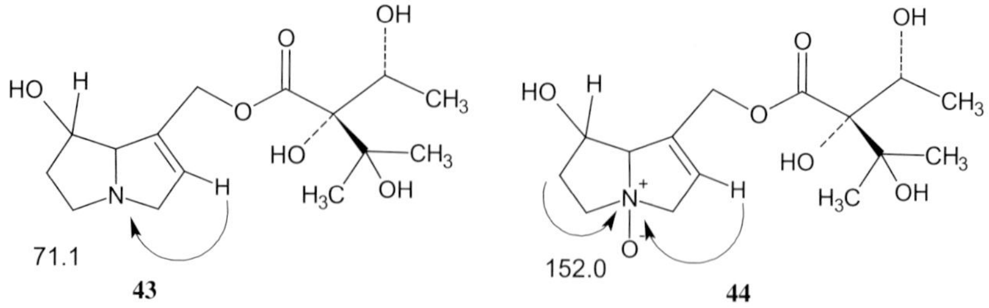

71.1 **43** 152.0 **44**

In an interesting report, the structure of tetapetalone-A (**45**), isolated from a *Streptomyces* sp., was revised through the use of long-range ^{1}H–^{15}N GHMBC data by Komoda *et al.* [76], who first reported the structure; the revised structure is shown by **46** [77].

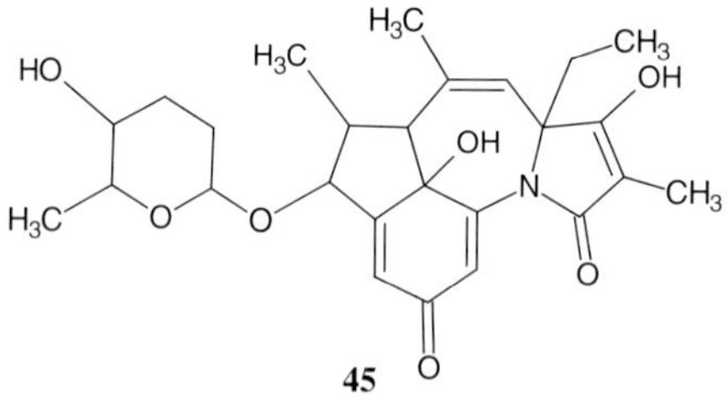

45

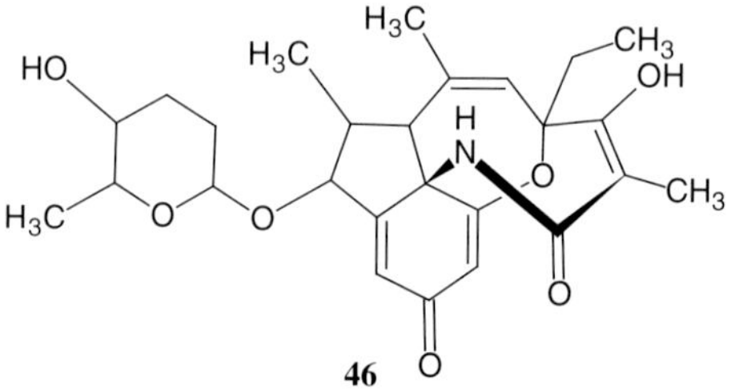

46

14.6.2
Tropane Alkaloids

In an interesting application, Fliniaux and coworkers [78] reported the use of long-range ^{1}H–^{15}N GHSQC and GHMBC data to monitor nitrogen metabolism leading to the formation of tropane alkaloids of *Datura stramonium* in transformed root culture. While there have not been many studies of this type, the increasing availability of cryogenic NMR probes, which offer greatly increased sensitivity over conventional NMR probes, will undoubtedly foster more such studies in the future.

Long-range ^{1}H–^{15}N 2D NMR methods were also used to examine the kinetics of the formation of calystegines (**47–50**) in root cultures of *Calystegia sepium* [79]. ^{15}N-Labeled tropinone was used as a precursor. The structures and assigned ^{15}N chemical shifts of some of the alkaloids that were studied are collected in Figure 14.10.

14.6.3 Indoles, Oxindoles, and Related Alkaloids

14.6.3.1 Strychnos Alkaloids

Members of the strychnos family of alkaloids were among the first compounds to which long-range ^1H–^{15}N 2D NMR methods were applied at natural abundance. Specifically, strychnine was studied both in the laboratory of one of the authors [33] and in the laboratory of Koshino and coworkers [31]. In addition, brucine (**51**) and holstiine (**52**) were studied in the author's laboratory. It is also quite probable that there have been more applications of long-range ^1H–^{15}N methods to indole alkaloids than to any other alkaloid class. The 1995 studies in the author's laboratories predated the development of accordion-optimized long-range methods by several years and hence employed conventional GHMBC experiments with multiple optimizations of the long-range ^1H–^{15}N delay in the pulse sequence.

The ^{15}N chemical shifts of the three alkaloids were quite similar. The N-9 amide shifts were observed toward the downfield end of the normal range for amide ^{15}N chemical shifts. The N-19 aliphatic nitrogen shifts were observed in the vicinity of 37 ppm, which is typical of secondary and tertiary aliphatic ^{15}N shifts. ^{15}N Chemical shifts for the three compounds are collected in Table 14.3.

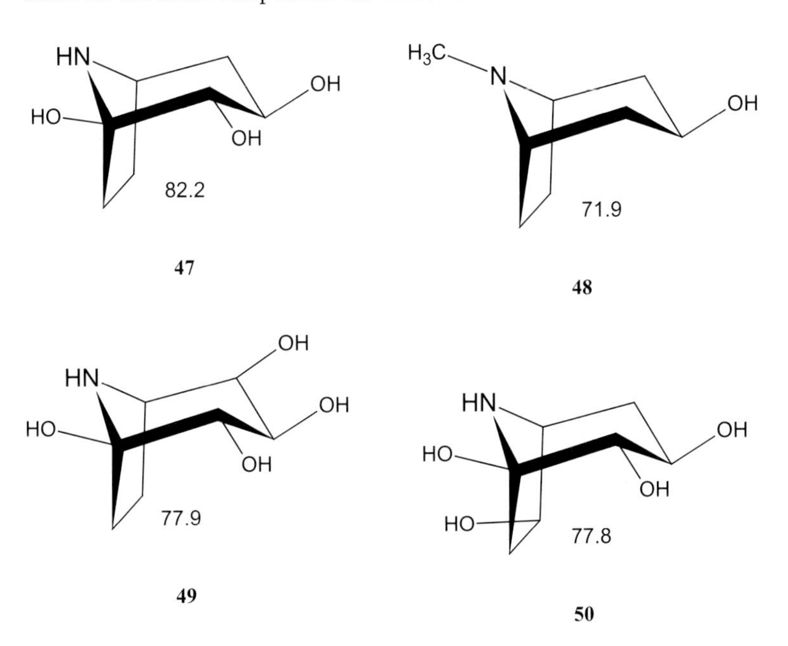

Fig. 14.10 Structures and ^{15}N chemical shift assignments of calystegins monitored in root cultures of *Calystegia sepium* [79]. ^{15}N chemical shifts were reported relative to potassium ^{15}N-nitrate. Chemical shifts in this figure have been referenced to liquid NH_3 using 376.5 ppm, the chemical shift of sodium ^{15}N-nitrate in water as the conversion factor.

Tab. 14.3 ^{15}N Chemical shift assignments for strychnine (**1**), brucine (**51**), holstiine (**52**), strychnine *N*-oxide (**54**), and brucine *N*-oxide (**55**) in deuterochloroform.

Compound	δ^{15}N N9 (ppm)	δ^{15}N N19 (ppm)
1	148.0 (D: 148.7/5.0)	35.0 (D: 34.8/5)
51	151.0 (D: 151.7)	37.0 (D: 36.8)
52	146.5 (135.2/10.7)	39.5 (31.7/19.8)
54	146.5 [−1.5]	136.3 [+101.3]
55	145.3 [−2.6]	135.5 [+99.6]

Values in square brackets are the shift relative to the parent molecule. Numbers in parentheses represent the predicted chemical shift and estimated error. The notation D indicates the structure is in the database. Occasionally a structure can be in the database multiple times with different chemical shifts (see Figure 14.11). The results will be the annotation of a record as contained in the database but a "prediction error" will be reported.

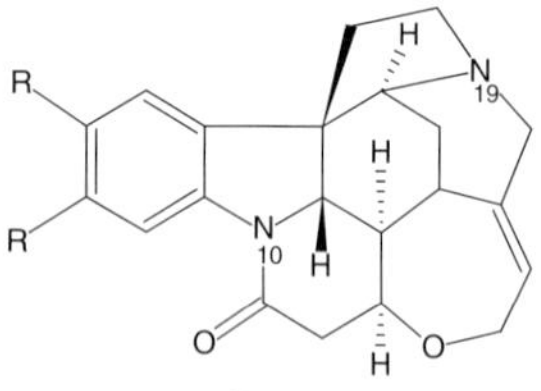

1R = -H
51R = -OCH_3

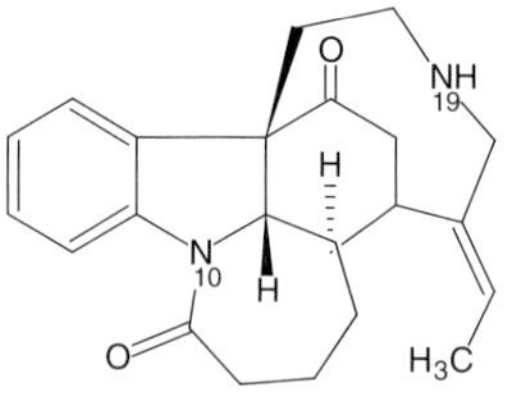

52

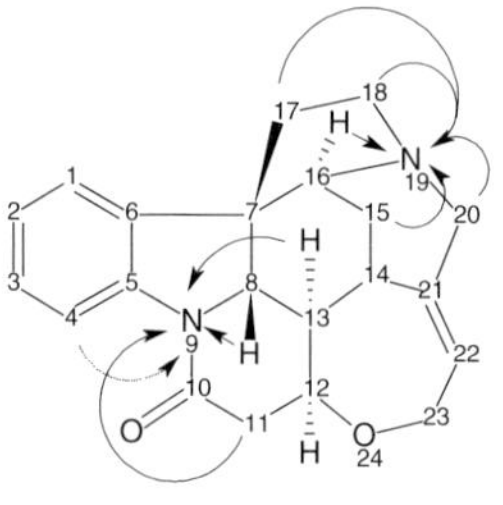

53

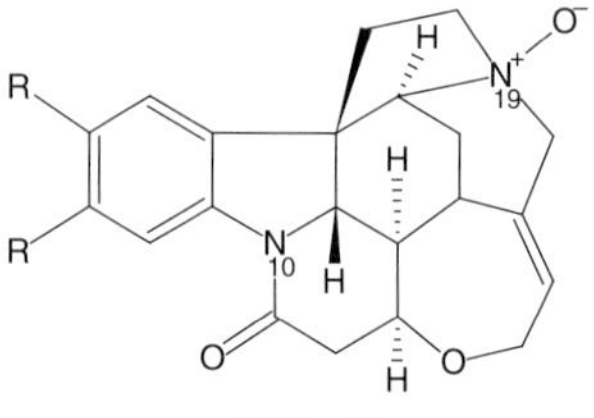

54R = -H
55R =-OCH_3

A complete summary of the long-range ^{1}H–^{15}N coupling pathways of strychnine is shown by **53**. The long-range proton–nitrogen couplings of brucine (**51**) and holstiine (**52**) were virtually identical to those of strychnine. In 1999, Hadden *et al.* [80] reported a study of the effects of N-oxidation at the N-19 position of strychnine (**54**) and brucine (**55**). As expected, the N-19 resonance of both compounds was shifted downfield by ~100 ppm relative to the chemical shift of N-19 in the parent alkaloid. In contrast, N-9 was shifted upfield slightly in both *N*-oxides.

14.6.3.2 Azaindoles

In 1996, a novel alkaloid, agrocybenine, **56**, with an azaindane skeleton was isolated from a Korean mushroom by Koshino and coworkers [81]. Long-range ^1H–^{15}N GHMBC data were used to locate the two annular nitrogens in the skeleton (Figure 14.11).

99.5

318.0

56

14.6.3.3 Indoloquinoline Alkaloids

The authors have also extensively studied the indoloquinoline families of alkaloids related to cryptolepine (indolo[3,2-*b*]quinoline **57**), applying direct and long-range

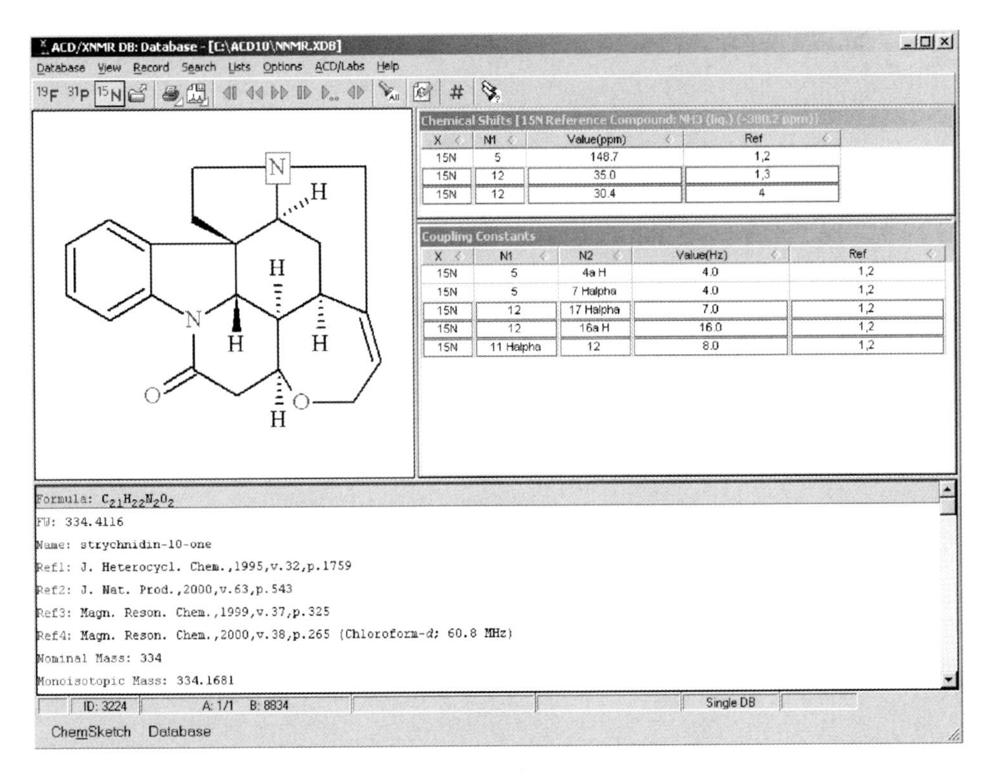

Fig. 14.11 The database record in ACD/NNMR associated with strychnine. Note that four references are reported with their associated chemical shifts and coupling constants.

^{1}H–^{15}N heteronuclear shift correlation methods during the study of a number of analogs.

In 1995, ^{1}H–^{15}N HSQC data were generated for an 800 μg sample of quindolinone (**58**) in ~2 h [82]. Both nitrogen resonances in the structure are protonated, and long-range ^{1}H–^{15}N heteronuclear shift correlation data for this alkaloid have never been reported.

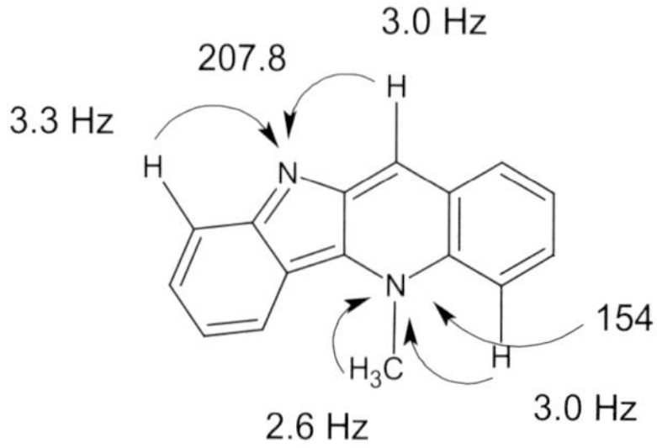

Cryptolepine (**57**), the parent alkaloid in the series, was studied in 1996 using long-range ^{1}H–^{15}N GHMBC data [83]. In order to observe a correlation to the N-10 nitrogen resonance, it was necessary to run GHMBC experiments with 4 and 10 Hz optimization to observe the weak *peri*-couplings to the N-5 and N-10 resonances. The ^{15}N shift of N-10 is in excellent agreement with the corresponding shift of an azacarboline with a similar exo double bond [84].

Cryptospirolepinone is the most complex member of the cryptolepis family of alkaloids, constituted as a spiro nonacyclic molecule **5** [85]. The molecule also required the acquisition of multiple long-range ^{1}H–^{15}N spectra to assign all of the ^{15}N resonances in the structure. Data were acquired with optimizations of 10, 6, 4, and 2.5 Hz. A single $^{4}J_{NH}$ correlation from H-13 to the N-3 resonance was observed only in the 2.5 Hz optimized spectrum, while correlations were readily observable to the other three nitrogen resonance in the structure [86].

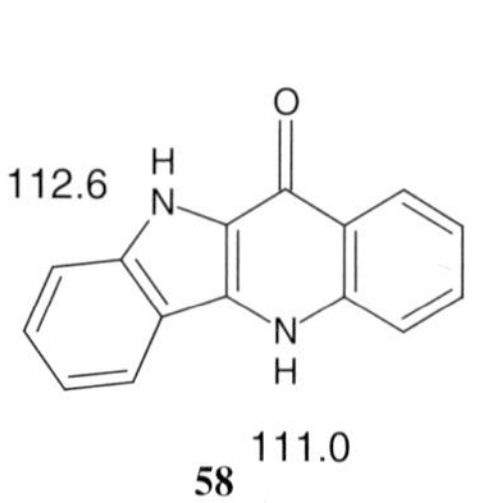

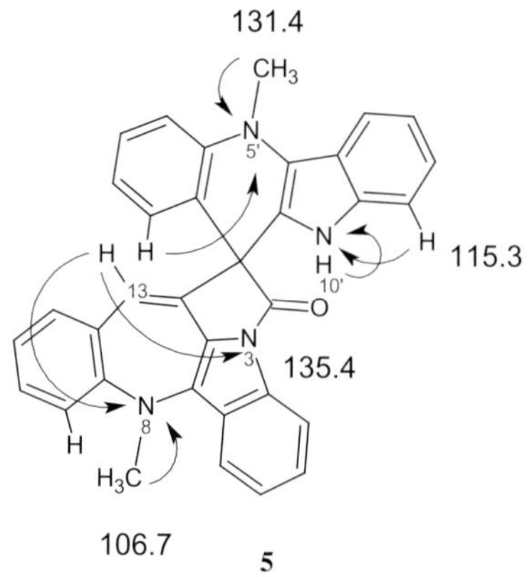

A 1999 study of the alkaloid cryptolepinone (**59**) also afforded ^{15}N chemical shift correlation data for the 5-oxide **60**, as this alkaloid undergoes facile air oxidation [87]. While transmitted electronic effects from the oxidized nitrogen to other nitrogen resonances are generally small [75] that was not the case for cryptolepinone 5-oxide, in which the N-10 nitrogen resonance was shifted downfield +23.9 ppm. Presumably, the effects of N-oxidation at the 5-position were transmitted to N-10 via the intervening C-5–C-10a double bond.

Later in 1999, an attempt was made to record long-range ^{1}H–^{15}N heteronuclear correlation data for 11-isopropylcryptolepine (**61**), using a <100 μg sample in a 1.7 mm submicro (SMIDG) 600 MHz NMR probe [88]. While correlations were observed from H-4 and the *N*-methyl resonance to the N-5 resonance at 147.8 ppm, no correlation was observed in the 3.3 Hz optimized spectrum to the N-10 resonance from H-9.

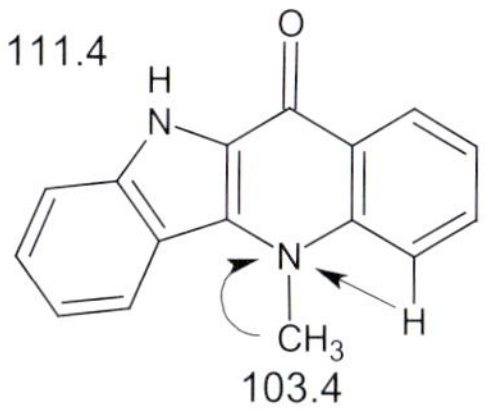

59 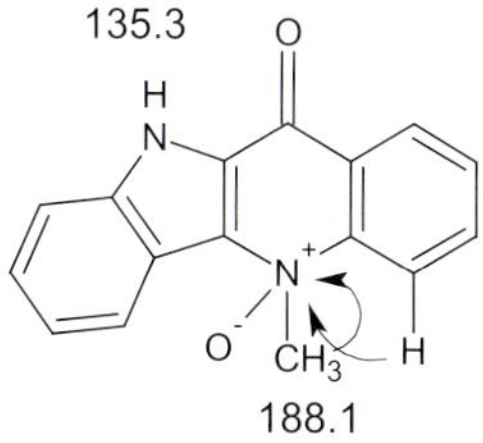 **60** 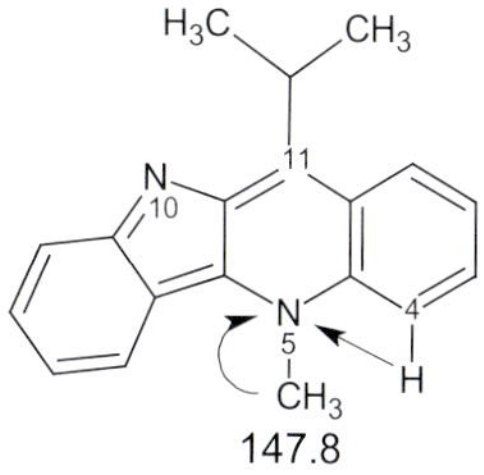 **61**

14.6.3.4 Vinca Alkaloids

To date, two studies have been reported in which long-range ^{1}H–^{15}N heteronuclear shift correlation data have been applied to vinca alkaloids. The first report in the literature was a 1995 application to the semisynthetic molecule vinorelbine (**62**) by one of the authors and a coworker [89]. Correlations to three of the four nitrogens in the structure of vinorelbine (**62**) were observed within a few hours of the initiation of data collection. In contrast, the single observable correlation to the azocine nitrogen in the structure (azocine ring is denoted in red) took a considerable period of time to observe and was then only weakly observed. The authors attributed this behavior to motion within the azocine ring. It is quite possible, however, that an IMPEACH-MBC ^{1}H–^{15}N spectrum would give better results because of the range of potential long-range coupling pathways being interrogated.

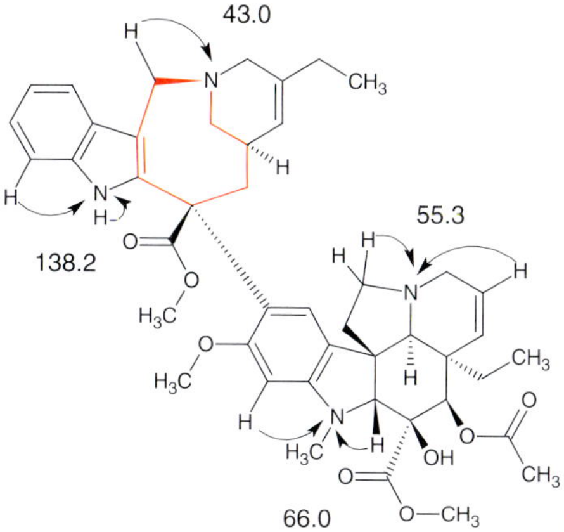

62 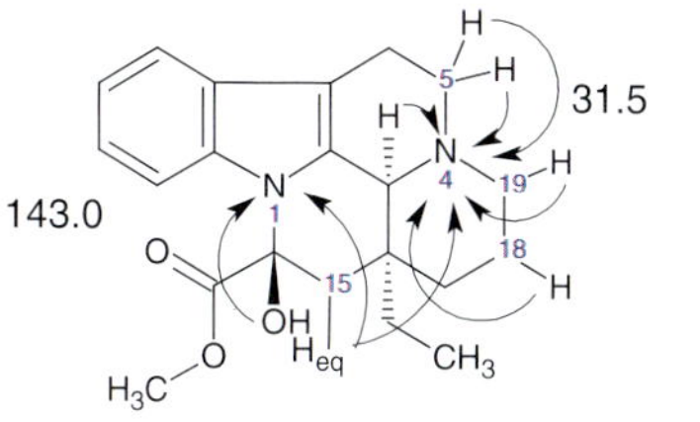 **63**

The only other reported application of long-range ^{1}H–^{15}N methods to a vinca alkaloid was an application to vincamine, **63** [90]. It is interesting to note that the hydroxyl proton exhibited a strong long-range correlation to the N-1 resonance observed at 143.0 ppm. The H-15e proton was long-range coupled to N-1 and also to N-4 via a $^{4}J_{NH}$ coupling pathway.

14.6.3.5 **Other Indole Alkaloids**

In addition to the molecules described above, a number of other reports have appeared in the literature detailing applications of direct and/or long-range ^{1}H–^{15}N 2D NMR methods to various indole alkaloids. One of the earliest of this group of studies was the 1997 report by Kim *et al.* [91] of the characterization of three indoline benzastatin alkaloids, E, F, and G that were isolated from a *Streptomyces nitrosporeus* culture broth. ^{1}H–^{15}N GHSQC data were reported only for benzastatin-E (**64**).

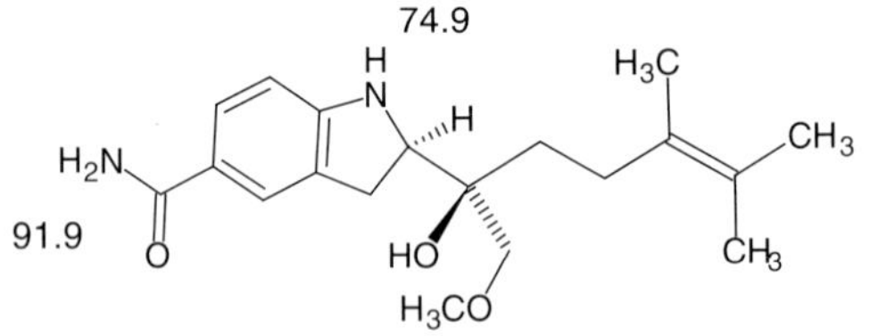

64

Wang and Ganesan [92], in the process of synthesizing a series of fumiquinazoline alkaloids, reported the extensive use of long-range ^{1}H–^{15}N correlation data. The

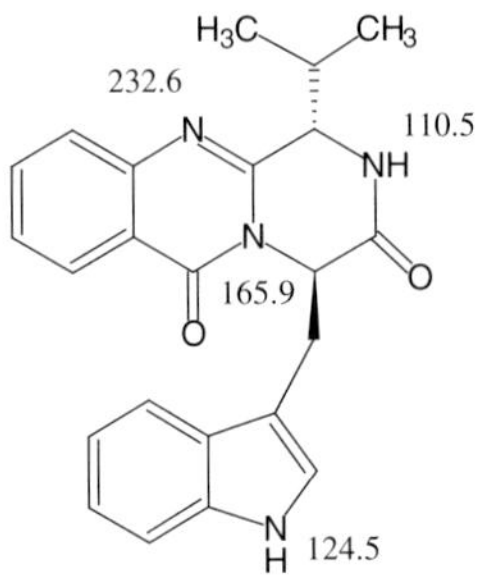

65

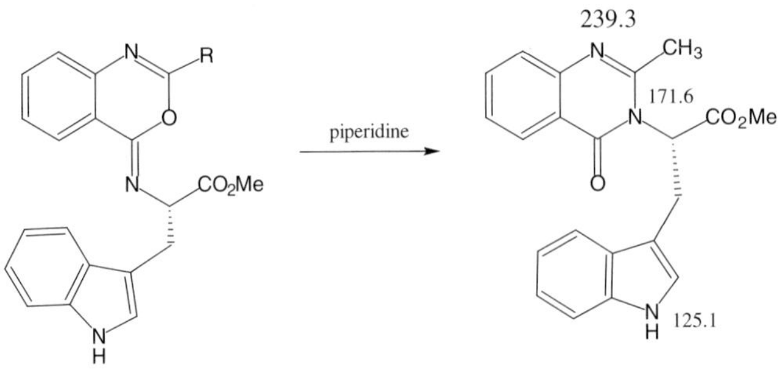

piperidine

66 **67**

Scheme 14.1

authors reported chemical shifts and long-range correlations observed for fiscalin-B (**65**) as well as for several of the substituted analogs shown in Scheme 14.1. ^{15}N Chemical shift data for analogs of **66** are collected in Table 14.4, while the ^{15}N chemical shifts for the one product, **67**, for which they were reported are shown on the structure. The original report also contains ^{15}N chemical shift data for a number of synthetic intermediates to which the interested reader is referred.

The authors also reported ^{15}N chemical shift assignments for one additional compound [92]. Refluxing the *t*-butyl substituted precursor of **66** in CH_2Cl_2 with triphenylphosphine, iodine, and triethylamine led to the formation of the substituted β-carboline, **68**

68

Seto and coworkers reported a study of the novel neuronal cell protecting substance mescengricin (**69**), a pyridoindole produced by *Streptomyces griseofulvis* [93]. There were only a few long-range correlations to the two nitrogen atoms, as shown on the structure. No details of the optimization of the long-range delays used in determining the long-range correlations to nitrogen were reported. This study is also interesting in that it is one of relatively few reports in the literature of the use of the D-HMBC experiment [94,95] The ^{15}N shifts of the indole and pyridine nitrogen resonance were reported as 122 and 207 ppm, respectively, the former in the normal range for an indole nitrogen resonance while the pyridine nitrogen is shifted considerably upfield relative to the normal pyridine nitrogen shift range of ~300–320 ppm.

69

Tab. 14.4 ^{15}N chemical shift assignments for several analogs of **66** [92].

Substituent	Indole NH	Exo C=N	Benzoxazine C=N
$R=CH_3$	122.2	224.2	223.0
$R=C_6H_5$	121.7	222.6	223.4
$R=CH_2CH(CH_3)_2$	123.4	221.9	225.0

Muhammad *et al.* [96] reported the isolation and structure characterization of a group of oxindole alkaloids **70–74**, from Una de Gato, or cat's claw, a plant indigenous to the rainforests of Peru and used by the native peoples for a variety of aliments. The plant has a wide range of secondary metabolites including tetra- and pentacyclic oxindole alkaloids among others. The ^{15}N chemical shifts reported for five oxindole alkaloids described in this report were interesting reporters of stereochemical differences between the alkaloids. The N-1 ^{15}N chemical shift was observed within a fairly narrow range of 136.9 to 135.1 ppm, while in contrast, the N-4 ^{15}N chemical shifts ranged more broadly from 54.6 to 64.8 ppm. These data are summarized in Table 14.5.

Later in 2001, Clark and coworkers [97] reported the isolation and structure characterization of an indolopyridoquinazoline alkaloid, 3-hydroxyrutaecarpine, (**75**) from *Leptothyrsa sprucei*. This study was particularly interesting in that the authors reported the substructural fragments **76–78**, which were assembled prior to the final determination of the structure of **75**. Long-range ^{1}H–^{13}C and ^{1}H–^{15}N correlations (GHMBC) were used to link the substructural fragments together to afford the final structure.

Tab. 14.5 Comparison of the stereochemical, N1 and N4 ^{15}N shifts and long-range couplings of a series of pentacyclic oxindole alkaloids isolated from the Peruvian plant Uña de Gato (Cat's Claw) [96].

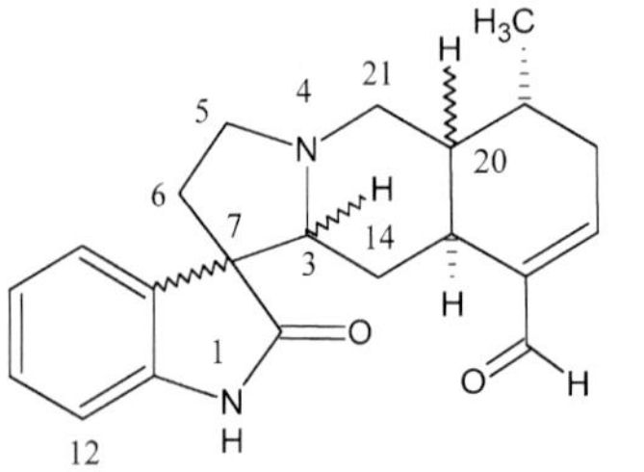

		Stereochemistry	N1 ^{15}N shift and long-range coupled protons	N4 ^{15}N shift and long-range coupled protons
70	Uncarine-C	3*S*, 7*R*, 15*S*, 19*S*, 20*S*	136.9 H-1, H-12	56.9 H-5, H-6, H-14, H-20, H-21
71	Uncarine-D	3*R*, 7*S*, 15*S*, 19*S*, 20*S*	135.1 H-1, H-12	64.8 H-3, H-6, H-14
72	Uncarine-E	3*S*, 7*S*, 15*S*, 19*S*, 20*S*	136.7 H-1	54.6 H-5, H-6, H-20, H-21
73	Mitraphylline	3*S*, 7*R*, 15*S*, 19*S*, 20*R*	135.1 H-1, H-12	64.7 H-3, H-5, H-6, H-14
74	Isomitraphylline	3*S*, 7*S*, 15*S*, 19*S*, 20*R*	135.8 H-1, H-12	64.7 H-5, H-6, H-21

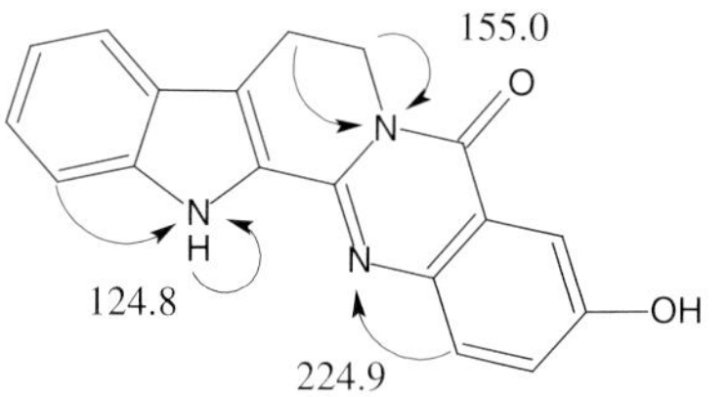

75

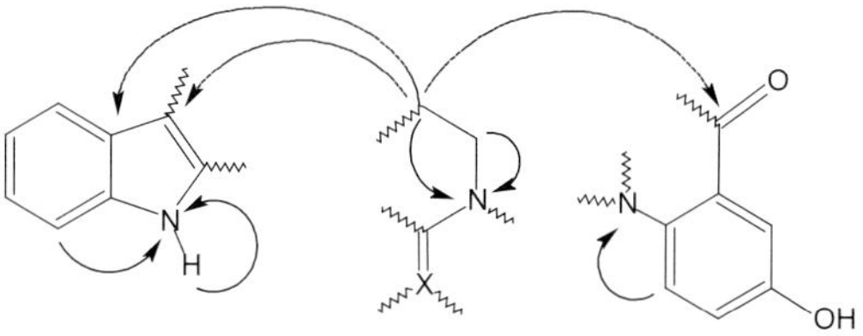

76 **77** **78**

Based on what might be considered the "normal" range of amide ^{15}N chemical shifts, the N-6 amide resonance of **75** might appear somewhat anomalous. However, it should be recalled (Table 14.3) that the N-9 amide chemical shifts of the strychnos alkaloids thus far studied are in the range of ~145–151 ppm; the ^{15}N shifts of oxoberberine (**128**, see Section 14.66) and nauclealine-A (**76**, following section) are assigned at 153.7 and 164.2 ppm, respectively. Hence, the N-6 ^{15}N chemical shift of 3-hydroxyrutaecarpine is quite reasonable.

In another study later in 2001, Clark and coworkers reported the application of long-range ^1H–^{15}N experiments during the isolation and structural characterization of another group of indole alkaloids, nauclealine-A (**76**), -B (**77**), and naucleaside-A and -B from the bark of *Nauclea orientalis* [98]. ^{15}N Chemical shifts were reported only for the former, although it is likely that the ^{15}N shifts of the latter would be quite similar.

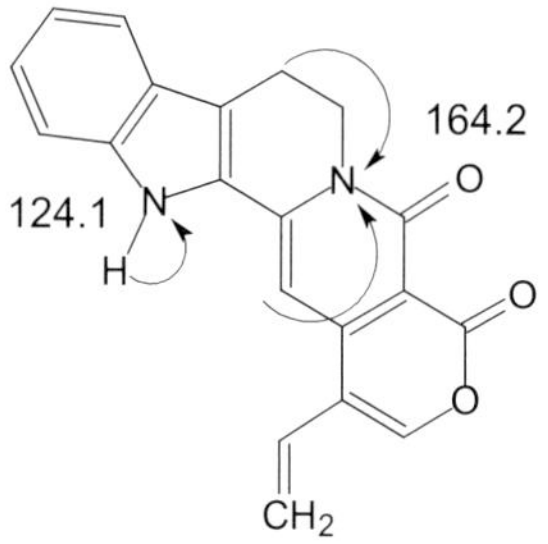

76 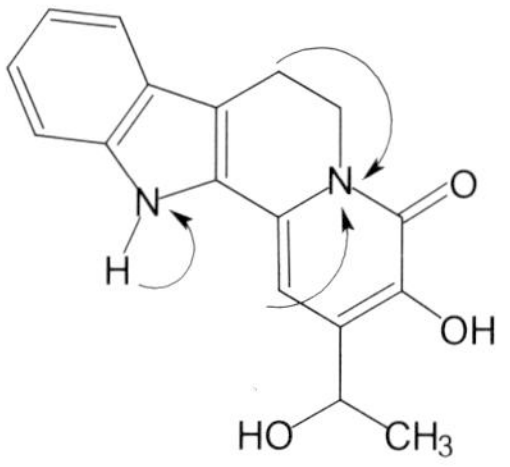 **77**

It is interesting to note that the ^{15}N chemical shift calculation of nauclealine-A (**76**) when performed using ACD/NNMR v10 gives a result for the indole nitrogen of 131.6 ± 6.8 ppm, which is quite reasonable and would readily allow the use of the predicted ^{15}N chemical shift to establish an F_1 window for the acquisition of long-range ^1H–^{15}N data. In stark contrast, the calculated ^{15}N chemical shift of the N-6 amide nitrogen was 84.9 ± 36.2 ppm, which is inconsistent with the observed and assigned ^{15}N chemical shift of 164.2 ppm for this resonance. On examining the compounds on which the chemical shift calculation is based, there is a strong bias toward bridgehead nitrogen-containing molecules with only scant representation of molecules such as oxoberberine, which contains a legitimate amide nitrogen. The bias inherent to the compounds used in the calculation, unfortunately, skewed the calculated ^{15}N chemical shift of N-6 unacceptably low. However, as ^{15}N chemical shift data for more compounds are added to expand the database, calculation problems of this type should diminish in frequency.

Another interesting report that appeared in 2002 dealt with the revision of the structure of porritoxin (**78**) reported by Horiuchi and coworkers [99]. When this alkaloid, which is produced by the fungus *Alternaria porri*, was first characterized [100], long-range ^{1}H–^{15}N 2D NMR methods were not available. The structure was originally reported as incorporating the eight-membered azoxacine as shown by **78**. When the structure was re-examined in 2002, it was necessary to revise the structure as shown by **79** on the basis of the long-range ^{1}H–^{15}N 2D NMR correlations observed in a GHMBC spectrum, which were inconsistent with the originally proposed azoxocine ring.

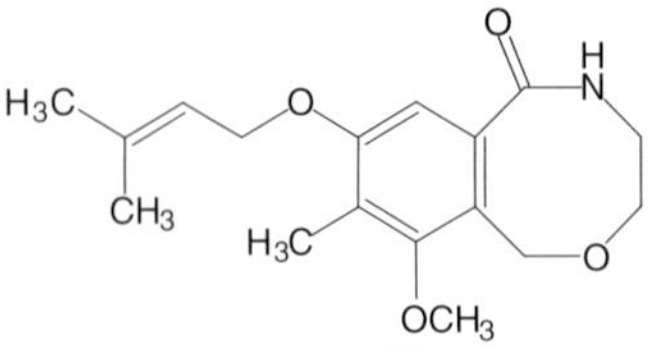

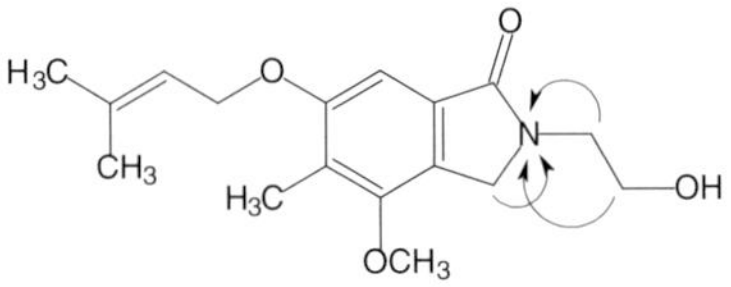

Copp and coworkers [101] made extensive use of both ^{1}H–^{15}N GHSQC and GHMBC data in determining the structures of kottamides A–D, novel indole-containing alkaloids from the New Zealand ascidian *Pycnoclavella kottae*. Reported ^{15}N chemical shifts and ^{1}H–^{15}N long-range coupling pathways are shown for kottamide-A (**80**, $R_1 = R_2 = Br$). ^{15}N data were unfortunately not reported for the other alkaloids in the series (**81–83**) whose structures were determined (Table 14.6). Key long-range ^{1}H–^{15}N correlations were observed from the methine protons of the

Tab. 14.6 Structural comparison of substituents and side chains of kottamides A–D[101].

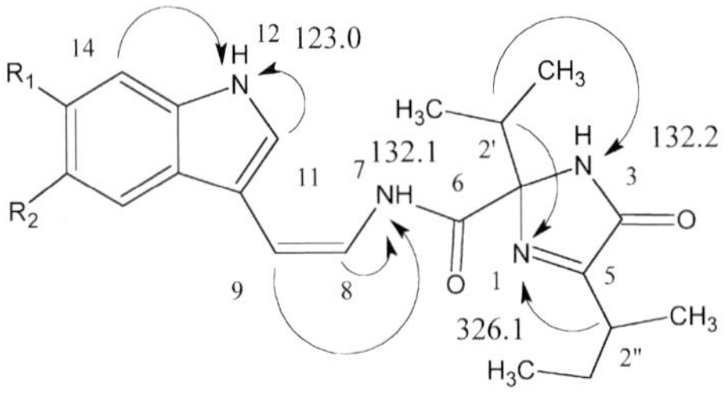

Structure	R_1	R_2	C-2 side chain	C-5 side chain
80	Br	Br	Isopropyl (as shown above)	*Sec*-butyl (as shown above)
81	Br	H	Isopropyl	*Sec*-butyl
82	H	Br	Isopropyl	*Sec*-butyl
83	Br	Br	*Sec*-butyl	Methyl

isopropyl and *sec*-butyl side chains to the imidazolone nitrogen, N-1, resonating at 326.1 ppm for **80**, which allowed the assembly of the imidazolone heterocyclic ring during the determination of the structures of these interesting alkaloids.

In 2003, Appleton and Copp [102] reported the isolation and characterization of another related alkaloid, kottamide-E (**84**) again making use of long-range ^1H–^{15}N heteronuclear chemical shift correlation data.

During the attempted total synthesis of the alkaloid roquefortine C, the synthetic step leading to the diketopiperazine went astray, leading to the formation of an unanticipated cyclization product. Hadden *et al.* [103] employed 6 Hz optimized long-range ^1H–^{15}N GHMBC data to establish the structure of the novel cyclization product as **85** shown in Scheme 14.2.

A novel pentacyclic oxindole, speradine A (**86**) was isolated from a marine-derived fungus, *Aspergillus tomarii* and characterized, in part using long-range ^1H–^{15}N 2D NMR data [104]. The only long-range correlations to the two nitrogens in the structure

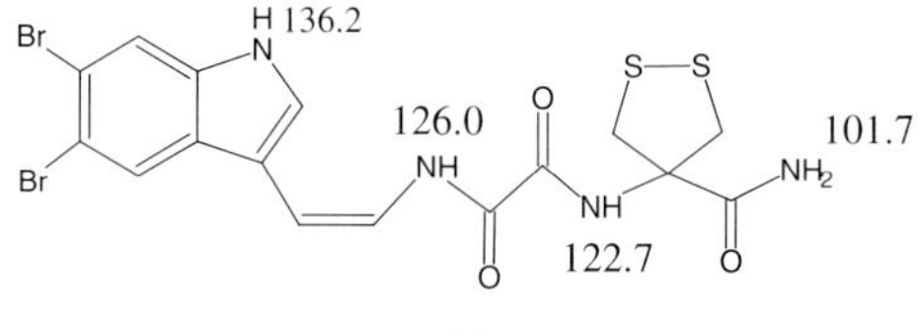

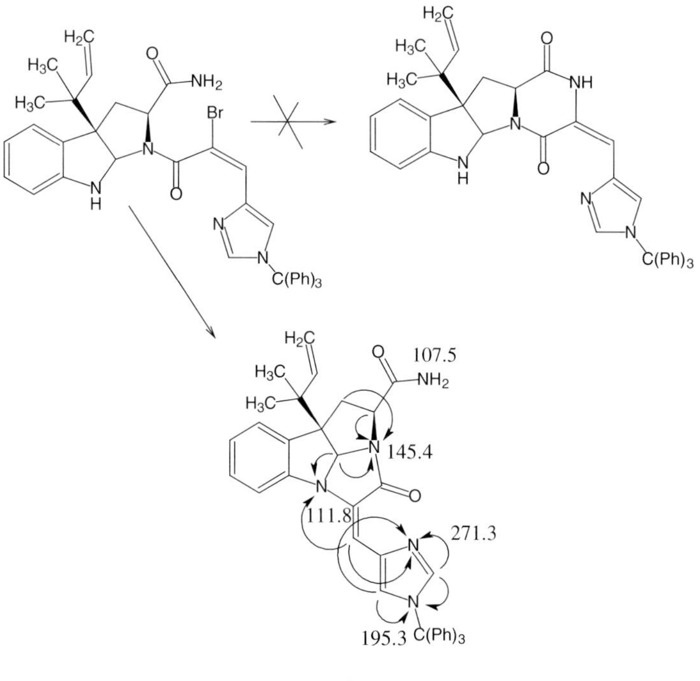

Scheme 14.2

of the alkaloid were from directly bound or nearby methyl groups, making the acquisition of the long-range correlation data a very simple undertaking.

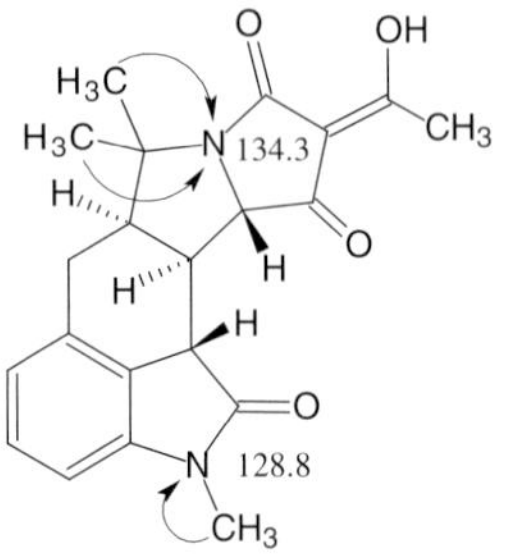

86

Another report in 2003 described the isolation and characterization of bacillamide (**87**) from a marine bacterium *Bacillus* sp. SY-1 [105]. This indole-containing alkaloid interestingly has both algicidal activity and potent activity against the harmful dinoflagellate *Cochlodinium polykiroides*

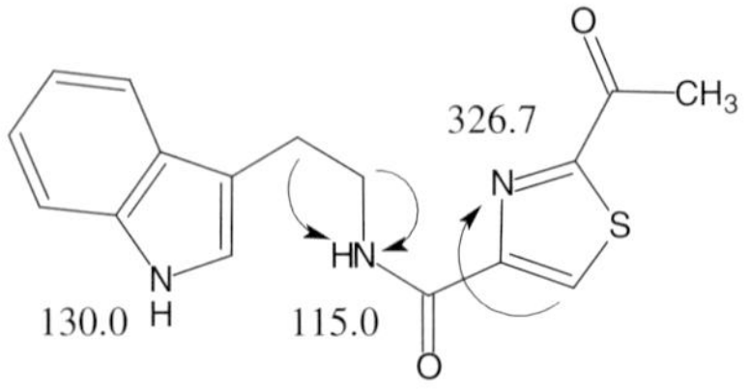

87

Despite the large numbers of reports that described the application of long-range ^{1}H–^{15}N 2D NMR methods to indole alkaloids during 2003, there were none reported in 2004. In 2005, Glover, Yoganathan, and Butler [106] reported the isolation and characterization of three indole-containing aspidofractinine alkaloids, kopsine, fruticosine, and fruticosamine (**88**). Long-range ^{1}H–^{15}N GHMBC data were acquired during the characterization of fruticosamine, although no details of the long-range coupling pathways observed were contained in their report.

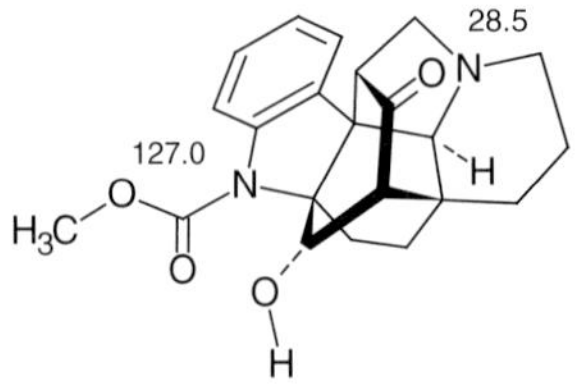

88

14.6.4
Carboline-derived Alkaloids

Several carboline-derived alkaloids have been studied using ^{1}H–^{15}N 2D NMR methods, the first in 1995 was ajmaline (**89**) [107]. Long-range ^{1}H–^{15}N couplings for ajmaline were measured and allowed the assignment of the resonances of the two nitrogens contained in the structure. Relatively few studies of alkaloids have reported the measurement of the actual long-range $^{1}H-^{15}N$ coupling constants. Instead, the vast majority of the reports in the literature have merely exploited the long-range coupling pathways to assign the ^{15}N chemical shifts, and/or to provide structure confirmation.

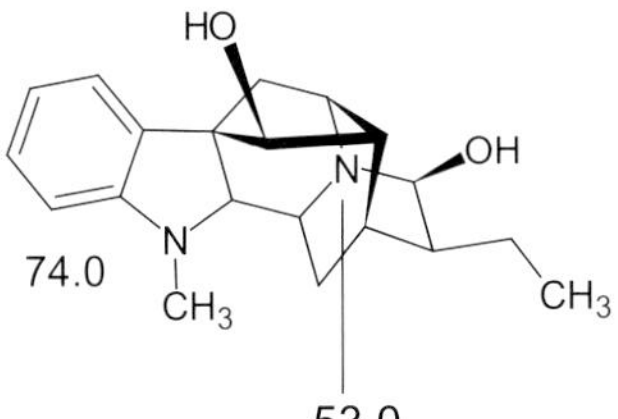

HO OH N N 74.0 CH_3 CH_3 **89** 53.0

Marek *et al.* [108] reported the ^{15}N chemical shifts and long-range correlation pathways for the alkaloid roemeridine (**90**) in 1996 in conjunction with data for a number of isoquinoline-derived alkaloids that are discussed below.

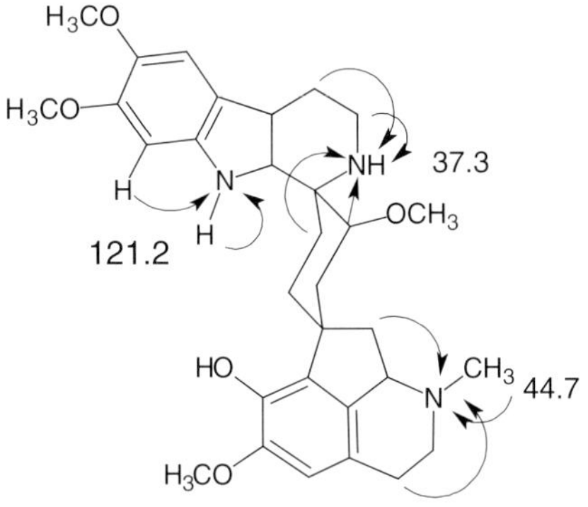

H_3CO H_3CO H N H 121.2 NH 37.3 OCH_3 HO CH_3 N 44.7 H_3CO **90**

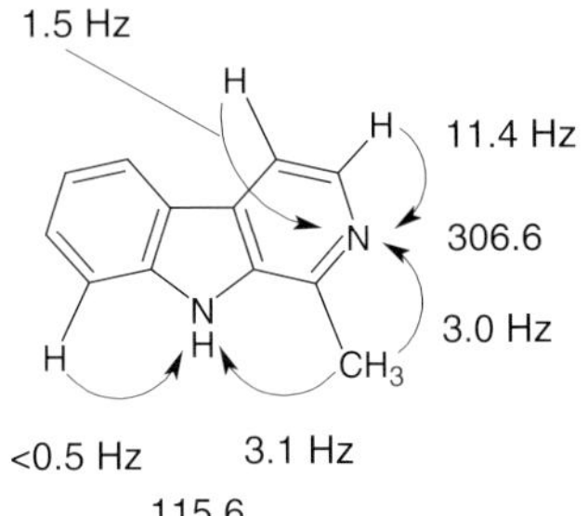

1.5 Hz H H 11.4 Hz N 306.6 3.0 Hz H N H CH_3 <0.5 Hz 3.1 Hz 115.6 **91**

A comprehensive study of the alkaloid harman (**91**) in which the long-range $^{1}H-^{15}N$ couplings were measured using modified $^{1}H-^{15}N$ J-HMBC methods has been reported [109]. The review of long-range $^{1}H-^{15}N$ 2D NMR applications by

Martin and Hadden [10] reported the ^{15}N chemical shifts and long-range coupling pathways of β-carboline (**92**).

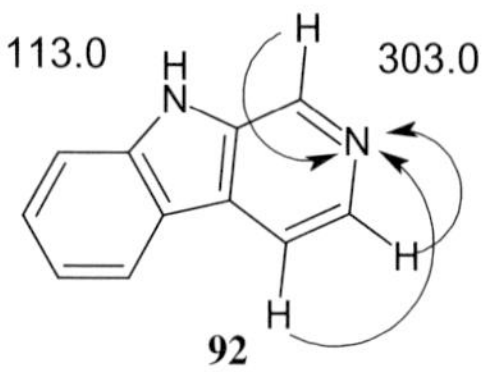

113.0 303.0

92

14.6.5
Quinoline, Isoquinoline, and Related Alkaloids

Marek and coworkers have reported extensively on the utilization of direct and long-range ^{1}H–^{15}N 2D NMR methods isoquinoline, benzo[*c*]phenanthridine, and related alkaloids. In a 1996 study, the ^{15}N chemical shift and long-range couplings of the alkaloid armepavine (**93**) were reported [108,110].

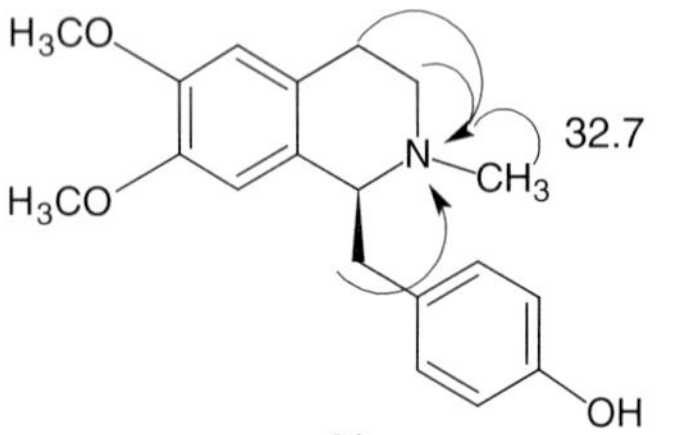

32.7

93

Two other closely related benzyl isoquinoline alkaloids are papaverine (**94**) and escholamine iodide (**95**). The ^{15}N resonance of the fully aromatic isoquinoline of papaverine, as would be expected, was observed at 297.2 ppm. In contrast, the quaternary methyl iodide salt form of escholamine iodide shifted the ^{15}N resonance upfield to 189.9 ppm [111].

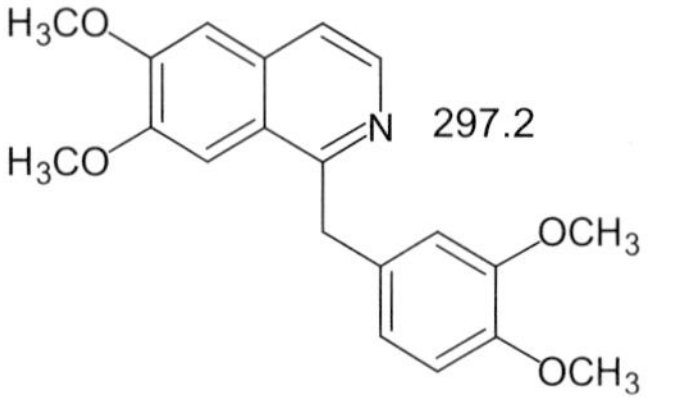

297.2 189.9

94 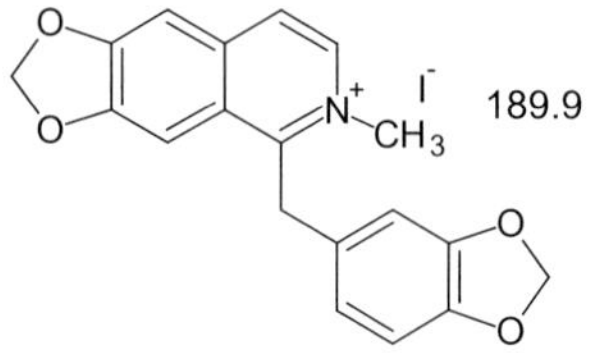 **95**

The investigation of bisbenzylisoquinolines has been an active area of natural product research that has been reviewed [112]. Frappier and coworkers [113] have studied and characterized some antiplasmodial bisbenzylisoquinoline alkaloids from *Isolona ghesquiereina* using long-range ^1H–^{15}N HMBC data in their study. The authors characterized (−)curine (**96**) and chondrofoline (**97**), for which ^{15}N chemical shift data were reported. Unfortunately, the observed long-range ^1H–^{15}N correlations were not reported. The N-2 and N-2′ resonances of **96** were assigned as 22.5 and 34.7 ppm, respectively. The corresponding resonance assignments for N-2 and N-2′ of **97** were comparable at 23.2 and 35.4 ppm.

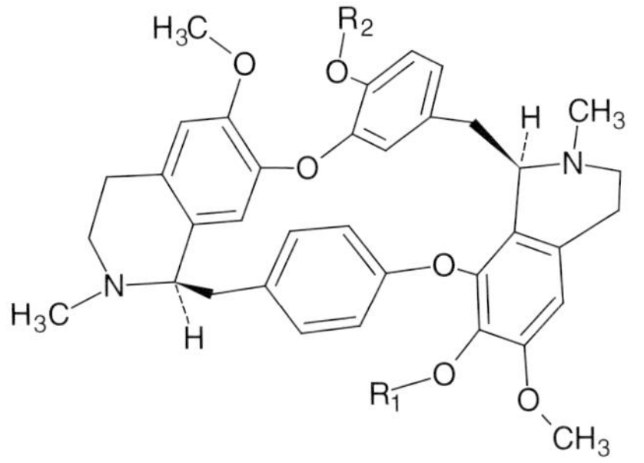

96 $R_1 = R_2 = -H$ **97** $R_1 = -CH_3$, $R_2 = -H$

Orabi and coworkers [114] reported the elucidation of the structures of several bis-1-oxaquinolizidine *N*-oxide alkaloids from Red Sea specimens of *Xestospongia exigua*, employing long-range ^1H–^{15}N HMBC data in the process. Two of the compounds were known, araguspongine A and araguspongine-C; however, their corresponding *N*-oxides, araguspongine-K (**98**) and araguspongine-L (**99**) were new compounds. Only partial ^{15}N data were reported for **98**, specifically the shift of the N-5-oxide nitrogen, which resonated at 117.1 ppm. Both shifts were reported for the symmetric **99**. In the case of the latter, N-5 resonated at 43.1 ppm, while the N-5′-oxide nitrogen resonated at 117.0 ppm, which represents a downfield shift of +73.4 ppm due to the N-oxidation. The downfield shift observed following N-oxidation is consistent with the downfield ^{15}N shifts following N-oxidation reported from these laboratories [75,80,88].

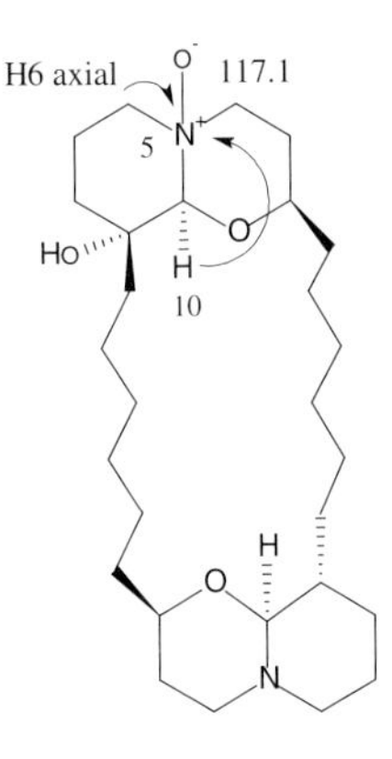

98 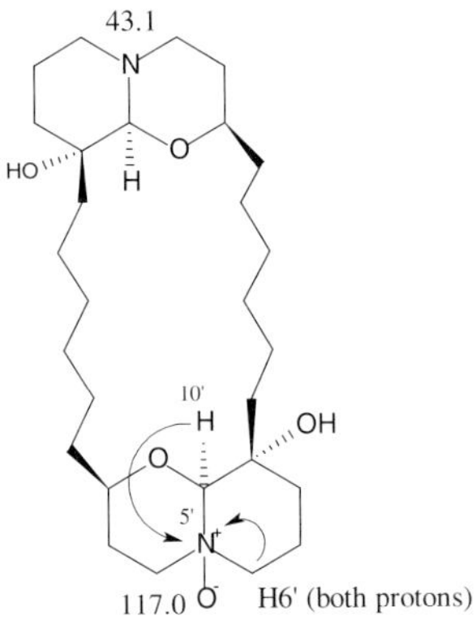 **99**

The new quinoline alkaloid sarcomejine (**100**) was isolated from the bark of *Sarcomelicope megistophylla* by Fokialakis *et al.* [115] The skeletal framework was assembled from NMR and mass spectrometric data, necessitating the authors to then differentiate between two alternative structures on the basis of a 5 Hz optimized long-range ^{1}H–^{15}N GHMBC spectrum. Correlations from the proton at the 8 position of the quinoline ring and from the side-chain methane proton resonating at 5.28 ppm allowed the authors to fix the locations of the side chain as shown by **100** rather than the alternative structure, **101**.

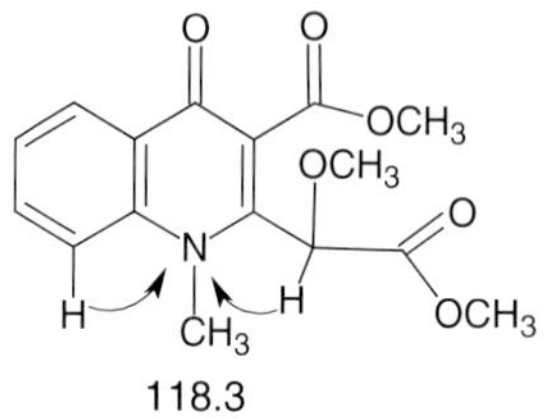

118.3

100 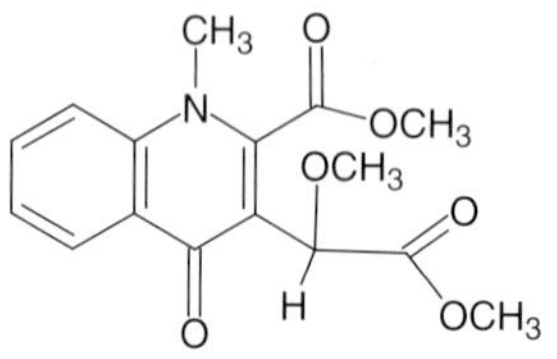 **101**

Muhammad and coworkers [116] reported the isolation and structural characterization of two alkaloids from *Duguetia hadrantha*, hadranthine-A and -B (**102**) that were related to the known alkaloid imbiline-1. To confirm the placement of the annular nitrogen atoms in the skeletal framework, the authors used long-range ^{1}H–^{15}N HMBC data. Thus, correlations from protons resonating at 9.11 and 8.22 ppm to a nitrogen resonating at 321.0 ppm were used to position N-1 of hadranthine-A (**102**, $R_1 = CH_3$; $R_2 = OCH_3$). A correlation from the amide methyl group to a nitrogen resonating at 135.6 ppm located N-6. To provide confirmation of the structural assignment, imbiline-1 (**102**, $R_1 = CH_3$, $R_2 = H$) was also examined by long-range ^{1}H–^{15}N methods. Identical long-range correlations were observed to nitrogens resonating at 321.9 and 135.4 ppm, confirming the structural assignment. Only partial ^{15}N data were reported by the authors for hadranthine-B (**102**, $R_1 = CH_3$; $R_2 = H$); the N-1 resonance was observed at 321.5 ppm. The ^{15}N chemical shift of the N-6 resonance was not reported.

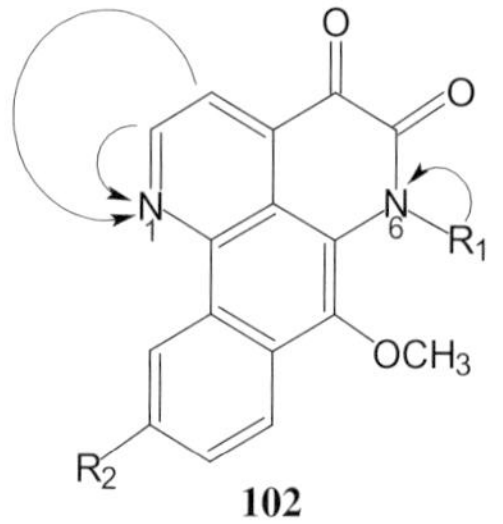

102

Francisco, Nasser, and Lopes [117] used long-range ^{1}H–^{15}N 2D NMR data in the characterization of tetrahydroisoquinoline alkaloids isolated from *Aristolochia arcuata*. The only compound for which ^{15}N data were reported was **103**.

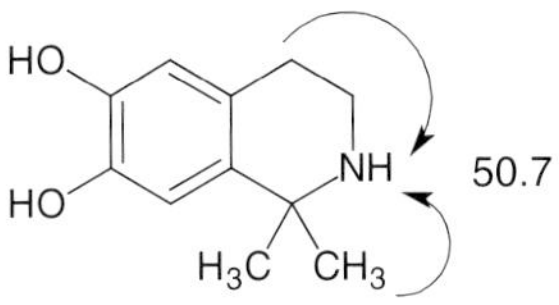

HO HO NH 50.7 H_3C CH_3

103

Yang *et al.* [118] reported the isolation and characterization of a novel naphthylisoquinoline alkaloid, ancisheynine (**104**) from the aerial part of *Ancistrocladus heyneanus*. The alkaloid is interesting in that it contains a previously undescribed N-2–C-8′ linkage between the naphthyl and isoquinoline subunits, the presence of which was established with long-range $^1H^{15}N$ correlation data.

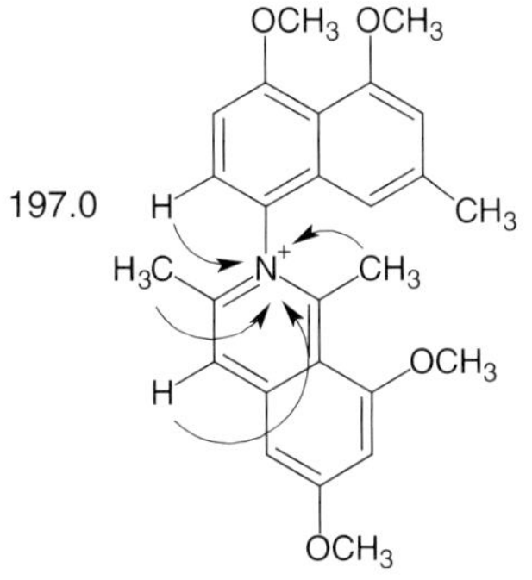

OCH_3 OCH_3 197.0 H CH_3 H_3C N^+ CH_3 OCH_3 H OCH_3

104

14.6.6 Benzo[*c*]phenanthridine Alkaloids

In their 1996 report cited above, in which long-range ^1H–^{15}N 2D NMR data were acquired in the study of roemeridine (**90**) and armepavine (**93**), Marek and coworkers also reported data for the benzo[*c*]phenanthridine alkaloids chelerythine (**105**) and sanguilutine (**106**), which had ^{15}N chemical shifts of 186.2 and 186.6 ppm, respectively [108].

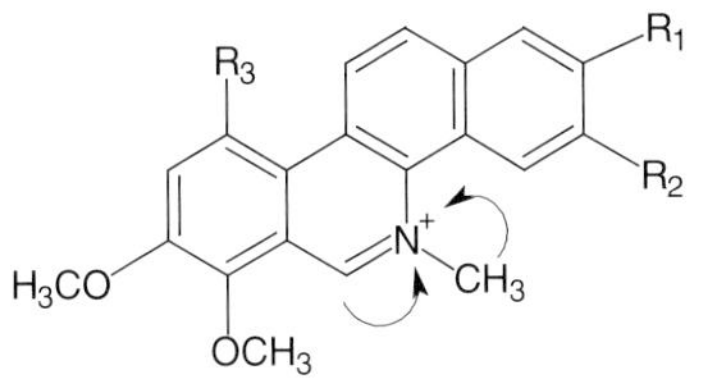

R_1 R_3 R_2 N^+ H_3CO CH_3 OCH_3

105 $R_1 + R_2$= -O-CH_2-O-, R_3= -H
106 $R_1 = R_2 = R_3$= -OCH_3

Marek and coworkers followed their initial report with a 1997 study of a group of dimeric benzo[*c*]phenanthridine alkaloids using ^1H–^{15}N GSQMBC data [119] (Table 14.7).

Tab. 14.7 ^{15}N Chemical shifts of selected benzo[*c*]phenanthridine alkaloids and their –*O*– and –NH– dimers.

Compound	X	$R_1 + R_2$	$R_3 + R_4$	R_5	$\delta^{15}N$ (ppm)
107a	—	O–CH_2–O	O–CH_2–O	OCH_3	49.7
107b	—	O–CH_2–O	O–CH_2–O	H	50.3
108a	–O–	O–CH_2–O	O–CH_2–O	OCH_3	40.1
108b	–O–	O–CH_2–O	O–CH_2–O	H	41.2
108c	–O–	O–CH_2–O	$R_3 = R_4 = OCH_3$	H	39.0
108d	–NH–	O–CH_2–O	O–CH_2–O	H	38.4
108e	–NH	O–CH_2–O	$R_3 = R_4 = OCH_3$	H	35.8

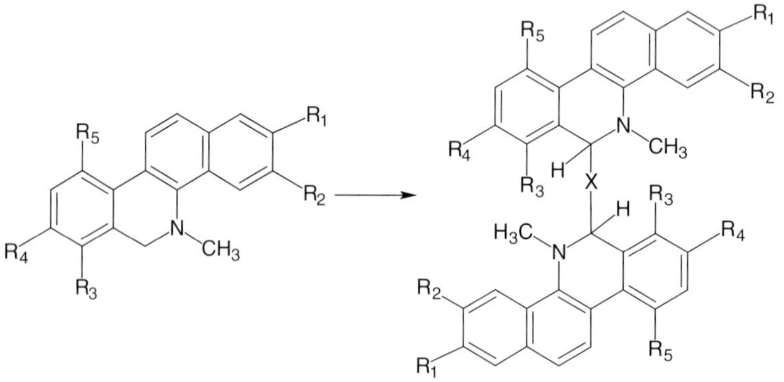

107 **108**

Sanguinarine pseudobase (**109**) was also re-examined by NMR, including long-range ^{1}H–^{15}N data [120]. The intent of the study was to unequivocally confirm the structure of sanguinarine pseudobase and to prove the existence of a hemiaminal OH group. The ^{13}C chemical shift of the C-6 resonance for **109a** (R = H) was 78.9 ppm, which is consistent with a hemiacetal. The ^{15}N shift of **107** (R_5 = H) was 50.3 ppm. In the case of **109b** (R = CH_3) and **109c** (R = C_2H_5), the ^{15}N shifts were 40.3 and 41.3 ppm, respectively.

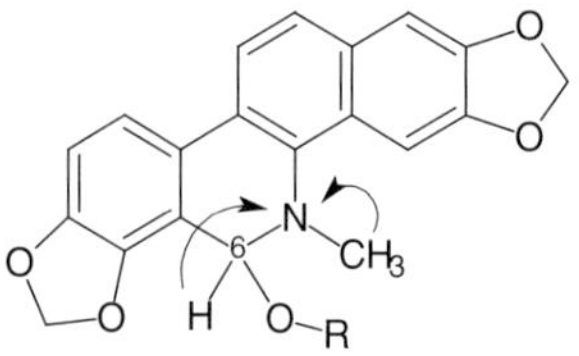

109

In addition to the study of the bis-benzo[*c*]phenanthridines just discussed above, Marek and coworkers have also extensively studied other benzo[*c*]phenanthridine alkaloids [111,121]. The structures of the alkaloids for which data have been reported and their respective ^{15}N chemical shifts are collected in Figure 14.12. These alkaloids include the methyl chloride salts, sanguinarine chloride (**110**), chelrythine chloride (**111**), sanguilutine chloride (**112**) and chelirubine chloride (**113**). 6-Substituted and

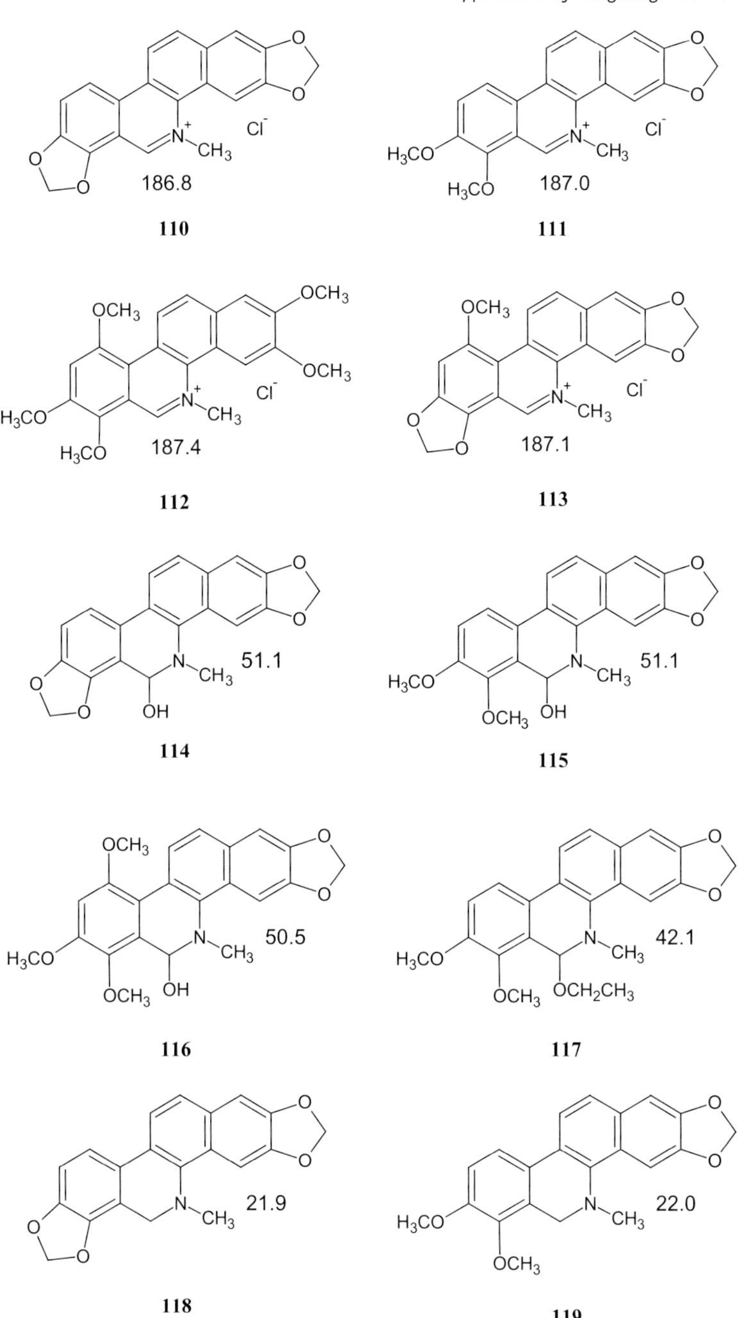

Fig. 14.12 ^{15}N Chemical shifts of benzo[*c*]phenanthridine alkaloids [111,121].

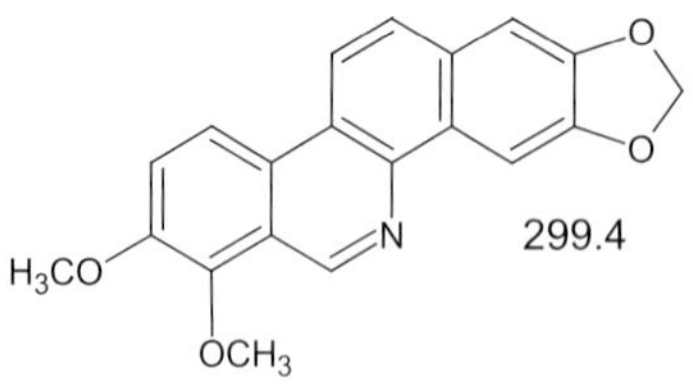

Fig. 14.12 *(Continued)*

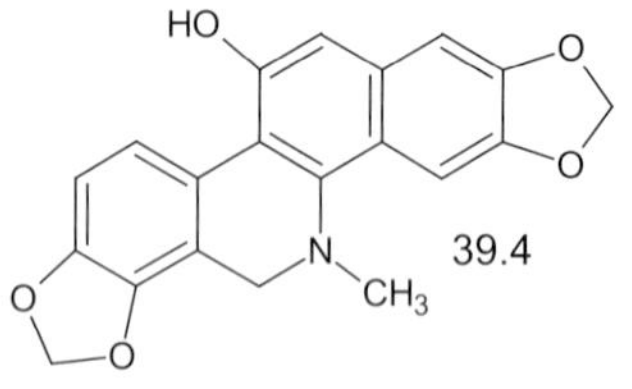

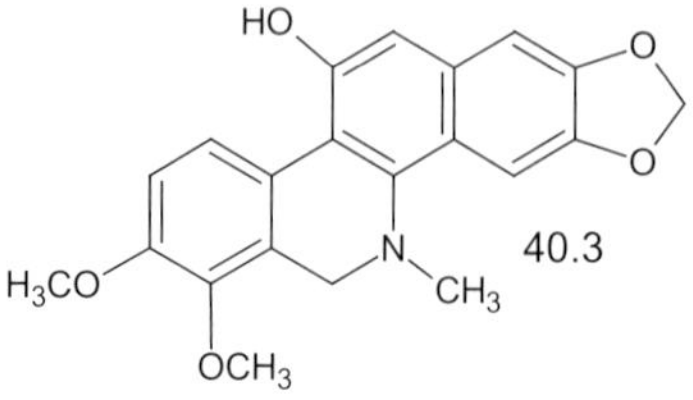

dihydro analogs included 6-hydroxydihydrosanguinarine (**109a**), 6-hydroxydihydro-chel-erythrine (**115**) 6-hydroxydihydrocelirubine (**116**) 6-methoxydihydrosanguinarine (**109b**), 6-ethoxydihydro-sanguinarine (**109c**), 6-ethoxydihyrochelerythrine (**117**), dehydro-sanguinarine (**118**) and dihydrochelerythrine (**119**). Related alkaloids included norchelerythrine (**120**), chelidone (**121**), and homochelidone, (**122**).

In addition to the benzo[*c*]phenanthridine alkaloids, Marek and coworkers [111,121] also reported on a number of protoberberine alkaloids which are collected in Figure 14.13. The protoberberines that have been studied include berberine chloride (**123**), coptisine chloride (**124**), canadine (**125**), stylopine (**126**), tetrahydro-palmatine (**127**), oxoberberine (**128**), dihydroberberine (**129**), 8-hydroxy-7,8-dihydro-berberine (**130**), corydaline (**131**), thalictricavine (**132**), palmitine chloride (**133**), jatrorhizine chloride (**134**), mecambridine (**135**), and cyclanoline iodide (**136**). These extensive reports also contain data for aporphine and proaporphine alkaloids, which are summarized in Figure 14.14. The aporphine and proaporphine alkaloids for which data are available include isothebaine (**137**), isocorydine (**138**), bulbocapnine (**139**), and mecambrine (**140**). Data have also been reported for several pavine analogs that are collected in Figure 14.15. The pavines include californidine perchlorate (**141**), eschscholtzine (**142**), and argemonine (**143**). The authors have also reported data for rhoeadine (**144**) and bicuculline (**145**).

14.6.7
Pyrazine Alkaloids

Chill, Aknin, and Kashman [122] used long-range $^{1}H-^{15}N$ data in the characterization of two pyrazine alkaloids with an unprecedented skeleton isolated from an unidentified tunicate collected at the Barren Islands, Madagascar. The assigned

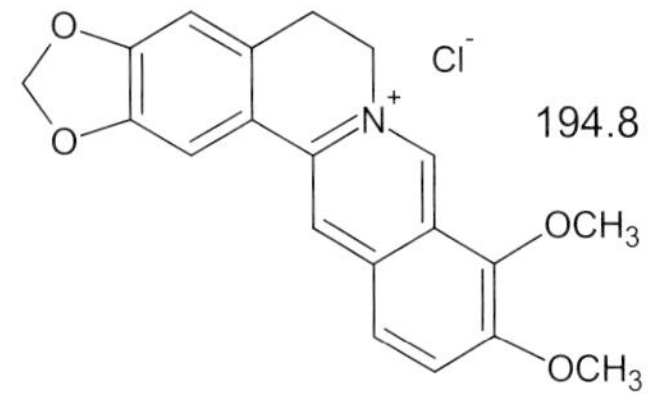

123

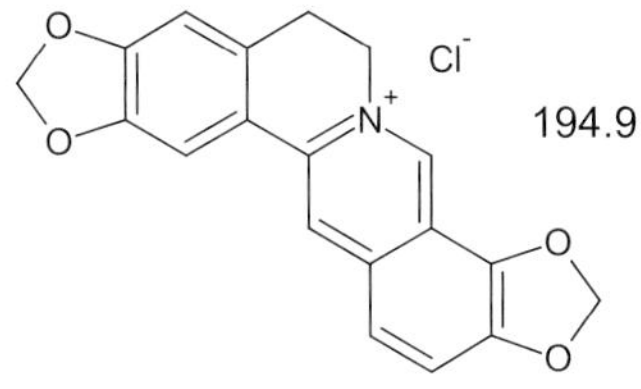

124

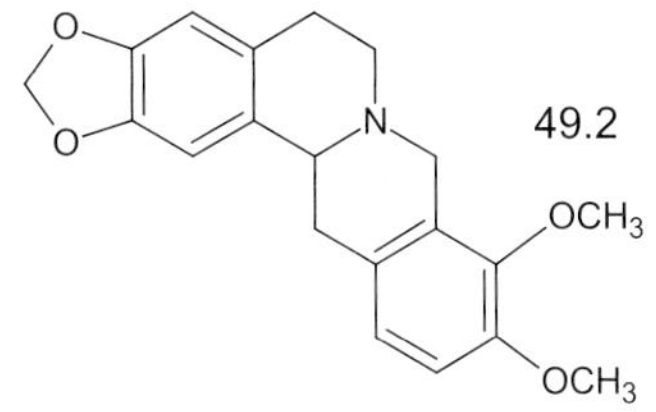

125

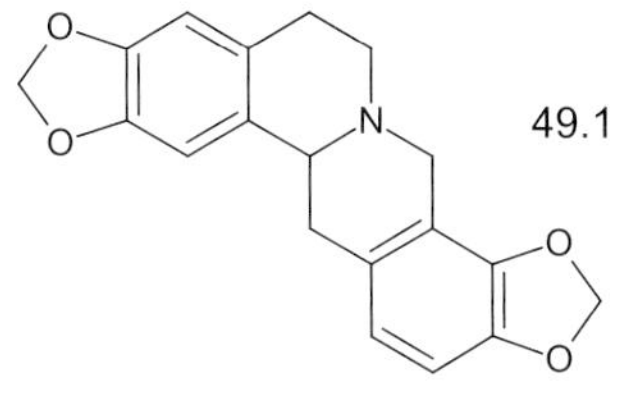

126

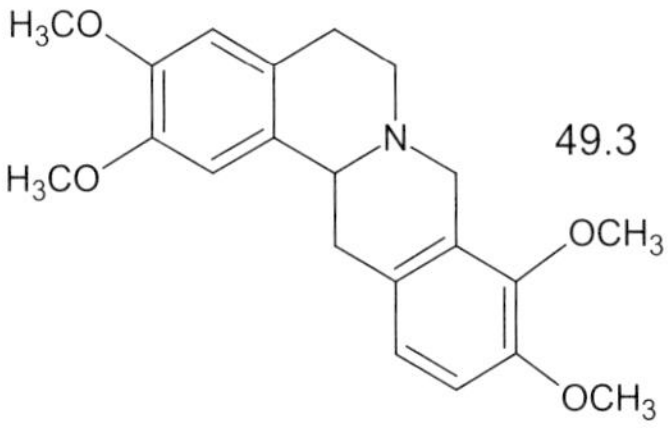

127

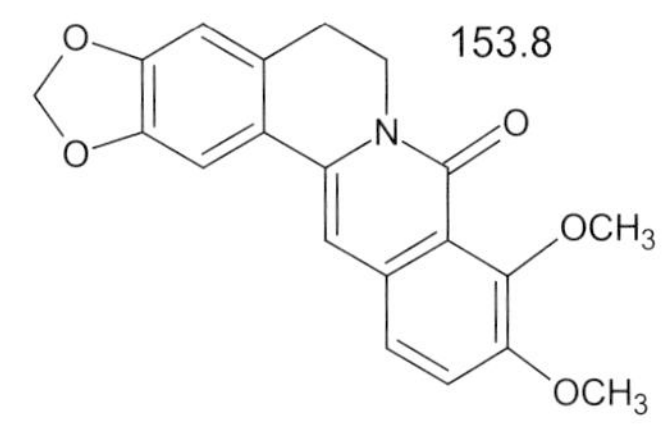

128

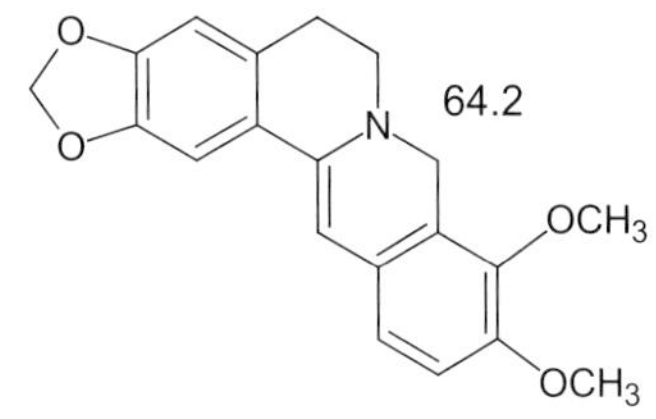

129

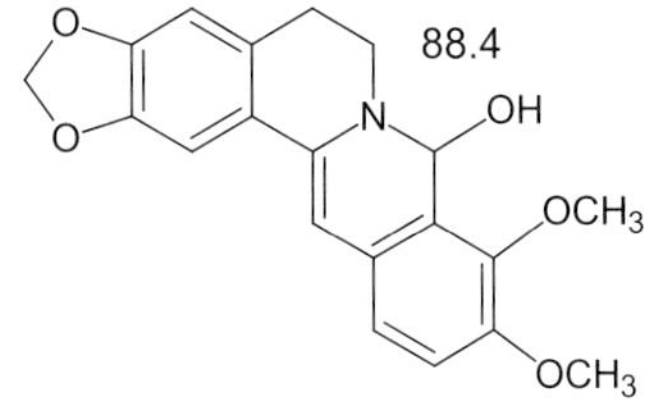

130

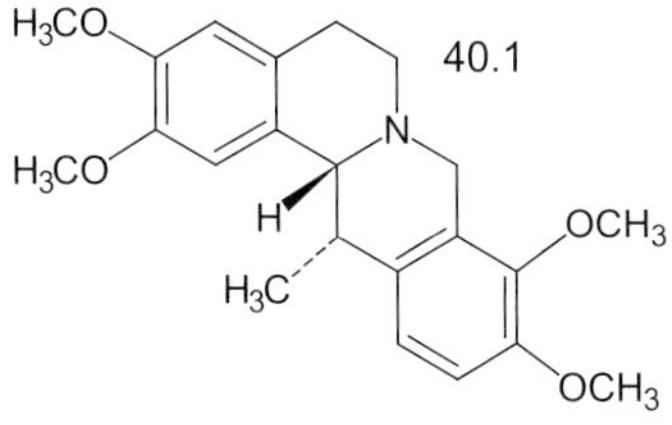

131

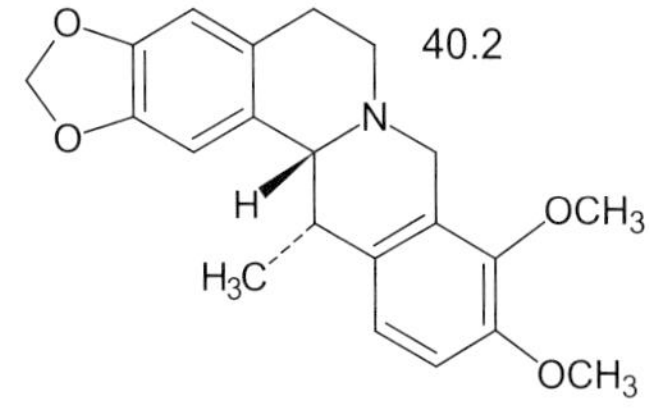

132

Fig. 14.13 Protoberberine Alkaloids. [111,121].

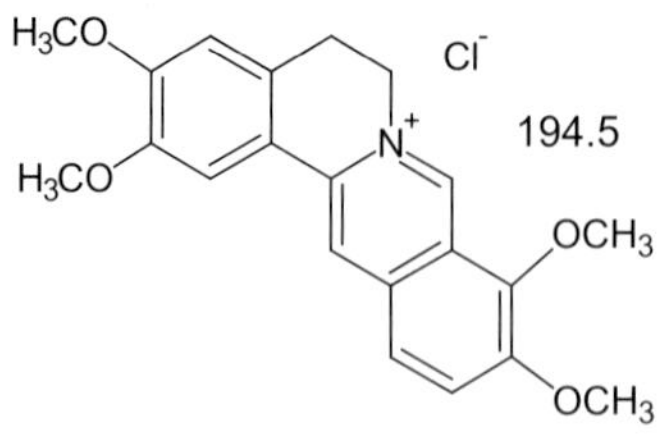

H3CO
H3CO
Cl-
N+
194.5
OCH3
OCH3

133

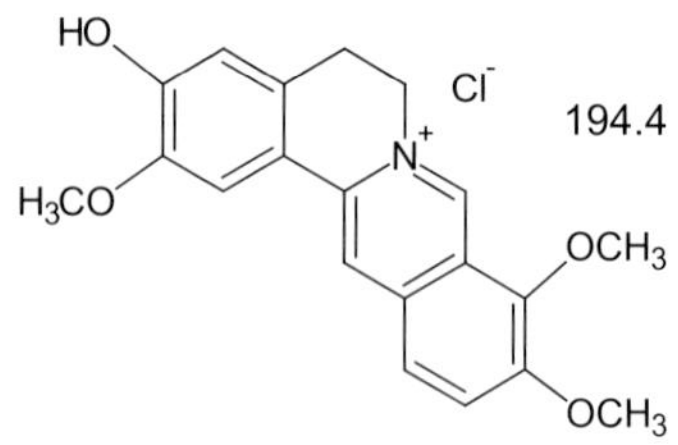

HO
H3CO
Cl-
N+
194.4
OCH3
OCH3

134

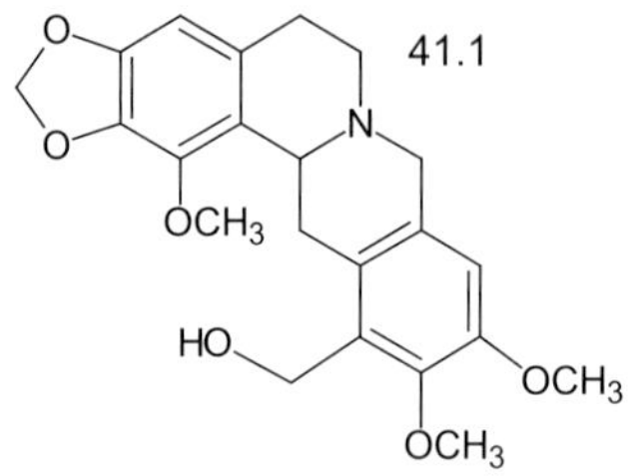

O
O
41.1
N
OCH3
HO
OCH3
OCH3

135

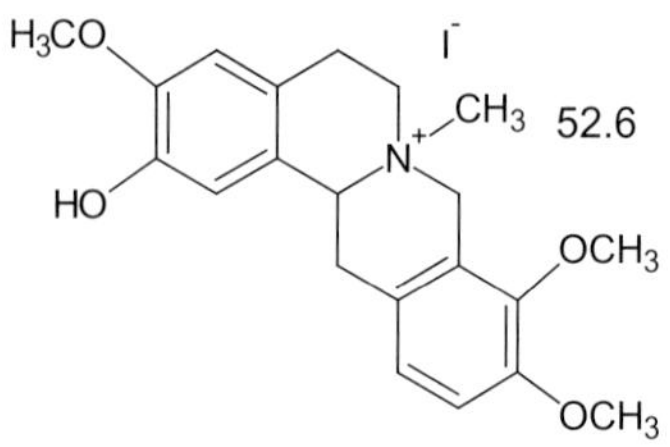

H3CO
HO
I-
N+
CH3
52.6
OCH3
OCH3

136

Fig. 14.13 (*Continued*)

H3CO
HO
H3CO
42.8
N
CH3

137

H3CO
H3CO
HO
H3CO
43.3
N
CH3

138

O
O
HO
H3CO
43.2
N
CH3

139

O
O
47.9
N
CH3
O

140

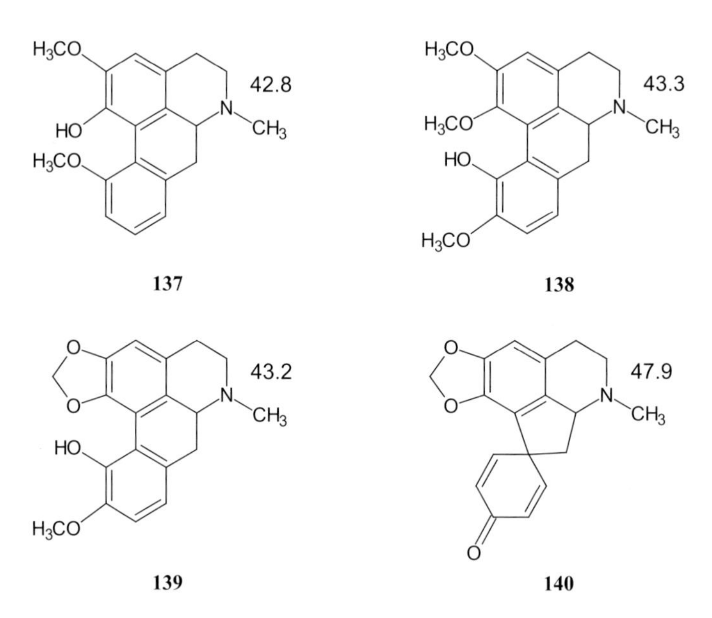

Fig. 14.14 Aporphine and proaporphine alkaloids [111,121].

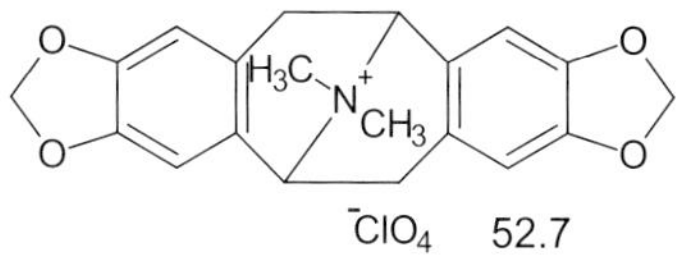

141

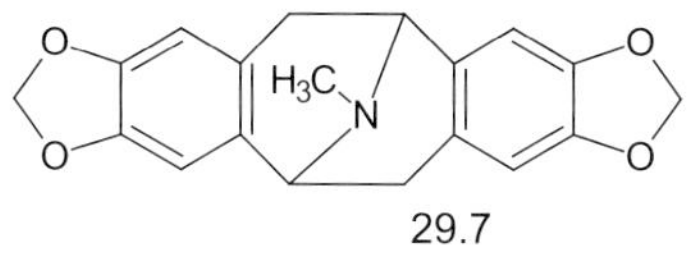

142

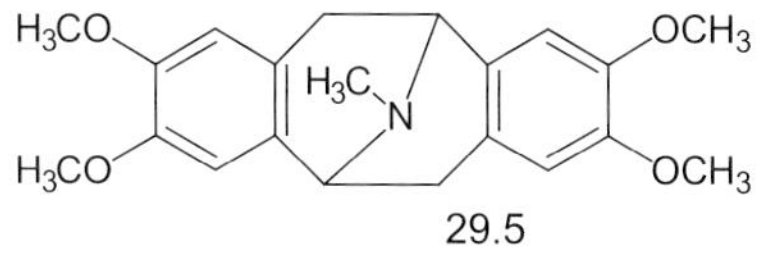

143

Fig. 14.15 Pavine alkaloids [111,121].

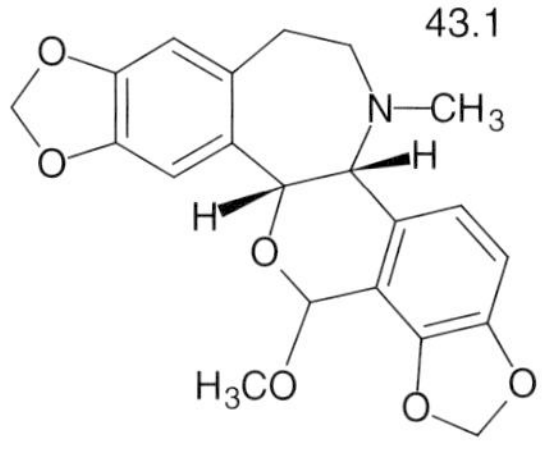

144

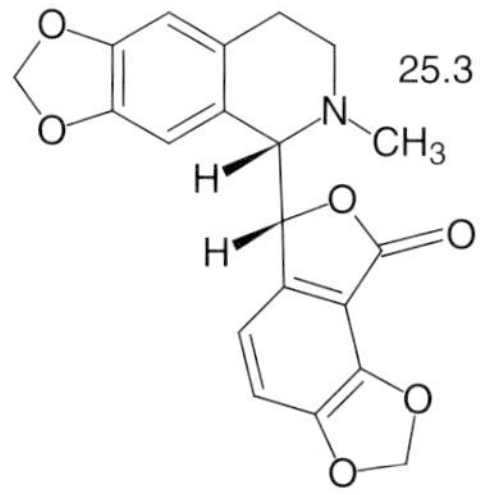

145

structure, **146**, was differentiated from an alternative isomeric structure, **147**, on the basis of the long-range ^1H–^{15}N correlation data.

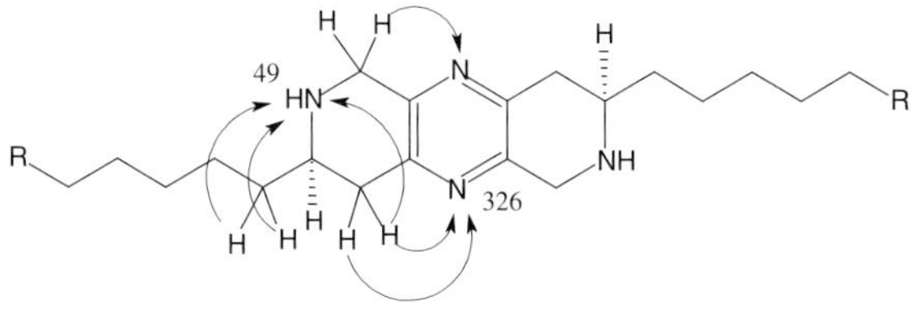

146

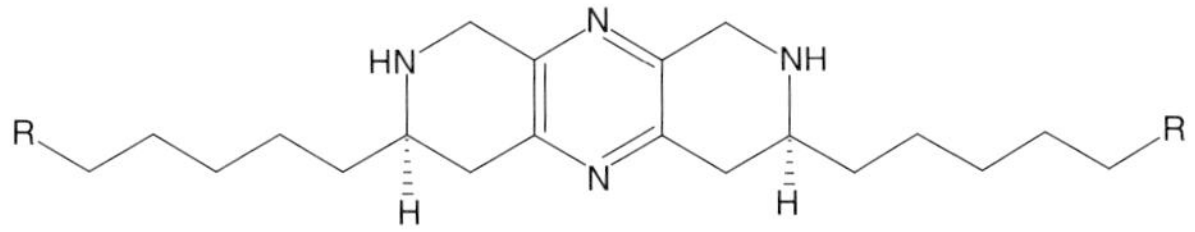

147

14.6.8
Diazepinopurine Alkaloids

Selective cytotoxicity of the marine sponge alkaloid asmarine-A (**148**) [123] triggered Kashman's laboratory to undertake the synthesis of this unique diazepinopurine

Tab. 14.8 ^{15}N Chemical shifts for asmarine-A (**148**), -F (**149**), -G (**150**), and -H (**151**) [125].

		^{15}N Chemical shifts				
Compound	**–R**	**N-1′**	**N-3′**	**N-7′**	**N-9′**	**N-10′**
148	–OH	233.3	247.0	245.8	153.9	166.5
149	$-OCH_3$ 5-epi, 8-oxo	224.0	239.2	120.0	113.5	186.9
150	$-OCH_3$	236.7	231.2	243.8	152.9	189.9
151	–H	243.0	231.2	243.8	155.9	116.6

alkaloid [124]. Long-range ^{1}H–^{15}N 2D NMR data were recorded for asmarine-A (**148**), -F (**149**), -G (**150**), and -H (**151**) [125]. The authors noted that the originally reported ^{15}N chemical shifts were misreferenced and reported the corrected assignments [124]. The corrected assignments are summarized in Table 14.8. Long-range ^{1}H–^{15}N coupling pathways in the asmarine series are shown by **152** [124]. The ^{15}N chemical shift assignments for a number of synthetic intermediates, **153–161**, are collected in Table 14.9. Further synthetic work reported by the same group of authors [126] resulted in ^{15}N chemical shift data for several additional compounds, which also afforded some insights into the effects of 9-substitution on the ^{15}N shift of the N-10 resonance. These data are presented in Table 14.10.

152

14.7
Pyridoacridine, Quinoacridine, and Related Alkaloids

There has been considerable interest in pyridoacridines and related compounds owing to their biological activity (cytotoxicity, mutagenicity, or antibacterial activity). Wawer and coworkers [127] reported the study of a pair of quinoacridinium salts from this

Tab. 14.9 ^{15}N Chemical shifts of tetrahydrodiazepinopurine analogs of asmarine-A (**148**) [124].

Compound	^{15}N Chemical shifts				
	N-1	N-3	N-7	N-9	N-10
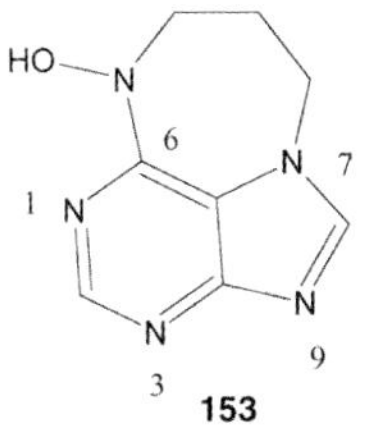 **153**	245.0	235.3	156.0	245.5	149.3
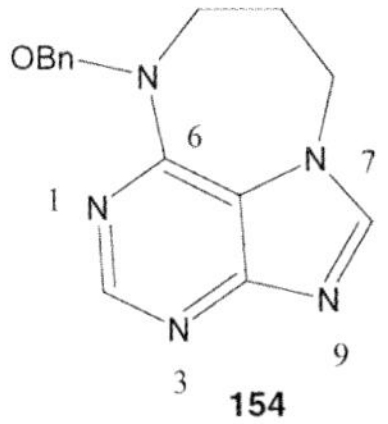 **154**	250.8	237.9	155.6	245.0	92.6
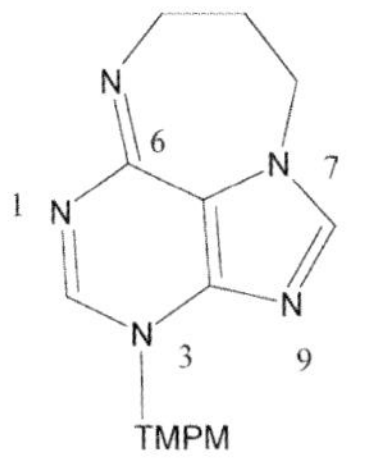 **155**	234.0	161.3	167.3	231.9	119.2
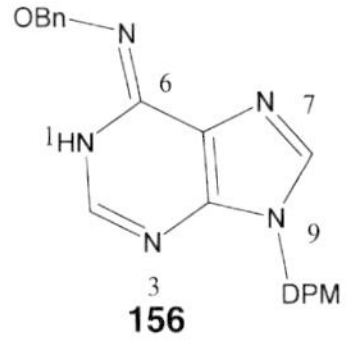 **156**	136.9	207.9	247.4	173.9	285.3
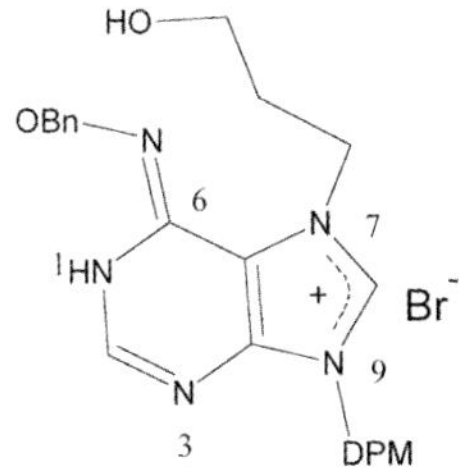 **157**	141.8	204.0	172.2	180.0	288.2

Tab. 14.9 *(Continued)*

Compound	^{15}N Chemical shifts				
	N-1	N-3	N-7	N-9	N-10
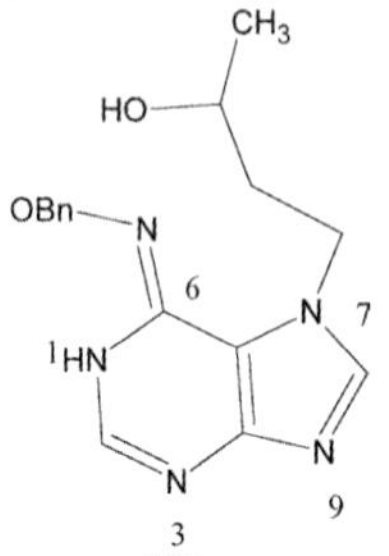 **158**	134.4	226.7	164.5	250.0	280.0
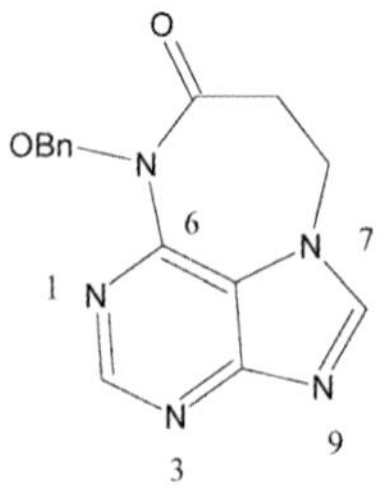 **159**	Not obs.	263.5	155.9	244.5	203.9
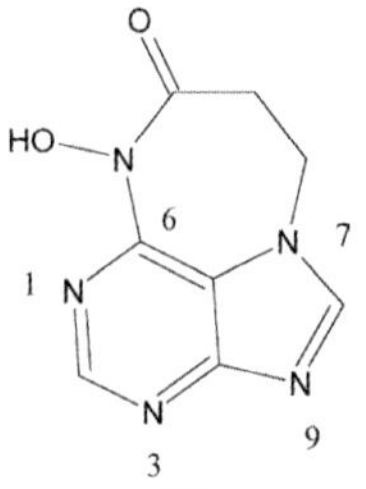 **160**	Not obs.	262.6	155.7	244.2	188.3
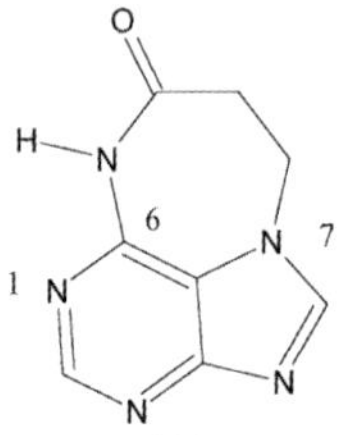 **161**	248.0	260.5	154.7	245.4	150.7 (HSQC)

TMPM, (3,4,5-timethoxyphenyl)methyl; DPM, diphenylmethyl.

Tab. 14.10 ^{15}N Chemical shifts of synthetic tetrahydrodiazepinopurine analogs of asmarine-A (**148**) [126].

Compound	^{15}N Chemical shifts				
	N-1	N-3	N-7	N-9	N-10
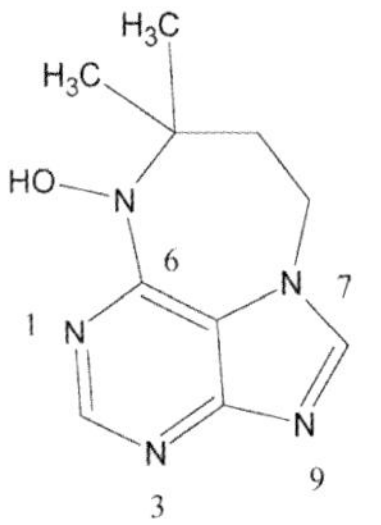 **162**	247.9	234.0	155.0	246.8	167.7
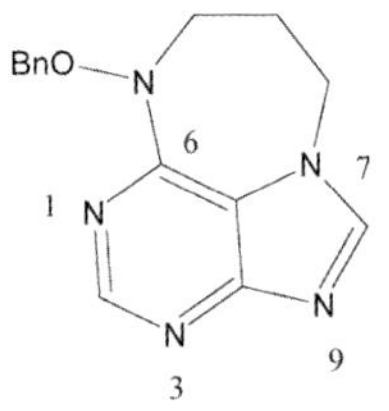 **163**	250.8	237.9	155.0	245.4	168.7
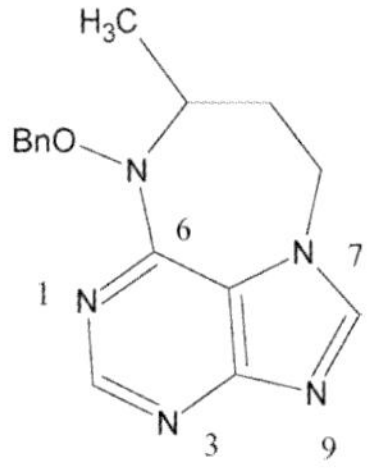 **164**	251.0	237.3	155.0	246.8	178.2
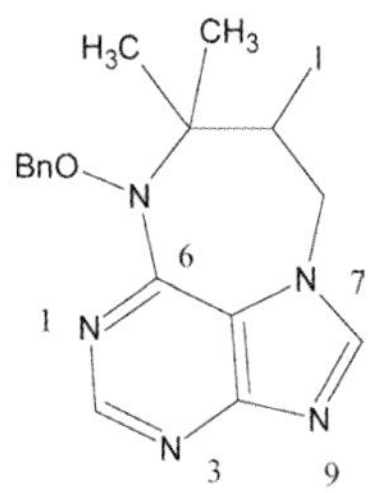 **165**	247.5	241.8	153.4	245.6	188.1

Tab. 14.10 *(Continued)*

Compound	^{15}N Chemical shifts				
	N-1	N-3	N-7	N-9	N-10
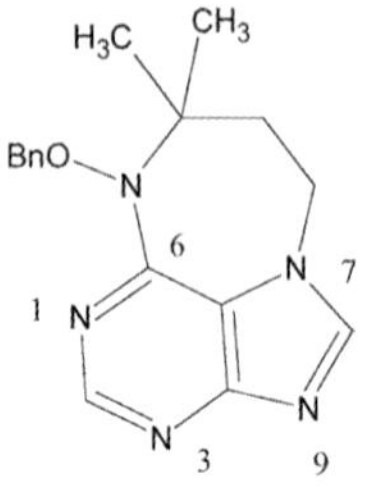 **166**	253.2	240.0	155.0	246.8	188.8

broad family of compounds using long-range ^{15}N heteronuclear shift correlation data in their study. There are numerous resonance structures, two of the 18 that the authors report as possible are shown as **167** and **168**. The authors were able to differentiate the chemical shifts of the two *N*-ethyl groups at N-8 and N-13 on the basis of calculated nitrogen shielding constants and by heteronuclear shift correlation spectra. Based on their work, the authors conclude that N-8 is significantly more shielded than N-13, giving rise to the chemical shift assignments contained in Table 14.11.

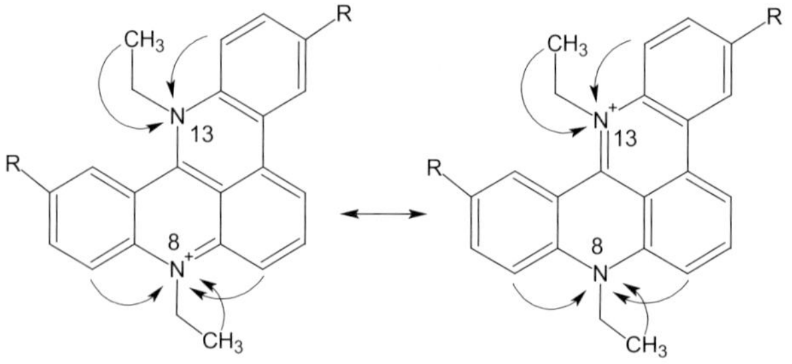

167 **168**

Copp and coworkers [128] reported the isolation of two new pyridoacridine analogs, isodiplamine (**169**) and lissoclinidine (**170**) from the New Zealand ascidian *Lissoclinum notti*. The authors used 6 Hz optimized ^{1}H–^{15}N HMBC data in the process of characterizing these novel compounds. The ^{15}N chemical shifts and observed long-range couplings are shown on the structures.

Tab. 14.11 ^{15}N Chemical shift assignments for two quinoacridinium salts [127].

Substituent	N-8	N-13
R = –H	135.6	143.7
R = –CH_3	134.9	142.7

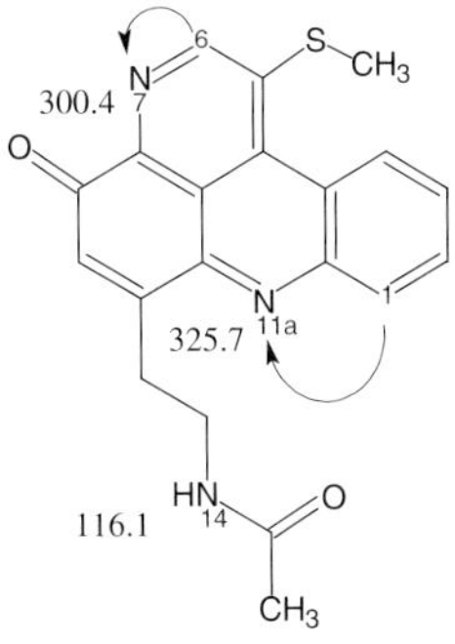

14.8 Conclusions

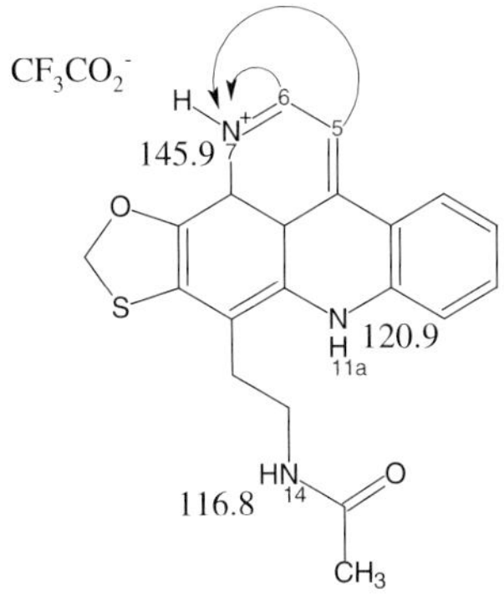

Alkaloids are widely distributed in both terrestrial and marine organisms. The feasibility of performing direct and long-range ^{1}H–^{15}N heteronuclear shift correlation experiments at natural abundance has provided investigators with an extremely powerful tool for alkaloid structure elucidation. Advances in the electronics employed in NMR spectrometers, the wider availability of very high field instruments, coupled with routine access to small-volume high-sensitivity NMR probes and, in growing numbers of laboratories, cryogenic NMR probes is also greatly facilitating the utilization of ^{15}N chemical shift and long-range coupling pathway information in the determination of the structures of novel alkaloids. We have tried to include in this chapter all the reports contained in the literature that involve natural abundance direct and long-range ^{1}H–^{15}N heteronuclear chemical shift correlation data. Doubtless, vast as the chemical literature is, we have overlooked some report(s). To the authors of those works, we apologize. As time goes on, we expect that there will growing numbers of reports in the literature describing the use of direct and long-range ^{1}H–^{15}N 2D NMR data for the elucidation of alkaloid structures.

We have also attempted to highlight the value of ^{15}N chemical shift calculations for establishing F_1 spectral windows when performing long-range ^{1}H–^{15}N heteronuclear chemical shift correlation experiments. Examples have also been presented illustrating the impact of ^{1}H–^{15}N heteronuclear shift correlation data in minimizing calculation times for computer-assisted structure elucidation (CASE) applications.

Finally, there are several very interesting papers currently in press. First, work by Kupče and Freeman [129] has demonstrated the possibility of obtaining natural abundance ^{15}N–^{13}C correlation spectra using vitamin B-12 as a model compound for their study. Their method uses a new reconstruction technique based on simultaneously acquired standard HMQC and HMBC spectra at ^{13}C and ^{15}N natural abundance. In contrast to standard 3D NMR methods, the authors claim the method they have described offers a two orders of magnitude improvement in sensitivity. It remains to be seen what impact this method will have on alkaloid chemical structure characterization. The authors have also reported work involving computation methods, specifically indirect covariance [130,131] and unsymmetrical indirect covariance NMR methods [132–134], that may facilitate the acquisition of hyphenated (HSQC-COSY, -TOCSY, -NOESY, -ROESY, and -HMBC) 2D NMR data that can be used in the determination of alkaloid structures.

Finally, Parella and coworkers [135] have reported a pulse sequence that allows for the simultaneous acquisition of ^{1}H–^{13}C and ^{1}H–^{15}N HMBC spectra. One of the model compounds that they used to demonstrate the technique, not surprisingly, was strychnine. How all of this new work will affect alkaloid structure characterization remains to be seen, but it does appear that the future is quite bright in terms of new spectroscopic techniques that may facilitate alkaloid structure characterization.

References

1 Müller, L. (1979) *Journal of the American Chemical Society*, **101**, 4481–4484.

2 Bodenhausen, G. and Ruben, D. J. (1980) *Chemical Physics Letters*, **69**, 185–189.

3 Bax, A., Griffey, R. H., Hawkins, B. L. (1983) *Journal of Magnetic Resonance*, **55**, 301–315.

4 Wrackmeyer, B., Garcia-Baez, E., Zuno-Cruz, F. J., Sanchez-Cabrera, G., Rosales, M. R. (2000) *Zeitschrift fur Naturforschung*, **55b**, 185–188.

5 Carbajo, R. J. and Lpez-Ortiz, F. (2001) *Journal of Magnetic Resonance*, **148**, 165–168.

6 Russell, D. J., Hadden, C. E., Martin, G. E., Krishnamurthy, V. V. (2002) *Magnetic Resonance in Chemistry*, **40**, 207–210.

7 Del Bene, J. E., Perera, S. A., Bartlett, R. J., Yanez, M., Mo, O., Elguero, J., Alkorta, I. (2003) *Journal of Physical Chemistry A*, **107**, 3121–3125.

8 Griesinger, C., Schwalbe, H., Schleucher, J., Sattler, M. (1994) Proton-Detected Heteronuclear and Multidimensional NMR, In Croasmun W. R. and Carlson R. M. K. (Eds.) *Two-Dimensional NMR Spectroscopy—Applications for Chemists and Biochemists*, 2nd edn VCH, New York, 458.

9 Hadden, C. E. (2005) *Magnetic Resonance in Chemistry*, **43**, 330–333.

10 Martin, G. E. and Hadden, C. E. (2000) *Journal of Natural Products*, **65**, 543–585.

11 Marek, R. and Lycka, A. (2002) *Current Organic Chemistry*, **6**, 35–66.

12 Martin, G. E. and Williams, A. J. (2005) Long-Range ^{1}H–^{15}N 2D NMR Methods, In Webb G. A. (Ed.) *Annual Reports on NMR Spectroscopy*, Vol. **55**, Elsevier, Amsterdam, 1–119.

13 Martin, G. E. and Crouch, R. C. (1994) Inverse-Detected 2D NMR Applications in Alkaloid Chemistry, In Linskens H. F.and Jackson J. F. (Eds.) *Modern Methods in Plant Analysis*, Vol. 15, Springer-Verlag, Berlin, 25–89.

14 Forgo, P., Hohmann, J., Dombi, G., Máthé, L. (2004) Advanced Multidimensional NMR Experiments as Tools for Structure Determination of *Amaryllidaceae* Alkaloids, In Acamovic, T. Steward, S. Pennycott T. W. (Eds.) *Poisonous Plants and Related Toxins*, CAB International, Wallingford, UK, 322–328.

15 Levy, G. C. and Lichter, R. L. (1979) *Nitrogen-15 NMR Spectroscopy*, Wiley, New York.

16 Witanowski, M. and Webb, G. A. (1973) *Nitrogen NMR*, Plenum Press, London.

17 Martin, G. J., Martin, M. L., Gouesnard, J. -P. (1981) *^{15}N-NMR Spectroscopy*, Springer-Verlag, Berlin.

18 Berger, S., Braun, S., Kalinowski, H.-O. (1997) *NMR Spectroscopy of the*

Non-Metallic Elements, Wiley, New York, pp. 111–318.

19 Mooney, E. F. and Winson, P. H. (1968) Nitrogen Magnetic Resonance Spectroscopy, In Mooney E. F. (Ed.) *Annual Reports on NMR Spectroscopy*, Vol. 2, Academic Press, New York, 125–152.

20 Witanowski, M., Stefaniak, L., Webb, G. A. (1972) Nitrogen NMR Spectroscopy, In Mooney E. F. (Ed.) *Annual Reports on NMR Spectroscopy*, Vol. 5a, Academic Press, New York, 395–457.

21 Witanowski, M., Stefaniak, L., Webb, G. A. (1977) Nitrogen NMR Spectroscopy, In Webb G. A. (Ed.) *Annual Reports on NMR Spectroscopy*, Vol. 7, Academic Press, New York, 118–239.

22 Witanowski, M., Stefaniak, L., Webb, G. A. (1981) Nitrogen NMR Spectroscopy, In Webb G. A. (Ed.) *Annual Reports on NMR Spectroscopy*, Vol. 11B, Academic Press, New York.

23 Witanowski, M., Stefaniak, L., Webb, G. A. (1986) Nitrogen NMR Spectroscopy, In Webb G. A. (Ed.) *Annual Reports on NMR Spectroscopy*, Vol. 18, Academic Press, New York.

24 Witanowski, M., Stefaniak, L., Webb, G. A. (1993) Nitrogen NMR Spectroscopy, In Webb G. A. (Ed.) *Annual Reports on NMR Spectroscopy*, Vol. 25, Academic Press, New York.

25 Bax, A. and Subramanian, S. (1986) *Journal of Magnetic Resonance*, **65**, 565–569.

26 Bax, A. and Summers, M. F. (1986) *Journal of the American Chemical Society*, **108**, 2093–2094.

27 Martin, G. E. and Zektzer, A. S. (1988) *Magnetic Resonance in Chemistry*, **26**, 631–652.

28 Martin, G. E., Crouch, R. C., Sharaf, M. H. M., and Schiff, P. L. Jr. (1993) 34th *Annual Meeting of the American Society of Pharmacognosy*, San Diego, CA, July 18–22 [Abstract P101].

29 Hurd, R. E. and John, B. K. (1991) *Journal of Magnetic Resonance*, **91**, 648–653.

30 Ruiz-Cabello, J., Vuister, G. W., Moonen, C. T. W., van Geldern, P., Cohen, J. S., van Zilj, P. C. M. (1992) *Journal of Magnetic Resonance*, **100**, 282–302.

31 Koshino, H. and Uzawa, J. (1995) *Kagaku to Seibutsu*, **33**, 252–258.

32 Crouch, R. C. and Martin, G. E. (1995) *Journal of Heterocyclic Chemistry*, **32**, 1665–1669.

33 Martin, G. E., Crouch, R. C., Andrews, C. W. (1995) *Journal of Heterocyclic Chemistry*, **32**, 1759–1766.

34 Martin, G. E. and Crouch, R. C. (1995) *Journal of Heterocyclic Chemistry*, **32**, 1839–1842.

35 Wagner, R. and Berger, S. (1998) *Magnetic Resonance in Chemistry*, **36**, S44–S46.

36 Hadden, C. E., Martin, G. E., Krishnamurthy, V. V. (1999) *Journal of Magnetic Resonance*, **140**, 274–280.

37 Martin, G. E. and Hadden, C. E. (2000) *Magnetic Resonance in Chemistry*, **38**, 251–256.

38 Kline, M. and Cheatham, S. (2003) *Magnetic Resonance in Chemistry*, **41**, 307–314.

39 Hadden, C. E., Martin, G. E., Krishnamurthy, V. V. (2000) *Magnetic Resonance in Chemistry*, **38** (143) 143–147.

40 Advanced Chemistry Development, Inc., 110 Yonge St., 14th Floor, Toronto, Ontario, M5C 1T4, Canada.

41 Bremser, W. (1978) *Analytica Chimica Acta*, **103**, 355–365.

42 Martin, G. E. and Williams, A. J. (April 18, 2004) Validation of ACD/NNMR predictor, ACD/Labs User Meeting, ENC 2004, Monterey.

43 Köck, M., Junker, J., Lindel, T. (1999) *Organic Letters*, **1**, 2041–2044.

44 Nuzillard, J. -M., Cormolly, J. D., Delavde, C., Richard, B., Zches-Hanrot, M., Le Me-Olivier, L. (1999) *Tetrahedron*, **55**, 11511–11518.

45 Martin, G. E., Hadden, C. E., Russell, D. J., Kaluzny, B. D., Guido, J. E., Duholke, W. K., Stiemsma, B. A., Thamann, T. J., Crouch, R. C., Blinov, K., Elyashberg, M. E.,

Martirosian, E. R., Molodtsov, S. G., Williams, A. J., Schiff, P. L., Jr. (2002) *Journal of Heterocyclic Chemistry*, **39**, 1241–1250.

46 Martin, G. E. and Williams, A. J. (March 7–8, 2002) Probes, Pulse *Sequences, and CASE Programs: Advances in the Characterization of Novel Structures Using NMR-Based Methods*. ACD/Labs North American User's Meeting, Princeton, NJ.

47 Grube, A. and Köck, M. (2006) *Journal of Natural Products*, **69**, 1212–1214.

48 Martin, G. E. and Williams, A. J. (March 11–16, 2001) Automated Structure Elucidation of Cryptolepine Derivatives, 42nd ENC, Orlando, FL.

49 Blinov, K. A., Elyashberg, M. E., Molodtsov, S. G., Williams, A. J., Martirosian, E. R. (2001) *Fresenius Journal of Analytical Chemistry*, **369**, 709–714.

50 Elyashberg, M. E., Blinov, K. A., Williams, A. J., Molodtsov, S. G., Martirosian, E. R. (2002) *Journal of Natural Products*, **65**, 693–703.

51 Blinov, K. A., Carlson, D., Elyashberg, M. E., Martin, G. E., Martirosian, E. R., Molodtsov, S. G., Williams, A. J. (2003) *Magnetic Resonance in Chemistry*, **41**, 359–372.

52 Martin, G. E., Russell, D. J., Blinov, K. A., Elyashberg, M. E., Williams, A. J. (2003) *Annales of Magnatic Resonance*, **2**, 1–31.

53 Blinov, K., Elyashberg, M., Martirosian, E. R., Molodtsov, S. G., Williams, A. J., Sharaf, M. H. M., Schiff, P. L., Jr., Crouch, R. C., Martin, G. E., Hadden, C. E., Guido, J. E. (2003) *Magnetic Resonance in Chemistry*, **41**, 577–584.

54 Elyashberg, M. E., Blinov, K. A., Martirosian, E. R., Molodtsov, S. G., Williams, A. J., Martin, G. E. (2003) *Journal of Heterocyclic Chemistry*, **40**, 1017–1029.

55 Elyashberg, M. E., Blinov, K. A., Williams, A. J., Molodtsov, S. G., Martin, G. E., Martirosian, E. R. (2004) *Journal of Chemical Information and Computer Science*, **44**, 771–792.

56 Molodtsov, S. G., Elyashberg, M. E., Blinov, K. A., Williams, A. J., Martirosian, E. R., Martin, G. E., Lefebvre, B. (2004) *Journal of Chemical Information and Computer Science*, **44**, 1737–1751.

57 Smurnyy, Y. D., Elyashberg, M. E., Blinov, K. A., Lefebvre, B. A., Martin, G. E., Williams, A. J. (2005) *Tetrahedron*, **61**, 9980–9989.

58 Elyashberg, M. E., Blinov, K. A., Molodtsov, S. G., Williams, A. J., Martin, G. E. (2007) *Journal of Chemical Information and Modeling*, **47**, 1053–1066.

59 Elyashberg, M. E., Blinov, K. A., Williams, A. J., Molodtsov, S. G., Martin, G. E. (2006) *Journal of Chemical Information and Modeling*, **46**, 1643–1656.

60 Günther, H. (1995) *NMR Spectroscopy: Basic Principles, Concepts, and Applications in Chemistry*, 2nd edn Wiley, New York.

61 Steinbeck, C. (2001) *Journal of Chemical Information and Computer Science*, **41**, 1500–1507.

62 Han, Y. and Steinbeck, C. (2004) *Journal of Chemical Information and Computer Science*, **44**, 489–498.

63 Steinbeck, C. (2004) *Natural Product Reports*, **21**, 512–518.

64 Steenberg, V. I., Narain, N. K., Singh, S. P., Obenauf, R. H., Albright, M. J. (1997) *Journal of Heterocyclic Chemistry*, **14**, 407–410.

65 Martin, G. E., Mills, K. A., and Williams, A. J. unpublished data.

66 Fanso-Free, S. N. Y., Furst, G. T., Srinivasan, P. R., Lichter, R. L., Nelson, R. B., Panetta, J. A., Gribble, G. W. (1979) *Journal of the American Chemical Society*, **101**, 1549–1553.

67 Koshino, H. and Uzawa, J. (1995) *Kagaku to Seibutsu*, **33**, 252–258.

68 Kato, Y., Koshino, H., Uzawa, J., Anzai, K. (1996) *Bioscience Biotechnology and Biochemistry*, **60**, 2081–2083.

69 Verotta, L., Pilati, T., Tatò, M., Elisabetsky, E., Amador, T. A., Nunes, D. S. (1998) *Journal of Natural Products*, **61**, 392–396.

70 El Sayed, K. A., Hamann, M. T., Abd El-Rahman, H. A., Zaghloul, A. M. (1998) *Journal of Natural Products*, **61**, 848–850.

71 Dunbar, D. C., Rinaldi, J. M., Clark, A. M., Kelly, M., Hamann, M. T. (2000) *Tetrahedron Letters*, **56**, 8795–8598.

72 Ford, J. and Capon, R. J. (2000) *Journal of Natural Products*, **63**, 1527–1528.

73 Ciminiello, P., Dell'Aversano, C., Fattorusso, E., Magno, S. (2001) *European Journal of Organic Chemistry*, 55–60.

74 Kretsi, O., Aligiannis, N., Skaltsounis, A. L., Chinou, I. B. (2003) *Helvetica Chimica Acta*, **86**, 3136–3140.

75 Farley, K. A., Bowman, P. B., Brumfield, J. C., Crow, F. W., Duholke, W. K., Guido, J. E., Robins, R. H., Sims, S. M., Smith, R. F., Thamann, T. J., Vonderwell, B. S., Martin, G. E. (1998) *Magnetic Resonance in Chemistry*, **36**, S11–S16.

76 Komoda, T., Sugiyama, Y., Abe, N., Imachi, M., Hirota, H., Koshiono, H., Hirota, A. (2003) *Tetrahedron Letters*, **44**, 7417–4719.

77 Komoda, T., Sugiyama, Y., Abe, N., Imachi, M., Hirota, H., Hirota, A. (2003) *Tetrahedron Letters*, **44**, 1659–1661.

78 Fliniaux, O., Mesnard, F., Raynaud, S., Baltora, S., Robins, R. J., Fliniaux, M.-A. (2001) *Comptes Rendus Academic des Sciences. Paris, Chemie*, **4**, 775–778.

79 Scholl, Y., Schneider, B., Dräger, B. (2003) *Phytochemistry*, **62**, 325–332.

80 Hadden, C. E., Kaluzny, B. D., Robins, R. H., Martin, G. E. (1999) *Magnetic Resonance in Chemistry*, **37**, 325–327.

81 Koshino, H., Lee, I.-K., Kim, J.-P., Kim, W.-G., Uzawa, J., Yoo, I.-D. (1996) *Tetrahedron Letters*, **37**, 4549–4550.

82 Crouch, R. C., Davis, A. O., Spitzer, T. D., Martin, G. E., Sharaf, M. H. M., Schiff, P. L., Jr.,Phoebe, C. H., Jr. (1995) *Journal of Heterocyclic Chemistry*, **32**, 1077–1080.

83 Martin, G. E., Crouch, R. C., Sharaf, M. H. M., Schiff, P. L. Jr. (1996) *Journal of Natural Products*, **59**, 2–4.

84 Kaczmarek, L., Nantka-Mamirski, P., Stefaniak, L., Webb, G. A., Davoust, D., Basselier, J. (1985) *Magnetic Resonance in Chemistry*, **23**, 853–855.

85 Tackie, A. N., Boye, G. L., Sharaf, M. H. M., Schiff, P. L., Jr.,Crouch, R. C., Spitzer, T. D., Johnson, R. L., Dunn, J., Minic, D., Martin, G. E. (1993) *Journal of Natural Products*, **56**, 653–670.

86 Hadden, C. E., Martin, G. E., Tackie, A. N., Schiff, P. L., Jr. (1999) *Journal of Heterocyclic Chemistry*, **36**, 1115–1117.

87 Martin, G. E., Hadden, C. E., Blinn, J. R., Sharaf, M. H. M., Tackie, A. N., Schiff, P. L., Jr. (1999) *Magnetic Resonance in Chemistry*, **37**, 1–6.

88 Hadden, C. E., Sharif, M. H. M., Guido, J. E., Robins, R. H., Tackie, A. N., Phobe, C. H., Jr.,Schiff, P. L., Jr.,Martin, G. E. (1999) *Journal of Natural Products*, **62**, 238–240.

89 Martin, G. E. and Crouch, R. C. (1997) *Journal of Heterocyclic Chemistry*, **32**, 1839–1842.

90 Martin, G. E. (1997) *Journal of Heterocyclic Chemistry*, **34**, 695–699.

91 Kim, W.-G., Kim, J.-P., Koshino, H., Shin-Ya, K., Seto, H., Yoo, I.-D. (1997) *Tetrahedron*, **53**, 4309–4316.

92 Wang, H. and Ganesan, A. (2000) *Journal of Organic Chemistry*, **65**, 1022–1030.

93 Shin-Ya, K., Kim, J.-K., Furihata, K., Hayakawa, Y., Seto, H. (2000) *Journal of Asian Natural Products Research*, **2**, 121–132.

94 Martin, G. E. (2002) Qualitative and Quantitative Exploitation of Heteronuclear Coupling Constants, In Webb G. A. (Ed.) *Annual Reports on NMR Spectroscopy*, Vol. 46, Academic Press, London, 37–100.

95 Seki, H., Tokunaga, T., Utsumi, H., Yamaguchi, K. (2000) *Tetrahedron*, **56**, 2935–2939.

96 Muhammad, I., Dunbar, D. C., Khan, R. A., Ganzera, M., Kahn, I. A. (2001) *Phytochemistry*, **57**, 781–785.

97 Li, X. -C., Dunbar, D. C., El Sohly, H. N., Walker, L. A., Clark, A. M. (2001) *Phytochemistry*, **58**, 627–629.
98 Zhang, Z., El Sohly, H. N., Jacob, M. R., Pasco, D. S., Walker, L. A., Clark, A. M. (2001) *Journal of Natural Products*, **64**, 1001–1005.
99 Horiuchi, M., Maoka, T., Iwase, N., Ohnishi, K. (2002) *Journal of Natural Products*, **65**, 1204–1205.
100 Suemitsu, R., Ohnishi, K., Horiuchi, M., Kitagichi, A., Odamura (1992) *Phytochemistry*, **31**, 2325–2326.
101 Appleton, D. R., Page, M. J., Lambert, G., Berridge, M. V., Copp, B. R. (2002) *Journal of Organic Chemistry*, **67**, 5402–5404.
102 Appleton, D. R. and Copp, B. R. (2003) *Tetrahedron*, **44**, 8963–8965.
103 Hadden, C. E., Richrd, D. J., Joullié, M. M., Martin, G. E. (2003) *Journal of Heterocyclic Chemistry*, **40**, 359–362.
104 Tsuda, M., Mugishima, T., Komatsu, K., Sone, T., Tanaka, M., Mikami, Y., Shiro, M., Hirai, M., Ohizumi, Y., Kobayashi, J. (2003) *Tetrahedron*, **59**, 3227–3230.
105 Jeong, S.-Y., Ishida, K., Ito, Y., Okada, S., Murakami, M. (2003) *Tetrahedron Letters*, **44**, 8005–8007.
106 Glover, R. P., Yoganathan, K., Butler, M. S. (2005) *Magnetic Resonance in Chemistry*, **43**, 483–485.
107 Crouch, R. C. and Martin, G. E. (1995) *Journal of Heterocyclic Chemistry*, **32**, 1665–1669.
108 Marek, R., Dostál, J., Slavk, J., Sklenář, V. (1996) *Molecules*, **1**, 166–169.
109 Seki, H., Tokunaga, T., Utsumi, H., Yamaguchi, K. (2000) *Tetrahedron*, **56**, 2935–2939.
110 Marek, R., Marek, J., Dostal, J., Slavk, J. (1997) *Collection of Czechoslovak Chemical Communications*, **62**, 1623–1230.
111 Marek, R., Humpa, O., Dostál, J., Slavk, J., Sklenář, V. (1999) *Magnetic Resonance in Chemistry*, **37**, 195–202.
112 Schiff, P. L., Jr. (1999) *Alkaloids: Chemical and Biological Perspectives*, **14**, 1–284.
113 Mambu, L., Martin, M.-T., Razafimahefa, D., Ramanitrahasimbola, D., Rasoanaivo, P., Frappier, F. (2000) *Planta Medica*, **66**, 537–540.
114 Orabi, K. Y., El Sayed, K. A., Harmann, M. T., Dunbar, C. D., Al-Said, M. S., Higa, T., Kelly, M. (2000) *Journal of Natural Products*, **65**, 1782–1785.
115 Fokialakis, N., Magiatis, P., Skaltsounis, A.-L., Tillequin, F., Sévenet, T. (2000) *Journal of Natural Products*, **63**, 1004–1005.
116 Muhammad, I., Dunbar, D. C., Takamatsu, S., Walker, L. A., Clark, A. M. (2001) *Journal of Natural Products*, **64**, 559–562.
117 Francisco, M. C., Nasser, A. L. M., Lopes, L. M. X. (2003) *Phytochemistry*, **62**, 1265–1270.
118 Yang, L. -K., Glover, R. P., Yogonathan, K., Sarnaik, J. P., Godbole, A. J., Soejarto, D. D., Buss, A. D., Butler, M. S. (2003) *Tetrahedron Letters*, **44**, 5827–5829.
119 Marek, R., Toušek, J., Králk, L., Dostál, J., Sklenář, V. (1997) *Chemistry Letters*, 369–370.
120 Dostál, J., Marek, R., Slavk, J., Táborská, E., Potácek, M., Sklenář, V. (1998) *Magnetic Resonance in Chemistry*, **36**, 869–872.
121 Marek, R., Marek, J., Dostál, J., Táborská, E., Slavk, J., Dommisse, R. (2002) *Magnetic Resonance in Chemistry*, **40**, 687–692.
122 Chill, L., Aknin, M., Kashman, Y. (2003) *Organic Letters*, **5**, 2433–2435.
123 Yosief, T., Rudi, A., Kashman, Y. (2000) *Journal of Natural Products*, **63**, 299–304.
124 Pappo, D. and Kashman, Y. (2003) *Tetrahedron*, **59**, 6493–6501.
125 Rudi, A., Shalom, H., Schleyer, M., Benayahu, Y., Kashman, Y. (2004) *Journal of Natural Products*, **67**, 106–109.
126 Pappo, D., Shimnlny, S., Kashman, Y. (2005) *Journal of Organic Chemistry*, **70**, 199–206.
127 Jaroszewska-Manaj, J., Maciejewska, D., Wawer, I. (2000) *Magnetic Resonance in Chemistry*, **38**, 482–485.

128 Appleton, D. R., Pearce, A. N., Lambert, G., Babcock, R. C., Copp, B. R. (2002) *Tetrahedron*, **58**, 9779–9783.

129 Kupče, E.,and Freeman, R. *Magnetic Resonance in Chemistry*, **45**, 103–105.

130 Zhang, F. and Brüschweiler, R. (2004) *Journal of the American Chemical Society*, **126**, 13180–13181.

131 Blinov, K. A., Larin, N. I., Kvasha, M. P., Moser, A., Williams, A. J., Martin, G. E. (2005) *Magnetic Resonance in Chemistry*, **43**, 999–1007.

132 Blinov, K. A., Larin, N. I., Williams, A. J., Zell, M., Martin, G. E. (2006) *Magnetic Resonance in Chemistry*, **44**, 107–109.

133 Blinov, K. A., Larin, N. I., Williams, A. J., Mills, K. A., Martin, G. E. (2006) *Journal of Heterocyclic Chemistry*, **43**, 163–166.

134 Blinov, K. A., Williams, A. J., Hilton B. D., Irish, P. A., Martin, G. E. *Magnetic Resonance in Chemistry*, **45**, 624–627.

135 Pérez-Trijillo, P. Nolis, Parella (2007) Organic letters, **9**, 29–32.

III
New Trends in Alkaloid Synthesis and Biosynthesis

Modern Alkaloids: Structure, Isolation, Synthesis and Biology. Edited by E. Fattorusso and O. Taglialatela-Scafati

ISBN: 978-3-527-31521-5

15
Synthesis of Alkaloids by Transition Metal-mediated Oxidative Cyclization

Hans-Joachim Knölker

The application of transition metals for selective C–H bond activation has emerged as a powerful tool for organic synthesis. We have demonstrated that the oxidative cyclization of appropriately substituted primary or secondary alkyl- and arylamines opens up simple and direct approaches to nitrogen-containing heterocyclic ring systems. This transformation can be efficiently induced using stoichiometric or even catalytic amounts of transition metals (e.g. iron, molybdenum, palladium, and silver). The advantages of this method are mild reaction conditions and toleration of a broad range of functional groups. Thus, the construction of nitrogen-heterocyclic frameworks by fusion of fully functionalized building blocks becomes feasible. Application of this chemistry in natural product total synthesis provides convergent and highly efficient short-step approaches to a variety of biologically active alkaloids [1]. The present chapter summarizes some of the advances in this area achieved since 2001 and emphasizes the utility of transition metal-mediated oxidative cyclizations.

15.1
Silver(I)-mediated Oxidative Cyclization to Pyrroles

The pyrrole ring system represents a pivotal substructure in naturally occurring alkaloids and pharmaceutical products. Following the classical Hantzsch, Knorr, and Paal–Knorr syntheses numerous alternative assemblies of pyrroles have been described [2]. Homopropargylamines and (allenylmethyl)amines represent easily available building blocks for pyrrole synthesis [3,4]. The cyclization of (allenylmethyl) amines to 2,5-dihydropyrroles can be induced by silver(I) salts and has already been investigated [4]. We have reported a novel pyrrole synthesis by silver(I)-mediated oxidative cyclization of homopropargylamines to pyrroles [5]. The required precursors are readily accessible by condensation of simple arylaldehydes to Schiff bases and subsequent Lewis acid-promoted addition of 3-trimethylsilylpropargylmagnesium bromide (Scheme 15.1). Under optimized reaction conditions, treatment with 1.1 equivalents of silver acetate at room temperature affords almost quantitative yields of the corresponding pyrroles. This procedure represents a versatile and simple synthetic route to 1,2-diarylpyrroles, which are of interest because of their biological activities [6].

Modern Alkaloids: Structure, Isolation, Synthesis and Biology. Edited by E. Fattorusso and O. Taglialatela-Scafati

ISBN: 978-3-527-31521-5

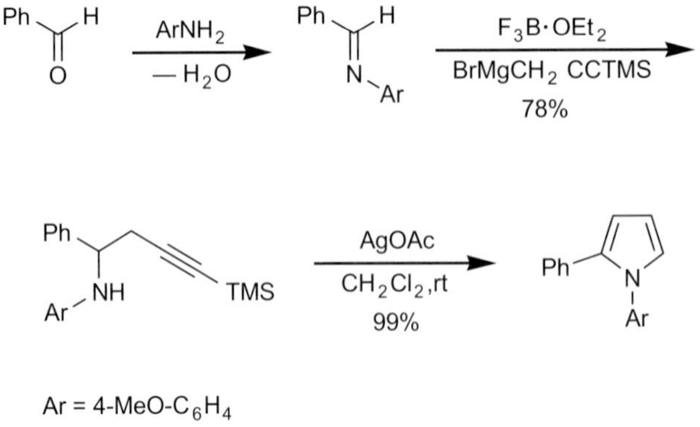

Ar = 4-MeO-C_6H_4

Scheme 15.1 Silver(I)-mediated oxidative cyclization of homopropargylamines to 1,2-diarylpyrroles.

Cyclic Schiff bases are easily prepared by Bischler–Napieralski cyclization of *N*-formyl-2-arylethylamines. Thus, the silver(I)-mediated oxidative cyclization of homopropargylamines can be exploited for the synthesis of polyheterocyclic ring systems. Annulation of a pyrrole ring on 3,4-dihydroisoquinoline provides simple access to the pyrrolo[2,1-*a*]isoquinoline framework (Scheme 15.2) [5]. Addition of the propargyl Grignard reagent affords the homopropargylamine as the major product along with the (allenylmethyl)amine resulting from attack at the γ-C atom. Silver(I)-mediated oxidative cyclization of the homopropargylamine to 5,6-dihydropyrrolo[2,1-*a*]isoquinoline proceeds at room temperature, whereas cyclization of the (allenylmethyl) amine requires elevated temperatures and gives lower yields. Protodesilylation of the silylacetylene and subsequent silver(I)-mediated oxidative

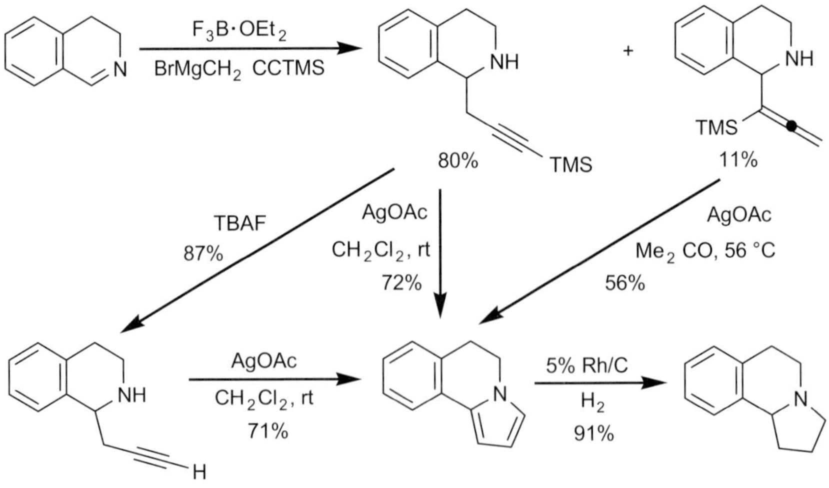

Scheme 15.2 Silver(I)-mediated oxidative cyclization to pyrrolo[2,1-*a*]isoquinolines.

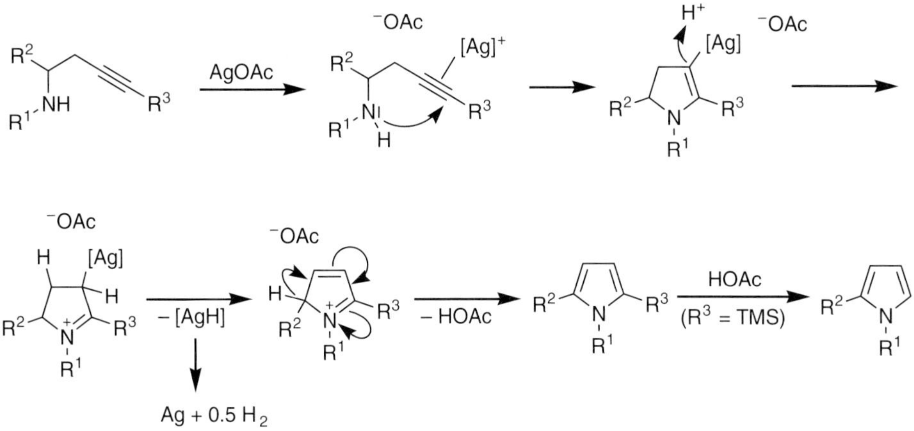

Scheme 15.3 Proposed mechanism for the silver(I)-

cyclization to 5,6-dihydropyrrolo[2,1-*a*]isoquinoline gives a yield similar to the cyclization of the silyl-protected precursor. Chemoselective hydrogenation of the pyrrole ring using 5 % rhodium on activated carbon as catalyst provides 1,2,3,5,6,10b-hexahydropyrrolo[2,1-*a*]isoquinoline [7], which represents a degradation product of the alkaloid norsecurinone [8] and is reported to show useful pharmacological activities [9].

Silver(I)-catalyzed cyclizations of substituted allenes to heterocyclic ring systems including 2,5-dihydropyrroles have been described previously [4,10]. Moreover, silver(I) salts are known to form stable π-complexes with terminal acetylenes [11]. On the other hand, on treatment with silver nitrate silylacetylenes were reported to afford silver acetylides [12]. Based on these considerations and additional experimental evidence [5,13], the following mechanism has been proposed for the silver(I)-mediated oxidative cyclization of homopropargylamines to pyrroles [5] (Scheme 15.3).

Activation of the acetylene by coordination of the triple bond to the silver cation enables a 5-endo-dig cyclization via nucleophilic attack of the amine [14]. Protonation of the resulting vinyl silver complex leads to an iminium ion. Subsequent β-hydride elimination affords metallic silver and a pyrrylium ion which aromatizes by proton loss to the pyrrole. For trimethylsilyl-substituted homopropargylamines (R^3 = $SiMe_3$), the resulting pyrrole (R^3 = $SiMe_3$) undergoes protodesilylation to the 1,2-disubstituted pyrrole.

15.1.1
Synthesis of the Pyrrolo[2,1-*a*]isoquinoline Alkaloid Crispine A

The hexahydropyrrolo[2,1-*a*]isoquinoline alkaloid crispine A was isolated in 2002 from *Carduus crispus* [15]. Extracts of this plant have been applied in Chinese folk medicine for the treatment of colds, stomach ache, and rheumatism; moreover they inhibit the growth of several human cancer cell lines. The useful biological activities induced a strong interest over the last few years in the synthesis of this alkaloid [7,16].

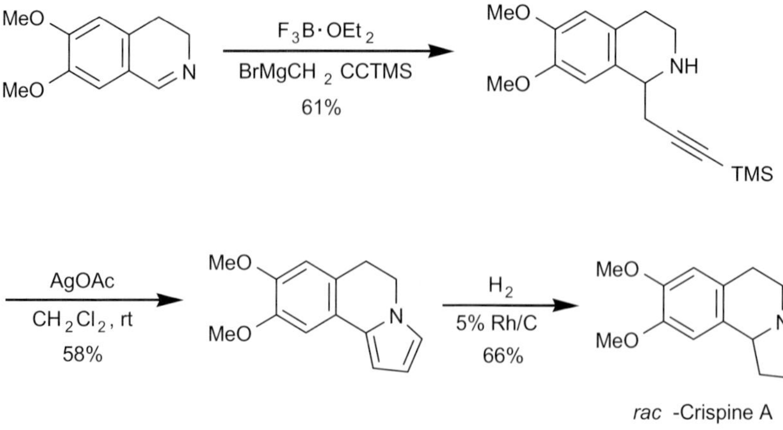

Scheme 15.4 Synthesis of (±)-crispine A via silver(I)-mediated oxidative cyclization.

We used the silver(I)-mediated oxidative cyclization of homopropargylamines to pyrroles for the synthesis of crispine A (Scheme 15.4) [7].

Reaction of 3,4-dihydro-6,7-dimethoxyisoquinoline with the propargyl Grignard reagent leads to the cyclic homopropargylamine, which on silver(I)-mediated oxidative cyclization affords 5,6-dihydro-8,9-dimethoxypyrrolo[2,1-*a*]isoquinoline. Chemoselective hydrogenation of the pyrrole ring provides racemic crispine A (three steps, 24 % overall yield).

15.1.2
Synthesis of the Indolizidino[8,7-*b*]indole Alkaloid Harmicine

The basic fraction of the ethanolic leaf extract of the Malaysian plant *Kopsia griffithii* exhibits a strong antileishmania activity. One of the two novel alkaloids isolated from this plant in 1998 was the indolizidino[8,7-*b*]indole harmicine [17]. Synthetic approaches to harmicine became attractive owing to its pharmacological potential [18,19]. We applied our three-step pyrrolidine annulation strategy to the synthesis of harmicine (Scheme 15.5) [19].

Addition of the propargyl Grignard reagent to 3,4-dihydro-β-carboline followed by silver(I)-mediated oxidative cyclization to the dihydroindolizino[8,7-*b*]indole and chemoselective hydrogenation provide racemic harmicine (three steps, 46 % overall yield).

15.2
Iron(0)-mediated Oxidative Cyclization to Indoles

Indole alkaloids represent a fundamental class of natural products and have important biological properties. Therefore, a large number of different methods for indole

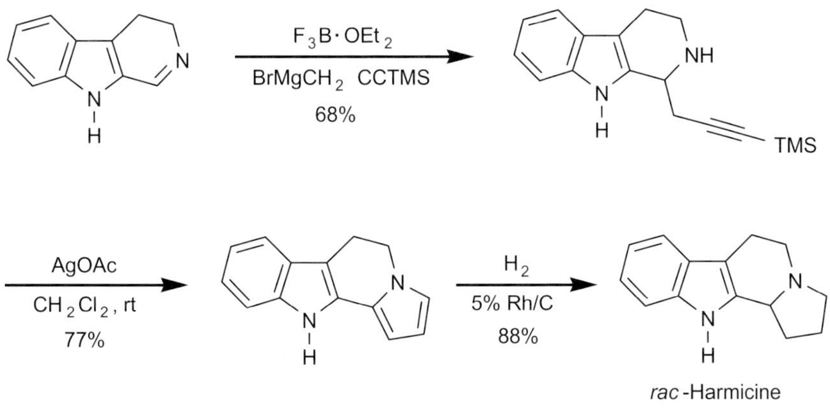

$F_3B \cdot OEt_2$, $BrMgCH_2$ CCTMS, 68%; AgOAc, CH_2Cl_2, rt, 77%; H_2, 5% Rh/C, 88%; *rac*-Harmicine

Scheme 15.5 Synthesis of (±)-harmicine via silver(I)-mediated oxidative cyclization.

annelation has been developed [20]. Taking advantage of the useful reactivity of tricarbonyl(η^4-cyclohexa-1,3-diene)iron complexes [21], we established a novel procedure for indole ring formation by an iron(0)-mediated oxidative cyclization of alkylamines (Scheme 15.6) [22].

The tricarbonyliron-coordinated cyclohexadienylium salts are readily available on a large scale by azadiene-catalyzed complexation of the corresponding cyclohexadienes with pentacarbonyliron [23] and subsequent hydride abstraction using trityl tetrafluoroborate [24]. Alkylation of methyl lithioacetate with the iron complex

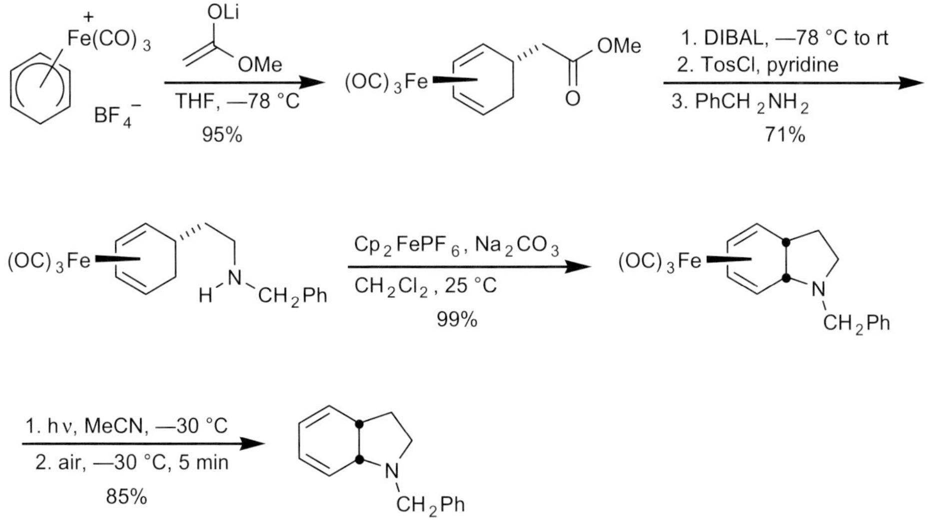

$Fe(CO)_3^+$ BF_4^-; OLi, OMe; THF, –78 °C, 95%; $(OC)_3Fe$; OMe; 1. DIBAL, –78 °C to rt 2. TosCl, pyridine 3. $PhCH_2NH_2$, 71%; $(OC)_3Fe$; CH_2Ph; Cp_2FePF_6, Na_2CO_3, CH_2Cl_2, 25 °C, 99%; $(OC)_3Fe$; CH_2Ph; 1. hν, MeCN, –30 °C 2. air, –30 °C, 5 min, 85%; CH_2Ph

Scheme 15.6 Iron(0)-mediated oxidative cyclization of alkylamines to indoles.

salt at low temperature proceeds almost quantitatively. Elaboration of the alkylamine side chain has been achieved by reduction to the primary alcohol using diisobutylaluminum hydride (DIBAL) followed by tosylation and nucleophilic substitution with benzylamine. Oxidative cyclization using ferricenium hexafluorophosphate in the presence of sodium carbonate provides the tricarbonyliron-coordinated 2,3,3a,7a-tetrahydroindole. Demetallation via a photolytically induced ligand exchange reaction with acetonitrile affords the corresponding free ligand [25].

This method can be applied to the total synthesis of natural products. The lycorine alkaloids anhydrolycorinone and hippadine have a pyrrolo[3,2,1-*de*]phenanthridine framework and have been isolated from Amaryllidaceae plants [26]. Hippadine exhibits reversible inhibition of fertility in male rats. Diverse synthetic approaches to the pyrrolophenanthridine alkaloids have been developed because of their potent biological activities [27]. We applied the iron(0)-mediated arylamine cyclization described above to a concise synthesis of anhydrolycorinone and hippadine (Scheme 15.7) [28].

Selective reduction of the methyl ester to the corresponding aldehyde using DIBAL at low temperature and subsequent reductive amination with iodopiperonylammonium chloride affords the tricarbonyliron–cyclohexadiene complex with the secondary alkylamine in the side chain. Iron(0)-mediated oxidative cyclization

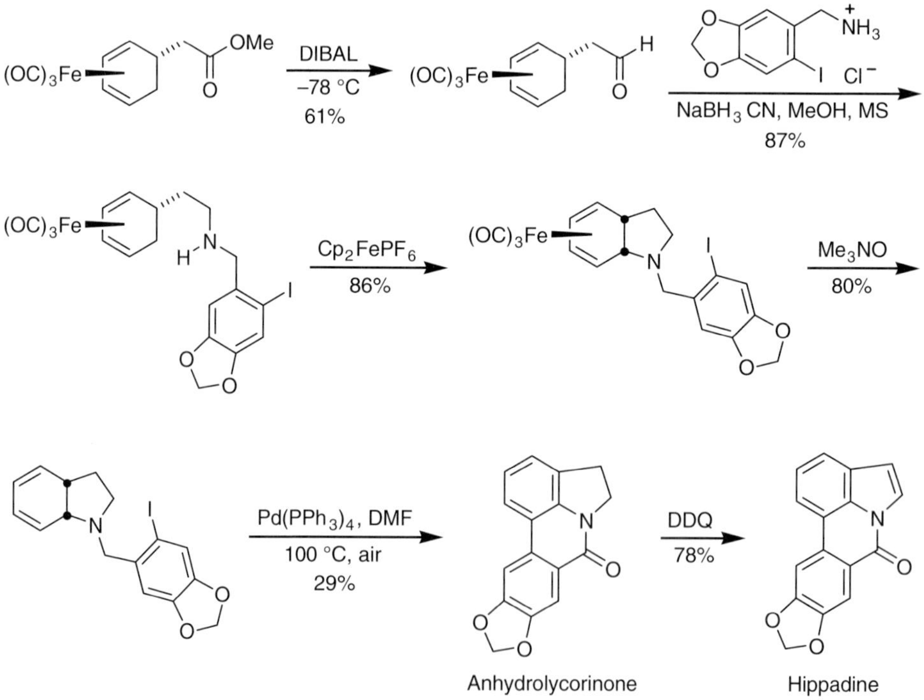

Scheme 15.7 Synthesis of anhydrolycorinone and hippadine via iron(0)-mediated oxidative cyclization.

followed by demetallation provides the tetrahydroindole. Heating the tetrahydroindole with a stoichiometric amount of tetrakis(triphenylphosphine)palladium in the presence of air leads to an intramolecular biaryl coupling with concomitant aromatization of the cyclohexadiene and oxidation at the benzylic position to the lactam, thus providing anhydrolycorinone directly. Finally, dehydrogenation leads to hippadine (seven steps, 8 % overall yield).

15.3 Iron(0)-mediated Oxidative Cyclization to Carbazoles

A broad structural variety of carbazole alkaloids with useful biological activities has been isolated from different natural sources. The pharmacological potential of this class of natural products led to the development of diverse methods for the synthesis of carbazoles [29,30]. We elaborated an efficient iron-mediated construction of the carbazole framework by consecutive C–C and C–N bond formation. This method provides highly convergent routes to carbazoles as demonstrated first for 4-deoxycarbazomycin B (Scheme 15.8) [31].

Electrophilic aromatic substitution of 2,3-dimethyl-4-methoxyaniline by reaction with the tricarbonyliron-coordinated cyclohexadienylium salt generates the aryl-substituted tricarbonyliron–cyclohexadiene complex. Treatment of this complex with very active manganese dioxide results in oxidative cyclization and aromatization with concomitant demetallation to afford directly 4-deoxycarbazomycin B, a degradation product of the antibiotic carbazomycin B [32]. Using ferricenium hexafluorophos-

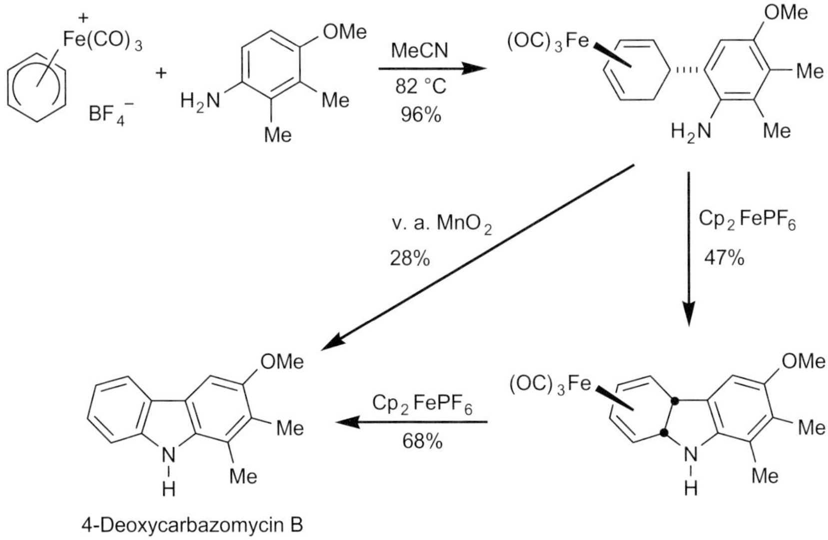

Scheme 15.8 Synthesis of 4-deoxycarbazomycin B via iron(0)-mediated oxidative cyclization.

phate as the oxidizing agent the two oxidations involved can be achieved sequentially: first, C–N bond formation by iron(0)-mediated oxidative cyclization to the tricarbonyliron-coordinated 4a,9a-dihydrocarbazole and second, aromatization followed by demetallation to 4-deoxycarbazomycin B.

15.3.1
3-Oxygenated Carbazole Alkaloids

Hyellazole and 6-chlorohyellazole have been isolated from the blue-green alga *Hyella caespitosa* [33]. They represent the first carbazole alkaloids that have been obtained from marine sources. Carazostatin has been isolated from *Streptomyces chromofuscus* and exhibits a strong inhibition of free-radical-induced lipid peroxidation [34]. Carazostatin and *O*-methylcarazostatin, a nonnatural derivative, are potent antioxidants [35]. We have developed a common approach to these 3-oxygenated carbazole alkaloids based on the iron-mediated carbazole synthesis (Scheme 15.9) [36,37].

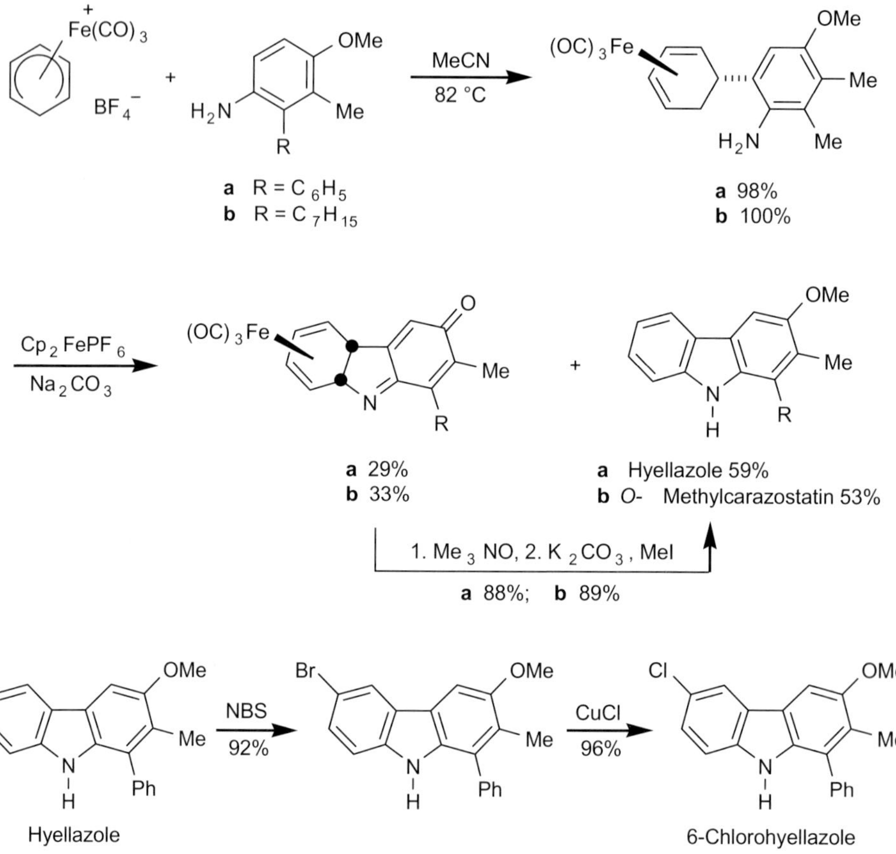

Scheme 15.9 Iron(0)-mediated synthesis of 3-oxygenated carbazole alkaloids.

Electrophilic substitution of the 4-methoxyanilines by reaction with the iron complex salt and subsequent oxidative cyclization using ferricenium hexafluorophosphate in the presence of sodium carbonate affords the 3-methoxycarbazoles hyellazole and *O*-methylcarazostatin along with their corresponding tricarbonyliron-coordinated 4b,8a-dihydrocarbazol-3-one complexes. Demetallation of the latter followed by *O*-methylation of the 3-hydroxycarbazoles resulting from tautomerization converts the iron complexes to the carbazole products, which become readily available (three steps and 82–83 % overall yield based on the iron complex salt for both compounds). Although three previous syntheses for 6-chlorohyellazole have been described, a transformation of hyellazole into 6-chlorohyellazole has not yet been reported. We found that halogenations of hyellazole occur with remarkable regiospecificity. While chlorination takes place selectively at the 4-position, bromination leads to the desired halogenation of the 6-position. Halogen exchange by treatment of 6-bromohyellazole with an excess of cuprous chloride provides 6-chlorohyellazole (five steps, 73 % overall yield based on the iron complex salt).

15.3.2
Carbazole-1,4-quinol Alkaloids

Carbazomycin G and H have been isolated from *Streptoverticillium ehimense* [38]. In nature carbazomycin G occurs as a racemic mixture and the same is assumed for carbazomycin H. Carbazomycin G was reported to show antifungal activity. The two compounds share a unique structure, namely the carbazole-1,4-quinol moiety. Starting from the appropriate iron complex salt, iron-mediated arylamine cyclization provides a simple route to both alkaloids (Scheme 15.10) [39].

Electrophilic aromatic substitution of 5-hydroxy-2,4-dimethoxy-3-methylaniline by reaction with the iron complex salts affords the corresponding aryl-substituted tricarbonyliron–cyclohexadiene complexes. *O*-Acetylation followed by iron-mediated arylamine cyclization with concomitant aromatization provides the substituted carbazole derivatives. Oxidation using cerium(IV) ammonium nitrate (CAN) leads to the carbazole-1,4-quinones. Addition of methyllithium at low temperature occurs preferentially at C-1, representing the more reactive carbonyl group, and thus provides in only five steps carbazomycin G (46 % overall yield) and carbazomycin H (7 % overall yield).

Carbazole-1,4-quinones, the key intermediates of this approach, can be even more efficiently prepared by the palladium(II)-catalyzed oxidative cyclization (see below, Scheme 15.17). In a screening for anti-TB active carbazoles, 3-methoxy-2-methylcarbazole-1,4-quinone showed significant inhibition of *Mycobacterium tuberculosis* strain $H_{37}Rv$, with an MIC_{90} value 2.2 μg/mL (9 μM) [40]. Thus, carbazoles may be developed as anti-TB drug candidates by structural modifications.

15.3.3
Furo[3,2-*a*]carbazole Alkaloids

Furostifoline, the first furocarbazole alkaloid obtained from natural sources, was isolated from the root bark of *Murraya euchrestifolia* (Rutaceae) [41]. The extracts of the

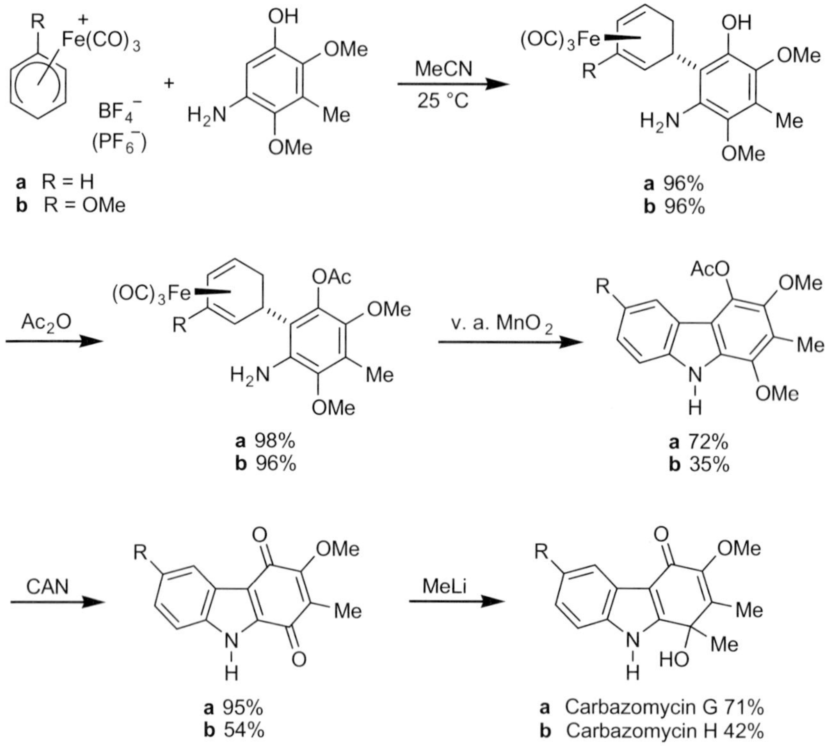

Scheme 15.10 Iron(0)-mediated synthesis of carbazomycin G and H.

leaves and bark of this shrub growing in Taiwan have been used in folk medicine. Furoclausine-A, another furo[3,2-*a*]carbazole alkaloid, has been isolated from the root bark of *Clausena excavata* [42]. In traditional Chinese folk medicine, extracts of this plant have been used for the treatment of various infections and poisonous snakebites. Because of their pharmacological potential the furocarbazole alkaloids became attractive synthetic targets [43]. Using an iron-mediated oxidative cyclization for the construction of the carbazole nucleus we established a simple three-step approach to the furo[3,2-*a*]carbazole framework, which was originally applied to the synthesis of furostifoline (Scheme 15.11) [44].

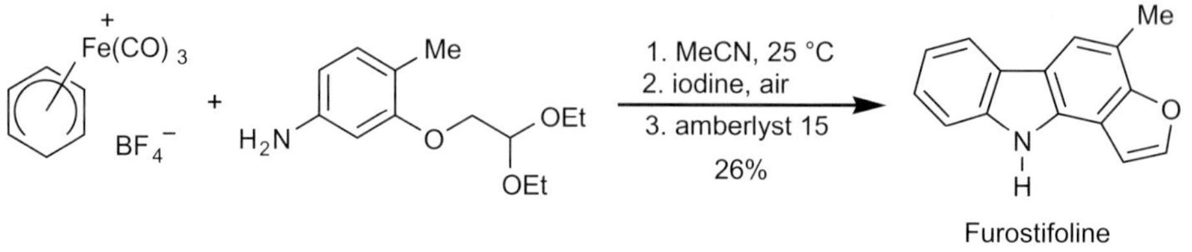

Scheme 15.11 Iron(0)-mediated synthesis of furostifoline.

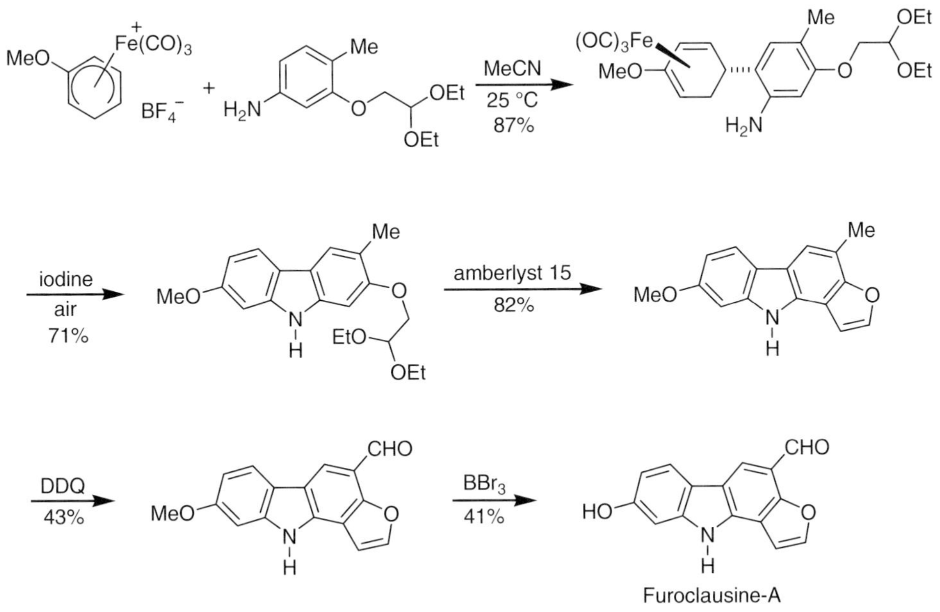

Scheme 15.12 Iron(0)-mediated synthesis of furoclausine-A.

Electrophilic substitution of the appropriately functionalized arylamine and subsequent iron-mediated oxidative cyclization with aromatization generates the carbazole skeleton. Annulation of the furan ring by treatment with catalytic amounts of amberlyst 15 affords furostifoline directly. Comparison of the six total syntheses reported so far for furostifoline demonstrates the superiority of the iron-mediated synthesis (Table 1 in ref. [43a]). Starting from the 2-methoxy-substituted tricarbonyliron-coordinated cyclohexadienylium salt this sequence has been applied to the synthesis of furoclausine-A (Scheme 15.12) [45].

The regioselective nucleophilic attack of the arylamine at the 2-methoxy-substituted iron complex salt is controlled by the methoxy group, which directs the arylamine to the *para*-position. Moreover, electrophilic attack takes place at the sterically less-hindered *ortho*-amino position. Iron-mediated oxidative cyclization of the resulting iron complex to the carbazole followed by proton-catalyzed annulation of the furan ring provides 8-methoxyfurostifoline. Oxidation with 2,3-dichloro-5,6-dicyano-1,4-benzoquinone (DDQ) to *O*-methylfuroclausine-A followed by cleavage of the methyl ether provides furoclausine-A (five steps, 9 % overall yield).

15.3.4 2,7-Dioxygenated Carbazole Alkaloids

The family of 2,7-dioxygenated carbazole alkaloids has been isolated from different terrestrial plants. Prior to our work this group of natural products was synthetically completely unexplored. 7-Methoxy-*O*-methylmukonal has been isolated from the roots of *Murraya siamensis*[46]. Clausine H, also named clauszoline-C, and clausine K, also named clauszoline-J, have been obtained from the stem bark of *Clausena excavata* [47]. Clausine K (clauszoline-J) was also isolated from the roots of *Clausena harmandiana* [48]. Clausine O was found in the root bark of *Clausena excavata* [49]. It has been reported that clausine H (clauszoline-C) shows antiplasmodial activity against *Plasmodium falciparum* [48] and clausine K (clauszoline-J) exhibits antimycobacterial activity against *Mycobacterium tuberculosis* [50]. These promising pharmacological activities prompted us to develop an efficient iron-mediated synthetic route to this class of carbazole alkaloids (Scheme 15.13) [51].

Electrophilic aromatic substitution of 3-methoxy-4-methylaniline using the 2-methoxy-substituted iron complex salt followed by oxidative cyclization with concomitant aromatization of the resulting iron complex affords 2,7-dimethoxy-3-methylcarbazole. Oxidation of the methyl substituent with DDQ leads to 7-methoxy-*O*-methylmukonal, which subsequently can be used as a relay compound to the other 2,7-dioxygenated carbazole alkaloids. Oxidation of 7-methoxy-*O*-methylmukonal using manganese dioxide in methanol in the presence of potassium cyanide provides clausine H (clauszoline-C) quantitatively. On ester cleavage this gives clausine K (clauszoline-J). Cleavage of the methyl ethers of 7-methoxy-*O*-methylmukonal leads to clausine O.

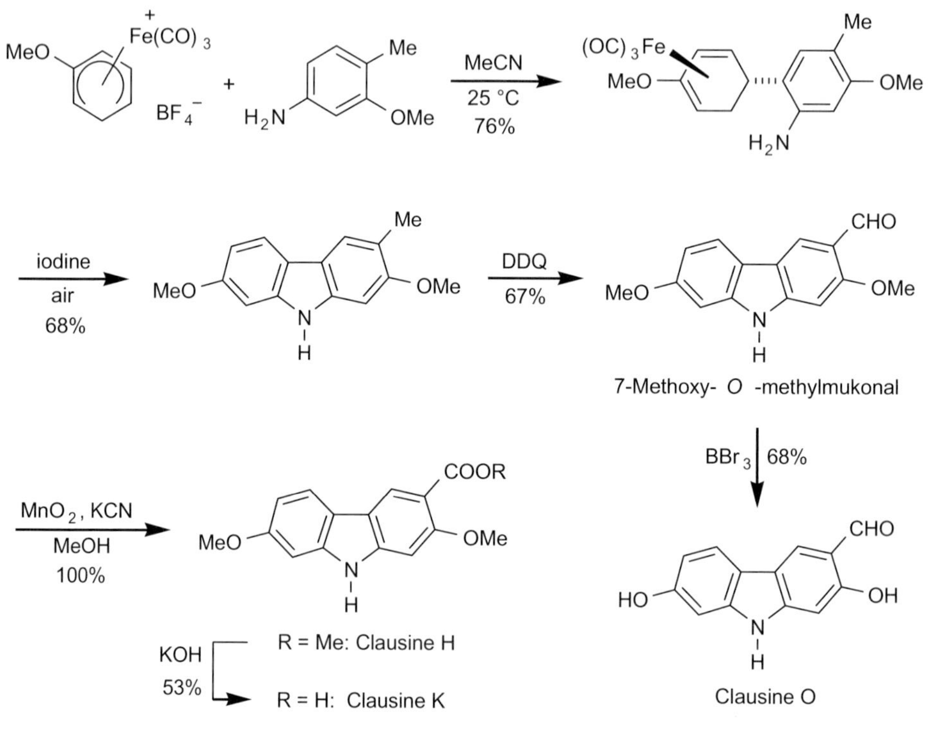

Scheme 15.13 Iron(0)-mediated synthesis of 2,7-dioxygenated carbazole alkaloids.

15.3.5
3,4-Dioxygenated Carbazole Alkaloids

In former applications, the iron-mediated oxidative cyclization has proven to provide highly efficient synthetic routes to 3,4-dioxygenated carbazole alkaloids [30]. In 1991 the neocarazostatins were isolated from *Streptomyces* sp. strain GP 38 on screening for novel free-radical scavengers [52]. They show a strong inhibition of free-radical-induced lipid peroxidation in rat brain homogenate. The absolute configuration for neocarazostatin B was not determined. Two years later, carquinostatin A, a strong antioxidant, was isolated from *Streptomyces exfoliatus* [53]. Although it contains an *ortho*-quinone, carquinostatin A is structurally related to neocarazostatin B. For carquinostatin A an (*R*)-configuration has been assigned to the stereogenic center of the side chain at C-1. Based on the structural analogy, the same absolute configuration has been assumed for neocarazostatin B. This hypothesis led us to develop an iron-mediated enantioselective synthesis for both natural products (Scheme 15.14) [54].

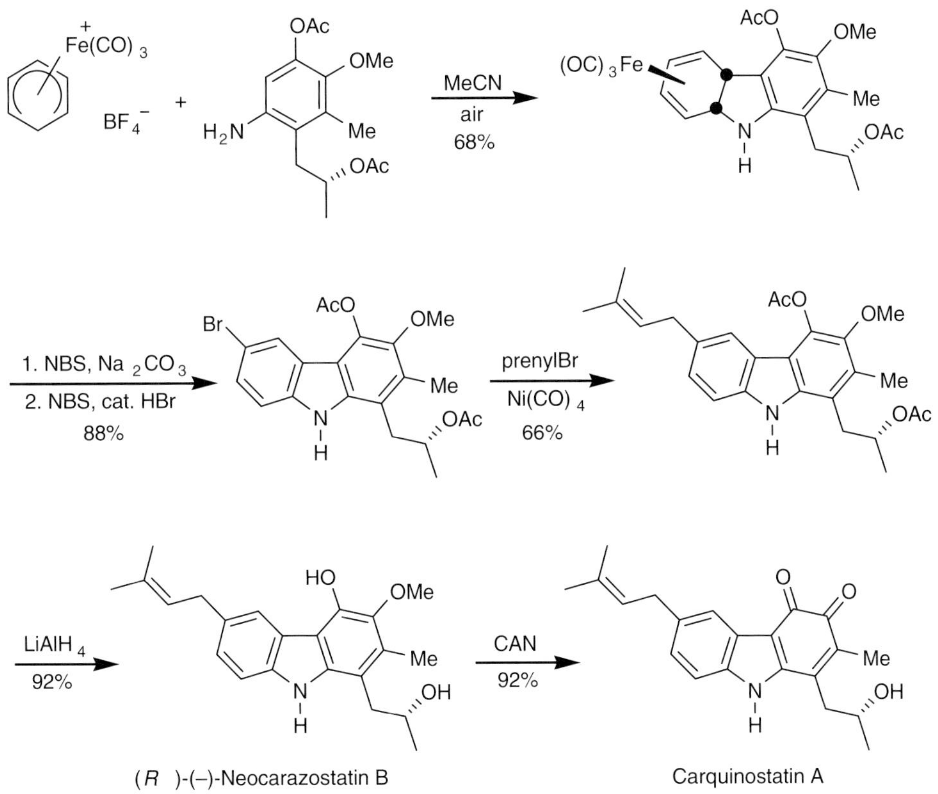

Scheme 15.14 Iron(0)-mediated synthesis of (*R*)-(–)-neocarazostatin B and carquinostatin A.

Reaction of the iron complex salt with the fully functionalized arylamine in air proceeds via sequential C–C and C–N bond formation to afford the tricarbonyliron-coordinated 4a,9a-dihydrocarbazole complex. This one-pot annulation is the result of an electrophilic aromatic substitution and subsequent iron-mediated oxidative cyclization, with air as the oxidizing agent. Aromatization with concomitant demetallation followed by regioselective electrophilic bromination leads to a 6-bromocarbazole that has been submitted to nickel-mediated prenylation. At the stage of the 6-bromocarbazole the absolute configuration of the stereogenic center in the side chain at C-1 has been confirmed by an X-ray crystal structure determination. Finally, removal of the acetyl groups by reduction using lithium aluminum hydride provides (*R*)-(−)-neocarazostatin B (five steps, 36 % overall yield). Smooth oxidation with cerium(IV) ammonium nitrate generates carquinostatin A. The identity of the absolute configurations of both alkaloids and also the enantiomeric purity of neocarazostatin B (>99 % ee) has been additionally confirmed by transformation of carquinostatin A to the corresponding Mosher ester [54].

15.4
Palladium(II)-catalyzed Oxidative Cyclization to Carbazoles

The oxidative cyclization of *N,N*-diarylamines represents a straightforward alternative route to the carbazole framework [55]. However, application of the classical thermally, photolytically or radical-induced process provides only moderate yields. Much higher yields are obtained for this transformation by palladium(II)-mediated oxidative cyclization, first reported by Åkermark (Scheme 15.15) [56].

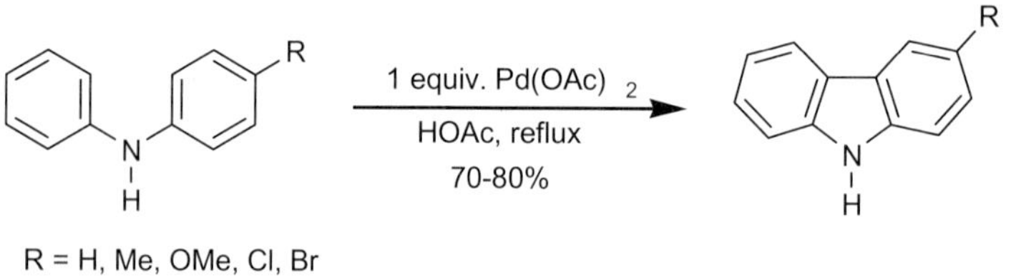

1 equiv. $Pd(OAc)_2$
HOAc, reflux
70-80%

R = H, Me, OMe, Cl, Br

Scheme 15.15 Palladium(II)-mediated oxidative cyclization of *N,N*-diarylamines to carbazoles.

Heating of the *N,N*-diarylamines with palladium(II) acetate in acetic acid at reflux results in smooth oxidative cyclization to the corresponding carbazole derivatives. A variety of substituents are tolerated in different positions. Thus, this procedure has found many applications in organic syntheses [30,55]. However, the drawback is that stoichiometric amounts of palladium(II) are required, as one equivalent of palladium(0) is formed in the final reductive elimination step. In the Wacker process, regeneration of the catalytically active palladium(II) species is achieved by oxidation of palladium(0) to palladium(II) with a copper(II) salt [57]. We were the first to demonstrate that oxidative regeneration of the catalytically active palladium(II)

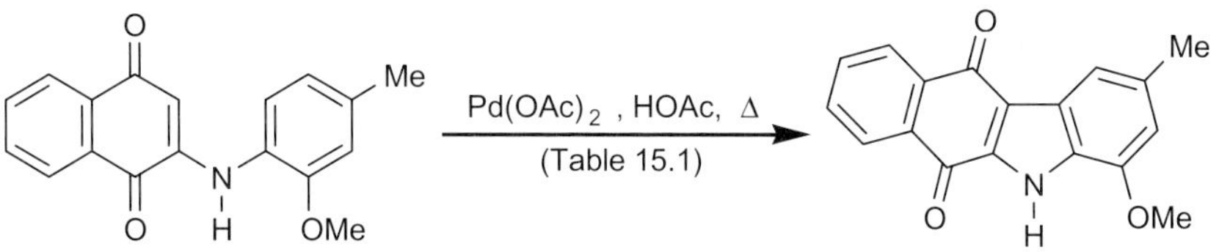

Scheme 15.16 Palladium(II)-catalyzed oxidative cyclization to 4-methoxy-2-methyl-5*H*-benzo[*b*]carbazole-6,11-dione.

Tab. 15.1 Palladium(II)-catalyzed oxidative cyclization to 4-methoxy-2-methyl-5*H*-benzo[*b*]carbazole-6,11-dione.

$Pd(OAc)_2$ (equiv.)	$Cu(OAc)_2$ (equiv.)	Reaction conditions	Start. mat., yield (%)[a]	Product, yield [%][b]	TON[c]
1.0	—	HOAc, 117 °C, 0.5 h	0	84	—
0.12	—	HOAc, 117 °C, 1 h	80	11	—
0.12	1.1	HOAc, 117 °C, 17 h	56	34	2.8

[a] Recovered starting material (*N*-[2-methoxy-4-methylphenyl]-2-amino-1,4-naphthoquinone).
[b] Isolated yield of 4-methoxy-2-methyl-5*H*-benzo[*b*]carbazole-6,11-dione.
[c] Turnover number.

species can be exploited for palladium(II)-catalyzed cyclizations of arylaminoquinones to carbazole derivatives (Scheme 15.16, Table 15.1) [58,59].

Oxidative cyclization of the 2-arylamino-1,4-naphthoquinone by reaction with stoichiometric amounts of palladium(II) acetate in refluxing acetic acid leads to the corresponding benzo[*b*]carbazole-6,11-dione in high yield. Using catalytic amounts (12 mol%) of palladium(II) acetate in the presence of equimolar amounts of copper(II) acetate provides the benzo[*b*]carbazolequinone in 34 % yield and 56 % of the starting material, demonstrating a catalytic cycle with a turnover number of 2.8 [58]. In a blank experiment (reaction with 12 mol% of palladium(II) acetate in the absence of copper(II) salt) the product was obtained in only 11 % yield, indicating the necessity for reoxidizing the palladium(0). Subsequently, other reoxidants for palladium(0) have also been applied in this reaction [60]. We have applied the palladium(II)-catalyzed oxidative cyclization to efficient syntheses of biologically active carbazole alkaloids.

15.4.1 Carbazolequinone Alkaloids

The carbazoquinocins are carbazole-3,4-quinone alkaloids and have been isolated from *Streptomyces violaceus* 2942-SVS3 [61]. They are strong antioxidative agents and thus represent potential drugs for the treatment of diseases initiated by oxygen-derived free radicals. We have developed an efficient synthesis of carbazoquinocin C using palladium(II)-catalyzed oxidative cyclization as the key step (Scheme 15.17, Table 15.2) [62].

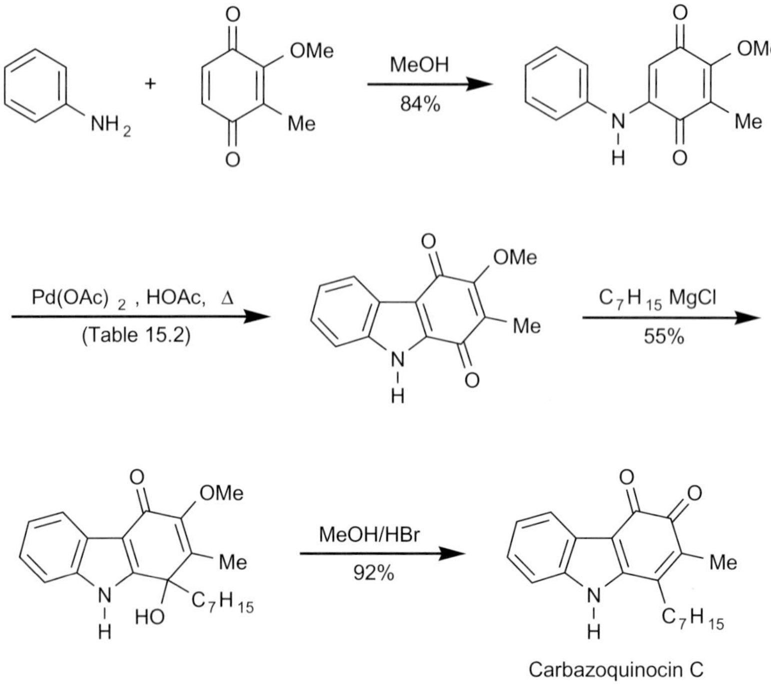

Scheme 15.17 Palladium(II)-catalyzed synthesis of carbazoquinocin C.

Addition of aniline to 2-methoxy-3-methyl-1,4-benzoquinone affords the anilino-substituted benzoquinone with complete regioselectivity. Optimization of the palladium(II)-catalyzed oxidative cyclization to 3-methoxy-2-methylcarbazol-1,4-quinone was achieved by varying the different reaction parameters. The best yield of the product was obtained using 30 mol% of the palladium(II) catalyst. Using 5 mol%

Tab. 15.2 Palladium(II)-catalyzed oxidative cyclization to 3-methoxy-2-methylcarbazole-1,4-quinone.

$Pd(OAc)_2$ (equiv.)	$Cu(OAc)_2$ (equiv.)	Reaction conditions	Product, yield [%][a]	TON[b]
1.0	—	HOAc, 117 °C, 5 h, Ar	83	—
0.3	2.5	HOAc, 117 °C, 72 h, air	91	3.0
0.2	2.5	HOAc, 117 °C, 22 h, air	73	3.7
0.1	2.5	HOAc, 117 °C, 96 h, air	78	7.8
0.1	1.0	HOAc, 117 °C, 17 h, air	54	5.4
0.1	0.25	HOAc, 117 °C, 18 h, air	24	2.4
0.05	2.5	HOAc, 117 °C, 19 h, air	56	11.2

[a] Isolated yield of 3-methoxy-2-methylcarbazole-1,4-quinone.
[b] Turnover number.

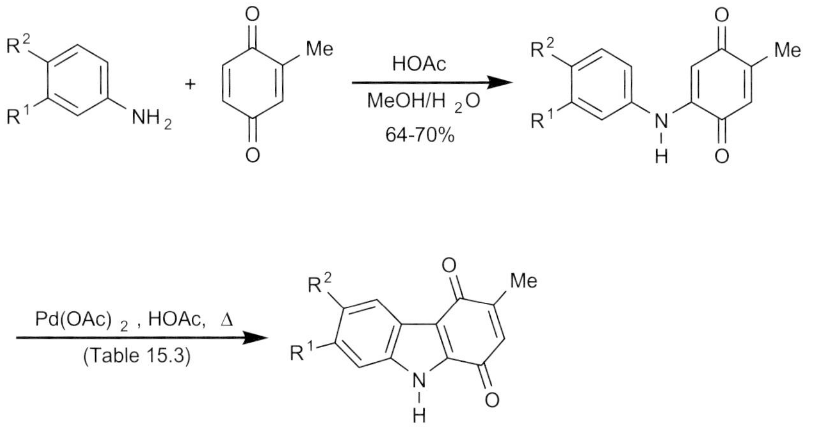

Scheme 15.18 Palladium(II)-catalyzed synthesis of carbazole-1,4-quinone alkaloids.

Tab. 15.3 Palladium(II)-catalyzed oxidative cyclization of 5-arylamino-2-methyl-1,4-benzoquinones.

R^1	R^2	$Pd(OAc)_2$ (equiv.)	$Cu(OAc)_2$ (equiv.)	Reaction conditions	Product, yield (%)[a]	TON[b]
H	H	1.0	—	HOAc, 117 °C, 4 h, Ar	68	—
H	H	0.1	2.5	HOAc, 95 °C, 48 h, air	73	7.3
OMe	H	1.0	—	HOAc, 117 °C, 4.5 h, air	71	—
OMe	H	0.1	2.5	HOAc, 117 °C, 12 h, air	44	4.4
OMe	OMe	1.0	—	HOAc, 117 °C, 4 h, air	73	—
OMe	OMe	0.2	2.5	HOAc, 117 °C, 6 h, air	69	3.5
OMe	OMe	0.1	2.5	HOAc, 117 °C, 10.5 h, air	52	5.2

[a]Isolated yield of murrayaquinone A (R^1, R^2 = H), koeniginequinone A (R^1 = OMe, R^2 = H) and koeniginequinone B (R^1, R^2 = OMe).
[b]Turnover number.

of the catalyst we observed the best turnover with respect to palladium(II). In contrast to the Wacker oxidation, copper(II) cannot be used in catalytic amounts under the present reaction conditions. Addition of heptylmagnesium chloride at low temperature takes place preferentially at the more reactive carbonyl group (C-1) to provide the corresponding carbazolequinol. On treatment with hydrogen bromide in methanol, this intermediate is smoothly converted into carbazoquinocin C (four steps, 39 % overall yield). Moreover, 3-methoxy-2-methylcarbazol-1,4-quinone also represents a synthetic precursor for carbazomycin G and exhibits significant anti-TB activity (see Section 15.3.2, Scheme 15.10) [39,40].

Because of their promising pharmacological properties, the carbazole-1,4-quinone alkaloids became attractive synthetic targets [30,63]. Murrayaquinone A has been isolated from the root bark of the Chinese medicinal plant *Murraya euchrestifolia* and shows cardiotonic activity [64]. The koeniginequinones A and B have been obtained from the stem bark of *Murraya koenigii* [65]. Palladium(II)-catalyzed oxidative cyclization opens up a simple two-step route to these natural products (Scheme 15.18, Table 15.3) [66].

Addition of the appropriate arylamines to 2-methyl-1,4-benzoquinone provides the desired 2-arylamino-5-methyl-1,4-benzoquinones as major products (ratios: 2.3–2.9 : 1). Oxidative cyclization of these intermediates leads directly to murrayaquinone A (51 % overall yield), koeniginequinone A (46 % overall yield) and koeniginequinone B (47 % overall yield). Using copper(II) acetate under the optimized reaction conditions described above (Table 15.2), these transformations have also been carried out with catalytic amounts of palladium(II).

15.4.2
Carbazomadurins and Epocarbazolins

The carbazomadurins have been isolated from the microorganism *Actinomadura madurae* 2808-SV1 and exhibit a strong neuronal cell-protecting activity against L-glutamate-induced cell death [67]. High extracellular concentrations of the excitatory amino acid L-glutamate occurring after brain ischemic attack lead to destruction of cerebral tissue. Thus, radical scavengers are considered to be potential drugs for the treatment of brain ischemia injury. The epocarbazolins have been obtained by a research group of Bristol–Myers from the actinomycete strain *Streptomyces anulatus* T688-8 on screening for 5-lipoxygenase inhibitors [68]. Inhibitors of 5-lipoxygenase represent potential therapeutics for inflammatory and allergic processes, such as psoriasis, asthma, and hypersensitivity. A common approach to the carbazomadurin and epocarbazolin alkaloids has been established by using sequentially three different palladium-catalyzed cross-coupling reactions (Scheme 15.19) [69,70].

A palladium(0)-catalyzed Buchwald–Hartwig coupling of the appropriately substituted aryl triflate, available in two steps and 91 % yield from isovanillic acid, with 2-bromo-4-methoxy-3-methylaniline followed by a palladium(II)-mediated oxidative cyclization of the resulting *N,N*-diarylamine, leads to the 1,2,3,5,8-penta-substituted carbazole nucleus. Cleavage of both methyl ethers and subsequent double silylation effect a switch of the protecting group. Introduction of either side chain at C-1 of the carbazole framework is achieved using a palladium(0)-catalyzed Stille coupling. The required 1-(*E*)-alkenylstannanes are readily prepared from the appropriate alkynes by application of Negishi's zirconium-catalyzed carboalumination as the key step. Reduction of the methyl ester with diisobutylaluminum hydride (DIBAL) affords the benzylic alcohols as common precursors for the carbazomadurins and the epocarbazolins (Scheme 15.20).

Removal of both silyl protecting groups from the disilyl-protected precursors using tetrabutylammonium fluoride (TBAF) provides carbazomadurin A (nine steps, 11 % overall yield) and carbazomadurin B (nine steps, 13 % overall yield). Based on our enantioselective synthesis of carbazomadurin B (>99 % ee) an (*S*)-configuration has been assigned to the stereogenic center in the side chain of the natural product [70]. Because direct epoxidation of the carbazomadurins and of the disilyl-protected precursors has been unsuccessful, the latter have been transformed quantitatively into the corresponding trisilyl-protected intermediates. Epoxidation of the fully

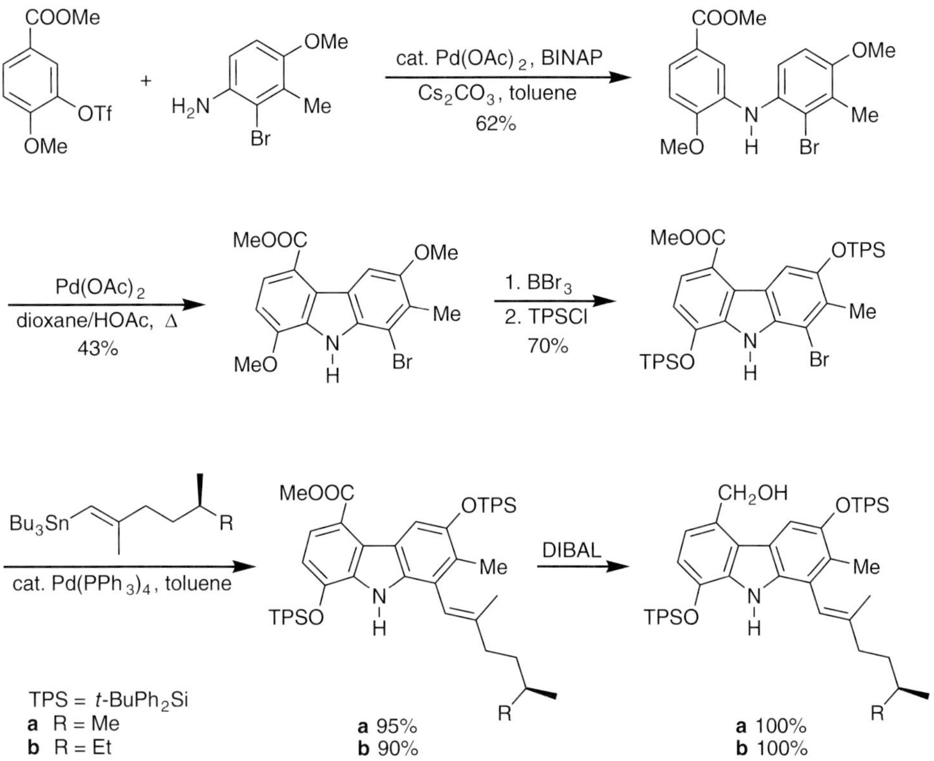

TPS = *t*-BuPh$_2$Si
a R = Me
b R = Et

Scheme 15.19 Palladium(II)-catalyzed synthesis of the disilyl-protected precursors.

protected compounds with dimethyldioxirane at −20 °C and subsequent desilylation provide racemic epocarbazolin A and epocarbazolin B [71].

15.4.3 7-Oxygenated Carbazole Alkaloids

Clauszoline-K and clausine C (clauszoline-L) have been isolated from the stem bark of the Chinese medicinal plant *Clausena excavata* [72]. Clausine M and clausine N have been found in the root bark of the same plant [73]. The bioassay-guided fractionation of the organic extract of *Murraya siamensis*, collected in Thailand, led to the isolation of siamenol, which shows HIV-inhibitory activity [74]. A simple route to these 7-oxygenated carbazole alkaloids has been developed, based on a highly efficient palladium-catalyzed approach (Scheme 15.21) [75] (Table 15.4).

Buchwald–Hartwig amination of *p*-bromotoluene with *m*-anisidine affords quantitatively the corresponding diarylamine. While oxidative cyclization using stoichiometric amounts of palladium(II) acetate provides only 36 % yield of 7-methoxy-3-methylcarbazole, up to 72 % yield is obtained using catalytic amounts of palladium(II). The highest turnover for the catalytic cycle is obtained with

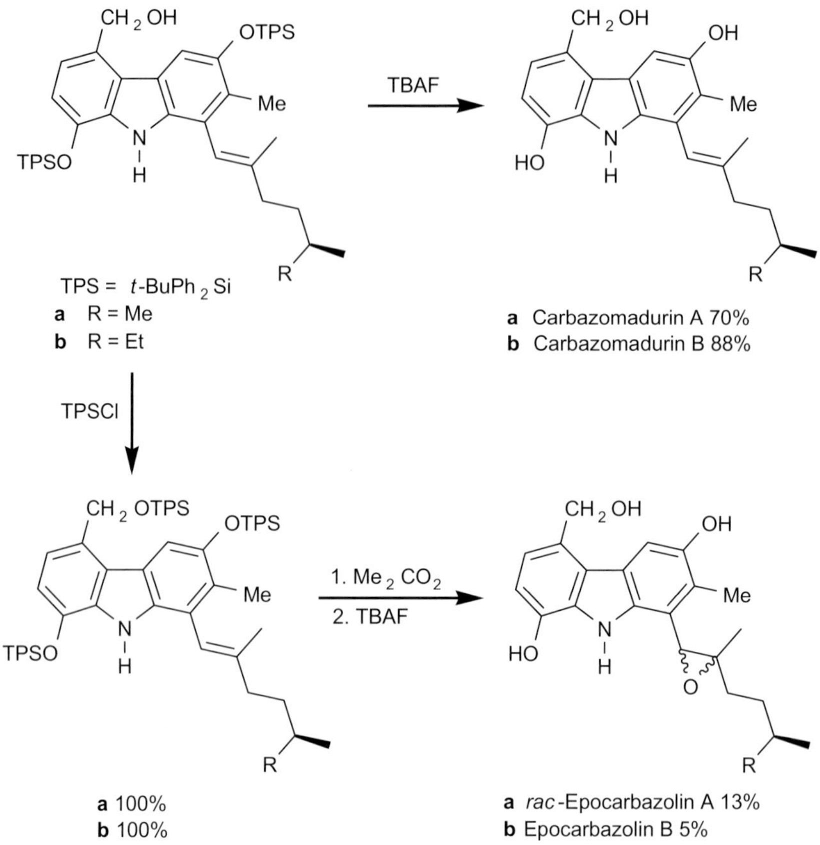

TBAF

TPS = t-BuPh$_2$Si
a R = Me
b R = Et

a Carbazomadurin A 70%
b Carbazomadurin B 88%

TPSCl

1. Me$_2$CO$_2$
2. TBAF

a 100%
b 100%

a *rac*-Epocarbazolin A 13%
b Epocarbazolin B 5%

Scheme 15.20 Syntheses of the carbazomadurins and epocarbazolins.

2 mol% of the palladium(II) catalyst. Obviously, stoichiometric amounts of palladium(II) lead to significant decomposition by oxidation of the electron-rich carbazole. Support for this hypothesis comes from a control experiment by treating 7-methoxy-3-methylcarbazole under the conditions for the stoichiometric reaction (1.2 equiv. Pd[OAc]$_2$, HOAc, 117 °C, 1 h). Only 65 % of 7-methoxy-3-methylcarbazole has been reisolated, because of decomposition. In the following, 7-methoxy-3-methylcarbazole serves as a relay compound to approach the other 7-oxygenated carbazole alkaloids. Oxidation with DDQ affords clauszoline-K. Further oxidation of clauszoline-K by treatment with manganese dioxide and potassium cyanide in methanol provides clausine C (clauszoline-L) quantitatively. Starting from clausine C, cleavage of the ester gives clausine N and cleavage of the ether leads to clausine M.

Regioselective electrophilic bromination of 7-methoxy-3-methylcarbazole followed by ether cleavage and introduction of the prenyl group in a nickel-mediated coupling reaction provides siamenol (five steps, 34 % overall yield) [75] (Scheme 15.22).

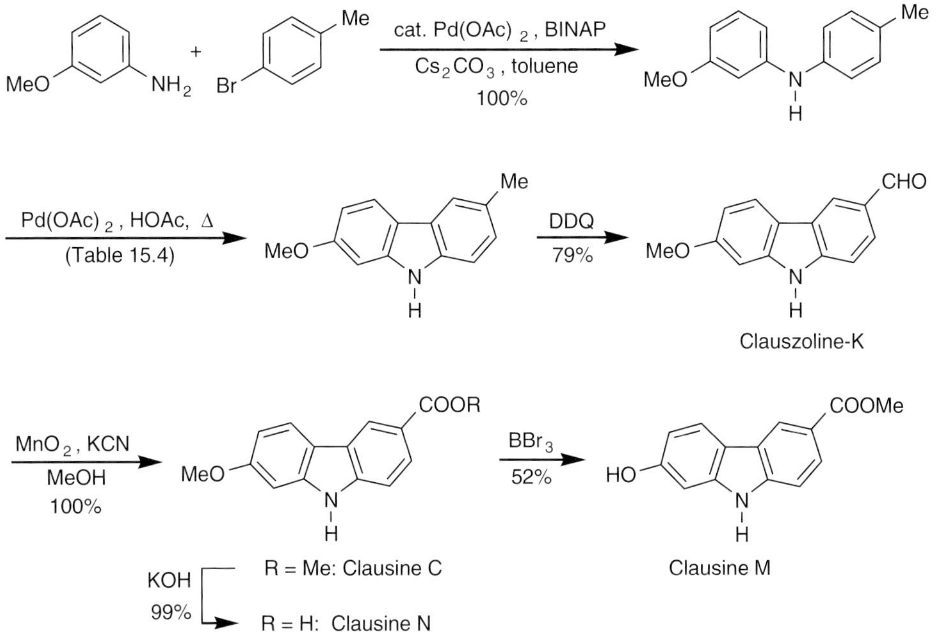

Scheme 15.21 Palladium(II)-catalyzed synthesis of clauszoline-K and clausine C, M, and N.

15.4.4
6-Oxygenated Carbazole Alkaloids

The 6-oxygenated carbazole alkaloids glycozoline and glycozolinine (glycozolinol) were first isolated from the Indian medicinal plant *Glycosmis pentaphylla* [76]. The alcohol glycozolinine (glycozolinol) shows a much stronger antibiotic activity than glycozoline [77]. Glycomaurrol, the 5-prenylated derivative of glycozolinine, has been obtained from the stem bark of *Glycosmis mauritiana* [78]. 3-Formyl-6-methoxycarbazole has been isolated from the roots of *Clausena lansium* [79]. The corresponding 5-prenylated derivative, micromeline, has been identified along with 3-formyl-

Tab. 15.4 Palladium(II)-catalyzed oxidative cyclization to 7-methoxy-3-methylcarbazole.

$Pd(OAc)_2$ (equiv.)	$Cu(OAc)_2$ (equiv.)	Reaction conditions	Product, yield (%)[a]	TON[b]
1.2	—	HOAc, 117 °C, 2 h, Ar	36	—
0.1	2.5	HOAc, 117 °C, 23 h, air	64	6.4
0.1	2.5	HOAc, 117 °C, 48 h, air	72	7.2
0.05	2.5	HOAc, 117 °C, 40 h, air	61	12.2
0.02	2.5	HOAc, 117 °C, 6 d, air	53	26.5

[a] Isolated yield of 7-methoxy-3-methylcarbazole.
[b] Turnover number.

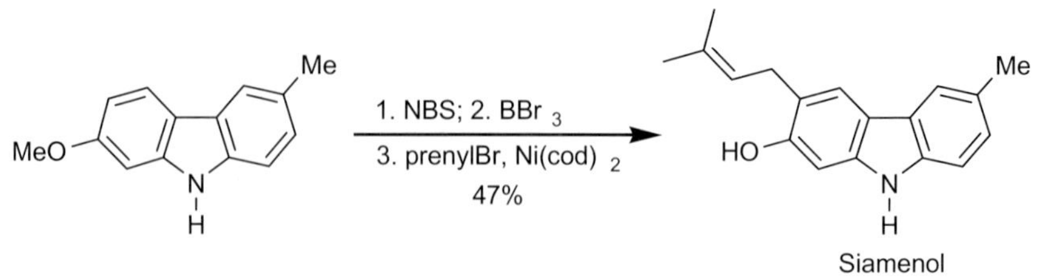

Me
MeO
N
H
1. NBS; 2. BBr_3
3. prenylBr, $Ni(cod)_2$
47%
Me
HO
N
H
Siamenol

Scheme 15.22 Nickel-mediated prenylation to siamenol.

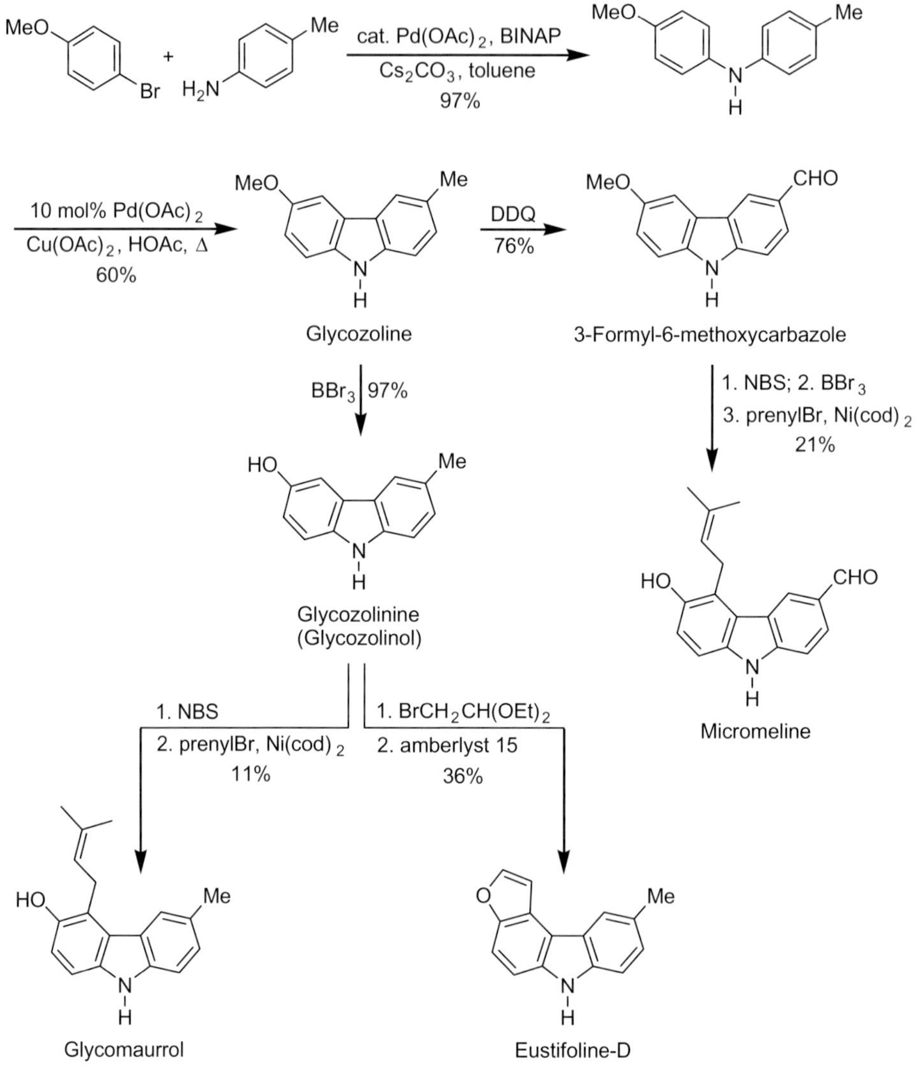

MeO
Br
+
H_2N
Me
cat. $Pd(OAc)_2$, BINAP
Cs_2CO_3, toluene
97%
MeO
Me
N
H

10 mol% $Pd(OAc)_2$
$Cu(OAc)_2$, HOAc, Δ
60%
MeO
Me
N
H
Glycozoline
DDQ
76%
MeO
CHO
N
H
3-Formyl-6-methoxycarbazole

BBr_3 97%
HO
Me
N
H
Glycozolinine
(Glycozolinol)

1. NBS; 2. BBr_3
3. prenylBr, $Ni(cod)_2$
21%
HO
CHO
N
H
Micromeline

1. NBS
2. prenylBr, $Ni(cod)_2$
11%
1. $BrCH_2CH(OEt)_2$
2. amberlyst 15
36%

HO
Me
N
H
Glycomaurrol

O
Me
N
H
Eustifoline-D

Scheme 15.23 Palladium(II)-catalyzed synthesis of 6-oxygenated carbazole alkaloids.

6-methoxycarbazole in an antituberculosis bioassay-directed fractionation of the stem bark extract of *Micromelum hirsutum* [80]. Both compounds show inhibitory activity against *Mycobacterium tuberculosis* $H_{37}Rv$. Eustifoline-D has been isolated along with furostifoline from the root bark of the Chinese medicinal plant *Murraya euchrestifolia*[41]. Easy access to the 6-oxygenated carbazole alkaloids is provided by the palladium-catalyzed construction of the carbazole framework (Scheme 15.23) [81].

Palladium(0)-catalyzed coupling of *p*-bromoanisole and *p*-toluidine followed by palladium(II)-catalyzed oxidative cyclization provides glycozoline directly. In this series glycozoline has been exploited as a relay compound to give access to the other 6-oxygenated carbazole alkaloids. Cleavage of the methyl ether affords glycozolinine (glycozolinol). Regioselective bromination of glycozolinine at C-5 and subsequent nickel-mediated prenylation leads to glycomaurrol. Williamson ether synthesis by alkylation of glycozolinine with 2-bromo-1,1-diethoxyethane and subsequent proton-catalyzed annulation of the furan ring provide eustifoline-D (five steps, 20 % overall yield). Oxidation of glycozoline with DDQ affords 3-formyl-6-methoxycarbazole. Regioselective bromination of the latter followed by ether cleavage and nickel-mediated prenylation provides micromeline (six steps, 9 % overall yield).

References

1 (a) Agarwal, S., Cämmerer, S., Filali, S., Fröhner, W., Knöll, J., Krahl, M. P., Reddy, K. R., Knölker, H. -J. (2005) *Current Organic Chemistry*, **9**, 1601. (b) Agarwal, S., Knöll, J., Krahl, M. P., Knölker, H. -J. (2005) *Journal of Fudan University (Natural Science)*, **44**, 699. (c) Agarwal, S., Filali, S., Fröhner, W., Knöll, J., Krahl, M. P., Reddy, K. R., Knölker, H. -J. (2006) In Kartsev, V. G. (Ed.) *The Chemistry and Biological Activity of Synthetic and Natural Compounds—Nitrogen-Containing Heterocycles*, Vol. 1, ICSPF Press, Moscow, 176.

2 (a) Sundberg, R. J. (1984) In Katritzky, A. R. and Rees, C. W. (Eds.) *Comprehensive Heterocyclic Chemistry*, Vol. 4, Pergamon Press, Oxford, 313. (b) Bean, G. P. (1990) In Jones, R. A. (Ed.) *Pyrroles*, Wiley, New York, 105. (c) Sundberg, R. J. (1996) In Katritzky, A. R., Rees, C. W. and Scriven, E. F. V. (Eds.) *Comprehensive Heterocyclic Chemistry II*, Vol. 2, Elsevier, Oxford, 119. (d) Gribble, G. W. (1996) In Katritzky, A. R., Rees, C. W. and Scriven, E. F. V. (Eds.) *Comprehensive Heterocyclic Chemistry II*, Vol. 2, Elsevier, Oxford, 207.

3 (a) Knight, D. W. and Sharland, C. M. (2003) Synlett 2258. (b) Knight, D. W. and Sharland, C. M. (2004) *Synlett*, 119.

4 (a) Claesson, A., Sahlberg, C., Luthman, K. (1979) *Acta Chemica Scandinavica B*, **33**, 309. (b) Prasad, J. S. and Liebeskind, L. S. (1988) *Tetrahedron Letters*, **29**, 4253. (c) Amombo, M. O., Hausherr, A., Reissig, H. -U. (1999) *Synlett*, 1871. (d) Ohno, H., Toda, A., Miwa, Y., Taga, T., Osawa, E., Yamaoka, Y., Fujii, N., Ibuka, T. (1999) *Journal of Organic Chemistry*, **64**, 2992. (e) Dieter, R. K. and Yu, H. (2001) *Organic Letters*, **3**, 3855.

5 Agarwal, S. and Knölker, H. -J. (2004) *Organic and Biomolecular Chemistry*, **2**, 3060.

6 Bellina, F. and Rossi, R. (2006) *Tetrahedron*, **62**, 7213.

7 Knölker, H. -J. and Agarwal, S. (2005) *Tetrahedron Letters*, **46**, 1173.

8 Saito, S., Tanaka, T., Kotera, K., Nakai, H., Sugimoto, N., Hori, Z., Ikeda, M., Tamura, Y. (1965) *Chemical and Pharmaceutical Bulletin*, **13**, 786.

9 (a) Chung, S. -H., Yook, J., Min, B. J., Lee, J. Y., Lee, Y. S., Jin, C. (2000) *Archives of Pharmacal Research*, **23**, 353. (b) Maryanoff, B. E., McComsey, D. F., Gardocki, J. F., Shank, R. P., Costanzo, M. J., Nortey, S. O., Schneider, C. R., Setler, P. E. (1987) *Journal of Medicinal Chemistry*, **30**, 1433.

10 (a) Marshall, J. A. and Wang, X. (1991) *Journal of Organic Chemistry*, **56**, 4913. (b) Marshall, J. A. and Bartley, G. S. (1994) *Journal of Organic Chemistry*, **59**, 7169.

11 Ginnebaugh, J. P., Maki, J. W., Lewandos, G. S. (1980) *Journal of Organometallic Chemistry*, **190**, 403.

12 (a) Schmidt, H. M. and Arens, J. F. (1967) *Recueil des Travaux Chimiques des Pays-Bas*, **86**, 1138. (b) Orsini, A., Vitérisi, A., Bodlenner, A., Weibel, J. -M., Pale, P. (2005) *Tetrahedron Letters*, **46**, 2259. (c) Vitérisi, A., Orsini, A., Weibel, J. -M., Pale, P. (2006) *Tetrahedron Letters*, **47**, 2779.

13 Roy, S. and Gribble, G. W. (2005) *Tetrahedron Letters*, **46**, 1325.

14 Baldwin, J. E. (1976) *Journal of the Chemical Society, Chemical Communications*, 734.

15 Zhang, Q., Tu, G., Zhao, Y., Cheng, T. (2002) *Tetrahedron*, **58**, 6795.

16 (a) Szawkalo, J., Zawadzka, A., Wojtasiewicz, K., Leniewski, A., Drabowicz, J., Czarnocki, Z. (2005) *Tetrahedron: Asymmetry*, **16**, 3619. (b) Meyer, N. and Opatz, T. (2006) *European Journal of Organic Chemistry*, 3997.

17 Kam, T. -S. and Sim, K. -M. (1998) *Phytochemistry*, **47**, 145.

18 (a) Itoh, T., Miyazaki, M., Nagata, K., Yokoya, M., Nakamura, S., Ohsawa, A. (2002) *Heterocycles*, **58**, 115. (b) Itoh, T., Miyazaki, M., Nagata, K., Nakamura, S., Ohsawa, A. (2004) *Heterocycles*, **63**, 655.

19 Knölker, H. -J. and Agarwal, S. (2004) *Synlett*, 1767.

20 (a) Gribble, G. W. (2000) *Journal of the Chemical Society, Perkin Transactions 1*, 1045. (b) Humphrey, G. R. and Kuethe, J. T. (2006) *Chemical Reviews*, **106**, 2875.

21 (a) Pearson, A. J. (1980) *Accounts of Chemical Research*, **13**, 463. (b) Pearson, A. J. (1982) In Wilkinson, G., Stone, F. G. A., and Abel, E. W. (Eds.) *Comprehensive Organometallic Chemistry*, Vol. 8, Pergamon Press, Oxford, (Chapter 58). (c) Pearson, A. J. (1985) *Metallo-Organic Chemistry*, Wiley, Chichester, (Chapters 7 and 8). (d) Knölker, H. -J. (1998) In Beller, M. and Bolm, C. (Eds.) *Transition Metals for Organic Synthesis*, Vol. 1, Wiley-VCH, Weinheim, (Chapter 3.13). (e) Knölker, H. -J. (1999) *Chemical Society Reviews*, **28**, 151. (f) Donaldson, W. A. (2000) *Current Organic Chemistry*, **4**, 837. (g) Knölker, H. -J. (2004) In Beller, M. and Bolm, C. (Eds.) *Transition Metals for Organic Synthesis*, 2nd edn, Vol. 1, Wiley-VCH, Weinheim, (Chapter 3.11).

22 Knölker, H. -J., El-Ahl, A. -A., Weingärtner, G. (1994) *Synlett*, 194.

23 (a) Knölker, H. -J., Baum, G., Foitzik, N., Goesmann, H., Gonser, P., Jones, P. G., Röttele, H. (1998) *European Journal of Inorganic Chemistry*, 993. (b) Knölker, H. -J., Baum, E., Gonser, P., Rohde, G., Röttele, H. (1998) *Organometallics*, **17**, 3916. (c) Knölker, H. -J., Ahrens, B., Gonser, P., Heininger, M., Jones, P. G. (2000) *Tetrahedron*, **56**, 2259. (d) Knölker, H. -J. (2000) *Chemical Reviews*, **100**, 2941.

24 Fischer, E. O. and Fischer, R. D. (1960) *Angewandte Chemie*, **72**, 919.

25 Knölker, H. -J., Goesmann, H., Klauss, R. (1999) *Angewandte Chemie*, **111**, 727. Knölker, H. -J., Goesmann, H., Klauss, R. (1999) *Angewandte Chemie International Edition*, **38**, 702.

26 (a) Ghosal, S., Rao, H. P., Jaiswal, D. K., Kumar, Y., Frahm, W. A. (1981) *Phytochemistry*, **20**, 2003. (b) Chattopadhyay, S., Chattopadhyay, U., Marthur, P. P., Saini, K. S., Ghosal, S. (1983) *Planta Medica*, **49**, 252.

27 See for examples (a) Wolkenberg, S. E. and Boger, D. L. (2002) *Journal of Organic Chemistry*, **67**, 7361. (b) Harayama, T., Hori, A., Abe, H., Takeuchi, Y. (2004) *Tetrahedron*,

60, 1611. (c) Torres, J. C., Pinto, A. C., Garden, S. J. (2004) *Tetrahedon*, **60**, 9889 and references cited therein.
28 Knölker, H. -J. and Filali, S. (2003) *Synlett*, 1752.
29 (a) Chakraborty, D. P. and Roy, S. (1991) In Herz, W., and Grisebach, H. Kirby, G. W., Steglich, W., Tamm, C. (Eds.) *Progress in the Chemistry of Organic Natural Products*, Vol. 57, Springer-Verlag, Wien, 71. (b) Chakraborty, D. P. (1993) In Cordell, G. A. (Ed.) *The Alkaloids*, Vol. 44, Academic Press, New York, 257.
30 (a) Knölker, H. -J. and Reddy, K. R. (2002) *Chemical Reviews*, **102**, 4303. (b) Knölker, H. -J. (2005) *Topics in Current Chemistry*, **244**, 115.
31 (a) Knölker, H. -J., Bauermeister, M., Bläser, D., Boese, R., Pannek, J. -B. (1989) *Angewandte Chemie*, **101**, 225. Knölker, H. -J., Bauermeister, M., Bläser, D., Boese, R., Pannek, J. -B. (1989) *Angewandte Chemie International Edition*, **28**, 223. (b) Knölker, H. -J. (1992) *Synlett*, 371. (c) Knölker, H. -J., Bauermeister, M., Pannek, J. -B., Bläser, D., Boese, R. (1993) *Tetrahedron*, **49**, 841.
32 Sakano, K. and Nakamura, S. (1980) *The Journal of Antibiotics*, **33**, 961.
33 Cardellina, J. H., Kirkup, M. P., Moore, R. E., Mynderse, J. S., Seff, K., Simmons, C. J. (1979) *Tetrahedron Letters*, 4915.
34 Kato, S., Kawai, H., Kawasaki, T., Toda, Y., Urata, T., Hayakawa, Y. (1989) *The Journal of Antibiotics*, **42**, 1879.
35 Jackson, P. M., Moody, C. J., Mortimer, R. J. (1991) *Journal of the Chemical Society, Perkin Transactions* **1**, 2941.
36 Knölker, H. -J., Fröhner, W., Heinrich, R. (2004) *Synlett*, 2705.
37 Knölker, H. -J. and Hopfmann, T. (2002) *Tetrahedron*, **58**, 8937.
38 Kaneda, M., Naid, T., Kitahara, T., Nakamura, S., Hirata, T., Suga, T. (1988) *The Journal of Antibiotics*, **41**, 602.
39 Knölker, H. -J., Fröhner, W., Reddy, K. R. (2003) *European Journal of Organic Chemistry*, 740.
40 Choi, T. A., Czerwonka, R., Fröhner, W., Krahl, M. P., Reddy, K. R., Franzblau, S. G., Knölker, H. -J. (2006) *Chem Med Chem*, **1**, 812.
41 Ito, C. and Furukawa, H. (1990) *Chemical and Pharmaceutical Bulletin*, **38**, 1548.
42 Wu, T. -S., Huang, S. -C., Wu, P. -L. (1997) *Heterocycles*, **45**, 969.
43 (a) Fröhner, W., Krahl, M. P., Reddy, K. R., Knölker, H. -J. (2004) *Heterocycles*, **63**, 2393. (b) Knölker, H. -J. and Reddy, K. R. (2005) In Kartsev, V. G. (Ed.) *Selected Methods for Synthesis and Modification of Heterocycles–The Chemistry and Biological Activity of Natural Indole Systems (Part 1)*, Vol. 4, ICSPF Press, Moscow, 166.
44 Knölker, H. -J. and Fröhner, W. (2000) *Synthesis*, 2131.
45 Knölker, H. -J. and Krahl, M. P. (2004) *Synlett*, 528.
46 Ruangrungsi, N., Ariyaprayoon, J., Lange, G. L., Organ, M. G. (1990) *Journal of Natural Products*, **53**, 946.
47 (a) Wu, T. -S., Huang, S. -C., Wu, P. -L., Teng, C. -M. (1996) *Phytochemistry*, **43**, 133. (b) Ito, C., Ohta, H., Tan, H. T. -W., Furukawa, H. (1996) *Chemical and Pharmaceutical Bulletin*, **44**, 2231. (c) Ito, C., Katsuno, S., Ohta, H., Omura, M., Kajiura, I., Furukawa, H. (1997) *Chemical and Pharmaceutical Bulletin*, **45**, 48.
48 Yenjai, C., Sripontan, S., Sriprajun, P., Kittakoop, P., Jintasirikul, A., Tanticharoen, M., Thebtaranonth, Y. (2000) *Planta Medica*, **66**, 277.
49 Wu, T. -S., Huang, S. -C., Wu, P. -L., Kuoh, C. -S. (1999) *Phytochemistry*, **52**, 523.
50 Sunthitikawinsakul, A., Kongkathip, N., Kongkathip, B., Phonnakhu, S., Daly, J. W., Spande, T. F., Nimit, Y., Rochanaruangrai, S. (2003) *Planta Medica*, **69**, 155.
51 Kataeva, O., Krahl, M. P., Knölker, H. -J. (2005) *Organic and Biomolecular Chemistry*, **3**, 3099.
52 Kato, S., Shindo, K., Kataoka, Y., Yamagishi, Y., Mochizuki, J. (1991) *The Journal of Antibiotics*, **44**, 903.

53 Shin-ya, K., Tanaka, M., Furihata, K., Hayakawa, Y., Seto, H. (1993) *Tetrahedron Letters*, **34**, 4943.

54 Czerwonka, R., Reddy, K. R., Baum, E., Knölker, H. -J. (2006) *Chemical Communications*, 711.

55 Knölker, H. -J. (1995) In Moody, C. J. (Ed.) *Advances in Nitrogen Heterocycles*, Vol. 1, JAI Press, Greenwich, CT, 173.

56 Åkermark, B., Eberson, L., Jonsson, E., Petersson, E. (1975) *Journal of Organic Chemistry*, **40**, 1365.

57 (a) Henry, P. M. (2002) In Negishi, E. (Ed.) *Handbook of Organopalladium Chemistry for Organic Synthesis*, Vol. 2, Wiley, New York (Chap. V.3.1.1). (b) Takacs, J. M. and Jiang, X. (2003) *Current Organic Chemistry*, **7**, 369. (c) Tsuji, J. (2004) *Palladium Reagents and Catalysts–New Perspectives for the 21st Century*, Wiley, Chichester, (Chapter 2).

58 Knölker, H. -J. and O'Sullivan, N. (1994) *Tetrahedron*, **50**, 10893.

59 Li, J. J. and Gribble, G. W. (2000) *Palladium in Heterocyclic Chemistry—A Guide for the Synthetic Chemist*, Pergamon, Oxford, (Chapters 1.1 and 3.2).

60 (a) Åkermark, B., Oslob, J. D., Heuschert, U. (1995) *Tetrahedron Letters*, **36**, 1325. (b) Hagelin, H., Oslob, J. D., Åkermark, B. (1999) *Chemistry—A European Journal*, **5**, 2413.

61 Tanaka, M., Shin-ya, K., Furihata, K., Seto, H. (1995) *The Journal of Antibiotics*, **48**, 326.

62 Knölker, H. -J., Fröhner, W., Reddy, K. R. (2002) *Synthesis*, 557.

63 (a) Furukawa, H. (1994) *Journal of the Indian Chemical Society*, **71**, 303. (b) Bouaziz, Z., Nebois, P., Poumaroux, A., Fillion, H. (2000) *Heterocycles*, **52**, 977.

64 (a) Wu, T. -S., Ohta, T., Furukawa, H., Kuoh, C. -S. (1983) *Heterocycles*, **20**, 1267. (b) Furukawa, H., Wu, T. -S., Ohta, T., Kuoh, C. -S. (1985) *Chemical and Pharmaceutical Bulletin*, **33**, 4132. (c) Takeya, K., Itoigawa, M., Furukawa, H. (1989) *European Journal of Pharmacology*, **169**, 137.

65 Saha, C. and Chowdhury, B. K. (1998) *Phytochemistry*, **48**, 363.

66 Knölker, H. -J. and Reddy, K. R. (2003) *Heterocycles*, **60**, 1049.

67 Kotada, N., Shin-ya, K., Furihata, K., Hayakawa, Y., Seto, H. (1997) *The Journal of Antibiotics*, **50**, 770.

68 Nihei, Y., Yamamoto, H., Hasegawa, M., Hanada, M., Fukagawa, Y., Oki, T. (1993) *The Journal of Antibiotics*, **46**, 25.

69 Knölker, H. -J. and Knöll, J. (2003) *Chemical Communications*, 1170.

70 Knöll, J. and Knölker, H. -J. (2006) *Synlett*, 651.

71 Knöll, J. and Knölker, H. -J. (2006) *Tetrahedron Letters*, **47**, 6079.

72 (a) Wu, T. -S., Huang, S. -C., Wu, P. -L. (1996) *Phytochemistry*, **43**, 1427. (b) Ito, C., Katsuno, S., Ohta, H., Omura, M., Kajiura, I., Furukawa, H. (1997) *Chemical and Pharmaceutical Bulletin*, **45**, 48.

73 Wu, T. -S., Huang, S. -C., Wu, P. -L., Kuoh, C. -S. (1999) *Phytochemistry*, **52**, 523.

74 Meragelman, K. M., McKee, T. C., Boyd, M. R. (2000) *Journal of Natural Products*, **63**, 427.

75 Krahl, M. P., Jäger, A., Krause, T., Knölker, H. -J. (2006) *Organic and Biomolecular Chemistry*, **4**, 3215.

76 (a) Chakraborty, D. P. (1966) *Tetrahedron Letters*, 661. (b) Chakraborty, D. P. (1969) *Phytochemistry*, **8**, 769. (c) Mukherjee, S., Mukherjee, M., Ganguly, S. N. (1983) *Phytochemistry*, **22**, 1064. (d) Bhattacharyya, P., Sarkar, T., Chakraborty, A., Chowdhury, B. K. (1984) *Indian Journal of Chemistry*, **23B**, 49.

77 (a) Chakraborty, D. P., Das, K., Das, B. P., Chowdhury, B. K. (1975) *Transactions of Bose Research Institute*, **38**, 1. (b) Chowdhury, D. N., Basak, S. K., Das, B. P. (1978) *Current Science*, **47**, 490.

78 Kumar, V., Reisch, J., Wickramasinghe, A. (1989) *Australian Journal of Chemistry*, **42**, 1375.

79 (a) Li, W. -S., McChesney, J. D., El-Feraly, F. S. (1991) *Phytochemistry*,

30, 343. (b) Wu, S. -L. and Li, W. -S. (1999) *Chinese Pharmacology Journal*, **51**, 227.

80 Ma, C., Case, R. J., Wang, Y., Zhang, H. -J., Tan, G. T., Hung, N. V., Cuong, N. M., Franzblau, S. G., Soejarto, D. D., Fong, H. H. S., Pauli, G. F. (2005) *Planta Medica*, **71**, 261.

81 Forke, R., Krahl, M. P., Krause, T., Schlechtingen, G., Knölker, H. -J. (2007) *Synlett*, 268.

16
Camptothecin and Analogs: Structure and Synthetic Efforts

Sabrina Dallavalle, Lucio Merlini

16.1
Introduction: Structure and Activity

Camptothecin (CPT, **1**) is a natural compound isolated for the first time [1] from the wood of *Camptotheca acuminata* Decne (Nyssaceae), a deciduous plant (xi shu, happy tree) of Southeastern China, but produced also by the Indian Icacinacea *Nothapodytes foetida* (Wight) Sleumer (formerly *Mappia foetida* Miers) [2], and by some other plants [3], the two former being the major sources of the compound.

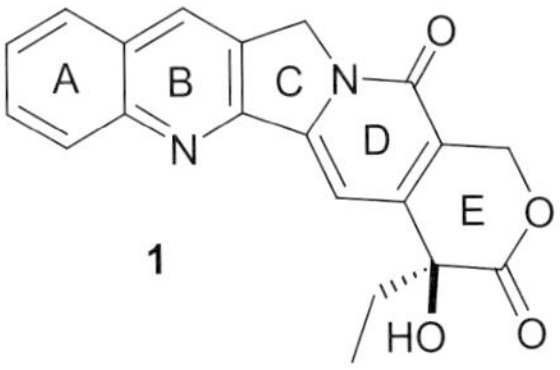

A B C N D E O O HO O

1

Although CPT is not basic, it certainly belongs to the alkaloid family, as its structure clearly shows the derivation from the basic precursors of monoterpenoid indole alkaloids, tryptamine and secologanin. The well-known intermediate of this pathway, strictosamide, has been shown to be a precursor of CPT, by incorporation of a radiolabeled sample [4]. The subsequent steps in the rearrangement of the indole to the quinoline nucleus most probably involve oxidation and recyclization of the C and D rings, oxidation of the D ring and removal of the C-21 glucose moiety, and oxidation of ring E. In agreement with this hypothesis is the isolation of 3-(*S*)- and 3-(*R*) deoxypumiloside and 3-(*S*)-pumiloside from *Ophiorrhiza pumila*, another plant producing CPT. (See ref. [3] for a detailed review of CPT biosynthesis.) (Scheme 16.1). ^{13}C NMR studies have established that the secologanin moiety is formed via the plastidic nonmevalonate (MEP) pathway [5], but details of the last steps of the biosynthesis remain hypothetical.

Camptothecin was discovered during a program of screening plant extracts for antitumor activity, launched by NCI in 1955. The unusual activity of the extracts of *Camptotheca acuminata* against some leukemia cellular lines prompted a study of the

Modern Alkaloids: Structure, Isolation, Synthesis and Biology. Edited by E. Fattorusso and O. Taglialatela-Scafati

ISBN: 978-3-527-31521-5

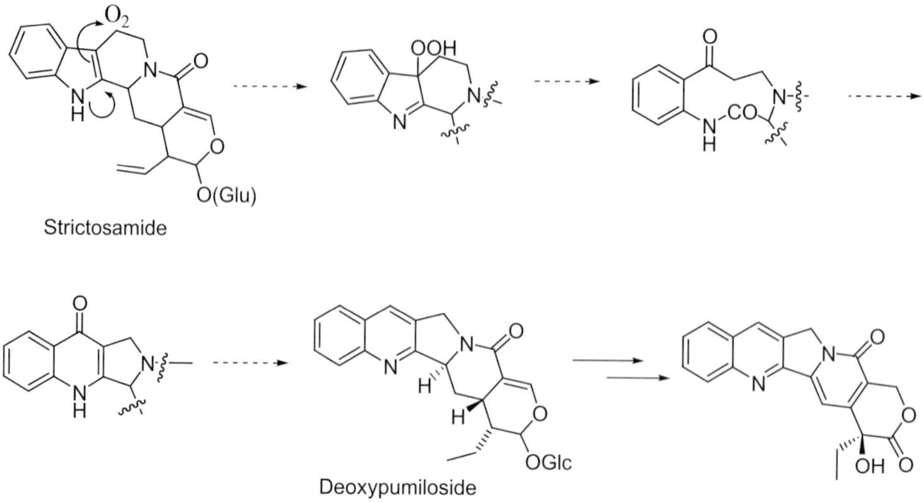

Strictosamide

Deoxypumiloside

Scheme 16.1 Putative last steps of camptothecin biosynthesis.

components, which led to the bio-guided fractionation, isolation, and structural elucidation of CPT in 1966 [1]. As soon as sufficient material became available, further *in vitro* and *in vivo* assays were conducted, culminating in Phase I and II clinical trials in 1970–1972 [6]. Owing to its extremely low solubility in water, CPT had to be administered as the sodium salt of the hydroxycarboxylic acid **2** (Scheme 16.2). However, shifting of the equilibrium toward the lactone form in tissutal compartments with acid pH caused precipitation of crystals of CPT, which caused severe hemorrhagic cystitis. This effect, together with other toxicities, led to the termination of clinical trials in 1972.

Interest in possible applications of CPT declined. However, renewed interest in CPT emerged when, as a result of a cooperative effort between Johns Hopkins University and SKB, it was found that DNA damage, which is the main reason for the antitumor activity, was due to inhibition of the ubiquitous nuclear enzyme topoisomerase I [7]. Elucidation of the mechanism of action of CPT and, therefore, of a definite biological target at which to aim new drugs gave rise to a fresh wave of research aimed at finding new more active and less toxic camptothecin derivatives.

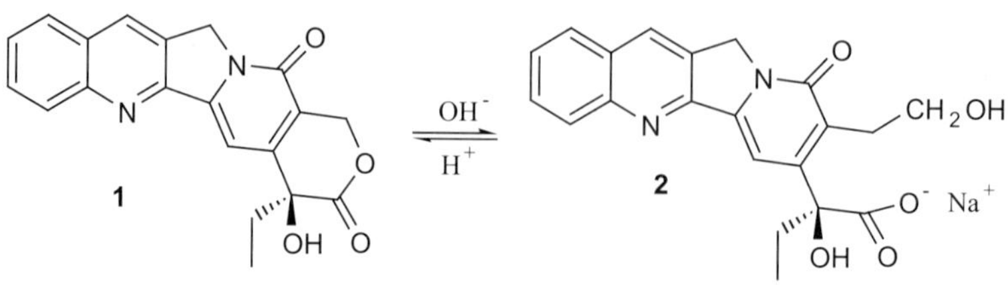

Scheme 16.2 Lactone—hydroxyacid equilibrium in camptothecin.

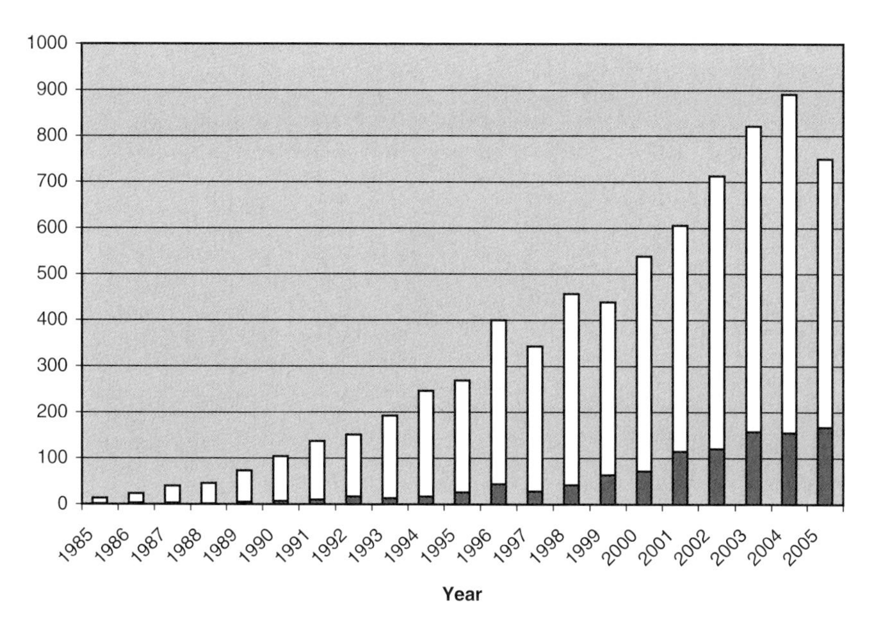

Fig. 16.1 Trend of publications and patents on camptothecins from 1985 to 2005.

This is clearly shown by the sharp increase in the number of publications and patents that followed Liu's paper (Figure 16.1).

To avoid the problems encountered with CPT itself, the introduction of functional groups able to make the compounds sufficiently water soluble to allow intravenous administration was a main issue.

The results of this effort were a detailed pattern of structure–activity relationships (Figure 16.2), and the production of two compounds, topotecan **3**[8] and irinotecan **4**[9], which were approved for clinical use in 1996, the main indications being ovarian and small-cell lung cancer for the former and metastatic colorectal cancer for the latter. Irinotecan is a water-soluble prodrug of the active compound SN-38 (**5**) Figure 16.3). Several reviews of this phase of research have been published [10–12].

Together with the synthesis and screening of new derivatives and analogs, research continued unabated to unveil the details of the mechanism of action of CPT at the molecular level. The decade 1995–2005 brought new exciting results and some changes in the perspective of research in the CPT field [13].

Camptothecin acts by forming a reversible ternary complex (''cleavable complex'') with DNA and topoisomerase I, preventing the re-ligation of the DNA strand cut by topoisomerase to allow relaxation, and thus inducing apoptosis [14]. The X-ray structure of crystals of such a complex of a 22-base DNA fragment with topoisomerase I and topotecan has been reported [15], and molecular models of the interaction have been proposed [16–18]. This kind of information should be of help in

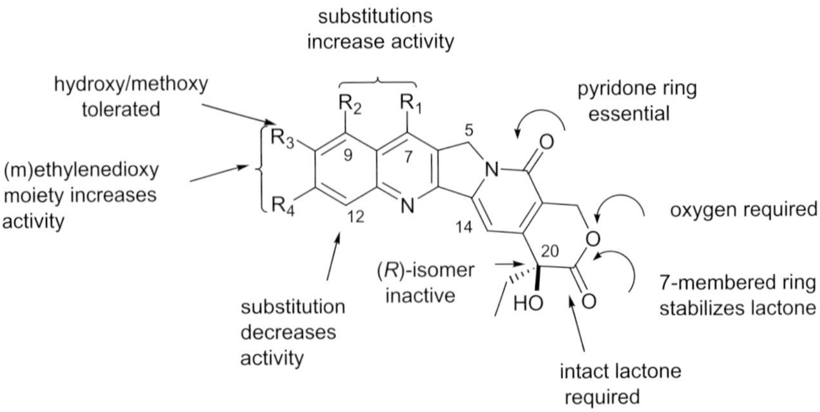

Fig. 16.2 Structure–activity relationships for antitumor activity in camptothecins as known around 1995.

designing new active compounds, but so far no breakthrough substance seems to have been obtained on such a basis, and discussion on which feature of ring E of camptothecin is essential for activity is still lively [19].

Over the years, the feature of interest for pharmacologists in camptothecins has progressively shifted, so that water solubility is no longer an essential requisite. Lipophilic compounds have the advantage of compartmentation in tissues, thus assuring the stabilization and enhanced persistence of the active lactone form, and allowing oral administration of the drug, with increased compliance by the patients. A seminal paper in this respect was published by Burke in 1993 [20], and now this trend is largely accepted [21]. These changes had important consequences in the design and synthesis of new analogs. In fact a series of lipophilic analogs of CPT are in preclinical development at the time of writing (2006) (Figure 16.4).

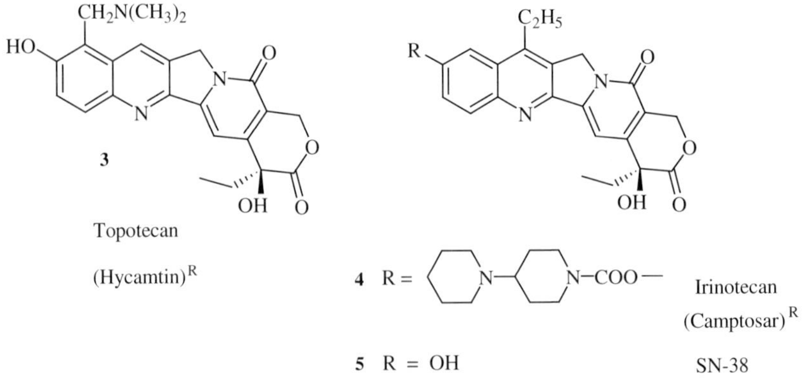

Fig. 16.3 Structures of topotecan and irinotecan.

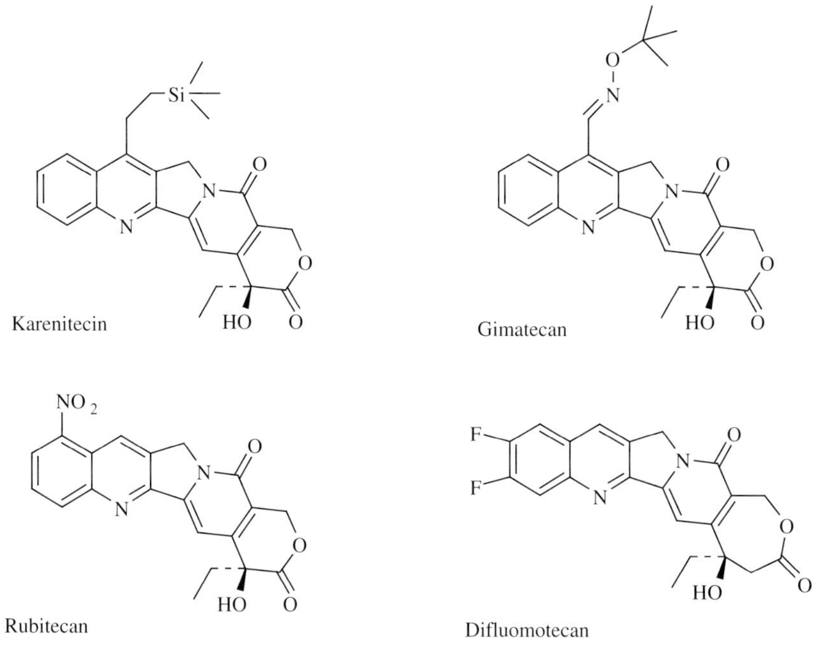

Karenitecin

Gimatecan

Rubitecan

Difluomotecan

Fig. 16.4 Lipophilic analogs of camptothecin in clinical development.

Another aspect of the progress toward the development of a camptothecin drug candidate concerns the study of proper formulations, such as liposomes [22], and the finding of innovative drug delivery systems [23].

16.2
Synthetic Efforts

For the synthesis of a new camptothecin derivative, the first choice is between a semisynthesis starting from the natural compound CPT, or a total synthesis. Camptothecin is a chiral compound, with only one asymmetric center, carbon 20, the active compounds possessing the natural configuration (*S*). A semisynthesis has the advantages of starting from a compound that possesses all the necessary structural features, including the required 20-(*S*) configuration. The drawbacks of this approach can be the limited reactivity of the quinoline nucleus and the sensitivity of the lactone ring. For the development of a drug, difficulties could derive from the possible failure of an adequate and constant supply of the natural material, and from an unpredictable pattern of impurities in the different batches. Owing to the high potency of the drugs, doses are rather low, so that the amount of camptothecin required has so far been within the capacity of the Chinese and Indian producers, although some concern has been raised on the conservation of *Camptotheca acuminata*, which grows only in an area of China south of the Yangtze river. However, the plant has been shown to grow in other areas of the world, and considerable effort has already been spent toward the production of camptothecin by cell cultures [3].

On the other hand, a total synthesis offers the possibility of substitutions and structural modifications that depend only on the manageability of the synthetic scheme, so enlarging the diversity of the target compounds, and is free from the constraints indicated above. However, an asymmetric synthesis is required, with several steps, and so far most of the total syntheses appear too expensive. Actually, the two drugs currently in clinical practice and most of the candidates presently (2006) in an advanced stage of development are produced by semisynthesis.

As Figure 16.2, 16.3 and 16.4 show, so far the most fruitful modifications of CPT to obtain an active antitumor compound have been the introduction of substituents in positions 7,9, and 10.

The electron-deficient ring of quinoline is not very reactive to electrophilic substitution, the preferred sites of attack being position 5 and 9 [24]. Nitration of CPT (best yields 70 % [25]) gives in fact a mixture of 12- and 9-nitrocamptothecin (**6**). The latter is itself a compound (Rubitecan) endowed with potent antitumor activity [26], and is a precursor of many derivatives, as it can be easily reduced to 9-amino-CPT (**7**), in turn convertible into 9-hydroxy- and 9-methoxycamptothecin, minor components of the plant extract (Scheme 16.3).

The accessibility of position 9 becomes much higher when an activating group, such as an OH, is present in position 10. Although 10-hydroxycamptothecin (**8**) is available in small amounts from the plant material, two efficient preparations of this compound were developed, via catalytic reduction of CPT in acid medium to a tetrahydroquinoline, followed by selective oxidation with lead tetraacetate [8], or phenyliodonium diacetate [27], or via a photochemical rearrangement of camptothe-

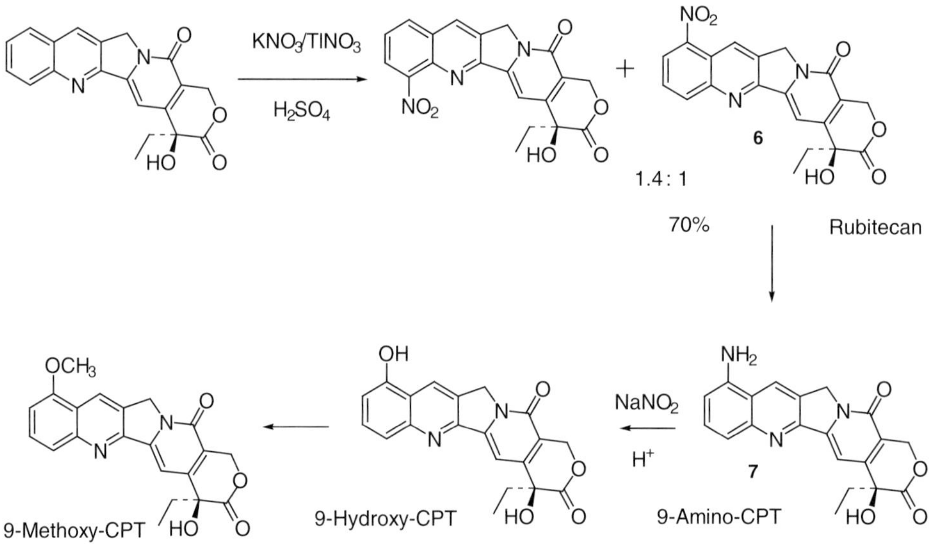

Scheme 16.3 Nitration of CPT and synthesis of 9-substituted CPTs.

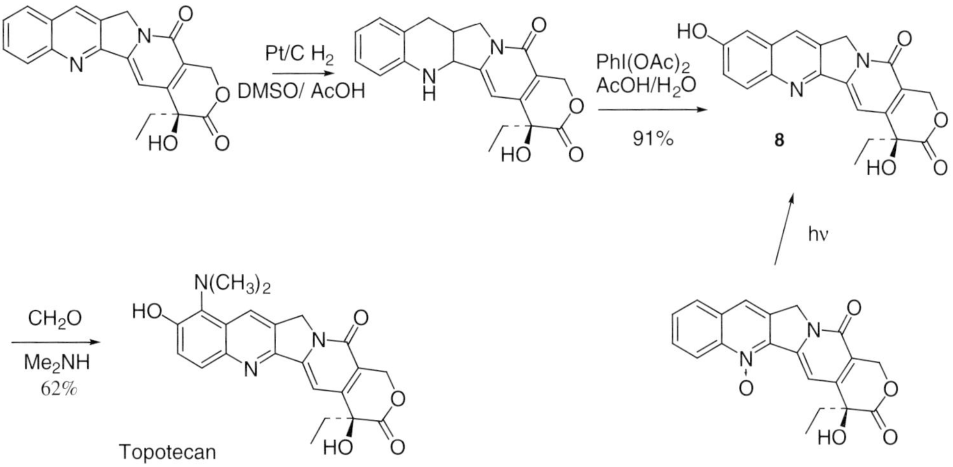

Scheme 16.4 Semisynthesis of topotecan.

cin *N*-oxide [28]. Thus activated, the nucleus smoothly undergoes the Mannich reaction to give topotecan (**3**) (Scheme 16.4).

The 10-hydroxy group can facilitate the alkylation of C-9 via a Claisen rearrangement, as in the case of 7-ethyl-10-hydroxy-CPT [29], or nitration in the same position, possibly followed by removal of the OH and reduction of the nitro group by palladium-catalyzed deoxygenation to give 9-aminoCPT (**7**)[30], another drug candidate (Scheme 16.5).

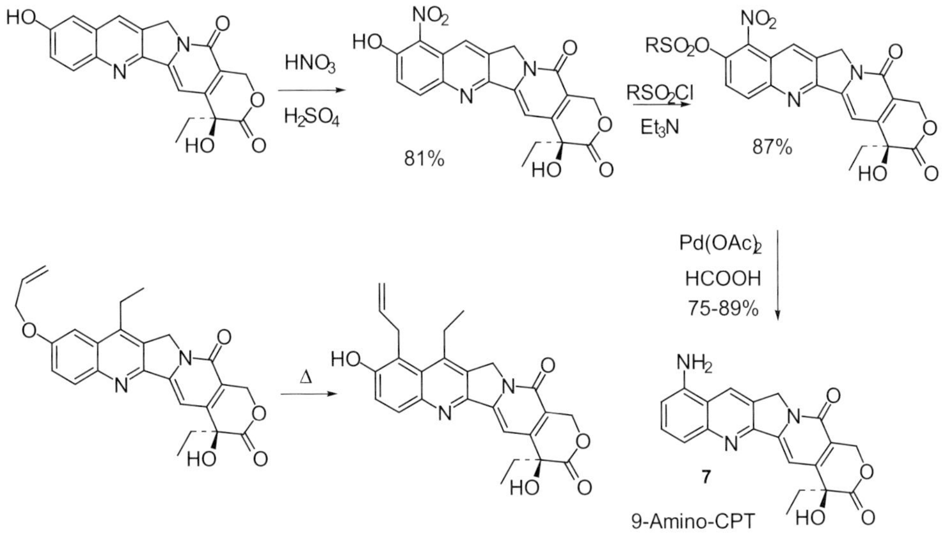

Scheme 16.5 Transformations of 10-hydroxycamptothecin.

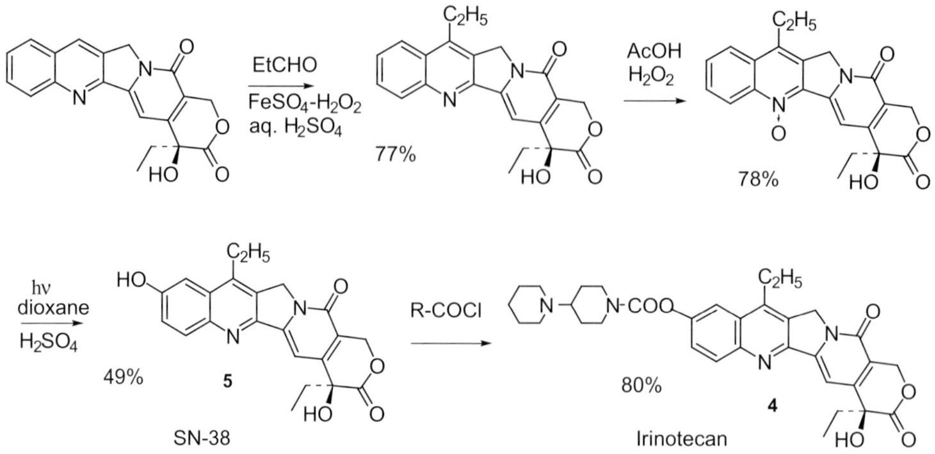

Scheme 16.6 Semisynthesis of irinotecan.

By contrast, substitution in position 7 is much easier thanks to the well-known Minisci reaction, which involves a nucleophilic radical attack on a protonated quinoline [31]. Moreover, due to the unavailability of position 2 of the quinoline nucleus, the reaction shows complete regioselectivity. Minisci alkylation with an ethyl radical produced *in situ* by decarbonylation of propionaldehyde is a crucial step in the process of preparation of irinotecan (**4**) (Scheme 16.6) [32], whereas the same kind of reaction led to the semisynthesis (Scheme 16.7) of gimatecan (**9**)[33], silatecan (**10**)[34], and belotecan (**11**)[35]. This last compound entered clinical practice in Korea in 2005.

A semisynthetic approach was also followed in the first synthesis of a camptothecin with a 7-membered lactone ring (**12**). This was indeed the first and so far the only modification of the E ring to give a strongly active compound. Lavergne and Bigg [36,37] reasoned that the reactivity of the lactone ring could be reduced by shifting the OH group from the α to the β position with respect to the lactone carbonyl. The modification was accomplished by reduction of CPT to a lactol, dehydration, and periodate oxidation followed by a Reformatzky reaction (Scheme 16.8).

As soon as the structure of camptothecin was published, the interest of many chemists, including some famous names, was directed toward this synthetic goal, encouraged by the relevance of the unusual antitumor activity. Later, when the compound had lost its novelty value, such studies were stimulated by the desire to achieve a process of production of the drugs derived from CPT and the preparation of new derivatives. Although at first sight the synthesis of CPT might not appear, by modern standards, a difficult task, the array of functional groups on ring E, not easily compatible with many synthetic procedures, has often required a number of steps and some detours to overcome the difficulties of a total synthesis. In some cases, the problem has been solved by the invention of new synthetic methods, so that the approaches have led to the addition of new tools to the arsenal of the synthetic organic chemist.

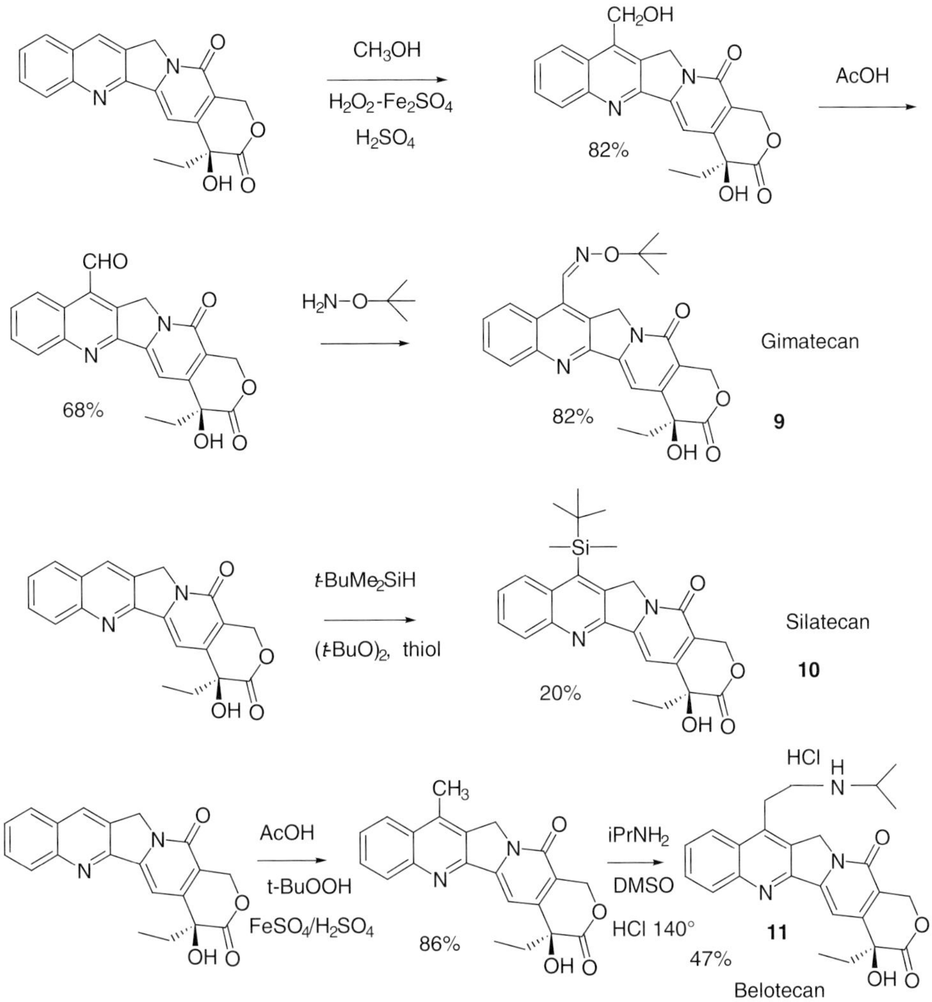

Scheme 16.7 Minisci reactions in the semisynthesis of camptothecin-derived drug candidates.

The early syntheses have been reviewed by Schultz [38] and Hutchinson [39]. Other reviews, more medicinally oriented, have appeared [40]. One of the most recent and detailed, covering work from 1990 onward, is that of Du [41].

As it is not possible to review here the large number of different syntheses of CPT, we will only attempt to call the attention of the reader to some particular or relevant, in our biased view, aspects of the large portfolio of synthetic approaches to camptothecins.

Some of the best organic chemists of the time, such as Stork, Danishefsky, and Corey, developed the early syntheses. The Stork synthesis of the racemate [42] was the first to use one of the most fruitful and popular approaches to the CPT skeleton, that is the building of ring B with a Friedländer synthesis (Scheme 16.9), but which

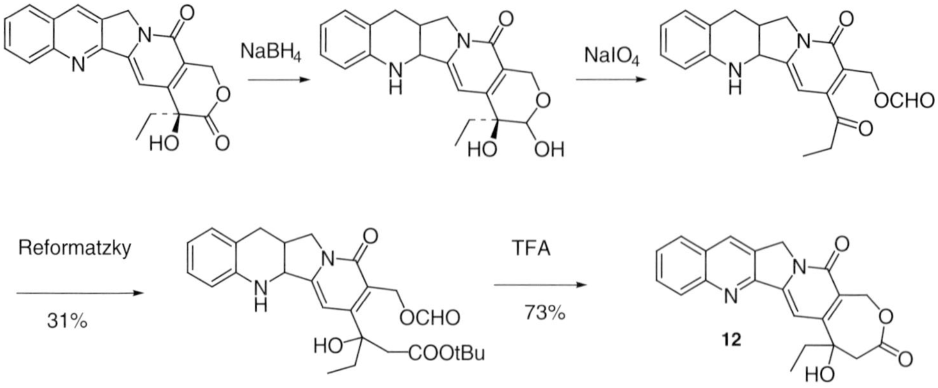

Scheme 16.8 Semisynthesis of racemic homocamptothecin.

encountered the problem of the conversion of a five-membered to a six-membered ring E, a difficulty experienced later by others.

The Corey synthesis [43] is worth revisiting for the originality of the approach in the construction both of ring C and of the D–E ring moiety, although it is flawed by the

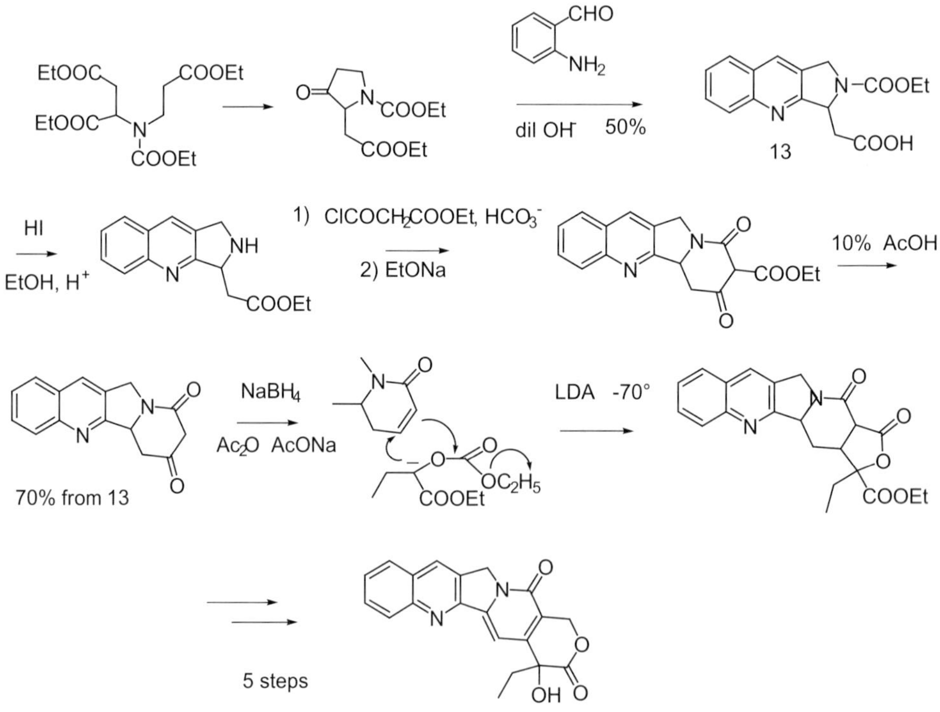

Scheme 16.9 Stork synthesis of racemic CPT.

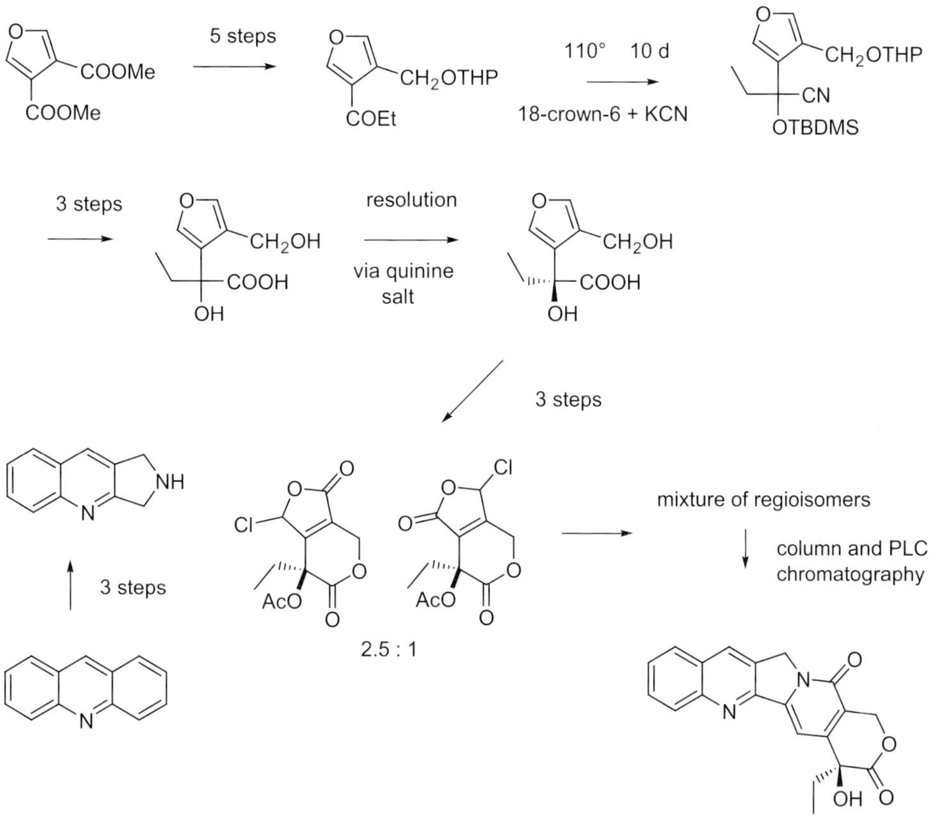

Scheme 16.10 Corey synthesis of 20(*S*)-camptothecin.

length of the preparation of the latter, and by lack of regioselectivity in the joining step (Scheme 16.10).

The Winterfeldt synthesis [44] of racemic camptothecin is remarkable for being the first to follow a biomimetic pathway, that is of taking advantage of the wealth of synthetic methods for indole alkaloids to synthesize the intermediate pyrido[3,4-b]indole intermediate to be converted by a biosynthetic-like oxidation into the expected pyrrolo[3,4-b]quinoline ring system. Moreover, this synthesis used simple and cheap reagents throughout (Scheme 16.11). Here, the last step could easily be made enantioselective by the use of a chiral hydroxylating reagent, such as Davis oxaziridines. Another truly biomimetic synthesis, but mostly only of academic interest, starting from strictosidine lactam, was reported by Brown [45].

As early as 1986, both Wall and coworkers [46] and a Chinese group [47] recognized the potentiality of a Friedländer synthesis approach from 2-aminobenzaldehyde with the synthon **14** and developed an approach to racemic **14**, based on the extremely efficient condensation of ethyl acetoacetate with cyanacetamide by Henecka [48], which provides in one step a pyridone intermediate **15** with three different sub-

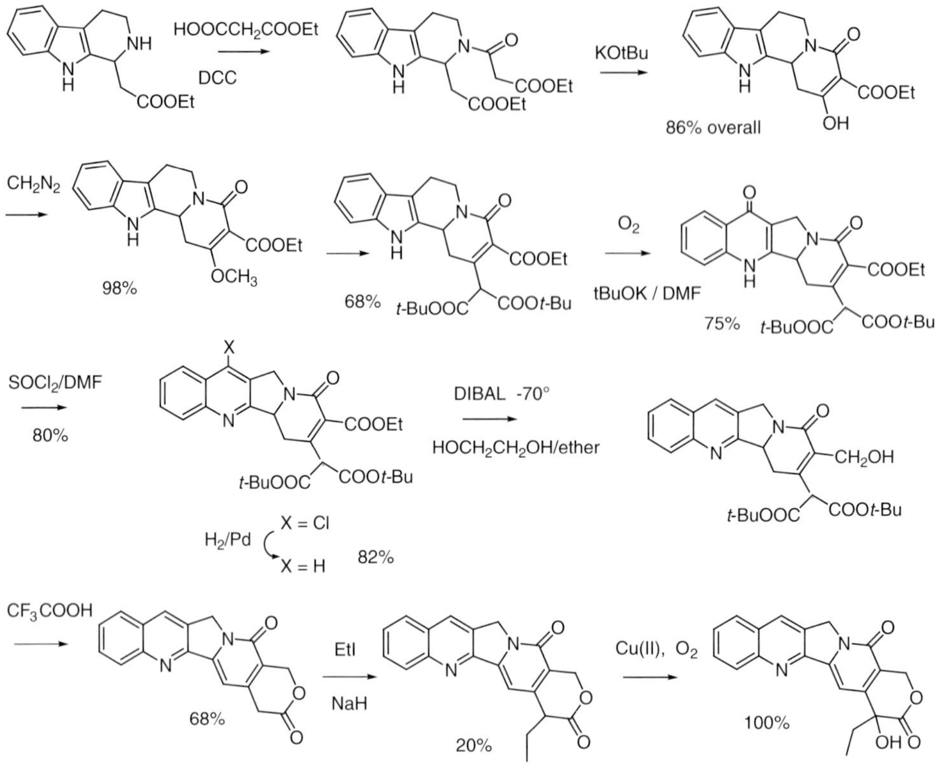

NH
N H
COOEt
$HOOCCH_2COOEt$
DCC
N
O
N H
COOEt
COOEt
KOtBu
N H
N
O
COOEt
OH
86% overall

CH_2N_2
N H
N
O
COOEt
OCH_3
98%
N H
N
O
COOEt
68% t-BuOOC COOt-Bu
O_2
tBuOK / DMF
O
N H
N
O
COOEt
75% t-BuOOC COOt-Bu

$SOCl_2$/DMF
80%
X
N
N
O
COOEt
t-BuOOC COOt-Bu
H_2/Pd
X = Cl
X = H
82%
DIBAL -70°
$HOCH_2CH_2OH$/ether
N
N
O
CH_2OH
t-BuOOC COOt-Bu

CF_3COOH
N
N
O
O
O
68%
EtI
NaH
N
N
O
O
O
20%
Cu(II), O_2
N
N
O
O
OH O
100%

Scheme 16.11 Winterfeldt synthesis.

stituent in the strategic positions. Elaboration of **15** and condensation with ethyl acrylate afforded **14** (Scheme 16.12).

Subsequent effort by various groups was dedicated to improvement of the scheme to provide an efficient synthesis of chiral **14**, via chemical [49] or enzymatic resolution [50] or, as in Tagawa's synthesis (Scheme 16.13), the use of a chiral auxiliary [51]. A procedure to recycle the otherwise wasted (*R,R*)-diastereoisomer of **16** via conversion to the mesylate of *ent*-**17** and inversion with CsOAc was also reported [52], as well as a variant to obtain the desired enantiomer via Sharpless dihydroxylation [53].

Among the more recent achievements, the Comins approach capitalized on progress in the formation of sp^2–sp^2 C—C bonds with palladium chemistry to

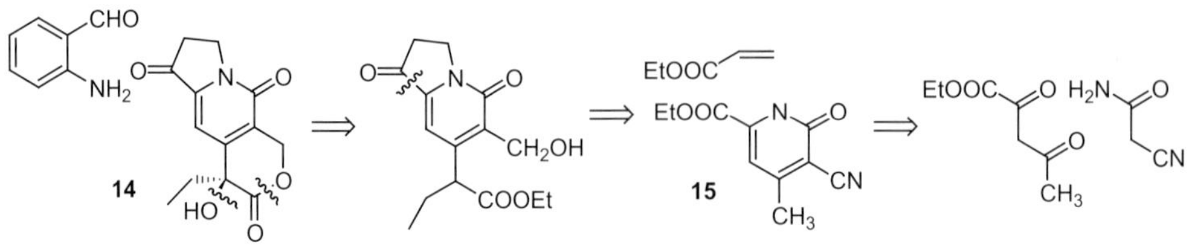

CHO
NH_2
O
N
O
14
HO
O
O
⟹
O
N
O
CH_2OH
COOEt
⟹
EtOOC
EtOOC
N
O
CN
15
CH_3
⟹
EtOOC
O
O
CH_3
H_2N
O
CN

Scheme 16.12 Approach to CPT synthesis via Friedländer condensation.

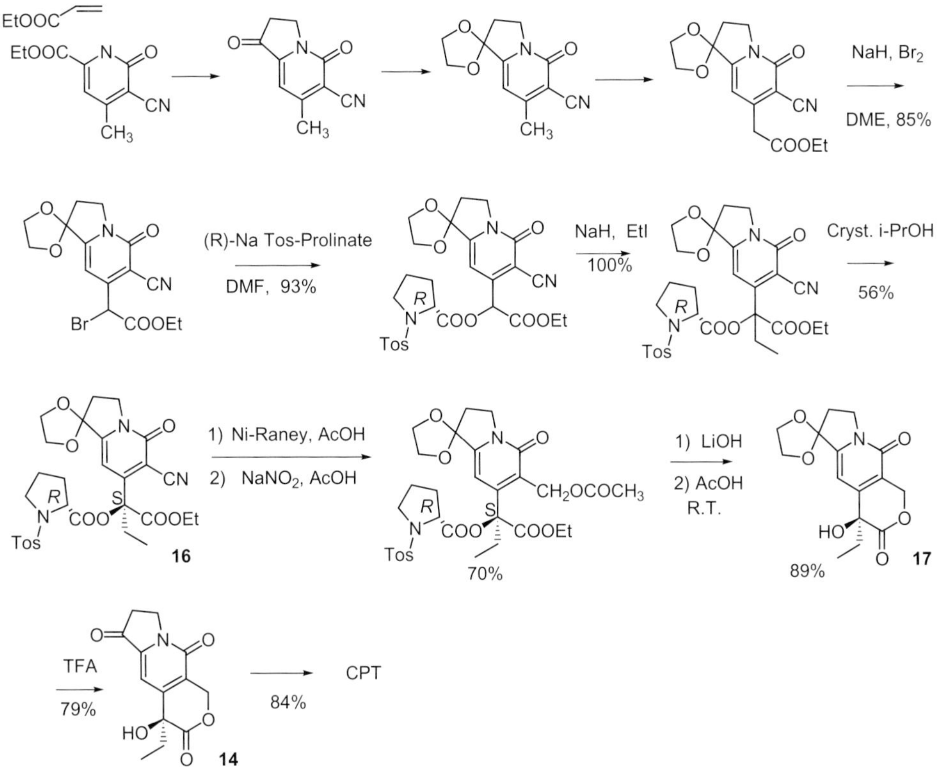

Scheme 16.13 Tagawa asymmetric synthesis of intermediate 14.

build ring C and on developments in the functionalization of pyridine by metallation [54]. For sake of brevity, Scheme 16.14 reports the final achievement, that is a six-step synthesis of CPT [55], but the reader is heartily invited to follow the masterly refinement and simplification of the synthesis across the series of Comins' papers [55–59]. It is a very instructive and enjoyable path.

On the basis of preceding experience in the synthesis of methylenecyclopentanes, Curran discovered a cascade reaction proceeding via a 4 + 1 radical annulation mechanism that led to a new synthesis of cyclopenta-fused quinolines [60] (Scheme 16.15).

The extension of this route to the case of (±)-camptothecin [61] was followed by a series of improvements [62,63], where the key intermediate **21** was obtained via the Sharpless dihydroxylation previously proposed by Fang [64] or via an asymmetric cyanosilylation reaction [65] (Scheme 16.16).

From a medicinal chemistry point of view, this approach can provide a wealth of camptothecins diversely substituted both in ring A, owing to the availability of anilines, immediate precursors of isonitriles, and at position 7, working on thepropargyl intermediates. Whereas para- and ortho-substituted isonitriles gave a regioselective cyclization, 3-substituted isonitriles gave a mixture of 9- and

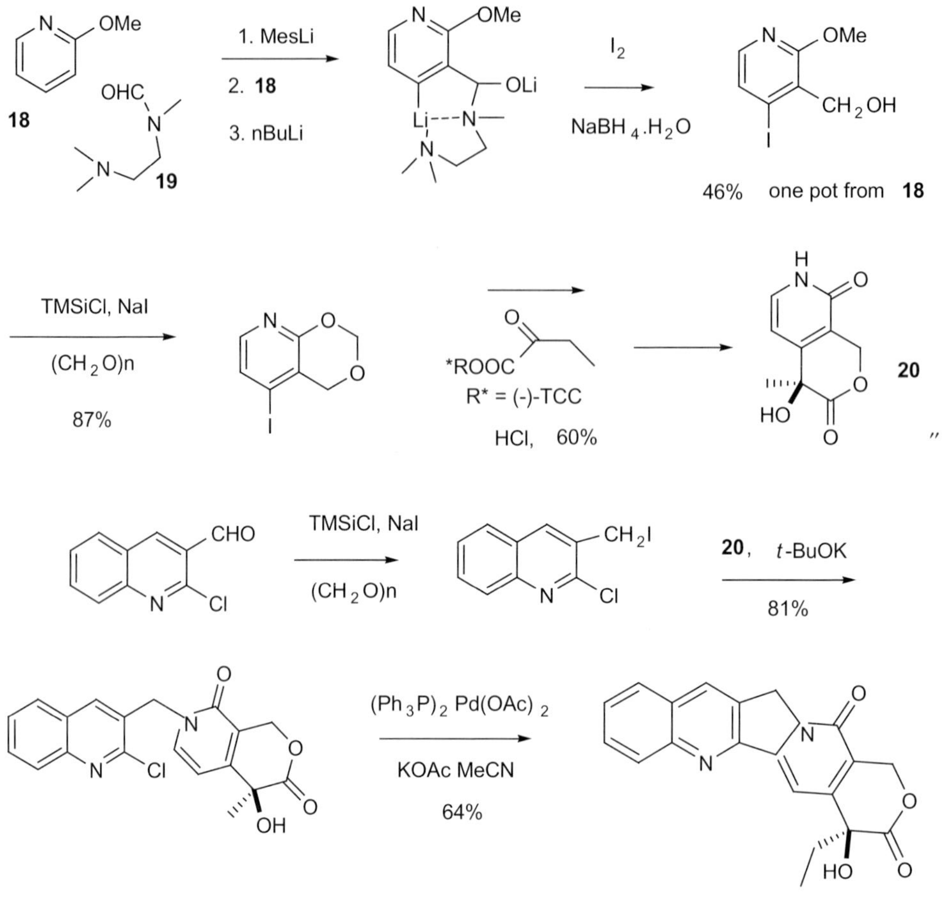

Scheme 16.14 Comins shortest synthesis of CPT.

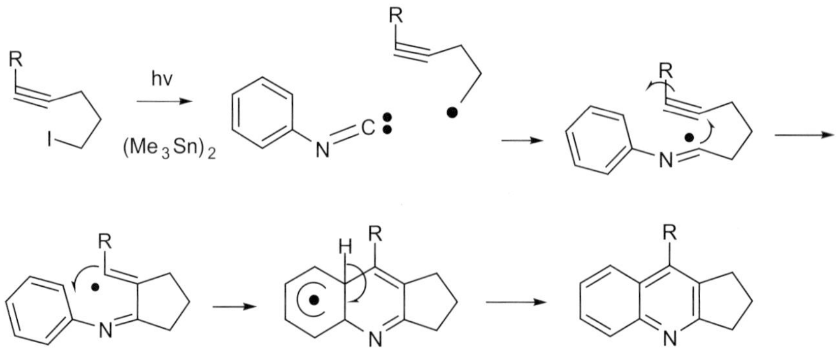

Scheme 16.15 Curran radical annulation to cyclopenta [2,3-b]quinolines.

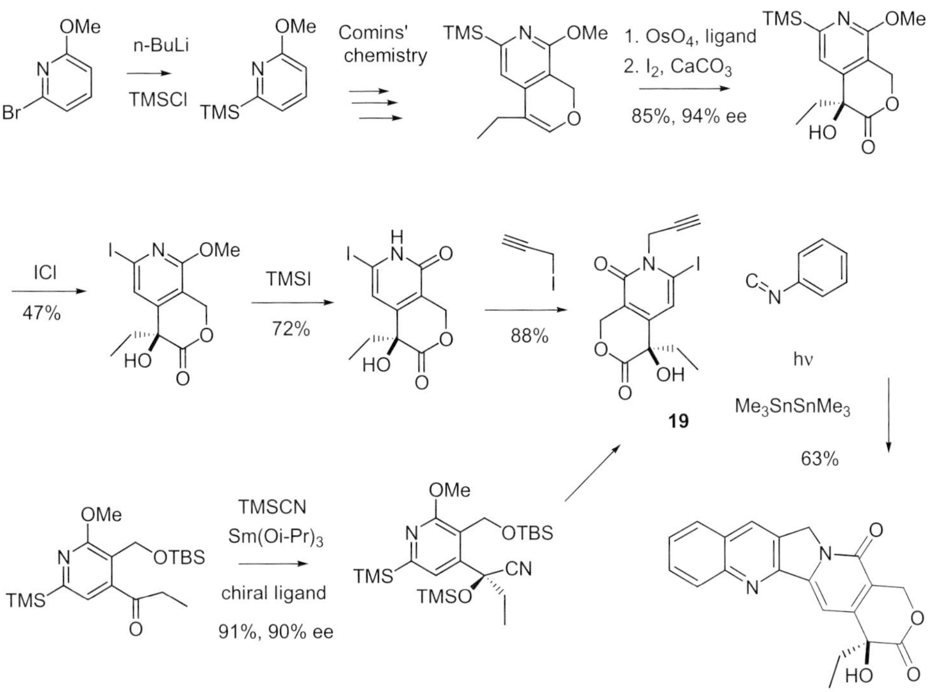

OMe
N
Br
n-BuLi
TMSCl
OMe
N
TMS
Comins' chemistry
TMS N OMe
O
1. OsO_4, ligand
2. I_2, $CaCO_3$
85%, 94% ee
TMS N OMe
HO O
O
ICl
47%
I N OMe
HO O
O
TMSI
72%
I H N O
HO O
O
I
88%
O N I
O OH
O
19
C N
hν
$Me_3SnSnMe_3$
63%
OMe
N OTBS
TMS O
TMSCN
$Sm(Oi\text{-}Pr)_3$
chiral ligand
91%, 90% ee
OMe
N OTBS
TMS CN
TMSO
N N O
O
HO O

Scheme 16.16 Curran synthesis of (+)-camptothecin.

11-substituted camptothecins. This problem was circumvented by using the easily removable trimethylsilyl group as a temporary protection [66] (Scheme 16.17).

The radical cascade synthesis was applied to the preparation of drugs such as irinotecan [62], and drug candidates such as lurtotecan [66], silatecan DB-67 [67] and homosilatecans [68]. Moreover, a convergent synthesis could be applied to a combinatorial synthesis, in which over one hundred homosilatecans were prepared by parallel synthesis and automated purification [69].

The years since 1985 have seen an enormous amount of work aimed at unravelling many facets of the reactivity of camptothecin and developing fast and ingenious syntheses. Although many of the synthetic issues concerning camptothecin have

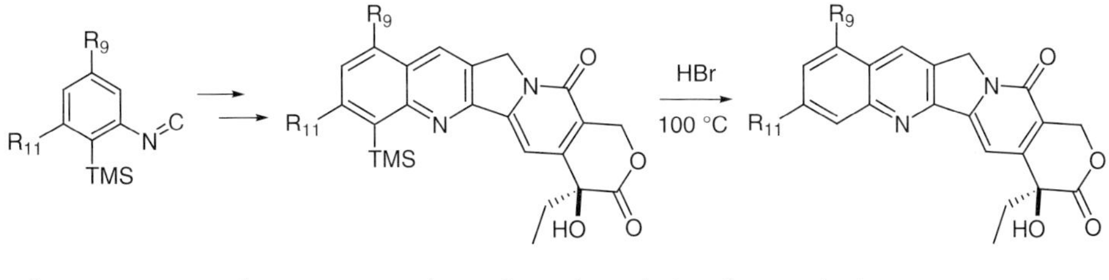

R_9
R_{11} N C
TMS
R_9
R_{11} N N O
TMS O
HO O
HBr
100 °C
R_9
R_{11} N N O
O
HO O

Scheme 16.17 Regioselective Curran synthesis of 9- and 11-substituted camptothecins.

been addressed, there is still room for the discovery of new straightforward and efficient methods of building the core ring system and of obtaining more specifically targeted derivatives and analogs. Future years will certainly bring exciting results toward these goals.

References

1 Wall, M. E., Wani, M. C., Cooke, C. E., Palmer, K. T., McPhail, A. T., Sim, G. A. (1966) *Journal of the American Chemical Society*, **88**, 3888–3890.

2 Govindachari, T. R. and Viswanathan, N. (1972) *Indian Journal of Chemistry*, **10**, 53–454.

3 Lorence, A. and Nessler, C. L. (2004) *Phytochemistry*, **65**, 2735–2749.

4 Hutchinson, C. R., Heckerdorf, A. H., Straughn, J. L., Daddona, P. E., Cane, D. E. (1979) *Journal of the American Chemical Society*, **101**, 3358–3366.

5 Yamazaki, Y., Kitajima, M., Arita, M., Takayama, H., Sudo, H., Yamazaki, M., Aimi, N., Saito, K. (2004) *Plant Physiology*, **134**, 161–170.

6 Moertel, C. G., Schutt, A. J., Reitemeyer, R. J., Hahn, R. G. (1972) *Cancer Chemotherapy Reports*, **56**, 95–101.

7 Hsiang, Y. -H., Hertzberg, R., Hecht, S., Liu, L. F. (1985) *Journal of Biological Chemistry*, **260**, 14873–14878.

8 Kingsbury, W. D., Boehm, J. C., Jakas, D. R., Holden, K. G., Hecht, S. M., Gallagher, G., Caranfa, M. J., McCabe, F. L., Faucette, L. F., Johnson, R. K., Hertzberg, R. P. (1991) *Journal of Medicinal Chemistry*, **34**, 98–107.

9 Kunimoto, T., Nitta, K., Tanaka, T., Uehara, N., Baba, H., Takeuchi, M., Yokokura, T., Sawada, S., Miyasaka, T., Mutai, M. (1987) *Cancer Research*, **47**, 5944–5947.

10 Potmesil, M.and Pinedo, H. (1995) *Camptothecins: New Anticancer Agents*, CRC Press, USA.

11 Pantaziz, P. and Giovanella, B. C. (1996) *The Camptothecins: From Discovery to Patient*, Annals of the New York Academy of Sciences, Vol. **803**.

12 Liehr, J. G., Giovanella, B. C., Verschraegen, C. F. Eds. (2000) *The Camptothecins: Unfolding Their Anticancer Potential*, Annals of the New York Academy of Sciences, Vol. 922.

13 Thomas, C. J., Rahier, N. J., Hecht, S. M. (2004) *Bioorganic and Medicinal Chemistry*, **12**, 1585–1604.

14 Liu, L. F., Desai, S. D., Li, T. -K., Mao, Y., Sun, M., Sim, S. -P., Liehr, J. G., Giovanella, B. C., Verschraegen, C. F. (eds) (2000) Annals of the New York Academy of Sciences, 922, 1.

15 Staker, B. L., Hjerrild, K., Feese, M. D., Behnke, C. A., Burgin, A. B., Jr., Stewart, L. (2002) *Proceedings of the National Academy of Sciences*, **99**, 15387–15392.

16 Redinbo, M. R., Stewart, L., Kuhn, P., Champoux, J. J., Hol, W. G. (1998) *Science*, **279**, 1504–1513.

17 Fan, Y., Weinstein, J., Kohn, K. W., Shi, L. M., Pommier, Y. (1998) *Journal of Medicinal Chemistry*, **41**, 2216–2226.

18 Kerrigan, J. E. and Pilch, D. S. (2001) *Biochemistry*, **40**, 9792–9798.

19 Hecht, S. M. (2005) *Current Medicinal Chemistry: Anti-Cancer Agents*, **5**, 353–362.

20 Burke, T. G., Mishra, A. K., Wani, M. C., Wall, M. C. (1993) *Biochemistry*, **32**, 5352–5364.

21 Zunino, F., Dallavalle, S., Laccabue, D., Beretta, G., Merlini, L., Pratesi, G. (2002) *Current Pharmaceutical Design*, **8**, 99–110.

22 Hofheinz, R. -D., Gnad-Vogt, S. U., Beyer, U., Hochhaus, A. (2005) *Anti-Cancer Drugs*, **16**, 691–707.

23 Haag, R. and Kratz, F. (2006) *Angewandte Chemie, International Edition*, **45**, 1198–1215.

24 Schofield, K., Crout, D. H. G., Penton, J. R. (1971) *Journal of the Chemical Society B*, 1254–1256.

25 Cao, Z., Armstrong, K., Shaw, M., Petry, E., Harris, N. (1998) *Synthesis*, 1724–1730.

26 Clark, J. W. (2006) *Expert Opinion on Investigational Drugs*, **15**, 71–79Z.

27 Wood, J. L., Fortunak, J. M., Mastrocola, A. R., Mellinger, M., Burk, P. L. (1995) *Journal of Organic Chemistry*, **60**, 5739–5740.

28 Sawada, S., Matsuoka, S., Nokata, K., Nagata, H., Furuta, T., Yokokura, T., Miyasaka, T. (1991) *Chemical and Pharmaceutical Bulletin*, **39**, 3183–3188.

29 Gao, H., Zhang, X., Chen, Y., Shen, H., Sun, J., Huang, M., Ding, J., Li, C., Lu, W. (2005) *Bioorganic and Medicinal Chemistry Letters*, **15**, 2003–2006.

30 Cabri, W., Candiani, I., Zarini, F., Penco, S., Bedeschi, A. (1995) *Tetrahedron Letters*, 9197–9200.

31 Reviews on the subject: Minisci, F. (1973) *Synthesis*, 1–24. Topics in Current Chemistry (1976) 62, 1–48. Minisci, F., Vismara, E., Fontana, F. (1989) *Heterocycles*, **28**, 489–519. Minisci, F., Fontana, F., Vismara, E., *Journal of Heterocyclic Chemistry*, **27**, (1990) 79–96.

32 Sawada, S., Okajima, S., Aiyama, R., Nokata, K., Furuta, T., Yokokura, T., Sugino, E., Yamaguchi, K., Miyasaka, T. (1991) *Chemical and Pharmaceutical Bulletin*, **39**, 1446–1450.

33 Dallavalle, S., Ferrari, A., Biasotti, B., Merlini, L., Penco, S., Gallo, G., Marzi, M., Tinti, M. O., Martinelli, R., Pisano, C., Carminati, P., Carenini, N., Beretta, G., Perego, P., De Cesare, M., Pratesi, G., Zunino, F. (2001) *Journal of Medicinal Chemistry*, **44**, 3264–3274.

34 Du, W., Kaskar, B., Blumbergs, P., Subramanian, P. K., Curran, D. P. (2003) *Bioorganic and Medicinal Chemistry*, **11**, 451–458.

35 Ahn, S. K., Choi, N. S., Jeong, B. S., Kim, K. K., Journ, D. J., Kim, J. K., Lee, S. J., Kim, J. W., Hong, C., Jew, S. -S. (2000) *Journal of Heterocyclic Chemistry*, **37**, 1141.

36 Lavergne, O., Lesueur-Ginot, L., Rodas, F. P., Bigg, D. C. H. (1997) *Bioorganic and Medicinal Chemistry Letters*, **7**, 2235–2238.

37 Lavergne, O., Lesueur-Ginot, L., Rodas, F. P., Kasprzyk, P. G., Pommier, J., Demarquay, D., Prevost, G., Ulibarri, G., Rolland, A., Schiano-Liberatore, A., Harnett, J., Pons, D., Camara, J., Bigg, D. C. H. (1998) *Journal of Medicinal Chemistry*, **41**, 5410–5419.

38 Schultz, A. G. (1973) *Chemical Reviews*, **73**, 385–405.

39 Hutchinson, C. R. (1981) *Tetrahedron*, **37**, 1047–1065.

40 (a) Wall, M.E. and Wani, M. C. (1994) Camptothecin. In Saxton, J.E. *The Monoterpenoid Indole Alkaloids*, Wiley, London. (b) Kawato, Y. and Terasawa, H. (1997) *Progress in Medicinal Chemistry*, **34**, 69–109 (c) Baurle, S. and Koert, U. (2000) Camptothecin –Synthesis of an Antitumor Agent. In *Organic Synthesis Highlights* IV (ed. H. G. Schmalz), pp. 232–240.

41 Du, W. (2003) *Tetrahedron*, **59**, 8649–8687.

42 Stork, G. and Schultz, A. G. (1971) *Journal of the American Chemical Society*, **93**, 4074–4075.

43 Corey, E. J., Crouse, D. N., Anderson, J. E. (1975) *Journal of Organic Chemistry*, **40**, 2140–2141.

44 Boch, M., Korth, T., Nelke, J. M., Pike, D., Radunz, H., Winterfeldt, E. (1972) *Chemische Berichte*, **105**, 2126–2142. Krohn, K. and Winterfeldt, E. (1975) *Chemische Berichte*, **108**, 3030. Krohn, K., Ohlendorf, H. -W., Winterfeldt, E. (1976) *Chemische Berichte*, **109**, 1389–1394.

45 Brown, R. T., Liu, J., Santos, C. A. M. (2000) *Tetrahedron Letters*, **41**, 859–862.

46 Wall, M. E., Wani, M. C., Natschke, S. M., Nicholas, A. W. (1986) *Journal of Medicinal Chemistry*, **29**, 1553–1555.

47 Kexue Tongbao, 21, 40–42 (1976) (Chemical Abstracts, 84, 122100n (1976)). Scientia Sinica, 21, 87–98 (1978).

48 Henecka, H. (1949) *Chemische Berichte*, **82**, 41–46.

49 Wani, M. C., Nicholas, A. W., Wall, M. E. (1987) *Journal of Medicinal Chemistry*, **30**, 1774–1779.

50 Imura, A., Itoh, M., Miyadera, A. (1998) *Tetrahedron: Asymmetry*, **9**, 2285–2291.

51 Ejima, A., Terasawa, H., Sugimori, M., Tagawa, H. (1990) *Journal Of The Chemical Society-Perkin Transactions 1*, 27–31.

52 Terasawa, H., Sugimori, M., Ejima, A., Tagawa, H. (1989) *Chemical and Pharmaceutical Bulletin*, **37**, 3382–3385.

53 Jew, S. -S., Ok, K. -D., Kim, H. -J., Kim, M. G., Kim, J. M., Hah, J. M., Cho, Y. -S. (1995) *Tetrahedron: Asymmetry*, **6**, 1245–1248.

54 Rocca, P., Cochennec, C., Marsais, F., Thomas-dit-Dumont, L., Mallet, M., Godard, A., Queguiner, G. (1993) *Journal of Organic Chemistry*, **58**, 7832–7838.

55 Comins, D. L. and Nolan, J. M. (2001) *Organic Letters*, **3**, 4255–4257.

56 Comins, D. L., Baevsky, M. F., Hong, H. (1992) *Journal of the American Chemical Society*, **114**, 10971–10972.

57 Comins, D. L., Hong, H., Gao, J. (1994) *Tetrahedron Letters*, **30**, 5331–5334.

58 Comins, D. L. and Saha, J. K. (1995) *Tetrahedron Letters*, **36**, 7995–77998.

59 Comins, D. L., Hong, H., Saha, J. K., Gao, J. (1994) *Journal of Organic Chemistry*, **59**, 5120–5121.

60 Curran, D. P. and Liu, H. (1991) *Journal of the American Chemical Society*, **113**, 2127–2132.

61 Curran, D. P. and Liu, H. (1992) *Journal of the American Chemical Society*, **114**, 5863–5864.

62 Curran, D. P., Ko, S. -B., Josien, H. (1996) *Angewandte Chemie, International Edition*, **34**, 2683–2684.

63 Curran, D. P., Liu, H., Josien, H., Ko, S. -B. (1996) *Tetrahedron*, **52**, 11385–11404.

64 Fang, F. G., Xie, S., Lowery, M. W. (1994) *Journal of Organic Chemistry*, **59**, 6142–6143.

65 Yabu, K., Masumoto, S., Kanai, M., Du, W., Curran, D. P., Shibasaki, M. (2003) *Heterocycles*, **59**, 369–385. and preceding papers

66 Josien, H., Ko, S. -B., Bom, D., Curran, D. P. (1998) *Chemistry—A European Journal*, **4**, 67–83.

67 Bom, D., Curran, P. D., Chavan, A. J., Kruszewski, S., Zimmer, S. G., Fraley, K. A., Bingcang, A. L., Latus, L. J., Pommier, Y., Burke, T. G. (2000) *Journal of Medicinal Chemistry*, **43**, 3970–3980.

68 Bom, D., Curran, P. D., Chavan, A. J., Kruszewski, S., Zimmer, S. G., Fraley, K. A., Burke, T. G. (1999) *Journal of Medicinal Chemistry*, **42**, 3018–3022.

69 Gabarda, A. E. and Curran, D. P. (2003) *Journal of Combinatorial Chemistry*, **5**, 617–624.

17
Combinatorial Synthesis of Alkaloid-like Compounds In Search of Chemical Probes of Protein–Protein Interactions

Michael Prakesch, Prabhat Arya, Marwen Naim, Traian Sulea, Enrico Purisima, Aleksey Yu. Denisov, Kalle Gehring, Trina L. Foster, Robert G. Korneluk

17.1
Introduction

There is a growing interest in the use of small molecules as chemical probes for understanding complex cellular networks [1]. There are several advantages in the use of small molecules. They include the ability of the small molecules (i) to interact with the biological targets in a reversible manner, (ii) to selectively modulate only one of the multiple interactions being made by the biological target, and (iii) to interfere with the cellular machinery in a nondestructive manner [2]. Another advantage is that, in addition to developing useful chemical probes for understanding cellular machinery, the study of these probes also has the potential, in certain cases, to develop into new therapeutic approaches in drug discovery with the probes functioning as starting candidates for drug design. The postgenomics chemical biology age has brought challenges to the biomedical research community trying to develop a better understanding of signaling networks based on protein–protein interactions [3,4]. It is becoming clear that signaling networks are central to both normal and dysfunctional cellular processes [5]. In most cases, these networks involve multiple dynamic, highly complex, protein–protein interactions. These confounding features have resulted in a lack of thorough understanding of normal and disease-related cellular signaling networks, which has severely limited our ability to develop therapeutic approaches that exploit these networks. Because of this, most therapeutic approaches developed to date do not involve signaling networks based on protein–protein interactions. At the same time, however, several possible solutions do exist for modulating the functions of signaling networks in a controlled and reversible manner. It is hoped that having a wide arsenal of *relevant* small-molecule chemical modulators of protein–protein interactions will result in more informative probing of such dynamic processes, in both normal and dysfunctional cellular processes. In the long run, it will allow biomedical researchers to explore the biological space of proteins, something which has not currently been considered highly attractive. Thus, tremendous opportunities do exist within the challenges of developing new therapeutic approaches, if one includes protein–protein interactions as biological targets [6,7].

Modern Alkaloids: Structure, Isolation, Synthesis and Biology. Edited by E. Fattorusso and O. Taglialatela-Scafati

ISBN: 978-3-527-31521-5

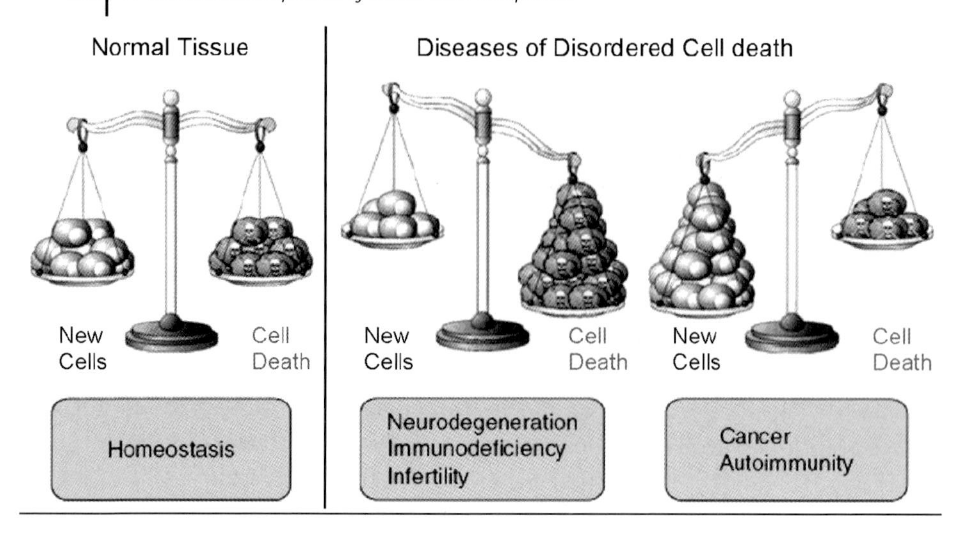

Fig. 17.1 Dysfunction in apoptosis leads to various diseases.

There is a growing desire to obtain a better understanding of the programmed cell death (apoptosis) machinery, as it is highly regulated, very well organized, and involves multiple protein–protein interactions and various other factors that contribute to its dysfunction in disease-related cells [8,9]. For example, as shown in Figure 17.1, it is now widely accepted that the apoptosis machinery slows down in a wide spectrum of cancer cells. On the other hand, immunodeficient or neurodegenerative cells have the tendency to exhibit enhanced apoptosis. The use of small molecules to study apoptosis is highly advantageous because, in addition to their use as chemical probes for enhancing our understanding of these processes, they also offer the possibility of correcting this machinery in disease-related disorders [10].

The initiation of apoptosis occurs by signals from two distinct but convergent pathways: (i) the extrinsic death receptor pathway, and (ii) the intrinsic mitochondrial apoptosome pathway. These two pathways consist of largely distinct molecular interactions, use different upstream initiator caspases, and are interconnected at numerous steps but ultimately converge at the level of downstream effector caspase activation. In the intrinsic pathway (also known as the mitochondrial pathway, see Figure 17.2), apoptosis involves two protein families around mitochondria. These two families are the proapoptotic proteins (Bak, Bad, Bid, Bax) and the antiapoptotic proteins (Bcl2, Bcl-XL, Mcl-1). The extrinsic pathway (also known as the Fas pathway) is initiated through the Death receptor (e.g. Fas, TNFR-1, or TRAIL receptors DR4 and DR5). The Inhibitors of Apoptosis Proteins (IAPs) are important proteins that bind to caspases and, through these protein–protein interactions, inhibit the caspase cascade – leading to the retardation or prevention of apoptosis [11].

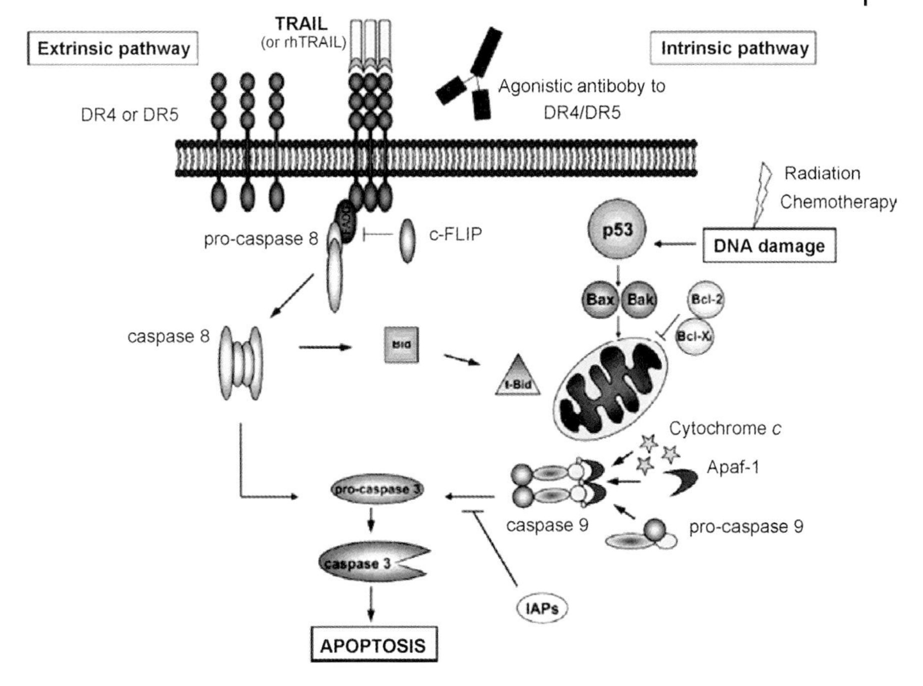

Fig. 17.2 Two major pathways leading to apoptosis.

17.2 Protein–Protein Interactions

Several factors have contributed to the limited success of small molecules as modulators of protein–protein interactions. In general, protein–protein interactions involve shallow surfaces and cover a relatively large surface area. Proteins also have areas known as "hot spots." In several cases, it has been observed that these "hot spots" contribute significantly to the overall binding, and the targeting of these "focused surface areas" has successfully led to the design of small-molecule binders. There are a few examples in the literature where structural information of a protein–protein complex has led to the design of small molecules that exploit the "hot spots." One such example of a rationally designed molecule involved the synthesis of small-molecule modulators of the antiapoptotic Bcl-2 protein family, which is known to interfere with Bcl-2 (and Bcl-XL)/BAK interactions [12,13]. As shown in Figure 17.3, information on the Bcl-XL protein structure (Figures 17.3a) and its partner, the BAK protein, led to the design of a peptide sequence that could bind to the extended hydrophobic grove of Bcl-XL (Figures 17.3b and c). The quest to obtain small-molecule mimics of this peptide sequence, which could promote apoptosis, spurred a deep interest in exploring various design strategies that include (i) the stapled approach [14] and (ii) the fragment-based approach [15]. In addition to these two approaches, which have been very successful in producing small-molecule probes

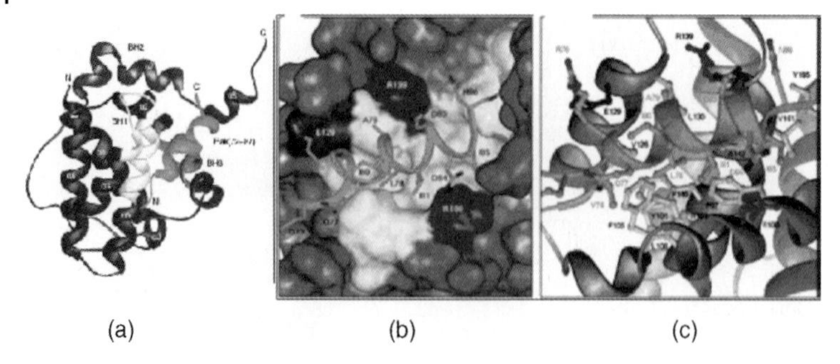

Fig. 17.3 (a) X-ray structure of Bcl-XL. (b) and (c) The interactions of the antiapoptotic protein, Bcl- XL with a peptide sequence from the proapoptotic BAK protein.

with enhanced apoptosis potential, an *in silico* approach [16] has also been widely used. However, despite having structural information on the protein surface(s), there are very few cases where this information has successfully led to the rational design of small-molecule modulators of protein–protein interactions. Thus, rapid access to small molecules by high-throughput organic synthesis still remains an attractive strategy for the identification of small-molecule modulators of protein–protein interactions. This leads to two important questions: (1) in the absence of any structural information, what types of small molecules are likely to be successful in disrupting signaling networks based on dynamic protein–protein interactions, and (2) does mother Nature already have an answer to this question?

17.3
Alkaloid Natural Products as Chemical Probes of Protein–Protein Interactions

Over the years, three-dimensional (3D), architecturally-complex natural products have been used as small-molecule probes for understanding protein function. Imbedded in these natural products are a number of highly diverse chiral functional groups, which are potential sites for protein binding. Although natural products from a variety of sources (e.g. plants, soil, sea, etc.) are very useful candidates for identifying lead compounds, the major disadvantage with natural products is the difficulty of the follow-up organic synthesis/medicinal chemistry efforts. In many cases, there is insufficient material for the various desired biological assays, thereby limiting the exploration of their full potential. One solution to this, developing relatively simple structural analogs with comparative biological responses to the natural product, is a challenging undertaking and it is at this point that a diversity-oriented synthesis (DOS) program [17–19] developed for the specific natural product class is extremely useful. Diversity-oriented synthesis is aimed at populating the unexplored natural product-based chemical space that is currently unoccupied by conventional combinatorial chemistry [20,21]. The combinatorial chemistry program in DOS utilizes stereo- and enantio-selective organic synthesis reactions, and is designed to produce small

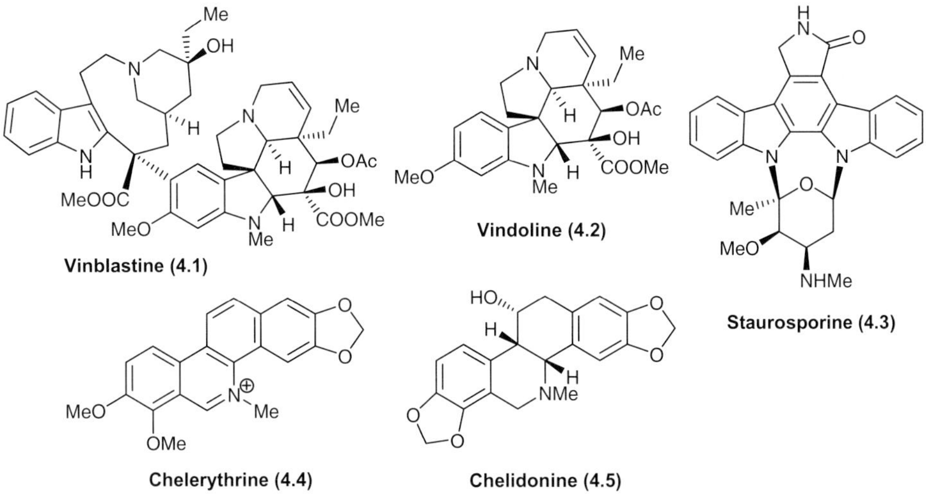

Vinblastine (4.1)

Vindoline (4.2)

Staurosporine (4.3)

Chelerythrine (4.4)

Chelidonine (4.5)

Fig. 17.4 A few examples of alkaloid natural products as modulators of protein–protein interactions.

molecules that are rich in (i) stereochemically defined polyfunctional groups and (ii) conformationally diverse natural product-like skeletons. Several alkaloid natural products are known to interfere with protein surfaces involved in protein–protein interactions. Some of the bioactive alkaloid natural products are shown in Figure 17.4. Vinca alkaloids are known as antimitotic agents and inhibit microtubule assembly, promoting tubulin self-association into spiral aggregates [22]. Vinca alkaloids (vinblastine (**1**) and vindoline (**2**), Figure 17.4) are important anticancer agents known to disrupt microtubule dynamics and inhibit assembly, resulting in the arrest of cell division at the metaphase. At sub-stoichiometric levels *in vitro*, vinca alkaloids are known to stabilize microtubules, possibly by binding to microtubule ends and inhibiting the hydrolysis of GTP.

Isolated in 1977 from the bacterium *Streptomyces staurosporeus*, staurosporine (**3**) is a natural product that inhibits most protein kinases at low nanomolar concentrations [23]. Through small-molecule/protein complex cocrystallization, it was shown that staurosporine binds tightly to the adenosine binding pocket of the catalytic subunit of the cAMP-dependent protein kinase. Chelerythrine (**4**) was identified as an inhibitor of Bcl-XL-Bak BH3 peptide binding with an IC50 of 1.5 μM, and also displaced Bax from Bcl-XL [24]. Chelidonine (**5**) is another example of an alkaloid natural product that inhibits the taxol-mediated polymerization of tubulin in the micromolar range (∼24.0 μM).

17.4
Indoline Alkaloid Natural Product-inspired Chemical Probes

Although the development of high-throughput methods for producing natural product-inspired chemical probes is at an early stage, significant progress has been made in recent years. By discussing a few selected examples from our group, we hope that this chapter will provide the readers with a flavor of what it takes to develop natural product-inspired small-molecule chemical modulators of protein–protein interactions involving Bcl-2/Bcl-XL and XIAP.

17.4.1
Indoline Alkaloid-inspired Chemical Probes

The indoline substructure is considered a privileged scaffold and is found in a wide variety of common alkaloid natural products. Based on this, we launched a synthesis program that was aimed at designing functionalized indoline derivatives that could further be used for building different natural product-inspired polycyclic architectures. Our first- and second-generation synthetic targets, **6** (racemic) and **7** (enantioenriched) are shown in Figure 17.5. The indoline scaffold 6 contains an amino alcohol functionality that could be further utilized in the design and syntheses of tricyclic structures. Through functioning as an anchoring site, the presence of the phenolic hydroxyl group allows development of solid-phase synthetic methods. Our second-generation indoline scaffold, **7** is densely functionalized and can be easily obtained in an enantio-enriched manner. In addition to having the functional groups that were present in the first-generation design, **7** also contains an amino group that is orthogonally protected from the indoline secondary amine. The presence of multiple functional groups on this scaffold make it highly attractive for developing modular approaches to obtaining three-dimensionally different tricyclic architectures.

Our first initiative, based on using an intermolecular Mitsunobu reaction to obtain tricyclic derivatives in solution and on solid phase, is shown in Figure 17.6 [25]. To test the scope of this method, **9** was obtained from **8** in a number of steps using solution chemistry. Compound **9** is an amino acid conjugate, with a free primary hydroxyl group in which the N-terminal amine is protected as the *o*-nosyl. The successful synthesis of this compound led to the production of the tricyclic compound, **10** which has two potential diversity sites for library generation. This synthetic method also worked well on solid phase using the bromo Wang resin, and a 100-membered library (**13**) was obtained from the solid-phase starting material, **11**.

With the goal of producing different indoline-based functionalized polycyclic architectures and using such architectures in library generation, we developed a practical, enantio-controlled synthesis of an aminoindoline scaffold, (**14**) (Figure 17.7) [26]. This scaffold has several attractive features that make it extremely

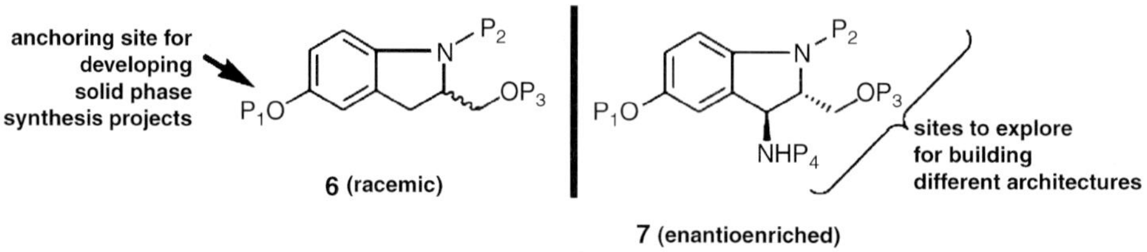

Fig. 17.5 First- and second-generation functionalized indoline scaffolds.

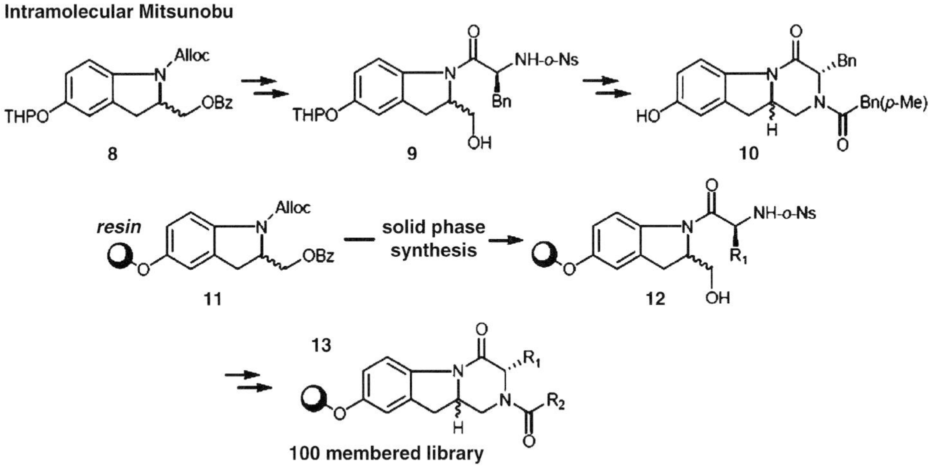

Intramolecular Mitsunobu

8 9 10

resin — solid phase synthesis → 11 12

13

100 membered library

Fig. 17.6 Solution and solid phase synthesis from the racemic first-generation indoline scaffold to obtain a 100-membered library using an intramolecular Mitsunobu approach.

versatile for producing different polycyclic 3D-architectures including the presence of four orthogonally protected functional groups, and the phenolic hydroxyl group, which could serve as an anchoring site during solid-phase synthesis. In one study, the aminoindoline scaffold **14** was converted into **15**, the starting material for developing the solid-phase synthesis project. The Broad Institute alkylsilyl linker-based

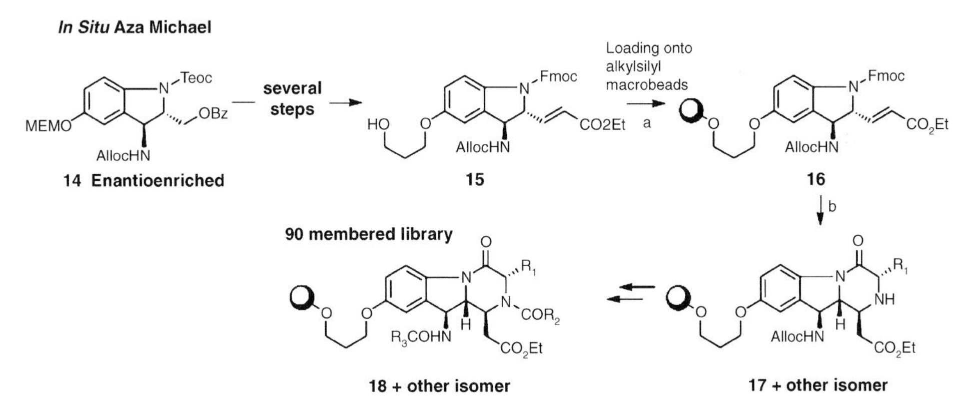

Fig. 17.7 Solid phase synthesis to obtain aminoindoline alkaloid-like tricyclic compounds by *in situ* aza Michael approach - (a) alkylsilyl linker-based polystyrene macrobeads (1.0 equiv), TfOH (6.0 equiv), 2,6-lutidine (10.0 equiv), **14** (0.5 equiv); (b) (i) 20% piperidine (ii) *N*-Fmoc amino acid chloride, collidine (iii) 20% piperidine.

polystyrene macrobeads were used to immobilize compound **15** on the solid support, giving **16**. In general, the loading using the Broad Institute protocol was observed to be in the range of 75–90 %, as determined after cleavage from the solid support. Following *N*-Fmoc removal in compound **16**, and amino acid coupling with the free indoline secondary amine, the key reaction in this approach was the formation of the third ring using an *in situ* stereocontrolled, aza-Michael reaction. Mirroring the original solution-phase results, this reaction worked nicely on solid support, producing compound **17** as the major diastereomer. The stereochemical outcome of this cyclization was dependent on the choice of the amino acid. Using the three diversity sites, as shown in compound **18**, a 90-membered library was obtained by employing an IRORI split-and-mix-type technology. The biological evaluation of this library is under investigation.

Another project was undertaken with the goal of obtaining two different tricyclic architectures from a common starting material, **20** (Figure 17.8) [27]. Compound **20** was obtained from **19** using orthogonally protected amines (i.e. *N*-Teoc and *N*H-Alloc) to develop a modular ring-closing metathesis (RCM). Two different unsaturated lactams with seven- (**21**) and eight-membered rings (**23**) were then obtained. An attractive feature of this approach is that by simply choosing one of the two amine moieties, it was possible to obtain a different, functionalized, indoline-based, tricyclic architecture. Following successful method development in solution, this synthesis was successfully undertaken on solid phase, where compound **24** (obtained from **20**) was anchored onto the alkylsilyl linker-based polystyrene macrobeads, giving product **25**. As observed in solution synthesis, subsequent reaction of **25** could lead to either of two different, tricyclic architectures. Further work is needed to employ this modular methodology for the generation of two libraries of indoline-derived compounds with different architectures.

17.4.2
Tetrahydroquinoline Alkaloid-inspired Chemical Probes

Tetrahydroquinoline and tetrahydroisoquinoline are two highly privileged substructures that are commonly found in a wide variety of alkaloid natural products. With the objective of exploring the chemical space around the tetrahydroquinoline substructure, we developed a highly practical, enantioselective synthesis of the functionalized tetrahydroaminoquinoline scaffold **28** (Figure 17.9) [28]. Several features of this scaffold make it versatile and amenable to the production of a wide variety of very different polycyclic architectures. The key features include the presence of: (i) the β-amino acid moiety, (ii) the δ-amino acid moiety, (iii) the γ-hydroxy carboxyl ester functionality, and (iv) the phenolic hydroxyl group, which could be used as an anchoring site during solid-phase synthesis. As an example, three different tricyclic structures (**31**, **32**, and **33**) were produced using a RCM strategy. Compound **31** is unique as it contains a bridged, 10-membered ring, unsaturated lactam moiety. The second compound, **32**, has a bridged 12-membered ring with the cis-olefin that was obtained from the RCM. Finally, compound **33** was obtained with trans-fused ring skeletons. The structures of all the lactams were determined by NMR experiments.

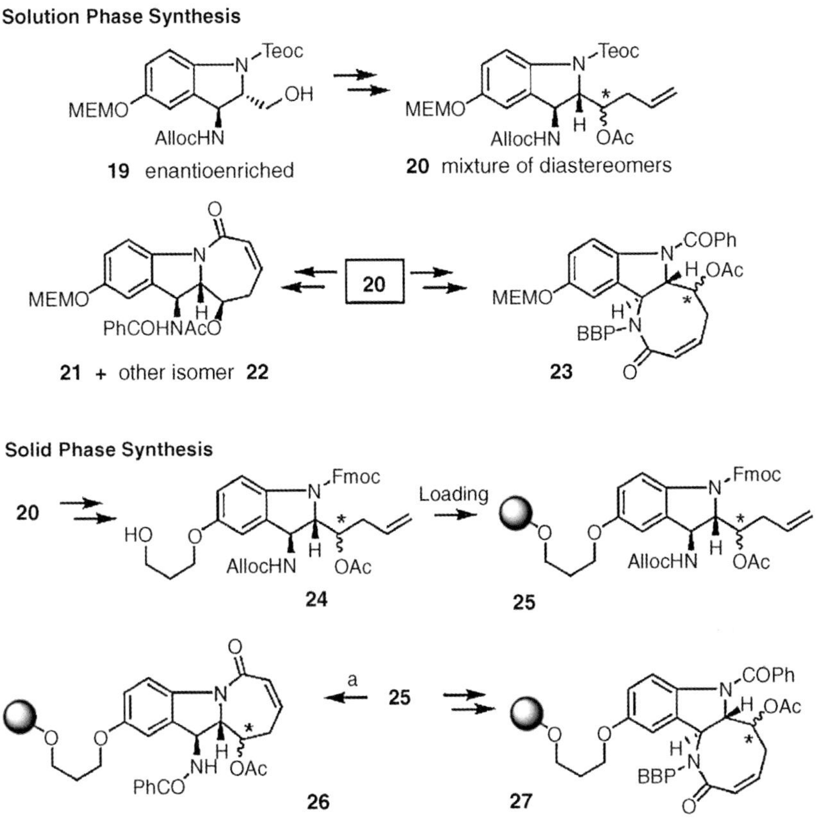

Solution Phase Synthesis

19 enantioenriched

20 mixture of diastereomers

21 + other isomer 22

23

Solid Phase Synthesis

24

25

26

27

Fig. 17.8 A modular solid phase approach to obtain aminoindoline-based two different tricyclic architectures having medium sized unsaturated lactams. (a): (i): $Pd(PPh_3)_4$, PPh_3, *N*-methyl morpholine, AcOH, CH_2Cl_2; (ii): benzoyl chloride, 2,6-collidine, CH_2Cl_2; (iii): 20% piperidine, DMF; (iv): acryloyl chloride, 2,6-collidine, CH_2Cl_2; (v) second-generation Grubbs'catalyst (40–50 mol%).

The development of the solid-phase synthesis of these compounds involved generating compound **29** in a number of steps from **28**. This was then successfully loaded onto the alkylsilyl linker-based polystyrene macrobeads with full regiocontrol, giving **30**. The ring-closing methods developed in solution were then applied to the solid-phase approach and three products, **34**, **35** and **36**, were successfully obtained in a modular manner. This approach to obtaining different functionalized macrocyclic-based architectures is highly attractive as it allows the divergent production of three libraries based on different polycyclic structures. Work toward this objective is progressing and the applications of these libraries will be reported as they become available.

Further use of the above scaffold to produce compound **37**, which has an allylic group at the C-2 position, is shown in Figure 17.10. Using this compound as the

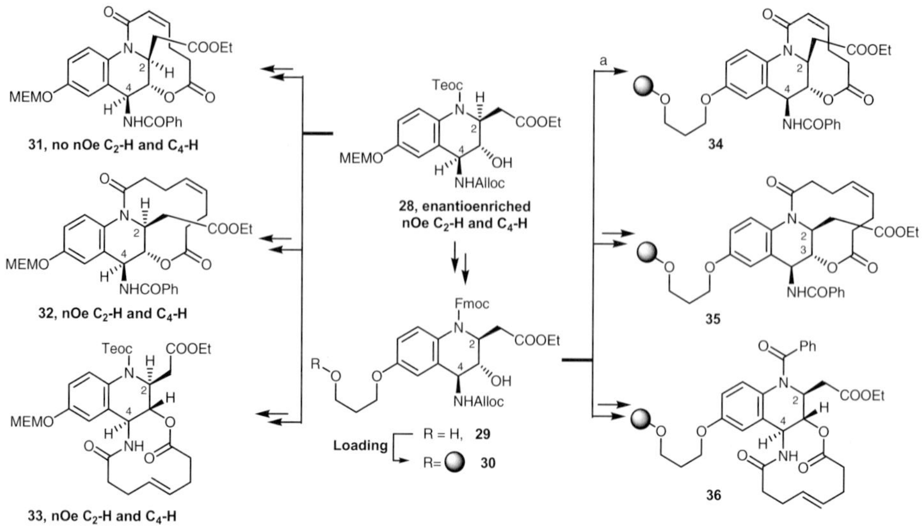

Fig. 17.9 Modular solid phase approaches to obtain tetrahydroquinoline-based different tricyclic compounds containing macrocyclic rings. (a): (i): 4-pentenoic acid, DMAP, DIC; (ii): 20% piperidine; (iii): *trans*-crotonoyl chloride, 2,4,6-collidine; (iv): $Pd(PPh_3)_4$, PPh_3, 4-methyl morpholine, CH_3CO_2H; (v): PhCOCl, 2,4,6-collidine; (vi): second-generation Grubbs'catalyst.

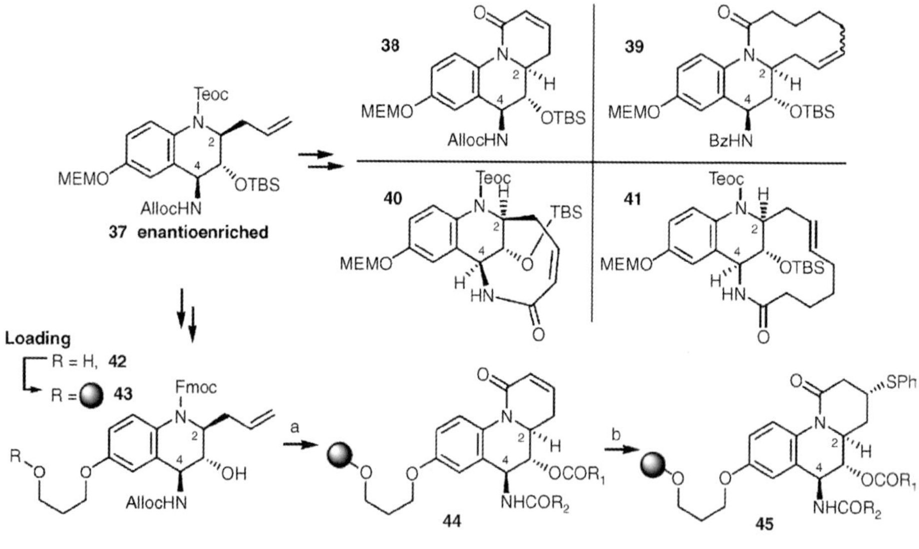

Fig. 17.10 Modular solution and solid phase approaches to obtain tetrahydroquinoline based tricyclic compounds containing different unsaturated lactam rings and macrocyclic rings. (a): (i): 20% piperidine, CH_2Cl_2; (ii): Et_3N, acryloyl chloride, −10 °C, CH_2Cl_2; (iii): second-generation Grubbs' catalyst (30 mol%), CH_2Cl_2; (iv): NaHMDS, acyl chloride, THF; (v): piperidine, $Pd(PPh_3)_4$ (10 mol%), CH_2Cl_2; (vi): Et_3N, acyl chloride, CH_2Cl_2; (b): Et_3N, benzenethiol, CH_2Cl_2.

starting material, we reported the synthesis of four different tricyclic architectures (**38–41**) that were obtained using RCM [29]. For the solid-phase synthesis, compound **42** was obtained from **37** and was successfully loaded onto the alkylsilyl linker-based polystyrene macrobeads, providing **43**. In one study, compound **43** was successfully transformed using RCM into the six-membered-ring unsaturated lactam **44**. Using this method for library generation remains to be undertaken.

By introducing an unsaturated eight-membered lactam, it was possible to explore the chemical space around the tetrahydroquinoline scaffold. To this end, compound **47** (Figure 17.11) was obtained from **46** [29]. As in the previous study, the successful implementation of the RCM reaction led to the synthesis of compound **48**. This approach was also tried on solid phase, using compound **49**, obtained from **46**, as the starting material. Loading **49** onto the alkylsilyl linker-based polystyrene macrobeads provided compound **50**, which was then successfully subjected to RCM conditions, yielding the final product **51**. This method, developed on solid phase, opens an attractive opportunity for producing a library based on a tricyclic tetrahydroquinoline architecture with an unsaturated eight-membered lactam ring.

As a final illustration, an unprecedented approach was discovered, wherein an *in situ* aza-Michael reaction was used that allowed the development of bridged tetrahydroquinoline-based tricyclic architectures under very mild reaction conditions [30]. In a typical example, enantio-enriched compound **52** (Figure 17.12) was converted to **53** in a series of steps. Following the acetonide and the *N*-Alloc removal, to our surprise there was no sign of the free amine derivative **54**. Instead, compound **55** was obtained as single diastereomer. Under these mild reaction conditions, the *in situ* aza-Michael cyclization produced the bridged architectures. The conditions for this transformation were highly attractive for developing this method on solid phase. Indeed, this reaction was found to work equally well in

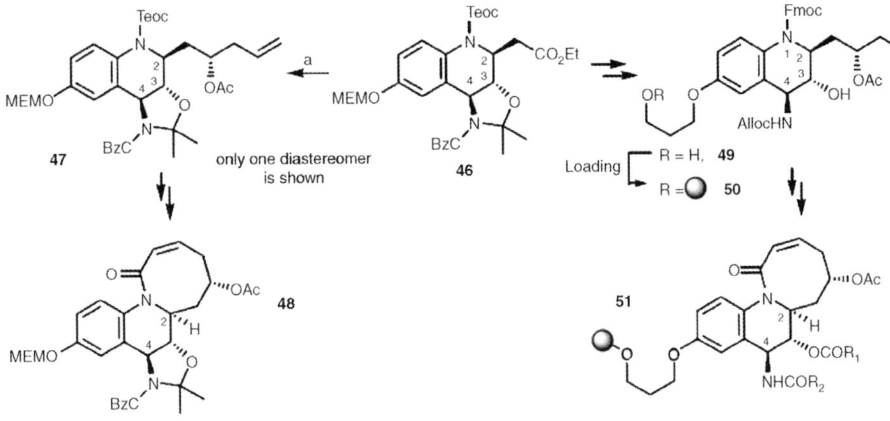

Fig. 17.11 Solution and solid phase approaches to obtain tetrahydroquinoline-based tricyclic compounds having medium sized unsaturated lactams. (a): (i): $LiBH_4$, THF, RT; (ii): Dess–Martin priodinane, RT: (iii): $ZnCl_2$, AllylMgBr, $-78\,^{\circ}C$; (iv): Ac_2O, DMAP, CH_2Cl_2, $0\,^{\circ}C$ to RT.

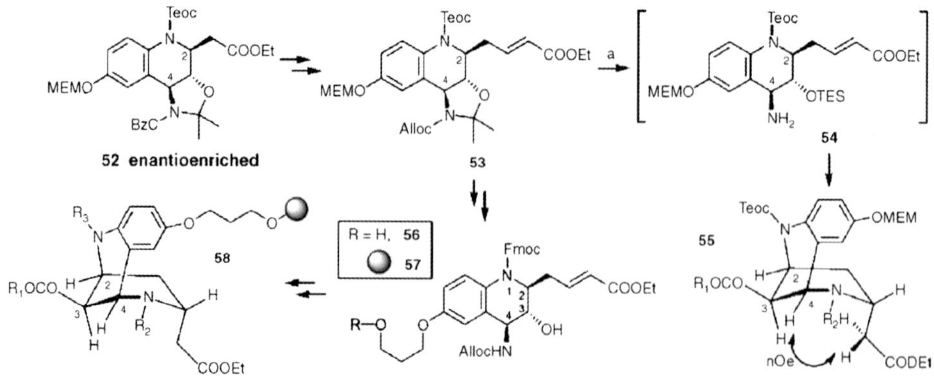

Fig. 17.12 *In situ* bridged aza Michael approach in solution and on solid phase to obtain tetrahydroquinoline-based tricyclic compounds. (a): (i): acetic acid/THF/H_2O, RT; (ii): TESOTf, pyridine, CH_2Cl_2, $-40\,^{\circ}C$; (iii): $Pd(PPh_3)_4$, morpholine, CH_2Cl_2, RT; (c): Et_3N, benzoyl chloride or cinnamoyl chloride, CH_2Cl_2, $0\,^{\circ}C$ to RT.

solution and on solid phase. Compound **56** was therefore loaded onto the alkylsilyl linker-based polystyrene macrobeads and produced the expected bridged tricyclic product **58** with complete diastereocontrol. Further work is ongoing in generating a library using this method.

17.5
Alkaloid Natural Product-inspired Small-molecule Binders to Bcl-2 and Bcl-XL and *In Silico* Studies

Figure 17.13 shows two libraries that were generated by our group. These two libraries (i.e. 105 and 200 compounds) use the tetrahydroaminoquinoline and aminoindoline scaffolds (Figure 17.13, **59** and **62**) and were designed to test these small molecules in several assays based on protein–protein interactions. With the objective of mapping the extended shallow hydrophobic domains of protein surfaces, a wide variety of aromatic hydrophobic groups were chosen in the library planning. These libraries are being tested in a variety of biological screening studies, including the search for small-molecule probes of Bcl-2/Bcl-XL-based protein–protein interactions. These two libraries were also tested, *in silico*, for binding to the hydrophobic domains of the Bcl-2 and Bcl-XL sites that are known to interact with the BAK protein. It was interesting to note that several members from these two libraries showed the potential for interacting with the Bcl-2 and Bcl-XL protein surface (e.g. three library members are shown in Figure 17.13 as binders to Bcl-2 and Bcl-XL). Further experimental studies are ongoing to confirm the scope of these library members as potential modulators of Bcl-2/Bcl-XL-BAK interactions and validate these *in silico* findings.

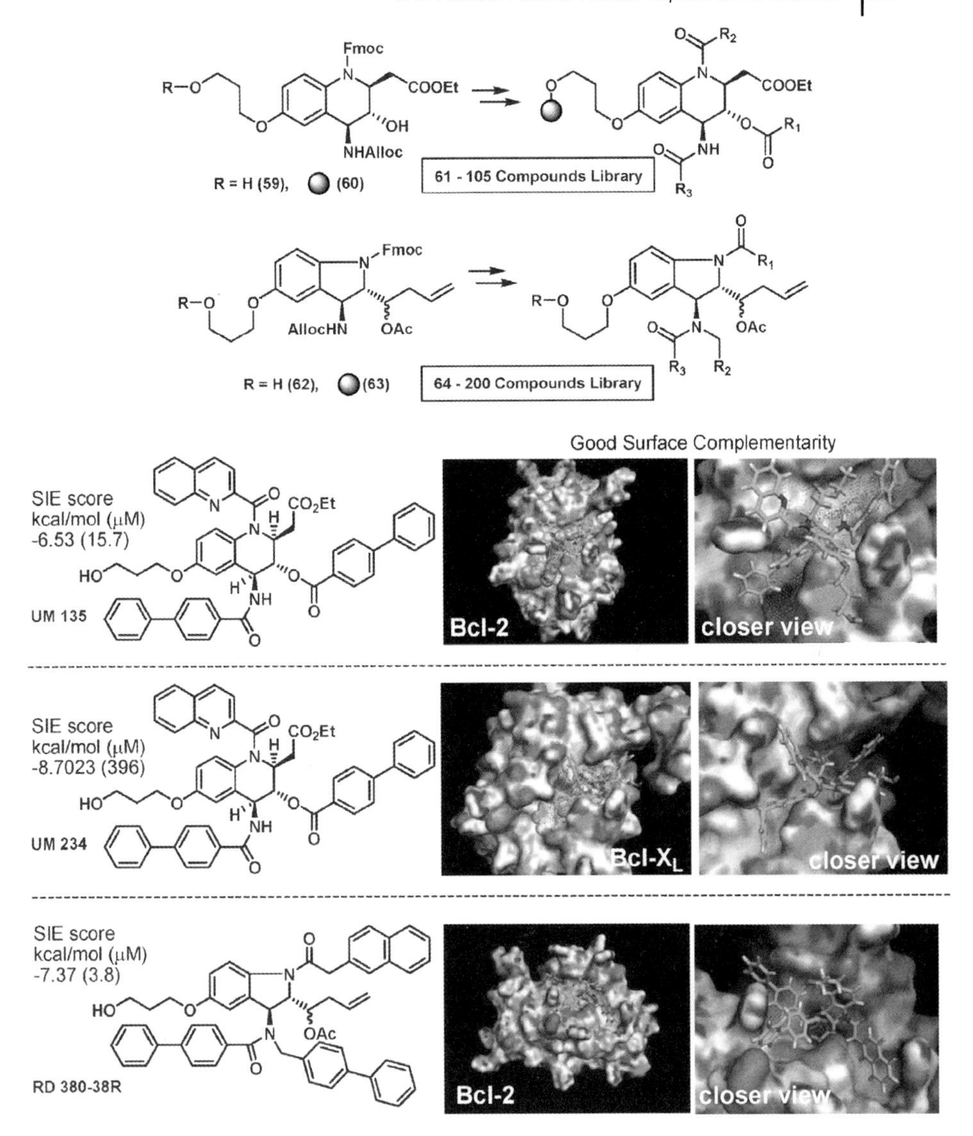

Fig. 17.13 The use of tetrahydroaminoquinoline and aminoindoline scaffolds in the library generation of natural product-like compounds and *in silico* testing of these two libraries as small molecule binders to Bcl-2/Bcl-XL.

17.5.1
Alkaloid Natural Product-inspired Small-molecule Binders to Bcl-XL and NMR Studies

In collaboration with the Gehring laboratory, the tetrahydroaminoquinoline and aminoindoline-derived natural product-like libraries, and several related intermediates synthesized by our group, were used in an NMR study to examine their binding with Bcl-XL. Using the fully ^{15}N labeled Bcl-XL protein, Gehring and coworkers are studying several known lead compounds from the literature with the goal of obtaining a better understanding of the interactions [31,32]. To date, two compounds with the tetrahydroaminoquinoline scaffold (**65** and **66**, Figure 17.14) have been identified as small-molecule binders to Bcl-XL. Combining this observation with molecular modeling studies, it has been possible to determine the nature of the interactions of **66** with Bcl-XL. Although it is a weak binder to Bcl-XL ($K_d = 0.2$ mM), **66** contains great functional group diversity and offers an excellent opportunity for designing second-generation compounds, by either focused medicinal or combinatorial chemistry [33]. Work is ongoing in both directions to

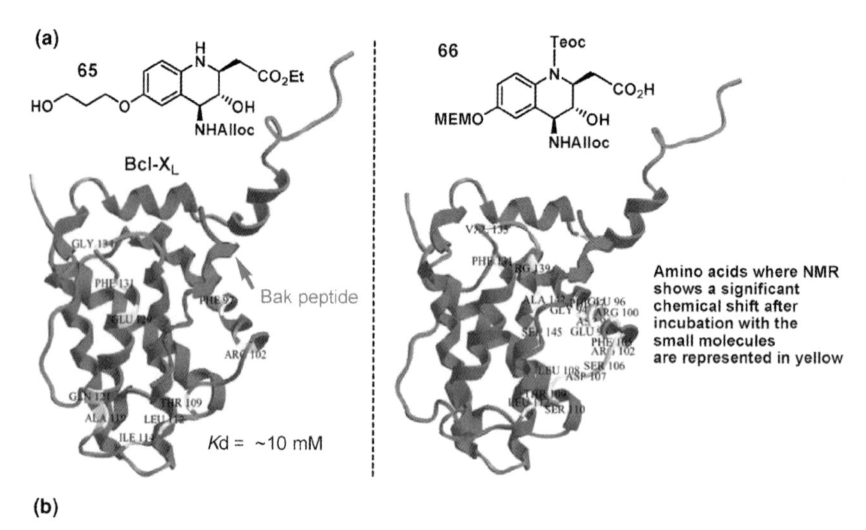

(b)

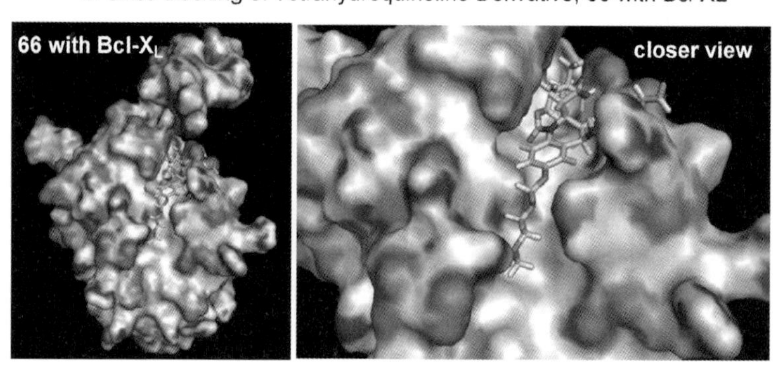

Fig. 17.14 (a) The binding of tetrahydroaminoquinoline scaffold-based small molecules (**65** and **66**) with Bcl-XL using fully ^{15}N-labeled NMR experiments. (b) The interactions tetrahydroquinoline derivative, **66** with Bcl-XL (NMR and in silico)

achieve this objective, and may lead to a new family of natural product-inspired small-molecule binders to Bcl-XL.

17.5.2 Alkaloid Natural Product-inspired Small-molecule Probes for XIAP

In recent years, the Inhibitors of Apoptosis Proteins (IAPs) field has emerged as an exciting area of research, from both from the academic and the commercial perspective [8,11]. The IAPs have been shown to interact with, and inhibit the activity of, a specific subset of caspases – a family of cysteine proteases that are the principle executioners of the apoptotic process in the cell. The IAPs play a pivotal role in the regulation of the apoptotic cascade, as they are the only known endogenous proteins that function as direct physiological repressors of both initiator and effector of caspases. XIAP is the most potent IAP with respect to its antiapoptotic functions; therefore, it is not surprising that XIAP is the best studied and the best characterized IAP. XIAP possesses three characteristic NH_2-terminal 70- to 80-amino-acid BIR domains and a COOH-terminal RING zinc finger domain. The BIR1 of XIAP also contains an Akt phosphorylation site at residue Ser87 that is involved in protein stabilization. The anticaspase activities of XIAP can be ascribed to the BIR domains and their linker regions: BIR3 is an inhibitor of the initiator caspase-9 and the linker preceding BIR2 functions as an inhibitor of the effector caspases-3 and -7 [34–36].

With the goals of finding small-molecule promoters of apoptosis signaling by (i) enhancing the release of caspase-3 and (ii) disrupting XIAP/caspase-9 interactions, a collaborative study with Korneluk and MacKenzie Laboratories at the Apoptosis Research Centre, Children's Hospital of Eastern Ontario (CHEO) in Ottawa yielded natural product-inspired compounds that were tested in the following three assays [37].

17.5.2.1 **Cell Death Assay** (Figure 17.15)

MDA-U6E1 and MDA-XG4 cell lines were used to examine the ability of various compounds to induce cell death in both a TRAIL-mediated and a non-TRAIL-mediated fashion. MDA-XG4 is a known XIAP-suppressed cell line, which allows for an initial validation (against the XIAP expressing cell line MDA-U6E1) of XIAP targeting by the complexes. XIAP-suppressed cell lines exhibit nearly 100 % cell death upon introduction of 50 ng/mL TRAIL. In contrast, the XIAP-expressing cell line shows nearly 100 % survival at this concentration of TRAIL. The sensitization to killing upon exposure to the compound was looked at and contrasted in each cell line. Similarly, introduction of a XIAP binding compound to the MDA-U6E1 cell line should result in cell death in a non-TRAIL-mediated environment. This assay provided a secondary means of looking at cell death induction by the compounds. After incubation with the compounds, with or without TRAIL, cells were harvested and then analyzed for viability by flow cytometry using Annexin V/PI staining or by a CellTiter Blue Fluorometric Viability Assay (Novagen). In all experiments, here and below, fluorescence readings were taken using a BMG

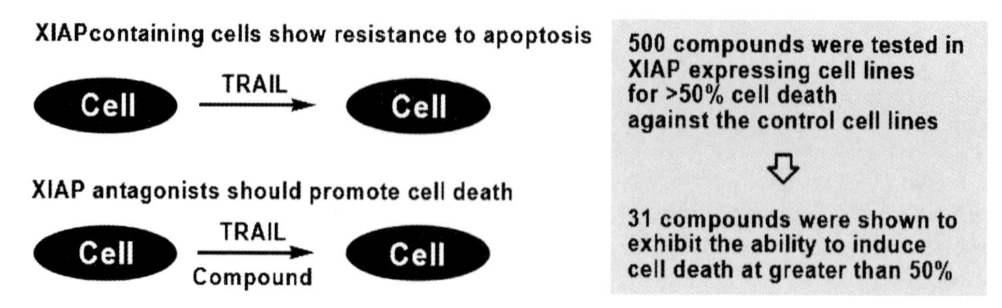

Fig. 17.15 The cell viability assay to explore the function of small molecules as promoters of apoptosis.

Laboratories, PolarStar Fluorometer. Using this assay, out of 500 compounds tested, 31 compounds were shown to exhibit the ability to introduce cell death greater than 50%.

17.5.2.2 **Caspase-3 Activation Assay** (Figure 17.16)
Thirty-one compounds showing activity in the above assay were then replicated using MDA-U6E1 and MDA-XG4 cells. A secondary analysis for caspase-3 activity was then carried out using a CaspaseGlo Fluorometric Assay (Promega). This confirms the presence of a caspase-3-mediated death pathway, which is a traditional marker for apoptotic (as opposed to necrotic) cell death.

17.5.2.3 **Caspase-9 Release Assay** (Figure 17.17)
In a final study, XIAP-BIR3 was expressed as a GST fusion protein using standard protocols established in the Korneluk laboratory. XIAP and caspase-9 were then incubated in the presence of CaspaseGlo 9 Reagent (Promega), with and without the lead targets identified in the above experiments from the caspase-3 activation assay. To our delight, of the three compounds identified as disruptors of XIAP-BIR 3/caspase-9 protein–protein interactions, RD-6 showed significant activity at low concentration (123 μM). RD-6 is an aminoindoline-derived natural product-inspired compound from the Arya laboratory. These finding are highly novel and further work is ongoing to obtain a better understanding of how this compound functions to disrupt a tightly bound XIAP-BIR 3/caspase-9 complex. These studies will be reported as they become available.

17.5.3 Summary and Future Outlook

There is growing interest in the biomedical research community in using small molecules to dissect signaling networks based on protein–protein interactions. The

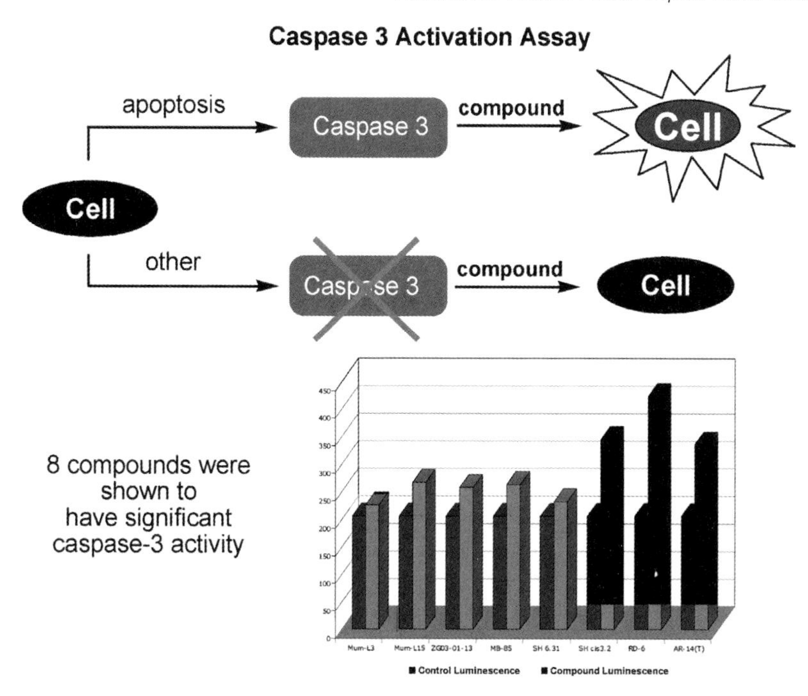

Fig. 17.16 *In vitro* cellular assay to explore the specificity of small molecules that interfere with caspase 3-related apoptosis signaling.

chemical probes that fulfill these criteria are not easy to find, and are in great demand. In addition to using these chemical probes for obtaining a better understanding of such interactions, in both normal and disease-related processes, these chemical entities could also provide useful starting points in probe-discovery research. Inspired by bioactive natural products that have a proven track record in this arena, the need to develop methods for generating natural product-like compounds to chart the natural product chemical space has also grown. The examples discussed in this concept chapter are indicative of a growing research community that is committed to the young field of "exploring the natural product chemical territory." In many cases, high-throughput synthesis methods have been successful in generating complex natural product-like architectures and, in a few cases, their applications are emerging in probe-discovery research. Although the examples covered are taken from our group only and, in many cases, the biological evaluation of these chemical probes has not been fully realized, this is an area that should be watched carefully in the coming years. Developing newer methods with the potential of generating natural product-like compounds in a high-throughput manner is the first major step toward reaching this dream: small molecule dissectors for all the signaling networks!

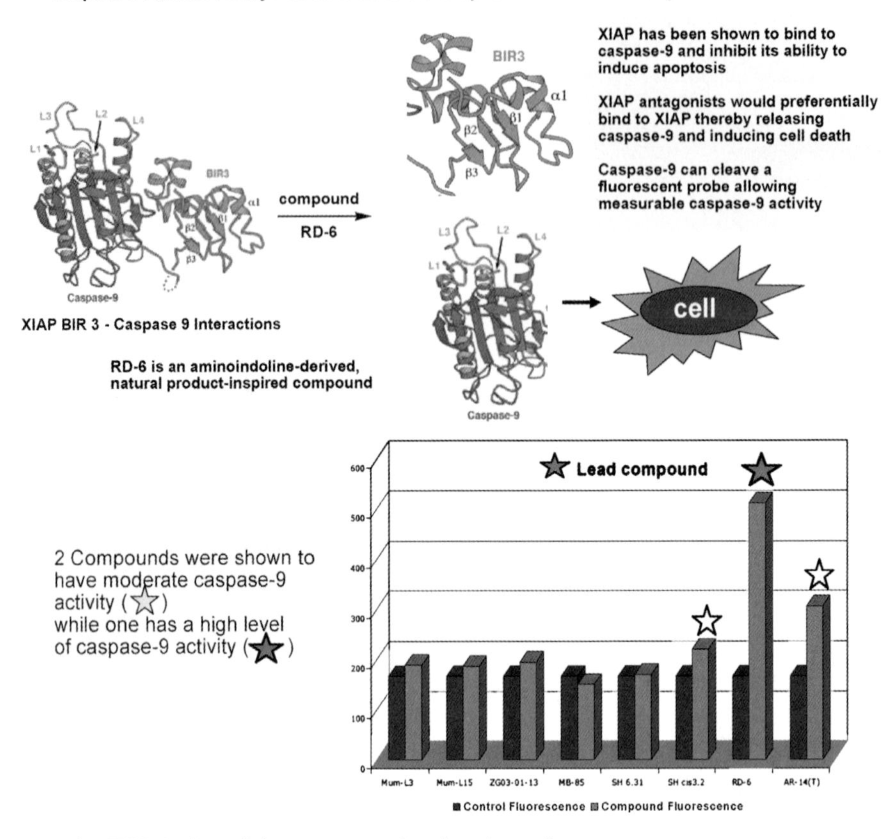

Fig. 17.17 *In vitro* cellular assay to monitor the release of caspase 9 from the XIAP BIR 3—caspase 9 protein–protein interactions.

17.6 Acknowledgments

We sincerely thank the Steacie Institute for Molecular Sciences (SIMS), the National Research Council of Canada's Genomics and Health Initiative (NRCC-GHI), the Canadian Institutes of Health Research (CIHR), and the National Cancer Institute of Canada (NCIC) for their continued support of this research. Our thanks to several former and current group members and collaborators for their numerous highly valuable contributions made to the work discussed in this chapter.

References

1 Schreiber, S. L. (2005) Small molecules: the missing link in the central dogma. *Nature Chemical Biology*, **1**, 64–66.

2 Schreiber, S. L. (2003) The small-molecule approach to biology. *Chemical and Engineering News*, 51–61.

3 Fishman, M. C. and Porter, J. A. (2005) Pharmaceuticals: a new grammar for drug discovery. *Nature*, **437**, 491–493.

4 Tate, E. W. (2006) Chemical intervention in signalling networks: recent advances and applications. *Signal Transduction*, **6**, 144–159.

5 Pawson, T. and Scott, J. D. (2005) Protein phosphorylation in signaling—50 years and counting. *Trends in Biochemical Sciences*, **30**, 286–290.

6 Arkin, M. R. and Wells, J. A. (2004) Small-molecule inhibitors of protein–protein interactions: progressing towards the dream. *Nature Reviews Drug Discovery*, **3**, 301–317.

7 Arkin, M. (2005) Protein-protein interactions and cancer: small molecules going in for the kill. *Current Opinion in Chemical Biology*, **9**, 317–324.

8 Fesik, S. W. (2005) Promoting apoptosis as a strategy for cancer drug discovery. *Nature Reviews Cancer*, **5**, 876–885.

9 Cheung, H. H., LaCasse, E. C., Korneluk, R. G. (2006) X-linked inhibitor of apoptosis antagonism: strategies in cancer treatment. *Clinical Cancer Research*, **12**, 3238–3242.

10 Reed, J. C. (2002) Apoptosis-based therapies. *Nature Reviews Drug Discovery*, **1**, 111–121.

11 Liston, P., Fong, W. G., Korneluk, R. G. (2003) The inhibitors of apoptosis: there is more to life than Bcl2. *Oncogene*, **22**, 8568–8580.

12 Walensky, L. D. (2006) BCL-2 in the crosshairs: tipping the balance of life and death. *Cell Death and Differentiation*, **13**, 1339–1350.

13 Oltersdorf, T., Elmore, S. W., Shoemaker, A. R., Armstrong, R. C., Augeri, D. J., Belli, B. A., Bruncko, M., Deckwerth, T. L., Dinges, J., Hajduk, P. J., Joseph, M. K., Kitada, S., Korsmeyer, S. J., Kunzer, A. R., Letai, A., Li, C., Mitten, M. J., Nettesheim, D. G., Ng, S., Nimmer, P. M., O'Connor, J. M., Oleksijew, A., Petros, A. M., Reed, J. C., Shen, W., Tahir, S. K., Thompson, C. B., Tomaselli, K. J., Wang, B. L., Wendt, M. D., Zhang, H. C., Fesik, S. W., Rosenberg, S. H. (2005) An inhibitor of Bcl-2 family proteins induces regression of solid tumours. *Nature*, **435**, 677–681.

14 Walensky, L. D., Kung, A. L., Escher, I., Malia, T. J., Barbuto, S., Wright, R. D., Wagner, G., Verdine, G. L., Korsmeyer, S. J. (2004) Activation of apoptosis in vivo by a hydrocarbonstapled BH3 helix. *Science*, **305**, 1466–1470.

15 Petros, A. M., Dinges, J., Augeri, D. J., Baumeister, S. A., Betebenner, D. A., Bures, M. G., Elmore, S. W., Hajduk, P. J., Joseph, M. K., Landis, S. K., Nettesheim, D. G., Rosenberg, S. H., Shen, W., Thomas, S., Wang, X., Zanze, I., Zhang, H., Fesik, S. W. (2006) Discovery of a potent inhibitor of the antiapoptotic protein Bcl-XL from NMR and parallel synthesis. *Journal of Medicinal Chemistry*, **49**, 656–663.

16 Leone, M., Zhai, D., Sareth, S., Kitada, S., Reed, J. C., Pellecchia, M. (2003) Cancer prevention by tea polyphenols is linked to their direct inhibition of antiapoptotic Bcl-2 family proteins. *Cancer Research*, **63**, 8118–8121.

17 Schreiber, S. L., Nicolaou, K. C., Davies, K. (2002) Diversity-oriented organic synthesis and proteomics: new frontiers for chemistry & biology. *Chemistry and Biology*, **9**, 1–2.

18 Tan, D. S. (2005) Diversity-oriented synthesis: exploring the intersections between chemistry and biology. *Nature Chemical Biology*, **1**, 74–84.

19 Arya, P., Joseph, R., Gan, Z. H., Rakic, B. (2005) Exploring new chemical space by stereocontrolled diversity-oriented synthesis. *Chemistry and Biology*, **12**, 163–180.

20 Reayi, A. and Arya, P. (2005) Natural product-like chemical space: search for chemical dissectors of macromolecular interactions. *Current Opinion in Chemical Biology*, **9**, 240–247.

21 Haggarty, S. J. (2005) The principle of complementarity: chemical versus biological space. *Current Opinion in Chemical Biology*, **9**, 296–303.

22 Duflos, A., Kruczynski, A., Barret, J. (2002) Novel aspects of natural and

modified vinca alkaloids. *Current Medicinal Chemistry: Anti-Cancer Agents*, **2**, 55–70.

23 Schimmer, A. D., Thomas, M. P., Hurren, R., Gronda, M., Pellecchia, M., Pond, G. R., Konopleva, M., Gurfinkel, D., Mawji, I. A., Brown, E., Reed, J. C. (2006) Identification of small molecules that sensitize resistant tumor cells to tumor necrosis factor-family death receptors. *Cancer Research*, **66**, 2367–2375.

24 Chan, S. L., Lee, M. C., Tan, K. O., Yang, L. K., Lee, A. S., Flotow, H., Fu, N. Y., Butler, M. S., Soejarto, D. D., Buss, A. D., Yu, V. C. (2003) Identification of chelerythrine as an inhibitor of BclXL function. *Journal of Biological Chemistry*, **278**, 20453–20456.

25 Arya, P., Wei, C.-Q., Barnes, M. L., Daroszewska, M. (2004) A solid phase library synthesis of hydroxyindoline-derived tricyclic derivatives by Mitsunobu approach. *Journal of Combinatorial Chemistry*, **6**, 65–72.

26 Gan, Z., Reddy, P. T., Quevillon, S., Couve-Bonnaire, S., Arya, P. (2005) Stereocontrolled solid-phase synthesis of a 90-member library of indoline-alkaloid-like polycyclics from an enantioriched aminoindoline scaffold. *Angewandte Chemie, International Edition*, **44**, 1366–1368.

27 Reddy, P. T., Quevillon, S., Gan, Z., Forbes, N., Leek, D. M., Arya, P. (2006) Solution- and solid-phase, modular approaches for obtaining different natural product-like polycyclic architectures from an aminoindoline scaffold for combinatorial chemistry. *Journal of Combinatorial Chemistry*, **8**, 856–871.

28 Prakesch, M., Sharma, U., Sharma, M., Khadem, S., Leek, D. M., Arya, P. (2006) Part 1. Modular approach to obtaining diverse tetrahydroquinoline-derived polycyclic skeletons for use in high-throughput generation of natural-product-like chemical probes. *Journal of Combinatorial Chemistry*, **8**, 715–734.

29 Sharma, U., Srivastava, S., Prakesch, M., Sharma, M., Leek, D. M., Arya, P. (2006) Part 2: Building diverse natural-product-like architectures from a tetrahydroaminoquinoline scaffold. modular solution- and solid-phase approaches for use in high-throughput generation of chemical probes. *Journal of Combinatorial Chemistry*, **8**, 735–761.

30 Prakesch, M., Srivastava, S., Leek, D. M., Arya, P. (2006) Part 3. A novel stereocontrolled, in situ, solution- and solid-phase, aza michael approach for high-throughput generation of tetrahydroaminoquinoline-derived natural-product-like architectures. *Journal of Combinatorial Chemistry*, **8**, 762–773.

31 Denisov, A. Y., Madiraju, M. S., Chen, G., Khadir, A., Beauparlant, P., Attardo, G., Shore, G. C., Gehring, K. (2003) Solution structure of human BCL-w: modulation of ligand binding by the C-terminal helix. *Journal of Biological Chemistry*, **278**, 21124–21128.

32 Denisov, A. Y., Chen, G., Sprules, T., Moldoveanu, T., Beauparlant, P., Gehring, K. (2006) Structural model of the BCL-w-BID peptide complex and its interactions with phospholipids micelles. *Biochemistry*, **45**, 2250–2256.

33 Prakesch, M. Denisov, A. Yu., Naim, M., Gehring, K., Arya, P. (2007) The Discovery of Natural Product–Inspired Chemical Probes of Bcl-XL and Mcl-l. Submitted.

34 Holcik, M. and Korneluk, R. G. (2001) XIAP, the guardian angel. *Nature Reviews Molecular Cell Biology*, **2**, 550–556.

35 Riedl, S. J. and Shi, Y. (2004) Molecular mechanisms of caspase regulation during apoptosis. *Nature Reviews Molecular Cell Biology*, **5**, 897–907.

36 Yan, N. and Shi, Y. (2005) Mechanisms of apoptosis through structural biology. *Annual Review of Cell and Developmental Biology*, **21**, 35–56.

37 Foster, T. L., Korneluk, R. G., MacKenzie, A. E., Prakesch, M., Reddy, P. T., Reddy, R. R., Arya, P. (2006) Unpublished results [This work was presented as a poster at the Quebec-Ontario Minisymposium, University of Western Ontario, London, Ontario, November 3–6, 2006.]

18
Daphniphyllum alkaloids: Structures, Biogenesis, and Activities

Hiroshi Morita, Jun'ichi Kobayashi

18.1
Introduction

Daphniphyllum alkaloids are a structurally diverse group of natural products, which are elaborated by the oriental tree "Yuzuriha" (*Daphniphyllum macropodum*; Daphniphyllaceae), dioecious evergreen trees, and shrubs native to central and southern Japan. *Daphniphyllum* comes from the Greek and refers to "Daphne" and "leaf." "Yuzurisha" means that the old leaf is replaced by a new leaf in the succeeding season. That is, to take over or take turns, with the old leaf dropping after the new leaf emerges, thus there is no interruption of the foliage. "Yuzurisha" in Japan is used as an ornamental plant for the New Year to celebrate the good relationships of the old and new generations. There are three *Daphniphyllum* species in Japan, *D. macropodum*, *D. teijsmanni*, and *D. humile*. Several other species, such as *D. calycinum*, *D. gracile*, *D. longeracemosum*, *D. yunnanense*, *D. longistylum*, *D. paxianum*, *D. oldhami*, and *D. glaucescens* are widely distributed in New Guinea, China, and Taiwan.

Since Hirata *et al.* began research into daphniphyllum alkaloids in 1966, a number of new alkaloids have been discovered. As a result, the number of known daphniphyllum alkaloids has grown markedly in recent years to a present count of 118 (compounds **1–118**). These alkaloids, isolated chiefly by Yamamura and Hirata *et al.* are classified into six different types of backbone skeletons [1–3]. These unusual ring systems have attracted great interest as challenging targets for total synthesis or biosynthetic studies. This chapter covers the reports on daphniphyllum alkaloids that have been published between 1966 and 2006. Since the structures and stereochemistry of these alkaloids are quite complex and the representation of the structure formula has not been unified, all the natural daphniphyllum alkaloids (**1–118**) are listed. Classification of the alkaloids basically follows that of the previous reviews [1,2], but sections on the newly found skeletons have been added.

It was of substantial interest when Heathcock and coworkers proposed a biogenetic pathway for daphniphyllum alkaloids [4,5] and demonstrated a biomimetic total synthesis of several of them. This review describes the recent studies on alkaloids isolated from the genus *Daphniphyllum*, the proposed biogenetic pathway, syntheses of daphniphyllum alkaloids based on these biogenetic proposals, and their activities.

Modern Alkaloids: Structure, Isolation, Synthesis and Biology. Edited by E. Fattorusso and O. Taglialatela-Scafati

ISBN: 978-3-527-31521-5

Section 18.2 surveys all the daphniphyllum alkaloids isolated so far, including our recent work, while Sections 18.3–18.5 mainly deal with biogenetic pathways, total syntheses, the biomimetic synthesis, and the activities of these compounds.

18.2
Structures of Daphniphyllum alkaloids

18.2.1
Daphnane-type Alkaloids

In 1909, Yagi isolated daphnimacrine, the parent compound of the C_{30}-type Daphniphyllum alklaloids [6]. The structure elucidation of daphniphylline (**1**), one of the major C_{30}-type daphniphyllum alkaloids, was carried out by X-ray crystallographic analysis of the hydrobromide [7–10]. Yamamura and Hirata also reported the structures of codaphniphylline (**2**) [10–12] and daphniphyllidine (**3**) [13], both of which are closely related to daphniphylline. Furthermore, they isolated methyl homodaphniphyllate (**7**), and the structure lacking the C_8 unit of daphniphylline was elucidated by chemical correlations from daphniphylline [19,20,37]. On the other hand, Nakano isolated daphnimacropine (**4**) [14], daphmacrine (**5**) [15–17], and daphmacropodine (**6**) [15,18], the structures of which were elucidated by X-ray crystallographic analysis of their methiodides. The structures of these alkaloids are listed in Figure 18.1.

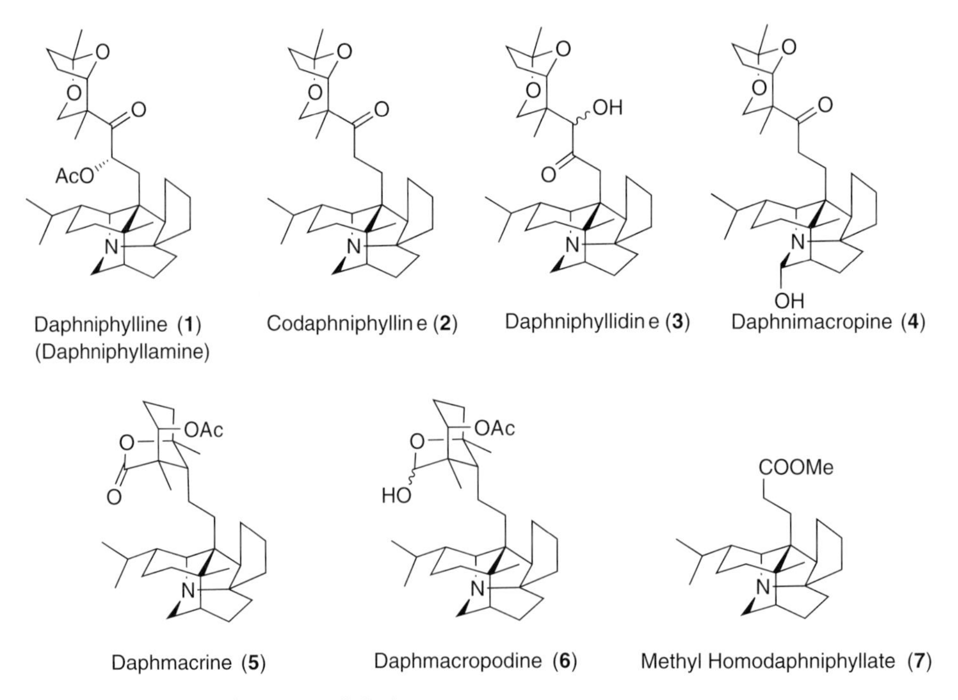

Fig. 18.1 Daphnane-type alkaloids.

18.2.2
Secodaphnane-type Alkaloids

Two secodaphnane-type alkaloids (Figure 18.2), secodaphniphylline (**8**) [19,21,22] and methyl homosecodaphniphyllate (**11**) [19,21,22], were isolated by Yamamura and Hirata, and their structures were elucidated by X-ray analysis of methyl *N*-bromoacetyl homosecodaphniphyllate and chemical correlations between **8** and **11**. The structures of the two related alkaloids, daphniteijsmine (**9**) [23] and daphniteijsmanine (**10**) [24], were elucidated by spectroscopic analysis coupled with an exhaustive comparison of the NMR data of secodaphniphylline and methyl homosecodaphniphyllate (**11**).

18.2.3
Yuzurimine-type Alkaloids

In 1966, Hirata *et al.* isolated yuzurimine (**12**) as one of the major alkaloids from *D. macropodum* and reported the crystal structure of yuzurimine hydrobromide [25]. They also isolated the two related alkaloids yuzurimines A (**13**) and B (**14**), whose structures were elucidated through spectroscopic data and chemical evidence in 1972. At almost the same time, Nakano *et al.* isolated macrodaphniphyllamine (**16**), macrodaphniphyllidine (**17**), and macrodaphnine (**18**) [15,29], whose structures were identical with deacetyl yuzurimine A, acetyl yuzurimine B, and dihydroyuzurimine, respectively. Yamamura *et al.* isolated deoxyyuzurimine (**19**) from *D. humile* [30], and daphnijsmine (**20**) and deacetyl daphnijsmine (**21**) from the seeds of *D. teijsmanni* [23]. Calycinine A (**22**) was isolated from *D. calycinum* distributed in China, together with deacetyl daphnijsmine, deacetyl yuzurimine, and the zwitterionic alkaloid **26** [31] (Figure 18.3).

18.2.4
Daphnilactone A-type Alkaloids

Hirata and Sasaki isolated daphnilactone A (**23**) as one of the minor alkaloids from *D. macropodum*, and the structure was determined by X-ray analysis [32,33]. The skeleton of daphnilactone A is considered to be constructed by the insertion of a C_1

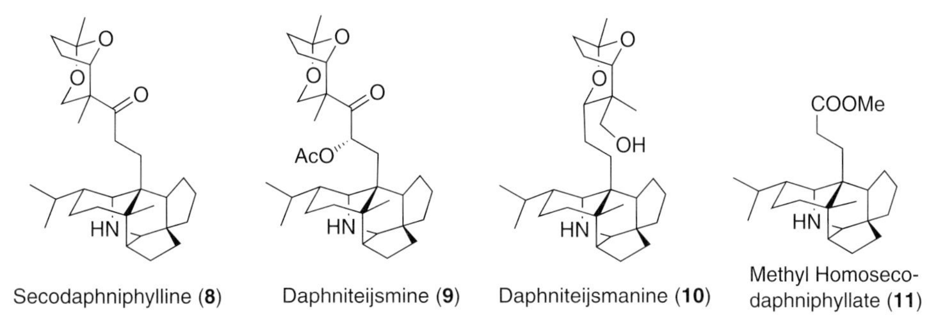

Fig. 18.2 Secodaphnane-type alkaloids.

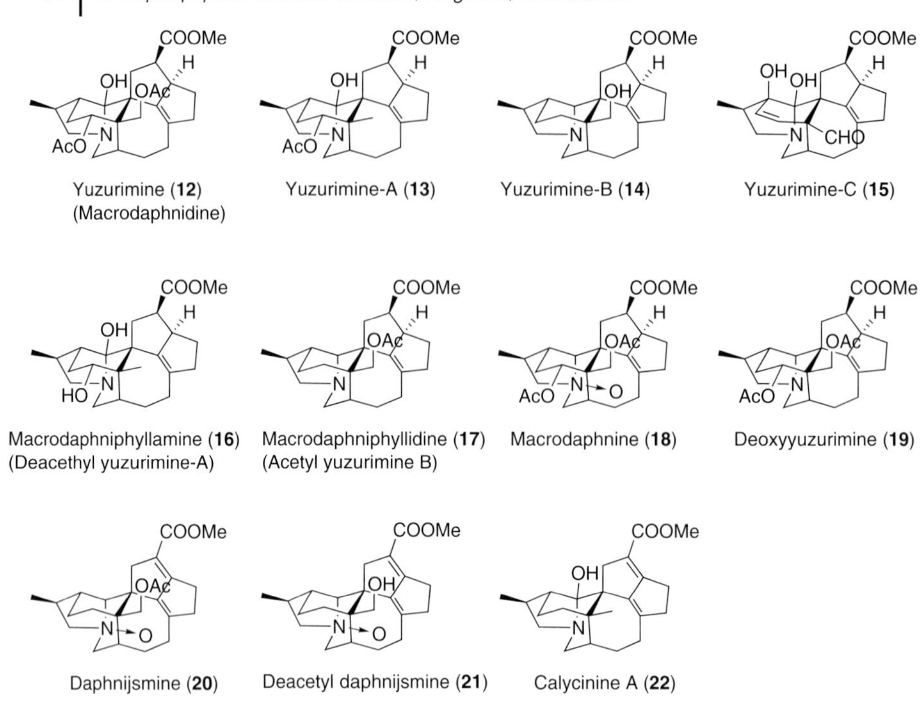

Fig. 18.3 Yuzurimine-type alkaloids.

unit into a nitrogen–carbon bond in the daphnane type skeleton, such as methyl homodaphniphyllate, followed by lactonization (Figure 18.4).

18.2.5
Daphnilactone B-type Alkaloids

Daphnilactone B (**24**) was isolated as one of the major alkaloids from the fruits of three *Daphniphyllum* species in Japan, *D. macropodum*, *D. teijsmanni*, and *D. humile*, and the structure was deduced by extensive spectral analysis, as well as by chemical evidence, and finally assigned by X-ray crystallographic analysis [34,35,37]. Isodaphnilactone B (**25**) was isolated from the leaves of *D. humile* and the structure was analyzed by spectroscopic methods [30]. A zwitterionic alkaloid **26**, the hydration product of daphnilactone B, was isolated from the fruits of *D. teijsmanni*, and the structure determined on the basis of its spectral and chemical properties [38].

18.2.6
Yuzurine-type Alkaloids

Nine alkaloids belonging to the yuzurine group were isolated from *D. macropodum* and *D. gracile* distributed in New Guinea. The structure of yuzurine (**27**) was

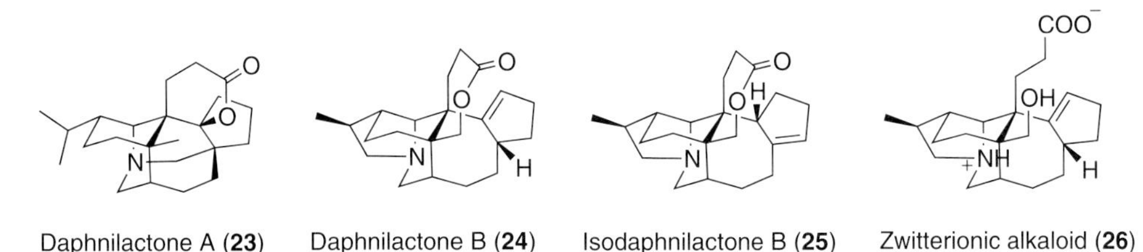

Fig. 18.4 Daphnilactones A- and B-type alkaloids.

established by X-ray crystallographic analysis of yuzurine methiodide [39]. The other alkaloids belonging to this group, daphnigracine (**28**), daphnigraciline (**29**), oxodaphnigracine (**30**), oxodaphnigraciline (**31**), epioxodaphnigraciline (**32**), daphgracine (**33**), daphgraciline (**34**), and hydroxydaphgraciline (**35**), were isolated from *D. gracile* and their structures were assigned by spectroscopic methods.

A new skeletal alkaloid, bukittinggine (**36**), was isolated from *Sapium baccatum* (Euphorbiaceae). The structure, which is closely related to both methyl homosecodaphniphyllate (**11**) and daphnilactone B, was determined by X-ray analysis of its hydrobromide [43]. The presence of bukittinggine indicates that *Sapium baccatum* may be closely related to the genus *Daphniphyllum* (Figure 18.5).

18.2.7
Daphnezomines

During the course of our studies for biogenetic intermediates of the daphniphyllum alkaloids, a project was initiated on the alkaloids of *D. humile*. A series of new daphniphyllum alkaloids, daphnezomines A–S (**37–55**), which were isolated from the leaves, stems, and fruits of *D. humile*, are of considerable interest from a biogenetic point of view. All these structures are listed in Figure 18.6.

The leaves, stems, and fruits of *D. humile* collected in Sapporo afforded daphnezomines A (**37**, 0.01 % yield), B (**38**, 0.008 %), C (**39**, 0.0001 %), D (**40**, 0.00007 %), E (**41**, 0.001 %), F (**42**, 0.0002 %), G (**43**, 0.0001 %), H (**44**, 0.0002 %), I (**45**, 0.005 %), J (**46**, 0.002 %), and K (**47**, 0.002 %), as unspecified salts [44–47]. The structures of daphnezomines A–G (**37–43**) were elucidated mainly on the basis of extensive spectroscopic studies, including several types of 2D NMR experiments.

From the IR absorptions, daphnezomine A (**37**), $C_{22}H_{35}NO_3$, was suggested to possess OH (3600 cm^{-1}), NH (3430 cm^{-1}), and carboxylate (1570 and 1390 cm^{-1}) functionalities. Detailed analysis of the ^1H–^1H COSY, HOHAHA, HMQC-HOHAHA and HMBC correlations defined the gross structure of **37**, consisting of an aza-adamantane core (N-1, C-1, and C-5–C-12) fused to a cyclohexane ring (C-1–C-5 and C-8) and another cyclohexane ring (C-9–C-11 and C-15–C-17), and three substituents at C-2, C-5, and C-8 [44].

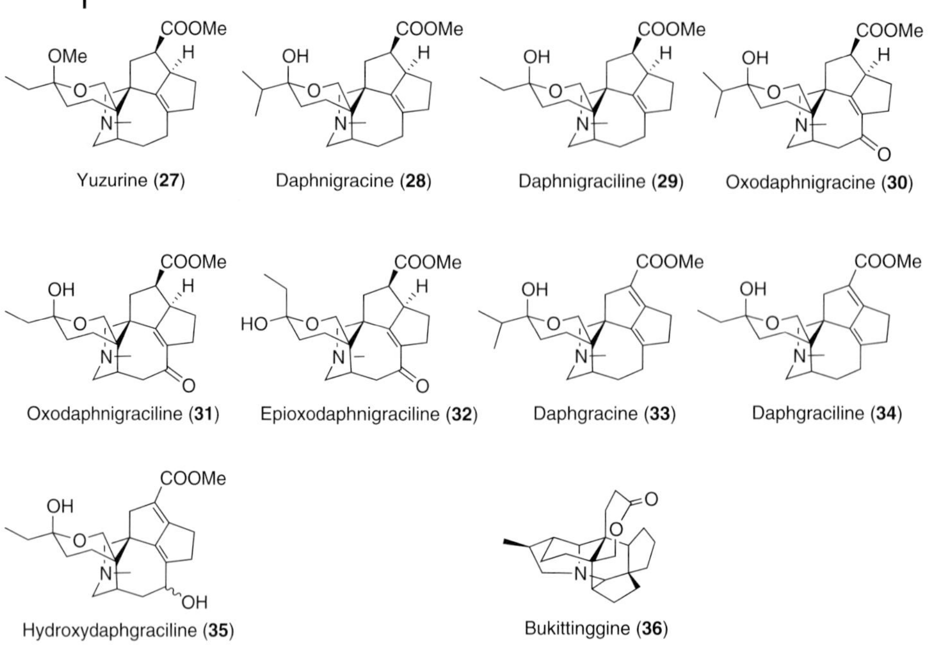

Fig. 18.5 Yuzurine-type alkaloids.

37

38: R=H
119: R=Ac

Daphnezomine B (**38**), $C_{23}H_{37}NO_3$, differs by a methoxy signal absent in **37**. Acetylation of **38** afforded the monoacetate **119**, in which the axially oriented tertiary hydroxyl group was acetylated. When daphnezomine B was treated with aqueous Na_2CO_3, it was converted into its free base **120**, showing spectroscopic anomalies as follows (Scheme 18.1). The ^{13}C signals (C-9, C-12, C-16, C-17, and C-18) of the free base showed extreme broadening, while the quaternary carbon (C-11) was not

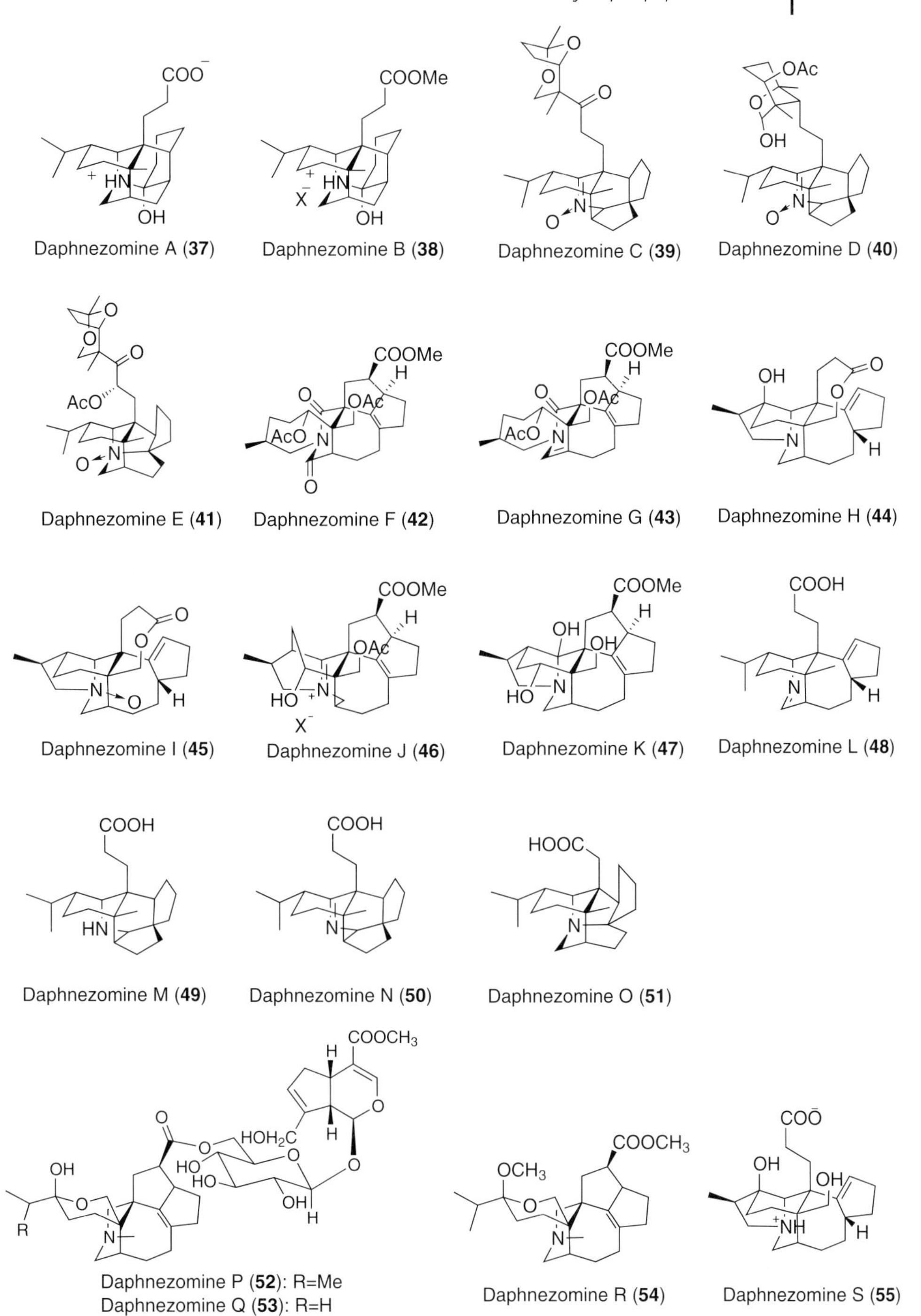

Fig. 18.6 Structures of daphnezomines A–S (**37**–**55**).

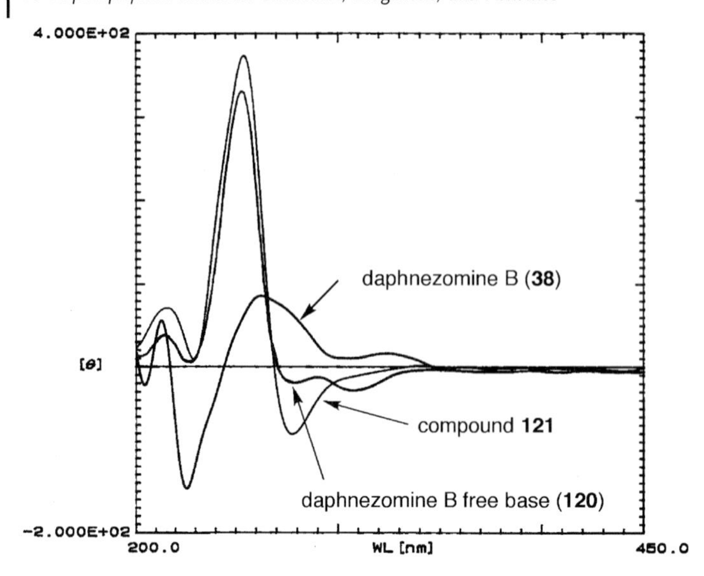

Fig. 18.7 CD spectra of daphnezomine B (**38**), daphnezomine B free base (**120**) and compound **121**[44].

observed. NMR evidence indicating a differently directed nitrogen lone pair was obtained by the ^{15}N NMR spectra (**38**: δ_N 93.7; free base of **38**: δ_N 63.0). On the other hand, in the IR spectrum, the close proximity of the carbonyl and the nitrogen permits pronounced interaction of these two functions to result in a lower shift of the carbonyl peak for **121**, whereas it was not observed for the free base of **38**. In order to examine these spectroscopic anomalies, yuzurimine free base with the similar amino ketal functionality was prepared under the same conditions, but the spectroscopic anomalies were not observed. When the free base was treated with acetic acid, it was easily converted into **38** again. Compound **121** was obtained by treatment of **38** with CH_3I/K_2CO_3 (Scheme 18.1) [40]. Production of **121** could have resulted from N-methylation of the free base followed by the alkaline-induced N–C-11 bond cleavage to generate a ketone at C-11. In the CD spectra (Figure 18.7), the structural similarity between the free bases of **38** showing spectroscopic anomalies and **121** were obtained by the CD spectra (MeOH) [free base of **38**: λ_{max} 255 (θ + 400) and 280 (−80) nm; **121**: λ_{max} 260 (θ + 350), 280 (−20), and 305 (−30) nm], showing a different CD curve from that of **38** [λ_{max} 225 (θ − 150) and 265 (+100) nm]. These data indicated that the balance of the amino ketal in the free base of **38** declined in the keto form in solution. Thus, the structure of daphnezomine B was elucidated to be **38**. The spectral data and the $[\alpha]_D$ value of the methyl ester of **37**, which was obtained by treatment of **37** with trimethylsilyldiazomethane, were in complete agreement with those of natural daphnezomine B [44].

In order to determine the absolute configurations of **37** and **38**, a crystal of the hydrobromide of **38** generated from MeOH/acetone (1 : 9) was submitted to X-ray crystallographic analysis [44]. The crystal structure containing the absolute

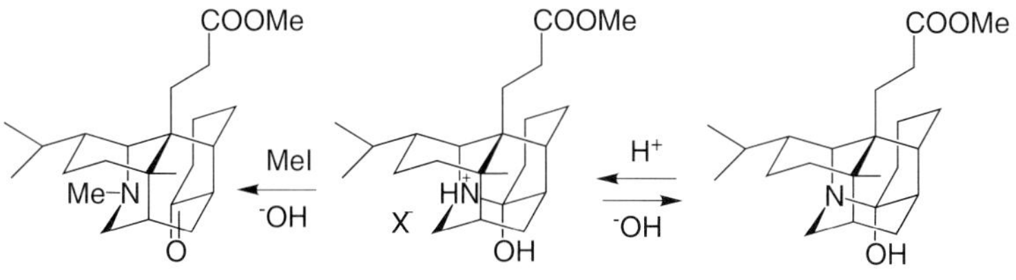

Scheme 18.1

configuration, which was determined through the Flack parameter [52], $\chi = -0.02(2)$, is shown in Figure 18.8. Daphnezomines A and B consisting of all six-membered rings are the first natural products containing an aza-adamantane core [53–56] with an amino ketal bridge, although there are reports on a number of daphniphyllum alkaloids containing five-membered rings, which may be generated from a nitrogen-involved squalene intermediate via the secodaphnane skeleton [1,2].

Daphnezomines C (**39**) and D (**40**), possess the secodaphniphylline-type skeleton with a nitrone functionality, while daphnezomine E (**41**) is the first *N*-oxide of a daphniphylline-type alkaloid, though the *N*-oxides of several yuzurimine-type alkaloids have been reported [45].

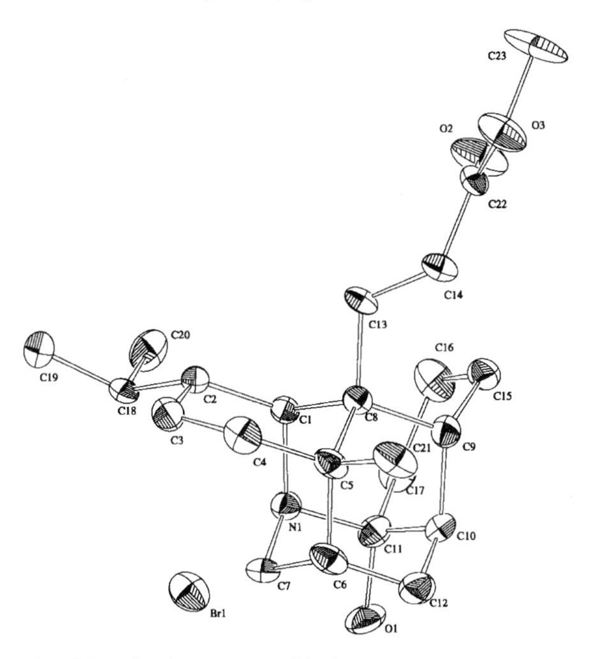

Fig. 18.8 Molecular structure of daphnezomine B (**38**) hydrobromide obtained by X-ray analysis (ORTEP drawing; ellipsoids are drawn at the 30 % probability level). Hydrogen atoms are omitted for clarity [44].

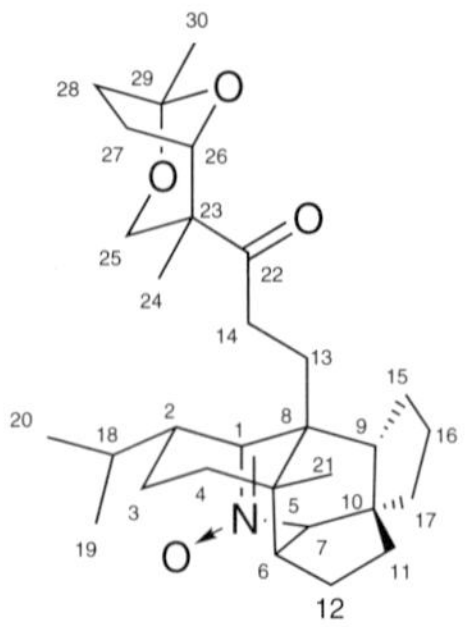

39 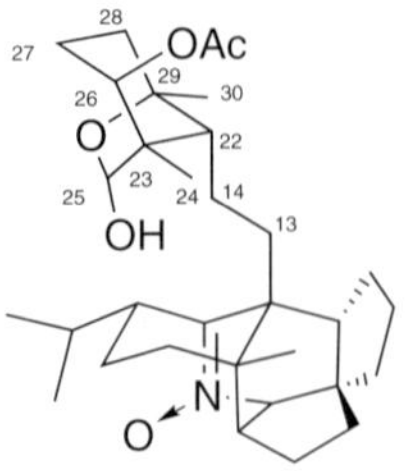 40 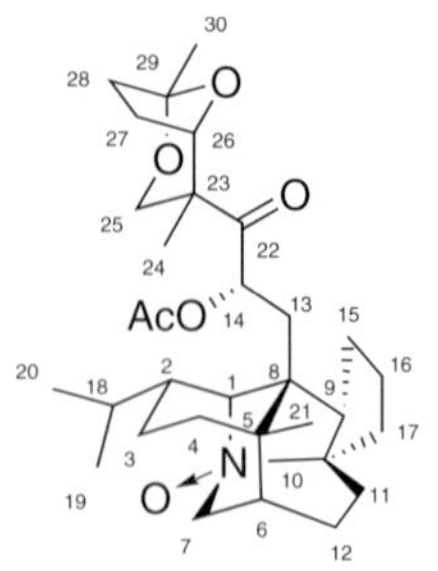 41

Spectral investigations of daphnezomines F (**42**) and G (**43**), whose molecular formulas are $C_{27}H_{35}NO_8$ and $C_{27}H_{35}NO_7$, respectively, revealed that they are structurally related and possess a 1-azabicyclo[5.2.2]undecane moiety. The conformation of the 1-azabicyclo[5.2.2]undecane ring in **43** was elucidated by a low-temperature NMR study and computational analysis [46].

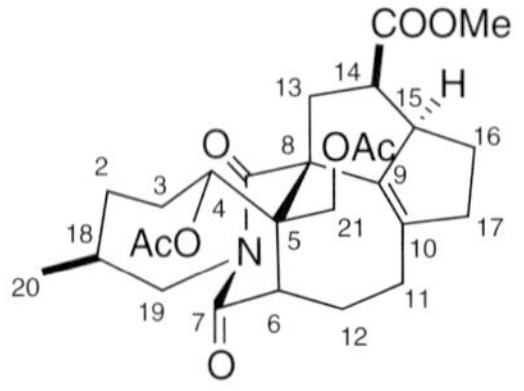

42 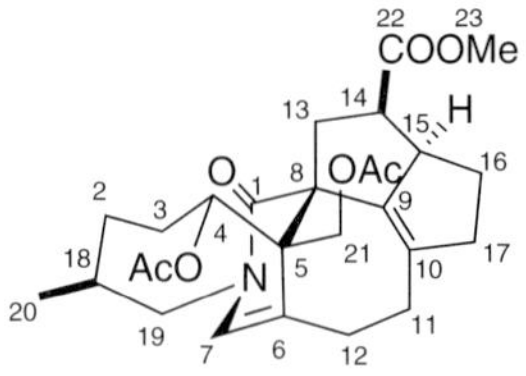 43

The structures, including relative stereochemistry, of daphnezomines H (**44**), I (**45**), J (**46**), and K (**47**), four new alkaloids possessing a daphnilactone-type (**44** and **45**) or a yuzurimine-type skeleton (**46** and **47**) were elucidated on the basis of spectroscopic data [47]. Daphnezomine I is the first *N*-oxide alkaloid having a daphnilactone-type skeleton, while daphnezomine J is the first alkaloid possessing a yuzurimine-type skeleton with an anti-Bredt-rule imine [57,58].

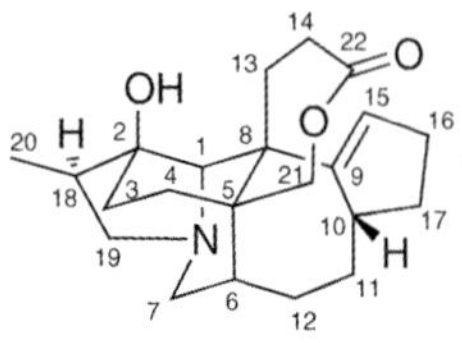

44 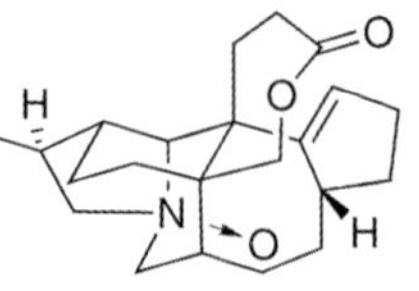 45 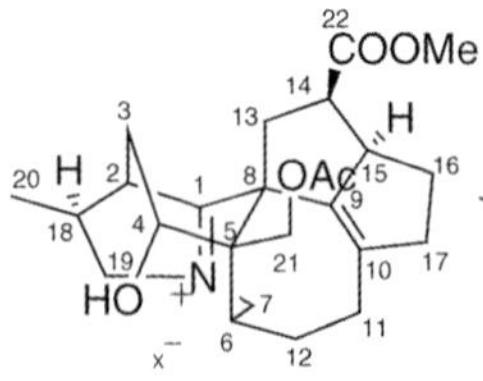 46 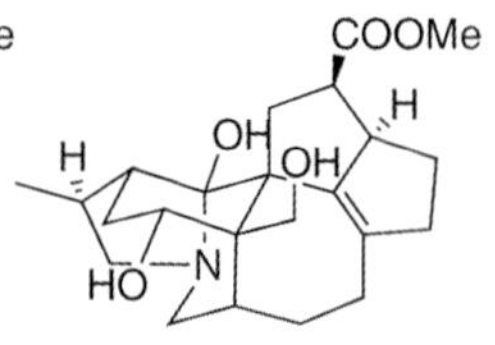 47

Relatively polar fractions prepared from the stems of *D. humile* afforded daphnezomines L (**48**, 0.0001 %), M (**49**, 0.00007 %), N (**50**, 0.00007 %), and O (**51**, 0.001 %) as colorless solids, together with the known zwitterionic alkaloid (**26**) (0.0005 %) [48].

Daphnezomine L (**48**) was close structurally to a biogenetic intermediate between the secodaphnane and daphnane skeletons.

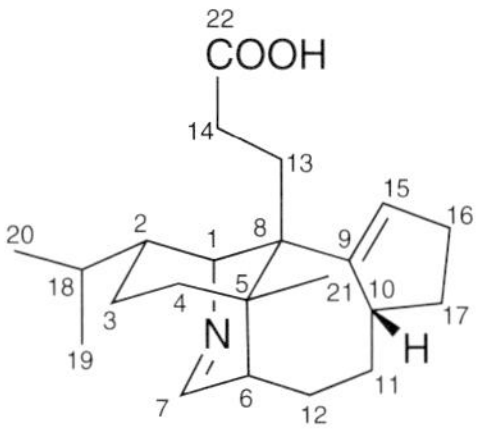

48 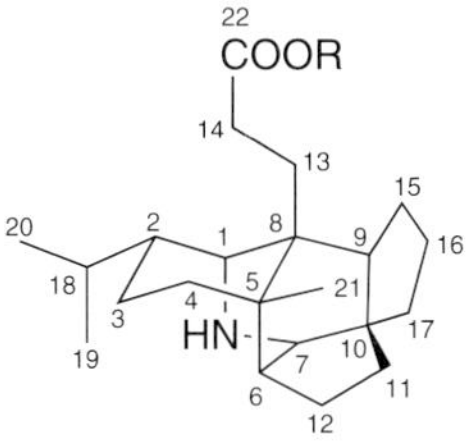 **49**: R=H **11**: R=Me **50** 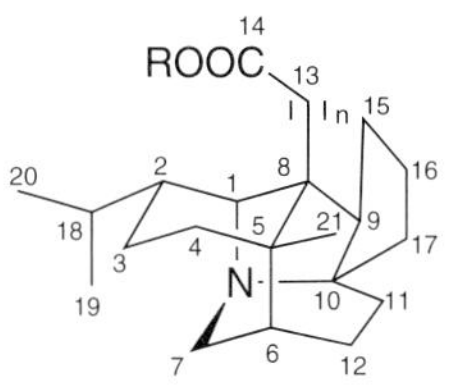 **51**: R=H, n=1 **7**: R=Me, n=2

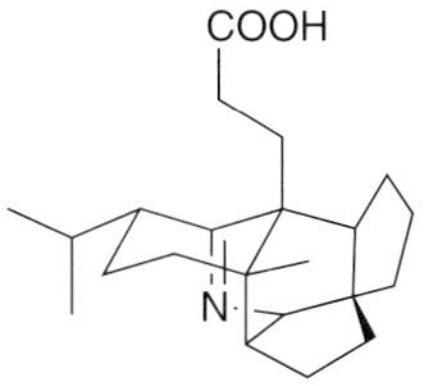

Four new alkaloids, daphnezomines P–S (**52**–**55**) have been isolated from the fruits of *D. humile* and daphnezomines P (**52**) and Q (**53**) were the first daphniphyllum alkaloids with an iridoid glycoside moiety [78].

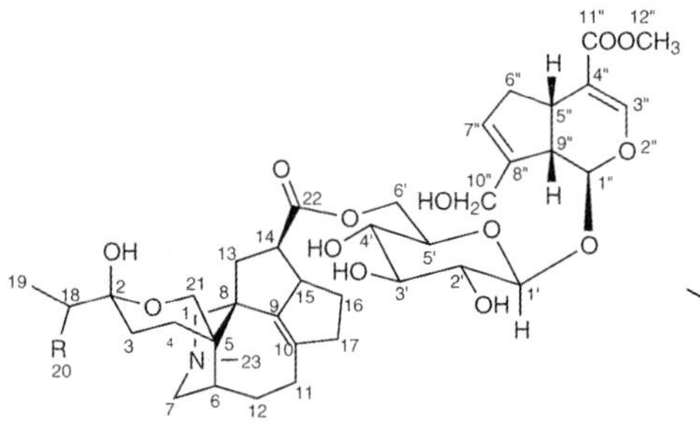

52: R=Me **53**: R=H **54** 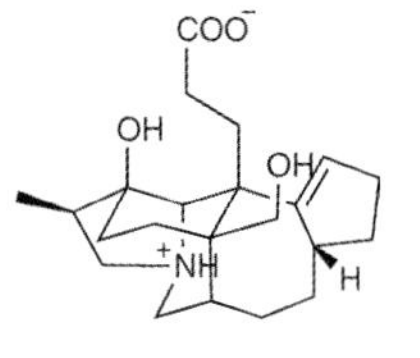 **55**

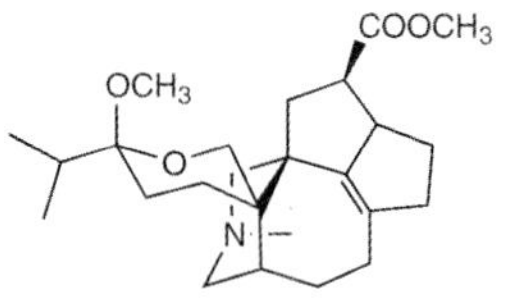

18.2.8
Daphnicyclidins

Eight highly modified daphniphyllum alkaloids with unprecedented fused hexa- or pentacyclic skeletons, daphnicyclidins A (**56**, 0.003 % yield), B (**57**, 0.0003 %), C (**58**, 0.001 %), D (**59**, 0.002 %), E (**60**, 0.001 %), F (**61**, 0.001 %), G (**62**, 0.001 %), and H (**63**, 0.004 %) were isolated from the stems of *D. teijsmanni* and *D. humile* [49] (Figure 18.9).

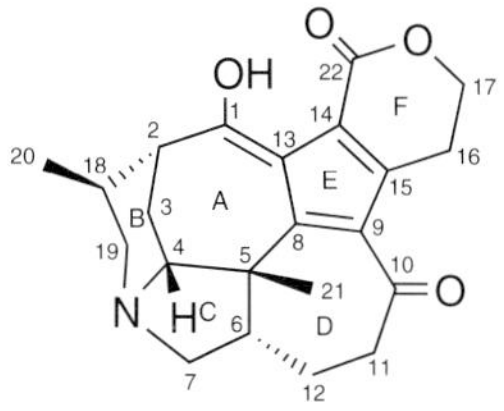

56 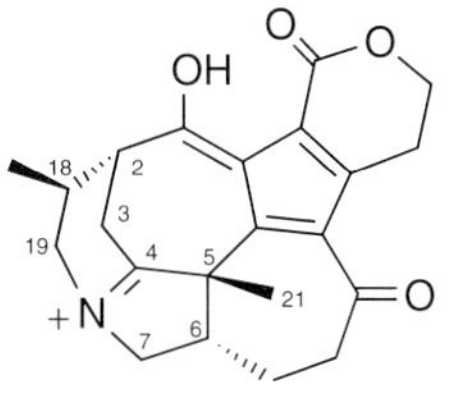 **57** 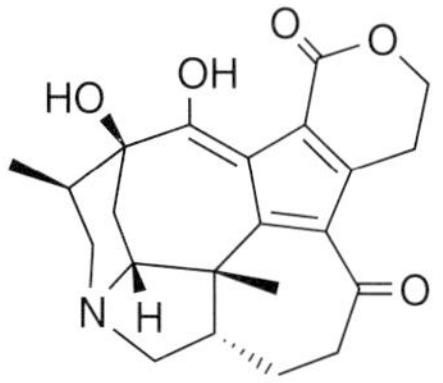 **58**

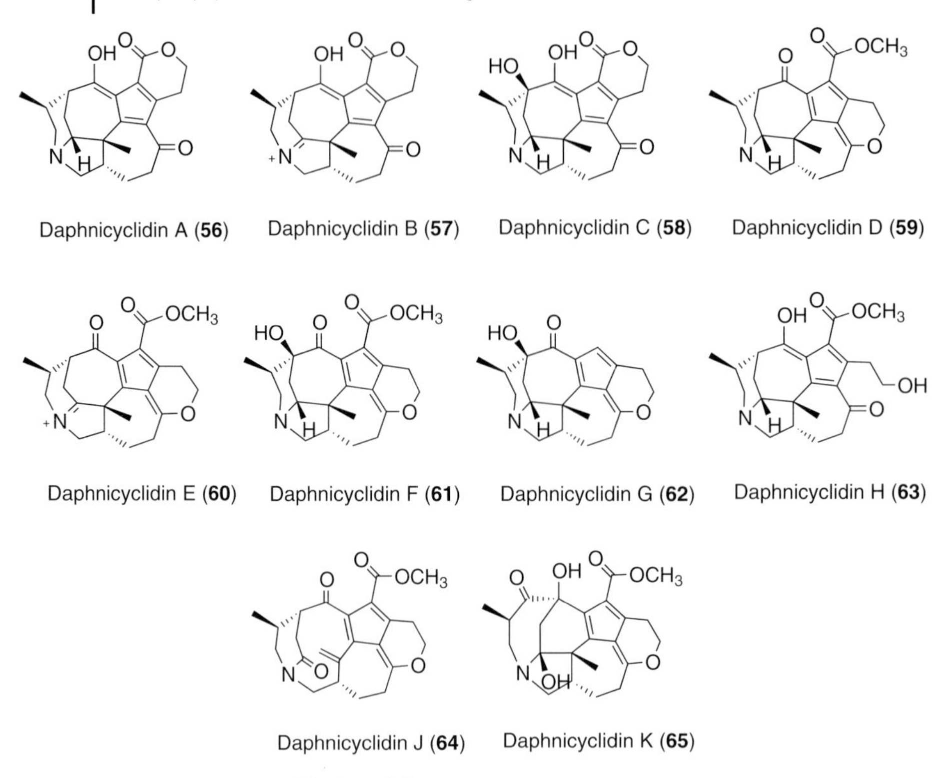

Fig. 18.9 Structures of daphnicyclidins A–K (**56**–**65**).

Daphnicyclidin A, $C_{22}H_{25}NO_4$, showed IR absorptions that implied the presence of OH and/or NH (3440 cm^{-1}) and conjugated carbonyl (1680 cm^{-1}) functionalities. Three partial structures, **a** (from C-2 to C-4 and from C-18 to C-19 and C-20), **b** (from C-6 to C-7 and C-12 and from C-11 to C-12), and **c** (from C-16 to C-17) were deduced from extensive analyses of 2D NMR data, including the ^1H–^1H COSY, HOHAHA, HMQC, and HMBC spectra in $CDCl_3$–CD_3OD (9 : 1). The connections of the three partial structures through a nitrogen atom (N-1) and also through a quaternary carbon (C-5) was established by the ^1H–^{13}C long-range (two- and three-bond) couplings detected in the HMBC spectrum to afford a proposed structure. The X-ray crystal structure (Figure 18.10) of daphnicyclidin A TFA salt revealed a unique fused-hexacyclic ring system consisting of two each of five-, six-, and seven-membered rings containing a nitrogen atom and two methyls at C-5 and C-18, in which an intramolecular hydrogen bond was observed between the C-1 hydroxyl proton and the C-22 carbonyl oxygen. The relative configurations at C-4, C-5, C-6, and C-18 were deduced from NOESY correlations, together with a stable chair conformation of ring B as depicted in the computer-generated 3D drawing (Figure 18.11).

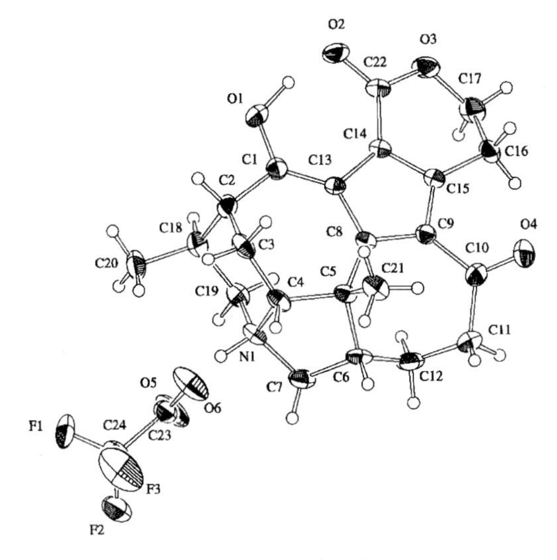

Fig. 18.10 Molecular structure of daphnicyclidin A (**56**) TFA salt obtained by X-ray analysis (ORTEP drawing; ellipsoids are drawn at the 30% probability level) [49].

The FABMS spectrum of daphnicyclidin B showed the molecular formula $C_{22}H_{24}NO_4$. The 2D NMR data of **57** were similar to those of the imine (C-4 and N-1) form of daphnicyclidin A. Spectral investigation of daphnicyclidin C, whose molecular formula is $C_{22}H_{25}NO_5$, revealed that it was the 2-hydroxy form of daphnicyclidin A [49].

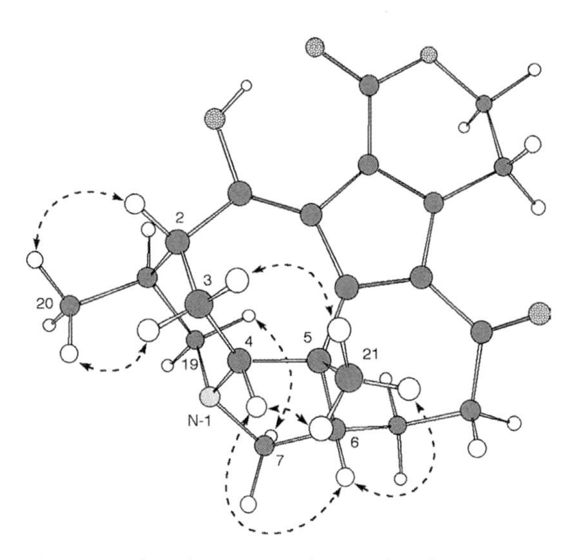

Fig. 18.11 Selected NOESY correlations (dotted arrows) and relative configurations for daphnicyclidin A (**56**)[49].

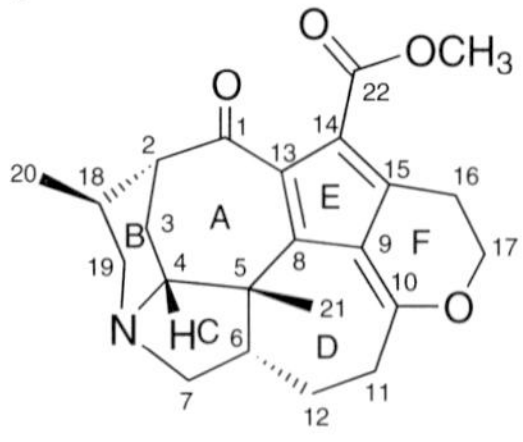

59

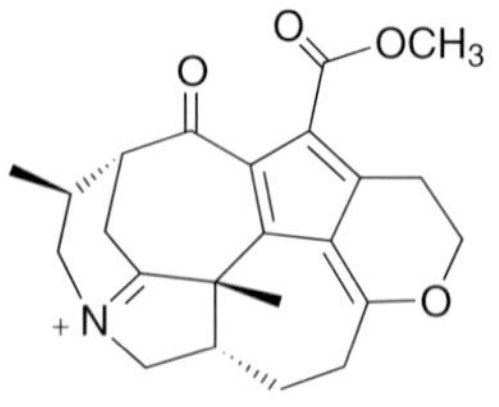

60

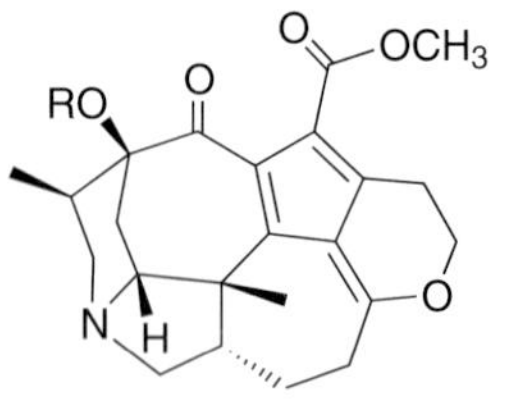

61: R=H
122: R=*p*-Br-Bz

The molecular formula, $C_{23}H_{27}NO_4$ of daphnicyclidin D (**59**), was larger than that of daphnicyclidin A by a CH_2 unit. ^{1}H and ^{13}C NMR data of **59** were similar to those of **56** in the left-hand part consisting of rings A to C with a nitrogen atom, except that a methoxy signal absent for **56** was observed for **59**. The structure of **59** was elucidated by 2D NMR data. HMBC cross-peaks indicated that C-10 and C-17 were connected through an oxygen atom to form a dihydropyran ring. In addition, the presence of a conjugated cyclopentadiene moiety (C-8–C-9 and C-13–C-15) as in **56** was suggested by HMBC correlations. Treatment of **56** with methanolic *p*-TsOH gave daphnicyclidin D (Scheme 18.2). Thus, the structure of daphnicyclidin D was assigned as **59** [49].

The IR spectrum of daphnicyclidin E (**60**), $C_{23}H_{25}NO_4$ was indicative of the presence of conjugated carbonyl and/or imine ($1680\,cm^{-1}$) functionalities. The presence of an iminium carbon (C-4) was indicated by HMBC correlations for H_3-21 and H_b-3 to C-4, and H_2-7 and H_a-19 to C-4 through a nitrogen atom. Treatment of

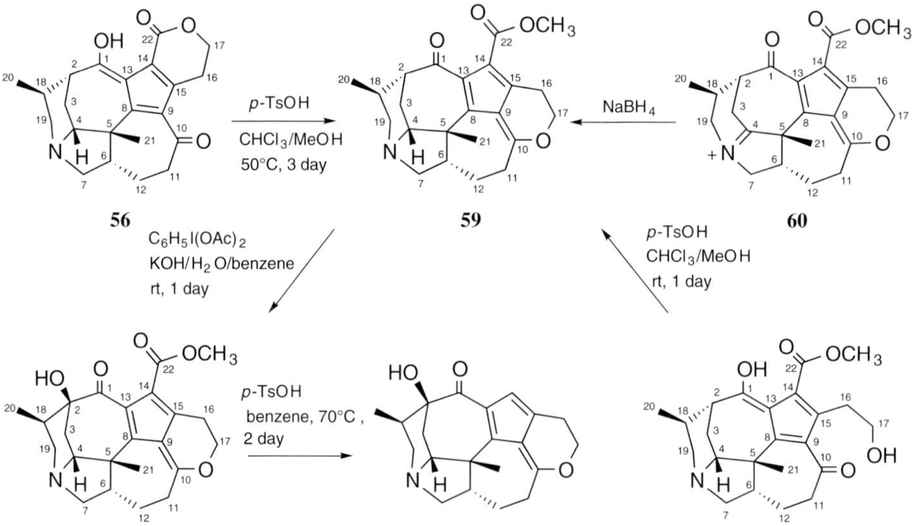

56 → 59: *p*-TsOH, $CHCl_3$/MeOH, 50°C, 3 day

60 → 59: $NaBH_4$

56 → 61: $C_6H_5I(OAc)_2$, KOH/H_2O/benzene, rt, 1 day

63 → 59: *p*-TsOH, $CHCl_3$/MeOH, rt, 1 day

61 → 62: *p*-TsOH, benzene, 70°C, 2 day

61 62 63

Scheme 18.2 Chemical correlations for daphnicyclidins A (**56**) and D–H **(59–63)** [49].

60 with $NaBH_4$ afforded daphnicyclidin D (Scheme 18.2). Thus, daphnicyclidin E was concluded to be the imine form at C-4 of daphnicyclidin D [49].

2D NMR analysis of daphnicyclidin F (**61**), $C_{23}H_{27}NO_5$, and chemical correlation of daphnicyclidin D with **61**, indicated that **61** is the 2-hydroxy form of daphnicyclidin D [49].

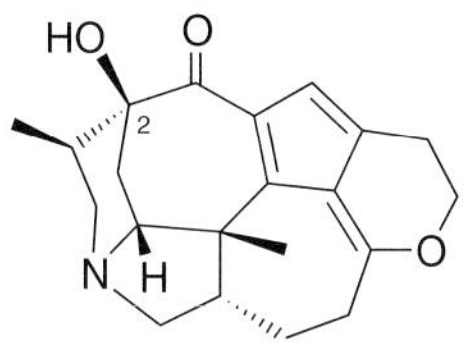

62

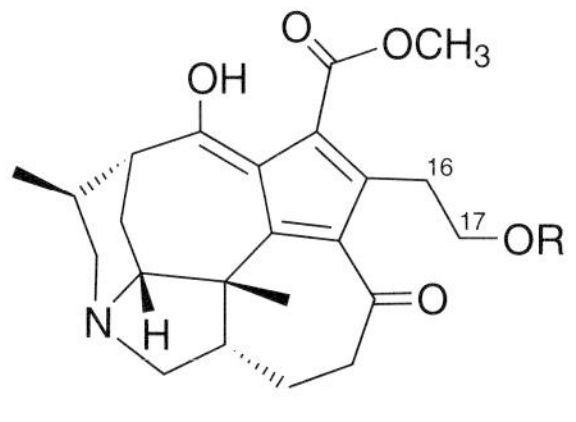

63 : R=H
123 : R=Ac

Daphnicyclidin F possesses a methoxy carbonyl moiety at C-14, while the ^1H and ^{13}C NMR data of **62** showed signals due to an sp^2 methine. Treatment of **61** with *p*-TsOH at 70 °C for two days gave daphnicyclidin G (**62**) (Scheme 18.2). Therefore, daphnicyclidin G was elucidated to be the 14-demethoxycarbonyl form of daphnicyclidin F [49].

Daphnicyclidin H (**63**), $C_{23}H_{29}NO_5$, showed IR absorptions at 3435 and 1680 cm^{-1} indicating the presence of hydroxyl and conjugated carbonyl groups, respectively. ^1H NMR signals assignable to H_2-17 were observed to be equivalent. Treatment of **63** with acetic anhydride afforded the monoacetate **123**, in which the hydroxyl group at C-17 was acetylated. On the other hand, the presence of a methoxy carbonyl group at C-14 and rings A–E with a ketone at C-10 was deduced from the 2D NMR analysis. The 2D NMR data indicated that the conjugated keto-enol moiety of **63** was the same as that of daphnicyclidin A. Treatment of daphnicyclidin H with *p*-TsOH gave daphnicyclidin D (Scheme 18.2) [49].

The absolute configuration of daphnicyclidin F was analyzed by applying the exciton chirality method [61] after introduction of a *p*-bromobenzoyl chromophore into the hydroxyl group at C-2. As the sign of the first Cotton effect [λ_{max} 280 (θ + 20 000) and 225 (−16 000) nm] was positive, the chirality between the cyclopentene moiety and the benzoate group of the *p*-bromobenzoyl derivative **122** of **61** was assigned as shown in Figure 18.12 (right-handed screw), indicating that the absolute stereochemistry at C-2 was (*S*) [49].

Daphnicyclidins J (**64**) and K (**65**), two alkaloids with unprecedented fused penta- or hexacyclic skeletons, respectively, were isolated from the stems of *D. humile* [50]. Daphnicyclidin J, $C_{23}H_{25}NO_5$, showed IR absorptions at 1690 and 1660 cm^{-1}, corresponding to ketone and amide carbonyl functionalities, respectively. The ^1H–^1H COSY and HOHAHA spectra proved information on the proton-connectivities for three partial structures **a** (C-2 to C-3 and C-18, and C-18 to C-19 and C-20), **b** (C-6 to C-7 and C-12, and C-11 to C-12), and **c** (C-16 to C-17). Long range ^1H–^{13}C

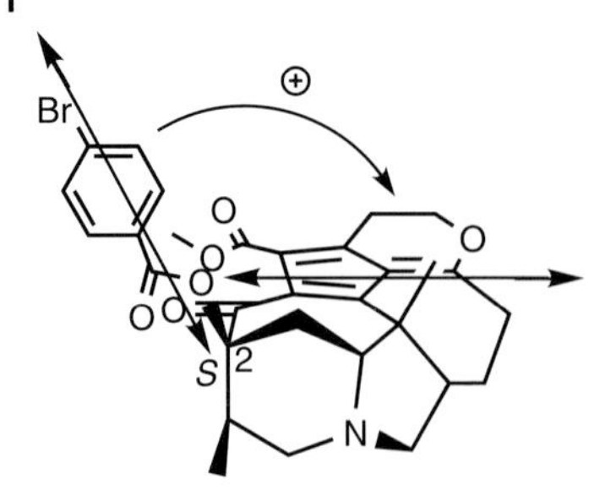

Fig. 18.12 Stereostructure of the *p*-bromobenzoate (**122**) of daphnicyclidin F (**61**). Arrows denote the electric transition dipole of the chromophore [49].

correlations showed that the partial structures are linked. The presence of a fulvene functionality (C-8–C-10 and C-13–C-15), which was conjugated with two carbonyl groups (C-1 and C-22) and an exo-methylene group (C-5 and C-21), was deduced by comparison of the carbon chemical shifts with those of daphnicyclidin D. UV absorptions (245, 320, and 330 nm) also supported the existence of the conjugated fulvene functionality. Thus, the structure of daphnicyclidin J was assigned as **64**, which has a uniquely fused-pentacyclic ring system (one five-, two six-, one seven-, and one ten-membered rings) containing a δ-lactam and a pyran ring. The absolute configuration was established by chemical correlation with a known related alkaloid, daphnicyclidin D, through a modified Polonovski reaction (Scheme 18.3) [50].

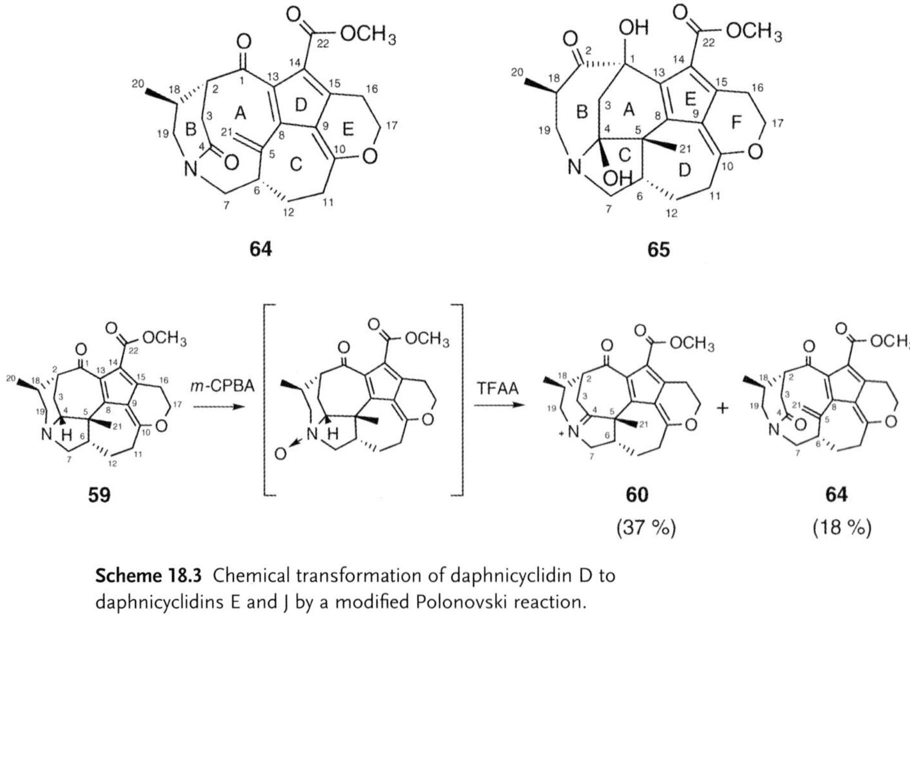

64 **65**

59 → *m*-CPBA → [N-oxide] → TFAA → **60** (37 %) + **64** (18 %)

Scheme 18.3 Chemical transformation of daphnicyclidin D to daphnicyclidins E and J by a modified Polonovski reaction.

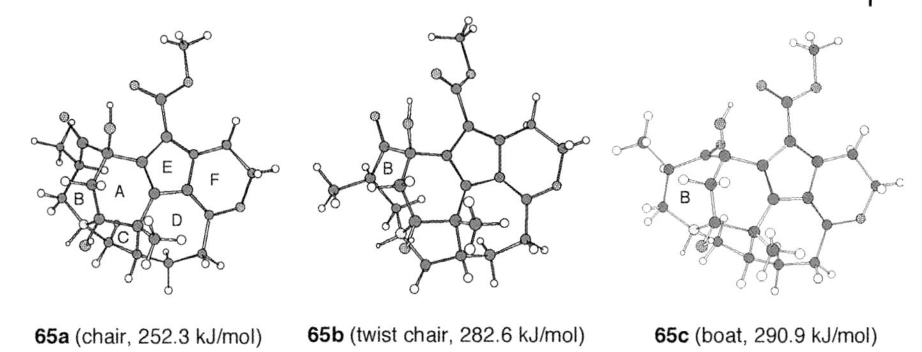

Fig. 18.13 Three representative stable conformers (**65a–65c**) for daphnicyclidin K (**65**) analyzed by Monte Carlo simulation followed by minimization and clustering analysis [50].

Daphnicyclidin K was shown to have the molecular formula $C_{23}H_{27}NO_6$. IR absorptions implied the presence of hydroxyl (3600 cm^{-1}), ester carbonyl (1700 cm^{-1}), and conjugated carbonyl (1650 cm^{-1}) functionalities. Analysis of 2D NMR data showed that the structure of **65** has an unusual skeleton consisting of a 6/7/5/7/5/6 hexacyclic ring system [50].

The relative configuration was assigned from NOESY correlations and conformational calculations by Monte Carlo simulation [59], which suggested that the seven-membered ring (ring B) with a chair conformation (**65a**) was the most stable, whereas those with twist chair (**65b**) and boat (**65c**) conformations had considerably higher energy (Figure 18.13). In addition, the NOESY correlations indicated that another seven-membered ring (ring D) assumed a twist-boat conformation similar to the crystal structure of daphnicyclidin A.

18.2.9
Daphmanidins

Further investigation of extracts of the leaves of *D. teijsmanii* resulted in the isolation of daphmanidin A (**66**, 0.0001 % yield), an alkaloid with an unprecedented fused-hexacyclic ring system, and daphmanidin B (**67**, 0.00003 %) with a pentacyclic ring system [51] (Figure 18.14).

66 **67**

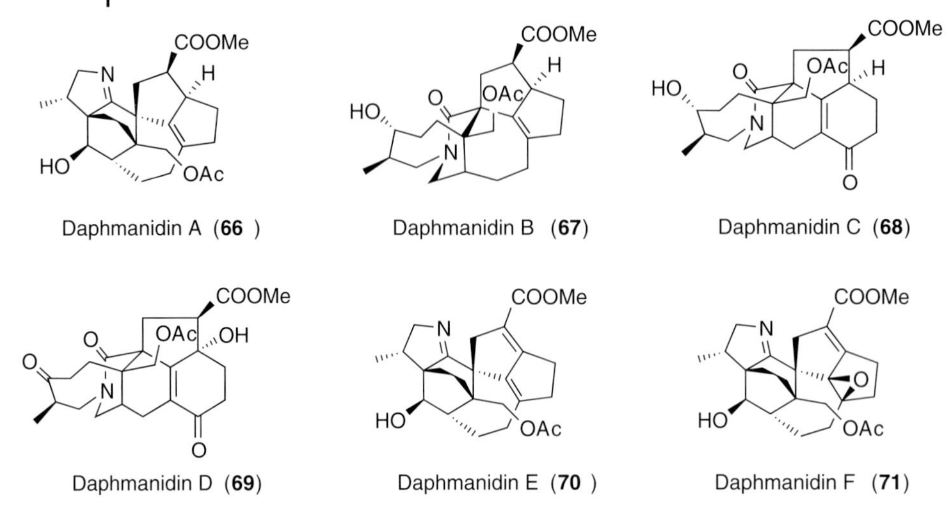

Fig. 18.14 Structures of daphmanidins A–F (**66–71**).

The IR absorptions of daphmanidin A (**66**), $C_{25}H_{33}NO_5$, implied the presence of hydroxyl (3616 cm^{-1}), ester carbonyl (1730 cm^{-1}), and imine (1675 cm^{-1}) functionalities. Detailed spectroscopic analysis revealed that the gross structure of daphmanidin A possesses a fused-hexacyclic ring system consisting of a dihydropyrrole ring (N-1, C-1, C-2, C-18, and C-19) with a methyl group at C-18, a bicyclo[2.2.2]octane ring (C-1–C-8) with a hydroxyl at C-7, and a decahydrocyclopenta [*cd*] azulene ring (C-5, C-6, C-8–C-17) with a methoxy carbonyl group at C-14 and an acetoxy methyl group at C-5. The relative and absolute stereochemistry of **66** was determined by a combination of NOESY correlations (Figure 18.15) and the modified Mosher method.

The structure of daphmanidin B (**67**), $C_{25}H_{36}NO_6$, was elucidated by 2D NMR data to possess a 1-azabicyclo[5.2.2]undecane moiety, like daphnezomines F and G [46]. The relative stereochemistry was deduced from NOESY correlations. The conformation of the unit (C-2–C-5, C-18 to C-2, C-19, and N) in the 1-azabicyclo[5.2.2]undecane moiety, with a twist-chair form as shown in Figure 18.16, was consistent with the results of a conformational search using MMFF force field [60] implemented in the Macromodel program [59].

Two novel alkaloids with an unprecedented fused-pentacyclic skeleton, daphmanidins C (**68**) and D (**69**), consisting of 1-azabicyclo[5.2.2]undecane, hexahydronaphthalen-1-one, and cyclopentane rings, have been isolated from the leaves of *D. teijsmanii* [79]. Daphmanidin C elevated the activity of NGF biosynthesis. New daphniphyllum alkaloids, daphmanidins E (**70**) and F (**71**), have also been isolated from the leaves of *D. teijsmannii*, and Daphmanidins E and F showed a moderate vasorelaxant effect on rat aorta [80].

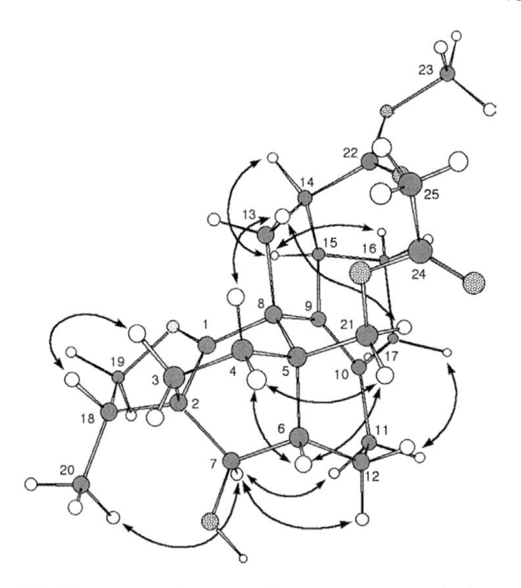

Fig. 18.15 Key NOESY correlations (arrows) and relative stereochemistry for daphmanidin A (**66**) [51].

18.2.10
Daphniglaucins

Two cytotoxic quaternary daphniphyllum alkaloids with an unprecedented fused-polycyclic skeleton containing a 1-azoniatetracyclo[5.2.2.0.^{1,6}0.^{4,9}]undecane ring

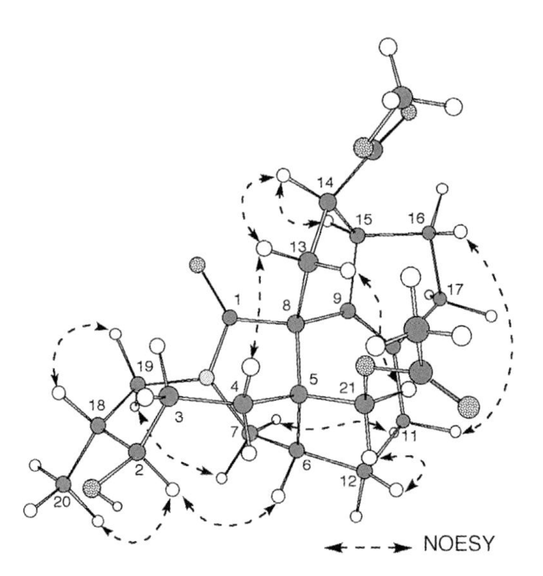

Fig. 18.16 Selected 2D NMR correlations and relative stereochemistry for daphmanidin B (**67**) [51].

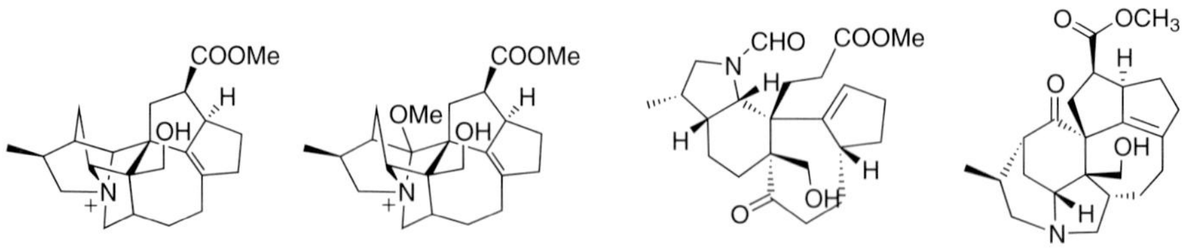

Daphniglaucin A (72) Daphniglaucin B (73) Daphniglaucin C (74) Daphniglaucin D (75)

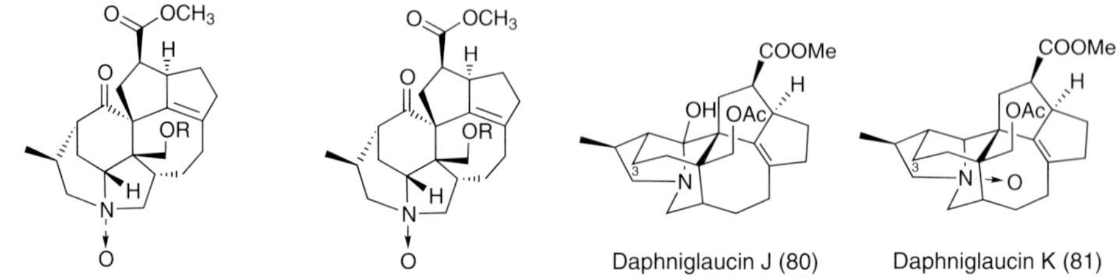

Daphniglaucin E (76): R=H
Daphniglaucin F (77): R=Ac
Daphniglaucin G (78): R=H
Daphniglaucin H (79): R=Ac
Daphniglaucin J (80) Daphniglaucin K (81)

Fig. 18.17 Structures of daphniglaucins A–K (**72–81**).

system, daphniglaucins A (**72**) and B (**73**), have been isolated from the leaves of *Daphniphyllum glaucescens* [81]. A novel daphniphyllum alkaloid with an unprecedented tetracyclic ring system consisting of octahydroindole and hexahydroazulene rings, daphniglaucin C (**74**), has been isolated from the leaves of *Daphniphyllum glaucescens* as a tubulin polymerization inhibitor [82]. Five new fused-hexacyclic alkaloids, daphniglaucins D (**75**), E (**76**), F (**77**), G (**78**), and H (**79**), and two new yuzurimine-type alkaloids, daphniglaucins J (**80**) and K (**81**), have also been isolated from the leaves of *D. glaucescens* [83] (Figure 18.17).

18.2.11 Calyciphyllines

Two types of daphniphyllum alkaloids with unprecedented fused-hexacyclic ring systems, calyciphyllines A (**82**) and B (**83**), have been isolated from the leaves of *Daphniphyllum calycinum* (Daphniphyllaceae) [84]. The structure of calyciphylline A was assigned as **82**, with a fused-hexacyclic ring system (three five-, two six-, and one seven-membered rings) containing an *N*-oxide group, and that of calyciphylline B was assigned as **83**, with a hexacyclic ring system consisting of a hexahydroindene ring and an octahydroindolizine ring fused to a cyclopentane ring with a δ-lactone ring at C-5 and C-8 as shown in Figure 18.18.

18.2.12 Daphtenidines

Daphtenidines A (**84**)–D (**87**) were isolated from the leaves of *D. teijsmannii* [85]. Daphtenidines A (**84**) and B (**85**) possess the daphnilactone A-type skeleton. This is

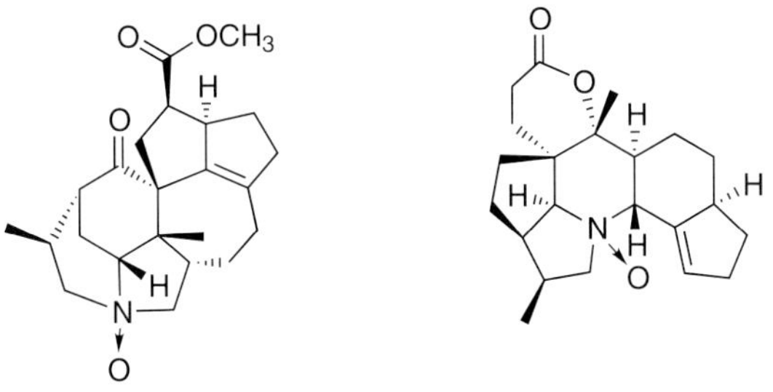

Calyciphylline A (**82**) Calyciphylline B (**83**)

Fig. 18.18 Structures of calyciphyllines A (**82**) and B (**83**).

the second isolation of daphnilactone A-type alkaloids from natural sources. Daphtenidine C (**86**) is the 4-acetoxy form of daphmanidin A, while daphtenidine D (**87**) is the 14-dehydro form of yuzurimine (Figure 18.19).

18.2.13 Other Related Alkaloids

Jossang *et al.* and Bodo *et al.* investigated the various parts of *D. calycinum* collected in Vietnam and isolated daphcalycine (**88**) [86], daphcalycinosidines A (**89**), B (**90**), and C (**92**), and the related alkaloids shown in Figure 18.20 [87,88]. Daphcalycinosidine A (**89**), B (**90**), and C (**92**) are characterized by an iridoid glucoside moiety linked to daphniphyllum alkaloid moieties such as daphnezomines P and Q [78]. Yue *et al.* also isolated the related alkaloids caldaphnidines A (**95**)–F (**100**) from *D. calycinum* [89] (Figure 18.20). Yue *et al.* also isolated various related daphniphyllum alkaloids (**101**–**118**) from various species distributed in China, such as *D. subverticillatum* [90], *D. paxianum* [91,92], *D. oldhami* [93], *D. longistylum* [94], *D. longeracemosum* [95], and *D. yunnanense* [96] as shown in Figure 18.21. Most of them belong to the categories that have already been isolated [3].

Daphtenidine A (**84**) Daphtenidine B (**85**) Daphtenidine C (**86**) Daphtenidine D (**87**)

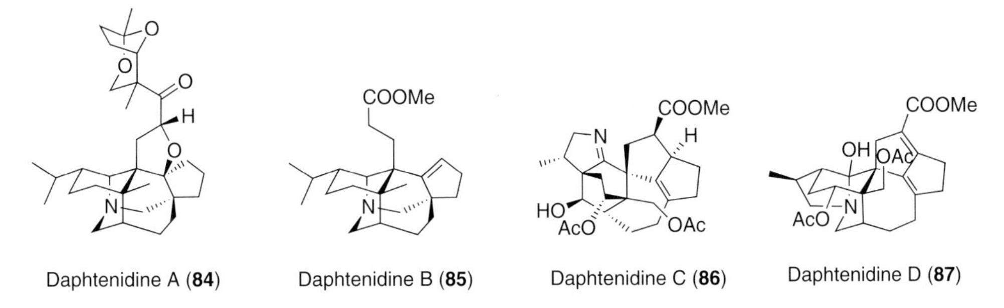

Fig. 18.19 Structures of daphtenidines A–D (**84**–**87**).

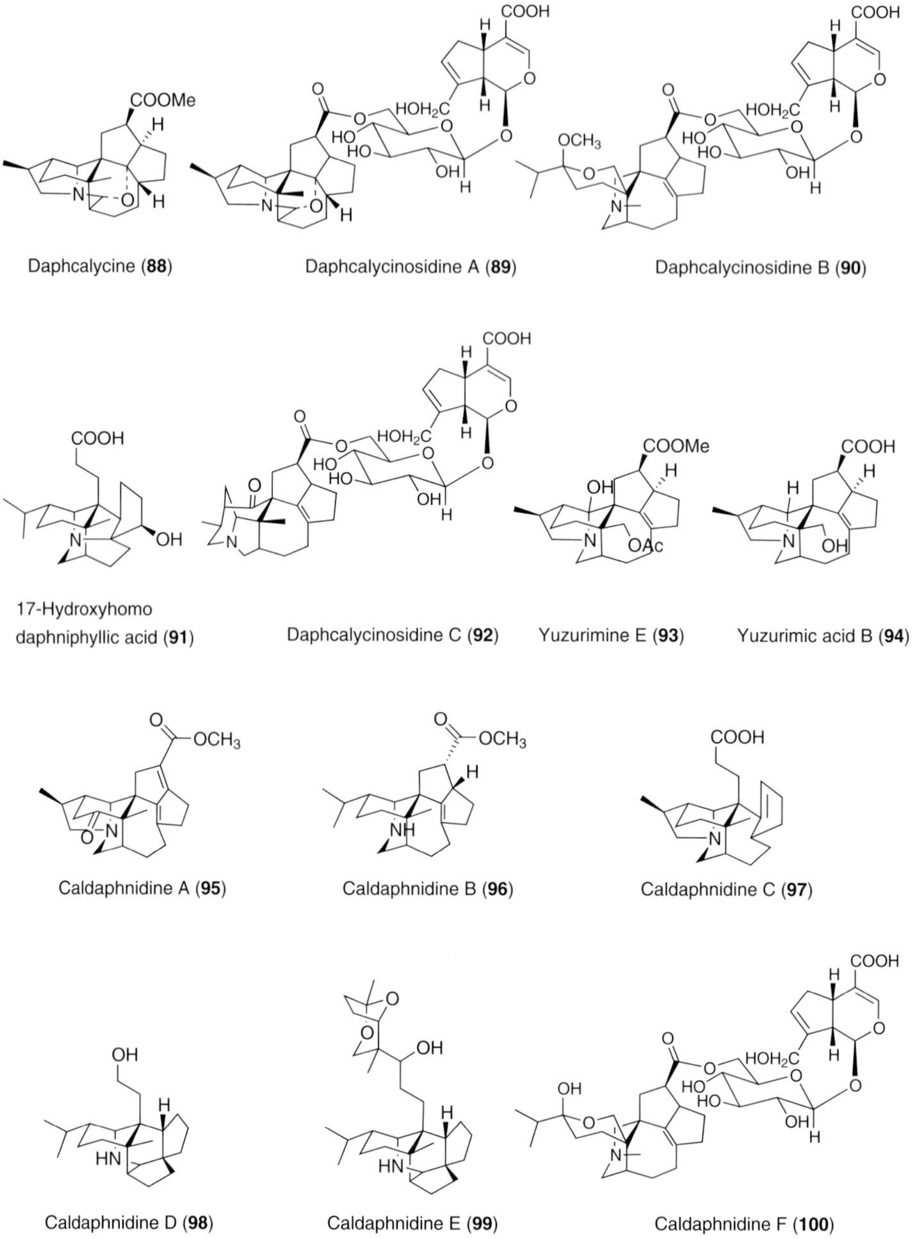

Daphcalycine (**88**)

Daphcalycinosidine A (**89**)

Daphcalycinosidine B (**90**)

17-Hydroxyhomo daphniphyllic acid (**91**)

Daphcalycinosidine C (**92**)

Yuzurimine E (**93**)

Yuzurimic acid B (**94**)

Caldaphnidine A (**95**)

Caldaphnidine B (**96**)

Caldaphnidine C (**97**)

Caldaphnidine D (**98**)

Caldaphnidine E (**99**)

Caldaphnidine F (**100**)

Fig. 18.20 Structures of other related alkaloids (**88–100**).

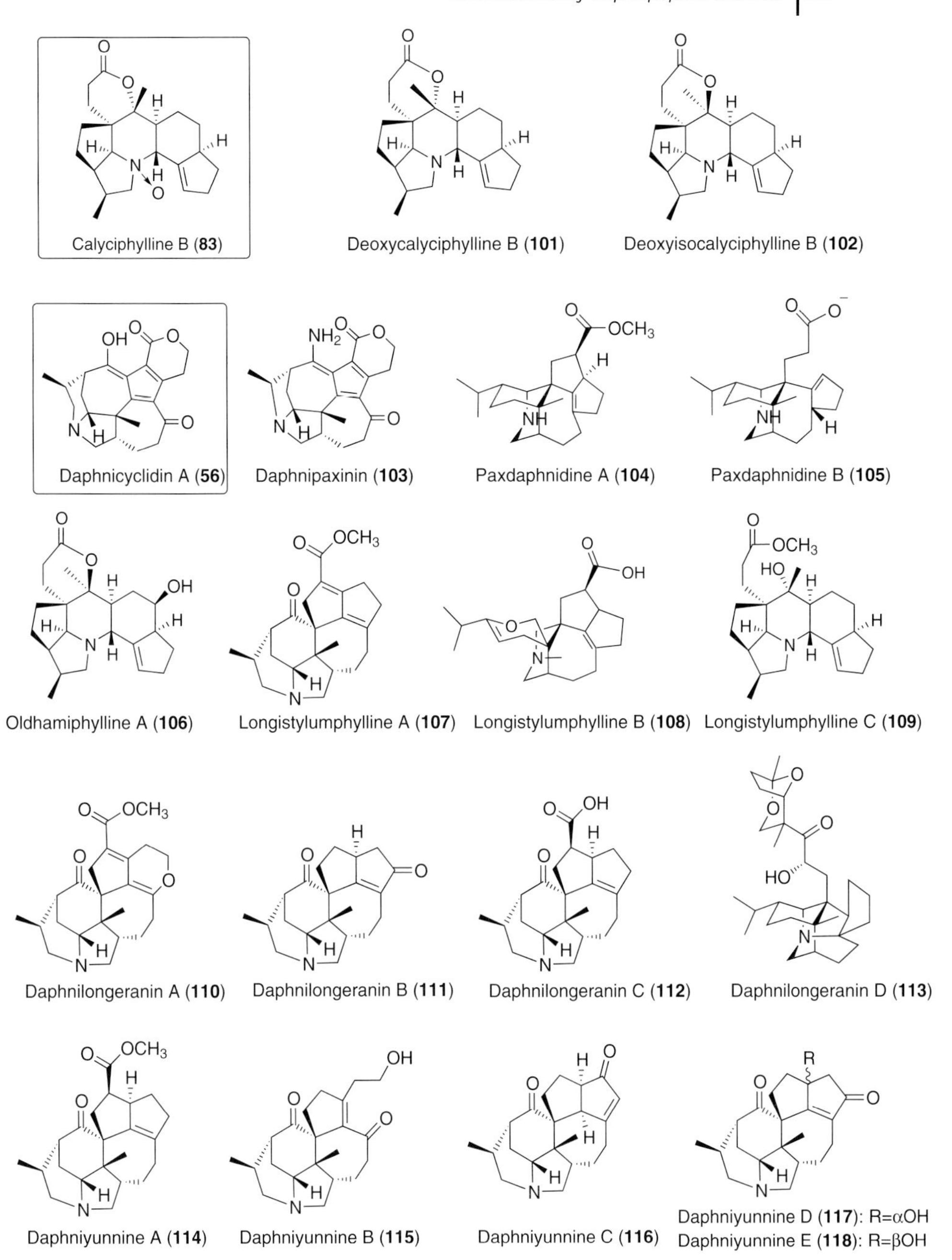

Fig. 18.21 Structures of other related alkaloids (**101–118**).

18.3
Biosynthesis and Biogenesis

18.3.1
Biosynthesis of Daphniphyllum Alkaloids

Suzuki and Yamamura conducted feeding experiments on the daphniphyllum alkaloids, using the leaves of *D. macropodum* [62]. The alkaloids present, as well as their amounts, varied with season, and the highest incorporation of DL-mevalonic acid (**124**, MVA) and squalene (**125**) into daphniphylline was recorded in June and July. From the feeding experiments, followed by degradation studies, daphniphylline and codaphniphylline were biosynthesized from six moles of MVA (**124**) through a squalene-like intermediate (Figure 18.22). In addition, feeding experiments using the fruits of *D. teijsmanni* resulted in the incorporation of four moles of MVA (**124**) into one of the major C_{22}-type daphniphyllum alkaloids, daphnilactone B [36].

18.3.2
Biogenesis of the Daphnane and Secodaphnane Skeletons

Heathcock proposed a biosynthetic pathway to the daphniphyllum alkaloids [4,5]. The linear squalene (**125**) molecule may be traced in the pentacyclic domain of the skeleton of secodaphniphylline. To convert squalene into secodaphniphylline, four C–C bonds must be formed: C-10 to C-14; C-6 to C-15; C-3 to the C-15 methyl group; and C-7 to the C-10 methyl group. In addition, the nitrogen atom is inserted between C-7 and the C-15 methyl group. For daphniphylline, however, the nitrogen seems to have been inserted between C-10 and its methyl group, which is also connected to C-7. Thus, it is likely that secodaphniphylline precedes daphniphylline in the biosynthetic pathway, and that an unsaturated amine such as compound **126** provides a biogenetic link between the two skeletons [5] (Scheme 18.4). The hypothetical

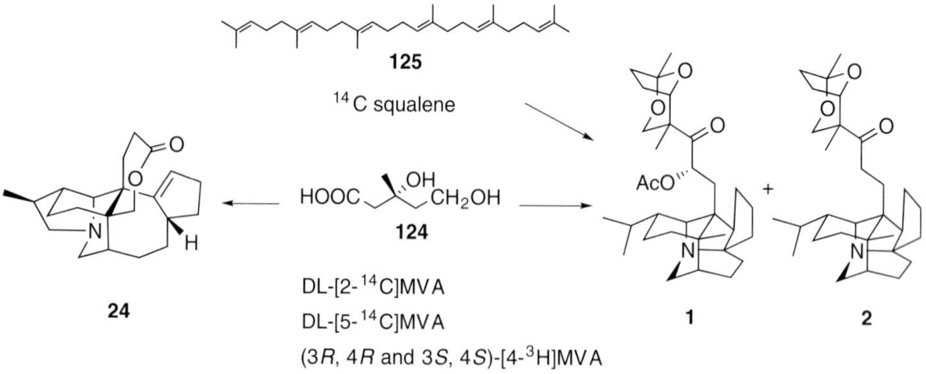

125
^{14}C squalene
HOOC–C(OH)(CH$_3$)–CH$_2$–CH$_2$OH
124
DL-[2-^{14}C]MVA
DL-[5-^{14}C]MVA
(3*R*, 4*R* and 3*S*, 4*S*)-[4-^{3}H]MVA
24
1
2

Fig. 18.22 Feeding experiments with labeled mevalonic acid (**124**) and squalene (**125**) into daphniphylline (**1**), codaphniphylline (**2**), and daphnilactone B (**24**) [62].

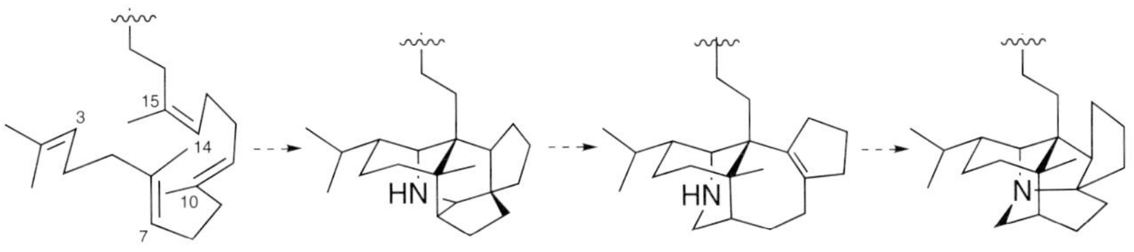

squalene (**125**) secodaphnane **126** daphnane

Scheme 18.4

unsaturated amine **126** also contains the bicyclo[4.4.1]undecane feature that is seen in yuzurimine, and could account for the extra carbon that is found in daphnilactone A (**23**) (Scheme 18.5).

This hypothesis led to the postulation of various scenarios whereby squalene (**125**) might acquire a nitrogen atom and be transformed into the pentacyclic secodaphniphylline skeleton. The outline of this proposal is shown in Scheme 18.6. Step 1 is an oxidative transformation of squalene **125** into a dialdehyde, **127**. In step 2, it is proposed that some primary amine, perhaps pyridoxamine or an amino acid, condenses with one of the carbonyl groups of compound **127**, affording the imine **128**. Step 3 is the prototopic rearrangement of a 1-azadiene **128** to a 2-azadiene **129**. A nucleophilic species adds to the imine bond of **129** in step 4 to give the product **130**, followed by subsequent cyclization to give compound **131**. In steps 6–9, the resulting bicyclic dihydropyran derivative **131** is transformed into a dihydropyridine derivative **133** by a sequence of proton-mediated addition and elimination processes. Alkaloid **133** would then be converted into **134** by a catalyzed Diels–Alder reaction, and the final ring would result from an ene-like cyclization, giving alkaloid **135**. Because **135** is the first pentacyclic alkaloid to occur in the biogenesis of the daphniphyllum alkaloids, it was named proto-daphniphylline (**135**).

18.3.3
Biogenesis of the Daphnezomines

Daphnezomines A and B consisting of all six-membered rings are the first natural products containing an aza-adamantane core with an amino ketal bridge. A biogenetic pathway for daphnezomine B is proposed in Scheme 18.7. Daphnezomines A

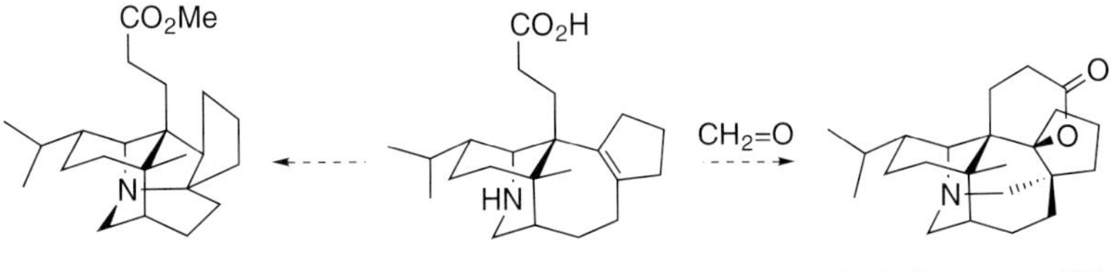

CO_2Me CO_2H $CH_2{=}O$

methyl homodaphniphyllate (**7**) **126** daphnilactone A (**23**)

Scheme 18.5

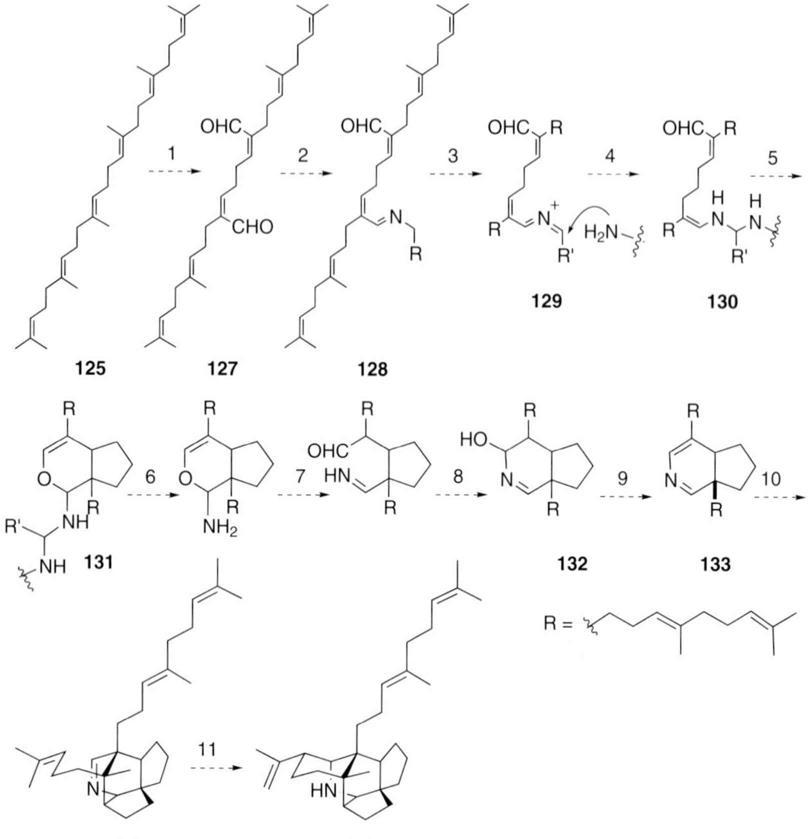

Scheme 18.6 Biogenesis of proto-daphniphylline (135).

and B might be generated through ring expansion accompanying backbone rearrangement of a common fragmentation intermediate.

Daphnezomines C and D are the first alkaloids possessing the secodaphniphylline-type skeleton with a nitrone functionality, while daphnezomine E is the first *N*-oxide of a daphniphylline-type alkaloid, although the *N*-oxides of yuzurimine-type alkaloids have been reported [23,29]. Heathcock offered a biogenetic conversion of the secodaphniphylline-type to the daphniphylline-type skeleton, in which an initial oxidation of the secodaphniphylline-type skeleton occurs on the nitrogen atom, followed by transformation into the daphniphylline-type skeleton through a ring-opened intermediate such as **B** (Schemes 18.4 and 18.8) [4,5]. The structures of daphnezomines C and D are very similar to that of a nitrone intermediate synthesized by Heathcock *et al.*[73]. Biogenetically, the daphniphylline-type skeleton (e.g. **1**) may be generated from the secodaphniphylline-type skeleton (e.g. **8**) through N-oxidation to generate an intermediate (**A**) or a nitrone such as **39**. Cleavage of the C-7–C-10

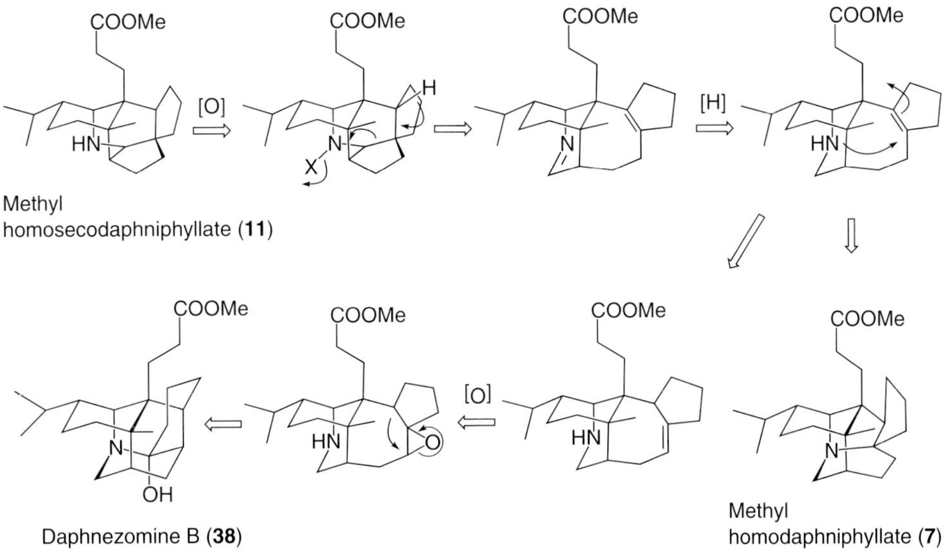

Methyl homosecodaphniphyllate (**11**)

Daphnezomine B (**38**)

Methyl homodaphniphyllate (**7**)

Scheme 18.7 Biogenesis of daphnezomine B (**38**).

bond, generation of a ring-opened imine intermediate (**B**), and formation of another C–N bond between N-1 and C-10, follows Heathcock's proposal (Figures 18.36 and 18.40).

The structures of daphnezomines F and G are similar to that of yuzurimine, but they lack the C-1–C-2 bond. A biogenetic pathway for daphnezomines F and G is proposed in Scheme 18.9. Daphnezomine G might be generated through oxidation

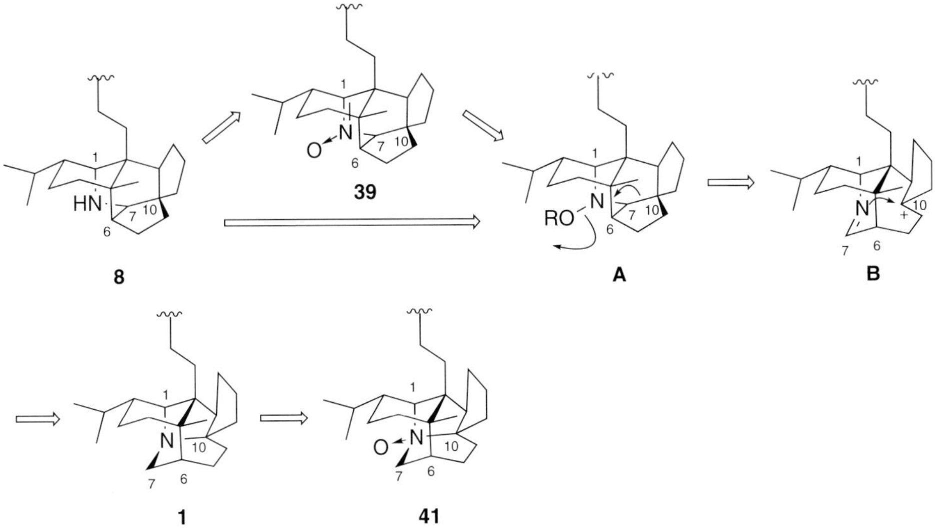

Scheme 18.8 Biogenesis of the daphniphylline skeleton of **1**.

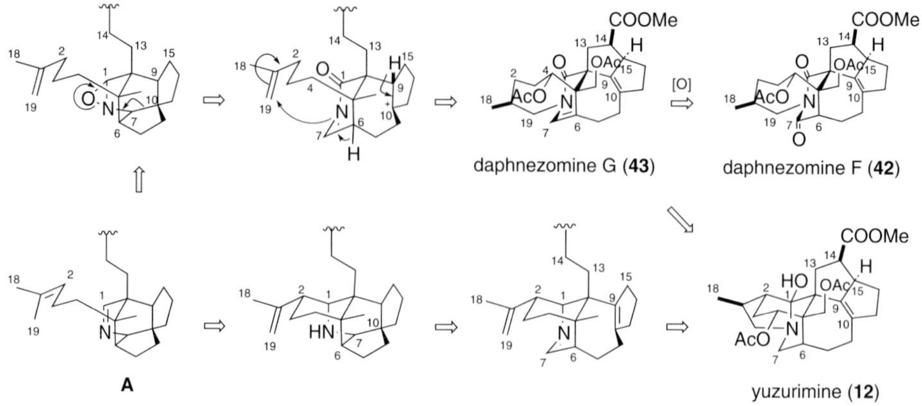

Scheme 18.9 Biogenesis of daphnezomines F (42) and G (43).

of a common imine intermediate **A** (proposed as a precursor of the secodaphniphylline-type skeleton by Heathcock *et al.*), and subsequent cleavage of the C-7–C-10 bond, followed by formation of the C-19–N-1 and C-14–C-15 bonds to give daphnezomine G. Daphnezomine F may be derived from daphnezomine G through oxidation of the C-7–C-6 bond. On the other hand, yuzurimine might be generated from the intermediate **A** through the secodaphniphylline-type skeleton, although an alternative pathway through **43** is also possible.

Biogenetically, daphnezomine I may be derived from daphnilactone B through oxidation at N-1, while daphnezomine J may be generated from yuzurimine through dehydroxylation at C-1. Daphnezomine L is structurally close to a biogenetic intermediate on the pathway from the secodaphnane to the daphnane skeleton [48]. Yamamura *et al.* suggested that a pentacyclic skeleton such as **48** is a biogenetic intermediate to the daphnane skeleton **51** [77], while Heathcock *et al.* proposed a biogenetic route from the secodaphnane to the daphnane skeletons through intermediates **A** and **B** (Scheme 18.10) [73]. Daphnezomines L and O might be biosynthesized through intermediates **A** and **B**, while daphnezomine N might be generated through intermediate **A** [48].

18.3.4
Biogenesis of the Daphnicyclidins

Daphnicyclidins A–G (**56**–**62**) and H (**63**) are novel alkaloids consisting of fused hexa- or pentacyclic ring systems, respectively. A biogenetic pathway for daphnicyclidins A–H is proposed in Scheme 18.11. The biogenetic origin of these alkaloids seems to be yuzurimine-type alkaloids, such as yuzurimine A and macrodaphniphyllamine, with an appropriate leaving group at C-4 and a methyl group at C-21. Rings B and C might be constructed by loss of the leaving group at C-4 followed by N-1–C-4 bond formation. Subsequently, cleavage of the C-1–C-8 bond followed by formation of the C-1–C-13 bond would result in enlargement of ring A, and aromatization of ring E to generate an intermediate **A**. Furthermore, oxidative cleavage of the C-10–C-17 bond

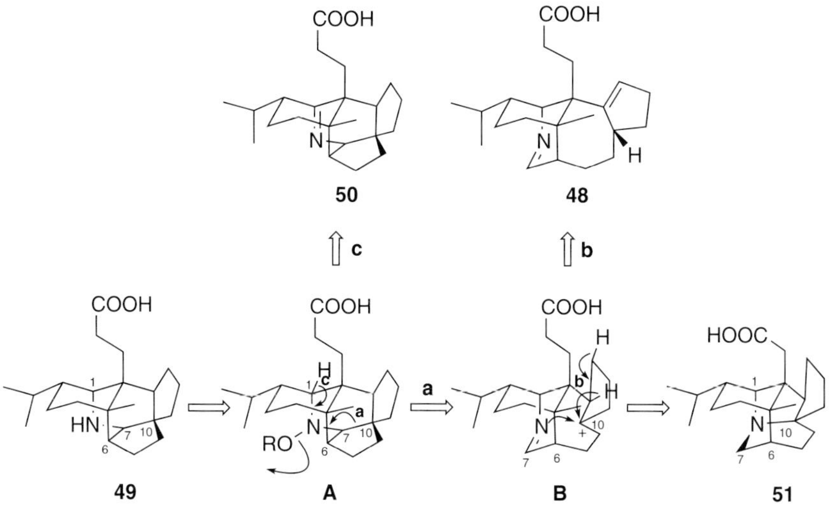

Scheme 18.10 Biogenesis of daphnezomines L–O (**48–51**).

could lead to daphnicyclidin H, followed by cyclization and dehydration to produce daphnicyclidin D, which may be oxidized to give daphnicyclidins E and F. On the other hand, cyclization of the 17-OH to C-22 in **63** to form ring F would generate daphnicyclidins A, B, and C.

A biogenetic pathway for daphnicyclidins J and K is proposed in Scheme 18.12. Daphnicyclidins J and K, as well as daphnicyclidins A–H reported more recently, might be derived from the yuzurimine-type alkaloids such as yuzurimine A and macrodaphniphyllamine. Daphnicyclidin J might be generated through N-oxidation of daphnicyclidin D, while daphnicyclidin K might be derived from an imine form **60** of daphnicyclidin D through introduction of hydroxy groups at C-2 and C-4, followed by acyloin rearrangement (Scheme 18.12).

18.3.5
Biogenesis of the Daphmanidins

A biogenetic pathway for daphmanidins A and B is proposed in Scheme 18.13. Daphmanidin A might be generated from a common imine intermediate **A**, which has been proposed as a precursor of the secodaphniphylline-type skeleton **B** by Heathcock *et al.* [4,5]. Cleavage of the C-7–C-10 bond in **B** will afford an intermediate with the yuzurimine-type skeleton, such as macrodaphniphyllidine, while subsequent cleavage of the N-1–C-7 bond, followed by formation of the C-7–C-2 bond will afford daphmanidin A. On the other hand, daphmanidin B might be derived from the imine intermediate **A** through formation of the N-1–C-19 bond.

Daphmanidins C and D might be derived through oxidative C–C bond fission followed by aldol-type condensation from daphmanidin B as shown in Scheme 18.14 [79].

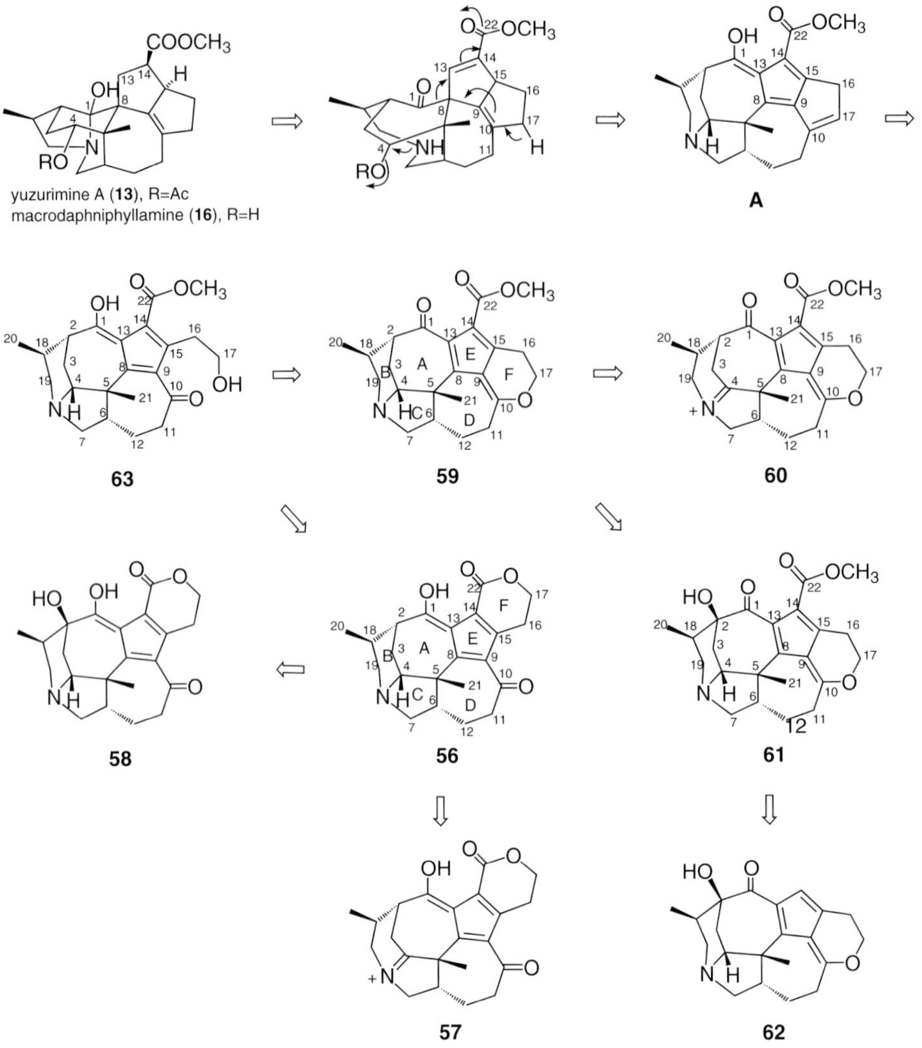

yuzurimine A (**13**), R=Ac
macrodaphniphyllamine (**16**), R=H

Scheme 18.11 Biogenetic pathway of daphnicyclidins A–H (**56–63**).

18.3.6
Biogenesis of the Daphniglaucins

A plausible biogenetic pathway for daphniglaucins A and D is proposed as shown in Scheme 18.15 [81,83]. Daphniglaucin A might be generated from the yuzurimine-type alkaloids such as yuzurimine A and macrodaphniphyllamine through a common imine intermediate **A**, which has been proposed as a precursor of the secodaphniphylline-type skeleton **B** by Heathcock *et al.* Loss of the leaving group at C-4 by attack of the nitrogen to form the N-1–C-4 bond will give daphniglaucin A [81].

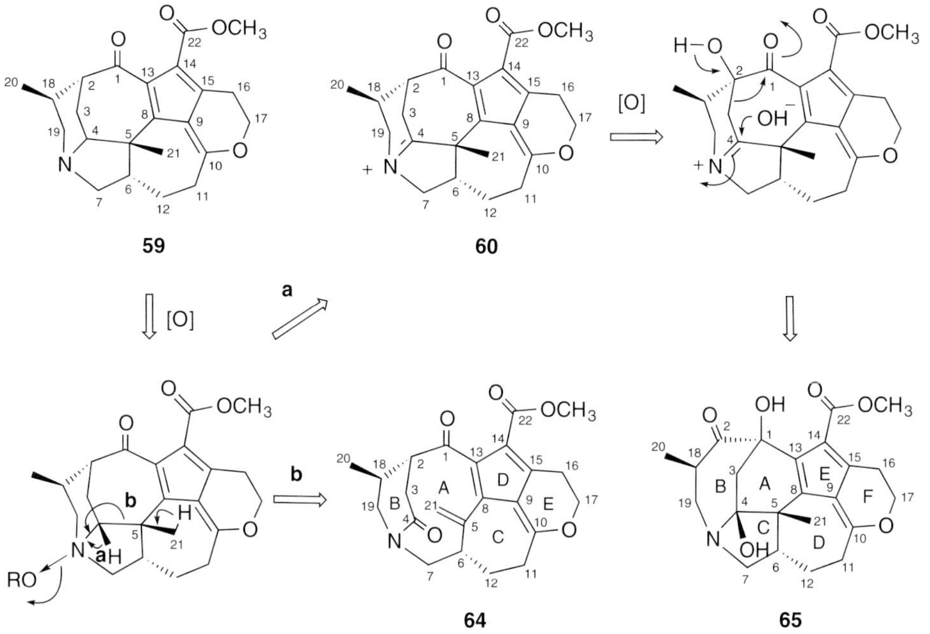

Scheme 18.12 Biogenetic pathway of daphnicyclidins J (**64**) and K (**65**).

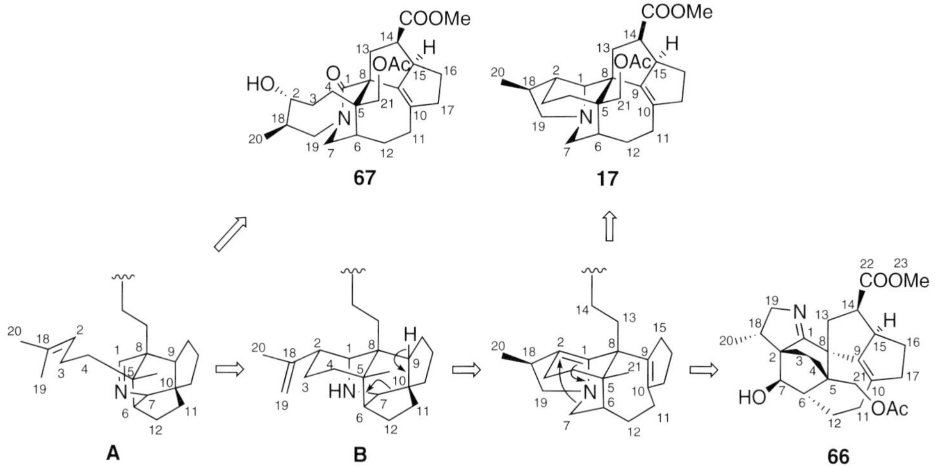

Scheme 18.13 Biogenesis of daphmanidins A (**66**) and B (**67**).

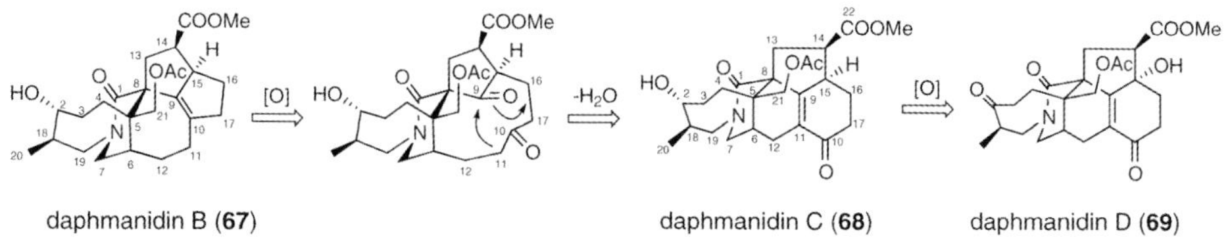

Scheme 18.14 Biogenesis of daphmanidins C (**68**) and D (**69**).

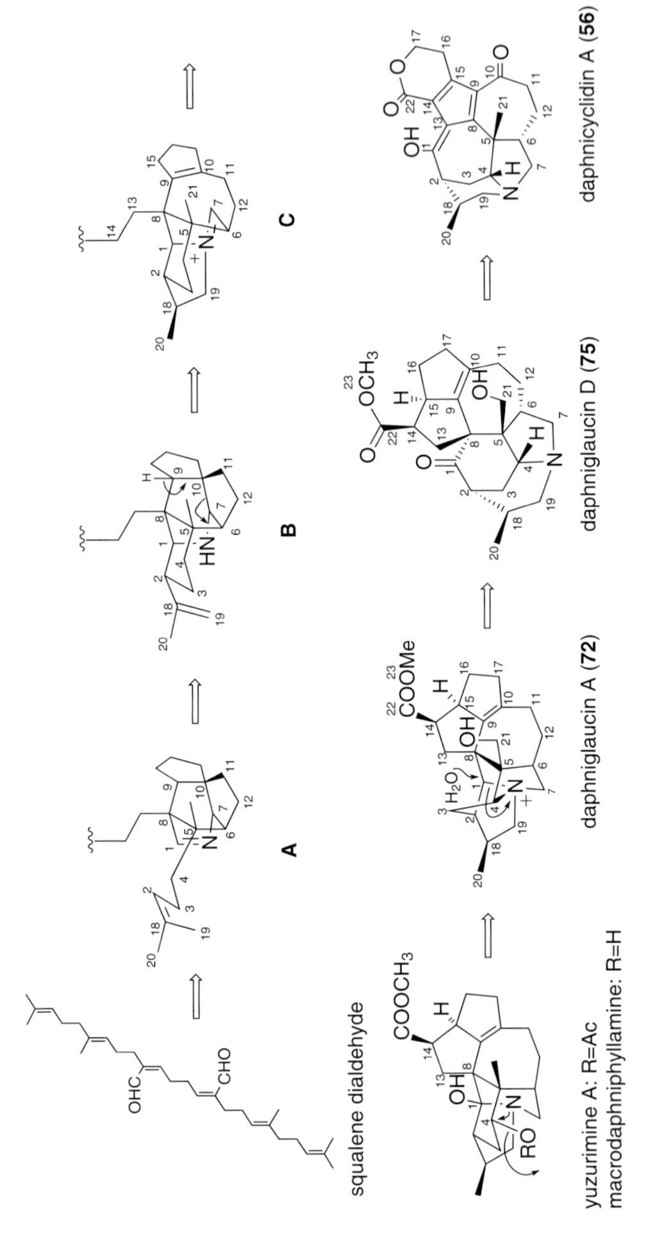

Scheme 18.15 Biogenesis of daphniglaucins A (**72**) and D (**75**).

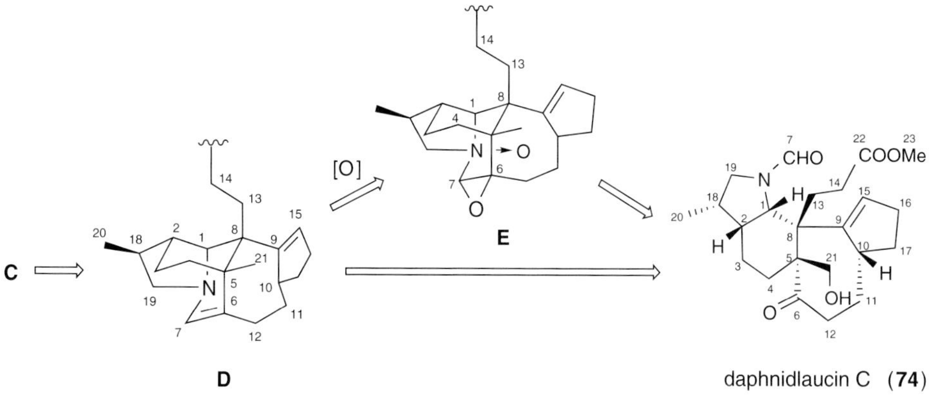

Scheme 18.16 Biogenesis of daphniglaucin C (**74**).

Cleavage of the C-1–N-1 bond of daphniglaucin A will give the skeleton of daphniglaucin D [83]. Furthermore, daphniglaucins A and D may be biogenetically related to daphnicyclidin A.

A plausible biogenetic pathway for daphniglaucin C is proposed in Scheme 18.16. The biogenetic origin of daphniglaucin C seems to be an imine intermediate **C** in Scheme 18.15. Oxidation of N-1, C-6, and C-7 of the intermediate **D** and cleavage of the C-6–C-7 bond of an intermediate **E** by Polonovski-type reaction will give the skeleton of daphniglaucin C, although an alternative path through oxidative cleavage of C-6–C-7 bond is also possible [82].

18.3.7
Biogenesis of the Calyciphyllines

A plausible biogenetic pathway for calyciphyllines A (**82**) and B (**83**) is shown in Scheme 18.17 [84]. Calyciphylline A (**82**) might be generated from the yuzurimine-type alkaloids such as daphniglaucin D. On the other hand, the biogenetic origin of calyciphylline B (**83**) seems to be an imine intermediate **C**, which might be produced through fragmentation reaction of the secodaphniphylline-type skeleton (**B**) derived from an imine intermediate **A**. Calyciphylline B (**83**) might be generated from attack of the carbonyl group to C-5 of the intermediate **C** and cleavage of the C-4–C-5 and C-8–C-9 bonds followed by C-7–C-9 bond formation. The stereochemistry at C-6 was suggested to epimerize through enamine formation during these backbone rearrangements.

18.3.8
Biogenesis of the Daphtenidines

Biogenetically, daphtenidines A (**84**) and B (**85**) might be generated through an intermediate **C** from secodaphnane-type alkaloid **B**, followed by the formation of daphnilactone A (**23**) in Scheme 18.18 [85].

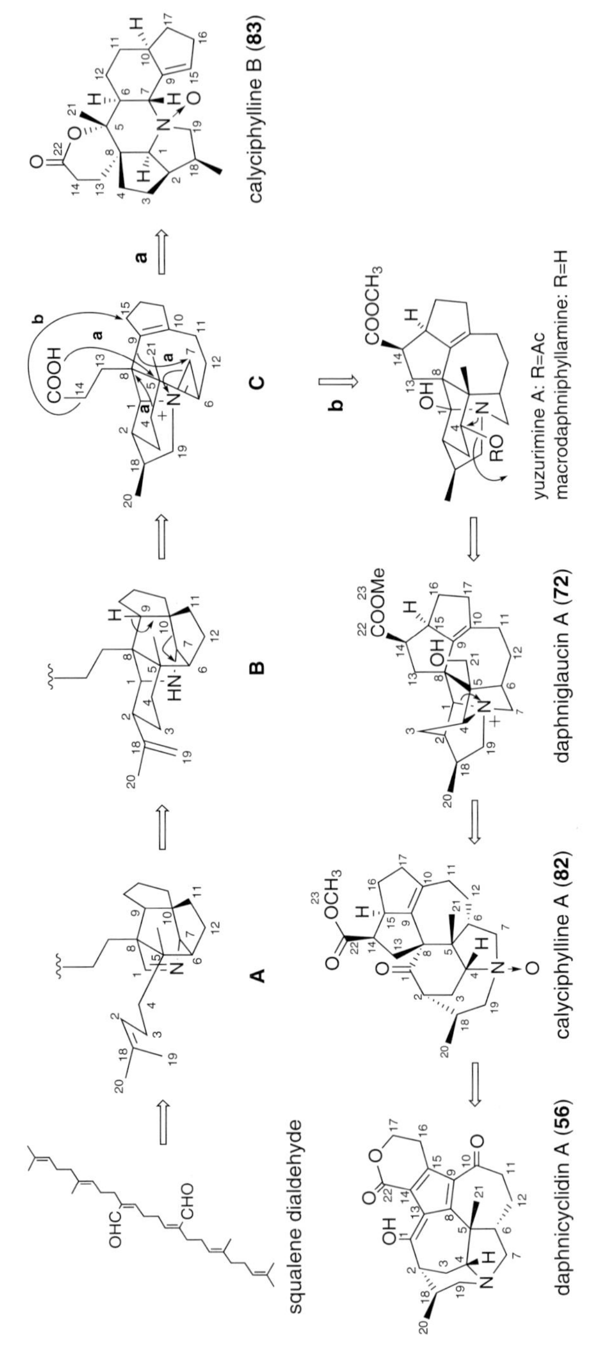

Scheme 18.17 Biogenesis of calyciphyllines A (**82**) and B (**83**).

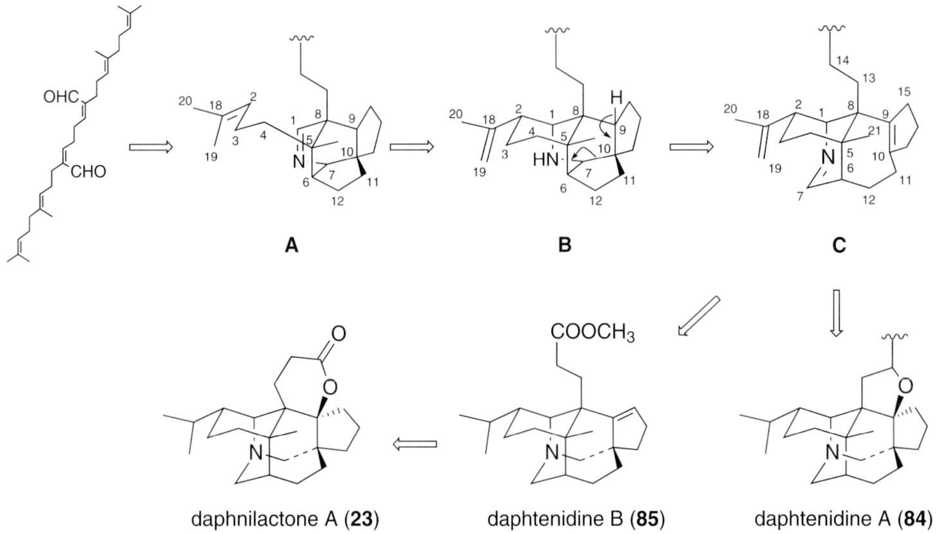

Scheme 18.18 Biogenesis of daphtenidines A (**84**) and B (**85**).

18.4
Synthesis

18.4.1
Biomimetic Chemical Transformations

18.4.1.1 Transformation of an Unsaturated Amine to the Daphnane Skeleton

Heathcock *et al.* suggested that the daphnane skeleton, such as methyl homodaphniphyllate, might arise by the cyclization of an unsaturated amine **136** [63]. Failure of this transformation under various acidic conditions presumably results from preferential protonation of the amine. In contrast, the bis-carbamoyl derivative **137**, obtained by treatment of the amino alcohol **136** with phenyl isocyanate, cyclizes smoothly in refluxing formic acid to provide the carbamate **138** (Scheme 18.19) [63]. The ease of cyclization of **137** raises the interesting question of whether a similar process might also be involved in the biosynthetic formation of the daphnane skeleton. The biogenetic carbamoylating agent could be carbamoyl phosphate.

18.4.1.2 Transformation of Daphnicyclidin D to Daphnicyclidins E and J

Daphnicyclidin J was obtained together with daphnicyclidin E from daphnicyclidin D through a modified Polonovski reaction [64] as shown in Scheme 18.20. Treatment of **59** with *m*-chloroperbenzoic acid (*m*-CPBA) followed by reaction with trifluoroacetic anhydride (TFAA) gave two compounds in 37 % and 18 % yields, whose spectral data were identical with those of natural daphnicyclidins E and J, respectively [50]. This result indicated that daphnicyclidin J might be generated through N-oxidation of daphnicyclidin D.

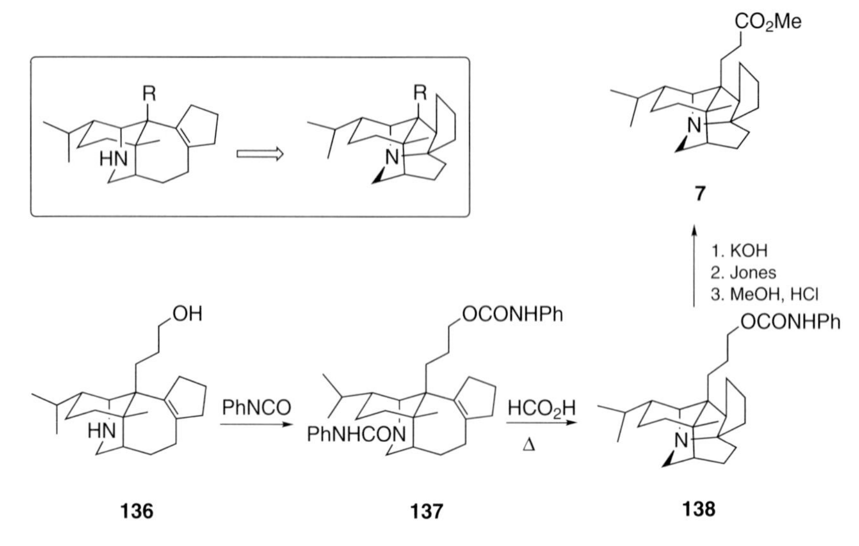

Scheme 18.19 Chemical transformation of **136** methyl homodaphniphyllate (**7**)[63].

18.4.2
Biomimetic Total Synthesis

18.4.2.1 Methyl Homosecodaphniphyllate and Protodaphniphylline

Heathcock *et al.* have embarked on a program to establish experimental methods to accomplish their proposals for the transformations of these alkaloids [4,5]. They initially focused their attention on the final stages of the polycyclization reaction leading to the secodaphniphylline skeleton [65,66]. Three simple building blocks, amide **139**, unsaturated ester **140**, and unsaturated iodide **141**, were combined in a highly convergent conjugate addition/enolate alkylation process to obtain the ester amide **142** in high yield. Straightforward methods were then employed to convert this substance into the dialdehyde **145**. Compound **145** was treated with ammonia, and then buffered acetic acid, to obtain the unsaturated amine **146** in excellent yield (64 % overall from **142** to **146**). The additional functional groups are used to convert **146** into racemic methyl homosecodaphniphyllate (**11**) [65,66].

The transformation of compound **145** to **146** involves a cascade of reactions and the two intermediates can be isolated. Thus, treatment of compound **145** with ammonia causes almost instantaneous transformation of the nonpolar dialdehyde to a complex mixture of polar materials, from which the dihydropyridine **147** can be isolated in about 45 % yield. This compound reacts rapidly on being treated with ammonium acetate in acetic acid at room temperature to give compound **148**, as the result of a formal intramolecular Diels–Alder reaction. Continued treatment

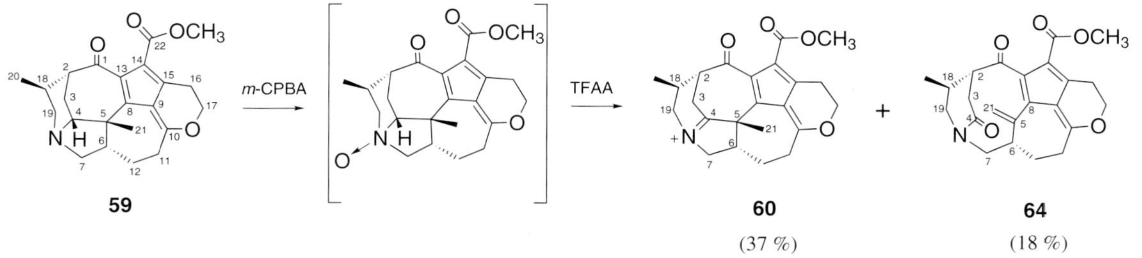

Scheme 18.20 Chemical transformation of daphnicyclidin D (**59**) to daphnicyclidins E (**60**) and J (**64**)[50].

with warm acetic acid converts compound **148** into compound **146** [65,66] (Scheme 18.21).

In addition, Heathcock and coworkers have intervened at an earlier stage in the biogenetic pathway depicted in Scheme 18.6. They prepared the dihydrosqualene

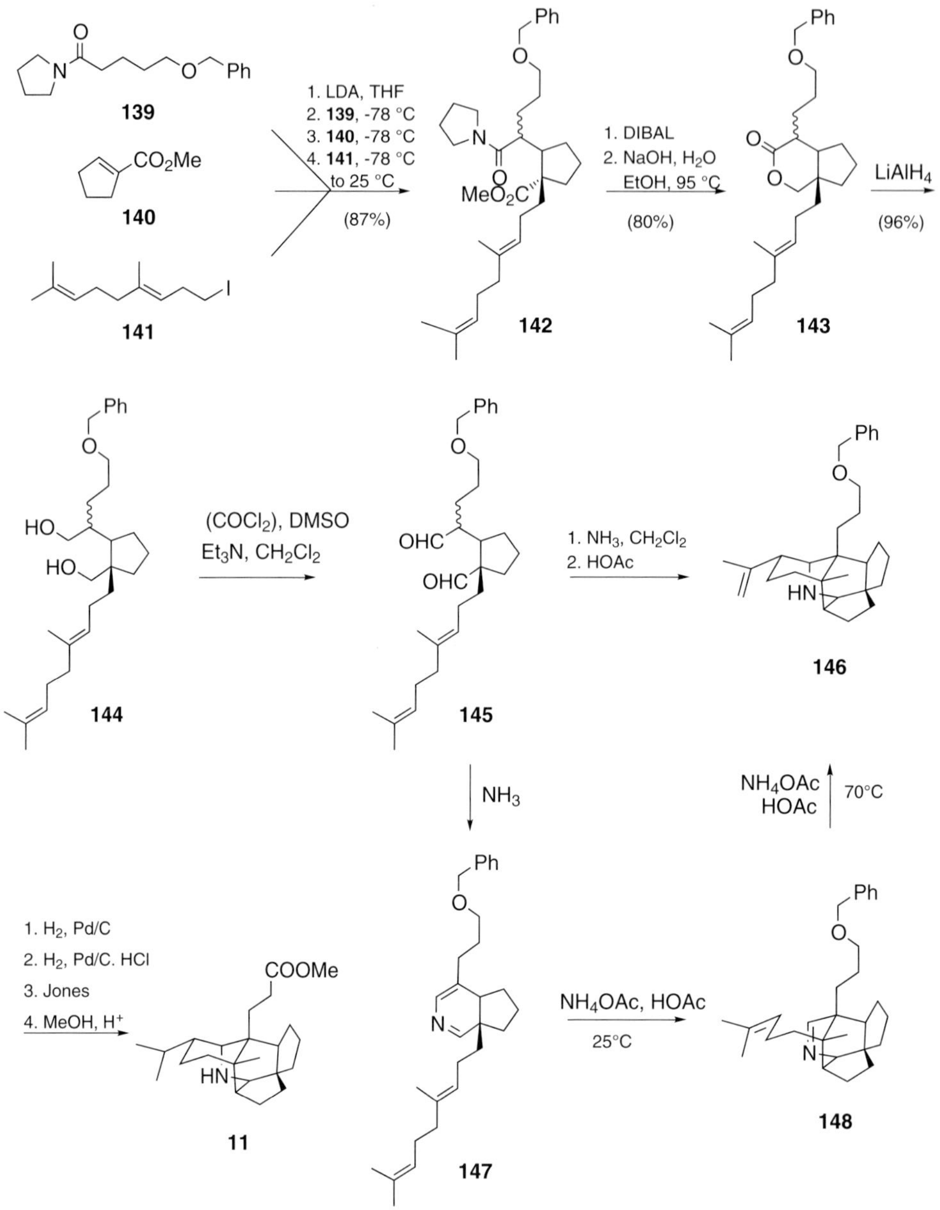

Scheme 18.21

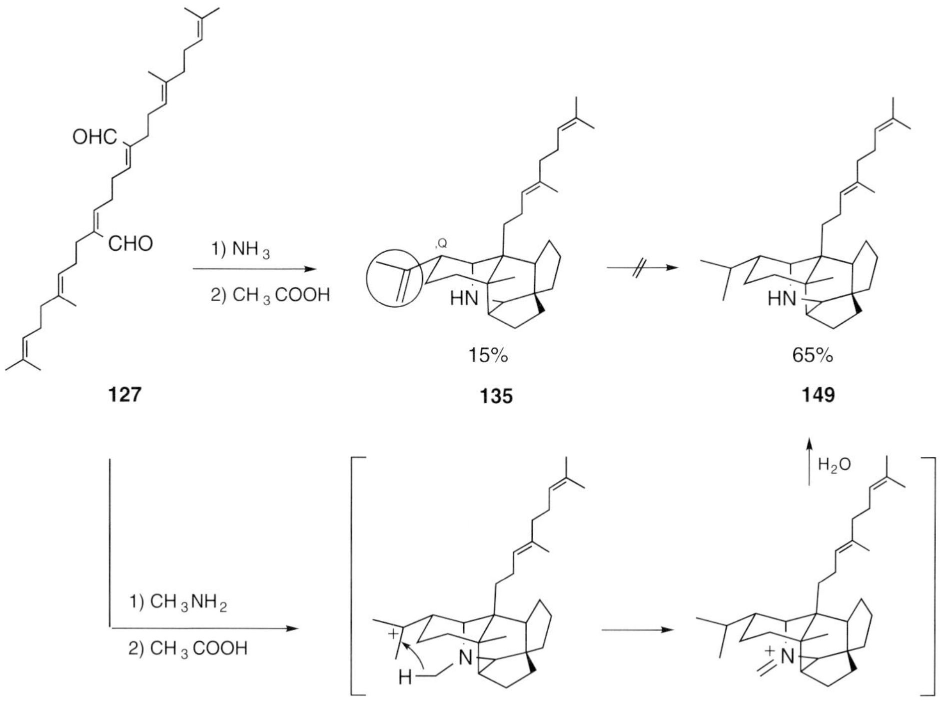

Scheme 18.22 Synthesis of dihydroprotodaphniphylline (**149**) (**79**).

dialdehyde **127** and treated it sequentially with ammonia and warm acetic acid. It was gratifying to find proto-daphniphylline **135** among the products of this reaction [67]. Although the isolation yield of **135** was only modest (15 %), a great deal has been accomplished, theoretically and practically, by the use of the simple reaction conditions. The fortuitous use of methylamine in place of ammonia suggested a possible solution to the problem of a low yield in the pentacyclization process with dihydrosqualene dialdehyde **127**. When compound **127** was treated successively with methylamine and warm acetic acid, dihydroprotodaphniphylline **149** was formed in 65 % yield (Scheme 18.22) [67].

This marvelous transformation results in the simultaneous formation of seven new sigma bonds and five rings. It is fully diastereoselective, and a necessary consequence of the reaction mechanism is that one of three similar carbon–carbon double bonds is regioselectively saturated.

18.4.2.2 **Secodaphniphylline**

An asymmetric total synthesis of (−)-secodaphniphylline was carried out using a mixed Claisen condensation between (−)-methyl homosecodaphniphyllate (**11**) and a carboxylic acid derivative **154** with the characteristic 2,8-dioxabicyclo[3.2.1]octane structure commonly found in the daphniphyllum alkaloids (Scheme 18.23) [68,69]. The necessary chirality was secured by an asymmetric Michael addition reaction of

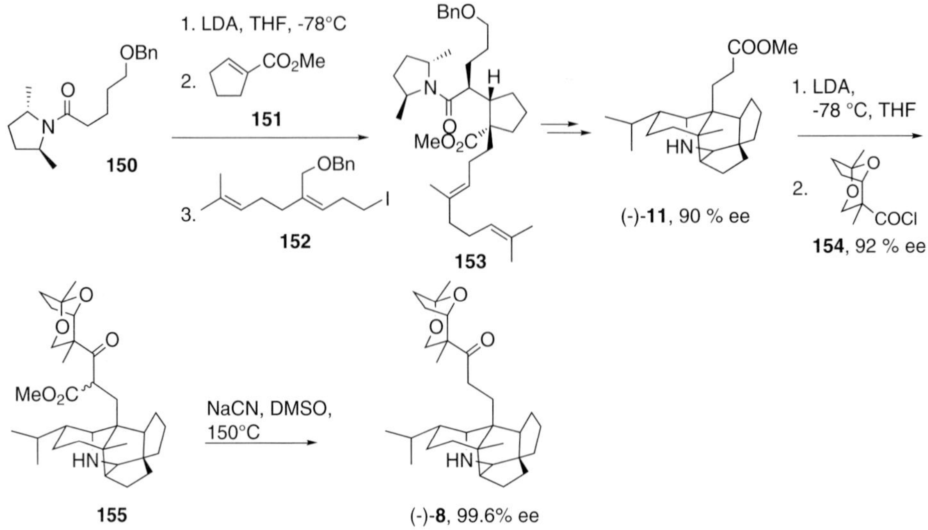

Scheme 18.23 Synthesis of (−)-secodaphniphylline (**8**) [68,69].

the lithium enolate of the C_2-symmetric amide **150** to the α,β-unsaturated ester **151** to give ester amide **153**. The conversion of **153** to (−)-**11** was performed by the same route as in the racemic series. Ester (−)-**11** and acid chloride **154** were joined by a mixed Claisen condensation and the resulting diastereomeric β-keto ester was demethylated and decarboxylated by treatment with NaCN in hot DMSO to obtain (−)-secodaphniphylline.

18.4.2.3 Methyl Homodaphniphyllate and Daphnilactone A

Synthetic work on the daphniphyllum alkaloids has been dominated by the versatile biomimetic synthesis developed by Heathcock and his collaborators. The first total synthesis of daphniphyllum alkaloids was achieved for methyl homodaphniphyllate (Scheme 18.24) [70,71]. The overall yield was about 1.1%. They employed network analysis outlined by Corey and chose an intramolecular Michael reaction for the strategic bond formation, since examination of molecular models of the hypothetical intermediate showed that there are conformations in which the indicated carbon in the tetrahydropyridone ring is within easy bonding distance of the β carbon of the cyclohexenone ring. The pentacyclic intermediate **167**, synthesized from the known keto acid **156**, was treated with a mixture of HCl and H_2SO_4 in aqueous acetone for two days to give two isomers in a ratio of 3 : 1. The major isomer was in full agreement with the expected Michael cyclization product **168**. Finally, racemic methyl homodaphniphyllate was obtained by reduction of **172** with hydrogen in the presence of Pearlman's catalyst, $Pd(OH)_2$ in ethanol at 120 °C and 1800 psi hydrogen pressure for 20 h, together with its isomer **173** at C-2 in the ratio of 1 : 1.

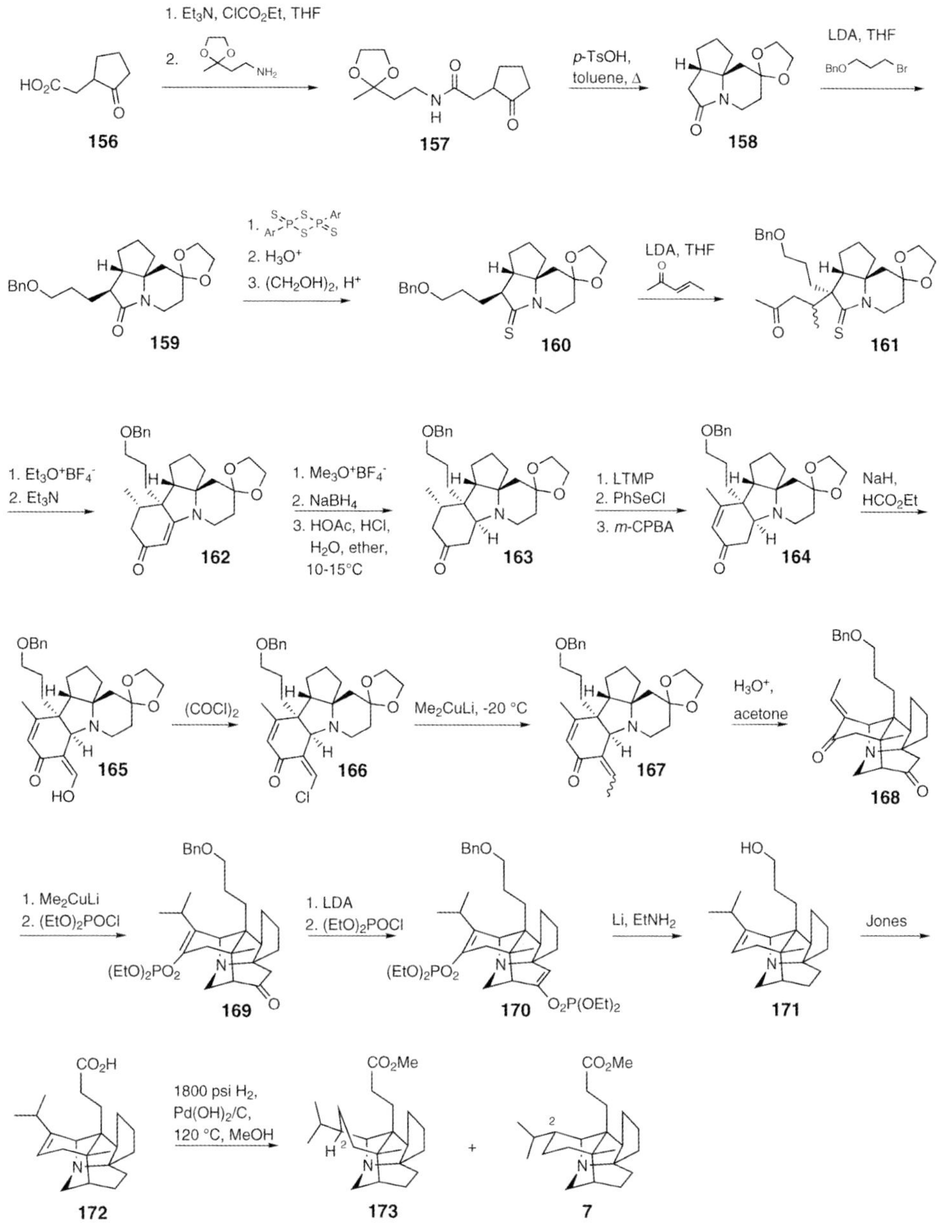

Scheme 18.24 Synthesis of methyl homodaphniphyllate (**7**). [71]

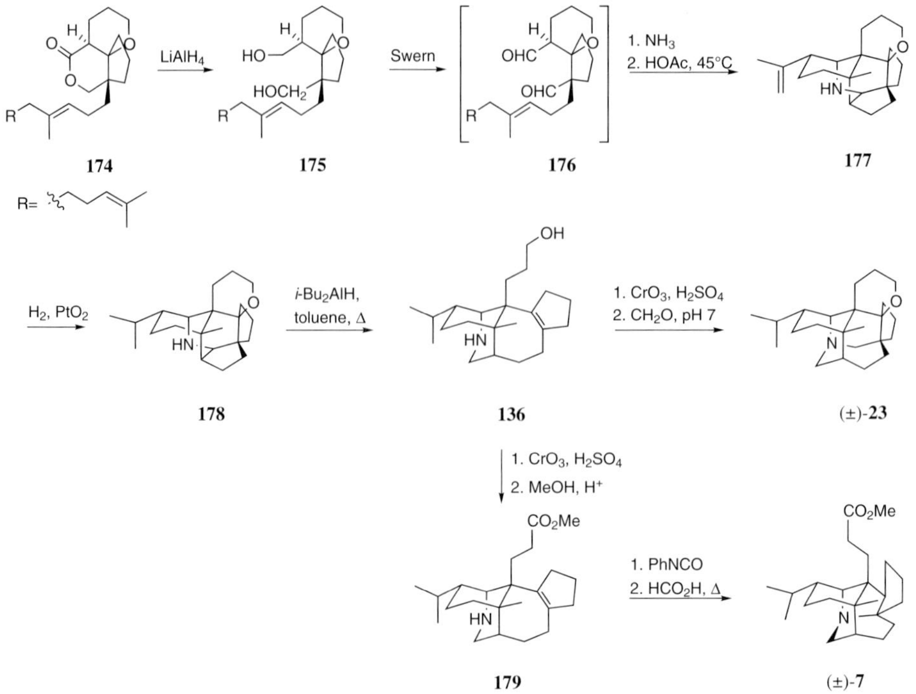

Scheme 18.25 Synthesis of (±)-methyl homodaphniphyllate (**7**) [63,72].

In addition, biomimetic total synthesis of (±)-methyl homodaphniphyllate has been carried out [63,72]. The synthesis began with the preparation of the tricyclic lactone ether **174**, which was reduced to the diol **175** with $LiAlH_4$. Oxidation of **175** gave a sensitive dialdehyde **176**, which was treated sequentially with ammonia and warm acetic acid to obtain the hexacyclic amino ether **177**. The tetracyclization process leading from **175** to **177** proceeded in 47% yield and resulted in the formation of five new sigma bonds and four new rings. Unsaturated amino alcohol **136** derived from **177** was converted into (±)-methyl homodaphniphyllate by a biomimetic process using a urea derivative as described previously. Furthermore, (±)-daphnilactone A (**23**) was synthesized from the unsaturated amino alcohol **136** by oxidation to the unsaturated amino acid, which was cyclized by treatment with aqueous formaldehyde at pH 7 [72] (Scheme 18.25).

A possibly biomimetic transformation of the secodaphnane to the daphnane skeleton with various Lewis acids has been investigated (Scheme 18.26) [73].

18.4.2.4 Codaphniphylline

(+)-Codaphniphylline, one of the C_{30} daphniphyllum alkaloids, was synthesized by a modification of Heathcock's biomimetic approach [74]. Modification was carried out by changing the tetrahydropyran to a tetrahydrofuran as in **189** (Scheme 18.27).

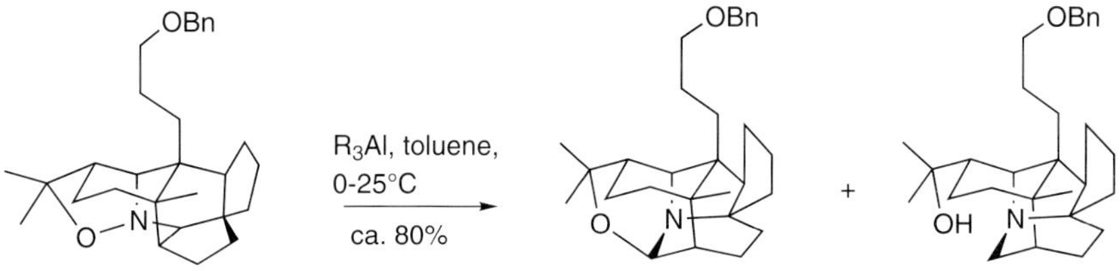

Scheme 18.26

This modification resulted in a yield improvement for the pentacyclization process from 47% to 66%. Treatment of the amino ether **192** with diisobutylaluminum hydride in refluxing toluene accomplished Eschenmoser–Grob fragmentation and reduction of the initially formed immonium ion, to give the unsaturated amino alcohol **193** in 86% yield. It was gratifying to find that **193** was the only product formed in this reaction. In the tetrahydropyran derivative, reduction of **192** to **193** is accompanied by about 15% simple elimination. Displacement of the tosyl group in **196** gives sulfide **197**, which is oxidized to sulfone **198**. This material is metallated and coupled with enantiomerically pure aldehyde to secure the codaphniphylline skelton [74].

18.4.2.5 **Bukittinggine**

Bukittinggine possesses key structural elements of both secodaphniphylline and yuzurimine. Consequently, the biogenesis of the heptacyclic alkaloid bukittinggine, isolated from *Sapium baccatum*, may be similar to that of the daphniphyllum alkaloids. The basic secodaphnane nucleus was synthesized in one step by application of the tetracyclization process to produce dihydroxy diether **205**. The pyrrolidine ring was formed by a Pd(II)-catalyzed oxidative cyclization of **206** to give the hexacyclic amine **207**. Hydrogenation of **207** proceeded with little diastereoselectivity in establishing the final stereocenter. However, the sequence of hydroboration/oxidation, tosylation, and reduction of **207** gave **209** under excellent stereocontrol. Debenzylation of **209**, followed by regiospecific oxidative lactonization of the diol, afforded (±)-bukittinggine (Scheme 18.28) [75].

18.4.2.6 **Polycyclization Cascade**

The scope of the 2-azadiene intramolecular Diels–Alder cyclization, employed for the synthesis of the daphniphyllum alkaloids, has been further investigated by Heathcock *et al.*[76]. The protocol involves Moffatt–Swern oxidation of the 1,5-diol to the dialdehyde, and treatment of the crude methylene chloride solution with ammonia followed by solvent exchange from methylene chloride to a buffered acetic acid solution. The cyclopentyl ring, quaternary carbon and tertiary carbon centers in

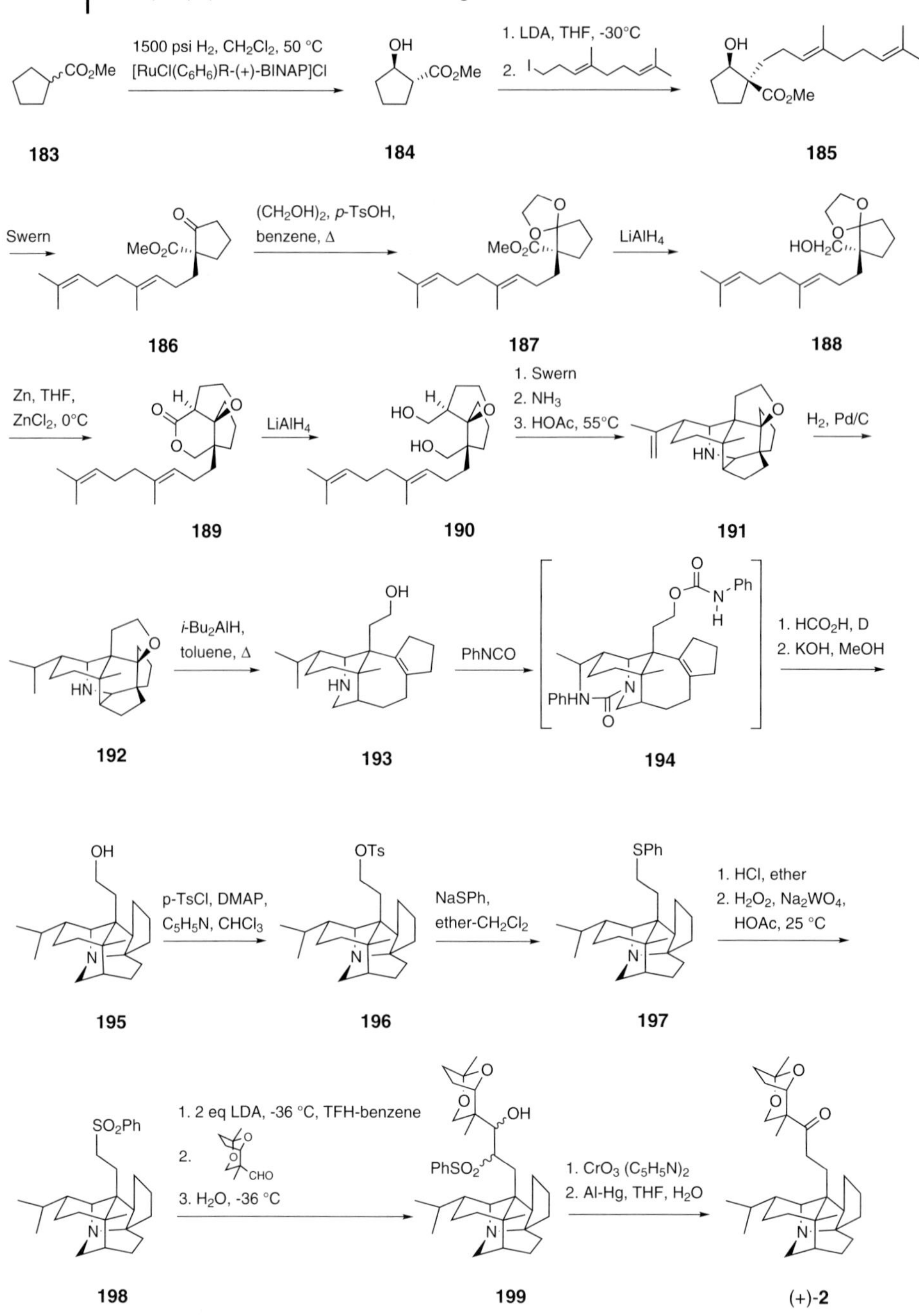

Scheme 18.27 Synthesis of (+)-codaphniphylline (**2**) [74].

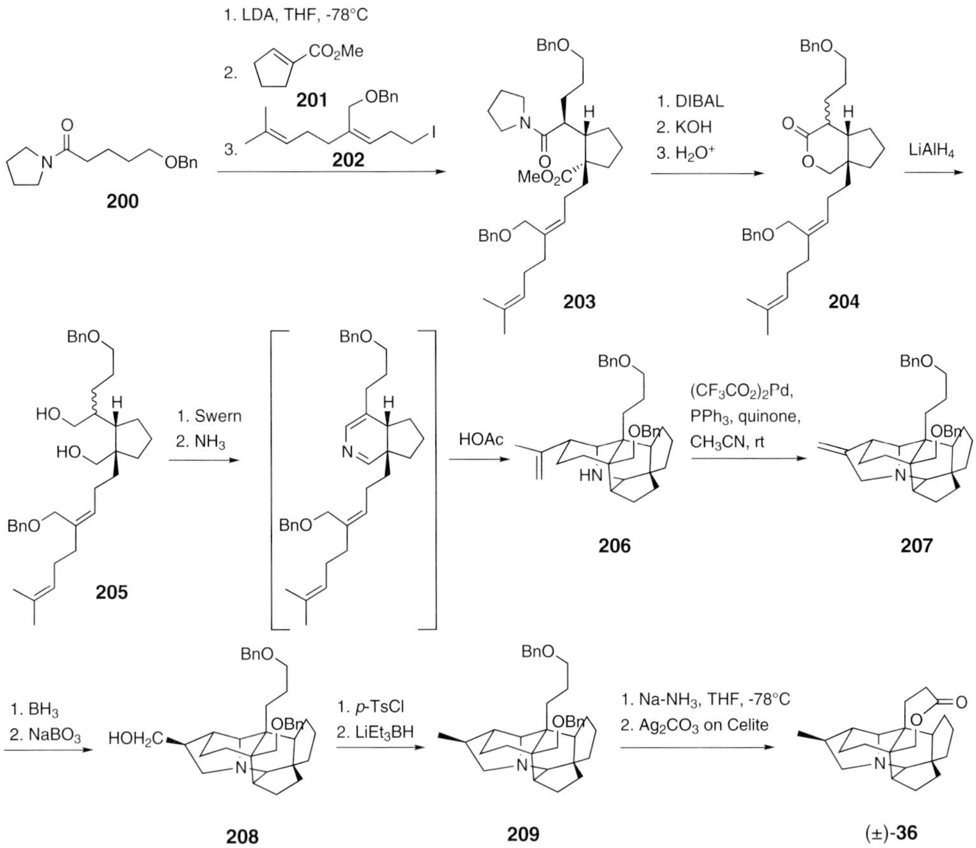

Scheme 18.28 Synthesis of (±)-bukittingine (**6**) [75].

the diol starting material all play a role in providing a selective and high-yielding cyclization (Scheme 18.29) [76].

18.5 Activities

Some daphniphyllum alkaloids, such as calyciphyllines A (**82**) and B (**83**), exhibited moderate cytotoxicity against murine lymphoma L1210 cells *in vitro* [84]. Daphniglaucin C showed inhibition of polymerization of tubulin at IC_{50} 25 mg/mL [82]. Recently, some daphniphyllum alkaloids such as daphmanidins E and F showed moderate vasorelaxant activity on rat aorta [80]. However, since the pharmacological activity of the daphniphyllum alkaloids is poorly studied, this area should be developed in future.

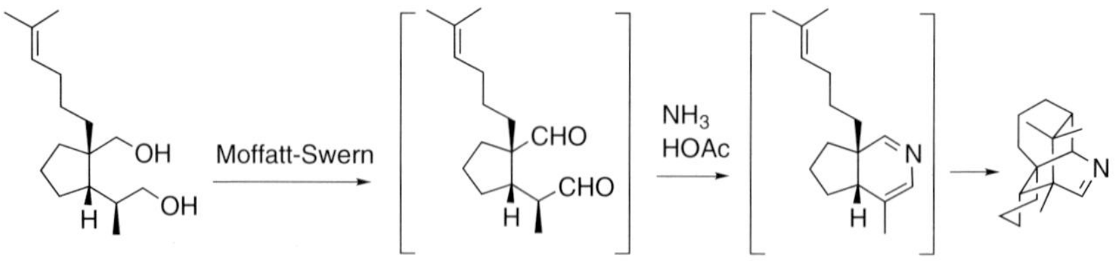

Scheme 18.29

18.6 Conclusions

Studies on the daphniphyllum alkaloids from 1966 to 2006 have been reviewed, with a particular focus on developments in the biomimetic synthesis of these alkaloids, and the structures of the new alkaloid types, such as daphnezomines, daphnicyclidins, daphmanidins, daphniglaucins, calyciphyllines, and daphtenidines. There are currently more than 100 daphniphyllum alkaloids of known structure. Further phytochemical investigations will bring increasing structural variation to this alkaloid group. Although the total syntheses of some of the daphnane and secodaphnane skeletons have been accomplished, the other skeletal variants remain an attractive subject. Similarly, the biosynthesis of daphniphyllum alkaloids has been only preliminarily studied, and the pathways have not been characterized with respect to the intermediates and the relevant enzymes. Widespread efforts for understanding the properties of these complex and fascinating alkaloids will result in further developments in this field.

References

1. Yamamura, S. and Hirata, Y. (1975) In Manske, R.H.F. (Ed.) *The Alkaloids*, Vol. 15, Academic Press, New York, 41.
2. Yamamura, S. (1986) In Brossi, A. *The Alkaloids*, Vol. 29, Academic Press, New York, 265.
3. Kobayashi, J. and Morita, H. (2003) In Cordell, G. A. *The Alkaloids*, Vol. 60, Academic Press, New York, 165.
4. Piettre, S. and Heathcock, C. H. (1990) *Science*, **248**, 1532–1534.
5. Heathcock, C. H. (1996) *Proceedings of the National Academy of Sciences of the United States of America*, **93**, 14323–14327.
6. Yagi, S. (1991) *Kyoto Igaku Zassi*, **6**, 208–223.
7. Sakabe, N., Irikawa, H., Sakurai, H., Hirata, Y. (1966) *Tetrahedron Letters*, 963–964.
8. Sakabe, N. and Hirata, Y. (1966) *Tetrahedron Letters*, 965–968.
9. Yamamura, S., Irikawa, H., Hirata, Y. (1967) *Tetrahedron Letters*, 3361–3364.
10. Irikawa, H., Sakabe, N., Yamamura, S., Hirata, Y. (1968) *Tetrahedron*, **24**, 5691–50.
11. Irikawa, H., Sakurai, H., Sakabe, N., Hirata, Y. (1966) *Tetrahedron Letters*, 5363–5368.
12. Irikawa, H., Yamamura, S., Sakabe, N., Hirata, Y. (1967) *Tetrahedron Letters*, 553–555.

13 Toda, M., Niwa, H., Hirata, Y., Yamamura, S. (1973) *Tetrahedron Letters*, 797–798.

14 Kamijo, N., Nakano, T., Terao, S., Osaki, K. (1966) *Tetrahedron Letters*, 2889–2892.

15 Nakano, T. and Saeki, Y. (1967) *Tetrahedron Letters*, 4791–4797.

16 Nakano, T., Saeki, Y., Gibbons, C. S., Trotter, J. (1968) *Chemical Communications*, 600–601.

17 Gibbons, C. S. and Trotter, J. (1969) *Journal of the Chemical Society B*, 840–843.

18 Nakano, T., Hasegawa, M., Saeki, Y. (1973) *Journal of Organic Chemistry*, **38**, 2404–2405.

19 Irikawa, H., Toda, M., Yamamura, S., Hirata, Y. (1969) *Tetrahedron Letters*, 1821–1824.

20 Toda, M., Yamamura, S., Hirata, Y. (1969) *Tetrahedron Letters*, 2585–2586.

21 Sasaki, K. and Hirata, Y. (1971) *Journal of the Chemical Society B*, 1565–1568.

22 Toda, M., Hirata, Y., Yamamura, S. (1972) *Tetrahedron*, **28**, 1477–1484.

23 Yamamura, S. and Hirata, Y. (1974) *Tetrahedron Letters*, 2849–2852.

24 Yamamura, S. and Hirata, Y. (1974) *Tetrahedron Letters*, 3673–3676.

25 Sakurai, H., Sakabe, N., Hirata, Y. (1966) *Tetrahedron Letters*, 6309–6314.

26 Irikawa, H., Yamamura, S., Hirata, Y. (1972) *Tetrahedron*, **28**, 3727–3738.

27 Sakurai, H., Irikawa, H., Yamamura, S., Hirata, Y. (1967) *Tetrahedron Letters*, 2883–2888.

28 Yamamura, S., Irikawa, H., Okumura, Y., Hirata, Y. (1975) *Bulletin of the Chemical Society of Japan*, **48**, 2120–2123.

29 Nakano, T. and Nilsson, B. (1969) *Tetrahedron Letters*, 2883–2884.

30 Yamamura, S. and Terada, Y. (1976) *Chemical Letters*, 1381–1382.

31 Hao, X.-J., Zhou, J., Node, M., Fuji, K. (1993) *Acta Botanica Yunnanica*, **15**, 205–207.

32 Sasaki, K. and Hirata, Y. (1972) *Journal of the Chemical Society, Perkin Transactions, 2*, 1411–1415.

33 Sasaki, K. and Hirata, Y. (1972) *Tetrahedron Letters*, 1275–1278.

34 Sasaki, K. and Hirata, Y. (1972) *Tetrahedron Letters*, 1891–1894.

35 Niwa, H., Toda, M., Hirata, Y., Yamamura, S. (1972) *Tetrahedron Letters*, 2697–20.

36 Niwa, H., Hirata, Y., Suzuki, K. T., Yamamura, S. (1973) *Tetrahedron Letters*, 2129–2132.

37 Toda, M., Niwa, H., Irikawa, H., Hirata, Y., Yamamura, S. (1974) *Tetrahedron*, **30**, 2683–2688.

38 Yamamura, S., Toda, M., Hirata, Y. (1976) *Bulletin of the Chemical Society of Japan*, **49**, 839.

39 Yamamura, S., Sasaki, K., Toda, M., Hirata, Y. (1974) *Tetrahedron Letters*, 2023–2026.

40 Yamamura, S., Lamberton, J. A., Irikawa, H., Okumura, Y., Hirata, Y. (1975) *Chemical Letters*, 923–926.

41 Yamamura, S., Lamberton, J. A., Irikawa, H., Okumura, Y., Toda, M., Hirata, Y. (1977) *Bulletin of the Chemical Society of Japan*, **50**, 1836–1840.

42 Yamamura, S., Lamberton, J. A., Niwa, M., Endo, K., Hirata, Y. (1980) *Chemical Letters*, 393–396.

43 Arbain, D., Byrne, L. T., Cannon, J. R., Patrick, V. A., White, A. H. (1990) *Australian Journal of Chemistry*, **43**, 185–190.

44 Morita, H., Yoshida, N., Kobayashi, J. (1999) *Journal of Organic Chemistry*, **64**, 7208–7212.

45 Morita, H., Yoshida, N., Kobayashi, J. (1999) *Tetrahedron*, **55**, 12549–12556.

46 Morita, H., Yoshida, N., Kobayashi, J. (2000) *Journal of Organic Chemistry*, **65**, 3558–3562.

47 Morita, H., Yoshida, N., Kobayashi, J. (2000) *Tetrahedron*, **56**, 2641–2646.

48 Morita, H. and Kobayashi, J. (2002) *Tetrahedron*, **58**, 6637–6641.

49 Kobayashi, J., Inaba, Y., Shiro, M., Yoshida, N., Morita, H. (2001) *Journal of the American Chemical Society*, **123**, 11402–11408.

50 Morita, H., Yoshida, N., Kobayashi, J. (2002) *Journal of Organic Chemistry*, **67**, 2278–2282.

51 Kobayashi, J., Ueno, S., Morita, H. (2002) *Journal of Organic Chemistry*, **67**, 6546–6549.

52 Flack, H. D. (1983) *Acta Crystallographica Section A*, **39**, 876–881.

53 Goto, T., Kishi, Y., Takahashi, S., Hirata, Y. (1964) *Tetrahedron Letters*, 779–786. Goto, T., Kishi, Y., Takahashi, S., Hirata, Y. (1965) *Tetrahedron*, **21**, 2059–2088. Woodward, R. B. (1964) *Pure and Applied Chemistry*, **9**, 49–74.

54 Roll, D. M., Biskupiak, J. E., Mayne, C. L., Ireland, C. M. (1986) *Journal of the American Chemical Society*, **108**, 6680–6682.

55 Morita, Y., Hesse, M., Schmid, H., Banerji, A., Banerji, J., Chatterjee, A., Oberhansli, W. E. (1977) *Helvetica Chimica Acta*, **60**, 1419–1434.

56 Borschberg, H.-J. (1984) *Helvetica Chimica Acta*, **67**, 1878–1882.

57 Kobrich, G. (1973) *Angewandte Chemie International Edition in English*, **12**, 464–473.

58 Toda, M., Hirata, Y., Yamamura, S. (1970) *Journal of the Chemical Society, Chemical Communications*, 1597–1598.

59 Mohamadi, F., Richards, N. G. J., Guida, W. C., Liskamp, R., Lipton, M., Caufield, C., Chang, G., Hendrickson, T., Still, W. C. (1990) *Journal of Computational Chemistry*, **11**, 440–467.

60 Halgren, T. (1990) *Journal of the American Chemical Society*, **112**, 4710–4723.

61 Harada, N., Nakanishi, K., Tatsuoka, S. (1969) *Journal of the American Chemical Society*, **91**, 5896–5898.

62 Suzuki, K. T., Okuda, S., Niwa, H., Toda, M., Hirata, Y., Yamamura, S. (1973) *Tetrahedron Letters*, 799–802.

63 Ruggeri, R. B. and Heathcock, C. H. (1990) *Journal of Organic Chemistry*, **55**, 3714–3715.

64 Grierson, D. (1990) *Organic Reactions*, **39**, 85–295.

65 Ruggeri, R. B., Hansen, M. M., Heathcock, C. H. (1988) *Journal of the American Chemical Society*, **110**, 8734–8736.

66 Heathcock, C. H., Hansen, M. M., Ruggeri, R. B., Kath, J. C. (1992) *Journal of Organic Chemistry*, **57**, 2544–2553.

67 Heathcock, C. H., Piettre, S., Ruggeri, R. B., Ragan, J. A., Kath, J. C. (1992) *Journal of Organic Chemistry*, **57**, 2554–2566.

68 Stafford, J. A. and Heathcock, C. H. (1990) *Journal of Organic Chemistry*, **55**, 5433–5434.

69 Heathcock, C. H. and Stafford, J. A. (1992) *Journal of Organic Chemistry*, **57**, 2566–2574.

70 Heathcock, C. H., Davidsen, S. K., Mills, S., Sanner, M. A. (1986) *Journal of the American Chemical Society*, **108**, 5650–5651.

71 Heathcock, C. H., Davidsen, S. K., Mills, S., Sanner, M. A. (1992) *Journal of Organic Chemistry*, **57**, 2531–2544.

72 Heathcock, C. H., Ruggeri, R. B., McClure, K. F. (1992) *Journal of Organic Chemistry*, **57**, 2585–2594.

73 Heathcock, C. H. and Joe, D. (1995) *Journal of Organic Chemistry*, **60**, 1131–1142.

74 Heathcock, C. H., Kath, J. C., Ruggeri, R. B. (1995) *Journal of Organic Chemistry*, **60**, 1120–1130.

75 Heathcock, C. H., Stafford, J. A., Clark, D. L. (1992) *Journal of Organic Chemistry*, **57**, 2575–2585.

76 Wallace, G. A. and Heathcock, C. H. (2001) *Journal of Organic Chemistry*, **66**, 450–454.

77 Niwa, H., Toda, M., Ishimaru, S., Hirata, Y., Yamamura, S. (1974) *Tetrahedron*, **30**, 3031–3036.

78 Morita, H., Takatsu, H., Kobayashi, J. (2003) *Tetrahedron*, **59**, 3575–3579.

79 Morita, H., Ishioka, N., Takatsu, H., Shinzatom, T., Obara, Y., Nakahata, N., Kobayashi, J. (2005) *Organic Letters*, **7**, 459–462.

80 Morita, H., Ishioka, N., Takatsu, H., Iizuka, T., Kobayashi, J. (2006) *Journal of Natural Products*, **69**, 418–420.

81 Kobayashi, J., Takatsu, H., Shen, Y.-C., Morita, H. (2003) *Organic Letters*, **5**, 1733–1736.

82 Morita, H., Takatsu, H., Shen, Y.-C., Kobayashi, J. (2004) *Tetrahedron Letters*, **45**, 901–904.

83 Takatsu, H., Morita, H., Shen, Y.-C., Kobayashi, J. (2004) *Tetrahedron*, **60**, 6279–6284.

84 Morita, H. and Kobayashi, J. (2003) *Organic Letters*, **5**, 2895–2898.

85 Kubota, T., Matsuno, Y., Morita, H., Shinzato, T., Sekiguchi, M., Kobayashi, J. (2006) *Tetrahedron*, **62**, 4743–4748.
86 Jossang, A., Bitar, H. E., Pham, V. C., Sévenet, T. (2003) *Journal of Organic Chemistry*, **68**, 300–304.
87 Bitar, H. E., Nguyen, V. H., Gramain, A., Sévenet, T., Bodo, B. (2004) *Tetrahedron Letters*, **45**, 515–518.
88 Bitar, H. E., Nguyen, V. H., Gramain, A., Sévenet, T., Bodo, B. (2004) *Journal of Natural Products*, **67**, 1094–1099.
89 Zhan, Z.-J., Zhang, C.-R., Yue, J.-M. (2005) *Tetrahedron*, **61**, 11038–11045.
90 Yang, S.-P., Yue, J.-M. (2003) *Journal of Organic Chemistry*, **68**, 7961–7966.
91 Yang, S.-P. and Yue, J.-M. (2004) *Organic Letters*, **6**, 1401–1404.
92 Zhan, Z.-J., Yang, S.-P., Yue, J.-M. (2004) *Journal of Organic Chemistry*, **69**, 1726–1729.
93 Chen, X., Zhan, Z.-J., Yue, J.-M. (2004) *Chemistry and Biodiversity*, **1**, 1513–1518.
94 Chen, X., Zhan, Z.-J., Yue, J.-M. (2005) *Helvetica Chimica Acta*, **88**, 854–860.
95 Yang, S.-P., Zhang, H., Zhang, C.-R., Cheng, H.-D., Yue, J.-M. (2006) *Journal of Natural Products*, **69**, 79–82.
96 Zhang, H., Yang, S.-P., Fan, C.-Q., Ding, J., Yue, J.-M. (2006) *Journal of Natural Products*, **69**, 553–557.

19
Structure and Biosynthesis of Halogenated Alkaloids

Gordon W. Gribble

19.1
Introduction

Of the more than 4500 known naturally occurring organohalogen compounds, a large fraction are alkaloids [1,3]. Most of these halogenated pyrroles, indoles, carbazoles, carbolines, tyrosines, and others have a marine origin. The present chapter surveys the occurrence, structure, and biosynthesis of these fascinating natural products. However, given their sheer number, this review focuses mainly on recent examples.

19.2
Structure of Halogenated Alkaloids

19.2.1
Indoles

Because the amino acid tryptophan is ubiquitous in nature, it is not surprising that a plethora of halogenated tryptophans, tryptamines, and simple indoles occur naturally [1–4]. In addition to the simple 3-chloroindole, 3-bromoindole, 3-chloro-6-bromoindole, 3,6-dibromoindole, 2,3,6,7-tetrabromoindole, and several other polyhalogenated indoles found in seaweeds, acorn worms, and ascidians [1–4], newer examples include di- and tribromoindoles from the common oyster (*Crassostrea virginica*) [5], sulfur-containing polybromoindoles (e.g. **11**) from the Formosan red alga *Laurencia brongniartii* [6], ancorinolates A (**2**) and C (**3**) from the sponge *Ancorina* sp. [7], and the novel oxindole matemone (**4**) from the Indian Ocean sponge *Iotrochota purpurea* [8]. The marine ascidian *Stomoza murrayi* has furnished several brominated indole-3-carbaldehydes (e.g. **5**) that prevent larval settlement or overgrowth by other species [9]. The Morocco tunicate *Cynthia savignyi* produces the novel chlorinated alkaloid cynthichlorine (**6**)[10].

Modern Alkaloids: Structure, Isolation, Synthesis and Biology. Edited by E. Fattorusso and O. Taglialatela-Scafati

ISBN: 978-3-527-31521-5

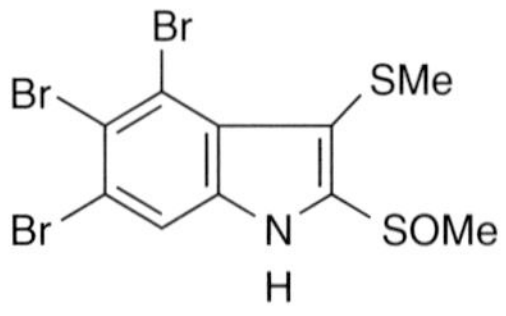

1

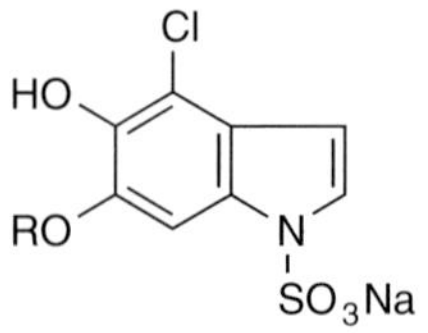

2, R = SO_3Na (ancorinolate A)
3, R = H (ancorinolate C)

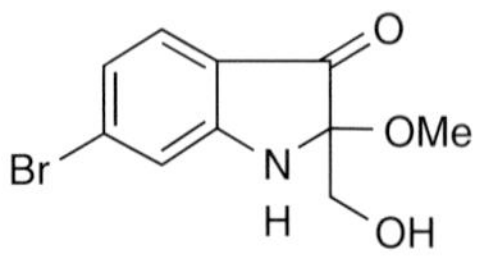

4 (ma temone)

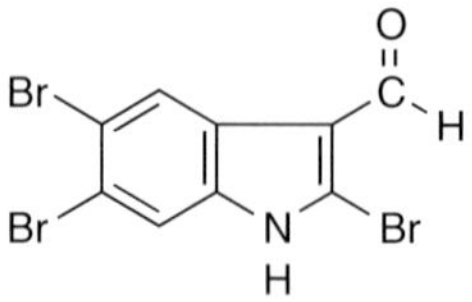

5

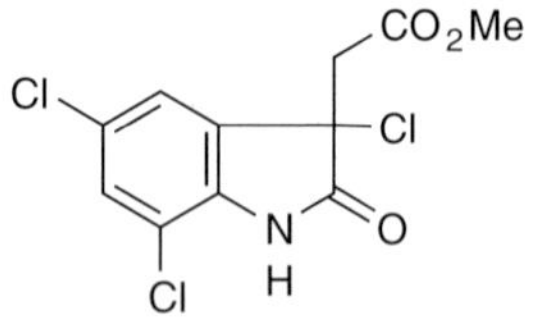

6 (cynthichlorine)

Many peptides from the marine snail *Conus* sp. contain 6-bromotryptophan [11], and a study of the sponges *Thorectandra* and *Smenospongia* has uncovered six novel brominated tryptophans (e.g., **7**) [12]. The Caribbean sponge *Plakortis simplex* has yielded six novel iodinated alkaloids, plakohypaphorines A–F [e.g., E (**8**) and F (**9**)] [13,14]. Plakohypaphorine E (**8**) is the first naturally occurring triiodinated indole. The California tunicate *Didemnum candidum* contains 6-bromotryptamine along with 2,2-bis(6′-bromo-3′-indolyl)ethylamine (**10**) [15]. The marine snail *Calliostoma canaliculatum* produces disulfide **11** as an apparent chemical weapon to thwart attack by the predatory starfish *Pycnopodia helianthoides* and *Pisaster giganteus* [16,17].

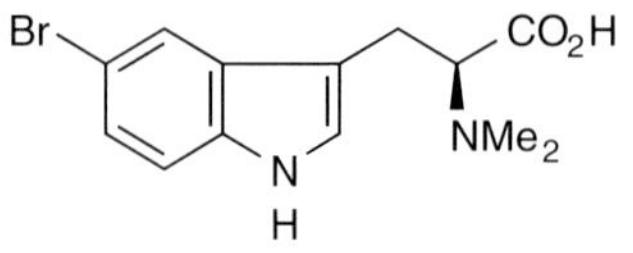

7

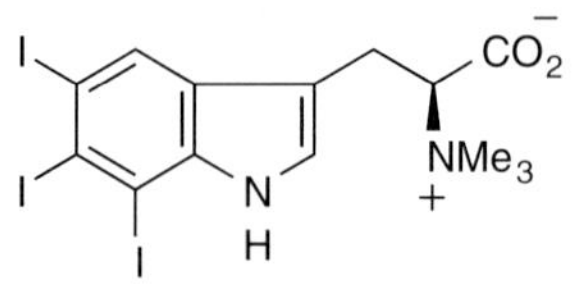

8 (plakohypaphorine E)

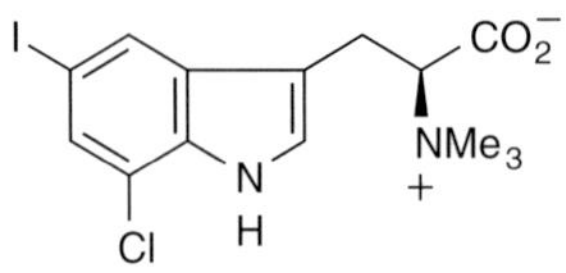

9 (plakohypaphorine F)

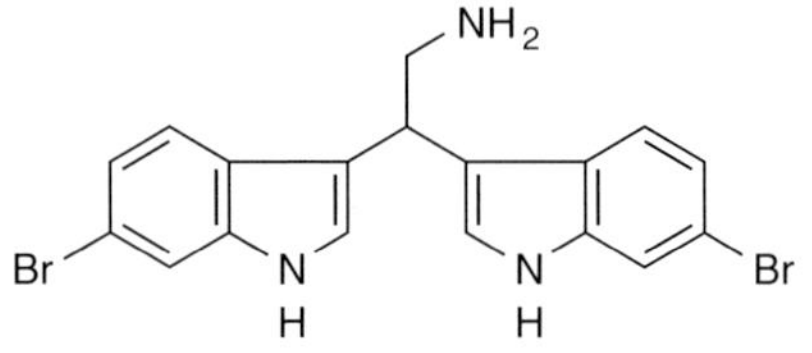

10

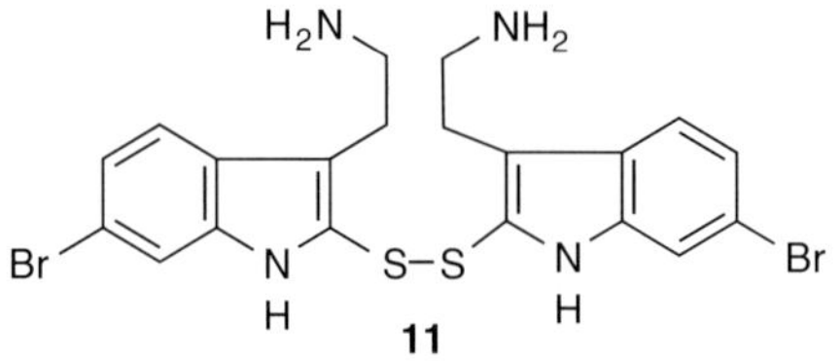

11

The marine bryozoan *Flustra foliacea* is a prolific source of brominated indoles [18], including several prenylated examples, such as **12** [19], **13** [20,21], **14** [21], and hexahydropyrrolo[2,3-*b*]indol-7-ol (**15**) [21]. An excellent review of marine alkaloids including those from bryozoa is available [18].

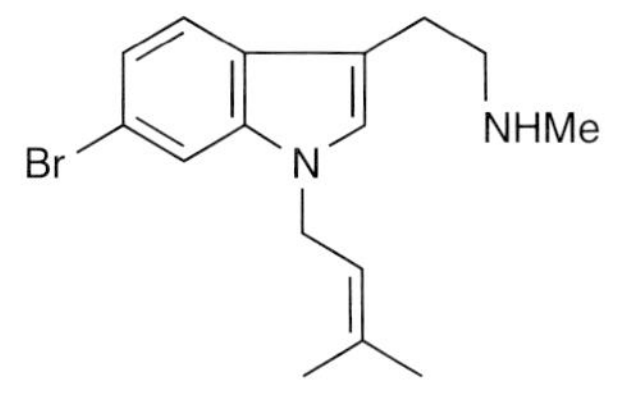

12

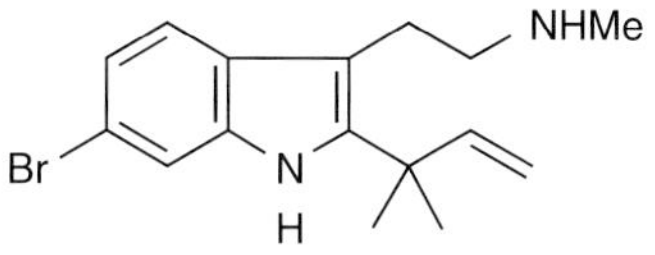

13

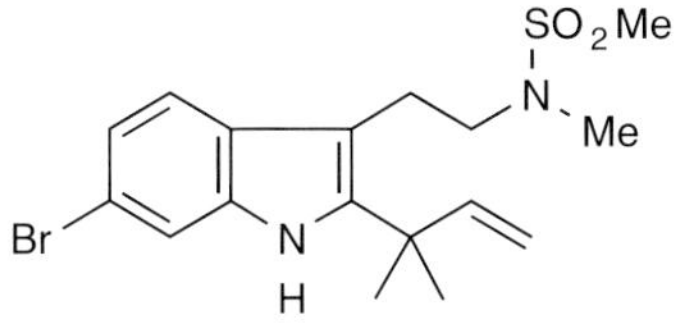

14

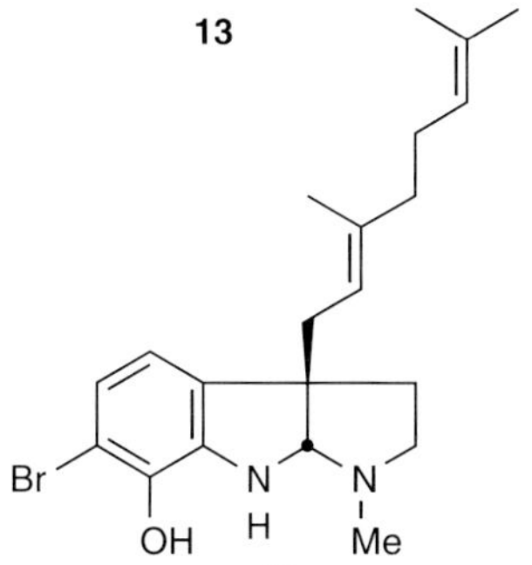

15

Dendridine A (**16**) is a novel bis-indole alkaloid isolated recently from the sponge *Dictyodendrilla* sp. [22]. The deep-water New Caledonian sponge *Orina* sp. (or *Gellius* sp.) has yielded several brominated bis- and tris-indoles, such as gelliusine (**17**), as a mixture of diastereomers [23,24]. The sponge *Spongosorites* sp. has yielded several cytotoxic bisindole alkaloids of the hamacanthin class ((*S*)-6″-debromohamacanthin B (**18**)) and the topsentin class (dibromodeoxytopsentin (**19**)) [25].

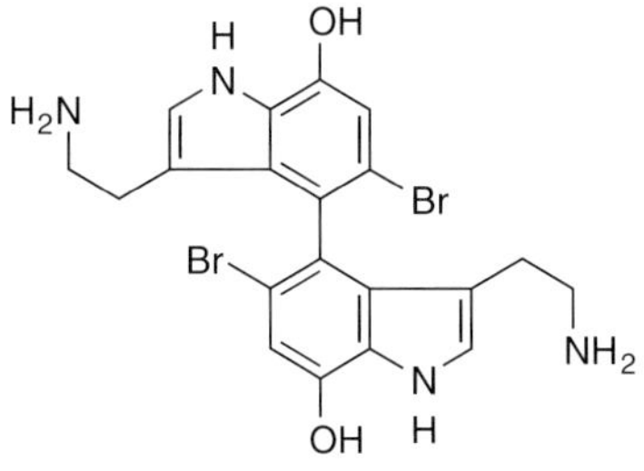

16 (dendridine A)

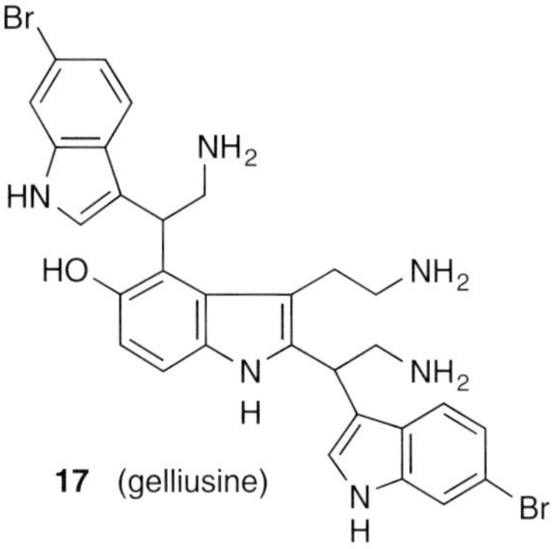

17 (gelliusine)

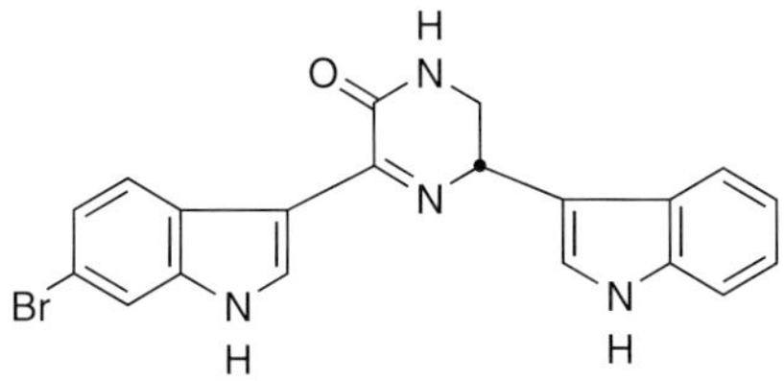

18 (*S* -6"-debromohamacanthin B)

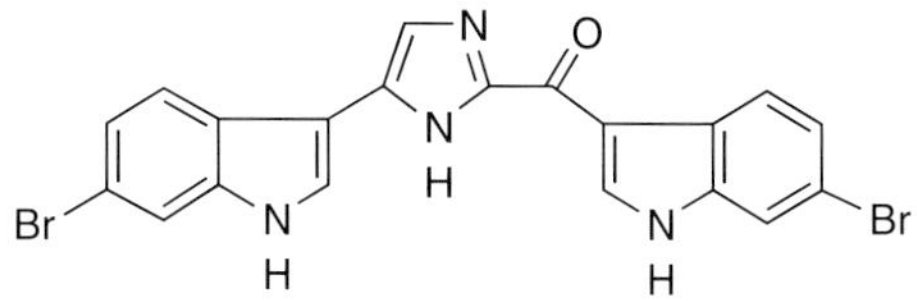

19 (dibromodeoxytopsentin)

Another large group of brominated tryptophan-derived marine alkaloids are the aplysinopsins and several new examples have been reported. The sponge *Hyrtios erecta* has furnished **20** and **21**[26], while a New Zealand ascidian is the source of kottamides A–E, for example A (**22**) [27] and E (**23**) [28]. The stony coral *Tubastraea* sp. contains the structurally complex and unprecedented bisindole tubastrindole A (**24**) [29].

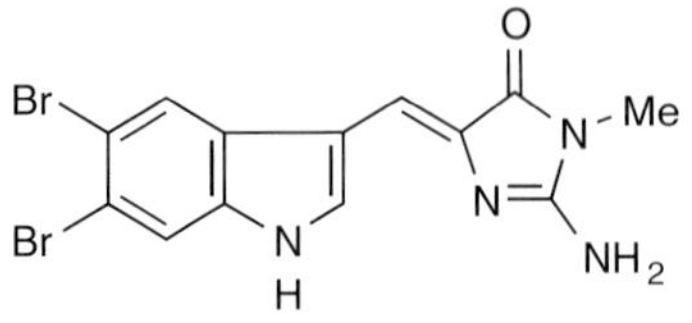

20

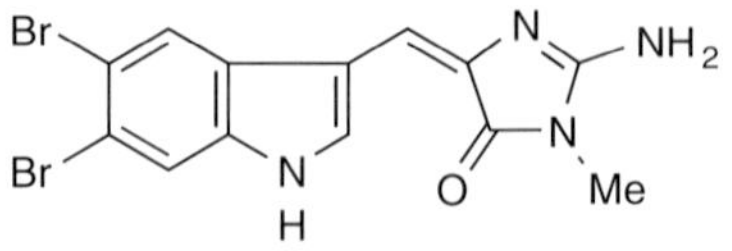

21

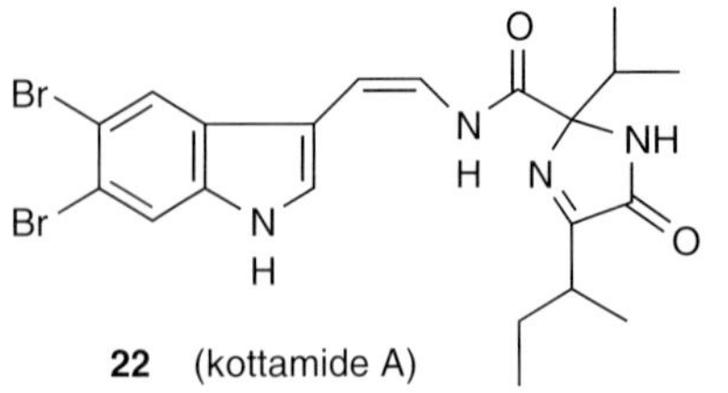

22 (kottamide A)

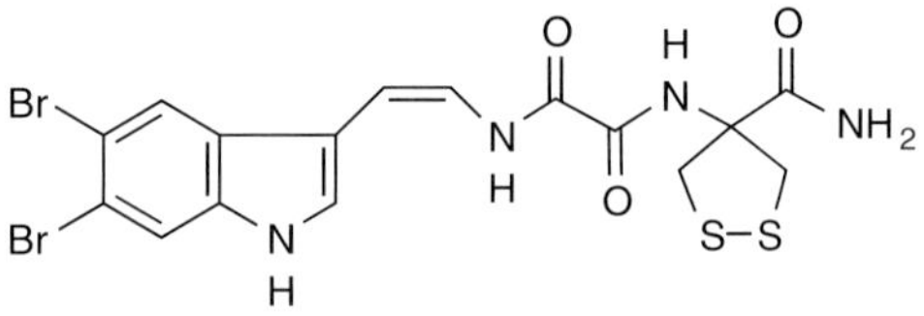

23 (kottamide E)

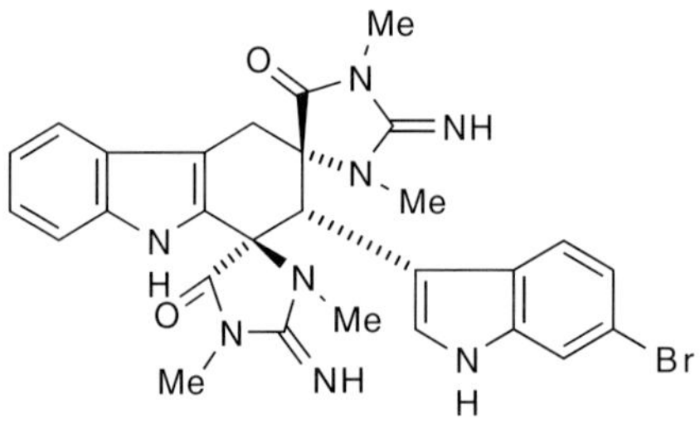

24 (tubastrindole A)

The pyrroloiminoquinone marine alkaloids, for example discorhabdins and batzellines, represent a large and diverse group of halogenated compounds [1,30], and several novel examples have been reported [31–34]. An Indo-Pacific collection of the sponge *Zyzzya fuliginosa* has revealed isobatzelline E (**25**) and batzelline D (**26**) [31]. The deep-water Caribbean sponge of genus *Batzella* has yielded discorhabdins S (**27**), T, and U [32]. An exhaustive study of four new species of South African latrunculid sponges *Tsitsikamma pedunculata, Tsitsikamma favus, Latrunculia bellae,* and *Strongylodesma algoaensis* afforded eight new discorhabdin alkaloids, including four brominated ones (e.g. 14-bromo-3-dihydro-7,8-dehydrodiscorhabdin C (**28**)) [33]. A New Zealand *Latrunculia* sp. sponge has produced the first dimeric discorhabdin (W, **29**) [34].

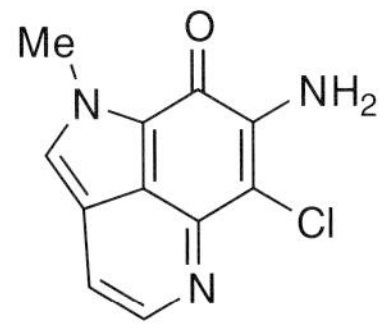

25 (isobatzelline E) 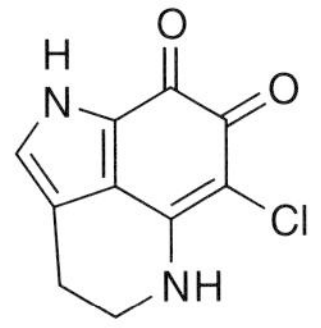 **26** (batzelline D) 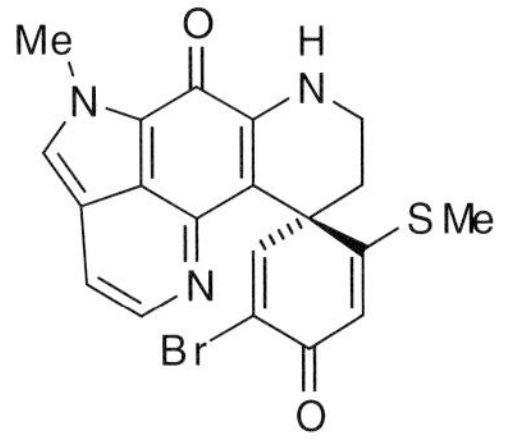 **27** (discorhabdin S)

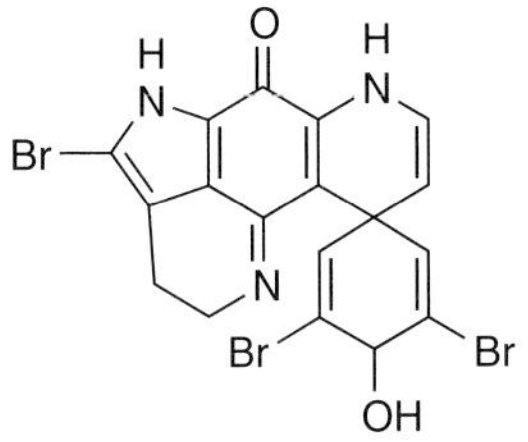

28 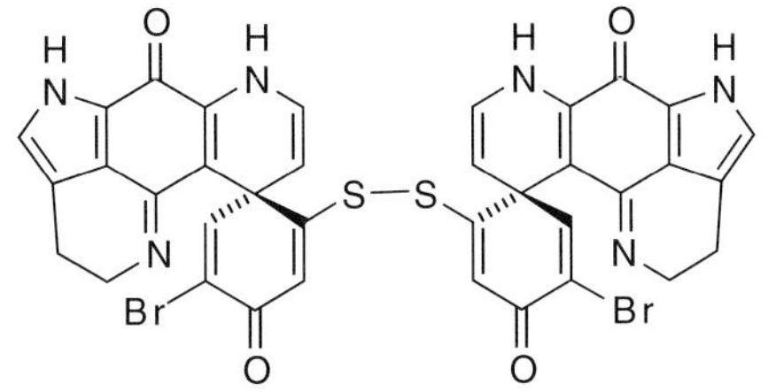 **29** (discorhabdin W)

The Philippine ascidian *Perophora namei* has yielded the polycyclic alkaloid perophoramidine (**30**), which is the first reported metabolite from the genus *Perophora* [35]. A Far Eastern *Eudistoma* ascidian was found to contain two brominated ergoline alkaloids pibocins A (**31**) [36] and B (**32**) [37].

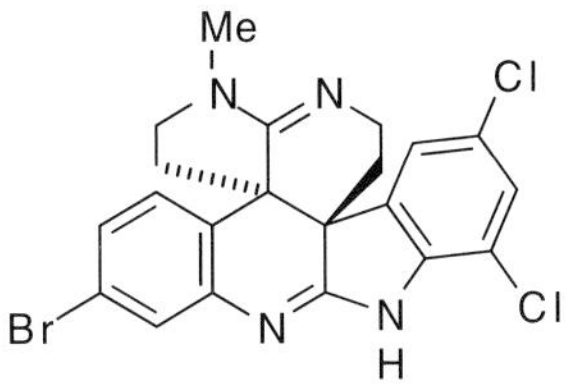

30 (perophoramidine)

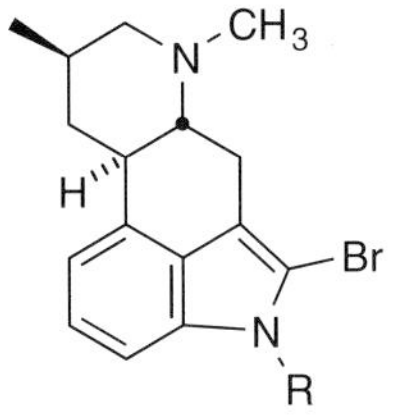

31 R = H (pibocin A)
32 R = OMe (pibocin B)

Blue-green algae have yielded a variety of related chlorine-containing indole alkaloids, fischerindoles, hapalindoles, and welwitindolinones [1,38]. A new example in this class is ambiguine G (**33**) from the terrestrial *Hapalosiphon delicatulus* [39]. Although there are precious few halogenated alkaloids from plants [1], one example is thomitrem A (**34**) from *Penicillium crustosum* [40], which is closely related to the well-known penitrems from fungi of the genera *Penicillium, Aspergillus,* and *Claviceps,* most examples from which are not halogenated.

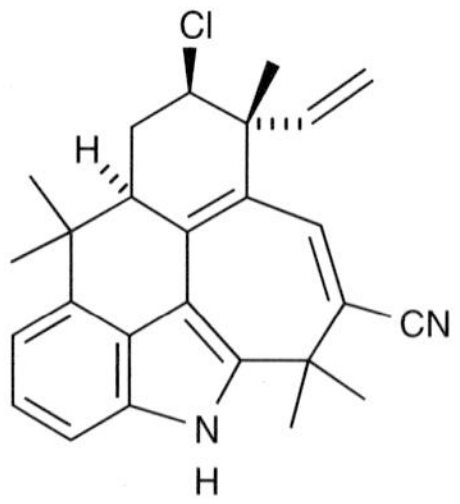

33 (ambiguine G nitrile)

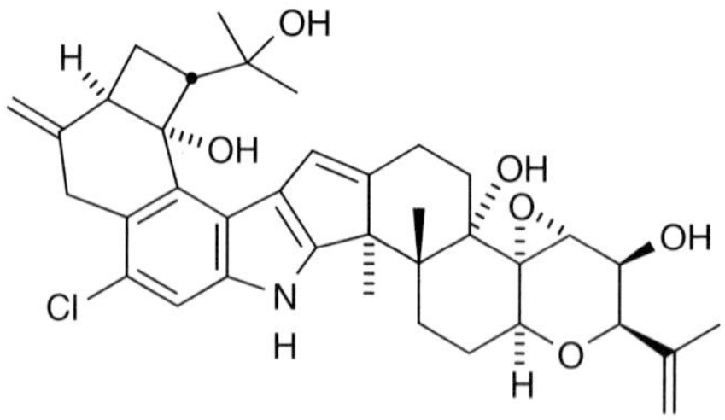

4 (thomitrem A)

19.2.2
Carbazoles

Although only a handful of naturally occurring halogenated carbazoles are known, these few examples are of great interest. Chlorohyellazole (**35**) is produced by the blue-green *Hyella caespitosa* [41], and 3-chlorocarbazole (**36**), from bovine urine, is a potent monoamine oxidase inhibitor [42]. The encrusting bacterium *Kyrtuthrix maculans* has yielded the three halogenated carbazoles **37–39** [43].

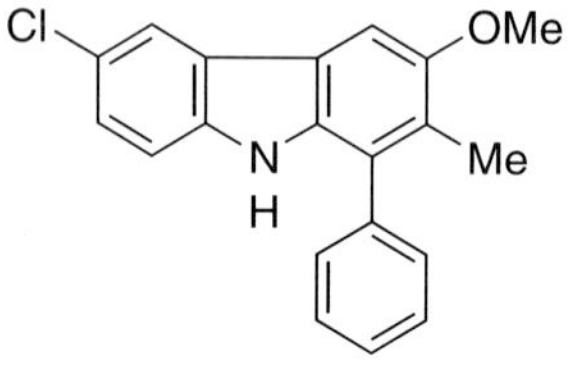

35 (chlorohyellazole)

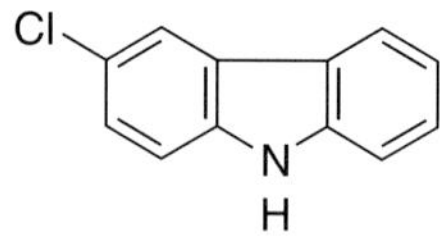

36

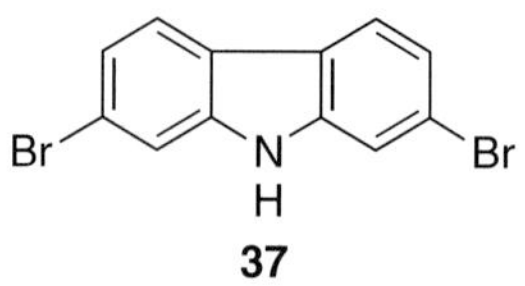

37

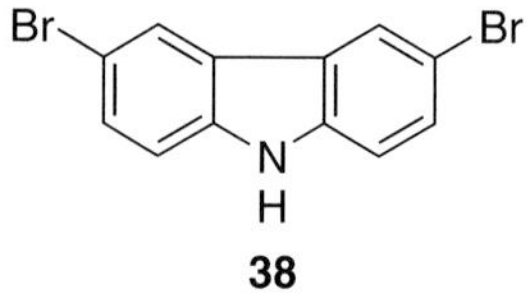

38

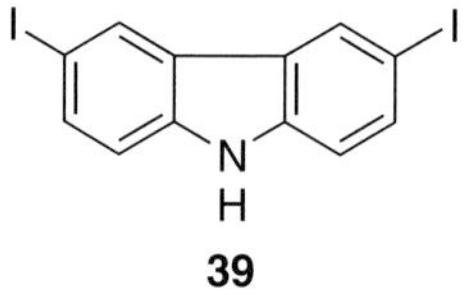

39

19.2.3
β-Carbolines

Unlike carbazoles, halogenated β-carbolines are abundant in nature [1]. For example, the simple eudistomin O (7-bromo-β-carboline) (**40**) is a ubiquitous marine ascidian metabolite [44]. Indeed, the tunicate genus *Eudistoma* has furnished most of the extant halogenated β-carbolines, some of which have significant antiviral (polio, herpes) and microbial activity. The Caribbean *Eudistoma olivaceum* produces at least 15 brominated carbolines [1]. A study of *Eudistoma gilboverde* uncovered the new eudistomins **41–43**[45], and the Australian ascidian *Pseudodistoma aureum* has yielded eudistomin V (**44**) [46]. The deep-water Palauan sponge *Plakortis nigra* contains plakortamines A–D (**45–48**) [47]. These β-carbolines exhibit activity against the HCT-116 human colon cancer cell line, with plakortamine B (**46**) being the most active. The freshwater cyanobacterium *Nostoc* 78-12A produces nostocarboline (**49**), which is a novel cholinesterase inhibitor comparable to galanthamine, an improved drug for the treatment of Alzheimer's disease [48].

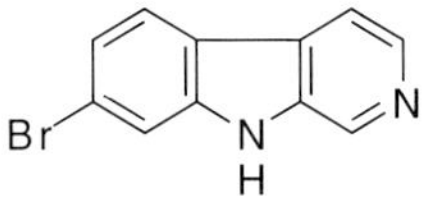

40

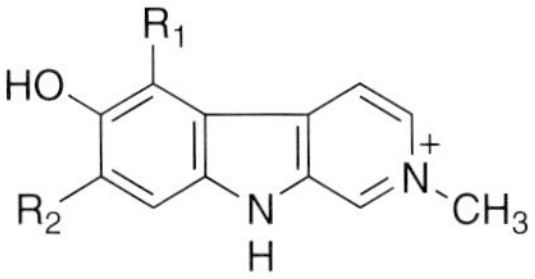

41, R_1 = Br, R_2 = H (2-methyleudistomin D)
42, R_1 = H, R_2 = Br (2-methyleudistomin J)

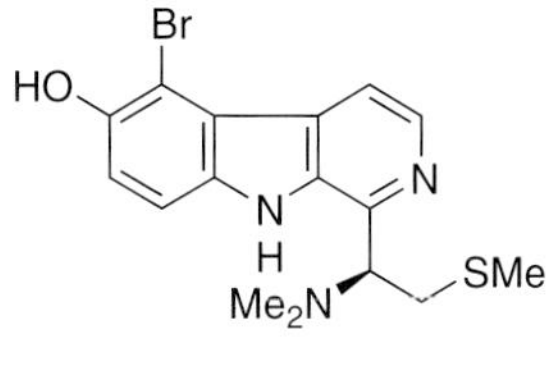

43

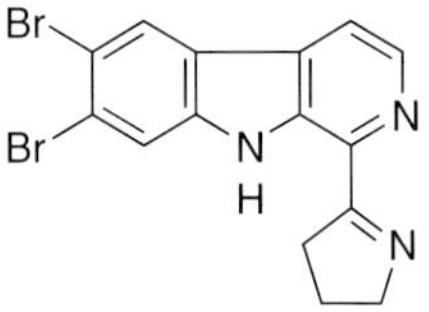

44 (eudistomin V)

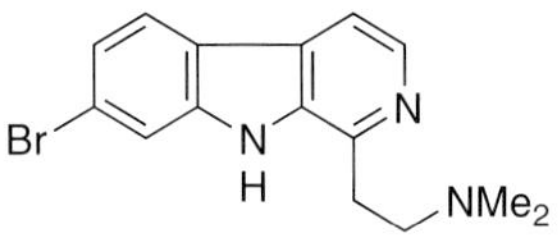

45 (plakortamine A)

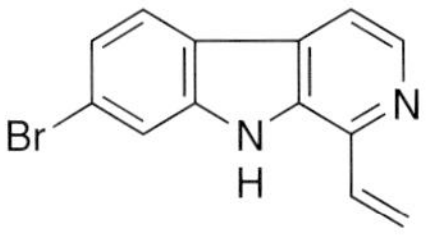

46 (plakortamine B)

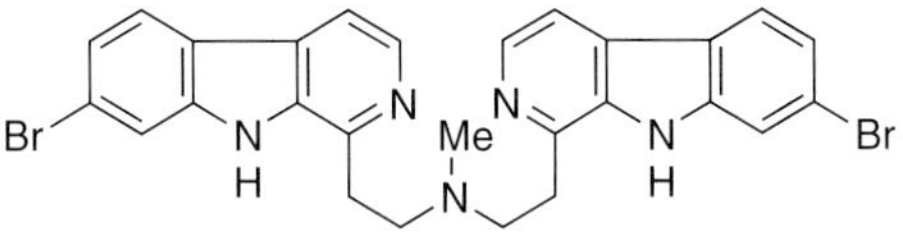

47 (plakortamine C)

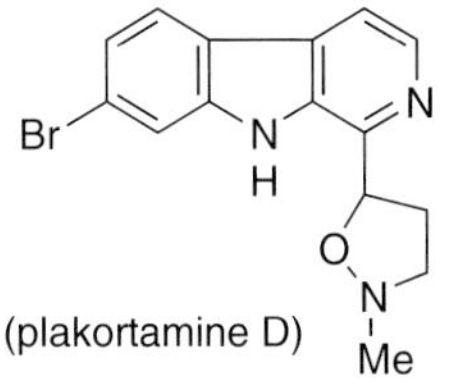

48 (plakortamine D)

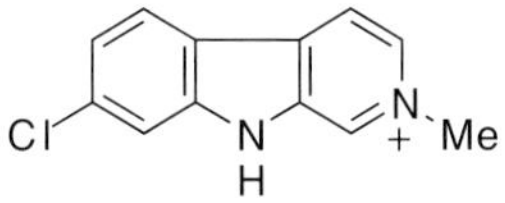

49 (nostocarboline)

Bauerines A–C (**50–52**) are chlorinated β-carbolines from the blue-green alga *Dichothrix baueriana* that show activity against herpes simplex virus type 1 [49].

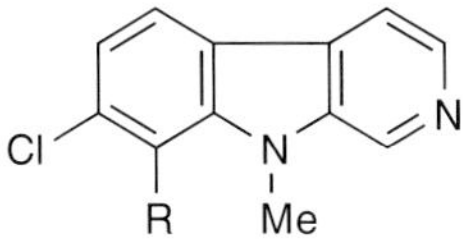

50, R = H (bauerine A)
51, R = Cl (bauerine B)

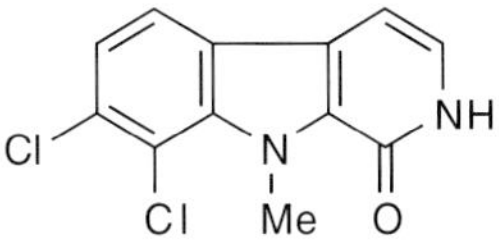

52 (bauerine C)

Amongst the more complex brominated β-carboline alkaloids are the fascaplysins and related metabolites from the tunicate *Didemnum* sp. and the sponge *Fascaplysinopsis reticulata* [50,51]. Several examples of each type are illustrated as **53–63**.

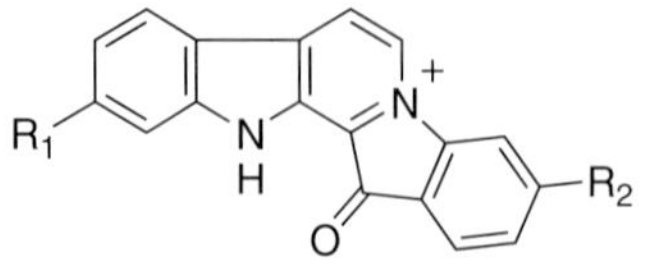

53, R_1 = H, R_2 = Br (3-bromofascaplysin)
54, R_1 = Br, R_2 = H (10-bromofascaplysin)
55, R_1 = R_2 = Br (3,10-dibromofascaplysin)

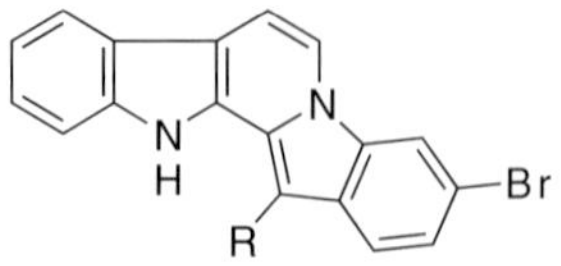

56, R = $COCO_2Me$ (3-bromohomofascaplysin B)
57, R = $COCO_2Et$ (3-bromohomofascaplysin B-1)
58, R = CHO (3-bromohomofascaplysin C)

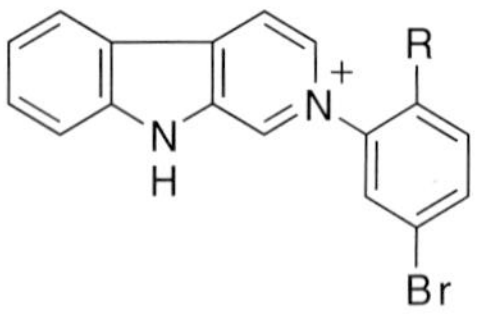

59, R = CO_2^- (14-bromoreticulatate)
60, R = CO_2Me (14-bromoreticulatine)
61, R = OH (14-bromoreticulatol)

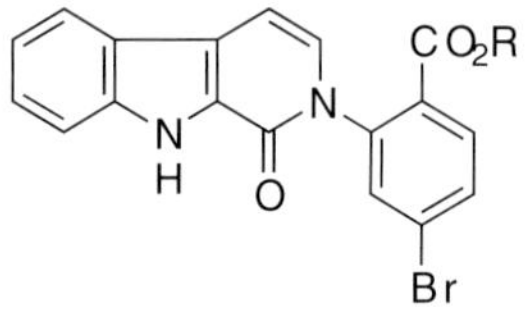

62, R = Me (3-bromosecofascaplysin A)
63, R = H (3-bromosecofascaplysin B)

Finally, it might be noted that several brominated tetrahydro- and dihydro-β-carbolines are known, for example woodinine (**64**) [52,53] and 19-bromoisoeudistomin U (**65**) [54], both from *Eudistoma* spp. The tunicate *Pseudodistoma arborescens* has yielded arborescidine A (**66**) a brominated derivative of the well-known indolo[2,3-*a*]quinolizidine alkaloid ring system [55].

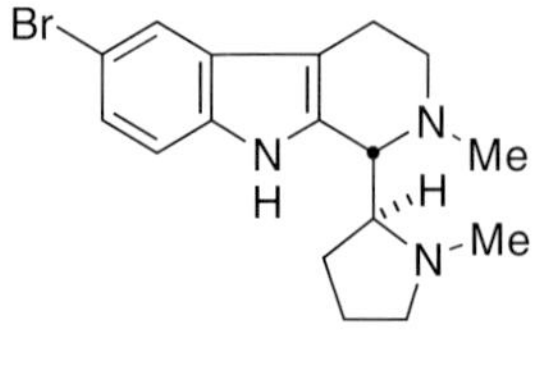

64 (woodinine) 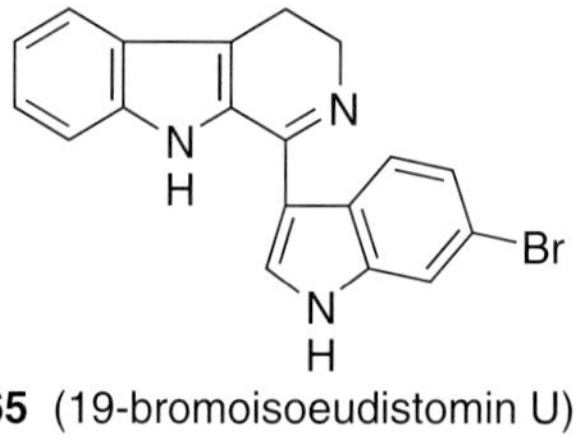 **65** (19-bromoisoeudistomin U) 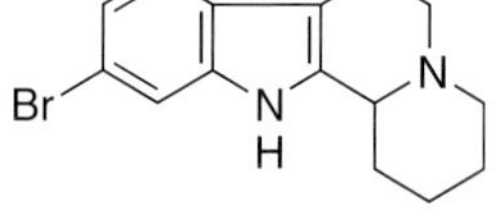 **66** (arborescidine A)

19.2.4
Tyrosines

Halogenated alkaloids derived from tyrosine represent an enormous and extraordinarily diverse collection of compounds, mostly sponge metabolites [1]. As expected, the electron-rich phenolic ring in tyrosine is very susceptible to electrophilic biohalogenation (mainly bromination). Indeed, 3,5-dibromotyrosine itself was isolated from verongid sponges as early as 1941 [56]. Furthermore, a large number of transformed halogenated tyrosines are also sponge metabolites [1]. Amongst the recently discovered simple bromotyrosine sponge metabolites are **67** from the Caribbean *Verongula gigantea* [57], mololipids **68** from a Hawaiian *Verongida* sp. that are derivatives of moloka'iamine (**68**, $R_1 = R_2 = H$) [58], as are **69** and **70** from the Japanese *Hexadella* sp. [59], and nakirodin A (**71**) from an Okinawan *Verongida* sp. [60].

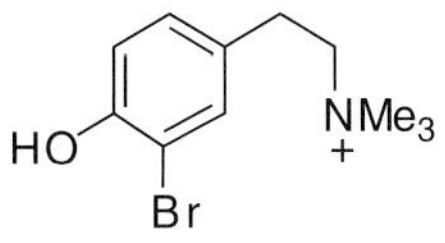

67

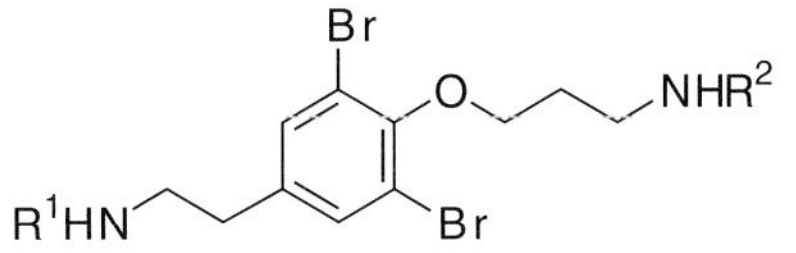

68 (mololipids)

R^1, R^2 = myristic, palmitic, margaric, oleic, stearic, arachidic, pentadecanoic, hexadecanoic (2), heptadecanoic, nondecanoic groups

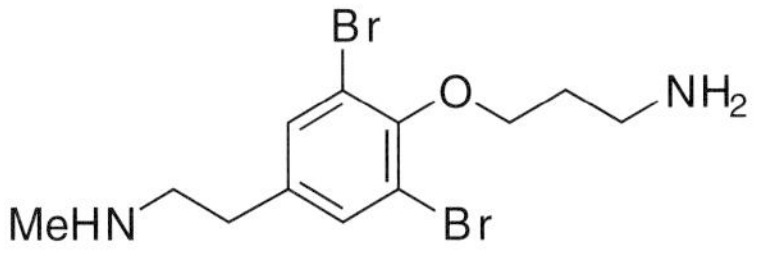

69

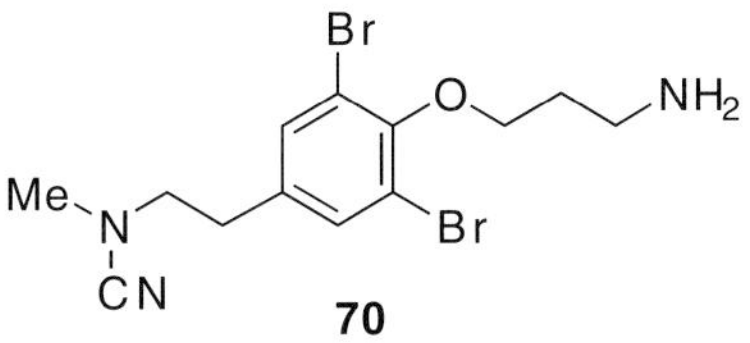

70

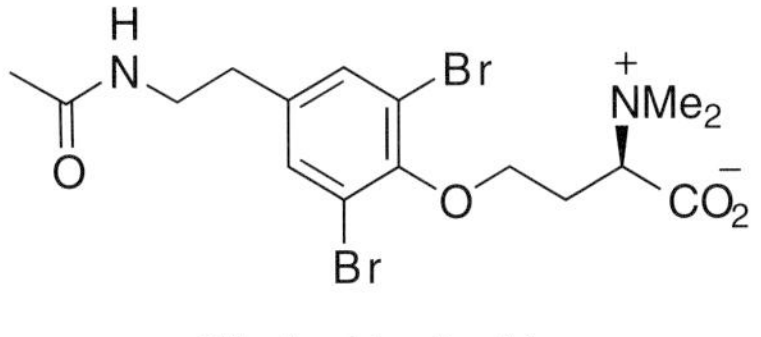

71 (nakirodin A)

The marine bryozoan *Amathia convoluta* produces convolutamine H (**72**) [61], and the Japanese gastropod *Turbo marmorata* has yielded the highly toxic (LD_{99} 1–4 mg/kg, mice) iodinated turbotoxins A (**73**) and B (**74**) [62]. The Palauan ascidian *Botrylloides tyreum* contains botryllamide G (**75**) [63], and several new psammaplins, for example psammaplin E (**76**), have been isolated from the sponge *Pseudoceratina purpurea* [64].

Numerous sponge metabolites that incorporate multiple tyrosine units are known [1], and some recent discoveries of compounds containing two tyrosines include tokaradine A (**77**) from *Pseudoceratina purpurea* [65], the novel bromotyrosine

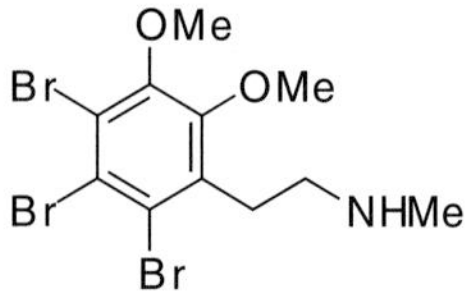

72 (convolutamine H)

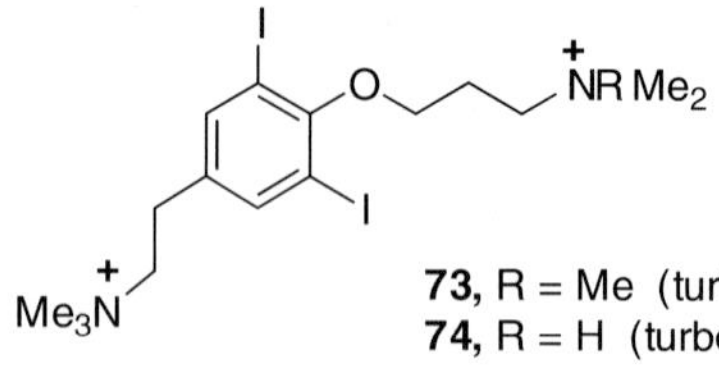

73, R = Me (turbotoxin A)
74, R = H (turbotoxin B)

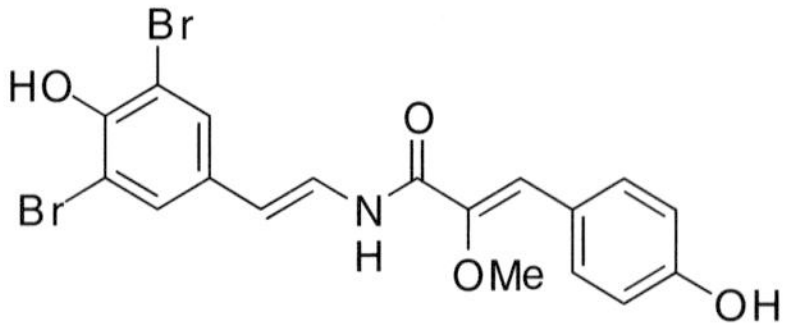

75 (botryllamide G)

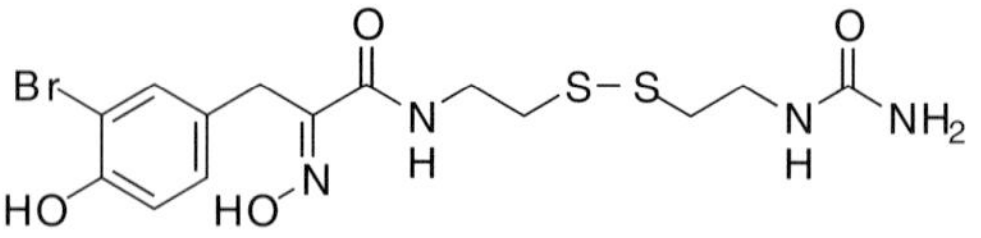

76 (psammaplin E)

N-oxide alkaloid purpuramine J (**78**) from the Fijian sponge *Druinella* sp. [66], aplyzanzine A (**79**) from the Indo-Pacific sponge *Aplysina* sp. [67], the structurally related suberedamines A and B from an Okinawan sponge *Subera* sp. [68], and the presumed biogenetically derived ma'edamines A (**80**) and B (**81**), which are also found in *Subera* sp. [69].

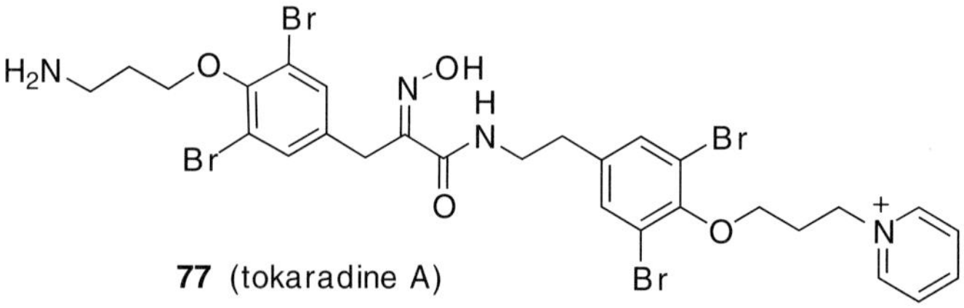

77 (tokaradine A)

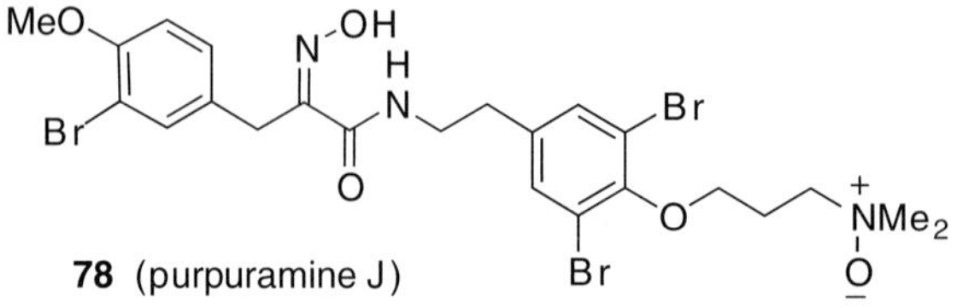

78 (purpuramine J)

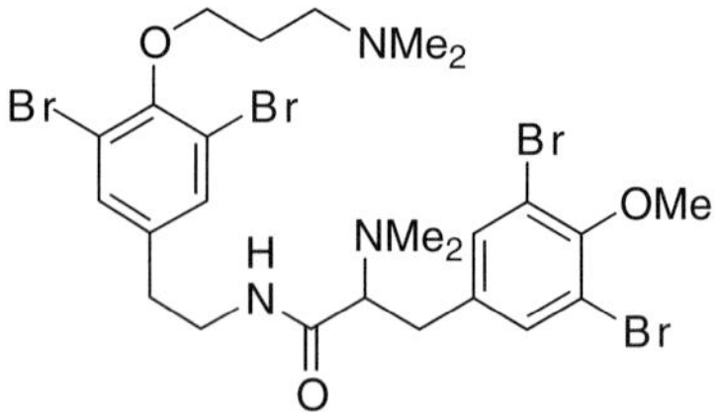

79 (aplyzanzine A)

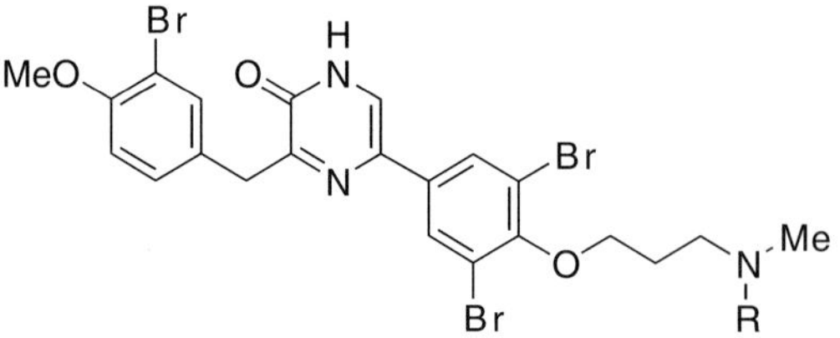

80, R = Me (ma'edamine A)
81, R = H (ma'edamine B)

The novel trisulfide derivative **82** of psammaplin A was isolated from the two associated sponges *Jaspis wondoensis* and *Poecillastra wondoensis* [70]. Compound **82** exhibits potent antibacterial activity, higher than meropenem. Among the more interesting multiple bromotyrosines are the bastidins. One example is bastidin 21 (**83**) from the Great Barrier Reef sponge *Iantella quadrangulata* [71].

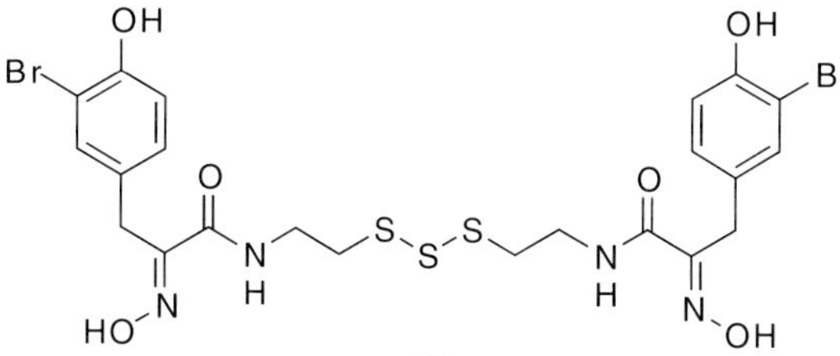

82

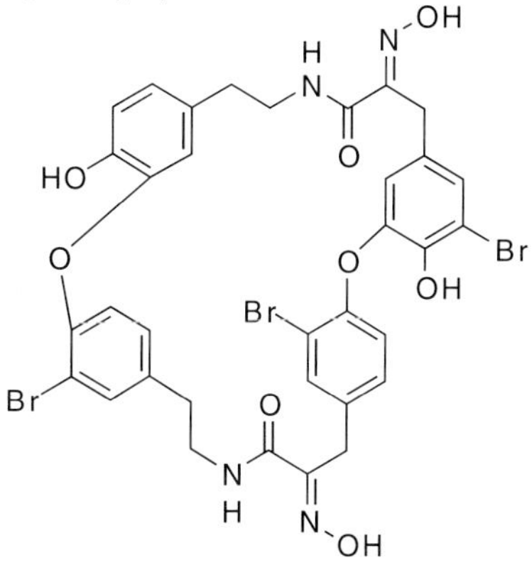

83 (bastadin 21)

Undoubtedly the most structurally and biogenetically intriguing bromotyrosines are the many spirocyclohexadienyl isoxazoles, which are too numerous to describe fully here [1]. Of those examples containing a single spiro unit, recently isolated compounds include ianthesine A (**84**) from the Australian sponge *Ianthella* sp. [72], caissarine A (**85**) from a Brazilian *Aplysina caissara* [73], and the unnamed **86** from an Australian non-verongid sponge *Oceanapia* sp. [74].

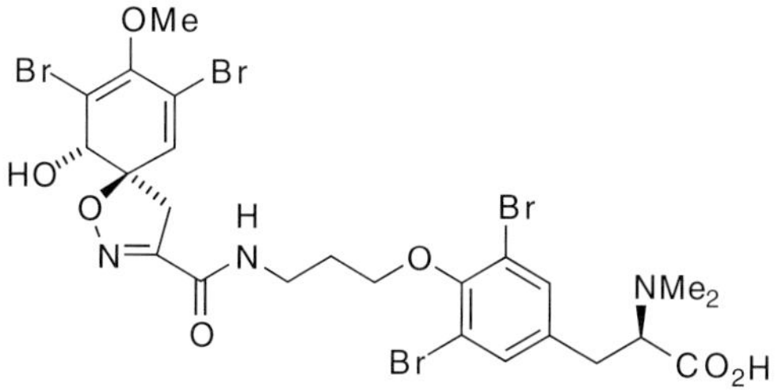

84 (ianthesine A)

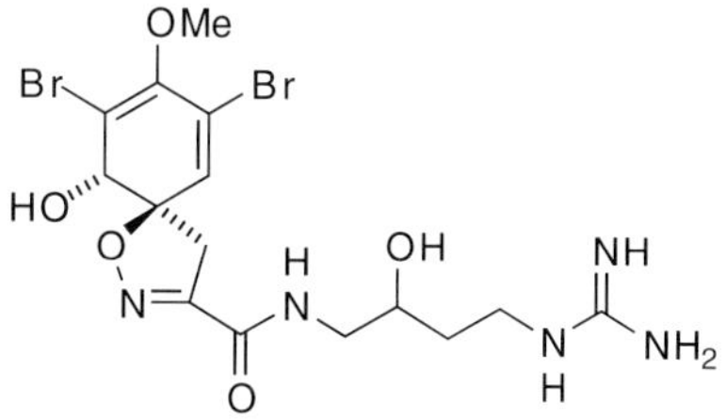

85 (caissarine A)

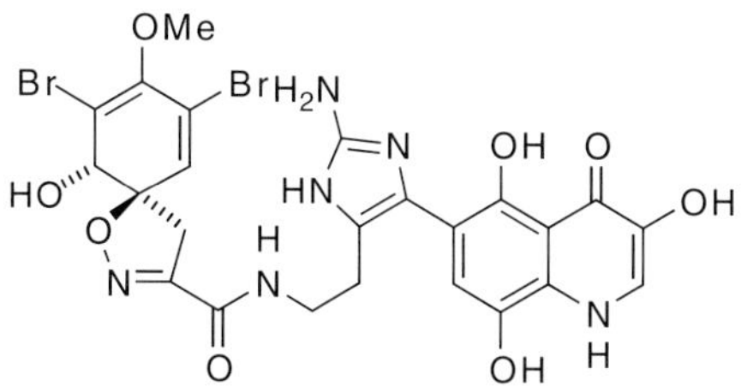

86

Several metabolites in this class consist of two spirocyclohexadienyl isoxazole units tethered by a linking chain [1]. Archerine (**87**) is a novel anti-histaminic bromotyrosine isolated from the Caribbean sponge *Aplysina archeri* [75], and calafianin (**88**) was found in the Mexican sponge *Aplysina gerardogreeni* [76] (structure revised in ref. [77]). Desoxyagelorin A (**89**) has been characterized in *Subera* aff. *praetensa* from the gulf of Thailand [78].

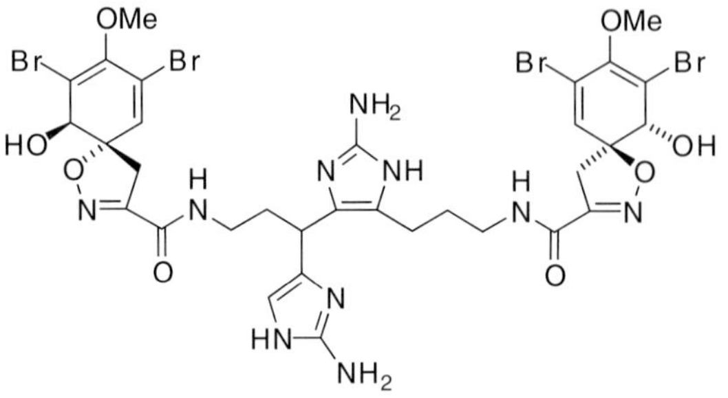

87 (archerine)

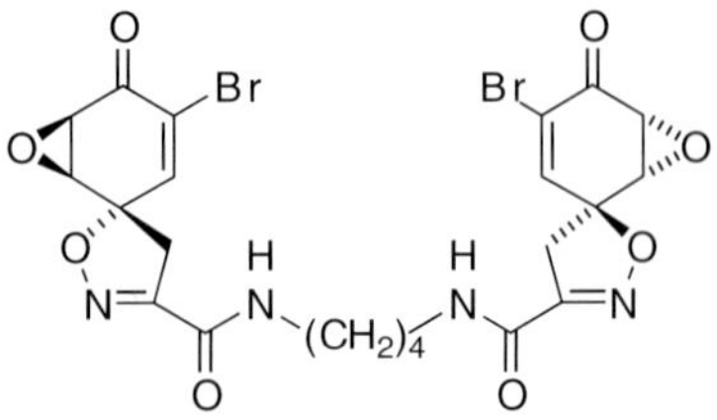

88 (calafianin)

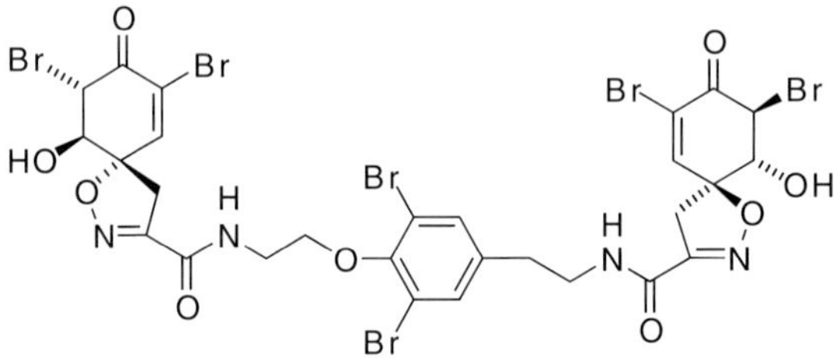

89 (desoxyagelorin A)

The structure of zamamistatin (**90**) previously isolated from the Okinawan *Pseudoceratina purpurea* [79] has been revised to that shown [80]. Another novel bromotyrosine sponge metabolite is kuchinoenamine (**91**) from *Hexadella* sp. [59].

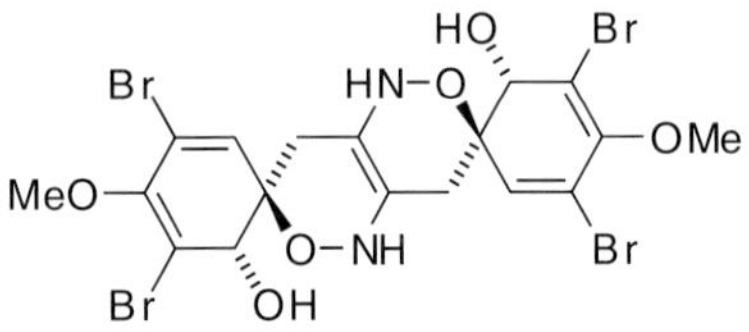

90 (zamamistatin)

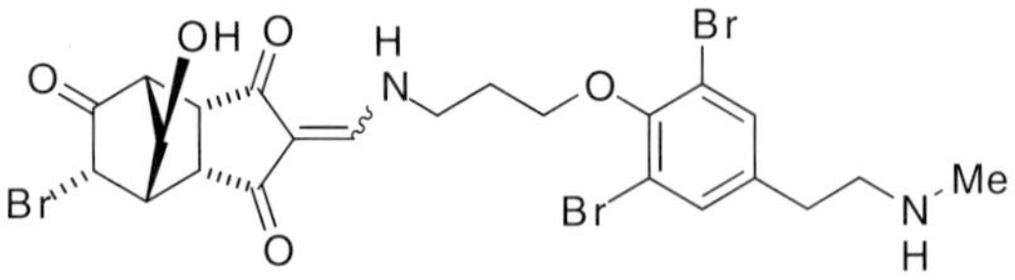

91 (kuchinoenamine)

19.2.5 Miscellaneous Halogenated Alkaloids

In addition to the myriad known indole, carbazole, and tyrosine-halogenated alkaloids, a number of other halogenated alkaloids of diverse structure exist [1]. Recent examples are presented here. The bryozoan *Euthyroides episcopalis* contains several euthyroideones, for example, **92**, which are novel brominated quinone methides [81]. Another bryozoan *Caulibugula intermis* from Palau has afforded caulibugulones A–F, for example **93** and **94**, which have potent cytotoxicity [82]. Interestingly the simple tribromoacetamide (**95**), which was found in the Okinawan alga *Wrangelia* sp., also displays potent cytotoxic activity against P388 leukemia (IC_{50} 0.02 μg/mL) [83]. Ceratamines A (**96**) and B (**97**) are novel heterocyclic alkaloids isolated from the Papuan sponge *Pseudoceratina* sp. [84]. The Australian sponge *Oceanopia* sp. has yielded petrosamine B (**98**), an inhibitor of a *Helicobacter pylori* key enzyme [85].

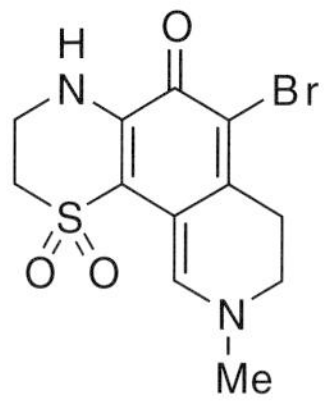

92 (euthyroideone A)

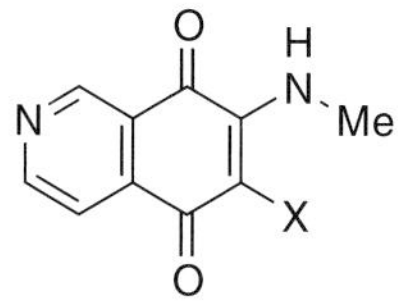

93, X = Cl (caulibugulone B)
94, X = Br (caulibugulone C)

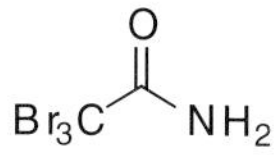

95

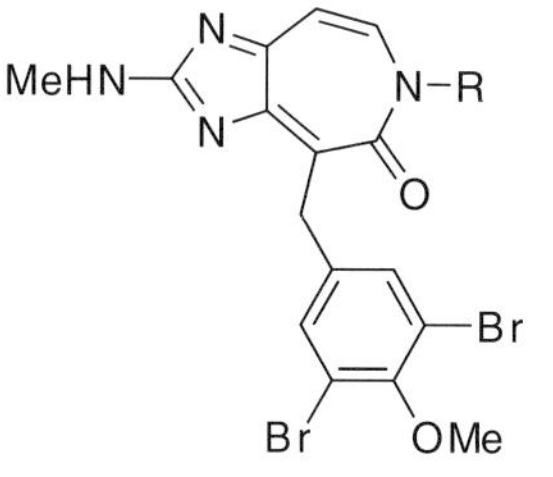

96, R = Me (ceratamine A)
97, R = H (ceratamine B)

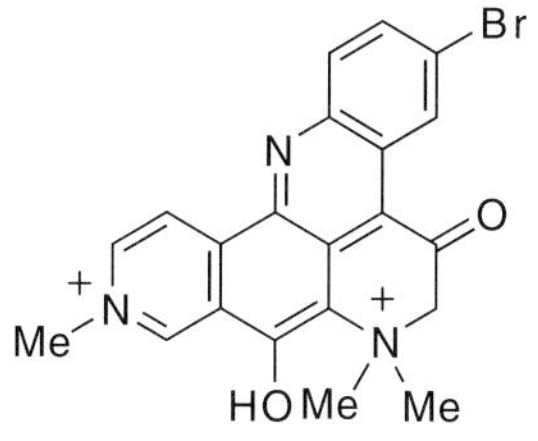

98 (petrosamine B)

The Okinawan ascidian *Lissoclinum* sp. produces the cytotoxic diterpene alkaloids haterumaimides J (**99**) and K (**100**) [86]. The incredibly prolific cyanobacterium *Lyngbya majuscula* has furnished the quinoline alkaloid **101** [87]. The new quinolones **102** and **103** were isolated from the sponge *Hyrtios erecta* [26].

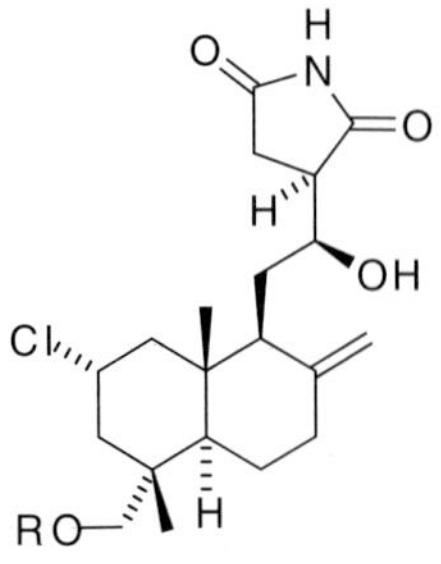

99, R = H (haterumaimide J)
100, R = Ac (haterumaimide K)

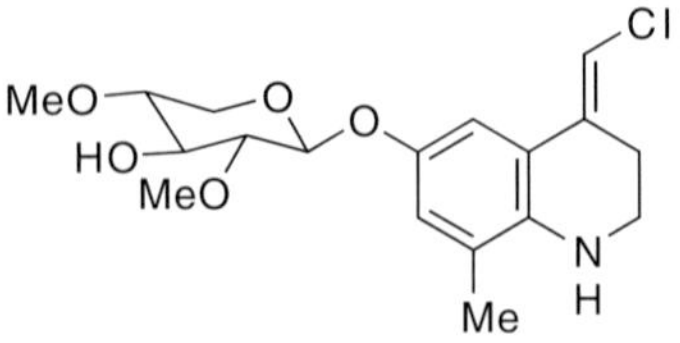

101

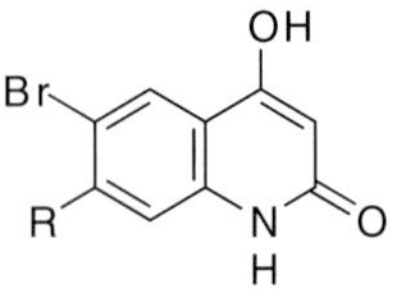

102, R = H
103, R = Br

Several noteworthy halogenated terrestrial alkaloids are known, with epibatidine (**104**) at the top of the list [88]. This apparent frog (*Epipedobates tricolor*) metabolite has powerful analgesic activity [89] and an intensive search is underway for a clinically useful drug [90]. A few chlorinated plant alkaloids have also been discovered. Romucosine F (**105**) is present in *Annora purpurea*, a South American bushy tree [91], and the closely related romucosine B (**106**) is found in the stems of *Rollinia mucosa* [92]. The furoquinoline alkaloid chlorodesnkolbisine (**107**) was isolated from the African folk medicine plant *Teclea nobilis* [93]. The authors of this latter study provide convincing evidence that **107** is not an isolation artifact (e.g. from the corresponding epoxide with HCl).

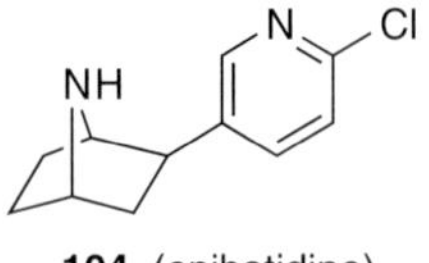

104 (epibatidine)

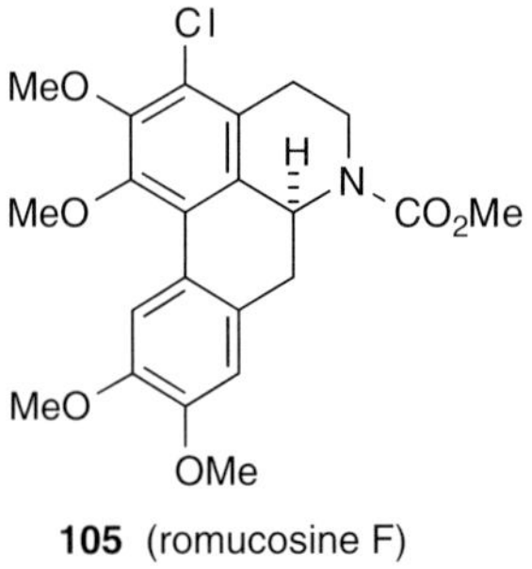

105 (romucosine F)

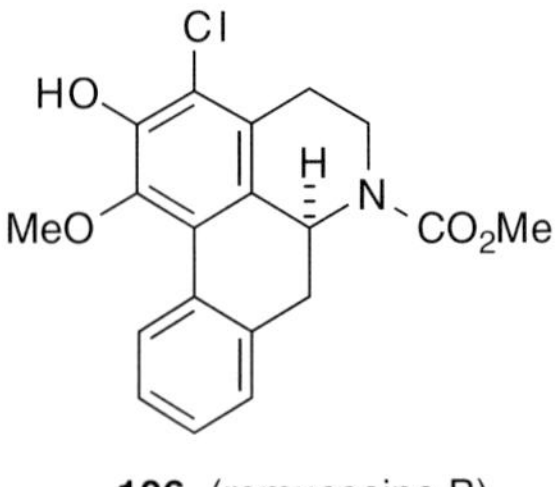

106 (romucosine B)

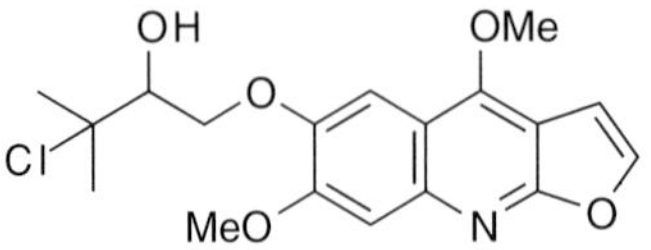

107 (chlorodesnkolbisine)

Root cultures of *Menispermum dauricum* DC. have yielded several chlorinated alkaloids, the most recent of which are dauricumine (**108**) and dauricumidine (**109**) [94,95]. The tetrahydroquinoline alkaloid virantmycin (**110**) is produced by *Streptomyces nitrosporeus* [96,97], as is benzastatin C (**111**) [98].

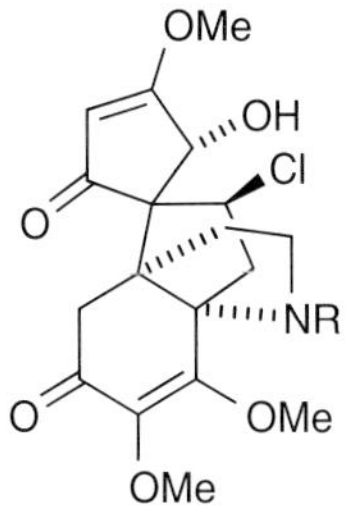

108, R = Me (dauricumine)
109, R = H (dauricumidine)

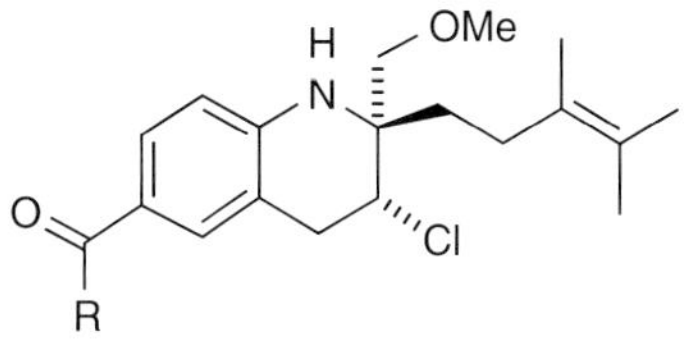

110, R = OH (virantmycin)
111, R = NH_2 (benzastatin C)

The fatty acid-derived alkaloids halichlorine (**112**) [99,100], pinnaic acid (**113**), and tauropinnaic acid (**114**) [101,102] have novel structures and unique anti-inflammatory activity.

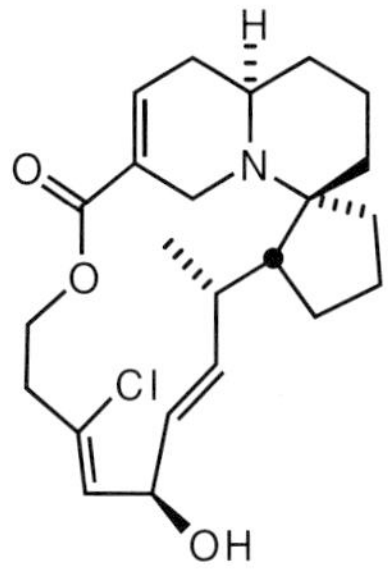

112 (halichlorine)

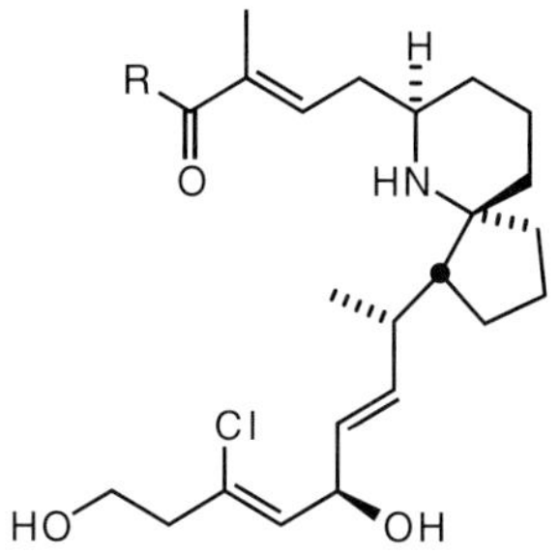

113, R = H (pinnaic acid)
114, R = $NHCH_2CH_2SO_3H$ (tauropinnaic acid)

19.3 Biosynthesis of Halogenated Alkaloids

Understanding of the biogenesis of halogenated alkaloids lags far behind our knowledge of the sources, biological properties, and characterization of these natural products. Nevertheless, this section illustrates proposed biogenetic routes to these alkaloids and the supporting experimental evidence.

19.3.1 Halogenation Enzymes

Several enzymes that halogenate organic substrates are well known and these enzymes have been studied extensively, especially those involving alkenes, alkynes, active methylene compounds, electron-rich heterocycles (pyrroles, indoles), and phenols [1,103–105]. Both chloroperoxidase and bromoperoxidase are widespread in the terrestrial and marine environments. For example, the vanadium haloperoxidases have been investigated thoroughly [106–108], as has white blood cell myeloperoxidase, the enzyme involved in the biohalogenation of invading microorganisms to fight infection [1,109]. Recent years have seen the discovery of a new class of halogenases [103], and a fluorinase that can introduce fluorine into acetate [104,110].

19.3.2
Indoles

While it is generally assumed that the simple naturally occurring halogenated indoles are tryptophan derived, no biogenetic labeling studies have been reported for simple haloindoles. However, it is clear from labeling studies that the bromine-containing eudistomin H (**121**), which is found in the tunicate *Eudistoma olivaceum*, is produced from tryptophan and proline [111]. Thus, both L-[side chain 3-^{14}C]tryptophan (**115**) and L-[2,3-^{3}H]proline (**117**) label both eudistomin H (**121**) and I (**120**), while [ethyl-^{3}H] tryptamine (**116**) is incorporated into eudistomin I but not into eudistomin H. Neither L-[2,3-^{3}H]ornithine or L-[2,3-^{3}H]arginine are utilized by this tunicate [111a]. These results reveal that decarboxylation of **118** to **119** must follow the biohalogenation of tryptophan (Scheme 19.1). A subsequent labeling study by Shen and Baker showed that both **118** and **119** are precursors to eudistomin H [111b].

A biosynthetic study of discorhabdin B (**123**) from the New Zealand sponge *Latrunculia* sp. demonstrates that [U-^{14}C]-L-phenylalanine is incorporated into **123** (via tyramine), and a biogenesis has been postulated (Scheme 19.2) [112]. It is not known when the phenolic ring is brominated. For a more general biogenetic proposal for the pyrroloquinolines see reference [110].

The pyrroloquinone imine **122** is considered to be a key branching point for the biogenesis of damirones, batzellines, isobatzellines, makaluvamines [110], and

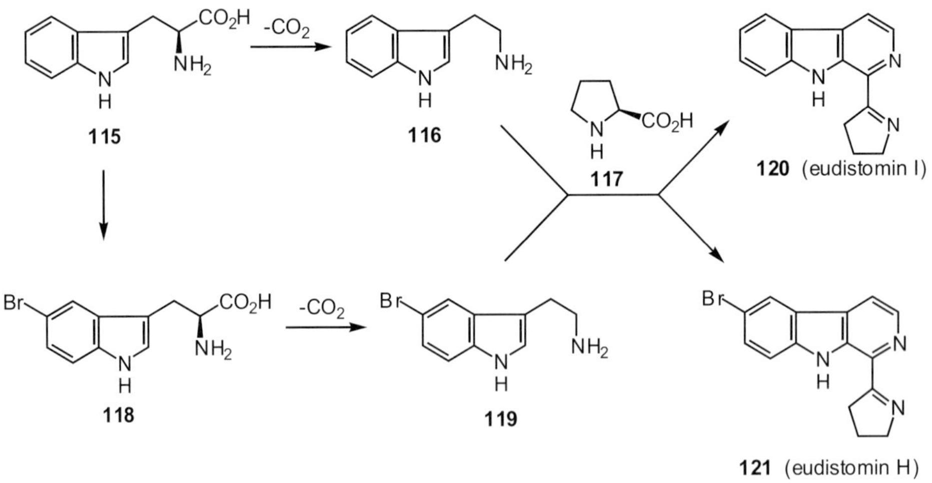

Fig. 19.1 Proposed biosynthesis of eudistomins H and I [111].

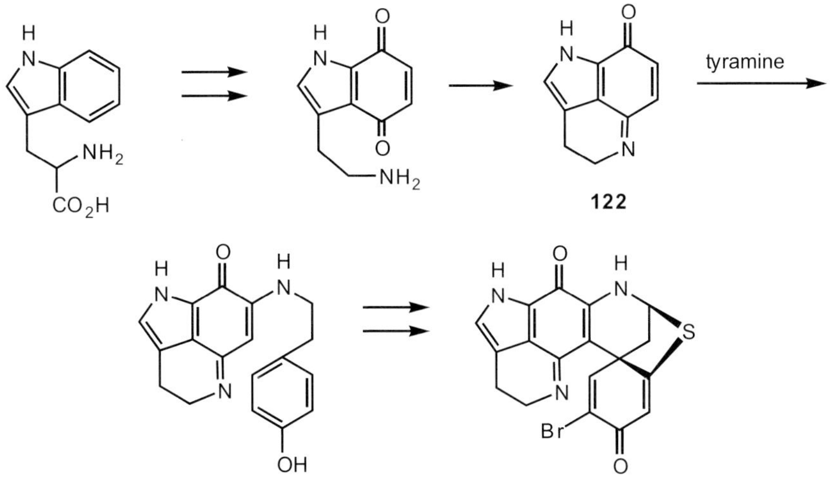

123 (discorhabdin B)

Fig. 19.2 Abbreviated suggested biosynthesis of discorhabdin B [112].

makaluvic acids (**125**)[113]. For example, a possible route to the last metabolites has been suggested (Scheme 19.3) [113], although these compounds are nonhalogenated.

Other proposed biogeneses of halogenated indole alkaloids for which little or no labeling evidence is available include schemes for the hapalindoles [38,114], penitrems [115], fascaplysins [51], and tubastrindoles [29]. With regard to the hapalindoles it has been determined that the isonitrile unit originates from a one-carbon donor related to tetrahydrofolate metabolism as [2-^{14}C]glycine, L-[3-^{14}C]serine, L-[methyl-^{14}C]methionine, [^{14}C]formate, and [^{14}C]cyanide are each incorporated into this indole alkaloid [114]. The timing for the biochlorination step is unknown but it is suggested to occur at the terpenoid level for related indole isonitrile alkaloids [116]. With regard to the penitrems, which are structurally similar to thomitrem A (**34**), it has been reported that the nonchlorinated [^{14}C]paxilline is incorporated into the chlorine-containing penitrem A [115a], apparently suggesting a late-stage biochlorination.

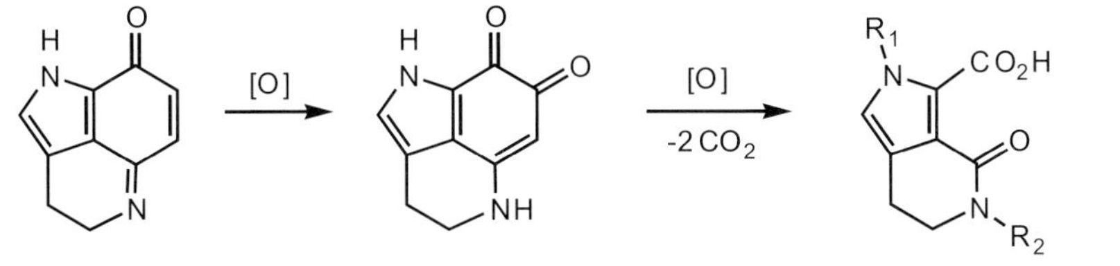

122　　124　　makaluvic acids (**125**)

Fig. 19.3 Suggested biosynthetic pathway to makaluvic acids [113].

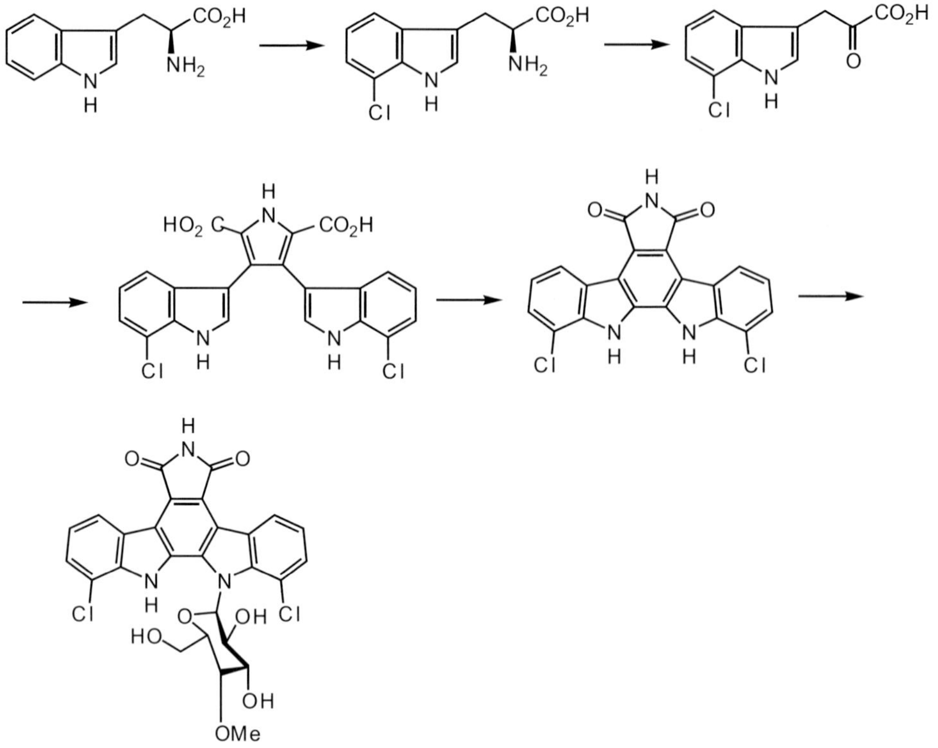

126 (rebeccamycin)

Scheme 19.4 Proposed biosynthesis of rebeccamycin [119]

The introduction of chlorine into tryptophan at C-7 is proposed as the initial step in the biosynthesis of rebeccamycin (**126**), the powerful indolocarbazole antitumor antibiotic from *Lechevalieria aerocolonigenes* [117–119]. The proposed overall biosynthesis of rebeccamycin is shown in Scheme 19.4 [119].

That chlorination occurs early in the biosynthesis of rebeccamycin is shown by the identification of two genes in the biosynthetic cluster and their ability to chlorinate tryptophan at C-7 as shown in Scheme 19.5 [117].

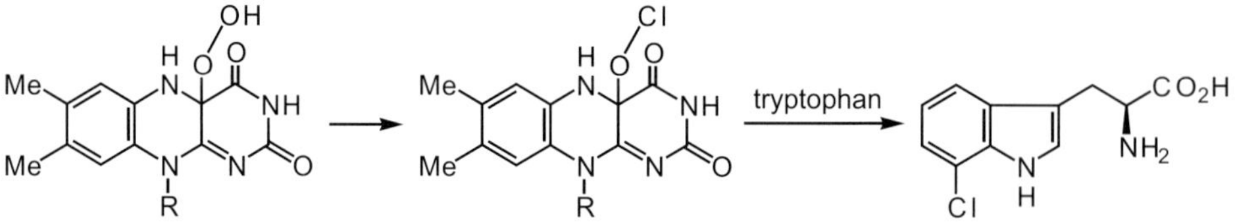

Fig. 19.5 Proposed C-7 chlorination of tryptophan [117].

19.3.3
Biosynthesis of Halogenated Tyrosines

The fantastic structural diversity of the natural brominated tyrosines has led to equally ingenious biosynthesis proposals, but only a few definitive labeling studies have been described. The early study by Tymiak and Rinehart on the biosynthesis of dibromotyrosine metabolites by the sponge *Aplysina fistularis* supports the incorporation of both phenylalanine (**127**) and tyrosine (**128**) into dienone **133** and dibromohomogentisamide (**134**) (Scheme 19.6) [120]. Metabolites **131**, **132**, **135**, and **136** were also identified along with **133** and **134** in this study, which utilized

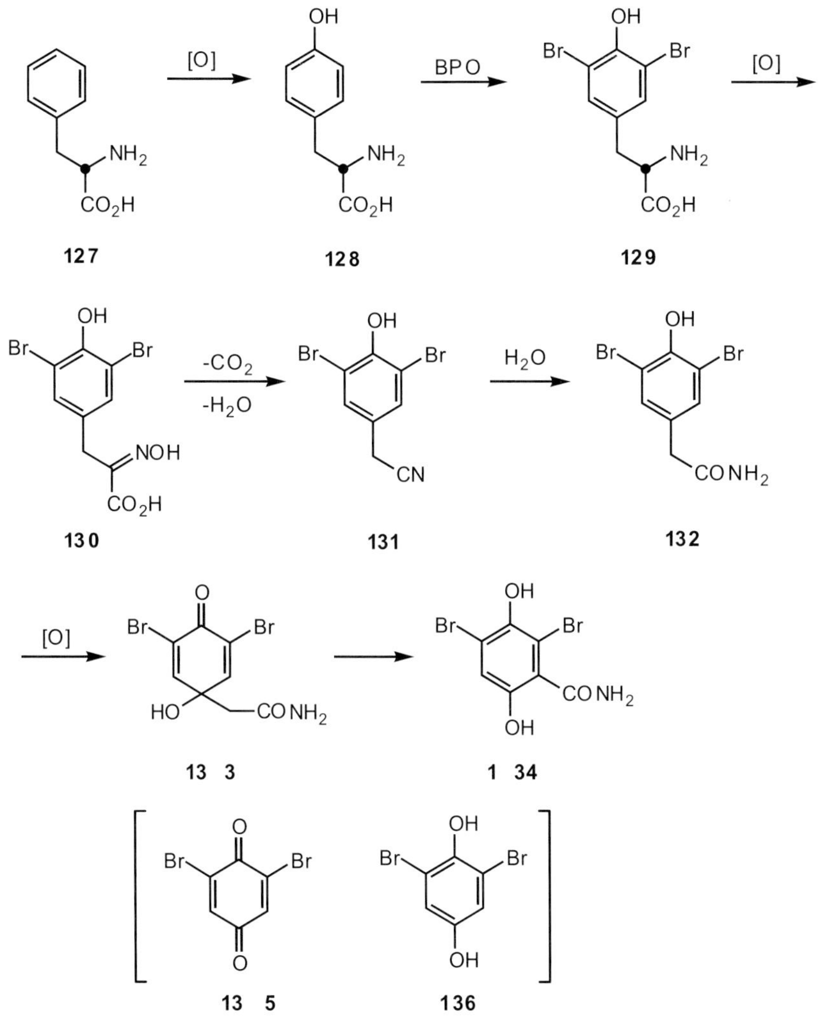

Scheme 19.6 Proposed biosynthetic pathway to dienone 133 and dibromohomogentisamide (134) [120].

[^{15}N]phenylalanine, [U-^{14}C]phenylalanine, and [U-^{14}C]tyrosine. The similarity of radioactivity in **133** and **134** suggests that the latter forms from the former.

A subsequent study by Carney and Rinehart provided additional evidence for the biosynthesis of brominated tyrosine metabolites [121]. Thus, [U-^{14}C]-L-tyrosine (**128**), [U-^{14}C]-L-3-bromotyrosine (**137**), and [U-^{14}C]-L-3,5-dibromotyrosine (**129**) are incorporated into both dibromoverongiaquinol (**133**) and aeroplysinin-1 (**142**). Moreover, [methyl-^{14}C]methionine is specifically incorporated into the *O*-methyl group of aeroplysinin-1 (**142**). These experiments led to the proposed biogenetic pathway shown in Scheme 19.7 [121].

Interestingly, a study of the bromide-dependent chloroperoxidase bromination of tyrosine reveals that the active brominating agent is free bromine and not a bromine-enzyme complex [122]. The situation is very different for bromoperoxidase- and vanadium peroxidase-catalyzed brominations [106–108].

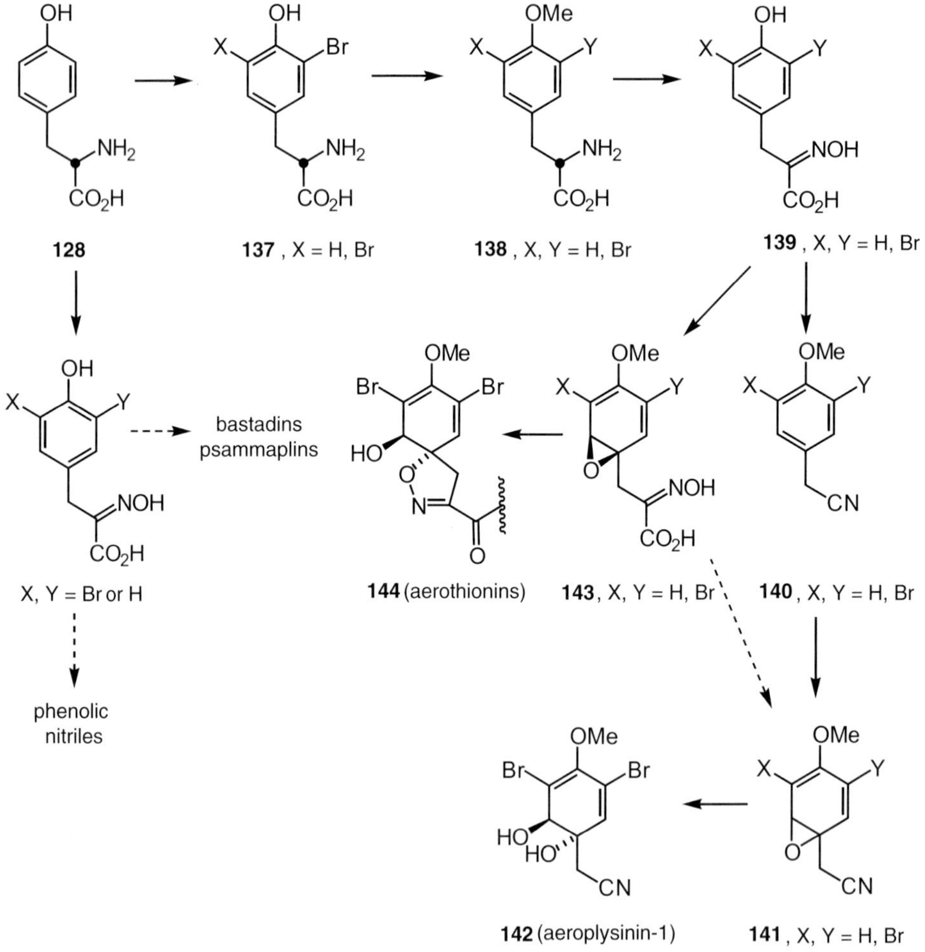

Fig. 19.7 Proposed biosynthetic pathways to brominated tyrosines [121].

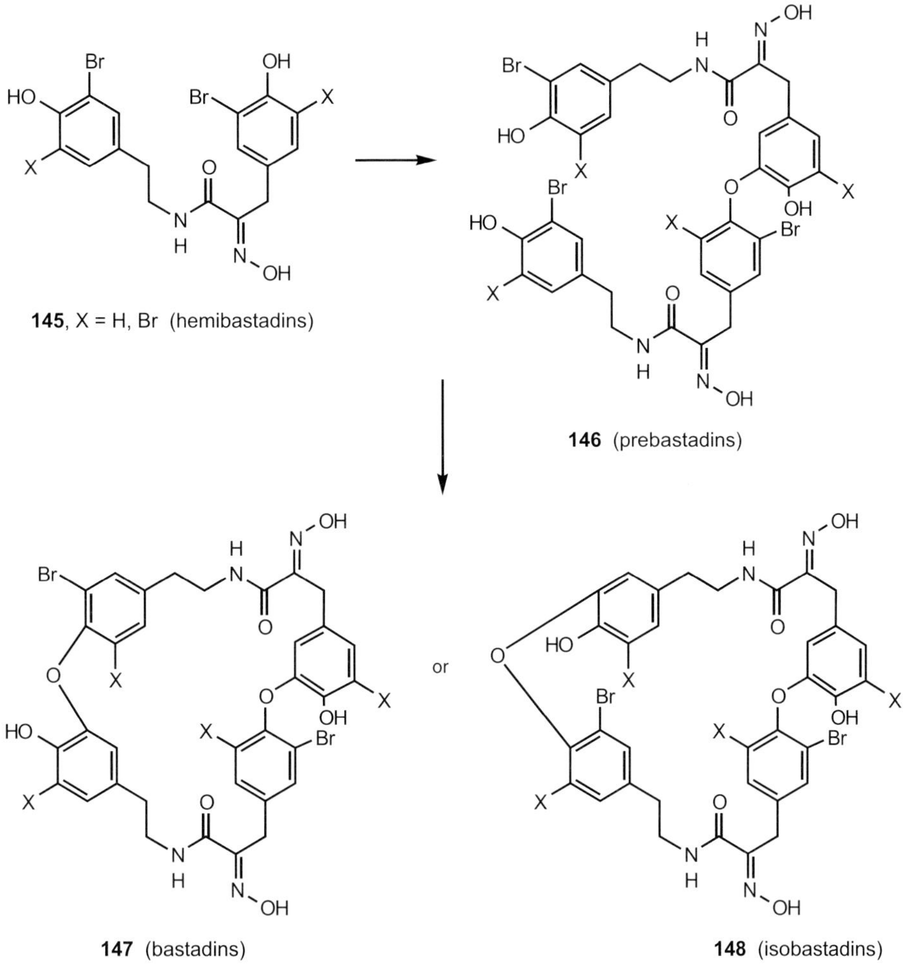

145, X = H, Br (hemibastadins)

146 (prebastadins)

147 (bastadins)

or

148 (isobastadins)

Fig. 19.8 Proposed biosynthetic pathway to bastadins [123].

Crews has formulated a collective scheme for the biosynthesis of the sponge metabolite bastadins based on the known structures (Scheme 19.8) [123]. Thus, dimerization of two brominated tyrosines can lead to hemibastadins **145**, which in turn can couple to form prebastadins **146**. Final ring closure can afford bastadins **147** or isobastadins **148**. The final cyclization is similar to the formation of polybrominated dibenzo-*p*-dioxins from polybrominated diphenyl ethers, which are ubiquitous in sponges [124,125].

Crews has also formulated a biosynthetic pathway for the disulfide bromotyrosine psammaplins from the sponge *Pseudoceratina purpurea* [64,126]. The structurally unique ma'edamines (**80**, **81**) may be derived from a dehydro derivative of aplysamine-2 or purpuramine H via cyclization and dehydroxylation [69]. The numerous bromotyrosine alkaloids that contain an aminopropanol unit, such as nakirodin A (**71**), may be derived from a homoserine unit through decarboxylation [60]. The

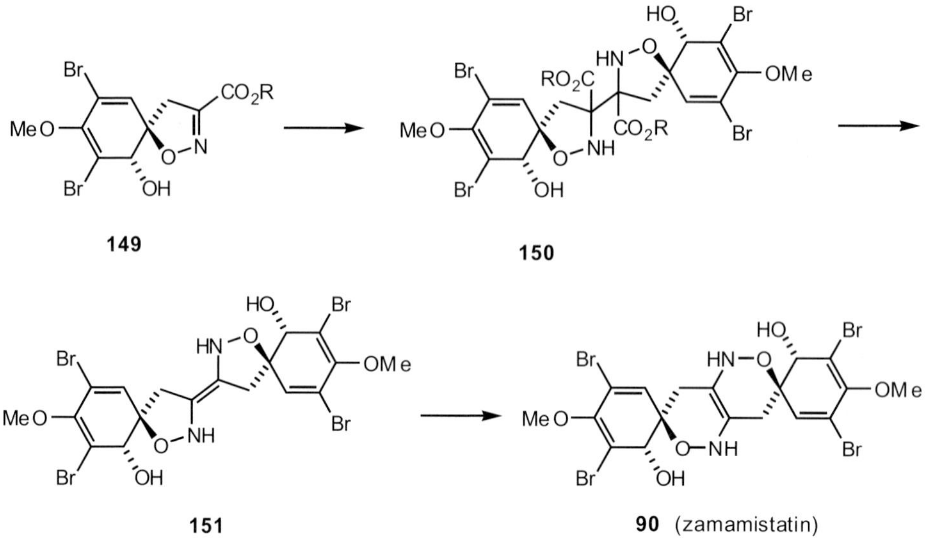

Scheme 19.9 Proposed biosynthesis of zamamistatin [80].

extraordinary tricyclo[5.2.1.$0^{2,6}$]decane skeleton in kuchinoenamine (**91**) may arise via a Diels–Alder cycloaddition between a cyclopenta-2,4-dienol and a 2-aminomethylenecyclopent-4-ene-1,3-dione. Subsequent formation of a bromohydrin and oxidation of the hydroxyl group to the ketone carbonyl completes the sequence [59]. Another dimerization reaction has been suggested to account for the biosynthesis of archerine (**87**). Thus, oxidative coupling of two molecules of aerophobin-2, which is also found in the sponge *Aplysina archeri* with archerine, could provide the latter metabolite [75]. Similarly, a reductive dimerization of isoxazoline **149** followed by oxidative decarboxylation of the resulting **150** could afford **151**. Hydrolysis and recyclization leads to zamamistatin (**90**) (Scheme 19.9) [80]. The novel ceratamines A (**96**) and B (**97**) are suggested to arise from the union of histidine and tyrosine similar to 5-bromoverongamine and ianthelline [84].

19.3.4
Biosynthesis of Miscellaneous Alkaloids

Very few halogenated terrestrial plant alkaloids are known [1]. A notable exception is the group of hasubanan-type alkaloids, dauricumine (**108**), dauricumidine (**109**), acutumine and acutumidine, and clolimalongine (**152**), each of which contains a single chlorine atom as shown for **108** and **109**[94,95,127–130]. The early proposed biosynthesis of acutumine by Barton [127], and later endorsed by Cavé for the related clolimalongine [129], is shown in Scheme 19.10. The only experimental support for this scheme is the observed incorporation of ^{36}Cl into these alkaloids using feeding experiments [94,127], and the incorporation of L-[U-^{14}C]tyrosine into acutumine [130].

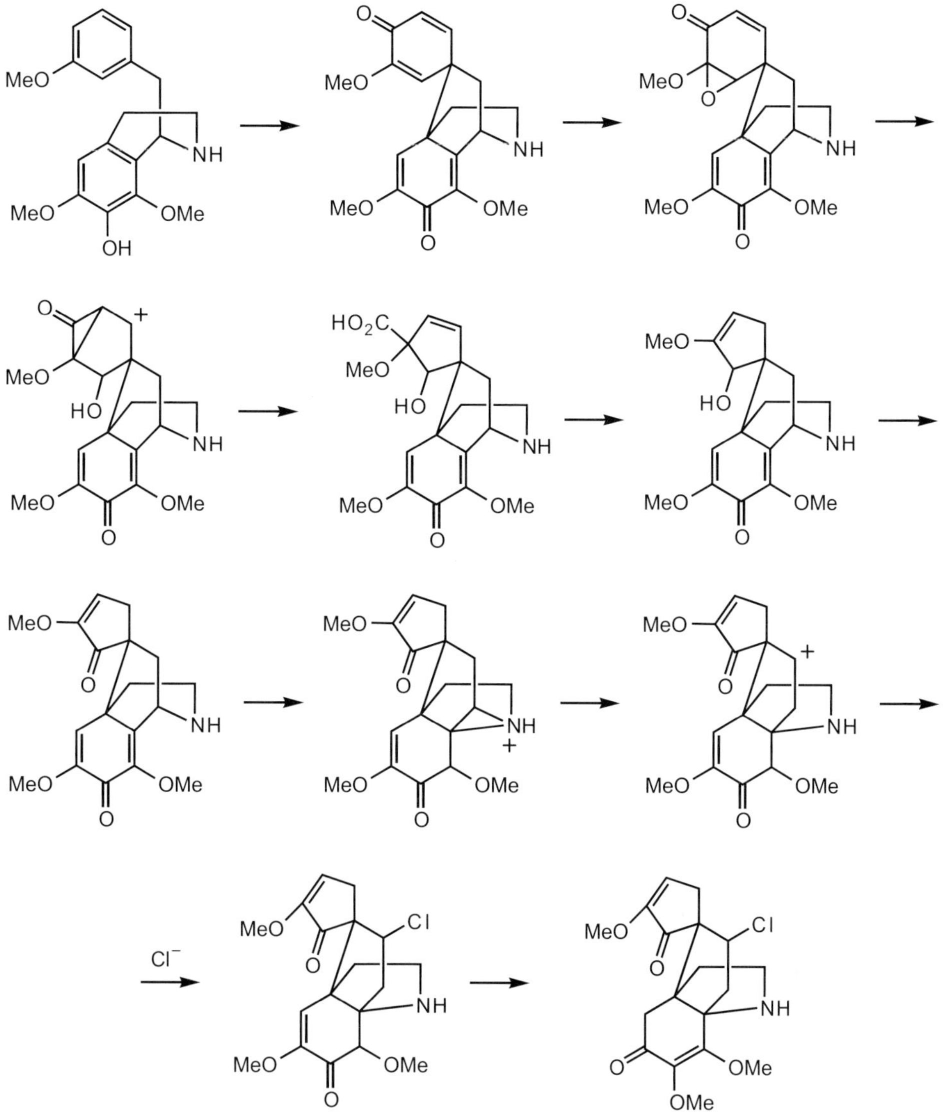

152 (clolimalongine)

Fig. 19.10 Proposed biosynthesis of acutumine (**152**)[127,129].

References

1 Gribble, G. W. (1996) *Progress in the Chemistry of Organic Natural Products*, **68**, 1–423.

2 Gribble, G. W. (2004) Natural organohalogens. *Euro Chlor Science Dossier*, 1–77.

3 G. W. Gribble, unpublished compilation.
4 Gribble, G. W. (2003) *Progress in Heterocyclic Chemistry*, **15**, 58–74.
5 Maruya, K. A. (2003) *Chemosphere*, **52**, 409–413.
6 El-Gamel, A. A., Wang, W.-L., Duh, C. -Y. (2005) *Journal of Natural Products*, **68**, 815–817.
7 Meragelman, K. M., West, L. M., Northcote, P. T., Pannell, L. K., McKee, T. C., Boyd, M. R. (2002) *Journal of Organic Chemistry*, **67**, 6671–6677.
8 Carletti, I., Banaigs, B., Amade, P. (2000) *Journal of Natural Products*, **63**, 981–983.
9 Moubax, I., Bontemps-Subielos, N., Banaigs, B., Combaut, G., Huitorel, P., Girard, J.-P., Pesando, D. (2001) *Environmental Toxicology and Chemistry*, **20**, 589–596.
10 Abourriche, A., Abboud, Y., Maoufoud, S., Mohou, H., Seffaj, T., Charrouf, M., Chaib, N., Bennamara, A., Bontemps, N., Francisco, C. (2003) *Il Farmaco*, **58**, 1351–1354.
11 Myers, R. A., Cruz, L. J., Rivier, J. E., Olivera, B. M. (1993) *Chemical Reviews*, **93**, 1923–1936.
12 Segraves, N. L. and Crews, P. (2005) *Journal of Natural Products*, **68**, 1484–1488.
13 Campagnuolo, C., Fattorusso, E., Taglialatela-Scafati, O. (2003) *European Journal of Organic Chemistry*, 284–287.
14 Borrelli, F., Campagnuolo, C., Capasso, R., Fattorusso, E., Taglialatela-Scafati, O. (2004) *European Journal of Organic Chemistry*, 3227–3232.
15 Fahy, E., Potts, B. C. M., Faulkner, D. J. (1991) *Journal of Natural Products*, **54**, 564–569.
16 Kelley, W. P., Wolters, A. M., Sack, J. T., Jockusch, R. A., Jurchen, J. C., Williams, E. R., Sweedler, J. V., Gilly, W. F. (2003) *Journal of Biological Chemistry*, **278**, 34934–34942.
17 Wolters, A. M., Jayawickrama, D. A., Sweedler, J. V. (2005) *Journal of Natural Products*, **68**, 162–167.
18 Christophersen, C. (1985) In Brossi, A. (Ed.) *The Alkaloids*, Academic Press, New York, Chapter 2.
19 Peters, L., Wright, A. D., Kehraus, S., Gündisch, D., Tilotta, M. C., König, G. M. (2004) *Planta Medica*, **70**, 883–886.
20 Lysek, N., Rachor, E., Lindel, T. (2002) *Zeitschrift fuer Naturforschung*, **57c**, 1056–1061.
21 Peters, L., König, G. M., Terlau, H., Wright, A. D. (2002) *Journal of Natural Products*, **65**, 1633–1637.
22 Tsuda, M., Takahashi, Y., Fromont, J., Mikami, Y., Kobayashi, J. (2005) *Journal of Natural Products*, **68**, 1277–1278.
23 Bifulco, G., Bruno, I., Minale, L., Riccio, R., Calignano, A., Debitus, C. (1994) *Journal of Natural Products*, **57**, 1294–1299.
24 Bifulco, G., Bruno, I., Riccio, R., Lavayre, J., Bourdy, G. (1995) *Journal of Natural Products*, **58**, 1254–1260.
25 Bao, B., Sun, Q., Yao, X., Hong, J., Lee, C. -O., Sim, C. J., Im, K. S., Jung, J. H. (2005) *Journal of Natural Products*, **68**, 711–715.
26 Aoki, S., Ye, Y., Higuchi, K., Takashima, A., Tanaka, Y., Kitagawa, I., Kobayashi, M. (2001) *Chemical and Pharmaceutical Bulletin*, **49**, 1372–1374.
27 Appleton, D. R., Page, M. J., Lambert, G., Berridge, M. V., Copp, B. R. (2002) *Journal of Organic Chemistry*, **67**, 5402–5404.
28 Appleton, D. R. and Copp, B. R. (2003) *Tetrahedron Letters*, **44**, 8963–8965.
29 Iwagawa, T., Miyazaki, M., Okamura, H., Nakatani, M., Doe, M., Takemura, K. (2003) *Tetrahedron Letters*, **44**, 2533–2535.
30 (a) Antunes, E. M., Copp, B. R., Davies-Coleman, M. T., Samaai, T. (2005) *Natural Product Reports*, **22**, 62–72. (b) Harayama, Y. and Kita, Y. (2005) *Current Organic Chemistry*, **9**, 1567–1588.
31 Chang, L. C., Otero-Quintero, S., Hooper, J. N. A., Bewley, C. A. (2002) *Journal of Natural Products*, **65**, 776–778.
32 Gunasekera, S. P., Zuleta, I. A., Longley, R. E., Wright, A. E., Pomponi, S. A. (2003) *Journal of Natural Products*, **66**, 1615–1617.
33 Antunes, E. M., Beukes, D. R., Kelly, M., Samaai, T., Barrows, L. R.,

Marshall, K. M., Sincich, C., Davies-Coleman, M. T. (2004) *Journal of Natural Products*, **67**, 1268–1276.

34 Lang, G., Pinkert, A., Blunt, J. W., Munro, M. H. G. (2005) *Journal of Natural Products*, **68**, 1796–1798.

35 Verbitski, S. M., Mayne, C. L., Davis, R. A., Concepcion, G. P., Ireland, C. M. (2002) *Journal of Organic Chemistry*, **67**, 7124–7126.

36 Makarieva, T. N., Ilyin, S. G., Stonik, V. A., Lyssenko, K. A., Denisenko, V. A. (1999) *Tetrahedron Letters*, **40**, 1591–1594.

37 Makarieva, T. N., Dmitrenok, A. S., Dmitrenok, P. S., Grebnev, B. B., Stonik, V. A. (2001) *Journal of Natural Products*, **64**, 1559–1561.

38 Stratmann, K., Moore, R. E., Bonjouklian, R., Deeter, J. B., Patterson, G. M. L., Shaffer, S., Smith, C. D., Smitka, T. A. (1994) *Journal of the American Chemical Society*, **116**, 9935–9942.

39 Huber, U., Moore, R. E., Patterson, G. M. L. (1998) *Journal of Natural Products*, **61**, 1304–1306.

40 Rundberget, T. and Wilkins, A. L. (2002) *Phytochemistry*, **61**, 979–985.

41 Cardellina, J. H., II,Kirkup, M. P., Moore, R. E., Mynderse, J. S., Seff, K., Simmons, C. J. (1979) *Tetrahedron Letters*, **20**, 4915–4916.

42 Luk, K.-C., Stern, L., Weigele, M., O'Brien, R. A., Spirt, N. (1983) *Journal of Natural Products*, **46**, 852–861.

43 Lee, S.-C., Williams, G. A., Brown, G. D. (1999) *Phytochemistry*, **52**, 537–540.

44 Schumacher, R. W. and Davidson, B. S. (1995) *Tetrahedron*, **51**, 10125–10130.

45 Rashid, M. A., Gustafson, K. R., Boyd, M. R. (2001) *Journal of Natural Products*, **64**, 1454–1456.

46 Davis, R. A., Carroll, A. R., Quinn, R. J. (1998) *Journal of Natural Products*, **61**, 959–960.

47 Sandler, J. S., Colin, P. L., Hooper, J. N. A., Faulkner, D. J. (2002) *Journal of Natural Products*, **65**, 1258–1261.

48 Becher, P. G., Beuchat, J., Gademann, K., Jüttner, F. (2005) *Journal of Natural Products*, **68**, 1793–1795.

49 Larsen, L. K., Moore, R. E., Patterson, G. M. L. (1994) *Journal of Natural Products*, **57**, 419–421.

50 Segraves, N. L., Lopez, S., Johnson, T. A., Said, S. A., Fu, X., Schmitz, F. J., Pietraszkiewicz, H., Valeriote, F. A., Crews, P. (2003) *Tetrahedron Letters*, **44**, 3471–3475.

51 Segraves, N. L., Robinson, S. J., Garcia, D., Said, S. A., Fu, X., Schmitz, F. J., Pietraszkiewicz, H., Valeriote, F. A., Crews, P. (2004) *Journal of Natural Products*, **67**, 783–792.

52 Debitus, C., Laurent, D., Païs, M. (1988) *Journal of Natural Products*, **51**, 799–801.

53 Mahboobi, S., Dove, S., Bednarski, P. J., Kuhr, S., Burgemeister, T., Schollmeyer, D. (1997) *Journal of Natural Products*, **60**, 587–591.

54 Kang, H. and Fenical, W. (1996) *Natural Product Letters*, **9**, 7–12.

55 Chbani, M., Païs, M., Delauneux, J.-M., Debitus, C. (1993) *Journal of Natural Products*, **56**, 99–104.

56 Ackermann, D. and Müller, E. (1941) *Hoppe-Seyler's Zeitschrift für Physiologische Chemie*, **269**, 146–157.

57 Ciminiello, P., Dell'Aversano, C., Fattorusso, E., Magno, S., Pansini, M. (2000) *Journal of Natural Products*, **63**, 263–266.

58 Ross, S. A., Weete, J. D., Schinazi, R. F., Wirtz, S. S., Tharnish, P., Scheuer, P. J., Hamann, M. T. (2000) *Journal of Natural Products*, **63**, 501–503.

59 Matsunaga, S., Kobayashi, H., van Soest, R. W. M., Fusetani, N. (2005) *Journal of Organic Chemistry*, **70**, 1893–1896.

60 Tsuda, M., Endo, T., Watanabe, K., Fromont, J., Kobayashi, J. (2002) *Journal of Natural Products*, **65**, 1670–1671.

61 Narkowicz, C. K., Blackman, A. J., Lacey, E., Gill, J. H., Heiland, K. (2002) *Journal of Natural Products*, **65**, 938–941.

62 Kigoshi, H., Kanematsu, K., Yokota, K., Uemura, D. (2000) *Tetrahedron*, **56**, 9063–9070.

63 Rao, M. R. and Faulkner, D. J. (2004) *Journal of Natural Products*, **67**, 1064–1066.

64 Piña, I. C., Gautschi, J. T., Wang, G.-Y.-S., Sanders, M. L., Schmitz, F. J., France, D., Cornell-Kennon, S., Sambucetti, L. C., Remiszewski, S. W., Perez, L. B., Bair, K. W., Crews, P. (2003) *Journal of Organic Chemistry*, **68**, 3866–3873.

65 Fusetani, N., Masuda, Y., Nakao, Y., Matsunaga, S., van Soest, R. W. M. (2001) *Tetrahedron*, **57**, 7507–7511.

66 Tabudravu, J. N. and Jaspars, M. (2002) *Journal of Natural Products*, **65**, 1798–1801.

67 Evan, T., Rudi, A., Ilan, M., Kashman, Y. (2001) *Journal of Natural Products*, **64**, 226–227.

68 Tsuda, M., Sakuma, Y., Kobayashi, J. (2001) *Journal of Natural Products*, **64**, 980–982.

69 Hirano, K., Kubota, T., Tsuda, M., Watanabe, K., Fromont, J., Kobayashi, J. (2000) *Tetrahedron*, **56**, 8107–8110.

70 Park, Y., Liu, Y., Hong, J., Lee, C.-O., Cho, H., Kim, D.-K., Im, K. S., Jung, J. H. (2003) *Journal of Natural Products*, **66**, 1495–1498.

71 Coll, J. C., Kearns, P. S., Rideout, J. A., Sankar, V. (2002) *Journal of Natural Products*, **65**, 753–756.

72 Okamoto, Y., Ojika, M., Kato, S., Sakagami, Y. (2000) *Tetrahedron*, **56**, 5813–5818.

73 Saeki, B. M., Granato, A. C., Berlinck, R. G. S., Magalhães, A., Schefer, A. B., Ferreira, A. G., Pinheiro, U. S., Hajdu, E. (2002) *Journal of Natural Products*, **65**, 796–799.

74 Nicholas, G. M., Newton, G. L., Fahey, R. C., Bewley, C. A. (2001) *Organic Letters*, **3**, 1543–1545.

75 Ciminiello, P., Dell'Aversano, C., Fattorusso, E., Magno, S. (2001) *European Journal of Organic Chemistry*, 55–60.

76 (a) Encarnación, R. D., Sandoval, E., Malmstrøm, J., Christophersen, C. (2000) *Journal of Natural Products*, **63**, 874–875.
(b) Ogamino, T. and Nishiyama, S. (2005) *Tetrahedron Letters*, **46**, 1083–1086.

77 Ogamino, T., Obata, R., Tomoda, H., Nishiyama, S. (2006) *Bulletin of the Chemical Society of Japan*, **79**, 134–139.

78 Kijjoa, A., Watanadilok, R., Sonchaeng, P., Silva, A. M. S., Eaton, G., Herz, W. (2001) *Zeitschrift fuer Naturforschung*, **56c**, 1116–1119.

79 Takada, N., Watanabe, R., Suenaga, K., Yamada, K., Ueda, K., Kita, M., Uemura, D. (2001) *Tetrahedron Letters*, **42**, 5265–5267.

80 Hayakawa, I., Teruya, T., Kigoshi, H. (2006) *Tetrahedron Letters*, **47**, 155–158.

81 Morris, B. D. and Prinsep, M. R. (1998) *Journal of Organic Chemistry*, **63**, 9545–9547.

82 Milanowski, D. J., Gustafson, K. R., Kelley, J. A., McMahon, J. B. (2004) *Journal of Natural Products*, **67**, 70–73.

83 Kigoshi, H., Ichino, T., Takada, N., Suenaga, K., Yamada, A., Yamada, K., Uemura, D. (2004) *Chemistry Letters*, **33**, 98–99.

84 Manzo, E., van Soest, R., Matainaho, L., Roberge, M., Andersen, R. J. (2003) *Organic Letters*, **5**, 4591–4594.

85 Carroll, A. R., Ngo, A., Quinn, R. J., Redburn, R., Hooper, J. N. A. (2005) *Journal of Natural Products*, **68**, 804–806.

86 Uddin, M. J., Kokubo, S., Ueda, K., Suenaga, K., Uemura, D. (2002) *Chemistry Letters*, 1028–1029.

87 Nogle, L. M. and Gerwick, W. H. (2003) *Journal of Natural Products*, **66**, 217–220.

88 Spande, T. F., Garraffo, H. M., Edwards, M. W., Yeh, H. J. C., Pannell, L., Daly, J. W. (1992) *Journal of the American Chemical Society*, **114**, 3475–3478.

89 Sullivan, J. P. and Bannon, A. W. (1996) *CNS Drug Reviews*, **2**, 21–39.

90 (a) Carroll, F. I. (2004) *Bioorganic and Medicinal Chemistry Letters*, **14**, 1889–1896.
(b) Olivo, H. F. and Hemenway, M. S. (2002) *Organic Preparations and Procedures International*, **34**, 1–26.

91 Chang, F. -R., Chen, C. -Y., Wu, P. -H., Kuo, R.-Y., Chang, Y.-C., Wu, Y.-C. (2000) *Journal of Natural Products*, **63**, 746–748.

92 Kuo, R.-Y., Chang, F.-R., Chen, C.-Y., Teng, C.-M., Yen, H.-F., Wu, Y.-C. (2001) *Phytochemistry*, **57**, 421–425.

93 Al-Rehaily, A. J., Ahmad, M. S., Muhammad, I., Al-Thukair, A. A., Perzanowski, H. P. (2003) *Phytochemistry*, **64**, 1405–1411.

94 Sugimoto, Y., Babiker, H. A. A., Saisho, T., Furumoto, T., Inanaga, S., Kato, M. (2001) *Journal of Organic Chemistry*, **66**, 3299–3302.

95 Sugimoto, Y., Matsui, M., Takikawa, H., Sasaki, M., Kato, M. (2005) *Phytochemistry*, **66**, 2627–2631.

96 Morimoto, Y., Matsuda, F., Shirahama, H. (1996) *Tetrahedron*, **52**, 10609–10630.

97 Morimoto, Y. and Shirahama, H. (1996) *Tetrahedron*, **52**, 10631–10652.

98 Kim, W.-G., Kim, J.-P., Yoo, I.-D. (1996) *Journal of Antibiotics*, **49**, 26–30.

99 Kuramoto, M., Tong, C., Yamada, K., Chiba, T., Hayashi, Y., Uemura, D. (1996) *Tetrahedron Letters*, **37**, 3867–3870.

100 Arimoto, H., Hayakawa, I., Kuramoto, M., Uemura, D. (1998) *Tetrahedron Letters*, **39**, 861–862.

101 Chou, T., Kuramoto, M., Otani, Y., Shikano, M., Yazawa, K., Uemura, D. (1996) *Tetrahedron Letters*, **37**, 3871–3874.

102 Carson, M. W., Kim, G., Danishefsky, S. J. (2001) *Angewandte Chemie, International Edition*, **40**, 4453–4456.

103 (a) van Pée, K. -H. and Unversucht, S. (2003) *Chemosphere*, **52**, 299–312.
(b) Unversucht, S., Hollmann, F., Schmid, A., van Pée, K. -H. (2005) *Advanced Synthesis and Catalysis*, **347**, 1163–1167.
(c) Zehner, S., Kotzsch, A., Bister, B., Süssmuth, R. D., Méndez, C., Salas, J. A., van Pée, K. -H. (2005) *Chemistry and Biology*, **12**, 445–452.

104 Murphy, C. D. (2003) *Journal of Applied Microbiology*, **94**, 539–548.

105 Murphy, C. D. (2006) *Natural Product Reports*, **23**, 147–152.

106 Butler, A. (1998) *Current Opinion in Chemical Biology*, **2**, 279–285.

107 Butler, A. (1999) *Coordination Chemistry Reviews*, **187**, 17–35.

108 Butler, A. and Carter-Franklin, J. N. (2004) *Natural Product Reports*, **21**, 180–188.

109 (a) Takeshita, J., Byun, J., Nhan, T. Q., Pritchard, D. K., Pennathur, S., Schwartz, S. M., Chait, A., Heinecke, J. W. (2006) *Journal of Biological Chemistry*, **281**, 3096–3104.
(b) Fu, X., Wang, Y., Kao, J., Irwin, A., d'Avignon, A., Mecham, R. P., Parks, W. C., Heinecke, J. W. (2006) *Biochemistry*, **45**, 3961–3971.

110 (a) Deng, H., Cobb, S. L., McEwan, A. R., McGlinchey, R. P., Naismith, J. H., O'Hagan, D., Robinson, D. A., Spencer, J. B. (2006) *Angewandte Chemie, International Edition*, **45**, 759–762.
(b) Cobb, S. L., Deng, H., McEwan, A. R., Naismith, J. H., O'Hagan, D., Robinson, D. A. (2006) *Organic and Biomolecular Chemistry*, **4**, 1458–1460.
(c) Huang, F., Haydock, S. F., Spiteller, D., Mironenko, T., Li, T. -L., O'Hagan, D., Leadlay, P. F., Spencer, J. B. (2006) *Chemistry and Biology*, **13**, 475–484.

111 (a) Shen, G. Q. and Baker, B. J. (1994) *Tetrahedron Letters*, **35**, 1141–1144.
(b) Shen, G. Q. and Baker, B. J. (1994) *Tetrahedron Letters*, **35**, 4923–4926.

112 Lill, R. E., Major, D. A., Blunt, J. W., Munro, M. H. G., Battershill, C. N., McLean, M. G., Baxter, R. L. (1995) *Journal of Natural Products*, **58**, 306–311.

113 Keyzers, R. A., Arendse, C. E., Hendricks, D. T., Samaai, T., Davies-Coleman, M. T. (2005) *Journal of Natural Products*, **68**, 506–510.

114 (a) Bornemann, V., Patterson, G. M. L., Moore, R. E. (1988) *Journal of the American Chemical Society*, **110**, 2339–2340.
(b) Karuso, P. and Scheuer, P. J. (1989) *Journal of Organic Chemistry*, **54**, 2092–2095.

115 (a) Mantle, P. G. and Penn, J. (1989) *Journal of the Chemical Society, Perkin Transactions 1*, 1539–1540.
(b) Hosoe, T., Nozawa, K., Udagawa, S., Nakajima, S., Kawai, K. (1990) *Chemical and Pharmaceutical Bulletin*, **38**, 3473–3475.

116 Park, A., Moore, R. E., Patterson, G. M. L. (1992) *Tetrahedron Letters*, **33**, 3257–3260.

117 (a) Dong, C., Flecks, S., Unversucht, S., Kaupt, C., van Pée, K. -H., Naismith, J. H. (2005) *Science*, **309**, 2216–2219.
(b) Yeh, E., Garneau, S., Walsh, C. T. (2005) *Proceedings of the National Academy of Sciences of the United States of America*, **102**, 3960–3965.

118 Sánchez, C., Butovich, I. A., Braña, A. F., Rohr, J., Méndez, C., Salas, J. A. (2002) *Chemistry and Biology*, **9**, 519–531.

119 Onaka, H., Taniguchi, S., Igarashi, Y., Furumai, T. (2003) *Bioscience, Biotechnology, and Biochemistry*, **67**, 127–138.

120 Tymiak, A. A. and Rinehart, K. L., Jr. (1981) *Journal of the American Chemical Society*, **103**, 6763–6765.

121 Carney, J. R. and Rinehart, K. L. (1995) *Journal of Natural Products*, **58**, 971–985.

122 Yang, Z. P., Shelton, K. D., Howard, J. C., Woods, A. E. (1995) *Comparative Biochemistry and Physiology*, **111B**, 417–426.

123 Jaspars, M., Rali, T., Laney, M., Schatzman, R. C., Diaz, M. C., Schmitz, F. J., Pordesimo, E. O., Crews, P. (1994) *Tetrahedron*, **50**, 7367–7374.

124 Utkina, N. K., Denisenko, V. A., Scholokova, O. V., Virovaya, M. V., Gerasimenko, A. V., Popov, D. Y., Krasokhin, V. B., Popov, A. M. (2001) *Journal of Natural Products*, **64**, 151–153.

125 Utkina, N. K., Denisenko, V. A., Virovaya, M. V., Scholokova, O. V., Prokofeva, N. G. (2002) *Journal of Natural Products*, **65**, 1213–1215.

126 Jiménez, C. and Crews, P. (1991) *Tetrahedron*, **47**, 2097–2102.

127 Barton, D. H. R., Kirby, A. J., Kirby, G. W. (1968) *Journal of the Chemical Society (C)*, 929–936.

128 Okamoto, Y., Yuge, E., Nagai, Y., Katsuta, R., Kishimoto, A., Kobayashi, Y., Kikuchi, T., Tomita, M. (1969) *Tetrahedron Letters*, **10**, 1933–1935.

129 Berthou, S., Leboeuf, M., Cavé, A., Mahuteau, J., David, B., Guinaudeau, H. (1989) *Journal of Organic Chemistry*, **54**, 3491–3493.

130 Sugimoto, Y., Uchida, S., Inanaga, S., Kimura, Y., Hashimoto, M., Isogai, A. (1996) *Bioscience, Biotechnology, and Biochemistry*, **60**, 503–505.

20
Engineering Biosynthetic Pathways to Generate Indolocarbazole Alkaloids in Microorganisms

César Sánchez, Carmen Méndez, José A. Salas

20.1
Introduction

The indolocarbazole alkaloids and the biosynthetically related bisindolylmaleimides constitute an important class of natural products, which have been isolated from actinomycetes, cyanobacteria, slime molds, and marine invertebrates [1–3]. They display a wide range of biological activities, including antibacterial, antifungal, antiviral, hypotensive, antitumor, and/or neuroprotective properties. The antitumor and neuroprotective activities of indolocarbazoles are the result of one, or several, of the following mechanisms: (a) inhibition of different protein kinases, (b) inhibition of DNA topoisomerases, or (c) direct DNA intercalation [3–6]. Hundreds of indolocarbazole derivatives have been produced by chemical synthesis or semisynthesis [1,2,6], and several of them have entered clinical trials for the treatment of diverse types of cancer, Parkinson's disease or diabetic retinopathy [3,7].

Structurally, most members of this family are characterized by a core consisting of an indolo[2,3-*a*]carbazole, which is frequently fused to a pyrrolo[3,4-*c*] unit. The chromophore is often decorated with a carbohydrate that is attached either through a single *N*-glycosidic bond, as in rebeccamycin **1** (REB), or through two C–N bonds, as in staurosporine **2** (STA). Another key structural difference between REB and STA resides at the pyrrole moiety, consisting of an imide function in the former but an

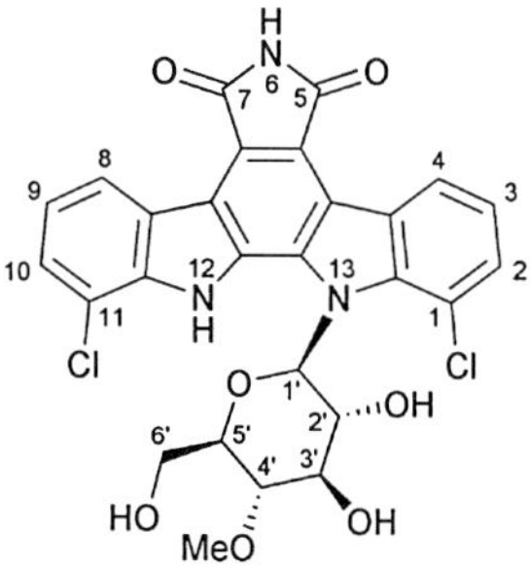

1
rebeccamycin (REB)

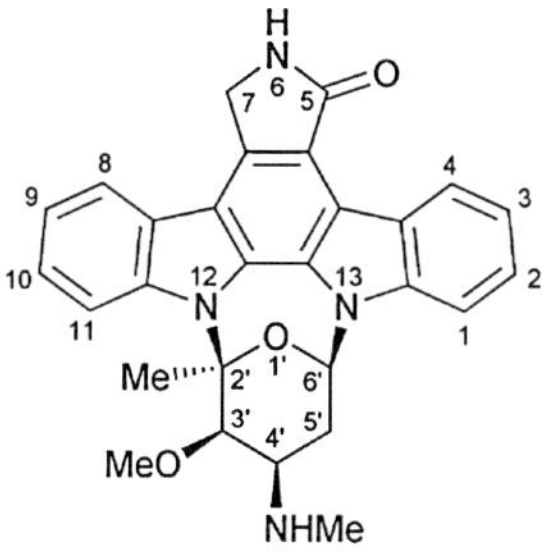

2
staurosporine (STA)

Modern Alkaloids: Structure, Isolation, Synthesis and Biology. Edited by E. Fattorusso and O. Taglialatela-Scafati

ISBN: 978-3-527-31521-5

amide function in the latter. These differences seem to be essential for target selectivity, since REB inhibits topoisomerase I while STA is a protein kinase inhibitor.

As a complementary approach to the methods of organic chemistry, manipulation of indolocarbazole biosynthesis offers a promising alternative means to prepare these compounds [3,8]. Here we summarize the findings concerning the generation of novel indolocarbazole derivatives in microorganisms, including the identification of genes involved in indolocarbazole biosynthesis in actinomycetes.

20.2
Studies Made Before the Identification of Biosynthetic Genes

The biosyntheses of STA and REB were studied by feeding radiolabeled precursors to *Lentzea albida* (formerly *Streptomyces staurosporeus*) [9–13] and *Lechevalieria aerocolonigenes* (formerly *Saccharothix aerocolonigenes*) [14,15], respectively. The results established that the indolocarbazole core was derived from two units of tryptophan (with the carbon skeleton incorporated intact), while the sugar moiety was derived from glucose and methionine (for methylations).

Before the identification of the biosynthesis genes involved, different strategies were successful for the production of novel analogs by indolocarbazole-producing microorganisms. Addition of potassium bromide in the fermentation of *Lech.*

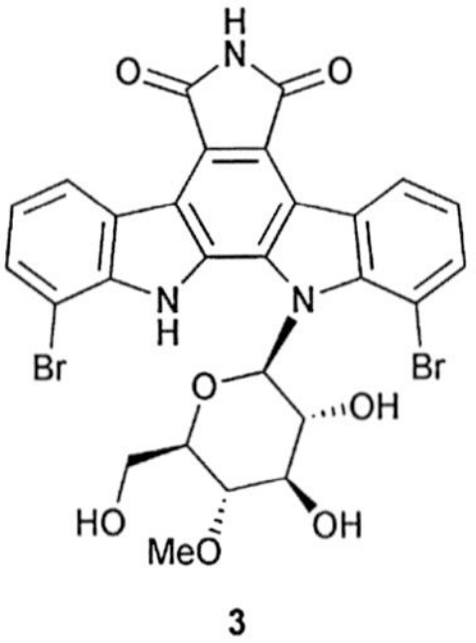

3 bromorebeccamycin

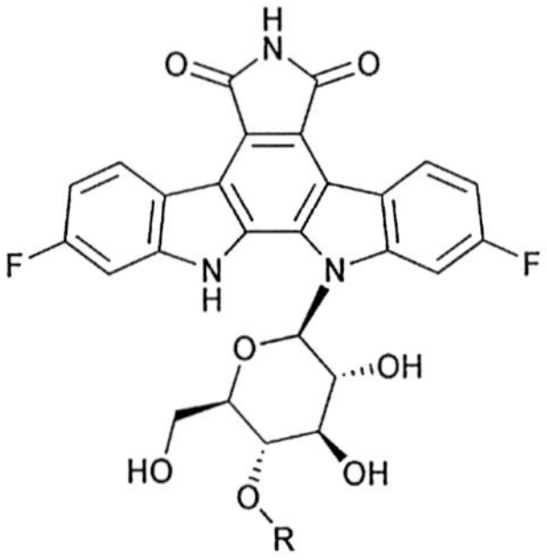

4 (R=Me)
5 (R=H)

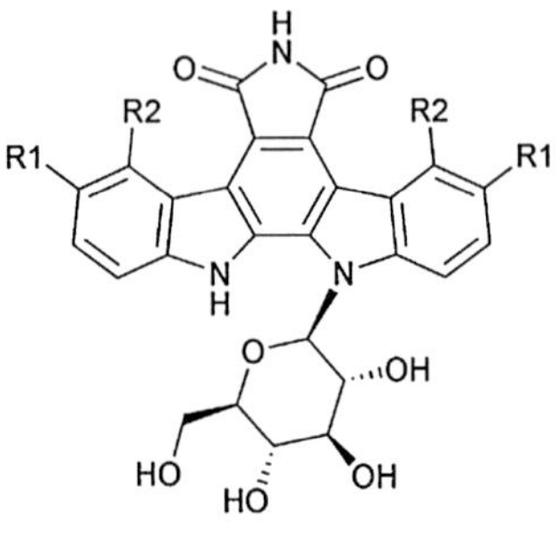

6 (R1=F, R2=H)
7 (R1=H, R2=F)

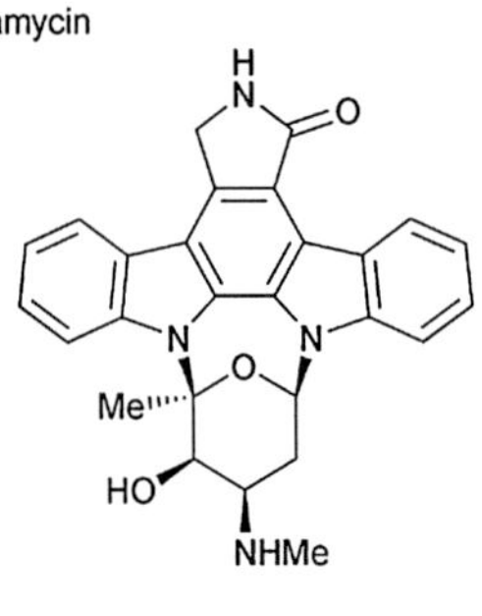

8

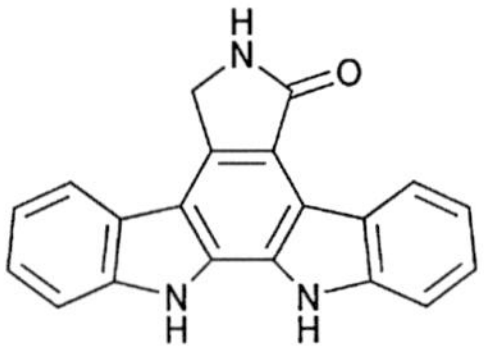

9 K252c (STA aglycone)

aerocolonigenes resulted in production of bromorebeccamycin (**3**) [16]. Moreover, addition of various types of fluorotryptophan to the fermentation of the same microorganism successfully produced a number of fluoroindolocarbazoles **4–7** [5,17]. The use of enzyme inhibitors was also evaluated for production of novel derivatives or intermediates. Sinefungin (an inhibitor of methyltransferases) affected the biosynthesis of STA in *Lentzea albida*, by blocking O-methylation of the sugar moiety. As a result, 3′-demethoxy-3′-hydroxy-STA **8** (or *O*-demethyl-STA) was efficiently accumulated in the medium [13].

A different STA-producing strain, *Strep. longisporoflavus*, was subjected to a mutagenic treatment by UV irradiation. This work led to the isolation of mutants M13 and M14, which accumulated STA aglycone (**9**, also known as K252c) and 3′-demethoxy-3′-hydroxy-STA (**8**), respectively [18,19]. The latter compound was converted into STA by cultures of the M13 mutant, and hence it was proposed as the last intermediate in STA biosynthesis [20].

20.3 Identification of Genes Involved in Indolocarbazole Biosynthesis

20.3.1 Genes Involved in Rebeccamycin Biosynthesis

A gene (*ngt*) encoding an indolocarbazole *N*-glycosyltransferase was cloned from the REB producer *Lech. aerocolonigenes* [21]. Heterologous expression of *ngt* in either *Strep. lividans* or *Strep. mobaraensis* allowed the introduction of a D-glucose moiety into two different indolocarbazoles (Scheme 20.1). These results suggested that *ngt* was probably involved in REB biosynthesis.

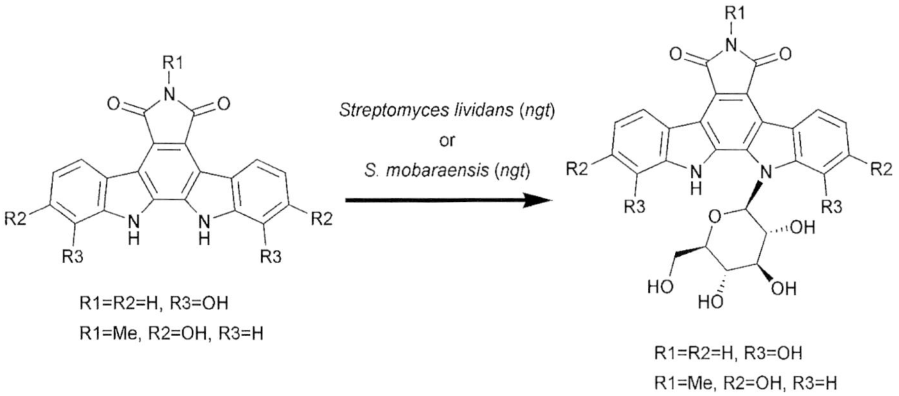

R1=R2=H, R3=OH
R1=Me, R2=OH, R3=H

Streptomyces lividans (ngt) or *S. mobaraensis (ngt)*

R1=R2=H, R3=OH
R1=Me, R2=OH, R3=H

Scheme 20.1 Glycosylation of indolocarbazoles by feeding appropriate aglycones to the fermentation broth of *ngt*-expressing streptomycetes.

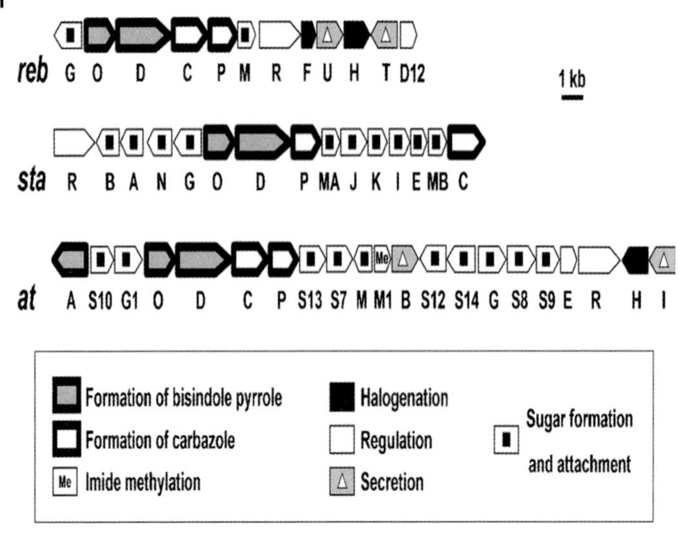

Fig. 20.1 Gene clusters for biosynthesis of rebeccamycin, staurosporine, and AT2433.

Given that biosynthetic genes for secondary metabolites usually occur as a cluster in actinomycetes, *ngt* served as the basis for the identification of the *reb* gene cluster, which was responsible for REB biosynthesis in *Lech. aerocolonigenes*. The biosynthetic locus was independently isolated from the same strain by several laboratories [22–25]. In a first approach [22], a *Lech. aerocolonigenes* genomic library was constructed in a shuttle cosmid vector (which was able to replicate in both *E. coli* and *Streptomyces* sp.), and the library was screened using an internal fragment of *ngt* as a probe. The result was the isolation of several overlapping cosmids containing *ngt*, which were subsequently introduced into an actinomycete host, *Strep. albus*. Two of the cosmids conferred the ability to produce REB in good yields. This result indicated that the cosmids carried all the genes needed for the biosynthesis, and the complete DNA sequence of the insert in one of these cosmids was determined. The *reb* gene cluster appeared to consist of 11 genes spanning 17.6 kb, with *ngt* (renamed as *rebG*) located at one end of the cluster (Figure 20.1). A number of functions were proposed for the *reb* genes, participating in a hypothetical pathway for REB biosynthesis (Table 20.1).

Some of these functions were supported by heterologous expression of subsets of *reb* genes in *Strep. albus*, which resulted in production of three nonchlorinated derivatives **10–12** [22]. Therefore, the heterologous expression of indolocarbazole biosynthetic genes in an actinomycete host appeared to be a practical solution for the production of derivatives.

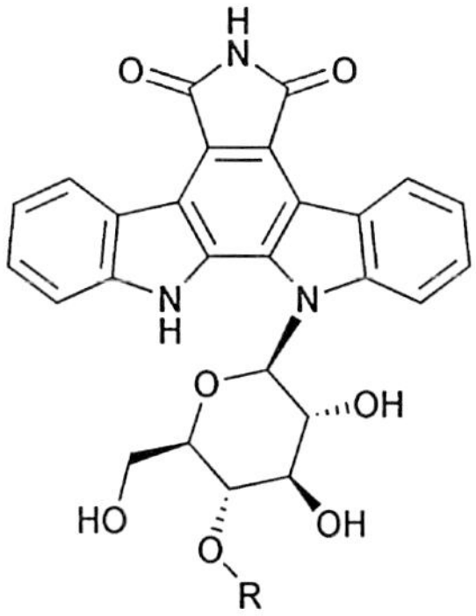

10 (R=Me)
11 (R=H)

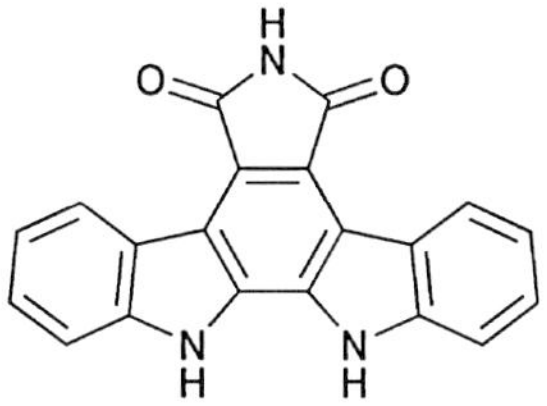

12
arcyriaflavin A

Tab. 20.1 Genes involved in the biosynthesis of rebeccamycin (*reb*), staurosporine (*sta*), and AT2433 (*at*).

	Genes			
Biosynthetic step	**reb**	**sta**	**at**	**Proposed function**
Tryptophan halogenation	*rebF*	–	–	Flavin reductase
	rebH	–	*atH*	Tryptophan halogenase
Formation of bisindole pyrrole	*rebO*	*staO*	*atO*	Amino oxidase
	rebD	*staD*	*atD*	Chromopyrrolic acid synthase
	–	–	*atA*	Amidotransferase
Formation of carbazole	*rebC*	*staC*	*atC*	Flavin-binding monooxygenase
	rebP	*staP*	*atP*	Cytochrome P450 monooxygenase
Sugar formation	–	*staA*	*atS7*	Phosphosugar nucleotidyltransferase
	–	*staB*	–	NDP-D-glucose 4,6-dehydratase
	–	–	*atS8*	NDP-hexose dehydrogenase
	–	–	*atS9*	NDP-hexose decarboxylase/4-epimerase
	–	*staJ*	*atS14*	NDP-hexose 2,3-dehydratase
	–	*staK*	*atS12*	NDP-hexose reductase
	–	*staI*	*atS13*	Aminotransferase
	–	*staE*	–	NDP-hexose 3,5-epimerase
	rebM	*staMA*	*atM*	Methyltransferases
		staMB	*atS10*	
Sugar attachment	*rebG*	*staG*	*atG*	Glycosyltransferases
			atG1	
	–	*staN*	–	Cytochrome P450 monooxygenase
Imide methylation	–	–	*atM1*	Methyltransferase
Secretion or self-resistance	*rebT*	–	*atI*	Major facilitator superfamily
	rebU	–	*atB*	Antiporter
Transcriptional regulation	*rebR*	*staR*	*atR*	LAL regulatory protein
	orfD12	–	*atE*	MarR regulatory protein

Independently, another laboratory also used the *ngt* gene to isolate the *reb* locus [23]. In order to confirm their possible role in REB biosynthesis, several *reb* genes were inactivated (by gene disruption experiments), and eight REB derivatives (**11**, **13–19**) were identified from the mutant strains. A key intermediate in the pathway consisted of 11,11′-dichlorochromopyrrolic acid **13** (11,11′-dichloro-CPA) [23]. REB production was reconstituted by feeding 11,11′-dichloro-CPA to a blocked mutant (unable to produce REB by itself), which confirmed compound **13** as an intermediate. Therefore, it appeared that decarboxylation occurred after the coupling of two tryptophan-derived units and the formation of the pyrrolo[3,4-*c*] ring. Additionally, a new genomic library of *Lech. aerocolonigenes* was prepared, but this time using a shuttle cosmid vector, and a clone was selected that contained the complete gene cluster. When this cosmid clone was transformed into *Strep. lividans*, REB production was detected [26].

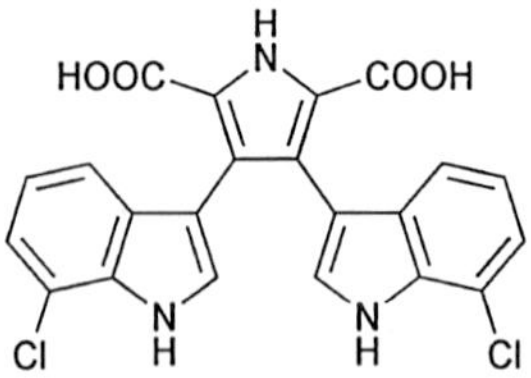

13
11,11'-dichloro-CPA

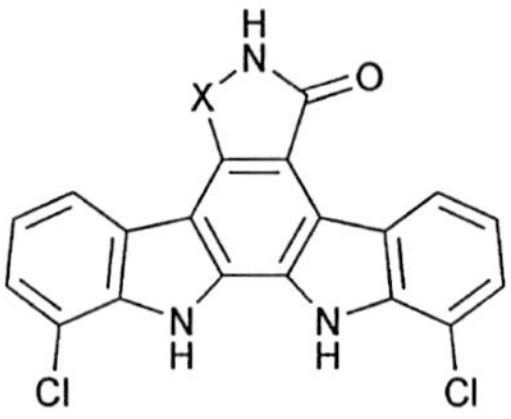

14 REB aglycone (X=CO)
15 (X=CHOH)
16 ($X=CH_2$)

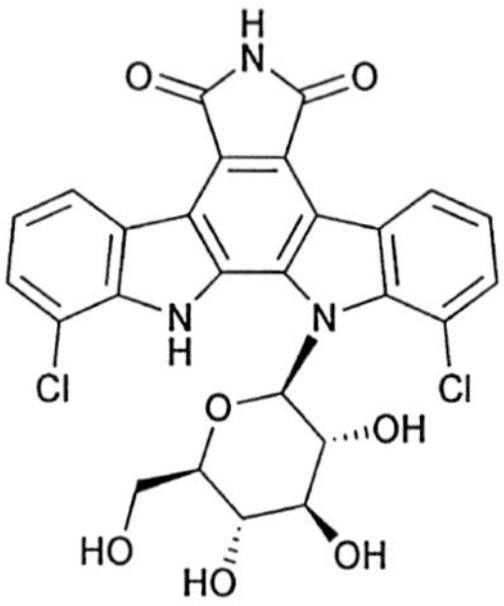

17

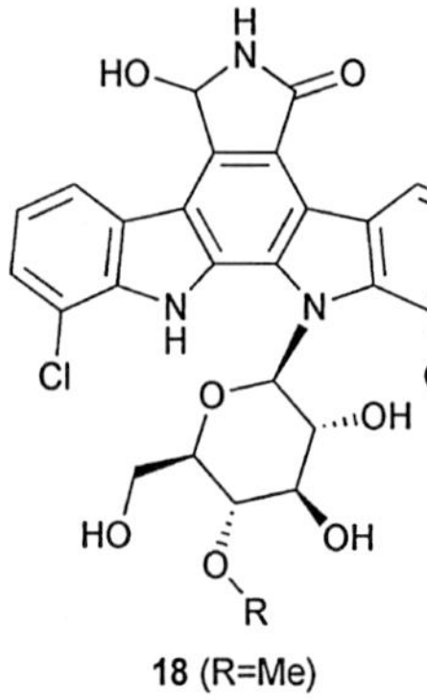

18 (R=Me)
19 (R=H)

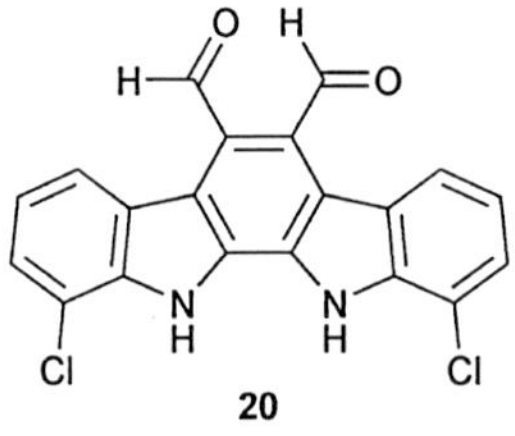

20

Finally, another research team isolated the complete gene cluster in a single cosmid clone that conferred the ability to produce REB in *E. coli*, although at low levels [24]. An *E. coli* strain carrying this cosmid produced REB (**1**, major component), REB aglycone (**14**), and metabolite **20** whose proposed structure lacked the pyrrolo[3,4-*c*] ring. The authors suggested that the first ring closure and aromatization in REB biosynthesis apparently occurred prior to forming the final pyrrolocarbazole ring. However, given that 11,11′-dichloro-CPA (**13**) has been confirmed as a true biosynthetic intermediate [23], it is more likely that metabolite **20** might be produced by some unidentified enzyme from the *E. coli* host, acting on a REB intermediate.

20.3.2
Genes Involved in Staurosporine Biosynthesis

Some genes needed for formation of the STA sugar moiety in *Strep. longisporoflavus* were disclosed in a patent application [27]. By transforming a genomic library of this microorganism into mutant M14 (which accumulated *O*-demethyl-STA **8**), a 2.1-kb DNA fragment was found to restore the ability to produce STA. The fragment was sequenced, and a gene putatively coding for a SAM-dependent methyltransferase was identified. Therefore, it was assumed that this gene coded for the *O*-methyltransferase responsible for the last step in STA formation. A DNA region of about 10 kb, which included the 2.1-kb fragment, was sequenced, and several additional genes were identified.

The complete STA biosynthetic gene cluster (*sta* genes) was later cloned from a different strain, *Streptomyces* sp. TP-A0274 [23,28]. A genomic library was constructed in a shuttle cosmid vector and was screened using as a probe a fragment of one of the *reb* genes (*rebD*). One of the positive clones was confirmed to include all the required biosynthetic genes by heterologous expression in *Strep. lividans*, resulting in STA production [28]. After the DNA region was sequenced, putative functions were proposed for the *sta* genes (Figure 20.1 and Table 20.1). Based upon the sequence information previously reported from *Strep. longisporoflavus* [27], the complete STA biosynthetic gene cluster from this strain was also isolated in a single cosmid clone, which directed STA production upon introduction in *Strep. albus* [29].

20.3.3
Genes Involved in Biosynthesis of Other Indolocarbazoles

Recently, a gene cluster (*at* genes) has been described for the production of AT2433 complex **21–24** in *Actinomadura melliaura* [30]. The *at* locus was identified by searching

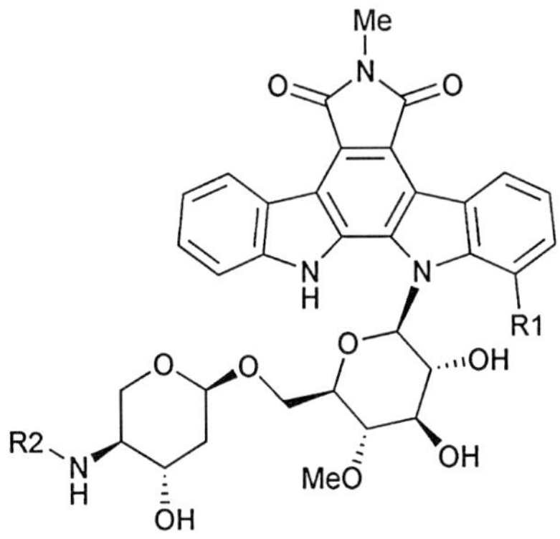

21 AT2433-A1 (R1=Cl, R2=Me)
22 AT2433-A2 (R1=Cl, R2=H)
23 AT2433-B1 (R1=H, R2=Me)
24 AT2433-B2 (R1=R2=H)

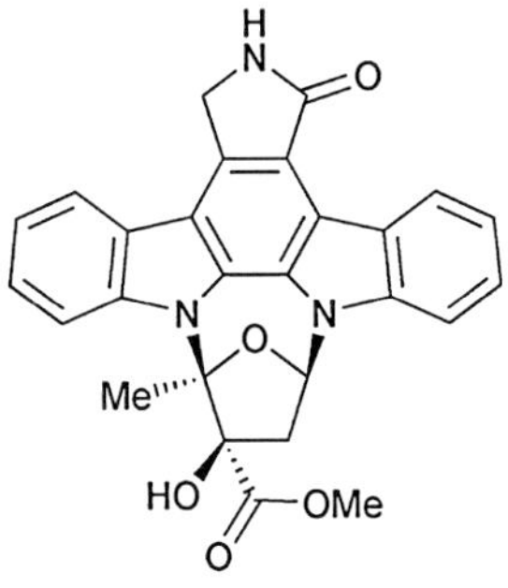

25
K252a

for genes similar to *rebD* and *rebP* in a genomic library of *A. melliaura*. The reported gene cluster spanned 35 kb and consisted of 21 genes (Figure 20.1). In agreement with the structural resemblance between AT2433 and REB, the gene cluster contained some genes that were likely homologs of their *reb* counterparts (Table 20.1). Additionally, there were other genes encoding proteins responsible for specific structural features, such as imide N-methylation and formation of the aminosugar moiety.

A biosynthetic locus (*ink*) responsible for production of K252a **25** has been reported from *Nonomuraea longicatena* (nucleotide sequence with accession number DQ399653) [31], although detailed information has not yet been published. Many of the identified *ink* genes were likely homologs of corresponding *sta* genes. Apparently, only one methyltransferase gene was found, probably for methylation at the furanose. A number of *ink* genes had no counterpart in the *sta* locus, and some of them might be responsible for those steps specific for K252a formation.

20.4 Indolocarbazole Biosynthetic Pathways and Their Engineering

20.4.1 Tryptophan Modification

The first step in REB biosynthesis is the conversion of L-tryptophan to 7-chloro-L-tryptophan (Scheme 20.2). *In vitro* halogenation by RebH, an $FADH_2$-dependent halogenase, requires molecular oxygen and flavin reductase RebF, which catalyzes the NADH-dependent reduction of FAD to provide $FADH_2$ for the halogenase [32]. Chloride salts are the source of chlorine for halogenation. Bromotryptophan could be produced by RebF/RebH in the presence of bromide ions, but neither fluoride nor iodide ions were competent for halogenation [32]. These results were in agreement with previous reports on the production of bromorebeccamycin **3** by cultures of *Lech. aerocolonigenes* [16].

Similarly, several bromoindolocarbazoles (**3**, **26–31**) were obtained by replacing chloride with bromide in the fermentation medium of *Strep. albus* strains expressing *reb* genes [33]. Nevertheless, in the absence of *rebH*, a full pathway for dideschloro-REB (**10**) could be reconstituted, indicating that the rest of the enzymes were able to use nonhalogenated intermediates [22,23,33]. Downstream enzymes could also utilize tryptophans carrying nonnatural halogenations, as demonstrated by production of fluoroindolocarbazoles **4–7** through addition of fluorotryptophans to fermentations of *Lech. aerocolonigenes* [5,17]. Moreover, the heterologous expression of different tryptophan halogenases (*pyrH*, *thal*), in addition to *reb* genes, resulted in formation of new halogenated derivatives **32–36** in *Strep. albus*, presumably by incorporation of 5- and 6-chlorotryptophan [33].

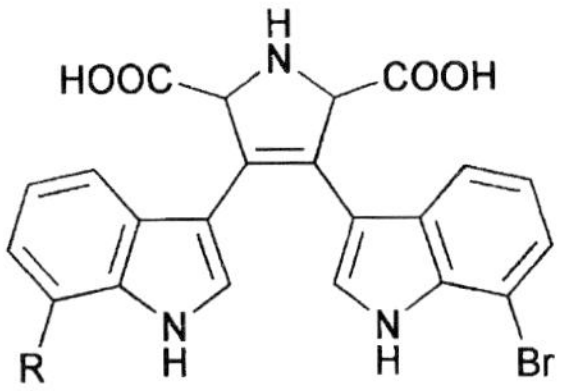

26 (R=Br)
27 (R=H)
28 (R=Cl)

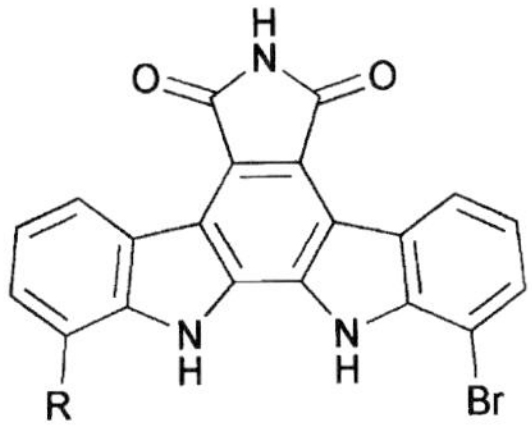

29 (R=Br)
30 (R=H)

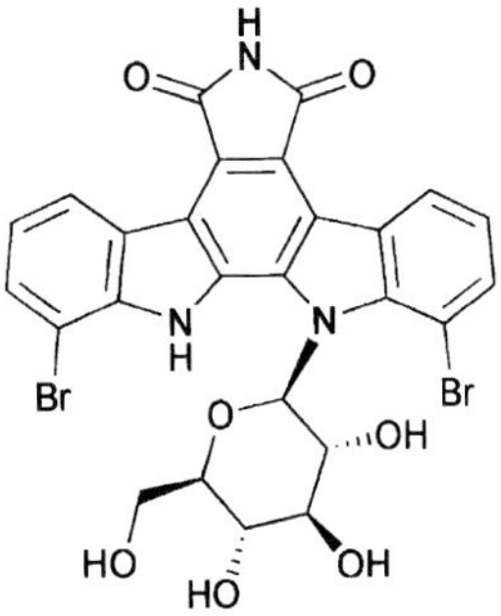

31

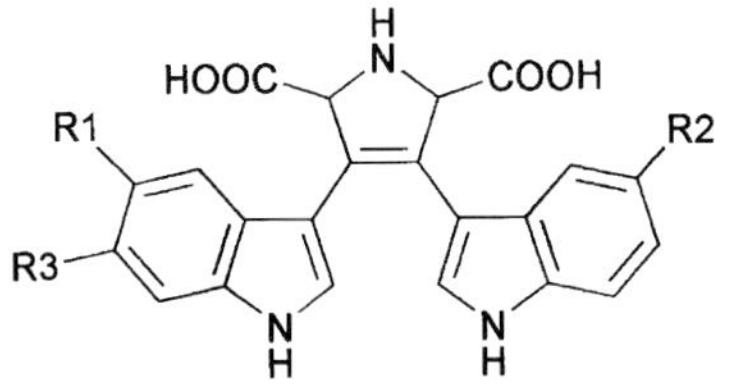

32 (R1=Cl, R2=R3=H)
33 (R1=R2=Cl, R3=H)
34 (R1=R2=H, R3=Cl)

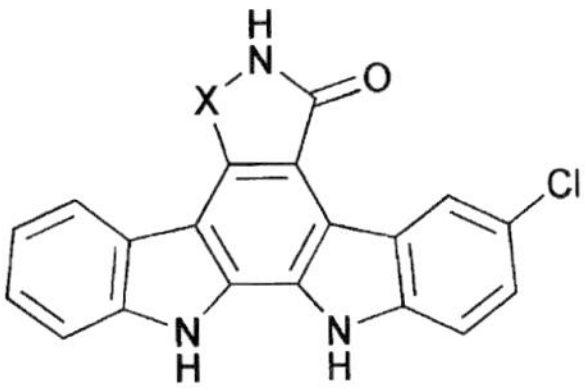

35 (X=CO)
36 (X=CH_2)

20.4.2
Formation of Bisindole Pyrrole

Two genes appear to be essential for indolocarbazole production in actinomycetes. One of the genes encodes a tryptophan oxidase (*rebO, staO*), while the second one codes for a heme-containing oxidase (*rebD, staD*). In their absence, no bisindole intermediates can be formed [23,33].

The enzyme RebO, from *Lech. aerocolonigenes*, is an FAD-dependent L-tryptophan oxidase that converts 7-chloro-L-tryptophan into 7-chloroindole-3-pyruvic acid imine, with production of hydrogen peroxide [25,34] (Scheme 20.2). Among the 20 natural amino acids, RebO showed high oxidase activity only against L-tryptophan [25]. Nevertheless, 1-methyl-L-tryptophan, 5-methyl-DL-tryptophan, and 5-fluoro-L-tryptophan were found to be substrates for RebO. The enzyme showed significant preference for 7-chloro-L-tryptophan over L-tryptophan, further supporting the role of the former as the natural early pathway intermediate [25]. Presumably, StaO might have activities similar to that of RebO, but acting on L-tryptophan to yield indole-3-pyruvic acid imine (in equilibrium with its ketone, Scheme 20.3).

Studies made with recombinant strains of *Strep. albus* showed that formation of the simplest bisindole intermediate during indolocarbazole biosynthesis required the coexpression of two genes: *rebO* and *rebD* [33]. The said metabolite consisted of chromopyrrolic acid (CPA, **37**), which seemed to be an intermediate during

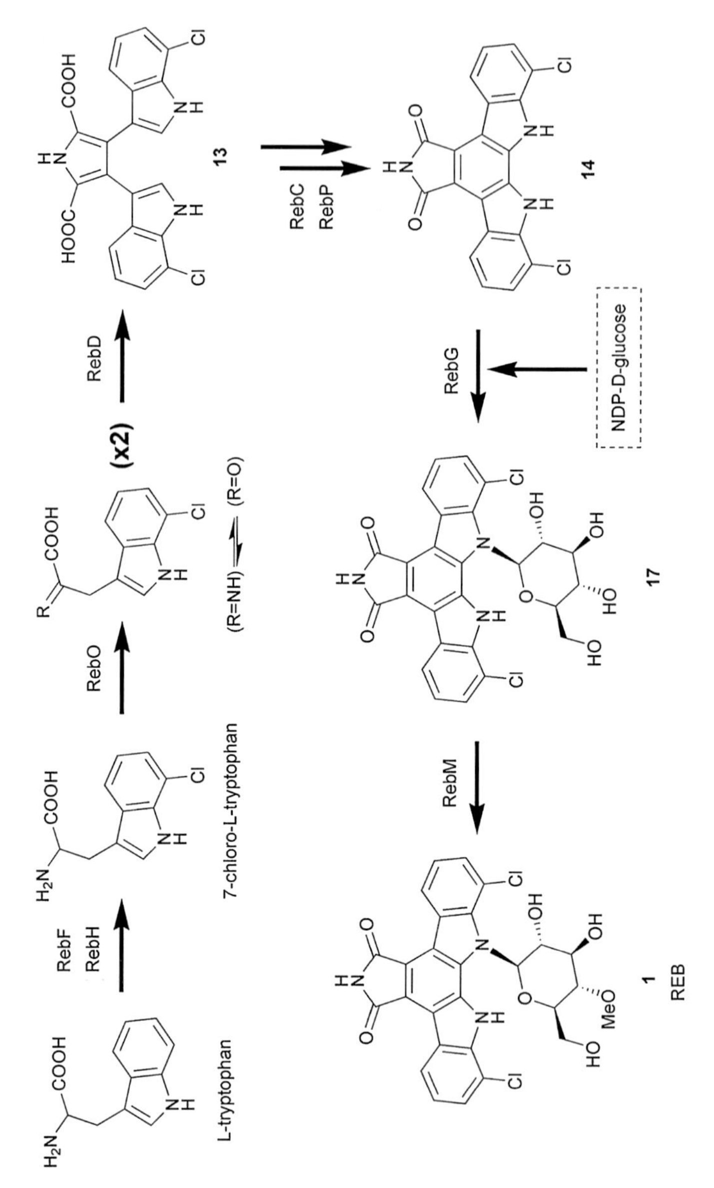

Scheme 20.2 Proposed pathway for biosynthesis of rebeccamycin (**1**).

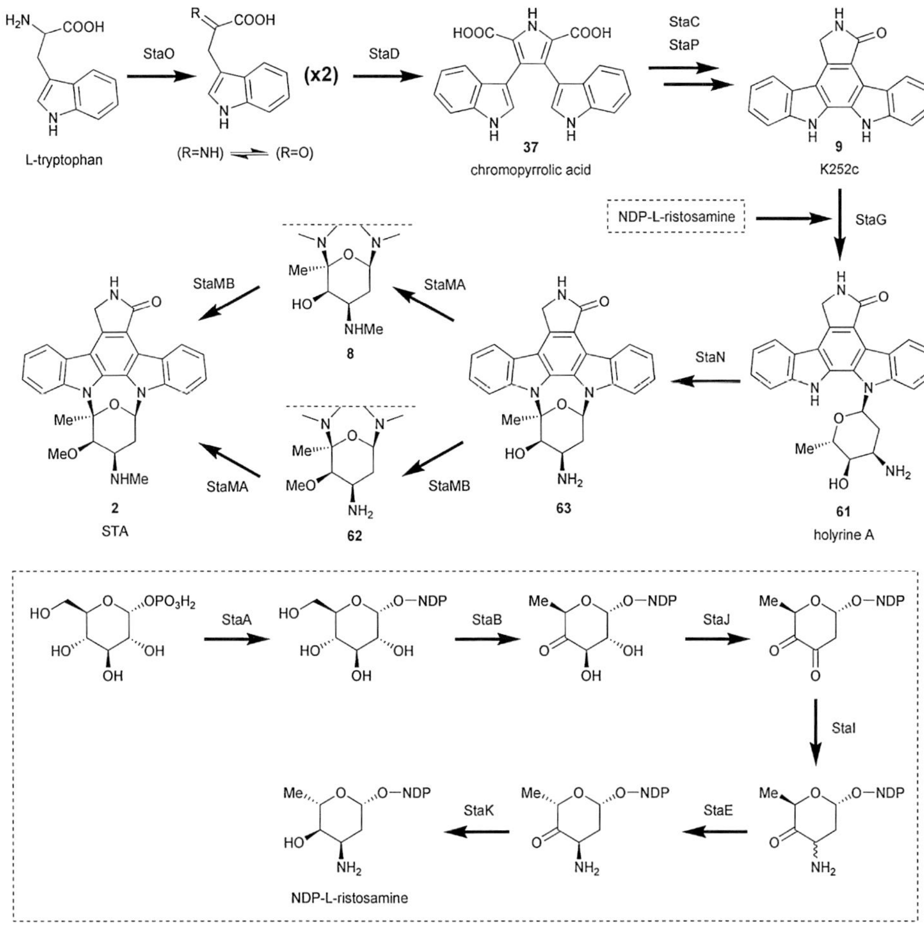

Scheme 20.3 Proposed pathway for biosynthesis of staurosporine (2).

production of nonchlorinated REB (**12**) [33] and STA (**2**) [35]. The role of a chromopyrrolic scaffold as a central intermediate in indolocarbazole formation was first discovered with the identification of 11,11′-dichloro-CPA (**13**), which was accumulated by a *rebP*-disrupted mutant of *Lech. aerocolonigenes* [23]. Metabolite **13** was also produced by coexpression of *rebH, rebO,* and *rebD* in *Strep. albus* [33].

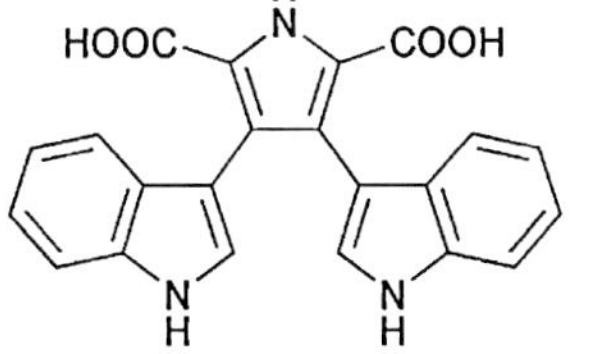

37
chromopyrrolic acid

The RebD enzyme was characterized as the first member of a new subfamily of heme-containing oxidases [34,36]. The enzyme acted as both a catalase and a CPA synthase, apparently converting two molecules of 7-chloroindole-3-pyruvic acid imine into 11,11′-dichloro-CPA. Formation of CPA by StaD, an enzyme homolog to RebD, was also confirmed for STA biosynthesis [35].

Recently, the relationship between the biosynthesis of indolocarbazoles and that of violacein (**38**, a bisindole pigment from *Chromobacterium violaceum*) was revealed [37]. A pair of genes (*vioA, vioB*), responsible for the earliest steps in violacein formation, was found to be functionally equivalent to their homolog pair in the indolocarbazole pathway (*rebO, rebD*), directing the formation of CPA. However, in contrast to indolocarbazole biosynthesis, CPA appeared to be a shunt product during violacein formation. A particular gene, *vioE*, was essential for production of asymmetrical intermediates instead of CPA.

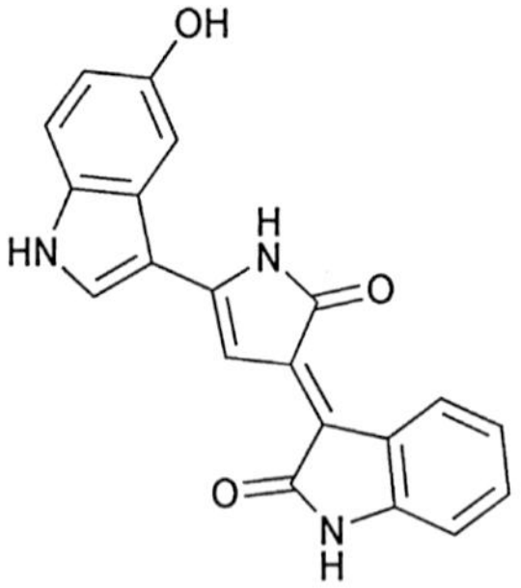

38
violacein

20.4.3
Formation of Carbazole

In order to complete the indolopyrrolocarbazole ring system, CPA intermediates undergo a decarboxylative ring closure. This biosynthetic step is performed by a cytochrome P450 enzyme (*rebP, staP*) and a flavin-dependent monooxygenase (*rebC, staC*).

The involvement of *rebP* and *rebC* in the ring-closure reaction was first indicated by the fact that *rebP*- and *rebC*-disrupted mutants of *Lech. aerocolonigenes* accumulated, respectively, 11,11′-dichloro-CPA (**13**) and a mixture of REB derivatives differing at C-7 [23]. Among the compounds accumulated by the *rebC* mutant, two of them were 7-deoxo-7-hydroxy-REB (**18**) and its 4′-*O*-demethyl analog (**19**). NMR spectroscopy results showed that each peak consisted of a mixture of 12-*N*- and 13-*N*-glycosides not separable on HPLC. A *rebG* and *rebC* double disruptant produced three peaks, identified as REB aglycone (**14**), 7-deoxo-7-hydroxy-REB aglycone (**15**) and 7-deoxo-REB aglycone (**16**), respectively. The 7-deoxo or 7-hydroxy derivatives did not seem to be REB intermediates, as they were not converted into REB in bioconversion experiments [23].

In agreement with those observations, efficient production of nonchlorinated REB aglycone (**12**, arcyriaflavin A) in *Strep. albus* required the coexpression of four genes:

rebO, rebD, rebC, and *rebP* [33]. If the *rebH* gene was also included, REB aglycone (**14**) was produced. Although it was possible to obtain **12** in the absence of *rebC* (by coexpression of *rebO, rebD*, and *rebP*), the result consisted of a mixture of three analogs differing at the C-7 position: (**12**), K252c (**9**), and 7-hydroxy-K252c (**39**).

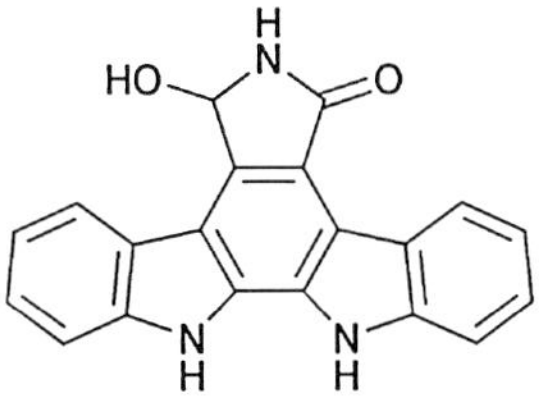

39

The involvement of *rebC, rebP*, and their homologs *staC* and *staP* in the ring-closure reaction was tested using recombinant *Strep. albus* strains [33]. The *rebO* and *rebD* genes were coexpressed together with different combinations of *rebC, rebP, staC*, and *staP*. The results showed that *rebP* and *staP* were functionally equivalent, and any of them was sufficient for processing CPA (**37**) into a mixture of the three analogs (**9**, **12**, **39**) differing at the C-7 position. When either *rebC* or *staC* was added to these gene combinations, a single product was obtained. Remarkably, the single product consisted of K252c (**9**) when using *staC*, or arcyriaflavin A (**12**) when using *rebC*. Therefore, a cytochrome P450 enzyme (RebP or StaP) appears to be responsible for the decarboxylative oxidations needed to convert a CPA intermediate into an indolopyrrolocarbazole. However, an additional monooxygenase (RebC or StaC) is needed for determination of the C-7 oxidation state of the indolocarbazole. Interestingly, RebC and StaC determine different oxidation states in the final product. This finding allowed the generation of a series of 7-deoxo derivatives (**9**, **16**, **40**–**46**) in *Strep. albus* strains expressing *reb* genes, by replacing *rebC* with *staC* [33]. Glycosylated derivatives **41**–**46** were probably a mixture of 12-*N*- and 13-*N*-glycosides not separable on HPLC, although this was not confirmed.

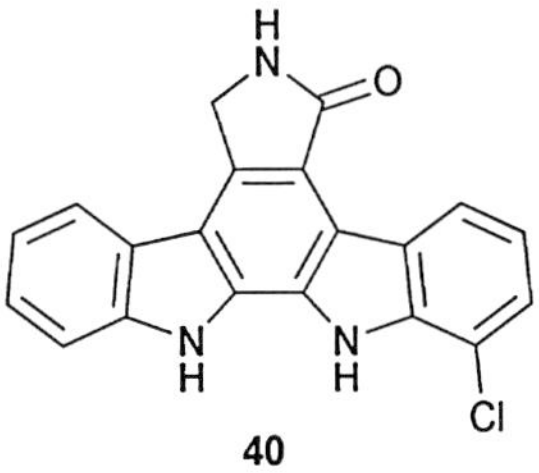

40

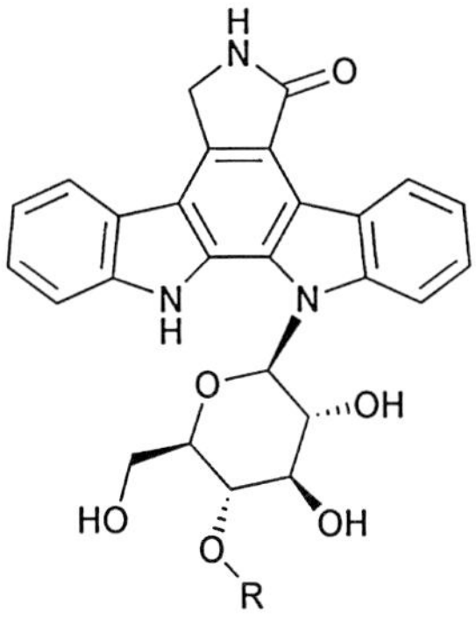

41 (R=H)
42 (R=Me)

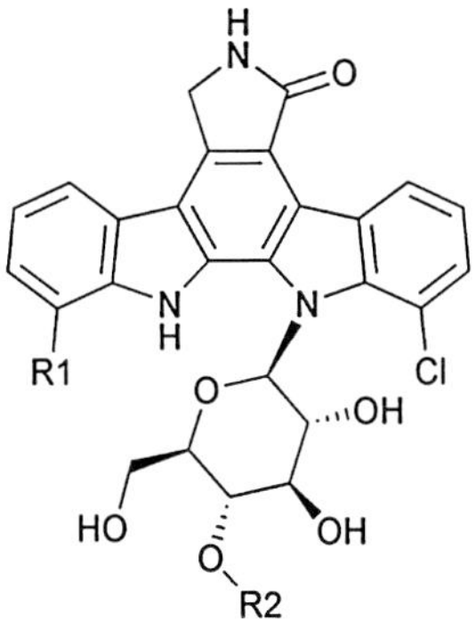

43 (R1=R2=H)
44 (R1=H, R2=Me)
45 (R1=Cl, R2=H)
46 (R1=Cl, R2=Me)

20.4.4
Formation of the Sugar Moiety

20.4.4.1 Sugar Moieties in Rebeccamycin and AT2433

REB (**1**) and AT2433 (**21–24**) contain a 4′-*O*-methyl-β-D-glucopyranosyl moiety. The *rebG*/*atG* and *rebM*/*atM* genes, respectively coding for a glycosyltransferase and a methyltransferase, appear to be responsible for formation of the carbohydrate (Table 20.1). RebG glycosylation takes place on a planar aromatic ring system, but not on chromopyrrolic intermediates. A *rebG* mutant of *Lech. aerocolonigenes* accumulated REB aglycone (**14**), which is the likely substrate for glycosylation [23] (Scheme 20.2). On the other hand, blocking the ring-closure step by disrupting *rebP* in *Lech. aerocolonigenes* resulted in accumulation of a nonglycosylated metabolite **13** [23]. Glycosylation on CPA **37** was not detected either by coexpression of *rebO*, *rebD*, and *rebG*, or by feeding **37** to a *rebG*-expressing *Strep. albus* [33]. Synthetic compounds structurally similar to CPA also failed as RebG substrates when fed to *rebG*-expressing strains [38].

A variety of *in vivo* experiments indicated that the RebG glycosyltransferase was able to add a D-glucose moiety to a variety of indolo[2,3-*a*]pyrrolo[3,4-*c*]carbazoles [5,16,17,21–23,33,38]. Additionally, indolo[2,3-*a*]carbazoles (lacking the pyrrole ring) were bioconverted by *rebG*-expressing strains to yield D-glucosyl derivatives **47–49**[38]. Glycosylation of asymmetrical aglycones resulted in a mixture of two regioisomers (such as **48** and **49**), not separable by HPLC, after indiscriminate N-glucosylation of either indole nitrogen [38].

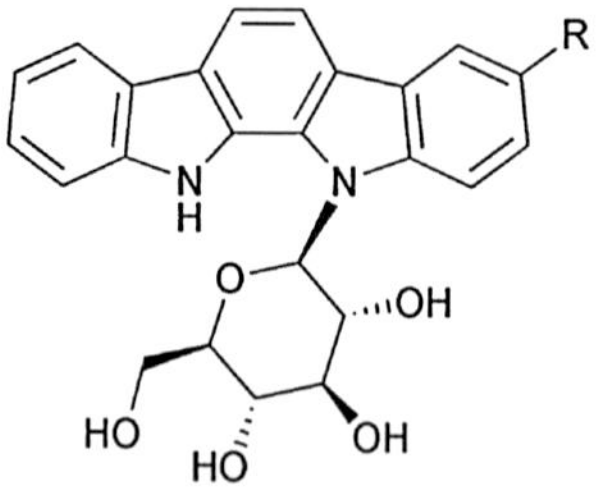

47 (R=H)
48 (R=Br)

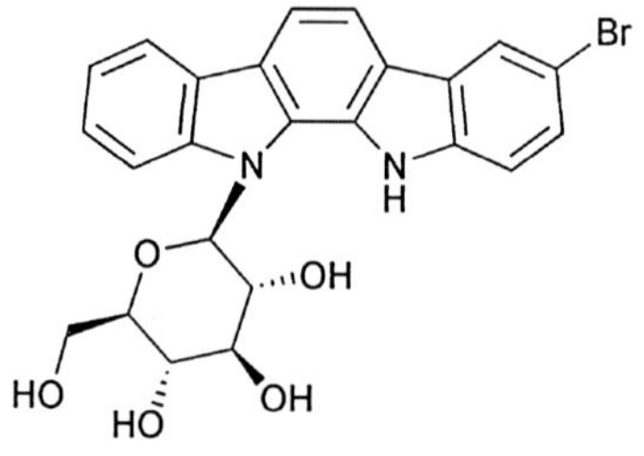

49

So far, no sugar moieties other than D-glucose have been reported for RebG glycosylations. Presumably, the sugar substrate directly used by RebG might consist of a nucleotide-activated form, that is, NDP-D-glucose. The biosynthesis of such an intermediate probably requires a glucose-1-phosphate nucleotidyltransferase, but a gene putatively encoding this activity was not found in the *reb* cluster. Such a gene is likely encoded in a different region of the *Lech. aerocolonigenes* genome. Production of RebG-glycosylated indolocarbazoles in *Strep. albus* [22,33], *Strep. lividans* [26], and *E. coli* [38] indicated that these hosts possessed nucleotidyltransferases able to supply the required nucleotide-activated D-glucose.

Sugar methylation is the last step during REB biosynthesis (Scheme 20.2). A *rebM*-disrupted mutant of *Lech. aerocolonigenes* accumulated 4′-*O*-demethyl-REB (**17**),

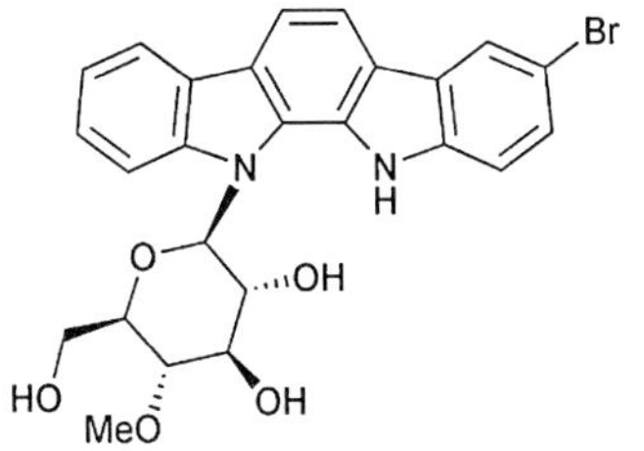

Br N H N O OH HO MeO OH

50

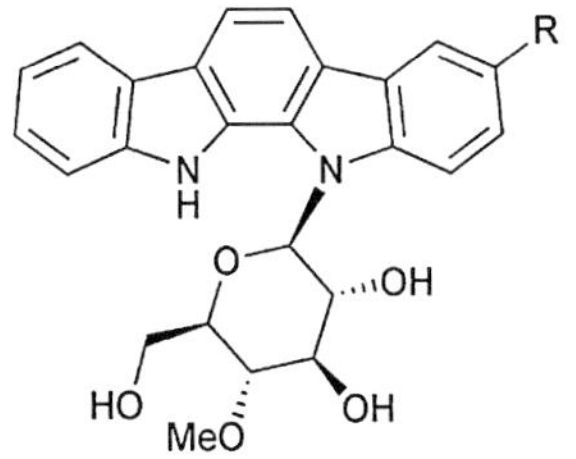

R N H N O OH HO MeO OH

51 (R=H)

52 (R=Br)

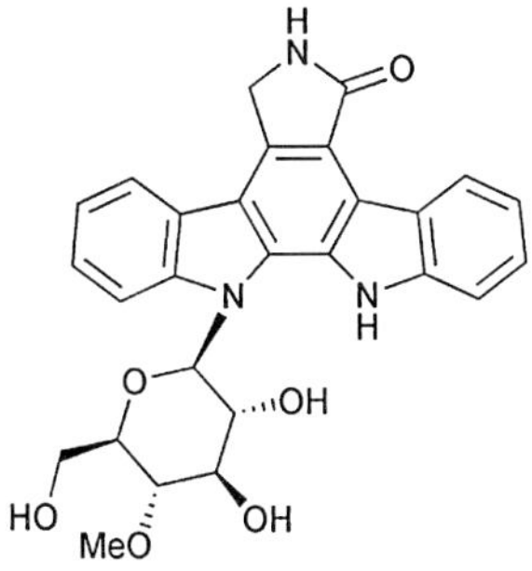

H N O N H N O OH HO MeO OH

53

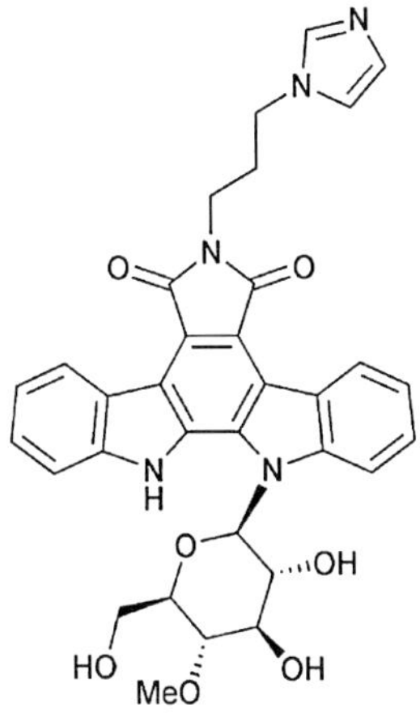

N N

O N O N H N O OH HO MeO OH

54

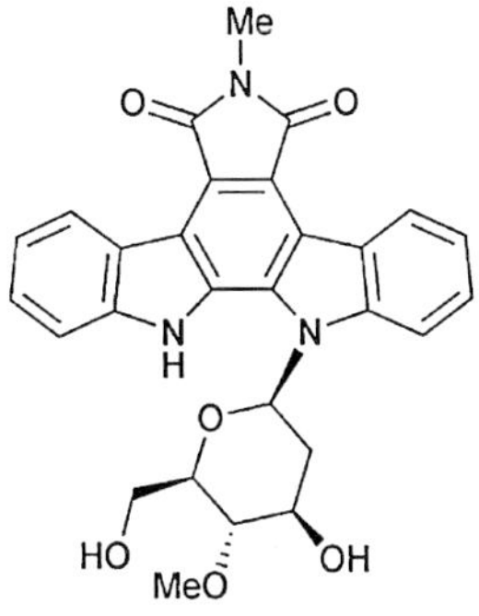

Me O N O N H N O HO MeO OH

55

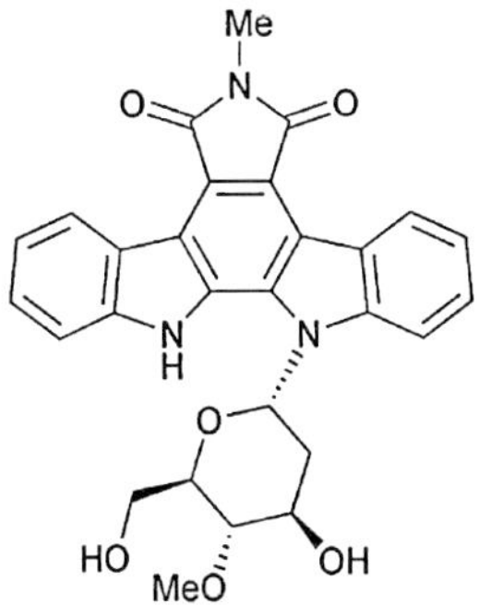

Me O N O N H N O HO MeO OH

56

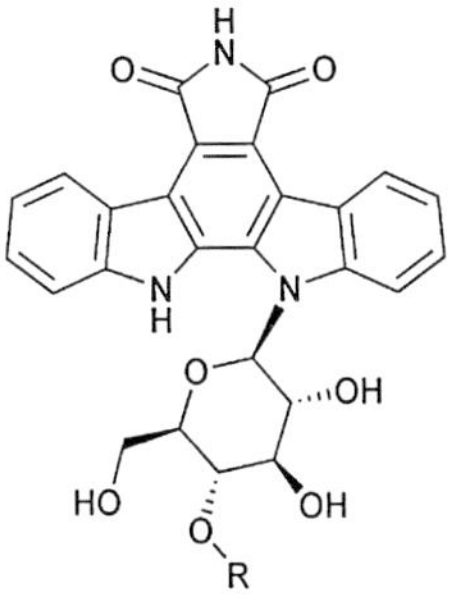

H O N O N H N O OH HO O R OH

57

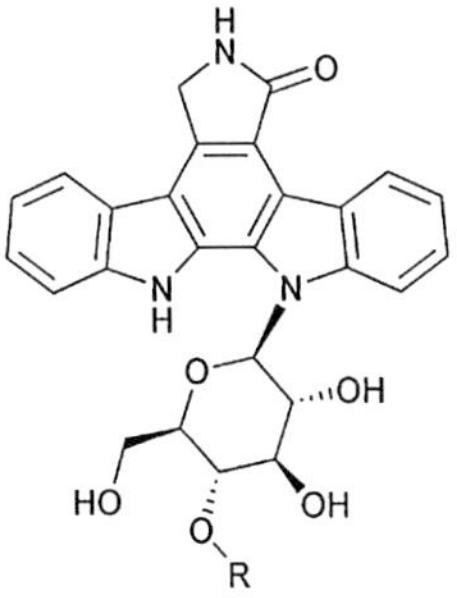

H N O N H N O OH HO O R OH

58

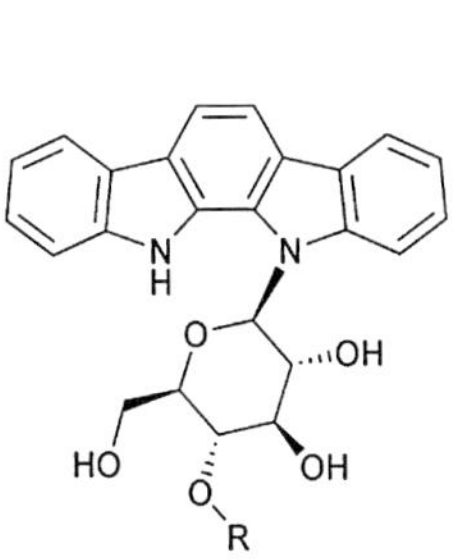

N H N O OH HO O R OH

59

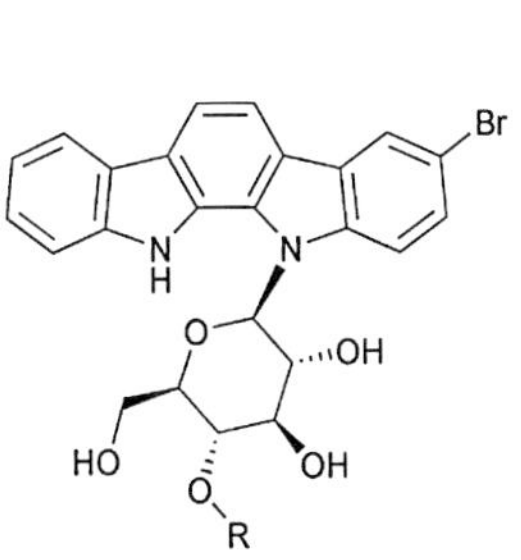

Br N H N O OH HO O R OH

60

R = $^{+}H_3N$ COO^- N^+ H O N N NH_2 N N HO OH

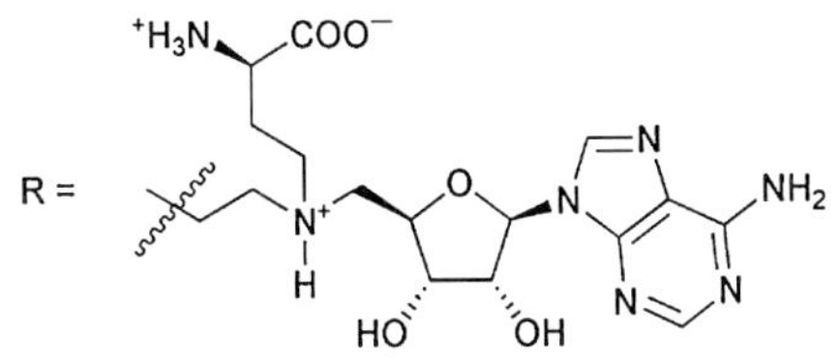

which was converted to REB by a *rebH* disruptant [23]. The RebM enzyme was characterized as the expected sugar 4′-*O*-methyltransferase [38,39]. Several glycosylated indolocarbazoles, including both α- and β-glycosidic analogs, could be modified *in vitro* by RebM-catalyzed methylation, to yield derivatives **10**, **11**, **50–56**. Glycosides containing L-deoxysugars were not methylated. Additionally, RebM was able to use an *N*-mustard analog of *S*-adenosyl-L-methionine for *in vitro* modification of several substrates, resulting in formation of novel derivatives **57–60**[39].

In comparison to REB (**1**), the sugar moiety of AT2433 (**21–24**) is remarkable as containing an aminodideoxypentose, in addition to the methylglucose. Accordingly, the *at* locus contained eight genes coding for enzymes putatively involved in formation and attachment of the aminopentose [30], which were not found in the *reb* gene cluster (Table 20.1). A proposed biosynthetic pathway for the aminodideoxypentose is shown in Scheme 20.4.

20.4.4.2 **The Staurosporine Sugar Moiety**

The carbohydrate present in STA is an aminodeoxysugar, which is linked to the aglycone through a pair of C–N bonds. Up to ten genes appear to be involved in formation of the sugar moiety (Table 20.1). Reconstitution of the STA pathway in *Strep. albus*, by heterologous expression of *sta* and *reb* genes, shed light on the nature of this process [29]. STA production in *Strep. albus* was achieved by coexpression of genes for K252c formation (*rebO, rebD, rebP, staC*) together with genes for sugar formation (*staG, staN, staMA, staJ, staK, staI, staE, staMB*). It was found that the *staA* and *staB* genes were not needed for STA formation in *Strep. albus*, suggesting that this host was providing the first two enzymatic activities required for deoxysugar biosynthesis. Deletion of *staG* abolished glycosylation, and only K252c (**9**) was produced. On the other hand, removal of *staN* (while keeping *staG*) resulted in production of holyrine A (**61**), which had been previously isolated from an STA-producing marine actinomycete [40] (Scheme 20.3). The carbohydrate in holyrine A was attached to the aglycone through a single glycosidic bond, and it lacked the two methylations found in STA. Holyrine A could be converted into STA when fed to a strain that expressed *staN, staMA*, and *staMB* [29]. In these experiments, independent removal of *staMA, staMB*, or both, caused holyrine A to be converted into 4′-*N*-demethyl-STA (**62**), 3′-*O*-demethyl-STA (**8**), or 3′-*O*-demethyl-4′-*N*-demethyl-STA (**63**), respectively (Scheme 20.3). Additionally, TP-A0274, a *staN*-deleted mutant of *Streptomyces* sp., accumulated holyrine A, which could be converted into STA when fed to a *staD* mutant of the same strain [41]. Therefore, the first sugar–aglycone linkage seems to be made by StaG, while the second linkage would be catalyzed by StaN acting on holyrine A (Scheme 20.3). Each one of the methylation steps can occur in the absence of the other, but only after the second linkage is formed. Presumably, most of the reactions needed for the biosynthesis of L-ristosamine might take place on nucleotide-activated sugar intermediates, before StaG-catalyzed attachment of L-ristosamine (or a previous intermediate) to the aglycone (Scheme 20.3).

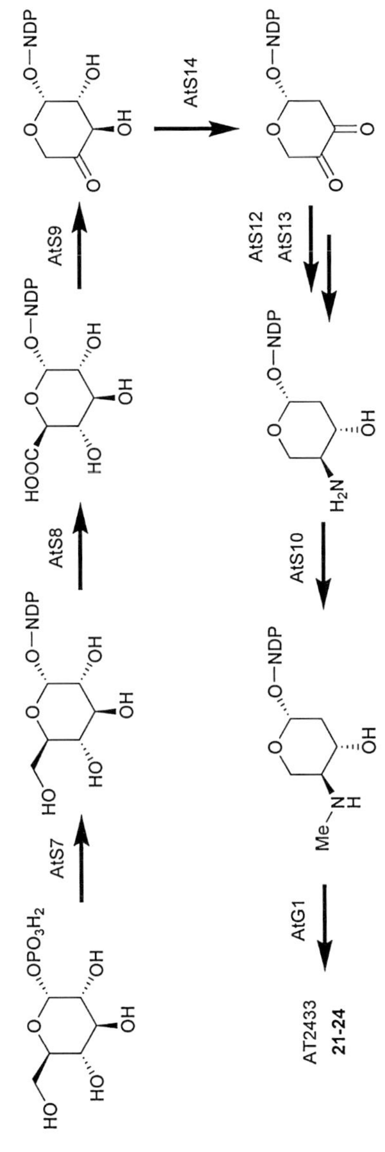

Scheme 20.4 Proposed pathway for biosynthesis of the aminopentose in AT2433. N-methylation might alternatively occur after the sugar is transferred to the aglycone.

In contrast to RebG, the StaG glycosyltransferase seems to possess a noteworthy regiospecificity toward its aglycone substrate: StaG appears to discriminate between the two indole nitrogens, and only one of them is chosen for the first glycosidic attachment. Accordingly, StaG-glycosylated indolocarbazoles produced in recombinant strains appeared to exist as single regioisomers (glycosidic linkage at N-13) [29], and not as mixtures of two regioisomers (12-*N*- and 13-*N*-glycosides) as occurred with RebG-glycosylated compounds [23,38].

On the other hand, StaG accepts a variety of sugar derivatives, as has been shown through the production of novel glycosylated indolocarbazoles in recombinant *Strep. albus* strains [29]. This was done by replacing the *sta* genes involved in formation of the STA sugar with any one of various sets of genes, each set directing formation of a different sugar. As a result, a number of derivatives were produced, with a sugar moiety consisting of L-rhamnose (**64–65**), L-olivose (**66–67**), L-digitoxose (**68–69**), or D-olivose (**70**). With the exception of D-olivose, which yielded only the single-linkage derivative, the sugar could be attached through either one or two linkages, indicating that StaN also showed some substrate flexibility.

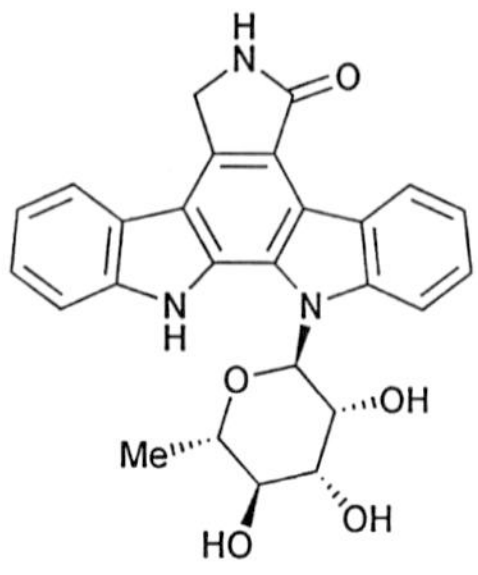

64 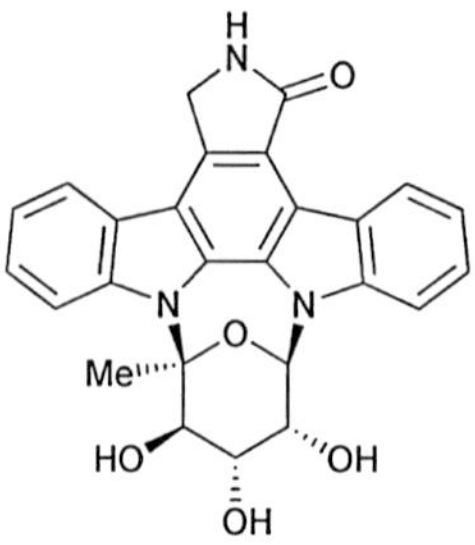 **65** 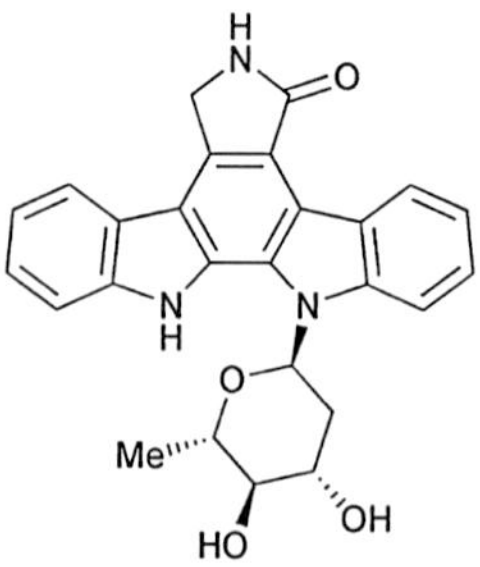 **66** 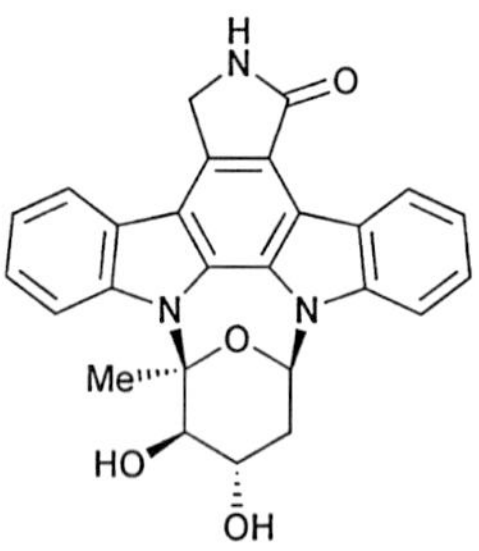 **67**

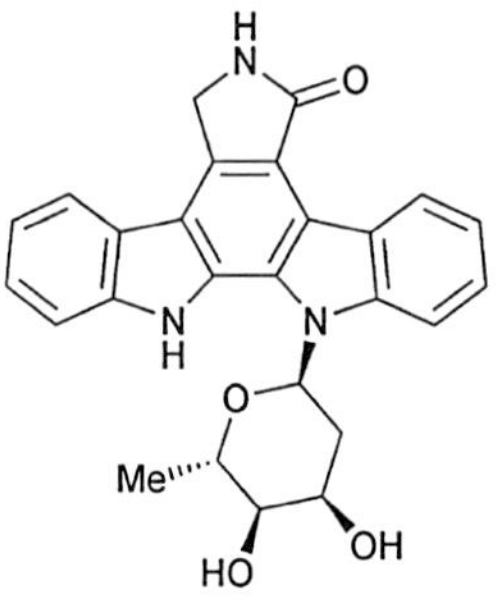

68 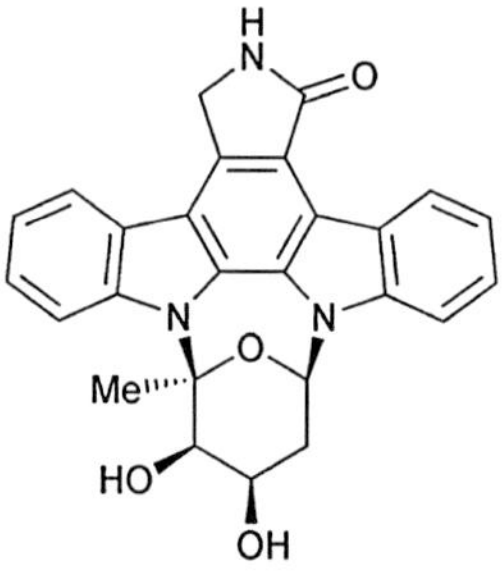 **69** 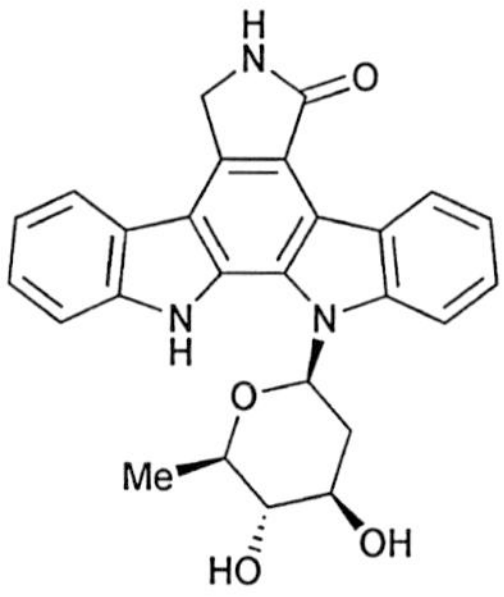 **70**

20.4.5
Regulation and Self-resistance

A gene (*rebR*/*staR*/*atR*/*inkR*) coding for a putative LAL transcriptional regulatory protein was found in the respective loci for the biosynthesis of REB [22–24], STA [28],

AT2433 [30], and K252a [31]. Experiments made with recombinant *Strep. albus* strains suggested that *rebR* was needed for expression of at least some of the *reb* genes [22]. This was further supported by the fact that a *rebR*-truncated mutant of *Lech. aerocolonigenes* did not produce REB or related compounds [23]. Additionally, the *reb* and *at* loci also included a gene (*orfD12/atE*) putatively encoding a MarR regulatory protein (Table 20.1).

The *reb* gene cluster contained two genes (*rebT, rebU*) coding for putative transmembrane transporter proteins, which might be involved in self-resistance, that is, protection of the microorganism against its own toxic product [22,23]. Expression of *rebT* conferred resistance to exogenously added REB in *Strep. albus* (which was otherwise sensitive to REB) [22]. Further studies made with recombinant *Strep. albus* strains showed that production of either REB (**1**) or 4′-*O*-demethyl-REB (**17**) (but not earlier intermediates or nonchlorinated analogs) was only possible if *rebT* was expressed [33]. However, disruption of *rebT* in *Lech. aerocolonigenes* did not decrease the production level of REB, or affect growth of the microorganism [23]. Therefore, despite the proved role of *rebT* as a protecting system in the *Strep. albus* host, the gene was not essential for REB production in *Lech. aerocolonigenes* in the conditions tested. RebT might protect *Lech. aerocolonigenes* cells from an excess of REB accumulation in certain circumstances. Additionally, other resistance mechanisms may exist in *Lech. aerocolonigenes* that are not found in *Strep. albus*. For instance, the possible role of *rebU* in any aspect related to REB biosynthesis remains unknown. The *at* locus from *A. melliaura* contained two genes, *atI* and *atB* [30], which were putative homologs of *rebT* and *rebU*, respectively (Table 20.1).

While REB inhibits the growth of several bacteria, STA is generally found to lack significant antibacterial activity [3]. This might explain the absence of candidate genes coding for specific self-resistance mechanisms in the *sta* locus. STA did not inhibit the growth of *Strep. albus*; therefore, STA production was achieved in this strain by coexpression of a defined set of *sta* and *reb* genes, in the absence of *rebT* or other dedicated resistance mechanisms [29]. However, STA appears to affect cell differentiation processes in streptomycetes [3].

20.5 Perspectives and Concluding Remarks

The indolocarbazole family of natural products is a source of lead compounds with potential therapeutic applications in the treatment of cancer and other diseases [3]. As an addition to chemical synthesis, the introduction of biological processes might help in the production of useful derivatives in a cost-effective way and with lower environmental impacts. The identification of novel producing organisms and the genes responsible for indolocarbazole biosynthesis provide the necessary "toolkit" for such purpose. Combining fermentation of microorganisms (either wild type or genetically manipulated) with *in vitro* enzymatic bioconversions and chemical synthesis will greatly expand the available repertory of analogs. Given the proven utility of targeting protein kinases and DNA topoisomerases for treating a range of important human diseases, it is hoped that some indolocarbazole analogs will provide new medicines in the near future.

20.6 Acknowledgments

We are indebted to Aaroa P. Salas and Alfredo F. Braña at the University of Oviedo, and to Lili Zhu, Igor A. Butovich and Jürgen Rohr at the University of Kentucky, who contributed to our research on indolocarbazole biosynthesis. C.M. and J.A.S. wish to acknowledge the generous financial support of the Spanish Ministry of Education and Science (BIO2000-0274 and BMC2003-00478), Plan Regional de I + D + i del Principado de Asturias (Grant GE-MED01-05), and Red Temática de Investigación Cooperativa de Centros de Cáncer (Ministerio de Sanidad y Consumo, Spain).

References

1 Gribble, G. W. and Berthel, S. J. (1993) Studies in natural product chemistry. In Atta-ur-Rahman (Ed.): *Stereoselective Synthesis, Part H*, Vol. 12, Elsevier, Amsterdam.

2 Knolker, H. J. and Reddy, K. R. (2002) *Chemical Reviews*, **102**, 4303–4427.

3 Sánchez, C., Méndez, C., Salas, J. A. (2006) *Natural Product Reports*, **23** 1007–1045.

4 Akinaga, S., Sugiyama, K., Akiyama, T. (2000) *Anticancer Drug Design*, **15**, 43–52.

5 Long, B. H., Rose, W. C., Vyas, D. M., Matson, J. A., Forenza, S. (2002) *Current Medicinal Chemistry: Anti-Cancer Agents*, **2**, 255–266.

6 Prudhomme, M. (2003) *European Journal of Medicinal Chemistry*, **38**, 123–140.

7 Butler, M. S. (2005) *Natural Product Reports*, **22**, 162–195.

8 Sánchez, C., Méndez, C., Salas, J. A. (2006) *Journal of Industrial Microbiology and Biotechnology*, **33**, 560–568.

9 Meksuriyen, D. and Cordell, G. A. (1988) *Journal of Natural Products*, **51**, 893–899.

10 Yang, S.-W. and Cordell, G. A. (1996) *Journal of Natural Products*, **59**, 828–833.

11 Yang, S.-W. and Cordell, G. A. (1997) *Journal of Natural Products*, **60**, 788–790.

12 Yang, S.-W. and Cordell, G. A. (1997) *Journal of Natural Products*, **60**, 236–241.

13 Yang, S. W., Lin, L. J., Cordell, G. A., Wang, P., Corley, D. G. (1999) *Journal of Natural Products*, **62**, 1551–1553.

14 Pearce, C. J., Doyle, T. W., Forenza, S., Lam, K. S., Schroeder, D. R. (1988) *Journal of Natural Products*, **51**, 937–940.

15 Lam, K. S., Forenza, S., Doyle, T. W., Pearce, C. J. (1990) *Journal of Industrial Microbiology and Biotechnology*, **6**, 291–294.

16 Lam, K. S., Schroeder, D. R., Veitch, J. M., Matson, J. A., Forenza, S. (1991) *Journal of Antibiotics (Tokyo)*, **44**, 934–939.

17 Lam, K. S., Schroeder, D. R., Veitch, J. M., Colson, K. L., Matson, J. A., Rose, W. C., Doyle, T. W., Forenza, S. (2001) *Journal of Antibiotics (Tokyo)*, **54**, 1–9.

18 Goeke, K., Hoehn, P., Ghisalba, O. (1995) *Journal of Antibiotics (Tokyo)*, **48**, 428–430.

19 Hoehn, P., Ghisalba, O., Moerker, T., Peter, H. H. (1995) *Journal of Antibiotics (Tokyo)*, **48**, 300–305.

20 Weidner, S., Kittelmann, M., Goeke, K., Ghisalba, O., Zahner, H. (1998) *Journal of Antibiotics (Tokyo)*, **51**, 679–682.

21 Ohuchi, T., Ikeda-Araki, A., Watanabe-Sakamoto, A., Kojiri, K., Nagashima, M., Okanishi, M., Suda, H. (2000) *Journal of Antibiotics (Tokyo)*, **53**, 393–403.

22 Sánchez, C., Butovich, I. A., Braña, A. F., Rohr, J., Méndez, C., Salas, J. A. (2002) *Chemistry and Biology*, **9**, 519–531.

23 Onaka, H., Taniguchi, S., Igarashi, Y., Furumai, T. (2003) *Bioscience, Biotechnology, and Biochemistry*, **67**, 127–138.

24 Hyun, C. G., Bililign, T., Liao, J., Thorson, J. S. (2003) *Chembiochem*, **4**, 114–117.

25 Nishizawa, T., Aldrich, C. C., Sherman, D. H. (2005) *Journal of Bacteriology*, **187**, 2084–2092.

26 Onaka, H., Taniguchi, S., Ikeda, H., Igarashi, Y., Furumai, T. (2003) *Journal of Antibiotics (Tokyo)*, **56**, 950–956.

27 Schupp, T., Engel, N., Bietenhader, J., Toupet, C., Pospiech, A. (1997) Worldwide Patent WO9708323.

28 Onaka, H., Taniguchi, S., Igarashi, Y., Furumai, T. (2002) *Journal of Antibiotics (Tokyo)*, **55**, 1063–1071.

29 Salas, A. P., Zhu, L., Sánchez, C., Braña, A. F., Rohr, J., Méndez, C., Salas, J. A. (2005) *Molecular Microbiology*, **58**, 17–27.

30 Gao, Q., Zhang, C., Blanchard, S., Thorson, J. S. (2006) *Chemistry and Biology*, **13**, 733–743.

31 Kim, S. Y., Park, J. S., Lee, J. Y., Oh, K. B. (2005) *International Meeting of the Federation of Korean Microbiological Societies*, Seoul, Korea.

32 Yeh, E., Garneau, S., Walsh, C. T. (2005) *Proceedings of the National Academy of Sciences of the United States of America*, **102**, 3960–3965.

33 Sánchez, C., Zhu, L., Braña, A. F., Salas, A. P., Rohr, J., Mendez, C., Salas, J. A. (2005) *Proceedings of the National Academy of Sciences of the United States of America*, **102**, 461–466.

34 Howard-Jones, A. R. and Walsh, C. T. (2005) *Biochemistry*, **44**, 15652–15663.

35 Asamizu, S., Kato, Y., Igarashi, Y., Furumai, T., Onaka, H. (2006) *Tetrahedron Letters*, **47**, 473–475.

36 Nishizawa, T., Gruschow, S., Jayamaha, D. H., Nishizawa-Harada, C., Sherman, D. H. (2006) *Journal of the American Chemical Society*, **128**, 724–725.

37 Sánchez, C., Braña, A. F., Méndez, C., Salas, J. A. (2006) *Chembiochem*, **7**, 1231–1240.

38 Zhang, C., Albermann, C., Fu, X., Peters, N. R., Chisholm, J. D., Zhang, G., Gilbert, E. J., Wang, P. G., Van Vranken, D. L., Thorson, J. S. (2006) *Chembiochem*, **7**, 795–804.

39 Zhang, C., Weller, R. L., Thorson, J. S., Rajski, S. R. (2006) *Journal of the American Chemical Society*, **128**, 2760–2761.

40 Williams, D. E., Bernan, V. S., Ritacco, F. V., Maiese, W. M., Greenstein, M., Andersen, R. J. (1999) *Tetrahedron Letters*, **40**, 7171–7174.

41 Onaka, H., Asamizu, S., Igarashi, Y., Yoshida, R., Furumai, T. (2005) *Bioscience, Biotechnology, and Biochemistry*, **69**, 1753–1759.

Index

Modern Alkaloids: Structure, Isolation, Synthesis and Biology. Edited by E. Fattorusso and O. Taglialatela-Scafati

ISBN: 978-3-527-31521-5

d

i

m

n

u